The ARRL Antenna Book

Published by
The American Radio Relay League
Newington, CT USA 06111

Editor
Gerald (Jerry) Hall, K1TD

Associate Editor
James W. Healy, NJ2L

Assistant Editors
Bruce S. Hale, KB1MW
Charles L. Hutchinson, K8CH
Mark J. Wilson, AA2Z
Larry Wolfgang, WA3VIL

Contributing Authors
John Almeida, KA1AIR
Wilfred N. Caron
Steve Cerwin, WA5FRF
Roger A. Cox, WBØDGF
Philip V. D'Agostino, W1KSC
Richard Guski, KC2MK
Robert (Ted) Hart, W5QJR
F. Robert Hawk, KØYEH
John Kraus, W8JK
Roy W. Lewallen, W7EL
Domenic M. Mallozzi, N1DM
Jim McKim, WØCY
William T. Schrader, K2TNO
Ralph Shaw, K5CAV
Allan White, W1EYI
Thomas Willeford, N8ETU

Contributors
Chip Angle, N6CA
Bob Atkins, KA1GT
Thomas A. Beery, WD5CAW
Dennis Bodson, W4PWF
Bruce Brown, W6TWW
Warren Bruene, W5OLY
Robert E. Cowan, K5QIN
Dave Fisher, WØMHS
Donald K. Johnson, W6AAQ
M. Walter Maxwell, W2DU
Charles J. Michaels, W7XC
Bill Myers, K1GQ
Jack Sobel, WØSVM
John J. Uhl, KV5E
Brian L. Wermager, KØEOU
Frank J. Witt, AI1H
Deane J. Yungling, KI6O

Production
Leslie Bartoloth, KA1MJP
Laird Campbell, W1CUT
Michelle Chrisjohn, WB1ENT
Sue Fagan, cover
Alison Halapin
Mark Kajpust
Joel Kleinman, N1BKE
Steffie Nelson, KA1IFB
David Pingree
Jean Wilson

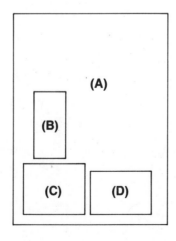

A—Photo of 120-foot tower at W1AW by Meyers Studio.
  Moon photo courtesy Chuck Hutchinson, K8CH

B—Photo courtesy Bob Cutter, KIØG

C—30-foot polar-mount dish at K5AZU

D—12 17-element long-boom 2-meter Yagis at N5BLZ

Foreword

The antenna launches energy from a transmitter into space or pulls it in from a passing wave for a receiver. The antenna is the vital link between a radio station and space. Without a suitable, properly installed antenna, the best transmitter and receiver are useless.

Antennas come in many shapes and sizes, from a simple straight wire to complex arrays, but all perform according to the same basic principles. These are clearly explained in this book, which then goes on to help one choose an antenna and to explain how to build, install and adjust it. Examples are given of hundreds of antennas with full constructional details. It's a very practical book, enabling the novice to build an antenna with confidence, while the more experienced will find innumerable ways of constructing new antennas or improving a present one.

Amateur Radio provides an important supplement to our educational system in that it gives hands-on, real-world experience that is lacking in most academic curricula. A great many professional radio engineers got their first practical experience as amateurs, and many have continued the hobby lifelong.

My love affair with antennas began 62 years ago as an amateur, and still continues. In 1950 I published a basic text and reference book, *Antennas,* based on courses I gave at the Ohio State University. That was about the time the fifth edition of *The ARRL Antenna Book* appeared. This year I published a new thoroughly revised and much enlarged second edition of *Antennas,* closely coinciding with the appearance of this bigger 15th edition of *The ARRL Antenna Book*. While my text goes more deeply into antenna theory, this book covers the practical and constructional aspects in more detail. The two books are complementary, the one supplementing the other. Anyone who wants to know what steps to take to build or improve an antenna will find this enlarged edition of *The ARRL Antenna Book* indispensable.

John Kraus, W8JK
Director, Radio Observatory
The Ohio State University
July, 1988

Introduction

We are pleased to offer this 15th edition of *The ARRL Antenna Book*. Since the first edition appeared in September 1939, each new edition has been dedicated to providing more and better information about the fascinating subject of antennas. The book has been well received over the years by amateurs and professionals alike, as indicated by the distribution of 717,222 copies since 1948.

With thirty chapters, this edition is certainly bigger, and I think you'll find better than any previous edition. Every page of the book has been set in new type and prepared in a layout that we feel is appealing. While most of the content of previous editions has been retained, the text has been totally reedited. Much new material has been added. Part of the new material is based on computer analyses of antennas, using state-of-the-art software. This, coupled with practical data, allows a more detailed look at the vagaries of antenna operation than has been possible heretofore, and therefore a more meaningful presentation.

Within these covers you'll also find dozens of references to earlier source material, and more bibliography references than ever before at the end of most chapters. You'll find an entire chapter devoted to the listing of significant antenna and related articles appearing in *QST* and other publications over the past quarter century. These are provided to guide you to additional information on topics presented here. Further, you'll find a lengthy list of suppliers of antenna products, with addresses, to aid you in locating materials for your own construction projects as well as locating ready-made antenna systems to fit your needs.

There are also a few items you will not find in this edition that have appeared in earlier versions. These items are of an operating nature, and do not pertain directly to antenna hardware, such as tables of latitudes and longitudes of DX locations, information on pointing your antenna, and so on. This information is now contained in *The ARRL Operating Manual,* which I highly recommend.

In a publishing effort of this magnitude, errors are inevitable. We've attempted to eliminate the errors, and would appreciate your letting us know about any which may have gone undetected. We'd also appreciate receiving your suggestions for making the next edition even better than this. A form for mailing your comments is included at the back of the book.

David Sumner, K1ZZ
Executive Vice President
Newington, Connecticut
July, 1988

Contents

Metric Equivalents, Gain Reference

Throughout this book distances and dimensions are usually expressed in English units—the mile, the foot, and the inch. Conversions to metric units may be made by using the following equations:

$$km = mi \times 1.609$$

$$m = ft\ (') \times 0.3048$$

$$mm = in.\ ('') \times 25.4$$

An inch is 1/12 of a foot. Tables in Chapter 21 provide information for accurately converting inches and fractions to decimal feet, and vice versa, without the need for a calculator.

Also throughout this book, gain is referenced to a dipole antenna in decibels (dB or dBd) unless indicated otherwise. Other references include an isotropic radiator, designated as dBi. A dipole has a gain of 2.14 dBi.

Chapter 1

Safety First

Safety begins with your attitude. If you make it a habit to plan your work carefully and to consider the safety aspects of a project before you begin the work, you will be much safer than "Careless Carl," who just jumps in, proceeding in a haphazard manner. Learn to have a positive attitude about safety. Think about the dangers involved with a job before you begin the work. Don't be the one to say, "I didn't think it could happen to me."

Having a good attitude about safety isn't enough, however. You must be knowledgeable about common safety guidelines and follow them faithfully. Safety guidelines can't possibly cover all the situations you might face, but if you approach a task with a measure of "common sense," you should be able to work safely.

This chapter offers some safety guidelines and protective measures for you and your Amateur Radio station. You should not consider it to be an all-inclusive discussion of safety practices, though. Safety considerations will affect your choice of materials and assembly procedures when building an antenna. Other chapters of this book will offer further suggestions on safe construction practices. For example, Chapter 22 includes some very important advice on constructing a proper base for your tower installation.

PUTTING UP SIMPLE WIRE ANTENNAS

No matter what type of antenna you choose to erect, you should remember a few key points about safety. If you are using a slingshot or bow and arrow to get a line over a tree, make sure you keep everyone away from the "downrange" area. Hitting one of your helpers with a rock or fishing sinker is considered not nice, and could end up causing a serious injury.

Make sure the ends of the antenna are high enough to be out of reach of passersby. Even when you are transmitting with low power there may be enough voltage at the ends of your antenna to give someone nasty "RF burns." If you have a vertical antenna with its base at ground level, build a wooden safety fence around it at least 4 feet away from it. Do not use metal fence, as this will interfere with the proper operation of the antenna. Be especially certain that your antenna is not close to any power wires. That is the *only* way you can be sure it won't come in contact with them!

Antenna work often requires that one person climb up on a tower, into a tree or onto the roof of a house. Never work alone! Work slowly, thinking out each move before you make it. The person on the ladder, tower, tree or rooftop should wear a safety belt, and keep it securely anchored. It is helpful (and safe!) to tie strings or light-weight ropes to all tools. If your tools are tied on, you'll save time getting them back if you drop them, and you'll greatly reduce the risk of injuring a helper on the ground. (There are more safety tips for climbing and working on towers later in this chapter. Those tips apply to any work that you must do above the ground to install even the simplest antenna.)

Tower Safety

Working on towers and antennas is dangerous, and possibly fatal, if you do not know what you are doing. Your tower and antenna can cause serious property damage and personal injury if any part of the installation should fail. Always use the highest quality materials in your system. Follow the manufacturer's specifications, paying close attention to base pier and guying details. Do not overload the tower. If you have any doubts about your ability to work on your tower and antennas safely, contact another amateur with experience in this area or seek professional assistance.

Chapter 22 provides more detailed guidelines for constructing a tower base and putting up a tower. It also explains how to properly attach guy wires and install guy anchors in the ground. These are extremely important parts of a tower installation, and you should not take shortcuts or use second-rate materials. Otherwise the strength and safety of your entire antenna system may be compromised.

Any mechanical job is easier if you have the right tools. Tower work is no exception. In addition to a good assortment of wrenches, screwdrivers and pliers, you will need some specialized tools to work safely and efficiently on a tower. You may already own some of these tools. Others may be purchased or borrowed. Don't start a job until you have assembled all of the necessary tools. Shortcuts or improvised tools can be fatal if you gamble and lose at 70 feet in the air. The following sections describe in detail the tools you will need to work safely on a tower.

CLOTHING

The clothing you wear when working on towers and antennas should be selected for maximum comfort and safety. Wear clothing that will keep you warm, yet allow complete freedom of movement. Long denim pants and a long-sleeve shirt will protect you from scrapes and cuts. (A pull-on shirt, like a sweat shirt with no openings or buttons to snag on tower parts, is best.) Wear work shoes with heavy soles, or better yet, with steel shanks (steel inserts in the soles), to give your feet the support they need to stand on a narrow tower rung.

Gloves are necessary for both the tower climber and all ground-crew members. Good quality leather gloves will protect hands from injury and keep them warm. They also offer protection and a better grip when you are handling rope. In cooler weather, a pair of gloves with light insulation will help keep your hands warm. The insulation should not be so bulky as to inhibit movement, however.

Ground-crew members should have hard hats for protection in case something falls from the tower. It is not uncommon for the tower climber to drop tools and hardware. A wrench dropped from 100 feet will bury itself several inches in soft ground; imagine what it might do to an unprotected skull.

SAFETY BELT AND CLIMBING ACCESSORIES

Any amateur with a tower *must* own a high-quality safety belt, such as the one shown in Fig 1. Do not attempt to climb a tower, even a short distance, without a belt. The climbing belt is more than just a safety device for the experienced climber. It is a tool to free up both hands for work. The belt allows the climber to lean back away from the tower to reach bolts or connections. It also provides a solid surface to lean against to exert greater force when hoisting antennas into place.

A climber must trust his life to his safety belt. For this reason, nothing less than a professional quality, commercially made, tested and approved safety belt is acceptable. Check the suppliers' list in Chapter 21 and ads in *QST* for suppliers of climbing belts and accessories. Examine your belt for defects *before each use*. If the belt or lanyard (tower strap) are cracked, frayed or worn in any way, destroy the damaged piece and replace it with a new one. You should never have to wonder if your belt will hold.

Along with your climbing belt, you should seriously consider purchasing some climbing accessories. A canvas bucket is a great help for carrying tools and hardware up the tower. Two buckets, a large one for carrying tools and a smaller one for hardware, make it easier to find things when needed. A few extra snap hooks like those on the ends of your belt lanyard are useful for attaching tool bags and equipment to the tower at convenient spots. These hooks are better than using rope and tying knots because in many cases they can be hooked and unhooked with one hand.

Gorilla hooks, shown in Fig 2, are especially useful for ascending and descending the tower. They attach to the belt and are hooked to the tower on alternating rungs as the climber progresses. With these hooks, the climber is secured to the tower at all times. Gorilla hooks were specially designed for amateur climbers by Ron Williams, W9JVF, 1408 W Edgewood, Indianapolis, IN 46217-3618.

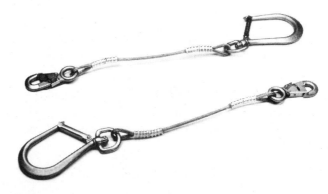

Fig 2—Gorilla hooks are designed to keep the climber attached to the tower at all times when ascending and descending.

Fig 1—Bill Lowry, W1VV, uses a good quality safety belt, a requirement for working on a tower. The belt should contain large steel loops for the strap snaps. Leather loops at the rear of the belt are handy for holding tools. *(photo by K1WA)*

Rope and Pulley

Every amateur who owns a tower should also own a good quality rope at least twice as long as the tower height. The rope is essential for safely erecting towers and installing antennas and cables. For most installations, a good quality ½-inch diameter manila hemp rope will do the job, although a thicker rope is stronger and may be easier to handle. Some types of polypropylene rope are acceptable also; check the manufacturer's strength ratings. Nylon rope is not recommended because it tends to stretch and cannot be knotted securely without difficulty.

Check your rope *before each use* for tearing or chafing. Do not attempt to use damaged rope; if it breaks with a tower section or antenna in mid-air, property damage and personal injury are likely results. If your rope should get wet, let it air dry thoroughly before putting it away.

Another very worthwhile purchase is a pulley like the one shown in Fig 3. Use the right size pulley for your rope. Be sure that the pulley you purchase will not jam or bind as the rope passes through it.

THE GIN POLE

A gin pole, like the one shown in Fig 4, is a handy device for working with tower sections and masts. This gin pole is designed to clamp onto one leg of Rohn No. 25 or 45 tower.

Fig 3—A good quality rope and pulley are essential for anyone working on towers and antennas. This pulley is encased in wood so the rope cannot jump out of the pulley wheel and jam.

The tubing, which is about 12 feet long, has a pulley on one end. A rope is routed through the tubing and over the pulley. When the gin pole is attached to the tower and the tubing extended into place, the rope may be used to haul tower sections or the mast into place. Fig 5 shows the basic process.

A gin pole can be expensive for an individual to buy, especially for a one-time tower installation. Some radio clubs own a gin pole for use by their members. Stores that sell tower sections to amateurs and commercial customers frequently will rent a gin pole to erect the tower. If you attempt to make your own gin pole, use materials heavy enough for the job. Provide a means for securely clamping the pole to the tower. There are many cases on record where homemade gin poles have failed, sending tower sections crashing down amidst the ground crew.

When you use a gin pole, make every effort to keep the load as vertical as possible. Although gin poles are strong, you are asking for trouble if you apply too much lateral force.

INSTALLING ANTENNAS ON THE TOWER

All antenna installations are different in some respects. Therefore, thorough planning is the most important first step in installing any antenna. At the beginning, before anyone climbs the tower, the whole process should be thought through. The procedure should be discussed to be sure each crew member understands what is to be done. Plan how to work out all bugs. Consider what tools and parts must be assembled and what items must be taken up the tower. Extra

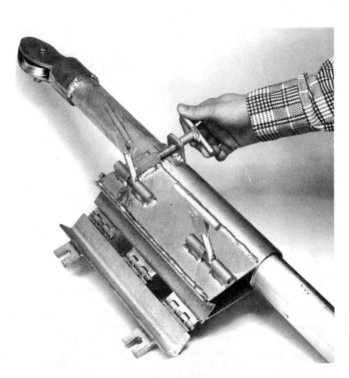

Fig 4—A gin pole is a mechanical device that can be clamped to a tower leg to aid in the assembly of sections as well as the installation of the mast. The aluminum tubing extends through the clamp and may be slipped into position before the tubing clamp is tightened. A rope should be routed through the tubing and over the pulley mounted at the top.

Fig 5—The assembly of tower sections is made simple when a gin pole is used to lift each one into position. Note that the safety belts of both climbers are fastened below the pole, thereby preventing the strap from slipping over the top section. (photo by K1WA)

trips up and down the tower can be avoided by using forethought.

Getting ready to raise a beam requires planning. Done properly, the actual work of getting the antenna into position can be accomplished quite easily with only one person at the top of the tower. The trick is to let the ground crew do all the work and leave the person on the tower free to guide the antenna into position.

Before the antenna can be hoisted into position, the tower and the area around it must be prepared. The ground crew should clear the area around the base while someone climbs the tower to remove any wire antennas or other objects that might get in the way. The first person to climb the tower should also rig the rope and pulley that will be used to raise the antenna. The time to prepare the tower is before the antenna leaves the ground, not after it becomes hopelessly entwined with your 3.5-MHz dipole.

SOME TOWER CLIMBING TIPS

The following tower climbing safety tips were compiled by Tom Willeford, N8ETU. The most important safety factor in any kind of hazardous endeavor is the right attitude. Safety is important and worthy of careful consideration and implementation. The right attitude toward safety is a requirement for tower climbers. Lip service won't do, however; safety must be practiced.

The safe ham's safety attitude is simple: *Don't take any unnecessary chances.* There are no exceptions to this plain and simple rule. It is the first rule of safety and, of course, of climbing. The second rule is equally simple: *Don't be afraid to terminate an activity* (climbing, in this case) at any time if things don't seem to be going well.

Take time to plan your climb; this time is never wasted, and it's the first building block of safety. Talk the climb over with friends who will be helping you. Select the date and alternative dates to do the work. Choose someone to be responsible for all activities on the ground and for all communication with the climbers. Study the structure to be climbed and choose the best route to your objective. Plan emergency ascent and descent paths and methods.

Make a list of emergency phone numbers to keep by your phone, even though they may never be used. Develop a plan for rescuing climbers from the structure, should that become necessary.

Give careful thought to how much time you will need to complete the project. Allow enough time to go up, do the work, and then climb down during daylight hours. Include time for resting during the climb and for completing the work in a quality fashion. Remember that the temperature changes fast as the sun goes down. Climbing up or down a tower with cold hands and feet is very difficult—and dangerous.

Give careful consideration to the weather, and climb only in good weather. Investigate wind conditions, the temperature and the weather forecast. The weather can change quickly, so if you're climbing a really tall tower it may be a good idea to have a weather alert radio handy during the climb. *Never climb a wet tower.*

The person who is going to do the climbing should be the one to disconnect and tag all sources of power to the structure. All switches or circuit breakers should be labeled clearly with DO NOT TOUCH instructions. Use locks on any switches designed to accept them. (See Fig 6.) Only the climber should reconnect power sources.

An important part of the climbing plan is to review notes on the present installation and any previous work. It's a good idea to keep a notebook, listing every bolt and nut size on your tower/antenna installation. Then, when you have to go up to make repairs, you'll be able to take the minimum number of tools with you to do the job. If you take too many tools up the tower, there is a much greater chance of dropping something, risking injury to the ground crew and possibly damaging the tool.

It is also a good idea to review the instruction sheets and take them with you. In other words, plan carefully what you are going to do, and what you'll need to do it efficiently and safely.

It's better to use a rope and pulley to hoist tools. Climbing is hard work and there's no sense making it more difficult by carrying a big load of tools. Always rig the pulley and rope so the ground crew raises and lowers tools and equipment.

Fig 6—If the switch box feeding power to equipment on your tower is equipped with a lock-out hole, use it. With a lock through the hole on the box, the power cannot be accidentally turned back on. *(Photos courtesy of American ED-CO®, at left, and Osborn Mfg Corp, at right.)*

Climbing Equipment

Equipment is another important safety consideration. By equipment, we don't just mean tools. We mean safety equipment. Safety equipment should be selected and cared for as if your life depends on it—because it does!

The list of safety equipment essential to a safe climb and safe work on the tower should include:

1) A first class safety belt,

2) Safety glasses,

3) Hard hat,

4) Long-sleeved, pullover shirt with no buttons or openings to snag (long sleeves are especially important for climbing wooden poles),

5) Long pants without cuffs,

6) Firm, comfortable, steel-shank shoes with no-slip soles and well-defined heels, and

7) Gloves that won't restrict finger movement (insulated gloves if you *must* work in cold weather).

Your safety belt should be approved for use on the structure you are climbing. Different structures may require different types of safety hooks or straps. The belt should be light weight, but strength should not be sacrificed to save weight. It should fit you comfortably. All moving parts, such as snap hooks, should work freely. You should inspect safety belts and harnesses carefully and thoroughly before each climb, paying particular attention to stitching, rivets and weight-bearing mechanical parts.

Support belt hooks should always be hooked to the D rings in an outward configuration. That is, the opening part of the hook should face away from the tower when engaged in the D rings (see Fig 7). Hooks engaged this way are easier to unhook deliberately but won't get squeezed open by a part of the tower or engage and snag a part of the tower while you are climbing. The engagement of these hooks should always be checked visually. A snapping hook makes the same sound whether it's engaged or not. Never check by sound—look to be sure the hook is engaged properly before trusting it.

Fig 7—Tom Miller, NK1P, shows the proper way to attach a safety hook, with the hook opening facing away from the tower. That way the hook can't be accidentally released by pressing it against a tower leg.

Remember that the D rings on the safety belt are for support hooks *only*. No tools or lines should be attached to these hooks. Such tools or lines may prevent the proper engagement of support belt hooks, or they may foul the hooks. At best, they could prevent the release of the hooks in an emergency. No one should have to disconnect a support hook to get a tool and then have to reconnect the support hook before beginning to work again. That's foolish.

Equipment you purchase new is best. Homemade belts or home-spliced lines are dangerous. Used belts may have worn or defective stitching, or other faulty components. Be careful of "bargains" that could cost you your life.

Straps, lanyards and lines should be as short as possible. Remember, in general knots reduce the load strength of a line by approximately 50%.

Before actually climbing, check the structure visually. Review the route. Check for obstacles, both natural (like wasp's nests) and man-made. Check the structure supports and add more if necessary. Guy wires can be obstacles to the climb, but it's better to have too many supports than not enough. Check your safety belt, support belts and hooks at the base of the tower. Really test them before you need them. Never leave the ground without a safety belt—even 5 or 10 feet. After all of this, the climb will be a "cakewalk" if you are careful.

Climb slowly and surely. Don't overreach or overstep. Patience and watchfulness is rewarded with good hand and foot holds. Take a lesson from rock climbers. Hook on to the tower and rest periodically during the climb. Don't try to rest by wedging an arm or leg in some joint; to rest, hook on. Rests provide an opportunity to review the remainder of the route and to make sure your safety equipment feels good and is working properly. Rest periods also help you conserve a margin of energy in case of difficulty.

Finally, keep in mind that the most dangerous part of working on a tower occurs when you are actually climbing. Your safety equipment is not hooked up at this time, so be extra careful during the ascent or descent.

You must climb the tower to install or work on an antenna. Nevertheless, any work that can be done on the ground should be done there. If you can do any assembly or make any adjustments on the ground, that's where you should do the work! The less time you have to spend on the tower, the better off you'll be.

When you arrive at the work area, hook on to the tower and review what you have to do. Determine the best position to do the work from, disconnect your safety strap and move to that position. Then reconnect your safety strap at a safe spot, away from joints and other obstacles. If you must move around an obstacle, try to do it while hooked on to the tower. Find a comfortable position and go to work. Don't overreach—move to the work.

Use the right tool for the task. If you don't have it, have the ground crew haul it up. Be patient. Lower tools, don't drop them, when you are finished with them. Dropped tools can bounce and cause injury or damage, or can be broken or lost. It's a good idea to tie a piece of string or light rope to the tools, and to tie the other end to the tower or some other point so if you do drop a tool, it won't fall all the way to the ground. Don't tie tools to the D ring or your safety belt, however!

Beware of situations where an antenna may be off balance. It's hard to obtain the extra leverage needed to handle even a small beam when you are holding it far from the balance point. Leverage can apply to the climber as well as the device being levered. Many slips and skinned knuckles result from such situations. A severely injured hand or finger can be a real problem to a climber.

Before descending, be sure to check all connections and the tightness of all the bolts and nuts that you have worked with. Have the ground crew use the rope and pulley to lower your tools. Lighten your load as much as possible. Remember, you're more tired coming down than going up. While still hooked on, wiggle your toes and move a little to get your senses working again. Check your downward route and begin to descend slowly and even more surely than you went up. Rest is even more important during the descent.

The ground captain is the director of all activities on the ground, and should be the only one to communicate with a person on a tower. Hand-held transceivers can be very helpful for this communication, but no one else should transmit to the workers on the tower. Even minor confusion or misunderstanding about a move to be made could be very dangerous.

"Antenna parties" can be lots of fun, but the joking and fooling around should wait until the job is done and everyone is down safely. Save the celebrating until after the work is completed, even for the ground crew.

These are just a few ideas on tower climbing safety; no list can include everything you might run into. You can't be too careful when climbing. Keep safety in mind while doing antenna work, and help ensure that after you have fallen *for* ham radio, you don't fall *from* ham radio.

THE TOWER SHIELD

A tower can be legally classified as an "attractive nuisance" that could cause injuries. You should take some precautions to ensure that "unauthorized climbers" can't get hurt on your tower. This tower shield was originally described by Baker Springfield, W4HYY, and Richard Ely, WA4VHM, in September 1976 *QST*, and should eliminate the worry.

Generally, the "attractive nuisance" doctrine applies to your responsibility to trespassers on your property. (The law is much stricter with regard to your responsibility to an invited guest.) You should expect your tower to attract children, whether they are already technically trespassing or whether the tower itself lures them onto your property. A tower is dangerous to children, especially because of their inability to appreciate danger. (What child could resist trying to climb a tower once they see one?) Because of this danger, you have a legal duty to exercise reasonable care to eliminate the danger or otherwise protect children against the perils of the attraction.

The tower shield is composed simply of panels that enclose the tower and make climbing practically impossible. These panels are 5 feet in height and are wide enough to fit snugly between the tower legs and flat against the rungs. A height of 5 feet is sufficient in almost every case. The panels are constructed from 18-gauge galvanized sheet metal obtained and cut to proper dimensions from a local sheet-metal shop. A lighter gauge could probably be used, but the extra physical weight of the heavier gauge is an advantage if no additional means of securing the panels to the tower rungs are used. The three types of metals used for the components of the shield are supposedly rustproof and nonreactive. The panels are galvanized sheet steel, the brackets aluminum, and the screws and nuts are brass. For a triangular tower, the shield consists of three panels, one for each of the three sides, supported by two brackets. Construct these brackets from 6-inch pieces of thin aluminum angle stock. Bolt two of the pieces together to form a Z bracket (see Figs 8, 9 and 10). The Z brackets are bolted together with

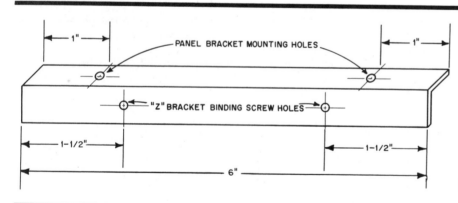

Fig 8—Z-bracket component pieces.

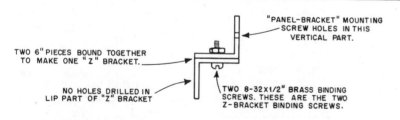

Fig 9—Assembly of the Z bracket.

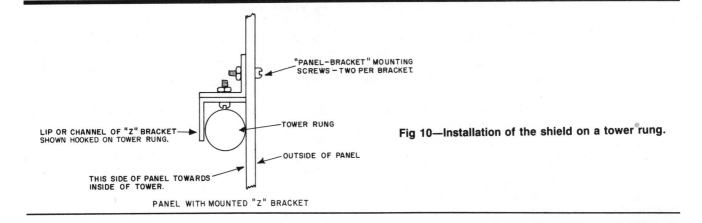

"PANEL-BRACKET" MOUNTING
SCREWS — TWO PER BRACKET.

LIP OR CHANNEL OF "Z" BRACKET
SHOWN HOOKED ON TOWER RUNG.

TOWER RUNG

OUTSIDE OF PANEL

THIS SIDE OF PANEL TOWARDS
INSIDE OF TOWER.

PANEL WITH MOUNTED "Z" BRACKET

Fig 10—Installation of the shield on a tower rung.

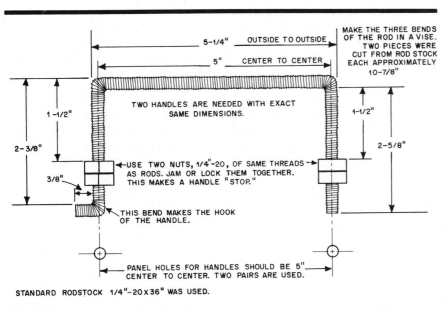

5-1/4" OUTSIDE TO OUTSIDE

5" CENTER TO CENTER

MAKE THE THREE BENDS
OF THE ROD IN A VISE.
TWO PIECES WERE
CUT FROM ROD STOCK
EACH APPROXIMATELY
10-7/8"

TWO HANDLES ARE NEEDED WITH EXACT
SAME DIMENSIONS.

1-1/2"

2-3/8"

1-1/2"

2-5/8"

3/8"

USE TWO NUTS, 1/4"-20, OF SAME THREADS
AS RODS. JAM OR LOCK THEM TOGETHER.
THIS MAKES A HANDLE "STOP."

THIS BEND MAKES THE HOOK
OF THE HANDLE.

PANEL HOLES FOR HANDLES SHOULD BE 5"
CENTER TO CENTER. TWO PAIRS ARE USED.

STANDARD RODSTOCK 1/4"-20 X 36" WAS USED.

Fig 11—Removable handle construction.

Fig 12—Installed tower shield. Note the holes for using the handles.

binding head brass machine screws.

Lay the panels flat for measuring, marking and drilling. First measure from the top of the upper mounting rung on the tower to the top of the bottom rung. (Mounting rungs are selected to position the panel on the tower.) Then mark this distance on the panels. Use the same size brass screws and nuts throughout the shield. Bolt the top vertical portion of each Z bracket to the panel. Drill the mounting-screw holes about 1 inch from the end of the Z brackets so there is an offset clearance between the Z-bracket binding-screw holes and the panel-bracket mounting-screw holes. Drill holes in each panel to match the Z-bracket holes.

The panels are held on the tower by their own weight. They are not easy to grasp because they fit snugly between the tower legs. If you feel a need for added safety against deliberate removal of the panels, this can be accomplished by means of tie wires. Drill a small hole in the panel just above, just below, and in the center of each Z bracket. Run a piece of heavy galvanized wire through the top hole, around the Z bracket, and then back through the hole just below the Z bracket. Twist together the two ends of the wire. One tie wire should be sufficient for each panel, but use two if desired.

The completed panels are rather bulky and difficult to handle. A feature that is useful if the panels have to be removed often for tower climbing or accessibility is a pair of removable handles. The removable handles can be constructed from one threaded rod and eight nuts (see Fig 11). Drill two pair of handle holes in the panels a few inches below the top Z bracket and several inches above the bottom Z bracket. For panel placement or removal, you can hook the handles in these panel holes. The hook, on the top of the handle, fits into the top hole of each pair of the handle holes. The handle is optional, but for the effort required it certainly makes removal and replacement much safer and easier.

Fig 12 shows the shield installed on a tower. This relatively simple device could prevent an accident.

Electrical Safety

Although the RF, ac and dc voltages in most amateur stations pose a potentially grave threat to life and limb, common sense and knowledge of safety practices will help you avoid accidents. Building and operating an Amateur Radio station can be, and is for almost all amateurs, a perfectly safe pastime. However, carelessness can lead to severe injury, or even death. The ideas presented here are only guidelines; it would be impossible to cover all safety precautions. Remember, there is no substitute for common sense.

A fire extinguisher is a requirement for the well-equipped amateur station. The fire extinguisher should be of the carbon-dioxide type to be effective in electrical fires. Store it in an easy-to-reach spot and check it at recommended intervals.

Family members should know how to turn the power off in your station. They should also know how to apply artificial respiration. Many community groups offer courses on cardiopulmonary resuscitation (CPR).

AC AND DC SAFETY

The primary wiring for your station should be controlled by one master switch, and other members of your household should know how to kill the power in an emergency. All equipment should be connected to a good ground. All wires carrying power around the station should be of the proper size for the current to be drawn and should be insulated for the voltage level involved. Bare wire, open-chassis construction and exposed connections are an invitation to accidents. Remember that high-current, low-voltage power sources are just as dangerous as high-voltage, low-current sources. Possibly the most-dangerous voltage source in your station is the 120-V primary supply; it is a hazard often overlooked because it is a part of everyday life. Respect even the lowliest power supply in your station.

Whenever possible, kill the power and unplug equipment before working on it. Discharge capacitors with an insulated screwdriver; don't assume the bleeder resistors are 100% reliable. In a power amplifier, always short the tube plate cap to ground just to be sure the supply is discharged. If you must work on live equipment, keep one hand in your pocket. Avoid bodily contact with any grounded object to prevent your body from becoming the return path from a voltage source to ground. Use insulated tools for adjusting or moving any circuitry. Never work alone. Have someone else present; it could save your life in an emergency.

National Electrical Code

The National Electrical Code® is a comprehensive document that details safety requirements for all types of electrical installations. In addition to setting safety standards for house wiring and grounding, the Code also contains a section on Radio and Television Equipment—Article 810. Sections C and D specifically cover Amateur Transmitting and Receiving Stations. Highlights of the section concerning Amateur Radio stations follow. If you are interested in learning more about electrical safety, you may purchase a copy of *The National Electrical Code* or *The National Electrical Code Handbook*, edited by Peter Schram, from the National Fire Protection Association, Batterymarch Park, Quincy, MA 02269.

Antenna installations are covered in some detail in the Code. It specifies minimum conductor sizes for different length wire antennas. For hard-drawn copper wire, the Code specifies no. 14 wire for open (unsupported) spans less than 150 feet, and no. 10 for longer spans. Copper-clad steel, bronze or other high-strength conductors may be no. 14 for spans less than 150 feet and no. 12 wire for longer runs. Lead-in conductors (for open-wire transmission line) should be at least as large as those specified for antennas.

The Code also says that antenna and lead-in conductors attached to buildings must be firmly mounted at least 3 inches clear of the surface of the building on nonabsorbent insulators. The only exception to this minimum distance is when the lead-in conductors are enclosed in a "permanently and effectively grounded" metallic shield. The exception covers coaxial cable.

According to the Code, lead-in conductors (except those covered by the exception) must enter a building through a rigid, noncombustible, nonabsorbent insulating tube or bushing, through an opening provided for the purpose that provides a clearance of at least 2 inches or through a drilled window pane. All lead-in conductors to transmitting equipment must be arranged so that accidental contact is difficult.

Transmitting stations are required to have a means of draining static charges from the antenna system. An antenna discharge unit (lightning arrester) must be installed on each lead-in conductor (except where the lead-in is protected by a continuous metallic shield that is permanently and effectively grounded, or the antenna is permanently and effectively grounded). An acceptable alternative to lightning arrester installation is a switch that connects the lead-in to ground when the transmitter is not in use.

Grounding conductors are described in detail in the Code. Grounding conductors may be made from copper, aluminum, copper-clad steel, bronze or similar erosion-resistant material. Insulation is not required. The "protective grounding conductor" (main conductor running to the ground rod) must be as large as the antenna lead-in, but not smaller than no. 10. The "operating grounding conductor" (to bond equipment chassis together) must be at least no. 14. Grounding conductors must be adequately supported and arranged so they are not easily damaged. They must run in as straight a line as practical between the mast or discharge unit and the ground rod.

The Code also includes some information on safety inside the station. All conductors inside the building must be at least 4 inches away from conductors of any lighting or signaling circuit except when they are separated from other conductors by conduit or a nonconducting material. Transmitters must be enclosed in metal cabinets, and the cabinets must be grounded. All metal handles and controls accessible by the operator must be grounded. Access doors must be fitted with interlocks that will disconnect all potentials above 350 V when the door is opened.

Ground

An effective ground system is necessary for every amateur station. The mission of the ground system is twofold. First, it reduces the possibility of electrical shock if something in a piece of equipment should fail and the chassis or cabinet becomes "hot." If connected properly, three-wire electrical systems ground the chassis, but older amateur equipment may use the ungrounded two-wire system. A ground system to prevent shock hazards is generally referred to as "dc ground."

The second job the ground system must perform is to provide a low-impedance path to ground for any stray RF current inside the station. Stray RF can cause equipment to malfunction and contributes to RFI problems. This low-impedance path is usually called "RF ground." In most stations, dc ground and RF ground are provided by the same system.

The first step in building a ground system is to bond together the chassis of all equipment in your station. Ordinary hookup wire will do for a dc ground, but for a good RF ground you need a low-impedance conductor. Copper strap, sold as "flashing copper," is excellent for this application, but it may be hard to find. Braid from coaxial cable is a popular choice; it is readily available, makes a low-impedance conductor, and is flexible.

Grounding straps can be run from equipment chassis to equipment chassis, but a more convenient approach is illustrated in Fig 13. In this installation, a ½-inch diameter copper water pipe runs the entire length of the operating bench. A thick braid (from discarded RG-8 cable) runs from each piece of equipment to a clamp on the pipe. Copper water pipe is available at most hardware stores and home centers. Alternatively, a strip of flashing copper may be run along the rear of the operating bench.

After the equipment is bonded to a common ground bus, the ground bus must be wired to a good earth ground. This run should be made with a heavy conductor (braid is a popular choice, again) and should be as short and direct as possible. The earth ground usually takes one of two forms.

In most cases, the best approach is to drive one or more ground rods into the earth at the point where the conductor from the station ground bus leaves the house. The best ground rods to use are those available from an electrical supply house. These rods are 8 to 10 feet long and are made from steel with a heavy copper plating. Do not depend on shorter, thinly plated rods sold by some home electronics suppliers. These rods begin to rust almost immediately after they are driven into the soil, and they become worthless within a short time. Good ground rods, while more expensive initially, offer long-term protection.

If your soil is soft and contains few rocks, an acceptable alternative to "genuine" ground rods is ½-inch diameter copper water pipe. A 6- to 8-foot length of this material offers a good ground, but it may bend while being driven into the

earth. Some people have recommended that you make a connection to a water line and run water down through the copper pipe so that it forces its own hole in the ground. There may be a problem with this method, however. When the ground dries, it may shrink away from the pipe and not make proper contact with the ground rod. This would provide a rather poor ground.

Once the ground rod is installed, clamp the conductor from the station ground bus to it with a clamp that can be tightened securely and will not rust. Copper-plated clamps made especially for this purpose are available from electrical supply houses, but a stainless-steel hose clamp will work too. Alternatively, drill several holes through the pipe and bolt the conductor in place. If a torch is available, solder the connection.

Another popular station ground is the cold water pipe system in the building. To take advantage of this ready made ground system, run a low-impedance conductor from the station ground bus to a convenient cold water pipe, preferably somewhere near the point where the main water supply enters the house. Avoid hot water pipes; they do not run directly into the earth. The advent of PVC (plastic) plumbing makes it mandatory to inspect the cold water system from your intended ground connection to the main inlet. PVC is an excellent insulator, so any PVC pipe or fittings rule out your cold water system for use as a station ground.

For some installations, especially those located above the first floor, a conventional ground system such as that just described will make a fine dc ground but will not provide the necessary low-impedance path to ground for RF. The length of the conductor between the ground bus and the ultimate ground point becomes a problem. For example, the ground wire may be about ¼ λ (or an odd multiple of ¼ λ) long on some amateur band. A ¼-λ wire acts as an impedance inverter from one end to the other. Since the grounded end is at a very low impedance, the equipment end will be at a high impedance. The likely result is RF hot spots around the station while the transmitter is operating. A ground system like this may be worse than having no ground at all.

An alternative RF ground system is shown in Fig 14. Connect a system of ¼-λ radials to the station ground bus. Install at least one radial for each band used. You should still be sure to make a connection to earth ground for the ac power wiring. Try this system if you have problems with RF in the shack. It may just solve a number of problems for you.

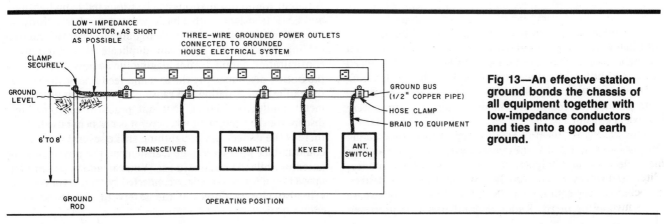

Fig 13—An effective station ground bonds the chassis of all equipment together with low-impedance conductors and ties into a good earth ground.

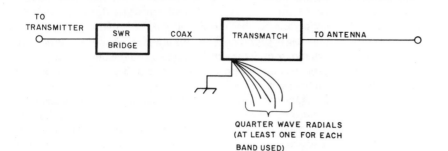

Fig 14—Here is an alternative to earth ground if the station is located far from the ground point and RF in the station is a problem. Install at least one ¼-λ radial for each band used.

Ground Noise

Noise in ground systems can affect sensitive radio equipment. It is usually related to one of three problems:
1) Insufficient ground conductor size,
2) Loose ground connections, or
3) Ground loops.

These matters are treated in precise scientific research equipment and some industrial instruments by paying attention to certain rules. The ground conductor should be at least as large as the largest conductor in the primary power circuit. Ground conductors should provide a solid connection to both ground and to the equipment being grounded. Liberal use of lock washers and star washers is highly recommended. A loose ground connection is a tremendous source of noise, particu-larly in a sensitive receiving system.

Ground loops should be avoided at all costs. A short discussion of what a ground loop is and how to avoid them may lead you down the proper path. A ground loop is formed when more than one ground current is flowing in a single conductor. This commonly occurs when grounds are "daisy-chained" (series linked). The correct way to ground equipment is to bring all ground conductors out radially from a common point to either a good driven earth ground or to a cold water system.

Ground noise can affect transmitted as well as received signals. With the low audio levels required to drive amateur transmitters, and with the ever-increasing sensitivity of our receivers, correct grounding is critical.

Lightning and EMP Protection

The National Fire Protection Association (NFPA) publishes a booklet called *Lightning Protection Code* (NFPA no. 78-1983) that should be of interest to radio amateurs. For information about obtaining a copy of this booklet, write to the National Fire Protection Association, Batterymarch Park, Quincy, MA 02269. Two paragraphs of particular interest to amateurs are presented here:

"3-26 Antennas. Radio and television masts of metal, located on a protected building, shall be bonded to the lightning protection system with a main size conductor and fittings.

"3-27 Lightning arresters, protectors or antenna discharge units shall be installed on electric and telephone service entrances and on radio and television antenna lead-ins."

The best protection from lightning is to disconnect all antennas from equipment and disconnect the equipment from the power lines. Ground antenna feed lines to safely bleed off static buildup. Eliminate the possible paths for lightning strokes. Rotator cables and other control cables from the antenna location should also be disconnected during severe electrical storms.

In some areas, the probability of lightning surges entering homes via the 120/240-V line may be high. Lightning produces both electrical and magnetic fields that vary with distance. These fields can be coupled into power lines and destroy electronic components in equipment that is miles from where the lightning occurred. Radio equipment can be protected from these surges to some extent by using transient-protective devices.

ELECTROMAGMETIC PULSE AND THE RADIO AMATEUR

The following material is based on a 4-part *QST* article by Dennis Bodson, W4PWF, that appeared in the August through November 1986 issues of *QST* (see the bibliography at the end of this chapter). The series was condensed from the National Communications System report NCS TIB 85-10.

An equipment test program demonstrated that most Amateur Radio installations can be protected from lightning and EMP transients with a basic protection scheme. Most of the equipment is not susceptible to damage when all external cabling is removed. You can duplicate this stand-alone configuration simply by unplugging the ac power cord from the outlet, disconnecting the antenna feed line at the rear of the radio, and isolating the radio gear from any other long metal conductors. Often it is not practical to completely disconnect the equipment whenever it is not being used. Also, there is the danger that a lightning strike several miles away could induce a large voltage transient on the power lines or antenna while the radio is in use. You can add two transient-protection devices to the interconnected system, however, that will also closely duplicate the safety of the stand-alone configuration.

The ac power line and antenna feed line are the two important points that should be outfitted with transient protection. This is the minimum basic protection scheme recommended for all Amateur Radio installations. (For fixed installations, consideration should also be given to the rotator connections—see Fig 15.) Hand-held radios equipped with a "rubber duck" require no protection at the antenna jack. If a larger antenna is used with the hand-held, however, a protection device should be installed.

General Considerations

Because of the unpredictable energy content of a nearby lightning strike or other large transient, it is possible for a metal-oxide varistor (MOV) to be subjected to an energy surge in excess of its rated capabilities. This may result in the destruction of the MOV and explosive rupture of the package. These fragments can cause damage to nearby components or operators and possibly ignite flammable material. Therefore, the MOV should be physically shielded.

A proper ground system is a key factor in achieving protection from lightning and EMP transients. A low-impedance ground system should be installed to eliminate transient paths through radio equipment and to provide a good physical ground for the transient-suppression devices. A single-point ground system is recommended (see Fig 16). Inside the station, single-point grounding can be had by installing a ground panel or bus bar. All external conductors going to the radio equipment should enter and exit the station through this panel. Install all transient-suppression devices directly on the panel. Use the shortest length(s) possible of no. 6 solid wire to connect the radio equipment case(s) to the ground bus.

Ac Power-Line Protection

Tests have indicated that household electrical wiring inherently limits the maximum transient current that it will pass to approximately 120 A. Therefore, if possible, the amateur station should be installed away from the house ac entrance panel and breaker box to take advantage of these limiting effects.

Ac power-line protection can be provided with easy-to-install, plug-in transient protectors. Ten such devices were tested (see Table 1). Six of these can be plugged directly into an ac outlet. Four are modular devices that require more extensive installation and, in some cases, more than one module.

The plug-in-strip units are the best overall choice for the typical amateur installation. They provide the protection needed, they're simple to install and can be easily moved to other operating locations with the equipment. The modular devices are second choices because they all require some installation, and none of the units tested provided full EMP protection for all three wires of the ac power system.

NCS considers the TII model 428 Plug-In Powerline Protector to be the best overall protector. It provides transient paths to ground from the hot and neutral lines (common mode) as well as a transient path between the hot and neutral lines (normal mode). The model 428 uses three MOVs and a 3-electrode gas-discharge-tube arrester to provide fast operation and large power dissipation capabilities. This unit was tested repeatedly and operated without failure.

Several other plug-in transient protectors provide 3-wire

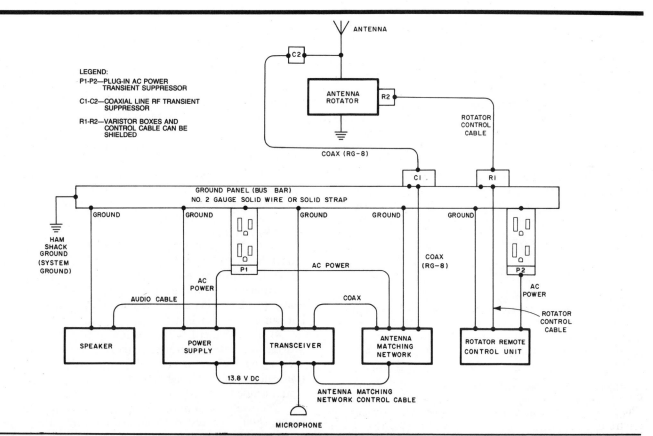

Fig 15—Transient suppression techniques applied to an Amateur Radio station.

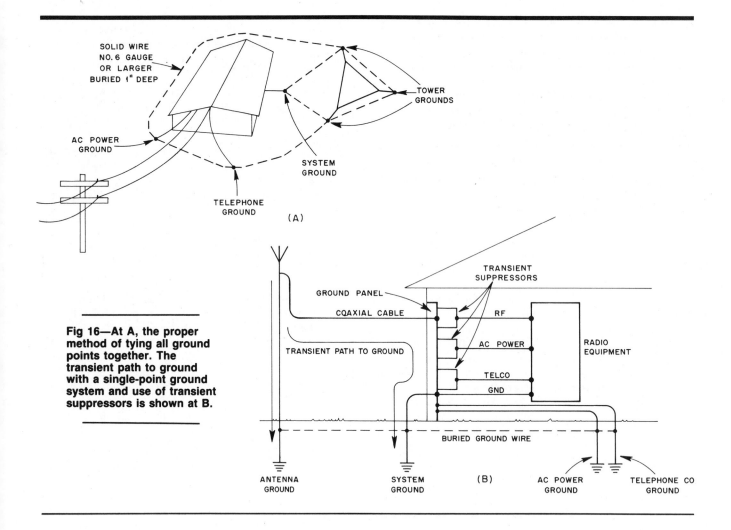

Fig 16—At A, the proper method of tying all ground points together. The transient path to ground with a single-point ground system and use of transient suppressors is shown at B.

Table 1
Ac Power-Line Protection Devices

Manufacturer	Device	Approximate Cost (US Dollars)	Measured High-Z Clamping Voltage (Volts)
(Modules)			
Fischer	FCC-120F-P	55	420
Joslyn	1250-32	31	940
General Semiconductor	587B051	56	600
General Semiconductor	PHP 120	50	400
(Plug-Ins)			
Joslyn	1270-02	49	600
TII	428	45	410
Electronic Protection Devices	Lemon	45	580
Electronic Protection Devices	Peach	60	1000
S. L. Waber	LG-10	13	600
Archer	61-2785*	22	300

*No longer available.

protection, but all operate at higher clamping voltages. Other low-cost plug-in devices either lack the 3-wire protection capability or have substantially higher clamping voltages. Some of these are the:

1) Joslyn 1270-02. It provides full 3-wire (common and normal mode) transient-path protection but at a slightly higher cost and at a higher clamping voltage.

2) Lemon and Peach protection devices, manufactured by Electronic Protection Devices, Inc. The Lemon provides full (common and normal mode) 3-wire protection, but at a higher clamping voltage; the Peach has a dangerously high (1000 V) clamping voltage.

3) Archer (Radio Shack) 61-2785 [replaced by a new model that wasn't tested—Ed.]. This unit provides excellent clamping performance at low cost, but it offers normal-mode protection only (a transient path between the hot and neutral leads). It will provide some protection for lightning transients, but not enough for EMP.

4) S. L. Waber LG-10. The lowest-cost device does not provide full three-wire protection (normal mode only) and has a clamping voltage of 600. This unit can provide limited transient protection for lightning, but not the 3-wire protection recommended for EMP transients.

The transient suppressors require a 3-wire outlet; the outlet should be tested to ensure all wires are properly connected. In older houses, an ac ground may have to be installed by a qualified electrician. The ac ground must be available for the plug-in transient suppressor to function properly. The ac ground of the receptacle should be attached to the station ground bus, and the plug-in receptacle should be installed on the ground panel behind the radio equipment.

Emergency Power Generators

Emergency power generators provide two major transient-protection advantages. First, the station is disconnected from the commercial ac power system. This isolates the radio equipment from a major source of damaging transients. Second, tests have shown that the emergency power generator may not be susceptible to EMP transients.

When the radio equipment is plugged directly into the generator outlets, transient protection may not be needed. If an extension cord or household wiring is used, transient protection should be employed.

An emergency power generator should be wired into the household circuit only by a qualified electrician. When connected properly, a switch is used to disconnect the commercial ac power source from the house lines before the generator is connected to them. This keeps the generator output from feeding back into the commercial power system. If this is not done, death or injury to unsuspecting linemen can result.

Feed Line Protection

Coaxial cable is recommended for use as the transmission line because it provides a certain amount of transient surge protection for the equipment to which it is attached. The outer conductor shields the center conductor from the transient field. Also, the cable limits the maximum conducted transient voltage on the center by arcing the differential voltage from the center conductor to the grounded cable shield.

By providing a path to ground ahead of the radio equipment, the gear can be protected from the large currents impressed upon the antenna system by lightning and EMP. A single protection device installed at the radio antenna jack will protect the radio, but not the transmission line. To protect the transmission line, another transient protector must be installed between the antenna and the transmission line. (See Fig 15.)

RF transient protection devices from three manufacturers were tested (see Table 2) using RG-8 cable equipped with UHF connectors. All of the devices shown can be installed in a coaxial transmission line. Recall that during the tests the RG-8 cable acted like a suppressor; damaging EMP energy arced from the center conductor to the cable shield when the voltage level approached 5.5 kV.

Low price and a low clamping-voltage rating must be considered in the selection of an RF transient-protection device. The lower-cost devices have the higher clamping voltages, however, and the higher-cost devices have the lower clamping voltages. Because of this, medium-priced devices manufactured by Fischer Custom Communications were selected for testing. The Fischer Spikeguard Suppressors ($55 price class) for coaxial lines can be made to order to operate at a specific clamping voltage. The Fischer devices satisfactorily suppressed the damaging transient pulses, passed the transmitter RF output power without interfering with the signal, and operated effectively over a wide frequency range.

Polyphaser Corporation devices are also effective in providing the necessary transient protection. The devices available limited the transmitter RF output power to 100 W or less, however. These units cost approximately $83 each.

The Alpha Delta Transi-Traps tested were low-cost items, but not suitable for EMP suppression because of their high (over 700-V) clamping levels. [New Alpha Delta "EMP" units have clamping voltages rated to be about one-third that of the older units tested here.—Ed.]

RF coaxial protectors should be mounted on the station ground bus bar. If the Fischer device is used, it should be attached to a grounded UHF receptacle that will serve as a hold-down bracket. This creates a conductive path between the outer shield of the protector and the bus bar. The Polyphaser device can be mounted directly to the bus bar with the bracket provided.

Attach the transceiver or antenna matching network to the grounded protector with a short (6 foot or less) piece of coaxial cable. Although the cable provides a ground path to the bus bar from the radio equipment, it is not a satisfactory transient-protection ground path for the transceiver. Another ground should be installed between the transceiver case and the ground bus using solid no. 6 wire. The coaxial cable shield should be grounded to the antenna tower leg at the tower base. Each tower leg should have an earth ground connection and be connected to the single-point ground system as shown in Fig 16.

Antenna Rotators

Antenna rotators can be protected by plugging the control box into a protected ac power source and adding protection to the control lines to the antenna rotator. When the control lines are in a shielded cable, the shield must be grounded at both ends. MOVs of the proper size should be installed at both ends of the control cable. At the station end, terminate the control cable in a small metal box that is connected to the station ground bus. Attach MOVs from each conductor to ground inside the box. At the antenna end of the control cable, place the MOVs inside the rotator case or in a small metal box that is properly grounded.

For example, the Alliance HD73 antenna rotator uses a

Table 2
RF Coaxial-Line Protectors

Manufacturer	Device	Approximate Cost (US Dollars)	Measured High-Z Clamping Voltage (Volts)
Fischer	FCC-250-300-UHF	55	393
Fischer	FCC-250-350-UHF	55	260
Fischer	FCC-250-150-UHF	55	220
Fischer	FCC-250-120-UHF	55	240
Fischer	FCC-450-120-UHF	55	120
Polyphaser	IS-NEMP	83	140
Polyphaser	IS-NEMP-1	83	150
Polyphaser	IS-NEMP-2	83	160
Alpha Delta	LT	20	700*
Alpha Delta	R-T	30	720*

Note: The transmitter output power, frequency of operation, and transmission line SWR must be considered when selecting any of these devices.

*The newer Alpha Delta LT and R-T "EMP" models have clamping voltages rated to be one-third of those shown here.

6-conductor unshielded control cable with a maximum control voltage of 24.7 V dc. Select an MOV with a clamping voltage level 10% higher (27 V or more) so the MOV won't clamp the control signal to ground. If the control voltage is ac, be sure to convert the RMS voltage value to peak voltage when considering the clamping voltage level.

Mobile Power Supply Protection

The mobile amateur station environment exposes radio equipment to other transient hazards in addition to those of lightning and EMP. Currents as high as 300 A are switched when starting the engine, and this can produce voltage spikes of over 200 V on the vehicle's electrical system. Lightning and EMP are not likely to impact the vehicle's electrical system as much as they would that of a fixed installation because the automobile chassis is not normally grounded. This would not be the case if the vehicle is inadvertently grounded; for example, when the vehicle is parked against a grounded metal conductor. The mobile radio system has two advantages over a fixed installation: Lightning is almost never a problem, and the vehicle battery is a natural surge suppressor.

Mobile radio equipment should be installed in a way that takes advantage of the protection provided by the battery. See Fig 17. To do this, connect the positive power lead of the radio directly to the positive battery post, not to intermediate points in the electrical system such as the fuse box or the auxiliary contacts on the ignition switch. To prevent equipment damage or fire, should the positive lead short to ground, an in-line fuse should be installed in the positive lead where it is attached to the battery post.

Connect the negative power lead to the chassis on the battery side of the quick-disconnect connector. Although it would help prevent alternator whine, connecting the negative power lead directly to the battery post is not recommended from an EMP standpoint.

An MOV should be installed between the two leads of the equipment power cord. A GE MOV (V36ZA80) is recommended for this application. This MOV provides the lowest measured clamping voltage (170 V) and is low in cost.

Mobile Antenna Installation

Although tests indicate that mobile radios can survive an EMP transient without protection for the antenna system, protection from lightning transients is still required. A coaxial-line transient suppressor should be installed on the vehicle chassis between the antenna and the radio's antenna connector.

A Fischer suppressor can be attached to a UHF receptacle that is mounted on, and grounded to, the vehicle chassis. The Polyphaser protector can be mounted on, and grounded to, the vehicle chassis with its flange. Use a short length of coaxial cable between the radio and the transient supressor.

Clamping Voltage Calculation

When selecting any EMP-protection device to be used at the antenna port of a radio, several items must be considered. These include transmitter RF power output, the SWR, and the operating frequency. The protection device must allow the outgoing RF signal to pass without clamping. A clamping voltage calculation must be made for each amateur installation.

The RF-power input to a transmission line develops a corresponding voltage that becomes important when a voltage-surge arrester is in the line. SWR is important because of its influence on the voltage level. The maximum voltage developed for a given power input is determined by:

$$V = \sqrt{2 \times P \times Z \times SWR} \qquad \text{(Eq 1)}$$

where

P = peak power in W
Z = impedance of the coaxial cable (ohms)
V = peak voltage across the cable

Eq 1 should be used to determine the peak voltage present across the transmission line. Because the RF transient-protection devices use gas-discharge tubes, the voltage level at which they clamp is not fixed; a safety margin must be added to the calculated peak voltage. This is done by multiplying the calculated value by a factor of three. This added safety margin is required to ensure that the transmitter's RF output power will pass through the transient suppressor without causing the device to clamp the RF signal to ground. The final clamping voltage obtained is then high enough to allow normal operation of the transmitter while providing the lowest practical clamping voltage for the suppression device. This ensures the maximum possible protection for the radio system.

Here's how to determine the clamping voltage required. Let's assume the SWR is 1.5:1. The power output of the

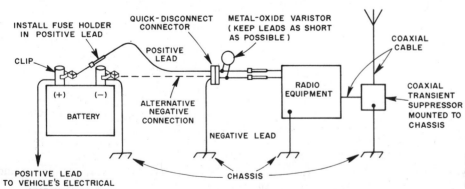

Fig 17—Recommended method of connecting mobile radio equipment to the vehicle battery and antenna.

transceiver is 100 W PEP. RG-8 coaxial cable has an impedance of 52 ohms. Therefore:

P = 100 W
Z = 52 ohms
SWR = 1.5

Substituting these values into Eq 1:

$$V = \sqrt{2 \times 100 \times 52 \times 1.5} = 124.89$$

Note that the voltage, V, is the peak value at the peak of the RF envelope. The final clamping voltage (FCV) is three times this value, or 374.7 V. Therefore, a coaxial-line transient suppressor that clamps at or above 375 V should be used.

The cost of a two-point basic protection scheme is estimated to be $100 for each fixed amateur station. This includes the cost of one TII model 428 plug-in power-line protector ($45) and one Fischer coaxial-line protector ($55).

Inexpensive Transient-Protection Devices

Here are two low-cost protection devices you can assemble. They performed flawlessly in the tests.

SIOV AC Box

The SIOV (SIemens metal-Oxide Varistor) power-line protection device shown in Fig 18 is fabricated by installing a duplex receptacle in a metal electrical box. Power is brought into the box through a 6-foot-long, 3-conductor power cord.

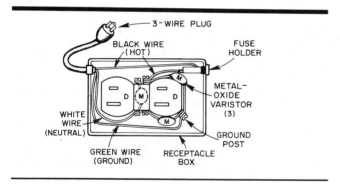

Fig 18—Pictorial diagram of an inexpensive, homemade ac power-line transient protector. This approach may be applied to multiple outlets; see text.

A fuse is installed in the incoming hot wire to guard against harmful effects if one of the protective devices shorts. MOVs (Siemens S14K130) are installed—with the shortest possible lead lengths—between the hot and neutral, hot and ground, and neutral and ground leads. The estimated cost of this unit is $11.

UHF Coaxial T

The radio antenna connection can be protected by means of another simple device. As shown in Fig 19, two spark gaps (Siemens BI-A350) are installed in series at one end of a coaxial-cable T connector. Use the shortest practical lead length (about ¼ inch) between the two spark gaps. One lead is bent forward and forced between the split sections of the inner coaxial connector until the spark gaps approach the

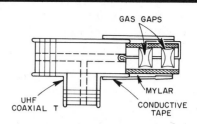

Fig 19—Pictorial diagram of an inexpensive, homemade transmission-line transient protector. See text for description of assembly.

body of the connector. A short length of insulating material (such as Mylar) is placed between the spark gaps and the connector shell. The other spark-gap lead is folded over the insulator, then conductive (metallic) tape is wrapped around the assembly. This construction method proved durable enough to allow many insertions and removals of the device during testing. Estimated cost of this assembly is $9. Similar devices can be built using components from Joslyn, General Electric, General Semiconductor or Siemens.

Summary

Amateurs should be aware of which components in their radio system are most likely to be damaged by EMP. They should also know how to repair the damaged equipment. Amateurs should know how to reestablish communications after an EMP event, taking into consideration its adverse effects on the earth's atmosphere and radio equipment. One of the first things that would be noticed, providing the radio equipment is operative, is a sudden silence in radio transmissions across all frequencies below approximately 100 MHz. This silence would be caused in part by damage to unprotected radio gear from the EMP transient. Transmissions from one direction, the direction of the nuclear blast, would be completely out. RF signal loss by absorption and attenuation by the nuclear fireball are the reasons for this.

After an EMP event, the amateur should be prepared to operate CW. CW gives the most signal power under adverse conditions. It also provides a degree of message security from the general public.

Amateurs should develop the capability and flexibility to operate in more than one frequency band. The lower ground-wave frequencies should be useful for long-distance communications immediately after an EMP event. Line-of-sight VHF would be of value for local communications.

What can be done to increase the survivability of an Amateur Radio station? Here are some suggestions:

1) If you have spare equipment, keep it disconnected; use only the primary station gear. The spare equipment would then be available after an EMP event.

2) Keep equipment turned off and antenna and power lines disconnected when the equipment is not in use.

3) Connect only those external conductors necessary for the current mode of operation.

4) Tie all fixed equipment to a single-point earth ground to prevent closed loops through the ground.

5) Obtain schematic diagrams of your equipment and tools for repair of the equipment.

6) Have spare parts on hand for sensitive components of the radio equipment and antenna system.

7) Learn how to repair or replace the sensitive components of the radio equipment.

8) Use nonmetallic guy lines and antenna structural parts where possible.

9) Obtain an emergency power source and operate from it during periods of increased world political tension. The power source should be completely isolated from the commercial power lines.

10) Equipment power cords should be disconnected when the gear is idle. Or the circuit breaker for the line feeding the equipment should be kept in the OFF position when the station is off the air.

11) Disconnect the antenna lead-in when the station is off the air. Or use a grounding antenna switch and keep it in the GROUND position when the equipment is not in use.

12) Have a spare antenna and transmission line on hand to replace a damaged antenna system.

13) Install EMP surge arresters and filters on all primary conductors attached to the equipment and antenna.

14) Retain tube-type equipment and spare components; keep them in good working order.

15) Do not rely on a microprocessor to control the station after an EMP event. Be able to operate without microprocessor control.

The recommendations contained in this section were developed with low cost in mind; they are not intended to cover all possible combinations of equipment and installation methods found in the amateur community. Amateurs should examine their own requirements and use this report as a guideline in providing protection for the equipment.

RF Radiation Safety

An often overlooked safety precaution is the avoidance of unnecessary exposure to RF energy. Amateur Radio is basically a safe activity, but accidents can always occur if we don't use common sense. This is where a general awareness of some basic safety precautions can help.

Is RF Power Hazardous?

Body tissues that are subjected to large amounts of RF energy may suffer serious heat damage. These effects depend upon the wavelength of the radiation, the energy density of the RF field that strikes the body, and even on factors like the polarization of the wave.

At frequencies where the body's length is around 0.4 wavelength, RF energy is absorbed most efficiently; this occurs in the VHF range, between 30 and 300 MHz. (The frequency range is so large partly because of the great differences in human heights.)

Most Amateur Radio operations use relatively low RF power, and the operations are highly intermittent. Hams spend more time listening than transmitting, and actual transmissions—like keyed CW and SSB—are inherently intermittent. The exception is RTTY, because the RF is present continuously at its maximum level when transmitting in this mode.

There has been considerable investigation of and discussion about the biological effects of RF radiation in recent years. Several government agencies have studied the issue, and the American National Standards Institute conducted extensive research on the matter. The ARRL even established a committee to gather information and prepare an educated opinion on the effects of radiation exposure from typical Amateur Radio operation.

The main purpose of this section is to suggest where some protection from RF radiation may be necessary. The RF protection problem comes in two parts: determining safe exposure levels, and estimating the local RF field strengths produced by a given transmitter power and antenna system. If the RF power density levels are greater than or even roughly equal to the maximum levels determined to be safe, then some protection or precaution is required.

Safe Exposure Levels

In recent years, scientists have devoted a great deal of effort to determining safe RF-exposure limits. This is a very complex problem, so it should come as no surprise that some changes in the recommended safe levels have occurred. In July of 1982 the American National Standards Institute (ANSI) released a standard for exposure limits to RF radiation, ANSI C95.1-1982 (see bibliography). The standard presents some guidelines, known as radio frequency protection guides, abbreviated RFPG. That standard took nearly five years to develop, and underwent repeated critical review by the scientific community. The standard recognizes the phenomenon of whole-body or geometric resonance, and recommends a frequency-dependent maximum permissible RF exposure level.

As mentioned earlier, the body absorbs a maximum amount of radiation when the height is about 0.4 wavelength at the frequency of the incident radiation. Because of this, the lowest safe exposure levels occur at frequencies between

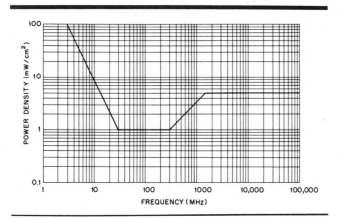

Fig 20—ANSI RF protection guidelines for body exposure of humans.

30 and 300 MHz. Outside of this range, the body can safely receive higher levels of radiation. The ANSI standard sets the maximum exposure limits between 30 and 300 MHz at 1 mW/cm².

The safe exposure limits rise gradually on either side of this frequency range. At 3 MHz the maximum permissible exposure level is 100 mW/cm² and at 1500 MHz and above the limit is 5 mW/cm². Fig 20 is a graph depicting these exposure limits. In specifying a constant 5 mW/cm² above 1500 MHz, the ANSI standard considers that there is very little penetration of the energy into body tissue with these extremely short wavelengths.

One other important point needs to be made with regard to the ANSI RFPG. On page 15 of its report, the ANSI committee that studied the biological effects noted, "...many low-power devices that are used by a large segment of the general population, such as citizen's band radio, and amateur, public safety, land mobile and marine transmitters, may generate localized fields that appear to exceed the RFPG but result in a significantly lower rate of whole-body averaged energy absorption as a result of the limited area of exposed tissue." Part of this lower rate of energy absorption also takes into account the limited, intermittent transmissions from these radios as compared to a commercial installation.

The report continues: "... the only practical way to cope with the problem of low-power devices was to enter an exclusion clause in the RFPG that would allow the power density (and local strengths) of incident fields to be exceeded under certain conditions."

The report details several reasons why such an exclusion is valid. Amateur Radio stations are excluded from the radiation limits set by the standard. Still, amateurs should be aware that the standard exists, and that they can take precautions to limit their exposure (and that of their family members and neighbors).

ESTIMATING POWER DENSITY

How can the power density of the radio signals transmitted from an amateur station be determined? Or more directly, how will you know if your radiation exposure exceeds the recommended limits? Although special equipment exists that accurately measures RF electric fields, most amateurs don't have access to such equipment. Power density will have

to be estimated with a few calculations.

These estimates involve some approximations, but they should serve as a prediction of the upper limit of exposure for a particular arrangement. The procedures outlined in this section should offer a conservative estimate of the dangers. Keep in mind that the results should be used only as a guideline, however.

These calculations should not be taken as *proof* that a particular setup is safe. For example, this section considers only radiation from an antenna. You can be exposed to RF radiation directly from a power amplifier if it is operated without proper shielding. Transmission lines can also radiate energy under some conditions. You can take precautions to ensure that your antenna is the only part of your station radiating RF energy, however.

The calculations shown here use a free-space propagation model as a first approximation of power density. You should probably include as much as a 4 to 6 dB safety factor to allow for situations where a reinforcing reflection might occur.

Antenna engineers generally split the region around an antenna into a *near-field* region and a *far-field* region. Fig 21 shows what is meant by near-field and far-field regions. Although a dipole antenna is shown in the drawing, the information applies to other common antennas, such as Yagi beams. The power density in the far-field region of an antenna can be calculated from

$$\rho = \frac{PG}{4\pi R^2} \qquad \text{(Eq 2)}$$

where
- ρ = estimated power density at a distance R from an antenna. (Units will be W/m² if P is in watts and R is in meters.)
- R = distance from observation point to closest point on antenna (in meters).
- P = *average* power at antenna feed point (in watts).
- G = antenna gain over a dipole as a power ratio (this is a numerical gain, not gain expressed in decibels)

Eq 2 is valid only if two requirements are met. The free-space radiation model must be appropriate, and the distance from the antenna must be sufficient to be in the far field.

To ensure an upper-bound estimate of power density, use

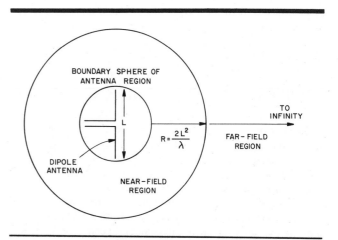

Fig 21—Diagram showing the antenna region, the near-field region and the far-field region around an antenna. The text gives an example for calculating power density of the antenna radiation in the far-field region.

the free-space antenna gain if you don't know the actual value. A textbook value will generally be useful. If you are given an antenna gain in dB, convert it to numerical gain from

$$\text{Gain ratio} = 10^{dB/10} \qquad \text{(Eq 3)}$$

Let's work through a sample power density calculation. You should be able to apply a similar technique to estimate the radiation exposure produced by your station. Suppose you are operating on the 14-MHz band, using a 3-element Yagi antenna with an approximate gain of 6 dB over a dipole. The average power delivered to the antenna feed point is 500 watts (this is probably a reasonable average power for a 1500-W PEP SSB transmission). What is the radiation exposure for a person 15 meters from the antenna?

First, we calculate the gain ratio for this antenna using Eq 3.

$$\text{Gain ratio} = 10^{6/10} = 10^{0.6} = 3.98 \approx 4$$

Next we calculate the minimum distance from the antenna that can be considered to be in the far field of the antenna. The equation for this calculation comes from Fig 21.

$$R_{min} = \frac{2\,L^2}{\lambda} \qquad \text{(Eq 4)}$$

where L = length of longest antenna element (in meters)

For our example, we'll use a longest element length of 35 feet. Convert this value to meters by multiplying by 0.3048. Thus, L = 10.7 m. Then we have

$$R_{min} = \frac{2 \times (10.7 \text{ m})^2}{20 \text{ m}} = \frac{228.98 \text{ m}^2}{20 \text{ m}} = 11.4 \text{ m}$$

Now we can be assured that Eq 2 is valid for the distance we want to do the calculation for, since 15 m is farther than the minimum distance. From Eq 2,

$$\rho = \frac{500 \text{ W} \times 4}{4 \pi \times (15 \text{ m})^2} = \frac{2000 \text{ W}}{2827 \text{ m}^2} = 0.707 \text{ W/m}^2$$

To compare this power density with the RFPG standard, it must be converted to units of mW/cm^2. That is a rather simple exercise in metric unit conversions.

$$\rho = \frac{0.707 \text{ W}}{\text{m}^2} \times \frac{1000 \text{ mW}}{1 \text{ W}} \times \frac{1 \text{ m}^2}{10,000 \text{ cm}^2}$$
$$= 0.070 \text{ mW/cm}^2$$

From the graph of Fig 20, we find that the recommended maximum power density at 14 MHz is about 4.6 mW/cm^2. From our calculation, the station seems to be well within the limits prescribed by the ANSI standard.

SOME FURTHER RF EXPOSURE GUIDELINES

Potential exposure situations should be taken seriously. Observing these "RF awareness" guidelines will help make your Amateur Radio operation safe.

1) Confine RF radiation to antenna radiating elements themselves. Provide a single, good station ground (earth), and eliminate radiation from transmission lines.

2) In high-power operation (several hundred watts and up) in the HF and VHF region, and particularly with RTTY, try to avoid human presence near antenna ends. With vertical monopole antennas, humans should go no closer than 10-15 feet during high-power, nonintermittent operation.

3) Don't operate RF power amplifiers, especially at VHF/UHF, with the covers removed.

4) In the UHF/SHF region, never look into the open end of an activated length of waveguide or point it toward anyone. Never point a high-gain, narrow-beamwidth antenna (a paraboloid, for instance) toward people.

5) With mobile rigs of 10 W or more RF power, do not transmit if anyone is standing near the antenna.

6) With hand-held transceivers RF power output above several watts, maintain an inch or so between forehead and antenna. Keep the antenna tip away from your head.

7) Don't work on antennas that have RF power applied.

8) Amateur antennas on towers, well away from people, pose no exposure problem. Make sure, however, that transmission lines are not radiating.

QST carries information regarding the latest developments for RF environmental regulations at the local and federal levels, and tells how these regulations may affect amateur operations. You can find additional information about the biological effects of RF radiation in the publications listed in the bibliography.

BIBLIOGRAPHY

Source material and more extended discussion of topics covered in this chapter can be found in the references given below and in the textbooks listed at the end of Chapter 2.

Q. Balzano, O. Garay and K. Siwiak, "The Near Field of Dipole Antennas, Part I: Theory," *IEEE Transactions on Vehicular Technology* (VT) 30, p 161, Nov 1981. Also "Part II: Experimental Results," same issue, p 175.

D. Bodson, "Electromagnetic Pulse and the Radio Amateur," in four parts, *QST*, Aug through Nov 1986. The series was condensed from the National Communications System report (NCS TIB 85-10) *Electromagnetic Pulse/Transient Threat Testing of Protection Devices for Amateur/Military Affiliate Radio System Equipment.* [A copy of the unabridged report is available from the NCS. Write to Dennis Bodson, Acting Assistant Manager, Office of Technology and Standards, National Communications System, Washington, DC 20305-2010.]

A. W. Guy and C. K. Chou, "Thermographic Determination of SAR in Human Models Exposed to UHF Mobile Antenna Fields," *Paper F-6, Third Annual Conference,* Bioelectromagnetics Society, Washington, DC, Aug 9-12, 1981.

D. L. Lambdin, "An Investigation of Energy Densities in the Vicinity of Vehicles with Mobile Communications Equipment and Near a Hand-Held Walkie Talkie," *EPA Report ORP/EAD 79-2,* Mar 1979.

D. I. McRee, *A Technical Review of the Biological Effects of Non-Ionizing Radiation,* Office of Science and Technology Policy, Washington, DC, 1978.

W. W. Mumford, "Heat Stress Due to RF Radiation," *Proceedings of the IEEE,* 57, 1969, pp 171-178.

R. J. Spiegel, "The Thermal Response of a Human in the Near-Zone of a Resonant Thin-Wire Antenna," *IEEE Transactions on Microwave Theory and Technology* (MTT) 30(2), pp 177-185, Feb 1982.

B. Springfield and R. Ely, "The Tower Shield," *QST,* Sep 1976, p 26.

American National Standards Institute (ANSI), "Safety Levels with Respect to Human Exposure to Radio Frequency Electromagnetic Fields (300 kHz to 100 GHz)," *ANSI C95.1-1982,* Jul 1982.

Chapter 2

Antenna Fundamentals

Antennas are electric circuits of a special kind. In ordinary circuits, the dimensions of coils, capacitors and connections usually are small compared with the wavelength that corresponds to the frequency in use. When this is the case, most of the electromagnetic energy stays in the circuit itself, and is either used up in performing useful work or is converted into heat. But when the dimensions of wiring or components become appreciable compared with the wavelength, some of the energy escapes by *radiation* in the form of electromagnetic waves. When the circuit is intentionally designed so that the major portion of the energy is radiated, the circuit is an *antenna*.

Usually, an antenna is a straight section of conductor, or a combination of such conductors. Frequently the conductor is a wire, although rods and tubing are also used. In this chapter the term "wire" is used to mean any type of conductor having a cross section that is small compared to its length.

The strength of the electromagnetic field radiated from a section of wire carrying RF current depends on the length of the wire and the amount of current flowing in it. The field strength also depends on the voltage across the section of wire, but it is generally more convenient to measure current. The electromagnetic field consists of both magnetic and electric energy, with the total energy equally divided between the two; one cannot exist without the other in an electromagnetic wave. The voltage in an antenna is just as much a measure of the field intensity it will produce as the current. Other factors being equal, field strength is directly proportional to current. It is therefore desirable to make the current as large as possible for a given amount of transmitter power. In any circuit that contains both resistance and reactance, the largest current flows (again, for a given amount of power) when the reactance is "tuned out"—in other words, when the circuit is made *resonant* at the operating frequency. This is the case with the common types of antenna; the current flowing in the antenna is largest, and the radiation therefore greatest, when the antenna is resonant.

In an ordinary circuit the inductance is usually concentrated in a coil, the capacitance in a capacitor, and the resistance is principally concentrated in resistors, although some resistance is distributed around the circuit wiring and coil conductors. Such circuits are said to have *lumped constants*. On the other hand, in an antenna the inductance, capacitance and resistance are distributed along the wire. Such a circuit is said to have *distributed constants*. Circuits with distributed constants are so frequently straight-line conductors that they are customarily called linear circuits.

RESONANCE IN LINEAR CIRCUITS

The shortest length of wire that resonates at a given frequency is one just long enough to permit an electric charge to travel from one end to the other and back in the time of one RF cycle. If the speed at which the charge travels is equal to the velocity of light, 299,793,077 meters per second (983,573,087 feet per second), the distance it covers in one cycle or period is equal to this velocity divided by the frequency in hertz, or approximately

$$\lambda = \frac{299,800,000}{f \text{ (hertz)}} \qquad \text{(Eq 1)}$$

in which λ is the wavelength, in meters for this case. (The concept of wavelength is further discussed in Chapter 23.) Because the charge traverses the wire *twice*, the length of wire needed to permit the charge to travel a distance λ in one cycle is $\lambda/2$, or one-half wavelength. The shortest *resonant* wire is therefore $\frac{1}{2} \lambda$ long.

The reason for this length can be made clear by a simple example. Imagine a trough with barriers at each end. If an elastic ball is started along the trough from one end, it will strike the far barrier, bounce back, travel along to the near barrier, bounce again, and continue until all the energy originally imparted to it is dissipated. If, however, each time the ball returns to the near barrier it is given a new push just as it starts away, this back and forth motion can be kept up indefinitely. The impulses, however, must be *timed* properly; in other words, the rate or frequency of the impulses must be adjusted to the length of travel and the rate of travel. Or, if the timing of the impulses and the speed of the ball are fixed, the length of the trough must be adjusted to "fit."

In the case of the antenna, the speed is essentially constant, so the alternatives of adjusting the frequency to a given length of wire, or adjusting the length of wire to a given operating frequency, are both available. (Adjusting the wire length is usually the practical condition.)

By changing the units in the equation just given and dividing by 2, the equation

$$\ell = \frac{491.8}{f(\text{MHz})} \qquad \text{(Eq 2)}$$

is obtained. In this case, ℓ is the length *in feet* of a *half* wavelength for a frequency f (in megahertz), when the wave travels at the velocity of light. This equation is the basis upon which several significant lengths in antenna work lie. It represents the length of a half wavelength in free space, when no factors that modify the speed of propagation exist. To determine a half wavelength in meters, the relationship is

$$\ell = \frac{149.9}{f(\text{MHz})} \qquad \text{(Eq 3)}$$

Current and Voltage Distribution

If the wire in an antenna was infinitely long, the charge

(voltage) and the current (an electric charge in motion) would both slowly decrease in amplitude with distance from the source. The slow decrease would result from dissipation of energy in the form of radiated electromagnetic waves and in wire heating resulting from conductor resistance. However, when the wire is short, the charge is reflected when it reaches the far end, just as the ball bounces back from the barrier.

With RF excitation of a ½-λ antenna, there is, of course, not just a single charge, but a continuous supply of energy, varying in voltage according to a sine wave cycle. This can be considered a series of charges, each of slightly different amplitude than the preceding one. When a charge reaches the end of the antenna and is reflected, the direction of current flow reverses. This is because the charge is now traveling in the opposite direction. The next charge is just reaching the end of the antenna, however, so two currents of essentially the same amplitude, but flowing in opposite directions, exist at that point on the wire. The resultant current at the end of the antenna therefore is zero.

Farther back from the end of the antenna, the magnitudes of the outgoing and returning currents are no longer the same, because the charges causing them have been supplied to the antenna at different parts of the RF cycle. There is less cancellation, therefore, and a measurable current exists. The greatest difference—that is, the largest resultant current—exists ¼ λ away from the end of the antenna. Back still farther from this point the current decreases until, a half wavelength away from the end of the antenna, it reaches zero again. Thus, in a ½-λ antenna, the current is zero at the ends and maximum at the center.

This resultant current distribution along a ½-λ wire is shown in Fig 1. The distance measured vertically from the antenna wire to the curve marked "current" at any point along the wire represents the relative amplitude of the current as measured by an ammeter at that point. This is called a *standing wave* of current. The *instantaneous* current at any point varies sinusoidally at the applied frequency, but its amplitude is different at every point along the wire, as shown by the curve. The standing-wave curve itself has the approximate shape of half of a sine wave.

The voltage along the wire behaves differently than the current; it is greatest at the end because at this point, two practically equal charges add together. Back along the wire,

however, the outgoing and returning charges are not equal and their sum is smaller. At the ¼-λ point the returning charge is equal in magnitude but opposite in phase with the outgoing charge. This is because at this time the polarity of the voltage wave from the source has reversed (½ cycle). The two voltages therefore cancel each other and the resultant voltage is zero. Beyond the ¼-λ point away from the end of the wire, the voltage again increases, but this time with the opposite polarity.

The voltage is therefore maximum at every point where the current is minimum, and vice versa. The polarity of the current or voltage reverses every ½ λ along the wire, but the reversals do not occur at the same points for both current and voltage; the respective reversals occur, in fact, at points ¼ λ apart. A maximum point on a standing wave is called a *loop* or *antinode*; a minimum point is called a *node*.

Harmonic Operation

If there is reflection from the end of a wire, the number of standing waves on the wire is equal to the length of the wire in half wavelengths. Thus, if the wire is two half waves long, there are two standing waves; if three half waves long, three standing waves, and so on. These longer wires, each multiples of ½ λ long, are therefore also resonant at the same frequency as the single ½-λ wire. When an antenna is two or more half waves in length at the operating frequency, it is said to be *harmonically resonant*, or to operate at a harmonic. The number of the harmonic is the number of standing waves on the wire. For example, a wire two half waves long is said to be operating on its *second harmonic*, one three half waves long on its *third harmonic*, and so on.

Harmonic operation of a wire is illustrated in Fig 2. Such operation is often used in antenna work because it permits operating the same antenna on several harmonically related amateur bands. It is also an important principle in the operation of certain types of directive antennas.

Electrical Length

The *electrical* length of a linear circuit such as an antenna wire is not necessarily the same as its *physical* length in wavelengths. Rather, the electrical length is measured by the *time* taken for the completion of a specified phenomenon.

For instance, imagine two linear circuits having such different characteristics that the speed at which a charge travels is not the same in both. Suppose both circuits are to resonate at the same frequency, and for that purpose the physical length of each is adjusted until a charge started at

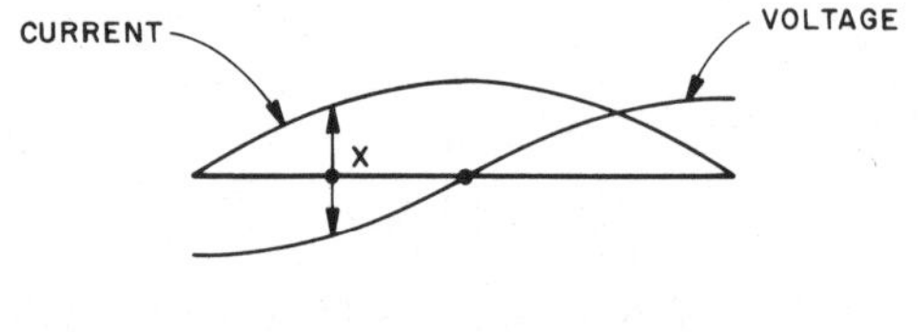

Fig 1—Current and voltage distribution on a ½-λ (half-wavelength) wire. The wire is represented by the heavy line. In this conventional representation the distance at any point (X, for instance) from the wire to the curve gives the relative intensity of current or voltage at that point on the wire. The relative direction of current flow (or polarity of the voltage) is indicated by drawing the curve either above or below the line representing the antenna. The voltage curve here, for example, indicates that the instantaneous polarity in one half of the antenna is opposite that in the other half.

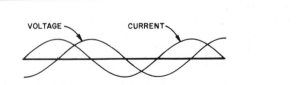

Fig 2—Harmonic operation of a long wire. The wire is long enough to contain several half waves—in this case, three. The current and voltage curves cross the heavy line representing the wire to indicate that there is reversal in the direction of the current, and a reversal in the polarity of the voltage, at intervals of a half wavelength. The reversals of current and voltage do not coincide, but occur at points ¼ λ apart.

one end travels to the far end, is reflected and completes its return journey to the near end in exactly the time of one cycle of the applied frequency. In such a case, the *physical* length of the circuit with the lower velocity of propagation is shorter than the physical length of the other. The *electrical* lengths, however, are identical, each being a half wavelength.

In ac circuits, the instantaneous values of current or voltage are determined by the instant during the cycle at which the measurement is made (assuming, of course, that such a measurement could be made rapidly enough). If the current and voltage follow a sine curve—which is the usual case—the time (for any instantaneous value) can be specified in terms of an angle. The sine of the angle gives the instantaneous value when multiplied by the *peak* value of the current or voltage. A complete sine curve occupies the 360° of a circle and represents one cycle of alternating current or voltage. Thus, a half cycle is equal to 180°, a quarter cycle 90°, and so on.

It is often convenient to use this same form of representation for linear circuits. When the electrical length of a circuit is such that a charge *traveling in one direction* takes the time of one cycle or period to traverse it, the length of the circuit is said to be 360°. This corresponds to one wavelength. On a wire that is a half wave in electrical length, the charge completes a one-way journey in one-half cycle, and its length is said to be 180°. The angular method of measurement is quite useful for lengths that are not simple fractions or multiples of such fractions. To convert fractional wavelengths to angular lengths, multiply the fraction by 360°.

Velocity of Propagation

The velocity at which electromagnetic waves travel through a medium depends upon the dielectric constant of the medium. At RF the dielectric constant of air is practically unity, so the waves travel at essentially the same velocity as light in a vacuum. This is also very close to the velocity of the charge traveling along a wire.

If the dielectric constant is greater than 1, the velocity of propagation is lowered. Thus, the introduction of insulating material having a dielectric constant greater than unity causes RF energy to travel more slowly. This effect is encountered in practice in connection with antennas and transmission lines.

It causes the electrical length of the line or antenna to be greater than the actual physical length indicates.

Length of a "Half-Wave" Antenna

Even if the antenna could be supported by insulators that did not cause the electromagnetic fields traveling along the wire to slow down, the physical length of a practical antenna is always somewhat less than its electrical length. That is, a "half-wave" antenna is not one having the same length as a half wavelength in space. It is one having an *electrical* length equal to 180°. Or to put it another way, it is one with a length which has been adjusted to "tune out" any reactance, so it is a *resonant* antenna.

The antenna length required to resonate at a given frequency (independently of any dielectric effects) depends on the ratio of the length of the conductor to its diameter. The smaller this ratio (or the "thicker" the wire), the shorter the antenna must be for a given electrical length. This effect is shown in Fig 3 as a factor K, by which a free-space half wavelength must be multiplied to find the resonant length. K is a function of the ratio of the free-space wavelength to conductor diameter, abbreviated λ/dia. The curve is based on theoretical considerations and is useful as a guide to the probable antenna length for a given frequency. It applies only to conductors of uniform diameter—tapered elements such as those used in some types of beam antennas are generally longer, for the same frequency, as discussed in detail later in this chapter. Neither does the curve of Fig 3 include any effects introduced by the method of supporting the conductor. (The effects of end loading are discussed in the next section.)

A λ/dia value of 20,000 is roughly average for wire antennas (it is approximately the ratio for a 7-MHz ½-λ antenna made of no. 12 wire). In this region K changes rather slowly, and a half-wave antenna made of wire is about 2% shorter than ½ λ in space.

The shortening effect is most pronounced when the λ/dia ratio is 200 or less. A resonant antenna constructed of 1-inch diameter tubing for use on 144 MHz for example, has a λ/dia ratio of about 80, and is almost 5% shorter than a free-space half wavelength.

If the antenna is made of rod or tubing and is not

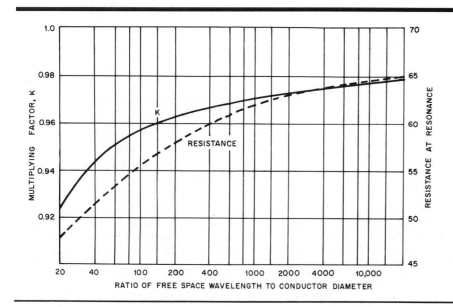

Fig 3—The solid curve shows the factor, K, by which the length of a half wave in free space should be multiplied to obtain the physical length of a resonant half-wave antenna versus the wavelength to diameter ratio. This curve does not take end effect or element tapering into account. The broken curve shows how the radiation resistance of a half-wave antenna varies with the wavelength to diameter ratio.

supported near the ends by insulators, the following equations give the required physical length of a ½-λ antenna based on Fig 3.

$$\text{Length (feet)} = \frac{491.8 \times K}{f(MHz)} \qquad \text{(Eq 4)}$$

$$\text{Length (inches)} = \frac{5902 \times K}{f(MHz)} \qquad \text{(Eq 5)}$$

where K is taken from Fig 3 for the particular λ/dia ratio of the conductor used.

End Effect

If Eqs 4 and 5 are used to determine the length of a wire antenna, the antenna will resonate at a somewhat lower frequency than is desired. The reason is that an additional "loading" effect is caused by the insulators used at the ends of the wires to support the antenna. These insulators and the wire loops that tie the insulators to the antenna add a small amount of capacitance to the system. This capacitance helps to tune the antenna to a slightly lower frequency, in much the same way that additional capacitance in any tuned circuit lowers the resonant frequency. In an antenna it is called *end effect*. The current at the ends of the antenna does not quite reach zero because of the end effect, as there is some current flowing into the end capacitance.

End effect increases with frequency and varies slightly with different installations. However, at frequencies up to 30 MHz (the frequency range over which wire antennas are most commonly used), experience shows that the length of a ½-λ antenna is on the order of 5% less than the length of a half wave in space. As an average, then, the physical length of a resonant ½-λ wire antenna can be found from

$$\ell = \frac{491.8 \times 0.95}{f(MHz)} \approx \frac{467}{f(MHz)} \qquad \text{(Eq 6)}$$

Eqs 4 through 6 are reasonably accurate for finding the physical length of a ½-λ antenna for a given frequency, but do not apply to antennas longer than a half wave in length. In the practical case, if the antenna length must be adjusted to exact frequency (not all antenna systems require it) the length should be "pruned" to resonance.

ANTENNA IMPEDANCE

In the simplified description given earlier of voltage and current distribution along an antenna, it was stated that the voltage is zero at the center of a ½-λ antenna (or at any current loop along a longer antenna). It is more accurate to say that the voltage reaches a *minimum*, rather than *zero*. Zero voltage with a finite value of current implies that the circuit is entirely without resistance. It also implies that no energy is radiated by the antenna, because a circuit without resistance could take no real power from the driving source.

Antennas, like any other circuit, consume power. The current that flows into the antenna therefore must be supplied at a finite voltage. The *impedance* of the antenna is simply equal to the voltage applied to its terminals divided by the current flowing into those terminals. If the current and voltage are exactly in phase the impedance is purely resistive. This is the case when the antenna is resonant. If the antenna is not exactly resonant, the current is somewhat out of phase with the applied voltage and the antenna shows reactance along with resistance.

Most amateur transmitting antennas are operated at or

close to resonance so that reactive effects are comparatively small. They are nevertheless present, and must be taken into account whenever an antenna is operated at other than the exact design frequency.

In the following discussion it is assumed that power is applied to the antenna by opening the conductor at the center and applying the driving voltage across the gap. This is shown in Fig 4. While it is possible to supply power to the antenna by other methods, the selection of different driving points leads to different values of impedance; this can be appreciated after study of Fig 1, which shows that the ratio of voltage to current (which by definition is the impedance) is different at every point along an antenna. To avoid confusion, it is desirable to use the conditions at the center of the antenna as a reference.

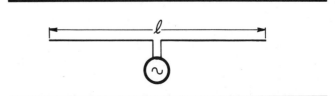

Fig 4—The center-fed antenna. It is assumed that the source of power is directly at the antenna feed point.

The Antenna as a Circuit

If the frequency applied at the center of a ½-λ antenna is varied above and below the resonant frequency, the antenna exhibits much the same characteristics as a conventional series-resonant circuit. Exactly at resonance, the current at the input terminals is in phase with the applied voltage. If the frequency is below resonance, the phase of the current leads the voltage; that is, the reactance of the antenna is capacitive. When the frequency is above resonance, the opposite occurs; the current lags the applied voltage and the antenna exhibits inductive reactance.

The following example illustrates this phenomenon. Consider the antennas shown in Fig 5—one resonant, one too short for the applied frequency, and one too long. In each case the applied voltage is shown at A, and the instantaneous current going *into* the antenna because of the applied voltage is shown at B. Note that this current is always in phase with the applied voltage, regardless of the antenna length. For the sake of simplicity only the current flowing in one leg of the antenna is considered; conditions in the other leg are similar.

In the case of the resonant antenna, the current travels out to the end and back to the driving point in one half cycle, because one leg of the antenna is 90° long and the total path out and back is therefore 180°. This makes the phase of the *reflected* current component differ from that of the outgoing current by 180°, because the outgoing current has gone through a half cycle in the meantime. Remember that there is, however, a phase shift of 180° at the end of the antenna, because the direction of current reverses at the end. The *total* phase shift between the outgoing and reflected currents is therefore 360°. In other words, the reflected component arrives at the driving point exactly in phase with the outgoing component. The reflected component, shown at C, adds to the outgoing component to form the resultant or total current

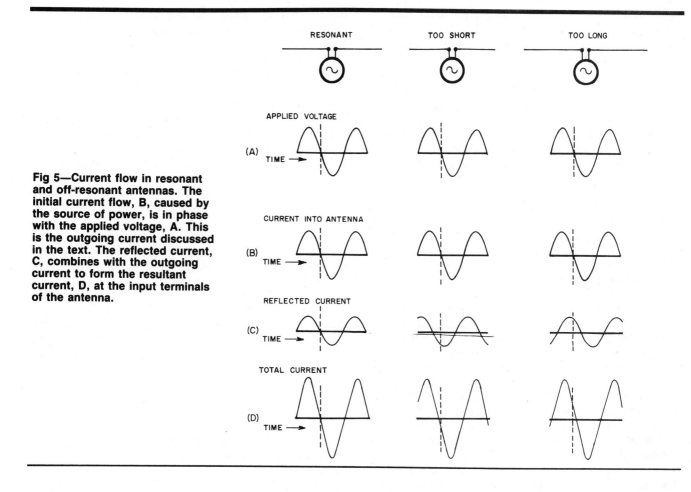

RESONANT   TOO SHORT   TOO LONG

APPLIED VOLTAGE

(A) TIME →

CURRENT INTO ANTENNA

(B) TIME →

REFLECTED CURRENT

(C) TIME →

TOTAL CURRENT

(D) TIME →

Fig 5—Current flow in resonant and off-resonant antennas. The initial current flow, B, caused by the source of power, is in phase with the applied voltage, A. This is the outgoing current discussed in the text. The reflected current, C, combines with the outgoing current to form the resultant current, D, at the input terminals of the antenna.

at the driving point. The resultant current is shown at D, and in the case of the resonant antenna it is easy to understand that the resultant current is *exactly in phase* with the applied voltage. This being the case, the load seen by the source of power is a pure resistance.

Now consider the antenna that is too short to be resonant. The outgoing component of current is still in phase with the applied voltage, as shown at B. The reflected component, however, gets back to the driving point *too soon*, because it travels over a path less than 180°, out and back. This means that the maximum value of the reflected component occurs at the driving point before (in time) the maximum value of the outgoing component, because that charge took less than a half cycle to get back. Including the 180° reversal at the end of the antenna, the total phase shift is therefore less than 360°. This is shown at C, and the resultant current is the combination of the outgoing and reflected components, as drawn at D. The resultant current leads the applied voltage, so the antenna looks like a resistance in series with a capacitance. The shorter the antenna, the greater the phase shift between voltage and current; that is, the capacitive reactance increases as the antenna is shortened.

When the antenna is too long for the applied frequency, the reflected current component arrives too late to be exactly in phase with the outgoing component because it must travel over a path more than 180° long. The maximum value of the reflected component therefore occurs later (in time) than the maximum value of the outgoing component, as shown at C. The resultant current at the antenna input terminals therefore

lags the applied voltage. The phase lag increases as the antenna is made longer. That is, an antenna that is too long shows inductive reactance along with resistance, and this reactance increases with an increase in antenna length over the length required for resonance.

If the antenna length is increased to 180° on each leg, the go-and-return path length for the current becomes 360°. This, plus the 180° reversal at the end, makes the total phase shift 540°, which is the same as a 180° shift. In this case the reflected current arrives at the input terminals exactly *out of phase* with the outgoing component, so the resultant current is very small. The resultant is in phase with the applied voltage, so the antenna impedance is again purely resistive. The resistance under these conditions is very high, and the antenna has the characteristics of a parallel-resonant circuit. A *voltage* loop, instead of a *current* loop, appears at the input terminals when each leg of the antenna is 180° long.

The amplitude of the reflected component is less than that of the component of current going into the antenna. This is the result of energy loss by radiation as the current travels along the wire. It is perhaps easier to understand if, instead of thinking of the electromagnetic fields as resulting from the current flow, we adopt the more fundamental viewpoint that *current flow along a conductor is caused by a moving electromagnetic field*. When some of the energy escapes from the system because the field travels out into space, it is not hard to understand why the current decreases as it travels farther. There is simply less energy left to cause it. The difference between the outgoing and reflected current amplitudes accounts for the fact that the current does not go to zero at a voltage loop, and a similar difference between the applied and reflected voltage components explains why the voltage does not go to zero at a current loop.

Resistance

The energy supplied to an antenna is principally dissipated in two ways: radiation of radio waves, and heat losses in the wire and nearby dielectrics. The radiated energy is the useful part, but it represents a loss just as much as the energy used in heating the wire is a loss. In either case, the dissipated power is equal to I^2R. In the case of heat losses, R is a real resistance. In the case of radiation, however, R is a "virtual" resistance which, if replaced with an actual resistor of the same value, would dissipate the power that is actually radiated from the antenna. This resistance is called the *radiation resistance*. The total power loss in the antenna is therefore equal to $I^2(R_0 + R)$, where R_0 is the radiation resistance and R the real resistance, or ohmic resistance.

In ordinary $\frac{1}{2}$-λ antennas operated at amateur frequencies, the power lost as heat in the conductor does not exceed a few percent of the total power supplied to the antenna. This is because the RF resistance of copper wire even as small as no. 14 is very low compared with the radiation resistance of an antenna that is reasonably clear of surrounding objects and is not too close to the ground. It can therefore be assumed that the ohmic loss in a reasonably well located antenna is negligible, and that the total resistance shown by the antenna (the feed-point resistance) is radiation resistance. As a radiator of electromagnetic waves, such an antenna is a highly efficient device.

The value of radiation resistance, as measured at the center of a $\frac{1}{2}$-λ antenna, depends on a number of factors. One is the location of the antenna with respect to other objects, particularly the earth. Another is the length to diameter ratio (λ/dia) of the conductor used. In "free space"—with the antenna remote from everything else—the radiation resistance of a resonant antenna made of an infinitely thin conductor is approximately 73 Ω. The concept of a free-space antenna forms a convenient basis for calculation because the modifying effect of the ground can be taken into account separately. If the antenna is at least several wavelengths away from ground and other objects, it can be considered to be in free space for the purposes of its electrical properties. This condition can be met with relative ease with antennas in the VHF and UHF range.

The way in which free-space radiation resistance varies with the λ/dia ratio of a $\frac{1}{2}$-λ antenna is shown by the broken curve in Fig 3. As the antenna is made thicker, the radiation resistance decreases. For most wire antennas the resistance is close to 65 Ω. For antennas constructed of rod or tubing, the radiation resistance usually is between 55 and 60 Ω.

The actual value of the radiation resistance—at least if it is 50 Ω or more—has no appreciable effect on the radiation efficiency of the antenna. This is because the ohmic resistance is on the order of only 1 Ω with the conductors used for "thick" antennas. The ohmic resistance does not become important unless the radiation resistance drops to very low values—say less than 10 Ω—as may be the case when several antennas are coupled together in the form of an array of elements.

The radiation resistance of a resonant antenna is the "load" for the transmitter or for the RF transmission line connecting the transmitter and antenna. Its value is important,

therefore, in determining the way in which the antenna and transmitter or line are coupled together.

The radiation resistance of an antenna varies with its length as well as with the λ/dia ratio. When the antenna is approximately $\frac{1}{2}$ λ long, the resistance changes rather slowly with length. This is shown by the curves of Fig 6, where the change in resistance as the length is varied on either side of resonance is shown by the broken lines. The resistance decreases somewhat when the antenna is slightly short, and increases when it is slightly long.

These curves also illustrate the effect of changing the frequency applied to an antenna of fixed length. This is because (as discussed earlier) increasing the frequency above resonance is the same as having an antenna that is too long, and vice versa.

The range covered by the curves in Fig 6 is representative of the frequency range over which a fixed antenna is operated between the limits of an amateur band. At greater departures from the resonant length, the resistance continues to decrease somewhat uniformly as the antenna is shortened, but tends to increase rapidly as the antenna is made longer. The resistance increases very rapidly when the length of a leg exceeds about 135° (3/8 λ), and reaches a maximum when the length of one side is 180°. This is discussed in more detail later in this chapter.

Reactance

The rate at which the reactance of the antenna increases as the length is varied from resonance also depends on the λ/dia ratio of the conductor. The change in reactance with λ/dia ratio is much more pronounced than the resistance variation. The thicker the conductor, the smaller the reactance variation for a given change in length. This is shown by the reactance curves in Fig 6. Curves for three values of λ/dia ratio are shown; λ represents a wavelength in free space, and dia is the diameter of the conductor in the same units as the wavelength.

The point where each curve crosses the zero axis (indicated by an arrow in each case) is the length at which an antenna of that particular λ/dia ratio is resonant. The effect of the wavelength to diameter ratio on the resonant

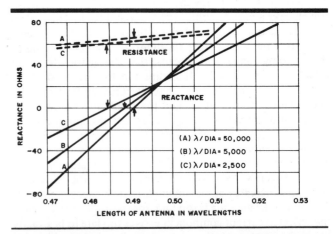

Fig 6—Resistance and reactance at the input terminals of a center-fed antenna as a function of its length. As shown by curves A, B and C, the reactance is affected more by the λ/dia ratio of the conductor than is the radiation resistance.

length is also illustrated by these curves; the smaller the ratio, the shorter the length at which the reactance is zero.

The reactance changes much more rapidly in the antenna with the smallest diameter (A) than it does in the antenna with the largest diameter (C). With still larger diameters, the rate at which the reactance changes is even smaller. As a practical matter, it is advantageous to keep the reactance change with a given change in length as small as possible. This means that when such an antenna is operated over a small frequency band centered on its resonant frequency, the reactance is comparatively low and the impedance change with frequency is small. This simplifies the problem of supplying power to the antenna when it must be used at frequencies somewhat different from its resonant frequency.

At lengths considerably different from the resonant length, the reactance changes more rapidly than it does close to resonance. As is the case with resistance, the reactance change is most rapid when the length exceeds 135° (3/8 λ) and approaches 180° (½ λ) per side. In this case the reactance is inductive and reaches a maximum at a length somewhat less than 180°. Between this maximum point and 180° of electrical length the reactance decreases very rapidly, reaching zero when the length is that for parallel resonance.

Very short antennas have a large capacitive reactance. In the preceding section, it was mentioned that with antennas shorter than 90° per side, the radiation resistance decreases at a fairly uniform rate as length is decreased. This is not true of reactance. It increases rather rapidly when the length of a side is shortened below about 45°.

The behavior of antennas with different λ/dia ratios corresponds to the behavior of ordinary resonant circuits having different values of Q. When the Q of a circuit is low, the reactance is small and changes rather slowly as the applied frequency is varied on either side of resonance. If the Q is high, the converse is true. The response curve of the low Q circuit is "broad"; that of the high Q circuit "sharp." So it is with antennas; a thick antenna works well over a comparatively wide band of frequencies while a thin antenna is rather sharp in response. The Q of the thick antenna is low; the Q of the thin antenna is high, assuming the radiation resistances are essentially the same in both cases. Antenna Q is discussed further in Chapter 26.

Coupled Antennas

A conventional tuned circuit far enough away from all other circuits so that no external coupling exists can be likened to an antenna in free space. That is, the circuit characteristics are unaffected by its surroundings. It will have a Q and resonant impedance determined by the inductance, capacitance and resistance of which it is composed, and those quantities alone. But if this circuit is coupled to another, its Q and impedance will change (depending on the characteristics of the other circuit and the degree of coupling).

A similar situation arises when two or more antennas (or antenna elements) are coupled together. This coupling takes place when the two antennas are in close proximity to each other (within a few wavelengths). The sharpness of resonance and the radiation resistance of each "element" of the system are affected by the mutual interchange of energy between the coupled elements. The extent of these changes depends on the degree of coupling (that is, how close the antennas are to each other in wavelengths, and whether or not the wires are parallel). The tuning condition (whether tuned to resonance or slightly off resonance) of each element also plays an

important role. Examples of antennas using coupled elements are Yagi beams and driven arrays.

Analysis of antenna systems with coupled elements is extremely difficult without a computer, and even with a computer must be based on some simplifying assumptions. Antenna systems consisting of coupled elements are covered in later chapters. At this point it is sufficient to know that the free-space values that have been discussed in this chapter are drastically modified when more than one antenna element is involved in the system.

The presence of the ground, as well as nearby conductors and dielectrics, also modifies the free-space values of an antenna. The free-space characteristics of the elementary ½-λ dipole are merely the point of departure for a practical antenna system. In other words, they give the basis for understanding antenna principles, but should not be applied too literally in practice.

The comparison between an isolated tuned circuit and an antenna in free space also cannot be taken too literally. In one sense, the comparison is wholly misleading; the tuned circuit is usually so small, physically (in comparison with the wavelength), that practically no energy escapes from it by radiation. Antennas, on the other hand, must be so large in comparison with the wavelength that practically *all* the energy supplied is radiated. Thus, the antenna can be said to be very tightly coupled to space, while the tuned circuit is not tightly coupled to anything. This very fundamental difference is one reason why antenna systems cannot be analyzed as readily, and with as satisfactory results in the form of simple equations, as ordinary electrical circuits.

With the proliferation of personal computers, there have recently been significant strides in computerized antenna system analysis. Using programs incorporating some simplifying assumptions, it is now possible for the amateur with a relatively inexpensive computer to evaluate rather complicated antenna systems. Software is available that takes into account the most significant effects of real ground and other factors that can be determined with reasonable accuracy. Amateurs are thus able to obtain a greater grasp of the operation of antenna systems—a subject that has been a great mystery to many in the past.

HARMONICALLY OPERATED ANTENNAS

An antenna operated at a harmonic of its fundamental frequency has considerably different properties than the ½-λ dipole previously discussed. It is important to realize that harmonic operation implies there is a reversal of the direction of current flow in alternate ½-λ sections of the antenna, as shown in Fig 2 and in Fig 7A. In Fig 7A, the curve shows the standing wave of current on the wire; the curve is above the line to indicate current flow in one direction (assumed to be to the right, in the direction of the arrow) and below the line to indicate current flow in the opposite direction in the other ½-λ section. (During the next radio-frequency half cycle the current flow in the left ½-λ section would be toward the left, and in the right ½-λ section to the right; this alternation in direction takes place in each succeeding half cycle. The direction of current flow in adjacent ½-λ sections, however, is opposite at all times.) The antenna in this drawing is 1 λ long and is operating on its second harmonic.

Now consider the ½-λ antenna shown at Fig 7B. It is opened in the center and fed directly by a source of RF power.

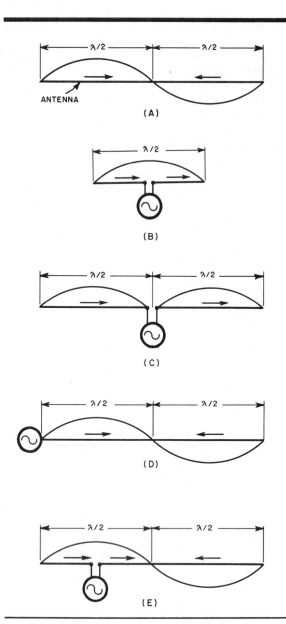

Fig 7—Moving the feed point makes a difference in current distribution along the antenna. With center feed, increasing the length of each side of the antenna keeps the current flowing in the same direction in the two halves, up to the point where each side is ½ λ long. For harmonic operation, the antenna must be fed in such a way that the current direction reverses in alternate ½-λ sections. Suitable methods are shown at D and E.

Because one terminal of the generator is positive at the same instant that the other terminal is negative, current flows *into* one side of the generator while it is flowing *out of* the other terminal. Consequently, the current flows in the same direction in both sections of the ½-λ antenna. It has the amplitude distribution shown by the curve over the antenna wire.

If the length of the wire on each side of the generator in Fig 7B is increased to ½ λ, the situation shown in Fig 7C exists. At the instant shown, current flows into the generator from the left ½-λ section, and out of the generator into the right ½-λ section. Thus the currents in the two sections are in the same direction, just as in Fig 7B. The current

distribution in this case is not the same as in Fig 7A. Although the overall lengths of the antennas shown at A and C are the same, the antenna at A is operating on a harmonic, but the one in C is not.

For true harmonic operation it is necessary that power be fed into the antenna at an appropriate point. Two methods that result in the proper current distribution are shown at D and E in Fig 7. If the source of power is connected to the antenna at one end, as in D, the direction of current flow is reversed in alternate ½-λ sections. Or if the power is inserted at the center of a ½-λ section, as in E, there is a similar reversal of current in the next ½-λ section. For harmonic operation, therefore, the antenna should be fed either at the end or at a current loop. If the feed point is at a current node, the current distribution is different than that expected on a properly fed harmonic antenna.

Length of a Harmonic Wire

The physical length of a harmonic antenna is not exactly the same as its electrical length, for the same reason discussed earlier in connection with the ½-λ antenna. The physical length is somewhat shorter than the length of the same number of half waves in space because of the λ/dia ratio of the conductor and the end effects. However, end effects are appreciable only where insulators introduce additional capacitance at a high-voltage point on the wire, and harmonic antennas usually have such insulation only at the ends. Therefore, the end-effect applies only to the ½-λ sections at each end of the antenna. The following equation for the length of a harmonic antenna works out well in practice for commonly used wire sizes:

$$\text{Length (feet)} = \frac{492 \, (N - 0.05)}{f(\text{MHz})} \tag{Eq 7}$$

where N is the number of *half* waves on the antenna.

Because the number of half waves varies with the harmonic on which the antenna is operated, consideration of the length equation together with that for the ½-λ antenna (the fundamental frequency) shows that the relationship between the antenna fundamental frequency and its harmonics is not exactly integral. That is, the "second-harmonic" frequency to which a given length of wire is resonant is not exactly twice its fundamental frequency; the "third-harmonic" resonance is not exactly three times its fundamental and so on. The actual resonant frequency of a harmonic antenna is always slightly higher than the exact multiple of the fundamental. A full-wave (second-harmonic) antenna, for example, must be slightly longer than twice the length of a ½-λ antenna.

Frequently it is desired to determine the electrical length of a harmonically operated wire antenna of fixed physical length for a given frequency. The following equation (a simple rearrangement of the one above) is useful for making these determinations:

$$\lambda = \frac{fL}{984} + 0.025 \tag{Eq 8}$$

where

λ = the length of the wire in wavelengths
f = frequency in megahertz
L = the physical length of the wire, feet

Impedance of Harmonic Antennas

Harmonic antennas can be looked upon as a series of ½-λ sections placed end to end (collinear) and supplied with power in such a way that the currents in alternate sections are out of phase. There is some coupling between adjacent ½-λ sections. Because of this coupling and the effect of radiation from the additional sections, the impedance measured at a current loop in a ½-λ section is not the same as the impedance at the center of a ½-λ antenna.

Just as in the case of a ½-λ antenna, the impedance consists of two main components: radiation resistance and reactance. The ohmic or loss resistance is low enough to be ignored in the practical case. If the antenna is exactly resonant, there is no reactance at the input terminals, and the impedance consists only of the radiation resistance.

Radiation resistance depends on the number of half waves in the wire, and is modified by the presence of nearby conductors and dielectrics, particularly the earth (as is the case with the ½-λ antenna). As a point of departure, however, it is of interest to know the *order of magnitude* of the radiation resistance of a theoretical harmonic antenna consisting of an infinitely thin conductor in free space, with its length adjusted to exact harmonic resonance. The radiation resistance of such an antenna having a length of 1 λ, measured at a current loop, is approximately 90 Ω. As the antenna length is increased, the resistance increases. At 10 λ it is approximately 160 Ω. The way in which the radiation resistance of a theoretical harmonic wire varies with length is shown by curve A in Fig 8. Radiation resistance is always measured at a current loop. (Curve B of Fig 8 is discussed in a later section of this chapter.)

If the antenna is operated at a frequency slightly off its exact resonant frequency, reactance and resistance appear at its input terminals. In a general way, the reactance varies with applied frequency in much the same fashion as in the case of the ½-λ antenna already described. However, the reactance varies *at a more rapid rate* as the applied frequency is varied; on a harmonic antenna a given percentage change in applied frequency causes a greater change in the phase of the reflected current as related to the outgoing current than is the case with a ½-λ antenna. This is because, in traveling the greater length of wire in a harmonic antenna, the reflected current gains the same amount of time in *each* ½-λ section, if the antenna is

too short for resonance, and these gains add up as the current travels back to the driving point.

If the antenna is too long, the reverse occurs and the reflected current progressively drops behind in phase as it travels back to the point at which the voltage is applied. This effect increases with the length of the antenna, and the change of phase can be quite rapid when the frequency applied to an antenna operated on a high order harmonic is varied.

Another way of looking at it is this. Consider the antenna of Fig 9A, driven at the end by a source of power having a frequency of f/2, where f is the fundamental or ½-λ resonant frequency of the antenna. When the frequency f/2 is applied, the wire is ¼ λ long, and has the current distribution shown. At this frequency the antenna is resonant, and it appears as a low, pure resistance to the source of power. This is because the current is large and the voltage is small at the feed point.

If the frequency is increased slightly, the antenna becomes too long, and the resultant current at the input terminals lags the applied voltage (as shown in Fig 5). The antenna has inductive reactance along with resistance. As the frequency is raised further, the inductive reactance increases to a maximum and then decreases. It reaches zero when the frequency is f, where the wire is ½ λ long, shown in Fig 9B.

With further frequency increase, the reactance becomes capacitive, increases to a maximum, and then decreases, reaching zero when the wire is an odd number of quarter wavelengths long. As the frequency is increased further still, the reactance again becomes inductive, reaches a maximum, and again goes to zero at 2f. At this point there are two complete standing waves of current (two half waves, or one wavelength) and the wire is exactly resonant on its second harmonic. This condition is shown in Fig 9C.

In varying the frequency from f/2 to 2f, the resistance seen by the source of power also varies. This resistance increases as the frequency is raised above f/2 and reaches a maximum when the wire is ½ λ long, decreases as the frequency is raised above f, reaching a minimum when the wire is an odd number of quarter wavelengths long. It then rises as the frequency is increased, reaching another maximum

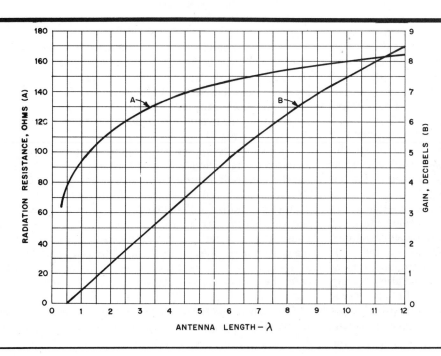

Fig 8—The variation in radiation resistance and power in the major lobe of harmonic (long-wire) antennas. Curve A shows the change in radiation resistance with antenna length, as measured at a current loop, while curve B shows the power gain in the lobes of maximum radiation for long-wire antennas as a ratio to the maximum of a ½-λ antenna.

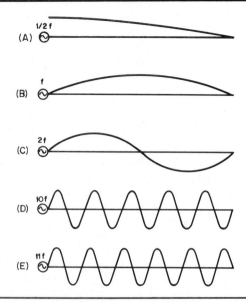

Fig 9—The percentage frequency change from one high-order harmonic to the next (for example, between the 10th and 11th harmonics shown at C and D) is much smaller than between the fundamental and second harmonic (A and B). This makes impedance variations more rapid as the wire becomes longer in wavelengths.

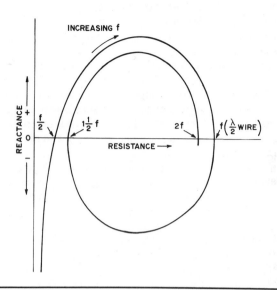

Fig 10—This drawing shows qualitatively the way in which the reactance and resistance of an end-fed antenna vary as the frequency is increased from one half the fundamental (f/2) to the second harmonic (2f). Relative resistance values are shown on the horizontal axis, and reactance values on the vertical axis; increasing frequency appears in a clockwise direction on the curve. Actual values of resistance and reactance and the frequencies at which the reactances are maximum depends on the size of the conductor and the height of the antenna above ground.

when the frequency is 2f.

This variation in reactance and resistance with frequency is shown in Fig 10. A similar change in reactance and resistance occurs when the frequency is moved from any harmonic to the *next adjacent* one, as well as between the fundamental and second harmonic as shown in Figs 9 and 10. That is, the impedance goes through a cycle, starting with a high value of pure resistance at f. It then becomes capacitive and decreases, passing through a low, pure resistance, and then becomes inductive and increases until it again reaches a high value of pure resistance at 2f. This cycle occurs as the frequency is continuously varied from any harmonic to the next higher one.

Look now at Fig 9D. The frequency has been increased to 10f, 10 times its original value, so the antenna is operated on its 10th harmonic. Raising the frequency to 11f causes the impedance of the antenna to go through the complete cycle described above. But 11f is only 10% higher in frequency than 10f, so a 10% change in frequency has caused a complete impedance cycle. In contrast, changing from f to 2f is a 100% increase in frequency—for the same impedance cycle. The impedance therefore changes 10 times faster when the frequency is varied about the 10th harmonic than it does when the frequency is varied by the same percentage about the fundamental.

To offset this, the actual impedance change—that is, the ratio of the maximum to the minimum impedance through the impedance cycle—is not as great at the higher harmonics as it is near the fundamental. This is because the radiation resistance increases with the order of the harmonic, raising the minimum point on the resistance curve and also lowering the maximum point. This is because the reflected current returning to the input end of a long harmonic wire is not as great as the outgoing current, as energy has been radiated.

The overall result, then, is that the *magnitude* of the impedance variations decrease as the wire is operated at increasingly higher harmonics. Nevertheless, the impedance reaches a maximum at each adjacent harmonic and a minimum halfway between, regardless of the actual impedance values.

OTHER ANTENNA PROPERTIES

Polarization

The polarization of a ½-λ dipole is the same as the direction of its axis. That is, at distances far enough from the antenna for the waves to be considered as plane waves, the direction of the electric component of the field is the same as the direction of the antenna wire. Vertical and horizontal polarization, the two most commonly used for antennas, are illustrated in Fig 11.

Antennas composed of a number of ½-λ elements arranged so that their axes lie in the same or parallel directions has the same polarization as that of any one of the elements. A system composed of a group of horizontal dipoles, for example, is horizontally polarized. If both horizontal and vertical elements are used in the same plane and radiate in phase, the polarization is the *resultant* of the contributions made by each set of elements to the total electromagnetic field at a given point some distance from the antenna. In such a case the resultant polarization is *linear*, tilted between horizontal and vertical.

If vertical and horizontal elements in the same plane are fed out of phase (where the beginning of the RF period applied

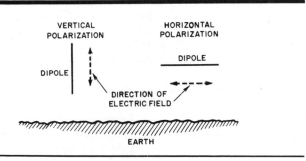

Fig 11—Vertical and horizontal polarization of a dipole. The direction of polarization is the direction of the electric field with respect to earth.

to the feed point of the vertical element is not in time phase with that applied to the horizontal), the resultant polarization is *elliptical*. Circular polarization is a special case of elliptical polarization. The wave front of a circularly polarized signal appears (in passing a fixed observer) to rotate every 90° between vertical and horizontal, making a complete 360° rotation once every period. Field intensities are *equal* at all instantaneous polarizations. Circular polarization is frequently used for space communications, and is discussed further in Chapter 19.

Harmonic antennas also are polarized in the direction of the wire axis. In some combinations of harmonic wires such as the V and rhombic antennas described in Chapter 13, however, the polarization becomes elliptical in most directions with respect to the antenna.

The polarization of a ½-λ sloping-wire dipole, with one end appreciably higher than the other, is linear. Polarization changes with compass direction, however. Broadside to the wire the polarization is tilted, but in the direction of the wire it is vertical.

Sky-wave transmission usually changes the polarization of traveling waves. (This is discussed in Chapter 23.) The polarization of receiving and transmitting antennas in the 3- to 30-MHz range, where almost all communication is by means of sky wave, need not be the same at both ends of a communication circuit (except for distances of a few miles). In this range the choice of polarization for the antenna is usually determined by factors such as the height of available antenna supports, polarization of man-made RF noise from nearby sources, probable energy losses in nearby objects, the likelihood of interfering with neighborhood broadcast or TV reception, and general convenience.

Reciprocity in Receiving and Transmitting

The basic conditions existing when an antenna is used for radiating power are not the same as when it is used for receiving a distant signal. In the transmitting case the electromagnetic field originates with the antenna, and the waves are not plane-polarized in the immediate vicinity. In the receiving case, the antenna is far enough away from the transmitter that the waves intercepted by the antenna are plane-polarized. This causes the current distribution in a receiving antenna to be different than in a transmitting antenna, except in a few special cases. These special cases, however, are those of most interest in amateur practice, because they occur when the antenna is resonant and is delivering power to a receiver.

For all practical purposes, then, the properties of a resonant antenna used for reception are the same as its properties in transmission. It has the same directive pattern in both cases, and delivers maximum signal to the receiver when the signal comes from a direction in which the antenna has its best response. The impedance of the antenna is the same, at the same point of measurement, in receiving as in transmitting.

In the receiving case, the antenna is the *source* of power delivered to the receiver, rather than the *load* for a source of power (as in transmitting). Maximum output from the receiving antenna is obtained when the load to which the antenna is connected is matched to the impedance of the antenna. Under these conditions half of the total power picked up by the antenna from the passing waves is delivered to the receiver and half is reradiated. (Under mismatched conditions, less than half the power is delivered to the receiver, and the remainder is reradiated.)

The *power gain* (defined later in this chapter) in receiving is the same as the gain in transmitting, assuming that certain conditions are met. One such condition is that both antennas (usually ½-λ antennas) must work into load impedances matched to their own impedances, so maximum power is transferred in both cases. In addition, the comparison antenna should be oriented so it gives maximum response to the signal used in the test; that is, it should have the same polarization as the incoming signal and should be placed so its direction of maximum gain is toward the signal source.

In long-distance transmission and reception via the ionosphere, the relationship between receiving and transmitting may not be exactly reciprocal. This is because the waves do not take exactly the same paths at all times and so may show considerable variation in alternate transmission and reception. Also, when more than one ionospheric layer is involved in the wave travel (see Chapter 23), it is sometimes possible for reception to be good in one direction and poor in the other, over the same path. Wave polarization is shifted in the ionosphere, as also discussed in Chapter 23. The tendency is for the arriving wave to be elliptically polarized, regardless of the polarization of the transmitting antenna. Vertically polarized antennas can be expected to show no more difference between transmission and reception than horizontally polarized antennas. On the average, an antenna that transmits well in a certain direction also gives favorable reception from the same direction, despite ionospheric variations.

Pickup Efficiency

Although the transmitting and receiving properties of an antenna are, in general, reciprocal, there is another fundamental difference between the two cases that is of great practical importance. In the transmitting case, virtually all the power supplied to an antenna is radiated (assuming negligible resistive losses). This is true regardless of the physical size of the antenna system. For example, a 300-MHz ½-λ radiator, which is only about 19 inches long, radiates just as efficiently as a 3.5-MHz ½-λ antenna, which is about 134 feet long. But in receiving, the 300-MHz antenna does not extract anything close to the amount of energy from passing waves that the 3.5-MHz antenna does.

This is because the section of the wave front from which an antenna can draw energy extends only about ¼ λ from the conductor. At 3.5 MHz this represents an area roughly ½ λ or 142 feet in diameter, but at 300 MHz the diameter of the area is only about 1.6 feet. Because the energy is evenly distributed throughout the wave front regardless of the wavelength, the effective area that the receiving antenna can utilize varies directly with the *square* of the wavelength. A 3.5-MHz ½-λ antenna therefore picks up something on the order of *7000 times* more energy than a 300-MHz ½-λ antenna, *the field strength being the same in both cases.*

The higher the frequency, consequently, the less energy a receiving antenna has to work with. This, it should be noted, does not affect the *gain* of the antenna. In making gain measurements, both the antenna under test and the comparison antenna are working at the same frequency. Both therefore suffer the same handicap with respect to the amount of energy that can be intercepted. The effective area of an antenna at a given frequency is therefore directly proportional to its gain. Although pickup efficiency decreases rapidly with increasing frequency, the smaller dimensions of resonant elements in the VHF and UHF regions make it relatively easy to combine elements to obtain high gain. This helps to overcome the lack of pickup efficiency.

The Induction Field

Throughout this chapter the fields discussed are those forming the traveling electromagnetic waves—the waves that go long distances from the antenna. These are the radiation fields. They are distinguished by the fact that their intensity is inversely proportional to the distance and that the electric and magnetic components, although perpendicular to each other in the wave front, are in time phase. Beyond several wavelengths from the antenna, these are the only fields that need to be considered.

Close to the antenna, however, the situation is much more complicated. In an ordinary electric circuit containing pure inductance or capacitance the magnetic field is 90° out of phase with the electric field. The intensity of these fields decreases in a complex way with distance from the source. These are the induction fields. The induction field exists about an antenna along with the radiation field, but dies away with much greater rapidity as the distance from the antenna is increased. For a short dipole (less than 0.1 λ long), at a distance equal to λ/2π or slightly less than 1/6 λ, the two types of fields have equal intensity.

Although the induction field has no consequential effects at any distance from an antenna, it *does* have significance when antenna elements are coupled together, particularly when the spacing between elements is small. Also, the induction field must be considered when making field-strength measurements about an antenna. Errors may occur if the measuring equipment is so close to the antenna system that the induction field has a major effect on the readings.

Radiation Patterns, Gain and Directivity

A graph showing the actual or relative field intensity at a fixed distance, as a function of the direction from the antenna system, is called a radiation pattern. To understand the basis of such a graph, see Fig 12. This drawing represents a flashlight shining in a totally darkened area.

Assume we have a sensitive light meter of the type used in photography, and that its scale is graduated in units from 0 to 10. We place the meter directly in front of the flashlight and adjust its distance so the meter reads 10, exactly full scale. We note this distance carefully. Then, *always keeping the meter the same distance* from the lamp in the flashlight, we move the light meter around the flashlight as indicated by the arrow, and take light readings at 16 different positions.

Next, after all the readings have been taken and recorded, we plot those values on a sheet of polar graph paper, like that shown in Fig 13. The values read on the meter are plotted at an angular position corresponding to that for which the meter reading was taken. Following this, we connect the plotted points with a smooth curve, also shown in Fig 13. When this is finished, we have completed a radiation pattern for the flashlight.

Antenna radiation patterns can be constructed in a similar manner. Power is fed to the antenna under test, and a field-strength meter indicates the amount of signal. For convenience, the antenna under test is rotated, rather than moving the measuring equipment to numerous positions about the antenna. However, a more common method of measuring antenna radiation patterns makes use of antenna reciprocity—its pattern while receiving is the same as that while transmitting. A source antenna that is fed from a low power transmitter illuminates the antenna under test, and the signal it intercepts is fed to a receiver and measuring equipment. Additional information on the mechanics of measuring antenna patterns is contained in Chapter 27.

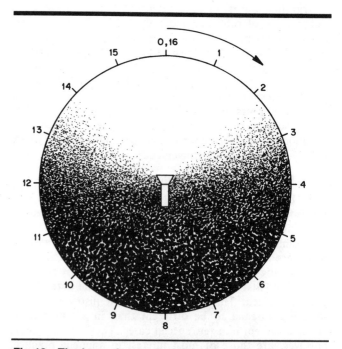

Fig 12—The beam from a flashlight illuminates a totally darkened area as shown here. Readings taken with a photographic light meter at the 16 points around the circle may be used to plot the radiation pattern of the flashlight.

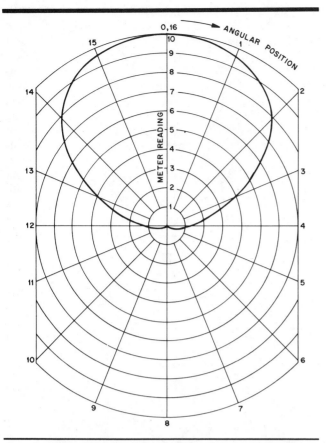

Fig 13—The radiation pattern of the flashlight of Fig 12. The measured values are plotted and connected with a smooth curve.

Patterns obtained as above represent the antenna radiation in just a one plane. Actually, the pattern for any antenna is three dimensional, and therefore cannot be represented in a single-plane drawing. The "solid" radiation pattern of an antenna in free space would be found by measuring the field strength at every point on the surface of an imaginary sphere having the antenna at its center. The information so obtained would then be used to construct a solid figure where the distance from a fixed point (representing the antenna) to the surface of the figure is proportional to the field strength from the antenna in any given direction.

For amateur work, *relative* values of field strength (rather than absolute) are quite adequate in pattern plotting. In other words, it is not necessary to know how many microvolts per meter a particular antenna will produce at a distance of 1 mile when excited with a specified power level. For whatever data is collected (or calculated from theoretical equations), it is common to "normalize" the plotted values so the field strength in the direction of maximum radiation coincides with the outer edge of the chart. On a given system of polar coordinate scales, the *shape* of the pattern is not altered by proper normalization, only its size.

COORDINATE SCALES FOR RADIATION PATTERNS

A number of different systems of coordinate scales or "grids" are in use for plotting antenna patterns. In current professional literature, almost universal use is made of rectangular coordinates. This may arise in part from the extensive use of small computer systems to develop the pattern plots, and the relative ease with which such computer systems can prepare rectangular graphs. Perhaps a disadvantage of rectangular coordinate plots is that they do not vividly portray the antenna response in various compass directions, or in directions relative to the direction of maximum response.

Antenna patterns published for amateur audiences are seldom placed on rectangular grids. Instead, polar coordinate systems are used almost universally. Polar coordinate systems may be divided generally into three classes: linear, logarithmic and log periodic.

A very important point to remember is that the shape of a pattern (its general appearance) is highly dependent on the grid system used for the plotting. This is exemplified in Fig 14, where patterns of three different antennas are presented on the four coordinate systems discussed in the paragraphs that follow. The antennas selected for plotting in Fig 14 run the gamut of the pattern types usually encountered in Amateur Radio. These are (1) a half-wave dipole, which has no minor lobes, (2) the extended double Zepp (see Chapter 8), which has four minor lobes down about 8 decibels, and (3) an array of four 19-element Cushcraft Yagi antennas for 144 MHz (see the Product Review column in November 1980 *QST*, p 48), a system having several minor lobes from 10 to 40 decibels down. (Decibel or dB units are discussed in detail later in this chapter.)

Linear Coordinate Systems

The polar coordinate system for the flashlight radiation pattern, Fig 13, is one of linear coordinates. The concentric circles are equally spaced, and are graduated from 0 to 10. Such a grid may be used to prepare a linear plot of the *power* contained in the signal. Parts A, E and I of Fig 14 show the three antenna patterns plotted on this *linear power grid*. However, for ease of comparison, the equally spaced concentric circles have been replaced with appropriately placed circles representing the decibel response, referenced to 0 dB at the outer edge of the plot.

Note that in these plots the minor lobes are suppressed. Lobes with peaks more than 15 dB or so below the main lobe disappear completely because of their small size. If the intent is to show the pattern of an array having high directivity (discussed later) and small minor lobes, this grid enhances those features.

The *voltage* in the signal, rather than the power, can also be plotted on a linear coordinate system. This *linear voltage grid* coordinate system is used by some manufacturers of amateur antennas in their literature, and was used in all editions of *The ARRL Antenna Book* prior to the 14th (1982). Parts B, F and J of Fig 14 show the three antennas patterns plotted on the linear voltage grid. Here again, however, the equally spaced circles have been replaced with circles showing the decibel response. Minor lobes that are 30 dB or so below the main lobe are not evident. This grid, too, enhances the appearance of directivity and suppresses minor lobes, but not to the same degree as the linear power grid.

Log Periodic Coordinate System

The log periodic grid has a system of concentric grid lines spaced periodically according to the logarithm of the voltage in the signal. Different values may be used for the constant of periodicity, and this choice, too, will have an effect on the appearance of plotted patterns. ARRL publications use a log periodic grid with a periodicity constant of 0.89 for 2-dB

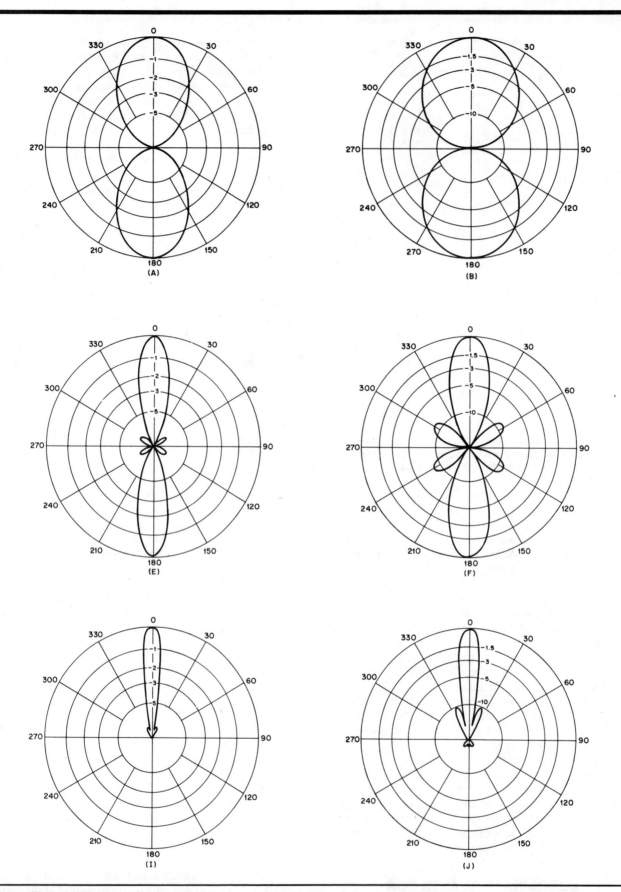

Fig 14—Radiation pattern plots for three antennas on four different grid coordinate systems. At A through D are plots for a ½-λ dipole; E through H for an extended double Zepp, and I through L for an array of four 19-element Yagis. At A, E and I are patterns on a linear power grid; at B, F and J on a linear voltage grid; at C, G and K on a log periodic grid, and at D, H and L on a logarithmic (linear decibel) grid. The concentric circles in all grids appearing here are graduated in decibels referenced to 0 dB at the outer edge of the chart.

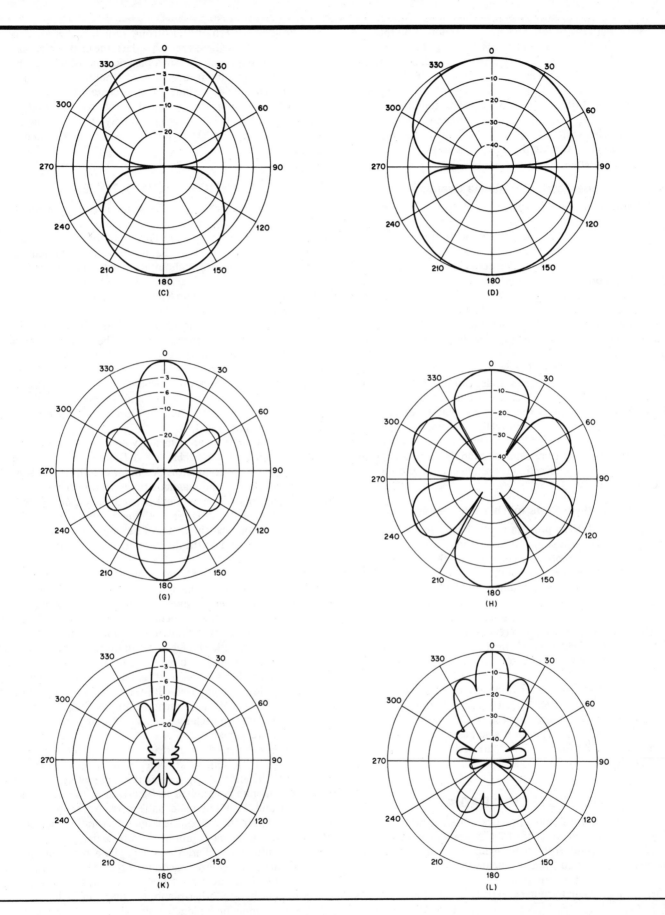

intervals. Patterns of the three antennas plotted on this grid appear in Fig 14C, 14G and 14K. With this grid, minor lobes that are 30 and 40 dB down from the main lobe are distinguishable. Such lobes are of concern in VHF and UHF work. Also on this grid, the spacing between plotted points at 0 dB and −3 dB is significantly greater than the spacing between −20 and −23 dB, which is significantly greater than the spacing between −50 and −53 dB. The spacings thus correspond generally to the relative significance of such changes in antenna performance.

Logarithmic Coordinate Systems

Another coordinate system used by several antenna manufacturers is the logarithmic grid, where the concentric grid lines are spaced according to the logarithm of the voltage in the signal. If the logarithmically spaced concentric circles are replaced with appropriately placed circles representing the decibel response, it ends up that the decibel circles are graduated linearly. In that sense, the logarithmic grid might be termed a linear grid, one having linear divisions calibrated in decibels. It is in this form that the grid usually appears. Parts D, H and L of Fig 14 show the three antenna patterns plotted on this *linear decibel grid*.

This grid enhances the appearance of the minor lobes. Note that in the plot for the extended double Zepp, Fig 14H, the minor lobes look almost as prominent as the major lobes. If the intent is to show the radiation pattern of an array supposedly having an omnidirectional response, this grid enhances that appearance. An antenna having a difference of 8 or 10 dB in pattern response around the compass appears to be closer to omnidirectional on this grid than on any of the others in Fig 14.

Pattern Plots in This Publication

Antenna pattern plots in this publication are made on the log periodic grid shown in basic form in Fig 14 at C, G and K. However, more concentric circles than in Fig 14 are provided, to offer greater resolution in interpreting the plots. The azimuth scale is indicated in degrees around the outside edge of the chart, and shows the angle of departure from the reference or starting point, usually 0°. Unless specifically stated otherwise, the 0-dB reference (outer edge of the chart) is taken as the field strength in the direction of maximum radiation in a given plane for the antenna system under consideration.

Work sheets for plotting antenna patterns on this grid are available from ARRL HQ, 100 for $3 at the time of this writing. A work sheet measures 8½ × 11 inches, with a 6-inch diameter 0-dB circle. The sheet also has blocks for entering related antenna information. (Chart design is by Jerry Hall, K1TD; drafting by Sue Fagan.)

THE ISOTROPIC RADIATOR

The radiation from a practical antenna never has the same intensity in all directions. The intensity may even be zero in some directions from the antenna; in others it may be greater than expected from an antenna that *did* radiate equally in all directions. Even though no actual antenna simultaneously radiates equally intense signals in all directions, it is useful to assume that such an antenna exists. It can be used as a "measuring stick" for comparing the properties of actual antenna systems. Such a hypothetical antenna is called an *isotropic radiator.*

The solid pattern of an isotropic radiator, therefore, is a sphere, because the field strength is the same in all directions. In any plane containing the isotropic antenna (which may be considered as a point in space, or a "point source") the pattern is a circle with the antenna at its center. The isotropic antenna has the simplest possible directive pattern; that is, it has no directivity at all. An infinite variety of pattern shapes, some quite complicated, is possible with actual antenna systems.

RADIATION FROM DIPOLES

In the analysis of antenna systems it is convenient to make use of another reference antenna called an *elementary doublet* or *elementary dipole*. This is simply a very short length of conductor, so short that it can be assumed that the current is the same throughout its length. (In an actual antenna, the current is different all along its length.) The radiation intensity from an elementary dipole is greatest at right angles to the line of the conductor, and decreases as the direction becomes more nearly in line with the conductor. Exactly off the ends, the intensity is zero. The directive pattern in a single plane (one containing the conductor) is shown in Fig 15A. If the pattern were drawn on a linear coordinate system showing field intensity, it would consist of two tangent circles. The solid pattern is the doughnut-shaped figure which results when the plane shown in the drawing is rotated about the conductor axis, Fig 15B.

The radiation from an elementary doublet is not uniform in all directions, because there is a definite direction to the current flow on the conductor. A similar condition exists in the ordinary electric and magnetic fields set up when current flows on any conductor; the field strength near a single-layer coil, for example, is greatest at the ends and least on the outside of the coil near the center of its length. Field strength should, then, depend on the direction in which it is measured from the radiator.

When an antenna has appreciable length (so the current in every part is not the same at any given instant), the shape of the radiation pattern changes. The pattern is the algebraic sum of the fields from *each* elementary dipole of which the antenna is "made." If the antenna is short compared with ½ λ, there is very little change in the pattern, but at ½ λ the pattern takes the shape shown in Fig 16. The intensity decreases somewhat more rapidly as the angle with the wire is made smaller, as compared with the elementary dipole of Fig 15. In the case of the elementary dipole, the half-power (or −3 dB) points on each lobe occur 90° apart; for the ½-λ dipole they are 78° apart.

As the wire length is increased farther, this tendency continues, with a somewhat wider null appearing in the pattern off the ends of the wire as the antenna approaches 1 λ. (The antenna is assumed to be driven at the center, as in Fig 7B and 7C.) The solid pattern from a ½-λ wire is formed, just as in the case of the dipole, by rotating the plane diagram shown in Fig 16 about the wire axis.

As mentioned earlier, the single-plane diagrams just discussed are actually *cross sections of the solid pattern*, cut by planes in which the axis of the antenna lies. If the solid pattern is cut by any other plane, the diagram is different. For instance, imagine a plane passing through the center of the wire at right angles to it. The cross section of the pattern for either the elementary dipole or the ½-λ antenna is simply a circle in that case. This is shown in Fig 17, where the dot

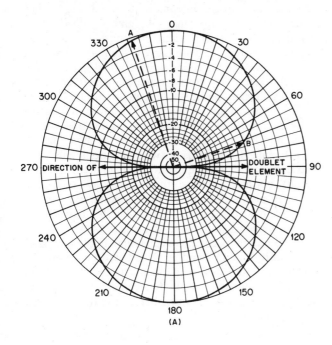

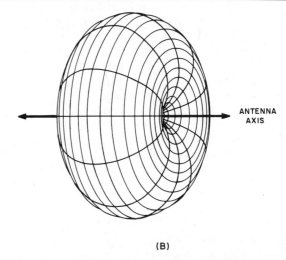

(B)

Fig 15—At A, directive diagram of an elementary doublet in the plane containing the wire axis. The length of each dash-line arrow represents the relative field strength in that direction, referenced to the direction of maximum radiation. At B, the solid pattern of the same antenna. These same diagrams apply to any center-fed antenna considerably less than a half wavelength long.

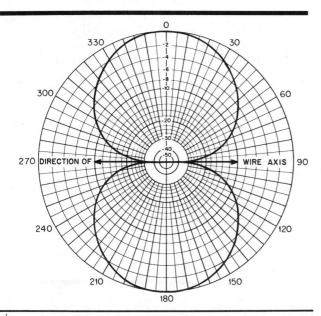

Fig 16—E-plane directive diagram of a ½-λ antenna. The solid line shows the direction of the wire, although the antenna itself is considered to be merely a point at the center of the diagram. As explained in the text, a diagram such as this is simply a cross section of the solid figure that corresponds to the relative radiation in all directions.

at the center represents the antenna as viewed "end on" (as if looking into the side of the doughnut of Fig 15B). In other words, the antenna is perpendicular to the page.

In any direction in a plane at right angles to the wire, the field intensity is exactly the same at the same distance from the antenna. At right angles to the wire, then, an antenna less than ½ λ in length is *nondirectional*. Also, at every point on

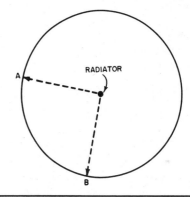

Fig 17—Directive diagram of a doublet or dipole in the plane perpendicular to the wire axis (H plane). The direction of the wire is into or out of the page.

such a circle the field is in the *same phase*.

E- AND H-PLANE PATTERNS

The solid pattern of an antenna cannot adequately be shown with field-strength data on a flat sheet of paper. For this purpose, two cross-sectional or plane diagrams as we've been discussing are very useful. Two such diagrams (as in Figs 16 and 17), one in the plane containing the axis of the antenna and one in the plane perpendicular to the axis, can give a great deal of information. The pattern in the plane containing the axis of the antenna is called the E-plane pattern, and the one in the plane perpendicular to the axis is called the H-plane pattern. These designations are used because they represent the planes in which the electric (symbol E), and the magnetic (symbol H) lines of force lie, respectively. The E lines are taken to represent the polarization of the antenna, consistent with the description of antenna polarization given earlier. The electromagnetic field pictured in Fig 1 of Chapter 23, as an example, is the field that would be radiated from a vertically polarized antenna; that is, an antenna in which the conductor is mounted perpendicular to the earth.

After a little practice, and with the exercise of some imagination, the complete solid pattern can be visualized with fair accuracy from inspection of the two diagrams. Plane diagrams are plotted on polar coordinate paper, as described earlier. The points on the pattern where the radiation is zero are called *nulls*. The curved section from one null to the next on the plane diagram, or the corresponding section on the solid pattern, is called a *lobe*.

HARMONIC-ANTENNA PATTERNS

Earlier was discussed the change in radiation patterns as the length of the antenna is increased from the elementary dipole to the ½-λ dipole. Further pattern changes occur as the antenna is made still longer. As a matter of fact, the patterns of harmonic antennas differ very considerably from the pattern of the ½-λ dipole.

As explained earlier in this chapter, a harmonic antenna consists of a series of ½-λ sections with the currents in adjacent sections always flowing in opposite directions at a given instant in time. This type of current flow causes the pattern to be split into a number of lobes. If there is an *even* number of ½-λ in the harmonic antenna, there is always a null in the plane at right angles to the wire; this is because the radiation from one ½-λ section cancels the radiation from the next in that particular direction.

If there is an *odd* number of ½ λ in the antenna, the radiation from all but one of the sections cancels itself in the plane perpendicular to the wire. The "left over" section radiates like a ½-λ dipole, so a harmonic antenna with an odd number of ½ λ does have some radiation at right angles to its axis.

The greater the number of ½ λ in a harmonic antenna, the larger the number of lobes into which the pattern splits. A feature of all such patterns is the fact that the "main" lobe—the one that gives the largest field strength at a given distance—always is the one that makes the smallest angle with

the antenna wire. Furthermore, this angle becomes smaller as the length of the antenna is increased. Fig 18 shows how the angle which the main lobe makes with the axis of the antenna varies with the antenna length in wavelengths. The angle shown by the solid curve is the maximum point of the lobe; that is, the direction in which field strength is greatest. The broken curve shows the angle at which the first null (the one that occurs at the smallest angle with the wire) appears. There is also a null in the direction of the wire itself (0°), so the total width of the main lobe is the angle between the wire and the first null. The curves of Fig 18 show that the width of the lobe decreases as the wire becomes longer. At 1 λ, for example, it has a width of 90° (from adjacent null to adjacent null), but at 8 λ, the width is slightly less than 30°.

A plane diagram of the radiation pattern of a 1 λ harmonic wire is shown in Fig 19. This is a free-space diagram in the plane containing the wire axis (E plane), corresponding to the diagrams for the elementary dipole and ½-λ dipole shown in Figs 15A and 16. It is based on an infinitely thin antenna conductor with ideal current distribution. In a practical antenna system the current is modified by the presence of the earth and other effects that are discussed in later chapters.

HOW PATTERNS ARE FORMED

The radiation pattern of the ½-λ dipole is found by algebraically adding, at every point on the surface of a sphere with the antenna at its center, the field components of all the elementary dipoles that can be considered to make up the dipole. Antenna systems often are composed of a group of ½-λ dipoles arranged in various ways, in which case each ½-λ dipole is called an antenna element. An antenna having two or more such dipoles is called a multielement antenna. (A harmonic antenna can be considered to be constructed of a number of such elements connected in series and fed power appropriately, as described earlier, but is not usually classed as a multielement antenna.)

In a multielement antenna system the radiation pattern is formed by the combination of the field components from the separate antenna elements. With two antenna elements, for example, the field strength at a given point depends on the amplitudes and phase relationship of the fields from each

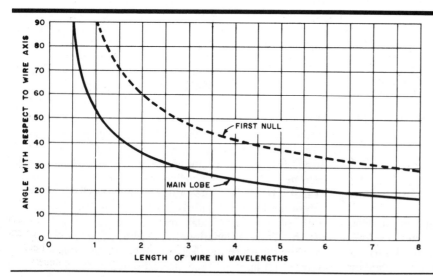

Fig 18—Angle at which the field intensity from the main lobe of a harmonic antenna is maximum, as a function of the wire length in wavelengths. The curve labeled "First Null" indicates the angle at which the intensity of the main lobe decreases to zero. The null marking the other boundary of the main lobe is always at 0° with the wire axis.

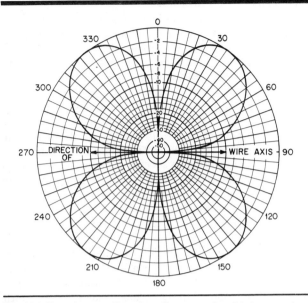

Fig 19—Free-space directive diagram of a 1-λ harmonic antenna in the plane containing the wire axis (E plane).

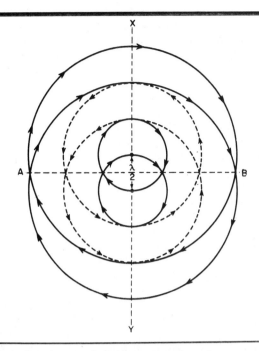

Fig 20—Interference between waves from two separate radiators causes the resultant directional effects to differ from those of either radiator alone. The two radiators shown here are separated ½ λ. The radiation fields of the two cancel along the line XY but, at distances which are large compared with the separation between the radiators, add together along line AB. The resultant field decreases uniformly as the line is swung through intermediate positions from AB to XY.

antenna. A requirement in calculating radiation patterns is that the field strength must be measured or calculated at a *distant* point—distant enough so that, if the elements carry equal currents, the field strength from each is exactly the same. This must be done, even though the size of the antenna system may be such that one antenna element is somewhat nearer the measuring point than another. On the other hand, this slight difference in distance, even though it may be only a small fraction of a wavelength, is very important in determining the *phase* relationships between the fields from the various elements.

The principle on which the radiated fields combine to produce directive patterns, in the case of multielement antennas, is illustrated in the simple example shown in Fig 20. In this case it is assumed that there are two antenna elements, each having a circular directive pattern. The two elements, therefore, could be ½-λ dipoles oriented perpendicular to the page (each having the plane pattern shown in Fig 17. The separation between the two elements is assumed to be ½ λ, and the currents in them are assumed to be equal. Furthermore, the two currents are in phase; that is, they reach their maximum values in the same polarity at the same instant.

Under these conditions the fields from the two antennas are in phase at any point that is an equal distance from both antenna elements. At the instant of time selected for the drawing of Fig 20 the solid-line circles having the upper antenna at their centers represent the location of all points at which the field intensity is maximum and has the direction indicated by the arrowheads. The *distance* between each pair of concentric solid circles, measured along a radius, is equal to 1 λ because, as described earlier in this chapter, it is only at intervals of this distance that the fields are in phase. The broken-line circle locates the points at which the field intensity is the same as in the case of the solid circles, but is *oppositely* directed. It is, therefore, 180° out of phase with the field denoted by the solid circles, and the distance between the solid and broken circles is therefore ½ λ.

Similarly, the solid circles centered on the lower antenna

locate all points at which the field intensity from that antenna is maximum and has the same direction as the solid circles about the upper antenna. In other words, these circles represent points in the same phase as the solid circles around the upper antenna. The broken circle having the lower antenna at its center likewise locates the points of opposite phase.

Considering the fields from both antennas, it can be seen that along the line AB, the fields from the two are always exactly in phase, because every point along AB is equally distant from both antenna elements. Along line XY, however, the field from one antenna is always *out of phase* with the other. This is because every point along XY is ½ λ nearer one element than the other. It takes one-half cycle longer, therefore, for the field from the more distant element to reach the same point as the field from the nearer antenna. One field thus arrives at that point 180° out of phase with the other.

Because we have assumed the points considered are sufficiently distant so that the amplitudes of the fields from the two antennas are the same, the *resultant* field at any point along XY is zero, and the antenna combination shown has a null in that direction. However, the two fields add together along AB and the field strength in that direction is twice the amplitude of the field in that direction from either antenna alone.

The drawing of Fig 20 is not quite accurate because it cannot be made large enough. Actually, the two fields along AB do not have exactly the same direction until the distance to the measuring point is large enough, compared with the dimensions of the antenna system, so the waves become planar. In a drawing of limited size the waves are necessarily represented as circles—that is, as representations of a *spherical* wave. Imagine Fig 20 as being so much enlarged that the circles crossing AB are essentially straight lines in the region under discussion.

Pattern Construction

The drawing of Fig 20 does not give much information about what happens to the field strength at points that do not lie on either AB or XY. It would be reasonable to guess that the field strength at intermediate points probably would decrease as the point was moved along the arc of a circle farther away from AB and nearer XY. To construct an actual pattern, it is necessary to use a different method. It is simple in principle and can be done with a ruler, protractor and pencil, or by trigonometry.

In Fig 21 the two antennas, A and B, are assumed to have circular radiation patterns, and to carry equal currents in the same phase. (In other words, the conditions are the same as in Fig 20.) The relative field strength at a distant point P is to be determined. Here again the limitations of the printed page make it necessary to use the imagination, because P is assumed to be far enough from A and B that the lines AP and BP are, for practical purposes, parallel. When this is so, the distance d, between B and a perpendicular line dropped to BP from A, is equal to the difference in length between the distance from A to P and the distance from B to P. The distance d thus measures the *difference* in distance the waves from A and B must cover to reach P. Therefore, d is a measure of the difference in the *time of arrival* or *phase* of the waves at P.

Under the assumed conditions, the relative field strengths can be easily combined graphically. The phase angle in degrees between the two fields at P is equal to

$$\frac{d}{\lambda} \times 360$$

where λ is the wavelength and d is found by constructing a figure similar to that shown in Fig 21 for P in any desired direction. The angle θ is the angle between a line to P and the line drawn between the two antenna elements, and is used simply to identify the direction of P from the antenna system;

λ and d must be expressed in the same length units.

For example, let us assume that θ is 40°. We then arbitrarily choose a scale such that 4 inches is equal to 1 λ. This scale is large enough for reasonable accuracy but not so large as to be unwieldy. Because the two antenna elements are assumed to be ½ λ apart, the drawing is started by placing two points 2 inches apart and connecting them by a line, as shown in Fig 22. Then, with B as a center and using the protractor, we lay off an angle of 40 degrees and draw the line BC. The next step is to drop a perpendicular from A to BC. This may be done with the 90° mark on the protractor, but the corner of an ordinary sheet of paper will do just about as well. The distance BD is then measured, preferably with a ruler graduated in tenths of inches rather than the more usual eighths. By actual measurement distance BD is found to be 1.53 inches. The phase difference is therefore d/λ × 360 = 1.53/4 × 360 = 138°.

The relative field strength in the direction given by θ (40° in this example) is found by arbitrarily selecting a line length to represent the field strength from each antenna. One inch is a convenient length. These line lengths are then combined "vectorially." Line XY, Fig 22, is such a line, representing the field strength from antenna element A. An angle of 138° is then measured from XY (using Y as a center), and line YZ is drawn one inch long to represent the strength and phase of the field from antenna element B.

The angle is measured clockwise from XY because the field from B lags that from A. The distance from X to Z then represents the relative field strength resulting from the combination of the separate fields from the two antennas, and measurement shows it to be approximately 0.72 inch. In the direction θ, therefore, the field strength is 72% as great as the field from either antenna alone. Using trigonometry, the determination may be made using the equation

$$\text{Field strength} = 2 \cos \left(\frac{S}{2} \cos \theta \right) \qquad \text{(Eq 1)}$$

where S is the spacing between elements in electrical degrees.

By selecting different values for θ and proceeding as above in each case, the complete pattern can be determined. When θ is 90°, the phase difference is zero and YZ and XY

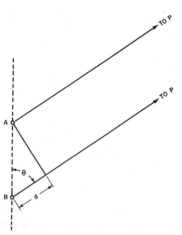

Fig 21—Graphical construction to determine the relative phase, at a distant point, of waves originating at two antennas, A and B. The phase is determined by the additional distance, d, that the wave from B must travel to reach the distant point. This distance will vary with the angle that the direction to P makes with the axis of the antenna system, θ.

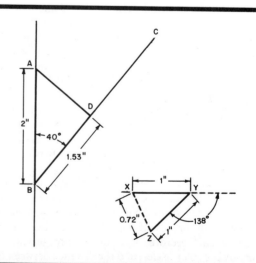

Fig 22—Graphical construction in the example discussed in the text.

are simply end-to-end along the same line. The maximum field strength is therefore twice that of either antenna alone. When θ is zero, YZ lies on top of XY (phase difference 180°) and the distance XZ is therefore zero; in other words the radiation from B cancels that from A at this angle.

The patterns of more complex antenna systems can be calculated by this method, although more work is required if the number of elements is increased. Whether or not actual patterns are calculated, an understanding of the method makes it clear why certain combinations of antenna elements result in specific directive patterns.

The illustration above is a very simple case, but it is only a short step to systems in which the antenna elements do not carry equal currents or currents in the same phase. A difference in current amplitude is easily handled by making the lengths of lines XY and YZ proportional to the current in the respective elements; if the current in B is one half that in A, for example, YZ would be drawn one half as long as XY. If the current in element B leads the current in A by 25°, then after the angle determined by the distance d is found, the line YZ is simply rotated 25° in the *counterclockwise* direction before measuring the distance XZ. The rotation would be *clockwise* for any line representing a lagging current.

The current lead or lag in the elements in the system always must be referred to the current in *one* element in the system, but any desired element can be chosen as the reference. For two elements fed out of phase but having equal currents, the relationship

$$\text{Field strength} = 2 \cos \left(\frac{\phi - S \cos \theta}{2} \right) \qquad \text{(Eq 2)}$$

may be used, where ϕ is the phase difference between the two fed elements. Trigonometric equations for determining array patterns become more complex when the currents in the elements are unequal or when the array consists of more than two elements.

The simple methods described above for determining pattern shapes do not compensate for mutual coupling between elements—that is, the fact that current flowing in each element induces voltages and causes current to flow in the others. If mutual coupling is taken into account, the shape of the pattern remains the same for a given condition of element spacing and phasing, but the *magnitudes* of the resultant vectors used in plotting points for various values of θ are altered by a fixed factor. The "fixed" factor varies with changes in spacing and phasing of the elements. Therefore, a *direct* comparison of the *sizes* of different patterns obtained by these simple procedures cannot be used for determining, say, the gain of one antenna system over another, even though both patterns were derived by using the same scale. Mutual coupling is covered in more detail in Chapter 8.

Practical Considerations

The theoretical radiation patterns of the antennas discussed so far are developed on the basis of sinusoidal distribution of current along the antenna, and on the assumption (in harmonic antennas) that the value of the current is the same at every current loop. Neither is strictly true. In particular, the current in a long harmonic antenna is not the same at every loop because energy is radiated along the length of the antenna. This affects both the forward and reflected currents. The result is that the radiation pattern does not have the perfect symmetry indicated in Fig 19. The lobes pointing away from the end at which the antenna is fed are tilted slightly toward the direction of the antenna wire. Similarly, the lobes pointing toward the fed ends are tilted away from the wire, and are of smaller amplitude. Typical patterns are shown in Fig 23 for antennas having lengths of 1, 1½ and 2 λ. There is even a slight tilt to the pattern of a ½-λ antenna when it is fed at one end; however, when such an antenna is fed at the center with a balanced line, the pattern is symmetrical.

The effect of nearby conductors and dielectrics cannot be included in the theoretical patterns. Conductors such as power and telephone lines, house wiring, piping, etc close to the antenna can cause considerable distortion of the pattern if induced voltages cause currents of appreciable magnitude to flow in them. Under similar conditions they can also have a marked effect on the radiation resistance. Poor dielectrics such as green foliage near the antenna introduce losses, and may make a noticeable difference between summer and winter performance.

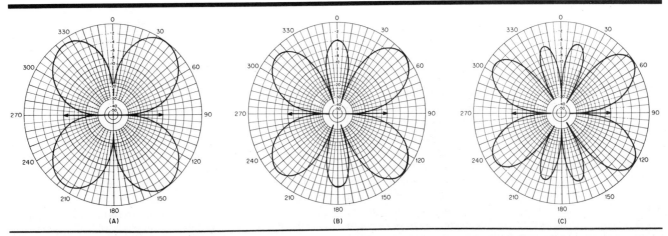

Fig 23—Patterns for three harmonic (long-wire) antennas. The arrows represent the axis of the antennas, which are fed at the left end. The antenna lengths are: A, 1 λ; B, 1½ λ; and C, 2 λ. The result of feeding a harmonic antenna at the end is smaller lobes at the feed-point end and tilted lobes, as shown in these patterns. Note also that (except off the ends of the wires) the nulls are only partial, indicated by their not extending to the centers of the charts.

The directional effects of an antenna conform more closely to theory if the antenna is located in a clear space, at least ½ λ from anything that might affect its properties. In cities, it may be difficult to find such a space at low frequencies. The worst condition arises when nearby wires or pipes happen to be resonant, or nearly so, at the operating frequency. Such resonances can often be eliminated by bonding pipes or BX coverings at trial points and checking with a diode detector wavemeter or RF current probe to determine the measures necessary to reduce the current flowing as a result of induced voltages. A "resonant breaker," as described in Chapter 6, is quite effective for eliminating the resonance effects of nearby conductors.

Metal masts or guy wires can cause distortion of the pattern unless detuned by grounding or by breaking up the wires with insulators, as discussed in Chapter 22. Masts and guy wires usually have relatively little effect on the performance of horizontal antennas because, being vertical or nearly so, they do not pick up much energy from a horizontally polarized wave. In considering nearby conductors the transmission line that feeds the antenna should not be overlooked. Under some conditions that are rather typical with amateur antennas, voltages are induced in the line by the antenna, leading to some undesirable effects on the radiation pattern. This is discussed in more detail in Chapter 26.

Radiation Resistance and Gain

The field strength produced at a distant point by a given antenna system is directly proportional to the current flowing in the antenna. In turn, the amount of current that flows, when a fixed amount of power is applied, is inversely proportional to the square root of the radiation resistance. Decreasing the radiation resistance increases the field strength, and raising the radiation resistance decreases it.

This statement is not to be interpreted broadly to mean that a low value of radiation resistance is good and a high value is bad, regardless of circumstances; that is far from the truth. For an antenna of given dimensions, a change that reduces the radiation resistance *in the right way* is accompanied by a change in the directive pattern that in turn increases the field strength in some directions, at the expense of reduced field strength in other directions. This principle is used in certain types of directive antenna systems described in detail in Chapter 8.

GAIN AND DIRECTIVITY

All antennas, even the simplest types, exhibit directive effects in that the intensity of radiation is not the same in all directions from the antenna. This property of radiating more strongly in some directions than in others is called the *directivity* of the antenna. Directivity can be expressed quantitatively by comparing the 3-dimensional pattern of the antenna under consideration with the spherical 3-dimensional pattern of the isotropic antenna. The field strength (and thus power per unit area, or "power density") are the same everywhere on the surface of an imaginary sphere having a radius of many wavelengths and having an isotropic antenna at its center. At the surface of the same imaginary sphere around an actual antenna radiating the same total power, the directive pattern results in greater power density at some points on this sphere and less at others. The ratio of the maximum power density to the average power density taken

over the entire sphere (which is the same as from the isotropic antenna under the specified conditions) is the numerical measure of the directivity of the antenna. That is,

$$D = \frac{P}{P_{av}} \qquad \text{(Eq 3)}$$

where

D = directivity
P = power density at its maximum point on the surface of the sphere
P_{av} = average power density

Gain

The gain of an antenna is closely related to its directivity. Because directivity is based solely on the *shape* of the directive pattern, it does not take into account any power losses that may occur in an actual antenna system. To determine gain, these losses must be subtracted from the power supplied to the antenna. The loss is normally a constant percentage of the power input, so the antenna gain is

$$G = k \frac{P}{P_{av}} \qquad \text{(Eq 4)}$$

where

G = gain (expressed as a power ratio)
k = efficiency (power radiated divided by power input) of the antenna
P and P_{av} are as above

For many of the antenna systems used by amateurs, the efficiency is quite high (the loss amounts to only a few percent of the total). In such cases the gain is essentially equal to the directivity.

The more the directive diagram is compressed—or, in common terminology, the "sharper" the lobes—the greater the power gain of the antenna. This is a natural consequence of the fact that as power is taken away from a larger and larger portion of the sphere surrounding the radiator, it is added to the volume represented by the narrow lobes. Power is therefore concentrated in some directions at the expense of others. In a general way, the smaller the volume of the solid radiation pattern, compared with the volume of a sphere having the same radius as the length of the largest lobe in the actual pattern, the greater the power gain.

As stated above, the gain of an antenna is related to its directivity, and directivity is related to the shape of the directive pattern. A commonly used index of directivity, and therefore the gain of an antenna, is a measure of the width of the major lobe (or lobes) of the plotted pattern. The width is expressed in degrees at the half-power or −3 dB points, and is often called the *beamwidth*.

This information provides only a general idea of relative gain, rather than an exact measure. This is because an absolute measure involves knowing the power density at every point on the surface of a sphere, while a single diagram shows the pattern shape in only one plane of that sphere. It is customary to examine at least the E-plane and the H-plane patterns before making any comparisons between antennas.

Gain referred to an isotropic radiator is necessarily theoretical; that is, it must be calculated rather than measured, because the isotropic radiator is only theoretical in existence. In practice, measurements on the antenna being tested are usually compared with measurements made on a ½-λ dipole. The dipole should be at the same height and have the same polarization as the antenna under test, and the reference field—that from the ½-λ dipole comparison antenna—should be measured in the most favored direction of the dipole. The data can be obtained either by measuring the field strengths produced at the same distance from both antennas when the same power is supplied to each, or by measuring the power required in each antenna to produce the same field strength at the same distance.

An infinitely thin ½-λ dipole has a theoretical gain of 2.14 dB over an isotropic radiator (dBi). Thus, the gain of an actual antenna over a ½-λ dipole can be referred to isotropic by adding 2.14 dB to the measured gain. Similarly, if the gain is expressed over an isotropic antenna, it can be referred to a ½-λ dipole by subtracting 2.14 dB.

It should be noted that the field strength (voltage) produced by an antenna at a given point is proportional to the square root of the power. That is, when the two are expressed as ratios (the usual case),

$$\frac{E1}{E2} = \sqrt{\frac{P1}{P2}} \qquad \text{(Eq 5)}$$

THE DECIBEL

As a convenience, the power gain of an antenna system is usually expressed in decibels. The decibel is an excellent practical unit for measuring power ratios because it is more closely related to the actual effect produced than the power ratio itself. One decibel represents a just detectable change in signal strength, regardless of the actual value of the signal voltage. A 20-decibel (20-dB) increase in signal, for example, represents 20 observable "steps" in increased signal. The power ratio (100 to 1) corresponding to 20 dB gives an entirely exaggerated idea of the improvement in communication to be expected. The number of decibels corresponding to any power ratio is equal to *10 times the common logarithm* of the power ratio, or

$$dB = 10 \log \frac{P1}{P2} \qquad \text{(Eq 6)}$$

If the *voltage* ratio is given, the number of decibels is equal to *20 times the common logarithm* of the ratio. That is,

$$db = 20 \log \frac{E1}{E2} \qquad \text{(Eq 7)}$$

When a voltage ratio is used, both voltages must be measured across the same value of impedance. Unless this is done the decibel figure is meaningless, because it is fundamentally a measure of a *power* ratio.

Even though the antenna patterns published in this book appear on a grid marked in decibels, the patterns themselves are in terms of relative field strength (voltage), referenced to the strength in the direction of maximum radiation (0 dB). The plotting of the patterns therefore necessarily involves the use of the second equation above, where E1 is the strength in the direction under consideration and E2 is the strength in the direction of maximum radiation. The two equations given above, if worked in reverse, yield field strength and power ratios for the published pattern information.

Table 1 shows the number of decibels corresponding to various power ratios, and Table 2 for various voltage ratios. One advantage of the decibel is that successive power gains expressed in decibels may simply be added together. Thus a gain of 3 dB followed by a gain of 6 dB gives a total gain of 9 dB. In ordinary power ratios, the ratios must be multiplied together to find the total gain.

A *reduction* in power is handled simply by subtracting the requisite number of decibels. Thus, reducing the power to ½ is the same as *subtracting* 3 decibels. For example, a power gain of 4 in one part of a system and a reduction to ½ in another part gives a total power gain of 4 × ½ = 2. In decibels, this is 6 − 3 = 3 dB. A power reduction or "loss" is simply indicated by including a negative sign in front of the appropriate number of decibels.

Power Gains of Harmonic Antennas

In splitting off into a series of lobes, the solid radiation pattern of a harmonic antenna is compressed into a smaller volume as compared with the single-lobed pattern of the ½-λ dipole. This means that there is a concentration of power in certain directions with a harmonic antenna, particularly in the main lobe. The result is that a harmonic antenna produces

Table 1
Power Ratio to Decibel Conversion

| | | | | | *Decimal Increments* | | | | | |
Ratio	0.0	0.1	0.2	0.3	0.4	0.5	0.6	0.7	0.8	0.9
1	0.00	0.41	0.79	1.14	1.46	1.76	2.04	2.30	2.55	2.79
2	3.01	3.22	3.42	3.62	3.80	3.98	4.15	4.31	4.47	4.62
3	4.77	4.91	5.05	5.19	5.31	5.44	5.56	5.68	5.80	5.91
4	6.02	6.13	6.23	6.33	6.43	6.53	6.63	6.72	6.81	6.90
5	6.99	7.08	7.16	7.24	7.32	7.40	7.48	7.56	7.63	7.71
6	7.78	7.85	7.92	7.99	8.06	8.13	8.20	8.26	8.33	8.39
7	8.45	8.51	8.57	8.63	8.69	8.75	8.81	8.86	8.92	8.98
8	9.03	9.08	9.14	9.19	9.24	9.29	9.34	9.40	9.44	9.49
9	9.54	9.59	9.64	9.68	9.73	9.78	9.82	9.87	9.91	9.96
10	10.00	10.04	10.09	10.13	10.17	10.21	10.25	10.29	10.33	10.37
× 10	+ 10									
× 100	+ 20									
× 1000	+ 30									
× 10,000	+ 40									
× 100,000	+ 50									

Table 2

Voltage Ratio to Decibel Conversion

Ratio	Decimal Increments									
	0.0	*0.1*	*0.2*	*0.3*	*0.4*	*0.5*	*0.6*	*0.7*	*0.8*	*0.9*
1	0.00	0.83	1.58	2.28	2.92	3.52	4.08	4.61	5.11	5.58
2	6.02	6.44	6.85	7.23	7.60	7.96	8.30	8.63	8.94	9.25
3	9.54	9.83	10.10	10.37	10.63	10.88	11.13	11.36	11.60	11.82
4	12.04	12.26	12.46	12.67	12.87	13.06	13.26	13.44	13.62	13.80
5	13.98	14.15	14.32	14.49	14.65	14.81	14.96	15.12	15.27	15.42
6	15.56	15.71	15.85	15.99	16.12	16.26	16.39	16.52	16.65	16.78
7	16.90	17.03	17.15	17.27	17.38	17.50	17.62	17.73	17.84	17.95
8	18.06	18.17	18.28	18.38	18.49	18.59	18.69	18.79	18.89	18.99
9	19.08	19.18	19.28	19.37	19.46	19.55	19.65	19.74	19.82	19.91
10	20.00	20.09	20.17	20.26	20.34	20.42	20.51	20.59	20.67	20.75
× 10	+20									
× 100	+40									
× 1000	+60									
× 10,000	+80									
× 100,000	+100									

an increase in field strength in its most favored direction over a ½-λ dipole in *its* most favored direction, when both antennas are supplied with the same amount of power.

The power gain from harmonic operation is small when the antenna is small in terms of wavelengths, but is quite appreciable when the antenna is fairly long. The theoretical power gain of harmonic antennas or "long wires" is shown by curve B in Fig 8, using the ½-λ dipole as a reference. A 1-λ or "second harmonic" antenna has only a slight power gain, but an antenna 9 λ long shows a power gain of nearly 7 dB over the dipole. This gain occurs in one direction by reducing or eliminating the power radiated in other directions; thus the longer the wire, the more directive the antenna becomes. Curve A in Fig 8 shows how the radiation resistance (as measured at a current loop) varies with the length of a harmonic antenna.

Antenna Frequency Scaling

Any antenna design can be scaled in size for use on another frequency or on another amateur band. The dimensions of the antenna may be scaled with Eq 1.

$$D = \frac{f1}{f2} \times d \qquad \text{(Eq 1)}$$

where

D = scaled dimension
d = original design dimension
f1 = original design frequency
f2 = scaled frequency (frequency of intended operation)

From this equation, a published antenna design for, say, 14 MHz, can be scaled in size and constructed for operation on 18 MHz, or any other desired band. Similarly, an antenna design could be developed experimentally at VHF or UHF and then scaled for operation in one of the HF bands. For example, from Eq 1, an element of 39.0 inches length at 144 MHz would be scaled to 14 MHz as follows: D = 144/14 × 39 = 401.1 inches, which is the same as 33.43 feet.

To scale an antenna properly, *all* physical dimensions must be scaled, including element lengths, element spacings, boom diameters, and *element diameters*. Lengths and spacings may be scaled in a straightforward manner as in the above example, but element diameters are often not as conveniently scaled. For example, assume a 14-MHz antenna is modeled at 144 MHz and perfected with 3/8-inch cylindrical elements. For proper scaling to 14 MHz, the elements should be cylindrical, of 144/14 × 3/8 or 3.86 inches diameter. From a realistic standpoint, a 4-inch diameter might be acceptable, but cylindrical elements of 4-inch diameter in lengths of 33 feet or so would be quite unwieldy (and quite expensive). Choosing another, more suitable diameter is the only practical answer.

DIAMETER (RADIUS) SCALING

Simply changing the diameter of dipole type elements during the scaling process is not satisfactory without making a corresponding element length correction. This is because changing the diameter results in a change in the λ/dia ratio from the original design, and this alters the corresponding resonant frequency of the element. In effect, the element length must be corrected by applying a different K factor, as discussed in connection with Fig 3 early in this chapter.

To be more precise, however, the purpose of diameter scaling is not to maintain the same resonant frequency for the element, but to maintain the same *reactance* at the *operating frequency*. As a matter of fact, for elements that are not resonant at the operating frequency in the original design, the ratio of the two resonant frequencies (before and after scaling) will not equal the scaling ratio, f1/f2, as defined for Eq 1. This is because the reactance varies at a different rate with frequency changes for different λ/dia elements. In other words, looking at the two elements as two resonant circuits, the Qs are different.

Necessary length corrections may be determined from Eqs 2 through 7 below. The calculations yield the proper length for a given element with a newly assigned diameter. For simplification, all dimensions are treated in wavelengths, rather than in physical units. Therefore, some values will be handled as quite small decimal fractions. The information which follows is adapted from Chapter 7 of the book by Jim Lawson, W2PV, *Yagi-Antenna Design* (see bibliography at the end of this chapter). The procedure may be performed with a scientific electronic calculator, but such calculations are somewhat tedious. A programmable calculator or a personal computer relieves the tedium. The following equations are used in the calculations. Their use is explained after all equations and definitions are presented. In the equations and procedure that follows, the suffix designator 1 indicates the *original* design, and 2 the *scaled* design.

$$M = \log \frac{2}{d} \qquad \text{(Eq 2)}$$

$$A = 430.8\,M - 339 \qquad \text{(Eq 3)}$$

$$\ell R = 0.5 - \frac{33.25 + 3.19M - 0.35M^2}{861.6M - 678} \qquad \text{(Eq 4)}$$

$$FR1 = \frac{\ell R1}{\ell 1} \qquad \text{(Eq 5)}$$

$$FR2 = 1 - \frac{A1\,(1 - FR1)}{A2} \qquad \text{(Eq 6)}$$

$$\ell 2 = \frac{\ell R2}{FR2} \qquad \text{(Eq 7)}$$

where

M = a constant related to the λ/dia of the element
d = element diameter in wavelengths
A corresponds to the slope of the reactance curve of the element
ℓR = approximate resonant length of element in wavelengths
FR = approximate resonant frequency of element, normalized to the design frequency
ℓ = element length in wavelengths
λ = free-space wavelength

The first few steps of the scaling procedure determine constants for the original design. Then the scaled length is calculated for the new diameter. Step by step instructions follow, after which an example is given. Remember, the suffix designator 1 indicates the original design, and the designator 2 the scaled design.

1) Determine d1, the original element diameter in free-space wavelengths. A free-space wavelength is $983.6/f_{MHz}$ feet or $11803/f_{MHz}$ inches.
2) Determine ℓ1, the element length in free-space wavelengths.
3) Determine M1 from Eq 2.
4) Determine A1 from Eq 3.
5) Determine ℓR1 from Eq 4.
6) Determine FR1 from Eq 5.
7) Assign d2 in free-space wavelengths.
8) Determine M2 from Eq 2.

9) Determine A2 from Eq 3.
10) Determine ℓR2 from Eq 4.
11) Determine FR2 from Eq 6. Use algebraic subtraction.
12) Determine ℓ2 from Eq 7. This is the scaled element length for the new diameter, in wavelengths. Convert this value to a physical length for the scaled frequency.

A Diameter Scaling Example

Earlier in this section we saw that a 3/8-inch diameter element of length 39 inches for 144 MHz, scaled to 14 MHz, would have a diameter of 3.86 inches and length 401.1 inches. A more practical element diameter would be 7/8 inch. The correct element length for this diameter is found by following the 12 steps outlined above. Five significant digits are used in working through this example; the results are rounded to four.

From step 1, determine d1, the original (144 MHz) element diameter in wavelengths. The free-space wavelength for this frequency is 11803/144 or 81.965 inches. For a 3/8-inch dia element, d1 = (3/8)/81.965 = 0.0045751 λ. From step 2, the element length ℓ1 = 39/81.965 = 0.47581 λ.

From step 3 and Eq 2, M1 = log (2/0.0045751) = 2.6406. From step 4 and Eq 3, A1 = 430.8 × 2.6406 − 339 = 798.57.

From step 5 and Eq 4,

$$\ell R1 = 0.5 - \frac{33.25 + 3.19 \times 2.6406 - 0.35 \times 2.6406^2}{861.6 \times 2.6406 - 678}$$

$$= \frac{39.233}{1597.1} = 0.47543$$

Next, from step 6 and Eq 5, FR1 = 0.47543/0.47581 = 0.99920. This completes the calculation of constants for the original design.

For the scaled design, from step 7 we determine d2 in wavelengths. A free-space wavelength at 14 MHz is 11803/14 or 843.07 inches. From this, for a 7/8-inch diameter, d2 = (7/8)/843.07 = 0.0010378 λ. From step 8 and Eq 2, M2 = log (2/0.0010378) = 3.2849. From step 9 and Eq 3, A2 = 430.8 × 3.2849 − 339 = 1076.1.

From step 10 and Eq 4,

$$\ell R2 = 0.5 - \frac{33.25 + 3.19 \times 3.2849 - 0.35 \times 3.2849^2}{861.6 \times 3.2849 - 678}$$

$$= 0.48144$$

From step 11 and Eq 6,

$$FR2 = 1 - \frac{798.57\,(1 - 0.99920)}{1076.1} = 0.99941$$

From step 12 and Eq 7, ℓ2 = 0.48144/0.99941 = 0.48172 λ. This is the scaled element length in wavelengths. Multiplying by the dimension of one wavelength at 14 MHz obtains the physical length; 843.07 × 0.48172 = 406.1 inches or 33.84 feet. (As a matter of interest, this is 5.0 inches longer than that calculated earlier for a diameter of 3.86 inches.)

Consequences of Diameter Scaling

An antenna for which *all* dimensions are scaled up or down to operate on a new frequency should behave exactly the same way as the original. In other words, the feed-point impedance, gain, and F/B ratio should be identical at the new design frequency. However, if diameter scaling is done for the elements, antenna performance will vary slightly with frequency departures because the Q of the elements is different.

TAPERED ELEMENTS

Rotatable beam antennas are usually constructed with elements made of metal tubing. Other than being held in place by attachment to the boom, the elements are self-supporting. This means the tubing selected for the center of the element must be strong enough to support the entire element weight without bending or breaking. This also means the material throughout the element must be rigid enough to avoid undue droop at the element ends.

Meeting the mechanical requirements for conductor selection at VHF and UHF is no problem, as the longest elements are seldom more than a few feet in length. In such antennas, a single length of tubing (or of solid rod at the higher frequencies) suffices for each element. These elements are cylindrical, of uniform diameter. At frequencies below 30 MHz, however, the situation is different. Resonant element lengths are many feet long, and for proper support, large diameter tubing with a thick wall becomes a requirement for the center section. But using the same material for the entire element contributes to unnecessary weight in the assembled antenna. This becomes a significant consideration especially at the lower frequencies, where large arrays often weigh in excess of 100 pounds.

The general practice at HF is to taper the elements with lengths of telescoping tubing. The center section has a large diameter, but the ends are relatively small. This reduces not only the weight, but also the cost of materials for the elements. Information on aluminum tubing sizes and suggested arrangements (schedules) for tapering the elements for various HF bands are given in Chapter 21.

Length Correction for Tapered Elements

The effect of tapering an element is to alter its electrical length. That is to say, two elements of the same length, one cylindrical and one tapered but with the same average diameter as the cylindrical element, will not be resonant at the same frequency. The tapered element must be made longer than the cylindrical element for the same resonant frequency. This is because the thicker portion at the center of the tapered element has less inductance than the cylinder, and must therefore be made longer. And at the ends, the smaller tapered sections have less end capacitance to load the element, also requiring a greater length. Replacing a cylindrical element with a tapered element (and vice versa) therefore requires a length correction for identical performance at the central design frequency.

A procedure for calculating the length correction for tapered elements has been worked out by Jim Lawson and is presented in his book, *Yagi-Antenna Design* (see bibliography). The information that follows is based on Lawson's procedure. The results are a very good approximation of the electrical effects of tapering an element.

As with diameter scaling, working through a few equations is required for determining taper corrections. Calculations are made for only one half of an element, assuming the element is symmetrical about the point of boom attachment.

In concept, the procedure is simple. Assume we have a tapered element and wish to know its equivalent length in a cylindrical element. The lengths for each telescoping section in the tapered element are first converted into equivalent lengths as sections of the cylindrical element. Then the individual cylindrical-equivalent lengths are simply added up to get the total equivalent length of the tapered element. That completes the conversion. As with the process of diameter scaling, a programmable calculator or a personal computer relieves the tedium of making the calculations.

Assume now that we want to work the other way, to determine the necessary length of a tapered element to replace one of cylindrical design. This procedure involves some iteration—some trial and error. First we determine the taper schedule for our new element, choosing the lengths and diameters of the various telescoping sections. (See Chapter 21 for suggestions.) Then, as before, we convert individual sections in the tapered design to equivalent lengths in a cylindrical element. The sum of these lengths is then compared to the length of the original cylindrical element. If the two are not in agreement, an adjustment is made to the design of the tapered element and the calculations are repeated. Two or three tries are often sufficient to find the schedule required for the tapered element.

At first it might seem logical to use the same set of equations for this procedure as those presented earlier to calculate diameter scaling corrections, just working with short sections of an element and making length corrections for each individual diameter in the tapered element. There are differences in the two situations, however. For example, inner sections of a tapered element have no end capacitance. Instead, impedance "lumps" exist in the element, caused by the transition from one diameter to another in the taper schedule. For reasons of this nature, different equations are necessary.

Eqs 8 through 12 apply to determining length corrections for element taper. In these equations, subscripted suffix designators T and C indicate the tapered and cylindrical designs, respectively. Note that all calculations are made for a *single frequency*. This procedure does not perform frequency scaling. Note also that all dimensions are expressed in *inches*. Wavelengths are not used in these calculations.

$$N = 4.373 - \log (f \times d) \qquad \text{(Eq 8)}$$

$$m = \frac{N_T - 0.7869}{N_C - 0.7869} \qquad \text{(Eq 9)}$$

$$\theta_2 = \frac{90S}{H} \qquad \text{(Eq 10)}$$

$$f_\theta = 28.648 \, \frac{\sin 2\theta_2 - \sin 2\theta_1}{\theta_2 - \theta_1} \qquad \text{(Eq 11)}$$

$$\ell_C = \frac{\ell_T}{2} \left[m + \frac{1}{m} + f_\theta \left(m - \frac{1}{m} \right) \right] \qquad \text{(Eq 12)}$$

where

f = frequency, megahertz
d = diameter of section, inches
S = distance from element center to outside end of telescoping section under consideration, inches
H = total half length of tapered element, inches
θ_1 and θ_2 represent angles in degrees
θ_1 for any section is the same as θ_2 for the adjacent inside section. (For the innermost section, $\theta_1 = 0$.)
ℓ = length of section, inches
All others are intermediate variables.

The procedure for determining a length correction to account for element taper is as follows. Calculations are made for just half of the antenna, from the center to one outside end.

1) If the conversion is being made from a cylindrical element, assign a taper schedule. In this case, the telescoping section lengths must initially be estimated.

2) If the conversion is being made from a cylindrical element, determine the corrected length of that element for the diameter *assigned at the center* of the half element in the taper schedule, using Eqs 2 through 7 above. If the conversion is being made from a tapered element, the diameter of the cylindrical element should initially be chosen as that at the center of the half element in the existing taper schedule.

3) From Eq 8 calculate N_C, and calculate N_T for each telescoping section.

4) From Eq 9, calculate m for each telescoping section.

5) From Eq 10, calculate θ_2 for each telescoping section.

6) From Eq 11, calculate f_θ for each telescoping section.

7) From Eq 12, calculate ℓ_C for each telescoping section.

8) Sum the ℓ_C values for the telescoping sections. This sum is the total equivalent half-length of the cylindrical element for the tapered element.

9) If the conversion is being made from a tapered element, the calculations for the cylindrical element length are complete. Its diameter may now be altered, if desired, and a length correction determined with Eqs 2 through 7 above. If the conversion is being made from a cylindrical element, compare the sum of the ℓ_C values with the cylindrical element half-length determined in step 2. If necessary, adjust the taper schedule and repeat this procedure from step 5. If the initial lengths were chosen with care, adjustment of only the outside end section should suffice.

A Taper Length-Correction Example

In the previous sections on antenna frequency and element diameter scaling, we saw as an example the case of an element for 144 MHz redesigned for 14 MHz. The original element was 39 inches long and 3/8 inch dia. Calculations showed that this element, scaled to 14 MHz, would have a length of 401.1 inches and a dia of 3.86 inches. To avoid this unwieldy diameter, we performed a diameter scaling operation and learned that an equivalent element could be made from a 406.1-inch length of tubing having 7/8-inch dia.

Let us assume now that instead of using a cylindrical element, we prefer a tapered element of telescoping sections of tubing. The calculation procedure follows. Recall that the subscripted suffix designator C indicates the cylindrical element, and T the tapered element.

Begin with step 1 by developing a taper schedule for the new element. Assume we choose the schedule shown in Fig 24. Initially, an estimate is required for the lengths of the various sections.

As given in step 2, we need to know the equivalent length of the cylindrical element for a diameter equal to that at the center of the tapered half-element. This diameter in our chosen taper schedule, Fig 24, is 7/8 inch. For this example we have already determined that length to be 406.1 inches.

From step 3 and Eq 8 we first calculate N_C, a constant for the cylindrical element. The frequency is 14 MHz. $N_C = 4.373 - \log(14 \times 7/8) = 4.373 - \log 12.25 = 3.2849$. Next we calculate N_T for each telescoping section. It is convenient to write the results of calculations in tabular form, as shown in Fig 24. For section 1, the innermost telescoping section, the diameter is 1-1/8 inches. From Eq 8, N_T for this section is $4.373 - \log(14 \times 1.125) = 3.1757$. In a similar manner, calculate values for N_T for sections 2, 3 and 4 in the taper

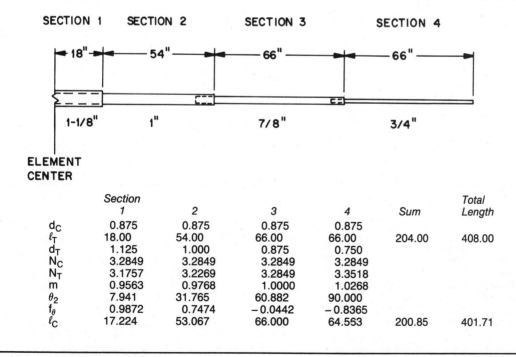

	Section 1	2	3	4	Sum	Total Length
d_C	0.875	0.875	0.875	0.875		
ℓ_T	18.00	54.00	66.00	66.00	204.00	408.00
d_T	1.125	1.000	0.875	0.750		
N_C	3.2849	3.2849	3.2849	3.2849		
N_T	3.1757	3.2269	3.2849	3.3518		
m	0.9563	0.9768	1.0000	1.0268		
θ_2	7.941	31.765	60.882	90.000		
f_θ	0.9872	0.7474	−0.0442	−0.8365		
ℓ_C	17.224	53.067	66.000	64.553	200.85	401.71

Fig 24—Half of a tapered element following a schedule chosen for the conversion from a 14-MHz cylindrical element. The tabulated data contains the results of calculations. This schedule represents an estimation; the final schedule must be determined by iteration.

schedule, sections having different diameters. The results of these calculations are tabulated in Fig 24.

Next, from step 4 and Eq 9, calculate values of m for each telescoping section. For section 1,

$$m = \frac{3.1757 - 0.7869}{3.2849 - 0.7869} = \frac{2.3888}{2.4980} = 0.9563$$

In a similar manner, calculate the value of m for sections 2, 3 and 4. The resulting values are shown in Fig 24.

Step 5 and Eq 10 are used to calculate θ_2 for each telescoping section. For section 1,

$$\theta_2 = \frac{90 \times 18}{204} = 7.941°$$

Note that S in Eq 10 is the spacing from the center of the element to the *outside* end of the telescoping section under consideration. For section 2, therefore,

$$\theta_2 = \frac{90\,(18 + 54)}{204} = 31.765°$$

In a similar way, calculate θ_2 for section 3. For the outermost section (section 4 in this example), the value of θ_2 is always 90°.

From step 6 and Eq 11 we calculate f_θ for each telescoping section. For section 1, the value of θ_1 is zero, and

$$f_\theta = 28.648 \frac{\sin (2 \times 7.941°)}{7.941°}$$

$$= 28.648 \frac{\sin 15.882°}{7.941°} = 0.9872$$

For succeeding sections, the value of θ_1 is the same as that for θ_2 for the adjacent inside section. Thus, for section 2,

$$f_\theta = 28.648 \frac{\sin (2 \times 31.765°) - \sin (2 \times 7.941°)}{31.765° - 7.941°}$$

$$= 28.648 \frac{0.62151}{23.824} = 0.7474$$

The values for f_θ are calculated in a similar manner for sections 3 and 4. In this example the results are negative values.

From step 7 and Eq 12, calculate ℓ_C for each telescoping section. For section 1,

$$\ell_C = \frac{18}{2}\left[0.9563 + \frac{1}{0.9563} + 0.9872\left(0.9563 - \frac{1}{0.9563}\right)\right]$$

$$= 9 \times 1.9138 = 17.224 \text{ in.}$$

Values of ℓ_C are calculated similarly for sections 2 through 4. For each telescoping section, these values are the electrically equivalent lengths in the cylindrical element.

Following step 8, we sum the ℓ_C values, 200.85 for this example. By doubling this value, we see that our chosen taper schedule of Fig 24 produces the equivalent of a 7/8-inch dia cylindrical element that is 401.7 inches long. From step 9, we compare this against our designed cylindrical element, which is 406.1 inches long.

At this point we still have more work to do; our taper schedule produces an element which is equivalent to being 4.4 inches too short. A half-length correction of 2.2 inches is required. This length adjustment can be made entirely in section 4 of our taper schedule, the outside end section.

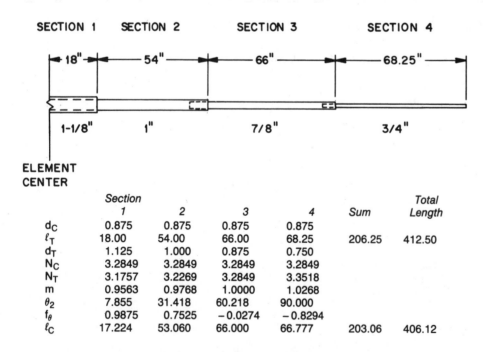

	Section					Total
	1	2	3	4	Sum	Length
d_C	0.875	0.875	0.875	0.875		
ℓ_T	18.00	54.00	66.00	68.25	206.25	412.50
d_T	1.125	1.000	0.875	0.750		
N_C	3.2849	3.2849	3.2849	3.2849		
N_T	3.1757	3.2269	3.2849	3.3518		
m	0.9563	0.9768	1.0000	1.0268		
θ_2	7.855	31.418	60.218	90.000		
f_θ	0.9875	0.7525	−0.0274	−0.8294		
ℓ_C	17.224	53.060	66.000	66.777	203.06	406.12

Fig 25—The adjusted taper schedule for the text example and the results of calculations. Twice the length of the half element shown is 412.5 inches, which is equivalent to a 7/8-inch-diameter cylindrical element of length 406.12 inches.

Experience with these calculations shows that a length added to the end section usually "shrinks" when it comes to the change in the equivalent cylindrical element. For this reason we'll want to add just a bit more than 2.2 inches to the length of section 4. Let's try 2¼ inches, making the end section now 68.25 inches long. This new taper schedule is shown in Fig 25.

As indicated in step 9, we repeat the procedure from step 5. Because the total half-length of the tapered element has now changed, we need to repeat the calculations for θ_2 for each telescoping section. This, in turn, requires subsequent calculations to be repeated.

The results of calculations for this new taper schedule are tabulated in Fig 25. Success! Our length adjustment to section 4 has produced a tapered element that is 412½ inches long, having an equivalent length of 406.12 inches for a cylindrical element of 7/8 inch dia. From a practical standpoint, this is the same as the 406.1-inch design value we are seeking. At the central design frequency, the tapered element of Fig 25 will have the same electrical characteristics as our 7/8-inch dia cylindrical element.

Calculations by Computer

A simple home-computer program can be used to calculate the correct taper schedule for a cylindrical element. TAPER is a program written for this purpose by Bill Myers, K1GQ. The program performs any necessary diameter scaling for the element (equivalent to step 2 of this procedure) and iterates within itself to find the correct length of the end section. The program listing was originally published in October 1986 *QST*, and is included here as Table 3.

TAPER uses the variables described in *Yagi-Antenna Design*, which are slightly different from those presented in this chapter. TAPER is written in Microsoft BASIC, so it should be easily adaptable to most home computers. Dimensions entered at the keyboard for the cylindrical element must be in wavelengths. With values for the above example entered, the computer program running in an IBM PC agrees within approximately 1/16 inch for the length of section 4 as determined above.

Table 3
Program Listing for TAPER

```
10 '*********************************************************************
20 '                          T A P E R                                *
30 '                                                                    *
40 '       Convert a cylindrical half-element to an equivalent tapered  *
50 '       half-element by computing the length of the end piece using  *
60 '       Lawson's method (Yagi Antenna Design).                       *
70 '                                                                    *
80 ' Microsoft BASIC Version 5.28                         Feb 86 K1GQ   *
90 '*********************************************************************
100 '                                                                   '
110 '   Reserve arrays.  MAXPARTS% establishes the maximum number of    '
120 '   pieces per half-element.                                        '
130 '                                                                   '
140 ''''''''''''''''''''''''''''''''''''''''''''''''''''''''''''''''''''''
150 MAXPARTS% = 9
160 DIM PARTD(MAXPARTS%), PARTL(MAXPARTS%), LP(MAXPARTS%), F(MAXPARTS%)
170 DIM M(MAXPARTS%), THETA(MAXPARTS%+1)
180 ''''''''''''''''''''''''''''''''''''''''''''''''''''''''''''''''''''''
190 '   Define functions for differential reactance, DELTAX, and        '
200 '   total reactance, X.  Coefficients are changed to use natural    '
210 '   logarithm instead of base-10 logarithm.  CAPK (CAPital K) is    '
220 '   the ratio of wavelength to radius.                              '
230 '                                                                   '
240 ''''''''''''''''''''''''''''''''''''''''''''''''''''''''''''''''''''''
250 DEF FN DELTAX(CAPK) = -18.7 * LOG(CAPK) + 33.9
260 DEF FN X(CAPK) = 33.25 + 1.385 * LOG(CAPK) - .066 * LOG(CAPK) ^ 2
270 ''''''''''''''''''''''''''''''''''''''''''''''''''''''''''''''''''''''
280 '   Constants.                                                      '
290 '                                                                   '
300 ''''''''''''''''''''''''''''''''''''''''''''''''''''''''''''''''''''''
310 C = 11802.85           ' Speed of light in inches/microsecond
320 PIO2 = 2 * ATN(1)      ' pi / 2
330 ''''''''''''''''''''''''''''''''''''''''''''''''''''''''''''''''''''''
340 '   Get design data and element tubing dimensions.                  '
350 '                                                                   '
360 ''''''''''''''''''''''''''''''''''''''''''''''''''''''''''''''''''''''
370 INPUT "Frequency (MHz)"; FREQ
380 INPUT "Cylinder halflength (wavelengths)"; HOL
390 INPUT "Cylinder diameter (wavelengths)"; DOL
400 PRINT "Number of pieces per half-element ( <="; MAXPARTS%; ")" ;
410 INPUT NPARTS%
420 IF NPARTS% > MAXPARTS% GOTO 400
430 FOR I% = 1 TO NPARTS% - 1
```

```
440      PRINT "Part"; I%; " Length (inches), diameter (eighth-inches)";
450      INPUT PARTL(I%), PARTD(I%)
460      PARTD(I%) = PARTD(I%)/8
470      NEXT I%
480 PRINT "part"; NPARTS%; "                        diameter (eighth-inches)";
490 INPUT PARTD(NPARTS%)
500 PARTD(NPARTS%) = PARTD(NPARTS%) / 8
510 LAMBDA = C / FREQ
520 '''''''''''''''''''''''''''''''''''''''''''''''''''''''''''''''''''''''''
530 '   Alter halflength to scale from design diameter to the            '
540 '   geometric average of the root and end piece diameters.           '
550 '                                                                    '
560 '''''''''''''''''''''''''''''''''''''''''''''''''''''''''''''''''''''''''
570 AVGDIA = SQR(PARTD(1) * PARTD(NPARTS%))
580 ADOL = AVGDIA / LAMBDA
590 CAPK = 2 / DOL
600 ACAPK = 2 / ADOL
610 SML = 2 * HOL
620 ASML = .5 + ( FNX(ACAPK) - FNX(CAPK) - 20 * FNDELTAX(CAPK) *
                  ( .5 - SML ) ) / ( 20 * FNDELTAX(ACAPK) )
630 HAOL = ASML / 2
640 HA = HAOL * LAMBDA
650 '''''''''''''''''''''''''''''''''''''''''''''''''''''''''''''''''''''''''
660 '   Set up Lawson's M functions for each piece.                      '
670 '                                                                    '
680 '''''''''''''''''''''''''''''''''''''''''''''''''''''''''''''''''''''''''
690 FOR I% = 1 TO NPARTS%
700      PDIA = PARTD(I%) / LAMBDA
710      CAPI = 2 / PDIA
720      M(I%) = FNDELTAX(CAPI) / FNDELTAX(ACAPK)
730      NEXT I%
740 '''''''''''''''''''''''''''''''''''''''''''''''''''''''''''''''''''''''''
750 ' Set up initial guess for the length of the end part.              '
760 '                                                                    '
770 '''''''''''''''''''''''''''''''''''''''''''''''''''''''''''''''''''''''''
780 PARTL(NPARTS%) = HA
790 FOR I% = 1 TO NPARTS% - 1
800      PARTL(NPARTS%) = PARTL(NPARTS%) - PARTL(I%)
810      NEXT I%
820 THETA(NPARTS%+1) = PIO2
830 '''''''''''''''''''''''''''''''''''''''''''''''''''''''''''''''''''''''''
840 '   Compute the cylindrical element which is equivalent to the       '
850 '   assumed tapered element, adjust the end piece length proportionally '
860 '   to the error between the computed cylinder length and target length '
870 '   (HA), iterate until the error is small.                          '
880 '                                                                    '
890 '''''''''''''''''''''''''''''''''''''''''''''''''''''''''''''''''''''''''
900 DELTA = 1
910 WHILE ABS(DELTA) > .00001*HA
920      ''''''''''''''''''''''''''''''''''''''''''''''''''''''''''''''''''''.
930      ' Find the total half-length of the tapered element.            '
940      '                                                                '
950      '''''''''''''''''''''''''''''''''''''''''''''''''''''''''''''''''''
```

```
960        S = 0
970        FOR I% = 1 TO NPARTS%
980             S = S + PARTL(I%)
990             THETA(I%) = 0
1000           NEXT I%
1010       SRAD = S / PIO2
1020       ''''''''''''''''''''''''''''''''''''''''''''''''''''''''''''''''''''''''''
1030       ' Compute the positions of the joints in radians.                        '
1040       '                                                                        '
1050       ''''''''''''''''''''''''''''''''''''''''''''''''''''''''''''''''''''''''''
1060       FOR I% = 2 TO NPARTS%
1070            THETA(I%) = THETA(I%-1) + PARTL(I%-1) / SRAD
1080           NEXT I%
1090       ''''''''''''''''''''''''''''''''''''''''''''''''''''''''''''''''''''''''''
1100       ' Evaluate Lawson's F function and determine the                         '
1110       ' equivalent length of each piece.                                       '
1120       '                                                                        '
1130       ''''''''''''''''''''''''''''''''''''''''''''''''''''''''''''''''''''''''''
1140       FOR I% = 1 TO NPARTS%
1150            F(I%) = ( SIN(2*THETA(I%+1)) - SIN(2*THETA(I%)) )
                        / ( 2 * ( THETA(I%+1) - THETA(I%) ) )
1160            LP(I%) = PARTL(I%) * ( M(I%) + 1/M(I%)
                         + ( M(I%) - 1/M(I%) ) * F(I%) ) / 2
1170           NEXT I%
1180       ''''''''''''''''''''''''''''''''''''''''''''''''''''''''''''''''''''''''''
1190       ' Find the error between the sum of the equivalent                       '
1200       ' piece lengths and the target length.                                   '
1210       '                                                                        '
1220       ''''''''''''''''''''''''''''''''''''''''''''''''''''''''''''''''''''''''''
1230       DELTA = HA
1240       FOR I% = 1 TO NPARTS%
1250            DELTA = DELTA - LP(I%)
1260           NEXT I%
1270       ''''''''''''''''''''''''''''''''''''''''''''''''''''''''''''''''''''''''''
1280       ' Add the error to the end piece and loop back.                          '
1290       '                                                                        '
1300       ''''''''''''''''''''''''''''''''''''''''''''''''''''''''''''''''''''''''''
1310       PARTL(NPARTS%) = PARTL(NPARTS%) + M(NPARTS%) * DELTA
1320       WEND
1330 ''''''''''''''''''''''''''''''''''''''''''''''''''''''''''''''''''''''''''
1340 ' Show the results, then go back to do another case with the same           '
1350 ' design parameters except halflength, and the same tubing schedule.        '
1360 '                                                                           '
1370 ''''''''''''''''''''''''''''''''''''''''''''''''''''''''''''''''''''''''''
1380 PRINT USING ">> End piece length = ###.#### inches <<"; PARTL(NPARTS%)
1390 INPUT "Another case (y or n)"; ANS$
1400 IF ANS$ = "n" GOTO 1430
1410     INPUT "Cylinder halflength (wavelengths)"; HOL
1420     GOTO 610
1430 END
```

Special Antenna Types

So far in this chapter, the underlying principles of antenna operation have been discussed primarily in terms of the single-conductor, ½-λ dipole, which is the elementary form from which more elaborate antenna systems are built. There are a number of other types of antennas that find application in amateur work, particularly when space limitations do not permit using a full-size dipole. The more common of these antennas are discussed in the sections that follow.

FOLDED DIPOLES

In the diagram shown in Fig 26, suppose for the moment that the upper conductor between points B and C is disconnected and removed. The system is then a simple center-fed dipole, and the direction of current flow along the antenna and line at a given instant is as shown by the arrows. If the upper conductor between B and C is restored, the current in it will flow away from C and toward B, in accordance with the rule for reversal of direction in alternate half-wave sections along a wire. However, the fact that the second wire is "folded" makes the currents in the two conductors of the antenna flow in the *same* direction. Although the antenna physically resembles a transmission line, it is not actually a line from the standpoint of antenna currents, but is merely two conductors in parallel. The connections at the ends of the two are assumed to be of negligible length.

A ½-λ dipole formed in this way has the same directional properties and total radiation resistance as an ordinary dipole. However, the feed line is connected to only *one* of the conductors. It is therefore to be expected that the antenna will "look" different with respect to its input impedance as viewed by the line.

The effect on the impedance at the antenna input terminals can be easily visualized. The center impedance of the dipole *as a whole* is the same as the impedance of a single-conductor dipole—that is, approximately 70 Ω. A given amount of power therefore causes a definite value of current, I. In the ordinary ½-λ dipole this current flows at the junction of the line and the antenna. In the folded dipole the same total current also flows, but is equally divided between two conductors in parallel. The current in each conductor is therefore I/2. Consequently, the line "sees" a higher impedance, because it is delivering the same power at only half the current. The new impedance value is equal to four times the impedance of a simple dipole ($Z = P/I^2$). If more wires are added in parallel, the current continues to divide between them and the terminal impedance is raised even higher. (This explanation is a simplified one based on the assumption that the conductors are close together and have the same diameter.)

The two-wire system in Fig 27A is an especially useful one because the input impedance is very close to 300 Ω. The antenna can be fed directly with 300-ohm twin-lead or open-wire line without any other matching arrangement, and the line will operate with a low standing wave ratio. The antenna itself can be built like an open-wire line—that is, the two conductors can be held apart by regular feeder spreaders. TV "ladder" line is suitable.

In the folded dipole there is also a transmission-line effect; the impedance of the dipole appears in parallel with the impedance of the shorted transmission-line sections. The value of 468 appearing in Fig 27A results in an antenna length

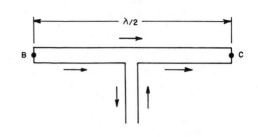

Fig 26—Direction of current flow in a folded dipole.

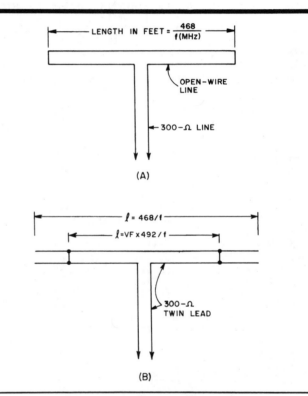

Fig 27—Folded half-wave dipoles. If solid-dielectric transmission line is used for the radiating element, the shorting connections should be placed inward from the ends of the antenna, as shown at B. VF = velocity factor of the line.

which is 95% of a half wave in free space. If the velocity factor of the antenna, looking at it as a transmission line, is of approximately this value, the shorting connections may be made at the ends of the antenna. (Velocity factors of transmission lines are discussed in detail in Chapter 24.)

Solid-dielectric lines, such as 300-Ω twin-lead, have a significantly lower velocity factor than open-wire line. In such cases the position of the shorting connections should be moved toward the center of the antenna, as shown in Fig 27B, to avoid introducing reactance at the feed point when the antenna is cut to a resonant physical length.

The folded dipole has a somewhat "flatter" impedance versus frequency characteristic than a simple dipole. That is, the reactance varies less rapidly, as the frequency is varied on either side of resonance, than with a single-wire antenna. The transmission-line effect mentioned above accounts for this phenomenon. At a frequency away from resonance, the reactance of the dipole is of the opposite "type" from that of the shorted line, and there is some reactance cancellation at the feed point as a result.

Harmonic Operation

A folded dipole will not accept power at twice the fundamental frequency, nor any even multiples of the fundamental. At such multiples the folded section simply acts as a continuation of the transmission line. No other current distribution is possible if the currents in the two conductors of the actual transmission lines are to flow in opposite directions.

On the third and other odd multiples of the fundamental, the current distribution is correct for operation of the system as a folded antenna. Because the radiation resistance of a $3/2\ \lambda$ antenna is not greatly different from that of a $\frac{1}{2}$-λ antenna, a folded dipole can be operated on its third harmonic.

Multi- and Unequal-Conductor Folded Dipoles

Impedance ratios larger than 4 to 1 are frequently desirable when the folded dipole is used as the driven element in a directive array, because the radiation resistance of such an array is usually quite low. A wide choice of impedance step-up ratios is available by varying the relative size and spacing of the conductors, and by using more than two conductors. Fig 28 gives design information of this nature for two-conductor folded dipoles and Fig 29 is a similar chart for three-conductor dipoles. Fig 29 assumes that the three conductors are in the same plane and that the two not directly connected to the transmission line are equally spaced from the driven conductor.

In computing the length of a folded dipole using thick conductors—that is, tubing such as is used in rotary beam antennas—the resonant length may be appreciably less than that of a single-wire antenna cut for the same frequency. Aside from the shortening required with thick conductors, the parallel conductors tend to act like the boundaries of a conducting sheet of the same width as the spacing between the conductors. The "effective diameter" of the folded dipole lies somewhere between the actual conductor diameter and the maximum distance between conductors. The relatively large effective thickness of the antenna reduces the rate of change of reactance with frequency, so the tuning becomes relatively broad and the antenna length is not too critical for a given frequency. Further information on the folded dipole, as pertains to feeding and matching, is contained in Chapters 4 and 15.

THE GROUNDED ANTENNA

In cases where vertical polarization is required—for example, when a low wave angle is desired at frequencies below 4 MHz—the antenna must be vertical. At these low frequencies, the height of a vertical $\frac{1}{2}$-λ antenna is beyond the constructional reach of most amateurs. A 3.5-MHz $\frac{1}{2}$-λ vertical is 133 feet high, for instance.

If the lower end of the vertical antenna is grounded, the

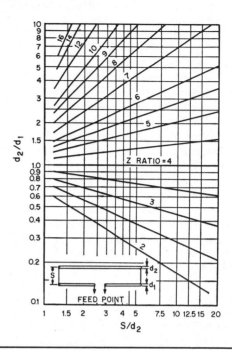

Fig 28—Impedance step-up ratio for the two-conductor folded dipole, as a function of conductor diameters and spacing. Dimensions d_1, d_2 and S are shown in the inset drawing. The step-up ratio, r, may also be determined by calculations:

$$r = \left(1 + \frac{\log \dfrac{2S}{d1}}{\log \dfrac{2S}{d2}}\right)^2$$

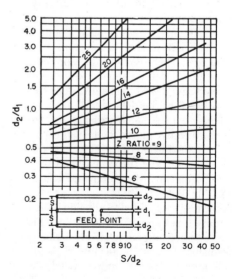

Fig 29—Impedance step-up ratio for the three-conductor folded dipole. The conductors that are not directly driven must have the same diameter, but this diameter need not be the same as that of the driven conductor. Dimensions are indicated in the inset.

antenna height need be only ¼ λ to resonate at the same frequency as an ungrounded ½-λ antenna. The reason for this is that ground having high conductivity acts as an electrical mirror, and the missing half of the antenna is supplied by the mirror image. This is shown in Fig 30. (The effects of ground are discussed in detail in Chapter 3.)

The directional characteristic of the grounded ¼-λ antenna are the same as that of a ½-λ antenna in free space. Thus, a grounded ¼-λ vertical antenna has a circular or omnidirectional radiation pattern in the horizontal or azimuth plane. In the vertical plane, assuming the earth is a perfect conductor, radiation decreases from maximum along the ground to zero directly overhead.

The current in a grounded ¼-λ vertical wire varies practically sinusoidally (as is the case with a ½-λ wire), and is highest at the ground connection. The RF voltage, however, is highest at the open (top) end and minimum at the ground. The current and voltage distribution are shown in Fig 31A.

The grounded antenna may be much smaller than a quarter wavelength and still be made resonant by "loading" it with inductance at the base, as shown in Fig 31 at B and C. By adjusting the inductance of the loading coil, even very short wires can be tuned to resonance. When the antenna is shorter than ¼ λ but is loaded to resonance, the current and voltage distribution are partial sine waves along the antenna wire. If the loading coil is essentially free from distributed capacitance, the voltage across it increases uniformly from minimum at the ground, as shown at B and C of Fig 31, while the current is the same throughout.

Of necessity, extremely short antennas are used in mobile work on the lower frequencies, such as the 3.5-MHz band. These may be "base loaded" as shown in Fig 31B and C, but there is an advantage to be realized by placing the loading coil at the center of the antenna. In neither case, however, is the current uniform throughout the coil. This is because the inductance required is so large that the coil acts as a linear circuit rather than as a lumped inductance.

If the antenna height is greater than ¼ λ but less than ½ λ, the antenna shows inductive reactance at its terminals and can be tuned to resonance by means of a capacitance of the proper value. This is shown in Fig 32A. As the length is increased progressively from ¼ λ to ½ λ, the current loop moves up the antenna, always being at a point ¼ λ from the top. When the height is ½ λ, the current distribution is as shown in Fig 32B. There is a voltage loop (current node) at the base, and power can be applied to the antenna through a parallel-tuned circuit, resonant at the same frequency as the antenna, as shown in the drawing.

Up to just over ½ λ, increasing the antenna height compresses the directive pattern in the vertical plane, resulting in an increase in field strength for a given power input at very low radiation angles. The theoretical improvement is about 1.7 dB for a ½-λ antenna when compared with a ¼-λ antenna, as shown in Fig 33A.

When the height of a vertical antenna is increased beyond ½ λ, secondary lobes appear in the pattern, Fig 33A and B. These become major lobes at relatively high angles when the length approaches ¾ λ. At 1 λ the low-angle lobe disappears and a single lobe remains at approximately 35° elevation.

Radiation Resistance

The radiation resistance of a grounded vertical antenna, as measured between the base of the antenna and ground, varies as a function of the antenna height, as shown in Fig 34. The word "height" used in this context has the same

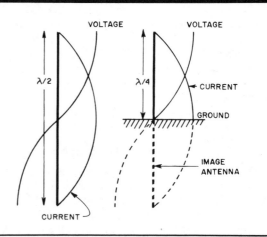

Fig 30—The ½-λ antenna and its grounded ¼-λ counterpart. The missing quarter wavelength can be considered to be supplied as an image in the ground, if it is of good conductivity.

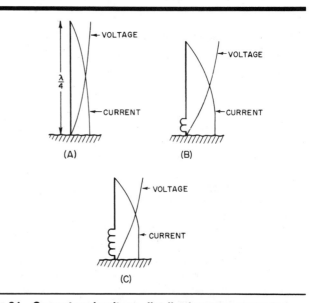

Fig 31—Current and voltage distribution on a grounded ¼-λ antenna (A) and on successively shorter antennas loaded to resonate at the same frequency.

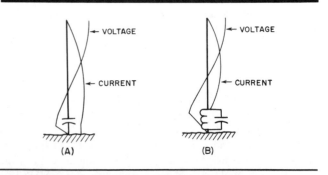

Fig 32—Current and voltage distribution on grounded antennas longer than 1/4 λ. At A, for lengths between 1/4 λ and approximately 3/8 λ; at B, 1/2 λ.

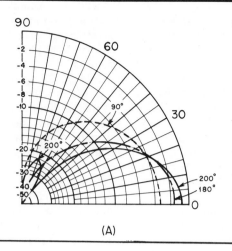

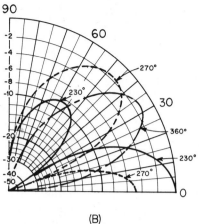

Fig 33—Vertical-plane radiation patterns of vertical antennas for several values of antenna height, ℓ. The amplitude responses in both A and B are to the same scale, relative to the peak response of the 230° (0.64 λ) antenna plotted at B. A perfect conductor beneath the antennas and zero loss resistances are assumed.

(A)   (B)

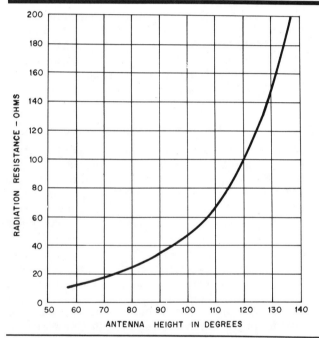

Fig 34—Radiation resistance of a vertical monopole as a function of free-space antenna height in degrees over a perfectly conducting ground (or a highly conducting ground plane). This curve may be used for center-fed antennas by multiplying the resistance values by two; the height in this case is half the actual antenna length.

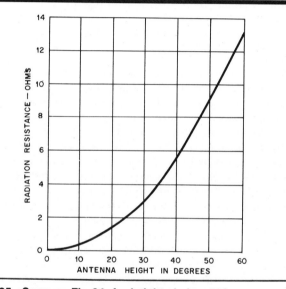

Fig 35—Same as Fig 34, for heights below 60°.

meaning as "length" when applied to horizontal antennas. This curve is for an antenna erected over (but not directly connected to) ground of perfect conductivity. The height is given in degrees, referenced to a free-space wavelength as 360°. The 60° to 135° range shown corresponds to heights from 1/6 λ to 3/8 λ. The antenna is approximately self-resonant at a height of 90° (¼ λ). The actual resonant length will be somewhat less because of the λ/dia ratio mentioned earlier in this chapter. The variation in radiation resistance with heights below 60° is shown in Fig 35. The values in this range are very low.

In the range of heights covered by Figs 34 and 35, the radiation resistance is practically independent of the λ/dia ratio. At greater heights the λ/dia ratio is important in determining the actual value of radiation resistance. At a height of ½ λ, the radiation resistance may be as high as several thousand ohms.

A very approximate curve of reactance vs height is given in Fig 36. The actual reactance depends highly on the λ/dia ratio of the conductor, so this curve should be used only as a rough guide. The curve is based on a conductor having a λ/dia ratio of about 2000 to 1. Thicker antennas can be expected to show less reactance at a given height, and thinner antennas should show more. At heights below and above the range covered by the curve, large reactance values are encountered, except for heights in the vicinity of ½ λ. In this region the reactance decreases, reaching zero when the antenna is resonant.

Efficiency

The *efficiency* of an antenna is the ratio of the radiation resistance to the total resistance of the system. The total

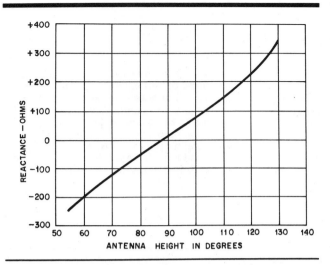

Fig 36—Approximate reactance of a vertical antenna over perfectly conducting ground. The λ/dia ratio is about 2000. Actual values vary considerably with λ/dia ratio. The remarks under Fig 34 also apply to this curve.

resistance includes radiation resistance, resistance in conductors and dielectrics (including the resistance of loading coils, if used) and the resistance of the ground system, usually referred to as "ground resistance."

It was stated earlier in this chapter that a ½-λ antenna operates at a very high efficiency because the conductor resistance is negligible compared with the radiation resistance. In the case of the grounded vertical antenna, ground resistance usually is *not* negligible, and if the antenna is short (compared with ¼ λ), the resistance of the necessary loading coil may become appreciable. To attain an efficiency comparable with that of a ½-λ antenna in a grounded antenna having a height of ¼ λ or less, great care must be used to reduce both ground resistance and the resistance of any required loading inductors. Without a fairly elaborate grounding system, the efficiency is not likely to exceed 50%, and it may be much less, particularly at heights below ¼ λ.

Grounding Systems

Based on the results of a study published in 1937 by Brown, Lewis and Epstein (see bibliography), the ideal grounding system for a grounded vertical antenna consists of about 120 wires, each at least ½ λ long, extending radially from the base of the antenna and spaced equally around a circle. Such a system is the practical equivalent of perfectly conducting ground and has negligible resistance. The wires can either be laid directly on the surface of the ground or buried a few inches below.

Such a system is not practical for most amateur installations. Unfortunately, ground-loss resistance increases rapidly when the number of radials is reduced. At least 15 radials should be used if at all possible. Experimental measurements show that even with this number, the resistance is such as to decrease the antenna efficiency to about 50% if its height is ¼ λ.

It has also been found that as the number of radials is reduced, the *length* required for optimum results with a particular *number* of radials also decreases; in other words, if only a small number of radials can be used, there is no point

in extending them out ½ λ. With 15 radials, for example, a length of 1/8 λ is sufficient. With as few as two radials the length is almost unimportant, but the efficiency of a ¼-λ antenna with such a grounding system is only about 25%. (It is considerably lower with shorter antennas.)

In general, a large number of radials (even though some or all of them must be short) is preferable to a few long radials. The conductor size is relatively unimportant; no. 12 to no. 28 copper wire is suitable. Chapter 3 contains specific information on appropriate radial lengths based on the number of wires, so that optimum use can be made of a given amount of wire (or real estate) for ground radials.

The measurement of ground resistance at the operating frequency is difficult. The power loss in the ground depends on the current concentration near the base of the antenna, and this depends on the antenna height. Typical values for small radial systems (15 or less) have been measured to be from about 5 to 30 Ω, for antenna heights from 1/16 λ to ¼ λ.

Counterpoise Systems

Recent studies have indicated that a counterpoise system is far more efficient than several radial wires on or beneath the surface of the earth. A counterpoise system is a collection of radial wires or a grid network of wires elevated above and insulated from the earth. The groundplane antenna, discussed in a later section of this chapter, uses a form of counterpoise system.

Top or End Loading

As Fig 35 indicates, the radiation resistance of a vertical antenna less than 1/8 λ (45°) high is less than 10 Ω. Because of the difficulty of obtaining a very low resistance ground system, it is always desirable to make a grounded vertical antenna as high as possible, as radiation resistance increases with height. (There is no point in increasing the height beyond ½ λ, however, as the radiation resistance decreases with further increases in height.) At the low frequencies where grounded antennas are generally used, the heights required to obtain high radiation resistances are usually impractical for amateur work. (In this section, "high" resistances are taken to be those greater than about 15 ohms.) It is desirable in the design of grounded vertical antennas that are necessarily ¼ λ (or less) high to bring the current loop as close to the top of the antenna as possible. The current throughout the length of the antenna must also be maximized. This requires *top loading*, which means replacing the missing height by some form of electrical circuit having the same characteristics as the missing part of the antenna.

One method of top loading is shown in Fig 37A. The vertical section of the antenna terminates in a "flat-top," which supplies a capacitance at the top into which current can flow. The simple single-conductor system shown at A is more easily visualized as a continuation of the antenna—so that the dimension X is essentially the overall length of the antenna. If this dimension is ½ λ, the resistance at the antenna terminals (indicated by the small circles, one being grounded) is high.

A disadvantage of this system is that the horizontal portion also radiates to some extent, although there is cancellation of radiation in the direction at right angles to the wire direction. (This is because the currents in the two portions flow in opposite directions at a given instant in time.)

A multiwire system such as the one shown in Fig 37B

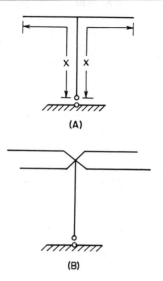

(A)

(B)

Fig 37—Simple top loading of a vertical antenna. The antenna terminals, indicated by the small circles, are the base of the antenna and ground, and should not be taken to include the length of any lead-ins or connecting wires.

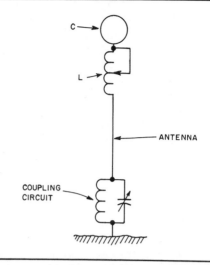

Fig 38—Top loading with lumped constants. The inductance, L, should be adjusted to give maximum field strength with constant power input to the antenna. A parallel-tuned circuit, independently resonant at the operating frequency, is required for coupling to the transmitter when the top loading is adjusted to bring a current node at the lower end of the antenna.

has more capacitance than the single-conductor arrangement, and thus does not need to be as long to resonate at a given frequency. This design does, however, require extra supports for the additional wires. Ideally, an arrangement of this sort should be in the form of a cross, but parallel wires separated by several feet give a considerable increase in capacitance over a single wire. With either system shown in Fig 37, dimension X (the length from the base of the antenna along one conductor to the end) should not be more than ½ λ, nor less than ¼ λ.

Instead of a flat top, it is possible to use a simple vertical wire with concentrated capacitance and inductance at its top to simulate the effect of the missing length. The capacitance used is not the usual type of capacitor (which is ineffective because the connection is one sided), but consists of a metallic structure large enough to have the necessary self-capacitance, Fig 38. Practically any sufficiently large metallic structure can be used for this purpose, but simple geometric forms such as the sphere, cylinder and disc are preferred because of the relative ease with which their capacitance can be calculated.

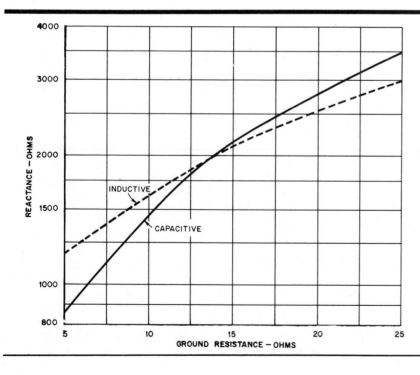

Fig 39—Inductive and capacitive reactance required for top loading a grounded antenna by the method shown in Fig 38. The reactance values should be converted to inductance and capacitance, using the usual formulas, at the operating frequency.

The inductance may be the usual type of RF coil, with suitable protection from the weather.

The minimum value of capacitive reactance required depends principally upon the ground resistance. Fig 39 is a set of curves giving the reactances required under representative conditions. These curves are based on obtaining 75% of the maximum possible increase in field strength over an antenna of the same height without top loading. The curves apply with sufficient accuracy to all antenna heights. An inductor of reasonably low loss construction is assumed. The general rule is to use as large a capacitance (low capacitive reactance) as the circumstances permit, because an increase in capacitance causes an increase in field strength. It is particularly important to do this when (as is usually the case) the ground resistance is not known and cannot be measured.

The capacitance of three geometric forms is shown by the curves of Fig 40 as a function of their size. For the cylinder, the length is specified equal to the diameter. The sphere, disc and cylinder can be constructed from sheet metal, if such construction is feasible, but the capacitance will be practically the same in each if a "skeleton" type of construction with screening or networks of wire or tubing are used.

Finding Capacitance Hat Size

In the broadcast industry the practical physical limit for top loading is considered to be approximately 30 electrical degrees if a disc alone is used. This means a vertical monopole that is 60° in physical height may be loaded with only a capacitance top hat to be 90° in electrical length. The loading efficiency is quite high, as inductor losses are eliminated.

The required size of a capacitance hat may be determined from the following procedure. The information in this section is based on a September 1978 *QST* article by Walter Schulz, K3OQF.

The physical length of a shortened antenna can be found from

$$h = \frac{32.8\ell}{f_{MHz}} \qquad \text{(Eq 1)}$$

where
    h = length in inches
    ℓ = length in degrees

Thus, using an example of 7 MHz and a shortened length of 60°, h = 32.8 × 60/7 = 281 inches, equivalent to 23.4 feet.

Now consider the vertical radiator as an open-ended transmission line, so the impedance and top loading may be determined. The characteristic impedance of a vertical antenna can be found from

$$Z_0 = 60\left[\ln\left(\frac{4h}{d}\right) - 1\right] \qquad \text{(Eq 2)}$$

where
    ln = natural logarithm
    h = length (height) of vertical radiator in inches (as above)
    d = diameter of radiator in inches

The vertical radiator for this example has a diameter of 1 inch. Thus, for this example,

$$Z_0 = 60\left(\ln \frac{4 \times 281}{1} - \right) = 361 \ \Omega$$

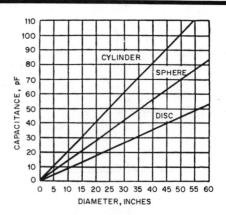

Fig 40—Capacitance of a sphere, disc and cylinder as a function of their diameters. The cylinder length is assumed equal to its diameter.

The capacitive reactance required for the amount of top loading can be found from

$$X = \frac{Z_0}{\tan \theta} \qquad \text{(Eq 3)}$$

where
    X = capacitive reactance, ohms
    Z_0 = characteristic impedance of antenna (from Eq 2)
    θ = amount of electrical loading, degrees

This value for a 30° hat is 361/tan 30° = 625 Ω. This capacitive reactance may be converted to capacitance with a slight rearrangement of the fundamental equation,

$$C = \frac{10^6}{2\pi f X_C} \qquad \text{(Eq 4)}$$

where
    C = capacitance in picofarads (pF)
    f = frequency, MHz
    X_C = capacitive reactance, ohms (from above)

For this example, the required C = $10^6/(2\pi \times 7 \times 625)$ = 36.4, which may be rounded to 36 pF. A disc is used in this example. The appropriate diameter for 36 pF of hat capacitance can be found from Fig 40. The disc diameter that yields 36 pF of capacitance is 40 inches.

The skeleton disc shown in Fig 41 is fashioned into a wagon wheel configuration. Six 20-inch lengths of 1.2-inch OD aluminum tubing are used as spokes. Each is connected to the hub at equidistant intervals. The outer ends of the spokes terminate in a loop made of no. 14 copper wire. Note that the loop increases the hat capacitance slightly, making a better approximation of a solid disc. The addition of this hat at the top of a 23.4-foot radiator makes it electrically 90° high at 7 MHz.

After construction, some slight adjustment in the radiator length or the hat size may be required if resonance at a specific frequency is desired. From Fig 35, the radiation resistance of a 60° high radiator is seen to be about 13 ohms. That graph does not take top loading into account, however. In practice the radiation resistance of top-loaded antennas will be somewhat higher than indicated by Figs 34 and 35. However, even though the shortened antenna is 90° long electrically, its radiation resistance will not be as high as if it were physically 90° long. For the antenna of this example, the actual radiation resistance is on the order of 20 ohms.

The 7-MHz antenna as designed above was constructed and placed over a system of 60 radials, each 0.2-λ long, at K3OQF. With modest power from the Philadelphia area, daily contacts have been made with stations in Europe, and on several occasions stations as far away as the Indian Ocean have been worked during periods of moderate sunspot activity.

GROUNDPLANE ANTENNAS

Instead of being actually grounded, a ¼-λ antenna can work against a simulated ground called a *ground plane*. Such a simulated ground can be formed from wires ¼ λ long radiating from the base of the antenna, as shown in Fig 42. With ¼-λ radials, the antenna and any one radial have a total length of ½ λ and therefore make up a resonant system. With only one radial, however, the directive pattern is that of a ½-λ antenna bent at a right angle at the center. If one section is vertical and the other horizontal, this results in equal components of horizontal and vertical polarization, and a nonuniform pattern in the horizontal plane. This can be overcome by using a ground plane in the shape of a disc with a radius of ¼ λ. The effect of the disc can be simulated, with simpler construction, by using at least three straight radials equally spaced around the circle. Four radials are more commonly used, as indicated in the drawing of Fig 42.

The groundplane antenna is widely used at VHF, for the purpose of establishing a "ground" for a vertical antenna mounted well above actual earth. The ground plane keeps a metallic antenna support from carrying currents that tend to turn the system into the equivalent of a vertical long-wire antenna, thereby raising the radiation angle.

At frequencies in the 14- to 30-MHz region, a ground plane of the type shown in Fig 42 permits using a ¼-λ vertical antenna for omnidirectional low-angle operation at a height allowing the antenna be clear of its surroundings. Such short antennas mounted on the ground itself are frequently so surrounded by energy-absorbing structures and trees that they are rather ineffective. Because ¼-λ radials are physically short at these frequencies, it is quite practical to mount the entire system on a roof top or pole.

At a groundplane height of ½ λ or more, earth losses become essentially negligible. This means that at such heights the antenna has high radiating efficiency, and that the radiation resistance curve of Fig 34 applies with reasonable accuracy. The vertical radiating element itself may be any desirable height, and does not have to be exactly ¼ λ long. The radials, however, should be ¼ λ long, taking their wavelength to diameter ratio into account.

A groundplane can also be of considerable benefit at lower frequencies, provided the radials and the base of the antenna are a moderate height above the ground. In order to take over the function of the actual ground connection, the ground plane must be constructed such that the field of the antenna tends to travel along the ground plane wires, rather than in the lossy earth. This is particularly important where the current is greatest, at the base of the vertical radiator.

If the groundplane must be near the earth, the number of wires should be increased, using as many as is practical and spacing them as evenly as possible in a circle around the

Fig 41—A close-up view of the capacitance hat for a 7-MHz vertical antenna. The ½-in. dia radial arms terminate in a loop of copper wire.

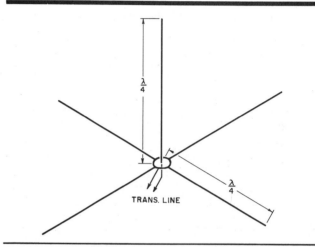

Fig 42—The groundplane antenna. Power is applied between the base of the vertical radiator and the center of the ground plane, as indicated in the drawing.

antenna. If the construction of a multiwire ground plane is impractical, bury as many radials as possible near the surface of the ground. (This is discussed in detail in Chapter 3.)

SHORT ANTENNAS

The earlier discussions in this chapter have principally been in terms of self-resonant antennas, particularly those ½ λ long (or ¼ λ long with a ¼ λ ground image). It is a mistake to assume that there is anything sacred (in effectiveness, at least) in using a resonant antenna. The resonant length happens to be one that is convenient to analyze. At other lengths, directive properties are different, radiation resistance is different, and the impedance looking into the terminals of the antenna contains reactance as well as resistance. The reactance presents no problem, because it can be "tuned out." The antenna *system* can thereby be resonated, even though the antenna itself is not resonant. The purpose of resonating either the antenna or the system as a whole is simply to allow power to be fed to the antenna easily. It is important to realize that such resonating or tuning *does not affect the radiating properties* of the antenna.

Physical conditions frequently make it necessary to use antennas shorter than ½ λ. The directive pattern of a short antenna does not differ greatly from that of a ½-λ antenna, and at the limit the pattern approaches that shown in Fig 15. The difference in the field strength caused by this shift in pattern shape is negligible. The most important difference is the decrease in radiation resistance and its effect on the efficiency of the antenna. (This is discussed in the preceding section on grounded antennas.)

The curves of Figs 34 and 35 can be used for any center-fed nongrounded antenna by using half the actual antenna length and multiplying the corresponding radiation resistance by two. For example, a center-fed dipole antenna having an actual length of 120° (⅓ λ) has a half length of 60° and a radiation resistance of about 13 Ω per side, or 26 Ω for the whole antenna. The reactance, which is capacitive and on the order of 400 Ω (from Fig 36, same technique as above), can be tuned out with a loading coil or coils. As described earlier, low resistance coils must be used if the antenna efficiency is to be kept reasonably high. Ground resistance loss can be neglected in a horizontal center-fed antenna of this type if the height is ¼ λ or more.

For optimum performance, a short antenna should not be made shorter than the physical circumstances require, because efficiency decreases rapidly as antenna length is reduced. For example, a center-fed antenna having an overall length of ¼ λ (half length 45°) has a radiation resistance of $2 \times 7 = 14$ Ω, as shown in Fig 35. Depending on the λ/dia ratio, the resistance of the loading coil necessary to cancel the reactance will likely be on the order of 3 to 6 Ω, so the probable efficiency will be 70 to 80%. This represents a loss of 1 to 1.5 dB (10 × log 100/efficiency %). While this is not intolerable, further shortening not only decreases the radiation resistance still more, but enters a length region where the reactance increases very rapidly. Compensating loading coil resistance quickly becomes larger than the radiation resistance.

Where the antenna must be short, a small λ/dia ratio (thick antenna) is definitely desirable as a means of keeping down the reactance and thus reducing the amount of loading inductance required. For quite short radiators (in terms of a wavelength), a combination of coil loading and capacitance hat loading is an efficient technique. End loading can also be increased with the addition of a second capacitance hat at some distance from the first. The loading of short antennas is also discussed in Chapter 6.

The ½-λ dipole and the few special types of antennas described in this chapter form the basis for practically all antenna systems in amateur use at frequencies from the VHF region down. Other fundamental types of radiators are applicable at microwaves, but they are not used at lower frequencies because the dimensions are such as to be wholly impractical when the wavelength is measured in meters rather than centimeters.

BIBLIOGRAPHY

Source material and more extended discussion of topics covered in this chapter can be found in the references given below.

J. S. Belrose, "Short Antennas for Mobile Operation," *QST*, Sep 1953.

G. H. Brown, "The Phase and Magnitude of Earth Currents Near Radio Transmitting Antennas," *Proc IRE*, Feb 1935.

G. H. Brown, R. F. Lewis and J. Epstein, "Ground Systems as a Factor in Antenna Efficiency," *Proc IRE*, Jun 1937, pp 753-787.

R. B. Dome, "Increased Radiating Efficiency for Short Antennas," *QST*, Sep 1934, pp 9-12.

A. C. Doty, Jr, J. A. Frey and H. J. Mills, "Efficient Ground Systems for Vertical Antennas," *QST*, Feb 1983, pp 20-25.

R. Fosberg, "Some Notes on Ground Systems for 160 Meters," *QST*, Apr 1965, pp 65-67.

G. Grammer, "More on the Directivity of Horizontal Antennas; Harmonic Operation—Effects of Tilting," *QST*, Mar 1937, pp 38-40, 92, 94, 98.

B. Myers, "The W2PV Four-Element Yagi," *QST*, Oct 1986, pp 15-19.

L. Richard, "Parallel Dipoles of 300-Ohm Ribbon," *QST*, Mar 1957.

W. Schulz, "Designing a Vertical Antenna," *QST*, Sep 1978, pp 19-21.

J. Sevick, "The Ground-Image Vertical Antenna," *QST*, Jul 1971.

J. Sevick, "The W2FMI 20-Meter Vertical Beam," *QST*, Jun 1972.

J. Sevick, "The W2FMI Ground-Mounted Short Vertical," *QST*, Mar 1973, pp 13-18, 41.

J. Sevick, "A High Performance 20-, 40- and 80-Meter Vertical System," *QST*, Dec 1973.

J. Sevick, "Short Ground-Radial Systems for Short Verticals," *QST*, Apr 1978, pp 30-33.

C. E. Smith and E. M. Johnson, "Performance of Short Antennas," *Proc IRE*, Oct 1947.

J. Stanley, "Optimum Ground Systems for Vertical Antennas," *QST*, Dec 1976.

R. E. Stephens, "Admittance Matching the Ground-Plane Antenna to Coaxial Transmission Line," Technical Correspondence, *QST*, Apr 1973, pp 55-57.

D. Sumner, "Cushcraft 32-19 'Boomer' and 324-QK Stacking Kit," Product Review, *QST*, Nov 1980, pp 48-49.

E. M. Williams, "Radiating Characteristics of Short-Wave Loop Aerials," *Proc IRE*, Oct 1940.

TEXTBOOKS ON ANTENNAS, TRANSMISSION LINES, AND PROPAGATION

C. A. Balanis, *Antenna Theory, Analysis and Design* (New York: Harper & Row, 1982).

D. S. Bond, *Radio Direction Finders*, 1st ed. (New York: McGraw-Hill Book Co).

K. Davies, *Ionospheric Radio Propagation—National Bureau of Standards Monograph 80* (Washington, DC: US Government Printing Office, April 1, 1965).

R. S. Elliott, *Antenna Theory and Design* (Englewood Cliffs, NJ: Prentice Hall, 1981).

A. E. Harper, *Rhombic Antenna Design* (New York: D. Van Nostrand Co, Inc, 1941).

H. Jasik, *Antenna Engineering Handbook*, 1st ed. (New York: McGraw-Hill, 1961).

W. C. Johnson, *Transmission Lines and Networks*, 1st ed.

(New York: McGraw-Hill Book Co, 1950).

R. C. Johnson and H. Jasik, *Antenna Engineering Handbook*, 2nd ed. (New York: McGraw-Hill, 1984).

E. C. Jordan and K. G. Balmain, *Electromagnetic Waves and Radiating Systems*, 2nd ed. (Englewood Cliffs, NJ: Prentice-Hall, Inc, 1968).

R. Keen, *Wireless Direction Finding*, 3rd ed. (London: Wireless World).

R. W. P. King, *Theory of Linear Antennas* (Cambridge, MA: Harvard Univ Press, 1956).

R. W. P. King, H. R. Mimno and A. H. Wing, *Transmission Lines, Antennas and Waveguides* (New York: Dover Publications, Inc, 1965).

King, Mack and Sandler, *Arrays of Cylindrical Dipoles* (London: Cambridge Univ Press, 1968).

M. G. Knitter, Ed., *Loop Antennas—Design and Theory* (Cambridge, WI: National Radio Club, 1983).

M. G. Knitter, Ed., *Beverage and Long Wire Antennas— Design and Theory* (Cambridge, WI: National Radio Club, 1983).

J. D. Kraus, *Electromagnetics* (New York: McGraw-Hill Book Co).

J. D. Kraus, *Antennas*, 2nd ed. (New York: McGraw-Hill Book Co, 1988).

E. A. Laport, *Radio Antenna Engineering* (New York: McGraw-Hill Book Co, 1952).

J. L. Lawson, *Yagi-Antenna Design*, 1st ed. (Newington: ARRL, 1986).

P. H. Lee, *The Amateur Radio Vertical Antenna Handbook*, 1st ed. (Port Washington, NY: Cowan Publishing Corp, 1974).

A. W. Lowe, *Reflector Antennas* (New York: IEEE Press, 1978).

G. M. Miller, *Modern Electronic Communication* (Englewood Cliffs, NJ: Prentice Hall, 1983).

V. A. Misek, *The Beverage Antenna Handbook* (Hudson, NH: V. A. Misek, 1977).

T. Moreno, *Microwave Transmission Design Data* (New York: McGraw-Hill, 1948).

Ramo and Whinnery, *Fields and Waves in Modern Radio* (New York: John Wiley & Sons).

V. H. Rumsey, *Frequency Independent Antennas* (New York: Academic Press, 1966).

P. N. Saveskie, *Radio Propagation Handbook* (Blue Ridge Summit, PA: Tab Books, Inc, 1980).

S. A. Schelkunoff, *Advanced Antenna Theory* (New York: John Wiley & Sons, Inc, 1952).

S. A. Schelkunoff and H. T. Friis, *Antennas Theory and Practice* (New York: John Wiley & Sons, Inc, 1952).

J. Sevick, *Transmission Line Transformers*, 1st ed. (Newington: ARRL, 1987).

H. H. Skilling, *Electric Transmission Lines* (New York: McGraw-Hill Book Co, Inc, 1951).

M. Slurzburg and W. Osterheld, *Electrical Essentials of Radio* (New York: McGraw-Hill Book Co, Inc, 1944).

G. Southworth, *Principles and Applications of Waveguide Transmission* (New York: D. Van Nostrand Co, 1950).

F. E. Terman, *Radio Engineers' Handbook*, 1st ed. (New York, London: McGraw-Hill Book Co, 1943).

F. E. Terman, *Radio Engineering*, 3rd ed. (New York: McGraw-Hill, 1947).

S. Uda and Y. Mushiake, *Yagi-Uda Antenna* (Sendai, Japan: Sasaki Publishing Co, 1954). [Published in English—Ed.]

P. P. Viezbicke, "Yagi Antenna Design," *NBS Technical Note 688* (US Dept. of Commerce/National Bureau of Standards, Boulder, CO), Dec 1976.

The GIANT Book of Amateur Radio Antennas (Blue Ridge Summit, PA: Tab Books, 1979), pp 55-85.

Radio Broadcast Ground Systems, available from Smith Electronics, Inc, 8200 Snowville Rd, Cleveland, OH 44141.

Radio Communication Handbook, 5th ed. (London: RSGB, 1976).

Radio Direction Finding, published by the Happy Flyer, 1811 Hillman Ave, Belmont, CA 94002.

Chapter 3

The Effects of the Earth

The earth acts as a huge reflector for those waves that are radiated from an antenna at angles lower than the horizon. These downcoming waves strike the surface and are reflected by a process very similar to that by which light waves are reflected from a mirror. As is the case with light waves, the angle of reflection is the same as the angle of incidence, so a wave striking the surface at an angle of, say, 15°, is reflected upward from the surface at 15°.

The reflected waves combine with direct waves (those radiated at angles above the horizon) in various ways. Some of the factors that influence this combining process are the orientation of the antenna with respect to the ground, the height of the antenna, its length, and the characteristics of the ground. At some vertical angles above the horizon the direct and reflected waves may be exactly in phase—that is, the maximum field strengths of both waves are reached at the same time at the same point in space, and the directions of the fields are the same. In such a case, the resultant field strength at that angle is simply the sum of the two. (This represents a theoretical increase in field strength of 6 dB over the free-space pattern at these angles.) At other vertical angles the two waves may be completely out of phase—that is, the fields are maximum at the same instant and the directions are opposite at the same spot. At still other angles, the resultant field will have intermediate values. Thus, the effect of the ground is to increase radiation intensity at some vertical angles and to decrease it at others.

It is often convenient to use the concept of an image antenna to show the effect of reflection. As Fig 1 shows, the reflected ray has the same path length (AD equals BD) that it would if it originated at a second antenna with the same

characteristics as the real antenna, but situated below the ground just as far as the actual antenna is above it. Like an image in a mirror, this image antenna is "in reverse," as shown in Fig 2.

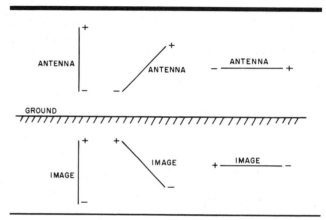

Fig 2—Horizontal, tilted and vertical half-wave antennas and their images.

If the real antenna is vertical, and is instantaneously charged so one end is positive and the other negative, then the image antenna, also vertical, is similarly charged; the top end is positively charged and the bottom end is negatively charged. In the transition from vertical to horizontal (the case with sloping antennas), the top end of both the antenna and the image are positively charged, and the bottom ends of the antenna and the image are negatively charged. In the case of horizontal antennas, one end of the antenna is positively charged and the same end of the image is negatively charged, and vice versa. Now if we look at the antenna and its image from a remote point on the surface of the ground, we will see that the currents in a horizontal antenna and its image are flowing in opposite directions, or are 180° out of phase, but the currents in a vertical antenna and its image are flowing in the *same* direction, or are *in* phase. Sloping antennas are subject to varying effects from their images based on the slope angle.

GROUND REFLECTION

A vertical monopole antenna provides a convenient means by which to understand the effects of ground reflection on radiation patterns, as azimuthal directivity is omni-

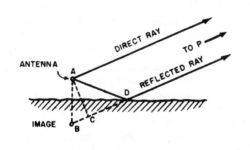

Fig 1—At any distant point, P, the field strength will be the vector sum of the direct ray and the reflected ray. The reflected ray travels farther than the direct ray by the distance BC, where the reflected ray is considered to originate at the "image" antenna.

directional. A quarter-wave vertical over ideal earth has the radiation pattern shown by the solid line in Fig 3. Over real earth, however, the pattern looks more like the shaded one in the same diagram. In this case, the low-angle radiation which might be expected from a vertical antenna is not achieved.

To understand why the desired low-angle radiation is not realized, examine Fig 4A. Radiation from each antenna segment reaches a point P in space by two paths; one directly from the antenna, path AP, and the other by reflection from the earth, path AGP. (Note that P is so far away that the slight difference in angles is insignificant—for practical purposes the waves are parallel to each other at point P.) If the earth was a perfectly conducting surface, there would be no phase shift of the vertically polarized wave upon reflection at point G; the two waves would add together with some phase difference because of the different path lengths. This is why the free-space radiation pattern differs from the pattern of the same antenna over ground. Now consider a point P that is close to the horizon, as in Fig 4B. The path lengths AP and AGP are almost the same, so the magnitudes of the two waves add together, producing a maximum at zero angle of radiation. The arrows on the waves point both ways since the process works similarly for transmitting and receiving.

With real earth, however, the reflected wave undergoes a change in *amplitude and phase* in the reflection process.

Indeed, at a low enough angle the phase of the reflected wave will actually change by 180°, and its magnitude will then *subtract* from that of the direct wave. At zero angle, it will be almost equal in amplitude, but 180° out of phase with the direct wave. Complete cancellation will result, inhibiting any radiation or reception at that angle.

THE PSEUDO-BREWSTER ANGLE

Much of the material presented here regarding pseudo-Brewster angle was prepared by Charles J. Michaels, W7XC, and first appeared in July 1987 *QST*. (See the bibliography at the end of this chapter.)

Most fishermen have noticed that when the sun is low, its light is reflected from the water's surface as glare, obscuring the underwater view. When the sun is high, however, the sunlight penetrates the water and it is possible to see objects below the surface of the water. The angle at which this transition takes place is known as the Brewster angle, named for the Scottish physicist Sir David Brewster (1781-1868).

A similar situation exists in the case of vertically polarized antennas; the RF energy behaves as the sunlight in the optical system, and the earth under the antenna acts as the water. The pseudo-Brewster angle (PBA) is the angle at which the reflected wave is 90° out of phase with respect to the direct wave. "Pseudo" is used here because the RF effect is similar to the optical effect from which the term gets its name. Below this angle, the reflected wave is between 90° and 180° out of phase with the direct wave, so some degree of cancellation takes place. The largest amount of cancellation occurs near zero degrees, and steadily less cancellation occurs as the PBA is approached from below.

The factors that determine the PBA for a particular location *are not related to the antenna itself, but to the ground around it.* The first of these factors is earth conductivity, G, which is a measure of the ability of the soil to conduct electricity. Conductivity is the inverse of resistance. The second factor is the dielectric constant, k, which is a unitless quantity that corresponds to the capacitive effect of the earth. For both of these quantities, the higher the number, the better the ground (for antenna purposes). The third factor determining the PBA for a given location is the frequency of operation. The PBA increases with increasing frequency, all other conditions being equal. Table 1 gives typical values of conductivity and dielectric constant for different types of soil. The map of Fig 5 shows the approximate conductivity values for different areas of the continental United States.

As the frequency is increased, the role of the dielectric constant in determining the PBA becomes more significant. Table 2 shows how the PBA varies with changes in ground conductivity, dielectric constant and frequency. The table shows trends in PBA dependency on ground constants and frequency. The constants chosen are not necessarily typical of any geographical area.

At angles below the PBA, the reflected vertically polarized wave subtracts from the direct wave, causing the radiation intensity to fall off rapidly. Similarly, above the PBA, the reflected wave adds to the direct wave, and the radiated pattern approaches the perfect-earth pattern. The PBA, usually labeled ψ_B, is shown in Fig 3.

In plotting vertical-antenna radiation patterns over real earth, the reflected wave from an antenna segment is multiplied by a factor called the reflection coefficient, and the product is then added vectorially to the direct wave to get

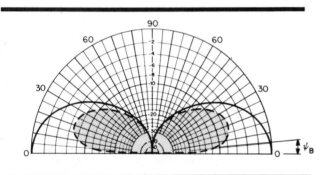

Fig 3—Vertical-plane radiation pattern for a ground-mounted quarter-wave vertical. The solid line is the pattern for perfect earth. The shaded pattern shows how the response is modified over average earth (k = 15, G = 0.005 S/m) at 14 MHz. ψ_B is the pseudo-Brewster angle (PBA), in this case 14°.

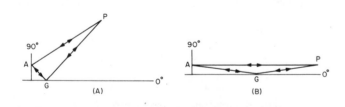

Fig 4—The direct wave and the reflected wave combine at point P to form the pattern (P is very far from the antenna). At A the two paths AP and AGP differ appreciably in length, while at B these two path lengths are nearly equal.

Table 1
Conductivities and Dielectric Constants for Common Types of Earth

Surface Type	Dielectric constant	Conductivity (S/m)	Relative quality
Fresh water	80	0.001	
Salt water	81	5.0	
Pastoral, low hills, rich soil typ Dallas, TX to Lincoln, NE areas	20	0.0303	Very good
Pastoral, low hills, rich soil typ OH and IL	14	0.01	
Flat country, marshy, densely wooded, typ LA near Mississippi River	12	0.0075	
Pastoral, medium hills and forestation, typ MD, PA, NY (exclusive of mountains and coastline)	13	0.006	
Pastoral, medium hills and forestation, heavy clay soil, typ central VA	13	0.005	Average
Rocky soil, steep hills, typ mountainous	12-14	0.002	Poor
Sandy, dry, flat, coastal	10	0.002	
Cities, industrial areas	5	0.001	Very Poor
Cities, heavy industrial areas, high buildings	3	0.001	Extremely poor

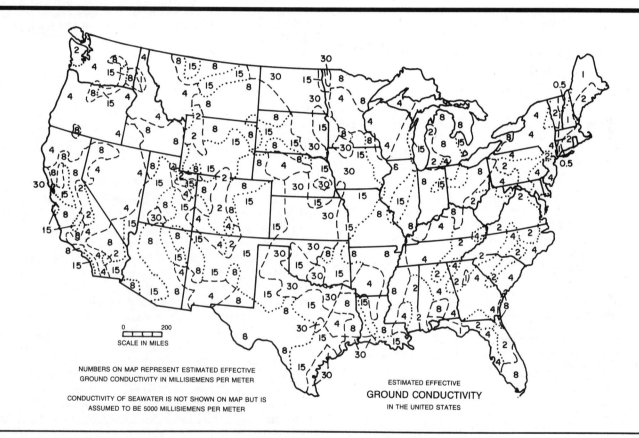

NUMBERS ON MAP REPRESENT ESTIMATED EFFECTIVE
GROUND CONDUCTIVITY IN MILLISIEMENS PER METER

CONDUCTIVITY OF SEAWATER IS NOT SHOWN ON MAP BUT IS
ASSUMED TO BE 5000 MILLISIEMENS PER METER

ESTIMATED EFFECTIVE
GROUND CONDUCTIVITY
IN THE UNITED STATES

Fig 5—Typical average soil conductivities for the continental United States. Numeric values indicate conductivities in millisiemens per meter (mS/m).

the resultant. The reflection coefficient consists of an attenuation factor, A, and a phase angle, ϕ, and is usually expressed as $A\angle\phi$. (ϕ is always a negative angle, because the earth acts as a lossy capacitor in this situation.) The following equation can be used to calculate the reflection coefficient for vertically polarized waves, for earth of given conductivity and dielectric constant at any frequency and wave angle.

$$A\angle\phi = \frac{k'\sin\psi - \sqrt{k' - \cos^2\psi}}{k'\sin\psi + \sqrt{k' - \cos^2\psi}} \qquad \text{(Eq 1)}$$

where

$A\angle\phi$ = reflection coefficient
ψ = wave angle

Table 2

Pseudo-Brewster Angle Variation with Frequency, Dielectric Constant, and Conductivity

Frequency (MHz)	Dielectric constant	Conductivity (S/m)	PBA (degrees)
7	15	0.01	10.5
	15	0.002	14.1
	12	0.005	13.5
	20	0.005	11.6
14	15	0.01	13.9
	15	0.002	14.4
	12	0.005	15.2
	20	0.005	12.3
21	15	0.01	13.6
	15	0.002	14.4
	12	0.005	15.7
	20	0.005	12.5

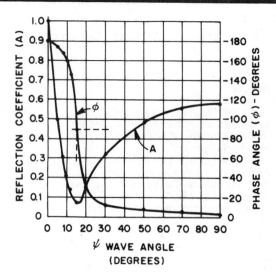

Fig 6—Reflection coefficient for vertically polarized waves. A and ϕ are magnitude and angle for wave angles ψ. This case is for average earth, (k = 13, G = 0.005 S/m), at 21 MHz.

$$k' = k - j\left(\frac{1.8 \times 10^4 \times G}{f}\right)$$

k = dielectric constant of earth (k for air = 1)
G = conductivity of earth in S/m
f = frequency in MHz
j = complex operator ($\sqrt{-1}$)

Solving this equation for several points indicates what effect the earth has on vertically polarized signals at a particular location for a given frequency range. Fig 6 shows the reflection coefficient as a function of wave angle at 21 MHz over average earth (G = 0.005 S/m, k = 13). Note that as the phase curve, ϕ, passes through 90°, the attenuation curve, A, passes through a minimum at the same wave angle, ψ. This is the PBA. At this angle, the reflected wave is not only at a phase angle of 90° with respect to the direct wave, but is so low in amplitude that it does not aid the direct wave by a significant amount. In the case illustrated in Fig 6 this wave angle is about 15°.

Variations In PBA With Earth Quality

From Eq 1, it is quite a task to search for either the 90° phase point or the attenuation curve minimum for a wide variety of earth conditions. Instead, the PBA can be calculated directly from the following equation.

$$\psi_B = \arcsin$$

$$\sqrt{\frac{k - 1 + \sqrt{(x^2+k^2)^2(k-1)^2 + x^2[(x^2+k^2)^2 - 1]}}{(x^2+k^2)^2 - 1}}$$

(Eq 2)

where

$$x = \frac{1.8 \times 10^4 \times G}{f}$$

k, G and f are as defined for Eq 1

Fig 7 contains curves that were calculated using Eq 2 for several different earth conditions, at frequencies between 1.8 and 30 MHz. As expected, poorer earths yield higher PBAs.

Unfortunately, at the higher frequencies (where low-angle radiation is most important for DX work), the PBAs are highest. The PBA is the same for both transmitting and receiving.

Relating PBA to Location and Frequency

Table 1 lists the physical descriptions of various kinds of earth with their respective conductivities and dielectric constants, as mentioned earlier. Note that in general, the dielectric constants and conductivities are higher for better earths. This enables the labeling of the earth characteristics as extremely poor, very poor, poor, average, very good, and so on, without the complications that would result from treating the two parameters independently.

Fresh water and salt water are special cases; in spite of high resistivity, the fresh-water PBA is 6.4°, and is nearly independent of frequency below 30 MHz. Salt water, because of its extremely high conductivity, has a PBA that never exceeds 1° in this frequency range. The extremely low conductivity listed for cities (last case) in Table 1 results more from the clutter of surrounding buildings and other obstructions than any actual earth characteristic. The PBA at any location can be found for a given frequency from the curves in Fig 7.

EFFECTS OF GROUND REFLECTION ON HORIZONTALLY POLARIZED WAVES

The situation for horizontal antennas is somewhat different from that of verticals. Fig 8 shows the reflection coefficient for horizontally polarized waves over average earth at 21 MHz. Note that in this case, the phase angle departure from 0° never gets very large, and the attenuation factor that causes the most loss for high-angle signals approaches unity for low angles. Attenuation increases with progressively poorer earth types.

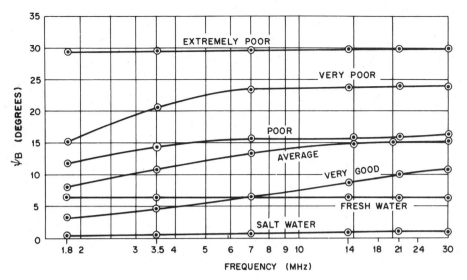

Fig 7—Pseudo-Brewster angle (ψ_B) for various qualities of earth over the 1.8- to 30-MHz frequency range. Note that the frequency scale is logarithmic. The constants used for each curve are given in Table 1.

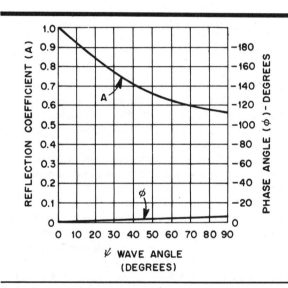

Fig 8—Reflection coefficient for horizontally polarized waves (magnitude A at angle ϕ), at 21 MHz over average earth (k = 13, G = 0.005 S/m).

In calculating the broadside radiation pattern of a horizontal half-wave dipole, the perfect-earth image current (equal to the true antenna current, but 180° out of phase with it) is multiplied by the horizontal reflection coefficient given by Eq 3 below. The product is then added vectorially to the direct wave to get the resultant at that wave angle. The reflection coefficient for horizontally polarized waves can be calculated using the following equation.

$$A\underline{/\phi} = \frac{\sqrt{k' - \cos^2\psi} - \sin\psi}{\sqrt{k' - \cos^2\psi} + \sin\psi} \qquad \text{(Eq 3)}$$

where

$A\underline{/\phi}$ = reflection coefficient
ψ = wave angle

$$k' = k - j\left(\frac{1.8 \times 10^4 \times G}{f}\right)$$

k = dielectric constant of earth
G = conductivity of earth in S/m
f = frequency in MHz
j = complex operator ($\sqrt{-1}$)

For a horizontal antenna near the earth, the resultant pattern is a modification of the free-space pattern of the antenna. Fig 9 shows how this modification takes place for a horizontal half-wave antenna over a perfectly conducting surface. The patterns at the left show the relative radiation when one views the antenna from the side; those at the right show the radiation pattern looking at the end of the antenna. Changing the height from ¼ to ½ λ makes a significant difference in the high-angle radiation.

Note that for an antenna height of ½ λ (lower part of Fig 9), the out-of-phase reflection from a perfectly conducting surface creates a null in the pattern at the zenith (90° wave angle). Over real earth, however, a "filling in" of this null occurs because of ground losses that prevent perfect reflection of high-angle radiation. This effect is examined again in a later section.

At a zero angle of radiation, horizontally polarized antennas develop a null because out-of-phase reflection cancels the direct wave. As the wave angle departs from zero degrees, however, there is a slight filling-in effect so that with other than perfect earth, radiation at lower-than-normal angles is enhanced. Therefore, a horizontal antenna will often outperform a vertical for low-angle DX work over some types of earth at the higher frequencies.

Reflection coefficients for vertically and horizontally polarized radiation differ considerably at most angles above ground, as can be seen by comparison of Figs 6 and 8. (Both sets of curves were plotted for the same ground constants and at the same frequency, so they may be compared directly.) This is because, as mentioned earlier, the image of a horizontally polarized antenna is out of phase with the antenna itself, and the image of a vertical antenna is in phase with the actual radiator.

The result is that the phase shifts and reflection magnitudes vary greatly at different angles for horizontal and vertical polarization. The magnitude of the reflection coefficient for vertically polarized waves is greatest (near unity)

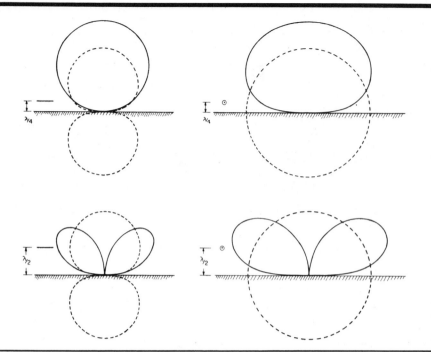

Fig 9—Effect of the ground on the radiation from a horizontal half-wave antenna, for heights of one-fourth and one-half wavelength. Broken lines show what the pattern would be if there were no reflection from the ground (free space).

at very low angles, and the phase angle is close to 180°. As mentioned earlier, this cancels nearly all radiation at very low angles. For the same range of angles, the magnitude of the reflection coefficient for horizontally polarized waves is also near unity, but the phase angle is near 0° for the specific conditions shown in Figs 6 and 8. This causes reinforcement of all low-angle horizontally polarized waves. At some relatively high angle, the reflection coefficients for horizontally and vertically polarized waves are equal in magnitude and phase. At this angle (approximately 81° for the example case), the effect of ground reflection on vertically and horizontally polarized signals will be exactly the same.

DEPTH OF RF CURRENT PENETRATION

In considering earth characteristics, questions about depth of RF current penetration often arise. For instance, if a given location consists of a 6-foot layer of soil overlying a highly resistive rock strata, which material dominates? The answer depends on the frequency, the soil and rock dielectric constants, and their respective conductivities. The following equation can be used to calculate the current density at any depth.

$$\frac{\text{Current Density at Depth D}}{\text{Current Density at Surface}} = e^{-pd} \qquad \text{(Eq 4)}$$

where

$$p = \left[\frac{X \times B}{2} \times \left(\sqrt{1 + \frac{G^2 \times 10^{-4}}{B^2}} - 1 \right) \right]^{1/2}$$

d = depth of penetration in cm
e = natural logarithm base (2.718)
X = 0.008 × π² × f
B = 5.56 × 10⁻⁷ × k × f
k = dielectric constant of earth
f = frequency in MHz
G = conductivity of earth in S/m

By some manipulation of this equation, it can be used to calculate the depth at which the current density is some fraction of that at the surface. The depth at which the current density is 37% ($1/e$) of that at the surface (often referred to as skin depth) is the depth at which the current density would be zero if it was distributed uniformly instead of exponentially. (This $1/e$ factor appears in many physical situations. For instance, a capacitor charges to within $1/e$ of full charge within one RC time constant.) At this depth, since the power loss is proportional to the square of the current, approximately 91% of the total power loss has occurred, as has most of the phase shift, and any current flow below this level is negligible.

Fig 10 shows the solutions to Eq 4 over the 1.8- to 30-MHz frequency range for various types of earth. For example, in very good earth, substantial RF currents flow down to about 3.3 feet at 14 MHz. This depth goes to 13 feet in average earth and as far as 40 feet in very poor earth. Thus, if the overlying soil is rich, moist loam, the underlying rock strata is of little concern. However, if the soil is only average, the underlying rock may constitute a major consideration in determining the PBA and the depth to which the RF current will penetrate.

This depth in fresh water is about 156 feet and is nearly independent of frequency in the amateur bands below 30 MHz. In salt water, the depth is about seven inches at 1.8 MHz and decreases rather steadily to about two inches at 30 MHz. Dissolved minerals in moist earth increase its conductivity.

The depth of penetration curves in Fig 10 illustrate a noteworthy phenomenon. While skin effect confines RF current flow close to the surface of a conductor, the earth is so lossy that RF current penetrates to much greater depths than in most other media. The depth of RF current penetration is a function of frequency as well as earth type. Thus, the only cases in which most of the current flows near the surface are with very highly conductive media (such as salt water), and at frequencies above 30 MHz.

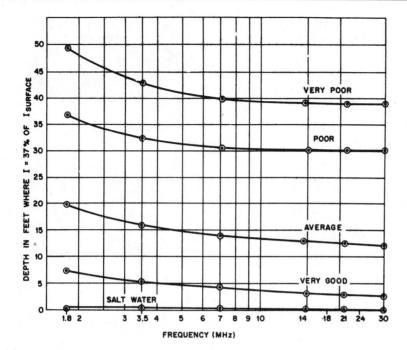

Fig 10—Depths at which the current density is 37% of that at the surface for different qualities of earth over the 1.8- to 30-MHz frequency range. The depth for fresh water, not plotted, is 156 feet and almost independent of frequency below 30 MHz. See text and Table 1 for ground constants.

Vertical Directivity Patterns

Because antenna radiation patterns are three-dimensional, it is helpful in understanding their operation to use a form of representation showing the *vertical* directional characteristic for different heights. Such patterns have already been considered, in Fig 9. It is possible to show selected vertical-plane patterns oriented in various directions with respect to the antenna axis. In the case of the horizontal half-wave dipole, a plane running in a direction along the axis and another broadside to the antenna will give a good deal of information.

The effect of reflection from the ground can be expressed as a pattern factor, given in decibels. For any given vertical angle, adding this factor algebraically to the value for that angle from the free-space pattern for that antenna gives the resultant radiation value at that angle. The limiting conditions are those represented by the direct ray and the reflected ray being exactly in phase and exactly out of phase, when both, assuming there are no ground losses, have equal amplitudes. Thus, the resultant field strength at a distant point may be either 6 dB greater than the free-space pattern (twice the field strength), or zero, in the limiting cases.

HORIZONTAL ANTENNAS

The way in which pattern factors vary with height for horizontal antennas is shown graphically in the plots of Fig 11. The solid-line plots are based on perfectly conducting ground, while the shaded plots are based on real earth conditions. These patterns apply to horizontal antennas of any length. While these graphs are, in fact, radiation patterns of horizontal single-wire antennas (dipoles) as viewed from the end of the wire, it must be remembered that the plots merely represent *pattern factors* for all other horizontal antennas.

Vertical radiation patterns in the directions off the ends of a horizontal half-wave dipole are shown in Fig 12 for various antenna heights. These patterns are scaled so they may be compared directly to those for the appropriate heights in Fig 11. Note that the perfect-earth patterns in Figs 12A and 11B are the same as those in the upper part of Fig 9. Note also that the perfect-earth patterns of Figs 12B and 11D are the same as those in the lower section of Fig 9. The reduction in field strength off the ends of the wire at the lower angles, as compared with the broadside field strength, is quite apparent. It is also clear from Fig 12 that, at some heights, the high-angle radiation off the ends is nearly as great as the broadside radiation.

In vertical planes making some intermediate angle between 0° and 90° with the wire axis, the pattern will have a shape intermediate between the broadside and end-on patterns. By visualizing a smooth transition from the end-on pattern to the broadside pattern as the horizontal angle is varied from 0° to 90°, a fairly good mental picture of the actual solid pattern may be formed. An example is shown in Fig 13. At A, the vertical pattern of a half-wave dipole at a height of ½ λ is shown through a plane 45° away from the favored direction of the antenna. At B and C, the vertical pattern of the same antenna is shown at heights of ¾ and

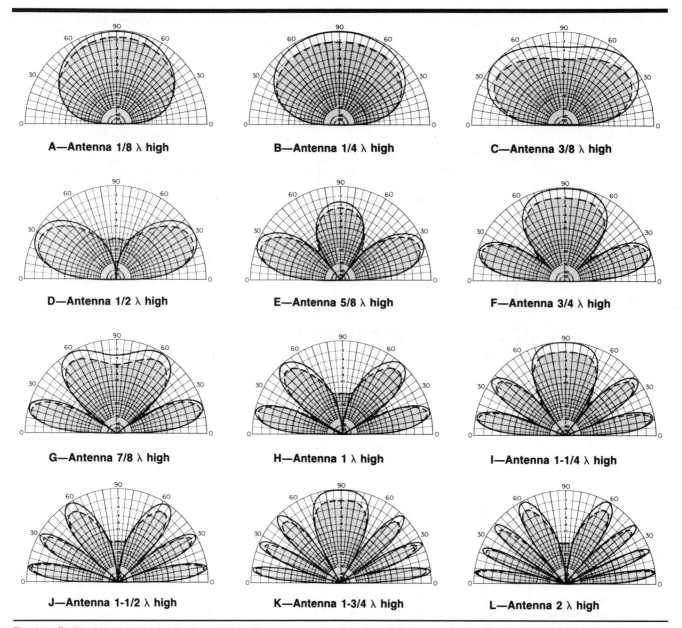

Fig 11—Reflection factors for horizontal antennas at various heights above ground. The solid-line curves are the perfect-earth patterns (broadside to the antenna wire); the shaded curves represent the effects of average earth (k = 15, G = 0.005 S/m) at 14 MHz. Add 6 dB to values shown.

1 λ (through the same 45° off-axis plane). These patterns are scaled so they may be compared directly with the broadside and end-fire patterns for the same antenna (at the appropriate heights) in Figs 11 and 12.

The curves presented in Fig 14 are useful for determining heights of horizontal antennas that give either maximum or minimum reinforcement at any given wave angle. For instance, if it is desired to place an antenna at such a height that it will have a null at 30°, the antenna should be erected at a height where a broken line crosses the 30° line on the horizontal scale. There are two heights (up to 2 λ) that will effect this null: 1 λ and 2 λ.

As a second example, it is desired to have the ground reflection give maximum reinforcement of the direct ray from a horizontal antenna at a 20° wave angle (angle of radiation). The antenna height should be 0.75 λ. The same height will

give a null at 42° and a second maximum at 90°.

Fig 14 is also useful for visualizing the vertical pattern of a horizontal antenna. For example, if an antenna is to be erected at 1.25 λ, it will have major lobes (solid-line crossings) at 12° and 37°, as well as at 90° (the zenith). The nulls in this pattern (dashed-line crossings) will appear at 24° and 53°. By using Fig 14 along with wave-angle information contained in Chapter 23, it is possible to calculate the antenna height that will best suit your needs.

VERTICAL ANTENNAS

In the case of a vertical half-wave dipole, the horizontal directional pattern is simply a circle at any wave angle (although the actual field strength will vary, at the different wave angles, with the height above ground). Hence, one

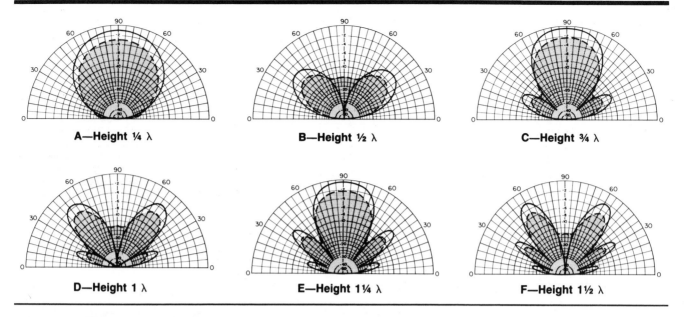

Fig 12—Vertical-plane radiation patterns of horizontal half-wave antennas in the plane of the antenna wire. The solid-line curves are the perfect-earth patterns, and the shaded curves represent the effects of average earth (k = 15, G = 0.005 S/m) at 14 MHz. The 0-dB reference in each plot corresponds to the peak of the main lobe in the favored direction of the antenna (the maximum gain). Add 6 dB to values shown.

vertical pattern is sufficient to give complete information (for a given antenna height) about the antenna in any direction with respect to the wire. A series of such patterns for various heights is given in Fig 15. These patterns are formed by using the procedure described in the early part of this chapter. The 3-dimensional radiation pattern in each case is formed by rotating the plane pattern about the zenith axis of the graph.

The solid-line curves represent the radiation patterns of the half-wave vertical dipole at different feed-point heights over perfectly conducting ground. The shaded curves show the patterns produced by the same antennas at the same heights over average ground (G = 0.005 S/m, k = 15) at 14 MHz. The PBA in this case is approximately 14°.

GROUND REFLECTION AND RADIATION RESISTANCE

Waves radiated from the antenna directly downward reflect vertically from the ground and, in passing the antenna

on their upward journey, induce a voltage in it. The magnitude and phase of the current resulting from this induced voltage depends on the height of the antenna above the reflecting surface.

The total current in the antenna thus consists of two components. The amplitude of the first is determined by the power supplied by the transmitter and the *free-space* radiation resistance of the antenna. The second component is induced in the antenna by the wave reflected from the ground. This second component, while considerably smaller than the first at most useful antenna heights, is by no means insignificant. At some heights, the two components will be in phase, so the total current is larger than is indicated by the free-space radiation resistance. At other heights, the two components are out of phase, and the total current is the difference between the two components.

Thus, merely changing the height of the antenna above ground will change the amount of current flow, assuming that the power input to the antenna is constant. A higher current

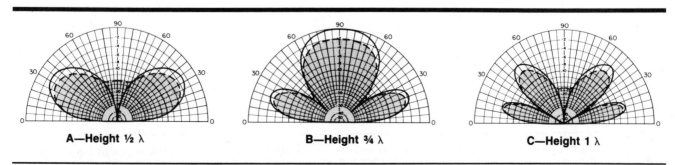

Fig 13—Vertical-plane radiation patterns of half-wave horizontal antennas at 45° from the antenna wire. The solid-line and shaded curves represent the same conditions as in Figs 11 and 12. These patterns are scaled so they may be compared directly with those of Figs 11 and 12.

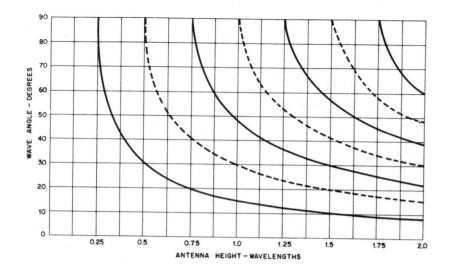

Fig 14—Angles at which nulls and maxima (factor = 6 dB) in the ground reflection factor appear for antenna heights up to two wavelengths. The solid lines are maxima, dashed lines nulls, for all horizontal antennas. See text for examples. Values may also be determined from the trigonometric relationship $\theta = \arcsin(A/4h)$, where θ is the wave angle and h is the antenna height in wavelengths. For the first maximum, A has a value of 1; for the first null A has a value of 2, for the second maximum 3, for the second null 4, and so on.

at the same power input means that the effective resistance of the antenna is lower, and vice versa. In other words, the radiation resistance of the antenna is affected by the height of the antenna above ground.

As discussed at length in the early part of this chapter, the electrical characteristics of the ground affect both the amplitude and the phase of reflected signals. For this reason, the electrical characteristics of the ground under the antenna will have some effect on the impedance of that antenna, the reflected wave having been influenced by the ground. In other words, different impedance values may be encountered when

an antenna is erected at identical heights but over different types of earth.

Fig 16 shows the way in which the radiation resistance of horizontal and vertical half-wave antennas vary with height (in wavelengths). For horizontal antennas, the differences between the effects of perfect ground and real earth are negligible if the antenna height is greater than 0.2 λ. At lesser heights, the radiation resistance decreases rapidly as the antenna is brought closer to a perfect conductor, but not so rapidly for actual ground. Over real earth, the resistance begins increasing at heights below about 0.08 λ. The reason

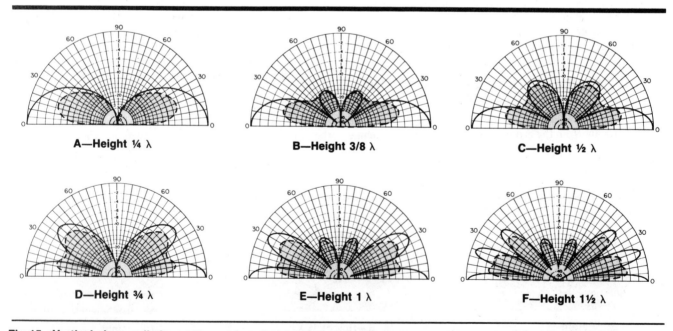

A—Height ¼ λ   B—Height 3/8 λ   C—Height ½ λ

D—Height ¾ λ   E—Height 1 λ   F—Height 1½ λ

Fig 15—Vertical-plane radiation patterns of vertical half-wave antennas above ground. The height is that of the center of the antenna. Solid lines are perfect-earth patterns; shaded curves show the effects of real earth. The patterns are scaled—that is, they may be directly compared to the solid-line ones for comparison of losses at any wave angle. These patterns were calculated for average ground (k = 15, G = 5 mS/m) at 14 MHz. The PBA for these conditions is 14°.

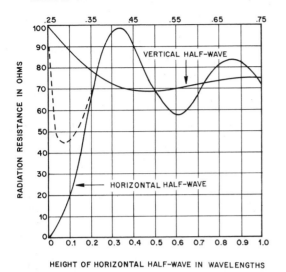

Fig 16—Variation in radiation resistance of vertical and horizontal half-wave antennas at various heights above ground. Solid lines are for perfectly conducting ground; the broken line is the radiation resistance of horizontal half-wave antennas at low heights over real ground.

for the increasing resistance at very low heights is that more and more of the induction field of the antenna is absor ͗d by the lossy ground.

For a half-wave vertical antenna, the center of which is ¼ λ or more above the surface, differences between the effects of perfect ground and real earth on the impedance is negligible. The resistance of a vertical dipole at various heights above ground is also shown in Fig 16. The antennas on which this chart is based are assumed to have infinitely thin conductors, and thus have somewhat higher free-space radiation resistances (73 ohms) than antennas constructed of wire or tubing.

Ground Screens

The effect of a perfectly conducting ground (for radiation-resistance purposes) can be simulated under an antenna by installing a metal screen or mesh such as poultry netting (chicken wire) or hardware cloth on or near the surface of the ground. The screen should extend at least a half wavelength in every direction from the antenna. Such a screen will effectively establish the height of the antenna as far as radiation resistance is concerned, as it substitutes for the actual earth underneath the antenna. However, such a screen will not have a significant effect on the vertical radiation pattern. This is because earth reflections as far as 10 to 20 or more wavelengths from the antenna, depending on the antenna height, play a part in forming the radiation pattern.

For current-fed (odd-multiple quarter-wave) vertical antennas, the screen also reduces ground losses near the antenna, because if the screen conductors are solidly bonded to each other, the resistance is much lower than that of the ground itself. With other types of antennas, such as horizontal dipoles at heights of a quarter wavelength or more, losses in the ground beneath the antenna are much less destructive to antenna efficiency.

DIRECTIVE PATTERNS AND THE WAVE ANGLE

Remember that radiation patterns are three dimensional, sometimes referred to as "solid" patterns. It is difficult to show more than a cross section of the solid pattern on a plane sheet of paper. The cross sections usually selected are the E plane, the plane containing the wire axis, and the H plane, the plane perpendicular to the wire axis. (These terms are discussed in detail in Chapter 2.)

If the antenna is horizontal, the E-plane pattern represents the horizontal or azimuthal radiation pattern of the antenna when the wave leaves the antenna (or arrives) at a zero-degree elevation angle. In the case of the vertical antenna, the horizontal radiation pattern at zero degrees is given by the H-plane pattern.

However, with two exceptions—surface waves at low frequencies and space waves at VHF and above—the wave angle used for communication is not zero. The subject of interest, then, is the directive pattern of the antenna at wave angles that are effective for the desired communications.

Directive Diagrams

The directive diagram for a wave angle of zero elevation (purely horizontal radiation) unfortunately does not give an accurate indication of the directive properties of a horizontal antenna at wave angles above zero. For example, consider the half-wave dipole pattern in Fig 17. It shows that there is no radiation directly in line with the antenna itself. This is true at zero wave angle. However, if the antenna is horizontal and some wave angle other than zero is considered, *it is not true at all.*

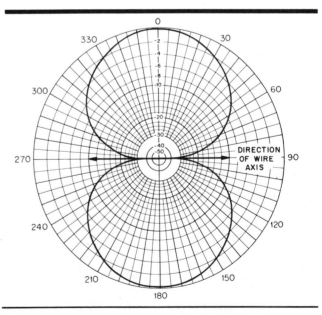

Fig 17—Horizontal radiation pattern of a half-wave dipole antenna. The arrow shows the direction of the wire, although the antenna itself is considered to be merely a point at the center of the diagram. This pattern is simply a cross-section of a three-dimensional solid figure that shows the radiation from the antenna in all directions.

The reason should become clear on inspection of Fig 18, which shows a horizontal half-wave antenna with a cross section of its free-space radiation pattern. The pattern is cut by a plane that is vertical with respect to the earth and which contains the axis of the antenna conductor. In other words, the view is looking broadside at the antenna wire. (For the moment, reflections from the ground are neglected.) The lines OA, OB and OC all point in the same *geographical* direction (the direction in which the wire itself points), but make different angles with the antenna in the vertical plane. Put another way, they correspond to different wave angles or angles of radiation, with all three rays aimed along the same line on the earth's surface. All three waves are leaving the *end of the antenna* (the same compass heading).

The purely horizontal wave OA has zero amplitude, but at a somewhat higher angle corresponding to line OB, the field strength is appreciable. At a still higher angle corresponding to line OC, the field strength is greater yet. In this pattern, the higher the wave angle, the greater the field strength in the compass direction OA. Thus, in plotting a directive pattern to show the behavior of the antenna in different compass directions, it is necessary to specify the angle of radiation at which the diagram applies. When the antenna is horizontal, the shape of the diagram will be altered considerably as the wave angle is changed.

As described in Chapter 23, the wave angles that are useful depend on two things—the distance over which communication is to be carried out, and the height of the ionospheric layer that does the reflecting. Whether the E or F2 layer (or a combination of the two) will be used depends on the operating frequency, the time of day, season, and the sunspot cycle. A horizontal antenna operating on a given frequency may be almost nondirectional for distances of a few hundred miles. It will, however, give substantially better results broadside than off the ends at distances on the order of 1000-1200 miles when propagation is by the E layer. When propagation is via the F layer, the directivity may be fairly well marked at long distances and not at all pronounced at 1000 miles or less.

From this it might seem that antenna directivity cannot be predicted. However, it is possible to get a very good idea of the directivity by choosing a few angles that are representative for different types of work. With patterns for

such angles available, it is fairly simple to interpolate for intermediate angles. Combined with some knowledge of the behavior of the ionosphere, a fairly good estimate of the directive characteristics of a particular antenna can be made for the particular time of day and distance of interest.

In the directive patterns given in Fig 19, the wave angles considered are 9, 15 and 30 degrees. These represent, respectively, the median values of a range of angles that have been found to be effective for communication at 28, 14 and 7 MHz. Because of the variable nature of ionospheric propagation, they should not be considered to be more than general guides to the sort of directivity to be expected.

The patterns of Fig 19 illustrate the trends that radiation from all horizontally polarized antennas follow as the wave angle is increased. Note that the rather sharp side rejection visible at a 9° wave angle becomes steadily less pronounced as the wave angle increases. At a 30° wave angle, the radiation is down only 8 dB from that in the broadside direction.

Because one S unit on the signal-strength scale is roughly 4 or 5 dB, it is easy to get an approximate idea of the operation of an antenna. For example, off the ends of a half-wave antenna (Fig 19) the signal can be expected to be down between 3 and 4 S units compared with its strength at right angles or broadside to the antenna, at a wave angle of 15°. This would be fairly representative of its performance on 14 MHz at distances of 500 miles or more. With a wave angle of 30°, the signal off the ends would be down only 1½ to 3 S units, while at an angle of 9° it would be down 4 to 5 S units. Because high wave angles become less useful as the frequency is increased, this illustrates the importance of running the antenna wire in the proper direction if best results

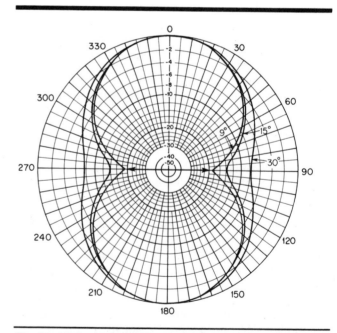

Fig 19—Horizontal radiation patterns for a horizontal half-wave antenna for vertical angles of 9°, 15° and 30°. The direction of the antenna itself is shown by the arrow. All three patterns are plotted to the same maximum, but the actual amplitudes at the various angles will depend on the antenna height, as described in the text. The patterns shown here indicate only the shape of the directive diagram as the angle is varied.

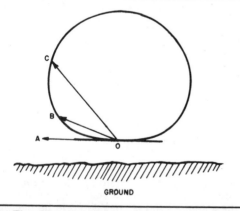

GROUND

Fig 18—The directive pattern of the antenna depends upon the angle of radiation considered. As shown by the arrows, the field strength in a given compass direction will be quite different at different vertical angles.

in a particular direction are to be realized at the higher frequencies.

Height Above Ground

The *shapes* of the directive patterns given in Fig 19 are not affected by the height of the antenna above the ground. However, the *amplitude* relationships between the patterns of a given antenna for various wave angles are modified by the height. The patterns in the figure are scaled so direct comparison at different wave angles is possible. To make best use of the patterns, the effect of the ground-reflection factor (Fig 11) should also be included.

Assume a horizontal half-wave antenna is placed a half wavelength above average ground. The graph of the ground-reflection factors for this height is given in Fig 11D. For angles of 9°, 15° and 30°, the values of the factor as read from the perfect-earth curve (with 6 dB added) are −0.5, 2.7 and 5 dB, respectively. These factors are applied to field strength. For convenience, take the 9° angle as reference. Then at a wave angle of 15° the field strength will be 3.2 dB greater than the field strength at 9°, *in any compass direction*. At a wave angle of 30° the field strength will be 5.5 dB greater than that at 9°, in any compass direction. Put another way, at a wave angle of 30°, the antenna is about 1 to 1½ S units better than it is at 9°. There is about ¾ S-unit difference between 9° and 15°, and about ½ an S unit between 15° and 30°.

PRACTICAL CONSIDERATIONS

A great deal of mystery and lack of information surrounds the vertical antenna ground system. The theory of ground effects has been covered in detail earlier in this chapter, but nothing has been said about radial systems. In the case of ground-mounted vertical antennas, many general statements such as "the more radials the better" and "lots of short radials are better than a few long ones" have served as rules of thumb to some, but many questions as to relative performance differences and optimum number for a given length remain unanswered. Most of these questions boil down to one, namely, how many radials, and how long, should be used in a vertical antenna installation. Table 3 answers these questions. John Stanley, K4ERO, first presented this material in December 1976 *QST*.

One source of information on ground-system design is *Radio Broadcast Ground Systems* (see the bibliography at the end of this chapter). Most of the data presented in Table 3 is taken from that source, or derived from the interpolation of data contained therein.

Table 3 gives numbers of radials and a corresponding optimum radial length for each case. Using radials considerably longer than suggested for a given number or using a lot more radials than suggested for a given length, while not adverse to performance, does not yield significant improvement either. That would represent a nonoptimum use of wire and construction time. Each suggested configuration represents an optimum relationship between length and number for a fixed amount of *total* radial wire. Obviously, the more total wire installed, the better the performance, provided the configuration is optimized in terms of number versus length.

The loss figures in Table 3 are calculated for a quarter-wave radiating element. A very rough approximation of loss when using shorter antennas can be obtained by doubling the loss in dB each time the antenna height is halved.

Table 3
Optimum Ground-System Configurations

| | Configuration Designation | | | | | |
	A	B	C	D	E	F
Number of radials	16	24	36	60	90	120
Length of each radial in wavelengths	0.1	0.125	0.15	0.2	0.25	0.4
Spacing of radials in degrees	22.5	15	10	6	4	3
Total length of radial wire installed, in wavelengths	1.6	3	5.4	12	22.5	48
Power loss in dB at low angles with a quarter-wave radiating element	3	2	1.5	1	0.5	0*
Feed-point impedance in ohms with a quarter-wave radiating element	52	46	43	40	37	35

Note: Configuration designations are indicated only for text reference.
*Reference. The loss of this configuration is negligible compared to a perfectly conducting ground.

For longer antennas the losses decrease, approaching 2 dB for configuration A of Table 3 for a half-wave radiator. Longer antennas yield correspondingly better performance.

The loss figures in the table are for the ground wave and are probably fairly accurate for low-angle sky wave signals. For higher angles, still useful for closer DX and other intermediate paths, the loss will be about half that shown.

The table is based on average ground conductivity. Variation of the loss values shown can be considerable, especially for configurations using fewer radials. Those building antennas over dry, sandy or rocky ground should expect more loss. On the other hand, higher than average soil conductivity and wet soils would make the "compromise" configurations (those with the fewest radials) even more attractive.

Keep in mind that the losses shown in Table 3 are significant only when transmitting. When receiving, the noise levels present below 30 MHz, as well as the received signal, are attenuated by ground losses, so that the signal-to-noise ratio is not affected. Hence, for reception, each configuration shown will be roughly equal for a given fixed radiator length.

When antennas are combined into arrays, either of parasitic or all-driven types, mutual impedances lower the radiation resistance of the elements, drastically increasing the effects of ground loss. For instance, an antenna with a 52-ohm feed-point impedance and 10 ohms of ground-loss resistance will have an efficiency of approximately 83%. An array of two similar antennas in a driven array with the same ground loss may have an efficiency of 70% or less. Special precautions must be taken in such cases to achieve satisfactory operation. Generally speaking, a wide-spaced broadside array presents little problem, but a close-spaced end-fire array should be avoided for transmission, unless the lower loss configurations are used or other precautions taken. Chapter 8 covers the subject of vertical arrays in great detail.

In cases where directivity is desirable or real estate dictates, longer, more closely spaced radials can be installed in one direction, and shorter, more widely spaced in another. Multiband ground systems can be designed using different optimum configurations for different bands. Usually it is most convenient to start at the lowest frequency with fewer radials and add more short radials for better performance on the higher bands.

There is nothing sacred about the exact details of the configurations given, and slight changes in the number of radials and lengths will not cause serious problems. Thus, a configuration with 32 or 40 radials of 0.14 λ or 0.16 λ will work as well as configuration C shown in the table.

If fewer than 90 radials are contemplated, there is no need to make then a quarter wavelength long. This differs rather dramatically from the case of a ground plane antenna where resonant radials are installed above ground. For the ground-mounted antenna, quarter-wave radials are far from optimum. Because the radials of a ground-mounted vertical are actually on, if not slightly below the surface, they are coupled by capacitance or conduction to the ground, and thus resonance effects are not important. The basic function of radials is to provide a low-loss return path for ground currents. The reason that short radials are sufficient when few are used is that at the perimeter of the circle to which the ground system extends, the few wires are so spread apart that most of the return currents are already in the ground between the wires rather than in the wires themselves. As more wires are added, the spaces between them are reduced and longer length helps to provide a path for currents still farther out.

Radio Broadcast Ground Systems states, "Experiments show that the ground system consisting of only 15 radial wires need not be more than 0.1 wavelength long, while the system consisting of 113 radials is still effective out to 0.5 wavelength." Many graphs in that publication confirm this statement. This is not to say that these two systems will perform equally well; they most certainly will not. However, if 0.1 wavelength is as long as the radials can be, there is little point in using more than 15 of them.

In most cases a "compromise" ground system actually turns out to be the "optimum" system when all factors are considered. Thus, the designer should (1) study the cost of various radial configurations versus the gain of each; (2) compare alternative means of improving transmitted signal and their cost (more power, etc); (3) consider increasing the physical antenna height or the electrical length of the vertical radiator, instead of improving the ground system; and (4) use multielement arrays for directivity and gain, observing the necessary precautions related to mutual impedances discussed in Chapter 8.

BIBLIOGRAPHY

Source material and more extended discussion of the topics covered in this chapter can be found in the references listed below and in the textbooks listed at the end of Chapter 2.

B. Boothe, "The Minooka Special," *QST*, Dec 1974.

Brown, Lewis and Epstein, "Ground Systems as a Factor in Antenna Efficiency," *Proc. IRE*, Jun 1937.

R. Jones, "A 7-MHz Vertical Parasitic Array," *QST*, Nov 1973.

C. J. Michaels, "Some Reflections on Vertical Antennas," *QST*, Jul 1987.

J. Sevick, "The Ground-Image Vertical Antenna," *QST*, Jul 1971.

J. Sevick, "The W2FMI 20-Meter Vertical Beam," *QST*, Jun 1972.

J. Sevick, "The W2FMI Ground-Mounted Short Vertical," *QST*, Mar 1973.

J. Sevick, "A High Performance 20-, 40- and 80-Meter Vertical System," *QST*, Dec 1973.

J. Sevick, "The Constant-Impedance Trap Vertical," *QST*, Mar 1974.

J. Stanley, "Optimum Ground Systems for Vertical Antennas," *QST*, Dec 1976.

F. E. Terman, *Radio Engineers' Handbook*, 1st ed. (New York, London: McGraw-Hill Book Co, 1943).

Radio Broadcast Ground Systems, available from Smith Electronics, Inc, 8200 Snowville Rd, Cleveland, OH 44141.

Chapter 4

Selecting Your Antenna System

Where should you start in putting together an antenna system? A newcomer to Amateur Radio, an amateur moving to a new location, or someone wanting to improve an existing "antenna farm" might ask this question. The answer: In a comfortable chair, with a pad and writing instrument.

The most important time spent in putting together an antenna system is that time spent in planning. It can save a lot of time, money and frustration. No one can tell you the exact steps you should take in developing your master plan. That plan and your method of deriving it will be different from those of others. This section, prepared by Chuck Hutchinson, K8CH, should help you with some ideas.

Begin planning by spelling out your communications desires. What bands are you interested in? Who (or where) do you want to talk to? When do you operate? How much time and money are you willing to spend on an antenna system? What physical limitations affect your master plan?

From the answers to the above questions, begin to formulate goals—short, intermediate and long range. Be realistic about those goals. Remember that there are three station effectiveness factors that are under your control. These are: operator skill, equipment in the shack, and the antenna system. There is no substitute for developing operating skills. Some trade-offs are possible between shack equipment and antennas. For example, a high power amplifier can compensate for a less than optimum antenna. By contrast, a better antenna has advantages for receiving as well as for transmitting.

Consider your limitations. Are there regulatory restrictions on antennas in your community? Are there any deed restrictions or covenants that apply to your property? Do other factors (finances, family considerations, other interests, and so forth) limit the type or height of antennas that you can erect? All of these factors must be investigated—they play a major role in determining the type of antennas you erect.

Chances are that you won't be able to do all you desire immediately. Think about how you can budget your resources over a period of time. Your resources are your money, your time available to work, materials you may have on hand, friends that are willing to help, etc. One way to budget is to concentrate your initial efforts on a given band or two. If your major interest is in chasing DX (working distant foreign stations), you might want to start with a very good antenna for the 14-MHz band. A simple multiband antenna could initially serve for other frequencies. Later you can add better antennas for those other bands.

SITE PLANNING

A map of your property or proposed antenna site can be of great help as you begin to consider alternative antennas. You'll need to know the size and location of buildings, trees and other major objects in the area. Be sure to note compass directions on your map. Graph paper or quadrille paper is very useful for this purpose. See Fig 1. It's a good idea to make a few photocopies of your site map so you can mark on the copies as you work on your plans.

Use your map to plan antenna layouts and locations of any supporting towers or masts. If your plan calls for more than one tower or mast, think about using them as supports for wire antennas. As you work on a layout, be sure to think in three dimensions even though the map shows only two.

Be sensitive to your neighbors. A 70-foot guyed tower in the front yard of a house in a residential neighborhood is not a good idea (and probably won't comply with local ordinances!).

ANALYSIS

Use the information in this book to analyze antenna patterns in both horizontal and vertical planes. If you want to work DX, you'll want antennas that radiate energy at low

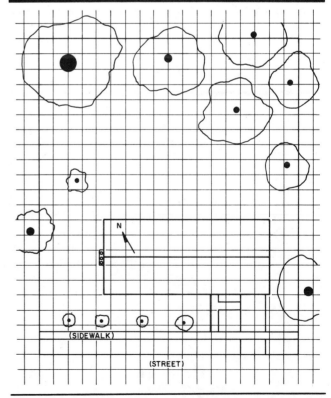

Fig 1—A site map such as the one shown here is a useful tool for planning your antenna installation.

angles. An antenna pattern is greatly affected by the presence of ground. Therefore, be sure to consider what effect ground will have on the antenna pattern at the height you are considering. An antenna at approximately 70 feet is on the order of ½, 1, 1½ and 2 wavelengths (λ) high on 7, 14, 21 and 28 MHz, respectively. Those heights are useful for long-distance communications. The same 70-foot height represents approximately ¼ λ at 3.5 MHz. Most of the radiated energy from a dipole at that height would be concentrated straight up. This condition is not outstanding for long-distance communication, but is still useful for DX work and excellent for short-range communications.

Lower heights can be useful for communications. However, it is generally true that "the higher, the better" as far as communications effectiveness is concerned.

There may be cases where is is not possible to install low-frequency dipoles at ¼ λ or more above the ground. A vertical antenna with many radials is a good choice for long-distance communications. You may want to install both a dipole and a vertical for the 3.5- or 7-MHz bands. You can then choose the antenna that performs best for a given set of conditions. The dipole will generally work better for shorter range communications, while the vertical will generally be the better performer over longer distances.

Consider the azimuthal pattern of fixed antennas. You'll want to orient those fixed antennas to favor the directions of greatest interest to you.

BUILDING THE SYSTEM

When the planning is completed, it is time to begin construction of the antenna system. Chances are that you can divide that construction into a series of phases or steps. Say, for example, that you have lots of room and that your long-range plan calls for a pair of 100-foot towers to support monoband Yagi antennas. The towers will also support a horizontal 3.5-MHz dipole at 100 feet, for DX work. On your map you've located them so the dipole will be broadside to Europe. Initially you decide to build a 60-foot tower with a triband beam and a 3.5-MHz inverted V dipole to begin the project. In your master plan, the 60-foot tower is really the bottom part of a 100-foot tower. The guys, anchors and all hardware are designed for use in the 100 footer. Initially you buy a heavy duty rotator and mast that will be needed for the monoband antennas later. Thus, you avoid having to buy, and then sell, a medium-duty rotator and lighter weight tower equipment. You could have saved money in the long run by putting up a monoband beam for your favorite band, but you decided that for now it is more important to have a beam on 14, 21 and 28 MHz.

The second step of your plan calls for installing the second tower. This time you've decided to wait until you can install all 100 feet of that second tower, and put a 7-MHz Yagi on top of it. Later you will remove the top section of the first (60-foot) tower and insert the sections and add the guys to bring it up to 100 feet. You decide that at that time you'll continue to use the tribander for a few months to see what difference the 60 foot to 100 foot height change makes.

COMPROMISES

Because of limitations, most amateurs are never able to build their "dream" antenna system. This means that some compromises must be made. Do not, under any circumstances, compromise the safety of an antenna installation. Follow the manufacturer's recommendations for tower assembly, installation and accessories. Make sure that all hardware is being used within its ratings.

Guyed towers are frequently used by radio amateurs because they cost less than more complicated tower types with similar ratings. Guyed towers are fine for those who can climb, or those with a friend who is willing to climb. But you may want to consider an antenna tower that folds over, or one that cranks up (and down). Some towers crank up (and down) and fold over too. See Fig 2. That makes for convenient access to antennas for adjustments and maintenance without

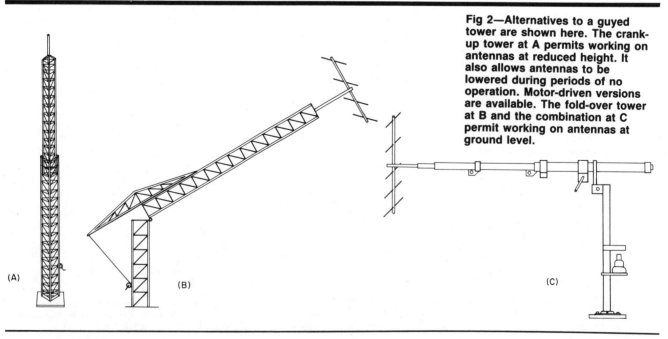

Fig 2—Alternatives to a guyed tower are shown here. The crank-up tower at A permits working on antennas at reduced height. It also allows antennas to be lowered during periods of no operation. Motor-driven versions are available. The fold-over tower at B and the combination at C permit working on antennas at ground level.

(A)　　(B)　　(C)

climbing. Crank-up towers also offer another advantage. They allow antennas to be lowered during periods of no operation, such as for aesthetic reasons or during periods of high winds.

A well designed monoband Yagi should out-perform a multiband Yagi. In a monoband design the best adjustments can be made for gain, front to back ratio (F/B ratio), and matching, *for a single band*. In a multiband design, there are always trade-offs in these properties for the ability to operate on more than one band. Nevertheless, a multiband antenna has many advantages over two or more single band antennas. A multiband antenna requires less heavy duty hardware, requires only one feed line, takes up less space, and it costs less.

Apartment dwellers face much greater limitations in their choice of antennas. For most, the possibility of a tower is only a dream. (One enterprising ham made arrangements to purchase a top-floor condominium from a developer. The arrangements were made before construction began, and the plans were altered to include a roof top tower installation.) For apartment and condominium dwellers, the situation is still far from hopeless. A later section presents ideas for consideration.

EXAMPLES

You can follow the procedure previously outlined to put together modest or very large antenna systems. What might a ham put together for antennas when he or she wants to try a little of everything, and has a modest budget? Let's suppose that the goals are (1) low cost, (2) no tower, (3) coverage of all HF bands and the repeater portion of one VHF band, and (4) the possibility of working some DX.

After studying the pages of this book, the station owner decides to first put up a 135-foot center-fed antenna. High trees in the back yard will serve as supports to about 50 feet. This antenna will cover all the HF bands by using a balanced feeder and a Transmatch. It should be good for DX contacts on 10 MHz and above, and will probably work okay for DX contacts on the lower bands. However, her plan calls for a vertical for 3.5 and 7 MHz to enhance the DX possibilities on those bands. For VHF, a chimney-mounted vertical is included.

Another Example

A licensed couple has bigger ambitions. Goals for their station are (1) a good setup for DX on 14, 21 and 28 MHz, (2) moderate cost, (3) one tower, (4) ability to work some DX on 1.8, 3.5 and 7 MHz, and (5) no need to cover the CW portion of the bands.

After considering the options, the couple decides to install a 65-foot guyed tower. A large commercial triband Yagi will be mounted atop the tower. The center of a trap dipole tuned for the phone portion of the 3.5- and 7-MHz bands will be supported by a wooden yard arm installed at the 60-foot level of the tower. An inverted L for 1.8 MHz starts near ground level and goes up to a similar yard arm on the opposite side of the tower. The horizontal portion of the inverted L runs away from the tower at right angles to the trap dipole.

Later, the husband will experiment with sloping antennas for 3.5 MHz. If those experiments are not successful, a ¼-λ vertical will be used on that band.

APARTMENT POSSIBILITIES

A complete and accurate assessment of antenna types, antenna placement, and feed line placement is very important for the apartment dweller. Among the many possibilities for types are balcony antennas, ''invisible'' ones (made of fine wire), vertical antennas disguised as flag poles, and indoor antennas.

A number of amateurs have been successful in negotiating with the apartment owner or manager for permission to install a short mast on the roof of the structure. Coaxial lines and rotator control cables might be routed through conduit troughs or through duct work. If you live in one of the upper stories of the building, routing the cables over the edge of the roof and in through a window might be the way to go. There is a story about one amateur who owns a triband beam mounted on a 10-foot mast. But even with such a short mast, he is the envy of all his amateur friends because of his superb antenna height. His mast stands atop a 22-story apartment building.

Usually the challenge is to find ways to install antennas that are unobtrusive. That means searching out antenna locations such as balconies, eaves, nearby trees, etc. For example, a simple but effective balcony antenna is a dangling vertical. Attach an ''invisible'' wire to the tip of a mobile whip or a length of metal rod or tubing. Then mount the rigid part of the antenna horizontally on the balcony rail, dangling the wire over the edge. The antenna is operated against the balcony railing or other metallic framework. A matching network is usually required at the antenna feed point. Metal in the building will likely give a directivity effect, but this may be of little consequence and perhaps even an advantage. The antenna may be removed and stored when not in use.

Frequently, the task of finding an inconspicuous route for a feed line is more difficult than the antenna installation itself. When Al Francisco, K7NHV, lived in an apartment, he used a tree-mounted vertical antenna. The coax feeder exited his apartment through a window and ran down the wall to the ground. Al buried the section of line that went from under the window to a nearby tree. At the tree, a section of enameled wire was connected to the coax center conductor. He ran the wire up the side of the tree away from foot traffic. A few short radials completed the installation. The antenna worked fine, and was never noticed by the neighbors.

See Chapters 6 and 15 for ideas about antennas that might fit into your available space. Your options are limited as much by your imagination and ingenuity as by your pocketbook.

Another option for apartment dwellers is to operate away from home. Some hams concentrate on mobile operation as an alternative to a fixed station. It is possible to make a lot of contacts on HF mobile. Some have worked DXCC that way.

Suppose that you like VHF contests. Because of other activities, you are not particularly interested in operating VHF outside the contests. Why not take your equipment and antennas to a hilltop for the contests? Many hams combine a love for camping or hiking with their interest in radio.

Working Out the Details

The preceding portions of this chapter are designed to make you think. They are not rich in detail, because you must supply the details for your situation. Much more could be said on these topics. Indeed, an entire book could be devoted to these matters.

When more than one antenna is to be used with the same transmitter or transceiver, some kind of switching arrangement is useful. The simplest arrangement is to change antennas by hand. That is not good for two reasons. First it is slow, and second it can cause wear, and even damage, to connectors. A better arrangement is to bring the transmission lines together near the operating position, and connect them to the equipment through an RF switch. See Fig 3A.

An alternative is to place a remotely controlled RF switch closer to the antennas as shown in Fig 3B. This can save money on transmission lines, and tends to make things neater near the operating position. When antennas and operating position are a long distance apart, it is a good idea to use the remotely controlled switch and a length of low loss transmission line.

INSTALLING TRANSMISSION LINES

When coaxial cable or 300-Ω ribbon line is connected to a dipole that does not have a support at the center, it is essential to avoid stressing the feed line conductors with the weight of the feed line. Fig 4 shows a method of accomplishing this with coaxial cable. The cable is looped around the center antenna insulator, and clamped before making connections to the antenna. In Fig 5, the weight of the ribbon line is supported by threading the line through a sheet of insulating material. The insulator is suspended from the antenna by threading the antenna through it. This arrangement is particularly suited to folded dipoles made of 300-Ω ribbon.

When open wire feed line is used, the conductors of the line should be anchored to the insulator by threading them through the eyes of the insulator two or three times, and twisting the wire back on itself before soldering. A slack tie wire should then be used between the feeder conductor and the antenna, as shown in Fig 6. (The tie wires may be extensions of the line conductors themselves.)

When using plastic-insulated open wire line, the tendency of the line to twist and short out close to the antenna can be counteracted by making the center insulator of the antenna longer than the spacing of the line, as shown in Fig 6. In this case, a heavier spreader insulator should be added just below the antenna insulator to prevent side stress from pulling the conductors away from the light plastic feeder spreaders.

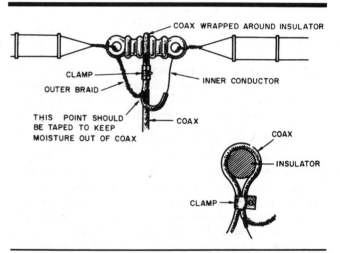

Fig 4—Method of relieving strain on conductors of coaxial cable when feeding a dipole. Six or eight wraps of solid copper bus wire may be used in place of the clamp.

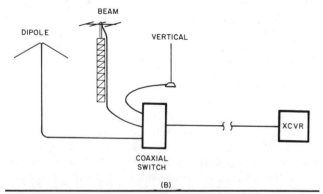

Fig 3—Two approaches to switching between antenna feed lines. The switching method at A can use a hand operated switch, but requires more line. When a remotely controlled switch is used, as at B, less line is required.

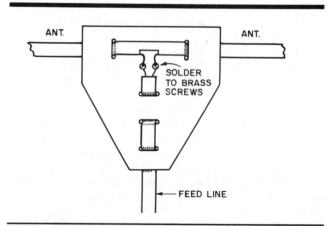

Fig 5—Strain reliever for conductors of 300-Ω ribbon line in a folded dipole. The strain reliever can be made from ¼-in. Lucite or Plexiglas sheet.

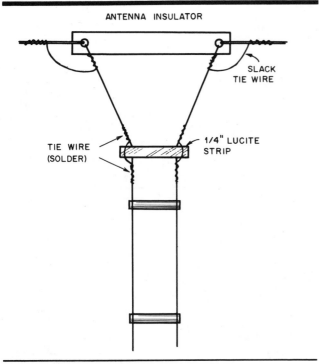

Fig 6—A method of connecting open wire line to an antenna center insulator. The Lucite strip keeps the feeder conductors from pulling away from the spreaders when plastic insulated open wire line is used.

Running the Feed Line from the Antenna to the Station

Chapter 24 contains some general guidelines for installing feed lines. More detailed information is contained in this section.

Coaxial cable requires no particular care in running from the antenna to the station entrance, other than protection from mechanical damage. If the antenna is not supported at the center, the line should be fastened to a post more than head high located under the center of the antenna, allowing enough slack between the post and the antenna to take care of any movement of the antenna in the wind. If the antenna feed point is supported by a tower or mast, the cable can be taped to the mast at intervals or to one leg of the tower.

Coaxial cable can also be buried a few inches in the ground to make the run from the antenna to the station. A deep slit can be cut by pushing a square-end spade full depth into the ground and moving the handle back and forth to widen the slit before removing the spade. After the cable has

been pushed into the slit with a piece of 1-inch board 3 or 4 inches wide, the slit can be tamped closed.

Ribbon line should be kept reasonably well spaced from other conductors running parallel to it for more than a few feet. TV-type standoff insulators with strap-clamp mountings can be used in running this type of line down a mast or tower leg. Similar insulators of the screw type can be used in supporting the line on poles for a long run.

Open wire lines require frequent supports to keep the lines from twisting and shorting out, as well as to relieve the strain. One method of supporting a long run of heavy open wire line is shown in Fig 7. The line must be anchored securely at a point under the feed point of the antenna. TV type line can be supported similarly by means of wire links fastened to the insulators. Fig 8 shows a method of supporting an open wire line from a tower.

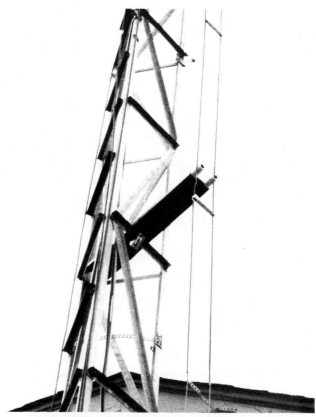

Fig 8—A board fitted with standoff insulators and clamped to the tower with U bolts keeps open wire line suitably spaced from a tower. (W4NML)

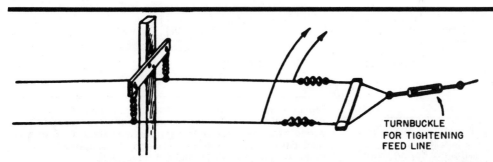

Fig 7—A support for open wire line. The support at the antenna end of the line must be sufficiently rigid to stand the tension of the line.

To keep the line clear of pedestrians and vehicles, it is usually desirable to anchor the feed line at the eave or rafter line of the station building (see Fig 9), and then drop it vertically to the point of entrance. The points of anchorage and entrance should be chosen so as to permit the vertical drop without crossing windows.

If the station is located in a room on the ground floor, one way of bringing coax transmission line into the house is to go through the outside wall below floor level, feed it through the basement or crawl space, and then up to the station through a hole in the floor. In making the entrance hole in the side of the building, suitable measurements should be made in advance to be sure the hole will go through the sill 2 or 3 inches above the foundation line (and between joists if the bore is parallel to the joists). The line should be allowed to sag below the entrance-hole level outside the building to allow rain water to drip off.

Open wire line can be fed in a similar manner, although it will require a separate hole for each conductor. Each hole should be insulated with a length of polystyrene or Lucite tubing. Drill the holes with a slight downward slant toward the outside of the building to prevent rain seepage. With TV type line, it will be necessary to remove a few of the spreader insulators, cut the line before passing through the holes (allowing enough length to reach the inside), and splice the remainder on the inside.

If the station is located above ground level, or if there is other objection to the procedure described above, entrance can be made at a window, using the arrangement shown in Fig 10. An Amphenol type 83-1F (UG-363) connector can be used as shown in Fig 11; ceramic feedthrough insulators can be used for open wire line. Ribbon line can be run through clearance holes in the panel, and secured by a winding of tape on either side of the panel, or by cutting the retaining rings and insulators from a pair of TV standoff insulators, and clamping one on each side of the panel.

LIGHTNING PROTECTION

Two or three types of lightning arresters for coaxial cable are available on the market. If the antenna feed point is at

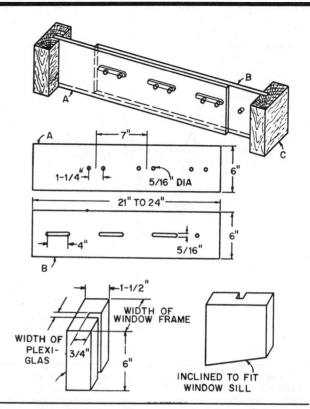

Fig 10—An adjustable window lead-in panel made of two sheets of Lucite or Plexiglas. A feedthrough connector for coax line can be made as shown in Fig 11. Ceramic feedthrough insulators are suitable for open wire line. (W1RVE)

the top of a well-grounded tower, the arrester can be fastened securely to the top of the tower for grounding purposes. A short length of cable, terminated in a coaxial plug, is then run from the antenna feed point to one receptacle of the arrester, while the transmission line is run from the other arrester receptacle to the station. Such arresters may also be placed at the entrance point to the station, if a suitable ground connection is available at that point (or arresters may be placed at both points for added insurance).

The construction of a homemade arrester for open wire line is shown in Fig 12. This type of arrester can be adapted to ribbon line an inch or so away from the center member of the arrester, as shown in Fig 13. Sufficient insulation should

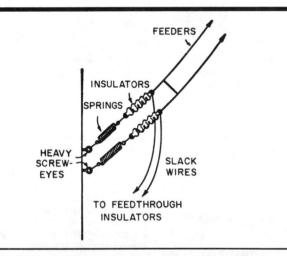

Fig 9—Anchoring open wire line at the station end. The springs are especially desirable if the line is not supported between the antenna and the anchoring point.

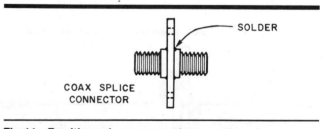

Fig 11—Feedthrough connector for coax line. An Amphenol 83-1J (PL-258) connector, the type used to splice sections of coax line together, is soldered into a hole cut in a brass mounting flange. An Amphenol bulkhead adapter 83-1F may be used instead.

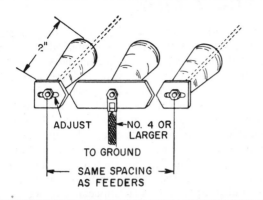

Fig 12—A simple lightning arrester for open-wire line made from three standoff or feedthrough insulators and sections of 1/8 × ½-in. brass or copper strap. It should be installed in the line at the point where the line enters the station. The heavy ground lead should be as short and direct as possible. The gap setting should be adjusted to the minimum width that will prohibit arcing when the transmitter is operated.

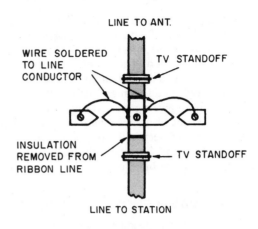

Fig 13—The lightning arrester of Fig 12 may be used with 300-Ω ribbon line in the manner shown here. The TV standoffs support the line an inch or so away from the grounded center member of the arrester.

be removed from the line where it crosses the arrester to permit soldering the arrester connecting leads.

Lightning Grounds

Lightning-ground connecting leads should be of conductor size equivalent to at least no. 10 wire. The no. 8 aluminum wire used for TV-antenna grounds is satisfactory.

Copper braid ¾ inch wide (Belden 8662-10) is also suitable. The conductor should run in a straight line to the grounding point. The ground connection may be made to a water pipe system (if the pipe is not plastic), the grounded metal frame of a building, or to one or more 5/8-inch ground rods driven to a depth of at least 8 feet. More detailed information on lightning protection is contained in Chapter 1.

Antennas for 3.5 and 7 MHz

Multiband antennas constructed as described in Chapter 7 obviously will be useful on 3.5 and 7 MHz, and, in fact, the end-fed and center-fed antennas are quite widely used for 3.5- and 7-MHz operation. The center-fed system is better because it is inherently balanced on both bands and there is less chance for feeder radiation and RF feedback troubles, but either system will give a good account of itself. On these frequencies the height of the antenna is not too important, and anything over 35 feet will work well for "average" operation. This section is concerned principally with antennas designed for use on one band only.

HALF-WAVELENGTH ANTENNAS

An untuned or "flat" feed line is a logical choice on any band because the losses are low, but this generally limits the use of the antenna to one band. Where only single-band operation is wanted, the ½-λ antenna fed with untuned line is one of the most popular systems on the 3.5- and 7-MHz bands. If the antenna is a single-wire affair (or two conductors in parallel, as shown in Fig 4), its impedance is in the vicinity of 60 Ω. The most logical way to feed the antenna is with 72-Ω twin-lead or else 52- or 75-Ω coaxial line. The heavy-duty twin-lead and the coaxial line present support problems, but these can be overcome by using a small auxiliary pole to take the weight of the line. The line should come away from the antenna at right angles, and it can be of any length.

A folded dipole shows an impedance of 300 Ω, and so it can be fed directly with any length of 300-Ω TV line. The line should come away from the antenna at as close to a right angle as possible. The folded dipole can be made of ordinary wire spaced by lightweight wooden or plastic spacers, 4 or

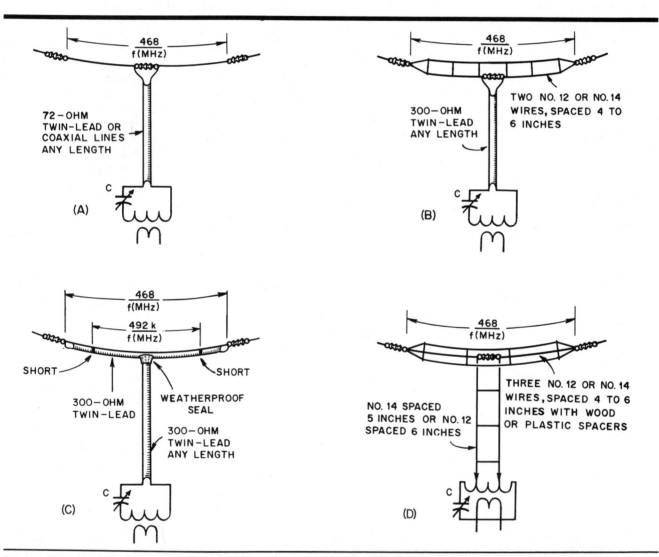

Fig 14—Half-wavelength antennas for single band operation. The multiwire types shown in B, C and D offer a better match to the feeder over a somewhat wider range of frequencies but otherwise the performances are identical. At C, k = velocity factor for the twin-lead. The feeder should run away from the antenna at a right angle for as great a distance as possible. In the coupling circuits shown, tuned circuits should resonate to the operating frequency. In the series-tuned circuits of A, B and C, high L and low C are recommended, and in D the inductance and capacitance should be similar to the output-amplifier tank, with the feeders tapped across at least ½ the coil. The tapped-coil matching circuit or the Transmatch, both shown in Chapter 25, can be substituted in each case.

6 inches long, or a piece of 300-Ω TV twin-lead.

A folded dipole can be fed with a 600-Ω open wire line with only a 2:1 SWR, but a nearly perfect match can be obtained with 600-Ω open line and a three-wire dipole.

The three types of ½-λ antennas just discussed are shown in Fig 14. One advantage of the two- and three-wire antennas over the single wire is that they offer a better match over a band. This is particularly important if full coverage of the 3.5-MHz band is contemplated.

While there are many other methods of matching lines to ½-λ antennas, the methods mentioned above are the most practical ones. It is possible, for example, to use a ¼-λ transformer of 150-Ω twin-lead to match a single-wire ½-λ antenna to 300-Ω feed line. But if 300-Ω feed line is to be used, a folded dipole offers an excellent match without the necessity for a matching section.

The formula shown above each antenna in Fig 14 can be used to compute the length at any frequency, or the length can be obtained directly from the charts in Fig 15.

Inverted-V Dipole

The halves of a dipole may be sloped to form an inverted V, as shown in Fig 16. This has the advantages of requiring only a single high support and less horizontal space. In theory, for sky-wave propagation the performance of a dipole in this form is essentially the same as a horizontal dipole. A number of amateurs report that an inverted-V dipole is more effective than a horizontal antenna, especially for frequencies of 7 MHz and lower.

Sloping of the wires results in a lowering of the resonant frequency and a decrease in feed-point impedance and bandwidth. Thus, for the same frequency, the length of the dipole must be decreased somewhat. The angle at the apex is not critical, although it should probably be made no smaller than 90°. Because of the lower impedance, a 52-Ω line should be used. For those who are dissatisfied with anything but a perfect match, the usual procedure is to adjust the angle for lowest SWR while keeping the dipole resonant by adjustment

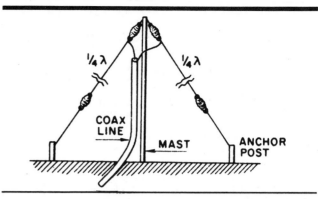

Fig 16—The inverted-V dipole. The length and apex angle should be adjusted as described in the text.

of length. Bandwidth may be increased by using multi-conductor elements, such as the cage configuration.

VERTICAL ANTENNAS

For 3.5-MHz work, the vertical can be ¼-λ long (if you can get the height), or it can be made shorter through the use of loading. The bottom of the antenna need clear the ground by only inches. Probably the cheapest construction of a ¼-λ vertical involves running copper or aluminum wire alongside a wooden mast. A metal tower can also be used as a radiator. If the tower is grounded, the antenna can be shunt-fed, as shown in B of Fig 17. The gamma matching system may also be used.

If the radiator is made of wire supported by non-conducting material, the approximate length for ¼-λ resonance can be found from

$$\ell_{ft} = 234/f_{MHz}$$

For tubing, the length for resonance must be shorter than

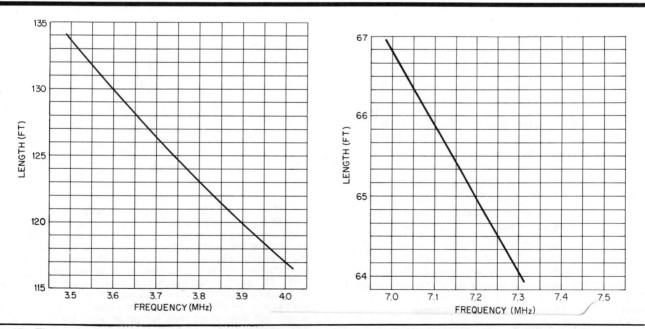

Fig 15—These charts can be used to determine the length of a ½-λ antenna of wire for the 3.5- or 7-MHz bands.

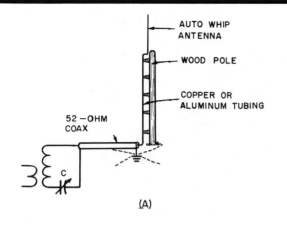

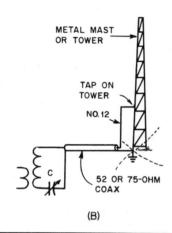

(A)

(B)

Fig 17—Vertical antennas are effective for 3.5- or 7-MHz work. The ¼-λ antenna shown at A is fed directly with 52-Ω coaxial line, and the resulting SWR is usually less than 1.5 to 1, depending on the ground resistance. If a grounded antenna is used as at B, the antenna can be shunt fed with either 52- or 75-Ω coaxial line. The tap for best match and the value of C will have to be found by experiment; the line running up the side of the antenna should be spaced 6 to 12 in. from the antenna. See text regarding the length (height) of the antenna.

given by the above equation, as the length to diameter ratio is lower than for wire (see Chapter 2). The following equation gives the approximate ¼-λ resonant length for tubing.

$$\ell_{ft} = 225/f_{MHz}$$

For a tower, the resonant length will be shorter still. In any case, after installation the antenna length (height) should be adjusted for resonance at the desired frequency.

A good ground system is essential in feeding a ¼-λ vertical antenna. The ground can be a number of radial wires extending out from the base of the antenna for about ¼ λ. Driven ground rods, while satisfactory for electrical safety and for lightning protection, are of little value as an RF ground for a vertical antenna except perhaps in marshy or beach areas or at 1.8 MHz. An elevated system of radials or a ground screen (counterpoise) may be used instead of radials at ground level, and will provide more efficient antenna.

The Groundplane Antenna

The size of a groundplane antenna makes it a little

impractical for 3.5-MHz work, but it can be used at 7 MHz to good advantage, particularly for DX work. This type of antenna can be placed higher above ground than an ordinary vertical without affecting the low-angle radiation.

The vertical member is a ¼-λ element and can be a length of self-supporting tubing at the top of a short mast. The resonant length depends on its length to diameter ratio, as described for verticals in the preceding section. The radials are also ¼-λ resonant elements. They can be lengths of wire used also to support the mast. Their length, too, depends on the length to diameter ratio. If the radiator is of tubing and the radials are of wire, the resonant radial length will be a few percent longer than that for the radiator. Probably this fact has given rise to the myth that radials should always be 5% longer than the radiator.

The radials do not have to be exactly horizontal, as shown in Fig 18. The ground-plane antenna can be fed directly with 52-Ω cable. With horizontal radials, the SWR will be in the neighborhood of 1.4 to 1. If the ends of the radials are lowered, as in Fig 18, the SWR will improve. However, the additional loss caused by an SWR as high as 2 to 1 will be inappreciable even in cable runs of several hundred feet when the frequency is as low as 7 MHz.

PHASED VERTICALS

Two or more vertical antennas spaced ½ λ apart can be operated as a single antenna system to obtain additional gain and a directional pattern. The following design for 7-MHz phased verticals is based on an April 1972 *QST* article by Gary Elliott, KH6HCM/W7UXP. A 3.5-MHz version can be constructed by proper scaling.

There are practical ways that verticals for 7 MHz can be combined, end-fire and broadside. In the broadside configuration, the two verticals are fed in phase, producing a figure-eight pattern that is broadside to the plane of the verticals. In an end-fire arrangement, the two verticals are

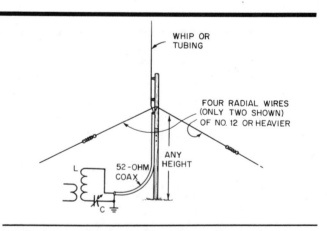

Fig 18—A groundplane antenna is effective for DX work on 7 MHz. Although its base can be any height above ground, losses in the ground underneath will be reduced by keeping the bottom of the antenna and the ground-plane as high above ground as possible. Feeding the antenna directly with 52-Ω coaxial cable will result in a low SWR. The vertical radiator and the radials are all ¼ λ long electrically. Contrary to popular myth, the radials need not necessarily be 5% longer than the radiator. Their physical length will depend on their length to diameter ratios, as discussed in text.

fed out of phase, and a figure-eight pattern is obtained that is in line with the two antennas, Fig 19. However, an end-fire pair of verticals can be fed 90° out of phase and spaced ¼ λ apart, and the resulting pattern will be unidirectional. The direction of maximum radiation is in line with the two verticals, and in the direction of the vertical receiving the lagging excitation; see Fig 20. This pattern, with a theoretically infinite F/B ratio, can be obtained only if the elements are excited by equal currents. (The feed method shown in Fig 20 is for illustration only, and does not produce the pattern shown there because the phases of the *induced* currents cause the feed impedances to be unequal. See Chapter 8 for detailed information on feeding phased verticals.) The feed system for this array is discussed in a subsequent section.

Construction

Physically, each vertical is constructed of telescoping aluminum tubing that starts off at 1½ inches dia and tapers down to ¼ inch dia at the top. The length of each vertical is 32 feet. Each vertical is supported on two standoff insulators set on a 2 × 4, 6 feet long, and strapped to a fence. An alternative method of mounting would be a 2 × 4 about 8 feet long and set about 2 feet in the ground.

Originally each vertical element was 32 feet, 6 inches long, 234/f (MHz). After one vertical was mounted on the 2 × 4 it was raised into position and the resonant frequency was checked with an antenna noise bridge. It was found that the vertical resonated too low in frequency, about 6.9 MHz. This was to be expected, as the equation for the ¼-λ vertical, 234/f, is only reasonably correct for very small diameter tubing or wire. When larger diameter tubing is used, the physical length will be shorter than this, as described earlier

in this chapter and in Chapter 2. While using the antenna noise bridge, an inch at a time was cut from the top until the resonant frequency was 7.1 MHz. This resulted in 6 inches being cut off, thus making the vertical exactly 32 feet long.

The ground system is very important in the operation of a vertical. The two usual methods of obtaining a ground system with verticals are shown in Fig 21. The radials are more effective than a single ground rod.

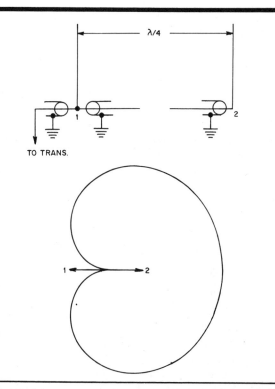

Fig 20—Pattern for two ¼-λ verticals spaced ¼ λ apart and fed 90° out of phase. The arrow represents the axis of the elements, with the element on the right being the one of lagging phase. The feed method shown here is for illustration only. See text and Fig 22.

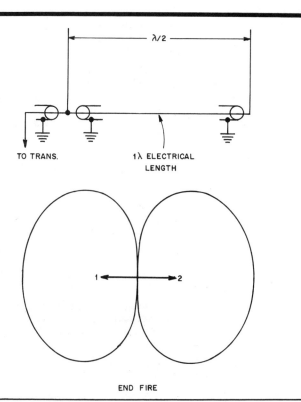

Fig 19—Pattern for two ¼-λ verticals spaced ½ λ apart and fed 180° out of phase. The arrow represents the axis of the elements.

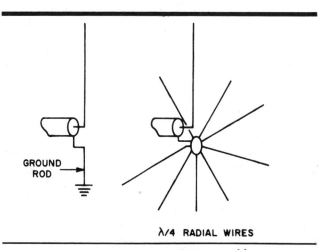

Fig 21—An 8- to 10-ft ground rod may provide a satisfactory ground system in marshy or beach areas, but in most locations a system of radial wires will be necessary.

Feed System

In order to obtain the unidirectional pattern shown in Fig 20, the two verticals must be separated by ¼ λ, and one vertical must be fed 90° delayed in phase as compared to the other. A suggested feed method is shown in Fig 22. The line lengths are calculated to compensate for the shifts in element impedances in the presence of one another, and will result in the proper current amplitude and phase at the base of each element to produce the pattern shown if a few radials are used. The length of an electrical line section in feet is based on the calculation

$$\ell = \frac{984 \times VF \times n}{7.1 \text{ MHz} \times 360°}$$

where n is the number of electrical degrees.

Using a velocity factor (VF) of 0.66, the calculated line length from the T connector to element 1 becomes 21 feet 4 inches, and 40 feet 11 inches to element 2. (Further information concerning velocity factor and transmission lines can be found in Chapter 24.) The SWR at 7.1 MHz is approximately 2:1, using 52-Ω coax and no matching network.

PHASED HORIZONTAL ARRAYS

Phased arrays with horizontal elements can be used to advantage at 7 MHz, if they can be placed at least 40 feet above ground. Any of the usual combinations will be effective. If a bidirectional characteristic is desired, the W8JK array, shown in Fig 23A, is a good one. If a unidirectional characteristic is required, two elements can be mounted about 20 feet apart and provision included for tuning one of the elements as either a director or reflector, as shown in Fig 23B. The parasitic element is tuned at the end of its feed line with a series- or parallel-tuned circuit (whichever would normally be required to couple power into the line), and the proper tuning condition can be found by using the system for receiving and listening to distant stations along the line of maximum radiation of the antenna. Tuning the feeder to the parasitic element will peak up the signal.

7-MHz "SLOPER" SYSTEM

One of the more popular antennas for 3.5 and 7 MHz

is the sloping dipole. David Pietraszewski, K1WA, made an extensive study of sloping dipoles at different heights with reflectors at the 3-GHz frequency range. From his experi-

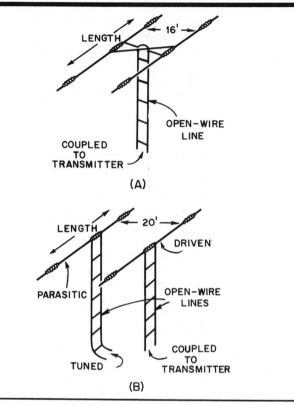

Fig 23—Directional antennas for 7 MHz. To realize any advantage from these antennas, they should be at least 40 feet high. The system at A is bidirectional, and that at B is unidirectional in a direction depending upon the tuning conditions of the parasitic element. The length of the elements in either antenna should be exactly the same, but any length from 60 to 150 feet can be used. If the length of the antenna at A is between 60 and 80 feet, the antenna will be bidirectional along the same line on both 7 and 14 MHz. The system at B can be made to work on 7 and 14 MHz in the same way, by keeping the length between 60 and 80 ft.

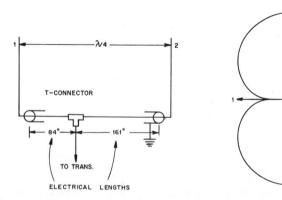

Fig 22—A method of feeding the phased verticals. The SWR in 50-Ω line is approximately 2:1.

ments, he developed the novel 7-MHz antenna system described here. With several sloping dipoles supported by a single mast and a switching network, an antenna with directional characteristics and forward gain can be simply constructed. This 7-MHz system uses several "slopers" equally spaced around a common center support. Each dipole is cut to ½ λ and fed at the center with 52-Ω coax. The length of each feed line is 36 feet. This length is just over 3/8 λ, which provides a useful quality. All of the feed lines go to a common point on the support (tower) where the switching takes place. At 7 MHz, the 36-foot length of coax looks inductive to the antenna when the end at the switching box is open circuited. This has the effect of adding inductance at the center of the sloping dipole element, which electrically lengthens the element. The 36-foot length of feed line serves to increase the length of the element about 5%. This makes any unused element appear to be a reflector.

The array is simple and effective. By selecting one of the slopers through a relay box located at the tower, the system becomes a parasitic array which can be electrically rotated. All but the driven element of the array become reflectors.

The physical layout is shown in Fig 24, and the basic materials required for the sloper system are shown in Fig 25. The height of the support point should be about 70 feet, but

Fig 25—The basic materials required for the sloper system. The control box appears at the left, and the relay box at the right.

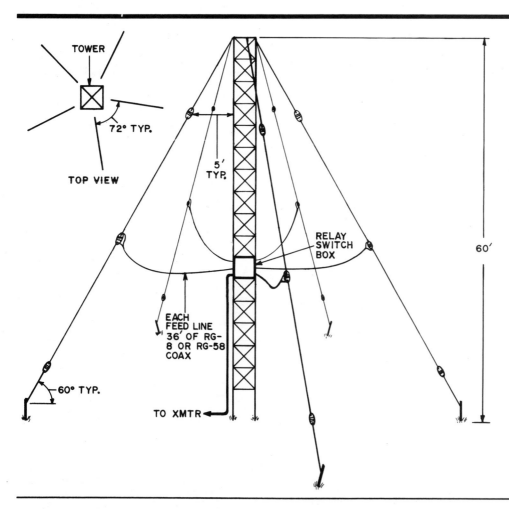

Fig 24—Five sloping dipoles suspended from one support. Directivity and forward gain can be obtained from this simple array. The top view shows how the elements should be spaced around the support.

can be less and still give reasonable results. The upper portion of the sloper is 5 feet from the tower, suspended by rope. The wire makes an angle of 60° with the ground. In Fig 26, the switch box is shown containing all the necessary relays to select the proper feed line for the desired direction. One feed line is selected at a time and the feed lines of those remaining are opened, Fig 27. In this way the array is electrically rotated. These relays are controlled from inside the shack with an appropriate power supply and rotary switch. For safety reasons and simplicity, 12 volt dc relays are used. The control line consists of a five-conductor cable, one wire used as a common connection; the others go to the four relays. By using diodes in series with the relays and a dual-polarity power supply, the number of control wires can be reduced, as shown in Fig 27B.

Measurements indicate that this sloper array provides up to 20 dB front-to-back ratio and forward gain of about 4 dB. If one direction is the only concern, the switching system can be eliminated and the reflectors should be cut 5 percent longer than the resonant frequency. The one feature which is worth noting is the good F/B ratio. By arranging the system properly, a null can be placed in an unwanted direction, thus making it an effective receiving antenna. In the tests conducted with this antenna, the number of reflectors used were as few as one and as many as five. The optimum combination appeared to occur with four reflectors and one driven element. No tests were conducted with more than five reflectors. This same array can be scaled to 3.5 MHz for similar results.

Fig 26—Inside view of relay box. Four relays provide control over five antennas. See text. The relays pictured here are Potter and Brumfeld type MR11D.

THE QUARTER-WAVELENGTH "HALF SLOPER"

Perhaps one of the easiest antennas to install is the ¼-λ sloper, Fig 28. A sloping ½-λ dipole is known among radio amateurs as a "full sloper" or "sloper." If only one half of

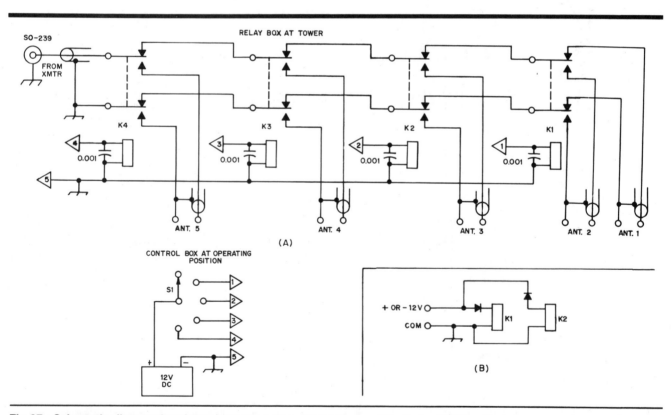

Fig 27—Schematic diagram for sloper control system. All relays are 12 volt dc, DPDT, with 8-A contact ratings. In A, the basic layout, excluding control cable and antennas. Note that the braid of the coax is also open circuited when not in use. Each relay is bypassed with 0.001-μF capacitors. The power supply is a low current type. In B, diodes are used to reduce the number of control wires when using dc relays. See text.

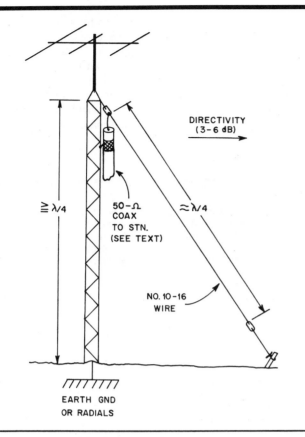

DIRECTIVITY
(3 - 6 dB)

$\geqq \lambda/4$

$\approx \lambda/4$

50-Ω
COAX
TO STN.
(SEE TEXT)

NO. 10-16
WIRE

EARTH GND
OR RADIALS

Fig 28—The ¼-λ half sloper antenna.

it is used it becomes a "half sloper." The performance of the two types of sloping antennas is similar: They exhibit some directivity in the direction of the slope and radiate energy at low angles respective to the horizon. The wave polarization is vertical. The amount of directivity will range from 3 to 6 dB, depending upon the individual installation, and will be observed in the slope direction.

The advantage of the half sloper over the full sloper is that the current portion of the antenna is higher. Also, only half as much wire is required to build the antenna for a given amateur band. The disadvantage of the half sloper is that it is sometimes impossible to obtain a low SWR when using coaxial-cable feed. This perplexing phenomenon is brought about by the manner in which the antenna is installed. Factors that affect the feed impedance are tower height, height of the attachment point, enclosed angle between the sloper and the tower, and what is mounted atop the tower (HF or VHF beams). Also the quality of the ground under the tower (ground conductivity, radials, etc) has a marked effect on the

antenna performance. The final SWR can vary (after optimization) from 1:1 to as high as 6:1. Generally speaking, the closer the low end of the slope wire is to ground, the more difficult it will be to obtain a good match.

Basic Recommendations

This excellent DX type of antenna is usually installed on a metal supporting structure such as a mast or tower. The support needs to be grounded at the lower end, preferably to a buried or on-ground radial system. If a nonconductive support is used, the outside of the coax braid becomes the return circuit and should be grounded at the base of the support. As a starting point one can attach the sloper so the feed point is approximately ¼ λ above ground. If the tower is not high enough to permit this, the antenna should be fastened as high on the supporting structure as possible. Start with an enclosed angle of approximately 45°, as indicated in Fig 28. The wire may be cut to the length determined from

$$\ell_{ft} = 260/f_{MHz}$$

This will allow sufficient extra length for pruning the wire for the lowest indicated SWR.

A metal tower or mast becomes an operating part of the half sloper system. In effect it and the slope wire function somewhat like an inverted-V dipole antenna. In other words, the tower operates as the missing half of the dipole. Hence its height and the top loading (beams) play a significant role.

The 52-Ω transmission line can be taped to the tower leg at frequent intervals to make it secure. The best method is to bring it to earth level, then route it to the operating position along the surface of the ground if it can't be buried. This will ensure adequate RF decoupling, which will help prevent RF energy from affecting the equipment in the station. Rotator cable and other feed lines on the tower or mast should be treated in a similar manner.

Adjustment of the half sloper is done with an SWR indicator in the 52-Ω transmission line. A compromise can be found between the enclosed angle and wire length, providing the lowest SWR attainable in the center of the chosen part of an amateur band. If the SWR "bottoms out" at 2:1 or lower, the system will work fine without using a Transmatch, provided the transmitter can work into the load. Typical optimum values of SWR for 3.5- or 7-MHz half slopers are between 1.3:1 and 2:1. A 100-kHz bandwidth is normal on 3.5 MHz, with 200 kHz being typical at 7 MHz. If the lowest SWR possible is greater than 2:1, the attachment point can be raised or lowered to improve the match. Readjustment of the wire length and enclosed angle may be necessary when the feed-point height is changed.

If the tower is guyed, the guy wires will need to be insulated from the tower and broken up with additional insulators to prevent resonance.

THE KØEOU BROADBAND SLOPER

Brian L. Wermager, KØEOU, described a broadband sloper antenna in April 1986 *QST*. This antenna covers the entire 3.5- to 4-MHz band, and is good for working DX. The feed arrangement is shown in Fig 29. A second wire is connected directly to the tower as shown in Fig 30. Wermager found that the SWR was below 1.2 across the entire band (see Fig 31).

You don't need to have a 70-foot tower to build this sloper. A 40-foot tower should be high enough. Kelly Davis, KD7XY built one of these antennas on his 50-foot tower. Measurements for his antenna are shown in Fig 32.

The antenna is nearly omnidirectional. An analysis done by Jerry Hall, K1TD, using the computer program MININEC (Mini-Numerical Electronics Code), yields the pattern diagrams in Fig 33. These close approximations show that the

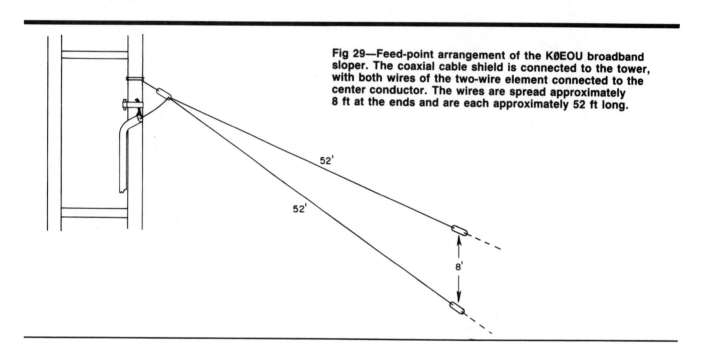

Fig 29—Feed-point arrangement of the KØEOU broadband sloper. The coaxial cable shield is connected to the tower, with both wires of the two-wire element connected to the center conductor. The wires are spread approximately 8 ft at the ends and are each approximately 52 ft long.

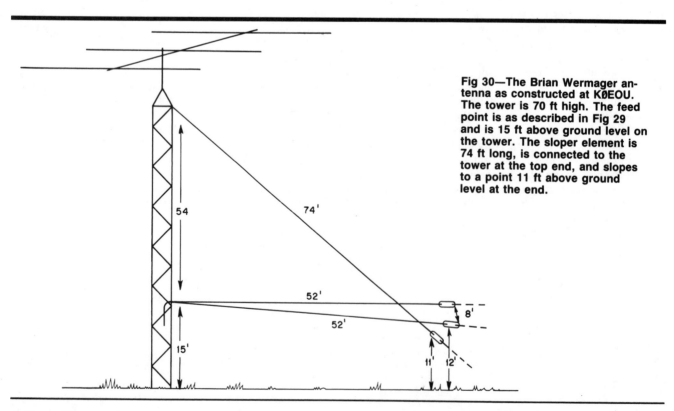

Fig 30—The Brian Wermager antenna as constructed at KØEOU. The tower is 70 ft high. The feed point is as described in Fig 29 and is 15 ft above ground level on the tower. The sloper element is 74 ft long, is connected to the tower at the top end, and slopes to a point 11 ft above ground level at the end.

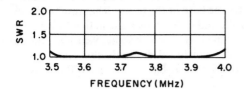

Fig 31—SWR measurements for the antenna at KØEOU. The highest SWR measurement between 3.5 and 4.0 MHz is 1.2, but this may be an indication of appreciable earth losses.

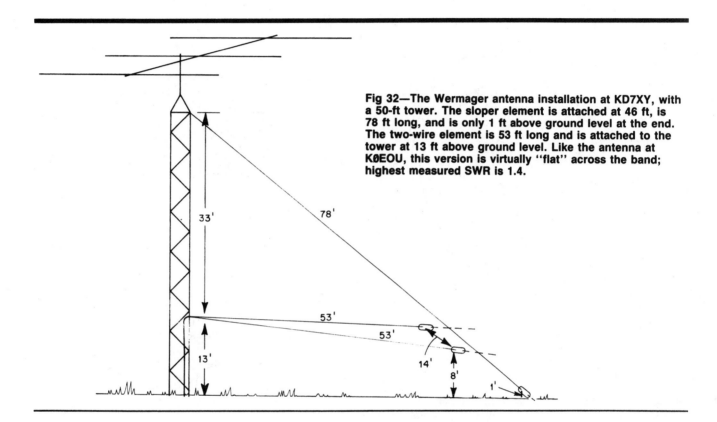

Fig 32—The Wermager antenna installation at KD7XY, with a 50-ft tower. The sloper element is attached at 46 ft, is 78 ft long, and is only 1 ft above ground level at the end. The two-wire element is 53 ft long and is attached to the tower at 13 ft above ground level. Like the antenna at KØEOU, this version is virtually "flat" across the band; highest measured SWR is 1.4.

radiation polarization is primarily vertical. The analysis also shows that the point of greatest current is between the tower and ground. This indicates that a good earth connection, and even a radial system, would offer highest efficiency.

A version of the Wermager sloper similar to the one shown in Fig 30 was built at K8CH where it was tested for a year. After the year-long test, 40 radials were extended from, and bonded to, the base of the tower. That resulted in a slight increase in SWR and a slight decrease in bandwidth after the antenna was retuned. However, another year-long test showed that performance improved dramatically. The antenna still covers the entire 3.5-MHz band with an SWR of less than 2.

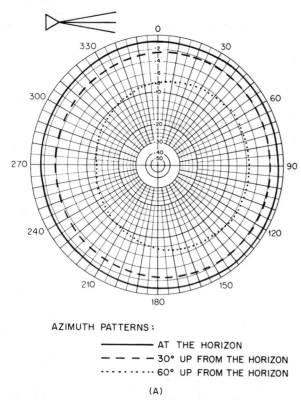

AZIMUTH PATTERNS:

——————— AT THE HORIZON

— — — — 30° UP FROM THE HORIZON

· · · · · · · 60° UP FROM THE HORIZON

(A)

Fig 33—Antenna radiation patterns for the KØEOU broadband sloper as calculated using MININEC. Values are in dBi; add 6 dB to the values shown. The insets with each pattern (upper left) portray the antenna in relation to the pattern. At A; the azimuth pattern. At B; the elevation pattern in the direction of the wires. At C; the elevation pattern broadside to the wires. All patterns were calculated for a perfectly conducting ground.

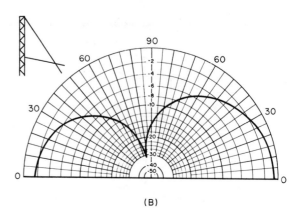

(B)

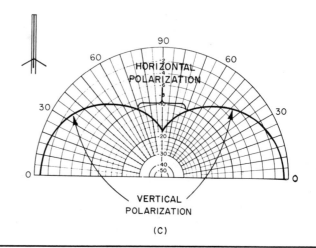

(C)

Antennas for 1.8 MHz

Any antenna or array commonly used on the higher frequencies is suitable for the 1.8-MHz band. However, practical considerations with regard to height and size usually limit the selection to a few basic types. These are the dipole, vertical, end-fed wire, loop and various combinations of these four. Further compromises are often necessary since even these antennas are still quite large. As the size and height decrease, so does the radiating effectiveness unless particular care is taken to reduce losses to a minimum. The most significant losses result from induced ground currents, conductor resistance, losses in matching networks and loading coils, and absorption of RF energy by surrounding objects.

The type of antenna installation finally selected is often dictated by those losses most easily eliminated. For example, vertical antennas are usually considered the most desirable ones to use on 1.8 MHz, but if a suitable ground system is not feasible the ground losses will be very high. In such a case, an ordinary dipole may give superior performance even though the angle of maximum radiation is farther from optimum than that of a vertical. Some experimentation is often necessary to find the best system, and the purpose of this section is to aid the reader in selecting the best one for his particular station.

PROPAGATION ON 1.8 MHz

While important, propagation characteristics on 1.8 MHz are secondary to system losses, as losses may offset any attempt to optimize for the angle of radiation. Generally speaking, the 1.8-MHz band has similar properties to those of the AM broadcast band (550 to 1600 kHz) but with greater significance of the sky wave. In this respect, it is not unlike the higher amateur frequencies such as 3.5 MHz, and most nighttime contacts over distances of a few hundred miles on 1.8 MHz will be by sky-wave propagation. During the daytime, absorption of the sky wave in the D region is almost complete, but reliable communication is still possible by means of the ground wave.

With respect to sky-wave transmission, 1.8-MHz waves entering the ionosphere, even vertically, are reflected to earth, so that there is no such phenomenon as skip distance on these frequencies. However, as at higher frequencies, to cover the greatest possible distance, the waves must enter the ionosphere at low angles.

POLARIZATION

It is mentioned in Chapter 23 that a ground wave must be vertically polarized, so the radiation from an antenna that is to produce a good ground wave likewise must be vertically polarized. This dictates the use of an antenna system of which the radiating part is mostly vertical. Horizontal polarization will produce practically no ground wave, and it is to be expected that such radiation will be ineffective for daytime communication. This is because absorption in the ionosphere in the daytime is so high at these frequencies that the reflected wave is too weak to be useful. At night a horizontal antenna will give better results than it will during the day. Ionospheric conditions permit the reflected wave to return to earth with less attenuation.

Some confusion over the term *ground wave* exists, since there are a number of propagation modes that go by this name. Here, only the type that travels over and near a conducting surface will be considered. If the surface is flat and has a very high conductivity, the attenuation of the wave follows a simple inverse-distance law. That is, every time the distance is doubled, the field strength drops by 6 dB. This law also holds for spherical surfaces for some distance, and then the field strength drops quite rapidly. For the earth, the break point is approximately 100 miles.

The conductivity of the surface is an important factor in ground-wave propagation. For example, sea water can be considered as almost a perfect conductor for this purpose, at frequencies well into the HF range. However, there may be as much as an additional 20 dB of attenuation for a 1- to 10-mile path over poorly conducting earth, compared to an equivalent path over the sea. The conductivity of sea water is roughly 400 times as great as good-conducting land (agricultural regions) and 4000 times better than poor land (cities and industrial areas).

After sundown, the propagation depends upon both the ground wave and sky wave. At the limit of the ground-wave region, the two may have equal field strengths and may either aid or cancel each other. The result is severe and rapid fading in this zone. While of less importance in amateur applications, this effect limits the useful nighttime range of broadcast stations. Antenna designs have been developed over the years which minimize sky-wave radiation and maximize the ground wave. For broadcast work, a vertical antenna of 0.528-λ height is optimum over a good ground system. However, caution should be exercised in applying this philosophy to amateur installations since effective antenna systems, even for DX work, are possible with relatively high angles of radiation.

REDUCING ANTENNA SYSTEM LOSSES

As the length of an antenna becomes small compared with the wavelength being used, the radiation resistance, R_a, drops to a very low value, as discussed in Chapter 2. The various losses can be represented by a resistance, R_L, in series with R_a. R_L may be larger than R_a in practical cases. Therefore, in an antenna system with high losses, most of the applied power is dissipated in the loss resistance and little is radiated in R_a. Since R_a is mostly dependent upon antenna construction, efforts to reduce the loss resistance will normally not affect the radiation resistance. Efficiency can be improved significantly by keeping the loss resistance as low as possible.

The simplest losses to reduce are the conductor losses. Since electrically short antennas exhibit a large series capacitive reactance, a loading coil is commonly used to tune out the reactance. If not part of the radiating system, the coil should have as high a Q as possible.

Incorporating the loading coil into the radiating system not only simplifies loading coil construction, but may actually increase the efficiency by redistributing the current in the antenna. Such loading coils are designed for low loss, rather than high Q. The reason for this is that one of the parameters resulting in low coil Q is radiation. But radiation is exactly the desired result in an antenna system. The radiation from a coil increases as its length to diameter ratio increases. In some instances, the entire antenna may consist of a single coil. A helically wound vertical is an example of this type. In any of the loading coils that are part of the radiating system, the conductor diameter should be as large as possible, and very close spacing between turns should be avoided.

The effect of the earth on antenna loss can best be seen by examination of Fig 34A. If a vertical radiator that is short compared with a wavelength is placed over a ground plane, the antenna current will consist of two components. Part of the current flows through C_W, which is the capacitance of the vertical to the radial wires, and part flows through C_E, the capacitance of the vertical to the earth. For a small number of radials, C_E will be much greater than C_W, and most of the current will flow through the circuit consisting of C_E and R_E (the earth resistance). Power will be dissipated in R_E which will not contribute to the radiation. The solution to the problem is to increase the number of radials. This will increase C_W, but, of more importance, will reduce R_E by providing more return paths through wire. Theory and experiments have shown that the ideal radial system with a 0.528-λ vertical consists of approximately 120 radials, each ½-λ long. If fewer radials are used (for example, 16), little is to be gained by running them out so far. The converse is also true. If space restricts the length of the radials, increasing the number much over 16 will have little effect for an antenna of this height. Since the current is greatest near the base of the antenna, a ground screen will also help if only a few radials are used.

Another method to reduce the ground currents is shown in Fig 34B. By raising the antenna and ground plane off the earth, C_W stays the same in value but C_E is considerably reduced (such a system is sometimes called a counterpoise). This decreased influence of the earth is also the reason why as few as two radials are sufficient for HF and VHF groundplane antennas that are several feet above the earth.

The simple lumped antenna-capacitance analysis is a good approximation to actual operation if the vertical is electrically short, but analysis becomes more complicated for greater antenna lengths. For instance, the maximum ground

loss for the 0.528-λ broadcast vertical mentioned earlier occurs at a point 0.35 λ away from the base.

Location of the antenna is perhaps more critical with regard to receiving applications than transmitting ones. Sources of strong local noise, such as TV sets and power lines, can cause considerable difficulty on 1.8 MHz. However, the proximity of RF-absorbing objects such as steel buildings may cut down on transmitting efficiency also. Since most installations are tailored to the space available, little can be done about the problem except to see that the other losses are kept to a minimum.

ANTENNA TYPES

A common misconception is that antennas for 1.8 MHz have to be much larger, higher and more elaborate than those for the higher bands. When one considers that even the gigantic antennas used for VLF work have radiating efficiencies of approximately 1%, it is not surprising that many contacts on 1.8 MHz can be made with little more than a piece of wire a few feet off the ground, or even from mobile installations. While it is true that a larger and more sophisticated system may perform better than a smaller one, the point here is that space restrictions should not discourage the use of the band.

Verticals

One of the most useful antennas for 1.8 MHz is a vertical radiator over a ground plane. A typical installation is shown in Fig 35. Some form of loading should be used, since economics would not justify a full-sized ¼-λ vertical. One of the disadvantages of the vertical is the necessity for a good ground system. For 1.8-MHz antennas, some improvement has been noted by using a combination of radials and ground

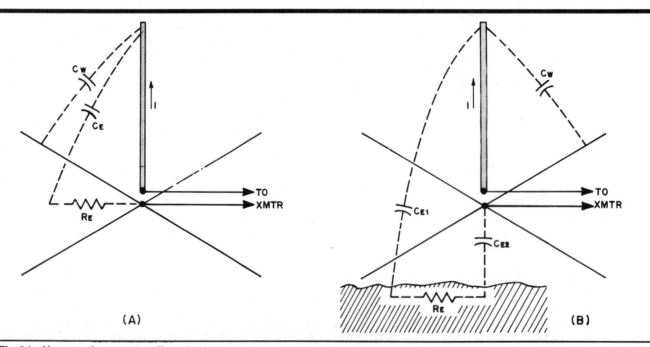

Fig 34—How earth currents affect the losses in a short vertical antenna system. At A, the current through the combination of C_E and R_E may be appreciable if C_E is much greater than C_W, the capacitance of the vertical to the ground wires. This ratio can be improved (up to a point) by using more radials. By raising the entire antenna system off the ground, C_E (which consists of the series combination of C_{E1} and C_{E2}) is decreased while C_W stays the same. The radial system shown at B is sometimes called a counterpoise.

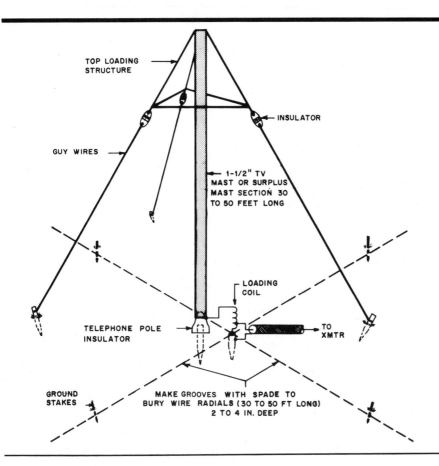

Fig 35—Physical layout of a typical vertical antenna suitable for 1.8-MHz operation. Without the top-loading structure or capacitance hat, the radiation resistance is approximately 1 Ω for a 30-ft vertical, and 3.3 Ω for a 50-ft height. The loading inductance for the 30-ft vertical is approximately 400 μH. Once the antenna is tuned approximately to resonance with the base loading coil, a suitable tap near the low end of the coil can be found which will give the best match for the transmitter. The radiation resistance can be increased by the use of a top-loading structure consisting of the guy wires (broken up near the top by insulators) which are interconnected by a horizontal wire, as shown. The radial system consists of wires buried a few inches underground.

rods where full-length radials were impractical. The exact configuration will vary from one installation to another, and the optimum placement of the ground rods will have to be determined by experimentation.

For verticals less than an electrical ¼ λ in height, the input reactance without loading will be capacitive. A simple series loading coil should be used to tune this reactance out and the coil may be the only matching network necessary.

Normally, matching to the feed line or transmitter can be accomplished with simple networks or a Transmatch. However, a unique method that also improves the radiating efficiency is used with certain VLF antennas. The technique, called multiple tuning, is illustrated in Fig 36A. A series of verticals is fed through a common flat-top structure with one of the "downleads" also acting as a feed point.

If the verticals are closely spaced (in comparison with a wavelength), the entire system can be considered to be one vertical with N times the current of one of the downleads taken alone. The result is that the radiation resistance is $N^2 \times R_a$, where R_a is the radiation resistance of a single vertical. This is the same principle as acquiring an impedance step-up in a multiconductor folded dipole. If the ground losses are also considered, the effective loss resistance (R_L) would also be transformed by the same amount. However, since the current distribution in the ground is usually improved by using this method, the ratio of R_a/R_L is also improved. The disadvantages of the system are increased complexity and difficulty of adjustment. While little if any use of this principle has been applied to amateur systems for 1.8 MHz, it offers some interesting possibilities where a good ground system is

impractical. The construction approach shown at B of Fig 36 may be used for the erection of an experimental antenna of this type.

Horizontal Antennas

In cases where a good ground system is not practical and when most of the operation will rely on sky-wave propagation, horizontal antennas can be used (see Fig 37). The relative simplicity of construction of an end-fed wire antenna makes it an attractive one for portable operation or where supporting structures are without much height.

As is the case with electrically short verticals, the input impedance of horizontal end-fed antennas less than ¼ λ long can be considered to be a resistance in series with a capacitive reactance. Matching networks for the end-fed wire are identical to those used for verticals.

Balanced center-fed antennas are also useful, even though they may be electrically short for 1.8 MHz and at heights typical of those used at the higher bands. For example, a 3.5-MHz dipole fed with open-wire line may also be used on 1.8 MHz with the appropriate matching network at the transmitter. Care should be taken to preserve the balanced configuration of the dipole in matching to this type. If one side of the feed line is connected to ground, part of the return circuit may be through the power line. This increases the chances of interference from appliances such as TV sets and fluorescent lamps on the same circuit. Also, since there are usually connections on the power service that are not soldered, rectification may take place. The result may be mixing of local broadcast stations with products on 1.8 MHz. Filtering will

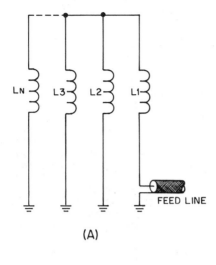

(A)

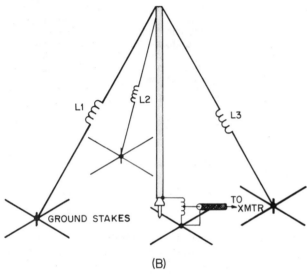

(B)

Fig 36—Possible configurations for a multiple-tuned vertical antenna for 1.8 MHz. Used extensively in VLF systems, little experimentation has been performed with it by amateurs. The principle is similar to that of the folded dipole where an impedance transformation occurs from a lower to higher value, simplifying matching. The ratio is equal to N^2, where N is the number of elements. In the system shown at B, the step-up ratio would be 16, since the total number of elements is four. The exact values of the loading inductors should be found experimentally, being such that the current in each leg is the same.

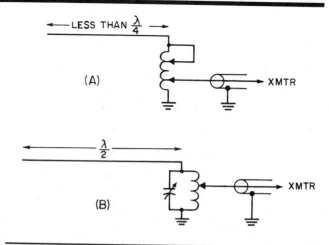

Fig 37—Two matching networks suitable for use with random-length horizontal (or vertical) wire antennas. If the electrical length is less than ¼ λ, the input impedance will be equivalent to a resistance in series with a capacitive reactance, and the circuit at A should be used. For lengths in the vicinity of ½ λ, the input impedance is fairly high and may have reactances that are either inductive or capacitive. For this case, the parallel-tuned circuit of B should be used.

not eliminate the problem because the products are in the same band with the desired signals. Problems of this type are usually less severe as the electrical length of the dipole approaches ½ λ.

Combinations of Vertical and Horizontal Antennas

The L and T antennas are the most common examples where combinations of horizontal and vertical radiators can be used to advantage. Various types are shown in Fig 38. Here, the philosophy is usually to run the vertical portion up as high as possible with the horizontal part merely acting as a top-loading structure. Such a system can be considered as equivalent to a vertical, and performance should be improved by the use of a ground system. Running the horizontal portion out to great distances may or may not improve the performance, unless the height is also increased.

A dipole fed with coaxial cable for a higher frequency band can be used as a T antenna, by tying the feed-line conductors together at the transmitter. This will also work with dipoles fed with open-wire line; however, they may work

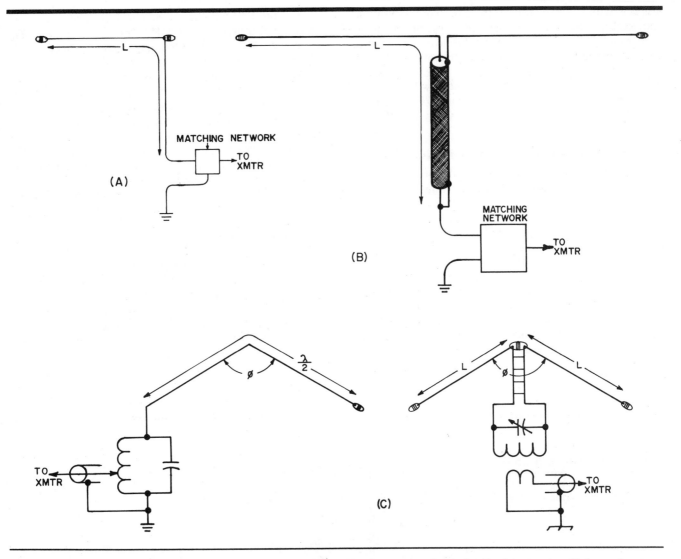

Fig 38—L, T and inverted-V dipole antennas. The type of matching network suitable for the L antenna will depend upon the length, L, and is the same for a straight horizontal antenna (see Fig 37). By tying the feed-line conductors together, an HF-band dipole can be used as a T for 1.8 MHz. The exact form that the matching network will take depends on the lengths of both the horizontal and vertical portions. At B, considering the length of only one leg should be sufficient for the majority of cases, and the equivalent L-antenna network can be used. The arrangement at C shows two different methods of feeding an inverted V. In either case, the apex angle, ϕ, should be greater than 90°.

just as well (or better) by using them in the more conventional manner discussed earlier. The inverted-V type of antenna has also given good results on 1.8 MHz. While the center of the antenna should be as high as possible, the total angle of the V should be greater than 90° at the apex. This will be determined by the height of the apex and how high the ends of the antenna are located above the ground. For angles less than 90°, the radiation efficiency drops very rapidly.

1.8 MHz ANTENNA SYSTEMS USING TOWERS

An existing metal tower used to support HF or VHF beam antennas can also be used as an integral part of a 1.8-MHz radiating system. The ¼-λ sloper or the half sloper discussed in other sections of this chapter will also perform well on 1.8 MHz. Those prominent 1.8-MHz operators who have had success with the half sloper antenna suggest a minimum tower height of 50 feet. Dana Atchley, W1CF, uses the configuration sketched in Fig 39. He reports that the uninsulated guy wires act as an effective counterpoise for the sloping wire. At Fig 40 is the feed system used by Doug DeMaw, W1FB, on a 50-foot self-supporting tower. The ground for the W1FB system is provided by buried radials connected to the tower base.

In April 1986 *QST*, Deane J. Yungling, KI6O, described a linear-loaded sloper for 1.8 MHz. This sloper, shown in Fig 41, has a 2:1 SWR bandwidth of approximately 70 kHz. This antenna is a derivative of the ¼ λ half sloper antenna.

A tower can also be used as a true vertical antenna, provided a good ground system is available. The shunt-fed tower is at its best on 1.8 MHz, where a full ¼-λ vertical antenna is rarely possible. Almost any tower height can be used. If the beam structure provides some top loading, so much the better—but anything can be made to radiate, if it is fed properly. W5RTQ uses a self-supporting, aluminum, crank-up, tilt-over tower, with a TH6DXX tribander mounted at 70 feet. Measurements showed that the entire structure has about the same properties as a 125-foot vertical. It thus works quite well as an antenna on 1.8 and 3.5 MHz for DX work requiring low-angle radiation.

Preparing the Structure

Usually some work on the tower system must be done before shunt-feeding is tried. Metallic guys should be broken up with insulators. They can be made to simulate top loading, if needed, by judicious placement of the first insulators. Don't overdo it; there is no need to "tune the radiator to resonance" in this way. If the tower is fastened to a house at a point more

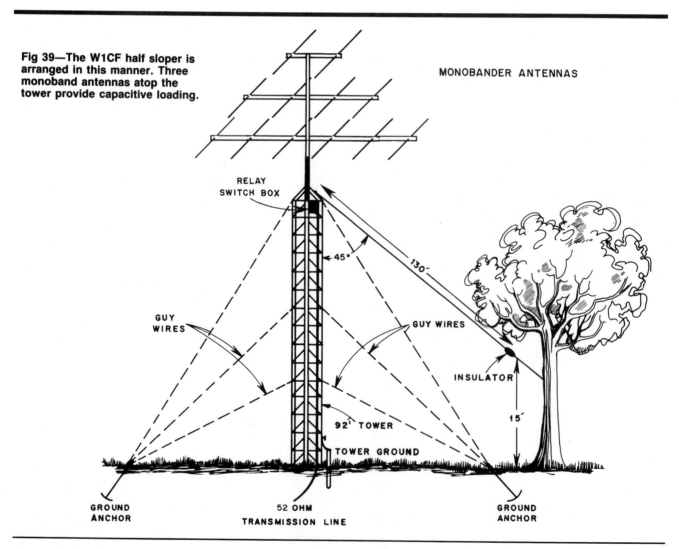

Fig 39—The W1CF half sloper is arranged in this manner. Three monoband antennas atop the tower provide capacitive loading.

MONOBANDER ANTENNAS

RELAY SWITCH BOX

45°

130'

GUY WIRES

GUY WIRES

INSULATOR

15'

92' TOWER

TOWER GROUND

GROUND ANCHOR

52 OHM TRANSMISSION LINE

GROUND ANCHOR

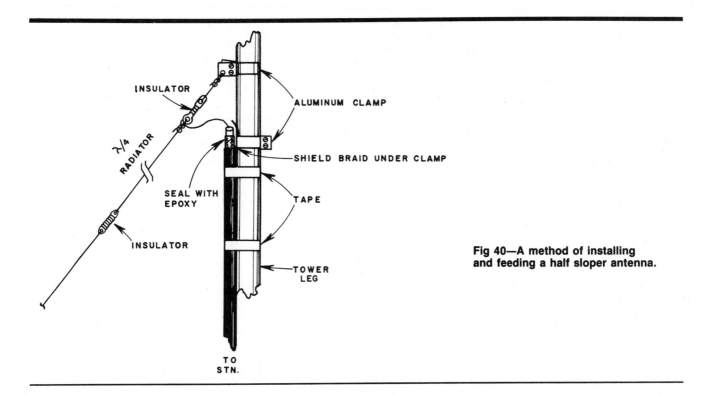

INSULATOR

ALUMINUM CLAMP

λ/4 RADIATOR

SHIELD BRAID UNDER CLAMP

SEAL WITH EPOXY

TAPE

INSULATOR

TOWER LEG

TO STN.

Fig 40—A method of installing and feeding a half sloper antenna.

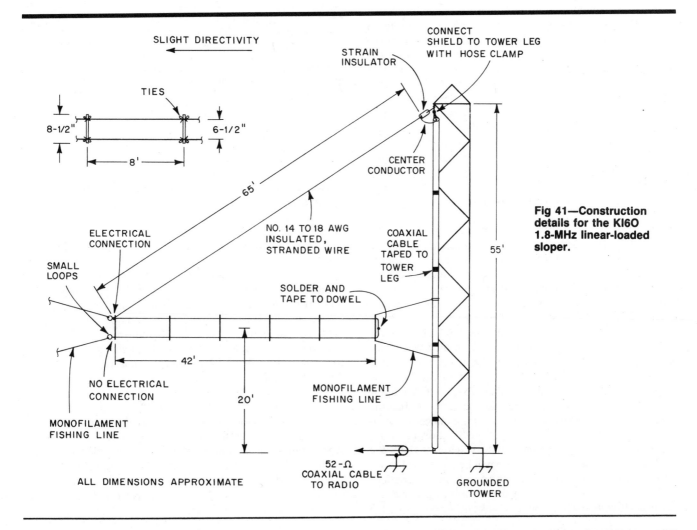

SLIGHT DIRECTIVITY

CONNECT SHIELD TO TOWER LEG WITH HOSE CLAMP

STRAIN INSULATOR

TIES

8-1/2"

6-1/2"

8'

CENTER CONDUCTOR

65'

ELECTRICAL CONNECTION

NO. 14 TO 18 AWG INSULATED, STRANDED WIRE

COAXIAL CABLE TAPED TO TOWER LEG

55'

SMALL LOOPS

SOLDER AND TAPE TO DOWEL

NO ELECTRICAL CONNECTION

42'

20'

MONOFILAMENT FISHING LINE

MONOFILAMENT FISHING LINE

52-Ω COAXIAL CABLE TO RADIO

GROUNDED TOWER

ALL DIMENSIONS APPROXIMATE

Fig 41—Construction details for the KI6O 1.8-MHz linear-loaded sloper.

than about one-fourth of the height of the tower, it may be desirable to insulate the tower from the building. Plexiglas sheet, ¼ inch or more thick, can be bent to any desired shape for this purpose, if it is heated in an oven and bent while hot.

All cables should be taped tightly to the tower, preferably on the inside, and run down to the ground level. It is not necessary to bond shielded cables to the tower electrically, but there should be no exceptions to the down-to-the-ground rule.

Although the effects of ground losses on feed-point impedance are less severe with the shunt-fed vertical than with the simple ¼-λ antenna, a good system of buried radials is very desirable. The ideal would be 120 radials, each 250 feet long, but fewer and shorter ones must often suffice. You can lay them around corners of houses, along fences or sidewalks, wherever they can be put a few inches under the surface, or even on the earth's surface. Aluminum clothesline wire may be used extensively in areas where it will not be subject to corrosion. Neoprene-covered aluminum wire will be better in highly acid soils. Contact with the soil is not important. Deep-driven ground rods and connection to underground copper water pipes are good, if available.

Installing the Shunt Feed

Principal details of the shunt-fed tower for 1.8 and 3.5 MHz are shown in Fig 42. Rigid rod or tubing can be used for the feed portion, but heavy-gauge aluminum or copper wire is easier to work with. Flexible stranded no. 8 copper wire is used at W5RTQ for the 1.8-MHz feed, because when the tower is cranked down, the feed wire must come down with it. Connection is made at the top, 68 feet, through a 4-foot length of aluminum tubing clamped to the top of the tower, horizontally. The wire is clamped to the tubing at the outer end, and runs down vertically through standoff insulators. These are made by fitting 12-inch lengths of PVC plastic water pipe over 3-foot lengths of aluminum tubing. These are clamped to the tower at 15- to 20-foot intervals, with the bottom clamp about 3 feet above ground. These lengths allow for adjustment of the tower to wire spacing over a range of about 12 to 36 inches, for impedance matching.

The gamma-match capacitor for 1.8 MHz is a 250-pF variable with about 1/6-inch plate spacing, which is adequate for power levels up to about 200 watts.

Tuning Procedure

The 1.8-MHz feed wire should be connected to the top of the structure if it is 75 feet tall or less. Mount the standoff insulators so as to have a spacing of about 24 inches between wire and tower. Pull the wire taut and clamp it in place at the bottom insulator. Leave a little slack below to permit adjustment of the wire spacing, if necessary.

Adjust the series capacitor in the 1.8-MHz line for minimum reflected power, as indicated on an SWR meter connected between the coax and the connector on the capacitor housing. Make this adjustment at a frequency near the middle of your expected operating range. If a high SWR is indicated, try moving the wire closer to the tower. Just the lower part of the wire need be moved for an indication as to whether reduced spacing is needed. If the SWR drops, move all insulators closer to the tower, and try again. If the SWR goes up, increase the spacing. There will be a practical range of about 12 to 36 inches. If going down to 12 inches does not give a low SWR, try connecting the top a bit farther down the tower. If wide spacing does not make it, the omega match

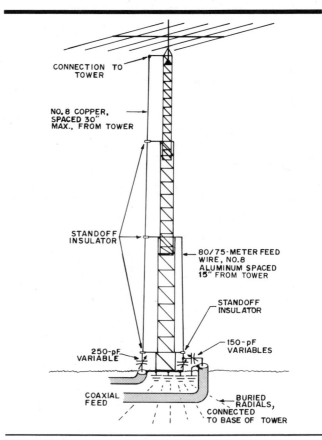

Fig 42—Principal details of the shunt-fed tower at W5RTQ. The 1.8-MHz feed, left side, connects to the top of the tower through a horizontal arm of 1-in. diameter aluminum tubing. The other arms have standoff insulators at their outer ends, made of 1-ft lengths of plastic water pipe. The connection for 3.5-4 MHz, right, is made similarly, at 28 ft, but two variable capacitors are used to permit adjustment of matching with large changes in frequency.

shown for 3.5-MHz work should be tried. No adjustment of spacing is needed with the latter arrangement, which may be necessary with short towers or installations having little or no top loading.

The two-capacitor arrangement is also useful for working in more than one 25-kHz segment of the 1.8-MHz band. Tune up on the highest frequency, say 1990 kHz, using the single capacitor, making the settings of wire spacing and connection point permanent for this frequency. To move to the lower frequency, say 1810 kHz, connect the second capacitor into the circuit and adjust it for the new frequency. Switching the second capacitor in and out then allows changing from one segment to the other, with no more than a slight retuning of the first capacitor.

A Different Approach

Fig 43 shows the method used by Doug DeMaw, W1FB, to gamma match his self-supporting 50-foot tower. A wire cage simulates a gamma rod of the proper diameter. The tuning capacitor is fashioned from telescoping sections of 1½ and 1¼ inch aluminum tubing with polyethylene tubing serving as the dielectric. This capacitor is more than adequate for power levels of 100 watts.

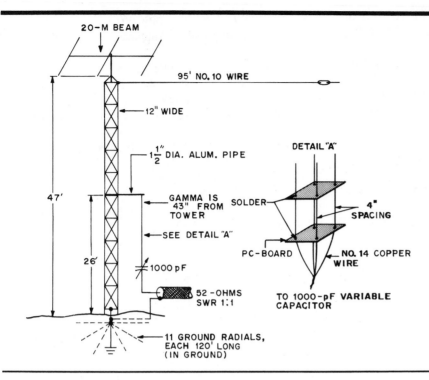

Fig 43—Details and dimensions for gamma-match feeding a 50-foot tower as a 1.8-MHz vertical antenna. The rotator cable and coaxial feed line for the 14-MHz beam is taped to the tower legs and run into the shack from ground level. No RF decoupling networks are necessary.

AN INVERTED L

The antenna shown in Fig 44 is simple and easy to construct. It is a good antenna for the beginner or the experienced 1.8-MHz DXer. Because the overall electrical length is greater than ¼ λ, the feed-point resistance is on the order of 50 ohms, with an inductive reactance. That reactance is canceled by a series capacitor.

A yardarm attached to a tower can be used to support the vertical section of the antenna. For best results the vertical section should be as long as possible. A good ground system is necessary for good results—the better the ground, the better the results.

BIBLIOGRAPHY

Source material and more extended discussion of topics covered in this chapter can be found in the references given below and in the textbooks listed at the end of Chapter 2.

D. Atchley, Jr., "Putting the Quarter-Wave Sloper to Work on 160," *QST*, Jul 1979, pp 19-20.

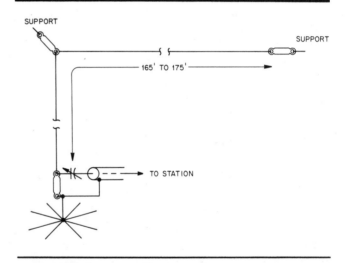

Fig 44—The 1.8-MHz inverted L. Overall wire length is 165 to 175 ft. The variable capacitor has a maximum capacitance of 500 to 800 pF. Adjust antenna length and variable capacitor for lowest SWR.

G. H. Brown, "The Phase and Magnitude of Earth Currents Near Radio Transmitting Antennas," *Proc IRE*, Feb 1935.

G. H. Brown, R. F. Lewis and J. Epstein, "Ground Systems as a Factor in Antenna Efficiency," *Proc IRE*, Jun 1937, pp 753-787.

A. Christman, "Feeding Phased Arrays: An Alternate Method," *Ham Radio,* May 1985, pp 58-59, 61-64.

D. DeMaw, "Additional Notes on the Half Sloper," *QST*, Jul 1979, pp 20-20.

A. C. Doty, Jr, J. A. Frey and H. J. Mills, "Efficient Ground Systems for Vertical Antennas," *QST*, Feb 1983, pp 20-25.

G. D. Elliott, "Phased Verticals for 40," *QST*, Apr 1972, pp 18-20.

R. Fosberg, "Some Notes on Ground Systems for 160 Meters," *QST*, Apr 1965, pp 65-67.

T. J. Gordon, "Invisible Antennas," *QST*, Nov 1965, pp 87-88.

G. Hubbell, "Feeding Grounded Towers as Radiators," *QST*, Jun 1960, pp 32-33, 140, 142.

P. H. Lee, *The Amateur Radio Vertical Antenna Handbook*, 1st ed. (Port Washington, NY: Cowan Publishing Corp, 1974).

J. Sevick, "The W2FMI Ground-Mounted Short Vertical," *QST*, Mar 1973, pp 13-18, 41.

B. L. Wermager, "A Truly Broadband Antenna for 80/75 Meters," *QST*, Apr 1986, pp 23-25.

D. J. Yungling, "The KI6O 160-Meter Linear-Loaded Sloper," *QST*, Apr 1986, p 26.

Chapter 5

Loop Antennas

A loop antenna is a closed-circuit antenna—that is, one in which a conductor is formed into one or more turns so its two ends are close together. Loops can be divided into two general classes, those in which both the total conductor length and the maximum linear dimension of a turn are very small compared with the wavelength, and those in which both the conductor length and the loop dimensions begin to be comparable with the wavelength.

A "small" loop can be considered to be simply a rather large coil, and the current distribution in such a loop is the same as in a coil. That is, the current has the same phase and the same amplitude in every part of the loop. To meet this condition, the total length of conductor in the loop must not exceed about 0.1 λ. Small loops are discussed later in this chapter, and further in Chapter 14.

A "large" loop is one in which the current is not the same either in amplitude or phase in every part of the loop. This change in current distribution gives rise to entirely different properties as compared with a small loop.

Half-Wave Loops

The smallest size of "large" loop generally used is one having a conductor length of ½ λ. The conductor is usually formed into a square, as shown in Fig 1, making each side 1/8 λ long. When fed at the center of one side, the current flows in a closed loop as shown at A. The current distribution is approximately the same as on a ½-λ wire, and so is maximum at the center of the side opposite the terminals X-Y, and minimum at the terminals themselves. This current distribution causes the field strength to be maximum in the plane of the loop and in the direction looking from the low-current side to the high-current side. If the side opposite the terminals is opened at the center as shown at B (strictly speaking, it is then no longer a loop because it is no longer

a closed circuit), the direction of current flow remains unchanged but the maximum current flow occurs at the terminals. This reverses the direction of maximum radiation.

The radiation resistance at a current antinode (which is also the resistance at X-Y in Fig 1B) is on the order of 50 Ω. The impedance at the terminals in A is a few thousand ohms. This can be reduced by using two identical loops side by side with a few inches spacing between them and applying power between terminal X on one loop and terminal Y on the other.

Unlike a ½-λ dipole or a small loop, there is no direction in which the radiation from a loop of the type shown in Fig 1 is zero. There is appreciable radiation in the direction perpendicular to the plane of the loop, as well as to the "rear"—the opposite direction to the arrows shown. The front-to-back (F/B) ratio is of the order of 4 to 6 dB. The small size and the shape of the directive pattern result in a loss of about 1 dB when the field strength in the optimum direction from such a loop is compared with the field from a ½-λ dipole in its optimum direction.

The ratio of the forward radiation to the backward radiation can be increased, and the field strength likewise increased at the same time to give a gain of about 1 dB over a dipole, by using inductive reactances to "load" the sides joining the front and back of the loop. This is shown in Fig 2. The reactances, which should have a value of approximately 360 Ω, decrease the current in the sides in which

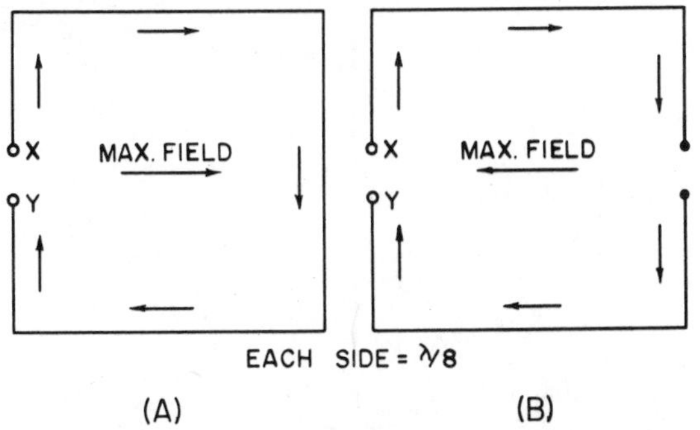

Fig 1—Half-wave loops, consisting of a single turn having a total length of ½ λ.

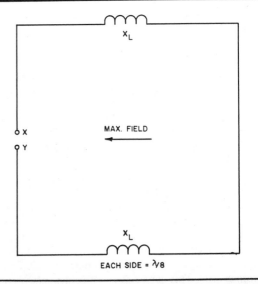

Fig 2—Inductive loading in the sides of a ½-λ loop to increase the directivity and gain. Maximum radiation or response is in the plane of the loop, in the direction shown by the arrow.

they are inserted and increase it in the side having terminals. This increases the directivity and thus increases the efficiency of the loop as a radiator.

One-Wavelength Loops

Loops in which the conductor length is 1 λ have different characteristics than ½-λ loops. Three forms of 1-λ loops are shown in Fig 3. At A and B the sides of the squares are equal to ¼ λ, the difference being in the point at which the terminals are inserted. At C the sides of the triangle are equal to ⅓ λ. The relative direction of current flow is as shown in the drawings. This direction reverses halfway around the perimeter of the loop, as such reversals always occur at the junction of each ½-λ section of wire.

The directional characteristics of loops of this type are opposite in sense to those of a small loop. That is, the radiation is maximum perpendicular to the plane of the loop and is minimum in any direction in the plane containing the loop. If the three loops shown in Fig 3 are mounted in a vertical plane with the terminals at the bottom, the radiation is horizontally polarized. When the terminals are moved to the center of one vertical side in A, or to a side corner in B, the radiation is vertically polarized. If the terminals are moved to a side corner in C, the polarization will be diagonal, containing both vertical and horizontal components.

In contrast to straight-wire antennas, the electrical length of the circumference of a 1-λ loop is *shorter* than the actual length. For loops made of wire and operating at frequencies below 30 MHz or so, where the ratio of conductor length to wire diameter is large, the loop will be close to resonance when

$$\text{Length}_{\text{feet}} = \frac{1005}{f_{\text{MHz}}}$$

The radiation resistance of a resonant 1-λ loop is

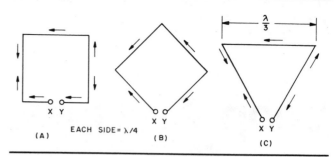

Fig 3—At A and B, loops having sides ¼ λ long, and at C having sides ⅓ λ long (total conductor length 1 λ). The polarization depends on the orientation of the loop and on the position of the feed point (terminals X-Y) around the perimeter of the loop.

approximately 100 Ω, when the ratio of conductor length to diameter is large. As the loop dimensions are comparable with those of a ½-λ dipole, the radiation efficiency is high.

In the direction of maximum radiation (that is, broadside to the plane of the loop, regardless of the point at which it is fed) the 1-λ loop will show a small gain over a ½-λ dipole. Theoretically, this gain is about 2 dB, and measurements have confirmed that it is of this order.

The 1-λ loop is more frequently used as an element of a directive antenna array (the quad and delta-loop antennas described in Chapter 12) than singly, although there is no reason why it cannot be used alone. In the quad and delta loop, it is nearly always driven so that the polarization is horizontal.

Small Loop Antennas

The electrically small loop antenna has existed in various forms for many years. Probably the most familiar form of this antenna is the ferrite loopstick found in portable AM broadcast-band receivers. Amateur applications of the small loop include direction finding, low-noise directional receiving antennas for 1.8 and 3.5 MHz, and small transmitting antennas. Because the design of transmitting and receiving loops requires some different considerations, the two situations are examined separately in this section. This information was written by Domenic M. Mallozzi, N1DM.

The Basic Loop

What is, and what is not a small loop antenna? By definition, the loop is considered to be electrically small when its total conductor length is less than 0.1 λ—0.085 is the number used in this section. This size is based on the fact that the current around the perimeter of the loop must be in phase. When the winding conductor is more than about 0.085 λ, this is no longer true. This constraint results in a very predictable figure-eight radiation pattern, shown in Fig 4.

The simplest loop is a 1-turn untuned loop with a load connected to a pair of terminals located in the center of one

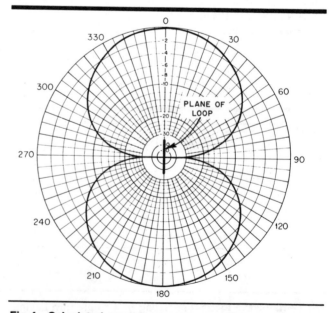

Fig 4—Calculated small loop antenna radiation pattern.

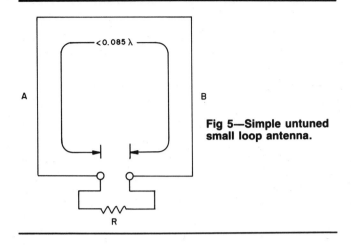

Fig 5—Simple untuned small loop antenna.

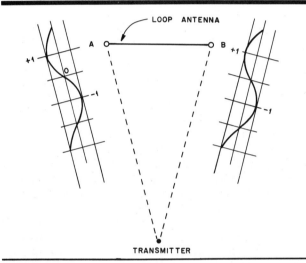

Fig 6—Example of orientation of loop antenna that does not respond to a signal source (null in pattern).

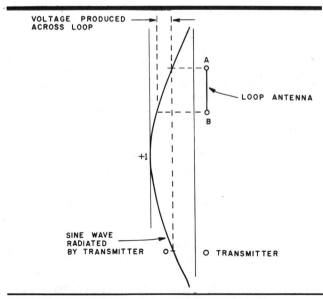

Fig 7—Example of orientation of loop antenna for maximum response.

of the sides, Fig 5. How its pattern is developed is easily pictured if we look at some "snapshots" of the antenna relative to a signal source. Fig 6 represents a loop from above, and shows the instantaneous radiated voltage wave. Note that points A and B of the loop are receiving the same instantaneous voltage. This means that no current will flow through the loop, because there is no current flow between points of equal potential. A similar analysis of Fig 7, with the loop turned 90° from the position represented in Fig 6, shows that this position of the loop provides maximum response. Of course, the voltage derived from the passing wave is small because of the small physical size of the loop. Fig 4 shows the ideal radiation pattern for a small loop.

The voltage across the loop terminals is given by

$$V = \frac{2\pi A N E \cos \theta}{\lambda} \qquad \text{(Eq 1)}$$

where

- V = voltage across the loop terminals
- A = area of loop in square meters
- N = number of turns in the loop
- E = RF field strength in volts per meter
- θ = angle between the plane of the loop and the signal source (transmitting station)
- λ = wavelength of operation in meters

This equation comes from a term called effective height. The effective height refers to the height (length) of a vertical piece of wire above ground that would deliver the same voltage to the receiver. The equation for effective height is

$$h = \frac{2\pi NA}{\lambda} \qquad \text{(Eq 2)}$$

where h is in meters and the other terms are as for Eq 1.

A few minutes with a calculator will show that, with the constraints previously stated, the loop antenna will have a very small effective height. This means it will deliver a relatively small voltage to the receiver, with even a large signal.

TUNED LOOPS

We can tune the loop by placing a capacitor across the antenna terminals. This causes a larger voltage to appear across the loop terminals because of the Q of the parallel resonant circuit that is formed.

The voltage across the loop terminals is now given by

$$V = \frac{2\pi A N E Q \cos \theta}{\lambda} \qquad \text{(Eq 3)}$$

where Q is the loaded Q of the tuned circuit, and other terms are as defined above.

Most amateur loops are of the tuned variety. For this reason, all comments that follow are based on tuned-loop antennas, consisting of one or more turns. The tuned-loop antenna has some particular advantages. For example, it puts high selectivity up at the "front" of a receiving system, where it can significantly help factors such as dynamic range. Loaded Q values of 100 or greater are easy to obtain with careful loop construction.

Consider a situation where the inherit selectivity of the loop is helpful. Assume we have a loop with a Q of 100 at 1.805 MHz. We are working a DX station on 1.805 MHz and are suffering strong interference from a local station 10-kHz

Table 1

Inductance Equations for Short Coils (Loop Antennas)

Triangle:

$$L(\mu H) = 0.006N^2s \left[\ln\left(\frac{1.1547\ sN}{(N+1)\ell} \right) + 0.65533 + \frac{0.1348(N+1)\ell}{sN} \right]$$

Square:

$$L(\mu H) = 0.008N^2s \left[\ln\left(\frac{1.4142\ sN}{(N+1)\ell} \right) + 0.37942 + \frac{0.3333(N+1)\ell}{sN} \right]$$

Hexagon:

$$L(\mu H) = 0.012N^2s \left[\ln\left(\frac{2\ sN}{(N+1)\ell} \right) + 0.65533 + \frac{0.1348(N+1)\ell}{sN} \right]$$

Octagon:

$$L(\mu H) = 0.016N^2s \left[\ln\left(\frac{2.613\ sN}{(N+1)\ell} \right) + 0.75143 + \frac{0.07153(N+1)\ell}{sN} \right]$$

where
  N = number of turns
  s = side length in cm
  ℓ = coil length in cm

Note: In the case of single-turn coils, the diameter of the conductor should be used for ℓ.

away. Switching from a dipole to a small loop will reduce the strength of the off-frequency signal by 6 dB (approximately one S unit). This, in effect, increases the dynamic range of the receiver. In fact, if the off-frequency station were further off frequency, the attenuation would be greater.

Another way the loop can help is by using the nulls in its pattern to null out on-frequency (or slightly off-frequency) interference. For example, say we are working a DX station to the north, and just 1 kHz away is another local station engaged in a contact. The local station is to our west. We can simply rotate our loop to put its null to the west, and now the DX station should be readable while the local will be knocked down by 60 or more dB. This obviously is quite a noticeable difference. Loop nulls are very sharp and are generally noticeable only on ground-wave signals (more on this later).

Of course, this method of nulling will be effective only if the interfering station and the station being worked are not in the same direction (or in exact opposite directions) from our location. If the two stations were on the same line from our location, both the station being worked and the undesired station would be nulled out. Luckily the nulls are very sharp, so as long as the stations are at least 10° off axis from each other, the loop null will be usable.

A similar use of the nulling capability is to eliminate local noise interference, such as that from a light dimmer in a neighbor's house. Just put the null on the offending light dimmer, and the noise should disappear.

Now that we have seen some possible uses of the small loop, let us look at a bit of detail about its design. First, the loop forms an inductor having a very small ratio of winding length to diameter. The equations for finding inductance given in most radio handbooks assume that the inductor coil is longer than its diameter. However, F. W. Grover of the US National Bureau of Standards has provided equations for inductors of common cross-sectional shapes and small

length-to-diameter ratios. (See the bibliography at the end of this chapter.) Grover's equations are shown in Table 1. Their use will yield relatively accurate numbers; results are easily worked out with a scientific calculator or home computer.

The value of a tuning capacitor for a loop is easy to calculate from the standard resonance equations. The only matter to consider before calculating this is the value of distributed capacitance of the loop winding. This capacitance shows up between adjacent turns of the coil because of their slight difference in potential. This causes each turn to appear as a charge plate. As with all other capacitances, the value of the distributed capacitance is based on the physical dimensions of the coil. An exact mathematical analysis of its value is a complex problem. A simple approximation is given by Medhurst (see bibliography) as

$$C = HD \qquad \text{(Eq 4)}$$

where
  C = distributed capacitance in pF
  H = a constant related to the length-to-diameter ratio of the coil (Table 2 gives H values for length-to-diameter ratios used in loop antenna work.)
  D = diameter of the winding in cm

Medhurst's work was with coils of round cross section. For loops of square cross section the distributed capacitance is given by Bramslev as

$$C = 60S \qquad \text{(Eq 5)}$$

where
  C = the distributed capacitance in pF
  S = the length of the side in meters

If you convert the length in this equation to centimeters you will find Bramslev's equation gives results in the same order of magnitude as Medhurst's equation.

Table 2

Table 2

Values of the Constant H for Distributed Capacitance

Length to Diameter Ratio	H
0.10	0.96
0.15	0.79
0.20	0.78
0.25	0.64
0.30	0.60
0.35	0.57
0.40	0.54
0.50	0.50
1.00	0.46

This distributed capacitance appears as if it were a capacitor across the loop terminals. Therefore, when determining the value of the tuning capacitor, the distributed capacitance must be subtracted from the total capacitance required to resonate the loop. The distributed capacitance also determines the highest frequency at which a particular loop can be used, because it is the minimum capacitance obtainable.

Electrostatically Shielded Loops

Over the years, many loop antennas have incorporated an electrostatic shield. This shield generally takes the form of a tube around the winding, made of a conductive but nonmagnetic material (such as copper or aluminum). Its purpose is to maintain loop balance with respect to ground, by forcing the capacitance between all portions of the loop and ground to be identical. This is illustrated in Fig 8. It is necessary to maintain electrical loop balance to eliminate what is referred to as the *antenna effect*. When the antenna becomes

unbalanced it appears to act partially as a small vertical antenna. This vertical pattern gets superimposed on the ideal figure-eight pattern, distorting the pattern and filling in the nulls. The type of pattern that results is shown in Fig 9.

Adding the shield has the effect of somewhat reducing the pickup of the loop, but this loss is generally offset by the increase in null depth of the loops. Proper balance of the loop antenna requires that the load on the loop also be balanced. This is usually accomplished by use of a balun transformer or a balanced input preamplifier. Two important points regarding the shield are that it cannot form a continuous electrical path around the loop perimeter, or it will appear as a shorted coil turn. Usually the insulated break is located opposite the feed point to maintain symmetry. Another point to be considered is that the shield should be of a much larger diameter than the loop winding, or it will lower the Q of the loop.

Various construction techniques have been used in making shielded loops. Genaille located his loop winding inside aluminum conduit, while True constructed an aluminum shield can around his winding. Others have used pieces

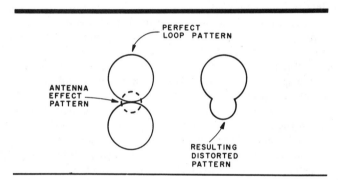

Fig 9—Distortion in loop pattern resulting from antenna effect.

Fig 8—At A, the loop is unbalanced by capacitance to its surroundings. At B, the use of an electrostatic shield overcomes this effect.

of Hardline to form a loop, using the outer conductor as a shield. DeMaw used flexible coax with the shield broken at the center of the loop conductor in a multiturn loop for 1.8 MHz. Goldman uses another shielding method for broadcast receiver loops. His shield is in the form of a barrel made of hardware cloth, with the loop in its center. (See bibliography for above references.) All these methods provide sufficient shielding to maintain the balance. It is possible, as Nelson shows, to construct an unshielded loop with good nulls (60 dB or better) by paying great care to symmetry.

LOOP Q

As previously mentioned, Q is an important consideration in loop performance because it determines both the loop bandwidth and its terminal voltage for a given field strength. The loaded Q of a loop is based on four major factors. These are (1) the intrinsic Q of the loop winding, (2) the effect of the load, (3) the effect of the electrostatic shield, and (4) the Q of the tuning capacitor.

The major factor is the Q of the winding of the loop itself. The ac resistance of the conductor caused by skin effect is the major consideration. The ac resistance for copper conductors may be determined from

$$R = \frac{0.996 \times 10^{-6} \sqrt{f}}{d} \qquad \text{(Eq 6)}$$

where

    R = resistance in ohms per foot
    f = frequency, Hz
    d = conductor diameter, inches

The Q of the inductor is then easily determined by taking the reactance of the inductor and dividing it by the ac resistance. If you are using a multiturn loop and are a perfectionist, you might also want to include the loss from conductor proximity effect. This effect is described in detail later in this chapter, in the section on transmitting loops.

Improvement in Q can be obtained in some cases by the use of Litz wire (short for Litzendraht). Litz wire consists of strands of individual insulated wires that are woven into bundles in such a manner that each conductor occupies each location in the bundle with equal frequency. Litz wire results in improved Q over solid or stranded wire of equivalent size, up to about 3 MHz.

Also the Q of the tuned circuit of the loop antenna is determined by the Q of the capacitors used to resonate it. In the case of air variables or dipped micas this is not usually a problem. But if variable-capacitance diodes are used to remotely tune the loop, pay particular attention to the manufacturer's specification for Q of the diode at the frequency of operation. The tuning diodes can have a significant effect on circuit Q.

Now we consider the effect of load impedance on loop Q. In the case of a directly coupled loop (as in Fig 5), the load is connected directly across the loop terminals, causing it to be treated as a parallel resistance in a parallel-tuned RLC circuit. Obviously, if the load is of a low value, the Q of the loop will be low. A simple way to correct this is to use a transformer to step up the load impedance that appears across the loop terminals. In fact, if we make this transformer a balun, it also allows us to use our unbalanced receivers with the loop and maintain loop symmetry. Another solution is to use what is referred to as an inductively coupled loop, such as DeMaw's four turn electrostatically shielded loop. A 1-turn link is connected to the receiver. This turn is wound with the four-turn loop. In effect, this builds the transformer into the antenna.

Another solution to the problem of load impedance on loop Q is to use an active preamplifier with a high impedance balanced input and unbalanced output. This method also has the advantage of amplifying the low-level output voltage of the loop to where it can be used with a receiver of even mediocre sensitivity. In fact, the Q of the loop when used with a balanced preamplifier having high input impedance may be so high as to be unusable in certain applications. An example of this situation would occur where a loop is being used to receive a 5 kHz wide AM signal at a frequency where the bandwidth of the loop is only 1.5 kHz. In this case the detected audio might be very distorted. The solution to this is to locate a Q-degrading resistor across the loop terminals.

FERRITE-CORE LOOP ANTENNAS

The ferrite-core loop antenna is a special case of the air-core receiving loops considered up to now. Because of its

use in every AM broadcast-band portable radio, the ferrite-core loop is, by quantity, the most popular form of the loop antenna. But broadcast-band reception is far from its only use; it is commonly found in radio direction finding equipment and low frequency receiving systems (below 500 kHz) for time and frequency standard systems. In recent years, design information on these types of antennas has been a bit sparse in the amateur literature, so the next few paragraphs are devoted to providing some details.

Ferrite loop antennas are characteristically very small compared to the frequency of use. For example, a 3.5-MHz version may be in the range of 15 to 30 cm long and about 1.25 cm in diameter. Earlier in this chapter, effective height was introduced as a measure of loop sensitivity. The effective height of an air-core loop antenna is given by Eq 2.

If an air-core loop is placed in a field, in essence it cuts the lines of flux without disturbing them (Fig 10A). On the other hand, when a ferrite (magnetic) core is placed in the field, the nearby field lines are redirected into the loop (Fig 10B). This is because the reluctance of the ferrite material is less than that of the surrounding air, so the nearby flux lines tend to flow through the loop rather than passing it by. (Reluctance is the magnetic analogy of resistance, while flux is analogous to current.) The reluctance is inversely proportional to the permeability of the rod core, μ_{rod}. (In some texts the rod permeability is referred to as effective permeability, μ_{eff}). This effect modifies the equation for effective height of a ferrite-core loop to

$$h = \frac{2\pi \, N \, A \, \mu_{rod}}{\lambda} \qquad \text{(Eq 7)}$$

where

    h = effective height (length) in meters
    N = number of turns in the loop
    A = RF field strength in volts/meter
    μ_{rod} = permeability of the ferrite rod
    λ = wavelength of operation in meters

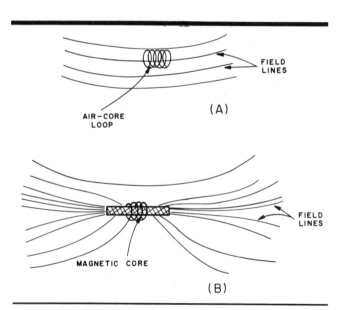

Fig 10—At A, an air-core loop has no effect on nearby field lines. B illustrates the effect of a ferrite core on nearby field lines. The field is altered by the reluctance of the ferrite material.

This obviously is a large increase in "collected" signal. If the rod permeability was 90, this would be the same as making the loop area 90 times larger with the same number of turns. For example, a 1.25 cm diameter ferrite-core loop would have an effective height equal to an air-core loop 22.5 cm in diameter (with the same number of turns).

By now you might have noticed we have been very careful to refer to *rod* permeability. There is a very important reason for this. The permeability that a rod of ferrite exhibits is a combination of the material permeability or μ, the shape of the rod, and the dimensions of the rod. In ferrite rods, μ is sometimes referred to as initial permeability, μ_i, or toroidal permeability, μ_{tor}. Because most amateur ferrite loops are in the form of rods, we will discuss only this shape.

The reason that μ_{rod} is different from μ is a very complex physics problem that is well beyond the scope of this book. For those interested in the details, books by Polydoroff and by Snelling cover this subject in considerable detail. (See bibliography.) For our purposes a simple explanation will suffice. The rod is in fact not a perfect director of flux, as is illustrated in Fig 11. Note that some lines impinge on the sides of the core and also exit from the sides. These lines therefore would not pass through all the turns of the coil if it were wound from one end of the core to the other. These flux lines are referred to as leakage flux, or sometimes as flux leakage.

Leakage flux causes the flux density in the core to be nonuniform along its length. From Fig 11 it can be seen that the flux has a maximum at the geometric center of the length of the core, and decreases as the ends of the core are approached. This causes some noticeable effects. As a short coil is placed at different locations along a long core, its inductance will change. The maximum inductance exists when the coil is centered on the rod. The Q of a short coil on a long rod is greatest at the center. On the other hand, if you require a higher Q than this, it is recommended that you spread the coil turns along the whole length of the core, even though this will result in a lower value of inductance. (The inductance can be increased to the original value by adding turns.) Fig 12 gives the relationship of rod permeability to material permeability for a variety of values.

The change in μ over the length of the rod results in an adjustment in the term μ_{rod} for its so called "free ends" (those not covered by the winding). This adjustment factor is given by

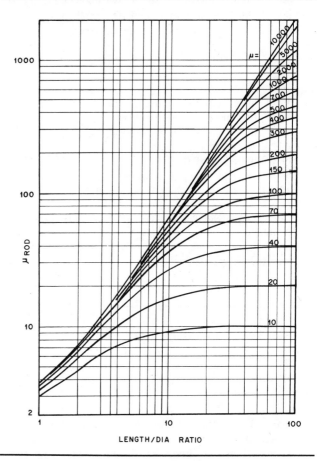

Fig 12—Rod permeability, μ_{rod}, versus material permeability, μ, for different rod length-to-diameter ratios.

$$\mu' = \mu_{rod} \sqrt[3]{\frac{a}{b}} \qquad \text{(Eq 8)}$$

where
μ' = the corrected permeability
a = the length of the core
b = the length of the coil

This value of μ' should be used in place of μ_{rod} in Eq 7 to obtain the most accurate value of effective height.

All these variables make the calculation of ferrite loop antenna inductance somewhat less accurate than for the air-core version. The inductance of a ferrite loop is given by

$$L = \frac{4\pi N^2 A \mu_{rod}}{\ell} \qquad \text{(Eq 9)}$$

where
L = inductance in henrys
N = number of turns
A = cross-sectional area of the core in square cm
ℓ = magnetic length of core in cm

Experiments indicate that the winding diameter should be as close to that of the rod diameter as practical in order to maximize both inductance value and Q.

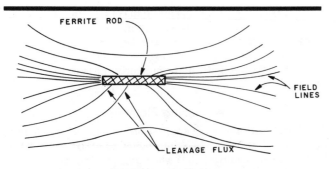

Fig 11—Example of magnetic field lines near a practical ferrite rod, showing leakage flux.

By using all this information, we may determine the voltage at the loop terminals and its signal-to-noise ratio (SNR). The voltage may be determined from

$$V = \frac{2\pi \, A \, N \, \mu' \, Q \, E}{\lambda} \qquad \text{(Eq 10)}$$

where

V = output voltage across the loop terminals
A = loop area in square meters
N = number of turns in the loop winding
μ' = corrected rod permeability
Q = loaded Q of the loop
E = RF field strength in volts per meter
λ = wavelength of operation in meters

Lankford's equation for the sensitivity of the loop for a 10 dB SNR is

$$E = \frac{1.09 \times 10^{-10} \, \lambda \, \sqrt{fLb}}{A \, N \, \mu' \, \sqrt{Q}} \qquad \text{(Eq 11)}$$

where

f = operating frequency in Hz
L = loop inductance in henrys
b = receiver bandwidth in Hz

Similarly, Belrose gives the SNR of a tuned loop antenna as

$$SNR = \frac{66.3 \, N \, A \, \mu_{rod} \, E}{\sqrt{b}} \sqrt{\frac{Qf}{L}} \qquad \text{(Eq 12)}$$

From this, if the field strength E, μ_{rod}, b, and A are fixed, then Q or N must increase (or L decrease) to yield a better SNR. Higher sensitivity can also be obtained (especially at frequencies below 500 kHz) by bunching ferrite cores together to increase the loop area over that which would be possible with a single rod. High sensitivity is important because loop antennas are not the most efficient collectors of signals, but they do offer improvement over other receiving antennas in terms of SNR. For this reason, you should attempt to maximize the SNR when using a small loop receiving antenna. In some cases there may be physical constraints that limit how large you can make a ferrite-core loop.

After working through Eq 11 or 12, you might find you still require some increase in antenna system gain to effectively use your loop. In these cases the addition of a low noise preamplifier may be quite valuable even on the lower frequency bands where they are not commonly used. Chapter 14 contains information on such preamplifiers.

The electrostatic shield discussed earlier with reference to air-core loops can be used effectively with ferrite-core loops. (Construction examples are presented in Chapter 14.) As in the air-core loop, a shield will reduce electrical noise and improve loop balance.

PROPAGATION EFFECTS ON NULL DEPTH

After building a balanced loop you may find it does not approach the theoretical performance in the null depth. This problem may result from propagation effects. Tilting the loop away from a vertical plane may improve performance under some propagation conditions, to account for the vertical angle of arrival. Basically, the loop performs as described above only when the signal is arriving perpendicular to the axis of rotation of the loop. At incidence angles other than perpendicular, the position and depth of the nulls deteriorate.

The problem can be even further influenced by the fact that if the loop is situated over less than perfectly conductive ground, the wave front will appear to tilt or bend. (This bending is not always detrimental; in the case of Beverage antennas, sites are chosen to take advantage of this effect.)

Another cause of apparent poor performance in the null depth can be from polarization error. If the polarization of the signal is not completely linear, the nulls will not be sharp. In fact, for circularly polarized signals, the loop might appear to have almost no nulls. Propagation effects are discussed further in Chapter 14.

SITING EFFECTS ON THE LOOP

The location of the loop has an influence on its performance that at times may become quite noticeable. For ideal performance the loop should be located outdoors and clear of any large conductors, such as metallic downspouts and towers. A VLF loop, when mounted this way, will show good sharp nulls spaced 180° apart if the loop is well balanced. This is because the major propagation mode at VLF is via ground wave. At frequencies in the HF region, a significant portion of the signals are propagated by sky wave, and nulls are often only partial.

For this reason most hams locate their loop antennas near their operating position. If you choose to locate a small loop indoors, its performance may show nulls of less than the expected depth, and some skewing of the pattern. For precision direction finding there may be some errors associated with wiring, plumbing, and other metallic construction members in the building. Also, a strong local signal may be reradiated from the surrounding conductors so that it cannot be nulled with any positioning of the loop. There appears to be no known method of curing this type of problem. All this should not discourage you from locating a loop indoors; this information is presented here only to give you an idea of some pitfalls. Many hams have reported excellent results with indoor mounted loops, in spite of some of the problems.

Locating a receiving loop in the field of a transmitting antenna may cause a large voltage to appear at the receiver antenna terminals. This may be sufficient to destroy sensitive RF amplifier transistors or front-end protection diodes. This can be solved by disconnecting your loop from the receiver during transmit periods. This can obviously be done automatically with a relay that opens when the transmitter is activated.

LOOP ANTENNA ARRAYS

Arrays of loop antennas, both in combination with each other and with other antenna types, have been used for many years. The arrays are generally used to cure some "deficiency" in the basic loop for a particular application, such as a 180° ambiguity in the null direction, low sensitivity, and so forth.

A Sensing Element

For direction finding applications the single loop suffers the problem of having two nulls which are 180° apart. This leads to an ambiguity of 180° when trying to find the direction to a transmitting station from a given location. A sensing element (often called a sense antenna) may be added to the loop, causing the overall antenna to have a cardioid pattern and only one null. The sensing element is a small vertical antenna whose height is equal to or greater than the loop effective height. This vertical is physically close to the loop, and when its omnidirectional pattern is adjusted so that its amplitude and phase are equal to one of the loop lobes, the patterns combine to form a cardioid. This antenna can be made quite compact by use of a ferrite loop to form a portable DF antenna for HF direction finding. Chapter 14 contains additional information and construction projects using sensing elements.

Arrays of Loops

A more advanced array which can develop more diverse patterns consists of two or more loops. Their outputs are combined through appropriate phasing lines and combiners to form a phased array. Two loops can also be formed into an array which can be rotated without physically turning the loops themselves. This method was developed by Bellini and Tosi in 1907 and performs this apparently contradictory feat by use of a special transformer called a goniometer. The goniometer is described in Chapter 14.

Aperiodic Arrays

The aperiodic loop array is a wide-band antenna. This type of array is useful over at least a decade of frequency, such as 2 MHz to 20 MHz. Unlike most of the loops discussed up to now, the loop elements in an aperiodic array are untuned. Such arrays have been used commercially for many years. One loop used in such an array is shown in Fig 13. This loop is quite different from all the loops discussed so far in this chapter because its pattern is not the familiar figure eight. Rather, it is omnidirectional.

The antenna is omnidirectional because it is purposely unbalanced, and also because the isolating resistor causes the antenna to appear as two closely spaced short monopoles. The loop maintains the omnidirectional characteristics over a frequency range of at least four or five to one. These loops, when combined into end-fire or broadside phased arrays, can provide quite impressive performance. A commercially made end-fire array of this type consisting of four loops equally spaced along a 25-meter baseline can provide gains in excess of 5 dBi over a range of 2 to 30 MHz. Over a considerable portion of this frequency range, the array can maintain F/B ratios of 10 dB. Even though the commercial version is very expensive, an amateur version can be constructed using the information provided by Lambert. One interesting feature of this type of array is that, with the proper combination of hybrids and combiners, the antenna can simultaneously feed two receivers with signals from different directions, as shown in Fig 14. This antenna may be especially interesting to one wanting a directional receiving array for two or more adjacent amateur bands.

TRANSMITTING LOOP ANTENNAS

The electrically small transmitting loop antenna involves some different design considerations from receiving loops. Unlike receiving loops, the size limitations of the antenna are not as clearly defined. For most purposes, any transmitting loop whose physical circumference (not total conductor length) is less than ¼ λ can be considered small. In most cases, as a consequence of their relatively large size (when compared to a receiving loop), transmitting loops have a nonuniform current distribution along their circumference. This leads to some performance changes from a receiving loop.

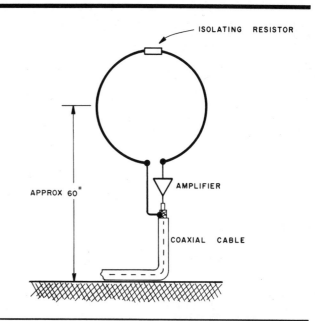

Fig 13—A single wide-band loop antenna used in an aperiodic array.

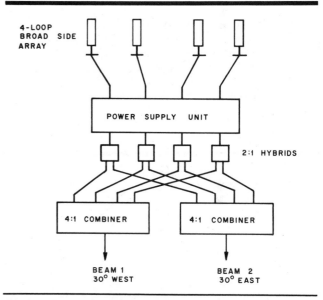

Fig 14—Block diagram of a four-loop broadside array with dual beams separated by 60° in azimuth.

The transmitting loop is a parallel-tuned circuit with a large inductor acting as the radiator. As with the receiving loop, the calculation of the transmitting loop inductance may be carried out with the equations in Table 1. Avoid equations for long solenoids found in most texts. Other fundamental equations for a transmitting loop are given in Table 3.

In recent years, two types of transmitting loops have been predominant in the amateur literature: the "army loop" by Lew McCoy, W1ICP, and the "high efficiency" loop by Ted Hart, W5QJR. The army loop is a version of a loop designed

Table 3

Transmitting Loop Equations

$X_L = 2\pi fL$ ohms

$Q = \dfrac{f}{\Delta f} = \dfrac{X_L}{2(R_R + R_L)}$

$R_R = 3.12 \times 10^4 \left[\dfrac{NA}{\lambda^2} \right]^2$ ohms

$V_c = \sqrt{PX_LQ}$

$I_L = \sqrt{\dfrac{PQ}{X_L}}$

where

X_L = inductive reactance, ohms
f = frequency, Hz
Δf = bandwidth, Hz
R_R = radiation resistance, ohms
R_L = loss resistance, ohms (see text)
N = number of turns
A = area enclosed by loop, square meters
λ = wavelength at operating frequency, meters
V_c = voltage across capacitor
P = power, watts
I_L = resonant circulating current in loop

for portable use in Southeast Asia by Patterson of the US Army. This loop is diagrammed in Fig 15A. It can be seen by examination that this loop appears as a parallel tuned circuit, fed by a tapped capacitance impedance-matching network. The Hart loop, shown in Fig 15B, has the tuning capacitor separate from the matching network. The matching

network is basically a form of gamma match. (Additional data and construction details for the Hart loop are presented later in this chapter.) Here we cover some matters which are common to both antennas.

The radiation resistance of a loop in ohms is given by

$$R_R = 3.12 \times 10^4 \left(\frac{N\,A}{\lambda^2} \right)^2 \qquad \text{(Eq 13)}$$

where

N = number of turns
A = area of loop in square meters
λ = wavelength of operation in meters

It is obvious that within the constraints given, the radiation resistance is very small. Unfortunately the loop has losses, both ohmic and from skin effect. By using this information, the radiation efficiency of a loop can be calculated from

$$\eta = \frac{R_R}{R_R + R_L} \times 100 \qquad \text{(Eq 14)}$$

where

η = antenna efficiency, %
R_R = radiation resistance, Ω
R_L = loss resistance, Ω

A simple ratio of R_R versus R_L shows the effects on the efficiency, as can be seen from Fig 16. The loss resistance is primarily the ac resistance of the conductor. This can be calculated from Eq 6. A transmitting loop generally requires the use of copper conductors of at least ¾ inch in diameter in order to obtain efficiencies that are reasonable. Tubing is as useful as a solid conductor because high-frequency currents flow only along a very small depth of the surface of the conductor; the center of the conductor has almost no effect on current flow.

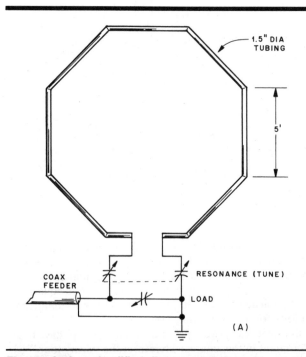
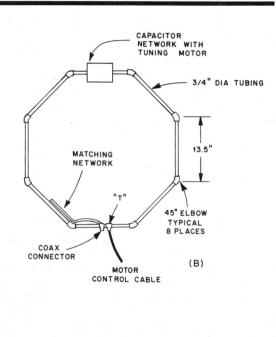

Fig 15—At A, a simplified diagram of the army loop. At B, the W5QJR Hart loop, which is described in more detail later in this chapter.

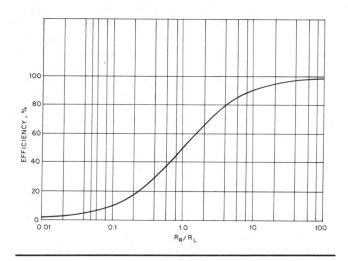

Fig 16—Effect of the ratio of R_R/R_L on loop efficiency.

efficient the loop, the worse the effect. For example, an 8-turn transmitting loop with an efficiency of 10% (calculated by the skin-effect method) actually only has an efficiency of 3% because of the additional losses introduced by the proximity effect. If you are contemplating construction of a multiturn transmitting loop, you might want to consider spreading the conductors apart to reduce this effect. G. S. Smith includes graphs that detail this effect in his 1972 IEEE paper.

The components in a resonated transmitting loop are subject to both high currents and voltages as a result of the large circulating currents found in the high-Q tuned circuit formed by the antenna. This makes it important that the capacitors have a high RF current rating, such as transmitting micas or the Centralab 850 series. Be aware that even a 100-watt transmitter can develop currents in the tens of amperes, and voltages across the tuning capacitor in excess of 10,000 volts. This consideration also applies to any conductors used to connect the loop to the capacitors. A piece of no. 14 wire may have more resistance than the rest of the loop conductor. It is therefore best to use copper strips or the braid from a piece of large coax cable to make any connections. Make the best electrical connection possible, using soldered or welded joints. Using nuts and bolts should be avoided, because at RF these joints generally have high resistance, especially after being subjected to weathering.

An unfortunate consequence of having a small but high-efficiency transmitting loop is high Q, and therefore limited bandwidth. This type of antenna may require retuning for frequency changes as little as 5 kHz. If you are using any wide-band mode such as AM or FM, this might cause fidelity problems and you might wish to sacrifice a little efficiency to obtain the required bandwidth.

A special case of the transmitting loop is that of the ferrite loaded loop. This is a logical extension of the transmitting loop if we consider the improvement that a ferrite core makes in receiving loops. The use of ferrites in a transmitting loop is still under development. (See the bibliography reference for DeVore and Bohley.)

In the case of multiturn loops there is an additional loss related to a term called proximity effect. The proximity effect occurs in cases where the turns are closely spaced (such as being spaced one wire diameter apart). As these current-carrying conductors are brought close to each other, the current density around the circumference of each conductor gets redistributed. The result is that more current per square meter is flowing at the surfaces adjacent to other conductors. This means that the loss is higher than a simple skin-effect analysis would indicate, because the current is bunched so it flows through a smaller cross section of the conductor than if the other turns were not present.

As the efficiency of a loop approaches 90%, the proximity effect is less serious. But unfortunately, the less

Small High Efficiency Loop Antennas for Transmitting

The ideal small transmitting antenna would have performance equal to a large antenna. A small loop antenna can approach that performance except for a reduction in bandwidth, but that effect can be overcome by retuning. This section was written by Robert T. (Ted) Hart, W5QJR. It includes information extracted from his book, *Small High Efficiency Antennas Alias the Loop.*

Small antennas are characterized by low radiation resistance. Typically, loading coils are added to small antennas to achieve resonance. However, the loss in the coils results in an antenna with low efficiency. If instead of coils a large capacitor is added to a low-loss conductor to achieve resonance, and if the antenna conductor is bent to connect the ends to the capacitor, a loop is formed. Based on this concept, the small loop is capable of high efficiency. In addition, the small loop, when mounted vertically, has the

unique characteristic of radiation at all elevation angles. Therefore it can replace both vertical and dipole antennas. Small size and high efficiency are advantages of using a properly designed and constructed loop on the lower frequency bands.

The only deficiency in a small loop antenna is narrow bandwidth; it must be tuned to the operating frequency. However, the use of a remote motor drive allows the loop to be tuned over a wide frequency range. For example, two loops could be constructed to provide continuous frequency coverage from 3.5 to 30 MHz.

The small transmitting loop has been around since 1957 (see the Patterson bibliography reference). Only recently has the small loop been developed into a practical antenna for amateurs. The most important aspect of the development was establishing a complete set of mathematical equations to define the loop. This was followed by designing a simple feed system, and finally a practical tuning capacitor was found. The results of this development program are presented here. Fig 17 presents computer-derived data for various size loop antennas for the HF amateur bands.

Loop Fundamentals

A small loop has the radiation pattern shown in Fig 18.

The pattern is easily conceived as a doughnut with a hole (null) in the pattern through the center of the loop at low elevation angles. When the circumference of the loop is less than ⅓ λ, regardless of the shape of the loop (round or square), that pattern will be obtained. In the majority of applications the loop will be mounted vertically. Mounted this way, it radiates at all vertical angles in the plane of the loop.

The loop has been defined mathematically by the

Loop No. 1

Frequency range, MHz	7.6-29.4					
Loop circumference, feet	8.5					
Conductor dia, inches	0.9					
Radials	No					
Frequency, MHz	10.1	14.2	18.0	21.2	24.0	29.0
Efficiency, dB below 100%	−6.5	−3.1	−1.6	−1.0	−0.7	−0.4
Bandwidth, kHz	5.5	9.9	18.2	30.2	46.0	91.4
Q	1552	1212	835	591	439	267
Tuning capacitance, pF	102.6	48.0	26.8	17.1	11.6	5.4
Capacitor voltage, kV P-P	38.21	40.03	37.40	34.16	31.32	26.86
Capacitor spacing, inches	0.255	0.267	0.249	0.228	0.209	0.179
Radiation resistance, ohms	0.009	0.034	0.088	0.170	0.279	0.594
Loss resistance, ohms	0.030	0.035	0.040	0.043	0.046	0.051

Loop No. 2

Frequency range, MHz	3.6-16.4			
Loop circumference, feet	20			
Conductor dia, inches	0.9			
Radials	No			
Frequency, MHz	4.0	7.2	10.1	14.2
Efficiency, dB below 100%	−8.9	−2.7	−1.0	−0.3
Bandwidth, kHz	3.3	8.4	22.1	73.8
Q	1356	965	515	217
Tuning capacitance, pF	310.5	86.1	36.8	11.6
Capacitor voltage, kV P-P	38.28	43.33	37.48	28.83
Capacitor spacing, inches	0.255	0.289	0.250	0.192
Radiation resistance, ohms	0.007	0.069	0.268	1.047
Loss resistance, ohms	0.044	0.059	0.070	0.083

Loop No. 3

Frequency range, MHz	2.1-10.0		
Loop circumference, feet	38		
Conductor dia, inches	0.9		
Radials	No		
Frequency, MHz	3.5	4.0	7.2
Efficiency, dB below 100%	−4.1	−3.0	−0.5
Bandwidth, kHz	4.2	5.6	33.2
Q	1014	880	265
Tuning capacitance, pF	192.3	142.4	29.9
Capacitor voltage, kV P-P	45.63	45.43	33.47
Capacitor spacing, inches	0.304	0.303	0.223
Radiation resistance, ohms	0.050	0.086	0.902
Loss resistance, ohms	0.079	0.084	0.113

Loop No. 4

Frequency range, MHz	0.9-4.1		
Loop circumference, feet	100		
Conductor dia, inches	0.9		
Radials	No		
Frequency, MHz	1.8	2.0	3.5	4.0
Efficiency, dB below 100%	−2.7	−2.1	−0.4	−0.2
Bandwidth, kHz	3.4	4.4	27.7	45.9
Q	663	565	156	108
Tuning capacitance, pF	215.7	166.4	24.9	8.8
Capacitor voltage, kV P-P	46.75	45.48	31.63	28.09
Capacitor spacing, inches	0.312	0.303	0.211	0.187
Radiation resistance, ohms	0.169	0.257	2.415	4.120
Loss resistance, ohms	0.148	0.157	0.207	0.221

Fig 17—Design data for loops to cover various frequency ranges. The information is calculated for an 8-sided loop, as shown in Fig 20. The capacitor specification data is based on 1000 W of transmitted power. See text for modifying these specifications for other power levels.

equations in Table 4. By using a computer and entering the circumference of the loop and the conductor diameter, all of the performance parameters can be calculated from these equations. Through such an analysis, it has been determined that the optimum size conductor is ¾-inch copper pipe, considering both performance and cost.

The loop circumference should be between 1/4 and 1/8 λ at the operating frequency. It will become self-resonant above ¼ λ, and efficiency drops rapidly below 1/8 λ. In the frequency ranges shown in Fig 17, the high frequency is for 5 pF of tuning capacitance, and the low frequency is that at which the loop efficiency is down from 100% by 10 dB.

Where smaller loops are needed, the efficiency can be increased by increasing the pipe size or by adding radials to form a ground screen under the loop (data are given in Fig 17). The effect of radials is to double the antenna area

Loop No. 5

Frequency range, MHz	5.1-29.4						
Loop circumference, feet	8.5						
Conductor dia, inches	0.9						
Radials	Yes						
Frequency, MHz	7.2	10.1	14.2	18.0	21.2	24.0	29.0
Efficiency, dB below 100%	−5.8	−2.7	−1.0	−0.5	−0.3	−0.2	−0.1
Bandwidth, kHz	4.9	9.2	24.4	55.7	102.4	164.6	344.1
Q	1248	925	490	272	174	123	71
Tuning capacitance, pF	209.7	102.6	48.0	26.8	17.1	11.6	5.4
Capacitor voltage, kV P-P	28.92	29.49	25.46	21.36	18.55	16.56	13.84
Capacitor spacing, inches	0.193	0.197	0.170	0.142	0.124	0.110	0.092
Radiation resistance, ohms	0.009	0.035	0.137	0.353	0.679	1.115	2.377
Loss resistance, ohms	0.025	0.030	0.035	0.040	0.043	0.046	0.051

Loop No. 6

Frequency range, MHz	2.4-16.4				
Loop circumference, feet	20				
Conductor dia, inches	0.9				
Radials	Yes				
Frequency, MHz	3.5	4.0	7.2	10.1	14.2
Efficiency, dB below 100%	−5.7	−4.3	−0.8	−0.3	−0.1
Bandwidth, kHz	3.7	4.6	21.9	74.5	278.7
Q	1061	976	369	152	57
Tuning capacitance, pF	409.8	310.5	86.1	36.8	11.6
Capacitor voltage, kV P-P	31.68	32.48	26.80	20.40	14.83
Capacitor spacing, inches	0.211	0.217	0.179	0.136	0.099
Radiation resistance, ohms	0.015	0.026	0.277	1.072	4.187
Loss resistance, ohms	0.041	0.044	0.059	0.070	0.083

Loop No. 7

Frequency range, MHz	1.4-10.0				
Loop circumference, feet	38				
Conductor dia, inches	0.9				
Radials	Yes				
Frequency, MHz	1.8	2.0	3.5	4.0	7.2
Efficiency, dB below 100%	−7.0	−5.8	−1.4	−1.0	−0.1
Bandwidth, kHz	2.3	2.6	9.2	14.0	121.8
Q	955	924	467	350	72
Tuning capacitance, pF	783.7	630.9	192.3	142.4	29.9
Capacitor voltage, kV P-P	31.74	32.92	30.97	28.64	17.48
Capacitor spacing, inches	0.212	0.219	0.206	0.191	0.117
Radiation resistance, ohms	0.014	0.021	0.201	0.344	3.607
Loss resistance, ohms	0.056	0.059	0.079	0.084	0.113

Loop No. 8

Frequency range, MHz	0.6-4.1			
Loop circumference, feet	100			
Conductor dia, inches	0.9			
Radials	Yes			
Frequency, MHz	1.8	2.0	3.5	4.0
Efficiency, dB below 100%	−0.9	−0.6	−0.1	−0.1
Bandwidth, kHz	8.7	12.5	104.2	176.4
Q	255	197	41	28
Tuning capacitance, pF	215.7	166.4	24.9	8.8
Capacitor voltage, kV P-P	29.01	26.87	16.30	14.32
Capacitor spacing, inches	0.193	0.179	0.109	0.095
Radiation resistance, ohms	0.676	1.030	9.659	16.478
Loss resistance, ohms	0.148	0.157	0.207	0.221

Fig 17 Continued.

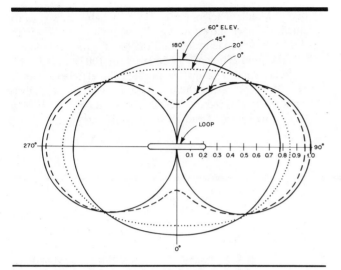

Fig 18—Azimuth patterns of a small vertical loop antenna at various elevation angles. The loop is bidirectional, with the greatest signal strength in the plane of the loop, as shown. In these directions the loop is vertically polarized.

because of the loop image. The length of each radial need be only twice the loop diameter. It should be noted that ¼ λ radials should be used for loops mounted over poor ground to improve performance. Data for Fig 17 was computed for ¾ inch copper water pipe (nominal OD of 0.9 inch). By comparing figures with radials (perfect screen assumed) and without, you will note that the effect of radials is greater for loops with a smaller circumference, for a given frequency. Also note the efficiency is higher and the Q is lower for loops having a circumference near ¼ λ. Larger pipe size will reduce the loss resistance, but the Q increases. Therefore the bandwidth decreases, and the voltage across the tuning capacitor increases.

The shape of the pipe forms a single-turn coil. The value of inductance and stray capacitance can be calculated and a corresponding value of tuning capacitance calculated to provide resonance for a given frequency. Fig 19 allows the selection of loop size versus tuning capacitance for any desired operating frequency range for the HF amateur bands. For example, a capacitor that varies from 5 to 50 pF, used with a loop 10 feet in circumference, tunes from 13 to 27 MHz (represented by the left dark vertical bar). A 25-150 pF

Table 4
Basic Equations for a Small Loop

Radiation resistance, ohms $\quad R_R = 3.38 \times 10^{-8} \, (f^2 A)^2$

Loss resistance, ohms $\quad R_L = 9.96 \times 10^{-4} \, \sqrt{f} \, \dfrac{S}{d}$

Efficiency $\quad \eta = \dfrac{R_R}{R_R + R_L}$

Inductance, henrys $\quad L = 1.9 \times 10^{-8} \, S \left(7.353 \log_{10} \dfrac{96 \, S}{\pi \, d} - 6.386 \right)$

Inductive reactance, ohms $\quad X_L = 2 \, \pi \, f L \times 10^6$

Tuning capacitor, farads $\quad C_T = \dfrac{1}{2 \, \pi \, f \, X_L \times 10^6}$

Quality factor $\quad Q = \dfrac{f \times 10^6}{\Delta f} = \dfrac{X_L}{2(R_R + R_L)}$

Bandwidth, hertz $\quad \Delta f = \dfrac{f \times 10^6}{Q} = (f_1 - f_2) \times 10^6$

Distributed capacity, pF $\quad C_D = 0.82S$

Capacitor potential, volts $\quad V_C = \sqrt{P X_L Q}$

where
    f = operating frequency, MHz
    A = area of loop, square feet
    S = conductor length, feet
    d = conductor diameter, inches
    η = decimal value; dB = $10 \log_{10} \eta$
    P = transmitter power, watts

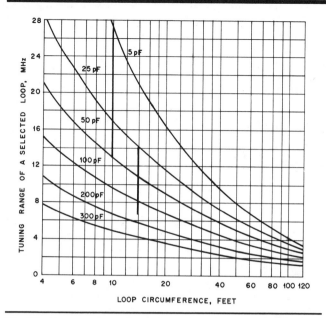

Fig 19—Frequency tuning range of an octagon-shaped loop using ¾-inch copper water pipe, for various values of tuning capacitance and loop circumference.

two, but the voltage rating is doubled.

The capacitor must be selected for transmitting-loop application; that is, all contacts must be welded, and no mechanical wiping contacts are allowed. For example, if the spacers between plates are not welded to the plates, there will be loss in each joint, and thus degraded loop efficiency. (Earlier transmitting loops exhibited poor efficiency because capacitors with wiping contacts were used.) There are two types of capacitors available for this application. A vacuum variable is an excellent choice, provided one is selected with adequate voltage rating. Unfortunately, those capacitors are very expensive. For this reason, a 170-pF variable—340 pF per section, with ¼-inch spacing—has been designed for transmitting loops (available from W5QJR Antenna Products; see Fig 20 caption). Another alternative is to obtain a large air variable, remove the aluminum plates, and replace them with copper or double-sided pc board material. Connect all plates together on the rotor and on the stators. Solder copper straps to the capacitor for soldering to the loop itself.

The spacing between plates determines the voltage-handling capability, rated at 75,000 volts per inch. Fig 17 includes the spacing required for each section of the split-stator capacitor for 1000 watts RF power. For other power ratings, multiply the spacing (and voltage) by the square root of the ratio of your power to 1000 watts. For example, for 100 watts, the ratio would be $\sqrt{100/1000} = 0.316$.

Remote Tuning

Because of the narrow bandwidth, the loop must be retuned each time the operating frequency is changed by more than a few kilohertz. A very high resolution motor and gear train is required. The use of a stepper motor with integral gear train provides an excellent drive. The preferred unit is available from Hurst Manufacturing Co, Princeton, IN 47670. Use motor no. 304-001 and controller no. 22001. The controller is an integrated circuit that provides all the functions of speed control and direction of rotation. Add a variable resistor for speed control, control switches and a 12 V dc source, and you have a complete drive. (The 1988 cost of the motor and controller is about $90.) For high RF power, it is advisable to add low-pass filters in the motor leads near the controller, to prevent RF from damaging the controller. Use 100 μH RF chokes (Radio Shack), with 0.011 μF disc capacitors from either side to ground.

CONSTRUCTION

After you select the loop design for your application, construct it as shown in Fig 20. The efficiency of a small loop is related to area, and therefore a round loop would provide the maximum area for a given circumference. The octagon shape is much easier to construct, with only a small difference in area. The third choice would be a square. The values presented in Fig 17 are for an octagon.

For a given loop circumference, divide the circumference by 8 and cut 8 equal-length pieces of ¾-inch copper pipe. Join the pieces with 45° elbows to form the octagon. With the loop lying on the ground, braze or solder all joints. In the center of one leg, cut the pipe and install a copper T. Adjacent to the T, install a mount for the coax connector. Make the mount from copper strap, which can be obtained by splitting a short piece of pipe and hammering it flat.

Make a box from clear plastic to house the variable

capacitor with a 13.5-foot loop covers the 7-14.4 MHz range, represented by the right vertical bar.

The equivalent electrical circuit for the loop is a parallel resonant circuit with a very high Q, and therefore a narrow bandwidth. The efficiency is a function of radiation resistance divided by the sum of the radiation plus loss resistance. The radiation resistance is much less than 1 Ω, so it is necessary to minimize the loss resistance, which is largely the skin effect loss of the conductor. However, if the loss is too low, the Q will be excessive and the bandwidth will be too narrow for practical use. These reasons dictate the need for a complete analysis to be performed before proceeding with the construction of a loop.

Additional Loss

When the loop is constructed, it will provide the performance determined from the equations of Table 4, provided no additional loss components are introduced. There are two sources for the additional losses. First, if the loop is mounted near lossy metallic conductors, the large magnetic field will induce currents into those conductors and be reflected as loss in the loop. Therefore the loop should be as far from other conductors as possible. If you use the loop inside a building constructed with large amounts of iron or near ferrous materials, you will simply have to live with the loss if the loop cannot otherwise be relocated.

The second source of loss is from poor construction, which can be avoided. All joints in the loop must be brazed or soldered. This also applies to the tuning capacitor. The use of a split-stator capacitor eliminates the resistance of wiper contacts, resistance that is inherent in a single-section capacitor. The loop ends are connected to the stators, and the rotor forms the variable coupling path between stators. With this arrangement the value of capacitance is divided by

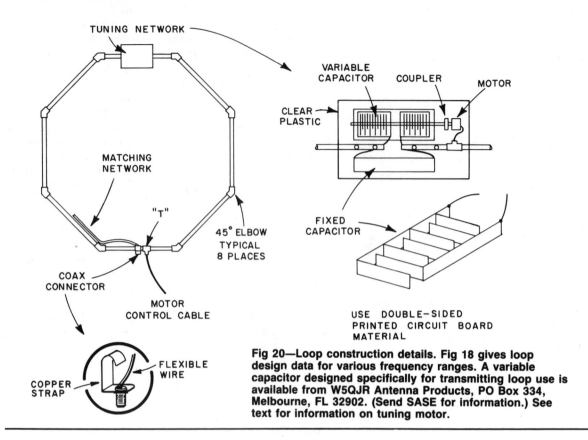

TUNING NETWORK

VARIABLE CAPACITOR   COUPLER   MOTOR

CLEAR PLASTIC

MATCHING NETWORK

"T"

45° ELBOW TYPICAL 8 PLACES

COAX CONNECTOR

MOTOR CONTROL CABLE

FIXED CAPACITOR

USE DOUBLE-SIDED PRINTED CIRCUIT BOARD MATERIAL

COPPER STRAP

FLEXIBLE WIRE

Fig 20—Loop construction details. Fig 18 gives loop design data for various frequency ranges. A variable capacitor designed specifically for transmitting loop use is available from W5QJR Antenna Products, PO Box 334, Melbourne, FL 32902. (Send SASE for information.) See text for information on tuning motor.

capacitor and drive motor. The side of the box that mounts to the loop and the capacitor should be at least 1/4 inch thick, preferably 3/8 inch. The remainder of the box can be 1/8-inch plastic sheet. Any good sign shop will cut the pieces to size for you. Mount the loop to the plastic using 1/4-inch bolts (two on either side of center). Remove the bolts and cut out a section of pipe 2 inches wide in the center. On the motor side of the capacitor, cut the pipe and install a copper T for the motor wiring.

The next step is to solder copper straps to the loop ends and to the capacitor stators, then remount the loop to the plastic. If you insert wood dowels, the pipe will remain round when you tighten the bolts.

Now you can install the motor drive cable through the loop and connect it to the motor. Antenna rotator cable is a good choice for this cable. Complete the plastic box using short pieces of aluminum angle and small sheet metal screws to join the pieces.

The loop is now ready to raise to the vertical position. Remember, no metal is allowed near the loop. Make a pole of 2 × 4-inch lumber with 1 × 4-inch boards on either side to form an I section. Hold the boards together with 1/4-inch bolts, 2 feet apart. Tie rope guys to the top. This makes an excellent mast up to 50 feet high. The pole height should be one foot greater than the loop diameter, to allow room for cutting grass or weeds at the bottom of the loop. By installing a pulley at the top, the loop can be raised and supported by rope. Support the bottom of the loop by tying it to the pole.

Tie guy ropes to the sides of the loop to keep it from rotating in the wind. By moving the anchor points, the loop can be rotated in the azimuth plane.

With the loop in the vertical position, cut a piece of 1/4-inch copper tubing the length of one of the sides of the loop. Flatten one end and solder a piece of flexible wire to the other. Wrap the tubing with electrical tape or cover with plastic tubing for insulation. Connect the flexible wire to the coax connector and install the tubing against the inside of the loop. Hold in place with tape. Solder the flat part to the loop. You have just constructed a form of gamma match, but without reactive components. This simple feed will provide better than 1.7:1 SWR over a 2:1 frequency range for the resonated loop. For safety, install a good ground rod under the loop and connect it to the strap for the coax connector, using large flexible wire.

TUNE-UP PROCEDURE

The resonant frequency of the loop can be readily found by setting the receiver to a desired frequency and rotating the capacitor (via remote control) until signals peak. The peak will be very sharp because of the high Q of the loop. Incidentally, the loop typically reduces electrostatic noise 26 dB compared to dipoles or verticals, thus allowing improved reception in noisy areas.

Turn on the transmitter in the TUNE mode and adjust either the transmitter frequency or the loop capacitor for maximum signal on a field strength meter, or for maximum forward signal on an SWR bridge. Adjust the matching network for minimum SWR by bending the matching line. Normally a small hump in the ¼-inch tubing line, as shown in Fig 20, will give the desired results. For a loop that covers two or more bands, adjust the feed to give equally low SWR at each end of the tuning range. The SWR will be very low in the center of the tuning range but will rise at each end.

If there is metal near the loop, the additional loss will reduce the Q and therefore the impedance of the loop. In those cases it will be necessary to increase the length of the matching line and tap higher up on the loop to obtain a 50-Ω match.

PERFORMANCE COMPARISON

As previously indicated, the loop will provide performance approaching full-size dipoles and verticals. To illustrate one case, a loop 100 feet in circumference would be 30 feet high for 1.8 MHz. However, a good dipole would be 240 feet (½ λ) in length and 120 feet high (¼ λ). A ¼-λ vertical would be 120 feet tall with a large number of radials, each 120 feet in length. The small loop would replace both of those

antennas. Since very few hams have full-size antennas on 1.8 MHz, it is easy for a loop to emanate the "big signal on the band."

On the higher frequencies, the same ratios apply, but the full-size antennas are less dramatic. However, very few city dwellers can erect good verticals even on 7 MHz with a full-size counterpoise. Even on 14 MHz a loop about 3 feet high can work the world.

Additional Comments

The loop should not be mounted horizontally except at great heights. The pattern for a horizontal loop will be horizontally polarized, but it will have a null overhead and be omnidirectional in the azimuth plane. The effect of the earth would be the same as on the pattern of a horizontal dipole at the same height.

It has taken a number of years to develop this small loop into a practical antenna for amateurs. Other than trading small size for narrow bandwidth, the loop is an excellent antenna and will find use where large antennas are not practical. It should be a useful antenna to a large number of amateurs.

The Loop Skywire

Are you looking for a multiband HF antenna that is easy to construct, costs nearly nothing and works great? Try this one. This information is based on a November 1985 *QST* article by Dave Fischer, W0MHS.

There is one wire antenna that performs exceptionally well on the lower HF bands, but relatively few amateurs use it. This is a full-size horizontal loop. The Loop Skywire antenna is that type. It is fundamental and simple, easy to construct, costs nearly nothing, and eliminates the need for multiple antennas to cover the HF bands. It is made only of wire and coaxial cable, and often needs no Transmatch. It is an efficient antenna that is omnidirectional over real earth. It is noticeably less susceptible than dipoles and verticals to man-made and atmospheric noise. The antenna can also be used on harmonics of the fundamental frequency, and fits on almost every amateur's lot.

It is curious that many references to this antenna are brief pronouncements that it operates best as a high-angle radiator and is good for only short-distance contacts. Such statements, in effect, dismiss this antenna as useless for most amateur work. This is not the case! Those who use the Loop Skywire know that its performance far exceeds the short haul. DX is easy to work.

THE DESIGN

The Loop Skywire is shown in Fig 21. This antenna is

a magnetic version of the open-wire, center-fed electric dipole that has performed extraordinarily well for many decades. Yet this one is less difficult to match and use. It is simply a loop antenna erected horizontal to the earth. Maximum enclosed area within the wire loop is the fundamental rule. The antenna has one wavelength of wire in its perimeter at the design or fundamental frequency. If you choose to calculate L_{total} in feet, the following equation should be used.

$$L_{total} = \frac{1005}{f}$$

where f equals the frequency in MHz

Given any length of wire, the maximum possible area the antenna can enclose is with the wire in the shape of a circle. Since it takes an infinite number of supports to hang a circular loop, the square loop (four supports) is the most practical. Further reducing the area enclosed by the wire loop (fewer supports) brings the antenna closer to the properties of the folded dipole, and both harmonic-impedance and feed-line voltage problems can result. Loop geometries other than a square are thus possible, but remember the two fundamental requirements for the Loop Skywire—its horizontal position and maximum enclosed area.

A little-known fact in the amateur community is that loops can be fed simply at all harmonics of the design frequency. There is another great advantage to this antenna system. It can be operated as a vertical antenna with top-hat

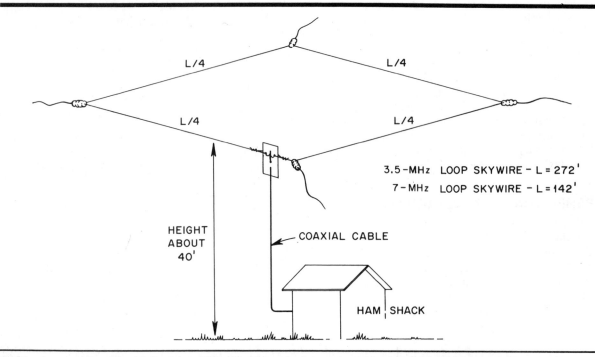

Fig 21—A complete view of the Loop Skywire. The square loop is erected horizontal to the earth.

loading on all bands as well. This is accomplished by simply keeping the feed-line run from the antenna to the shack as vertical as possible and clear of objects. Both feed-line conductors are then tied together (via a shorted SO-239 jack, for example), and the antenna is fed against a good ground.

CONSTRUCTION

Antenna construction is simple. Although the loop can be made for any band or frequency of operation, the following two Loop Skywires are star performers. The 10-MHz band can also be operated on both.

3.5-MHz Loop Skywire (3.5-28 MHz loop and 1.8-MHz vertical)

    Total loop perimeter: 272 feet
    Square side length: 68 feet

7-MHz Loop Skywire (7-28 MHz loop and 3.5-MHz vertical)

    Total loop perimeter: 142 feet
    Square side length: 35.5 feet

The actual total length can vary from the above by a few feet, as the length is not at all critical. Do not worry about tuning and pruning the loop to resonance. No signal difference will be detected on the other end when that method is used.

Copper wire is usually used in the loop. Lamp or "zip" cord and Copperweld can also be used. Several loops have even been constructed successfully with steel wire, but soldering is difficult. Fig 22 shows the placement of the insulators at the loop corners. Two common methods are used to attach the insulators. Either lock or tie the insulator in place with a loop wire tie, as shown in Fig 22A, or leave the insulator free to "float" or slide along the wire, Fig 22B. Most loop users float at least two insulators. This allows pulling the slack out of the loop once it is in the air, and eliminates the need to have all the supports exactly placed for proper tension in

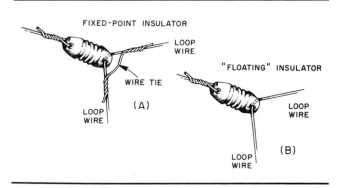

Fig 22—Two methods of installing the insulators at the loop corners.

each leg. Floating two opposite corners is recommended. The feed point can be positioned anywhere along the loop that you wish. However, most users feed the Skywire at a corner. Fig 23 depicts a method of doing this. It is advantageous to keep the feed point mechanicals away from the corner support. Feeding a foot or so from one corner allows the feed line to exit more freely. This method keeps the feed line free from the loop support.

Generally a minimum of four supports is required. If trees are used for supports, then at least two of the ropes or guys used to support the insulators should be counterweighted and allowed to move freely. The feed-line corner is almost always tied down, however. Very little tension is needed to support the loop (far less than that for a dipole). Thus, counterweights are light. Several loops have been constructed with bungie cords tied to three of the four insulators. This eliminates the need for counterweighting.

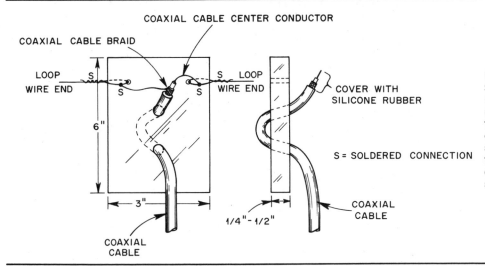

Fig 23—Most users feed the Skywire at a corner. A high-impedance weather-resistant insulant should be used for the feed-point insulator. Cover the end of the coaxial cable with silicone rubber for protection from the weather and added electrical insulation. Dimensions shown are approximate.

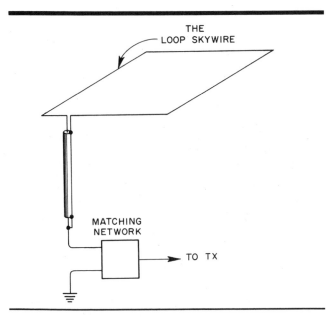

Fig 24—The feed arrangement for operating the loop as a vertical antenna.

on your operating frequency and the type of feed line used. Coaxial cable is sufficient. Open wire does not appear to make the loop perform any better or matching to it easier. Most users feed with RG-58, RG-59 or RG-62. RG-8 and RG-11 are generally too cumbersome to use. With full power and coaxial cable feeding these loops, feed-line problems have not been reported. The SWR from either of these loops is rarely over 3:1. If you are concerned about the SWR, use a Transmatch and eliminate all worries about power transfer and maximum signal strength. When constructing the loop, connect (solder) the coaxial feed line ends directly to the loop wire ends. Don't do anything else. Baluns or choke coils at the feed point are not to be used. They are unnecessary. Don't let anyone talk you into using them. The feed arrangement for operating the loop as a vertical antenna is shown in Fig 24.

The highest line SWR usually occurs at the second harmonic of the design frequency. The Loop Skywire is somewhat more broadband than corresponding dipoles, and the loop is efficient. Do not expect SWR curves that are "dummy load" flat!

Since the loop is high in the air and has considerable electrical exposure to the elements, proper methods should be employed to eliminate the chance of induced or direct lightning hazard to the shack and operator. Some users simply employ a three-connector (PL-259/PL-258/PL-259) weather-protected junction in the feed line outside the shack and completely disconnect the antenna from the rig and shack during periods of possible lightning activity.

Some skeptics have commented that the Loop Skywire is actually a vertical antenna in disguise. Yet when the loops have been used in on-the-air tests with both local and distant stations, the loop operating as a loop consistently "out-signals" the loop operating as a vertical.

Recommended height for the antenna is 40 feet or more. The higher the better, especially if you wish to use the loop in the vertical mode. However, successful local and DX operation has been reported in several cases with the antenna at 20 feet.

If you are preoccupied with SWR, the reading will depend

7-MHz Loop

An effective but simple 7-MHz antenna that has a theoretical gain of approximately 2 dB over a dipole is a full-wave, closed vertical loop. Such a loop need not be square, as illustrated in Fig 25. It can be trapezoidal, rectangular, circular, or some distorted configuration in between those shapes. For best results, however, the builder should attempt to make the loop as square as possible. The more rectangular the shape, the greater the cancellation of energy in the system, and the less effective it will be. The effect is similar to that of a dipole, its effectiveness becoming impaired as the ends of the dipole are brought closer and closer together. The practical limit can be seen in the inverted-V dipole antenna, where a 90° apex angle between the legs is the minimum value ordinarily used. Angles that are less than 90° cause serious cancellation of the RF energy.

The loop can be fed in the center of one of the vertical sides if vertical polarization is desired. For horizontal polarization, it is necessary to feed either of the horizontal sides at the center.

Optimum directivity occurs at right angles to the plane of the loop, or in more simple terms, broadside from the loop. One should try to hang the system from available supports which will enable the antenna to radiate the maximum amount in some favored direction.

Just how the wire is erected will depend on what is available in one's yard. Trees are always handy for supporting antennas, and in many instances the house is high enough to be included in the lineup of solid objects from which to hang a radiator. If only one supporting structure is available it should be a simple matter to put up an A frame or pipe mast to use as a second support. (Also, tower owners see Fig 25 inset.)

The overall length of the wire used in a loop is determined in feet from the formula 1005/f(MHz). Hence, for operation at 7.125 MHz the overall wire length will be 141 feet. The matching transformer, an electrical ¼ λ of 75 Ω coax cable, can be computed by dividing 246 by the operating frequency in MHz, then multiplying that number by the velocity factor of the cable being used. Thus, for operation at 7.125 MHz, 246/7.125 MHz = 34.53 feet. If coax with solid polyethylene insulation is used a velocity factor of 0.66 must be employed. Foam-polyethylene coax has a velocity factor of 0.80. Assuming RG-59 is used, the length of the matching transformer becomes 34.53 (feet) × 0.66 = 22.79 feet, or 22 feet, 9½ inches.

This same loop antenna may be used on the 14- and 21-MHz bands, although its pattern will be somewhat different than on its fundamental frequency. Also, a slight

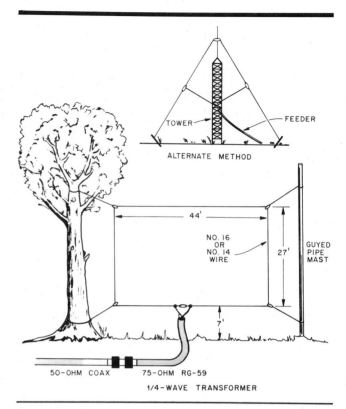

Fig 25—Details of the full-wave loop. The dimensions given are for operation at 7.05 MHz. The height above ground was 7 feet in this instance, although improved performance should result if the builder can install the loop higher above ground without sacrificing length on the vertical sides. The inset illustrates how a single supporting structure can be used to hold the loop in a diamond-shaped configuration. Feeding the diamond at the lower tip provides radiation in the horizontal plane. Feeding the system at either side will result in vertical polarization of the radiated signal.

mismatch will occur, but this can be overcome by a simple matching network. When the loop is mounted in a vertical plane, it tends to favor low-angle signals. If a high-angle system is desired, say for 3.5 MHz, the full-wave loop can be mounted in a horizontal plane, 30 or more feet above ground. This arrangement will direct most of the energy virtually straight up, providing optimum sky-wave coverage on a short-haul basis.

A Receiving Loop for 1.8 MHz

Small shielded loop antennas can be used to improve reception under certain conditions, especially at the lower amateur frequencies. This is particularly true when high levels of man-made noise are prevalent, when the second-harmonic energy from a nearby broadcast station falls in the 1.8-MHz band, or when interference exists from some other amateur station in the immediate area. A properly constructed and tuned small loop will exhibit approximately 30 dB of front-to-side response, the minimum response being at right angles to the plane of the loop. Therefore, noise and interference can be reduced significantly or completely nulled out, by rotating the loop so that it is sideways to the interference-causing source. Generally speaking, small shielded loops are far less responsive to man-made noise than are the larger antennas used for transmitting and receiving. But a trade-off in performance must be accepted when using the loop, for the strength of received signals will be 10 or 15 dB less than when using a full-size resonant antenna. This condition is not a handicap on 1.8 or 3.5 MHz, provided the station receiver has normal sensitivity and overall gain. Because a front-to-side ratio of 30 dB may be expected, a shielded loop can be used to eliminate a variety of receiving problems if made rotatable, as shown in Fig 26.

To obtain the sharp bidirectional pattern of a small loop, the overall length of the conductor must not exceed 0.1 λ. The loop of Fig 27 has a conductor length of 20 feet. At 1.81 MHz, 20 feet is 0.036 λ. With this style of loop, 0.036 λ is about the maximum practical dimension if one is to tune the element to resonance. This limitation results from the distributed capacitance between the shield and inner conductor of the loop. RG-59 was used for the loop element in this example. The capacitance per foot for this cable is 21 pF, resulting in a total distributed capacitance of 420 pF. An additional 100 pF was needed to resonate the loop at 1.810 MHz. Therefore, the approximate inductance of the loop is 15 μH. The effect of the capacitance becomes less pronounced at the higher end of the HF spectrum, provided the same percentage of a wavelength is used in computing the conductor length. The ratio between the distributed capacitance and the lumped capacitance used at the feed point becomes greater at resonance. These facts should be con-

templated when scaling the loop to those bands above 1.8 MHz.

There will not be a major difference in the construction requirements of the loop if coaxial cables other than RG-59 are used. The line impedance is not significant with respect to the loop element. Various types of coaxial line exhibit different amounts of capacitance per foot, however, thereby requiring more or less capacitance across the feed point to establish resonance.

Shielded loops are not affected noticeably by nearby objects, and therefore they can be installed indoors or out after being tuned to resonance. Moving them from one place to another does not significantly affect the tuning.

In the model shown here it can be seen that a supporting structure was fashioned from bamboo poles. The X frame is held together at the center by means of two U bolts. The loop element is taped to the cross-arms to form a square. It is likely that one could use metal cross arms without seriously degrading the antenna performance. Alternatively, wood can be used for the supporting frame.

A Minibox was used at the feed point of the loop to contain the resonating variable capacitor. In this model a 50- to 400-pF compression trimmer is used to establish resonance. It is necessary to weatherproof the box for outdoor installations.

The shield braid of the loop coax is removed for a length of one inch directly opposite the feed point. The exposed areas should be treated with a sealing compound once this is done.

In operation this receiving loop has been very effective in nulling out second-harmonic energy from local broadcast stations. During DX and contest operation on 1.8 MHz it helped prevent receiver overloading from nearby 1.8-MHz stations that share the band. The marked reduction in response to noise has made the loop a valuable station accessory when receiving weak signals. It is not used all of the time, but is available when needed by connecting it to the

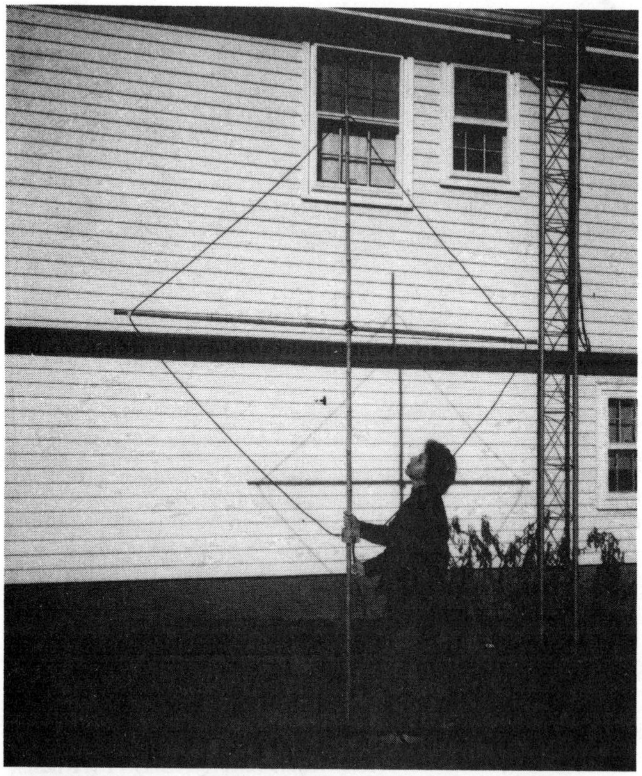

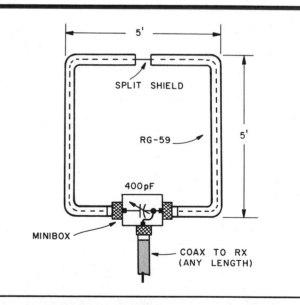

Fig 27—Schematic diagram of the loop antenna. The dimensions are not critical provided overall length of the loop element does not exceed approximately 0.1 λ. Small loops which are one half or less the size of this one will prove useful where limited space is a consideration.

Fig 26—Jeannie DeMaw, W1CKK, tests the 1.8-MHz shielded loop. Bamboo cross arms are used to support the antenna.

receiver through an antenna selector switch. Reception of European stations with the loop has been possible from New England at times when other antennas were totally ineffective because of noise.

It was also discovered that the effects of approaching storms (with attendant atmospheric noise) could be nullified considerably by rotating the loop away from the storm front. It should be said that the loop does not exhibit meaningful directivity when receiving sky-wave signals. The directivity characteristics relate primarily to ground-wave signals. This is a bonus feature in disguise, for when nulling out local noise

or interference, one is still able to copy sky-wave signals from all compass points!

For receiving applications it is not necessary to match the feed line to the loop, though doing so may enhance the performance somewhat. If no attempt is made to obtain an SWR of 1, the builder can use 50- or 75-Ω coax for a feeder, and no difference in performance will be observed. The Q of this loop is sufficiently low to allow the operator to peak it for resonance at 1.9 MHz and use it across the entire 1.8-MHz band. The degradation in performance at 1.8 and 2 MHz will be so slight that it will be difficult to discern.

The Bi-Square Antenna

A development of the lazy H, known as the *bi-square* antenna, is shown in Fig 28. The gain of the bi-square is somewhat less than that of the lazy H, but this array is attractive because it can be supported from a single pole. It has a circumference of 2 λ at the operating frequency, and is vertically polarized.

The bi-square antenna is not a true loop by strict definition because the ends opposite the feed point are open. However, it is included in this chapter because identical construction techniques can be used for the two antenna types. Indeed, with a means of remotely closing the connection at the top for lower frequency operation, the antenna can be operated on two harmonically related bands. As an example, an array with 17 feet per side can be operated as a bi-square at 28 MHz and as a full-wave loop at 14 MHz. (The polarization will be horizontal at 14 MHz). For two-band operation in this manner, the side length should favor the higher frequency, using the formula in the caption of Fig 28. The length of a closed loop is not as critical.

The bi-square antenna consists of two 1-λ radiators, fed 180° out of phase at the bottom of the array. The radiation resistance is 300 Ω, so it can be fed with either 300 or 600 Ω line. The gain usually claimed is 4 dBd, but in practice a gain of 3 dBd is probably more realistic because of the rather close spacing of the elements. The gain may be increased by adding a reflector or director. Two bi-square arrays can be mounted at right angles and switched to provide omnidirectional coverage. In this way, the antenna wires may be used as part of the guy system.

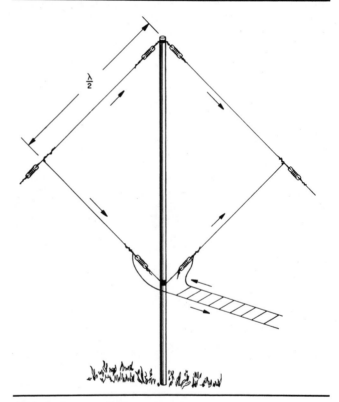

Fig 28—The bi-square array. It has the appearance of a loop, but is not a true loop because the conductor is open at the top. The length of each side, in feet, is 480/f (MHz).

BIBLIOGRAPHY

Source material and more extended discussion of topics covered in this chapter can be found in the references given below and in the textbooks listed at the end of Chapter 2.

C. F. W. Anderson, "A Crossed-Loop/Goniometer DF Antenna for 160 Meters," *The ARRL Antenna Compendium, Volume 1* (Newington: ARRL, 1985), pp 127-132.

J. S. Belrose, "Ferromagnetic Loop Aerials," *Wireless Engineer*, Feb 1955, pp 41-46.

D. S. Bond, *Radio Direction Finders*, 1st ed. (New York: McGraw-Hill Book Co, 1944).

G. Bramslev, "Loop Aerial Reception," *Wireless World*, Nov 1952, pp 469-472.

R. W. Burhans, "Experimental Loop Antennas for 60 kHz to 200 kHz," *Technical Memorandum (NASA) 71* (Athens, OH: Ohio Univ, Dept of Electrical Engr), Dec 1979.

R. W. Burhans, "Loop Antennas for VLF-LF," *Radio-Electronics*, Jun 1983, pp 83-87.

D. DeMaw, "Beat the Noise with a Scoop Loop," *QST*, Jul 1977, pp 30-34.

M. F. DeMaw, *Ferromagnetic-Core Design and Application Handbook* (Englewood Cliffs, NJ: Prentice-Hall Inc, 1981).

D. DeMaw and L. Aurick, "The Full-Wave Delta Loop at Low Height," *QST*, Oct 1984, pp 24-26.

R. Devore and P. Bohley, "The Electrically Small Magnetically Loaded Multiturn Loop Antenna," *IEEE Trans on Ant and Prop,* Jul 1977, pp 496-505.

T. Dorbuck, "Radio Direction-Finding Techniques," *QST*, Aug 1975.

R. G. Fenwick, "A Loop Array for 160 Meters," *CQ*, Apr 1986, pp 25-29.

D. Fischer, "The Loop Skywire," *QST*, Nov 1985, pp 20-22. Also Feedback, *QST*, Dec 1985, p 53.

R. A. Genaille, "V.L.F. Loop Antenna," *Electronics World*, Jan 1963, pp 49-52.

G. Gercke, "Radio Direction/Range Finder," *73*, Dec 1971, pp 29-30.

R. S. Glasgow, *Principles of Radio Engineering* (New York: McGraw-Hill Book Co, Inc, 1936).

S. Goldman, "A Shielded Loop for Low Noise Broadcast Reception," *Electronics*, Oct 1938, pp 20-22.

F. W. Grover, *Inductance Calculation-Working Formulas and Tables* (New York: D. VanNostrand Co, Inc, 1946).

J. V. Hagan, "A Large Aperture Ferrite Core Loop Antenna for Long and Medium Wave Reception," *Loop Antennas Design and Theory*, M. G. Knitter, Ed. (Cambridge, WI: National Radio Club, 1983), pp 37-49.

T. Hart, "Small, High-Efficiency Loop Antennas," *QST*, Jun 1986, pp 33-36.

T. Hart, *Small High Efficiency Antennas Alias The Loop* (Melbourne, FL: W5QJR Antenna Products, 1985).

F. M. Howes and F. M. Wood, "Note on the Bearing Error and Sensitivity of a Loop Antenna in an Abnormally Polarized Field," *Proc IRE*, Apr 1944, pp 231-233.

J. A. Lambert, "A Directional Active Loop Receiving Antenna System," *Radio Communication*, Nov 1982, pp 944-949.

D. Lankford, "Loop Antennas, Theory and Practice," *Loop Antennas Design and Theory*, M. G. Knitter, Ed. (Cambridge, WI: National Radio Club, 1983), pp 10-22.

D. Lankford, "Multi-Rod Ferrite Loop Antennas," *Loop Antennas Design and Theory*, M. G. Knitter, Ed. (Cambridge, WI: National Radio Club, 1983), pp 53-56.

G. Levy, "Loop Antennas for Aircraft," *Proc IRE*, Feb 1943, pp 56-66. Also see correction, *Proc IRE*, Jul 1943, p 384.

J. Malone, "Can a 7 foot 40m Antenna Work?" *73*, Mar 1975, pp 33-38.

L. G. McCoy, "The Army Loop in Ham Communications," *QST*, Mar 1968, pp 17, 18, 150, 152. (See also Technical Correspondence, *QST*, May 1968, pp 49-51 and Nov 1968, pp 46-47.)

R. G. Medhurst, "HF Resistance and Self Capacitance of Single Layer Solenoids," *Wireless Engineer*, Feb 1947, pp 35-43, and Mar 1947, pp 80-92.

G. P. Nelson, "The NRC FET Altazimuth Loop Antenna," *N.R.C. Antenna Reference Manual Vol. 1*, 4th ed., R. J. Edmunds, Ed. (Cambridge, WI: National Radio Club, 1982), pp 2-18.

K. H. Patterson, "Down-To-Earth Army Antenna," *Electronics*, Aug 21, 1967, pp 111-114.

R. C. Pettengill, H. T. Garland and J. D. Meindl, "Receiving Antenna Design for Miniature Receivers," *IEEE Trans on Ant and Prop*, Jul 1977, pp 528-530.

W. J. Polydoroff, *High Frequency Magnetic Materials— Their Characteristics and Principal Applications* (New York: John Wiley and Sons, Inc, 1960).

E. Robberson, "QRM? Get Looped," *Radio and Television News*, Aug 1955, pp 52-54, 126.

G. S. Smith, "Radiation Efficiency of Electrically Small Multiturn Loop Antennas," *IEEE Trans on Ant and Prop*, Sep 1972, pp 656-657.

E. C. Snelling, *Soft Ferrites—Properties and Applications* (Cleveland, OH: CRC Press, 1969).

G. Thomas, "The Hot Rod—An Inexpensive Ferrite Booster Antenna," *Loop Antennas Theory and Design*, M. G. Knitter, Ed. (Cambridge, WI: National Radio Club, 1983), pp 57-62.

J. R. True, "Low-Frequency Loop Antennas," *Ham Radio*, Dec 1976, pp 18-24.

E. G. VonWald, "Small-Loop Antennas," *Ham Radio*, May 1972, pp 36-41.

"An FET Loop Amplifier with Coaxial Output," *N.R.C. Antenna Reference Manual, Vol 2*, 1st ed., R. J. Edmunds, Ed. (Cambridge, WI: National Radio Club, Oct 1982), pp 17-20.

Aperiodic Loop Antenna Arrays (Hermes Electronics Ltd, Nov 1973).

Reference Data for Radio Engineers, 6th ed. (Indianapolis: Howard W. Sams & Co, subsidiary of ITT, 1977).

Chapter 6

Antennas for Limited Space

It is not always practical to erect full-size antennas for the HF bands. Those who live in apartment buildings may be restricted to the use of minuscule radiators because of house rules, or simply because the required space for full-size antennas is unavailable. Other amateurs may desire small antennas for aesthetic reasons, perhaps to keep peace with neighbors who do not share their enthusiasm about high towers and big antennas. There are many reasons why some amateurs prefer to use physically shortened antennas; this chapter discusses proven designs and various ways of building and using them effectively.

Few compromise antennas are capable of delivering the performance one can expect from the full-size variety. But the patient and skillful operator can often do as well as some who are equipped with high power and full-size antennas. One with an antenna of reduced size may not be able to "bore a hole" in the bands as often, and with the commanding dispatch enjoyed by those who are better equipped, but DX can be worked successfully when band conditions are suitable.

INVISIBLE ANTENNAS

We amateurs don't regard our antennas as eyesores; in fact, we almost always regard them as works of art! But there are occasions when having an outdoor or visible antenna can present problems.

When we are confronted with restrictions—self-imposed or otherwise—we can take advantage of a number of options toward getting on the air and radiating at least a moderately effective signal. In this context, a poor antenna is certainly better than no antenna at all! This section describes a number of techniques that enable us to use indoor antennas or "invisible" antennas outdoors. Many of these systems will yield good to excellent results for local and DX contacts, depending on band conditions at any given time. *The most important consideration is that of not erecting any antenna that can present a hazard (physical or electrical) to humans, animals and buildings. Safety first!*

Clothesline Antenna

Clotheslines are sometimes attached to pulleys (Fig 1) so that the user can load the line and retrieve the laundry from a back porch. Laundry lines of this variety are accepted parts of the neighborhood "scenery," and can be used handily as amateur antennas by simply insulating the pulleys from their support points. This calls for the use of a conducting type of clothesline, such as heavy-gauge stranded electrical wire with Teflon or vinyl insulation. A high-quality, flexible steel cable (stranded) is suitable as a substitute if one doesn't mind cleaning it each time clothing is hung on it.

A jumper wire can be brought from one end of the line to the ham shack when the station is being operated. If a good electrical connection exists between the wire clothesline and

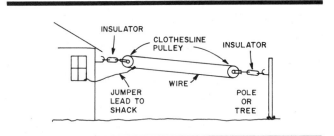

Fig 1—The clothesline antenna is more than it appears to be.

the pulley, a permanent connection can be made by connecting the lead-in wire between the pulley and its insulator. A Transmatch can be used to match the "invisible" random-length wire to the transmitter and receiver.

Invisible Long Wire

A wire antenna is not actually a "long wire" unless it is one wavelength or greater in length. Yet many amateurs refer to (relatively) long physical spans of conductor as "long wires." For the purpose of this discussion we will assume we have a fairly long span of wire, and refer to it as an "end-fed" wire antenna.

If we use small-diameter enameled wire for our end-fed antenna, chances are that it will be very difficult to see against the sky and neighborhood scenery. The smaller the wire, the more "invisible" the antenna will be. The limiting factor with small wire is fragility. A good compromise is no. 24 or no. 26 magnet wire for spans up to 130 feet; lighter gauge wire can be used for shorter spans, such as 30 or 60 feet. The major threat to the longevity of fine wire is icing. Also, birds may fly into the wire and break it. Therefore, this style of antenna may require frequent service or replacement.

Fig 2 illustrates how we might install an invisible end-fed wire. It is important that the insulators also be lacking in prominence. Tiny Plexiglas blocks perform this function well. Small-diameter clear plastic medical vials are suitable also. Some amateurs simply use rubber bands for end insulators, but they will deteriorate rapidly from sun and air pollutants. They are entirely adequate for short-term operation with an invisible antenna, however.

Rain Gutter and TV Antennas

A great number of amateurs have taken advantage of standard house fixtures when contriving inconspicuous antennas. A very old technique is the use of the gutter and downspout system on the building. This is shown in Fig 3, where a lead wire is routed to the operating room from one

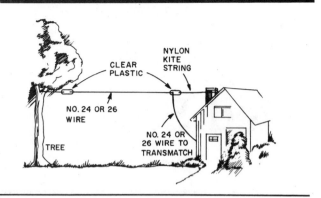

Fig 2—The "invisible" end-fed antenna.

end of the gutter trough. We must assume that the wood to which the gutter is affixed is dry and of good quality to provide reasonable electrical insulation. The rain-gutter antenna may perform quite poorly during wet weather or when there is ice and snow on it and the house roof.

All joints between gutter and downspout sections must be bonded electrically with straps of braid or flashing copper to provide good continuity in the system. Poor joints can permit rectification of RF and subsequently cause TVI and other harmonic interference. Also, it is prudent to insert a section of plastic downspout about 8 feet above ground to prevent RF shocks or burns to passersby while the antenna is being used. Improved performance may result if the front and back gutters of the house are joined by a jumper wire to increase the area of the antenna.

Fig 3 also shows a TV or FM antenna that can be employed as an invisible amateur antenna. Many of these antennas can be modified easily to accommodate the 144- or 220-MHz bands, thereby permitting the use of the 300-ohm line as a feeder system. Some FM antennas can be used on 6 meters by adding no. 10 bus-wire extensions to the ends of the elements, and adjusting the match for an SWR of 1:1. If 300-ohm line is used it will require a balun or Transmatch to interface the line with the station equipment.

For operation in the HF bands, the TV or FM antenna feeders can be tied together at the transmitter end of the span and the system treated as a random-length wire. If this is done, the 300-ohm line will have to be on TV standoff insulators and spaced well away from phone and power company service-entrance lines. Naturally, the TV or FM radio must be disconnected from the system when it is used for amateur work! Similarly, masthead amplifiers and splitters must be removed from the line if the system is to be used for amateur operation. If the system is mostly vertical, a good RF ground system with many radials around the base of the house should be used to improve performance.

Flagpole Antennas

We can exhibit our patriotism and have an invisible amateur antenna at the same time by disguising our antenna as shown in Fig 4. The vertical antenna is a wire that has been placed inside a plastic or fiberglass pole.

The flagpole antenna shown is structured for a single amateur band, and it is assumed that the height of the pole corresponds to a quarter wavelength for the chosen band. The radials and feed line can be buried in the ground as shown. In a practical installation, the sealed end of the coax cable would protrude slightly into the lower end of the plastic pole.

If a large-diameter fiberglass pole were available, a multiband trap vertical may be concealed inside it. Or we might use a metal pole and bury a water-tight box at its base, containing fixed-tuned matching networks for the bands of interest. The networks could then be selected remotely by means of relays inside the box. A 30-foot flagpole would provide good results in this kind of system, provided it was used in conjunction with a buried radial system.

Still another technique is one that employs a wooden flagpole. A small-diameter wire can be stapled to the pole and routed to the coax feeder or matching network. The halyard could by itself constitute the antenna wire if it were made from heavy-duty insulated hookup wire. There are countless variations for this type of antenna, and they are limited only by the imagination of the amateur. Detailed plans for a flagpole antenna can be found later in this chapter.

Other Invisible Antennas

Some amateurs have used the metal fence on apartment verandas as antennas, and have had good results on the upper HF bands (14, 21 and 28 MHz). We must presume that the fences were not connected to the steel framework of the

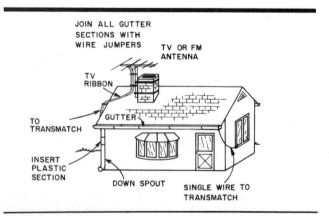

Fig 3—Rain gutters and TV antenna installations can be used as inconspicuous Amateur Radio antennas.

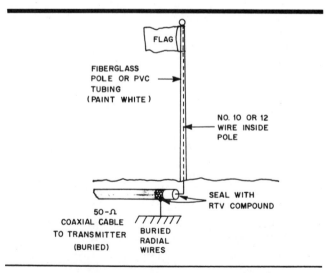

Fig 4—A flagpole antenna.

building, but rather were insulated by the concrete floor to which they were affixed. These veranda fences have also been used effectively as ground systems (counterpoises) for HF-band vertical antennas that were put in place temporarily after dark.

One New York City amateur uses the fire escape on his apartment building as a 7-MHz antenna, and reports good success in working DX stations with it. Another apartment dweller makes use of the aluminum frame on his living-room picture window as an antenna for 21 and 28 MHz. He works it against the metal conductors of the baseboard heater in the same room.

Many jokes have been told over the years about "bedspring antennas." The idea is by no means absurd. Bedsprings and metal end boards have been used to advantage as antennas by many apartment dwellers as 14, 21 and 28-MHz radiators. A counterpoise ground can be routed along the baseboard of the room and used in combination with the bedspring. It is important to remember that any independent (insulated) metal object of reasonable size can serve as an antenna if the transmitter can be matched to it. An amateur in Detroit once used his Shopsmith craft machine (about 5 feet tall) as a 28-MHz antenna. He worked a number of DX stations with it when band conditions were good.

A number of operators have used metal curtain rods and window screens for VHF work, and found them to be acceptable for local communications. Best results with any of these makeshift antennas will be had when the "antennas" are kept well away from house wiring and other conductive objects.

INDOOR ANTENNAS

Without question, the best place for your antenna is outdoors, and as high and in the clear as possible. Some of us, however, for legal, social, neighborhood, family or landlord reasons, are restricted to indoor antennas. Having to settle for an indoor antenna is certainly a handicap for the amateur seeking effective radio communication, but that is not enough reason to abandon all operation in despair.

First, we should be aware of the reasons why indoor antennas *do not* work well. Principal faults are (1) low height above ground—the antenna cannot be placed higher than the highest peak of the roof, a point usually low in terms of wavelength at HF; (2) the antenna must function in a lossy RF environment that involves close coupling to electrical wiring, guttering, plumbing and other parasitic conductors, besides dielectric losses in such nonconductors as wood, plaster and masonry; (3) sometimes the antenna must be made small in terms of a wavelength and (4) usually it cannot be rotated. These are appreciable handicaps. Nevertheless, global communication with an indoor antenna is still possible.

Some practical points *in favor* of the indoor antenna include (1) freedom from weathering effects and damage caused by wind, ice, rain and sunlight (the SWR of an attic antenna, however, can be affected somewhat by a wet or snow-covered roof); (2) indoor antennas can be made from materials that would be altogether impractical outdoors, such as aluminum foil and thread (the antenna need support only its own weight); (3) the supporting structure is already in place, eliminating the need for antenna masts and (4) the antenna is readily accessible in all weather conditions, simplifying pruning or tuning, which can be accomplished without climbing or tilting over a tower.

Empiricism

A typical house or apartment involves such a complex electromagnetic environment that it is impossible to predict theoretically which location or orientation of the indoor antenna will work best. This is where good old-fashioned cut-and-try, use-what-works-best empiricism pays off. But to properly determine what really is most suitable requires an understanding of some antenna-measuring fundamentals.

Unfortunately, many amateurs do not know how to evaluate performance scientifically or compare one antenna with another. Typically, they will put up one antenna and try it out on the air to see how it "gets out" in comparison with a previous antenna. This is obviously a very poor evaluation method because there is no way to know if the better or worse reports are caused by changing band conditions, different S-meter characteristics, or any of several other factors that could influence the reports received.

Many times the difference between two antennas or between two different locations for identical antennas amounts to only a few decibels, a difference that is hard to discern unless instantaneous switching between the two is possible. Those few decibels are not important under strong-signal conditions, of course, but when the going gets rough, as is often the case with an indoor antenna, a few dB can make the difference between solid copy and no possibility of real communication.

Very little in the way of test equipment is needed for casual antenna evaluation, other than a communications receiver. You can even do a qualitative comparison by ear, *if* you can switch antennas instantaneously. Differences of less than 2 dB, however, are still hard to discern. The same is true of S meters. Signal-strength differences of less than a decibel are usually difficult to see. If you want that last fraction of a decibel, you should use a good ac voltmeter at the receiver audio output (with the AGC turned off).

In order to compare two antennas, switching the coaxial transmission line from one to the other is necessary. No elaborate coaxial switch is needed; even a simple double-throw toggle or slide switch will provide more than 40 dB of isolation at HF. See Fig 5. Switching by means of manually connecting and disconnecting coaxial lines is not recommended because that takes too long. Fading can cause signal-strength changes during the changeover interval.

Whatever difference shows up in the strength of the received signal will be the difference in performance between the two antennas in the direction of that signal. For this test

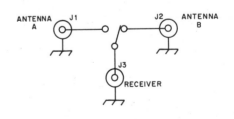

Fig 5—When antennas are compared on fading signals, the time delay involved in disconnecting and reconnecting coaxial cables is too long for accurate measurements. A simple slide switch will do well for switching coaxial lines at HF. The four components can be mounted in a tin can or any small metal box. Leads should be short and direct. J1 through J3 are coaxial connectors.

to be valid, both antennas must have nearly the same feed-point impedance, a condition that is reasonably well met if the SWR is below 2:1 on both antennas.

On ionospheric-propagated signals (sky wave) there will be constant fading, and for a valid comparison it will be necessary to take an average of the difference between the two antennas. Occasionally, the inferior antenna will deliver a stronger signal to the receiver, but in the long run the law of averages will put the better antenna ahead.

Of course with a ground-wave signal, such as that from a station across town, there will be no fading problems. A ground-wave signal will enable the operator to properly evaluate the antenna under test in the direction of the source. The results will be valid for ionospheric propagated signals at low elevation angles in that direction. On 28 MHz, all sky-wave signals arrive and leave at low angles. But on the lower bands, particularly 3.5 and 7 MHz, we often use signals propagated at high elevation angles, almost up to the zenith. For these angles a ground-wave test will not provide a proper evaluation of the antenna, and use of sky-wave signals becomes necessary.

Dipoles

At HF the most practical indoor antenna is usually the dipole. Any attempt to get more gain with parasitic elements will usually fail because of close proximity of the ground or coupling to house wiring. Beam antenna dimensions determined outdoors will not usually be valid for an attic antenna because the roof structure will cause dielectric loading of the parasitic elements. It is usually more worthwhile to spend time optimizing the location and performance of a dipole than to try to improve results with parasitic elements.

Most attics are not long enough to accommodate half-wave dipoles for 7 MHz and below. If this is the case, some folding of the dipole will be necessary. The final shape of the antenna will depend on the dimensions and configuration of the attic. Remember that the center of the dipole carries the most current and therefore does most of the radiating. This part should be as high and unfolded as possible. Because the dipole ends radiate less energy than the center, their orientation is not as important. They do carry a maximum voltage, nevertheless, so care should be taken to position the ends far enough from other conductors to avoid arcing. The dipole may end up being L shaped, Z shaped, U shaped or some

indescribable corkscrew shape, depending on what space is available, but reasonable performance can often be had even with such a nonlinear arrangement. Fig 6 shows some possible configurations. Multiband operation is possible with the use of open-wire feeders and a Transmatch. One alternative not shown here is the aluminum-foil dipole, which was conceived by Rudy Stork, KA5FSB. He suggests mounting the dipole behind wallpaper or in the attic, with portability, ease of construction and adjustment, and economy in design among its desirable features. This antenna should also display reasonably good bandwidth resulting from the large area of its conductor material.

If coaxial feed is to be used, some pruning of an attic antenna to establish minimum SWR at the band center will be required. Tuning the antenna outdoors and then installing it inside is usually not feasible since the behavior of the antenna will not be the same when placed in the attic. Resonance will be affected somewhat if the antenna is bent. Even if the antenna is placed in a straight line, parasitic conductors and dielectric loading by nearby wood structures will affect the impedance.

Trap and loaded dipoles are shorter than the full-sized versions, but are comparable performers. Trap dipoles are discussed in Chapter 7; loaded dipoles later in this chapter.

Dipole Orientation

Theoretically a vertical dipole is most effective at low radiation angles, but practical experience shows that the horizontal dipole is usually a better indoor antenna. A high horizontal dipole does exhibit directional effects at low radiation angles, but you will not be likely to see much, if any, directivity with an attic-mounted dipole. Some operators place two dipoles at right angles to each other with provisions at the operating position for switching between the two. Their reasoning is the radiation patterns will inevitably be distorted in an unpredictable manner by nearby parasitic conductors. There will be little coupling between the dipoles if they are oriented at right angles to each other as shown in Figs 7A and 7B. There will be some coupling with the arrangement shown in Fig 7C, but even this orientation is preferable to a single dipole.

With two antennas mounted 90° apart, you may find that one dipole is consistently better in nearly all directions, in which case you will want to remove the inferior dipole,

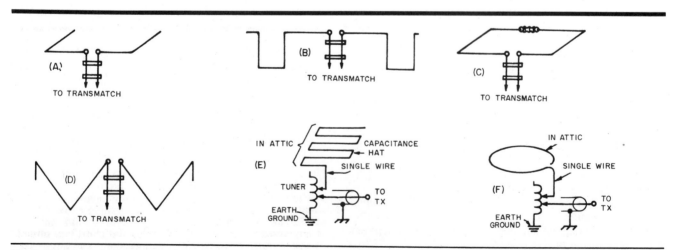

Fig 6—Various configurations for small indoor antennas. A discussion of installation is contained in the text.

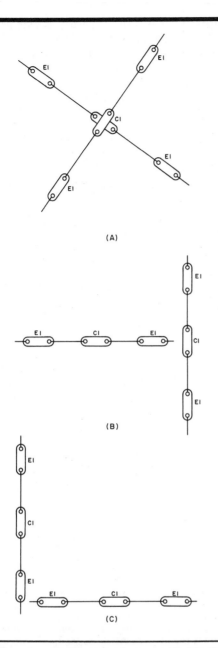

(A)

(B)

(C)

Fig 7—Ways to orient a pair of perpendicular dipoles. The orientations of A and B will result in no mutual coupling between the two dipoles, but there will be some coupling with the configuration shown at C. End (EI) and center (CI) insulators are shown.

perhaps placing it someplace else. In this manner the best spots in the house or attic can be determined experimentally.

Parasitic Conductors

Inevitably, any conductor in your house near a quarter wave in length or longer at the operating frequency will be parasitically coupled to your antenna. The word *parasitic* is particularly appropriate in this case because these conductors usually introduce losses and leave less energy for radiation into space. Unlike the parasitic elements in a beam antenna, conductors such as house wiring and plumbing are usually connected to lossy objects such as earth, electrical appliances, masonry or other objects that dissipate energy. Even where this energy is reradiated, it is not likely to be in the right phase

in the desired direction; it is, in fact, likely to be a source of RFI.

There are, however, some things that can be done about parasitic conductors. The most obvious is to reroute them at right angles to the antenna or close to the ground, or even underground—procedures that are usually not feasible in a finished home. Where these conductors cannot be rerouted, other measures can be taken. Electrical wiring can be broken up with RF chokes to prevent the flow of radio-frequency currents while permitting 60-Hz current (or audio, in the case of telephone wires) to flow unimpeded. A typical RF choke for a power line can be 100 turns of no. 10 insulated wire close-wound on a length of 1-inch dia plastic pipe. Of course one choke will be needed for each conductor. A three-wire line calls for three chokes.

THE RESONANT BREAKER

Obviously, RF chokes cannot be used on conductors such as metal conduit or water pipes. But it is still possible, surprising as it may seem, to obstruct RF currents on such conductors without breaking the metal. The resonant breaker was first described by Fred Brown, W6HPH, in Oct 1979 *QST*.

Fig 8 shows a method of accomplishing this. A figure-eight loop is inductively coupled to the parasitic conductor and is resonated to the desired frequency with a variable capacitor. The result is a very high impedance induced in series with the pipe, conduit or wire. This impedance will block the flow of radio-frequency currents. The figure-eight coil can be thought of as two turns of an air-core toroid and since the parasitic conductor threads through the hole of this core, there will be tight coupling between the two. Inasmuch as the figure-eight coil is parallel resonated, transformer action will reflect a high impedance in series with the linear conductor.

Before you bother with a "resonant breaker" of this type, be sure that there is a significant amount of RF current flowing in the parasitic conductor, and that you will therefore benefit from installing one. The relative magnitude of this current can be determined with an RF current probe of the type described in Chapter 27. According to the rule of thumb

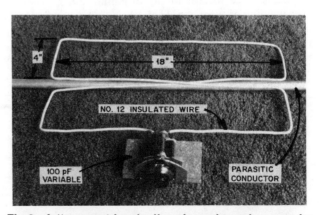

Fig 8—A "resonant breaker" such as shown here can be used to obstruct radio-frequency currents in a conductor without the need to break the conductor physically. A vernier dial is recommended for use with the variable capacitor because tuning is quite sharp. The 100-pF capacitor is in series with the loop. This resonant breaker tunes from 14 through 29.7 MHz. Larger models may be constructed for the lower frequency bands.

regarding parasitic conductor current, if it measures less than 1/10 of that measured near the center of the dipole, the parasitic current is generally not large enough to be of concern.

The current probe is also needed for resonating the breaker after it is installed. Normally, the resonant breaker will be placed on the parasitic conductor near the point of maximum current. When it is tuned through resonance, there will be a sharp dip in RF current, as indicated by the current probe. Of course, the resonant breaker will be effective only on one band. You will need one for each band where there is significant current as indicated by the probe.

Power-Handling Capability

So far, our discussion has been limited to the indoor antenna as a receiving antenna, except for the current measurements, where it is necessary to supply a small amount of power to the antenna. These measurements will not indicate the full power-handling capability of the antenna. Any tendency to flash over must be determined by running full power or, preferably, somewhat more than the peak power you intend to use in regular operation. The antenna should be carefully checked for arcing or RF heating before you do any operating. Bear in mind that attics are indeed vulnerable to fire hazards. A potential of several hundred volts exists at the ends of a dipole fed by the typical Amateur Radio transmitter. If a power amplifier is used, there could be a few thousand volts at the ends of the dipole. Keep your antenna elements well away from other objects. *Safety first!*

LOADED ANTENNAS

Physically shortened antennas, particularly dipoles, are practical and should be of interest to the indoor-antenna user. There are several ways to load antennas so they may be reduced in size without severe reduction in effectiveness. Loading is always a compromise; the best method is determined by the amount of space available and the band(s) to be worked. Most indoor dipole antennas can be conveniently loaded both inductively (coils in the legs of the dipole) and with capacitive end loading. The most serious drawback associated with inductive loading is high losses in the coils themselves. It is important that you use inductors made from reasonably large wire or tubing to minimize this problem. Close winding of turns should also be avoided if possible. A good compromise is to use some off-center inductive loading in combination with capacitive end loading, keeping the

inductor losses small and the efficiency as high as possible.

Some examples of off-center coil loading and capacitive-end loading are shown in Fig 9. This technique was described by Jerry Hall, K1TD (ex-K1PLP) in Sep 1974 *QST*. For the antennas shown, the longer the overall length (dimension A, Fig 9A) and the farther the loading coils are from the center of the antenna (dimension B), the greater the efficiency of the antenna. As dimension B is increased, however, the inductance required to resonate the antenna at the desired frequency increases. Approximate inductance values for single-band resonance (for the antenna in Fig 9A only) may be determined with the aid of Fig 10 or from Eq 1. The final

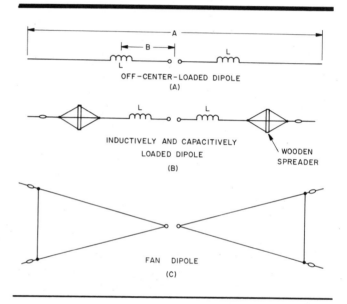

Fig 9—At A is a dipole antenna lengthened electrically with off-center loading coils. For a fixed dimension A, greater efficiency will be realized with greater distance B, but as B is increased, L must be larger in value to maintain resonance. If the two coils are placed at the ends of the antenna, in theory they must be infinite in size to maintain resonance. At B, capacitive loading of the ends, either through proximity of the antenna to other objects or through the addition of capacitance hats, will reduce the required value of the coils. At C, a fan dipole provides some electrical lengthening as well as broadbanding.

$$X_L = \frac{10^6}{34\pi f} \cdot \left\{ \frac{\left[\ln \frac{24\left(\frac{234}{f} - B\right)}{D} - 1\right]\left[\left(1 - \frac{fB}{234}\right)^2 - 1\right]}{\frac{234}{f} - B} - \frac{\left[\ln \frac{24\left(\frac{A}{2} - B\right)}{D} - 1\right]\left[\left(\frac{\frac{fA}{2} - fB}{234}\right)^2 - 1\right]}{\frac{A}{2} - B} \right\}$$

(Eq 1)

where

ln = natural log
f = frequency, megahertz
A = overall antenna length, feet

B = distance from center to each loading coil, feet
D = diameter of radiator, inches

values will depend on the proximity of surrounding objects in individual installations and must be determined experimentally. The use of high-Q low-loss coils is important for maximum efficiency.

A dip meter or SWR indicator is recommended for use during adjustment of the system. Note that the minimum inductance required is for a center-loaded dipole. If the inductive reactance is read from Fig 10 for a dimension B of zero, one coil having approximately twice this reactance can be used near the center of the dipole. Fig 11 illustrates this idea. This antenna was conceived by Jack Sobel, WØSVM, who dubbed the 7-MHz version the "Shorty Forty."

An alternative to inductive loading is linear loading. This little-understood method of shortening radiators can be applied to almost any antenna configuration—including parasitic arrays. Although commercial antenna manufacturers make use of linear loading in their HF antennas, relatively few hams have used it in their own designs. Linear loading can be used to our advantage in many antennas because it introduces very little loss, does not degrade directivity patterns, and has low enough Q to allow reasonably good bandwidth. Some examples of linear-loaded antennas are shown in Fig 12. Since the dimensions and spacing of linear-loading devices vary greatly from one antenna installation to

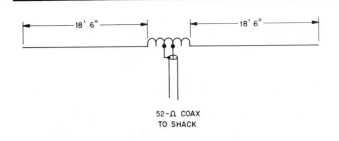

Fig 11—The WØSVM "Shorty Forty" center-loaded antenna. Dimensions given are for 7.0 MHz. The loading coil is 5 in. long and 2½ in. in diameter. It has a total of 30 turns of no. 12 wire wound at 6 turns per in. (Miniductor 3029 stock).

another, the best way to employ this technique is to try a length of conductor 10 to 20% longer than the difference between the shortened antenna and the full-size dimension for the linear-loading device. Then use the "cut and try" method, varying both the spacing and length of the loading

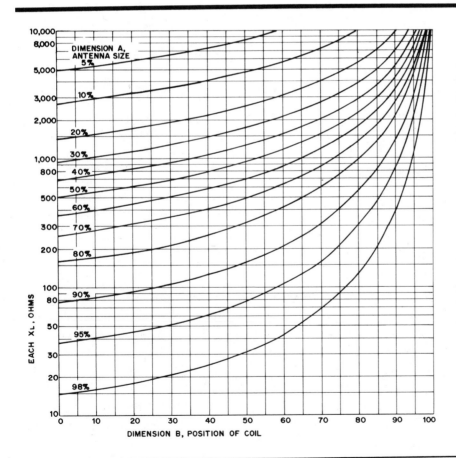

Fig 10—Chart for determining approximate inductance values for off-center-loaded dipoles. See Fig 9A. At the intersection of the appropriate curve from the body of the chart for dimension A and proper value for the coil position from the horizontal scale at the bottom of the chart, read the required inductive reactance for resonance from the scale at the left. Dimension A is expressed as percent length of the shortened antenna with respect to the length of a half-wave dipole of the same conductor material. Dimension B is expressed as the percentage of coil distance from the feed point to the end of the antenna. For example, a shortened antenna which is 50% or half the size of a half-wave dipole (one-quarter wavelength overall) with loading coils positioned midway between the feed point and each end (50% out) would require coils having an inductive reactance of approximately 950 ohms at the operating frequency for antenna resonance. (Based on Eq 1.)

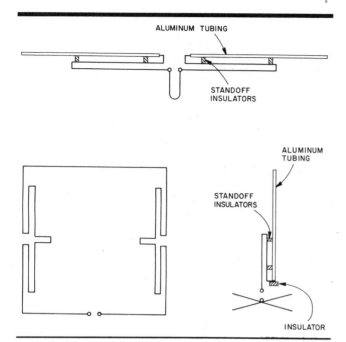

Fig 12—Some examples of linear loading. The small circles indicate the feed points of the antennas.

Fig 13—Jerry Sevick, W2FMI, adjusts the 6-ft, 40-meter vertical.

Fig 14—Construction details for the top hat. For a diameter of 7 ft, ½-in. aluminum tubing is used. The hose clamp is of stainless steel and available at Sears. The rest of the hardware is aluminum.

device to optimize the match. A hairpin at the feed point can be useful in achieving a 1:1 SWR at resonance.

Outdoor Antennas

It is possible to reduce the physical size of an antenna by 50% or more and still obtain good results. Use of an outdoor loaded dipole, as described in the previous section, will permit a city-size lot to accommodate a doublet antenna on the lower frequency HF bands, 7, 3.5 or even 1.8 MHz. Short vertical antennas can also be made effective by using lumped inductance to obtain resonance, and by using a capacitance hat to increase the feed-point impedance of the system. As is the case with full-size vertical quarter-wave antennas, the ground-radial system should be as extensive as possible for maximum effectiveness. Several articles by Jerry Sevick, W2FMI, and others have been published in *QST* showing the relative effectiveness of various radial systems with vertical radiators. Ground-mounted vertical radiators should be used in combination with several buried radials. Above-ground vertical antennas should be worked against at least four quarter-wavelength radials.

A 6-FOOT-HIGH 7-MHz VERTICAL ANTENNA

Figs 13 through 16 give details for building short, effective vertical quarter-wavelength radiators. The information gathered and presented here was provided by Jerry Sevick, W2FMI. (See the bibliography at the end of this chapter for reference to his *QST* articles on the subject of shortened antennas.)

A short vertical antenna, properly designed and installed, approaches the efficiency of a full-size resonant quarter-wave antenna. Even a 6-foot vertical on 7 MHz can produce an exceptional signal. Theory tells us that this should be possible, but the practical achievement of such a result requires an

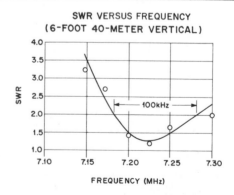

Fig 15—Standing-wave ratio of the 6-ft vertical using a 7-ft top hat and 14 turns of loading 6 in. below the top hat.

Fig 16—Base of the vertical antenna showing the 60 radial wires. The aluminum disc is 15 in. in diameter and ¼-in. thick. Sixty tapped holes for ¼-20 aluminum hex-head bolts form the outer ring and 20 form the inner ring. The inner bolts were used for performance comparisons with more than 60 radials. The insulator is polystyrene material (phenolic or Plexiglas suitable) with a 1-in. diameter. Also shown is the impedance bridge used for measuring input resistance.

understanding of the problems of ground losses, loading, and impedance matching, treated in the theory chapters of this book.

The key to success with shortened vertical antennas lies in the efficiency of the ground system with which the antenna is used. A system of at least 60 radial wires is recommended for best results, although the builder may want to reduce the number at the expense of some performance. The radials can be tensioned and pinned at the far ends to permit on-the-ground installation, which will enable the amateur to mow the lawn without the wires becoming entangled in the mower blades. Alternatively, the wires can be buried in the ground where they will not be visible. There is nothing critical about the wire size for the radials. No. 28, 22 or even 16 gauge, will provide the same results. The radials should be at least 0.2 wavelength long (27 feet or greater).

A top hat is formed as illustrated in Fig 14. The diameter is 7 feet, and a continuous length of wire is connected to the spokes around the outer circumference of the wheel. A loading coil consisting of 14 turns of B&W 3029 Miniductor stock (2½ inch dia, 6 TPI, no. 12 wire) is installed 6 inches below the top hat (see Fig 13). This antenna exhibits a feed-point impedance of 3.5 ohms at 7.21 MHz. For operation above or below this frequency, the number of coil turns must be decreased or increased, respectively. Matching is accomplished by increasing the feed-point impedance to 14 ohms through addition of a 4:1 transformer, then matching 14 ohms to 50 ohms (feeder impedance) by means of a pi network. The 2:1 SWR bandwidth for this antenna is approximately 100 kHz.

More than 200 contacts with the 6-foot antenna have indicated the efficiency and capability of a short vertical. Invariably at distances greater than 500 or 600 miles, the short vertical yields excellent signals. Similar antennas can be scaled and constructed for bands other than 7 MHz. The 7-foot-diameter top hat was tried on an 3.5-MHz vertical, with an

antenna height of 22 feet. The loading coil had 24 turns and was placed 2 feet below the top hat. On-the-air results duplicated those on 40 meters. The bandwidth was 65 kHz.

Short verticals such as these have the ability to radiate and receive almost as well as a full-size quarter-wave antenna. The differences are practically negligible. Trade-offs are in lowered input impedances and bandwidths. However, with a good image plane and a proper design, these trade-offs can be made entirely acceptable.

THE DDRR ANTENNA

Another physically small but effective antenna is the DDRR (directional discontinuity ring radiator), described in *Electronics*, January 1963. An in-depth mathematical analysis of this low-profile antenna was given by Robert Dome, W2WAM, in July 1972 *QST*. Fig 17 shows details for constructing a DDRR antenna.

In this example the radiating element of the antenna is made from 2-inch diameter automobile exhaust pipe. Most muffler shops can supply the materials as well as bend the pipe to specifications. Table 1 lists the dimensions required for operation in the amateur bands from 1.8 to 148 MHz, inclusive. The following technique illustrates how a 7-MHz model can be assembled (from an article by Ed English, W6WYQ, Dec 1971 *QST*).

In forming the ring to these dimensions, four 10-foot lengths of tubing are used. A 10-degree bend is made at 9-inch intervals in three of the lengths. The fourth length is similarly treated, except for the last 18 inches, which is bent at right angles to form the upright leg of the ring. One end of each section is flared so that the sections can be coupled together by slipping the end of one into the flare of its mate.

The required flares are easily made at the muffler shop with the aid of the forming tools. Another task which can best be completed at the shop is to weld a flange onto the end of the upright legs. This flange facilitates attaching the leg to the mounting plate that provides a chassis for the tuning mechanism and the coaxial-feed coupler. After bending

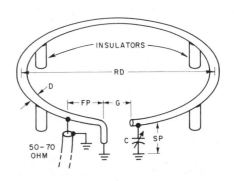

Fig 17—RD = 0.078 λ (28°); SP = 0.0069 λ (2.5°); FP = 0.25SP (see Note 1); C = (see Note 2); D = (see Note 3); G = (see Table 1).

Notes:
1) **Actual dimension must be found experimentally.**
2) **Value to resonate the antenna to the operating freq.**
3) **D ranges upward from ½″. The larger D is, the higher the efficiency is. Use largest practical size, such as ½″ for 28 MHz, 5″ or 6″ for 3.5 or 1.8 MHz.**

Table 1

Dimensions for ¼-Wavelength DDRR Elements

Frequency (MHz)	1.8	3.5	7	10	14	18	21	24	28	50	144
Feed Point (FP)	12"	6"	6"	3.5"	2"	1.5"	1.5"	1.8"	3"	1"	½"
Gap (G)	16"	7"	5"	4"	3"	2.8"	2.5"	2.2"	2"	1.5"	1"
Capacitor, pF (C)	150	100	70	50	35	20	15	15	15	10	5
Spacing (Height) (SP)	48"	24"	11"	7.5"	6"	5"	4¾"	4"	3"	1½"	1"
Tubing Diameter (D)	5"	4"	2"	1.5"	1"	7/8"	¾"	¾"	¾"	½"	¼"
Ring Diameter (RD)	36'	18'	9'	6'	4.5'	4'	3'4"	2.5'	2'4"	16¼"	6"

and flaring is completed, the ring is assembled and minor adjustments made to bring it into round and to the proper dimensions. This can best be done by drawing a circle on the floor with chalk and fitting the ring inside the circle. The circle must be slightly larger than the center-to-center ring diameter so that the reference line can be seen easily. For example, with two-inch tubing the diameter of the reference circle must be 9 feet, 2 inches. When a satisfactory fit is obtained between the tubing ring and the chalk ring, drill a ¼-inch hole through each of the joints to accept a ¼-inch bolt. These bolts clamp the sections together. They can also be used to attach the insulators which support the ring at a fixed height above the ground plane.

Insulators for the antenna are made from 11-inch lengths of 2-inch PVC pipe inserted into a standard cap of the same material. The PVC caps are first drilled through the center to accept the ¼-inch bolt previously installed at the joints. The caps are then slipped onto the bolts, and nuts are installed and tightened to secure the caps in place. The 11-inch length of pipe, when inserted into the cap and pressed firmly until it touches bottom, results in a total insulator length of 12 inches. Four insulators are required, one at each of the joints and one near the open end of the ring for support. It is wise to locate this insulator as far back from the end of the ring as possible because of the increasingly high RF voltage that develops as the end of the ring is approached. (Because of the danger of RF burns in the event of accidental contact with the antenna, precautions should be taken to prevent random access to the completed installation.) As a final measure, the bottom ends of the insulators are sealed to prevent moisture from forming on the inside surfaces. Standard PVC caps may be used here, but plastic caps from 15-ounce aerosol cans fit well.

A mounting plate is required to provide good mechanical and electrical connections for the grounded leg of the radiator, the coaxial feed-line connection, and the tuning mechanism. If you are using aluminum tubing, you should use an aluminum plate, and for steel tubing, a steel plate (to lessen corrosion from the junction of dissimilar metals). Dimensions for the plate are shown in Fig 18. The important consideration here is that good, solid mechanical and electrical connections be made between the ground side at the coaxial connector, the ring base and the tuning capacitor.

In the installation shown in Fig 19, the 9-foot ring resonated easily with approximately 20 pF of capacitance between the high-impedance end of the ring and the base plate or ground. Any variable capacitor which will tune the system to resonance and which will not arc under full power should be satisfactory. Remember, the RF voltage at the high

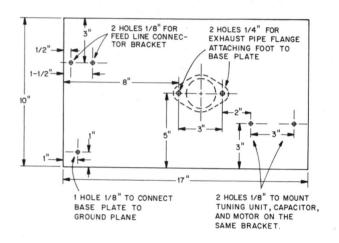

Fig 18—Drawing of the base plate, which can be made from either steel or aluminum, as described in the text. The lower right portion of the plate may be used for the mounting of the tuning capacitor (and motor, if used).

Fig 19—The chicken-wire ground plane is evident in the background. The base plate is visible at the lower right. Note the relative positions of the 52-ohm coaxial feed at the left end of the plate, the flange on the foot of the post, and the tuning unit at the right-hand end of the plate.

impedance end of this antenna can reach 20 to 30 kV with high power, so if you are using the maximum legal limit, you would do well to consider using a vacuum variable capacitor. To provide for full band coverage, a 35-pF variable capacitor was coupled to a reversible, slow-speed motor which allowed remote tuning of the antenna from the operating position. An indicated SWR of 1.1 to 1 was obtained easily over the entire 7-MHz band. The motor used was a surplus item made by Globe Industries of Dayton, Ohio. At 20 volts dc the shaft of this motor turns at about 1 RPM, which is ideal for DDRR tuning. The gears used were surplus items. If you cannot obtain gears, a string and pulley drive will do almost as well, or you can mount both the motor and the capacitor in line and use direct coupling. Of course, if you operate on a fixed frequency, or within a 40- to 50-kHz segment of the band, you can dispense with the motor entirely and simply tune the capacitor manually. In any case, the tuning unit must be protected from the weather. A plastic refrigerator box may be used to house the tuning capacitor and its drive motor.

Electrical Connections and the Ground Plane

The connection between the open end of the ring and the tuning capacitor is made with no. 12 wire or larger. On the end of the base plate opposite the tuning unit, and directly under the ring about 8 inches from the grounded post, install a bracket for a coaxial connector. The connector should be oriented so the feed line will lead away from the ring at close to 90°. Install a clamp on the ring directly above the coaxial connector. Connect a lead of no. 12 or larger wire from the coaxial connector to the clamp. This wire must have a certain amount of flexibility to accommodate the movement necessary when adjusting for a match. The matching point must be found experimentally. It will be affected by the nature and quality of the ground plane over which the antenna is operating. The antenna will function over earth ground, but a ground-plane surface of chicken wire (laid under the antenna and bonded to the base plate) will provide a constant ground reference and improved performance. In a roof-top location, sheet-metal roofing should provide an excellent ground plane. A poor ground usually results in a matching point for the feed line far out along the circumference of the circle. In the installation shown, a near-perfect match was obtained with the feed line connected to the ring about 12 inches from the grounded post. During testing, when the antenna was set up on a concrete surface without the ground plane, a match was found when the feed line was connected nearly 7 feet from the post!

As shown in the photo, the compactness of the antenna is readily apparent. The ground plane is made up of three 12-foot lengths of chicken wire, each 4 feet wide, which are bonded along the edges at about 6-inch intervals. In this installation the antenna, with the ground plane, could be dismantled in about 30 minutes. If portability is not important, it is best to bond all of the joints in the tubing so that good electrical continuity is assured.

After all construction is completed, the antenna should be given a coat of primer paint to minimize the possibility of rust formation. If it suits you, there is no reason why a final coat of enamel could not be applied.

Tuning Procedure

Once the mechanical construction is completed, the antenna should be erected in its intended operating location. Coupling to the station may be accomplished with either 52- or 75-ohm coaxial cable. Tune and load the transmitter as with any antenna. While observing an SWR meter in the line, operate the tuning motor. Indication of resonance is the noticeable decrease in indicated reflected power. At this point, note the loading of the transmitter; it will probably increase markedly as antenna resonance is approached. Retune the transmitter and move the feed-point tap on the antenna for a further reduction in indicated reflected power. There is interaction between the movement at the feed tap and the resonance point; therefore, it will be necessary to operate the tuning motor each time the tap is adjusted until the lowest SWR is found. Don't settle for anything above 1.1 to 1. With a good ground and proper tuning and matching, this ratio can be obtained and held over the entire band. Once the proper feed point has been located, the only adjustment necessary when changing frequency is retuning the antenna to resonance by means of the motor. If the antenna is to be fixed tuned, provide an insulated shaft extension of 18 inches or so to the tuning-capacitor shaft for manual adjustment. This not only provides insulation from the high RF voltage but also minimizes body capacitance effects during the tuning process.

THE INVERTED-L ANTENNA

The inverted L is a convenient low-band antenna for amateurs with space limited to one tree or tower and some other supporting structure. This antenna, shown in Fig 20, is a top-loaded vertical which requires a fairly good ground-radial system for efficient operation. Inverted Ls are popular on 1.8 and 3.5 MHz especially, because they can be put up easily and give good results for their modest space requirements. The overall dimensions of this antenna are roughly the same as a quarter-wave vertical, but the antenna element can be bent at any point along its length. Optimum DX performance can be expected when the vertical section is as long as possible, but a good compromise is to have equal-length vertical and horizontal sections. The dimensions of the antenna are dictated by the amount of real estate available to the builder.

Analysis shows that the horizontal pattern of an inverted L over average ground is essentially omnidirectional, although

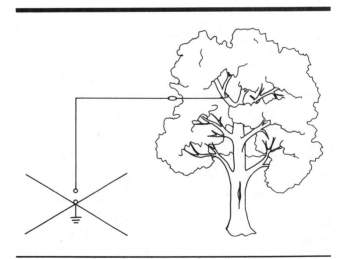

Fig 20—The inverted-L antenna can be supported by any convenient means. Matching is easiest when the total length is near 5/16 λ.

some amateurs claim directivity in the direction of the horizontal wire. If the horizontal wire is considerably longer than the vertical wire, however, there is some directivity (about 1.5 dB) in line with the horizontal wire, but in the *opposite* direction. The top-loading wire may be sloped up or down away from the vertical wire without drastically changing the feed-point impedance, and the antenna will still be essentially omnidirectional, if at least half of its length remains vertical.

The inverted L can be fed directly with 50-ohm coax if its total length is near a quarter wavelength. The end opposite the feed point can be pruned for resonance at the desired frequency. The easiest configuration for an inverted L from a tuning standpoint is a total length somewhat greater than a quarter wavelength (160 to 170 feet for 1.8 MHz), tuned with a series capacitance. This is also helpful to reduce interaction with other antennas in close proximity to the inverted L, as the total length is far from resonance on any of the HF amateur bands.

If the antenna is constructed over average earth and configured with, say, 65 feet vertical and 100 feet horizontal, the horizontal component of the signal in the direction broadside to the horizontal wire will be about 5 dB down from the vertical component. (The horizontal component is down about 10 dB from the vertical for an antenna with 65 feet vertical and 65 feet horizontal.) This can be a desirable situation, because as many 160-meter DXers have observed, there are times when conditions favor horizontal polarization over vertical, and vice versa. Since this antenna radiates a significant horizontal component broadside to the top-loading wire, the inverted L is a good compromise in either case. Also, since horizontal antennas are more immune to atmospheric noise pickup on the low bands, the inverted L is a somewhat quieter receiving antenna than a full-size vertical. Inverted-L antennas are quite common and can be very effective on 1.8 MHz. They are far easier to put up and maintain in an area without many high supports than any other antenna with comparable effectiveness for both local and DX work.

SHORTENED YAGI BEAM ANTENNA WITH LOADING COILS

At some sacrifice in bandwidth it is practical to shrink the element dimensions of Yagi antennas. Resonance can be established through the use of loading inductors in the elements, or by using inductors in combination with capacitance hats. Though not as effective in terms of gain, the short Yagi beam offers the advantage of small size, less weight, and lower wind loading when compared to a full-size array. When tower-mounted, the two-band Yagi of Figs 21 through 25 requires only 11.3 feet maximum turning radius for its 16-foot elements and boom. The antenna consists of interlaced elements for 14 and 21 MHz. A driven element and reflector are used for 21-MHz operation. The 14 MHz section is comprised of a driven element and a director. Both driven elements are gamma matched. A low-cost TV antenna rotator has sufficient torque to handle this lightweight array. (This information was written by R. Myers, W1XT, and C. Greene, K1JX, and originally appeared in September 1973 *QST*.)

A misconception among amateurs is that any element short of full size is no good in an antenna system. Reducing the size of an antenna by 50 percent does lower the efficiency somewhat, but the gain capability of a parasitic array outweighs this small loss in efficiency. Mounting the antenna above the interference-generating neighborhood can greatly

Fig 21—This short two-band Yagi can be turned by a light-duty rotator.

reduce susceptibility to man-made noise and certainly aids in the reduction of RF heating to trees, telephone poles and buildings. Placing the antenna above these energy-absorbing objects is very desirable.

The dual-band beam has four elements, the longest of which is 16 feet. All of the elements and the boom are made from 1¼-inch diameter aluminum tubing available at most hardware stores. A complete parts list is given in Table 2. Element sections and boom pieces are joined together by slotting a 10-inch length of 1¼-inch tubing with a nibbling tool and compressing it for a snug fit inside the element and boom tubing. Coupling details are shown in Fig 22.

The loading coils are wound on 1-1/8-inch diameter Plexiglas rod. The rod slips into the element tubing and is held in place with compression clamps. Be sure to slit the end of the aluminum where the compression clamps are placed. See Fig 24. The model shown in the photographs has coils made of surplus Teflon-insulated miniature audio coaxial cable with the shield braid and inner conductor shorted together. A suitable substitute would be no. 14 enameled copper wire wound to the same dimensions as those given in Fig 22.

All of the elements are secured to the boom with common TV U-bolt hardware. Plated bolts are desirable to prevent the formation of rust. A ¼-inch thick boom-to-mast plate is constructed from a few pieces of sheet aluminum cut into 10-inch square sheets and held together with no. 8 hardware. Several cookie tins could be used if sheet aluminum is not available. A plate from a large electrical box might even

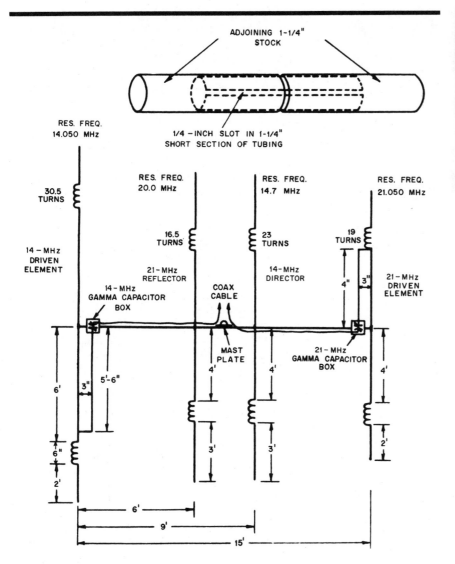

ADJOINING 1-1/4"
STOCK

1/4 -INCH SLOT IN 1-1/4"
SHORT SECTION OF TUBING

RES. FREQ.
14.050 MHz

30.5
TURNS

14 - MHz
DRIVEN
ELEMENT

14 - MHz
GAMMA CAPACITOR
BOX

RES. FREQ.
20.0 MHz

16.5
TURNS

21 - MHz
REFLECTOR

RES. FREQ.
14.7 MHz

23
TURNS

14 - MHz
DIRECTOR

COAX
CABLE

MAST
PLATE

RES. FREQ.
21.050 MHz

19
TURNS

4" 3"

21 - MHz
DRIVEN
ELEMENT

21 - MHz
GAMMA CAPACITOR
BOX

6'

3" 5'-6"

6"

2'

4'

3'

4'

3'

4'

2'

6'

9'

15'

Fig 22—Constructional details for the 14 and 21-MHz beam. The coils for each side of the element are identical. The gamma capacitors are each 140-pF variable units. The capacitors are insulated from ground within the container. Since the antenna is one-half size for each band, the tuning is somewhat critical. The builder is encouraged to carefully follow the dimensions given here.

Fig 23—The gamma assembly is held in place by means of a small U bolt. The capacitors are mounted on etched circuit board.

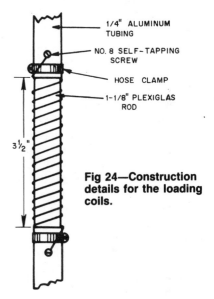

1/4" ALUMINUM
TUBING

NO. 8 SELF-TAPPING
SCREW

HOSE CLAMP

1-1/8" PLEXIGLAS
ROD

3½"

Fig 24—Construction details for the loading coils.

be used as a boom-to-mast bracket. The material for this plate should be chosen based upon the fact that it must withstand all of the wind and rotation forces on the antenna—it should be as rigid as possible. Galvanized material is best since it is quite resistant to harsh weather.

A boom strut (sometimes called a truss) is recommended because the weight of the elements is sufficient to cause the boom to sag a bit. A 1/8-inch diameter nylon line is plenty strong. A U-bolt clamp is placed on the mast several feet above the antenna and provides the attachment point for the center of the truss line. To reduce the possibility of water accumulating in the element tubing and subsequently freezing (rupture may be the end result), crutch caps are placed over the element ends. Rubber tips suitable for keeping steel-tubing furniture from scratching hardwood floors would serve the same purpose.

A heavy-duty steel mast should be used, such as a 1-inch

Table 2

Complete Parts List for the Short Beam

Qty	Material
9	Eight-foot lengths of aluminum tubing, 1¼" dia
11	U bolts
2	Variable capacitors, 140 pF (E. F. Johnson)
4'	Plexiglas cast rod, 1-1/8" dia
16	Stainless steel hose clamps, 1½" dia
1	Aluminum plate, 8" square
10'	Solid aluminum rod, ¼" dia
2	Refrigerator boxes, 4 × 4 × 4 in.
25'	Nylon rope, 1/8" diameter
16	No. 8 sheet metal screws
16	No. 8 solder lugs
8	Plastic (or rubber) end caps, 1¼-in.

Fig 25—The boom-to-mast plate.

diameter galvanized water pipe. Steel TV mast is also acceptable. Any conventional TV type antenna rotator should hold up under load conditions presented by this antenna. Nevertheless, certain precautions should be taken to assure continued trouble-free service. For instance, wherever possible, mount the rotator inside the tower and extend the mast through the tower top sleeve. This procedure relieves the rotator of lateral pressures during windy weather conditions. A thrust bearing is desirable to reduce downward forces on the rotator bearing.

The monoband nature of the beam requires the use of two coaxial feed lines. The coaxial cable is attached to the 21-MHz element (at the front of the beam) at the gamma-capacitor box. The other end of the cable is connected to a surplus 28-V dc single-pole coaxial switch. The cable for the 14-MHz element is connected in a similar fashion. The switch allows the use of a single feed line from the shack to a point just below the antenna where the switch is mounted. It is a simple matter to provide voltage to the switch for operation on one of the two bands. At the price of coaxial cable today, a double run of feed line represents a substantial investment and should be avoided if possible.

An etched circuit board was mounted inside an aluminum Minibox to provide support and insulation for each of the gamma tuning capacitors. Plastic refrigerator boxes available from most department stores would serve just as well. The capacitor housing is mounted to the boom by means of U bolts.

The builder is encouraged to follow the dimensions given in Fig 22 as a starting point for the position of the gamma rods and shorting bar. Placing the antenna near the top of the tower and then tilting it to allow the capacitors to be reached makes it possible to adjust the capacitors for minimum SWR as indicated by an SWR meter (or power meter) connected in the feed line at the relay. If the SWR cannot be reduced below some nominal figure of approximately 1.1:1, a slight repositioning of the gamma short

might be required. The dimensions given are for operation at 14.050 and 21.050 MHz. The SWR climbs above 2:1 about 50 kHz in either direction from the center frequency. Although tests were not conducted at more than 150 watts input to the transmitter, there is no reason why the system should not operate correctly with a kilowatt of power supplied to it.

After many months of testing this antenna, several characteristics were noted. During this period the antenna withstood several wind and ice storms. Performance is what can be expected from a two-element Yagi. The front-to-back ratio on 20 meters is a bit less than 10 dB. On 15 meters the front-to-back ratio is considerably better—on the order of 15 dB. Gain measurements were not made.

A YAGI ANTENNA WITH CONTINUOUSLY LOADED ELEMENTS

Another practical approach in building shortened Yagi antennas is the use of helically wound elements. Bamboo poles or fiberglass quad antenna spreaders are utilized as forms for the spirally wound elements. The 7-MHz beam illustrated in Figs 26 through 28 is only 28 percent of full size. The elements measure 18 feet tip to tip, and the boom is 16 feet long. The feed-point impedance is approximately 12 ohms, thereby permitting the use of a 4:1 broadband balun transformer to match the antenna to a 50-ohm coaxial feed line.

This antenna can be built for any 50-kHz segment of the 7-MHz amateur band and will operate with an SWR of less than 2.5:1 across that range. A 1:1 SWR can be obtained at the center of the 50-kHz range to which the beam is adjusted, and a gradual rise in SWR will occur as the frequency of

Fig 26—The short beam with helically wound elements for 7 MHz is shown here mounted on top of a 40-foot tower. A nylon-rope cross strut was not used with this installation and a slight amount of boom sag is noticeable.

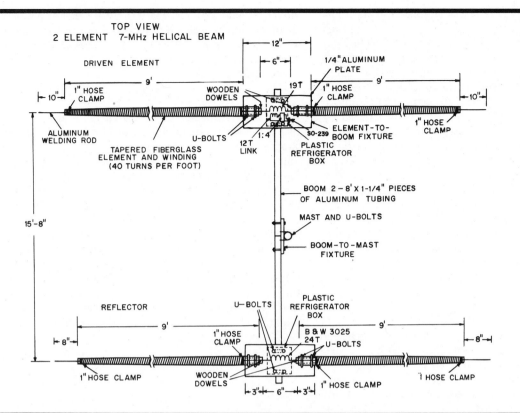

Fig 27—Overall dimensions for the 40-meter short beam. The boom consists of two pieces of standard 1¼-in. dia Do-It-Yourself aluminum tubing.

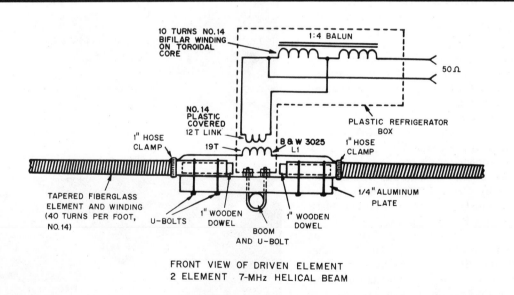

FRONT VIEW OF DRIVEN ELEMENT
2 ELEMENT 7-MHz HELICAL BEAM

Fig 28—Schematic diagram of the balun assembly mounted inside the plastic utility box. The core is a single T-200-2 Amidon. The 12-turn link is wound directly over L1.
L1—19 turns B&W 3205 Miniductor stock or equiv (no. 14 conductor, 2-in. dia, 6 TPI).

operation is changed toward the plus or minus 25-kHz points from center frequency.

Stubs of aluminum welding rod or no. 8 aluminum clothesline wire, 10 inches long, are used at the tips of each element to help lower the Q (in the interest of increased bandwidth). The stubs are useful in trimming the elements to resonance after the beam is elevated to its final height above ground. Coarse adjustment of the elements is effected by means of tapped inductors located at the center of each element. Plastic refrigerator boxes are mounted at the center of each element to protect the loading inductors and balun transformer from the effects of weather. Two coats of exterior spar varnish should be applied to the helically wound elements after they are adjusted to resonance. This will keep the turns in place and offer protection against moisture. Details for a 7-MHz version of this antenna are given here, but the same approach can be used in fabricating short beams for the other HF bands. Performance with the test model was excellent. The antenna was mounted 36 feet above ground (rotatable) on a steel tower.

Construction Details

The construction of the 40-meter beam is very simple and requires no special tools or hardware. Two fiberglass 21-MHz quad arm spreaders are mounted on an aluminum plate with U bolts, as shown in Fig 27. A wooden dowel is inserted approximately 6 inches into the end of each fiberglass arm to prevent the U bolts from crushing the poles. The aluminum mounting plate is equipped with U-bolt hardware for attachment to the 1¼-inch diameter boom.

A plastic refrigerator box is mounted on each element support plate and is used to house a Miniductor coil. No. 14 copper wire is used for the elements. The wire is wound directly onto the fiberglass poles at a density of *40 turns per foot* (not turns per inch) for a total of 360 evenly spaced turns. The wire is attached at each end with an automotive hose clamp of the proper size to fit the fiberglass spreader. Since the fiberglass is tapered, care must be taken to keep the turns from sliding in the direction of the end tips. Several pieces of plastic electrical tape were wrapped around the pole and wire at intervals of about every foot. All of the element half sections are identical in terms of wire and pitch. Coil dimensions and type are given in Figs 27 and 28.

The driven-element matching system consists of a 4:1 balun transformer and a tightly coupled link to the main-element Miniductor. Complete details are given in Fig 28.

Mounted at the end of each element and held in place by the hose clamp is a short section of stiff wire material used for final tuning of the system. Since the overall antenna is very small in relation to a full-sized array, the 2:1 SWR bandwidth is rather narrow. The bandwidth of the antenna shown in the photograph is about 60 kHz. This particular antenna was tuned for 7.040 MHz and can be used throughout the CW portion of the band. Tuning the antenna for phone-band operation should not be difficult, and the procedure outlined below should be suitable.

Tuning

The parasitic element was adjusted to be about four percent lower in frequency than the driven element. A dip oscillator was coupled to the center loading coil and the stiff-wire element tips were trimmed (a quarter of an inch at a time) until resonance was indicated at 6.61 MHz. For phone-band use, the ends could be snipped for 6.91 MHz. Adjusting the

driven element is simple. Place an SWR meter or power meter at the input connector and cut the end wires (or add some if necessary) to obtain the best match between the line and the antenna.

SMALL YAGI FOR 7-MHz

A 7-MHz antenna for most amateur installations consists of a half-wavelength dipole attached between two convenient supports and fed at the center with coaxial cable. When antenna gain is a requirement on this frequency, the dimensions of the system can become overwhelming. A full-size, three-element Yagi for 7 MHz would typically have 68-foot elements and a 36-foot boom. Accordingly, half-size elements present some distinct mechanical and economical advantages. Reducing the spacing between elements is not recommended since it would severely restrict the bandwidth of operation and make the tuning critical. The array shown in Fig 29 features good directivity and reasonable gain, yet the mechanical design allows the use of a "normal" heavy-duty rotator and a conventional tower support. Element loading is accomplished by lumped inductance and

Fig 29—The shortened 7-MHz Yagi is close in size to a standard three-element 14-MHz Yagi. It is shown on a 60-foot telephone pole.

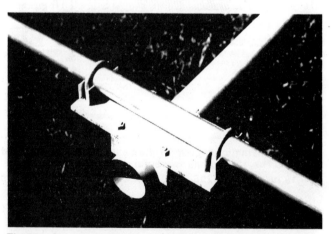

Fig 30—An aluminum plate and four automotive muffler clamps are used to affix the parasitic beam elements to the boom.

capacitance hats along the 38-foot elements. This design concept can be applied on any of the amateur HF bands.

Construction

The system described here uses standard sizes and lengths of aluminum tubing available through most aluminum suppliers. For best mechanical and electrical performance, 6061-T6 alloy should be used. All three elements are the same length; the tuning of the inductor is slightly different on each element, however. The two parasitic elements are grounded at the center with the associated boom-to-element hardware. A helical hairpin match is used to match the split and insulated driven element. Two sections of steel angle stock are used to reinforce the driven-element mounting plate since the Plexiglas center insulating material is not rigid and element sag might otherwise result. The parasitic element center sections are continuous sections of aluminum tubing, and additional support is not needed here. Figs 30 and 31 show the details clearly.

Fig 32—Each loading coil is wound on Plexiglas rod. The capacitance hats for the parasitic elements are mounted next to the coil, as shown here. The hose clamps compress the tubing against the Plexiglas rod. Each capacitance hat consists of two sections of tubing and associated muffler clamps.

Fig 31—The driven element of the antenna is insulated from the boom by means of PVC tubing, as shown.

The inductors for each element are wound on 1-1/8-inch diameter solid Plexiglas cast rod. Each end of the coil is secured in place with a solder lug, and the Plexiglas is held in position with an automotive compression clamp. The total number of turns needed to resonate the elements correctly is given in Fig 33. The capacitive hats consist of ½-inch tubing 3 feet long (two pieces used) attached to the element directly next to the coil on each parasitic element and 2 inches away from the coil for the driven element. Complete details are given in Figs 32 and 33.

The boom is constructed from three sections of aluminum tubing, each 2½ inches in diameter and 12 feet long. These pieces are joined together with inner tubes made from 2¼-inch stock shimmed with aluminum flashing. Long strips, approximately one inch wide, are wound on the inner tubing before it is placed inside the boom sections. A pair of 3/8 × 3½-inch steel bolts are placed at right angles to each other at every connection point to secure the boom. Caution: Do not overtighten the bolts as this will distort the tubing, making it impossible to pull apart sections, should the need arise. It

is much better to install locking nuts over the original ones to assure mechanical security.

The helical hairpin details are given in Fig 34. Quarter-inch copper tubing is formed into seven turns approximately 4 inches long and 2¼ inches ID.

Tuning and Matching

The builder is encouraged to follow carefully the dimensions given in Fig 33. Tuning the elements with the aid of a dip oscillator has proved to be unreliable; accordingly, no resonant frequencies are given.

The hairpin matching system may not resemble the usual form, but its operation and adjustment are essentially the same. For a detailed explanation of this network, see Chapter

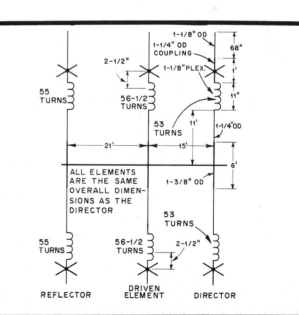

Fig 33—Mechanical details and dimensions for the 7-MHz Yagi. Each of the elements uses the same dimensions; the difference is only the number of turns on the inductors and the placement of the capacitive hats. See the text for more details.

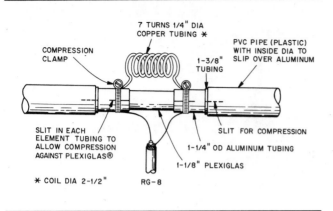

Fig 34—Details of the helical hairpin matching network on the driven element.

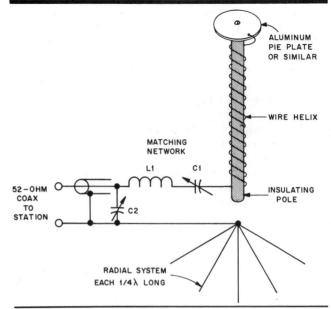

Fig 35—Details for building and connecting the helically wound short vertical antenna.

26. The driven element resonant frequency required for the hairpin match is determined by the placement of the capacitance hats with respect to the ends of the coils. Sliding the capacitance hats away from the ends of the coils increases the resonant frequency (capacitive reactance) of the element to cancel the effect of the hairpin inductive reactance. The model shown here had capacitance hats mounted 2½ inches out from the ends of the coils (on the driven element only). An SWR indicator or wattmeter should be installed in series with the feed line *at the antenna*. The hairpin coil may be spread or compressed with an insulated tool (or by hand if power is removed!) to provide minimum reflected power at 7.050 MHz.

The builder should not necessarily strive for a perfect match by changing the position of the capacitance hats since this may reduce the bandwidth of the matching system. An SWR of less than 2:1 was measured across the entire 7-MHz band with the antenna mounted atop an 80-foot tower.

The tuning of the array can be checked by making front to back ratio measurements across the band. With the dimensions given here, a front to back ratio of approximately 25 to 30 dB should be noticed in the CW portion of the band. Should the builder suspect the tuning is incorrect, or if the antenna is mounted at some height greatly different than 30 feet, retuning of the elements may be necessary.

SHORT CONTINUOUSLY LOADED VERTICAL ANTENNAS

The concept of size reduction can be applied to vertical antennas as well as to Yagi beams. One has the option of using lumped L and C to achieve resonance in a shortened system, or the antenna can be helically wound to provide a linear distribution of the required inductance, as shown in Fig 35. No added capacitance is necessary to establish resonance at the operating frequency. Shortened quarter-wavelength vertical antennas can be made by forming a helix on a long cylindrical insulator. The diameter of the helix must be very small in terms of wavelength in order to prevent the antenna from radiating in the axial mode. Acceptable form diameters for HF-band operation are from 1 inch to 10 inches when the practical aspects of antenna construction are considered. Insulating poles of fiberglass, PVC tubing, treated bamboo

or wood, or phenolic are suitable for use in building helically wound radiators. If wood or bamboo is used the builder should treat the material with at least two coats of exterior spar varnish prior to winding the antenna element. The completed structure should be given two more coats of varnish, regardless of the material used for the coil form. Application of the varnish will help weatherproof the antenna and prevent the coil turns from changing position.

No strict rule has been established concerning how short a helically wound vertical can be before a significant drop in performance is experienced. As a general recommendation, one should use the greatest amount of length consistent with available space. A guideline might be to maintain an element length of 0.05 wavelength or more for antennas which are electrically a quarter wavelength long. Thus, use 13 feet or more of stock for an 80-meter antenna, 7 feet for 40 meters, and so on.

A quarter wavelength helically wound vertical can be used in the same manner as a full-size vertical. That is, it can be worked against an above-ground wire radial system (four or more radials), or it can be ground-mounted with radials buried or lying on the ground. Some operators have reported good results when using antennas of this kind with four helically wound radials cut for resonance at the operating frequency. The latter technique should capture the attention of those persons who must use indoor antennas.

Winding Information

There is no hard and fast formula for determining the amount of wire needed to establish resonance in a helical antenna. The relationship between the length of wire needed for resonance and a full quarter wave at the desired frequency depends on several factors. Some of these are wire size, diameter of the turns, and the dielectric properties of the form material, to name a few. Experience has indicated that a section of wire approximately one half wavelength long, wound on an insulating form with a linear pitch (equal spacing between turns) will come close to yielding a resonant quarter wavelength. Therefore, an antenna for use on 160 meters would require approximately 260 feet of wire, spirally wound on the support. No specific rule exists concerning the size or type of wire one should use in making a helix. Larger wire sizes are, of course, preferable in the interest of minimizing I^2R losses in the system. For power levels up to 1000 watts it is wise to use a wire size of no. 16 or larger. Aluminum clothesline wire is suitable for use in systems where the spacing between turns is greater than the wire diameter. Antennas requiring close-spaced turns can be made from enameled magnet wire or no. 14 vinyl jacketed, single-conductor house wiring stock. Every effort should be made to keep the turn spacing as large as is practical to maximize efficiency.

A short rod or metal disc should be made for the top or high-impedance end of the vertical. This is a necessary part of the installation to assure reduction in antenna Q. This broadens the bandwidth of the system and helps prevent extremely high amounts of RF voltage from being developed at the top of the radiator. (Some helical antennas act like Tesla coils when used with high-power transmitters, and can actually catch fire at the high-impedance end when a stub or disc is not used.) Since the Q-lowering device exhibits some additional capacitance in the system, it must be in place before the antenna is tuned.

Tuning and Matching

Once the element is wound it should be mounted where it will be used, with the ground system installed. The feed end of the radiator can be connected temporarily to the ground system. Use a dip meter and check the antenna for resonance by coupling the dipper to the last few turns near the ground end of the radiator. Add or remove turns until the vertical is resonant at the desired operating frequency.

It is impossible to predict the absolute value of feed impedance for a helically wound vertical. The value will depend upon the length and diameter of the element, the ground system used with the antenna, and the size of the disc or stub atop the radiator. Generally speaking, the radiation resistance will be very low—approximately 3 to 10 ohms. An L network of the kind shown in Fig 36 can be used to increase the impedance to 50 ohms. Constants are given for operation at 7.0 MHz. The Q_L (loaded Q) of the network inductors is low to provide reasonable bandwidth, consistent with the bandwidth of the antenna. Network values for other operating bands and frequencies can be determined by using the reactance values listed below. The design center for the network is based on a radiation resistance of 5 ohms. If the exact feed impedance is known, the following equations can be used to determine precise component values for the

matching network. (See Chapter 25 for additional information on L-network matching.)

$$X_{C1} = QR_L$$

$$X_{C2} = 50\sqrt{\frac{R_L}{50 - R_L}}$$

$$X_{L1} = X_{C1} + \left(\frac{R_L 50}{X_{C2}}\right)$$

where

X_{C1} = Capacitive reactance of C1
X_{C2} = Capacitive reactance of C2
X_{L1} = Inductive reactance of L1
Q = Loaded Q of network
R_L = Radiation resistance of antenna

Example: Find the network constants for a helical antenna with a feed impedance of 5 ohms at 7 MHz, Q = 3:

$$X_{C1} = 3 \times 5 = 15 \text{ and}$$

$$X_{C2} = 50\sqrt{\frac{5}{50 - 5}} = 50\sqrt{0.111}$$

$$= 50 \times 0.333 = 16.666 \text{ and}$$

$$X_{L1} = 15 + \left(\frac{250}{16.66}\right)$$

$$= 15 + 15 = 30$$

Therefore, C1 = 1500 pF, C2 = 1350 pF, and L1 = 0.7 μH. The capacitors can be made from parallel or series combinations of transmitting micas. L1 can be a few turns of large Miniductor stock. At RF power levels of 100 W or less, large compression trimmers can be used at C1 and C2 because the maximum RMS voltage at 100 W (across 50 ohms) will be 50. At, say, 800 W there will be approximately 220 volts RMS developed across 50 ohms. This suggests the use of small transmitting variables at C1 and C2, possibly connected in parallel with fixed values of capacitance to constitute the required amount of capacitance for the network. By making some part of the network variable, it will be possible to adjust the circuit for an SWR of 1:1 without knowing precisely what the antenna feed impedance is. Actually, C1 is not required as part of the matching network. It is included here to bring the necessary value for L1 into a practical range.

Fig 35 illustrates the practical form a typical helically wound ground-plane vertical might take. Performance from this type antenna is comparable to that of many full-size quarter-wavelength vertical antennas. The major design trade-off is in usable bandwidth. All shortened antennas of this variety are narrow-band devices. At 7 MHz, in the example illustrated here, the bandwidth between the 2:1 SWR points will be on the order of 50 kHz, half that amount on 80 meters, and twice that amount on 20 meters. Therefore, the antenna should be adjusted for operation in the center of the frequency spread of interest.

THE FLAGPOLE DELUXE

The flagpole deluxe is a trap vertical that covers 7, 14,

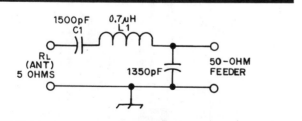

Fig 36—An L network suitable for matching the low feed-point impedance of the helically wound vertical to a 50-ohm coaxial cable at 7 MHz. The loaded Q of this network is 3.

21 and 28 MHz. From it you can fly a 4 × 6-foot flag. The antenna is built from aluminum tubing with slim, home-built traps, covered by standard PVC pipe, and topped with a toilet-tank ball. The PVC pipe has nothing to do with the antenna itself; it is the flagpole. The base is small and simple, thus easily concealed by brickwork, rocks or flowers.

The system described by F. Schnell, W6OZF, in Mar 1978 *QST* used two ground rods and the "skin" of a mobile home for a ground system. You should try to put down a system of ground radials if at all possible. A ground rod does not make a very good RF ground. Alternatively, a large metal structure, such as a mobile home, can serve as a counterpoise. Table 3 is a complete list of the necessary parts.

Preliminary Construction

Fig 37 shows details of the base mounting assembly. Drill and countersink a hole to accommodate a no. 8-32 × 9/16-inch flat-head screw. Drill and tap the insulator block and mount it inside the U channel. Drill eight 3/8-inch holes in the T bar as shown; the 2-inch U bolts will straddle the U channel. Assemble the U and T bars as shown in Fig 38, leaving the U bolts loose.

Fit the insulator sleeves on the 36-inch length of 1-1/8-inch, 0.125-inch wall tubing (Fig 38). These insulators should fit snugly and the U bolts which clamp them to the base assembly will hold them tight. Drill and tap the tubing

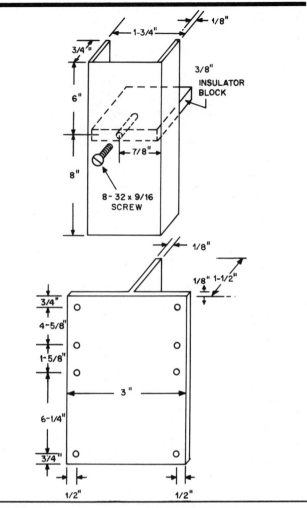

Fig 37—U- and T-bar stock details for the base assembly.

Table 3

Parts List for the Flagpole Deluxe Antenna

Qty	Material
1	Insulator block, 1 × 1½ × 3/8″ (Formica, Plexiglas, plastic, etc)
1	Length aluminum tubing, 1¼ × 0.058 × 144″ 6061-T6
1	Length aluminum tubing, 1-1/8 × 0.125 × 36″ 6061-T6
1	Piece of aluminum rod, 1 × 1″ round
1	Fiberglass round rod, 1 × 1″
4	2″ U bolts
5	25-pF Centralab transmitting capacitors, 850S series
1	Piece of construction aluminum channel stock, ¾ × 14″, 1/8″ wall thickness
1	Piece of construction aluminum T-bar stock, 3 × 1½ × 14″, 1/8″ wall thickness
8	1¼″ hose clamps
1	2-foot length 1¼″ EMT conduit
1	3-foot length 1¼″ EMT conduit
1	12-foot length of 2″ PVC schedule 40 water pipe
1	2″-to-2″ PVC coupling
1	1½″ PVC cap
1	12-foot length of 1½″ PVC schedule 40 water pipe
1	PVC reducer, 2″ to 1½″
1	Tank ball, 5″
1	10-foot length copper water pipe (for ground, if required)
1	Piece of copper or brass foil, 0.005 × 1 × 12″; this may be salvaged from old transformer trim stock, or may be purchased from auto or metal suppliers
2	Insulating sleeves made from plastic tubing 1-1/8″ inside diameter, ½″ wall thickness. These can be made from two PVC 1¼″-to-¾″ bushings. These are obtainable where PVC pipe is sold. The bushings will have to be reamed very slightly to have a press fit over the 1-1/8 × 0.125″ aluminum tubing. Either file or turn the flange down on a lathe.

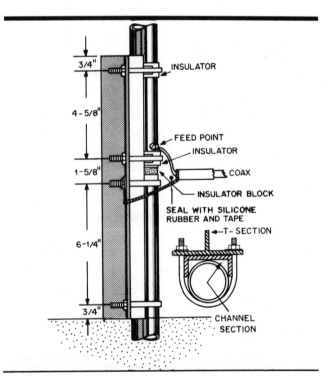

Fig 38—Construction details of the T bar and U channel for the base mounting assembly.

for a no. 10 screw, ¼ inch above the top insulator; this will be the feed point.

Traps

Refer to Fig 39. Cut one 7½-inch and two 8-inch lengths of 1-inch fiberglass rod, and through all three pieces drill a 1/8-inch hole lengthwise. In one end of the 7½-inch rod, drill a ¾-inch hole ¾ inch deep. In the other two pieces, drill a hole in one end 1½ inches deep. Use masonry drill bits for these holes.

In the 7½-inch rod, drill two 1/8-inch holes (3/16 inch deep) for pegs, as indicated in Fig 39. Insert 1/8-inch pegs in these holes. These pegs will be used to hold the windings in place.

Referring to Fig 40, wind the 28-MHz trap. Begin by soldering one end of a suitable length of no. 16 wire (Formvar or Thermaleze) to one end of a 25-pF capacitor. Push the wire through the rod from the large hole end and push the capacitor all the way in, then pull the wire snugly over the end of the rod to the nearest peg and wind five turns, close spaced. Coming off the next peg, dress the wire to the capacitor and solder to it by means of a no. 6 solder lug. This completes assembly of the 28-MHz trap.

Following the same procedure, wind the 14 and 21-MHz traps. Note that these traps are space wound and that the last turn is spaced a little more than the others; Figs 40 and 41 illustrate. The 21-MHz trap has 13 turns spaced to a length of 1-3/8 inches and the 14 MHz trap has 23 turns spaced to a length of 2-3/8 inches, with the last turn of each spaced slightly more than the rest. This helps in tuning of the traps.

Referring to Fig 42, scrape the enamel off both ends of each winding from the pegs to the ends of the fiberglass rods, and carefully tin each one with a hot soldering iron. Cut six foil tabs, each ¾ × 1½ inches and run a solder path lengthwise along each tab, on one side only. Now solder these tabs onto the previously tinned wires (Fig 42).

Cut six lengths of the 1-1/8-inch tubing, each 6 inches long. In one end only of two of the tubes cut a slot 3½ inches long, using two hacksaw blades together in the hacksaw (Fig 43). This will make a slot about 1/8 inch wide. On the other four tubes, make this slot 3 inches long. Remove one hacksaw blade and cut three more slots in the same end as the wide slot, spacing them 90° apart. Clean and deburr all cuts. These slots permit tight compression when the hose clamps are applied.

Carefully push these bushings (the slotted tubes) into place on the trap forms, up against the pegs, Fig 44. The longest slotted bushings go onto the 28-MHz trap. Then gently press the foil tabs down against the tubing. The bushings must be in place for the next step.

The traps should be checked with a dip meter and adjusted for resonance at the following frequencies: 28.0 MHz, 20.5 MHz and 14.0 MHz. Do not couple too tightly to the traps as the dipper can be "pulled," resulting in erroneous readings. Adjust the traps by carefully spreading or compressing the last turn on each trap.

Construction of Adjustable Section

Refer to Fig 45 and cut the 12-foot length of 1¼-inch tubing into four lengths, as follows: 5 feet 6 inches; 1 foot 9 inches;* 2 feet 10½ inches;* 1 foot 10½ inches. The starred (*) items may be cut in half and a bushing inserted to permit

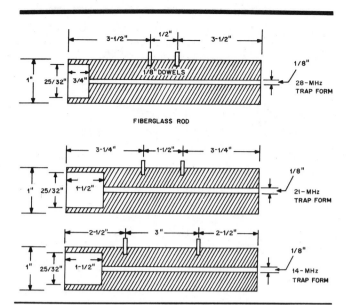

Fig 39—Cross-sectional views of the three traps.

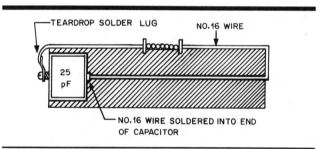

Fig 40—28-MHz trap winding detail.

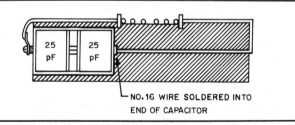

Fig 41—14 and 21-MHz trap details.

adjustment for the 14 and 21-MHz bands (Fig 44). If you choose this method, four more hose clamps will be required.

The tubes just cut should all be slotted and deburred on both ends, inside and out. If this deburring is not done the aluminum may seize or gall and it then becomes difficult (if not impossible) to separate or adjust. A stainless steel hose clamp is placed over each slotted end of each tube.

Cut and deburr a 6-foot length of the 1-1/8-inch tubing and insert the 1-inch piece of round aluminum rod, which should be drilled and tapped with ¼-20 threads all the way

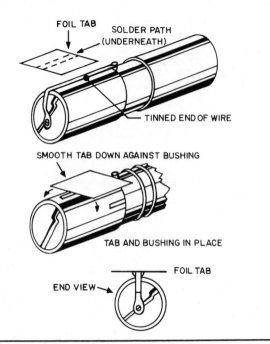

Fig 42—Further details of trap construction showing the foil tab and bushings.

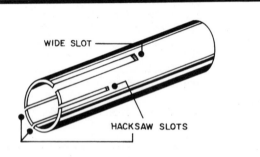

Fig 43—Method of slotting the tubing for use as a bushing.

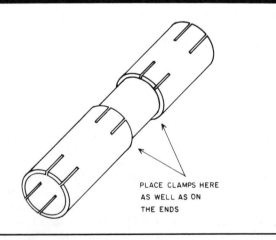

Fig 44—Bushings mounted in place on the 14-MHz trap.

through. Secure this in the end of the tubing by drilling two no. 36 holes and tapping for 6-32 screws.

Assembly

Figs 45 and 46 provide an overall picture of assembly of the vertical. Nothing is critical except that care must be exercised when attaching the traps to the 1¼-inch tubing sections. Start by spreading *one* of the slots in the tubing ends to expand the diameter slightly. The foil on each trap must be started smoothly between the trap bushing and the tubing sections. This is made easier by carefully pushing the foil tightly against the bushing before sliding the bushing onto the trap itself.

Start by sliding the 21-inch, 21-MHz tubing section onto the capacitor end of the 28-MHz trap, up against the peg, and tighten the hose clamp securely. In the same manner, clamp the 34½-inch, 14-MHz section onto the capacitor end of the 14-MHz trap. Tighten all clamps securely, then slide the 14 and 21-MHz sections onto the bottoms of their respective traps. See Fig 45. Next, clamp the 6-foot length of 1¼-inch tubing to the bottom of the 28-MHz trap, the 36-inch length of 1-1/8-inch tubing into the last tubing just installed and adjust the total length to 87¾ inches from the 28-MHz trap to the end of the tubing.

Install the ball on the end of the 6-foot length of 1-1/8-inch tubing that has the aluminum plug in it, using a 2-inch length of ¼-20 threaded stock. This can be cut from a ¼-20 bolt. Install this section of tubing into the end of the tubing on the 14-MHz trap and adjust to the dimension shown in Fig 45. Recheck all clamps for tightness.

Testing and Adjustment

Do your tuning in the final position if it is at all possible. Set the 3-foot length of 1¼-inch EMT conduit in concrete next to your ground system, leaving 7¾ inches above the concrete. Mount the antenna assembly on this conduit (Fig 37), and connect a strap from the ground system to the base assembly.

Once everything is in place and tightened, use a dip meter to determine the resonant frequencies of the vertical on each band. When making adjustments, remember that any change made on one band affects all *lower* frequency bands. Adjust the 28-MHz section first, and measure the SWR by attaching the coax and applying very low power to the antenna through an appropriate SWR indicator. Continue with the other bands; very little trimming or adding should be required. If adding length is necessary, the alternate method described earlier of cutting a section of tubing in half and adding a sleeve is recommended.

After the 7-MHz section is adjusted, remove the hose clamp and secure the tubing with four no. 6 sheet-metal screws. This is necessary because the 1½-inch PVC pipe will not clear the hose clamp. After all adjustments have been made, the SWR should not exceed 2:1 at any band edge.

Camouflaging

The time has come to turn the vertical into a flagpole, or vice versa. Lay the entire antenna (including the base mounting assembly) flat on the ground, with the 2-inch PVC alongside it. Carefully measure the distance from the top of the base assembly (Point A, Fig 45) to the top of the 14-MHz trap (Point B) and cut the PVC to length. Install the 2-inch to 1½-inch reducer assembly with PVC cement. Next measure the distance from the ridge inside the reducer to the top of the tubing of the 7-MHz section; make sure you have the end

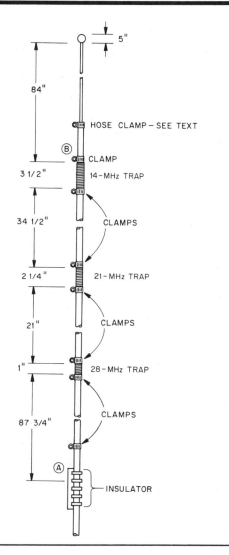

Fig 45—The complete Flagpole Deluxe vertical antenna.

Fig 46—Would you believe this flagpole is really a four-band antenna, covering 7, 14, 21 and 28 MHz? Sturdy enough to fly the flag in a stiff breeze, this trap vertical presents a good match to 50-ohm coaxial line.

of the 2-inch PVC even with the top of the base assembly. Cut the 1½-inch PVC to length and cement it into the reducer.

Drill a hole in the PVC cap to clear the ¼-20 stud in the end of the 40-meter section. Remove the base assembly and slide the antenna through the PVC pipe until the stud comes out the top; slip the cap onto the stud and screw the top ball on tight. Then slide the antenna back until the cap slides all the way on in place on the end of the 1½-inch PVC. Do *not* cement the cap in place. One word of caution—make certain all clamps are tight before sliding the antenna back and forth in the PVC, to prevent any change in dimensions. The end of the 2-inch PVC should now be even with the base mounting assembly when the antenna is reattached. A small pulley and rope should be added for flying a flag from the vertical.

BIBLIOGRAPHY

Source material and more extended discussion of topics covered in this chapter can be found in the references given below.

J. Belrose, "Transmission Line Low Profile Antennas," *QST*, Dec 1975.

C. Collinge, "Linear-Loaded 20-Meter Beam," *QST*, Jun 1976.

D. Courtier-Dutton, "Some Notes on a 7-MHz Linear-Loaded Quad," *QST*, Feb 1972.

R. B. Dome, "A Study of the DDRR Antenna," *QST*, Jul 1972.

W. E. English, "A 40-Meter DDRR Antenna," *QST*, Dec 1971.

J. Hall, "Off-Center-Loaded Dipole Antennas," *QST*, Sep 1974.

C. J. Michaels, "Some Reflections on Vertical Antennas," *QST*, May 1987.

R. M. Myers and D. DeMaw, "The HW-40 Micro Beam," *QST*, Feb 1974.

R. M. Myers and C. Greene, "A Bite Size Beam," *QST*, Sep 1973.

F. J. Schnell, "The Flagpole Deluxe," *QST*, Mar 1978, pp 29-32.

J. Sevick, "The Ground-Image Vertical Antenna," *QST*, Jul 1971.

J. Sevick, "The W2FMI Ground-Mounted Short Vertical," *QST*, Mar 1973.

J. Sevick, "The Constant-Impedance Trap Vertical," *QST*, Mar 1974.

J. Sevick, "Short Ground-Radial Systems for Short Verticals," *QST*, Apr 1978.

J. Tyskewicz, "The Heli-Rope Antenna," *QST*, Jun 1971.

Chapter 7

Multiband Antennas

For operation in a number of bands such as those between 3.5 and 30 MHz it would be impractical, for most amateurs, to put up a separate antenna for each band. But this is not necessary; a dipole, cut for the lowest frequency band to be used, can be operated readily on higher frequencies. To do so, one must be willing to accept the fact that such harmonic-type operation leads to a change in the directional pattern of the antenna (see Chapter 2). The user must also be willing to use "tuned" feeders. A center-fed single-wire antenna can be made to accept power and radiate it with high efficiency on any frequency higher than its fundamental resonant frequency and, with some reduction in efficiency and bandwidth, on frequencies as low as one half the fundamental.

In fact, it is not necessary for an antenna to be a full half-wavelength long at the lowest frequency. It has been determined that an antenna can be considerably shorter than ½ λ, as short as ¼ λ, and still be a very efficient radiator.

In addition, methods have been devised for making a single antenna structure operate on a number of bands while still offering a good match to a transmission line, usually of the coaxial type. It should be understood, however, that a "multiband antenna" is not *necessarily* one that will match a given line on all bands on which it is intended to be used. Even a relatively short whip type of antenna can be operated as a multiband antenna with suitable loading for each band. Such loading may be in the form of a coil at the base of the antenna on those frequencies where loading is needed, or it may be incorporated in the tuned feeders which run from the transmitter to the base of the antenna.

This chapter describes a number of systems that can be used on two or more bands. Beam antennas are treated separately in later chapters.

DIRECTLY FED ANTENNAS

The simplest multiband antenna is a random length of no. 12 or no. 14 wire. Power can be fed to the wire on practically any frequency by one or the other of the methods shown in Fig 1. If the wire is made either 67 or 135 feet long, it can also be fed through a tuned circuit, as in Fig 2. It is advantageous to use an SWR bridge or other indicator in the coax line at the point marked "X."

If a 28- or 50-MHz rotary beam has been installed, in many cases it will be possible to use the beam feed line as an antenna on the lower frequencies. Connecting the two wires of the feeder together at the station end will give a random-length wire that can be conveniently coupled to the transmitter as in Fig 1. The rotary system at the far end will serve only to "end load" the wire and will not have much other effect.

One disadvantage of all such directly fed systems is that

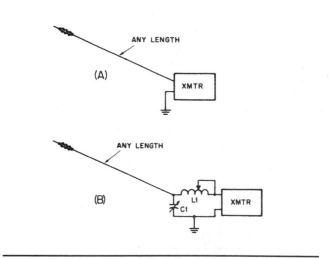

Fig 1—At A, a random-length wire driven directly from the pi-network output of a transmitter. At B, an L network for use in cases where sufficient loading cannot be obtained with the arrangement at A. C1 should have about the same plate spacing as the final tank capacitor in a vacuum-tube type of transmitter; a maximum capacitance of 100 pF is sufficient if L1 is 20 to 25 μH. A suitable coil would consist of 30 turns of no. 12 wire, 2½ in. dia, 6 turns per in. Bare wire should be used so the tap can be placed as required for loading the transmitter.

part of the antenna is practically within the station, and there is a good chance that trouble with RF feedback will be encountered. The RF within the station can often be minimized by choosing a length of wire so that a *current loop* occurs at or near the transmitter. This means using a wire length of ¼ λ (65 feet at 3.6 MHz, 33 feet at 7.1 MHz), or an odd multiple of ¼ λ (¾ λ is 195 feet at 3.6 MHz, 100 feet at 7.1 MHz). Obviously, this can be done for only one band in the case of even harmonically related bands, since the wire length that presents a current loop at the transmitter will present a voltage loop at two (or four) times that frequency.

When one is operating with a random-length wire antenna, as in Figs 1 and 2, it is wise to try different types of grounds on the various bands, to see which will give the best results. In many cases it will be satisfactory to return to the transmitter chassis for the ground, or directly to a convenient metallic water pipe. If neither of these works well (or the metallic water pipe is not available), a length of no. 12 or no. 14 wire (approximately ¼ λ long) can often be used to good advantage. Connect the wire at the point in the

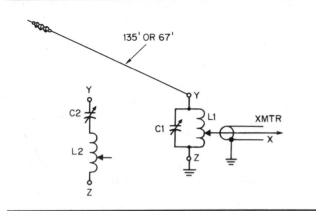

Fig 2—If the antenna length is 135 ft, a parallel-tuned coupling circuit can be used on each amateur band from 3.5 through 30 MHz, with the possible exception of the WARC 10, 18 and 24-MHz bands. C1 should duplicate the final tank tuning capacitor and L1 should have the same dimensions as the final tank inductor on the band being used. If the wire is 67 ft long, series tuning can be used on 3.5 MHz as shown at the left; parallel tuning will be required on 7 MHz and higher frequency bands. C2 and L2 will in general duplicate the final tank tuning capacitor and inductor, the same as with parallel tuning. The L network shown in Fig 1B is also suitable for these antenna lengths.

circuit that is shown grounded, and run it out and down the side of the house, or support it a few feet above the ground if the station is on the first floor or in the basement. It should not be connected to actual ground at any point.

END-FED ANTENNAS

When a straight-wire antenna is fed at one end by a two-wire line, the length of the antenna portion becomes critical if radiation from the line is to be held to a minimum. Such an antenna system for multiband operation is the "end-fed" or "Zepp-fed" antenna shown in Fig 3. The antenna length is made ½ λ long at the lowest operating frequency. The feeder length can be anything that is convenient, but feeder lengths that are multiples of ¼ λ generally give trouble with parallel currents and radiation from the feeder portion of the system. The feeder can be an open-wire line of no. 14 solid copper wire spaced 4 or 6 inches with ceramic or plastic spacers. Open-wire TV line (not the type with a solid web of dielectric) is a convenient type to use. This type of line is available in approximately 300- and 450-ohm characteristic impedances.

If one has room for only a 67-foot flat top and yet wants to operate in the 3.5-MHz band, the two feeder wires can be tied together at the transmitter end and the entire system treated as a random-length wire fed directly, as in Fig 1.

The simplest precaution against parallel currents that could cause feed-line radiation is to use a feeder length that is not a multiple of ¼ λ. A Transmatch can be used to provide multiband coverage with an end-fed antenna with any length of open-wire feed line, as shown in Fig 3.

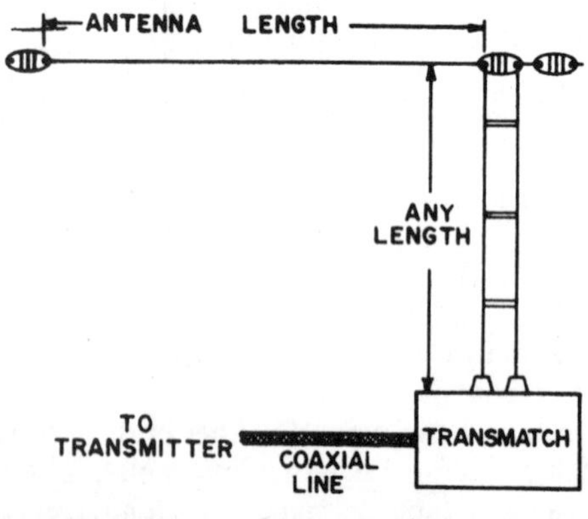

CENTER-FED ANTENNAS

The simplest and most flexible (and also least expensive) all-band antennas are those using open-wire parallel-conductor feeders to the center of the antenna, as in Fig 4. Because each half of the flat top is the same length, the feeder currents will be balanced at all frequencies unless, of course, unbalance is introduced by one half of the antenna being closer to ground (or a grounded object) than the other. For best results and to maintain feed-current balance, the feeder should run away at right angles to the antenna, preferably for at least ¼ λ.

Center feed is not only more desirable than end feed because of inherently better balance, but generally also results in a lower standing wave ratio on the transmission line, provided a parallel-conductor line having a characteristic impedance of 450 to 600 ohms is used. TV-type open-wire line is satisfactory for all but possibly high power installations (over 500 watts), where heavier wire and wider spacing is desirable to handle the larger currents and voltages.

The length of the antenna is not critical, nor is the length of the line. As mentioned earlier, the length of the antenna can be considerably less than ½ λ and still be very effective. It the overall length is at least ¼ λ at the lowest frequency, a quite usable system will result. The only difficulty that may exist with this type of system is the matter of coupling the antenna-system load to the transmitter. Most modern transmitters are designed to work into a 52-ohm coaxial load. With this type of antenna system a coupling network (a Transmatch) is required.

Feed-Line Radiation

The preceding sections have pointed out means of

Fig 3—An end-fed Zepp antenna for multiband use.

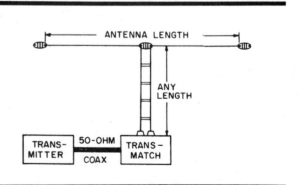

Fig 4—A center-fed antenna system for multiband use.

reducing or eliminating feed-line radiation. However, it should be emphasized that any radiation from a transmission line is not "lost" energy and is not necessarily harmful. Whether or not feed-line radiation is important depends entirely on the antenna system being used. For example, feed-line radiation is *not* desirable when a directive array is being used. Such radiation can distort the desired pattern of such an array, producing responses in unwanted directions. In other words, one wants radiation *only* from the directive array.

On the other hand, in the case of a multiband dipole where general coverage is desired, if the feed line happens to radiate, such energy could actually have a desirable effect. Antenna purists may dispute such a premise, but from a practical standpoint where one is not concerned with a directive pattern, much time and labor can be saved by ignoring possible transmission-line radiation.

MULTIPLE-DIPOLE ANTENNAS

The antenna system shown in Fig 5 consists of a group of center-fed dipoles, all connected in parallel at the point where the transmission line joins them. The dipole elements are stagger tuned. That is, they are individually cut to be ½ λ at different frequencies. Chapter 9 discusses the stagger tuning of dipole antennas to attain a low SWR across a broad range of frequencies. An extension of the stagger tuning idea is to construct multiwire dipoles cut for different bands.

In theory, the 4-wire antenna of Fig 5 can be used with a coaxial feeder on five bands. The four wires are prepared as parallel-fed dipoles for 3.5, 7, 14 and 28 MHz. The 7-MHz dipole can be operated on its 3rd harmonic for 21-MHz operation to cover the 5th band. However, in practice it has been found difficult to get a good match to coaxial line on all bands. The ½-λ resonant length of any one dipole in the presence of the others is not the same as for a dipole by itself, and attempts to optimize all four lengths can become a frustrating procedure. The problem is compounded because the optimum tuning changes in a different antenna environment, so what works for one amateur may not work for another. Even so, many amateurs with limited antenna space are willing to accept the mismatch on some bands just so they can operate on those frequencies.

Since this antenna system is balanced, it is desirable to use a balanced transmission line to feed it. The most desirable type of line is 75-ohm transmitting twin-lead. However, either 52-ohm or 75-ohm coaxial line can be used; coax line introduces some unbalance, but this is tolerable on the lower frequencies.

The separation between the dipoles for the various frequencies does not seem to be especially critical. One set of wires can be suspended from the next larger set, using insulating spreaders (of the type used for feeder spreaders) to give a separation of a few inches.

An interesting method of construction used successfully by Louis Richard, ON4UF, is shown in Fig 6. The antenna has four dipoles (for 7, 14, 21 and 28 MHz) constructed from 300-ohm ribbon transmission line. A single length of ribbon makes two dipoles. Thus, two lengths, as shown in the sketch, serve to make dipoles for four bands. Ribbon with copper-clad steel conductors (Amphenol type 14-022) should be used because all of the weight, including that of the feed line, must be supported by the uppermost wire.

Two pieces of ribbon are first cut to a length suitable for the two halves of the longest dipole. Then one of the conductors in each piece is cut to proper length for the next band higher in frequency. The excess wire and insulation is stripped away. A second pair of lengths is prepared in the same manner, except that the lengths are appropriate for the next two higher frequency bands.

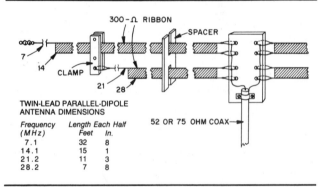

TWIN-LEAD PARALLEL-DIPOLE ANTENNA DIMENSIONS

Frequency (MHz)	Length Each Half	
	Feet	In.
7.1	32	8
14.1	15	1
21.2	11	3
28.2	7	8

Fig 6—Sketch showing how the twin-lead multiple-dipole antenna system is assembled. The excess wire and insulation are stripped away.

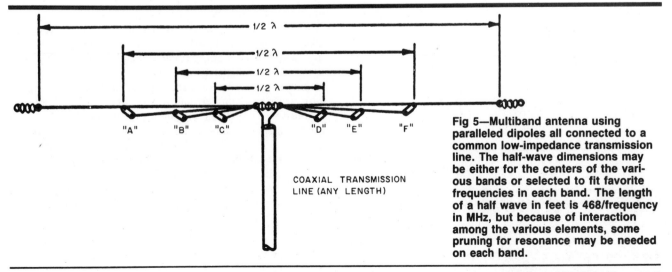

Fig 5—Multiband antenna using paralleled dipoles all connected to a common low-impedance transmission line. The half-wave dimensions may be either for the centers of the various bands or selected to fit favorite frequencies in each band. The length of a half wave in feet is 468/frequency in MHz, but because of interaction among the various elements, some pruning for resonance may be needed on each band.

A piece of thick polystyrene sheet drilled with holes for anchoring each wire serves as the central insulator. The shorter pair of dipoles is suspended the width of the ribbon below the longer pair by clamps also made of poly sheet. Intermediate spacers are made by sawing slots in pieces of poly sheet so they will fit the ribbon snugly.

The multiple-dipole principle can also be applied to vertical antennas. Parallel or fanned ¼-λ elements of wire or tubing can be worked against ground or tuned radials from a common feed point.

The Open-Sleeve Antenna

Although only recently adapted for the HF and VHF amateur bands, the open-sleeve antenna has been around since 1946. The antenna was invented by Dr. J. T. Bolljahn, of Stanford Research Institute. This section on sleeve antennas was written by Roger A. Cox, WBØDGF.

The basic form of the open-sleeve monopole is shown in Fig 7. The open-sleeve monopole consists of a base-fed central monopole with two parallel closely spaced parasites, one on each side of the central element, and grounded at each base. The lengths of the parasites are roughly one half that of the central monopole.

Impedance

The operation of the open sleeve can be divided into two modes, an antenna mode and a transmission line mode. This is shown in Fig 8.

The antenna mode impedance, Z_A, is determined by the length and diameter of the central monopole. For sleeve lengths less than that of the monopole, this impedance is essentially independent of the sleeve dimensions.

The transmission line mode impedance, Z_T, is deter-

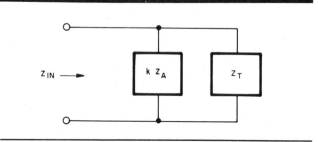

Fig 8—Equivalent circuit of an open-sleeve antenna.

mined by the characteristic impedance, end impedance, and length of the 3-wire transmission line formed by the central monopole and the two sleeve elements. The characteristic impedance, Z_c, can be determined by the element diameters and spacing if all element diameters are equal, and is found from

$$Z_c = 207 \log 1.59 \, (D/d)$$

where

D = spacing between the center of each sleeve element and the center of the driven element
d = diameter of each element

This is shown graphically in Fig 9. However, since the end impedance is usually unknown, there is little need to know the characteristic impedance. The transmission line mode impedance, Z_T, is usually determined by an educated guess and experimentation.

As an example, let us consider the case where the central monopole is ¼ λ at 14 MHz. It would have an antenna mode impedance, Z_A, of approximately 52 ohms, depending upon the ground conductivity and number of radials. If two sleeve elements were added on either side of the central monopole, with each approximately half the height of the monopole and at a distance equal to their height, there would be very little effect on the antenna mode impedance, Z_A, at 14 MHz.

Also, Z_T at 14 MHz would be the end impedance transformed through a 1/8-λ section of a very high characteristic impedance transmission line. Therefore, Z_T would be on the order of 500-2000 ohms resistive plus a large capacitive reactance component. This high impedance in parallel with

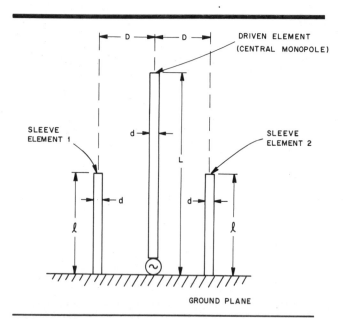

Fig 7—Diagram of an open-sleeve monopole.

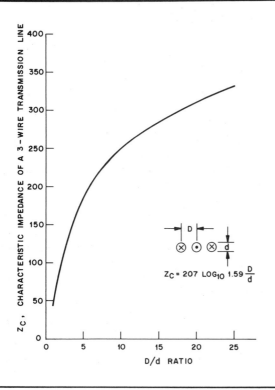

Fig 9—Characteristic impedance of transmission line mode in an open-sleeve antenna.

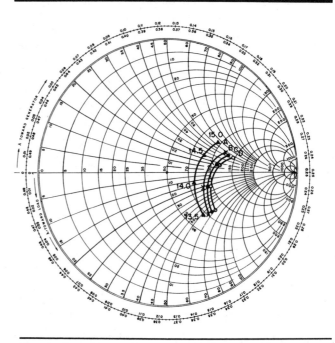

Fig 10—Impedance of an open-sleeve monopole for the frequency range 13.5-15 MHz. Curve A is for a 14-MHz monopole alone. For curves B, C and D, the respective spacings from the central monopole to the sleeve elements are 8, 6 and 4 in. See text for other dimensions.

52 ohms would still give a resultant impedance close to 52 ohms.

At a frequency of 28 MHz, however, Z_A is that of an end-fed half-wave antenna, and is on the order of 1000-5000 ohms resistive. Also, Z_T at 28 MHz would be on the order of 1000-5000 ohms resistive, since it is the end impedance of the sleeve elements transformed through a quarter-wave section of a very high characteristic impedance 3-wire transmission line. Therefore, the parallel combination of Z_A and Z_T would still be on the order of 500-2500 ohms resistive.

If the sleeve elements were brought closer to the central monopole such that the ratio of the spacing to element diameter was less than 10:1, then the characteristic impedance of the 3-wire transmission line would drop to less than 250 ohms. At 28 MHz, Z_A remains essentially unchanged, while Z_T begins to edge closer to 52 ohms as the spacing is reduced. At some particular spacing the characteristic impedance, as determined by the D/d ratio, is just right to transform the end impedance to exactly 52 ohms at some frequency. Also, as the spacing is decreased, the frequency where the impedance is purely resistive gradually increases.

The actual impedance plots of a 14/28 MHz open sleeve monopole appear in Figs 10 and 11. The length of the central monopole is 195.5 inches, and of the sleeve elements 89.5 inches The element diameters range from 1.25 inches at the bases to 0.875 inch at each tip. The measured impedance of the 14 MHz monopole alone, curve A of Fig 10, is quite high. This is probably because of a very poor ground plane

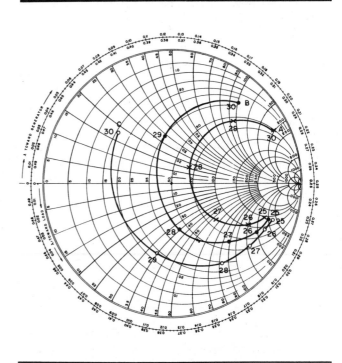

Fig 11—Impedance of the open-sleeve monopole for the range 25-30 MHz. For curves A, B and C the spacings from the central monopole to the sleeve elements are 8, 6 and 4 in., respectively.

under the antenna. The addition of the sleeve elements raises this impedance slightly, curves B, C and D.

As curves A and B in Fig 11 show, an 8 inch sleeve spacing gives a resonance near 27.8 MHz at 70 ohms, while a 6 inch spacing gives a resonance near 28.5 MHz at 42 ohms. Closer spacings give lower impedances and higher resonances. The optimum spacing for this particular antenna would be somewhere between 6 and 8 inches. Once the spacing is found, the lengths of the sleeve elements can be tweaked slightly for a choice of resonant frequency.

In other frequency combinations such as 10/21, 10/24, 14/21 and 14/24 MHz, spacings in the 6 to 10 inch range work very well with element diameters in the 0.5-1.25 inch range.

Bandwidth

The open-sleeve antenna, when used as a multiband antenna, does not exhibit broad SWR bandwidths unless, of course, the two bands are very close together. For example, Fig 12 shows the return loss and SWR of a single 10-MHz vertical antenna. Its 2:1 SWR bandwidth is 1.5 MHz, from 9.8-11.3 MHz. Return loss and SWR are related as given by the following equation.

$$SWR = \frac{1 + k}{1 - k}$$

where
$$k = 10^{-\frac{RL}{20}}$$

RL = return loss, dB

When sleeve elements are added for a resonance near 22 MHz, the 2:1 SWR bandwidth at 10 MHz is still nearly 1.5 MHz, as shown in Fig 13. The total amount of spectrum under 2:1 SWR increases, of course, because of the additional band, but the individual bandwidths of each resonance are virtually unaffected.

The open-sleeve antenna, however, can be used as a broadband structure, if the resonances are close enough to overlap. With the proper choices of resonant frequencies, sleeve and driven element diameters and sleeve spacing, the SWR ''hump'' between resonances can be reduced to a value less than 3:1. This is shown in Fig 14.

Current Distribution

According to H. B. Barkley (see bibliography at the end of this chapter), the total current flowing into the base of the

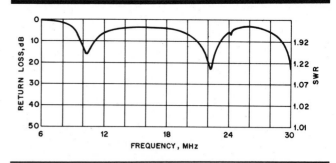

Fig 13—Return loss and SWR of a 10/22 MHz open-sleeve vertical antenna.

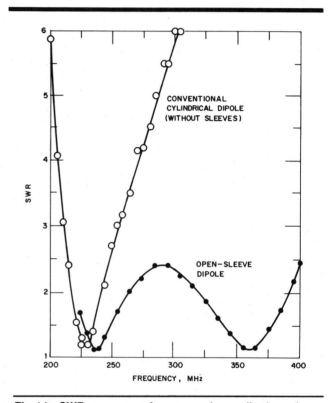

Fig 14—SWR response of an open-sleeve dipole and a conventional dipole.

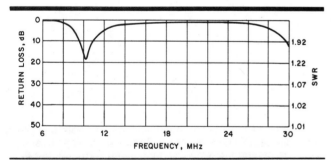

Fig 12—Return loss and SWR of a 10-MHz vertical antenna. A return loss of 0 dB represents an SWR of infinity. The text contains an equation for converting return loss to an SWR value.

open-sleeve antenna may be broken down into two components, that contributed by the antenna mode, I_A, and that contributed by the transmission line mode, I_T. Assuming that the sleeves are approximately half the height of the central monopole, the impedance of the antenna mode, Z_A, is very low at the resonant frequency of the central monopole, and the impedance of the transmission line mode, Z_T, is very high. This allows almost all of the current to flow in the antenna mode, and I_A is very much greater than I_T. Therefore, the current on the central ¼-λ monopole assumes the standard sinusoidal variation, and the radiation and gain characteristics are much like those of a normal ¼-λ vertical antenna.

However, at the resonant frequency of the sleeves, the impedance of the central monopole is that of an end fed half-wave monopole and is very high. Therefore I_A is small.

If proper element diameters and spacings have been used to match the transmission line mode impedance, Z_T, to 52 ohms; then I_T, the transmission line mode current, is high compared to I_A.

This means that very little current flows in the central monopole above the tops of the sleeve elements, and the radiation is mostly from the transmission line mode current, I_T, in all three elements below the tops of the sleeve elements. The resulting current distribution is shown in Figs 15 and 16 for this case.

Radiation Pattern and Gain

The current distribution of the open-sleeve antenna where all three elements are nearly equal in length is nearly that of a single monopole antenna. If, at a particular frequency, the elements are approximately ¼ λ long, the current distribution is sinusoidal.

If, for this and other length ratios, the chosen diameters and spacings are such that the two sleeve elements approach an interelement spacing of 1/8 λ; the azimuthal pattern will show directivity typical of two in-phase vertical radiators, approximately 1/8 λ apart. If a bidirectional pattern is needed, then this is one way to achieve it.

Spacings closer than this will produce nearly circular azimuthal radiation patterns. Practical designs in the 10-30 MHz range using 0.5-1.5 inch diameter elements will produce azimuthal patterns that vary less than plus or minus 1 dB.

If the ratio of the length of the central monopole to the length of the sleeves approaches 2:1, then the elevation pattern of the open-sleeve vertical antenna at the resonant frequency of the sleeves becomes slightly compressed. This is because of the in-phase contribution of radiation from the ½-λ central monopole.

As shown in Fig 17, the 10/21 MHz open-sleeve vertical antenna produces a lower angle of radiation at 21.2 MHz with a corresponding increase in gain of 0.66 dB over that of the 10-MHz vertical alone.

At length ratios approaching 3:1, the antenna mode and transmission line mode impedance become nearly equal again, and the central monopole again carries a significant portion of the antenna current. The radiation from the top ½ λ combines constructively with the radiation from the ¼ λ sleeve elements to produce gains of up to 3 dB more than just a quarter-wave vertical element alone.

Length ratios in excess of 3.2:1 produce higher level sidelobes and less gain on the horizon, except for narrow spots near the even ratios of 4:1, 6:1, 8:1, etc. These are where the central monopole is an even multiple of a half-wave, and the antenna mode impedance is too high to allow much antenna mode current.

Up to this point, it has been assumed that only ¼-λ resonance could be used on the sleeve elements. The third, fifth and seventh-order resonances of the sleeve elements and the central monopole element can be used, but their radiation patterns normally consist of high-elevation lobes, and the gain on the horizon is less than that of a ¼-λ vertical.

Practical Construction and Evaluation

The open-sleeve antenna lends itself very easily to home construction. For the open-sleeve vertical antenna, only a feed-point insulator and a good supply of aluminum tubing

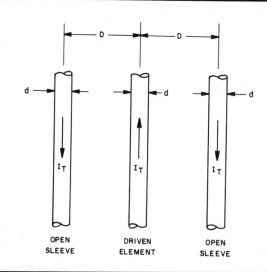

Fig 15—Current distribution in the transmission line mode. The amplitude of the current induced in each sleeve element equals that of the current in the central element but the phases are opposite, as shown.

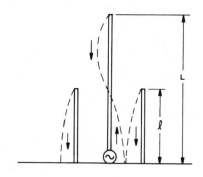

Fig 16—Total current distribution with ℓ = L/2.

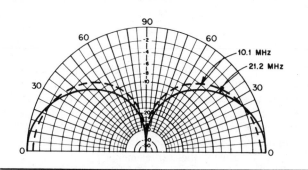

Fig 17—Vertical-plane radiation patterns of a 10/21 MHz open-sleeve vertical antenna on a perfect ground plane. At 10.1 MHz the maximum gain is 5.09 dBi, and 5.75 dBi at 21.2 MHz.

are needed. No special traps or matching networks are required. The open-sleeve vertical can produce up to 3 dB more gain than a conventional ¼-λ vertical. Further, there is no reduction in bandwidth, because there are no loading coils.

The open-sleeve design can also be adapted to horizontal dipole and beam antennas for HF, VHF and UHF. A good example of this is Telex/Hy-Gain's Explorer 14 triband beam which utilizes an open sleeve for the 10/15 meter driven element. The open-sleeve antenna is also very easy to model in computer programs such as NEC and MININEC, because of the open tubular construction and lack of traps or other intricate structures.

In conclusion, the open-sleeve antenna is an antenna experimenters delight. It is not difficult to match or construct, and it makes an ideal broadband or multiband antenna.

Trap Antennas

By using tuned circuits of appropriate design strategically placed in a dipole, the antenna can be made to show what is essentially fundamental resonance at a number of different frequencies. The general principle is illustrated by Fig 18.

Even though a trap-antenna arrangement is a simple one, an explanation of how a trap antenna works is elusive. For some designs, traps are resonated in our amateur bands, and for others (especially commercially made antennas) the traps are resonant far outside any amateur band.

A trap in an antenna system can perform either of two functions, depending on whether or not it is resonant at the operating frequency. A familiar case is where the trap is resonant in an amateur band. For the moment, let us assume that dimension A in Fig 18 is 33 feet and that each L/C combination is resonant in the 7-MHz band. Because of its resonance, the trap presents a high impedance at that point in the antenna system. The electrical effect at 7 MHz is that the trap behaves as an insulator. It serves to divorce the outside ends, the B sections, from the antenna. The result is easy to visualize—we have an antenna system that is resonant in the 7-MHz band. Each 33-foot section (labeled A in the drawing) represents ¼ λ, and the trap behaves as an insulator. We therefore have a full-size 7-MHz antenna.

The second function of a trap, obtained when the frequency of operation is *not* the resonant frequency of the trap, is one of electrical loading. If the operating frequency is below that of trap resonance, the trap behaves as an inductor; if above, as a capacitor. Inductive loading will electrically lengthen the antenna, and capacitive loading will electrically shorten the antenna.

Let's carry our assumption a bit further and try using the antenna we just considered at 3.5 MHz. With the traps resonant in the 7-MHz band, they will behave as inductors when operation takes place at 3.5 MHz, electrically lengthening the antenna. This means that the total length of sections A and B (plus the length of the inductor) may be something less than a physical ¼ λ for resonance at 3.5 MHz. Thus, we have a two-band antenna that is shorter than full size on the lower frequency band. But with the loading provided by the traps, the overall *electrical* length is ½ λ. The total antenna length needed for resonance in the 3.5-MHz band will depend on the L/C ratio of the trap elements.

The key to trap operation off resonance is its L/C ratio, the ratio of the value of L to the value of C. At resonance, however, within practical limitations the L/C ratio is immaterial as far as electrical operation goes. For example, in the antenna we've been discussing, it would make no difference for 7-MHz operation whether the inductor were

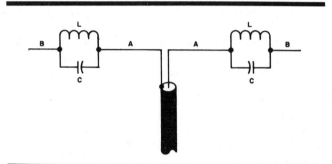

Fig 18—A trap dipole antenna. This antenna may be fed with 52-ohm coaxial line. Depending on the L/C ratio of the trap elements and the lengths chosen for dimensions A and B, the traps may be resonant either in an amateur band or at a frequency far removed from an amateur band for proper two-band antenna operation.

1 µH and the capacitor were 500 pF (the reactances would be just below 45 ohms at 7.1 MHz), or whether the inductor were 5 µH and the capacitor 100 pF (reactances of approximately 224 ohms at 7.1 MHz). But the choice of these values will make a significant difference in the antenna size for resonance at 3.5 MHz. In the first case, where the L/C ratio is 2000, the necessary length of section B of the antenna for resonance at 3.75 MHz would be approximately 28.25 feet. In the second case, where the L/C ratio is 50,000, this length need be only 24.0 feet, a difference of more than 15%.

The above example concerns a two-band antenna with trap resonance at one of the two frequencies of operation. On each of the two bands, each half of the dipole operates as an electrical ¼ λ. However, the same band coverage can be obtained with a trap resonant at, say, 5 MHz, a frequency quite removed from either amateur band. With proper selection of the L/C ratio and the dimensions for A and B, the trap will act to shorten the antenna electrically at 7 MHz and lengthen it electrically at 3.5 MHz. Thus, an antenna that is intermediate in physical length between being full size on 3.5 MHz and full size on 7 MHz can cover both bands, even though the trap is not resonant at either frequency. Again, the antenna operates with electrical ¼-λ sections.

Additional traps may be added in an antenna section to cover three or more bands. Or a judicious choice of dimensions and the L/C ratio may permit operation on three or more bands with just a pair of identical traps in the dipole.

An important point to remember about traps is this. *If the operating frequency is below that of trap resonance, the trap behaves as an inductor; if above, as a capacitor.* The above discussion is based on dipoles that operate electrically as 1/2 λ antennas. This is not a requirement, however. Elements may be operated as electrical 3/2 λ, or even 5/2 λ, and still present a reasonable impedance to a coaxial feeder. In trap antennas covering several HF bands, using electrical lengths that are odd multiples of ½ λ is often done at the higher frequencies.

To further aid in understanding trap operation, let's now choose trap L and C components which each have a reactance of 20 ohms at 7 MHz. Inductive reactance is directly proportional to frequency, and capacitive reactance is inversely proportional. When we shift operation to the 3.5-MHz band, the inductive reactance becomes 10 ohms, and the capacitive reactance becomes 40 ohms. At first thought, it may seem that the trap would become capacitive at 3.5 MHz with a higher capacitive reactance, and that the extra capacitive reactance would make the antenna electrically shorter yet. Fortunately, this is not the case. The inductor and the capacitor are connected in parallel *with each other*, but the *series equivalent* of this parallel combination is what affects the electrical operation of the antenna. The series equivalent of unlike reactances in parallel may be determined from the equation

$$Z = \frac{-j\,X_L\,X_C}{X_L - X_C}$$

where *j* indicates a reactive impedance component, rather than resistive. A positive result indicates inductive reactance, and a negative result indicates capacitive. In this 3.5-MHz case, with 40 ohms of capacitive reactance and 10 ohms of inductive, the equivalent series reactance is 13.3 ohms inductive. This inductive loading lengthens the antenna to an electrical ½ λ overall, assuming the B end sections in Fig 18 are of the proper length.

With the above reactance values providing resonance at 7 MHz, X_L equals X_C, and the theoretical series equivalent is infinity. This provides the insulator effect, divorcing the ends.

At 14 MHz, where $X_L = 40$ ohms and $X_C = 10$ ohms, the resultant series equivalent trap reactance is 13.3 ohms capacitive. If the total physical antenna length is slightly longer than 3/2 λ at 14 MHz, this trap reactance at 14 MHz can be used to shorten the antenna to an electrical 3/2 λ. In this way, 3-band operation is obtained for 3.5, 7 and 14 MHz with just one pair of identical traps. The design of such a system is not straightforward, however, for any chosen L/C ratio for a given total length affects the resonant frequency of the antenna on both the 3.5 and 14 MHz bands.

Trap Losses

Since the tuned circuits have some inherent losses, the efficiency of a trap system depends on the Q values of the tuned circuits. Low-loss (high-Q) coils should be used, and the capacitor losses likewise should be kept as low as possible. With tuned circuits that are good in this respect—comparable with the low-loss components used in transmitter tank circuits, for example—the reduction in efficiency compared with the

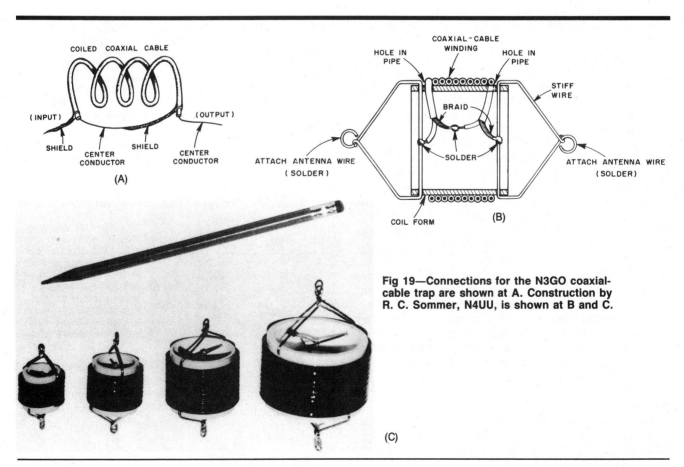

Fig 19—Connections for the N3GO coaxial-cable trap are shown at A. Construction by R. C. Sommer, N4UU, is shown at B and C.

efficiency of a simple dipole is small, but tuned circuits of low unloaded Q can lose an appreciable portion of the power supplied to the antenna.

The above commentary applies to traps assembled from conventional components. The important function of a trap that is resonant in an amateur band is to provide a high isolating impedance, and this impedance is directly proportional to Q. Unfortunately, high Q restricts the antenna bandwidth, because the traps provide maximum isolation only at trap resonance. A type of trap described by Gary O'Neil, N3GO, in October 1981 *Ham Radio* achieves high impedance with low Q, effectively overcoming the bandwidth problem. Shown in Fig 19, the N3GO trap is fabricated from a single length of coaxial cable. The cable is wound around a form as a single-layer coil, and the shield becomes the trap inductor.

The capacitance between the center conductor and shield resonates the trap. At each end of the coil the center conductor and shield are separated. At the "inside" end of the trap, nearer the antenna feed point, the shield is connected to the antenna wire. At the outside end, the center conductor is attached to the outside antenna wire. The center conductor from the inside end is joined to the shield from the outside end to complete the trap. Constructed in this way, the trap provides high isolation over a greater bandwidth than is possible with conventional traps.

Robert C. Sommer, N4UU, in December 1984 *QST* described how to optimize the N3GO trap. The analysis shows that best results are realized when the trap diameter is from 1 to 2.25 times greater than the length. Trap diameters toward the higher end of that range are better.

Five-Band Antenna

A trap antenna system has been worked out by C. L. Buchanan, W3DZZ, for the five pre-WARC amateur bands from 3.5 to 30 MHz. Dimensions are given in Fig 20. Only one set of traps is used, resonant at 7 MHz to isolate the inner (7-MHz) dipole from the outer sections, which cause the overall system to be resonant in the 3.5-MHz band. On 14, 21 and 28 MHz the antenna works on the capacitive-reactance principle just outlined. With a 75-ohm twin-lead feeder, the SWR with this antenna is under 2 to 1 throughout the three highest frequency bands, and the SWR is comparable with that obtained with similarly fed simple dipoles on 3.5 and 7 MHz.

Trap Construction

Traps frequently are built with coaxial aluminum tubes (usually with polystyrene tubing between them for insulation) for the capacitor, with the coil either self-supporting or wound on a form of larger diameter than the tubular capacitor. The coil is then mounted coaxially with the capacitor to form a unit assembly that can be supported at each end by the antenna wires. In another type of trap devised by William J. Lattin, W4JRW (see bibliography at the end of this chapter), the coil is supported inside an aluminum tube, and the trap capacitor is obtained in the form of capacitance between the coil and the outer tube. This type of trap is inherently weatherproof.

A simpler type of trap, easily assembled from readily available components, is shown in Fig 21. A small transmitting-type ceramic capacitor is used, together with a length of commercially available coil material, these being supported by an ordinary antenna strain insulator. The circuit constants and antenna dimensions differ slightly from those of Fig 20, in order to bring the antenna resonance points closer

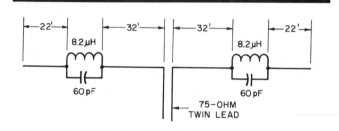

Fig 20—Five-band (3.5, 7, 14, 21 and 28 MHz) trap dipole for operation with 75-ohm feeder at low SWR (C. L. Buchanan, W3DZZ). The balanced (parallel-conductor) line indicated is desirable, but 75-ohm coax can be substituted with some sacrifice of symmetry in the system. Dimensions given are for resonance (lowest SWR) at 3.75, 7.2, 14.15 and 29.5 MHz. Resonance is very broad on the 21-MHz band, with SWR less than 2:1 throughout the band.

Fig 21—Easily constructed trap for wire antennas (A. Greenburg, W2LH). The ceramic insulator is 4¼ inches long (Birnback 688). The clamps are small service connectors available from electrical supply and hardware stores (Burndy KS90 servits).

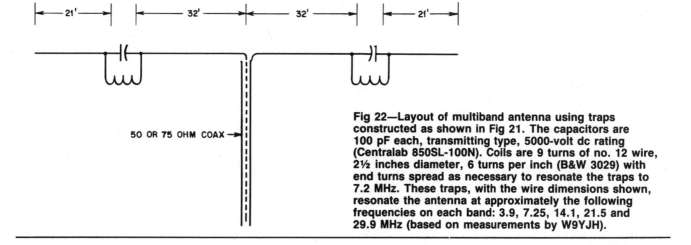

50 OR 75 OHM COAX →

Fig 22—Layout of multiband antenna using traps constructed as shown in Fig 21. The capacitors are 100 pF each, transmitting type, 5000-volt dc rating (Centralab 850SL-100N). Coils are 9 turns of no. 12 wire, 2½ inches diameter, 6 turns per inch (B&W 3029) with end turns spread as necessary to resonate the traps to 7.2 MHz. These traps, with the wire dimensions shown, resonate the antenna at approximately the following frequencies on each band: 3.9, 7.25, 14.1, 21.5 and 29.9 MHz (based on measurements by W9YJH).

to the centers of the various phone bands. Construction data are given in Fig 22. If a 10-turn length of inductor is used, a half turn from each end may be used to slip through the anchor holes in the insulator to act as leads.

The components used in these traps are sufficiently weatherproof in themselves so that no additional weatherproofing has been found necessary. However, if it is desired to protect them from the accumulation of snow or ice, a plastic cover can be made by cutting two discs of polystyrene slightly larger in diameter than the coil, drilling at the center to pass the antenna wires, and cementing a plastic cylinder on the edges of the discs. The cylinder can be made by wrapping two turns or so of 0.02-inch poly or Lucite sheet around the discs, if no suitable ready-made tubing is available. Plastic drinking glasses and soft 2-liter soft-drink bottles are easily adaptable for use as trap covers.

Four-Band Trap Dipole

In case there is not enough room available for erecting the 100-odd-foot length required for the five-band antenna just described, Fig 23 shows a four-band dipole operating on the same principle that requires only half the space. This antenna covers the 7- through 28-MHz bands but does not work on the WARC bands. The trap construction is the same as shown in Fig 21. With the dimensions given in Fig 23 the resonance points are 7.2, 14.1, 21.15 and 28.4 MHz. The capacitors are 27-pF transmitting type ceramic (Centralab type 857). The inductors are 9 turns of no. 12 wire, 2½ inches diameter, 6 turns per inch (B&W 3029), adjusted so that the trap resonates at 14.1 MHz before installation in the antenna.

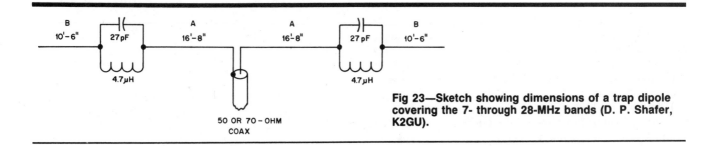

Fig 23—Sketch showing dimensions of a trap dipole covering the 7- through 28-MHz bands (D. P. Shafer, K2GU).

Vertical Antennas

There are two basic types of vertical antennas; either type can be used in multiband configurations. The first is the ground-mounted vertical and the second, the ground plane. These antennas are described in detail in Chapter 2.

The efficiency of any ground-mounted vertical depends a great deal on earth losses. As pointed out in Chapter 3, these losses can be reduced or eliminated with an adequate radial system. Considerable experimentation has been conducted on this subject by Jerry Sevick, W2FMI, and several important results were obtained. It was determined that a radial system consisting of 40 to 50 radials, 0.2 λ long, would reduce the earth losses to about 2 ohms when a ¼-λ radiator was being used. These radials should be on the earth's surface, or if buried, placed not more than an inch or so below ground. Otherwise, the RF current would have to travel through the lossy earth before reaching the radials. In a multiband vertical system, the radials should be 0.2 λ long for the lowest band, that is, 55 feet long for 3.5-MHz operation. Any wire size may be used for the radials. The radials should fan out in a circle, radiating from the base of the antenna. A metal plate, such as a piece of sheet copper, can be used at the center connection.

The other common type of vertical is the ground-plane antenna. Normally, this antenna is mounted above ground with the radials fanning out from the base of the antenna. The vertical portion of the antenna is usually an electrical ¼ λ, as is each of the radials. In this type of antenna, the system of radials acts somewhat like an RF choke, to prevent RF currents from flowing in the supporting structure, so the number of radials is not as important a factor as it is with a ground-mounted vertical system. From a practical standpoint, the customary number of radials is four or five. In a multiband configuration, ¼-λ radials are required for each band of operation with the ground-plane antenna. This is not so with the ground-mounted antenna, where the ground plane is relied upon to provide an image of the radiating section. In the ground-mounted case, as long as the ground-screen radials are approximately 0.2 λ long at the lowest frequency, the length will be more than adequate for the higher frequency bands.

Short Vertical Antennas

A short vertical antenna can be operated on several bands by loading it at the base, the general arrangement being similar to Figs 1 and 2. That is, for multiband work the vertical can be handled by the same methods that are used for random-length wires.

A vertical antenna should not be longer than about ¾ λ at the highest frequency to be used, however, if low-angle radiation is wanted. If the antenna is to be used on 28 MHz and lower frequencies, therefore, it should not be more than approximately 25 feet high, and the shortest possible ground lead should be used.

Another method of feeding is shown in Fig 24. L1 is a loading coil, tapped to resonate the antenna on the desired band. A second tap permits using the coil as a transformer for matching a coax line to the transmitter. C1 is not strictly necessary, but may be helpful on the lower frequencies, 3.5 and 7 MHz, if the antenna is quite short. In that case C1

makes it possible to tune the system to resonance with a coil of reasonable dimensions at L1. C1 may also be useful on other bands as well, if the system cannot be matched to the feed line with a coil alone.

The coil and capacitor should preferably be installed at the base of the antenna, but if this cannot be done a wire can be run from the antenna base to the nearest convenient location for mounting L1 and C1. The extra wire will of course be a part of the antenna, and since it may have to run through unfavorable surroundings it is best to avoid its use if at all possible.

This system is best adjusted with the help of an SWR indicator. Connect the coax line across a few turns of L1 and take trial positions of the shorting tap until the SWR reaches its lowest value. Then vary the line tap similarly; this should bring the SWR down to a low value. Small adjustments of both taps then should reduce the SWR to close to 1 to 1. If

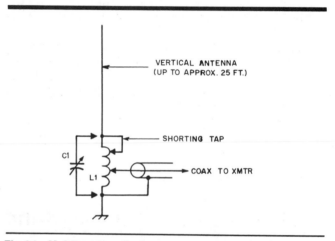

Fig 24—Multiband vertical antenna system using base loading for resonating on 3.5 to 28 MHz. L1 should be wound with bare wire so it can be tapped at every turn, using no. 12 wire. A convenient size is 2½ inches diameter, 6 turns per inch (such as B&W 3029). Number of turns required depends on antenna and ground lead length, more turns being required as the antenna and ground lead are made shorter. For a 25-ft antenna and a ground lead of the order of 5 ft, L1 should have about 30 turns. The use of C1 is explained in the text. The smallest capacitance that will permit matching the coax cable should be used; a maximum capacitance of 100 to 150 pF will be sufficient in any case.

not, try adding C1 and go through the same procedure, varying C1 each time a tap position is changed.

Trap Verticals

The trap principle described in Fig 18 for center-fed dipoles also can be used for vertical antennas. There are two principal differences. Only one half of the dipole is used, the ground connection taking the place of the missing half, and the feed-point impedance is one half the feed-point impedance of a dipole. Thus it is in the vicinity of 30 ohms (plus the ground-connection resistance), so 52-ohm cable should be used since it is the commonly available type that comes closest to matching.

As in the case of any vertical antenna, a good ground is essential, and the ground lead should be short. Some amateurs have reported successfully using a ground plane dimensioned for the lowest frequency to be used; for example, if the lowest frequency is 7 MHz, the ground-plane radials can be approximately 34 feet long.

A Trap Vertical For 21 and 28 MHz

Simple antennas covering the upper HF bands can be quite compact and inexpensive. The two-band vertical ground plane described here is highly effective for long-distance communication when installed in the clear.

Figs 25, 26 and 27 show the important assembly details. The vertical section of the antenna is mounted on a ¾-inch thick piece of plywood board that measures 7 × 10 inches. Several coats of exterior varnish or similar material will help protect the wood from inclement weather. Both the mast and the radiator are mounted on the piece of wood by means of TV U-bolt hardware. The vertical is electrically isolated from the wood with a piece of 1-inch diameter PVC tubing. A piece approximately 8 inches long is required, and it is of the schedule-80 variety. To prepare the tubing it must be slit along the entire length on one side. A hacksaw will work quite well. The PVC fits rather snugly on the aluminum tubing and will have to be "persuaded" with the aid of a hammer. The mast is mounted directly on the wood with no insulation. An SO-239 coaxial connector and four solder lugs are mounted on an L-shaped bracket made from a piece of aluminum sheet. A short length of test probe wire, or inner conductor of RG-58 cable, is soldered to the inner terminal of the connector. A UG-106 connector hood is then slid over the wire and onto the coaxial connector. The hood and connector are bolted to the aluminum bracket. Two wood screws are used to secure the aluminum bracket to the plywood as shown in the drawing and photograph. The free end of the wire coming from the connector is soldered to a lug which is mounted on the bottom of the vertical radiator. Any space between the wire and where it passes through the hood is filled with GE silicone glue and seal or similar material to keep moisture out. The eight radials are soldered to the four lugs on the aluminum bracket. The two sections of the vertical member are separated by a piece of clear acrylic rod. Approximately 8 inches of 7/8-inch OD material is required. The aluminum tubing must be slit lengthwise for several inches so the acrylic rod may be inserted. The two pieces of aluminum tubing are separated by 2¼ inches.

The trap capacitor is made from RG-8 coaxial cable and is 30.5 inches long. RG-8 cable has 29.5 pF of capacitance per foot and RG-58 has 28.5 pF per foot. RG-8 cable is recommended over RG-58 because of its higher breakdown-voltage characteristic. The braid should be pulled back 2 inches on one end of the cable, and the center conductor soldered to one end of the coil. Solder the braid to the other end of the coil. Compression type hose clamps are placed over the capacitor/coil leads and put in position at the edges of

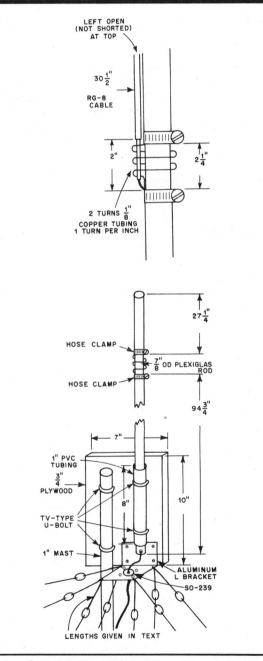

Fig 25—Constructional details of the 21- and 28-MHz dual-band antenna system.

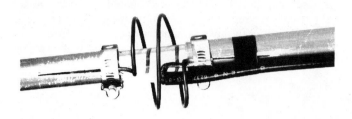

Fig 26—A close-up view of a trap. The coil is 3 inches in diameter. The leads from the coaxial-cable capacitor should be soldered directly to the pigtails of the coil. These connections should be coated with varnish after they have been secured under the hose clamps.

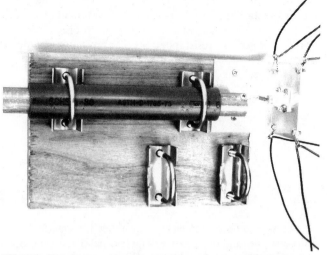

Fig 27—The base assembly of the 21- and 28-MHz vertical. The SO-239 coaxial connector and hood can be seen in the center of the aluminum L bracket. The U bolts are TV-type antenna hardware. The plywood should be coated with varnish or similar material.

the aluminum tubing. When tightened securely, the clamps serve a two-fold purpose—they keep the trap in contact with the vertical members and prevent the aluminum tubing from slipping off the acrylic rod. The coaxial-cable capacitor runs upward along the top section of the antenna. This is the side of the antenna to which the braid of the capacitor is connected. A cork or plastic cap should be placed in the very top of the antenna to keep moisture out.

Installation and Operation

The antenna may be mounted in position using a TV-type tripod, chimney, wall or vent mount. Alternatively, a telescoping mast or ordinary steel TV mast may be used, in which case the radials may be used as guys for the structure.

The 28-MHz radials are 8 feet 5 inches long, and the 21-MHz radials are 11 feet 7 inches.

Any length of 52-ohm cable may be used to feed the antenna. The SWR at resonance should be on the order of 1.2:1 to 1.5:1 on both bands. The reason the SWR is not 1 is that the feed-point resistance is something other than 52 ohms—closer to 35 or 40 ohms.

Adapting Manufactured Trap Verticals to the WARC Bands

The frequency coverage of a multiband HF vertical antenna can be modified simply by altering the lengths of the tubing sections and/or adding a trap. Several companies manufacture trap verticals covering 7, 14, 21 and 28 MHz. Many amateurs roof-mount these antennas for any of a number of reasons—because an effective ground radial system isn't practical, to keep children away from the antenna, or to clear metal-frame buildings. On the three highest frequency bands, the tubing and radial lengths are convenient for rooftop installations, but 7 MHz sometimes presents problems. Prudence dictates erecting an antenna with the assumption that it will fall down. When the antenna falls, it and the radial system must clear any nearby power lines. Where this consideration rules out 7-MHz operation, careful measurement may show that 10-MHz dimensions will allow adequate safety. The antenna is resonated by pruning the tubing above the 14-MHz trap and installing tuned radials.

Several new frequency combinations are possible. The simpler ones are 24/28, 18/21/28 and 7/10/14/21/28 MHz. These are shown in Fig 28 as applied to the popular ATV series of trap verticals manufactured by Cushcraft. Operation in the 10-MHz band requires an additional trap—use Fig 26 as a guide for constructing this component.

Combining Vertical and Horizontal Conductors

The performance of vertical antennas such as just described depends a great deal on the quality of the ground system. If you can eliminate the ground connection as a part of the antenna system, it simplifies things. Fig 29 shows how it can be done. Instead of a ground, the system is completed by a wire—preferably, but not necessarily, horizontal—of the same length as the antenna. This makes a center-fed system somewhat like a dipole.

It is desirable that the length of each conductor be on the order of 30 feet, as shown in the drawing, if the 3.5-MHz band is to be used. At 7 MHz, this length doesn't really represent a compromise, since the total length is almost ½ λ overall on that band. Because the shape of the antenna differs from that of a regular ½-λ dipole, the radiation

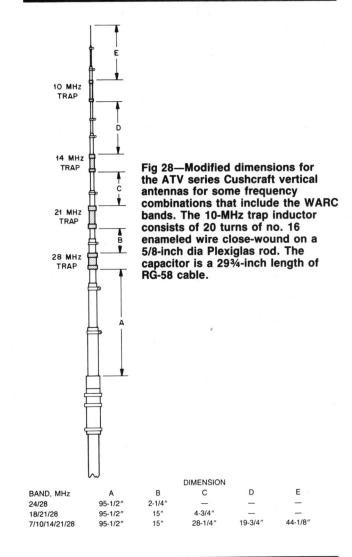

Fig 28—Modified dimensions for the ATV series Cushcraft vertical antennas for some frequency combinations that include the WARC bands. The 10-MHz trap inductor consists of 20 turns of no. 16 enameled wire close-wound on a 5/8-inch dia Plexiglas rod. The capacitor is a 29¾-inch length of RG-58 cable.

			DIMENSION		
BAND, MHz	A	B	C	D	E
24/28	95-1/2"	2-1/4"	—	—	—
18/21/28	95-1/2"	15"	4-3/4"	—	—
7/10/14/21/28	95-1/2"	15"	28-1/4"	19-3/4"	44-1/8"

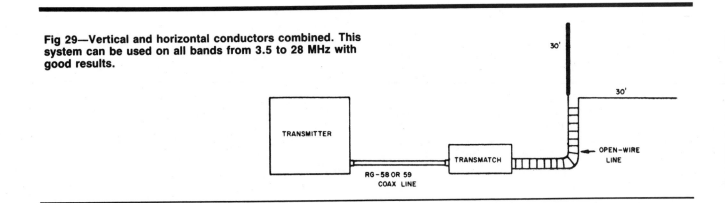

Fig 29—Vertical and horizontal conductors combined. This system can be used on all bands from 3.5 to 28 MHz with good results.

characteristics will be different, but the efficiency will be high on 7 MHz and higher frequencies. Although the vertical radiating part is only about 1/8 λ at 3.5 MHz, the efficiency on this band, too, will be higher than it would be with a grounded system. If one is not interested in 3.5 MHz and can't use the dimensions shown, the lengths can be reduced. Fifteen feet in both the vertical and horizontal conductors will not do too badly on 7 MHz and will not be greatly handicapped, as compared with a ½-λ dipole, on 14 MHz and higher.

The vertical part can be mounted in a number of ways. However, if it can be put on the roof of your house, the extra height will be worthwhile. Fig 30 suggests a simple base mount using a glass bottle as an insulator. Get one with a neck diameter that will fit into the tubing used for the vertical part of the antenna. To help prevent possible breakage, put a piece of some elastic material such as rubber sheet around the bottle where the tubing rests on it.

The lower wire conductor doesn't actually have to be horizontal. It can be at practically any angle that will let it run off in a straight line to a point where it can be secured. Use an insulator at this point, of course.

TV ladder line should be used for the feeder in this system. On most bands the standing wave ratio will be high, and you will lose a good deal of power in the line if you try to use coax, or even 300-ohm twin-lead. This system can be tuned up by using an SWR indicator in the coax line between the transmitter and a Transmatch.

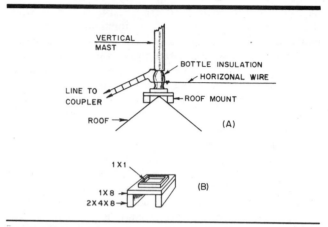

Fig 30—One method of mounting the vertical section on a rooftop. The mounting base dimensions can be adjusted to fit the pitch of the roof. The 1 × 1 pieces should fit snugly around the bottom of the bottle to keep it from shifting position.

The Multee Antenna

Two-band operation may be obtained on 1.8/3.5 MHz or on 3.5/7 MHz within the confines of the average city lot by using the multee antenna shown in Fig 31. Dimensions are given for either pair of bands in the drawing. If built for the lower frequencies, the top portion will do little radiating on 1.8 MHz; it acts merely as top loading for the 52-foot vertical section. On 3.5 MHz, the horizontal portion radiates and the vertical section acts as a matching stub to transform the high feed-point impedance to the coaxial cable impedance.

Since the antenna must work against ground on its lower frequency band, it is necessary to install a good ground system. Minimum requirements in this regard would include 20 radials, each 55 to 60 feet long for the 1.8/3.5-MHz version, or half that for the 3.5/7 MHz version. If not much area is available for the radial system, wires as short as 25 feet long (12 feet for 3.5/7 MHz) may be used if many are installed, but some reduction in efficiency will result.

With suitable corrections in length to account for the velocity factor, 300-ohm TV twin-lead may be substituted for the open wire. The velocity factor should be taken into account for both the vertical and horizontal portions, to preserve the impedance relationships.

HARMONIC RADIATION FROM MULTIBAND ANTENNAS

Since a multiband antenna is intentionally designed for operation on a number of different frequencies, any harmon-

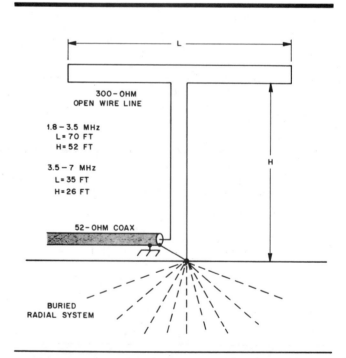

Fig 31—Two-band operation in limited space may be obtained with the multee antenna. The portion identified as H should remain as vertical as possible, as it does the radiating on the lower frequency band.

ics or spurious frequencies that happen to coincide with one of the antenna resonant frequencies will be radiated with very little, if any, attenuation. Particular care should be exercised, therefore, to prevent such harmonics from reaching the antenna.

Multiband antennas using tuned feeders have a certain inherent amount of built-in protection against such radiation, since it is nearly always necessary to use a tuned coupling circuit (Transmatch) between the transmitter and the feeder. This adds considerable selectivity to the system and helps to discriminate against frequencies other than the desired one.

Multiple dipoles and trap antennas do not have this feature, since the objective in design is to make the antenna show as nearly as possible the same resistive impedance in all the amateur bands the antenna is intended to cover. It is advisable to conduct tests with other amateur stations to determine whether harmonics of the transmitting frequency can be heard at a distance of, say, a mile or so. If they can, more selectivity should be added to the system since a harmonic that is heard locally, even if weak, may be quite strong at a distance because of propagation conditions.

BIBLIOGRAPHY

Source material and more extended discussion of topics covered in this chapter can be found in the references given below and in the textbooks listed at the end of Chapter 2.

H. B. Barkley, *The Open-Sleeve As A Broadband Antenna*, Technical Report No. 14, US Naval Postgraduate School, Monterey, CA, Jun 1955.

W. M. Bell, "A Trap Collinear Antenna," *QST*, Aug 1963, pp 30-31.

H. J. Berg, "Multiband Operation with Paralleled Dipoles," *QST*, Jul 1956.

E. L. Bock, J. A. Nelson and A. Dorne, "Sleeve Antennas," *Very High Frequency Techniques*, H. J. Reich, ed. (New York: McGraw-Hill, 1947), Chapter 5.

J. T. Bolljahn and J. V. N. Granger, "Omnidirectional VHF and UHF Antennas," *Antenna Engineering Handbook*, H. Jasik, ed. (New York: McGraw-Hill, 1961) pp 27-32 through 27-34.

G. H. Brown, "The Phase and Magnitude of Earth Currents Near Radio Transmitting Antennas," *Proc IRE*, Feb 1935.

G. H. Brown, R. F. Lewis and J. Epstein, "Ground Systems as a Factor in Antenna Efficiency," *Proc IRE*, Jun 1937, pp 753-787.

C. L. Buchanan, "The Multimatch Antenna System," *QST*, Mar 1955.

R. A. Cox, "The Open-Sleeve Antenna," *CQ*, Aug 1983, pp 13-19.

D. DeMaw, "Lightweight Trap Antennas—Some Thoughts," *QST*, Jun 1983, p 15.

W. C. Gann, "A Center-Fed 'Zepp' for 80 and 40," *QST*, May 1966.

A. Greenberg, "Simple Trap Construction for the Multiband Antenna," *QST*, Oct 1956.

G. L. Hall, "Trap Antennas," Technical Correspondence, *QST*, Nov 1981, pp 49-50.

W. Hayward, "Designing Trap Antennas," Technical Correspondence, *QST*, Aug 1976, p 38.

R. H. Johns, "Dual-Frequency Antenna Traps," *QST*, Nov 1983, p 27.

R. W. P. King, *Theory of Linear Antennas* (Cambridge, MA: Harvard Univ Press, 1956), pp 407-427.

W. J. Lattin, "Multiband Antennas Using Decoupling Stubs," *QST*, Dec 1960.

W. J. Lattin, "Antenna Traps of Spiral Delay Line," *QST*, Nov 1972, pp 13-15.

M. A. Logan, "Coaxial-Cable Traps," Technical Correspondence, *QST*, Aug 1985, p 43.

J. R. Mathison, "Inexpensive Traps for Wire Antennas," *QST*, Aug 1977, p 18.

L. McCoy, "An Easy-to-Make Coax-Fed Multiband Trap Dipole," *QST*, Dec 1964.

G. E. O'Neil, "Trapping the Mysteries of Trapped Antennas," *Ham Radio*, Oct 1981, pp 10-16.

L. Richard, "Parallel Dipoles of 300-Ohm Ribbon," *QST*, Mar 1957.

D. P. Shafer, "Four-Band Dipole with Traps," *QST*, Oct 1958.

R. R. Shellenbach, "Try the 'TJ'," *QST*, Jun 1982, p 18.

R. R. Shellenbach, "The JF Array," *QST*, Nov 1982, p 26.

R. C. Sommer, "Optimizing Coaxial-Cable Traps," *QST*, Dec 1984, p 37.

Chapter 8

Multielement Arrays

The gain and directivity offered by an array of elements represents a worthwhile improvement both in transmitting and receiving. Power gain in an antenna is the same as an equivalent increase in the transmitter power. But unlike increasing the power of one's own transmitter, antenna gain works equally well on signals received from the favored direction. In addition, the directivity reduces the strength of signals coming from the directions not favored, and so helps discriminate against a good deal of interference.

One common method of obtaining gain and directivity is to combine the radiation from a group of ½-λ dipoles in such a way as to concentrate it in a desired direction. The way in which such combinations affect the directivity has been explained in Chapter 2. A few words of additional explanation may help make it clear how power gain is obtained.

In Fig 1, imagine that the four circles, A, B, C and D, represent four dipoles so far separated from each other that the coupling between them is negligible. Also imagine that point P is so far away from the dipoles that the distance from P to each one is exactly the same (obviously P would have to be much farther away than it is shown in this drawing). Under these conditions the fields from all the dipoles will add up at P if all four are fed RF currents in the same phase.

Let us say that a certain current, I, in dipole A will produce a certain value of field strength, E, at the distant point P. The same current in any of the other dipoles will produce the same field at P. Thus, if only dipoles A and B are operating, each with a current I, the field at P will be 2E. With A, B and C operating, the field will be 3E, and with all four operating with the same I, the field will be 4E. Since the power received at P is proportional to the square of the field strength, the relative *power* received at P is 1, 4, 9 and 16, depending on whether one, two, three or four dipoles are operating.

Now, since all four dipoles are alike and there is no coupling between them, the same power must be put into each in order to cause the current I to flow. For two dipoles the relative power input is 2, for three dipoles it is 3, for four dipoles 4, and so on. The actual gain in each case is the relative received (or output) power divided by the relative input power. Thus we have the results shown in Table 1. The power ratio is directly proportional to the number of elements used.

It is well to have clearly in mind the conditions under which this relationship is true:

1) The fields from the separate antenna elements must be in phase at the receiving point.

2) The elements are identical, with equal currents in all elements.

3) The elements must be separated in such a way that the current induced in one by another is negligible; that is, the radiation resistance of each element must be the same as it would be if the other elements were not there.

Very few antenna arrays meet all these conditions exactly. However, it may be said that the power gain of a directive array using dipole elements with optimum values of element spacing is approximately proportional to the number of elements. Another way to say this is that a gain of approximately 3 dB will be obtained each time the number of elements is doubled, assuming the proper element spacing is maintained. It is possible, though, for an estimate based on this rule to be in error by a ratio factor of two or more (gain error of 3 dB or more).

DEFINITIONS

The "element" in a multielement directive array is usually a ½-λ radiator or a ¼-λ vertical element above ground. The length is not always an exact electrical half or quarter wavelength, because in some types of arrays it is desirable that the element show either inductive or capacitive reactance. However, the departure in length from resonance is ordinarily small (not more than 5%, in the usual case) and so has no appreciable effect on the radiating properties of the element.

Antenna elements in multielement arrays of the type

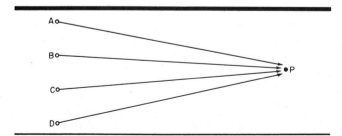

Fig 1—Fields from separate antennas combine at a distant point, P, to produce a field strength that exceeds the field produced by the same power in a single antenna.

Table 1

Comparison of Dipoles with Negligible Coupling (See Fig 1)

Dipoles	Relative Output Power	Relative Input Power	Power Gain	Gain in dB
A only	1	1	1	0
A and B	4	2	2	3
A, B and C	9	3	3	4.8
A, B, C and D	16	4	4	6

considered in this chapter are always either *parallel*, as at A in Fig 2, or *collinear* (end to end), Fig 2B. Fig 2C shows an array combining both parallel and collinear elements. The elements can be either horizontal or vertical, depending on whether horizontal or vertical polarization is desired. Except for space communications, there is seldom any reason for mixing polarization, so arrays are customarily constructed with all elements similarly polarized.

A *driven element* is one supplied power from the transmitter, usually through a transmission line. A *parasitic element* is one that obtains power solely through coupling to another element in the array because of its proximity to such an element.

A *driven array* is one in which all the elements are driven elements. A *parasitic array* is one in which one or more of the elements are parasitic elements. At least one element in a parasitic array must be a driven element, as it is necessary to introduce power into the array.

A *broadside array* is one in which the principal direction of radiation is perpendicular to the axis of the array and to the plane containing the elements, as shown in Fig 3. The elements of a broadside array may be collinear, as in Fig 3A, or parallel (two views in Fig 3B).

An *end-fire array* is one in which the principal direction of radiation coincides with the direction of the array axis. This definition is illustrated in Fig 4. An end-fire array must consist of parallel elements. They cannot be collinear, as ½ λ elements do not radiate straight off their ends.

A *bidirectional array* is one that radiates equally well in either direction along the line of maximum radiation. A bidirectional pattern is shown in Fig 5A. A *unidirectional* array is one that has only one principal direction of radiation, as illustrated by the pattern in Fig 5B.

The *major lobes* of the directive pattern are those in which the radiation is maximum. Lobes of lesser radiation intensity are called *minor lobes*. The *beamwidth* of a directive antenna is the width, in degrees, of the major lobe between the two directions at which the relative radiated power is equal to one half its value at the peak of the lobe. At these "half-power points" the field intensity is equal to 0.707 times its maximum value, or down 3 dB from maximum. Fig 6 shows a lobe having a beamwidth of 30°.

Unless specified otherwise, the term "gain" as used in this section is the power gain over a ½-λ dipole of the same orientation and height as the array under discussion, and having the same power input. Gain may either be measured

experimentally or determined by calculation. Experimental measurement is difficult and often subject to considerable error, for two reasons. First, errors normally occur in measurement because the accuracy of simple RF measuring equipment is relatively poor; even high quality instruments suffer in accuracy compared with their low-frequency and dc counterparts). And second, the accuracy depends considerably on conditions—the antenna site, including height, terrain characteristics, and surroundings—under which the measurements are made. Calculations are frequently based on the measured or theoretical directive patterns of the antenna (see Chapter 2). The theoretical gain of an array may be determined approximately from

$$G = 10 \log \frac{41,253}{\theta_H \theta_V}$$

where

G = decibel gain over a dipole in its favored direction
θ_H = horizontal half-power beamwidth in degrees
θ_V = vertical half-power beamwidth in degrees

This equation, strictly speaking, applies only to lossless antennas having approximately equal and narrow E- and H-plane beamwidths—up to about 20°—and no large minor lobes. (The E and H planes are discussed in Chapter 2.) The

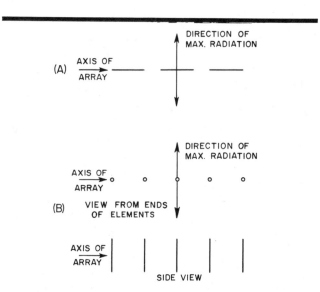

Fig 3—Representative broadside arrays. At A, collinear elements, with parallel elements at B.

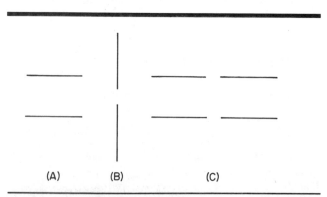

Fig 2—Parallel (A) and collinear (B) antenna elements. The array shown at C combines both parallel and collinear elements.

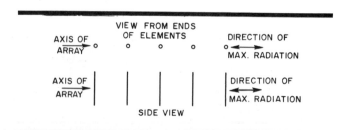

Fig 4—An end-fire array. Practical arrays may combine both broadside directivity (Fig 3) and end-fire directivity, including both parallel and collinear elements.

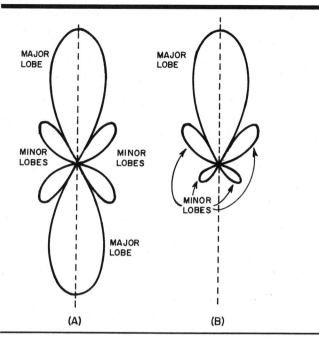

Fig 5—Typical bidirectional pattern (A) and unidirectional directive pattern (B). These drawings also illustrate the application of the terms "major" and "minor" to the pattern lobes.

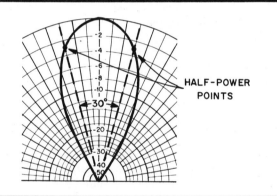

Fig 6—The width of a beam is the angular distance between the directions at which the received or transmitted power is half the maximum power (−3 dB). Each angular division of the pattern grid is 5°.

error may be considerable when the formula is applied to simple directive antennas having relatively large beamwidths. The error is in the direction of making the calculated gain larger than the actual gain.

Front-to-back or *F/B ratio* is the ratio of the power radiated in the favored direction to the power radiated in the opposite direction.

Phase

The term "phase" has the same meaning when used in connection with the currents flowing in antenna elements as it does in ordinary circuit work. For example, two currents are in phase when they reach their maximum values, flowing in the same direction, at the same instant. The direction of current flow depends on the way in which power is applied to the element.

This is illustrated in Fig 7. Assume that by some means an identical voltage is applied to each of the elements at the ends marked A. Assume also that the coupling between the elements is negligible, and that the instantaneous polarity of the voltage is such that the current is flowing away from the point at which the voltage is applied. The arrows show the assumed current directions. Then the currents in elements 1 and 2 are in phase, since they are flowing in the same direction in space and are caused by the same voltage. However, the current in element 3 is flowing in the *opposite* direction in space because the voltage is applied to the opposite end of the element. The current in element 3 is therefore 180° out of phase with the currents in elements 1 and 2.

The phasing of driven elements depends on the direction of the element, the phase of the applied voltage, and the point at which the voltage is applied. In many systems used by amateurs, the voltages applied to the elements are exactly in or exactly out of phase with each other. Also, the axes of the elements are nearly always in the same direction, since parallel or collinear elements are invariably used. The currents in driven elements in such systems are therefore always either exactly in or exactly out of phase with the currents in other elements.

It is possible to use phase differences of less than 180° in driven arrays. One important case is where the voltage applied to one set of elements differs by 90° from the voltage applied to another set. However, making provision for proper phasing in such systems is considerably more complex than in the case of simple 0° or 180° phasing, as described in a later section of this chapter.

In parasitic arrays the phase of the currents in the parasitic elements depends on the spacing and tuning, as described later.

Ground Effects

The effect of the ground is the same with a directive antenna as it is with a simple dipole antenna. The reflection

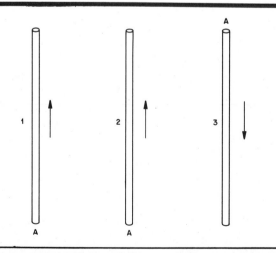

Fig 7—This drawing illustrates the phase of currents in antenna elements, represented by the arrows. The currents in elements 1 and 2 are in phase, while that in element 3 is 180° out of phase with 1 and 2.

Multielement Arrays   8-3

factors discussed in Chapter 3 may therefore be applied to the vertical pattern of an array, subject to the same modifications mentioned in that chapter. In cases where the array elements are not all at the same height, the reflection factor for the mean height of the array may be used for a close approximation. The mean height is the average of the heights measured from the ground to the centers of the lowest and highest elements.

MUTUAL IMPEDANCE

Consider two ½ λ elements that are fairly close to each other. Assume that power is applied to only one element, causing current to flow. This creates an electromagnetic field, which induces a voltage in the second element and causes current to flow in it as well. The current flowing in element no. 2 will in turn induce a voltage in element no. 1, causing additional current to flow there. The total current in no. 1 is then the sum (taking phase into account) of the original current and the induced current.

With element no. 2 present, the amplitude and phase of the resulting current in element no. 1 will be different than if element no. 2 was not there. This indicates that the presence of the second element has changed the impedance of the first. This effect is called mutual coupling. Mutual coupling results in a mutual impedance between the two elements. The mutual impedance has both resistive and reactive components. The actual impedance of an antenna element is the sum of its self-impedance (the impedance with no other antennas present) and its mutual impedances with all other antennas in the vicinity.

The magnitude and nature of the feed point impedance of the first antenna depends on the amplitude of the current induced in it by the second, and on the phase relationship between the original and induced currents. The amplitude and phase of the induced current depend on the spacing between the antennas and whether or not the second antenna is tuned to resonance.

In the discussion of the several preceding paragraphs, it was specified that power is applied to only one of the two elements. Do not interpret this to mean that mutual coupling exists only in parasitic arrays! It is important to remember that mutual coupling exists between *any two conductors that are located near one another*. The mutual coupling between two given elements is the same no matter if either, both, or neither is fed power from a transmission line.

Amplitude of Induced Current

The induced current will be largest when the two antennas are close together and are parallel. Under these conditions the voltage induced in the second antenna by the first, and in the first by the second, has its greatest value and causes the largest current flow. The coupling decreases as the parallel antennas are moved farther apart.

The coupling between collinear antennas is comparatively small, and so the mutual impedance between such antennas is likewise small. It is not negligible, however.

Phase Relationships

When the separation between the two antennas is an appreciable fraction of a wavelength, a measurable period of time elapses before the field from antenna no. 1 reaches antenna no. 2. There is a similar time lapse before the field set up by the current in no. 2 gets back to induce a current

in no. 1. Hence the current induced in no. 1 by no. 2 will have a phase relationship with the original current in no. 1 that depends on the spacing between the two antennas.

The induced current can range all the way from being completely in phase with the original current to being completely out of phase with it. If the currents are in phase, the total current is larger than the original current, and the antenna feed point impedance is reduced. If the currents are out of phase, the total current is smaller and the impedance is increased. At intermediate phase relationships the impedance will be lowered or raised depending on whether the induced current is mostly in or mostly out of phase with the original current.

Except in the special cases when the induced current is exactly in or out of phase with the original current, the induced current causes the phase of the total current to shift with respect to the applied voltage. Consequently, the presence of a second antenna nearby may cause the impedance of an antenna to be reactive—that is, the antenna will be detuned from resonance—even though its self-impedance is entirely resistive. The amount of detuning depends on the magnitude and phase of the induced current.

Tuning Conditions

A third factor that affects the impedance of antenna no. 1 when no. 2 is present is the tuning of no. 2. If no. 2 is not exactly resonant the current that flows in it as a result of the induced voltage will either lead or lag the phase it would have if the antenna were resonant. This causes an additional phase advance or delay that affects the phase of the current induced back in no. 1. Such a phase lag has an effect similar to a change in the spacing between self-resonant antennas. However, a change in tuning is not exactly equivalent to a change in spacing because the two methods do not have the same effect on the *amplitude* of the induced current.

MUTUAL IMPEDANCE AND GAIN

The mutual coupling between antennas is important because it can have a significant effect on the amount of current that will flow for a given amount of power supplied. And it is the amount of *current* flowing that determines the field strength from the antenna. Other things being equal, if the mutual coupling between two antennas is such that the currents are greater for the same total power than would be the case if the two antennas were not coupled, the power gain will be greater than that shown in Table 1. On the other hand, if the mutual coupling is such as to reduce the current, the gain will be less than if the antennas were not coupled. The term "mutual coupling," as used in this paragraph, assumes that the mutual impedance between elements is taken into account, along with the added effects of propagation delay because of element spacing, and element tuning or phasing.

The calculation of mutual impedance between antennas is a complex problem. Data for two simple but important cases are graphed in Figs 8 and 9. These graphs do not show the mutual impedance, but instead show a more useful quantity—the feed point resistance measured at the center of an antenna as it is affected by the spacing between two antennas.

As shown by the solid curve in Fig 8, the feed point resistance at the center of either antenna, when the two are self-resonant, parallel, and operated in phase, decreases rapidly as the spacing between them is increased until the

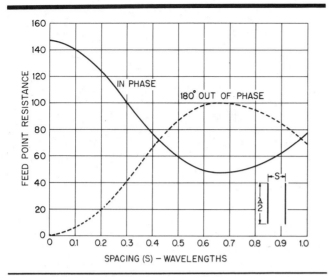

Fig 8—Feed point resistance measured at the center of one element as a function of the spacing between two parallel ½-λ self-resonant antenna elements. For ground mounted ¼-λ vertical elements, divide these resistances by two.

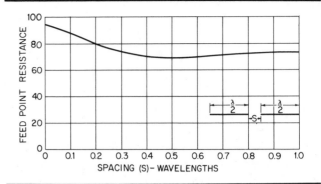

Fig 9—Feed point resistance measured at the center of one element as a function of the spacing between the ends of two collinear self-resonant ½-λ antenna elements operated in phase.

spacing is about 0.7 λ. This is a broadside array. The maximum gain is achieved from a pair of such elements when the spacing is in this region, because the current is larger for the same power and the fields from the two arrive in phase at a distant point placed on a line perpendicular to the line joining the two antennas.

The broken curve in Fig 8, representing two antennas operated 180° out of phase (end-fire), cannot be interpreted quite so simply. The feed point resistance decreases with decreasing spacing in this case. However, for the range of spacings considered, only when the spacing is ½ λ do the fields from the two antennas add up exactly in phase at a distant point in the favored direction. At smaller spacings the fields become increasingly out of phase, so the total field is less than the simple sum of the two. Smaller spacings thus decrease the gain at the same time that the reduction in feed point resistance is increasing it. The gain goes through a maximum when the

spacing is in the region of 1/8 λ.

The curve for two collinear elements in phase, Fig 9, shows that the feed point resistance decreases and goes through a broad minimum in the region of 0.4- to 0.6-λ spacing between the adjacent ends of the antennas. As the minimum is not significantly less than the feed point resistance of an isolated antenna, the gain will not exceed the gain calculated on the basis of uncoupled antennas. That is, the best that two collinear elements will give, even with the optimum spacing, is a power gain of about 2 (3 dB). When the separation between the ends is very small—the usual method of operation—the gain is reduced.

PARASITIC ARRAYS

The foregoing information in this chapter applies to multielement arrays of both types, driven and parasitic. However, there are special considerations for driven arrays that do not necessarily apply to parasitic arrays, and vice versa. Such considerations for parasitic arrays are presented in Chapter 11. The remainder of this chapter is devoted to driven arrays.

Driven Arrays

Driven arrays in general are either broadside or end-fire, and may consist of collinear elements, parallel elements, or a combination of both. From a practical standpoint, the maximum number of usable elements depends on the frequency and the space available for the antenna. Fairly elaborate arrays, using as many as 16 or even 32 elements, can be installed in a rather small space when the operating frequency is in the VHF range, and more at UHF. At lower frequencies the construction of antennas with a large number of elements is impractical for most amateurs.

Of course the simplest of driven arrays is one with just

two elements. If the elements are collinear, they are always fed in phase. The effects of mutual coupling are not great, as illustrated in Fig 9. Therefore, feeding power to each element in the presence of the other presents no significant problems. This may not be the case when the elements are parallel to each other. However, because the combination of spacing and phasing arrangements for parallel elements is infinite, the number of possible radiation patterns is endless. This is illustrated in Fig 10. When the elements are fed in phase, a broadside pattern always results. At spacings of less than 5/8 λ with the elements fed 180° out of phase, an end-

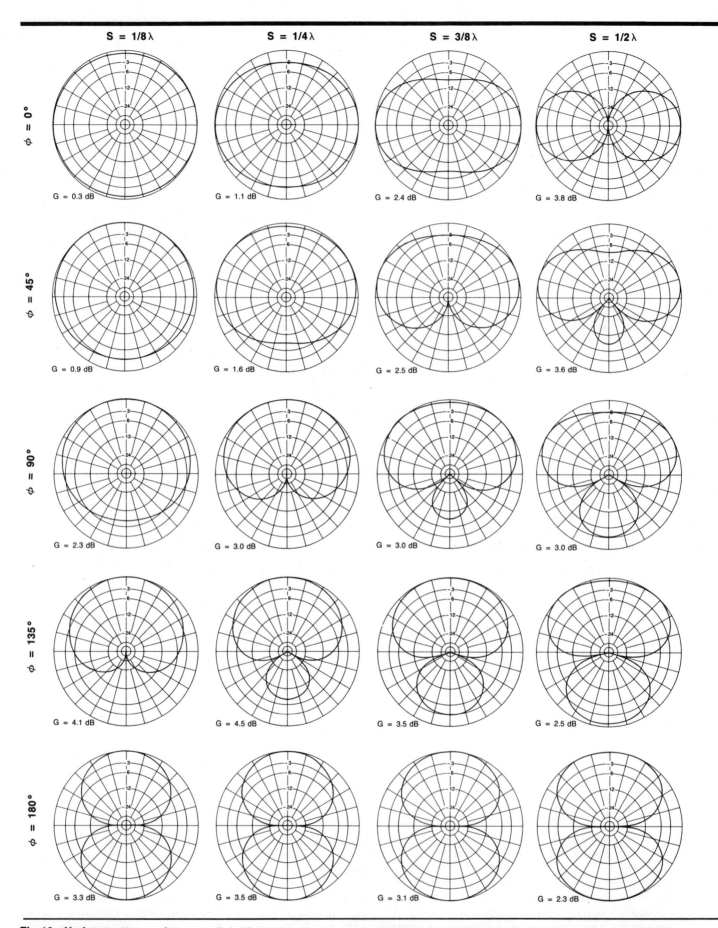

Fig 10—H-plane patterns of two parallel driven elements, spaced and phased as indicated (S = spacing, φ = phasing). In these plots the elements are aligned with the vertical axis, and the uppermost element is the one of lagging phase at angles other than 0°. The two elements are assumed to be the same length, with exactly equal currents. The gain figure

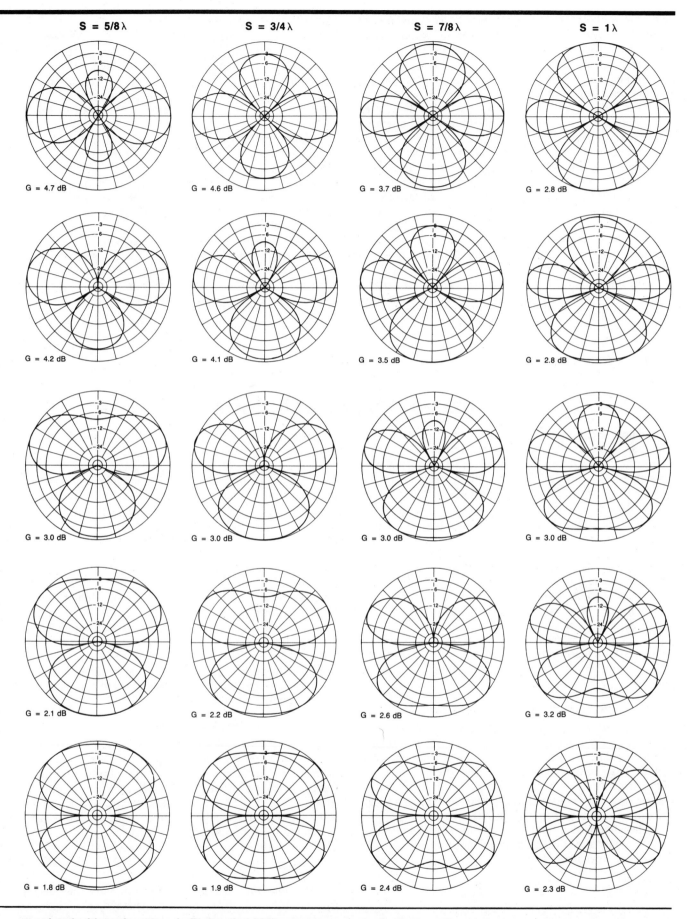

S = 5/8 λ   S = 3/4 λ   S = 7/8 λ   S = 1 λ

G = 4.7 dB   G = 4.6 dB   G = 3.7 dB   G = 2.8 dB

G = 4.2 dB   G = 4.1 dB   G = 3.5 dB   G = 2.8 dB

G = 3.0 dB   G = 3.0 dB   G = 3.0 dB   G = 3.0 dB

G = 2.1 dB   G = 2.2 dB   G = 2.6 dB   G = 3.2 dB

G = 1.8 dB   G = 1.9 dB   G = 2.4 dB   G = 2.3 dB

associated with each pattern indicates that of the array over a single element. The plots may be interpreted as the horizontal or azimuth pattern of two vertical elements at a 0° elevation angle, or the free-space pattern of two horizontal elements when viewed on end, with one element above the other.

Multielement Arrays   8-7

fire pattern always results. With intermediate amounts of phase difference, the results cannot be so simply stated. Patterns evolve which are not symmetrical in all four quadrants.

In the plots of Fig 10, the two elements are assumed to have the same length. In addition, equal currents are assumed to be flowing in each element, a condition that most often will not exist in practice without devoting special attention to the feeder system.

Because of the effects of mutual coupling between the two driven elements, greater or lesser currents will flow in each element with changes in spacing and phasing, as described in the previous section. This, in turn, affects the gain of the array in a way that cannot be shown merely by plotting the *shapes* of the patterns, as has been done in Fig 10. Therefore, supplemental gain information is also shown in Fig 10, adjacent to the pattern plot for each combination of spacing and phasing, referenced to a single element. For example, a pair of elements fed 90° apart at a spacing of ¼ λ will have a gain in the direction of maximum

radiation of 3.0 dB over a single element.

It is characteristic of broadside arrays that the power gain is proportional to the length of the array but is substantially independent of the number of elements used, provided the optimum element spacing is not exceeded. This means, for example, that a 5-element array and a 6-element array will have the same gain, provided the elements in both are spaced so that the overall array length is the same. Although this principle is seldom used for the purpose of reducing the number of elements, because of complications introduced in feeding power to each element in the proper phase, it does illustrate the fact that there is nothing to be gained by increasing the number of elements if the space occupied by the antenna is not increased proportionally.

Generally speaking, the maximum gain in the smallest linear dimensions will result when the antenna combines both broadside and end-fire directivity and uses both parallel and collinear elements. In this way the antenna is spread over a greater volume of space, which has the same effect as extending its length to a much greater extent in one linear direction.

Phased Array Techniques

Phased antenna arrays have become increasingly popular for amateur use, particularly on the lower frequency bands where they provide one of the few practical methods of obtaining substantial gain and directivity. This section on phased array techniques was written by Roy W. Lewallen, W7EL. The operation and limitations of phased arrays, how to design feed systems to make them work properly, and how to make necessary tests and adjustments are discussed in the pages that follow. The examples deal primarily with vertical HF arrays, but the principles apply to horizontal and VHF/UHF arrays as well.

The performance of a phased array is determined by several factors. Most significant among these are the characteristics of a single element, reinforcement or cancellation of the fields from the elements, and the effects of mutual coupling. To understand the operation of phased arrays, it is first necessary to understand the operation of a single antenna element.

Fundamentals of Phased Arrays

Of primary importance is the strength of the field produced by the element. Information in Chapter 2 explains that the field radiated from a linear (straight) element, such as a dipole or vertical monopole, is proportional to the sum of the elementary currents flowing in each part of the antenna element. For this discussion it is important to understand what determines the current in a single element.

The amount of current flowing at the base of a resonant ground-mounted vertical or groundplane antenna is given by the familiar formula

$$I = \sqrt{\frac{P}{R}} \qquad \text{(Eq 1)}$$

where

P is the power supplied to the antenna
R is the feed point resistance

R consists of two parts, the loss resistance and the radiation resistance. The loss resistance, R_L, includes losses in the conductor, in the matching and loading components, and dominantly (in the case of ground-mounted verticals), in ground losses. The power "dissipated" in the radiation resistance, R_R, is the power which is radiated, so maximizing the power "dissipated" by the radiation resistance is desirable. However, the power dissipated in the loss resistance truly is lost (as heat), so resistive losses should be made as small as possible.

The radiation resistance of an element may be derived from electromagnetic field theory, being a function of antenna length, diameter, and geometry. Graphs of radiation resistance versus antenna length are given in Chapter 2. The radiation resistance of a thin ¼-λ ground-mounted vertical is about 36 ohms. A ½-λ dipole in free space has a radiation resistance of about 73 ohms. Reducing the antenna lengths by one half drops the radiation resistances to approximately 7 and 14 ohms, respectively.

Radiation Efficiency

To generate a stronger field from a given radiator, it is necessary to increase the power P (the "brute force" solution), to decrease the loss resistance R_L (by putting in a more elaborate ground system for a vertical, for instance), or to somehow decrease the radiation resistance R_R so more current will flow with a given power input. This can be seen by expanding the formula for base current as

$$I = \sqrt{\frac{P}{R_R + R_L}} \qquad \text{(Eq 2)}$$

Dividing the feed point resistance into components R_R and R_L easily leads to an understanding of element efficiency. The efficiency of an element is the proportion of the total power that is actually radiated. The roles of R_R and R_L in determining efficiency can be seen by analyzing a simple equivalent circuit, shown in Fig 11.

The power "dissipated" in R_R (the radiated power) equals $I^2 R_R$. The total power supplied to the antenna system is

$$P = I^2(R_R + R_L) \qquad \text{(Eq 3)}$$

so the efficiency (the fraction of supplied power which is actually radiated) is

$$\text{Eff} = \frac{I^2 R_R}{I^2(R_R + R_L)} = \frac{R_R}{R_R + R_L} \qquad \text{(Eq 4)}$$

Efficiency is frequently expressed in percent, but expressing it in decibels relative to a 100% efficient radiator gives a better idea of what to expect in the way of signal strength. The field strength of an element relative to a lossless but otherwise identical element, in dB, is

$$\text{FSG} = 10 \log \frac{R_R}{R_R + R_L} \qquad \text{(Eq 5)}$$

where FSG = field strength gain, dB

For example, information presented by Sevick in March 1973 *QST* shows that a ¼-λ ground-mounted vertical antenna with four 0.2-λ radials has a feed point resistance of about 65 ohms (see the bibliography at the end of this chapter). The efficiency of such a system is 36/65 = 55.4%. It is rather disheartening to think that, of 100 watts fed to the antenna, only 55 watts are being radiated, with the remainder literally warming up the ground. Yet the signal will be only 10 log (36/65) = −2.57 dB relative to the same vertical with a *perfect* ground system. In view of this information, trading a small reduction in signal strength for lower cost and greater simplicity may become an attractive consideration.

So far, only the current at the base of a resonant antenna has been discussed, but the field is proportional to the sum of currents in each tiny part of the antenna. The field is a function of not only the magnitude of current flowing at the base, but also the distribution of current along the radiator and the length of the radiator. However, nothing can be done at the base of the antenna to change the current distribution, so *for a given element*, the field strength is proportional to the base current (or center current, in the case of a dipole). However, changing the radiator length or loading it at some point other than the feed point *will* change the current distribution. More information on shortened or loaded radiators may be found in Chapters 2 and 6, and in the bibliography references of this chapter. A few other important facts follow.

1) If there is *no loss*, the field from even an infinitesimally short radiator is less than ½ dB weaker than the field from a half-wave dipole or quarter-wave vertical. Without loss, all the supplied power is radiated regardless of the antenna length, so the only factor influencing gain is the slight difference in the patterns of very short and ½-λ antennas. The small pattern difference arises from different current distributions, as discussed in Chapter 2. A short antenna has a very low radiation resistance, resulting in a heavy current flow over its short length. In the absence of loss, this generates a field strength comparable to that of a longer antenna. Where loss *is* present—that is, in practical antennas—shorter radiators usually don't do so well, since the low radiation resistance leads to lower efficiency for a given loss resistance. If care is taken, short antennas can achieve good efficiency.

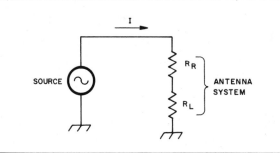

Fig 11—Simplified equivalent circuit for a single-element resonant antenna. R_R represents the radiation resistance, and R_L the ohmic losses in the total antenna system.

2) The feed point resistance of folded antennas isn't the radiation resistance as the term is used here. The act of folding an antenna only transforms the input impedance to a higher value, providing an easier match in some cases. The higher feed point impedance doesn't help the efficiency, since the resulting smaller currents flow through more conductors, for the same net loss. In a folded vertical, the same total current ends up flowing through the ground system, again resulting in the same loss.

3) The current flowing in an element with a given power input can be increased, or decreased, by mutual coupling to other elements. The effect is equivalent to changing the element radiation resistance. Mutual coupling is sometimes regarded as a "minor" effect, but most often it is not minor!

Field Reinforcement and Cancellation

Consider two elements which each produce a field strength of, say, exactly 1 millivolt per meter (mV/m) at some distance many wavelengths from the array. In the direction in which the fields are in phase, a total field of 2 mV/m results; in the direction in which they are out of phase, a zero field results. The ratio of maximum to minimum field strength of this array is 2/0, or infinite.

Now suppose, instead, that one field is 10% high and the other 10% low—1.1 and 0.9 mV/m, respectively. In the forward direction, the field strength is still 2 mV/m, but in the canceling direction, the field will be 0.2 mV/m. The front-to-back ratio has dropped from infinite to 2/0.2, or 20 dB. (Actually, slightly more power is required to redistribute the field strengths this way, so the forward gain is reduced—but only by a small amount, less than 0.1 dB.) For most arrays, unequal fields from the elements have a minor effect on forward gain, *but a major effect on pattern nulls*.

Even with perfect current balance, deep nulls aren't assured. Fig 12 shows the minimum spacing required for total field reinforcement or cancellation. If the element spacing isn't adequate, there may not be any direction in which the fields are completely out of phase (see curve B of Fig 12). Slight physical and environmental differences between elements will invariably affect null depths, and null depths will vary with elevation angle. However, a properly designed and fed array can, in practice, produce very impressive nulls. The key to achieving good performance is being able to control the fields from the elements. This, in turn, requires knowing how to control the currents in the elements, since the fields are proportional to the currents. Most phased arrays require the

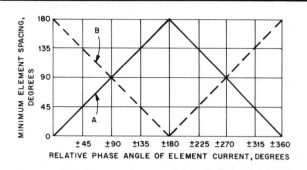

Fig 12—Minimum element spacing required for total field reinforcement, curve A, or total field cancellation, curve B. Total cancellation results in pattern nulls in one or more directions. Total reinforcement does not necessarily mean there is gain over a single element, as the effects of loss and mutual coupling must also be considered.

element currents to be equal in magnitude and different in phase by some specific amount. Just how this can be accomplished is explained in a subsequent section.

MUTUAL COUPLING

Mutual coupling refers to the effects which the elements in an array have on each other. Mutual coupling can occur intentionally or entirely unintentionally. For example, Lewallen has observed effects such as a quad coupling to an inverted-V dipole to form a single, very strange, antenna system. The current in the "parasitic element" (nondriven antenna) was caused entirely by mutual coupling, just as in the familiar Yagi antenna. The effects of mutual coupling are present regardless of whether or not the elements are driven.

Suppose that two driven elements are very far from each other. Each has some voltage and current at its feed point. For each element, the ratio of this voltage to current is the element self-impedance. If the elements are brought close to each other, the current in each element will change in amplitude and phase because of coupling with the field from the other element. Significant mutual coupling occurs at spacings as great as a wavelength or more. The fields change the currents, which change the fields. There is an equilibrium condition in which the currents in all elements (hence, their fields) are totally interdependent. The feed point impedances of all elements also are changed from their values when far apart, and all are dependent on each other. In a driven array, the changes in feed point impedances can cause additional changes in element currents, because the operation of many feed systems depends on the element feed point impedances.

Connecting the elements to a feed system to form a driven array does not eliminate the effects of mutual coupling. In fact, in many driven arrays the mutual coupling has a greater effect on antenna operation than the feed system does. All feed system designs must account for the impedance changes caused by mutual coupling if the desired current balance and phasing are to be achieved.

Several general statements can be made regarding phased array systems. Mutual coupling accounts for these characteristics.

1) The resistances and reactances of all elements of an array generally will change substantially from the values of an isolated element.

2) If the elements of a two-element array are identical, and have equal currents which are in phase or 180° out of phase, the feed point impedances of the two elements will be equal. But they will be different than for an isolated element. If the two elements are part of a larger array, their impedances can be very different from each other.

3) If the elements of a two-element array have currents which are neither in phase (0°) nor out of phase (180°), their feed point impedances will not be equal. The difference will be substantial in typical amateur arrays.

4) The feed point resistances of the elements in a closely spaced, 180° out-of-phase array will be very low, resulting in poor efficiency unless care is taken to minimize loss. This is also true for any other closely spaced array with significant predicted gain.

Gain

Gain is strictly a relative measure, so the term is completely meaningless unless accompanied by a statement of just what it is relative to. One useful measure for phased array gain is *gain relative to a single similar element*. This is the increase in signal strength which would be obtained by replacing a single element by an array made from elements just like it. All gain figures in this section are relative to a single similar element unless otherwise noted. In some instances, such as investigating what happens to array performance when *all* elements become more lossy, gain refers to a more absolute, although unattainable standard: a lossless element. Uses of this standard are explicitly noted.

Why does a phased array have gain? One way to view it is in terms of directivity. Since a given amount of radiated power, whether radiated from one or a dozen elements, must be radiated *somewhere*, field strength must be increased in some directions if it is reduced in others. There is no guarantee that the fields from the elements of an arbitrary array will completely reinforce or cancel in *any* direction; element spacing must be adequate for either to happen (see Fig 12). If the fields reinforce or cancel to only a slight extent, causing a pattern similar to that of a single element, the gain will also be similar to that of a single element.

To get a feel for how much gain a phased array can deliver, consider what would happen if there were no change in element feed point resistance from mutual coupling. This actually does occur at some spacings and phasings, but not in commonly used systems. It is a useful example, nevertheless.

In the fictitious array the elements are identical and there are no resistance changes from mutual coupling. The feed point resistance, R_F, equals $R_R + R_L$, the sum of radiation and loss resistances. If power P is put into a single element, the feed point current is

$$I_F = \sqrt{\frac{P}{R_F}} \tag{Eq 6}$$

At a given distance, the field strength is proportional to the current, so the field strength is

$$E = kI_F = k\sqrt{\frac{P}{R_F}} \tag{Eq 7}$$

where k is the constant relating the element current to the field strength at the chosen distance.

If, instead, the power is equally split between two elements,

$$I_{F1} = I_{F2} = \sqrt{\frac{P/2}{R_F}} \qquad \text{(Eq 8)}$$

From this,

$$E_1 = E_2 = k\sqrt{\frac{P/2}{R_F}} \qquad \text{(Eq 9)}$$

If the elements are spaced far enough apart to allow full field reinforcement, the total field in the favored direction will be

$$E_1 + E_2 = 2k\sqrt{\frac{P/2}{R_F}} = \sqrt{2}\ k\sqrt{\frac{P}{R_F}} \qquad \text{(Eq 10)}$$

This represents a field strength gain of

$$FSG = 20 \log \sqrt{2} = 3\ dB \qquad \text{(Eq 11)}$$

where FSG = field strength gain, dB

The power gain in dB equals the field strength gain in dB.

The above argument leading to Eq 11 can be extended to show that the gain in dB for an array of n elements, without resistance changes from mutual coupling and with sufficient spacing and geometry for total field reinforcement, is

$$FSG = 20 \log \sqrt{n} = 10 \log n \qquad \text{(Eq 12)}$$

That is, a 5-element array satisfying these assumptions would have a power gain of 5 times, or about 7 dB. Remember, the assumption was made that equal power is fed to each element. With equal element resistances and no resistance changes from mutual coupling, *equal currents* are therefore made to flow in all elements!

The gain of an array can be increased or decreased from 10 log n decibels by mutual coupling, but any loss will move the gain back toward 10 log n. This is because resistance changes from mutual coupling get increasingly swamped by the loss as the loss increases. Arrays designed to have substantially more gain than 10 log n decibels require heavy element currents. As designed gain increases, the required currents increase dramatically, resulting in power losses which partially or totally negate the expected gain. The net result is a practical limit of about 20 log n for the gain in dB of an n-element array, and this gain can be achieved only if extreme attention is paid to keeping losses very small. The majority of practical arrays, particularly arrays of ground-mounted verticals, have gains closer to 10 log n decibels.

The foregoing comments indicate that many of the claims about the gain of various arrays are exaggerated, if not ridiculous. But an honest 3 dB or so of gain from a two-element array can really be appreciated if an equally honest 3 dB has been attempted by other means. Equations for calculating array gain and examples of their use are given in a later section of this chapter.

FEEDING PHASED ARRAYS

The previous section explained why the currents in the elements must be very close to the ratios required by the array design. This section explains how to feed phased arrays to produce the desired current ratio and phasing. Since the desired current ratio is 1:1 for virtually all two-element and for most larger amateur arrays, special attention is paid to

methods of assuring equal element currents. Other current ratios are also examined.

Phasing Errors

For an array to produce the desired pattern, the element currents must have the required magnitude and the required phase relationship. On the surface, this sounds easy; just make sure that the difference in electrical lengths of the feed lines to the elements equals the desired phase angle. Unfortunately, this approach doesn't necessarily achieve the desired result. The first problem is that the phase shift through the line is not equal to its electrical length. The current (or, for that matter, voltage) delay in a transmission line is equal to its electrical length in only a few special cases—cases which do not exist in most amateur arrays! The impedance of an element in an array is frequently very different from the impedance of an isolated element, and the impedances of all the elements in an array can be different from each other. Consequently, the elements seldom provide a matched load for the element feed lines. The effect of mismatch on phase shift can be seen in Fig 13. Observe what happens to the phase of the current and voltage on a line terminated by a purely resistive impedance which is lower than the characteristic impedance of the line (Fig 13A). At a point 45° from the load, the current has advanced less than 45°, and the voltage more than 45°. At 90° from the load, both are advanced 90°. At 135°, the current has advanced more and the voltage less than 135°. This apparent slowing down and speeding up of the current and voltage waves is caused by interference between the forward and reflected waves. It occurs on any line not terminated with a pure resistance equal to its characteristic impedance. If the load resistance is greater than the characteristic impedance of the line, as shown in Fig 13B, the voltage and current exchange angles. Adding reactance to the load

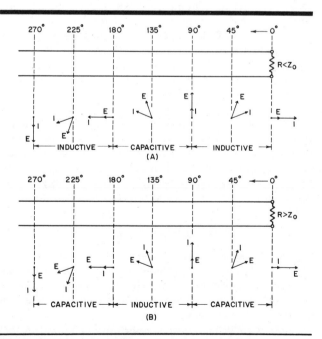

Fig 13—Resultant voltages and currents along a mismatched line. At A, R less than Z_0, and at B, R greater than Z_0.

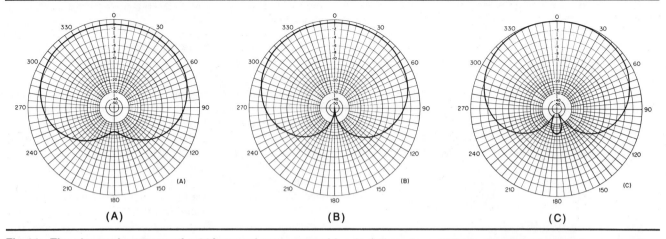

(A)   (B)   (C)

Fig 14—The change in pattern of a 90° spaced array caused by deviations from 90° phasing (equal currents assumed). At A, B and C the respective phase angles are 80°, 90° and 100°. Note the minor changes in gain as well as in pattern shapes with phase angle deviations. Gain is referenced to a single element; add 3.4 dB to the scale values shown for each plot.

causes additional phase shift. The *only* cases in which the current (or voltage) delay is equal to the electrical length of the line are

1) when the line is "flat," that is, terminated in a purely resistive load equal to its characteristic impedance;

2) when the line length is an integral number of half wavelengths;

3) when the line length is an odd number of quarter wavelengths *and* the load is purely resistive; and

4) when other specific lengths are used for specific load impedances.

Just how much phase error can be expected if two lines are simply hooked up to form an array? There is no simple answer. Some casually designed feed systems might deliver satisfactory results, but most will not. Later examples show just what the consequences of casual feeding can be.

The effect of phasing errors is to alter the basic shape of the radiation pattern. Nulls may be reduced in depth, and additional lobes added. Actual patterns can be calculated by using Eq 15 in a later section of this chapter. The effects of phasing errors on the shape of a 90° fed, 90° spaced array pattern are shown in Fig 14.

A second problem with simply connecting feed lines of different lengths to the elements is that the lines will change the *magnitudes* of the currents. The magnitude of the current (or voltage) out of a line does not equal the magnitude in, except in cases 1, 2 and 4 above. The feed systems presented here assure currents which are correct in both magnitude and phase.

The Wilkinson Divider

The Wilkinson divider, sometimes called the Wilkinson power divider, has been promoted in recent years as a means to distribute power among the elements of a phased array. It is therefore worthwhile to investigate just what the Wilkinson divider does.

The Wilkinson divider is shown in Fig 15. It is a very useful device for *splitting power* among several loads, or, in reverse, combining the outputs from several generators. If all loads are equal to the design value (usually 50 ohms), the power from the source is split equally among them, and no

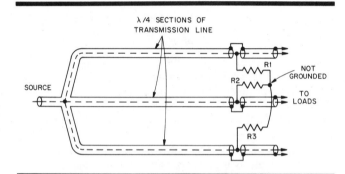

Fig 15—The Wilkinson divider. Three output ports are shown here, but the number may be reduced to two or increased as necessary. If (and only if) the source and all load impedances equal the design impedance, the power from the source will be split equally among the loads. The Z_0 of the ¼-λ sections is equal to the load impedance times the square root of the number of loads.

R1, R2, R3—Noninductive resistors having a value equal to the impedance of the loads.

power is dissipated in the resistors. If the impedance of one of the loads should change, however, the power which was being delivered to that load becomes shared between it and the resistors. The power to the other loads is unchanged, so they are not affected by the errant load. The network is also commonly used to combine the outputs of several transmitters to obtain a higher power than a single transmitter can deliver. The great value of the network becomes evident by observing what happens if one transmitter fails. The other transmitters continue working normally, delivering their full power to the load. The Wilkinson network prevents them, or the load, from "seeing" the failed transmitter, except as a reduction of total output power. Most other combining techniques would result in incorrect operation or failure of the remaining transmitters.

The Wilkinson divider is a port-to-port isolation device. It does *not* assure equal powers or currents in all loads. When connected to a phased array, it might make the system more broadband—by an amount directly related to the amount of power being lost in the resistors! Amateurs feeding a "four-square" array (reference Atchley, Steinhelfer and White—see bibliography) with this network have reported one or more resistors getting very warm, indicating lost power that would be used to advantage if radiated.

Incidentally, if the divider is to be used for its intended purpose, the *source* impedance must be correct for proper operation. Hayward and DeMaw have pointed out that amateur transmitters do not necessarily have a well defined output impedance (see bibliography).

In summary, if the Wilkinson divider is used for feeding a phased array, (1) it will *not* assure equal element powers (which are not wanted anyway). (2) It will *not* assure equal element currents (which *are* wanted). (3) It will waste power. The Wilkinson divider is an extremely useful device. But it is not what is needed for feeding phased antenna arrays.

The Broadcast Approach

Networks can be designed to transform the element base impedances to, say, 50 ohms resistive. Then another network can be inserted at the junction of the feed lines to properly divide the power among the elements (not necessarily equally!). And finally, additional networks must be built to correct for the phase shifts of the other networks! This general approach is used by the broadcast industry. Although this technique can be used to feed any type of array, design is difficult and adjustment is tedious, as all adjustments interact. When the relative currents and phasings are adjusted, the feed point impedances change, which in turn affect the element currents and phasings, and so on. A further disadvantage of using this method is that switching the array direction is generally impossible. Information on applying this technique to amateur arrays may be found in Paul Lee's book.

A PREFERRED FEED METHOD

The feed method introduced here has been used in its simplest form to feed television receiving antennas and other arrays, as presented by Jasik, pages 2-12 and 24-10. However, this feed method has not been widely applied to amateur arrays.

The method takes advantage of an interesting property of ¼-λ transmission lines. (All references to lengths of lines are electrical length, and lines are assumed to have negligible loss.) See Fig 16. The magnitude of the *current* out of a ¼-λ transmission line is equal to the *input* voltage divided by the characteristic impedance of the line, independent of the load

impedance. In addition, the phase of the output current lags the phase of the input voltage by 90°, also independent of the load impedance. This property can be used to advantage in feeding arrays with certain phasings between elements.

If any number of loads are connected to a common driving point through ¼-λ lines of equal impedance, the currents in the loads will be *forced* to be equal and in phase, regardless of the load impedances. So any number of in-phase elements can be correctly fed using this method. Arrays which require unequal currents can be fed through lines of unequal impedance to achieve other current ratios.

The properties of ½-λ lines also are useful. Since the current out of a ½-λ line equals the input current shifted 180°, regardless of the load impedance, any number of half wavelengths of line may be added to the basic ¼ λ, and the current and phase "forcing" property will be preserved. For example, if one element is fed through a ¼-λ line, and another element is fed from the same point through a ¾-λ line of the same characteristic impedance, the currents in the two elements will be forced to be equal in magnitude and 180° out of phase, regardless of the feed point impedances of the elements.

If an array of two identical elements is fed in phase or 180° out of phase, both elements have the same feed point impedance. With these arrays, feeding the elements through equal lengths of feed line (in phase) or lengths differing by 180° (out of phase) will lead to the correct current and phase match, regardless of what the line length is. Unless the lines are an integral number of half wavelengths long, the currents out of the lines will not be equal to the input currents, and the phase will not be shifted an amount equal to the electrical lengths of the lines. But both lines will produce the same transformation and phase shift because their load impedances are equal, resulting in a properly fed array. In practice, however, feed point impedances of elements frequently are different even in these arrays, because of such things as different ground systems (for vertical elements), proximity to buildings or other antennas, or different heights above ground (for horizontal elements). In many larger arrays, two or more elements must be fed either in phase or out of phase with equal currents, but coupling to other elements may cause their impedances to change unequally—sometimes extremely so. Using the current forcing method allows the feed system designer to ignore all these effects while guaranteeing equal and correctly phased currents in any combination of 0° and 180° fed elements.

Feeding Elements in Quadrature

Many popular arrays have elements or groups of elements which are fed in quadrature (90° relative phasing). A combination of the forcing method and a simple adjustable network can produce the correct current balance and element phasing.

Suppose that ¼-λ lines of the same impedance are connected to two elements. The magnitudes of the element currents equal the voltages at the feed line inputs, divided by the characteristic impedance of the lines. The currents are both shifted 90° relative to the input voltages. If the two input voltages can be made equal in magnitude but 90° different in phase, the element currents will also be equal and phased at 90°. Many networks will accomplish the desired function, the simplest being the L network. Either a high-pass or low-pass network can be used. A high-pass network will give a phase lead, and a low-pass network causes a phase lag. The

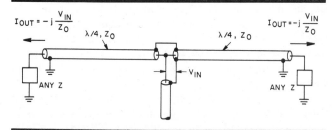

Fig 16—A useful property of ¼-λ transmission lines; see text. This property is utilized in the "current forcing" method of feeding an array of coupled elements.

low-pass network offers dc continuity, which can be beneficial by eliminating static buildup. Only low-pass networks are described here. The harmonic reduction properties of low-pass networks should not be a consideration in choosing the network type; antenna system matching components should not be depended upon to achieve an acceptable level of harmonic radiation. The quadrature feed system is shown in Fig 17.

For element currents of equal magnitude and 90° relative phase, equations for designing the network are

$$X_{ser} = \frac{Z_0^2}{R_2} \qquad \text{(Eq 13)}$$

$$X_{sh} = \frac{Z_0^2}{X_2 - R_2} \qquad \text{(Eq 14)}$$

where

X_{ser} = the reactance of the series component
X_{sh} = the reactance of the shunt component
Z_0 = the characteristic impedance of the ¼-λ lines
R_2 = the feed point resistance of element 2
X_2 = the feed point reactance of element 2

R_2 and X_2 may be calculated from Eqs 21 and 22, presented later. If X_{ser} or X_{sh} is positive, that component is an inductor; if negative, a capacitor. In most practical arrays, X_{ser} is an inductor, and X_{sh} is a capacitor.

Unlike the current forcing methods, the output-to-input voltage transformation and the phase shift of an L network *do* depend on the feed point impedances of the array elements. So the impedances of the elements, when coupled to each other and while being excited to have the proper currents, must be known in order to design a proper L network. Methods for determining the impedance of one element in the presence of others are presented in later sections. Suffice it to say here that the self-impedances of the elements and their mutual impedance must be known in order to calculate the element feed point impedances. In practice, if simple dipoles or verticals are used, a rough estimation of self- and mutual impedances is generally enough to provide a starting point for determining the component values. Then the components may be adjusted for the desired array performance.

The Magic Bullet

Two elements could be fed in quadrature without the necessity to determine self- and mutual impedances if a quadrature forcing network could be found. This passive network would have any one of the following characteristics, but the condition must be *independent of the network load impedance:*

1) The output voltage is equal in amplitude and 90° delayed or advanced in phase relative to the input voltage.

2) The output current is equal in amplitude and 90° delayed or advanced in phase relative to the input current.

3) The output voltage is in phase or 180° out of phase with the input current, and the magnitude of the output voltage is related to the magnitude of the input current by a constant.

4) The output current is in phase or 180° out of phase with the input voltage, and the magnitude of the output current is related to the magnitude of the input voltage by a constant.

Such a network would be the "magic bullet" to extend the forcing method to quadrature feed systems. Lewallen has looked long and hard for this magic bullet without success. Among the many unsuccessful candidates is the 90° hybrid coupler. Like the Wilkinson divider, the hybrid coupler is a useful port-to-port isolation device which does not accomplish the needed function for this application. The feeding of amateur arrays could be greatly simplified by use of a suitable network. Any reader who is aware of such a network is encouraged to publish it in amateur literature, or to contact Lewallen or the editors of this book.

PATTERN AND GAIN CALCULATION

The following equations are derived from those published by Brown in 1937. Findings from Brown's and later works are presented in concise form by Jasik. Equivalent equations may be found in other texts, such as *Antennas* by Kraus. (See the bibliography at the end of this chapter.) The equations in this part will enable the mathematically inclined amateur armed with a calculator or computer to determine patterns, actual gains, and front-to-back or front-to-side ratios of two-element arrays. Although only two-element arrays are presented in detail in this part, the principles hold for larger arrays.

The importance of equal element currents (assuming identical elements) in obtaining the best possible nulls was explained earlier. Maximum forward gain is obtained usually, if not always, for two-element arrays when the currents are equal. Therefore, most of the equations in this part have been simplified to assume that equal element currents are produced. Just how this can be accomplished for many common array types has already been described briefly, and is covered in more detail later in this chapter. Equations which include the effects of unequal currents are also presented later in this chapter.

The equations given below are valid for horizontal or vertical arrays. However, ground reflection effects must be taken into account when dealing with horizontal arrays, doubling the number of "elements" which must be dealt with. In fact, the impedance and vertical radiation patterns of horizontal arrays over a reflecting surface (such as the ground)

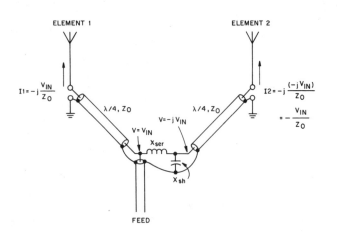

Fig 17—Quadrature feed system. Equations in the text permit calculation of values for the L network components, X_{ser} and X_{sh}.

can be derived by treating the images as additional array elements.

For two-element arrays of identical elements with equal element currents, the field strength gain at a distant point relative to a single similar element is

$$FSG = 10 \log \frac{(R_R + R_L)[1 + \cos (S \cos \theta + \phi_{12})]}{(R_R + R_L) + R_m \cos \phi_{12}} \quad \text{(Eq 15)}$$

where

FSG = field strength gain, dB
R_R = radiation resistance of a single isolated element
R_L = loss resistance of a single element
S = element spacing in degrees
θ = direction from array (see Fig 18)
ϕ_{12} = phase angle of current in element 2 relative to element 1. ϕ_{12} is negative if element 2 is delayed (lagging) relative to element 1.
R_m = mutual resistance between elements (see Fig 19)

The Gain Equation

The gain value from Eq 15 is the *power gain* in dB, which equals the *field strength gain* in dB. Eq 15 should not be confused with equations such as those in Chapter 2 for "relative" field strength, which are used to calculate only the *shape* of the pattern. The above equation gives not only the shape of the pattern, but also the actual gain at each angle, relative to a single element.

The quantity for which the logarithm is taken in Eq 15 is composed of two major parts,

$$1 + \cos(S \cos \theta + \phi_{12}) \quad \text{(Term 1)}$$

which relates to field reinforcement or cancellation, and

$$\frac{R_R + R_L}{(R_R + R_L) + R_m \cos \phi_{12}} \quad \text{(Term 2)}$$

which is the gain change caused by mutual coupling. It is informative to look at each of these terms separately, to see what effect they have on the overall gain.

If there were no mutual coupling at all, Eq 15 would reduce to

$$FSG = 10 \log [1 + \cos(S \cos \theta + \phi_{12})] \quad \text{(Eq 16)}$$

The term

$$\cos(S \cos \theta + \phi_{12}) \quad \text{(Term 3)}$$

can assume values from -1 to $+1$, depending on the element spacing, current phase angle, and direction from the array. In the directions in which the term is -1, the gain becomes zero; a null occurs. Where the term is equal to $+1$, a maximum gain of

$$FSG = 10 \log 2 = 3 \text{ dB} \quad \text{(Eq 17)}$$

occurs. This is the same conclusion reached earlier (Eq 11). If the element spacing is insufficient, the term will fail to reach -1 or $+1$ in any direction, resulting in incomplete nulls or reduced gain, or both. Analysis of the spacing required for the term to reach -1 and $+1$ results in the graphs of Fig 12.

Analyzing array operation without mutual coupling is not simply an intellectual exercise, even though mutual coupling is present in all arrays. There are some circumstances which

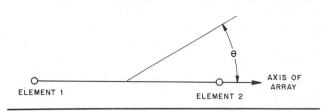

Fig 18—Definition of the angle θ for pattern calculation.

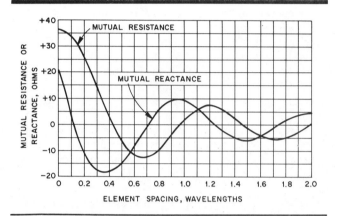

Fig 19—Mutual impedance between two parallel ¼-λ vertical elements. Multiply the resistance and reactance values by two for ½-λ dipoles. Values for vertical elements that are between 0.15 and 0.25 λ high may be approximated by multiplying the given values by $R_R/36$, where R_R is the radiation resistance of the vertical given by graphs in Chapter 2.

will make the mutual coupling portion of the gain equation equal, or very nearly equal, to one. Term 2 above will equal one if

$$R_m \cos \phi_{12} \quad \text{(Term 4)}$$

is equal to zero. This will happen if $R_m = 0$, which does occur at an element spacing of about 0.43 λ (see Fig 19). Arrays don't usually have elements spaced at 0.43 λ, but a much more common circumstance can cause the effect of mutual coupling on gain to be zero. Term 4 also equals zero if ϕ_{12}, the phase angle between the element currents, is $\pm 90°$. As a result, the gain of any two-element array with 90° phased elements is 3 dB in the favored directions, provided that the spacing is at least ¼ λ. The ¼ λ minimum is dictated by the requirement for full field reinforcement. If the elements are closer together, the gain will be less than 3 dB, as indicated in Fig 10.

Loss Resistance and Antenna Gain

A circumstance which reduces the gain effects of mutual coupling is the presence of high losses. If the loss resistance R_L becomes very large, the $R_R + R_L$ part of Term 2 above gets much larger than the $R_m \cos \phi_{12}$ part. Then Term 2, the mutual coupling part of the gain equation, becomes approximately

$$\frac{R_R + R_L}{R_R + R_L} = 1$$

Thus, the gain of any very lossy two-element array is 3 dB relative to a single similar element, providing that the spacing is adequate for full field reinforcement. Naturally, higher losses will always lower the gain relative to a single *lossless* element.

This principle can be used to obtain substantial gain if an inefficient antenna system is in use. The technique is to construct one or more additional closely spaced elements (each with its own ground system), and feed the resulting array with all elements in phase. The array won't have appreciable directivity, but it will have significant gain if the original system is very inefficient. As losses increase, the gain approaches 10 log n, where n is the number of elements—3 dB for two elements. This gain, of course, is relative to the original lossy element, so the system gain is unlikely to exceed that of a single lossless element.

Why does a close-spaced second element provide gain? An intuitive way to understand it is to note that two or more closely spaced in-phase elements behave almost like a single element, because of mutual coupling. However, the ground systems are not coupled, so they behave like parallel resistors. The result is a more favorable ratio of radiation to loss resistance. In an efficient system, which has a favorable ratio to begin with, the improvement is not significant, but it can be very significant if the original antenna is inefficient.

The following example illustrates the use of this technique to improve the performance of a 1.8-MHz antenna system. Suppose the original system consists of a single 50-foot-high vertical radiator with a 6-inch effective diameter. This antenna will have a radiation resistance R_R of 3.12 Ω at 1.9 MHz. A moderate ground system on a city lot will have a loss resistance R_L of perhaps 20 Ω. The efficiency of the antenna will be 3.12/(20 + 3.12) = 13.5%, or −8.7 dB relative to a perfectly efficient antenna.

If a second 50-foot antenna with a similar ground system is constructed just ten feet away from the first, the mutual resistance between elements will be 3.86 Ω. (Calculation of mutual resistance for very short radiators isn't covered in this chapter, but Brown shows that the mutual resistance between short radiators drops approximately in proportion to the self-resistance of each element.) Putting the appropriate values into Eq 15 shows an array gain of 2.34 dB relative to the original single element.

When the effects of mutual coupling are present, the gain in the favored direction can be greater or less than 3 dB, depending on the sign of Term 4. Analysis becomes easier if the element spacing is assumed to be sufficient for full field reinforcement. If this is true, the gain in the favored direction is

$$FSG = 10 \log \frac{2(R_R + R_L)}{(R_R + R_L) + R_m \cos \phi_{12}}$$

$$= 3 \text{ dB} + 10 \log \frac{R_R + R_L}{(R_R + R_L) + R_m \cos \phi_{12}} \quad \text{(Eq 18)}$$

Note that Term 4 above appears in the denominator of Eq 18. If maximum gain is the goal, this term should be made as negative as possible. One of the more obvious ways is to make ϕ_{12}, the phase angle, be 180°, so that $\cos \phi_{12} = -1$, and space the elements closely to make R_m large and positive (see Fig 19). Unfortunately, close spacing does not permit total field reinforcement, so Eq 18 is invalid for this approach. However, the very useful gain of just under 4 dB is still obtainable with this concept if the loss is kept very low. The highest gains for two-element arrays (about 5.6 dB) occur at close spacings with feed angles just under 180°. All close-spaced, moderate to high gain arrays are very sensitive to loss, so they generally will produce disappointing results when made with ground-mounted vertical elements.

Here are some examples which illustrate the use of Eq 15. Consider an array of two parallel, ¼ λ high, ground-mounted vertical elements, spaced ½ λ apart and fed 180° out of phase. For this array,

$R_R = 36$ ohms
$S = 180°$
$\phi_{12} = 180°$
$R_m = -6$ ohms (from Fig 19)

R_L must be measured or approximated, measurements being preferred for best accuracy. Suitable methods are described later. Alternatively, R_L can be estimated from graphs of ground system losses. Probably the most extensive set of measurements of vertical antenna ground systems was published by Brown et al in their classic 1937 paper. Their data have been republished countless times since, in amateur and other literature. Unfortunately, information is sparse for systems of only a few radials because Brown's emphasis is on broadcast installations. More recent measurements by Sevick nicely fill this void. From his data, we find that the typical feed point resistance of a ¼ λ vertical with four 0.2-0.4 λ radials is 65 ohms. (See Fig 23.) The loss resistance is 65 − 36 = 29 ohms. This value is used for the example.

Putting the values into Eq 15 results in

$$FSG = 10 \log \frac{65 \, [1 + \cos (180° \cos \theta + 180°)]}{65 + (-6 \cos 180°)}$$

Calculating the result for various values of θ reveals the familiar two-lobed pattern with maxima at 0° and 180°, and complete nulls at 90° and 270°. Maximum gain is calculated from Eq 15 by taking θ as 0°.

$$FSG = 10 \log \frac{65 \, (1 + 1)}{65 + 6} = 2.63 \text{ dB}$$

In this array, the mutual coupling decreases the gain slightly from the nominal 3 dB figure. The reader can confirm that if the element losses were zero ($R_L = 0$), the gain would be 2.34 dB relative to a similar, lossless element. If the elements were extremely lossy, the gain would approach 3 dB relative to a single similar and very lossy element. The efficiency of the original example elements is 36/65 = 55%, and a single isolated element would have a signal strength of 10 log 36/55 = −2.57 dB relative to a *lossless* element. As determined above, this phased array has a gain of 2.63 dB relative to a single 55% efficient element. Comparing the decibel numbers indicates the array performance in its favored directions is approximately the same as a single *lossless* element.

Changing the phasing of the array to 0° rotates the pattern 90°, and changes the gain to

$$FSG = 10 \log \frac{65 \times 2}{65 - 6} = 3.43 \text{ dB}$$

A system of very lossy elements would give 3 dB gain as before, and a lossless system would show 3.80 dB (each relative to a single similar element). In this case, the mutual coupling increases the gain above 3 dB, but the losses drop it back toward that figure. This effect can be generalized for larger arrays: Increasing loss in a system of n elements tends to move the gain toward 10 log n relative to a single similar (lossy) element, provided that spacing is adequate for full field reinforcement. If the spacing is closer, losses can reduce gain below this value.

MUTUAL COUPLING AND FEED POINT IMPEDANCE

The feed point impedances of the elements of an array are important to the design of some of the feed systems presented here. When elements are placed in an array, their feed point impedances change from the self-impedance values (impedances when isolated from other elements). The following information shows how to find the feed point impedances of elements in an array.

The impedance of element number 1 in a two-element array is given by Jasik as

$$R_1 = R_S + M_{12}(R_m \cos \phi_{12} - X_m \sin \phi_{12}) \qquad \text{(Eq 19)}$$

$$X_1 = X_S + M_{12}(X_m \cos \phi_{12} + R_m \sin \phi_{12}) \qquad \text{(Eq 20)}$$

where
   R_1 = the feed point resistance of element 1
   X_1 = the feed point reactance of element 1
   R_S = the self-resistance of a single isolated element = radiation resistance R_R + loss resistance R_L.
   X_S = the self-reactance of a single isolated element
   M_{12} = the magnitude of current in element 2 relative to that in element 1
   ϕ_{12} = the phase angle of current in element 2 relative to that in element 1
   R_m = the mutual resistance between elements 1 and 2
   X_m = the mutual reactance between elements 1 and 2

For element 2,

$$R_2 = R_S + M_{21}(R_m \cos \phi_{21} - X_m \sin \phi_{21}) \qquad \text{(Eq 21)}$$

$$X_2 = X_S + M_{21}(X_m \cos \phi_{21} + R_m \sin \phi_{21}) \qquad \text{(Eq 22)}$$

where

$$M_{21} = \frac{1}{M_{12}}$$

$$\phi_{21} = -\phi_{12}$$

and other terms are as defined above

Equations for the impedances of elements in larger arrays are given later.

Two Elements Fed Out of Phase

Consider the earlier example of a two-element array of ¼-λ verticals spaced ½ λ apart and fed 180° out of phase. To find the element feed point impedances, first the values of R_m and X_m are found from Fig 19. These are −6 and −15 Ω, respectively. Assuming that the element currents can

be balanced and that the desired 180° phasing can be obtained, the feed point impedance of element 1 becomes

$$R_1 = R_S + 1[-6\cos 180° - (-15)\sin 180° = R_S + 6 \, \Omega$$

$$X_1 = X_S + 1[-15\cos 180° + (-6)\sin 180° = X_S + 15 \, \Omega$$

Suppose that the elements, when not in an array, are resonant ($X_S = 0$) and that they have good ground systems so their feed point resistances (R_S) are 40 Ω. The feed point impedance of element 1 changes from 40 + j0 for the element by itself to 40 + 6 + j(0 + 15) = 46 + j15 Ω, because of mutual coupling with the second element. Such a change would be quite noticeable.

The second element in this array would be affected by the same amount, since the elements "look" the same to each other—there is no difference between 180° leading and 180° lagging. Mathematically, the difference in the calculation for element 2 involves changing +180° to −180° in the equations, leading to identical results. Elements fed in phase ($\phi_{12} = 0°$) also "look" the same to each other. So, for two-element arrays fed in phase (0°) or out of phase (180°), the feed point impedances of both elements change by the same amount and in the same direction because of mutual coupling. This is not generally true for a pair of elements which are part of a larger array, as a later example shows.

Two Elements with 90° Phasing

Now see what happens with two elements having a different relative phasing. Consider the popular vertical array with two elements spaced ¼ λ and fed with a 90° relative phase angle to obtain a cardioid pattern. Assuming equal element currents and ¼-λ elements, Fig 19 shows that $R_m = 20 \, \Omega$ and $X_m = -15 \, \Omega$. Use Eqs 19 and 20 to calculate the feed point impedance of the leading element, and Eqs 21 and 22 for the lagging element.

$$R_1 = R_S + 1[20\cos(-90°) - (-15)\sin(-90°)$$
$$= R_S - 15 \, \Omega$$

$$X_1 = X_S + 1[-15\cos(-90°) + 20\sin(-90°)$$
$$= X_S - 20 \, \Omega$$

And for the lagging element,

$$R_2 = R_S + 1[20\cos 90° - (-15)\sin 90°] = R_S + 15 \, \Omega$$

$$X_2 = X_S + 1[(-15)\cos 90° + 20\sin 90°] = X_S + 20 \, \Omega$$

These values represent quite a change in element impedance from mutual coupling. If each element, when isolated, is 50 Ω and resonant (50 + j0 Ω impedance), the impedances of the elements in the array become 35 − j20 and 65 + j20 Ω. These very different impedances can lead to current imbalance and serious phasing errors, if a casually designed or constructed feed system is used.

Close-Spaced Elements

Another example provides a good illustration of several principles. Consider an array of two parallel ½-λ dipoles fed 180° out of phase and spaced 0.1 λ apart. To avoid complexity in this example, assume these dipoles are a free-space ½-λ long, which is about 1.4% longer than a thin, resonant dipole. At this spacing, from Fig 19, $R_m = 67\ \Omega$ and $X_m = 7\ \Omega$. (Remember to double the values from the graph of Fig 19 for dipole elements.) For each element,

$$R_1 = R_2 = R_S + 1(67 \cos 180° - 7 \sin 180°)$$
$$= R_S - 67\ \Omega$$

$$X_1 = X_2 = X_S + 1(7 \cos 180° + 67 \sin 180°)$$
$$= X_S - 7\ \Omega$$

The feed point impedance of an isolated, free-space ½-λ dipole is approximately $74 + j44\ \Omega$. Therefore the elements in this array will each have an impedance of about $74 - 67 + j(44 - 7) = 7 - j37\ \Omega$! Aside from the obvious problem of matching the array to a feed line, there are some other consequences of such a radical change in the feed point impedance. Because of the very low feed point impedance, relatively heavy current will flow in the elements. Normally this would produce a larger field strength, but note from Fig 12 that the element spacing (36°) is far below the 180° required for total field reinforcement. What happens here is that the fields from the elements of this array partially or totally cancel in all directions; there is no direction in which they fully reinforce. As a result, the array produces only moderate gain. Even a few ohms of loss resistance will dissipate a substantial amount of power, reducing the array gain.

This type of array was first described in 1940 by Dr John Kraus, W8JK (see bibliography). At 0.1-λ spacing, the array will deliver just under 4 dB gain if there is no loss, and just over 3 dB if there is 1 Ω loss per element. The gain drops to about 1.3 dB for 5 Ω of loss per element, and to zero dB at 10 Ω. These figures can be calculated from Eq 15 or read directly from the graphs in Kraus's paper. The modern "8JK" array (presented later in this chapter) is based on the array just described, but it overcomes some of the above disadvantages by using four elements instead of two (two pairs of two half waves in phase). Doubling the size of the array provides a theoretical 3 dB gain increase over the above values, and feeding the array as pairs of half waves in phase increases the feed point impedance to a more reasonable value. However, the modern 8JK array is sensitive to losses, as described above, because of relatively high currents flowing in the elements.

LARGER ARRAYS

As mentioned earlier, the feed point impedance of any given element in an array of dipole or ground-mounted vertical elements is altered from its self-impedance by coupling to other elements in the array. Eqs 19 through 22 may be used to calculate the resistive and reactive components of the elements in a two-element array. In a larger array, however, mutual coupling must be taken into account between any given element and all other elements in the array.

Element Feed Point Impedances

The equations presented in this section may be used to calculate element feed point impedances in larger arrays. Jasik

gives the impedance of an element in an n-element array as follows. For element 1,

$$R_1 = R_{11} + M_{12}(R_{12} \cos \phi_{12} - X_{12} \sin \phi_{12}) +$$
$$M_{13}(R_{13} \cos \phi_{13} - X_{13} \sin \phi_{13}) + \ldots +$$
$$M_{1n}(R_{1n} \cos \phi_{1n} - X_{1n} \sin \phi_{1n}) \qquad \text{(Eq 23)}$$

$$X_1 = X_{11} + M_{12}(R_{12} \sin \phi_{12} + X_{12} \cos \phi_{12}) +$$
$$M_{13}(R_{13} \sin \phi_{13} + X_{13} \cos \phi_{13}) + \ldots +$$
$$M_{1n}(R_{1n} \sin \phi_{1n} + X_{1n} \cos \phi_{1n}) \qquad \text{(Eq 24)}$$

For element p,

$$R_p = R_{pp} + M_{p1}(R_{p1} \cos \phi_{p1} - X_{p1} \sin \phi_{p1}) +$$
$$M_{p2}(R_{p2} \cos \phi_{p2} - X_{p2} \sin \phi_{p2}) + \ldots +$$
$$M_{pn}(R_{pn} \cos \phi_{pn} - X_{pn} \sin \phi_{pn}) \qquad \text{(Eq 25)}$$

$$X_p = X_{pp} + M_{p1}(R_{p1} \sin \phi_{p1} + X_{p1} \cos \phi_{p1}) +$$
$$M_{p2}(R_{p2} \sin \phi_{p2} + X_{p2} \cos \phi_{p2}) + \ldots +$$
$$M_{pn}(R_{pn} \sin \phi_{pn} + X_{pn} \cos \phi_{pn}) \qquad \text{(Eq 26)}$$

And for element n,

$$R_n = R_{nn} + M_{n1}(R_{n1} \cos \phi_{n1} - X_{n1} \sin \phi_{n1}) +$$
$$M_{n2}(R_{n2} \cos \phi_{n2} - X_{n2} \sin \phi_{n2}) + \ldots +$$
$$M_{n(n-1)}(R_{n(n-1)} \cos \phi_{n(n-1)} - X_{n(n-1)} \sin \phi_{n(n-1)})$$
$$\text{(Eq 27)}$$

$$X_n = X_{nn} + M_{n1}(R_{n1} \sin \phi_{n1} + X_{n1} \cos \phi_{n1}) +$$
$$M_{n2}(R_{n2} \sin \phi_{n2} + X_{n2} \cos \phi_{n2}) + \ldots +$$
$$M_{n(n-1)}(R_{n(n-1)} \sin \phi_{n(n-1)} + X_{n(n-1)} \cos \phi_{n(n-1)})$$
$$\text{(Eq 28)}$$

where

R_{jj} = self resistance of element j
X_{jj} = self reactance of element j
M_{jk} = magnitude of current in element k relative to that in element j
R_{jk} = mutual resistance between elements j and k
X_{jk} = mutual reactance between elements j and k
ϕ_{jk} = phase angle of current in element k relative to that in element j

These are more general forms of Eqs 19 and 20. Examples of using these equations appear in a later section.

Quadrature Fed Elements in Larger Arrays

In some arrays, *groups of elements* must be fed in quadrature. Such a system is shown in Fig 20. The current in each element in the left-hand group equals

$$I1 = -j \frac{V_{in}}{Z_0} \qquad \text{(Eq 29)}$$

The current in the elements in the right-hand group equals

$$I2 = -j \frac{V_{out}}{Z_0} \qquad \text{(Eq 30)}$$

Thus, if $V_{out} = -jV_{in}$, the right-hand group will have currents equal in magnitude to and 90° delayed from the

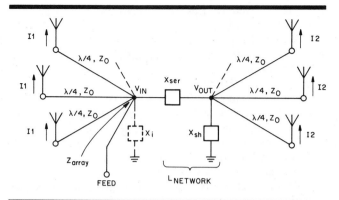

Fig 20—The L network applied to larger arrays. Coaxial cable shields and ground connections for the elements have been omitted for clarity. The text gives equations for determining the component values of X_{ser}, X_{sh} and X_i. X_i is an optional impedance matching component.

currents in the left-hand group. The feed point resistances of the elements have nothing to do with determining the current relationship, except that the relationship between V_{out} and V_{in} is a function of the impedance of the load presented to the L network. That load is determined by the impedances of the elements in the right-hand group.

Values of network components are given by

$$X_{ser} = \frac{Z_0^2}{\Sigma R_2} \qquad \text{(Eq 31)}$$

$$X_{sh} = \frac{Z_0^2}{\Sigma X_2 - \Sigma R_2} \qquad \text{(Eq 32)}$$

where

X_{ser} = the reactance of the series network element
X_{sh} = the reactance of the shunt network element (at the output side)
Z_0 = the characteristic impedance of the element feed lines
ΣR_2 = the sum of the feed point resistances of all elements connected to the output side of the network
ΣX_2 = the sum of the feed point reactances of all elements connected to the output side of the network

These are more general forms of Eqs 13 and 14. If the value of X_{ser} or X_{sh} is positive, that component is an inductor; if negative, a capacitor.

Array Impedance and Array Matching

Although the impedance matching of an array to the main feed line is not covered in any depth in this chapter, simply adding X_i to the L network, as shown in Fig 20, can improve the match of the array. X_i is a shunt component with reactance, added at the network input. With the proper X_i, the array common-point impedance is made purely resistive,

improving the SWR or allowing Q-section matching. X_i is determined from

$$X_i = \frac{Z_0^2}{\Sigma X_1 - \Sigma R_2} \qquad \text{(Eq 33)}$$

where

X_i = the reactance of the shunt network matching element (at the input side)
ΣX_1 = the sum of the feed point reactances of all elements connected to the input side of the network

and other terms are as defined above

If the value of X_i is positive, the component is an inductor; if negative, a capacitor.

With the added network element in place, the array common point impedance is

$$Z_{array} = \frac{Z_0^2}{\Sigma R_1 + \Sigma R_2} \qquad \text{(Eq 34)}$$

where

ΣR_1 = the sum of the feed point resistances of all elements connected to the input side of the network

and other terms are as described above

X_{ser} and X_{sh} should be adjusted for correct phasing and current balance as described later. They should *not* be adjusted for the best SWR. X_i, only, is adjusted for the best SWR, and has no effect on phasing or current balance.

CURRENT IMBALANCE AND ARRAY PERFORMANCE

The result of phase error in a driven array was discussed earlier. Changes in phase from the design value produce pattern changes such as shown in Fig 14. Now we turn our attention to the effects of current amplitude imbalance in the elements. This requires the introduction of more general gain equations to take the current ratio into account; the equations given earlier are simplified, based on equal element currents.

Gain, Nulls, and Null Depth

A more general form of Eq 15, taking the current ratios into account, is

$$FSG = 10 \log$$

$$\frac{(R_R + R_L)[1 + M_{12}^2 + 2M_{12} \cos (S \cos \theta + \phi_{12})]}{(R_R + R_L)(1 + M_{12}^2) + 2M_{12} R_m \cos \phi_{12}}$$

$$\text{(Eq 35)}$$

where

FSG = field strength gain relative to a single, similar element, dB
M_{12} = the magnitude of current in element 2 relative to the current in element 1.

and other symbols are as defined for Eq 15

Eq 35 may be used to determine the array field strength at a distant point relative to that from a single similar element for any spacing of two array elements.

Now consider arrays where the spacing is sufficient for total field reinforcement or total field cancellation, or both. Fig 12 shows the spacings necessary to achieve these conditions. The curves of Fig 12 show spacings which will allow the term

$$\cos(S \cos \theta + \phi_{12})$$

to equal its maximum possible value of $+1$ (total field reinforcement) and minimum possible value of -1 (total field cancellation). In reality, the fields from the two elements cannot add to zero unless this term is -1 *and* the element currents are equal. For a given set of element currents, the directions in which the term is $+1$ are those of maximum gain, and the directions in which the term is -1 are those of the deepest nulls.

The elements in many arrays are spaced at least as far apart as given by the two curves in Fig 12. Considerable simplification results in gain calculations for unequal currents if it is assumed that the elements are spaced to satisfy the conditions of Fig 12. Such simplified equations follow.

In the directions of maximum signal,

$$FSG =$$
$$10 \log \frac{(R_R + R_L)(1 + M_{12})^2}{(R_R + R_L)(1 + M_{12}^2) + 2M_{12} R_m \cos \phi_{12}}$$
$$\text{(Eq 36)}$$

This is a more general form of Eq 18, and is valid provided that the element spacing is sufficient for total field reinforcement.

In the directions of minimum gain (nulls),

$$FSG \text{ at nulls} =$$
$$10 \log \frac{(R_R + R_L)(1 - M_{12})^2}{(R_R + R_L)(1 + M_{12}^2) + 2M_{12} R_m \cos \phi_{12}}$$
$$\text{(Eq 37)}$$

This equation is valid if the spacing is enough for total field cancellation.

The "front-to-null" ratio can be calculated by combining the above two equations.

$$\text{Front-to-null ratio (dB)} = 10 \log \frac{(1 + M_{12})^2}{(1 - M_{12})^2} \quad \text{(Eq 38)}$$

This equation is valid if the spacing is sufficient for total field reinforcement *and* cancellation.

The equation for forward gain is further simplified for those special cases where

$$R_m \cos \phi_{12} \qquad \text{(Term 5)}$$

is equal to zero. (See the discussion of Eq 15 and Term 4 in the earlier section, "The Gain Equation.")

$$FSG = 10 \log \frac{(1 + M_{12})^2}{1 + M_{12}^2} \qquad \text{(Eq 39)}$$

This equation is valid if the element spacing is sufficient for total field reinforcement.

If an array is more closely spaced than indicated above, the gain will be less, the nulls poorer, or front-to-null ratio

worse than given by Eqs 36 through 39. Eq 35 is valid regardless of spacing.

Graphs of Eqs 38 and 39 are shown in Fig 21. Note that the "forward gain" graph applies only to arrays for which Term 5, above, equals zero (which includes all two-element arrays phased at 90° and spaced at least ¼ λ). The graph is useful, however, to get a "ballpark" idea of the gain of other arrays. The "front-to-null" graph applies to any two-element array, provided that spacing is wide enough for both full reinforcement and cancellation. Fig 21 clearly shows that current imbalance affects the front-to-null ratio much more strongly than it affects forward gain.

If the two elements have different loss resistances (for example, from different ground systems in a vertical array), gain relative to a single similar element becomes meaningless. However, gain relative to a single *lossless* element can still be calculated

$$\text{Gain} = 10 \log$$
$$\frac{R_R[1 + M_{12}^2 + 2M_{12} \cos (S \cos \theta + \phi_{12})]}{(R_R + R_{L1}) + M_{12}^2 (R_R + R_{L2}) + 2M_{12} R_m \cos \phi_{12}}$$
$$\text{(Eq 40)}$$

where
the gain is relative to a lossless element
R_{L1} = loss resistance of element 1
R_{L2} = loss resistance of element 2

Current Errors with Simple Feed Systems

It has already been said that casually designed feed systems can lead to poor current balance and improper phasing. To illustrate just how significant the errors can be, consider various arrays with "typical" feed systems.

The first array consists of two resonant, ¼-λ ground-mounted vertical elements, spaced ¼ λ apart. Each element has a feed point resistance of 65 Ω when the other element is open circuited. This is the approximate value when four radials per element are used. In an attempt to obtain 90° relative phasing, element 1 is fed with a line of electrical length L_1, and element 2 is fed with a line 90 electrical degrees longer (L_2). The results appear in Table 2.

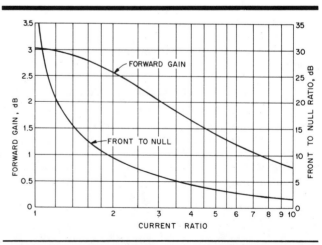

Fig 21—Effect of element current imbalance on forward gain and front-to-null ratio for certain arrays. See text.

Table 2

Two ¼-λ Vertical Elements with ¼-λ Spacing

Feeder system: Line lengths to elements 1 and 2 are given below as L_1 and L_2, respectively. The line length to element 2 is electrically 90° longer than to element 1.

| | Feed Lines | | | El. Feed Point Impedances | | El. Current Ratio | |
No.	Z_0, Ohms	L_1, Deg	L_2, Deg	Z_1, Ohms	Z_2, Ohms	Mag	Phase, Deg
1	50	90	180	$50.8 - j6.09$	$69.8 + j40.0$	0.620	-120
2	75	90	180	$45.1 - j14.0$	$73.3 + j24.3$	0.973	-108
3	50	180	270	$45.7 - j14.1$	$73.9 + j24.6$	0.956	-107
4	75	180	270	$51.5 - j11.4$	$79.4 + j32.4$	0.705	-103
5	50	45	135	$45.2 - j8.44$	$68.5 + j28.9$	0.859	-120
6	75	45	135	$50.2 - j14.9$	$79.4 + j26.1$	0.840	-98
7	Correctly fed			$50.0 - j20.0$	$80.0 + j20.0$	1.000	-90

Not only is the magnitude of the current ratio off by as much as nearly 40%, but the phase angle is incorrect by as much as 30°! The pattern of the array fed with feed system number 1 is shown in Fig 22, with a correctly fed array pattern for reference. Note that the example array has only a 9.0 dB front-to-back ratio, although the forward gain is only 0.1 dB more than the correctly fed array. This pattern was calculated from Eq 35.

Results will be different for arrays with different ground systems. For example, if the array fed with feed system no. 1 had elements with an initial feed point resistance of 40 instead of 65 Ω, the current ratio would be almost exactly 1—but the phase angle would still be $-120°$, resulting in poor nulls. The forward gain of the array is $+4.0$ dB, but the front-to-back ratio is only 11.5 dB.

The advantage of using the current forcing method to feed arrays of in-phase and 180° out-of-phase elements is shown by the following example. Suppose that the ground systems of two half-wave spaced, ¼-λ vertical elements are slightly different, so that one element has a feed point resistance of 50 Ω, the other 65 Ω. (Each is measured when the other element is open circuited.) What happens in this case is shown in Table 3.

The patterns of the nonforced arrays are only slightly distorted, with the main deficiency being imperfect nulls. The in-phase array fed with feed system number 1 exhibits a front-to-side ratio of 18.8 dB. The out-of-phase array fed with feed system number 6 has a front-to-side ratio of 17.0 dB. Both these arrays have forward gains very nearly equal to that of a correctly fed array.

Even when the ground systems of the two elements are only slightly different, a substantial current imbalance can

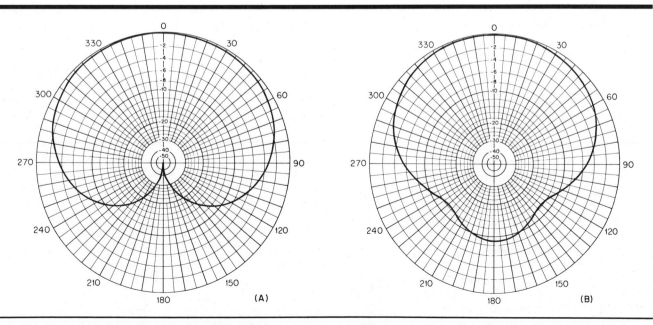

Fig 22—Patterns of an array when correctly fed, A, and when casually fed, B. (See text.) The difference in gain is about 0.1 dB. Gain is referenced to a single similar element; add 3.1 dB to the scale values shown.

Table 3

Two ¼-λ Vertical Elements with ½-λ Spacing and Different Self-Resistances

Self-resistances: Element 1—50 Ω; Element 2—65 Ω (difference caused by different ground losses)
Feeder system: Line lengths to elements 1 and 2 are given below as L_1 and L_2, respectively.

| | Feed Lines | | | El. Feed Point Impedances | | El. Current Ratio | |
No.	Z_0, Ohms	L_1, Deg	L_2, Deg	Z_1, Ohms	Z_2, Ohms	Mag	Phase, Deg
1	Any*	180	180	45.9 − j12.2	56.5 − j18.3	0.800	+ 3.1
2	50	135	135	43.8 − j11.9	59.7 − j18.6	0.834	− 5.8
3	75	135	135	43.2 − j12.5	60.3 − j17.7	0.883	− 6.8
4	Any*	270†	270†	44.0 − j15.0	59.0 − j15.0	1.000	0.0
5	50	45	225	53.2 + j12.9	74.8 + j17.1	0.820	− 172
6	Any*	180	360	55.6 + j11.0	71.1 + j20.2	0.764	− 185
7	Any*	90†	270†	56.0 + j15.0	71.0 + j15.0	1.000	− 180

*Both lines must have the same Z_0
†Current forced

occur in in-phase and 180° out-of-phase arrays if casually fed. Two elements with feed point resistances of 36 and 41 Ω (when isolated), fed with ½ and 1 λ of line, respectively, will have a current ratio of 0.881. This is a significant error for a small resistance difference that may be impossible to avoid in practice. As explained earlier, two horizontal elements of different heights, or two elements in many larger arrays, even when fed in phase or 180° out of phase, require more than a casual feed system for correct current balance and phasing.

Practical Array Design Examples

This section, also written by Roy Lewallen, W7EL, presents four examples of practical arrays using the design principles given in previous sections. All arrays are assumed to be made of ¼ λ vertical elements.

General Array Design Considerations

If the quadrature feed system (Fig 17) is used, the self-impedance of one or more elements must be known. If the elements are common types, such as plain vertical wires or tubes, the impedance can be estimated quite closely from the graphs in Chapter 2. Elements which are close to ¼ λ high will be near resonance, and calculations can be simplified by adjusting each element to exact resonance (with the other element open-circuited at the feed point) before proceeding. If the elements are substantially less than ¼ λ high, they will have a large amount of capacitive reactance. This should be reduced in order to keep the SWR on the feed lines to a value low enough to prevent large losses, possible arcing, or other problems. Any tuning or loading done to the elements at the feed point must be in *series* with the elements, so as not to shunt any of the carefully balanced current to ground. A loading coil in series with a short element is permissible, provided that all elements have identical loading coils, but any shunt component at the element feed point must be avoided.

For the following examples, it is assumed that the elements are close to ¼ λ high and that they have been adjusted for resonance. The radiation resistance of each element is then close to 36 Ω, and the self-reactance is zero because it is resonant.

In any real vertical array, there is ground loss associated with each element. The amount of loss depends on the length and number of ground radials, and on the type and wetness of the ground under and around the antenna. This resistance appears in series with the radiation resistance. The self-resistance is the sum of the radiation resistance and the loss resistance. Fig 23 gives resistance values for typical ground systems, based on measurements by Sevick (July 1971 and March 1973 *QST*). The values of quadrature feed system components based on Fig 23 will be reasonably close to correct, even if the ground characteristics are somewhat different than Sevick's.

Feed systems for the design example arrays to follow are based on the resistance values given below.

Number of Radials	Loss Resistance, Ohms
4	29
8	18
16	9
Infinite	0

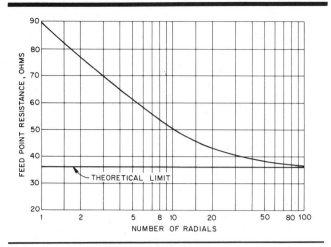

Fig 23—Approximate feed point resistance of a resonant ¼-λ ground-mounted vertical element versus the number of radials, based on measurements by Jerry Sevick, W2FMI. Moderate length radials (0.2 to 0.4 λ) were used for the measurements. The exact resistance, especially for only a few radials, will depend on the nature of the soil under the antenna.

The mutual impedance of the elements also must be known in order to calculate the impedances of the elements when in the array. The mutual impedance of parallel elements of near-resonant length may be taken from Fig 19. For elements of different lengths, or for unusual shape or orientation, the mutual impedance is best determined by measurement, using measurement methods as given later. Fig 19 suffices for the mutual impedance values in the example arrays.

The matter of matching the array for the best SWR on the feed line to the station is not discussed here. Many of the simpler arrays provide a match which is close to 50 or 75 Ω, so no further matching is required. If better matching is necessary, the appropriate network should be placed in the single feed line running to the station. Attempts to improve the match by adjustment of the phasing L network, antenna lengths, or individual element feeder lengths will ruin the current balance of the array. Information on impedance matching may be found in Chapters 25 and 26.

90° FED, 90° SPACED ARRAY

The feed system for a 90° fed, 90° spaced array is shown in Fig 17. The values of the inductor and capacitor must be calculated, at least approximately. The exact values can be determined by adjustment.

In this example the elements are assumed to be close to ¼ λ high, and each is assumed to have been adjusted for resonance with the other element open circuited. If each element has, say, four ground radials, the ground loss resistance is approximately 29 Ω. The self-resistance is 65 Ω. The self-reactance is zero, since the elements are resonant. From Fig 19, the mutual resistance of two parallel ¼-λ verticals spaced ¼ λ apart is 20 Ω, and the mutual reactance is −15 Ω. These values are used in Eqs 19 through 22 to calculate the feed point impedances of the elements. Element currents of equal magnitude are required, so $M_{12} = M_{21} = $

1. The L network causes the current in element 2 to lag that in element 1 by 90°, so $\phi_{12} = -90°$ and $\phi_{21} = 90°$. Summarizing,

$R_S = 65 \ \Omega$
$X_S = 0 \ \Omega$
$M_{12} = M_{21} = 1$
$\phi_{12} = -90°$
$\phi_{21} = 90°$
$R_m = 20 \ \Omega$
$X_m = -15 \ \Omega$

Putting these values into Eqs 19 through 22 results in the following values.

$R_1 = 50 \ \Omega$
$X_1 = -20 \ \Omega$
$R_2 = 80 \ \Omega$
$X_2 = 20 \ \Omega$

These are the actual impedances at the bases of the two elements when placed in the array and fed properly. It is necessary only to know the impedance of element 2 in order to design the L network, but the impedance of element 1 was calculated here to show how different the impedances are. Next, the impedance of the feed lines is chosen. Suppose the choice is 50 Ω. For Eqs 13 and 14,

$Z_0 = 50 \ \Omega$
$R_2 = 80 \ \Omega$
$X_2 = 20 \ \Omega$

From Eq 13,

$$X_{ser} = \frac{50^2}{80} = 31.3 \ \Omega$$

And from Eq 14,

$$X_{sh} = \frac{50^2}{20 - 80} = -41.7 \ \Omega$$

The signs show that X_{ser} is an inductor and X_{sh} is a capacitor. The actual values of L and C can be calculated for the desired frequency by rearranging and modifying the basic equations for reactance.

$$L = \frac{X_L}{2\pi f} \qquad \qquad \text{(Eq 41)}$$

$$C = \frac{-10^6}{2\pi f X_c} \qquad \qquad \text{(Eq 42)}$$

where

 L = inductance, μH
 C = capacitance, pF
 f = frequency, MHz
 X_L and X_C = reactance values, ohms

The negative sign in Eq 42 is included because capacitive reactance values are given here as negative.

A similar process is followed to find the values of X_{ser} and X_{sh} for different ground systems and different feed line impedances. The results of such calculations appear in Table 4.

To obtain correct performance, both network components must be adjustable. If an adjustable inductor is not convenient or available, a fixed inductor in series with a variable capacitor will provide the required adjustability. The

Table 4

L Network Values for Two Elements ¼ λ Apart, Fed 90° Out of Phase (Fig 17)

R_S, Ohms	No. of Radials per Element	Z_0, Ohms	X_{ser} Ohms	X_{sh} Ohms
65	4	50	31.3	−41.7
65	4	75	70.3	−93.8
54	8	50	36.2	−51.0
54	8	75	81.5	−114.8
45	16	50	41.7	−62.5
45	16	75	93.8	−140.6
36	∞	50	49.0	−80.6
36	∞	75	110.3	−181.5

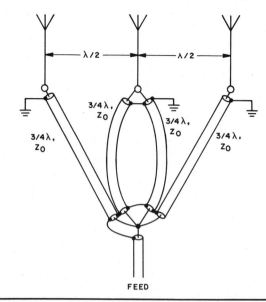

Fig 24—Feed system for the three-element 1:2:1 "binomial" array. All feed lines are ¾ electrical wavelength long and have the same characteristic impedance.

equivalent reactance should be equal to the value calculated for X_{ser}. For example, to use the above design at 7.15 MHz, $L_{ser} = 0.697 \ \mu H$, and $C_{sh} = 534$ pF. The 0.697 μH inductor (reactance = 31.3 Ω) can be replaced by a 1.39 μH inductor (reactance = approximately 62.6 Ω) in series with a variable capacitor capable of being adjusted on both sides of 711 pF (reactance = −31.3 Ω). The reactance of the series combination can then be varied on both sides of $62.6 − 31.3 = 31.3$ Ω. Actually, it might be preferable to use 75-Ω feed line instead of 50 for this array. Table 4 shows that the L network reactances are about twice as great if 75-Ω line is chosen. This means that the required capacitance would be one half as large. Smaller adjustable capacitors are more common, and more compact.

The voltages across the network components are relatively low. Components with breakdown voltages of a few hundred volts will be adequate for a few hundred watts of output power. If fixed capacitors are used, they should be good quality mica or ceramic units.

A THREE ELEMENT BINOMIAL BROADSIDE ARRAY

An array of three in-line elements spaced ½ λ apart and fed in phase gives a pattern which is generally bidirectional. If the element currents are equal, the resulting pattern has a forward gain of 5.7 dB (for lossless elements) but substantial side lobes. If the currents are tapered in a binomial coefficient 1:2:1 ratio (twice the current in the center element as in the two end elements), the gain drops slightly to 5.2 dB, the main lobes widen, and the side lobes disappear.

The array is shown in Fig 24. To obtain a 1:2:1 current ratio in the elements, each end element is fed through a ¾-λ line of impedance Z_0. Line lengths of ¾-λ are chosen because ¼-λ lines will not physically reach. The center element is fed from the same point through two parallel ¾ λ lines of the same characteristic impedance, which is equivalent to feeding it through a line of impedance $Z_0/2$. The currents are thus forced to be in phase and to have the correct ratio.

A FOUR-ELEMENT RECTANGULAR ARRAY

The four-element array shown with its pattern in Fig 25 has appeared numerous times in amateur publications. However, the accompanying feed systems invariably fail to

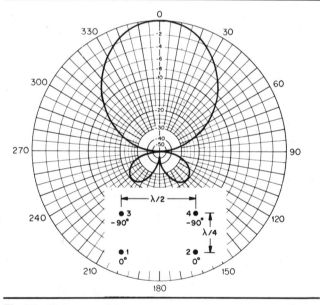

Fig 25—Pattern and layout of the four-element rectangular array. Gain is referenced to a single similar element; add 6.8 dB to the scale values shown.

deliver currents in the proper amounts and phases to the various elements. The array can be correctly fed using the principles discussed in this section.

Elements 1 and 2 can be forced to be in phase and to have equal currents by feeding them through ¾-λ lines. (Again, ¾-λ lines are chosen because ¼-λ lines won't physically reach.) Likewise, the currents in elements 3 and 4 can be forced to be equal and in phase. Elements 3 and 4 are made to have currents of equal amplitude but of 90° phase difference from elements 1 and 2 by use of the quadrature feed system shown in Fig 26. The phasing network is the type shown in Fig 17, but Eqs 31 and 32 must be used to calculate the network component values. For this array they are

$$X_{ser} = \frac{Z_0^2}{R_3 + R_4} = \frac{Z_0^2}{2R_3} \qquad \text{(Eq 43)}$$

$$X_{sh} = \frac{Z_0^2}{X_3 + X_4 - (R_3 + R_4)} = \frac{Z_0^2}{2(X_3 - R_3)} \qquad \text{(Eq 44)}$$

The impedances of elements 3 and 4 will change by the same amount because of mutual coupling. If their ground systems are identical, they will also have equal values of R_L. If the ground systems are different, an adjustment of network values must be made, but the currents in all elements will be equal and correctly phased once the network is adjusted.

Eqs 25 and 26 are used to calculate R_3 and X_3. For element 3, they become

$$R_3 = R_S + M_{31}(R_{31} \cos \phi_{31} - X_{31} \sin \phi_{31}) +$$
$$M_{32}(R_{32} \cos \phi_{32} - X_{32} \sin \phi_{32}) +$$
$$M_{34}(R_{34} \cos \phi_{34} - X_{34} \sin \phi_{34})$$

$$X_3 = X_S + M_{31}(R_{31} \sin \phi_{31} + X_{31} \cos \phi_{31}) +$$
$$M_{32}(R_{32} \sin \phi_{32} + X_{32} \cos \phi_{32}) +$$
$$M_{34}(R_{34} \sin \phi_{34} + X_{34} \cos \phi_{34})$$

where
$$M_{31} = M_{32} = M_{34} = 1$$
$$\phi_{31} = +90°$$
$$\phi_{32} = +90°$$
$$\phi_{34} = 0°$$
$R_{31} = 20 \ \Omega$ (from Fig 19, 0.25-λ spacing)
$X_{31} = -15 \ \Omega$ (0.25-λ spacing)
$R_{32} = -10 \ \Omega$ (0.56-λ spacing)
$X_{32} = -10 \ \Omega$ (0.56-λ spacing)
$R_{34} = -6 \ \Omega$ (0.50-λ spacing)
$X_{34} = -15 \ \Omega$ (0.50-λ spacing)

resulting in $R_3 = R_S + 19 \ \Omega$ and $X_3 = X_S - 5.0 \ \Omega$. R_S and X_S are the self-resistance and self-reactance of a single isolated element. In this example, they are assumed to be the same for all elements. Thus, element 4 will have the same impedance as element 3.

It is now possible to make a table of X_{ser} and X_{sh} values for this array for different ground systems and feed line impedances. The information appears in Table 5. Calculation of actual values of L and C are the same as for the earlier example.

THE FOUR-SQUARE ARRAY

A versatile array is one having four elements arranged in a square, commonly called the four-square array. The array layout and its pattern are shown in Fig 27. This array has several attractive properties:

1) 5.5 dB forward gain over a single similar element, for any value of loss resistance;

2) 3 dB or greater forward gain over a 90° angle;

3) 20 dB or better F/B ratio maintained over a 130° angle;

4) symmetry that allows directional switching in 90° increments.

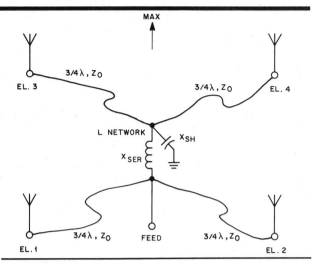

Fig 26—Feed system for the four-element rectangular array. Grounds and cable shields have been omitted for clarity.

Table 5

L Network Values for the Four-Element Rectangular Array (Fig 26)

R_S, Ohms	No. of Radials per Element	Z_0, Ohms	X_{ser} Ohms	X_{sh} Ohms
65	4	50	14.9	−14.0
65	4	75	33.5	−31.6
54	8	50	17.1	−16.0
54	8	75	38.5	−36.1
45	16	50	19.5	−18.1
45	16	75	43.9	−40.8
36	∞	50	22.7	−20.8
36	∞	75	51.1	−46.9

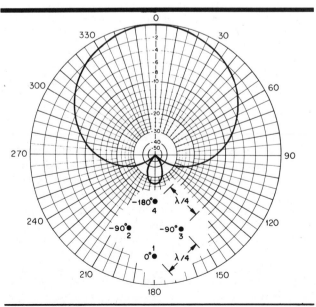

Fig 27—Pattern and layout of the four-square array. Gain is referenced to a single similar element; add 5.5 dB to the scale values shown.

Because of the large differences in element feed point impedances from mutual coupling, casual feed systems nearly always lead to poor performance of this array. Using the feed system described here, performance is very good, being limited chiefly by environmental factors. Such an array and feed system have been in use at W7EL for several years.

Although the impedances of only two of the four elements need to be calculated to design the feed system, all element impedances will be calculated to show the wide differences in value. This is done by using Eqs 23 through 28, with the following values for the variables.

$M_{jk} = 1$ for all j and k
$R_{12} = R_{21} = R_{13} = R_{31} = R_{24} = R_{42} = R_{34} = R_{43} = 20\ \Omega$ (from Fig 19, 0.25 λ spacing)
$X_{12} = X_{21} = X_{13} = X_{31} = X_{24} = X_{42} = X_{34} = X_{43} = -15\ \Omega$ (0.25 λ spacing)
$R_{14} = R_{41} = R_{23} = R_{32} = 8\ \Omega$ (0.354-λ spacing)
$X_{14} = X_{41} = X_{23} = X_{32} = -18\ \Omega$ (0.354-λ spacing)
$\phi_{12} = \phi_{13} = \phi_{24} = \phi_{34} = -90°$
$\phi_{21} = \phi_{31} = \phi_{42} = \phi_{43} = 90°$
$\phi_{14} = \phi_{41} = \pm 180°$
$\phi_{23} = \phi_{32} = 0°$

resulting in

$R_1 = R_S - 38\ \Omega$
$X_1 = X_S - 22\ \Omega$
$R_2 = R_3 = R_S + 8\ \Omega$
$X_2 = X_3 = X_S - 18\ \Omega$
$R_4 = R_S + 22\ \Omega$
$X_4 = X_S + 58\ \Omega$

where R_S and X_S are the resistance and reactance of a single element when isolated from the array.

If element 1 had a perfect ground system and were resonant (a self-impedance of $36 + j0\ \Omega$), in the array it would have a feed point impedance of $36 - 38 - j22 = -2 - j22\ \Omega$. The negative resistance means that it would be delivering power *into* the feed system. This can, and does, happen in some phased arrays, and is a perfectly legitimate result. The power is, of course, coupled into it from the other elements by mutual coupling. Elements having impedances of precisely zero ohms could have the feed line short circuited at the feed point without effect; that is what a "parasitic" element is. This is yet another illustration of the error of trying to deliver equal *powers* to the elements.

The basic system for properly feeding the four-square array is shown in Fig 28. Foamed-dielectric cable must be used for the ¼-λ lines. The velocity factor of solid dielectric cable is lower, making an electrical ¼ λ of that type physically too short to reach. Elements 2 and 3 are forced to have equal and in-phase currents regardless of differences in ground systems. Likewise, elements 1 and 4 are forced to have equal, 180° out-of-phase currents, in spite of extremely different feed point impedances. The 90° phasing between element pairs is accomplished, as before, by an L network.

Eqs 43 and 44 may be used directly to generate a table of network element values for this array. For this array the values of resistance and reactance for element 3 are as calculated above.

$R_3 = R_S + 8\ \Omega$

$X_3 = X_S - 18\ \Omega = -18\ \Omega$

(Because each element was resonated when isolated from the other elements, $X_S = 0$.) Table 6 shows values of L-network components for various ground systems and feed line impedances.

This array is more sensitive to adjustment than the two-element 90° fed, 90° spaced array. Adjustment procedures and a method of remotely switching the direction of this array are described in the section that follows.

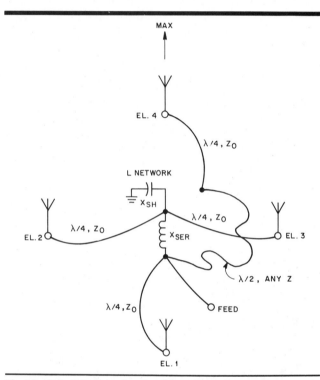

Fig 28—Feed system for the four-square array. Grounds and cable shields have been omitted for clarity.

Table 6

L Network Values for the Four-Square Array (Fig 28)

R_S, Ohms	No. of Radials per Element	Z_0, Ohms	X_{ser} Ohms	X_{sh} Ohms
65	4	50	17.1	-13.7
65	4	75	38.5	-30.9
54	8	50	20.2	-15.6
54	8	75	45.4	-35.2
45	16	50	23.6	-17.6
45	16	75	53.1	-39.6
36	∞	50	28.4	-20.2
36	∞	75	63.9	-45.4

Practical Aspects of Phased Array Design

With almost any type of antenna system, there is much that can be learned from experimenting with, testing and using various array configurations. In this section, Roy Lewallen, W7EL, shares the benefit of years of his experience from actually building, adjusting, and using phased arrays. There is much more work to be done in most of the areas covered here, and Roy encourages the reader to build on this work.

Adjusting Phased Array Feed Systems

If a phased array is constructed only to achieve forward gain, adjusting it is seldom worthwhile. This is because the forward gain of most arrays is quite insensitive to either the magnitude or phase of the relative currents flowing in the elements. If, however, good rejection of unwanted signals is desired, adjustment may be required.

The in-phase and 180° out-of-phase current-forcing methods supply very well-balanced and well-phased currents to the elements without adjustment. If the pattern of an array fed using this method is unsatisfactory, this is generally the result of environmental differences; the elements, furnished with correct currents, do not generate correct fields. Such an array can be optimized in a single direction, but a more general approach than the current forcing method must be taken. Some possibilities are described by Paul Lee and Forrest Gehrke (see bibliography).

Unlike the current forcing methods, the quadrature feed systems described earlier in this chapter are dependent on the element self- and mutual impedances. The required L network component values can be computed to a high level of precision, but the results are only as good as the knowledge of the relevant impedances. A practical approach is to estimate the impedances or measure them with moderate accuracy, and adjust the network for the best performance. Simple arrays, such as the two-element 90° fed and spaced array, may be adjusted as follows.

Place a low power signal source at a distance from the array (preferably several wavelengths), in the direction of the null. While listening to the signal on a receiver connected to the array, alternately adjust the two L-network components for the best rejection of the signal.

This has proved to be a very good way to adjust two-element arrays. However, variable results were obtained when a four-square array was adjusted using this technique. The probable reason is that more than one combination of current balance and phasing will produce a null in a given direction. But the overall array pattern is different for each combination. So a different method must be used for adjusting more complex arrays. This involves actually measuring the element currents one way or another, and adjusting the network until the currents are correct.

MEASURING ELEMENT CURRENTS

The element currents can be measured two ways. One way is to measure them directly at the element feed points, as shown in Fig 29. A dual-channel oscilloscope is required to monitor the currents. This method is the most accurate, and it provides a direct indication of the actual relative magnitudes and phases of the element currents. The current probe is shown in Fig 30.

Instead of measuring the element currents directly, they

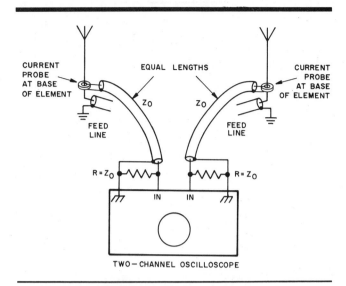

Fig 29—One method of measuring element currents in a phased array. Details of the current probe are given in Fig 30. Caution: Do not run high power to the antenna system for this measurement, or damage to the test equipment may result.

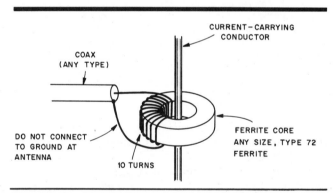

Fig 30—The current probe for use in the test setup of Fig 29. The ferrite core is of type 72 material, and may be any size. The coax line must be terminated at the opposite end with a resistor equal to its characteristic impedance.

may be indirectly monitored by measuring the voltages on the feed lines an electrical ¼ or ¾ λ from the array. The voltages at these points are directly proportional to the element currents. All the example arrays presented earlier (Figs 17, 20, 24, 26 and 28) have ¼ or ¾-λ lines from all elements to a common location, making this measurement method convenient. The voltages may be observed with a dual-channel oscilloscope, or, to adjust for equal magnitude currents and 90° phasing, the test circuit shown in Fig 31 may be used. The test circuit is connected to the feed lines of two elements which are to be adjusted for 90° phasing (such as elements 1 and 2, or 2 and 4 of the four-square array of Fig 28). Adjust the L-network components alternately until both meters read zero. Proper operation of the test circuit may be verified by

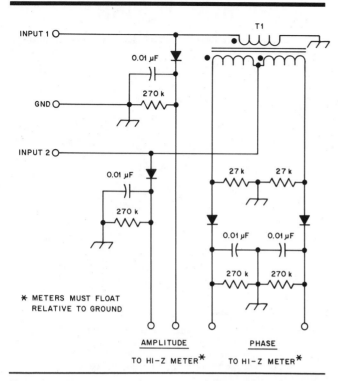

Fig 31—Quadrature test circuit. All diodes are germanium, such as 1N34A, 1N270, or equiv. All resistors are ¼ or ½ W, 5% tolerance. Capacitors are ceramic. Alligator clips are convenient for making the input and ground connections to the array.

T1—7 trifilar turns on an Amidon FT-37-72 or equiv ferrite toroid core.

disconnecting one of the inputs. The *phase* output should then remain close to zero. If not, there is an undesirable imbalance in the circuit, which must be corrected. Another means of verification is to first adjust the L network so the tester indicates correct phasing (zero volts at the *phase* output). Then reverse the tester input connections to the elements. The *phase* output should remain close to zero.

DIRECTIONAL SWITCHING OF ARRAYS

One ideal directional switching method would take the entire feed system, including the lines to the elements, and rotate them. The smallest possible increment of rotation depends on the symmetry of the array—the feed system would need to rotate until the array again "looks" the same to it. For example, any two-element array can be rotated 180° (although that wouldn't accomplish anything if the array was bidirectional to begin with). The four-element rectangular array of Figs 25 and 26 can also be reversed, and the four-square array of Figs 27 and 28 can be switched in 90° increments. Smaller increment switching can be accomplished only by reconfiguring the feed system, including the phase shift network, if used. Switching in smaller increments than dictated by symmetry will create a different pattern in some directions than in others, and must be thoughtfully done to maintain equal and properly phased element currents. The methods illustrated here will deal only with switching in

increments related to the array symmetry, except one, a two-element broadside/end-fire array.

In arrays containing quadrature fed elements, the success of directional switching depends on the elements and ground systems being identical. Few of us can afford the luxury of having an array many wavelengths away from all other conductors, so an array will nearly always perform somewhat differently in each direction. The array, then, should be adjusted when steered in the direction requiring the most signal rejection in the nulls. Forward gain will, for practical purposes, be equal in all the switched directions, since gain is much more tolerant of error than nulls are.

BASIC SWITCHING METHODS

Following is a discussion of basic switching methods, how to power relays through the main feed line, and other practical considerations. In diagrams, grounds are frequently omitted to aid clarity, but connections of the ground conductors must be carefully made. In fact, it is recommended that the ground conductors be switched just as the center conductors are. This is explained in more detail in subsequent text. In all cases, interconnecting lines must be very short.

A pair of elements spaced ½ λ apart can readily be switched between broadside and end-fire bidirectional patterns, using the current forcing properties of ¼-λ lines. The method is shown in Fig 32. The switching device can be a relay powered via a separate cable or by dc sent along the main feed line.

Fig 33 shows directional switching of a 90° fed, 90° spaced array. The rectangular array of Figs 25 and 26 can be switched in a similar manner, as shown in Fig 34.

Switching the direction of an array in increments of 90°, when permitted by its symmetry, requires at least two relays. A method of 90° switching of the four-square array is shown in Fig 35.

Powering Relays Through Feed Lines

All of the above switching methods can be implemented without additional wires to the switch box. A single-relay

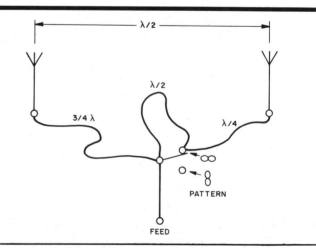

Fig 32—Two-element broadside/end-fire switching. All lines must have the same characteristic impedance. Grounds and cable shields have been omitted for clarity.

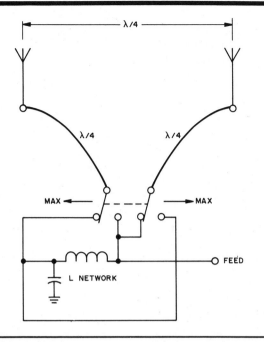

Fig 33—90° fed, 90° phased array reversal switching. All interconnections must be very short. Grounds and cable shields have been omitted for clarity.

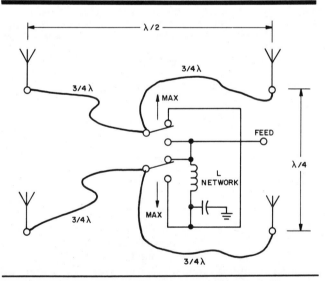

Fig 34—Directional switching of a four-element rectangular array. All interconnections must be very short. Grounds and cable shields have been omitted for clarity.

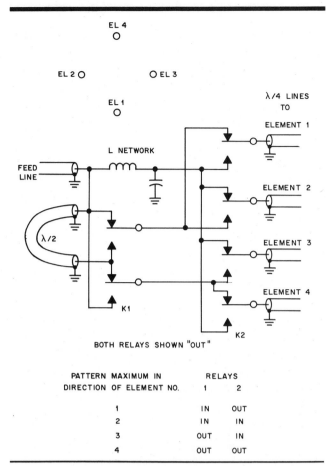

BOTH RELAYS SHOWN "OUT"

PATTERN MAXIMUM IN DIRECTION OF ELEMENT NO.	RELAYS 1	2
1	IN	OUT
2	IN	IN
3	OUT	IN
4	OUT	OUT

Fig 35—Directional switching of the four-square array. All interconnections must be very short.

C2 may be omitted if the antenna system is an open circuit at dc. C3 and C4 should be ceramic, 0.001 μF or larger.

In Fig 36B, capacitors C5 through C8 should be selected with the ratings of their counterparts in Fig 35A, as given above. Electrolytic capacitors across the relay coils, C9 and C10 in Fig 36B, should be large enough to prevent the relays from buzzing, but not so large as to make relay operation too slow. Final values for most relays will be in the range from 10 to 100 μF. They should have a voltage rating of at least double the relay coil voltage. Some relays do not require this capacitor. All diodes are 1N4001 or similar. A rotary switch may be used in place of the two toggle switches in the two-relay system to switch the relays in the desired sequence.

Although plastic food storage boxes are inexpensive and durable, using them to contain the direction switching circuitry might lead to serious phasing errors. If the circuitry is implemented as shown in Figs 32 through 35 and the feed-line grounds are simply connected together, the currents from more than one element share a single conductive path and get phase shifted by the reactance of the wire. As much as 30° of phase shift has been measured at 7 MHz from one side of a plastic box to the other, a distance of only four inches! No. 12 wire was connecting the two points. Since this experience, twice the number of relay contacts have been used, and the ground conductor of each coaxial cable has been switched right along with the center conductor. A solid metal

system is shown in Fig 36A, and a two-relay system in Fig 36B. Small 12- or 24-volt dc power relays can be used in either system at power levels up to at least a few hundred watts. Do not attempt to change directions while transmitting, however. Blocking capacitors C1 and C2 should be good quality ceramic or transmitting mica units of 0.01 to 0.1 μF. No problems have been encountered using 0.1-μF, 300-volt monolithic ceramic units at RF output levels up to 300 watts.

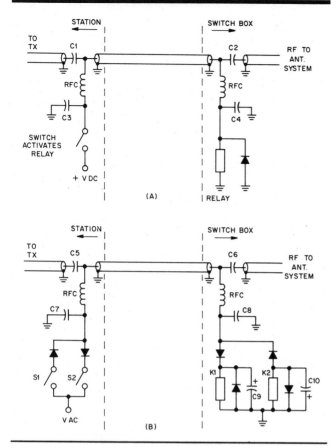

Fig 36—Remote switching of relays. See text for component information. A one-relay system is shown at A, and a two-relay system at B. In B, S1 activates K1, and S2 activates K2.

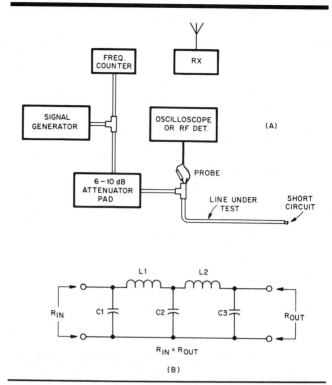

(A)

(B)

Fig 37—At A, the setup for measurement of the electrical length of a transmission line. The receiver may be used in place of the frequency counter to determine the frequency of the signal generator. The signal generator output must be free of harmonics; the half-wave harmonic filter at B may be used outboard if there is any doubt. It must be constructed for the frequency band of operation. Connect the filter between the signal generator and the attenuator pad.

C1, C3—Value to have a capacitive reactance = R_{IN}.
C2—Value to have a capacitive reactance = ½ R_{IN}.
L1, L2—Value to have an inductive reactance = R_{IN}.

box might present a path of low enough impedance to prevent the problem. If it does not, the best solution is to use a nonconductive box, and switch the grounds as described.

MEASURING THE ELECTRICAL LENGTH OF FEED LINES

When using the feed methods described earlier, the feed lines must be very close to the correct length. For best results, they should be correct within one percent or so. This means that a line which is intended to be, say, ¼ λ at 7 MHz, should actually be ¼ λ at some frequency within 70 kHz of 7 MHz. A simple but accurate method to determine at what frequency a line is ¼ or ½ λ is shown in Fig 37A. The far end of the line is short circuited with a *very* short connection. A signal is applied to the input, and the frequency is swept until the impedance at the input is a minimum. This is the frequency at which the line is ½ λ. Either the frequency counter or the receiver may be used to determine this frequency. The line is, of course, ¼ λ at one half the measured frequency. The detector can be a simple diode detector, or an oscilloscope may be used if available. A 6-10 dB attenuator pad is included to prevent the signal generator from looking into a short circuit at the measurement frequency. The signal generator output must be free of harmonics. If there is any doubt, an outboard low-pass filter, such as a half-wave harmonic filter,

should be used. The half-wave filter circuit is shown in Fig 37B, and must be constructed for the frequency band of operation.

Another satisfactory method is to use a noise or resistance bridge at the input of the line, again looking for a low impedance at the input while the output is short circuited. Simple resistance bridges are described in Chapter 27.

Dip oscillators have been found to be unsatisfactory. The required coupling loop has too great an effect on measurements.

MEASURING ELEMENT SELF-IMPEDANCE

The self-impedance of an unbalanced element, such as a vertical monopole, can be measured directly at the feed point using an impedance bridge. Commercial noise bridges are available, and noise and RLC bridges for home construction are described in Chapter 27.

When the measurement is being made, all other elements must be open-circuited. If the feed point is not readily accessible, the impedance can be measured remotely through one or more half wavelengths of transmission line. Other line lengths may also be used, but then an impedance conversion becomes necessary, such as with a Smith Chart (see Chapter

28). A balanced antenna, for example a dipole, must be measured through a transmission line to permit insertion of the proper type of balun (see below) unless the impedance meter can be effectively isolated from the ground and nearby objects, including the person doing the measurement. When measuring impedance through a transmission line, the following precautions must be taken to avoid substantial errors.

1) The characteristic impedance of the transmission line should be as close as possible to the impedance being measured. The closer the impedances, the less the sensitivity to feed line loss and length.

2) Do not use any more ½-λ sections of line than necessary. Errors are multiplied by the number of sections. Measurements made through lines longer than 1 λ should be suspect.

3) Use low-loss line. Lossy line will skew the measured value toward the characteristic impedance of the line. If the line impedance is close to the impedance being measured, the effect is usually negligible.

4) If a ½-λ section of line or multiple is being used, measure the line length using one of the methods described earlier. Do not try to make measurements at frequencies very far away from the frequency at which the line is the correct length. The sensitivity to electrical line length is less if the line impedance is close to the impedance being measured.

5) If the impedance of a balanced antenna such as a dipole is being measured, the correct type of balun must be used. (See Lewallen on baluns, listed in the bibliography). One way to make the proper type of balun is to use coaxial feed line, and pass the line through a large, high permeability ferrite core several times, near the antenna. Or a portion of the line may be wound into a flat coil of several turns, a foot or two in diameter, near the antenna. A third method is to string a large number of ferrite cores over the feed line, as Maxwell describes. The effectiveness of the balun can be tested by watching the impedance measurement while moving the coax about, and grasping it and letting go. The measurement should not change when this is done.

MEASURING MUTUAL IMPEDANCE

Various methods for determining the mutual impedance between elements have been devised. Each method has advantages and disadvantages. The basic difficulty in achieving accuracy is that the measurement of a small change in a large value is required. Two methods are described here. Both require the use of a calibrated impedance bridge. The necessary calculations require a knowledge of complex arithmetic. If measurements are made through feed lines, instead of directly at the feed points, the precautions listed above must be observed.

Method 1

1) Measure the self-impedance of one element with the second element open circuited at the feed point, or with the second element connected to an open-circuited feed line that is an integral number of ½ λ long. This impedance is designated Z_{11}.

2) Measure the self-impedance of the second element with the first element open circuited. This impedance is called Z_{22}.

3) Short circuit the feed point of the second element,

directly or at the end of an integral number of ½ λ of feed line. Measure the impedance of the first element. This is called Z_{1S}.

4) Calculate the mutual impedance Z_{12}.

$$Z_{12} = \pm \sqrt{Z_{22}(Z_{11} - Z_{1S})} \qquad \text{(Eq 45)}$$

where all values are complex.

Because the square root is extracted, there are two answers to this equation. One of these answers is correct and one is incorrect. There is no way to be sure which answer is correct except by noticing which one is closest to a theoretical value, or by making another measurement with a different method. This ambiguity is one disadvantage of using method 1. The other disadvantage is that the difference between the two measured values is small unless the elements are very closely spaced. This can cause relatively large errors in the calculated value of Z_{12} if small errors are made in the measured impedances. Useful results can be obtained using this method if care is taken, however. The chief advantage of method 1 is its simplicity.

Method 2

1) As in method 1, begin by measuring the self-impedance of one element, with the second element open circuited at the feed point, or with the second element connected to a ½-λ (or multiple) open-circuited line. Designate this impedance Z_{11}.

2) Measure the self-impedance of the second element with the first element open circuited. Call this impedance Z_{22}.

3) Connect the two elements together with ½ λ of transmission line, and measure the impedance at the feed point of one element. A ½-λ line may be added to both elements for this measurement if necessary. That is, the line to element 1 would be ½ λ, and the line to element 2 a full wavelength. Be sure to *read and observe* the precautions necessary when measuring impedance through a transmission line, enumerated earlier. This measured impedance is called Z_{1X}.

4) Calculate the mutual impedance Z_{12}.

$$Z_{12} = Z_{21} = -Z_{1X} \pm \sqrt{(Z_{1X} - Z_{11})(Z_{1X} - Z_{22})}$$

$$\text{(Eq 46)}$$

where all values are complex.

Again, there are two answers. But the correct one is generally easier to identify than when method 1 is used. For most systems, Z_{11} and Z_{22} are about the same. If they are, the *wrong* answer will be about equal to $-Z_{11}$ (or $-Z_{22}$). The *correct* answer will be about equal to $Z_{11} - 2Z_{1X}$ (or $Z_{22} - 2Z_{1X}$). The advantages of this method are that the correct answer is easier to identify, and that there is a larger difference between the two measured impedances. The disadvantage is that the ½-λ line adds another possible source of error.

The wrong answers from methods 1 and 2 will be different, but the correct answers should be the same. Measure with both methods, if possible. Accuracy in these measurements will enable the builder to determine more precisely the proper values of components for a phasing L network. And with precision in these measurements, the performance features of the array, such as gain and null depth, can be determined more accurately with methods given earlier in this chapter.

Collinear Arrays

Collinear arrays are always operated with the elements in phase. (If alternate elements in such an array are out of phase, the system simply becomes a harmonic type of antenna.) A collinear array is a broadside radiator, the direction of maximum radiation being at right angles to the line of the antenna.

Power Gain

Because of the nature of the mutual impedance between collinear elements, the feed point resistance is increased as shown earlier in this chapter (Fig 9). For this reason the power gain does not increase in direct proportion to the number of elements. The gain with two elements, as the spacing between them is varied, is shown by Fig 38. Although the gain is greatest when the end-to-end spacing is in the region of 0.4 to 0.6 λ, the use of spacings of this order is inconvenient constructionally and introduces problems in feeding the two elements. As a result, collinear elements are almost always operated with their ends quite close together—in wire antennas, usually with just a strain insulator between.

With very small spacing between the ends of adjacent elements the theoretical power gain of collinear arrays is approximately as follows:

2 collinear elements—1.9 dB
3 collinear elements—3.2 dB
4 collinear elements—4.3 dB

More than four elements are rarely used.

Directivity

The directivity of a collinear array, in a plane containing the axis of the array, increases with its length. Small secondary lobes appear in the pattern when more than two elements are used, but the amplitudes of these lobes are low enough so that they are not important. In a plane at right angles to the array the directive diagram is a circle, no matter what the number of elements. Collinear operation, therefore, affects only E-plane directivity, the plane containing the antenna. At right angles to the wire the pattern is the same as that of the individual ½-λ elements of which it is composed.

When a collinear array is mounted with the elements vertical, the antenna radiates equally well in all geographical directions. An array of such "stacked" collinear elements tends to confine the radiation to low vertical angles.

If a collinear array is mounted horizontally, the directive pattern in the vertical plane at right angles to the array is the same as the vertical pattern of a simple ½-λ antenna at the same height (Chapter 2).

TWO-ELEMENT ARRAY

The simplest and most popular collinear array is one using two elements, as shown in Fig 39. This system is commonly known as "two half-waves in phase." The manner in which the desired current distribution is obtained is described in Chapter 26. The directive pattern in a plane containing the wire axis is shown in Fig 40.

Depending on the conductor size, height, and similar factors, the impedance at the feed point can be expected to be in the range from about 4 to 6 kΩ, for wire antennas. If the elements are made of tubing having a low λ/dia

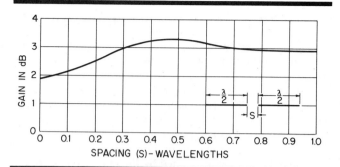

Fig 38—Gain of two collinear ½-λ elements as a function of spacing between the adjacent ends.

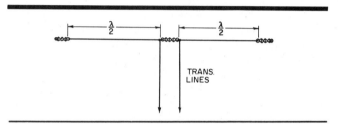

Fig 39—A two-element collinear array (two half-waves in phase). The transmission line shown would operate as a tuned line. A matching section can be substituted and a nonresonant line used if desired.

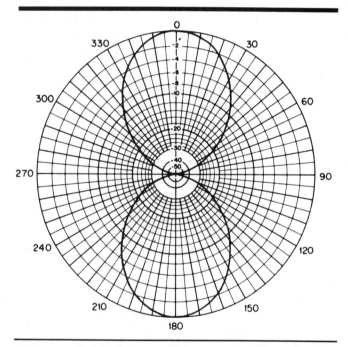

Fig 40—Free-space E-plane directive diagram for a two-element collinear array. The axis of the elements lies along the 90°-270° line. This is the horizontal pattern at low wave angles when the array is horizontal. The array gain is approximately 1.9 dBd.

(wavelength to diameter) ratio, values as low as 1 kΩ are representative. The system can be fed through an open-wire tuned line with negligible loss for ordinary line lengths, or a matching section may be used if desired.

THREE- AND FOUR-ELEMENT ARRAYS

When more than two collinear elements are used it is necessary to connect "phasing" stubs between adjacent elements in order to bring the currents in all elements in phase. It will be recalled from Chapter 2 that in a long wire the direction of current flow reverses in each ½-λ section. Consequently, collinear elements cannot simply be connected end to end; there must be some means for making the current flow in the same direction in all elements. In Fig 41A the direction of current flow is correct in the two left-hand elements because the transmission line is connected between them. The phasing stub between the second and third elements makes the instantaneous current direction correct in the third element. This stub may be looked upon simply as the alternate ½-λ section of a long-wire antenna folded back on itself to cancel its radiation. In Fig 41A the part to the right of the transmission line has a total length of three half wavelengths, the center half wave being folded back to form a ¼-λ phase-reversing stub. No data are available on the impedance at the feed point in this arrangement, but various considerations indicate that it should be over 1 kΩ.

An alternative method of feeding three collinear elements is shown in Fig 41B. In this case power is applied at the center of the middle element and phase-reversing stubs are used between this element and both of the outer elements. The impedance at the feed point in this case is somewhat over 300 Ω and provides a close match to 300-Ω line. The SWR will be less than 2 to 1 when 600-Ω line is used. Center feed of this type is somewhat preferable to the arrangement in Fig 41A because the system as a whole is balanced. This assures more uniform power distribution among the elements. In A, the right-hand element is likely to receive somewhat less power than the other two because a portion of the fed power is radiated by the middle element before it can reach the element located at the extreme right.

A four-element array is shown in Fig 41C. The system is symmetrical when fed between the two center elements as shown. As in the three-element case, no data are available on the impedance at the feed point. However, the SWR with a 600-Ω line should not be much over 2 to 1. Fig 42 shows the directive pattern of a four-element array. The sharpness of the three-element pattern is intermediate between Figs 40

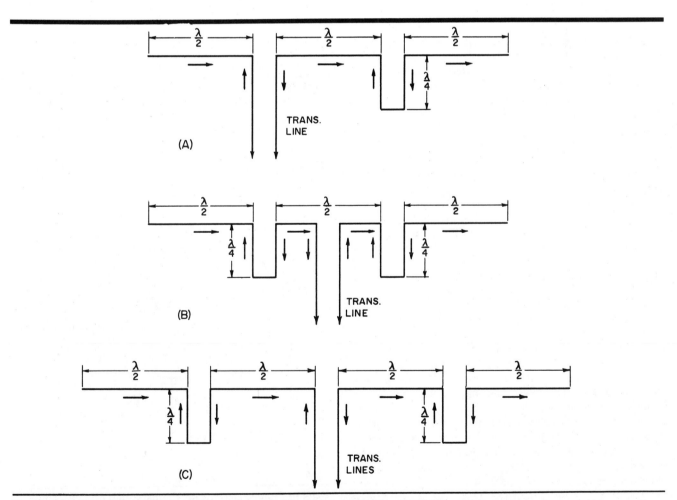

Fig 41—Three- and four-element collinear arrays. Alternative methods of feeding a three-element array are shown at A and B. These drawings also show the current distribution on the antenna elements and phasing stubs. A matched transmission line can be substituted for the tuned line by using a suitable matching section.

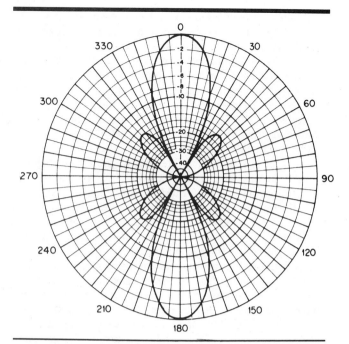

Fig 42—E-plane pattern for a four-element collinear array. The axis of the elements lies along the 90°-270° line. The array gain is approximately 4.3 dBd.

and 42, with four small minor lobes at 25° off the array axis.

Collinear arrays can be extended to more than four elements. However, the simple two-element collinear array is the type most used, since it lends itself well to multiband operation. More than two collinear elements are seldom used because more gain can be obtained from other types of arrays.

Adjustment

In any of the collinear systems described the lengths of the radiating elements in feet can be found from the formula $468/f_{MHz}$. The lengths of the phasing stubs can be found from the equations given in Chapter 26 for the type of line used. If the stub is open-wire line (500 to 600 Ω impedance) it is satisfactory to use a velocity factor of 0.975 in the formula for a ¼-λ line. On-the-ground adjustment is, in general, an unnecessary refinement. If desired, however, the following procedure may be used when the system has more than two elements.

Disconnect all stubs and all elements except those directly connected to the transmission line (in the case of feed such as is shown in Fig 41B leave only the center element connected to the line). Adjust the elements to resonance, using the still-connected element. When the proper length is determined, cut all other elements to the same length. Make the phasing stubs slightly long and use a shorting bar to adjust their length. Connect the elements to the stubs and adjust the stubs to resonance, as indicated by maximum current in the shorting bars or by the SWR on the transmission line. If more than three or four elements are used it is best to add elements two at a time (one at each end of the array), resonating the system each time before a new pair is added.

THE EXTENDED DOUBLE ZEPP

An expedient that may be adopted to obtain the higher gain that goes with wider spacing in a simple system of two collinear elements is to make the elements somewhat longer than ½ λ. As shown in Fig 43, this increases the spacing between the two in-phase ½-λ sections at the ends of the wires. The section in the center carries a current of opposite phase, but if this section is short the current will be small; it represents only the outer ends of a ½-λ antenna section. Because of the small current and short length, the radiation from the center is small. The optimum length for each element is 0.64 λ. At greater lengths the system tends to act as a long-wire antenna, and the gain decreases.

This system is known as the "extended double Zepp." The gain over a ½-λ dipole is approximately 3 dB, as compared with slightly less than 2 dB for two collinear ½-λ dipoles. The directional pattern in the plane containing the axis of the antenna is shown in Fig 44. As in the case of all other collinear arrays, the free-space pattern in the plane at right angles to the antenna elements is the same as that of a ½-λ antenna—circular.

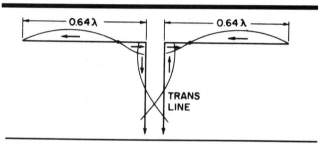

Fig 43—The extended double Zepp. This system gives somewhat more gain than two ½-λ collinear elements.

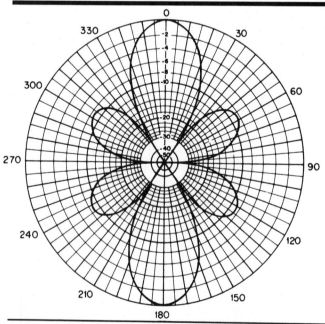

Fig 44—E-plane pattern for the extended double Zepp. This is also the horizontal directional pattern when the elements are horizontal. The axis of the elements lies along the 90°-270° line. The array gain is approximately 3 dBd.

Broadside Arrays

To obtain broadside directivity with parallel elements the currents in the elements must all be in phase. At a distant point lying on a line perpendicular to the axis of the array and also perpendicular to the plane containing the elements, the fields from all elements add up in phase. The situation is similar to that pictured in Fig 1 in this chapter.

Broadside arrays of this type theoretically can have any number of elements. However, practical limitations of construction and available space usually limit the number of broadside parallel elements to two, in the amateur bands below 30 MHz, when horizontal polarization is used. More than four such elements seldom are used even at VHF.

Power Gain

The power gain of a parallel-element broadside array depends on the spacing between elements as well as on the number of elements. The way in which the gain of a two-element array varies with spacing is shown in Fig 45. The greatest gain is obtained when the spacing is in the vicinity of ⅔ λ.

The theoretical gains of broadside arrays having more than two elements are approximately as follows:

No. of Parallel Elements	dB Gain with ½-λ Spacing	dB Gain with ¾-λ Spacing
3	5	7
4	6	8.5
5	7	10
6	8	11

The elements must, of course, all lie in the same plane and

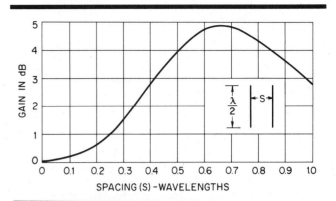

Fig 45—Gain as a function of the spacing between two parallel elements operated in phase (broadside).

all must be fed in phase.

Directivity

The sharpness of the directive pattern depends on spacing between elements and on the number of elements. Larger element spacing will sharpen the main lobe, for a given number of elements. The two-element array has no minor lobes when the spacing is ½ λ, but small minor lobes appear at greater spacings, as indicated in Fig 10 of this chapter. When three or more elements are used the pattern always has minor lobes.

End-Fire Arrays

The term "end-fire" covers a number of different methods of operation, all having in common the fact that the maximum radiation takes place along the array axis, and that the array consists of a number of parallel elements in one plane. End-fire arrays can be either bidirectional or unidirectional. In the bidirectional type commonly used by amateurs there are only two elements, and these are operated with currents 180° out of phase. Even though adjustment tends to be complicated, unidirectional end-fire driven arrays have also seen amateur use, primarily as a pair of phased, ground-mounted ¼-λ vertical elements. Extensive discussions of this array are contained in earlier sections of this chapter.

Horizontally polarized unidirectional end-fire arrays see little amateur use except in logarithmic periodic arrays (described in Chapter 10). Instead, horizontally polarized unidirectional arrays usually have parasitic elements (described in Chapter 11).

TWO-ELEMENT ARRAY

In the two-element array with equal currents out of

phase, the gain varies with the spacing between elements as shown in Fig 46. The maximum gain occurs in the neighborhood of 1/8-λ spacing. Below 0.05-λ spacing the gain decreases rapidly, since the system is approaching the spacing used for nonradiating transmission lines.

The feed point resistance for either element is very low at the spacings giving the greatest gain, as shown by Fig 8 earlier in this chapter. The spacings most frequently used are 1/8 and 1/4 λ, at which the resistances of center-fed 1/2-λ elements are about 8 and 32 Ω, respectively. When the spacing is 1/8 λ it is advisable to use good-sized conductors—preferably tubing—for the elements because with the feed point resistance so low the heat losses in the conductor can represent an appreciable portion of the power supplied to the antenna. Excessive conductor losses will mean that the theoretical gain cannot be realized, as discussed in earlier sections.

UNIDIRECTIONAL END-FIRE ARRAYS

Two parallel elements spaced ¼ λ apart and fed equal

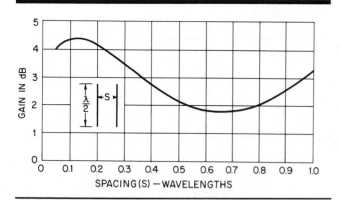

Fig 46—Gain of an end-fire array consisting of two elements fed 180° out of phase, as a function of the spacing between elements. Maximum radiation is in the plane of the elements and at right angles to them at spacings up to ½ λ, but the direction changes at greater spacings.

opposite direction the fields from the two elements cancel.

When the currents in the elements are neither in phase nor 180° out of phase, the feed point resistances of the elements are not equal. This complicates the problem of feeding equal currents to the elements, as treated in earlier sections. If the currents are not equal, one or more minor lobes will appear in the pattern and decrease the front-to-back ratio.

More than two elements can be used in a unidirectional end-fire array. The requirement for unidirectivity is that there must be a progressive phase shift in the element currents equal to the spacing, in electrical degrees, between the elements. The amplitudes of the currents in the various elements also must be properly related. This requires "binomial" current distribution—that is, the ratios of the currents in the elements must be proportional to the coefficients of the binomial series. In the case of three elements, this requires that the current in the center element be twice that in the two outside elements, for 90° (¼-λ) spacing and element current phasing. This antenna has an overall length of ½ λ. The directive diagram is shown in Fig 48. The pattern is similar to that of Fig 47, but the three-element binomial array has greater directivity, evidenced by the narrower half-power beamwidth (148° versus 180°).

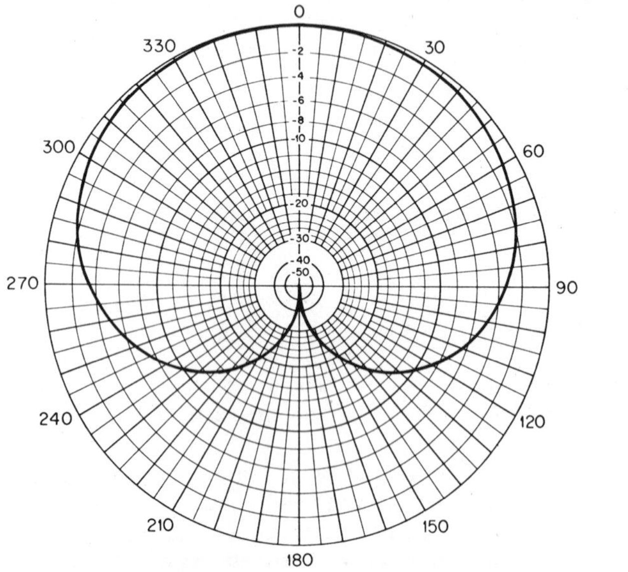

currents 90° out of phase will have a directional pattern, in the plane at right angles to the plane of the array, as represented in Fig 47. The maximum radiation is in the direction of the element in which the current lags. In the

Fig 47—Representative H-plane pattern for a two-element end-fire array with 90° spacing and phasing. The elements lie along the vertical axis, with the uppermost element the one of lagging phase.

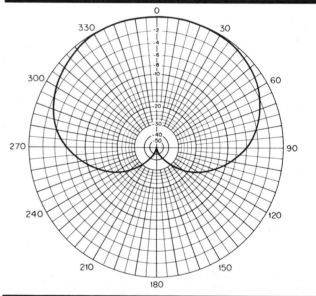

Fig 48—H-plane pattern for a three-element end-fire array with binomial current distribution (the current in the center element is twice that in each end element). The elements are spaced ¼ λ apart along the 0°-180° axis. The center element lags the lower element by 90°, while the upper element lags the lower element by 180° in phase.

Combination Driven Arrays

Broadside, end-fire and collinear elements can readily be combined to increase gain and directivity, and this is in fact usually done when more than two elements are used in an array. Combinations of this type give more gain, in a given amount of space, than plain arrays of the types just described. The combinations that can be worked out are almost endless, but in this section are described only a few of the simpler types.

The accurate calculation of the power gain of a multi-element array requires a knowledge of the mutual impedances between all elements, as discussed in earlier sections. For approximate purposes it is sufficient to assume that each *set* (collinear, broadside, end-fire) will have the gains as given earlier, and then simply add up the gains for the combination. This neglects the effects of cross-coupling between sets of elements. However, the array configurations are such that the mutual impedances from cross-coupling should be relatively small, particularly when the spacings are ¼ λ or more, so the estimated gain should be reasonably close to the actual gain.

FOUR-ELEMENT END-FIRE AND COLLINEAR ARRAY

The array shown in Fig 49 combines collinear in-phase elements with parallel out-of-phase elements to give both broadside and end-fire directivity. It is popularly known as a "two-section W8JK" or "two-section flat-top beam." The approximate gain calculated as described above is 6.2 dB with 1/8-λ spacing and 5.7 dB with 1/4-λ spacing. Directive patterns are given in Figs 50 and 51.

The impedance between elements at the point where the phasing line is connected is of the order of several thousand ohms. The SWR with an unmatched line consequently is quite high, and this system should be constructed with open-wire line (500 or 600 Ω) if the line is to be resonant. With ¼-λ element spacing the SWR on a 600-Ω line is estimated to be in the vicinity of 3 or 4 to 1.

To use a matched line, a closed stub 3/16 λ long can be connected at the transmission-line junction shown in Fig 49, and the transmission line itself can then be tapped on this matching section at the point resulting in the lowest line SWR. This point can be determined by trial.

This type of antenna can be operated on two bands having a frequency ratio of 2 to 1, if a resonant feed line is used. For example, if designed for 28 MHz with 1/4-λ spacing

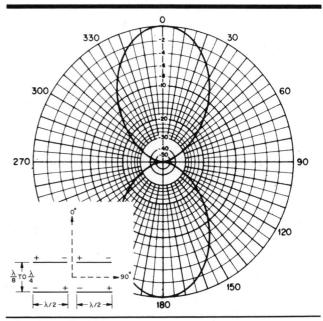

Fig 50—E-plane pattern for the antenna shown in Fig 49. The elements are parallel to the 90°-270° line in this diagram. Less than a 1° change in half-power beamwidth results when the spacing is changed from 1/8 to 1/4 λ.

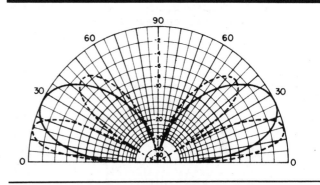

Fig 51—Vertical pattern for the four-element antenna of Fig 49 when mounted horizontally. Solid curve, height ½ λ; broken curve, height 1 λ above a perfect conductor. Fig 50 gives the horizontal pattern.

between elements it can be operated on 14 MHz as a simple end-fire array having 1/8-λ spacing.

FOUR-ELEMENT BROADSIDE ARRAY

The four-element array shown in Fig 52 is commonly known as the "lazy H." It consists of a set of two collinear elements and a set of two parallel elements, all operated in phase to give broadside directivity. The gain and directivity will depend on the spacing, as in the case of a simple parallel-element broadside array. The spacing may be chosen between the limits shown on the drawing, but spacings below 3/8 λ

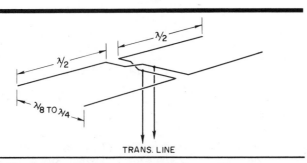

Fig 49—A four-element array combining collinear broadside elements and parallel end-fire elements, popularly known as the W8JK array.

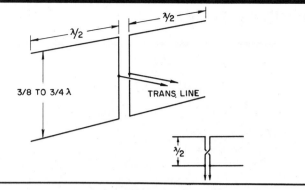

Fig 52—Four-element broadside array ("lazy H") using collinear and parallel elements.

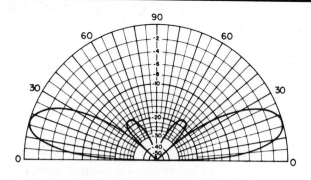

Fig 54—Vertical pattern of the four-element broadside antenna of Fig 52, when mounted with the elements horizontal and the lower set ½ λ above a perfect conductor. "Stacked" arrays of this type give best results when the lowest elements are at least ½ λ high. The gain is reduced and the wave angle raised if the lowest elements are close to ground.

are not worthwhile because the gain is small. Estimated gains are as follows

3/8-λ spacing—4.4 dB
1/2-λ spacing—5.9 dB
5/8-λ spacing—6.7 dB
3/4-λ spacing—6.6 dB

Half-wave spacing is generally used. Directive patterns for this spacing are given in Figs 53 and 54.

With ½-λ spacing between parallel elements, the impedance at the junction of the phasing line and transmission line is resistive and is in the vicinity of 100 Ω. With larger or smaller spacing the impedance at this junction will be reactive as well as resistive. Matching stubs are recommended in cases where a nonresonant line is to be used. They may be calcu-

lated and adjusted as described in Chapter 26.

The system shown in Fig 52 may be used on two bands having a 2-to-1 frequency relationship. It should be designed for the higher of the two frequencies, using 3/4-λ spacing between parallel elements. It will then operate on the lower frequency as a simple broadside array with 3/8-λ spacing.

An alternative method of feeding is shown in the small diagram in Fig 52. In this case the elements and the phasing line must be adjusted exactly to an electrical half wavelength. The impedance at the feed point will be resistive and of the order of 2 kΩ.

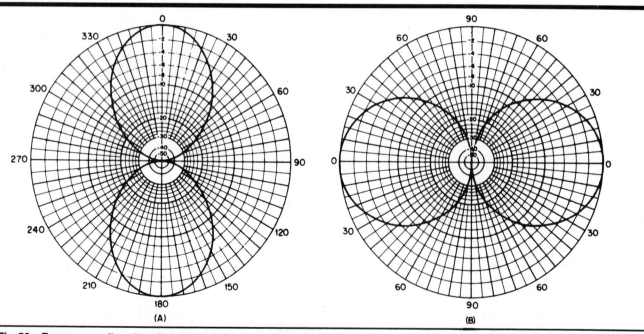

Fig 53—Free-space directive diagrams of the four-element antenna shown in Fig 52. At A is the E-plane pattern, the horizontal directive pattern at low wave angles when the antenna is mounted with the elements horizontal. The axis of the elements lies along the 90°-270° line. At B is the free-space H-plane pattern, viewed as if one set of elements is above the other from the ends of the elements.

Other Forms of Multielement Driven Arrays

For those who have the available room, multielement arrays based on the broadside concept have something to offer. The antennas are large but of simple design and noncritical dimensions; they are also very economical in terms of gain per unit of cost.

Arrays of three and four elements are shown in Fig 55. In the three-element array with ½-λ spacing at A, the array is fed at the center. This is the most desirable point in that it tends to keep the power distribution among the elements uniform. However, the transmission line could be connected at either point B or C of Fig 55A, with only slight skewing of the radiation pattern.

When the spacing is greater than ½ λ, the phasing lines must be 1 λ long and are not transposed between elements. This is shown at B in Fig 55. With this arrangement, any element spacing up to 1 λ can be used, if the phasing lines can be folded as suggested in the drawing.

The four-element array at C is fed at the center of the system to make the power distribution among elements as uniform as possible. However, the transmission line could be connected at either point B, C, D or E. In this case the section of phasing line between B and D must be transposed in order to make the currents flow in the same direction in all elements. The four-element array at C and the three-element array at B have approximately the same gain when the element spacing in the array at B is ¾ λ.

An alternative feeding method is shown in Fig 55D. This system can also be applied to the three-element arrays, and will result in better symmetry in any case. It is necessary only to move the phasing line to the center of each element, making connection to both sides of the line instead of one only.

The free-space pattern for a four-element array with ½-λ spacing is shown in Fig 56. This is also approximately the pattern for a three-element array with ¾-λ spacing.

Larger arrays can be designed and constructed by following the phasing principles shown in the drawings. No accurate figures are available for the impedances at the various feed points indicated in Fig 55. It can be estimated to be in the vicinity of 1 kΩ when the feed point is at a junction between the phasing line and a ½-λ element, becoming smaller as the number of elements in the array is increased. When the feed point is midway between end-fed elements as in Fig 55C, the impedance of a four-element array as seen by the transmission line is in the vicinity of 200 to 300 Ω, with 600-Ω open wire phasing lines. The impedance at the feed point with the antenna shown at D should be about 1.5 kΩ.

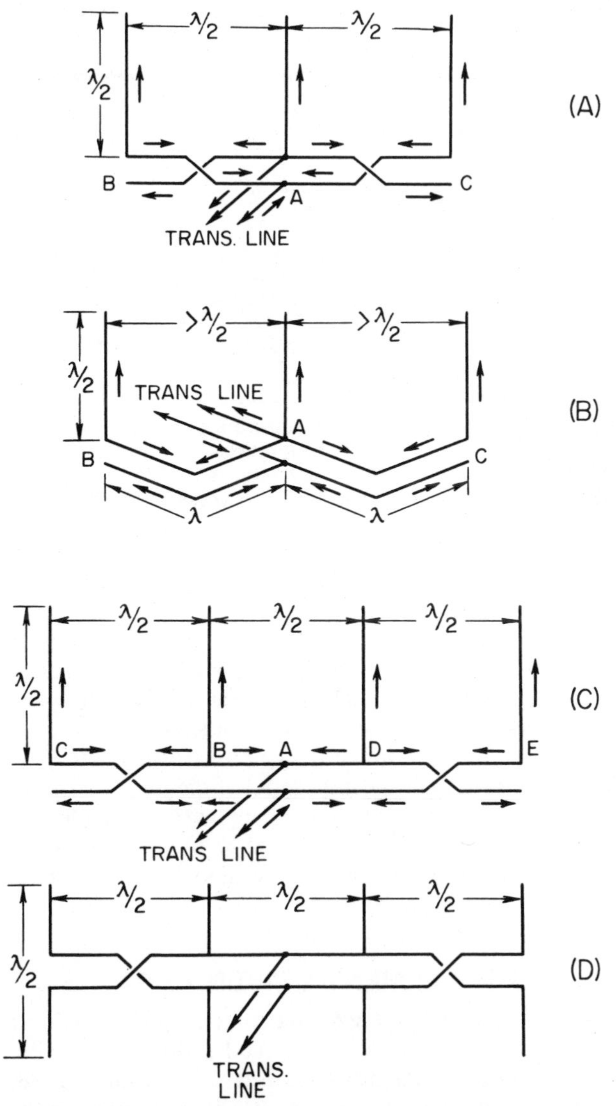

Fig 55—Methods of feeding three- and four-element broadside arrays with parallel elements.

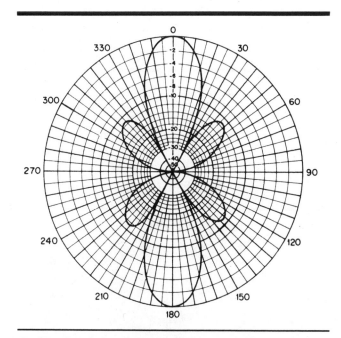

Fig 56—Free-space H-plane pattern of a four-element broadside array using parallel elements (Fig 55). This corresponds to the horizontal directive pattern at low wave angles for a vertically polarized array. The axis of the elements lies along the 90°-270° line.

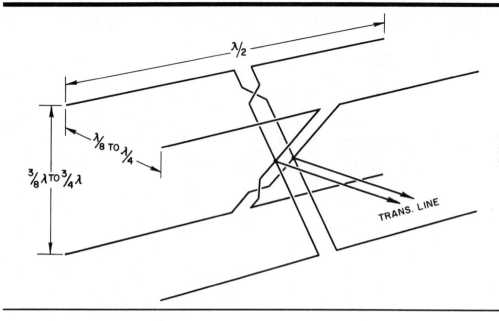

Fig 57—Four-element array combining both broadside and end-fire elements.

FOUR-ELEMENT BROADSIDE AND END-FIRE ARRAY

The array shown in Fig 57 combines parallel elements in broadside and end-fire directivity. Approximate gains can be calculated by adding the values from Figs 45 and 46 for the element spacings used. The smallest array (physically)—3/8-λ spacing between broadside and 1/8-λ spacing between end-fire elements—has an estimated gain of 6.8 dB and the largest—3/4- and 1/4-λ spacing, respectively—about 8.5 dB. The optimum element spacings are 5/8 λ broadside and 1/8 λ end-fire, giving an overall gain estimated at 9.3 dB. Directive patterns are given in Figs 58 and 59.

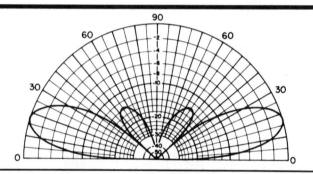

Fig 59—Vertical pattern of the antenna shown in Fig 57 at a mean height of ¾ λ (lowest elements ½ λ above a perfect conductor) when the antenna is horizontally polarized. For optimum gain and low wave angle the mean height should be at least ¾ λ.

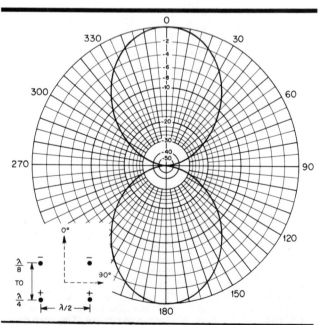

Fig 58—Free-space H-plane pattern of the four-element antenna shown in Fig 57.

The impedance at the feed point will not be purely resistive unless the element lengths are correct and the phasing lines are exactly a ½ λ long. (This requires somewhat less than ½-λ spacing between broadside elements.) In this case the impedance at the junction is estimated to be over 10 kΩ. With other element spacings the impedance at the junction will be reactive as well as resistive, but in any event the SWR will be quite large. An open-wire line can be used as a resonant line, or a matching section may be used for nonresonant operation.

EIGHT-ELEMENT DRIVEN ARRAY

The array shown in Fig 60 is a combination of collinear and parallel elements in broadside and end-fire directivity. The gain can be calculated as described earlier, using Figs 38, 45 and 46. Common practice is to use ½-λ spacing for the parallel broadside elements and 1/8-λ spacing for the end-fire elements. This gives an estimated gain of about 10 dB. Directive patterns for an array using these spacings are similar

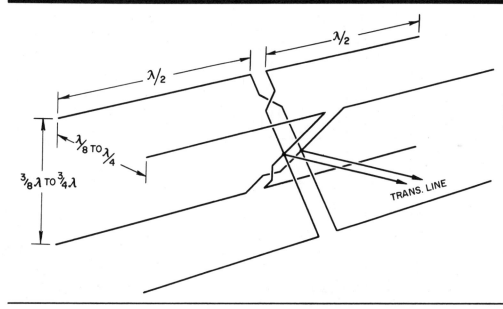

Fig 60—Eight-element driven array combining collinear and parallel elements for broadside and end-fire directivity.

to those of Figs 58 and 59, being somewhat sharper.

Although even approximate figures are not available, the SWR with this arrangement will be high. Matching stubs are recommended for making the line nonresonant. Their position and length can be determined as described in Chapter 26.

This system can be used on two bands related in frequency by a 2-to-1 ratio, providing it is designed for the higher of the two, with ¾-λ spacing between the parallel broadside elements and ¼-λ spacing between the end-fire elements. On the lower frequency it will then operate as a four-element antenna of the type shown in Fig 57, with 3/8-λ broadside spacing and 1/8-λ end-fire spacing. For two-band operation a resonant transmission line must be used.

OTHER DRIVEN SYSTEMS

Two other types of driven antennas are worthy of mention, although their use by amateurs has been rather limited. The Sterba array, shown at A in Fig 61, is a broadside radiator consisting of both collinear and parallel elements with ½-λ spacing between the latter. Its distinctive feature is the method of closing the ends of the system. For direct current and low-frequency ac, the system forms a closed loop, which is advantageous in that heating currents can be sent through the wires to melt the ice that forms in cold climates. There is comparatively little radiation from the vertical connecting wires at the ends because the currents are relatively small and are flowing in opposite directions with respect to the center (the voltage loops are marked with dots in this drawing).

The system obviously can be extended as far as desired. The approximate gain is the sum of the gains of one set of collinear elements and one set of broadside elements, counting the two ¼-λ sections at the ends as one element. The antenna shown, for example, is about equivalent to one set of four collinear elements and one set of two parallel broadside elements, so the total gain is approximately 4.3 + 4.0 = 8.3 dB. Horizontal polarization is the only practicable type at the

lower frequencies, and the lower set of elements should be at least ½ λ above ground for best results.

When fed at the point shown, the impedance is of the order of 600 Ω. Alternatively, this point can be closed and the system fed between any two elements, as at X. In this case a point near the center should be chosen so the power distribution among the elements will be as uniform as possible. The impedance at any such point will be 1 kΩ or less in systems with six or more elements.

The Bruce array is shown at B in Fig 61. It consists simply of a single wire folded so that the vertical sections carry large currents in phase while the horizontal sections carry small currents flowing in opposite directions with respect to the center of that section (indicated by the dots). The radiation consequently is vertically polarized. The gain is proportional to the length of the array but is somewhat smaller, because of the short radiating elements, than is obtainable from a broadside array of ½-λ parallel elements of the same overall length. The array should be two or more wavelengths long to achieve a worthwhile gain. The system can be fed at any current loop; these occur at the centers of the vertical wires.

Another form of the Bruce array is shown at C. Because the radiating elements have twice the height, the gain is increased. The system can be fed at the center of any of the connecting lines.

BOBTAIL CURTAIN

The antenna system of Fig 62 uses the principles of cophased verticals to produce a broadside, bidirectional pattern providing approximately 5 dB of gain over a single element. The antenna performs as three in-phase top-fed vertical radiators approximately ¼ λ in height and spaced approximately ½ λ. It is most effective for low-angle signals and makes an excellent long-distance antenna for either 3.5 or 7 MHz.

The three vertical sections are the actual radiating

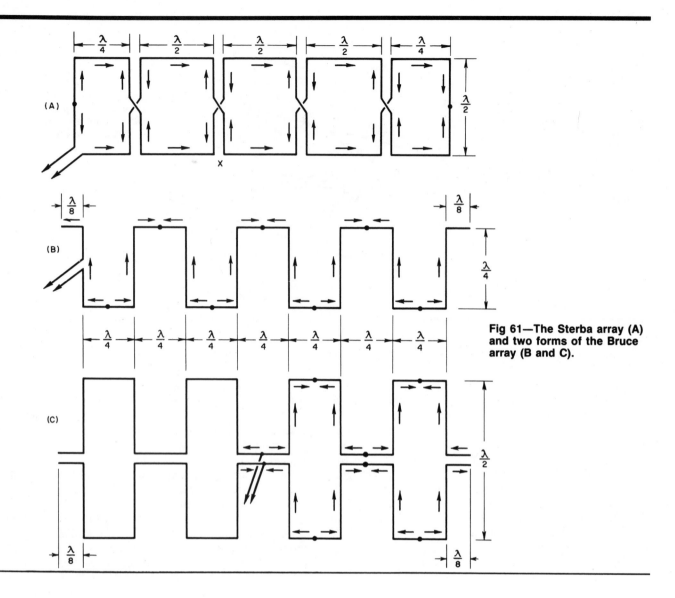

Fig 61—The Sterba array (A) and two forms of the Bruce array (B and C).

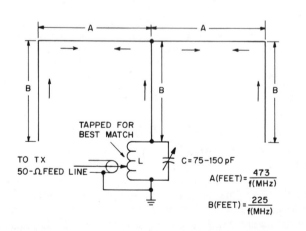

Fig 62—The bobtail curtain is an excellent low-angle radiator having broadside bidirectional characteristics. Current distribution is represented by the arrows. Dimensions A and B (in feet) can be determined from the equations.

$$A(FEET) = \frac{473}{f(MHz)}$$

$$B(FEET) = \frac{225}{f(MHz)}$$

components, but only the center element is fed directly. The two horizontal parts, A, act as phasing lines and contribute very little to the radiation pattern. Because the current in the center element must be divided between the end sections, the current distribution approaches a binomial 1:2:1 ratio. The radiation pattern is shown in Fig 63.

The vertical elements should be as vertical as possible. The height for the horizontal portion should be slightly greater than B, as shown in Fig 62. The tuning network is resonant at the operating frequency. The L/C ratio should be fairly low to provide good loading characteristics. As a starting point, a maximum capacitor value of 75 to 150 pF is recommended, and the inductor value is determined by C and the operating frequency. The network is first tuned to resonance and then the tap point is adjusted for the best match. A slight readjustment of C may be necessary. A link coil consisting of a few turns can also be used to feed the antenna.

PHASING ARROWS IN ARRAY ELEMENTS

In the antenna diagrams of preceding sections, the

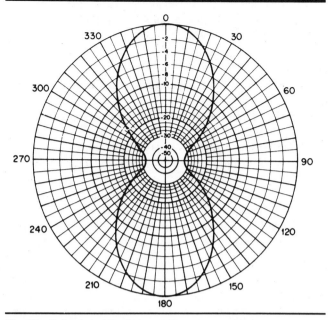

Fig 63—Calculated horizontal directive diagram of the antenna shown in Fig 62. The array lies along the 90°-270° axis.

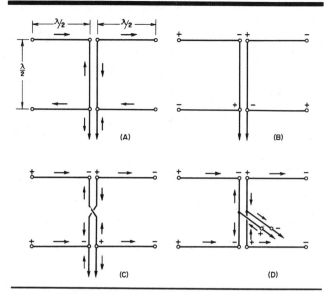

Fig 64—Methods of checking the phase of currents in elements and phasing lines.

relative direction of current flow in the various antenna elements and connecting lines is shown by arrows. In laying out any antenna system it is necessary to know that the phasing lines are properly connected; otherwise the antenna may have entirely different characteristics than anticipated. The phasing may be checked either on the basis of current direction or polarity of voltages. There are two rules to remember:

1) In every ½-λ section of wire, starting from an open end, the current directions reverse. In terms of voltage, the polarity reverses at each ½-λ point, starting from an open end.

2) Currents in transmission lines always must flow in opposite directions in adjacent wires. In terms of voltage, polarities always must be opposite.

Examples of the use of current direction and voltage polarity are given at A and B, respectively, in Fig 64. The ½-λ points in the system are marked by small circles. When current in one section flows toward a circle, the current in the next section must also flow toward it, and vice versa. In the four-element antenna shown at A, the current in the upper right-hand element cannot flow toward the transmission line, because then the current in the right-hand section of the phasing line would have to flow upward and thus would be flowing in the same direction as the current in the left-hand wire. The phasing line would simply act like two wires in parallel in such a case. Of course *all* arrows in the drawing could be reversed, and the net effect would be unchanged.

C shows the effect of transposing the phasing line. This transposition reverses the direction of current flow in the lower pair of elements, as compared with A, and thus changes the array from a combination collinear and end-fire arrangement into a collinear-broadside array.

The drawing at D shows what happens when the transmission line is connected at the center of a section of

phasing line. Viewed from the main transmission line, the two parts of the phasing line are simply in parallel, so the half wavelength is measured from the antenna element along the upper section of phasing line and thence along the transmission line. The distance from the lower elements is measured in the same way. Obviously the two sections of phasing line should be the same length. If they are not, the current distribution becomes quite complicated; the element currents are neither in phase nor 180° out of phase, and the elements at opposite ends of the lines do not receive the same power. To change the element current phasing at D into the phasing at A, simply transpose the wires in one section of the phasing line; this reverses the direction of current flow in the antenna elements connected to that section of phasing line.

BIBLIOGRAPHY

Source material and more extended discussion of topics covered in this chapter can be found in the references given below and in the textbooks listed at the end of Chapter 2.

D. W. Atchley, H. E. Stinehelfer and J. F. White, "360°-Steerable Vertical Phased Arrays," *QST*, Apr 1976, pp 27-30.

G. H. Brown, "Directional Antennas," *Proc IRE*, Vol 25, No. 1, Jan 1937, pp 78-145.

G. H. Brown, R. F. Lewis and J. Epstein, "Ground Systems as a Factor in Antenna Efficiency," *Proc IRE*, Jun 1937, pp 753-787.

G. H. Brown and O. M. Woodward, Jr, "Experimentally Determined Impedance Characteristics of Cylindrical Antennas," *Proc IRE*, Apr 1945.

A. Christman, "Feeding Phased Arrays: An Alternate Method," *Ham Radio*, May 1985, pp 58-59, 61-64.

F. Gehrke, "Vertical Phased Arrays," in six parts, *Ham Radio*, May-Jul, Oct and Dec 1983, and May 1984.

W. Hayward and D. DeMaw, *Solid State Design for the Radio Amateur* (Newington, CT: ARRL, 1977).

H. Jasik, *Antenna Engineering Handbook*, 1st ed. (New York: McGraw-Hill, 1961).

H. W. Kohler, "Antenna Design for Field-Strength Gain," *Proc IRE*, Oct 1944, pp 611-616.

J. D. Kraus, "Antenna Arrays with Closely Spaced Elements," *Proc IRE*, Feb 1940, pp 76-84.

J. D. Kraus, *Antennas* (New York: McGraw-Hill Book Co, 1950).

E. A. Laport, *Radio Antenna Engineering* (New York: McGraw-Hill Book Co, 1952).

J. L. Lawson, "Simple Arrays of Vertical Antenna Elements," *QST*, May 1971, pp 22-27.

P. H. Lee, *The Amateur Radio Vertical Antenna Handbook*, 1st ed. (Port Washington, NY: Cowan Publishing Corp, 1974).

R. W. Lewallen, "Baluns: What They Do and How They Do It," *The ARRL Antenna Compendium, Volume 1* (Newington, CT: ARRL, 1985).

M. W. Maxwell, "Some Aspects of the Balun Problem," *QST*, Mar 1983, pp 38-40.

J. Sevick, "The Ground-Image Vertical Antenna," *QST*, Jul 1971, pp 16-19, 22.

J. Sevick, "The W2FMI Ground-Mounted Short Vertical," *QST*, Mar 1973, pp 13-18, 41.

E. J. Wilkinson, "An N-Way Hybrid Power Divider," *IRE Transactions on Microwave Theory and Techniques*, Jan 1960.

Radio Broadcast Ground Systems, available from Smith Electronics, Inc, 8200 Snowville Rd, Cleveland, OH 44141.

Reference Data for Radio Engineers, 5th ed. (Indianapolis: Howard W. Sams & Co, subsidiary of ITT, 1968).

Chapter 9

Broadband Antennas

Antennas that provide a good impedance match over a wide frequency range have been a topic of interest to hams for many years. The advantages of a broadband antenna are obvious—fewer adjustments during tune-up. For some of the new broadband transceivers and amplifiers, a broadband antenna means no tune-up at all; after setting the band and frequency, one is "in business."

Bandwidth Factor

The traditional measure of antenna bandwidth in terms of impedance is the standing wave ratio or SWR in the feed line. Most modern amateur equipment is designed to work into a 52-Ω load with an SWR of better than 2:1. Therefore, the frequency range within which the SWR in 52-Ω line is less than 2:1 is often used for antenna bandwidth comparisons. However, any SWR range and any line impedance may be used for comparison, as long as they are clearly specified and are used consistently for the antennas being compared.

In making bandwidth comparisons with SWR, the operating frequency must also be taken into account, for a specified frequency bandwidth range in one amateur band will not apply if the antenna is scaled to another band. For instance, if a dipole antenna has a 2:1-SWR bandwidth of 500 kHz when cut for 14 MHz, it will not have a 500-kHz bandwidth when scaled for 3.5 MHz.

Expressing the 2:1-SWR bandwidth as a percentage is convenient, using the following relationship.

$$\text{SWR bandwidth} = \frac{f2 - f1}{f_c} \times 100\% \qquad \text{(Eq 1)}$$

where

    f1 is the lower frequency, above which the SWR is less than the specified limit

    f2 is the upper frequency, below which the SWR is less than the specified limit

    f_c is the center frequency, determined from

$$f_c = \sqrt{f1 \times f2} \qquad \text{(Eq 2)}$$

For example, the solid-line curve of Fig 1 shows the SWR versus frequency plot for a theoretical 3.75-MHz single-wire dipole in free space, fed with a 52-Ω line. The 2:1-SWR frequencies are 3.665 and 3.825 MHz. The 2:1-SWR bandwidth of this antenna is

$$\text{Bandwidth factor} = \frac{3.825 - 3.665}{\sqrt{3.825 \times 3.665}} \times 100 = 4.3\%$$

The bandwidth factor calculated in this way can be used if the antenna is scaled to another band and the same impedance feeder is used. If this 3.75-MHz dipole is scaled to operate at 14.2 MHz, the 2:1-SWR frequency range in that band should be $14.2 \times 4.3\% = 0.611$ MHz or 611 kHz.

It is important to note that the bandwidth percentage factor will not be the same if a different type of feeder is used. The broken-line curve of Fig 1 shows the SWR response of the very same antenna, but with a 75-Ω feeder. The reason for the different SWR response is that, while the antenna impedance is unchanged for each individual frequency, the degree of mismatch does change with another feeder impedance. In this case, the 2:1-SWR bandwidth factor becomes 5.5%.

The curves of Fig 1 and the information in the preceding paragraphs should not be interpreted to mean that 75-Ω line will necessarily provide a better match or broader SWR bandwidth with a dipole than 52-Ω line. A dipole near the earth will not have the same impedance as one in free space, and the curves indicate only that the free-space impedance is nearer 75 Ω than 52 Ω. The only conclusion that should be drawn from this presentation is that the SWR bandwidth of an antenna is dependent upon the type of feeder employed. Thus, comparison of bandwidths with different feed lines can be misleading.

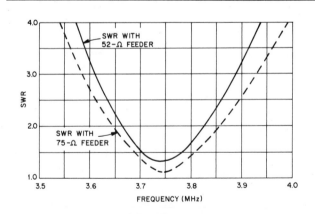

Fig 1—The SWR versus frequency plots for a hypothetical 3.75-MHz single-wire dipole in free space for 52- and 75-Ω feed lines. The 2:1-SWR bandwidth is 4.3% with a 52-Ω feeder and 5.5% with a 75-Ω feeder. These bandwidths will change as the antenna is brought near the earth.

Antenna Q

Another measure of antenna bandwidth is its Q. The method of determining antenna Q is discussed in Chapter 26. Briefly, however, the method requires knowing the values of antenna resistance and reactance at a known frequency percentage away from resonance. The Q of the dipole in the above example is approximately 13.

Antenna Q is independent of the feeder impedance and the band of operation. However, determining the Q of an antenna is somewhat nebulous, as the radiation resistance changes with frequency (but at a slower rate than the reactance changes). Calculating the Q from measurements at a frequency, say, 2% above resonance will generally not produce the same value as from a frequency 2% below resonance.

Another more serious difficulty arises in determining the Q of antenna systems which include some broadbanding schemes, such as those with resonating stubs or lumped LC constants at the antenna feed point. When such matching networks are used, the antenna may be resonant at more than one frequency within an amateur band. (Resonance is defined as the frequency or frequencies where the reactance at the feed point goes through zero.) With resonance ambiguities,

departure from resonance by a specific frequency percentage becomes meaningless.

THE CAGE DIPOLE

The bandwidth of a single-wire dipole may be increased by using a thick radiator, one with a large diameter. The radiator does not necessarily have to be solid; open construction such as shown in Fig 2 may be used.

The theoretical SWR response of a cage dipole having a 6-inch diameter is shown in Fig 3. The bandwidth factor of this antenna with 52-Ω line is 7.7%, and the Q is approximately 8. Its 2:1-SWR frequency range is 1.79 times broader than the antenna of Fig 1.

There are also other means of obtaining a thick radiator, thereby gaining greater bandwidth. The bow-tie and fan dipole make use of the same Q-lowering principle as the cage to obtain increased bandwidth.

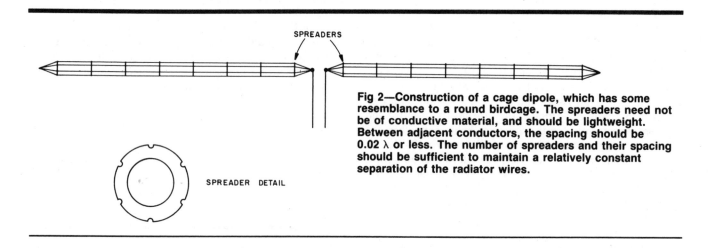

Fig 2—Construction of a cage dipole, which has some resemblance to a round birdcage. The spreaders need not be of conductive material, and should be lightweight. Between adjacent conductors, the spacing should be 0.02 λ or less. The number of spreaders and their spacing should be sufficient to maintain a relatively constant separation of the radiator wires.

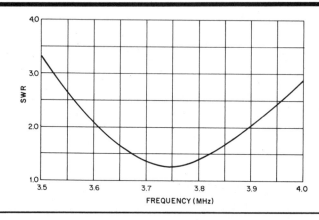

Fig 3—Theoretical SWR versus frequency response for a cage dipole of length 122 ft 6 in. and a spreader diameter of 6 in., fed with 52-Ω line. The 2:1-SWR bandwidth frequencies are 3.610 and 3.897 MHz, with a resulting bandwidth factor of 7.7%.

Efficiency Factor

The bandwidth factor as calculated in the preceding section does not include any indication of radiation efficiency. The efficiency factor is another very important consideration when it comes to broadbanded antenna systems. Unfortunate-

ly, most broadbanding schemes involve some loss of radiation efficiency.

Typically with broadbanding schemes, the efficiency falls off at the band edges as the SWR increases. Losses occur not only in the matching-system components, but also the losses increase in the transmission line as the SWR rises. These losses can amount to several dB. So in addition to the SWR versus frequency response, attention should be given to the efficiency versus frequency characteristic of any broadbanded antenna. Trade-offs must generally be made between SWR bandwidth and efficiency at the band edges.

QST for October 1986 contains an article by Frank Witt, AI1H, disclosing the results of computerized calculations of dipole bandwidth versus efficiency for various broadbanding schemes. (See the bibliography at the end of this chapter.) Information in this section is based on that article.

When the SWR at the antenna end of the transmission line is less than 2:1, the transmission-line losses are virtually the same as those from the length of a matched line. Thus, for the part of the band over which the SWR is less than 2:1, one need consider only losses in the matching network when computing efficiency.

Efficiency is related to resistive or ohmic losses in the matching network. The lower the losses, the higher the efficiency. However, ohmic losses in the matching network will broaden the response of a dipole system beyond that possible with a lossless or ideal matching network. Users must decide whether they are willing to accept the lower efficiency in trade for the increased bandwidth.

An extreme degree of bandwidth broadening is illustrated in Fig 4. The broadening is accomplished by adding resistive losses. One may resort to network theory and derive the RLC (resistor, inductor, capacitor) matching network shown. The network provides the complement of the antenna impedance. Note that the SWR is virtually 1:1 over the entire band, but the efficiency falls off dramatically away from resonance.

The efficiency loss may be converted to decibels from:

$$dB \; (loss) \; = \; -10 \; log \; \frac{Efficiency}{100} \qquad (Eq \; 3)$$

From this, the band-edge efficiency of 25 or 30% shown in Fig 4 means that the antenna has about 5 dB of loss relative to an ideal dipole. Also note that at the band edges, 70 to 75% of the power delivered down the transmission line from the transmitter is heating up the matching-network resistor. For a 1-kW output level, the resistor must have a power rating of at least 750 watts! Use of an RLC complementary network for broadbanding is not recommended, but it does illustrate how resistance (or losses) in the matching network can significantly increase the apparent antenna SWR bandwidth.

Resonators as Matching Networks

The most practical broadbanding network for a dipole is the parallel LC tuned circuit connected directly across the antenna terminals. This circuit may be constructed either with lumped constants, by placing a coil in parallel with a capacitor at the feed point, or by using one or more coaxial resonator stubs at the feed point.

THE DOUBLE BAZOOKA

The response of the somewhat controversial double bazooka antenna is shown in Fig 5. This antenna actually consists of a dipole with two quarter-wave coaxial resonator stubs connected in series.

Not much bandwidth enhancement is provided by this resonator connection because the impedance of the matching network is too high. With a 72-Ω feeder, this antenna offers a 2:1-SWR bandwidth frequency range that is only 1.14 times that of a simple dipole with the same feeder.

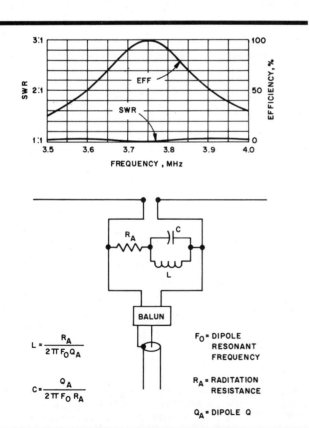

Fig 4—Matching the dipole with a complementary RLC network greatly improves the SWR characteristics, nearly 1:1 across the 3.5-MHz band. However, the relative loss at the band edges is greater than 5 dB.

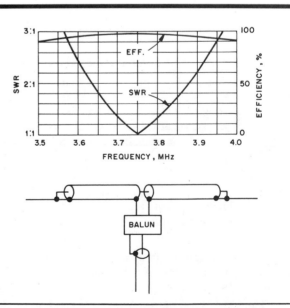

Fig 5—The double bazooka, sometimes called a coaxial dipole. The antenna is self-resonant at 3.75 MHz. The resonator stubs are 43.23-ft lengths of RG-58A coax.

THE CROSSED DOUBLE BAZOOKA

A modified version of the double-bazooka antenna is shown in Fig 6. In this case, the impedance of the matching network is reduced to one-fourth of the impedance of the standard double-bazooka network. The lower impedance provides more reactance correction, and hence increases the bandwidth frequency range noticeably, to 1.55 times that of a simple dipole. Notice, however, that the efficiency of the antenna drops to about 80% at the 2:1-SWR points. This amounts to a loss of approximately 1 dB. The broadbanding, in part, is caused by the resistive losses in the coaxial resonator stubs, which have a remarkably low Q (only 20).

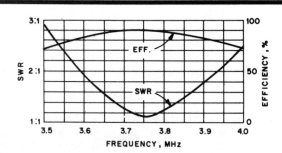

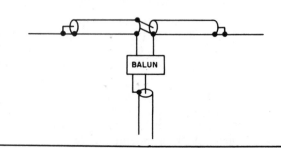

Fig 6—The crossed double bazooka yields bandwidth improvement by using two quarter-wave resonators, parallel connected, as a matching network.

The Q of Coaxial Resonators

The Q that can be acquired when resonators are made from coaxial cable is a parameter of interest. Table 1 summarizes the resonator Q that can be obtained from different types of coax at 1.9 and 3.75 MHz. If the cable loss is known, the Q of the resonator may be determined from

$$Q = \frac{278 \, f_c}{A \times VF} \qquad \text{(Eq 4)}$$

where
f_c = dipole resonant frequency, MHz
A = line attenuation per 100 ft, dB
VF = velocity factor of line

For example, RG-8 foam coax has a velocity factor of 80% and an attenuation of 0.3 dB at 3.75 MHz. The Q is calculated as

$$Q = \frac{278 \times 3.75}{0.3 \times 80} = 43.4$$

Table 1
Resonator Q for Various Types of Coaxial Cable

Cable Type	Resonator Q	
	3.75 MHz	1.9 MHz
RG-174	6.5	4.5
RG-58A	20.0	15.1
RG-141	22.5	16.1
RG-8	41.0	30.6
½-in. Hardline	75.5	53.9
¾-in. Hardline	109.1	77.1

Chebyshev Matching

It is possible to widen the bandwidth further by again resorting to network theory. However, in contrast to matching with a complementary network (Fig 4), no resistors are used. The matching network parameters are chosen to yield a Chebyshev (often called equi-ripple) approximation. The simplest way to make use of this theory for broadbanding the dipole is to deliberately mismatch the dipole at the center of the band by adding a transformer to the matching network. This transformer must provide a voltage step-up between the transmission line and the antenna. The result is a W-shaped SWR characteristic. Low SWR is sacrificed at the band center to obtain greater bandwidth.

A broadband dipole using a Chebyshev matching network with a step-up transformer is shown in Fig 7. The transformer can also serve as a balun. The SWR is better than

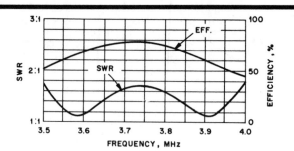

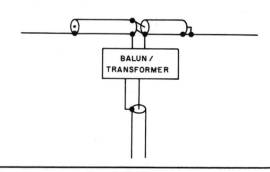

Fig 7—Chebyshev matching provides greater broadbanding by trading midband matching for increased bandwidth. The feed arrangement is 52-Ω line and a 1.91:1 step-up transformer balun. Different lengths of RG-58A resonator stubs are used, one of 30.75 ft and one of 12.49 ft. Note that the longer stub is left open at the outer end. (Design by Frank Witt, AI1H)

1.8:1 over the entire 3.5-MHz band—not bad for about 43 feet of RG-58 coax and a slightly modified balun.

Can this be true? Are we getting something for nothing? Not really. Notice that the efficiency in Fig 7 falls to only 45% and 52% at the 3.5-MHz band edges. Only half of the available power is radiated. This low efficiency is directly attributable to the low Q of the coaxial resonator.

LC MATCHING NETWORK

Efficiency can be improved by using lower-loss coax or by using a matching network made up of a high-Q inductor-capacitor parallel-tuned circuit. The SWR response and efficiency offered by a network of lumped constants is shown in Fig 8. The 2:1-SWR bandwidth with 52-Ω line is 460 kHz,

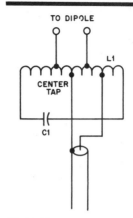

Fig 9—A practical LC matching network which provides reactance compensation, impedance transformation and balun action.

C1—400 pF transmitting mica rated at 3000 V, 4 A (RF).

L1—4.5 µH, 8½ turns of B&W coil stock, type 3029 (6 turns per in., 2½-in. dia, no. 12 wire). The primary and secondary portions of the coil have 1¾ and 3 turns, respectively.

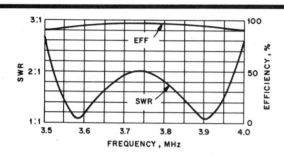

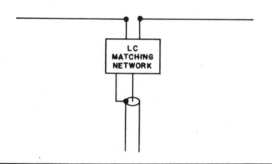

Fig 10—A method of constructing the AI1H LC matching network. See Figs 8 and 9. Components must be chosen for a high Q and must have adequate voltage and current ratings. The network is designed for use at the antenna feed point, and should be housed in a weatherproof package.

Fig 8—Efficient broadbanding with an LC matching network. The feeder is 52-Ω coax, and the matching network provides a step-up ratio of 2.8:1. See Fig 9 for details of the matching network. (Design by Frank Witt, AI1H)

not quite great as that provided by the coaxial resonator in Fig 7. The LC network uses deliberate mismatching at band center, resulting in the W-shaped SWR characteristic. The network also functions as a balun. The capacitor is connected across the entire coil in order to obtain practical element values.

The efficiency at the band edges for the antenna system shown in Fig 8 is 90%, compared to 45% and 52% for that of Fig 7. However, the increase in efficiency is obtained at the expense of bandwidth, as noted above. Unfortunately, the very low impedance required cannot be easily realized with practical inductor-capacitor values. It is for this reason that a form of impedance transformation is used.

A practical circuit for the LC matching network is shown in Fig 9, and Fig 10 shows a method of construction. The taps on L1 serve to reduce the impedance of the matching network,

while still permitting the use of practical element values. L1 is resonated at midband with C1.

The selection of a capacitor for this application must be made carefully, especially if high power is to be used. For the capacitor described in the caption of Fig 9, the allowable peak power (limited by the breakdown voltage) is 2450 watts. However, the allowable average power (limited by the RF current rating) is only 88 watts! These limits apply at the 1.75:1 SWR points.

Another Version

With slight alteration, the antenna system of Fig 8 can provide the performance indicated in Fig 11. Dubbed the 80-meter DXer's delight, this antenna has SWR minima near 3.5 and 3.8 MHz. A single antenna permits operation with a near-perfect match in the DX portions of the band, both CW and phone.

The modifications involve resonating the antenna and the LC network at a lower frequency, 3.67 MHz instead of 3.75. This requires 4.7 µH of inductance, rather than 4.5. In addition, the impedance step-up ratio is altered from 2.8:1 to 2:1. This is accomplished by setting the dipole taps on L1 for 2½ turns, rather than 3.

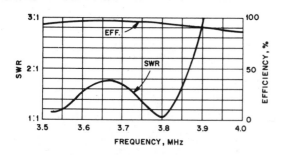

Fig 11—The 80-meter DXer's delight permits operation with a near-perfect match in the DX portions of the band, both CW and phone. See text for alterations required from the antenna system of Figs 8 and 9.

Bandwidth Versus Efficiency Trade-Off

As is apparent in the preceding section, there is clearly a trade-off between bandwidth enhancement and efficiency. This is true because the broadbanding results from two causes: reactance compensation and resistive loading. Pure reactance compensation would be achieved with resonators having infinite Q. The resistive loading caused by nonideal resonators further increases the bandwidth, but the price paid is that some of the output power heats up the resonator, leading to a loss in efficiency.

The best one can do with 100% efficiency is to double the bandwidth. Larger improvements are accompanied by efficiency loss. For example, a tripling of the bandwidth would be obtained with an efficiency of only 38% at the 2:1-SWR points.

THE SNYDER ANTENNA

A commercially manufactured antenna utilizing the principles described in the preceding section is the Snyder dipole. Patented by Richard D. Snyder in late 1984 (see bibliography), it immediately received much public attention through articles that Snyder published. Snyder's claimed performance for the antenna is a 2:1 SWR bandwidth of 20% with high efficiency.

The configuration of the Snyder antenna is like that of Fig 6, with 25-ohm line used for the resonators. The antenna is fed with 52-ohm line through a 2:1 balun, and exhibits a W-shaped SWR characteristic like that of Fig 7. The SWR at band center, based on information in the patent document, is 1.7 to 1. There is some controversy in professional circles regarding the claims for the Snyder antenna.

STAGGER TUNED DIPOLES

A single-wire dipole exhibits a relatively narrow bandwidth in terms of coverage for the 3.5-MHz band. A technique that has been used for years to cover the entire band is to have two dipoles, one cut for the CW portion and one for the phone portion. Of course separate antennas with separate feed lines may be used, but it is more convenient to connect the dipoles in parallel at the feed point and use a single feeder. This technique is known as stagger tuning.

Fig 12 shows the theoretical SWR response of a pair of stagger tuned dipoles fed with 52-Ω line. No mutual coupling

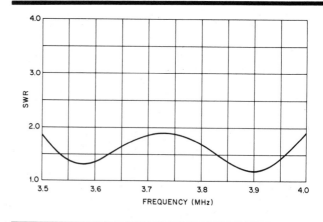

Fig 12—Theoretical SWR response of two stagger tuned dipoles. They are connected in parallel at the feed point and fed with 52-Ω line. The dipoles are of wire such as no. 12 or no. 14, with total lengths of 119 and 132 ft.

between the wires is assumed, a condition that would exist if the two antennas were at right angles to one another. As Fig 12 shows, the SWR response is less than 1.9 to 1 across the entire band.

A difficulty with crossed dipoles is that four supports are required if the antennas are to be horizontal. A more common arrangement is to use inverted V dipoles with just one support, at the apex of each element. The radiator wires can also act as guy lines for the supporting mast.

When the dipoles are crossed at something other than a right angle, mutual coupling between them comes into play. This causes interaction between the two elements—tuning of one by length adjustment will affect the tuning of the other. The interaction becomes most critical when the two dipoles are run parallel to each other, suspended by the same supports, and the wires are close together. Finding the optimum length for each dipole for total band coverage can become a tedious and frustrating process.

THE CONICAL MONOPOLE ANTENNA

A trapless vertical antenna which works well over a 4-to-1 frequency range should appeal to users of 3.5 and 7 MHz. It is operated at ground potential (which affords a measure of lightning protection). Such an antenna is the conical monopole shown in Fig 13. Its electrical representation is shown in Fig 13C, and dimensions for operation on 3.5, 7 and 14 MHz are given at B and C.

Although the length of a ¼-wave vertical antenna for 3.5 MHz would be on the order of 66 feet or so, the conical monopole requires only 43 feet (0.17 λ) of height. The shortening is a result of increased diameter, which also increases antenna bandwidth. Few antennas will allow an operator to use both the low and high end of the 3.5-MHz band with essentially the same low SWR; this is one that will.

Like vertical antennas in general, the conical monopole requires a ground system beneath it to reduce ground losses and increase radiation efficiency. At least 30 wires, each 62 feet long, should be used. Every third radial should be connected to a ground rod at its far end and all radials should be joined at their far ends.

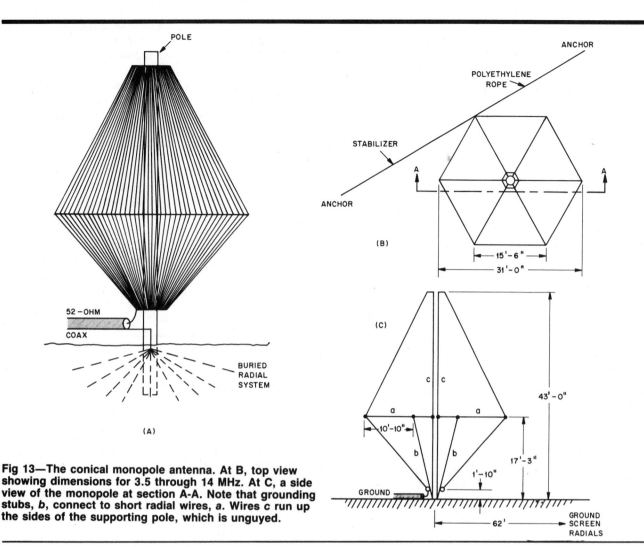

Fig 13—The conical monopole antenna. At B, top view showing dimensions for 3.5 through 14 MHz. At C, a side view of the monopole at section A-A. Note that grounding stubs, *b*, connect to short radial wires, *a*. Wires *c* run up the sides of the supporting pole, which is unguyed.

THE DISCONE ANTENNA

The discone is a vertically polarized broadband antenna which maintains an SWR of 1.5:1 or less, over several octaves in frequency. Fig 14 shows the configuration of the antenna. Dimension L of the equilaterally skirted bottom section is approximately equal to the free-space ¼-wavelength at the lowest frequency for which the antenna is built.

Below the design frequency, the SWR rises rapidly, but within its "resonant" region the antenna provides an excellent match to the popular 52-Ω coax.

Because of its physical bulk at HF, the antenna has not enjoyed much use by amateurs working in that part of the spectrum. However, the antenna has much to offer at VHF and UHF. If designed for 50 MHz, for example, the antenna will also work well on 144 and 220 MHz. Construction at HF would best be done by simulating the skirt with a grid of wires. On VHF and UHF there would be no problem in fashioning a solid skirt of some easily workable metal, such as flashing copper.

The disc-like top-hat section should be insulated from the skirt section. This is usually done with a block of material strong enough to support the disc. The inner conductor of the coax runs up through this block and is attached to the disc; the shield of the coax is connected to the skirt section.

The optimum spacing of the disc from the skirt varies as a function of the part of the spectrum for which the antenna is designed. At HF this spacing may be as much as 6 inches for 14 MHz, while at 144 MHz the spacing may be only 1 inch. It does not appear to be particularly critical.

The gain of the discone is essentially constant across its useful frequency range. The angle of radiation is very low, for the most part, rising only slightly at some frequencies.

AN HF DISCONE ANTENNA

The problem of covering all of the existing amateur HF allocations without complications or compromises seems formidable. A discone (a contraction of disc and cone) is one possible solution. Developed in 1945 by Armig G. Kandoian, this antenna can provide efficient radiation and low SWR over a decade of bandwidth. (See bibliography.) Thus, it should be possible to cover the 3.5- to 29.7-MHz spectrum with a single antenna and transmission line. However, this would require a 75-foot vertical structure and a clear circular area 65 feet in diameter on the ground. These dimensions are impractical for many amateurs, but a 7- through 29.7-MHz version should be practical at most locations. John Belrose, VE2CV, described the design presented here in July 1975 *QST*.

The antenna comprises a vertical cone beneath a horizontal disc (see Fig 15). For frequencies within the range

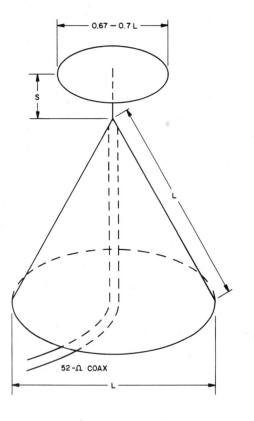

Fig 14—The discone antenna is a wideband, coaxially fed type best suited to VHF and UHF coverage because of its cumbersome size at HF. Dimension L is equal to a free-space quarter wavelength at the lowest operating frequency. The profile of the skirt is an equilateral triangle; the skirt itself can be of a cage type of construction, with adjoining wires separated by not more than 0.02 λ at the bottom of the cone. Dimension S varies from 1 to 6 inches, depending on the low-frequency cutoff of the design.

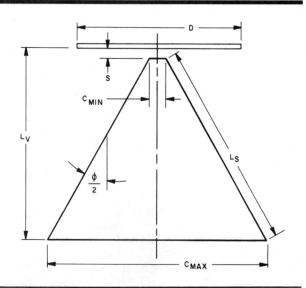

Fig 15—Cross-section sketch of the discone antenna. See text for definitions of terms.

of the antenna, radiation results from a resonance between the fields caused by current flow over the disc and over the surface of the cone, which is established by its flare angle. The apex of the cone, which is vertical, approaches and becomes common with the outer conductor of the coaxial feeder at its extremity. The center conductor of the coaxial feeder terminates at the center of the disc, which is perpendicular to the axis of the cone and the feed line. The discone can be thought of as an upside-down conical monopole.

The advantages of the discone are that it can be operated remote from and independent of ground. Furthermore, since the current maximum is at the top instead of at the bottom of the antenna, and since its structural configuration lends itself to mounting on a pole or on top of a building, the radiation characteristics of a practical discone antenna can approximate an ideal dipole antenna in free space. The change of impedance versus frequency is, however, very much less than for any ordinary dipole, even dipoles with rather small length to diameter ratios. The same is true for the radiation characteristics of the discone.

The antenna exhibits good impedance characteristics over a 10 to 1 frequency range and low-angle radiation with little change in the radiation pattern over a 3:1 or 4:1 frequency range. At the high-frequency end, the pattern begins to turn upward, with a resulting decrease in the radiation at low elevation angles. The discone antenna may be visualized as a radiator intermediate between a conventional dipole and a biconical horn. A biconical horn is essentially a conical dipole operated at frequencies for which the physical dimensions of the antenna become large compared with a wavelength. At the lower frequencies the antenna behaves very much like a dipole; at much higher frequencies it becomes essentially a horn radiator.

Design Considerations

Refer to the sketch of the discone radiator in Fig 15. The following nomenclature is used:

ϕ = cone flare angle (total)
L_s = slant height of cone
L_v = vertical distance from the disc to the base of the cone
C_{max} = maximum diameter of cone
C_{min} = minimum diameter of cone
D = diameter of disc
S = disc-to-cone spacing

The optimum parameters for discone antennas are as follows:

$S = 0.3 \, C_{min}$
$D = 0.7 \, C_{max}$

and typically, for an optimum design

$L_s/C_{min} > 22$
$\phi = 60°$

The performance of the antenna is not critical in regard to the value of flare angle ϕ, except there is less irregularity in the SWR versus frequency if ϕ is greater than 50°, although values of ϕ above 90° were not investigated. Since the bandwidth is inversely proportional to C_{min}, that dimension must be small. For a frequency range of 10 to 1, L_s/C_{min} should be greater than 22.

From the circuit standpoint, the discone antenna behaves essentially as a high-pass filter. It has an effective cutoff frequency, f_c, below which it becomes very inefficient, causing severe standing waves on the coaxial feed line. Above the cutoff frequency, little mismatch exists and the radiation pattern remains essentially the same over a wide range of frequencies (from some minimum frequency, f_{min}, to some maximum frequency, f_{max}). The slant height of the cone, L_s, is approximately equal to a quarter wavelength at the cutoff frequency, f_c, and the vertical height (or altitude) of the cone is approximately a quarter wavelength at the lowest operating frequency, f_{min}.

The radiation from the discone can be viewed in this somewhat oversimplified way. A traveling wave, excited by the antenna input between the apex of the cone and the disc, travels over the surface of the cone toward the base until it reaches a distance along the slant surface of the cone where the vertical dimension between that point and the disc is a quarter wavelength. The wave field therefore sees a resonant situation and is almost entirely radiated.

For f_{min} = 7.0 MHz and a velocity factor for propagation along the surface of the cone equal to 0.96,

$$L_v = \frac{2834}{f_{min}} = 405 \text{ in.}$$

If $\phi = 60°$, then $L_s = 456$ in. and

$$f_c = \frac{2834}{L_s} = 6.22 \text{ MHz.}$$

The disc diameter is D = 0.7 C_{max} = 0.7 × 456 = 319.2 in.

For C_{min} = 13.5 in. (a practical dimension, as we shall see later), S = 0.3 C_{min} = 0.3 × 13.5 = 4 in.

The ratio L_s/C_{min} = 456/13.5 = 33.7.

The frequency response of a discone antenna constructed with these dimensions is shown in Fig 16. Here we see that the SWR is 3.25:1 at f_c and decreases rapidly with increasing frequency to about 1.5:1 at f_{min}. The SWR is less than 1.5 over the frequency range 7 to 23 MHz. While this ratio increases for frequencies above 23 MHz, the SWR is less than

2.5:1 over the frequency range 6.5 to 30 MHz, except for the irregularity for frequencies 23.5 to 25.5 MHz. The SWR peak in the frequency range 23.5 to 25.5 MHz is thought to be caused by a resonance in the metal structure of a nearby part of the building on which the discone antenna was mounted. During these measurements the antenna was mounted on a flat roof, 70 feet from a penthouse that is 21.25 feet high (including the grounded metal rail around the top). This height is a resonant λ/2 at 24.4 MHz.

Practical Construction

At HF, the discone can be built using closely spaced wires to simulate the surface of the cone. The disc can be simulated by a structure consisting of eight spreaders with wires connected between them. It is important that a skirt wire connect the bottom ends of all slant wires simulating the cone, and another connect the outer ends of the spreaders which simulate the disc. These wires increase the effectiveness of the wire structures to a considerable extent. An antenna constructed in this way closely approximates the performance of a solid disc and a cone over the frequency range of the antenna.

The discone assembly and construction details are given in Fig 17. The antenna is supported by an 8-inch triangular aluminum mast (item 1) that is 36 feet high. The insulator separating the disc and the cone (item 2) is detailed in Fig 18. Basically it is two metal plates separated by an insulating section. The lower plate has a coaxial feedthrough connector

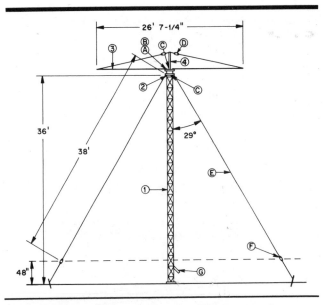

Fig 17—Construction details for the HF discone antenna.
A—Hex-head screw, ¼-20 × 2 in. long, 16 req'd.
B—Hex nut, ¼-20 thread, 16 req'd.
C—Hex-head screw, 3/8-16 × 1 in. long, 8 req'd.
D—6-in. turnbuckle, 8 req'd.
E—No. 12 Copperweld wire, 1400 ft req'd.
F—Porcelain or ceramic insulators, 24 req'd.
G—52-Ω coaxial feed line, length as required. Line is secured to mast and connected at feed point shown in Fig 18.
1—Antenna mast with cap.
2—Insulator subassembly. See Fig 18.
3—Spreaders, made from 1-in. aluminum tubing, 8 req'd.
4—Spreader support, 3-ft length of steel or aluminum pipe or tubing, flange mounted.

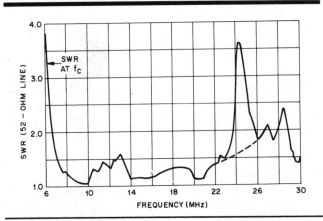

Fig 16—Standing-wave ratio versus frequency for the discone antenna designed for operation on 7 MHz and above. The "spike" in the curve at approximately 24 MHz is believed to be caused by an adjacent metal structure, as explained in the text.

mounted at its center, and the outer edge is drilled with 24 equally spaced holes, 5/32-inch diameter, on a 13.5-inch dia circle for the guy wires that simulate the cone. The end of each wire is soldered to a spade lug that is attached by a self-tapping screw to the plate. This plate is bolted to the top of the mast. Eight 1-inch dia disc spreaders (item 3) are bolted to the top plate. A short 3-foot rod (item 4) is flange mounted at the center of the upper plate. Supporting cables for the far ends of the spreaders are connected to this rod. The center conductor of the coaxial feed line is attached to the center of the top plate, as shown in Fig 18.

As shown in Fig 19, the antenna is mounted on the flat roof of a three-story building. The height of the lower edge of the cone is 4 feet above the roof. The 24 guy wires simulating the cone are broken by 12-inch porcelain insulators (item F) at their bottom ends. As previously mentioned, the ends of each wire are joined by a skirt, as shown in the drawing.

Performance

The discone antenna shown in the photograph has survived more than one freezing rain ice storm. The entire antenna and all supporting wires on at least one occasion were

Fig 19—The completed discone antenna, installed on the roof of a three-story building.

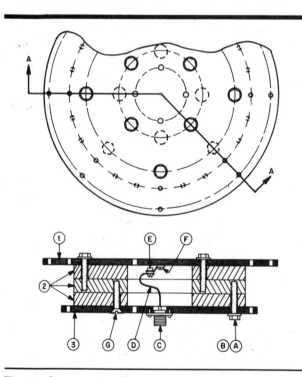

Fig 18—Construction details of insulator subassembly.
A—Hex-head screw, ½-13 × 2½ in. long, 12 req'd.
B—Flat washers, ½-in., 12 req'd.
C—RF connector, as required.
D—6-in. length of copper wire.
E—Wire lug, Emco 14-6 or equiv.
F—Round-head screw, 10-32 × 3/8 in. long.
G—Flat-head screw, ½-13 × 2½ in. long, 4 req'd.
1—Aluminum mounting plate for disc spreaders.
2—Phenolic insulator rings.
3—Guy mounting plate.

covered with a ½-inch radial thickness of ice. A 3-element triband amateur beam covered with this thickness of ice also survived the ice storm but it was unusable at the time; it was detuned too much by the ice sheath. The performance of the discone was unaffected by the ice. In fact, at an operating frequency of about 14 MHz, the SWR was marginally lower when the antenna was covered with ice compared to normal.

The antenna exhibits most of the usual characteristics of a vertical monopole. However, vertical monopole antennas have a characteristic overhead null in the radiation pattern, and for short-distance sky-wave communications a horizontal dipole is generally the better antenna. But communication has always been possible with the discone, to distances beyond that over which the ground wave could be received, provided of course that the ionosphere would reflect a frequency of 7 MHz (the lowest frequency for which the antenna could be used). While there is certainly a null overhead, it is not a very deep one.

A SIMPLIFIED DISCONE

The discone antenna structure can be reduced to a mere skeleton and still provide reasonably good radiation characteristics and a good impedance match over a single amateur band. Fig 20 shows the ultimate simplification of the discone principle as applied by Mike Wintzer, PAØMWI, in October 1974 *QST*.

If one has room for both a horizontal dipole and an inverted-V dipole, the configuration of Fig 20 is worth trying. It may outperform both the dipole and the inverted-V in terms of all-angle coverage. The bandwidth certainly is impressive; 550 kHz was obtained at 7 MHz. This suggests that a 3.5-MHz

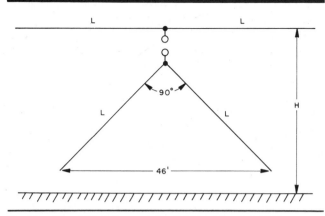

Fig 20—Simplified discone for 7 MHz. The 2:1-SWR bandwidth is 550 kHz.
H—25 ft, 1 in.
L—32 ft, 6 in.

version would exhibit a good match over the phone or CW portion of the band. The wire elements can be in a single plane or the "skirt" can be rotated 90° with respect to the flat top.

THE DISCONE—A VHF-UHF TRIBANDER

The broadband, vertically polarized antenna of Fig 21 can be assembled easily and inexpensively. The model described can be used on 144, 220 and 420 MHz. It was originally described by Dave Geiser, WA2ANU, in December 1978 *QST*.

The discone antenna functions as a wide-bandwidth, impedance-matching transformer, coupling a low-impedance transmission line to the higher impedance of free space. In the process, it radiates with a pattern similar to that of a quarter-wavelength vertical antenna above a ground plane. Waves form at the feed point (cone apex) and travel on the antenna surface to the edges of the cone and disc. The dimensions and geometry of the antenna are chosen so as to make the impedance at the edges similar to that of free space. We know that maximum energy transfer occurs when impedances are matched, so the antenna radiates.

A discone antenna acts as a high-pass filter. Below some cutoff frequency, the SWR will increase rapidly. Above this frequency, the antenna SWR remains low up to a maximum of 10 times the cutoff frequency value, depending on the design proportions. The antenna described here shows less than 2:1 SWR from 140 to 450 MHz. At 1300 MHz, the SWR measured 5:1. Fig 22 gives dimensional information for the antenna. The slant height and diameter of the cone are the same, about 110 percent of a quarter wavelength at the lowest operating frequency. Diameter of the disc is about 66% of a quarter wavelength.

Construction

The radiating surfaces are cut from hardware cloth. A 5-foot long piece of material, 24 inches wide, can be purchased at most hardware stores. The galvanized-steel wire that makes up the hardware cloth is spaced ¼ inch.

Cutting information for the discone is given in Fig 22. A felt-tip pen can be used to draw a pattern on the hardware cloth. Forming the cone may require some help from a second

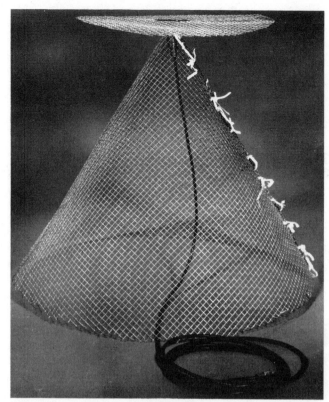

Fig 21—A completed discone antenna suitable for use on the 144, 220 and 420-MHz bands. This antenna wouldn't last long in the outdoors, but is fine for indoor use. For outside installation a more robust construction is required.

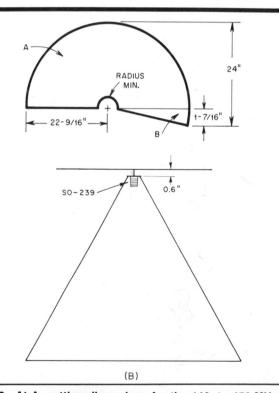

Fig 22—At A, cutting dimensions for the 140- to 450-MHz discone. The size of "radius min." will depend on the type of connector used at the apex of the cone. Not shown at A is the 15-in. diameter disc to be cut. At B, placement of the disc above the cone. Dimensions given may be scaled for other frequencies.

person. Leather-palmed gloves will protect your hands from the sharp ends of the wire. Use bread-wrapper ties to hold the edges together. To make sure it stays in place, solder the seam in a few locations. This is only for mechanical reasons—current flows down the cone, not around it, so electrical continuity isn't required. After the seam is soldered, the ties should be removed.

A 1 × 3-inch piece of sheet copper is supported by the SO-239 connector, and is soldered to the disc. The connector is soldered to the cone with its threaded end pointed down. The disc is supported about ½ inch above the cone.

The lower edge of the cone is at the same potential all the way around, allowing the antenna to be mounted on a metal surface, although this will change the radiation pattern somewhat. A support mast that extends into the cone will have little effect on the antenna performance. Lower frequency discones may be built using a number of individual wires to make the cone and disc. The disc may be approximated with metal rods if desired. If necessary, thin, nonconductive insulators may be used to support the disc.

BIBLIOGRAPHY

Source material and more extended discussion of the topics covered in this chapter can be found in the references given below and in the textbooks listed at the end of Chapter 2.

J. S. Belrose, "The HF Discone Antenna," *QST*, Jul 1975, pp 11-14, 56.

W. Conwell, "Broadband Antennas Employing Coaxial Transmission Line Sections," *QEX*, Apr 1985, pp 8-9.

R. S. Elliott, *Antenna Theory and Design* (Englewood Cliffs, NJ: Prentice Hall, 1981), pp 277-321.

R. M. Fano, "Theoretical Limitations on the Broadband Matching of Arbitrary Impedances," *Journal of the Franklin Institute,* Jan 1950, pp 57-83, and Feb 1950, pp 139-155.

D. Geiser, "An Inexpensive Multiband VHF Antenna," *QST*, Dec 1978, pp 28-29.

J. Hall, "The Search for a Simple, Broadband 80-Meter Dipole," *QST*, Apr 1983, pp 22-27.

J. Hall, "Maxcom Antenna Matcher and Dipole Cable Kit," Product Review, *QST*, Nov 1984, pp 53-54.

R. C. Johnson and H. Jasik, *Antenna Engineering Handbook*, 2nd ed. (New York: McGraw-Hill, 1984), pp 43-27 to 43-31.

A. G. Kandoian, "Three New Antenna Types and Their Applications," *Proc IRE*, Vol 34, Feb 1946, pp 70W-75W.

R. D. Snyder, "The Snyder Antenna," *RF Design*, Sep/Oct 1984, pp 49-51.

R. D. Snyder, "Broadband Antennae Employing Coaxial Transmission Line Sections," United States Patent no. 4,479,130, issued Oct 23, 1984.

M. Wintzer, "Dipole Passe?" *QST*, Oct 1974, pp 15-18, 21.

F. J. Witt, "Broadband Dipoles—Some New Insights," *QST*, Oct 1986.

Reference Data for Engineers, 7th ed. (Indianapolis: Howard W. Sams & Co, subsidiary of Macmillan, Inc, 1985), Chap 29.

Chapter 10

Log Periodic Arrays

A log periodic antenna is a system of driven elements, designed to be operated over a wide range of frequencies. Its advantage is that it exhibits essentially constant characteristics over the frequency range—the same radiation resistance (and therefore the same SWR), and the same pattern characteristics (approximately the same gain and the same front-to-back ratio). Not all elements in the system are active on a single frequency of operation; the design of the array is such that the active region shifts among the elements with changes in operating frequency. R. H. DuHamel and D. E. Isbell published the first information on log periodic arrays in professional literature in the late 1950s. The first log periodic antenna article to be published in amateur literature appeared in November 1959 *QST*, and was written by Carl T. Milner, W1FVY. (See the bibliography at the end of this chapter.)

Several varieties of log periodic antenna systems exist, such as the zig-zag, planar, trapezoidal, slot, V and the dipole. The type which is favored by amateurs is the log periodic dipole array, often abbreviated LPDA. The LPDA, shown in Fig 1, was invented by D. E. Isbell at the University of Illinois in 1958. Similar to a Yagi antenna in construction and appearance, a log periodic dipole array may be built as a rotatable system for all the upper HF bands, such as 18 to 30 MHz. The longest element, at the rear of the array, is a half wavelength at the lower design frequency.

Depending upon its design parameters, the LPDA can be operated over a range of frequencies having a ratio of 2:1 or higher. Over this range its electrical characteristics—gain, feed-point impedance, front-to-back ratio, and so forth—remain more or less constant. This is not true of any other type of antenna discussed in this book. With a Yagi or quad antenna, for example, either the gain factor or the front-to-back ratio, or both, deteriorate rapidly as the frequency of operation departs from the optimum design frequency of the array. And because those antennas are based upon resonant elements, off-resonance operation introduces reactance which causes the SWR in the feeder system to increase. Even terminated antennas such as a rhombic exhibit significant changes in gain over a 2:1 frequency ratio.

As may be seen in Fig 1, the log periodic array consists of several dipole elements which are each of different lengths and different relative spacings. A distributive type of feeder

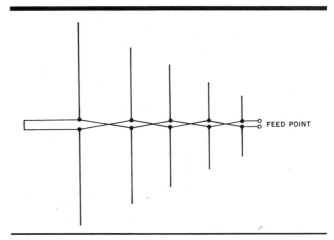

Fig 1—A log periodic dipole array. All elements are driven, as shown. The forward direction of the array as drawn here is to the right. Sometimes the elements are sloped forward, and sometimes parasitic elements are used to enhance the gain and front-to-back ratio.

system is used to excite the individual elements. The element lengths and relative spacings, beginning from the feed point for the array, are seen to increase smoothly in dimension, being greater for each element than for the previous element in the array. It is this feature upon which the design of the LPDA is based, and which permits changes in frequency to be made without greatly affecting the electrical operation. With changes in operating frequency, there is a smooth transition along the array of the elements which comprise the active region. The following information is based on a November 1973 *QST* article by Peter Rhodes, K4EWG.

A good LPDA may be designed for any single amateur band or for adjacent bands, HF to UHF, and can be built to meet the amateur's requirements at nominal cost: high forward gain, good front-to-back ratio, low SWR, and a boom length equivalent to a full sized three-element Yagi. The LPDA exhibits a relatively low SWR (usually not greater than 2 to 1) over a wide band of frequencies. A well-designed LPDA can yield a 1.3-to-1 SWR over a 1.8-to-1 frequency range with a typical gain of 9.5 dB gain over an isotropic

EDITOR'S NOTE (added in third printing): All gain figures appearing in this chapter for log periodic dipole arrays are based on information published by R. L. Carrel in 1961 (see bibliography at the end of the chapter). Subsequent work by others, recently brought to our attention, has revealed that Carrel erred in his calculations, resulting in gain figures that average 1.5 dB too high. To compensate for that error, subtract 1.5 dB from all LPDA gain values shown in this chapter. Note this correction especially for the family of curves in Fig 4 on page 10-3.

Reference: P. C. Butson and G. T. Thompson, "A Note on the Calculation of the Gain of Log-Periodic Dipole Antennas," *IEEE Trans on Antennas and Propagation.* Vol AP-24, No. 1, Jan 1976, pp 105-106.

radiator (dBi) assuming a lossless system. This equates to approximately 7.4 dB gain over a half-wave dipole (dBd).

BASIC THEORY

The LPDA is frequency independent in that the electrical properties vary periodically with the logarithm of the frequency. As the frequency f1 is shifted to another frequency f2 within the passband of the antenna, the relationship is

$$f2 = f1/\tau \qquad \text{(Eq 1)}$$

where

τ = a design parameter, a constant; $\tau < 1.0$. Also,
$f3 = f1/\tau^2$
$f4 = f1/\tau^3$

.
.
.

$f_n = f1/\tau^{n-1}$
$n = 1, 2, 3, \ldots n$
$f1$ = lowest frequency
f_n = highest frequency

The design parameter τ is a geometric constant near 1.0 that is used to determine the element lengths, ℓ, and element spacings, d, as shown in Fig 2. That is,

$\ell2 = \tau\ell1$
$\ell3 = \tau\ell2$

.
.
.

$$\ell_n = \tau\ell_{(n-1)} \qquad \text{(Eq 2)}$$

where

ℓ_n = shortest element length, and
$d_{23} = \tau d_{12}$
$d_{34} = \tau d_{23}$

.
.
.

$$d_{n-1,n} = \tau d_{n-2,n-1} \qquad \text{(Eq 3)}$$

where d_{23} = spacing between elements 2 and 3.

Each element is driven with a phase shift of 180° by switching or alternating element connections, as shown in Fig 2. At a median frequency the dipoles near the input, being nearly out of phase and close together, nearly cancel each other's radiation. As the element spacing d increases along the array, there comes a point where the phase delay in the transmission line combined with the 180° switch gives a total of 360°. This puts the radiated fields from the two dipoles in phase in a direction toward the apex. Hence, a lobe coming off the apex results.

This phase relationship exists in a set of dipoles known as the "active region." If we assume that an LPDA is designed for a given frequency range, then that design must include an active region of dipoles for the highest and lowest design frequency. It has a bandwidth which we shall call B_{ar}, bandwidth of the active region.

Assume for the moment that we have a 12-element LPDA. Currents flowing in the elements are both real and imaginary, the real current flowing in the resistive component

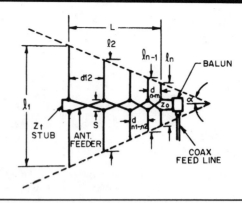

Fig 2—Schematic diagram of log periodic dipole array, with some of the design parameters indicated. Design factors are:

$$\tau = \frac{\ell_n}{\ell_{n-1}} = \frac{d_{n,n-1}}{d_{n-2,n-1}}$$

$$\sigma = \frac{d_{n,n-1}}{2\ell_{n-1}}$$

where

ℓ = element length
d = element spacing
τ = design constant
σ = relative spacing constant
S = feeder spacing
Z_0 = characteristic impedance of antenna feeder

of the impedance of a particular dipole, and the imaginary flowing in the reactive component. Assume that the operating frequency is such that element number 6 is near to being half-wave resonant. The imaginary parts of the currents in shorter elements 7 to 12 are capacitive, while those in longer elements 1 to 5 are inductive. The capacitive current components in shorter elements 9 and 10 exceed the conductive components; hence, these elements receive little power from the feeder and act as parasitic directors. The inductive current components in longer elements 4 and 5 are dominant and they act as parasitic reflectors. Elements 6, 7 and 8 receive most of their power from the feeder and act as driven elements. The amplitudes of the currents in the remaining elements are small and they may be ignored as primary contributors to the radiation field. Hence, we have a generalized Yagi array with seven elements comprising the active region. It should be noted that this active region is for a specific set of design parameters ($\tau = 0.93$, $\sigma = 0.175$). The number of elements making up the active region varies with τ and σ. Adding more elements on either side of the active region cannot significantly modify the circuit or field properties of the array.

This active region determines the basic design parameters for the array, and sets the bandwidth for the structure, B_s. That is, for a design-frequency coverage of bandwidth B, there exists an associated bandwidth of the active region such that

$$B_s = B \times B_{ar} \qquad \text{(Eq 4)}$$

where

$$B = \text{operating bandwidth} = \frac{f_n}{f1} \qquad \text{(Eq 5)}$$

$f1$ = lowest frequency, MHz
f_n = highest frequency, MHz

B_{ar} varies with τ and α as shown in Fig 3. Element lengths which fall outside B_{ar} play an insignificant role in the operation of the array. The gain of an LPDA is directly related to its directivity, and is determined by the design parameter τ and the relative element spacing constant σ. Fig 4 shows the relationship between these parameters. For each value of τ in the range $0.8 \leq \tau < 1.0$, there exists an optimum value for σ we shall call σ_{opt}, for which the gain is maximum. However, the increase in gain obtained by using σ_{opt} and τ near 1.0 (such as $\tau = 0.98$) is only 3 dB when compared with the minimum σ ($sigma_{min} = 0.05$) and $\tau = 0.98$, as may be seen in Fig 4.

An increase in τ means more elements, and optimum σ means a long boom. A high-gain (8.3 dBi) LPDA can be designed in the HF region with $\tau = 0.9$ and $\sigma = 0.05$. The relationship of τ, σ and α is as follows:

$$\sigma = (\tfrac{1}{4})(1 - \tau) \cot \alpha \qquad \text{(Eq 6)}$$

where

$\alpha = \frac{1}{2}$ the apex angle
$\tau = $ design constant
$\sigma = $ relative spacing constant

Also $\sigma = \dfrac{d_{n,n-1}}{2\ell_{n-1}}$ $\qquad$ (Eq 7)

$$\sigma_{opt} = 0.243\tau - 0.051 \qquad \text{(Eq 8)}$$

FEEDING THE LPDA

The method of feeding the antenna is rather simple. As shown in Fig 2, a balanced feeder is required for each element, and all adjacent elements are fed with a 180° phase shift by alternating element connections. In this section the term *antenna feeder* is defined as that line which connects each adjacent element. The *feed line* is that line between antenna and transmitter.

The input resistance of the LPDA, R_0, varies with frequency, exhibiting a periodic characteristic. The range of the feed point resistance depends primarily on Z_0, the characteristic impedance of the antenna feeder. R_0 may therefore

be selected to some degree by choosing Z_0, that is, by choosing the conductor size and the spacing of the antenna feeder conductors. Other factors that affect R_0 are the average characteristic impedance of a dipole, Z_{av}, and the mean spacing factor, σ'. As an approximation (to within about 10%), the relationship is as follows:

$$R_0 = \frac{Z_0}{\sqrt{1 + \dfrac{Z_0}{4\sigma' Z_{av}}}} \qquad \text{(Eq 9)}$$

where

$R_0 = $ mean radiation resistance level of the LPDA input impedance
$Z_0 = $ characteristic impedance of antenna feeder
$Z_{av} = $ average characteristic impedance of a dipole

$$= 120 \left[\ln\left(\frac{\ell_n}{d_n}\right) - 2.25 \right] \qquad \text{(Eq 10)}$$

$\ell_n/dia_n = $ length to diameter ratio of nth element

$$\sigma' = \text{mean spacing factor} = \frac{\sigma}{\sqrt{\tau}} \qquad \text{(Eq 11)}$$

The mean spacing factor, σ', is a function of τ and α (Eqs 6 and 11). For a fixed value of Z_0, R_0 decreases with increasing τ and increasing α.

If all element diameters are identical, then the element ℓ/dia ratios will increase along the array. Ideally the ratios should remain constant, but from a practical standpoint the SWR performance of a single-band LPDA will not be noticeably degraded if all elements are of the same diameter. But to minimize SWR variations for multiband designs, the LPDA may be constructed with progressively increasing element diameters from the front to the back of the array. This approach also offers structural advantages for self-supporting elements, as larger conductors will be in place for

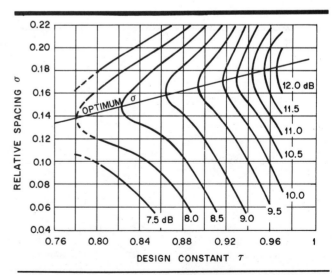

Fig 4—LPDA directivity (gain over isotropic, assuming no losses) as a function of τ and σ, for a length to diameter ratio of 125 for the element at the feed point. For each doubling of ℓ/dia, the directivity decreases by about 0.2 dB for ℓ/dia values in the range 50 to 10000. Gain relative to a dipole may be obtained by subtracting 2.14 dB from the values indicated. (After Carrel)

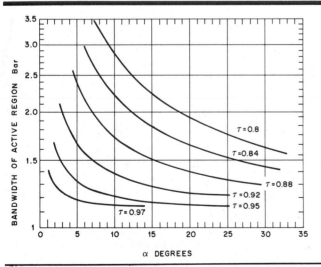

Fig 3—Design graph showing the relationships among α, τ and the bandwidth of the active region, B_{ar}. (After Carrel)

the longer elements.

The standing wave ratio varies periodically with frequency. The mean value of SWR, with respect to R_0, has a minimum of about 1.1:1 at σ_{opt} (Eq 8), and rises to a value of 1.8:1 at $\sigma = 0.05$. In other words, the periodic SWR variation (with frequency changes) swings over a wider range of SWR values with lower values of σ. These SWR ranges are acceptable when using standard 52-ohm and 72-ohm coax for the feed line. However, a 1:1 SWR match can be obtained at the transmitter end by using a coax-to-coax Transmatch. A Transmatch enables the transmitter low-pass filter to see a 52-ohm load on each frequency within the array passband. The Transmatch also eliminates possible harmonic radiation caused by the frequency-independent nature of the array.

R_0 should be chosen for the intended balun and feed line characteristics. For HF arrays, a value of 208 ohms for R_0 usually works well with a 4:1 balun and 52-Ω coax. Direct 52-Ω feed is usually not possible. (Attempts may result in smaller conductor spacing for the antenna feeder than the conductor diameter, a physical impossibility.)

For VHF and UHF designs, the antenna feeder may also serve as the boom. With this technique, element halves are supported by feeder conductors of tubing that are closely spaced. If R_0 is selected as 72 Ω, direct feed with 72-Ω cable is possible. An effective balun exists if the coax is passed through one of the feeder conductors from the rear of the array to the feed point. Fig 5 shows such a feed point arrangement.

If the design bandwidth of the array is fairly small (single band), another possible approach is to design the array for a 100-Ω R_0 and use a ¼-wave matching section of 72-Ω coax between the feed point and 52-Ω feed line. In any case, select the element feeder diameters based on mechanical considerations. The required feeder spacing may then be calculated.

The antenna feeder termination, Z_t, is a short circuit at a distance of $\lambda_{max}/8$ or less behind element no. 1, the longest

element. In his 1961 paper on LPDAs, Carrel reported satisfactory results in some cases by using a short circuit at the terminals of element no. 1. If this is done, the shorted element acts as a passive reflector at the lowest frequencies. Some constructors indicate that Z_t may be eliminated altogether without significant effect on the results. The terminating stub impedance tends to increase the front-to-back ratio for the lowest frequencies. If used, its length may be adjusted for the best results, but in any case it should be no longer than $\lambda_{max}/8$. For HF-band operation a 6-inch shorting jumper wire may be used for Z_t.

It might also be noted that one could increase the front-to-back ratio on the lowest frequency by moving the passive reflector (element no. 1) a distance of 0.15 to 0.25 λ behind element no. 2, as would be done in the case of an ordinary Yagi parasitic reflector. This of course would necessitate lengthening the boom. The front-to-back ratio increases somewhat as the frequency increases. This is because more of the shorter inside elements form the active region, and the longer elements become additional reflectors.

DESIGN PROCEDURE

The preceding section provides information on the fundamentals of a log periodic dipole array. From that discussion, some insights may be gained into the effects of changing the various design parameters. However, a thorough understanding of LPDA basic theory is not necessary in order to design your own array. A systematic step-by-step design procedure of the LPDA is presented in this section, with design examples. There are necessarily some mathematical calculations to be performed, but these may be accomplished with a four-function electronic calculator that additionally handles square-root and logarithmic functions. The procedure that follows may be used for designing any LPDA for any desired bandwidth.

1) Decide on an operating bandwidth B between f1,

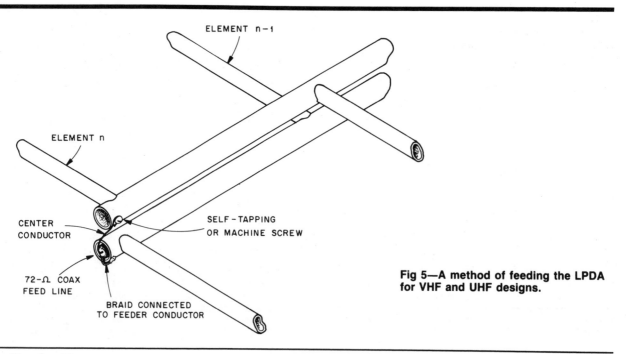

Fig 5—A method of feeding the LPDA for VHF and UHF designs.

lowest frequency and f_n, highest frequency, using Eq 5.

2) Choose τ and σ to give a desired gain (Fig 4).

$0.8 \leq \tau \leq 0.98$
$0.05 \leq \sigma \leq \sigma_{opt}$

The value of σ_{opt} may be determined from Fig 4 or from Eq 8.

3) Determine the value for the cotangent of the apex half-angle α from

$$\cot \alpha = \frac{4\sigma}{1 - \tau} \qquad \text{(Eq 12)}$$

Note: α, the apex half angle itself, need not be determined as a part of this design procedure, but the value for $\cot \alpha$ is used frequently in the steps that follow.

4) Determine the bandwidth of the active region B_{ar} either from Fig 3 or from

$$B_{ar} = 1.1 + 7.7(1 - \tau)^2 \cot \alpha \qquad \text{(Eq 13)}$$

5) Determine the structure (array) bandwidth B_s from Eq 4.

6) Determine the boom length L, number of elements N, and longest element length $\ell 1$.

$$L_{ft} = \left[\frac{1}{4} \left(1 - \frac{1}{B_s} \right) \cot \alpha \right] \lambda_{max} \qquad \text{(Eq 14)}$$

$$N = 1 + \frac{\log B_s}{\log \frac{1}{\tau}} = 1 + \frac{\ln B_s}{\ln \frac{1}{\tau}} \qquad \text{(Eq 15)}$$

$$\ell 1_{ft} = \frac{492}{f1} \qquad \text{(Eq 16)}$$

where λ_{max} = longest free-space wavelength = $984/f1$. Usually the calculated value for N will not be an integral number of elements. If the fractional value is significant, more than about 0.3, increase the value to the next higher integer. Doing this will also increase the actual value of L over that obtained from Eq 14.

Examine L, N and $\ell 1$ to determine whether or not the array size is acceptable for your needs. If the array is too large, increase f1 or decrease σ or τ and repeat steps 2 through 6. (Increasing f1 will decrease all dimensions. Decreasing σ will decrease primarily the boom length. Decreasing τ will decrease both the boom length and the number of elements.)

7) Determine the terminating stub Z_t. (Note: For HF arrays, short out the longest element with a 6-inch jumper. For VHF and UHF arrays use:

$$Z_t = \lambda_{max}/8 \qquad \text{(Eq 17)}$$

8) Solve for the remaining element lengths from Eq 2.
9) Determine the element spacing d_{12} from

$$d_{12} = \frac{1}{2} (\ell 1 - \ell 2) \cot \alpha \qquad \text{(Eq 18)}$$

and the remaining element-to-element spacings from Eq 3.

10) Choose R_0, the desired feed point resistance, to give the lowest SWR for the intended balun ratio and feed line impedance. From the following equation, determine the neces-

sary antenna feeder impedance, Z_0, using the definitions of terms for Eq 9.

$$Z_0 = \frac{R_0{}^2}{8\sigma' Z_{av}} + R_0 \sqrt{\left(\frac{R_0}{8\sigma' Z_{av}} \right)^2 + 1} \qquad \text{(Eq 19)}$$

11) Once Z_0 has been determined, select a combination of conductor size and spacing to provide that impedance from

$$S = \left(\frac{\text{dia}}{2} \right) \times 10^{Z_0/276} \qquad \text{(Eq 20)}$$

where

S = center-to-center distance between conductors
dia = outer diameter of conductor (in same units as S)
Z_0 = intended characteristic impedance for antenna feeder

Note: This equation assumes round feeder conductors.

If an impractical spacing results for the antenna feeder, select a different conductor diameter and repeat step 11. In severe cases it may be necessary to select a different R_0 and repeat steps 10 and 11. Once a satisfactory feeder arrangement is found, the LPDA design is completed.

Design Example—Short Four-Band Array

Suppose we wish to design a log periodic dipole array to cover the frequency range 18.06 to 29.7 MHz. Such an array will offer operation on any frequency in the 17, 15, 12 and 10-meter amateur bands. In addition, we desire for this to be a short array, constructed on a boom of no more than 10 feet in length.

To follow through this example, it is suggested that you write the parameter names and their values as they are calculated, in columns, on your worksheet. This will provide a ready reference for the values needed in subsequent calculations.

We begin the design procedure with step 1 and determine the operating bandwidth from Eq 5: f1 = 18.06, f_n = 29.7, and B = 29.7/18.06 = 1.6445. (Note: Because log periodics have reduced gain at the low-frequency end, some designers lower f1 by several percent to assure satisfactory gain at the lower operating frequencies. Increasing f_n, the design frequency at the high end, however, appears to offer no advantage other than extended frequency coverage.) Because we wish to have a compact design, we choose not to extend the lower frequency range.

Next, step 2, we examine Fig 4 and choose values for τ, σ and gain. Knowing from the basic theory section that larger values of σ call for a longer boom, we choose the not-too-large value of 0.06. Also knowing that a compact array will not exhibit high gain, we choose a modest gain, 8.0 dBi. For these values of σ and gain, Fig 4 shows the required τ to be 0.885.

From step 3 and Eq 12, we determine the value for $\cot \alpha$ to be $4 \times 0.06/(1 - 0.885) = 2.0870$. We need not determine α, the apex half angle, but if we wish to go the trouble we can use the relationship

$\alpha = \text{arc cot } 2.0870 = \text{arc tan } (1/2.0870) = 25.6°$

This means the angle at the apex of the array will be $2 \times 25.6 = 51.2°$.

From step 4 and Eq 13, we calculate the value for B_{ar} as $1.1 + 7.7(1 - 0.885)^2 \times 2.087 = 1.3125$.

Next, from step 5 and Eq 4, we determine the structure bandwidth B_s to be $1.6445 \times 1.3125 = 2.1584$.

From step 6 and the associated equations we determine the boom length, number of elements, and longest element length.

$$L = \left[\frac{1}{4}\left(1 - \frac{1}{2.1584}\right) \times 2.0870\right]\frac{984}{18.06} = 15.26 \text{ ft}$$

$$N = 1 + \frac{\log 2.1584}{\log (1/0.885)} = 1 + \frac{0.3341}{0.05306} = 7.30$$

(Because a *ratio* of logarithmic values is determined here, either common or natural logarithms may be used in the equation, so long as both the numerator and the denominator are the same type; the results are identical.)

$\ell 1 = 492/18.06 = 27.243$ ft

The 15.26-foot boom length is greater than the 10-foot limit we desired, so some change in the design is necessary. The 7.30 elements should be increased to 8 elements if we chose to proceed with this design, adding still more to the boom length. The longest element length is a function solely of the lowest operating frequency, so we do not wish to change that.

Decreasing either σ or τ will yield a shorter boom. Because σ is already close to the minimum value of 0.05, we decide to retain the value of 0.06 and decrease the value of τ. Let's try $\tau = 0.8$. Repeating steps 2 through 6 with these values, we calculate the following.

Gain = 6.8 dBi?   (outside curves of graph)
cot α = 1.2000
B_{ar} = 1.4696
B_s = 2.4168
L = 9.58 ft
N = 4.95
$\ell 1$ = 27.243 ft

These results nicely meet our requirement for a boom length not to exceed 10 feet. The 4.95 elements obviously must be increased to 5. The 6.8 dBi gain (4.7 dBd) is nothing spectacular, but it is near to what one would expect for a two-element Yagi. For four-band coverage with a short boom, we decide this gain and array dimensions are acceptable, and we choose to go ahead with the design. The variables summarized on our worksheet now should be those shown in the first portion of Table 1.

Continuing at step 7, we make plans to use a 6-inch

Table 1
Design Parameters for the 4-Band LPDA

f1 = 18.06 MHz	Element lengths:
f_n = 29.7 MHz	$\ell 1$ = 27.243 ft
B = 1.6445	$\ell 2$ = 21.794 ft
τ = 0.8	$\ell 3$ = 17.436 ft
σ = 0.06	$\ell 4$ = 13.948 ft
Gain = 6.8 dBi = 4.7 dBd	$\ell 5$ = 11.159 ft
cot α = 1.2000	Element spacings:
B_{ar} = 1.4696	d_{12} = 3.269 ft
B_s = 2.4168	d_{23} = 2.616 ft
L = 9.58 ft	d_{34} = 2.092 ft
N = 4.95 elements (increase to 5)	d_{45} = 1.674 ft
Z_t = 6-in. shorted jumper	Element diameters:
R_0 = 208 Ω	dia_5 = ½ in.;
Z_{av} = 400.8 Ω	$\ell 5/dia_5$ = 267.8
σ' = 0.06708	dia_4 = 5/8 in.;
Z_0 = 490.5 Ω	$\ell 4/dia_4$ = 267.8
Antenna feeder:	dia_3 = ¾ in.;
No. 12 wire spaced 2.4 in.	$\ell 3/dia_3$ = 279.0
Balun: 4 to 1	dia_2 = 1 in.;
Feed line: 52-Ω coax	$\ell 2/dia_2$ = 261.5
	dia_1 = 1¼ in.;
	$\ell 1/dia_1$ = 261.5

[Note: At this writing the 17-meter band is not scheduled to be available to US amateurs until July 1989.—Ed.]

shorted jumper for the terminating stub, Z_t.

From step 8 and Eq 2 we determine the element lengths:

$\ell 2 = \tau \ell 1 = 0.8 \times 27.243 = 21.794$ ft
$\ell 3 = 0.8 \times 21.794 = 17.436$ ft
$\ell 4 = 0.8 \times 17.436 = 13.948$ ft
$\ell 5 = 0.8 \times 13.948 = 11.159$ ft

From step 9 and Eq 18 we calculate the element spacing d_{12} as ½ (27.243 − 21.794) × 1.2 = 3.269 ft. Then from Eq 3 we determine the remaining element spacings:

$d_{23} = 0.8 \times 3.269 = 2.616$ ft
$d_{34} = 0.8 \times 2.616 = 2.092$ ft
$d_{45} = 0.8 \times 2.092 = 1.674$ ft

This completes the calculations of the array dimensions. The work remaining is to design the antenna feeder. From step 10, we wish to feed the LPDA with 52-Ω line and a 4:1 balun, so we select R_0 as $4 \times 52 = 208 \Omega$.

Before we calculate Z_0 from Eq 19 we must first determine Z_{av} from Eq 10. At this point we must assign a diameter to element no. 5. We wish to make the array rotatable with self-supporting elements, so we shall use aluminum tubing for all elements. For element no. 5, the shortest element, we plan to use tubing of ½-inch OD. We calculate the length to diameter ratio by first converting the length to inches:

$$\ell5/\text{dia}_5 = 11.159 \times 12/0.5 = 267.8$$

At this point in the design process we may also assign diameters to the other elements. To maintain an essentially constant ℓ/dia ratio along the array, we shall use larger tubing for the longer elements. (From a practical standpoint for large values of τ, 2 or 3 adjacent elements could have the same diameter. For a single-band design, they could all have the same diameter.) From data in Chapter 21 we see that, above ½ inch, aluminum tubing is available in diameter steps of 1/8 inch. We assign additional element diameters and determine ℓ/dia ratios as follows:

$\text{dia}_4 = 5/8$ in.; $\ell4/\text{dia}_4 = 13.948 \times 12/0.625 = 267.8$
$\text{dia}_3 = ¾$ in.; $\ell3/\text{dia}_3 = 17.436 \times 12/0.75 = 279.0$
$\text{dia}_2 = 1$ in.; $\ell2/\text{dia}_2 = 21.794 \times 12/1 = 261.5$
$\text{dia}_1 = 1¼$ in.; $\ell1/\text{dia}_1 = 27.243 \times 12/1.25 = 261.5$

Tapered elements with telescoping tubing at the ends may certainly be used. From a matching standpoint, the difference from cylindrical elements is of minor consequence.

In Eq 10 the required length to diameter ratio is that for element no. 5, or 267.8. Now we may determine Z_{av} as

$$Z_{av} = 120 \, [\ln 267.8 - 2.25] = 120 \, [5.590 - 2.25] = 400.8$$

Additionally, before solving for Z_0 from Eq 19, we must determine σ' from Eq 11.

$$\sigma' = \frac{0.06}{\sqrt{0.8}} = 0.06708$$

And now we use Eq 19 to calculate Z_0.

$$Z_0 = \frac{208^2}{8 \times 0.06708 \times 400.8}$$
$$+ \, 208\sqrt{\left(\frac{208}{8 \times 0.06708 \times 400.8}\right)^2 + 1}$$
$$= 201.1 + 208 \times \sqrt{1.935} = 490.5 \; \Omega$$

From step 11, we are to determine the conductor size and spacing for a Z_0 of 490.5 Ω for the antenna feeder. We elect to use no. 12 wire, and from data in Chapter 21 learn that its diameter is 80.8 mils or 0.0808 inch. We determine the spacing from Eq 20 as

$$S = \left(\frac{0.0808}{2}\right) \times 10^{490.5/276} = \frac{0.0808}{2} \times 10^{1.777}$$
$$= \frac{0.0808}{2} \times 59.865 = 2.42 \; \text{in.}$$

An open wire line of no. 12 wire with 2.4-inch spacers may be used for the feeder. This completes the design of the four-band LPDA. The design data are summarized in Table 1.

Wire Log-Periodic Dipole Arrays for 3.5 or 7 MHz

These log-periodic dipole arrays are simple and easy to build. They are designed to have reasonable gain, be inexpensive and lightweight, and they may be assembled with stock items found in large hardware stores. They are also strong—they can withstand a hurricane! These antennas were first described by John J. Uhl, KV5E, in *QST* for August 1986. Fig 6 shows one method of installation. You can use the information presented here as a guide and point of reference for building similar LPDAs.

If space is available, the antennas can be "rotated" or repositioned in azimuth after they are completed. A 75-foot tower and a clear turning radius of 120 feet around the base of the tower are needed. The task is simplified if only three anchor points are used, instead of the five shown in Fig 6.

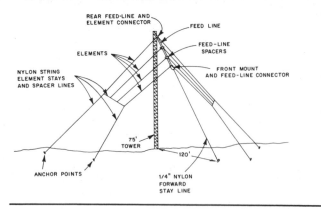

Fig 6—Typical 4-element log-periodic dipole array erected on a tower.

Table 2
Design Parameters for the 3.5-MHz Single-Band LPDA

f_1 = 3.3 MHz	Element lengths:
f_n = 4.1 MHz	ℓ_1 = 149.091 ft
B = 1.2424	ℓ_2 = 125.982 ft
τ = 0.845	ℓ_3 = 106.455 ft
σ = 0.06	ℓ_4 = 89.954 ft
Gain = 7.5 dBi = 6.4 dBd	Element spacings:
cot α = 1.5484	d_{12} = 17.891 ft
B_{ar} = 1.3864	d_{23} = 15.118 ft
B_s = 1.7225	d_{34} = 12.775 ft
L = 48.42 ft	Element diameters:
N = 4.23 elements (decrease to 4)	All = 0.0641 in.
Z_t = 6-in. shorted jumper	ℓ/dia ratios:
R_0 = 208 Ω	ℓ_4/dia$_4$ = 16840
Z_{av} = 897.8 Ω	ℓ_3/dia$_3$ = 19929
σ' = 0.06527	ℓ_2/dia$_2$ = 23585
Z_0 = 319.8 Ω	ℓ_1/dia$_1$ = 27911
Antenna feeder:	
No. 12 wire spaced 0.58 in.	
Balun: 4 to 1	
Feed line: 52-Ω coax	

Omit the two anchor points on the forward element, and extend the two nylon strings used for element stays all the way to the forward stay line.

DESIGN OF THE LOG-PERIODIC DIPOLE ARRAYS

Design constants for the two arrays are listed in Tables 2 and 3. The preceding sections of this chapter contain a more precise design procedure than that presented in earlier editions of *The ARRL Antenna Book*, resulting in slightly different feeder design values than those appearing in *QST*.

The process for determining the values in Tables 2 and 3 is identical to that given in the preceding example. The primary differences are the narrower frequency ranges and the use of wire, rather than tubing, for the elements. As additional design examples for the LPDA, you may wish to work through the step-by-step procedure and check your results against the values in Tables 2 and 3.

From the design procedure, the feeder spacings for the two arrays are slightly different, 0.58 inch for the 3.5-MHz array and 0.66 inch for the 7-MHz version. As a compromise toward the use of common spacers for both bands, a spacing of 5/8 inch is quite satisfactory. Surprisingly, the feeder spacing is not at all critical here from a matching standpoint, as may be verified from Z_0 = 276 log (2S/dia) and from Eq 9. Increasing the spacing to as much as ¾ inch results in an R_0 SWR of less than 1.1 to 1 on both bands.

Constructing the Arrays

The construction techniques are the same for both the 3.5 and the 7-MHz versions of the array. Once the designs are completed, the next step is to fabricate the fittings; see Fig 7 for details. Cut the wire elements and feed lines to the proper sizes and mark them for identification. After the wires are cut and placed aside, it will be difficult to remember which

Table 3
Design Parameters for the 7-MHz Single-Band LPDA

f_1 = 6.9 MHz	Element lengths:
f_n = 7.5 MHz	ℓ_1 = 71.304 ft
B = 1.0870	ℓ_2 = 60.252 ft
τ = 0.845	ℓ_3 = 50.913 ft
σ = 0.06	ℓ_4 = 43.022 ft
Gain = 7.5 dBi = 6.4 dBd	Element spacings:
cot α = 1.5484	d_{12} = 8.557 ft
B_{ar} = 1.3864	d_{23} = 7.230 ft
B_s = 1.5070	d_{34} = 6.110 ft
L = 18.57 ft	Element diameters:
N = 3.44 elements (increase to 4)	All = 0.0641 in.
Z_t = 6-in. shorted jumper	ℓ/dia ratios:
R_0 = 208 Ω	ℓ_4/dia$_4$ = 8054
Z_{av} = 809.3 Ω	ℓ_3/dia$_3$ = 9531
σ' = 0.06527	ℓ_2/dia$_2$ = 11280
Z_0 = 334.2 Ω	ℓ_1/dia$_1$ = 13349
Antenna feeder:	
No. 12 wire spaced 0.66 in.	
Balun: 4 to 1	
Feed line: 52-Ω coax	

is which unless they are marked. When you have finished fabricating the connectors and cutting all of the wires, the antenna can be assembled. Use your ingenuity when building one of these antennas; it isn't necessary to duplicate these LPDAs exactly.

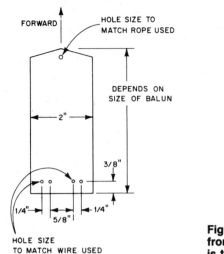

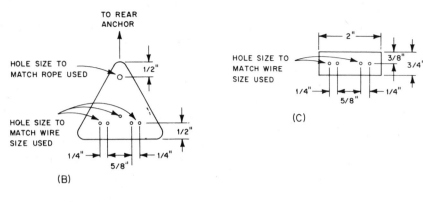

HOLE SIZE
TO MATCH WIRE USED

(A)

(B)

(C)

Fig 7—Pieces to be fabricated for the LPDA. At A, the forward connector, made from ½-in. Lexan. At B, the rear connector, also made from ½-in. Lexan. At C is the pattern for the feed-line spacers, made from ¼-in. Plexiglas. Two of these spacers will be required.

The elements are made of standard no. 14 stranded copper wire. The two parallel feed lines are made of no. 12 solid copper-coated steel wire, such as Copperweld®. This will not stretch when placed under tension. The front and rear connectors are cut from ½-inch thick Lexan® sheeting, and the feed-line spacers from ¼-inch Plexiglas® sheeting.

Study the drawings carefully and be familiar with the way the wire elements are connected to the two feed lines, through the front, rear and spacer connectors. Details are sketched in Figs 8 and 9. Connections made this way prevent the wire from breaking. All of the rope, string and connectors must be made of materials that can withstand the effects of tension and weathering. Use nylon rope and strings, the type that yachtsmen use. Fig 6 shows the front stay rope coming down to ground level at a point 120 feet from the base of a 75-foot tower. It may not be possible to do this in all cases. An alternative installation technique is to put a pulley 40 feet up in a tree and run the front stay rope through the pulley and down to ground level at the base of the tree. The front stay rope will have to be tightened with a block and tackle at ground level.

Putting an LPDA together is not difficult if it is assembled in an orderly manner. It is easier to connect the elements to the feeder lines when the feed-line assembly is stretched between two points. Use the tower and a block and tackle. Attaching the rear connector to the tower and assembling the LPDA at the base of the tower makes raising the antenna into place a much simpler task. Tie the rear connector securely to the base of the tower and attach the two feeder lines to it. Then thread the two feed-line spacers onto the feed line. The spacers will be loose at this time, but will be positioned properly when the elements are connected. Now connect the front connector to the feed lines. A word of

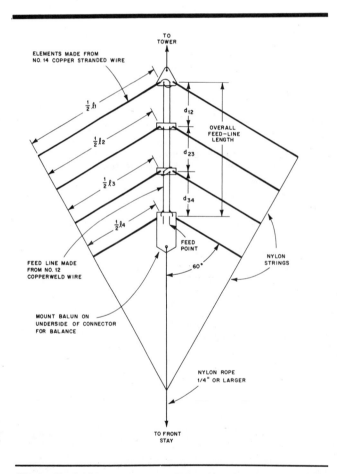

Fig 8—Typical layout for the LPDA. Use a 4:1 balun at the point indicated. See Tables 2 and 3 for dimensions.

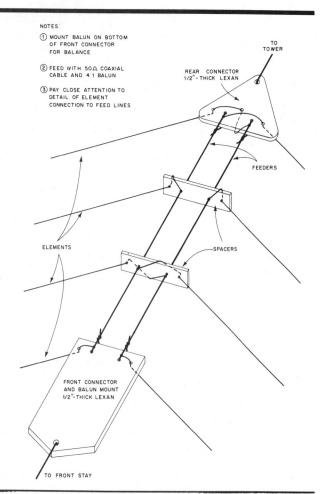

NOTES:
1. MOUNT BALUN ON BOTTOM OF FRONT CONNECTOR FOR BALANCE
2. FEED WITH 50 Ω COAXIAL CABLE AND 4:1 BALUN
3. PAY CLOSE ATTENTION TO DETAIL OF ELEMENT CONNECTION TO FEED LINES

TO TOWER

REAR CONNECTOR 1/2"-THICK LEXAN

FEEDERS

ELEMENTS

SPACERS

FRONT CONNECTOR AND BALUN MOUNT 1/2"-THICK LEXAN

TO FRONT STAY

Fig 9—Details of electrical and mechanical connections of the elements to the feed line. Knots in the nylon stay lines are not shown.

caution: Measure accurately and carefully! Double-check all measurements before you make permanent connections.

Connect the elements to the feeder lines through their respective plastic connectors, beginning with element 1, then element 2, and so on. Keep all of the element wires securely coiled. If they unravel, you will have a tangled mess of kinked wire. Check that the element-to-feeder connections have been made properly. (See Figs 8 and 9.) Once you have completed all of the element connections, attach the 4:1 balun to the underside of the front connector. Connect the feeder lines and the coaxial cable to the balun.

You will need a separate piece of rope and a pulley to raise the completed LPDA into position. First secure the eight element ends with nylon string, referring to Figs 6 and 8. The string must be long enough to reach the tie-down points. Connect the front stay rope to the front connector, and the completed LPDA is now ready to be raised into position. While raising the antenna, uncoil the element wires to prevent their getting away and tangling up into a mess. Use care! Raise the rear connector to the proper height and attach it securely to the tower, then pull the front stay rope tight and secure it. Move the elements so they form a 60-degree angle with the feed lines, in the direction of the front, and space them properly relative to one another. By adjusting the end positions of the elements as you walk back and forth, you will be able to align all the elements properly. Now it is time to hook your rig to the system and make some contacts.

Performance

The reports received from these LPDAs were compared with an inverted-V dipole. All of the antennas are fixed; the LPDAs radiate to the northeast, and the dipole to the northeast and southwest. The apex of the dipole is at 70 feet, and the 40- and 80-meter LPDAs are at 60 and 50 feet, respectively. The gain of the LPDAs is several dB over the dipole. This was apparent from many of the reports received. During pileups, it was possible to break in with a few tries on the LPDAs, yet it was impossible to break in the same pileups using the dipole.

During the CQ WW DX Contest some *big* pileups were broken after a few calls with the LPDAs. Switching to the dipole, it was found impossible to break in after many, many calls. Then, after switching back to the LPDA, it was easy to break into the same pileup and make the contact.

Think of the possibilities that these wire LPDA systems offer hams worldwide. They are easy to design and to construct, real advantages in countries where commercially built antennas and parts are not available at reasonable cost. The wire needed can be obtained in all parts of the world, and cost of construction is low! If damaged, the LPDAs can be repaired easily with pliers and solder. For those who travel on DXpeditions where space and weight are large considerations, LPDAs are lightweight but sturdy, and they perform well.

5-Band Log Periodic Dipole Array

A log periodic array designed to cover the frequency range from 13 to 30 MHz is pictured in Fig 10. This is a large array having a gain of 8.2 dBi or 6.1 dBd (approximately 1.2 dB below what one would expect with a full-size three-element Yagi array). This antenna system was originally described by Peter D. Rhodes, WA4JVE, in November 1973 *QST*. The radiation pattern, measured at 14 MHz, is shown in Fig 11.

The characteristics of the array are:
1) Half-power beamwidth, 43° (14 MHz)
2) Design parameter $\tau = 0.9$
3) Relative element spacing constant $\sigma = 0.05$
4) Boom length, L = 26 ft
5) Longest element $\ell 1 = 37$ ft 10 in. (a tabulation of element lengths and spacings is given in Table 4)
6) Total weight, 116 pounds
7) Wind-load area, 10.7 sq ft
8) Required input impedance (mean resistance), $R_0 = 72$ ohms, $Z_t = 6$-inch jumper no. 18 wire
9) Average characteristic dipole impedance,

Z_{av}: 337.8 ohms

   10) Impedance of the feeder, Z_0: 117.1 ohms

   11) Feeder: no. 12 wire, close spaced

   12) With a 1:1 toroid balun at the input terminals and a 72-ohm coax feed line, the maximum SWR is 1.4 to 1.

The mechanical assembly uses materials readily available from most local hardware stores or aluminum supply houses. The materials needed are given in Table 5. In the construction diagram, Fig 12, the materials are referenced by their respective material list number. The photograph shows the overall construction, and the drawings show the details. Table 6 gives the required tubing lengths to construct the elements.

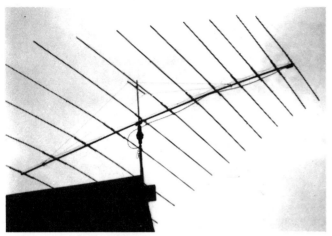

Fig 10—The 13-30 MHz log periodic dipole array.

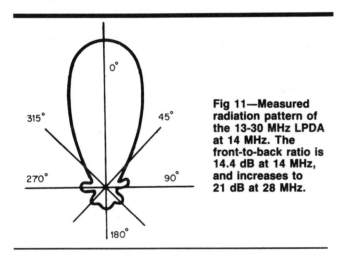

Fig 11—Measured radiation pattern of the 13-30 MHz LPDA at 14 MHz. The front-to-back ratio is 14.4 dB at 14 MHz, and increases to 21 dB at 28 MHz.

Table 5
Materials List, 13-30 MHz Log Periodic Dipole Array

	Material Description	Quantity
1)	Aluminum tubing—0.047″ wall thickness	
	1″—12′ or 6′ lengths	126 lineal feet
	7/8″—12′ lengths	96 lineal feet
	7/8″—6′ or 12′ lengths	66 lineal feet
	¾″—8′ lengths	16 lineal feet
2)	Stainless-steel hose clamps—2″ max	48 ea
3)	Stainless-steel hose clamps—1¼″ max	26 ea
4)	TV type U bolts	14 ea
5)	U bolts, galv. type	
	5/16″ × 1½″	4 ea
	¼″ × 1″	2 ea
6	1″ ID polyethylene water-service pipe— 160 psi test, approx. 1¼″ OD	20 lineal feet
A)	1¼″ × 1¼ × 1/8″ aluminum angle—6′ lengths	30 lineal feet
B)	1″ × ¼″ aluminum bar—6′ lengths	12 lineal feet
7)	1¼″ top rail of chain-link fence	26 lineal feet
8)	1:1 toroid balun	1 ea
9)	6-32 × 1″ stainless steel screws	24 ea
	6-32 stainless steel nuts	48 ea
	No. 6 solder lugs	24 ea
10)	No. 12 copper feeder wire	60 lineal feet
11A)	12″ × 8″ × ¼″ aluminum plate	1 ea
B)	6″ × 4″ × ¼″ aluminum plate	1 ea
12A)	¾″ galv. pipe	3 lineal feet
B)	1″ galv. pipe—mast	5 lineal feet
13)	Galv. guy wire	50 lineal feet
14)	¼″ × 2″ turnbuckles	4 ea
15)	¼″ × 1½″ eye bolts	2 ea
16)	TV guy clamps and eye bolts	2 ea

Table 4
13-30 MHz Array Dimensions, Feet

El. No.	Length	$d_{n-1,n}$ (spacing)	Nearest Resonant
1	37′ 10.2″	—	
2	34′ 0.7″	3′ 9.4″ = d_{12}	14 MHz
3	30′ 7.9″	3′ 4.9″ = d_{23}	
4	27′ 7.1″	3′ 0.8″ = d_{34}	
5	24′ 10.0″	2′ 9.1″ = d_{45}	18 MHz
6	22′ 4.2″	2′ 5.8″ = d_{56}	21 MHz
7	20′ 1.4″	2′ 2.8″ = d_{67}	
8	18′ 1.2″	2′ 0.1″ = d_{78}	24.9 MHz
9	16′ 3.5″	1′ 9.7″ = d_{89}	28 MHz
10	14′ 7.9″	1′ 7.5″ = $d_{9,10}$	
11	13′ 2.4″	1′ 5.6″ = $d_{10,11}$	
12	11′ 10.5″	1′ 3.8″ = $d_{11,12}$	

Table 6
Element Material Requirements, 13-30 MHz LPDA

El. No.	1″ tubing Lth	1″ tubing Qty	7/8″ tubing Lth	7/8″ tubing Qty	¾″ tubing Lth	¾″ tubing Qty	1¼″ angle Lth	1″ bar Lth
1	6′	2	6′	2	8′	2	3′	1′
2	6′	2	12′	2	—	—	3′	1′
3	6′	2	12′	2	—	—	3′	1′
4	6′	2	8.5′	2	—	—	3′	1′
5	6′	2	7′	2	—	—	3′	1′
6	6′	2	6′	2	—	—	3′	1′
7	6′	2	5′	2	—	—	2′	1′
8	6′	2	3.5′	2	—	—	2′	1′
9	6′	2	2.5′	2	—	—	2′	1′
10	3′	2	5′	2	—	—	2′	1′
11	3′	2	4′	2	—	—	2′	1′
12	3′	2	4′	2	—	—	2′	1′

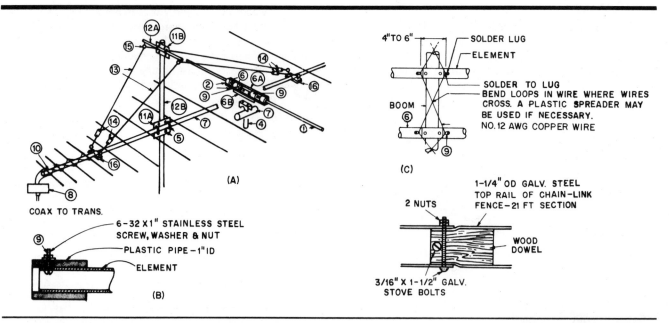

Fig 12—Construction diagram of the 13-30 MHz log periodic array. At B and C are shown the method of making electrical connection to each half element, and at D is shown how the boom sections are joined.

The Telerana

The Telerana (Spanish for "spider web") is a rotatable log periodic antenna that is lightweight, easy to construct and relatively inexpensive to build. Designed to cover 12.1 to 30 MHz, it was co-designed by George Smith, W4AEO, and Ansyl Eckols, YV5DLT, and first described by Eckols in *QST* for July 1981. Some of the design parameters are as follow.

1) $\tau = 0.9$

2) $\sigma = 0.05$

3) Gain = 8.2 dBi (6.1 dBd)

4) Feed arrangement: 400-ohm feeder line with 4:1 balun, fed with 52-Ω coax. The SWR is 1.5:1 or less in all amateur bands.

The array consists of 13 dipole elements, properly spaced and transposed, along an open wire feeder having an impedance of approximately 400 ohms. See Figs 13 and 14. The array is fed at the forward (smallest) end with a 4:1 balun and RG-8 cable placed inside the front arm and leading to the transmitter. An alternative feed method is to use open wire or ordinary TV cable and a tuner, eliminating the balun.

The frame (Fig 15) that supports the array consists of four 15-foot fiberglass vaulting poles slipped over short nipples at the hub, appearing like wheel spokes (Fig 16). Instead of being mounted directly into the fiberglass, short metal tubing sleeves are inserted into the outer ends of the arm and the necessary holes are drilled to receive the wires and nylon.

A shopping list is provided in Table 7. The center hub is made from a 1¼-inch galvanized four-outlet cross or X and four 8-inch nipples (Fig 16). A 1-inch dia X may be used alternatively, depending on the diameter of the fiberglass. A hole is drilled in the bottom of the hub to allow the cable to be passed through after welding the hub to the rotator mounting stub.

All four arms of the array must be 15 feet long. They should be strong and springy for maintaining the tautness of the array. If vaulting poles are used, try to obtain all of them with identical strength ratings.

The front spreader should be approximately 14.8 feet

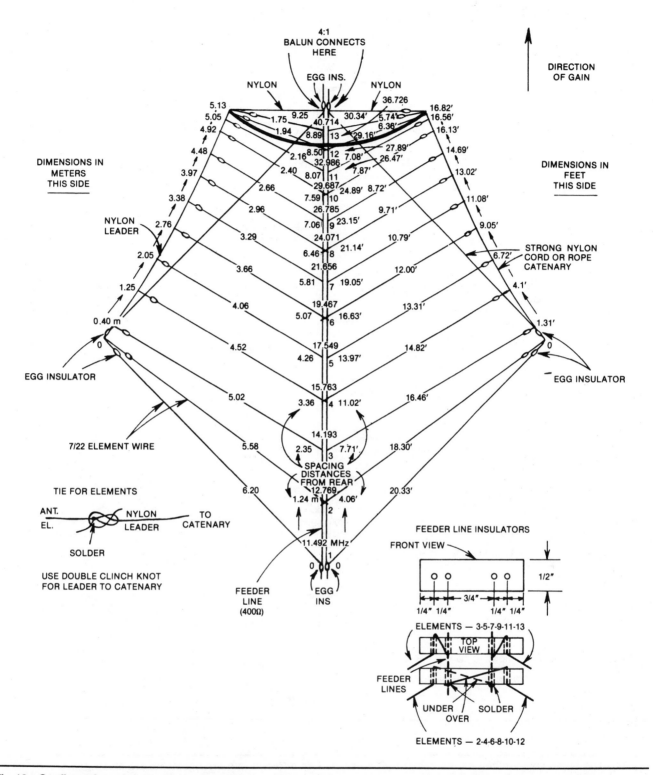

Fig 13—Configuration of the spider web antenna. Nylon monofilament line is used from the ends of the elements to the nylon cords. Solder all metal-to-metal connections. Use nylon line to tie every point where lines cross. The forward fiberglass feeder lies on the feeder line and is tied to it. Note that both metric and English measurements are shown except for the illustration of the feed-line insulator. Use soft-drawn copper or stranded wire for elements 2 through 12. Element 1 should have no. 7/22 flexible wire or no. 14 Copperweld.

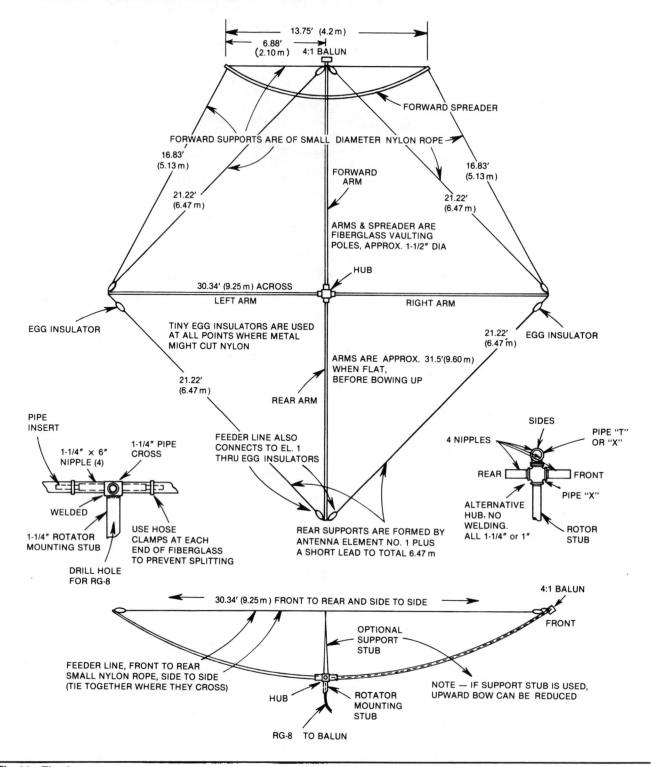

Fig 14—The frame construction for the spider web antenna. Two different hub arrangements are illustrated.

long. It can be much lighter than the four main arms, but must be strong enough to keep the lines rigid. If tapered, the spreader should have the same measurements from the center to each end. *Do not use metal for this spreader.*

Building the frame for the array is the first construction step. Once that is prepared, then everything else can be built onto it. Begin by assembling the hub and the four arms, letting

them lie flat on the ground with the rotator stub inserted into a hole in the ground. The tip-to-tip length should be about 31.5 feet each way. A hose clamp is used at each end of the arms to prevent splitting. Insert the metal inserts at the outer ends of the arms, with 1 inch protruding. The mounting holes should have been drilled at this point. If the egg insulators and nylon cords are mounted to these tube inserts, the whole

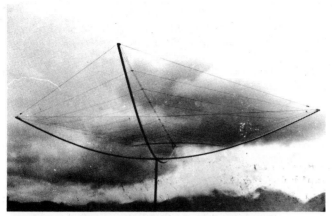

Fig 15—The spider web antenna, as shown in this somewhat deceptive photo, might bring to mind a rotatable clothesline. Of course it is much larger than a clothesline, as indicated by Figs 13 and 14. It can be lifted by hand.

Fig 16—The simple arrangement of the hub of the spider web. See Fig 13 and the text for details.

antenna can be disassembled simply by bending up the arms and pulling out the inserts with everything still attached.

Choose the arm to be at the front end. Mount two egg insulators at the front and rear to accommodate the interelement feeder. These insulators should be as close as possible to the ends.

At each end of the cross arm on top, install a small pulley and string nylon cord across and back. Tighten the cord until the upward bow reaches 3 feet above the hub. All cords will require retightening after the first few days because of stretching. The cross arm can be laid on its side while preparing the feeder line. For the front-to-rear bowstring it is important to use a wire that will not stretch, such as no. 14 Copperweld. This bowstring is actually the interelement transmission line. See Fig 17.

Secure the rear ends of the feeder to the two rear insulators, soldering the wrap. Before securing the fronts, slip the 12 insulators onto the two feed lines. A rope can be used temporarily to form the bow and to aid in mounting the feeder line. The end-to-end length of the feeder should be 30.24 feet.

Now, lift both bows to their upright position and tie the feeder line and the cross arm bowstring together where they cross, directly over and approximately 3 feet above the hub.

The next step is to install the no. 1 rear element from the rear egg insulators to the right and left cross arms using other egg insulators to provide the proper element length. Be sure to solder the element halves to the transmission line. Complete this portion of the construction by installing the nylon cord catenaries from the front arm to the cross arm tips. Use egg insulators where needed to prevent cutting the nylon cords.

In preparing the fiberglass front spreader, keep in mind that it should be 14.75 feet long before bowing and is approximately 13.75 feet when bowed. Secure the center of the bowstring to the end of the front arm. Lay the spreader on top of the feed line, then tie the feeder to the spreader with

Table 7

Shopping List for the Telerana

1—1¼-inch galvanized, 4-outlet cross or X.
4—8-inch nipples.
4—15-ft long arms. Vaulting poles suggested. These must be strong and all of the same strength (150 lb) or better.
1—Spreader, 14.8 ft long (must not be metal).
1—4:1 balun unless open-wire or TV cable is used.
12—Feed-line insulators made from Plexiglas or fiberglass.
36—Small egg insulators.
328 ft copper wire for elements; flexible 7/22 is suggested.
65.6 ft (20 m) no. 14 Copperweld wire for interelement feed line.
164 ft (50 m) strong 1/8-inch dia cord.
1—Roll of nylon monofilament fishing line, 50 lb test or better.
4—Metal tubing inserts go into the ends of the fiberglass arms.
2—Fiberglass fishing-rod blanks.
4—Hose clamps.

Fig 17—The elements, balun, transmission line and main bow of the spider web antenna.

nylon fish line. String the catenary from the spreader tips to the cross arm tips.

At this point of assembly, antenna elements 2 through 13 should be prepared. There will be two segments for each element. At the outer tip make a small loop and solder the

wrap. This will be for the nylon leader. Measure the length plus 0.4 inch for wrapping and soldering the element segment to the feeder. Seven-strand no. 22 antenna wire is suggested for use here. Slide the feed-line insulators to their proper position and secure them temporarily.

The drawings show the necessary transposition scheme. Each element half of elements 1, 3, 5, 7, 9, 11 and 13 is connected to its own side of the feeder, while elements 2, 4, 6, 8, 10 and 12 cross over to the opposite side of the transmission line.

There are four holes in each of the transmission-line insulators (see Fig 13). The inner holes are for the transmission line, and the outer ones are for the elements. Since the array elements are slanted forward, they should pass through the insulator from front to back, then back over the insulator to the front side and be soldered to the transmission line. The small drawings of Fig 13 show the details of the element transpositions.

Each place where lines cross, they are tied together with nylon line, whether copper/nylon or nylon/nylon. This makes the array much more rigid. All elements should be mounted loosely before you try to align the whole thing. Tightening any line or element affects all the others. There will be plenty of walking back and forth before the array is aligned properly. Do not expect it to be extremely taut.

The Pounder—A Single-Band 144-MHz LPDA

The 4-element Pounder LPDA pictured in Fig 18 was developed by Jerry Hall, K1TD, for the 144-148 MHz band. Because it started as an experimental antenna, it utilizes some unusual construction techniques. However, it gives a very good account of itself, exhibiting a theoretical gain of 8.7 dBi and a front-to-back ratio of 20 dB or better. The Pounder is small and light. It weighs just 1 pound, and hence its name. In addition, as may be seen in Fig 19, it can be disassembled and reassembled quickly, making it an excellent antenna for portable use. This array also serves well as a fixed station antenna, and may be changed easily to either vertical or horizontal polarization.

The antenna feeder consists of two lengths of $\frac{1}{2} \times \frac{1}{2}$ × 1/16-inch angle aluminum. The feeder also serves as the boom for the Pounder. In the first experimental model the array contained only two elements with a spacing of 1 foot, so a boom length of 1 foot was the primary design requirement for the 4-element version. Table 8 gives the design data for the 4-element array.

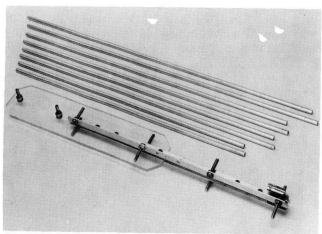

Fig 19—One end of each half element is tapped to fasten onto boom-mounted screws. Thus, disassembly of the array consists of merely unscrewing 8 half elements from the boom, and the entire array can be packaged in a small bundle of only 21 in. length.

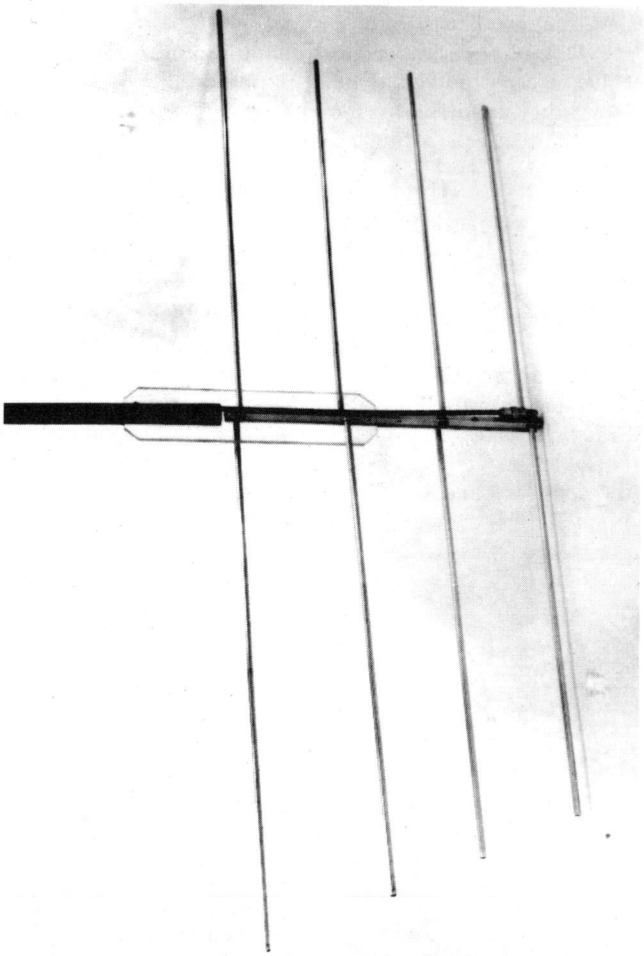

Fig 18—The 144-MHz Pounder. The boom extension running out of the picture is a 40-in. length of slotted PVC tubing, 7/8-in. OD. This tubing may be clamped to the side of a tower or attached to a mast with a small boom-to-mast plate. Rotating the tubing appropriately at the clamp will provide for either vertical or horizontal polarization.

Table 8
Design Parameters for the 144-MHz Pounder

f_1 = 143 MHz	Element lengths:
f_n = 148 MHz	ℓ_1 = 3.441 ft
B = 1.0350	ℓ_2 = 3.165 ft
τ = 0.92	ℓ_3 = 2.912 ft
σ = 0.053	ℓ_4 = 2.679 ft
Gain = 8.7 dBi = 6.6 dBd	Element spacings:
cot α = 2.6500	d_{12} = 0.365 ft
B_{ar} = 1.2306	d_{23} = 0.336 ft
B_s = 1.2736	d_{34} = 0.309 ft
L = 0.98 ft	Element diameters:
N = 3.90 elements (increase to 4)	All = 0.25 in.
Z_t = none	ℓ/dia ratios:
R_0 = 52 Ω	ℓ_4/dia$_4$ = 128.6
Z_{av} = 312.8 Ω	ℓ_3/dia$_3$ = 139.8
σ' = 0.05526	ℓ_2/dia$_2$ = 151.9
Z_0 = 75.1 Ω	ℓ_1/dia$_1$ = 165.1

Antenna feeder:
 $\frac{1}{2} \times \frac{1}{2} \times 1/16''$ angle aluminum spaced 1/4″
Balun: 1:1 (See text)
Feed line: 52-Ω coax (see text)

Construction

The general construction approach for the Pounder may be seen in the photographs. Drilled and tapped pieces of Plexiglas sheet, ¼-inch thick, serve as insulating spacers for the angle aluminum feeder. Two spacers are used, one near the front and one near the rear of the array. Four no. 6-32 × ¼ inch pan head screws secure each aluminum angle section to the Plexiglas spacers, Figs 20 and 21. Use flat washers with each screw to prevent it from touching the angle stock on the opposite side of the spacer. Be sure the screws are not so long as to short out the feeder! A clearance of about 1/16 inch has been found sufficient. If you have doubts about the screw lengths, check the assembled boom for a short with your ohmmeter on a megohms range.

Either of two mounting techniques may be used for the Pounder. As shown in Figs 18 and 19, the rear spacer measures 10 × 2½ inches, with 45° corners to avoid sharp points. This spacer also accommodates a boom extension of PVC tubing, which is attached with two no. 10-32 × 1-inch screws. This tubing provides for side mounting the Pounder away from a mast or tower.

An alternative support arrangement is shown in Fig 20. Two ½ × 3-inch Plexiglas spacers are used at the front and rear of the array. Each spacer has four holes drilled 5/8 inch apart and tapped with no. 6-32 threads. Two screws enter each spacer from either side to make a tight aluminum-Plexiglas-aluminum sandwich. At the center of the boom, secured with only two screws, is a 2 × 18-inch strip of ¼-inch Plexiglas. This strip is slotted about 2 inches from each end to accept hose clamps for mounting the Pounder atop a mast. As shown, the strip is attached for vertical polarization. Alternate mounting holes, visible on the now-horizontal lip of the angle stock, provide for horizontal polarization. Although sufficient, this mounting arrangement is not as sturdy as that shown in Fig 18.

The elements are lengths of thick-wall aluminum tubing, ¼-inch OD. The inside wall conveniently accepts a no. 10-32 tap. The threads should penetrate the tubing to a depth of at least 1 inch. Eight no. 10-32 × 1-inch screws are attached to the boom at the proper element spacings and held in place with no. 10-32 nuts, Fig 19. For assembly, the elements are then simply screwed into place.

Note that with this construction arrangement, the two halves of any individual element are not aligned with each other; they are offset by about ¾ inch. This offset does not seem to affect performance.

The Feed Arrangement

Use care in initially mounting and cutting the elements to length. To obtain the 180° crossover feed arrangement, the element halves from a single side of the feeder must alternate directions. That is, the halves of elements 1 and 3 will point to one side, and of elements 2 and 4 to the other. This arrangement may be seen by observing the element mounting screws in Fig 19. Because of this mounting scheme, the length of tubing for an element "half" is not simply half of the length given in Table 8. After final assembly, halves for elements 2 and 4 will have a slight overlap, while elements 1 and 3 are extended somewhat by the boom thickness. The best procedure is to cut each assembled element to its final

Fig 21—The feed arrangement. A right-angle chassis-mount BNC connector, modified by removing a portion of the flange, provides for ready connection of a coax feed line. A short length of bus wire connects the center pin to the opposite feeder conductor.

Fig 20—A close-up look at the boom, showing an alternative mounting scheme for the Pounder. This photo shows an earlier 2-element array, but the boom construction is unchanged with added elements. See text for details.

length by measuring from tip to tip.

The Pounder may be fed with RG-58 or RG-59 coax and a BNC connector. A modified right-angle chassis-mount BNC connector is attached to one side of the feeder/boom assembly for cable connection, Fig 21. The modification consists of cutting away part of the mounting flange that would otherwise protrude from the boom assembly. This leaves only two mounting-flange holes, but these are sufficient. A short length of small bus wire connects the center pin to the opposite side of the feeder, where it is secured under the mounting-screw nut for the shortest element.

For operation, the coax may be secured to the PVC boom extension or to the mast with electrical tape. It is also advisable to use a balun, especially if the Pounder is operated with vertical elements. A choke type of balun is satisfactory, formed by taping 6 turns of the coax into a coil of 3 inches diameter. This choke should be formed at the point where the coax is brought away from the boom. If the mounting arrangement of Fig 20 is used with vertical polarization, a second choke coil should be formed approximately ¼ wavelength down the coax line from the first. This will place it at about the level of the lower tips of the elements. For long runs of coax to the transmitter, a transition from RG-58 to RG-8 or from RG-59 to RG-9 is suggested, to reduce line losses. Make this transition at some convenient point near the array.

No shorting feeder termination is used with the array described here. In the basic theory section of this chapter, it is stated that direct feed of an LPDA is usually not possible with 52-ohm coax if a good match is to be obtained. The feeder Z_0 of this array is in the neighborhood of 120 ohms, and with this value, Eq 9 indicates R_0 to be 72.6 ohms. Thus, the theoretical mean SWR with 52 Ω line is 72.6/52 or 1.4 to 1. Upon array completion, the measured SWR (52-Ω line) was found to be relatively constant across the band, with a value of about 1.7 to 1. The Pounder offers a better match to 72-ohm coax.

Being an all-driven array, the Pounder is more immune to changes in feed-point impedance caused by nearby objects than is a parasitic array. This became obvious during portable use when the array was operated near trees and other objects...the SWR did not change noticeably with antenna rotation toward and away from those objects. This indicates the Pounder should behave well in a restricted environment, such as an attic. For weighing just one pound, this array indeed does give a good account of itself.

The Log Periodic V Array

The log periodic resonant V array is a modification of the LPDA, as shown in Fig 22. Dr Paul E. Mayes and Dr Robert L. Carrel published a report on the log periodic V array (LPVA) in the *IRE Wescon Convention Record* in 1961. (See the bibliography listing at the end of this chapter.) At the antenna laboratory of the University of Illinois, they found that by simply tilting the elements toward the apex, the array could be operated in higher resonance modes with an increase in gain (9 to 13 dBd total gain), yielding a pattern with negligible side lobes. The information presented here is based on an October 1979 *QST* article by Peter D. Rhodes, K4EWG.

A higher resonance mode is defined as a frequency that is an odd multiple of the fundamental array frequency. For example, the higher resonance modes of 7 MHz are 21 MHz, 35 MHz, 49 MHz and so on. The fundamental mode is called the $\lambda/2$ (half-wavelength) mode, and each odd multiple as follows: $3\lambda/2$, $5\lambda/2$, $7\lambda/2$, and so forth, to the $(2n - 1)\lambda/2$ mode.

The usefulness of such an array becomes obvious when one considers an LPVA with a fundamental frequency design of 7 to 14 MHz that can also operate in the $3\lambda/2$ mode at 21 to 42 MHz. A six-band array can easily be developed to yield 7 dBd gain at 7, 10 and 14 MHz, and 10 dBd gain at 21, 24.9 and 28 MHz, without traps. Also, using proper design parameters, the same array can be employed in the $5\lambda/2$ mode to cover the 35- to 70-MHz range.

A 7-30 MHz LPVA with minimum design parameters (fewest elements and shortest boom) is shown in Fig 23. This array was designed and built to test the LPVA theory under the most extreme minimum design parameters, and the results confirmed the theory.

Theory of Operation

The basic concepts of the LPDA also apply to the LPV array. That is, a series of interconnected "cells" or elements are constructed so that each adjacent cell or element differs by the design or scaling factor, τ (Fig 24). If ℓ_1 is the length of the longest element in the array and ℓ_n the length of the shortest, the relationship to adjacent elements is as follows:

$$\ell_1 = \frac{492}{f_1} \qquad \text{(Eq 1)}$$

$$\ell_2 = \tau\ell_1$$
$$\ell_3 = \tau\ell_2$$
$$\ell_4 = \tau\ell_3, \text{ and so on, to}$$
$$\ell_n = \tau\ell_{n-1} \qquad \text{(Eq 2)}$$

where

f_1 = lowest desired frequency and
n = total number of elements

Assume d_{12} is the spacing between elements ℓ_1 and ℓ_2. Then d_{n-1} is the spacing between the last or shortest elements ℓ_{n-1} and ℓ_n, where n is equal to the total number of elements.

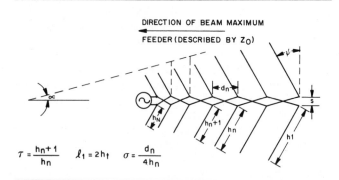

Fig 22—LPVA schematic diagram and definition of terms.

Fig 23—A pedestrian's view of the 5-element 7-30 MHz log periodic V array showing one of the capacitance hats on the rear element.

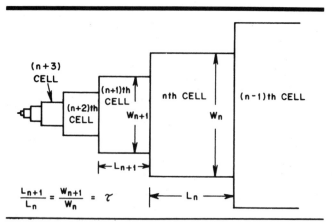

Fig 24—An interconnection of a geometric progression of cells.

The relationship to adjacent element spacings is as follows:

$$d_{12} = \tfrac{1}{2} (\ell 1 - \ell 2) \cot \alpha$$
$$d_{23} = \tau d_{12}$$
$$d_{34} = \tau d_{23}$$
$$d_{45} = \tau d_{34}$$
$$\cdot$$
$$\cdot$$
$$\cdot$$
$$d_{n-1,n} = \tau d_{n-2,n-1} \qquad \text{(Eq 3)}$$

where

$\alpha = \tfrac{1}{2}$ the apex angle

$$\sigma = \tfrac{1}{4}(1 - \tau) \cot \alpha \qquad \text{(Eq 4)}$$

The above information is no different than was presented earlier in this chapter for the LPDA. It becomes obvious that the elements, cells of elements and their associated spacings, differ by the design parameter τ. Each band of frequencies between any f and τf corresponds to one period of the structure. In order to be frequency independent (or nearly so), the variation in performance (impedance, gain, front-to-back ratio, pattern, and so forth) across a frequency period must be negligible.

The active region is defined as the radiating portion or cell within the array which is being excited at a given frequency, f, within the array passband. As the frequency decreases, the active cell moves toward the longer elements, and as the frequency increases, the active cell moves toward the shorter elements. With variations of the design constant τ, the apex half angle α (or relative spacing constant σ), and the element-to-element feeder spacing S, the following trends are found:

1) The gain increases as τ increases (more elements for a given f) and α decreases (wider element spacing).

2) The average input impedance decreases with increasing α (smaller element spacing) and increasing τ (more elements for a given f).

3) The average input impedance decreases with decreasing S, and increasing conductor size of the element-to-element feeder.

As described earlier, the LPVA operates at higher order resonance points. That is, energy is readily accepted from the feeder by those elements which are near any of the odd-multiple resonances ($\lambda/2$, $3\lambda/2$, $5\lambda/2$, and so on). The higher order modes of the LPVA are higher order space harmonics (see Mayes, Deschamps and Patton bibliography listing). Hence, when an LPVA is operated at a frequency whose half-wavelength is shorter than the smallest element, the energy on the feeder will propagate to the vicinity of the $3\lambda/2$ element and be radiated.

The elements are tilted toward the apex of the array by an angle, ψ, shown in Fig 22. The tilt angle, ψ, determines the radiation pattern and subsequent gain in the various modes. For each mode there is a different tilt angle that produces maximum gain. Mayes and Carrel did extensive experimental work with an LPVA of 25 elements with $\tau = 0.95$ and $\sigma = 0.0268$. The tilt angle, ψ, was varied from 0° to 65° and radiation patterns were plotted in the $\lambda/2$ through

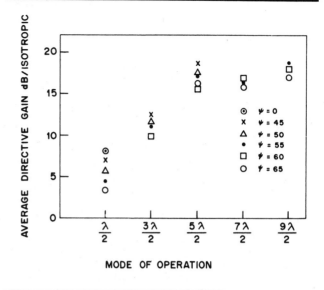

Fig 25—Average directive gain above isotropic (dBi). Subtract 2.1 from gain values to obtain gain above a dipole (dBd).

$7\lambda/2$ modes. Gain data are plotted in Fig 25. Operation in the higher modes is improved by increasing τ (more elements) and decreasing σ (closer element spacing).

When considering any single mode, the characteristic impedance is comparable with that of the LPDA; it is predominantly real and clustered around a central value, R_0. The central value R_0 for each mode increases with Z_0 (feeder impedance). Thus, as with the LPDA, control of the LPVA input impedance can be accomplished by controlling Z_0.

When multimode operation is desired, a compromise must be made in order to determine a fixed impedance level. The multimode array impedance is defined as the weighted mean resistance level, R_{wm}. Also, it can be shown that R_{wm} lies between the R_0 central values of two adjacent modes. For example,

$$R_{0_{1/2}} < R_{wm} < R_{0_{3/2}} \qquad \text{(Eq 5)}$$

where

$R_{0_{1/2}} = \lambda/2$ mode impedance, center value

$R_{0_{3/2}} = 3\lambda/2$ mode impedance, center value

and where

$$R_0 = \sqrt{R_{max} \times R_{min}} \qquad \text{(Eq 6)}$$

$$SWR = \sqrt{\frac{R_{max}}{R_{min}}} \qquad \text{(Eq 7)}$$

The weighted mean resistance level between the λ/2 and 3λ/2 modes is defined by

$$R_{wm} = \sqrt{R_{0_{1/2}} R_{0_{3/2}} \frac{SWR_{3/2}}{SWR_{1/2}}} \qquad \text{(Eq 8)}$$

where

$SWR_{1/2}$ = SWR in λ/2 mode

$SWR_{3/2}$ = SWR in 3λ/2 mode

Once Z_0 and ψ have been chosen, Fig 26 can be used to estimate the R_{wm} value for a given LPVA. Notice the dominant role that Z_0 (feeder impedance) plays in the array impedance.

It is apparent from the preceding data that the LPVA is useful for covering a number of different bands spread over a wide range of the spectrum. It is fortunate that most of the amateur bands are harmonically related. By choosing a large design parameter, $\tau = 0.9$, a small relative spacing constant, $\sigma = 0.02$, and a tilt angle of $\psi = 40°$, an LPVA could easily cover the amateur bands from 7 through 54 MHz!

DESIGN PROCEDURE

A step-by-step design procedure for the log periodic V array follows.

1) Determine the operational bandwidth, B, in the λ/2 (fundamental) mode:

$$B = \frac{f_n}{f1} \qquad \text{(Eq 9)}$$

where

f1 = lowest frequency, MHz
f_n = highest frequency, MHz

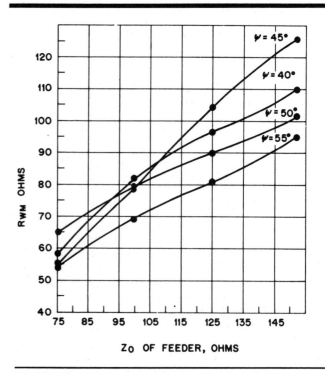

Fig 26—Weighted mean resistance level, R_{wm}, versus characteristic impedance of the feeder Z_0 for various ψ angles.

2) Determine τ for a desired number of elements, n, using Fig 27.

3) Determine element lengths ℓ1 to ℓ_n using Eqs 1 and 2 of this section.

4) Choose the highest operating mode desired and determine σ and ψ from Fig 28.

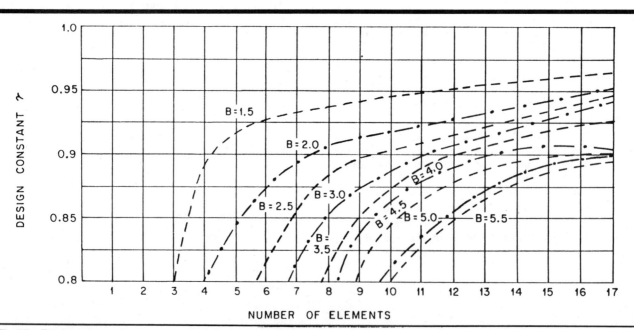

Fig 27—Design parameter, τ, versus number of elements, n, for various operational bandwidths, B.

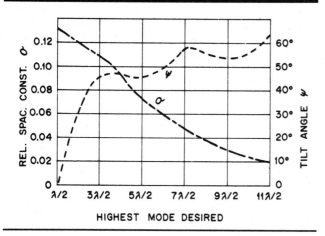

Fig 28—Optimum σ and ψ for an LPVA when the highest operating mode has been chosen.

Fig 29—The element-to-boom detail is depicted here. Aluminum angle brackets, U bolts, and sections of PVC tubing are shown securing each element to the boom at two points. The 300-Ω twin-lead, threaded through a piece of polystyrene and attached to the foremost element, may be seen entering the picture at the top left. The end of the linear loading line for $\ell 1$ is visible near the bottom.

5) Determine cell boom length, L, from

$$L = \frac{2\sigma(\ell 1 - \ell_n)}{1 - \tau} \qquad \text{(Eq 10)}$$

Note: If more than one LPVA cell is to be driven by a common feeder, the spacing between cells can be determined from

$$D_{12} = 2\sigma_1 \ell_{n1} \qquad \text{(Eq 11)}$$

where

$\quad D_{12}$ = element spacing between cell 1 (lower frequency cell) and cell 2 (higher frequency cell).

$\quad \sigma_1$ = relative spacing constant for cell 1

$\quad \ell_{n1}$ = shortest or last element within cell 1

6) Determine the mean resistance level, R_{wm}, using Fig 26.

7) Determine the element spacings using Eqs 3 and 4 of this section.

Construction Considerations

The 7-30 MHz LPVA shown in the photographs gives good results. The structural details can be seen in Figs 29 and 30, and additional data is presented in Tables 1 and 2.

Fig 30—A shot of the rearmost elements looking at an angle to the boom. The linear loading line may be seen supported at various points along the boom and at the rear element by pieces of polystyrene.

Table 1

Design Dimensions for the LPVA

Element Lengths, ft	Element Spacings, ft	Design Parameters
$\ell 1 = 56.22$*	$d_{12} = 9.15$	$\tau = 0.8$
$\ell 2 = 56.22$	$d_{23} = 7.32$	$\sigma = 0.05$
$\ell 3 = 45.00$	$d_{34} = 5.86$	$\alpha = 38.2°$
$\ell 4 = 36.00$	$d_{45} = 4.67$	$L = 27$ ft**
$\ell 5 = 28.79$		$\psi = 45°$

*$\ell 1$ is a shortened element; the full-size dimension is 70.28 ft.

**The total physical boom length is L plus the distance to the $\ell 5$ cross bracing. The cross braces are 3 ft long, and $\psi = 45°$; hence, the total boom length is 27 ft + 1.5 ft = 28.5 ft.

Table 2

Basic Materials for the LPVA

Elements	1½″, 6061-T6, 0.047″ wall aluminum tubing
Bracing	1¼″ × 1¼″ × 1/8″ aluminum angle
Boom	2½″ OD, 0.107″ wall aluminum tubing
U bolts	¼″ squared at loop to accommodate tilt angle ψ
Feeder	No. 12, solid copper
Cap. hat for $\ell 1$	No. 10 aluminum wire, 24″ dia
Linear loading for $\ell 1$	4′ loop, 3″ spacing each half of $\ell 1$

Although it performs well, it is likely that a more conservative design (two additional elements) would yield a narrower half-power (3 dB) beamwidth on 7 and 14 MHz.

It may be of interest to note that both linear and capacitive loading were used on $\ell 1$. The relationship in the next section may be used to estimate linear loading stub length and/or capacitance hat size if construction constraints prohibit a full sized array. However, performance in higher mode operations was less than optimum when shortened elements were used.

Linear Loading Stub Design

The following linear loading stub design equation may be used for approximating the stub length (one half of element, two stubs required).

$$L_s = \frac{2.734}{f} \arctan\left[\frac{33.9\left[\ln\frac{24h}{d} - 1\right]\left[1 - \left(\frac{fh}{234}\right)^2\right]}{fh \log\left(\frac{b}{a}\right)}\right]$$

(Eq 12)

where

L_s = linear loading stub length in feet required for each half element

h = element half length in feet

f = element resonant frequency in MHz

b = loading stub spacing in inches

a = *radius* (not diameter) of loading stub conductors in inches

d = average element dia in inches

Note: The resonant frequency, f, of an individual element of length, ℓ, can be found from:

$$f = \frac{468}{\ell}$$

(Eq 13)

The capacitance hat dimensions for each half element can be found from data in Chapter 2.

Log Periodic-Yagi Arrays

Several possibilities exist for constructing high-gain arrays that use the log periodic dipole as a basis. Tilting the elements toward the apex, for example, increases the gain by 3 to 5 dB on harmonic-resonance modes, as discussed in the previous section of this chapter. Another technique is to add parasitic elements to the LPDA to increase both the gain and the front-to-back ratio for a specific frequency within the passband. The LPDA-Yagi combination is simple in concept. It utilizes an LPDA group of driven elements, along with parasitic elements at normal Yagi spacings from the end elements of the LPDA.

The LPDA-Yagi combinations are endless. An example of a single-band high-gain design is a 2- or 3-element LPDA for 21.0 to 21.45 MHz with the addition of two or three parasitic directors and one parasitic reflector. The name Log-Yag array has been coined for these combination antennas. The LPDA portion of the array is of the usual design to cover the desired bandwidth, and standard Yagi design procedures are used for the parasitic elements. Information in this section is based on a December 1976 *QST* article by P. D. Rhodes, K4EWG, and J. R. Painter, W4BBP, "The Log-Yag Array."

THE LOG-YAG ARRAY

The Log-Yag array provides higher gain and greater directivity than would be realized with either the LPDA or Yagi array alone. The Yagi array requires a long boom and wide element spacing for wide bandwidth and high gain. This is because the Q of the Yagi system increases as the number of elements is increased and/or as the spacing between adjacent elements is decreased. An increase in the Q of the Yagi array means that the total bandwidth of that array is decreased, and optimum gain, front-to-back ratio and side lobe rejection are obtainable only over small portions of the band.

The Log-Yag system overcomes this difficulty by using a multiple driven element "cell" designed in accordance with the principles of the log periodic dipole array. Since this log cell exhibits both gain and directivity by itself, it is a more effective radiator than a simple dipole driven element. The front-to-back ratio and gain of the log cell can be improved with the addition of a parasitic reflector and director.

It is not necessary for the parasitic element spacings to be large with respect to wavelength, as in the Yagi array, since the log cell is the determining factor in the array bandwidth. In fact, the element spacings within the log cell may be small with respect to a wavelength without appreciable deterioration of the cell gain. For example, decreasing the relative spacing constant (σ) from 0.1 to 0.05 will decrease the gain by less than 1 dB.

A Practical Example

The photographs and figures show a Log-Yag array for the 14-MHz amateur band. The array design takes the form of a 4-element log cell, a parasitic reflector spaced at 0.085 λ_{max}, and a parasitic director spaced at 0.15 λ_{max} (where λ_{max} is the longest free-space wavelength within the array passband). It has been found that array gain is almost unaffected with reflector spacings from 0.08 λ to 0.25 λ, and the increase in boom length is not justified. The function of the reflector is to improve the front-to-back ratio of the log cell while the director sharpens the forward lobe and decreases the half-power beamwidth. As the spacing between the parasitic elements and the log cell decreases, the parasitic elements must increase in length.

The log cell is designed to meet upper and lower band limits with $\sigma = 0.05$. The design parameter τ is dependent on the structure bandwidth, B_s. When the log periodic design parameters have been found, the element length and spacings can be determined.

Array layout and construction details can be seen in Figs 31 through 34. Characteristics of the array are given in Table 1.

The method of feeding the antenna is identical to that of feeding the log periodic dipole array without the parasitic elements. As shown in Fig 31, a balanced feeder is required for each log-cell element, and all adjacent elements are fed with a 180° phase shift by alternating connections. Since the Log-Yag array will be covering a relatively small bandwidth, the radiation resistance of the narrow-band log cell will vary from 80 to 90 ohms (tubing elements) depending on the operating bandwidth. The addition of parasitic elements

---

Table 1

Log-Yag Array Characteristics

1) Frequency range	14-14.35 MHz
2) Operating bandwidth	B = 1.025
3) Design parameter	τ = 0.946657
4) Apex half angle	α = 14.92°; cot α = 3.753
5) Half-power beamwidth	42° (14-14.35 MHz)
6) Bandwidth of structure	B_s = 1.17875
7) Free-space wavelength	λ_{max} = 70.28 ft
8) Log cell boom length	L = 10.0 ft
9) Longest log element	$\ell1$ = 35.14 ft (a tabulation of element lengths and spacings is given in Table 2)
10) Forward gain over dipole	11.5 dB (theoretical)
11) Front-to-back ratio	32 dB (theoretical)
12) Front-to-side ratio	45 dB (theoretical)
13) Input impedance	Z_0 = 37 ohms
14) SWR	1.3 to 1 (14-14.35 MHz)
15) Total weight	96 pounds
16) Wind-load area	8.5 sq ft
17) Feed-point impedance	Z_0 = 37 ohms
18) Reflector length	36.4 ft at 6.0 ft spacing
19) Director length	32.2 ft at 10.5 ft spacing
20) Total boom length	26.5 ft

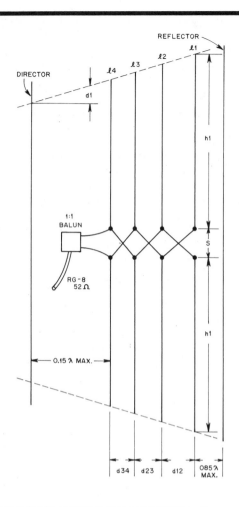

Fig 33—The attachment of the elements to the boom.

Fig 31—Layout of the Log-Yag array.

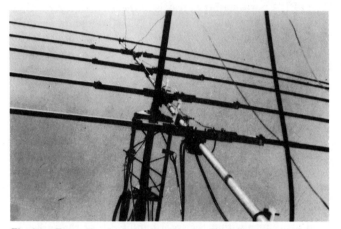

Fig 34—From the front to the back of the Log-Yag array. Note the truss provides lateral and vertical support.

Fig 32—Assembly details. The numbered components refer to Table 4.

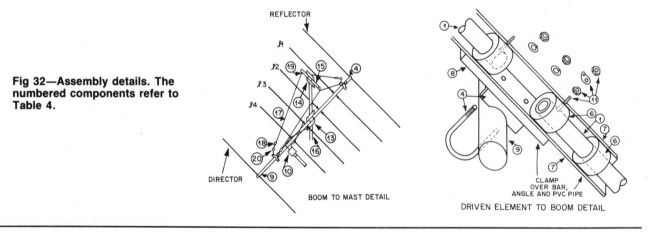

BOOM TO MAST DETAIL

DRIVEN ELEMENT TO BOOM DETAIL

CLAMP OVER BAR, ANGLE AND PVC PIPE

lowers the log-cell radiation resistance. Hence, it is recommended that a 1-to-1 balun be connected at the log-cell input terminals and 52-ohm coaxial cable be used for the feed line. The measured radiation resistance of the 14-MHz Log-Yag is 37 ohms, 14.0 to 14.35 MHz. It is assumed that tubing elements will be used. However, if a wire array is used then the radiation resistance R_o and antenna-feeder input impedance Z_o must be calculated so that the proper balun and coax may be used. The procedure is outlined in detail in the early part of this chapter.

Table 2 has array dimensions. Tables 3 and 4 contain lists of the materials necessary to build the Log-Yag array.

Table 2
Log-Yag Array Dimensions

Element	Length Feet	Spacing Feet
Reflector	36.40	6.00 (Ref. to ℓ1)
ℓ1	35.14	3.51 (d_{12})
ℓ2	33.27	3.32 (d_{23})
ℓ3	31.49	3.14 (d_{34})
ℓ4	29.81	10.57 (ℓ4 to dir.)
Director	32.20	

Table 3
Element Material Requirements, Log-Yag Array

	1-In. Tubing		7/8-In. Tubing		3/4-In. Tubing		1 1/4-In. Angle	1 × 1/4-In. Bar
	Lth Ft	Qty	Lth Ft	Qty	Lth Ft	Qty	Lth Ft	Lth Ft
Reflector	12	1	6	2	8	2	None	None
ℓ1	6	2	6	2	8	2	3	1
ℓ2	6	2	6	2	8	2	3	1
ℓ3	6	2	6	2	6	2	3	1
ℓ4	6	2	6	2	6	2	3	1
Director	12	1	6	2	6	2	None	None

Table 4
Materials List, Log-Yag Array

1) Aluminum tubing—0.047 in. wall thickness
   1 in.—12 ft lengths, 24 lin. ft
   1 in.—12 ft or 6 ft lengths, 48 lin. ft
   7/8 in.—12 ft or 6 ft lengths, 72 lin. ft
   3/4 in.—8 ft lengths, 48 lin. ft
   3/4 in.—6 ft lengths, 36 lin. ft
2) Stainless steel hose clamps—2 in. max, 8 ea
3) Stainless steel hose clamps—1 1/4 in. max, 24 ea
4) TV-type U bolts—1 1/2 in., 6 ea
5) U bolts, galv. type: 5/16 in. × 1 1/2 in., 4 ea
6) U bolts, galv. type: 1/4 in. × 1 in., 2 ea
7) 1 in. ID water-service polyethylene pipe 160 lb/sq in. test, approx. 1-3/8 in. OD, 7 lin. ft
8) 1 1/4 in. × 1 1/4 in. × 1/8 in. aluminum angle—6 ft lengths, 12 lin. ft
9) 1 in. × 1/4 in. aluminum bar—6 ft lengths, 6 lin. ft
10) 1 1/4 in. top rail of chain-link fence, 26.5 lin. ft
11) 1:1 toroid balun, 1 ea
12) No. 6-32 × 1 in. stainless steel screws, 8 ea
    No. 6-32 stainless steel nuts, 16 ea
    No. 6 solder lugs, 8 ea
13) No. 12 copper feed wire, 22 lin. ft
14) 12 in. × 6 in. × 1/4 in. aluminum plate, 1 ea
15) 6 in. × 4 in. × 1/4 in. aluminum plate, 1 ea
16) 3/4 in. galv. pipe, 3 lin. ft
17) 1 in. galv. pipe—mast, 5 lin. ft
18) Galv. guy wire, 50 lin. ft
19) 1/4 in. × 2 in. turnbuckles, 4 ea
20) 1/4 in. × 1 1/2 in. eye bolts, 2 ea
21) TV guy clamps and eyebolts, 2 ea

BIBLIOGRAPHY

Source material and more extended discussion of topics covered in this chapter can be found in the references given below and in the textbooks listed at the end of Chapter 2.

C. A. Balanis, *Antenna Theory, Analysis and Design* (New York: Harper & Row, 1982), pp 427-439.

R. L. Carrel, "The Design of Log-Periodic Dipole Antennas," *1961 IRE International Convention Record*, Part 1, Antennas and Propagation; also PhD thesis, "Analysis and Design of the Log-Periodic Dipole Antenna," Univ of Illinois, Urbana, 1961.

R. H. DuHamel and D. E. Isbell, "Broadband Logarithmically Periodic Antenna Structures," *1957 IRE National Convention Record*, Part 1.

A. Eckols, "The Telerana—A Broadband 13- to 30-MHz Directional Antenna," *QST*, Jul 1981, pp 24-27.

D. E. Isbell, "Log-Periodic Dipole Arrays," *IRE Transactions on Antennas and Propagation*, Vol AP-8, No. 3, May 1960.

D. A. Mack, "A Second-Generation Spiderweb Antenna," *The ARRL Antenna Compendium Vol 1* (Newington, CT: The American Radio Relay League, Inc, 1985), pp 55-59.

P. E. Mayes and R. L. Carrel, "Log Periodic Resonant-V Arrays," *IRE Wescon Convention Record*, Part 1, 1961.

P. E. Mayes, G. A. Deschamps, and W. T. Patton, "Backward Wave Radiation from Periodic Structures and Application to the Design of Frequency Independent Antennas," *Proc IRE*, Vol 49, No. 5, May 1961.

C. T. Milner, "Log Periodic Antennas," *QST*, Nov 1959.

P. D. Rhodes, "The Log-Periodic Dipole Array," *QST*, Nov 1973.

P. D. Rhodes and J. R. Painter, "The Log-Yag Array," *QST*, Dec 1976.

P. D. Rhodes, "The Log-Periodic V Array," *QST*, Oct 1979.

V. H. Rumsey, *Frequency Independent Antennas* (New York: Academic Press, 1966).

J. J. Uhl, "Construct a Wire Log-Periodic Dipole Array for 80 or 40 Meters," *QST*, Aug 1986.

The GIANT Book of Amateur Radio Antennas (Blue Ridge Summit, PA: Tab Books, 1979), pp 55-85.

Chapter 11

Yagi Arrays

Multielement arrays containing parasitic elements are called parasitic arrays, even though at least one of the elements is driven. A parasitic element obtains its power through coupling with a driven element, as opposed to receiving it by direct connection to the power source. A parasitic array with linear (dipole type) elements is frequently called a Yagi or Yagi-Uda antenna, after its inventors.

As explained in Chapter 8, the amplitude and phase of the current induced in an antenna element depends on the element tuning and the spacing between it and the driven element. The fact that the relative phases of the currents in driven and parasitic elements can be adjusted is very advantageous. For example, the spacing and tuning in a Yagi array can be adjusted to approximate the conditions that exist when two driven elements spaced ¼ λ apart are operated with a phase difference of 90°. This arrangement gives the unidirectional pattern shown in Fig 1. However, complete cancellation of radiation in the rear direction is not possible when a parasitic element is used, as it is not possible to make *both* the amplitude and phase of the current reach the desired values simultaneously. Nevertheless, a properly designed parasitic array can be adjusted to yield a large front-to-back

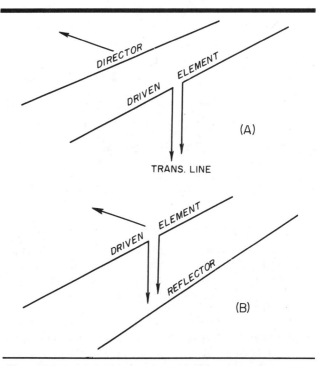

Fig 2—Antenna systems using a single parasitic element. At A the parasitic element acts as a director, and at B as a reflector. The arrows show the direction in which maximum radiation takes place.

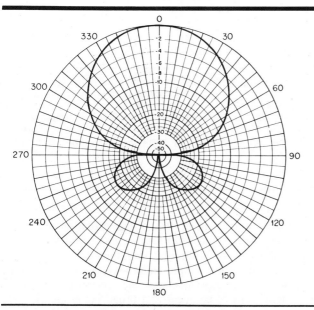

Fig 1—Azimuth pattern for two horizontal driven elements spaced ¼ λ apart and fed 90° out of phase. When parasitic elements are used, complete cancellation of radiation in the rear direction (toward the bottom of the drawing) is not possible. This is because the amplitude and the phase of the current in the parasitic element cannot simultaneously reach the desired values.

ratio. The essentially unidirectional pattern and the relatively simple electrical configuration of a parasitic array make it especially useful for antenna systems that are to be rotated.

REFLECTORS AND DIRECTORS

Although there are special cases where a parasitic array will have a bidirectional pattern, in most applications the pattern tends to be unidirectional. That is, the array will radiate more energy in one direction than it will in any other. As shown in Fig 2A, a parasitic element is called a director when it makes the radiation maximum along the perpendicular line from the driven to the parasitic element. When the maximum radiation is in the opposite direction—that is, from the parasitic element through the driven element, Fig 2B—the parasitic element is called a reflector.

Whether the parasitic element operates as a director or a reflector depends on the relative phases of the currents in the driven and parasitic elements. At the element spacings commonly used (between 0.06 and 0.25 λ), the parasitic element will act as a reflector when it is tuned below resonance (inductive reactance), and it will act as a director when it is tuned above resonance (capacitive reactance). The proper tuning is ordinarily accomplished by adjusting the lengths of the parasitic elements. Therefore, for the usual range of element spacings, a reflector is cut longer than a self-resonant length, and a director is cut shorter. As an alternative to changing the element lengths, the elements may be loaded at the center with lumped inductance or capacitance to obtain the same effect. This alternative is often used for beams with wire elements in a fixed position, where remote switching of reactances is used to switch the direction of maximum radiation.

If the parasitic element is self-resonant, the element spacing determines whether it will act as a reflector or director.

The Two-Element Beam

The simplest Yagi antenna is one with just two elements, as sketched in Fig 2. Based on an extensive study of the two-element Yagi with a method-of-moments computer analysis, Jerry Hall, K1TD, prepared the information in this section. For this study, the two elements were considered to be cylindrical (not tapered), with a diameter of $1/500 \lambda$ (dia = $\lambda/500$). This corresponds approximately to ½ inch tubing at 50 MHz.

Self-Resonant Parasitic Elements

The special case of the self-resonant parasitic element is of interest, as it gives a good idea of the overall performance of two-element systems, even though the results can be modified by detuning the parasitic element. Fig 3 shows the gain referenced to a dipole (dBd) and the radiation resistance as a function of the element spacing for this case. Relative field strength in the direction A on the small drawing is indicated by curve A; signal strength in direction B is shown by curve B. The front-to-back ratio at any spacing is the difference between the values given by curves A and B at that spacing. Whether the parasitic element is functioning primarily as a director or as a reflector is determined by whether curve A or curve B is on top; the function shifts at a spacing of about 0.11 λ. That is, at closer spacings the parasitic element acts as a director, while at greater spacings it acts as a reflector. At a spacing of 0.11 λ, the radiation is the same in both directions; the antenna is bidirectional and has a calculated gain of about 3.6 dB, referenced to a half-wave dipole.

The front-to-back ratios obtained with a self-resonant parasitic element are not very great except in the case of extremely close spacings. These close spacings (on the order of 0.04 λ) are not very practical with outdoor installations, as it is difficult to make the elements sufficiently stable mechanically. Another difficulty arises from the low radiation resistance at such spacings, on the order of 5 Ω. The low radiation resistance at the spacings giving highest gain tends to reduce the radiation efficiency, as discussed in subsequent paragraphs. Ordinary practice is to use spacings of at least 0.1 λ and tune the parasitic element for greatest attenuation in the rearward direction (greatest front-to-back ratio).

With the self-resonant parasitic element acting as a reflector, the radiation resistance increases rapidly for spacings greater than 0.13 λ, while the gain changes quite slowly. If front-to-back ratio is not an important consideration, a spacing as great as 0.25 λ can be used without much reduction

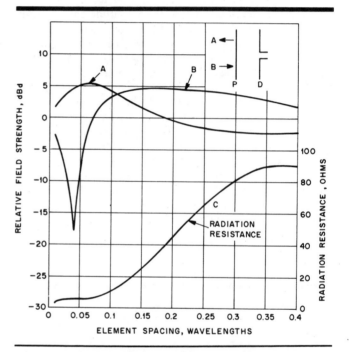

Fig 3—Curves A and B show the calculated gain of a two-element parasitic array over a half-wave dipole as a function of element spacing when the parasitic element is self-resonant. Curve C shows the radiation resistance at the center of the driven element. These curves assume cylindrical elements having a dia = λ/500. (From a study by K1TD)

in gain. At a spacing of 0.22 λ the radiation resistance approaches 52 Ω, suitable for feeding with common types of coaxial line. (Some form of balun should be employed.) Spacings such as these are particularly well suited to antennas using wire elements, such as multielement arrays consisting of combinations of collinear and broadside elements.

Low Radiation Resistance

With a low radiation resistance, and with a fixed loss resistance, more of the power supplied to the antenna is lost in heat. Less power is radiated as the radiation resistance approaches the loss resistance in magnitude. The loss

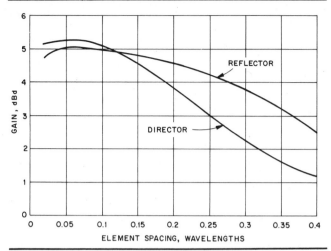

Fig 4—The maximum possible gain obtainable with a parasitic element over a half-wave antenna alone, assuming the parasitic element tuning is adjusted for the greatest gain at each spacing. These curves, based on calculations, assume cylindrical elements having a dia = λ/500, and no ohmic losses. In practical antennas having resistive losses the gain will be less, particularly at close spacings.

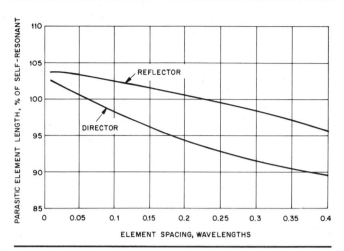

Fig 5—Element spacing versus the amount of length change from self-resonance required for the parasitic element in order to obtain maximum gain. At increased spacings the parasitic element must be made shorter to obtain maximum gain. This is true whether the element is tuned as a director or as a reflector. These curves assume the array is in free space. *(From a study by K1TD)*

resistance can be decreased by using low-resistance conductors for the antenna elements. This means large diameter conductors—usually aluminum, copper, or copper-plated steel tubing. Such conductors have mechanical advantages as well, in that it is relatively easy to provide adjustable sliding sections for changing length. Also, because they can be largely self-supporting, they are well suited for rotatable antenna construction. With ½ inch or larger tubing, the loss resistance in any two-element antenna should be small.

With low radiation resistance, the current and voltage standing waves on the antenna reach considerably higher maximum values than is the case with a simple dipole. For this reason, losses in insulators at the ends of the elements become more important. The use of tubing, rather than wire, helps reduce the end voltage. Further, the tubing does not require support at the ends, eliminating end insulators and their inherent losses.

MAXIMUM GAIN

As implied earlier, the amplitude and phase of the current in the parasitic element can be controlled by adjusting its spacing from the driven element and by adjusting its length (or inserting a lumped reactance at its center). Rather than using a self-resonant parasitic element, usual practice is to tune the parasitic element for either maximum forward gain or for the greatest possible front-to-back ratio.

The maximum gain theoretically obtainable with a single parasitic element is shown in Fig 4 as a function of element spacing. The two curves show the gain over a dipole versus element spacing when the parasitic element is tuned for optimum performance as a director or for optimum performance as a reflector. The selection of director or reflector performance is accomplished by tuning the parasitic element

appropriately—usually by changing its length.

Maximum gain is obtained when the element spacing is approximately 0.06 λ. This is true whether the parasitic element is tuned as a director or as a reflector. At spacings below 0.12 λ, the director provides slightly more gain than the reflector, but the difference is less than ½ dB. At spacings above 0.12 λ, greater gain is obtained when the parasitic element is tuned as a reflector. The difference increases with increased spacing, being ½ dB at a spacing of approximately 0.18 λ.

In only two cases are the gains shown in Fig 4 obtained when the parasitic element is self-resonant. These occur at approximately 0.06- and 0.25-λ spacings, with the parasitic element acting as director and reflector, respectively. For all other cases, the element must be tuned away from self-resonance at the operating frequency. Fig 5 shows the percent of change in element length required for the parasitic element in order to obtain the gains shown in Fig 4, for both the director and reflector cases. These curves are based on cylindrical elements having a diameter of λ/500, but the trends indicated should apply to any practical ratio. As a general statement, for maximum gain the parasitic element must be made correspondingly shorter for increased element spacing.

For reflector operation, the parasitic element must be tuned to a frequency below resonance to obtain maximum gain at all spacings less than about 0.25 λ. At spacings greater than 0.25 λ, the parasitic element must be *shorter* than a self-resonant length to operate as a reflector. The closer the spacing, the greater the detuning required.

Conversely, the director must be detuned toward a higher frequency (it must be shorter than the self-resonant length) at spacings greater than 0.06 wavelength in order to obtain maximum gain. The amount of detuning necessary becomes greater as the spacing is increased. At less than 0.06-wavelength spacing, the director must be tuned below resonance (*longer* than a self-resonant length) to obtain the maximum gain indicated by the Fig 4 curve.

Input Impedance

The mutual impedance between two parallel antenna elements contains reactance as well as resistance. Therefore, the presence of a director or reflector near the driven element affects not only the radiation resistance of the driven element, but also introduces a reactive component (assuming that the driven element is self-resonant). In other words, the parasitic element detunes the driven element. The amount of detuning depends on the spacing and tuning of the parasitic element, and also on the length to diameter ratios of the elements.

For the spacings and tuning conditions that give the gains indicated by the curves of Fig 4, the radiation resistances and reactances at the center of the driven element are shown in Fig 6. For these curves the driven element is assumed to be self-resonant, while the parasitic element is tuned for maximum gain. Fig 6A shows the values when the parasitic element is tuned as a director, and 6B for a reflector. The resistance values are essentially the same for both cases at spacings below 1/8 λ, while at greater spacings the resistance is higher when the parasitic element is tuned as a reflector. The reactive components of the impedances differ for the two cases at all but very close and very wide spacings.

Note from Fig 6 that at close spacings the radiation resistances are quite low, below 5 Ω for spacings of 0.06 λ and less. As discussed earlier, resistive losses contribute to inefficient operation with low values of radiation resistance.

Canceling Feed-Point Reactance

A Yagi array is normally fed at the center of the driven element, and the curves of Fig 6 therefore represent the feed-point resistances and reactances. The reactances shown in Fig 6 may be canceled effectively by altering the length of the *driven* element. (A slight change in radiation resistance will accompany the altered length, but this is normally inconsequential.)

The element should be shortened to cancel inductive reactance, and lengthened to cancel capacitive reactance. With the parasitic element tuned for maximum gain, the effect of the coupled reactance is to make the driven element look more inductive (or less capacitive) with increased parasitic element length. Therefore, the driven element should be slightly longer if the parasitic element is a director, compared to when the parasitic element is a reflector. These principles apply to all spacings of 0.4 λ and below.

FRONT-TO-BACK RATIO

The tuning conditions that give maximum forward gain do not give maximum signal reduction (attenuation) to the rear. The tuning condition, or element length, which gives maximum attenuation to the rear is considerably more critical than that for maximum gain, so a good front-to-back ratio can be obtained without sacrificing much gain. However, some gain must be sacrificed in order to get the highest possible front-to-back ratio. The reduction in rearward response is brought about by adjustment of the tuning of the parasitic element. Most amateurs prefer to adjust a Yagi array for maximum front-to-back (F/B) ratio, rather than for maximum gain. This is done because maximum F/B ratio is more advantageous for receiving, providing better rejection of signals off the rear of the array. Further, the sacrifice in

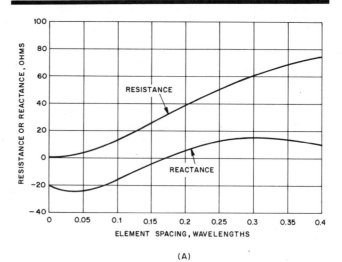

(A)

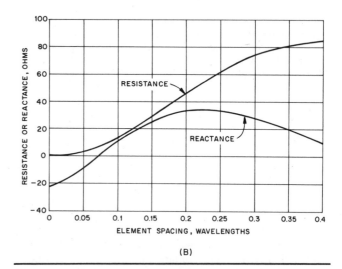

(B)

Fig 6—Calculated free-space radiation resistance and reactance (R and jX) at the center of the driven element as a function of element spacing when the parasitic element is adjusted for the gains given in Fig 4. The driven element is assumed to be self-resonant. Graph A is for the case when the parasitic element is tuned as a director, and graph B as a reflector. These curves assume cylindrical elements with dia = λ/500, and no ohmic losses.

gain is usually less than 1 dB when the array is adjusted for the maximum F/B ratio.

Based on the computer study, the highest possible F/B ratios obtainable for a two-element array are shown as a function of parasitic element spacing in Fig 7. Curves are shown for both cases, when the parasitic element is tuned for maximum F/B ratio as a director, and as a reflector. As may be seen from Fig 7, the F/B ratios attainable with a director are significantly better than with a reflector at spacings below 0.12 λ. Interestingly, it is this same range of spacings in which a director also provides more gain than a reflector, even though the parasitic element is tuned for the best possible F/B ratio.

As the element spacing is increased from 0.12 λ, the F/B ratio with a director falls off rapidly, while that for a reflector changes rather slowly. For a wide-spaced array, the reflector

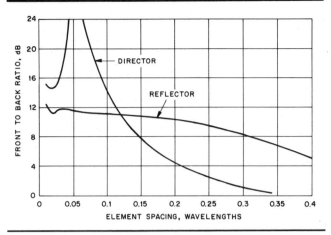

Fig 7—Maximum possible front-to-back ratio versus element spacing for a two-element Yagi having cylindrical elements with dia = λ/500. The parasitic element is assumed to be tuned for the maximum possible F/B ratio as a director or as a reflector at each spacing. The data for these curves were calculated for the array in free space, and in the plane of the array. The off-scale value at a spacing of 0.05 λ is near 30 dB.

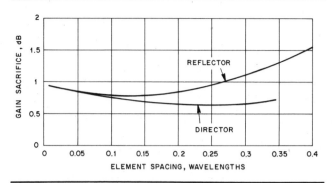

Fig 8—The sacrifice in gain when the parasitic element in a two-element Yagi is adjusted for the highest possible front-to-back ratio, as opposed to the highest possible gain. See text regarding convergence of the curves at spacings below 0.1 λ.

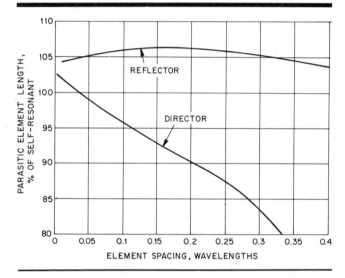

Fig 9—Element spacing versus the amount of length change from self-resonance required for the parasitic element in order to obtain the highest possible F/B ratio. These curves assume the array is in free space. (From a study by K1TD)

offers superior operation over a director for both F/B ratio and gain.

Earlier it was mentioned that when a Yagi array is tuned for maximum F/B ratio, there is a sacrifice in the gain obtainable from the array. Fig 8 shows just how much gain is sacrificed in a two-element array, for the case of a director and a reflector. These curves are based on the K1TD computer study, and show the average of rather widely separated plot points at spacings below 0.1 λ. Thus, it is uncertain without further computer analysis that the two curves actually converge at close spacings, as drawn. Nevertheless, the information should be useful for a builder who is considering the many possible ways of designing a 2-element array. At spacings greater than 0.1 λ, more gain is sacrificed when the parasitic element is adjusted for the maximum F/B ratio as a reflector, rather than as a director, but the difference is less than 1 dB.

When a two-element Yagi array is tuned for maximum F/B ratio by element length adjustment, a director must be made shorter than its self-resonant length at all spacings greater than 0.04 λ. Conversely, a reflector must always be made longer than self-resonant. Fig 9 shows the percent of change in element length required for the parasitic element in order to obtain the F/B ratios shown in Fig 7, for both the director and reflector cases. As with all graphs in this section, these curves are based on cylindrical elements having a diameter of λ/500, but the trends indicated should apply to any practical ratio. As a general statement, for maximum F/B ratio with a director, it must be made correspondingly shorter for increased element spacing. For a reflector, optimum operation is obtained at all spacings when the element is in the range of 5% to 6% longer than a self-resonant length.

The director curve of Fig 9 shows that as the element spacing is increased, the length requirement changes quickly. At a spacing of ⅓ λ, the director must be made approximately 20% shorter than its self-resonant length in order to obtain

the best possible F/B ratio. However, at this great a spacing and with that much element detuning, performance is far from optimum for a two-element array—the F/B ratio is less than 1 dB, and the array gain is only about 1 dBd. But this information clearly illustrates that stray coupling to other conductors in the vicinity of a Yagi array can have a decided effect on its performance, even if the conductors are detuned by an appreciable percentage. Such conductors might be elements of other antennas, guy wires, or even the feed line to the array in question, if it is not properly decoupled.

For the spacings and tuning conditions that give the F/B ratios indicated in Fig 7, the radiation resistances and reactances at the center of the driven element are shown in Fig 10. For these curves the driven element is assumed to be self-resonant, while the parasitic element is tuned for the maximum F/B ratio. The curves of A and B in Fig 10 compare

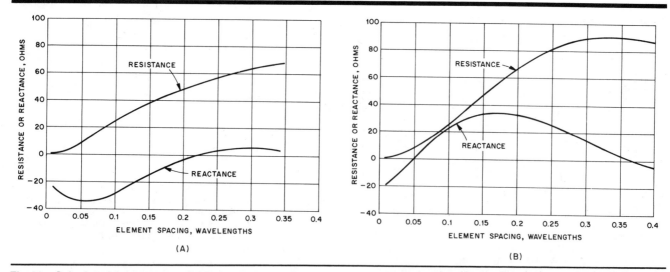

Fig 10—Calculated free-space radiation resistance and reactance (R and jX) at the center of the driven element as a function of element spacing when the parasitic element is adjusted for the F/B ratios given in Fig 7. The driven element is assumed to be self-resonant. Graph A is for the case when the parasitic element is tuned as a director, and graph B as a reflector. These curves assume cylindrical elements with dia = λ/500, and no ohmic losses.

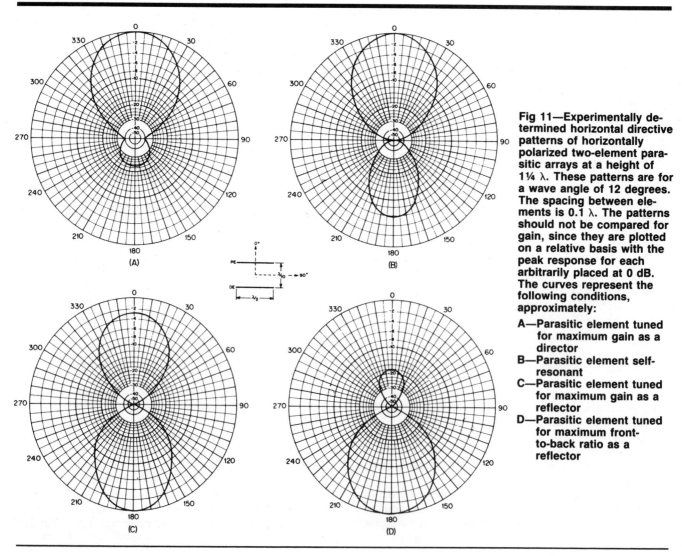

Fig 11—Experimentally determined horizontal directive patterns of horizontally polarized two-element parasitic arrays at a height of 1¼ λ. These patterns are for a wave angle of 12 degrees. The spacing between elements is 0.1 λ. The patterns should not be compared for gain, since they are plotted on a relative basis with the peak response for each arbitrarily placed at 0 dB. The curves represent the following conditions, approximately:

A—Parasitic element tuned for maximum gain as a director

B—Parasitic element self-resonant

C—Parasitic element tuned for maximum gain as a reflector

D—Parasitic element tuned for maximum front-to-back ratio as a reflector

in general in the same way as do the curves of Fig 6A and B, the case where parasitic element tuning was for maximum gain. Much of the earlier discussion about Fig 6 applies here as well.

When the parasitic element is a director and the tuning is from maximum gain to maximum F/B ratio, the radiation resistance at the feed point increases. The reactance values become less inductive (more capacitive) with this shift in tuning.

When the parasitic element is a reflector that is tuned from maximum gain to maximum F/B ratio, the radiation resistance increases, as with the director case. On the other hand, the reactance values at spacings below 0.2 λ become more inductive (less capacitive). At spacings greater than 0.2 λ, the reactance changes in the same way as for the director case at all spacings.

Directional Patterns

The directional patterns obtained with two-element arrays vary considerably with the tuning and spacing of the parasitic element. Typical patterns are shown in Figs 11 and 12. Four cases are represented: the parasitic element length is adjusted for optimum gain as a director, for self-resonance, for optimum gain as a reflector, and for optimum front-to-back ratio as a reflector. Over this range of adjustment, the width of the main beam does not change significantly. These patterns are based on experimental measurements by J. L. Gillson, W3GAU.

BANDWIDTH

The bandwidth of an antenna can be specified in various ways, such as the width of the band over which the gain is higher than some stated figure. Other common criteria are the band over which at least a given front-to-back ratio is obtained, or the band over which the SWR on the transmission line is below a given value. The last is probably the most useful, as the SWR affects the coupling between the transmitter and the line.

The SWR bandwidth depends on the Q of the antenna (see Chapter 26). The Q of close-spaced parasitic arrays is quite high, which results in a relatively narrow SWR bandwidth. Data for a driven element and close-spaced director from another set of experimental measurements, by J. P. Shanklin, are given in Table 1. The antenna elements used in these measurements had a diameter of λ/660.

The antenna with 0.075-λ spacing will, through a suitable matching device, operate with an SWR of less than 3:1 over

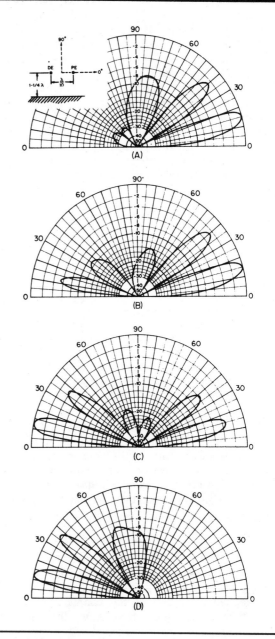

Fig 12—Vertical patterns of a horizontally polarized two-element array under the conditions given in Fig 11. These patterns are in the vertical plane at right angles to the antenna elements. (Patterns of Figs 11 and 12 are based on measurements by J. L. Gillson, W3GAU.)

Table 1
Feed Impedance and Front-to-Back Ratio of a Fed Dipole with One Director
Data based on measurements by J. P. Shanklin.

Element Spacing	Fed Dipole Length	Director Length	Input Resistance at Band Center	Q	Front-to-Back Ratio at Band Center
0.05 λ	0.509 λ	0.484 λ	13.2 Ω	53.2	26.0 dB
0.075	0.504	0.476	24.4	29.4	18.4
0.1	0.504	0.469	28.1	20.0	12.7

a bandwidth of about 3% of the center frequency. This corresponds to about 420 kHz for a 14-MHz antenna. The same antenna has a front-to-back ratio of approximately 10 dB or better over this frequency range. At greater element spacings than those shown in the table, the Q is lower and the bandwidth consequently greater, but the front-to-back ratio is smaller. This is to be expected from the trend shown by the resistance and reactance curves of Figs 3, 6 and 10. The gain is a nearly constant 5 dB for all spacings shown in Table 1.

The same series of experimental measurements showed that with the parasitic element tuned as a reflector for maximum front-to-back ratio, the optimum spacing was 0.2 λ. The maximum front-to-back ratio was 16 dB. In both the director and reflector cases, the front-to-back ratio decreased rather rapidly as the operating frequency was moved away from the frequency for which the system was tuned. With the reflector at 0.2-λ spacing and tuned for maximum front-to-back ratio, the input resistance was 72 Ω and the Q of the antenna was 4.7.

As stated earlier, the antenna elements used in these measurements had a diameter of λ/660. A larger diameter will decrease the rate of reactance change with length and hence decrease the Q, while a smaller diameter will increase the Q. The use of fairly thick elements is desirable when maximum bandwidth is desired.

The Three-Element Beam

It is possible to use more than one parasitic element in conjunction with a single driven element. With two parasitic elements, the optimum gain and directivity result when one is used as a reflector and the second as a director. Such a three-element antenna is shown in Fig 13.

As the number of parasitic elements is increased, the problem of determining the optimum element spacings and lengths to meet given specifications (maximum gain, maximum F/B ratio, maximum bandwidth, and so on) becomes extremely tedious because of the large number of variables involved. In general, when one of these quantities—gain, F/B ratio, or bandwidth—is maximized, the other two cannot be. Also, if it is desired to design the antenna with a specific input impedance, the other three cannot be maximized.

Power Gain

A theoretical investigation of the three-element case (director, driven element and reflector) by Uda and Mushiake has indicated a maximum gain of slightly more than 7 dB (see bibliography at the end of this chapter). A number of experimental investigations have shown that the optimum spacing between the driven element and reflector is between 0.15 and 0.25 λ, with 0.2 λ representing probably the best overall choice. With 0.2-λ reflector spacing, Fig 14 shows the variation in gain with director length, with the director also spaced 0.2 λ from the driven element. Fig 15 shows the gain variation with director spacing. The director spacing is not especially critical, and the overall length of the array (boom length in the case of a rotatable antenna) can be anywhere between 0.35 and 0.45 λ with no appreciable difference in gain.

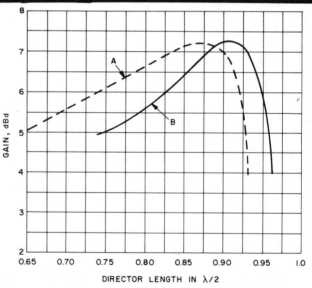

Fig 14—Gain of a three-element Yagi over a dipole as a function of the director length for 0.2-λ spacing between driven element and director and between driven element and reflector. These curves show how the element thickness affects the optimum length. Curve A is for an element diameter of λ/50, while Curve B is for a diameter of λ/500. (A diameter of λ/500 is approximately ½ in. at 50 MHz.) Where the relative diameter is smaller, as on the lower frequencies, the optimum director length will be somewhat greater.

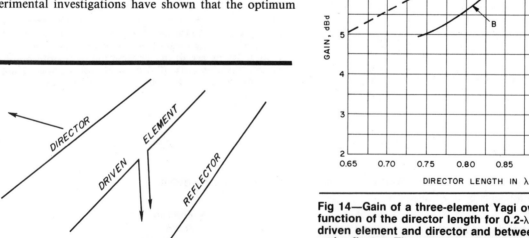

Fig 13—Antenna system using a driven element and two parasitic elements, one as a reflector and one as a director.

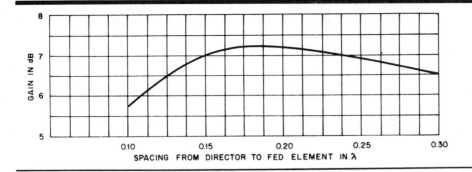

Fig 15—Gain of three-element Yagi as a function of director spacing; the reflector spacing is fixed at 0.2 λ. *(Curves of Figs 14 and 15 are from work of Carl Greenblum.)*

Wide spacing of both elements is desirable not only because it results in high gain, but also because adjustment of tuning or element length is less critical. Also, the input impedance of the driven element is higher than with close spacing. The higher feed-point impedance results in higher efficiency and makes a greater bandwidth possible. However, a total antenna length (director to reflector) of more than 0.3 λ at frequencies on the order of 14 MHz introduces considerable difficulty from a constructional standpoint, so lengths of 0.25 to 0.3 λ are frequently used for this band, even though they are less than optimum.

In general, the gain of a three-element antenna drops off less rapidly when the reflector length is increased beyond the optimum value than it does for a corresponding decrease below the optimum value. The opposite is true of a director, as shown by Fig 14. It is therefore advisable to err, if necessary, on the long side for a reflector and on the short side for a director. This also tends to make the antenna performance less dependent on the exact frequency at which it is operated, because an increase above the design frequency has the same effect as increasing the length of both parasitic elements, while a decrease in frequency has the same effect as shortening both elements. By making the director slightly short and the reflector slightly long, there will be a greater spread between the upper and lower frequencies at which the gain decreases rapidly.

Input Impedance

The radiation resistance (as measured at the center of the driven element) of a three-element array can vary over a fairly wide range, as it is a function of the spacing and tuning of the parasitic elements. There are, however, certain fairly well-defined trends:

1) The resistance reaches a minimum at the parasitic-element tuning condition that gives maximum gain, and increases as the element is detuned in either direction.

2) The resistance is lower for closer spacings between the parasitic and driven elements. Values on the order of 10 Ω are typical with a three-element beam having 0.1-λ director spacing when the director length is adjusted for maximum gain. This can be raised considerably—to 50 ohms or more—by sufficient change in director length, but some gain must be sacrificed. The minimum value of resistance increases with increased director spacing, and is on the order of 30 ohms at a spacing of 0.25 λ.

As in the case of the two-element beam, tuning and spacing of the parasitic elements affect the reactance of the driven element. (A change in the spacing or length of the parasitic elements will change the resonant frequency of the driven element.) Interestingly, the resonant length of the driven element is essentially the same whether the parasitic elements are present (and properly tuned) or not present at all. This is because the two parasitic elements reflect opposing reactances to the driven element and tend to cancel each other's effects at the feed point.

Fig 16 shows the input resistance of three-element arrays with an overall length (director to reflector) of 0.3 λ. The curves give resistance contours as a function of the spacing

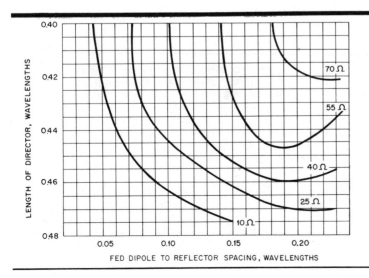

Fig 16—Resonant resistance of the driven element in a three-element parasitic antenna, overall length 0.3 λ. *(Based on measurements by J. P. Shanklin.)*

between the driven element and the reflector and the length of the director. If the reflector length is in the optimum region for a good F/B ratio (as described in the next section), small changes in reflector length have only a comparatively small effect on the input resistance. In Fig 16, the spacing between the director and driven element is equal to the difference between 0.3 λ and the selected driven-element to reflector spacing, as the length of the array is constant. The elements used in obtaining the data in Fig 16 have a diameter of λ/660.

Front-to-Back Ratio

The element lengths and spacings are more critical when a high F/B ratio is the objective than when the antenna is designed for maximum gain. Some gain must be sacrificed for the sake of a good F/B ratio, just as is the case with the two-element array. In general, a high F/B ratio requires fairly close spacing between the director and driven element, but considerably larger spacings are optimum for the reflector. The F/B ratio will change more rapidly than the gain when the operating frequency differs from that for which the antenna was adjusted.

Front-to-back ratio decreases with increased element spacing. However, with a director spacing of about 0.2 λ, it is possible to obtain a very good front-to-*side* ratio, which is useful in some situations. As is the case with F/B ratio, the reflector spacing is much less critical.

Bandwidth

The bandwidth with respect to input impedance (as evidenced by the change in SWR over a band of frequencies) is generally smaller for lower values of input resistance. In turn, this becomes smaller when the element spacing decreases. Hence, close spacings are usually associated with small bandwidths, especially when the element lengths are adjusted for maximum gain.

Fig 17 shows how the Q of a three-element antenna with a total length of 0.3 λ varies as a function of spacing and director length. Based on measurements by J. P. Shanklin, the data is for an element diameter of λ/660, but should hold for diameters between λ/400 and λ/800. From the standpoint

of impedance bandwidth, the upper right-hand region of the chart is best, since this region is associated with low Q values.

These measurements showed that the length of the reflector for optimum F/B ratio did not vary widely. In the low-Q region of Fig 17 it was 0.51 λ, increasing to 0.525 λ in the high-Q region. The proper driven element length was found to be 0.49 λ (at the center of the band) for all conditions.

Similar tests were made on antennas with overall lengths of 0.2 and 0.4 λ. These tests showed that the smaller length would give high F/B ratios but with high Q and consequently small bandwidth. The 0.4-λ case gives low Q and good bandwidth, but the F/B ratio was smaller. The Q values given by the chart can be used as described in Chapter 26 to find the bandwidth over which the SWR will not exceed a specified value.

The low radiation resistance values are accompanied by a high degree of selectivity in the antenna; that is, its impedance is constant over only a small frequency range. These changes in impedance make it difficult to couple power from the transmitter to the line. Such difficulties can be reduced by using wider spacing (on the order of 0.2 λ or more).

Directive Patterns

Directive patterns for three-element arrays, based on experimental measurements made at VHF by J. L. Gillson, W3GAU, show that the beam is somewhat sharper (as is to be expected) when the parasitic element is tuned for maximum gain. Raising of the antenna will lower the wave angle, as the shape and amplitude of the vertical lobes are determined by the ground reflection factors given in Chapter 3, as well as by the free space pattern of the antenna itself.

DESIGNING THREE-ELEMENT YAGIS

Perhaps the most popular type of antenna used on the 14- to 28-MHz amateur bands is the three-element array. Three elements offer the best compromise between gain, size, weight, wind loading and F/B ratio. Construction techniques are described later in this chapter.

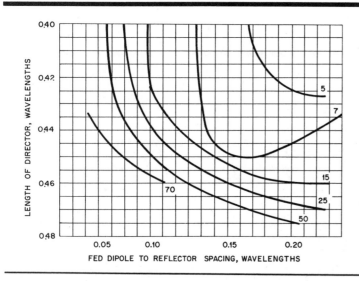

Fig 17—Q of input impedance of the driven element in a three-element parasitic antenna, overall length 0.3 λ. *(J. P. Shanklin)*

As stated earlier, the element tuning for maximum gain is not very critical, and the lengths given by the following equations have been found to work well in practice for three-element antennas:

$$\text{Driven element length (ft)} = \frac{475}{f(MHz)}$$

$$\text{Director length (ft)} = \frac{455}{f(MHz)}$$

$$\text{Reflector length (ft)} = \frac{500}{f(MHz)}$$

These are average lengths determined experimentally for elements having a diameter of $\lambda/400$ to $\lambda/800$, and with element spacings from 0.1 to 0.2 λ.

Fig 18 gives dimensions for a three-element antenna with attention given to the F/B ratio. The reflector length is the most critical for optimum F/B ratio, and it should be adjusted as part of the tune-up procedure for best performance.

As an example of the use of Fig 18, assume we wish to design a 0.1D-0.15R beam for 28.6 MHz. This array would have a director length of 465/28.6 = 16.259 ft = 16 ft 3 in. The reflector length would be 496/28.6 = 17.343 ft = 17 ft 4 in., and the driven-element length 475.2/28.6 = 16.615 ft = 16 ft 7 in. The radiation resistance to be expected is on the order of 21 Ω.

FOUR-ELEMENT ARRAYS

Parasitic arrays with a single driven element and three parasitic elements—reflector and two directors—are fre-

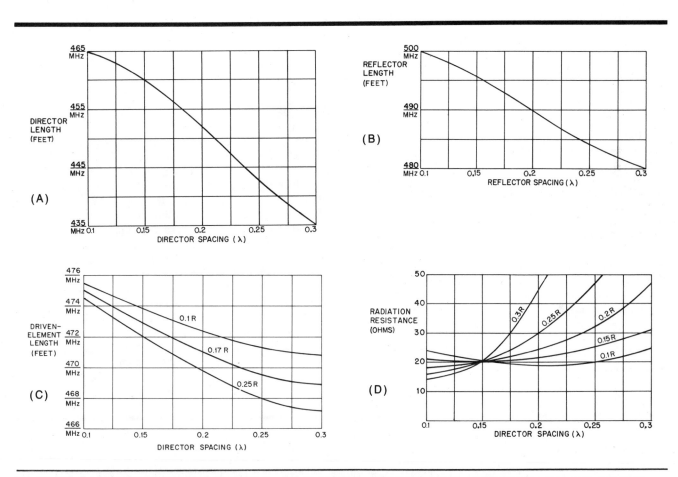

Fig 18—Element lengths for three-element Yagis. These lengths will hold closely for tubing elements supported at or near the center. The radiation resistance (D) is useful information in planning a matching system, but it varies with height above ground and must be considered an approximation. The driven-element length (C) may require modification if reactance is to be tuned out with a gamma- or hairpin-match feed system.

quently used at and above 14 MHz. This type of antenna is shown in Fig 19.

Close spacing is undesirable in a four-element antenna because of the low radiation resistance. An optimum design, based on an experimental determination at 50 MHz, uses the following spacings:

Driven element to reflector—0.2 λ
Driven element to first director—0.2 λ
First director to second director—0.25 λ

Using a diameter of about λ/200 for the elements, the element lengths for maximum gain were found to be

Reflector—0.51 λ
Driven element—0.47 λ
First director—0.45 λ
Second director—0.44 λ

The input resistance with these spacings and lengths is about 30 ohms, and the antenna gives useful gain over a total bandwidth equal to about 4% of the center frequency.

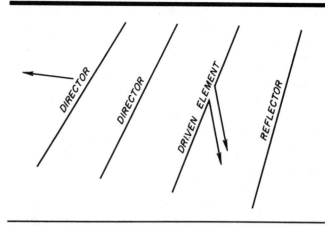

Fig 19—A four-element antenna system, using two directors and one reflector in conjunction with a driven element.

Long Yagis

Parasitic arrays are not limited as to the number of elements that can be used, although very large antennas (eight elements or more) are not very practical below 30 MHz. However, on the VHF bands an array that is long (in terms of wavelengths) is often physically practical. Several independent investigations of the properties of multielement Yagi antennas have shown that the gain of the antenna expressed as a power ratio is proportional to the length of the array, provided the number, lengths and spacings of the elements are chosen properly.

The results of one such study by Carl Greenblum are shown in terms of the number of elements in the antenna, Figs 20 and 21. In each case the antenna consists of a driven element, one reflector and a series of directors properly spaced and tuned. Thus, if the antenna is to have a gain of 12 dB, Fig 20 shows that eight elements—a driven, a reflector, and

six directors—will be required. Fig 21 shows that for such an eight-element antenna the array length required is 1.75 λ.

Table 2 shows the optimum element spacings determined from investigations by C. Greenblum. There is a fair amount of latitude in the placement of the elements along the length of the array, although the optimum tuning of each element will vary somewhat with the exact spacing chosen. Within the spacing ranges shown, the gain will not vary more than 1 dB provided the director lengths are appropriately adjusted.

The optimum director lengths are generally greater for closer director-to-driven element spacings, but the length does not uniformly decrease with increasing distance from the driven element. Fig 22 shows the experimentally determined lengths for various element diameters, based on cylindrical elements mounted through a cylindrical metal boom that is two or three times the element diameter. These curves

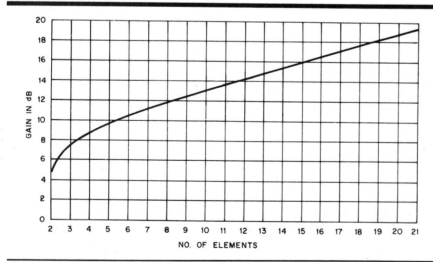

Fig 20—Gain in decibels over a half-wave dipole as a function of the number of elements in the Yagi, assuming the array length is as given in Fig 21. (C. Greenblum)

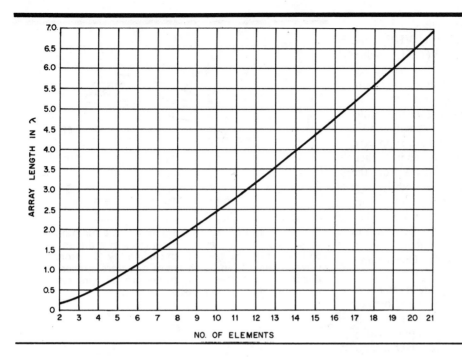

Fig 21—Optimum length of Yagi antenna as a function of the number of elements. (C. Greenblum)

Table 2

Optimum Element Spacings for Multielement Yagi Arrays

No. Elements	R-DE	$DE-D_1$	D_1-D_2	D_2-D_3	D_3-D_4	D_4-D_5	D_5-D_6
2	0.15-0.2 λ						
2		0.07-0.11 λ					
3	0.16-0.23	0.16-0.19					
4	0.18-0.22	0.13-0.17	0.14-0.18 λ				
5	0.18-0.22	0.14-0.17	0.15-0.20	0.17-0.23 λ			
6	0.16-0.20	0.14-0.17	0.16-0.25	0.22-0.30	0.25-0.32 λ		
8	0.16-0.20	0.14-0.16	0.18-0.25	0.25-0.35	0.27-0.32	0.27-0.33 λ	0.30-0.40 λ
8 to N	0.16-0.20	0.14-0.16	0.18-0.25	0.25-0.35	0.27-0.32	0.27-0.33	0.35-0.42

DE—Driven Element; R—Reflector; D—Director; N—any number; director spacings beyond D_6 should be 0.35-0.42 λ.

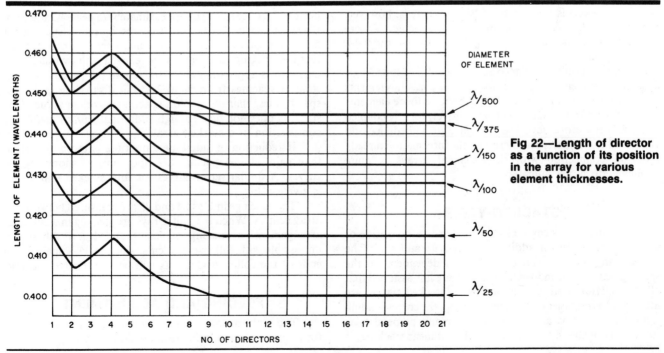

Fig 22—Length of director as a function of its position in the array for various element thicknesses.

probably would not be useful for other shapes.

In another study of long Yagi antennas at VHF, J. A. Kmosko, W2NLY, and H. G. Johnson, W6QKI, reached essentially the same general conclusions concerning the relationship between overall antenna length and power gain, although their gain figures differ from Greenblum's. The comparison is shown in Fig 23. The Kmosko-Johnson results are based on a somewhat different element spacing and a construction in which thin director elements are supported above the metal boom rather than running through it. In their optimum design the first director is spaced 0.1 λ from the driven element. The next two directors are slightly over 0.1 λ apart, the fourth director is approximately 0.2 λ from the third. Succeeding directors are spaced 0.4 λ apart. The Kmosko-Johnson figures are based on a simplified method of computing gain from the beamwidth of the antenna pattern (the beamwidths were measured experimentally). Greenblum's data is from experimental measurements.

Experimental gain figures, based on measurements at 9 GHz by H. W. Ehrenspeck and H. Poehler, are shown by a third curve in Fig 23. They indicate lower gain for a given antenna length but confirm the gain/length proportionality relationship. These measurements were made over a large ground plane using elements on the order of ¼-λ high. The general conclusions of this study were:

1) Reflector spacing and tuning is independent of the other antenna dimensions; the optimum fed-element to reflector spacing is about 0.25 λ, but is not critical.

2) For a given antenna length, gain is essentially independent of the number of directors provided the director-to-director spacing does not exceed 0.4 λ.

3) The optimum director tuning varies with director spacing, but that for constant spacing, all directors can be similarly tuned.

4) A slight improvement in gain results from using an extra director spaced about 0.1 λ from the driven element.

The agreement between these three sets of measurements is not as close as might be expected, confirming the difficulty of determining an optimum design where a multiplicity of elements is used. Measuring gain with a degree of accuracy that will permit reconciliation of the results obtained by various observers is comparably difficult. There is, however, agreement on the general principle that overall length is more important than the number of elements (provided the 0.4-λ element spacing is not exceeded).

The feed-point impedance and bandwidth of long Yagis depends almost entirely on the two or three parasitic elements closest to the driven element. This is because those elements farther from the driven element are relatively loosely coupled to it, and therefore have less influence on impedance and bandwidth. In this respect, therefore, the information already given in connection with three-element arrays is quite applicable to larger antennas.

STACKED YAGIS

Parasitic arrays can be stacked either in a broadside or a collinear fashion for additional directivity and gain. The increase in gain that can be realized is dependent on the spacing between the individual arrays. Here it is assumed that all the individual arrays making up the stacked system are identical. In the case of broadside stacking, the corresponding elements must be parallel and must lie in planes perpendicular to the axis of the individual arrays. In collinear stacking, it

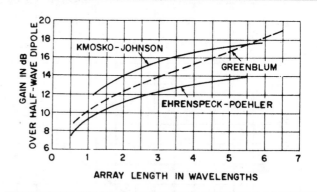

Fig 23—Gain of long Yagi antennas as a function of overall length. The antenna consists of a driven element, a single reflector spaced approximately ¼ λ from the driven element, and a series of directors spaced as described in the text. The three curves represent the results of three independent studies.

is assumed that the corresponding elements are collinear and all elements of the individual arrays lie in the same plane. In both cases the driven elements must be fed in phase.

The decrease in beamwidth of the main radiation lobe that accompanies stacking is usually accompanied by a "splitting off" of one or more sets of side lobes. Their amplitudes will depend on the shape of the directive pattern of the unit array, the number of unit arrays, and their spacing. An optimum spacing is one which gives as much gain as possible without allowing the side lobes to exceed some specified amplitude relative to the main lobe. Fig 24 shows the optimum spacings for three such conditions (no side lobes, side lobes down 20 dB and side lobes down 13 dB) as a function of the half-power beamwidth of the unit array. Calculations by H. W. Kaspar, K2GAL, show that maximum gain occurs when the side lobes are approximately 13 dB down, as indicated in Fig 24.

A single three-element array will have a half-power beamwidth (free space) of approximately 75 degrees, and from Fig 24 it can be determined that the optimum spacing for maximum gain is ¾ λ. Measurements by Greenblum show that the stacking gain that can be realized with two such Yagi antennas is approximately 3 dB, remaining practically constant at spacings from ¾ to 2 λ, but decreasing rapidly at spacings less than ¾ λ. With spacings less than about ½ λ, stacking does not yield enough additional gain to justify the construction of a stacked array.

If reduction of side-lobe amplitude is more important than gain, closer spacings are optimum, as shown by the curves. A similar set of curves for four stacked unit arrays is given in Fig 25.

The spacings in Figs 24 and 25 are measured between the array centers. When the unit arrays are stacked in a collinear arrangement, a spacing of less than ½ λ is physically impossible with full-size elements, since at ½-λ spacing the ends of the collinear elements will be practically touching.

FEEDING AND ADJUSTMENT

The problems of matching and adjusting parasitic arrays for maximum performance are the same in principle as with

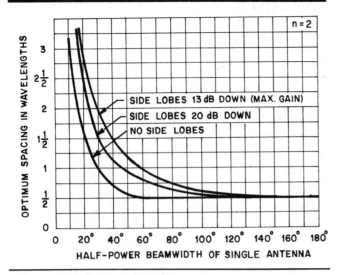

Fig 24—Optimum stacking spacing for two antennas. The spacing required to eliminate side lobes may result in almost no gain improvement with stacking, especially for narrow beamwidths.

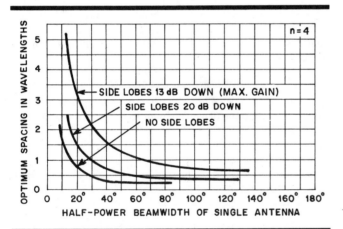

Fig 25—Optimum stacking distance for four antennas. Spacings less than ½ λ are physically possible only for shortened dipoles in the case of collinear elements, or for stacking in the plane perpendicular to the plane of polarization.

other antenna systems. Adjustment of element lengths for optimum performance usually necessitates measurement of relative field strength. However, the experience of a great many amateurs who have followed this rather laborious procedure (adjusting each element a little at a time and measuring the relative field after each such change) has accumulated a large amount of data on optimum lengths.

Depending on the objective in designing the antenna—maximum gain, maximum F/B ratio, etc—it is possible to predetermine the actual element lengths for a given center frequency and thus avoid the necessity for such adjustments. Charts giving proper element lengths for three-element beams are shown in Chapter 9 for the maximum-gain condition, and data for maximum F/B ratio were given earlier in this chapter. The principal adjustment required is proper matching of the antenna to the feed line.

Feeding Stacked Arrays

The load presented by the driven element in a parasitic array differs from the load presented by a simple dipole only in that its resistance may be considerably lower, especially if close spacings are used between elements. The rate of change of reactance as the operating frequency is moved away from the design frequency may also be greater in a parasitic array. With low input resistance, a fairly large impedance step-up is required for matching to the feed line, and the mismatch will increase more rapidly than with a simple dipole when the applied frequency is varied.

Practically any of the matching systems detailed in Chapter 26 are applicable. The use of a folded or multi-conductor dipole for the driven element provides a useful method of matching a low antenna resistance to a suitable transmission line. Design details are given in Chapter 26.

The choice of a matching system is affected by constructional considerations, because parasitic arrays are usually built to be rotated. The hairpin match, beta match and gamma match are favorites with many amateurs because they fit in well (physically) when the driven element is made of tubing. The T match is also a favorite for VHF and UHF arrays. Another matching system suitable for continuously rotatable antennas uses two large inductively coupled loops, one fixed and one rotatable. Open-wire line is preferred for such coupling, however, so this method is seldom used. Because of the inherent problems with installing open-wire lines, most amateurs prefer coaxial lines for carrying RF energy up a tower or other support, and use rotators with mechanical stops to avoid tangling the feed line around the tower.

Broadening the Response

As mentioned previously, the tuning conditions that give maximum gain with parasitic elements are not highly critical. However, the varying reactance coupled into the driven element and the low radiation resistance at the center of the driven element cause the impedance of the array to change rapidly when the applied frequency is varied.

This impedance variation can be made less rapid by using fairly wide spacing between elements, as mentioned. It is also beneficial to use elements with a fairly large diameter, as this reduces the impedance change with frequency. The use of a folded dipole driven element broadens the bandwidth of the antenna because of the greater effective diameter that results from using two or more conductors. The shorted-transmission-line effect of the conductors is also beneficial.

Yagi Construction

Chapter 21 discusses antenna materials and construction techniques that will meet the needs for most amateur installations. Constructing a rotatable antenna requires materials that are strong, lightweight and easy to obtain. The materials required to build a suitable rotatable antenna will vary, depending on many factors. Perhaps the most important factor that determines the type of hardware needed is the weather conditions normally encountered. High winds usually don't cause as much damage to an antenna as does ice—or even ice along with high winds. Aluminum element and boom sizes should be selected so that the various sections of tubing will telescope to provide the necessary total length.

While the maximum safe length of an antenna element depends to some extent on its diameter, the only laws that specify the minimum diameter of an element are the laws of nature. That is, the element must be rugged enough to survive whatever weather conditions it will encounter.

Adjusting Parasitic Arrays

There are two separate processes in adjusting an array with parasitic elements. One is the determination of the optimum element lengths, depending on whether maximum gain or maximum F/B ratio is desired. The other is matching the antenna to the transmission line. The second is usually dependent on the first, and the results observed on adjusting the element tuning can be meaningless unless the line is equally well matched under all tuning conditions.

Many amateurs have found that very satisfactory results can be obtained simply by cutting the elements to the lengths given by equations, graphs or tables. In fact, even after considerable time is spent optimizing dimensions, the final values are often found to be very close to the starting dimensions. Front-to-back ratio requires more careful adjustment than forward gain. Reflector tuning is generally more critical than any of the other adjustments in this respect.

If the array is constructed with dimensions from equations and so forth, the only adjustment needed is the matching of the driven element to the transmission line. The adjustment procedure for each type of matching arrangement is described in Chapter 26.

Test Setup

The only practical method of adjusting parasitic element lengths for best performance is to measure the field strength from the antenna as adjustments are made. Relative measurements are entirely satisfactory for determining the operating conditions that result in the maximum forward gain or greatest F/B ratio. For this purpose, the measuring equipment does not need to be calibrated; only relative readings are required.

A nearby friend who has a receiver equipped with an S meter can provide the necessary assistance in tuning the array. A few precautions must be taken if this method is to be reliable. The receiving antenna must have the same polarization as the transmitting antenna under test (usually horizontal) and should be reasonably high above its surroundings. The receiving system should be checked for pickup on the transmission line to make sure that the indications given by the receiver result only from signals picked up by the receiving antenna. This can be verified by temporarily disconnecting the line from the antenna (but leaving it in place) and observing the signal strength on the S meter. (The line should

be terminated in its characteristic impedance at the antenna end.) If the readings are not several S units below the readings with the antenna connected, the results from adjustment of the transmitting antenna cannot be relied upon. In checking the F/B ratio, stray pickup at the receiving location must be well below the smallest signal resulting from the transmitter antenna if the results are to be meaningful.

Another method of tuning the antenna is to use a field-strength indicator of the diode-detector variety. Such an indicator should be connected to a dipole antenna mounted some distance away from the transmitter antenna at a height at least equal to the transmitter antenna. There should be no obstructions between the two antennas, and both should have the same polarization. The receiving dipole need not be ½ λ long, although that length is desirable because it will increase the ratio of received energy picked up on the antenna to stray pickup. To eliminate coupling effects, the distance between the two antennas should be at least 3 λ. At shorter distances, the mutual impedance may be large enough to cause the receiving antenna to become part of the transmitting system, leading to inaccurate results.

A useful signal-strength indicating system is shown in Fig 26. The transmission line should drop vertically down to the indicator to eliminate stray pickup. This pickup can be checked as described in the preceding paragraph. If the distance between the two antennas is such that greater sensitivity is needed, a reflector may be placed ¼ λ behind the receiving dipole.

Adjustment Procedure

First set the element lengths to those given by the equations. Match the driven element to the transmission line,

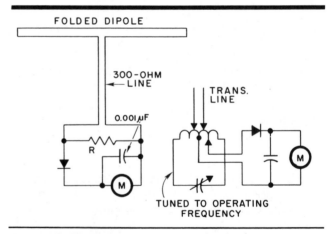

Fig 26—Field-strength measurement setup. The folded dipole should be at least as high as the antenna under test and should be three or more wavelengths away. R should be a 300-ohm noninductive resistor to provide a proper load for the line, so that a line of any desired length can be used. If the sensitivity is not high enough with this arrangement, the alternative connections at the right will result in increased meter readings. The taps are adjusted for maximum reading, keeping the transmission-line taps spaced equally on either side of the coil center tap. The indicating meter, M, may be either a micro-ammeter or 0-1 milliammeter.

obtaining as low an SWR as possible. In subsequent adjustments, a close watch should be kept on the SWR, and the transmitter power output should be kept constant. If the SWR changes enough to affect the coupling at the transmitter when an adjustment is made (but not enough to raise the line loss significantly), readjust the coupling to bring the input back to the same value. If the line loss increases more than a fraction of a decibel, *rematch at the driven element*. If this is not done, the results may be entirely misleading; it is absolutely necessary to maintain constant power input to the driven element if parasitic-element adjustments are to give meaningful results.

The experience of most amateurs in adjusting parasitic arrays indicates that the order in which the elements are tuned is not very important. There is, however, slightly less interaction if the director is first tuned for maximum gain and the reflector is then adjusted to give either maximum gain or maximum F/B ratio, whichever is desired. After the second parasitic element has been adjusted, check the tuning of the first to make sure that it has not been affected by mutual coupling. If there are three parasitic elements, the other two should be checked each time an appreciable change is made in one. The actual lengths should not be very far from those given by the equations when the optimum settings are finally determined. As mentioned previously, the reflector length may be somewhat greater when it is adjusted to give maximum F/B ratio.

Radiation from the transmission line must be eliminated, or at least reduced to a very low level compared to radiation from the antenna itself. Conditions are usually favorable to low line radiation in horizontally polarized parasitic arrays, because the line is usually symmetrical with respect to the antenna and is brought away perpendicular to it. Nevertheless, the line radiation can be appreciable unless the line is decoupled as described in Chapter 26. With coaxial line, some method of line balancing at the antenna should always be incorporated to avoid skewing the beam pattern or lowering the F/B ratio. Decoupling the feed line is necessary for both the transmitting and receiving antennas.

After arriving at the optimum adjustments at the frequency for which the antenna was designed, the performance should be checked over a frequency range either side of the design frequency. If the field strength falls off rapidly with frequency, it may be desirable to shorten the director a bit to increase the gain at frequencies above resonance, and lengthen the reflector slightly to increase gain at frequencies below resonance. Do not confuse the change in SWR with the change in antenna gain. The antenna itself may give good gain over a considerable frequency range, but the SWR may vary considerably in this range. To check the *antenna* behavior, keep the power input to the transmission line constant and rematch the driven element to the line whenever the line losses increase appreciably. If such rematching is found necessary over the band of frequencies to be used, it may be advisable to retune the system to give a higher input impedance. This decreases the Q of the antenna, even though some gain is sacrificed in so doing.

Adjustment by Reception

As an alternative to applying power to the array and checking the field strength, it is possible to adjust the array by measuring received signal strength. It is impractical to do this on distant signals because of fading. The most reliable method is to erect a temporary antenna of the type recommended for field-strength measurements (Fig 26) and drive it with a low-power transmitter. The same precautions as mentioned earlier with respect to distance between the two antennas apply.

Feed-line radiation must also be minimized with this method if the results are to be reliable. The same tests just discussed are valid, but it is not as easy to control the SWR with this method. In the receiving case, the SWR on the transmission line depends on the load presented by the *receiver* to the line. Under most conditions the SWR will be reasonably constant over an amateur band, although its value may not be known. However, the energy transfer from the antenna to the line depends on the mismatch between the driven element and the line. There is no convenient way to check this in the receiving case. The only practical method is to apply power to the array after a set of tuning conditions has been reached, and then rematch at the driven element if necessary. After rematching, the measurement must be repeated. (Double-checking is necessary if the results are to be comparable with those obtained by transmitting from the antenna under test.)

The PV4 Four-Element Monoband Yagi

In the 1970s, Jim Lawson, W2PV, began studying Yagi-Uda antenna performance with computer programs that he developed. (The results of his study were eventually published in the book, *Yagi Antenna Design*—see bibliography at the end of this chapter). One of the conclusions he reached was that designs with equally spaced elements and equal-length directors perform as well as any alternative design, for a fixed boom length. A second conclusion was that boom lengths near an odd number of quarter-wavelengths produce good patterns with high F/B ratios.

Also during that study, Lawson developed an exceptional design that violates both of these principles. The four-element Yagi described here has radically uneven element spacing, and the boom length is near 0.6 λ. The material presented in this section is condensed from an article in October 1986 *QST* by Bill Myers, K1GQ.

The PV4 antenna provides very good gain and excellent F/B on a physically manageable boom length. At 14 MHz, for example, the boom length is a convenient 40 feet. This length has two advantages: (1) the boom itself is nonresonant in the amateur bands (if the elements are isolated from the boom), minimizing the possibility of interaction between the boom and other antennas, and (2) irrigation tubing suitable for boom material is commonly available in 40-foot lengths.

Furthermore, the four-element arrangement has an open space at the center of the boom, so the antenna can easily be side-mounted on most towers.

The PV4 does have some disadvantages. The array has a pronounced weather-vane characteristic caused by the nonuniform element spacing. The driven element is mounted so far toward the reflector end of the boom that it is unreachable from the tower except in the 28-MHz design. Finally, the design is narrowbanded, both in F/B and in input impedance.

The bandwidth limitation is unimportant for the new 10, 18 and 24-MHz bands, since they are quite narrow. The PV4 bandwidth for a 20 dB F/B ratio or greater is about 1.5 percent. Thus, the PV4 is an excellent design choice for the new amateur bands, whereas compromises are necessary for any of the other bands below 30 MHz.

Free-Space Performance

Fig 27 is a view of the antenna from above. It shows all important dimensions and is drawn to scale. Note that the spacings and lengths are given as a fraction of a wavelength. All elements have the same uniform diameter, $\lambda/\text{dia} = 1000$.

The computed free-space gain and F/B ratio for this design are plotted as a function of normalized frequency in Fig 28. The central design frequency corresponds to normalized frequency = 1.0, so you can see how gain and F/B ratio vary over a range of frequencies above and below the design frequency.

The gain of the PV4 is about 9.6 dBi. The F/B curve in Fig 28 shows a very pronounced peak, and the frequency range over which the F/B remains high is much smaller than the bandwidth indicated for the gain curve. If we arbitrarily select a threshold of 20 dB as defining "F/B bandwidth," the PV4 performance bandwidth is about 1.5 percent. As mentioned, this bandwidth is less than the width of any of most amateur bands.

Of course, the central design frequency could be selected to place the F/B maximum at some favored frequency within any amateur band. The performance of the PV4 changes rapidly above the central design frequency, however, so a CW-band design is likely to perform poorly in the phone band. A phone-band design, though, will perform reasonably well on CW, since the gain doesn't drop very much. The F/B ratio will suffer, however. Thus, it is best to center the design in the middle or upper part of the wider ham bands, rather than in the bottom half.

Matching

Fig 29 charts the PV4 input impedance and SWR versus normalized frequency. The reactance is zero at $f = 1.0097$ because the driven element is resonant there; it is deliberately shortened to yield capacitive reactance at the central design frequency. The length of the driven element affects only the

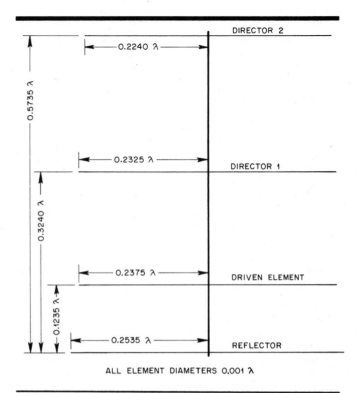

Fig 27—Diagram for the normalized PV4 design. This drawing is to scale.

input impedance—gain and pattern are not changed when the driven element length is modified. The match changes slowly below the design frequency, but it rises steeply above the design frequency. The 2:1 SWR bandwidth extends from 2.5 percent below the central design frequency to just 0.8 percent above. This is another reason to favor centering the PV4 higher in the band rather than lower.

The PV4 provides good results when fed via a hairpin match. (Details of hairpin matching are contained in Chapter 26.) The hairpin inductor can be made of a shorted section of "open-wire" transmission line constructed from two parallel pieces of tubing running along the boom, as shown in Fig 30. The support for the ends of these tubes can be connected to the boom, providing a path to ground for electrostatic charges built up on the driven element. As this feed arrangement is symmetric (that is, balanced), it is useful to isolate currents on the outside of the transmission line by inserting a balun. The choke balun shown in Fig 30 is made of sleeve beads at the connection to the driven element, as described by Walt Maxwell, W2DU. (See bibliography; also see Chapter 26.)

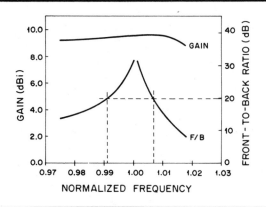

Fig 28—Gain and front-to-back ratio for the PV4 in free space.

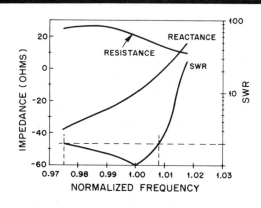

Fig 29—Input impedance and SWR for the PV4 antenna.

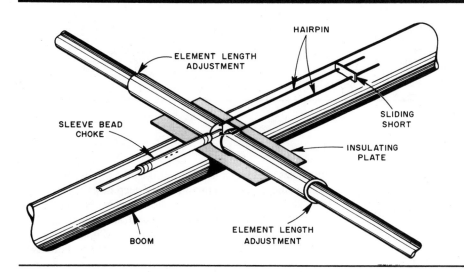

Fig 30—Details of the hairpin matching section for the PV4 driven element.

Radiation Patterns

Fig 31 contains a sequence of radiation patterns for different frequencies around the central design frequency. These patterns are taken in the vertical plane containing the boom, and are normalized to the peak gain at the corresponding frequency (which occurs along the boom for a Yagi in free space). The values of the peak gains are given in Table 3.

The high-angle backward lobe in these patterns is of no consequence because propagation at frequencies above

14 MHz rarely supports such high wave angles. The backward lobe directly along the boom is responsible for the behavior of the F/B ratio plotted in Fig 28. This lobe "tucks in" to some minimum value just above the central design frequency (f = 1.000), corresponding to the peak shown in Fig 28. When mounted over real ground, the wave angles of interest are between some very small value, say 2°, and 10° to 20°, depending on the frequency. The sequence of elevation radiation patterns suggests that F/B ratio is maximum between f = 1.000 and f = 1.005.

Azimuthal radiation patterns for the same sequence of

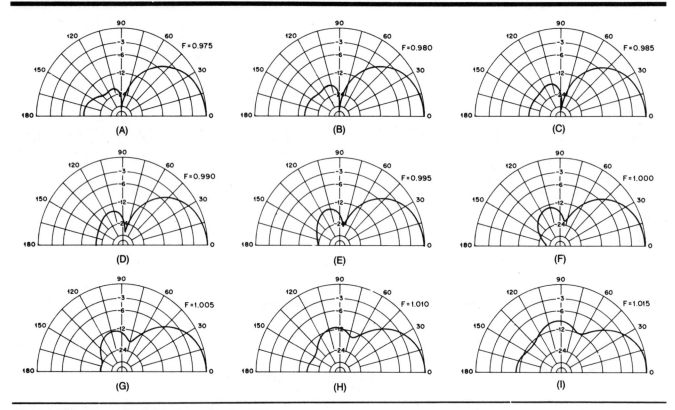

Fig 31—Elevation radiation patterns for the PV4.

Table 3

Variations in Gain, F/B Ratio and Input Impedance of the PV4 Array with Frequency

Normalized Frequency	Gain (dBi)	F/B Ratio	Input Impedance, Ω
0.9750	9.16	14.15 dB	24.96 − j37.82
0.9775	9.18	14.84	25.70 − j35.48
0.9800	9.21	15.57	26.25 − j33.29
0.9825	9.24	16.37	26.58 − j31.20
0.9850	9.28	17.28	26.65 − j29.19
0.9875	9.31	18.35	26.43 − j27.20
0.9900	9.36	19.68	25.90 − j25.17
0.9925	9.41	21.42	25.04 − j23.03
0.9950	9.46	23.82	23.88 − j20.71
0.9975	9.51	27.38	22.45 − j18.12
1.0000	9.56	31.74	20.79 − j15.20
1.0025	9.60	28.81	18.98 − j11.89
1.0050	9.63	23.30	17.12 − j8.16
1.0075	9.62	19.18	15.30 − j4.01
1.0100	9.57	15.95	13.60 + j0.54
1.0125	9.44	13.25	12.10 + j5.45
1.0150	9.21	10.88	10.86 + j10.66
1.0175	8.85	8.74	9.93 + j16.10

frequencies are shown in Fig 32. Like all Yagis with a single driven element, the antenna has very deep nulls off the ends of the elements. The back lobe along the boom diminishes as the frequency increases toward the central design frequency, until the entire lobe is contained within a −20 dB circle. As the frequency is increased still further, the lobe splits into two lobes. The null between these two lobes falls along the boom, producing the maximum in F/B ratio near the central design frequency. This null immediately begins to fill in with a third backward lobe above the central design frequency.

24-MHz DESIGN EXAMPLE

The steps for converting the design parameters into a plan for constructing a real antenna are:

1) Select the central operating frequency.

2) Convert element lengths and spacings from wavelengths to physical units.

3) Adjust the element lengths to account for radius scaling and tapering.

The 24-MHz band extends from 24.89 to 24.99 MHz. As pointed out earlier, the fractional bandwidth is very small, so any frequency within the band can be taken as the central

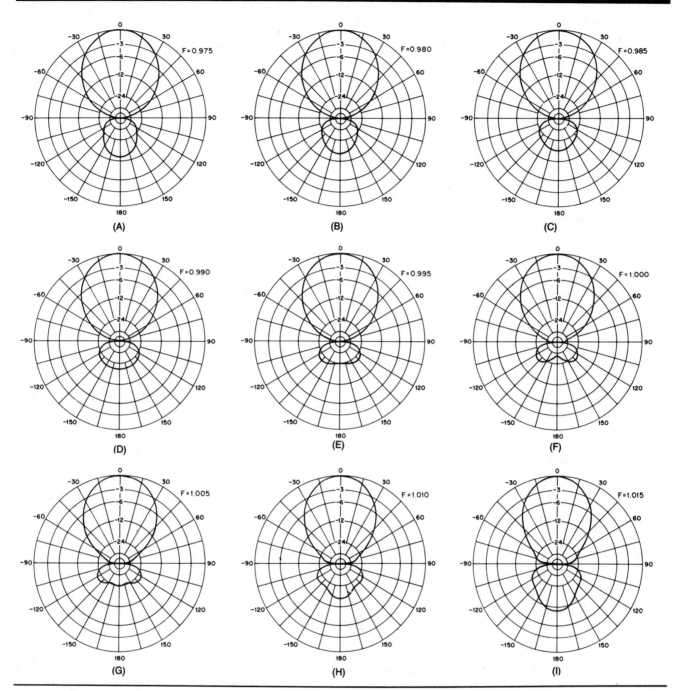

Fig 32—PV4 azimuth radiation patterns.

design frequency with no effect on the final antenna. For convenience, let's take 24.9 MHz as the central design frequency. The wavelength at this frequency, using speed of light = 299.79 meters per microsecond (983.56 feet per microsecond) is 39.50 feet. The dimensions given below are rounded to the nearest ¼ inch, which is about 0.0005 λ.

The boom length, 0.5735 λ, converts to 22 feet, 7¾ inches. The boom should be about 23 feet long to allow space for mounting the end elements. A single section of 3 inches OD irrigation tubing should be used. At 24.9 MHz, the elements are short enough that a boom made from this material wouldn't need any special mechanical supports or reinforcements. The element positions along the boom are:

Reflector: 0 ft 2 in.
Driven element: 5 ft ½ in.
Director 1: 12 ft 11½ in.
Director 2: 22 ft 9¾ in.

The element half-lengths (using 0.2330 λ for the driven element) convert from wavelengths to physical units as follows:

Reflector: 10 ft ¼ in.
Driven element: 9 ft 2½ in.
Director 1: 9 ft 2¼ in.
Director 2: 8 ft 10¼ in.

The lengths calculated above apply if the elements are cylindrical, with a diameter of 0.001 λ. For 24.9 MHz, this diameter calculates to be 0.474 inch. Tubing of ½-inch dia might be acceptable, although the diameter difference is in excess of 5%. Further, ½-inch tubing for the lengths required will not provide a sturdy array. If elements of a different diameter or tapered elements are used, a correction should be applied to these lengths.

Radius scaling accounts for the difference between the design element diameter and the actual average diameter. The taper correction accounts for the effect of variations in element diameter with telescoping tubing used in the construction. Radius scaling and element taper are treated in Chapter 2. Additional corrections can be made to account for the effects of the boom and boom-to-element mounting structures, but these effects are negligible at 24 MHz unless the elements pass through the boom.

A very rugged element for 24 MHz can be made by telescoping lengths of ¾-inch OD tubing into the ends of a single 12-foot length of 7/8-inch OD tubing. The 7/8-inch tubing must have a wall thickness of 0.058 inch (a standard size). Other thicknesses will not telescope properly. The elements for the PV4 24-MHz beam can be made from four 12-foot lengths of 7/8-inch and three 12-foot lengths of ¾-inch tubing.

With scaling and taper corrections, the lengths of the ¾-inch element end pieces are calculated to be:

Reflector: 49½ in.
Driven element: 38½ in.
Director 1: 38¼ in.
Director 2: 33¾ in.

Allowing for 4 to 5 inches of overlap at the telescoping joints, these end pieces can be cut from the three 12-foot lengths of ¾-inch tubing as follows:

1) From the first piece of tubing, cut one 54-inch piece and two 45-inch pieces. This yields one reflector tip and the two driven element tips.

2) Cut the second tube in the same way, yielding the second reflector tip and the two director 1 tips.

3) Cut two 38-inch pieces for director 2 from the third tube; you will have 68 inches left for your next project.

PV4 arrays have been built for various amateur bands. Measured performance, that is, patterns and SWR curves, agree well with the computer predictions.

A Linear-Loaded 14-MHz Beam

There are two reasons for considering a shortened antenna. The first has to do with considerations of space and appearance. The second is cost. The antenna of Fig 33 is a winner in both categories. It was first described by Cole Collinge, WØYNF, in June 1976 QST.

Aluminum for the elements consists of four 12-foot lengths of 7/8-inch diameter tubing. Two 2-foot lengths of aluminum channel are used for the element supports (see Fig 34). Each element section is attached using 1-inch ceramic standoff insulators. A ¾-inch birch dowel was used to stiffen and prevent crushing of the tubing where the attachment is made. Element supports are grooved and attached to a 10-foot TV mast section using a single 5/16-inch U bolt.

The loading wire is no. 12 copper wire. It is supported by two Bakelite blocks on each element section. The blocks are anchored to the tubing by small sheet-metal screws. Ends of the wire sections are formed into a loop to hold their position.

Dimensions for the antenna are given in Fig 35. Details of the driven element are shown in the photo.

Fig 33—The driven element of the linear-loaded beam.

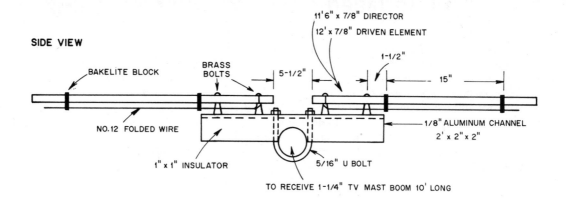

SIDE VIEW

BAKELITE BLOCK

BRASS BOLTS

11'6" x 7/8" DIRECTOR

12' x 7/8" DRIVEN ELEMENT

1-1/2"

5-1/2"

15"

NO.12 FOLDED WIRE

1" x 1" INSULATOR

5/16" U BOLT

1/8" ALUMINUM CHANNEL 2' x 2" x 2"

TO RECEIVE 1-1/4" TV MAST BOOM 10' LONG

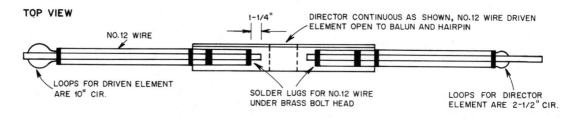

TOP VIEW

NO.12 WIRE

1-1/4"

DIRECTOR CONTINUOUS AS SHOWN, NO.12 WIRE DRIVEN ELEMENT OPEN TO BALUN AND HAIRPIN

LOOPS FOR DRIVEN ELEMENT ARE 10" CIR.

SOLDER LUGS FOR NO.12 WIRE UNDER BRASS BOLT HEAD

LOOPS FOR DIRECTOR ELEMENT ARE 2-1/2" CIR.

Fig 34—Construction details for the linear-loaded 14-MHz beam.

SET SCREW

2-1/8"

7/8"

1-3/4"

1-1/8"

BLOCK DETAIL

7/16"

FOR NO.12 WIRE SUPPORT (1/8" DRILL)

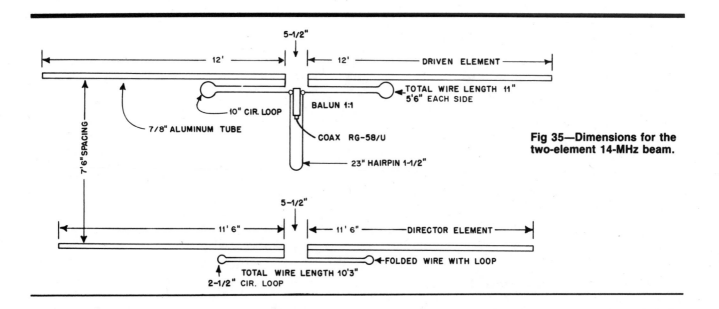

5-1/2"

12'

12'

DRIVEN ELEMENT

7'6" SPACING

7/8" ALUMINUM TUBE

10" CIR. LOOP

BALUN 1:1

COAX RG-58/U

23" HAIRPIN 1-1/2"

TOTAL WIRE LENGTH 11" 5'6" EACH SIDE

Fig 35—Dimensions for the two-element 14-MHz beam.

5-1/2"

11' 6"

11' 6"

DIRECTOR ELEMENT

FOLDED WIRE WITH LOOP

TOTAL WIRE LENGTH 10'3"

2-1/2" CIR. LOOP

An Interlaced Yagi For 14 and 21 MHz

It is often desirable to install more than one antenna on top of a single tower or mast. Stacking antennas one above the other creates a large stress on the mast and the rotator. With large arrays, it is desirable to reduce the weight and wind loading characteristics in every possible way to reduce damage possibilities from ice and wind. One simple solution to the problem is to mount two complete antennas on one boom.

Installing elements for two different antennas on one boom has been popular for many years. Most commercially manufactured triband antennas use this technique. The question which develops however, is whether or not interaction between elements for different bands causes detrimental effects. It is generally accepted that interaction, if any, is very minimal between bands which are not harmonically related. The example shown here is a Wilson Electronics Model DB-54 (see bibliography) designed to operate on both 14 and 21 MHz. Two driven elements are required and each is fed independently with separate transmission lines. The boom is 40 feet long and is 3 inches OD. Smaller boom sizes are not recommended.

This is a large array, with 5 elements for 14 MHz and 4 elements for 21 MHz. A heavy duty rotator equipped with

a brake is strongly recommended if the antenna is to be rotated.

The element center sections for 14 MHz begin with 1¼ inch tubing and telescope down in size. The 21-MHz elements begin with 1-1/8 inch tubing. All of the critical dimensions are given in Fig 36.

A long boom needs to have additional support if appreciable sag or droop is to be avoided. The truss can be made of any suitable steel wire and should be connected about 10 feet in each direction from the boom-to-mast plate. Turnbuckles should be used at the mast to create suitable tension for the wires. Each of the interlaced arrays can be treated as separate antennas for tune-up. As there is little (if any) interaction between elements, the 21-MHz section could be removed from the boom if only a 14-MHz monoband system is needed.

As mentioned, the array requires separate transmission lines for each band. Of course a single line could be used with a remotely controlled switching arrangement at the antenna. Separate gamma matching networks are used for each band. Suggested dimensions are given in Fig 37. The measured SWR versus frequency for the two bands is plotted in Fig 38.

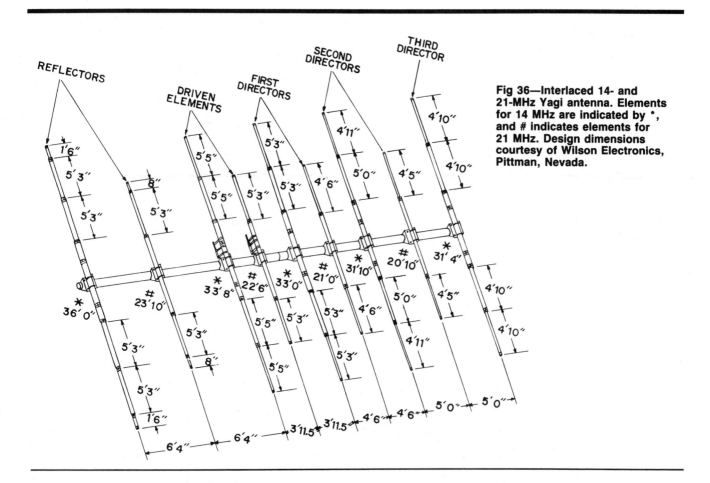

Fig 36—Interlaced 14- and 21-MHz Yagi antenna. Elements for 14 MHz are indicated by *, and # indicates elements for 21 MHz. Design dimensions courtesy of Wilson Electronics, Pittman, Nevada.

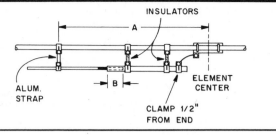

Fig 37—Gamma match arrangement for the interlaced 14-
and 21-MHz Yagi. Dimensions for the two bands are:

	14 MHz	21 MHz
A	41 in.	36 in.
B	7 in.	5 in.

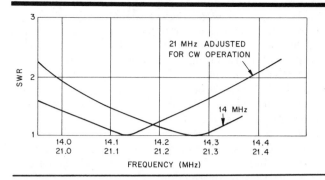

Fig 38—Measured SWR versus frequency for the
interlaced 14- and 21-MHz array.

A Beam Antenna For 21 MHz

When the amateur is interested in constructing an array that can be rotated, aluminum tubing is generally used for the elements. The mechanical problems encountered are usually not much greater for several elements than with single-element rotatable antenna systems.

This four-element Yagi antenna provides appreciable gain and exhibits significant back-rejection characteristics. Overall dimensions are given in Fig 39. Construction is straight-forward, using commonly available tubing material which normally is 12 feet long. The mechanical dimensions for this antenna were developed by Wilson Electronics, Pittman, Nevada.

The center of each element is made from a 12-foot length of 6061-T6 aluminum alloy of 1-1/8 inches diameter. The overall length of the element is determined by the distance the telescoping section is extended beyond the end of the center piece. Each of these telescoping sections is 1-inch OD and 6 feet long to provide the proper fit. Data in Chapter 21 provides a guide for determining proper sizes for element material diameters. When telescoping sections are needed, the

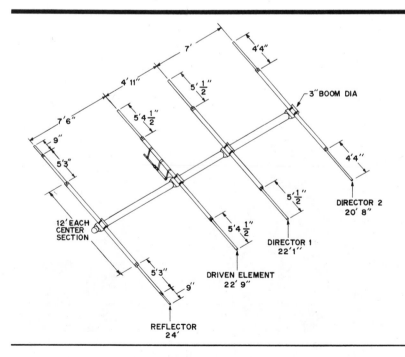

Fig 39—Overall dimensions for the four-
element 21-MHz array.

difference between joining pieces (in terms of diameters) should be about 0.009 inch. An additional section of 7/8-inch OD material is used at the tips of the reflector element to meet the dimensions specified. The two 7/8-inch pieces extend about 9 inches beyond the 1-inch diameter stock.

Each element is held in place with two U bolts that clamp it to a 6-inch long piece of angle aluminum stock. These pieces of angle material are then fastened to the boom with automotive muffler clamps. The size of the muffler clamp depends on the size of the boom. For this model, a 2-inch diameter boom size should be satisfactory for all but the roughest climate conditions. A 3-inch diameter boom does offer very rugged construction.

Matching a feed line to the driven element can be accomplished by using the gamma match dimensions given in Fig 40. Final adjustment of the gamma system should be made after the antenna is mounted in place atop the mast by placing a power meter (or SWR indicator) in series with the feed line at the input connector and adjusting the capacitor along with the tap point for minimum reflected power, as described in Chapter 26.

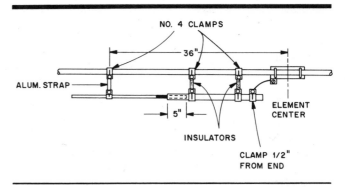

Fig 40—Gamma match dimensions for the four-element 21-MHz array.

While gain measurements are impossible without a test facility, the estimated power gain of this system is on the order of 9 dB. The F/B ratio is typically 20 to 25 dB.

A DL-Style Yagi For 28 MHz

Guenter Schwarzbeck, DL1BU, has made extensive comparisons between various Yagi and quad antenna designs on an antenna test range and has documented his work in the pages of *CQ-DL*. He has been especially interested in bandwidth, F/B ratio and gain. He has designed seven-element Yagis for the 28- and 50-MHz bands that perform well in all three respects. The design has been copied using sizes of aluminum tubing commonly available in the U.S. The 28-MHz version requires a 40-foot boom; the 50-MHz version is described in Chapter 18. Fig 41 shows both arrays.

A gamma match is used, but the constructor may substitute a favorite matching system. It is worth noting that the dimensions of the driven element are not especially critical on any Yagi design, and reasonable adjustments to the length of the driven element can be made in the interests of impedance matching without affecting the other characteristics of the antenna. The same does not hold true for the parasitic elements, as departures from the design can result in significant differences in performance.

Building the 28-MHz version of this seven-element antenna is an ambitious project. The minimum requirement for boom material is 2 inches OD, and additional bracing is essential (see Chapter 21). If the elements are secured to the boom by a single U bolt, the suggested double thickness of tubing should be used at the center of the element. See Figs 42 and 43 for details. While the dimensions given are to cover the bottom half of the 28.0-MHz band, the antenna has reasonable gain but a reduced F/B ratio in the top half.

The 3-dB beamwidth of this seven-element antenna is on the order of 40° in the horizontal plane, which means that a pointing error of as little as 20° will result in a 3-dB loss in signal strength. Operators accustomed to antennas with shorter booms and broader patterns will find that their antenna rotators are getting more of a workout when they switch to an antenna like this.

Fig 41—A 7-element 50-MHz Yagi of the "DL design" is mounted 4 ft above a 7-element 28-MHz Yagi of similar design.

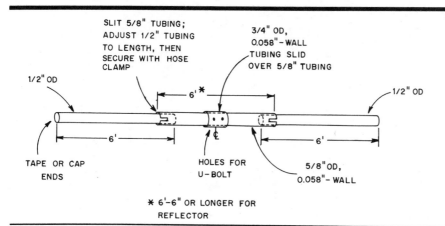

Fig 42—Construction of a typical element for the DL-style 28-MHz Yagi.

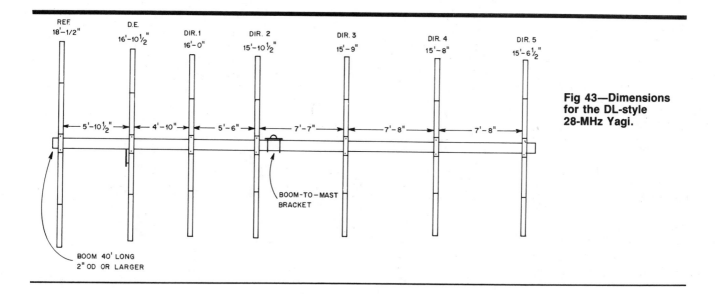

Fig 43—Dimensions for the DL-style 28-MHz Yagi.

BIBLIOGRAPHY

Source material and more extended discussion of topics covered in this chapter can be found in the references given below and in the textbooks listed at the end of Chapter 2.

G. H. Brown, "Directional Antennas," *Proc IRE*, Vol 25, No. 1, Jan 1937, pp 78-145.

C. Collinge, "Linear Loaded 20-Meter Beam," *QST*, Jun 1976, pp 18-19.

H. W. Ehrenspek and H. Poehler, "A New Method for Obtaining Maximum Gain from Yagi Antennas," *IRE Transactions on Antennas and Propagation*, Vol AP-7, No. 4, Oct 1959, pp 379-386.

P. C. Erhorn, "Element Spacing in 3-Element Beams," *QST*, Oct 1947, pp 37-42, 116, 118.

C. Greenblum, "Notes on the Development of Yagi Arrays," Parts 1 and 2, *QST*, Aug and Sep 1956.

H. W. Kasper, "Array Design with Optimum Antenna Spacing," *QST*, Nov 1960, p 23.

King, Mack and Sandler, *Arrays of Cylindrical Dipoles* (London: Cambridge Univ Press, 1968).

J. A. Kmosko and H. G. Johnson, "LONG Long Yagis," *QST*, Jan 1956.

E. A. Laport, *Radio Antenna Engineering* (New York: McGraw-Hill Book Co, 1952).

J. L. Lawson, *Yagi Antenna Design*, 1st edition (Newington: ARRL, 1986).

W. Maxwell, "Some Aspects of the Balun Problem," *QST*, Mar 1983, pp 38-40.

B. Myers, "The W2PV Four-Element Yagi," *QST*, Oct 1986, pp 15-19.

G. C. Southworth, "Certain Factors Affecting the Gain of Directive Antennas," *Proc IRE*, Vol 18, No. 9, Sep 1930, pp 1502-1536.

F. E. Terman, *Radio Engineering*, 3rd edition (New York: McGraw-Hill Book Co, 1947).

S. Uda and Y. Mushiake, *Yagi-Uda* Antenna (Sendai, Japan: Sasaki Publishing Co, 1954). [Published in English—Ed.]

P. P. Viezbicke, "Yagi Antenna Design," *NBS Technical Note 688* (US Dept of Commerce/National Bureau of Standards, Boulder, CO), Dec 1976.

"Wilson Electronics DB-54 20- and 15-Meter Duo-Band Beam," *Recent Equipment, QST*, Jun 1974, pp 40-41.

Chapter 12

Quad Arrays

In the previous chapter it was assumed that the various antenna arrays were assemblies of linear half-wave (or approximately half-wave) dipole elements. However, other element forms may be used according to the same basic principles. For example, loops of various types may be combined into directive arrays. A popular type of parasitic array using loops is the quad antenna, in which loops having a perimeter of one wavelength are used in much the same way as dipole elements in the Yagi antenna.

The quad antenna was designed by Clarence Moore, W9LZX, in the late 1940s. Since its inception, there has been extensive controversy whether the quad is a better performer than a Yagi. This argument continues, but over the years several facts have become apparent. For example, J. Lindsay, W7ZQ, has made many comparisons between quads and Yagis. His data show that the quad has a gain of approximately 2 dB over a Yagi for the same array length. Another argument that has existed is that for a given array height, the quad has a lower angle of radiation than a Yagi. Even among authorities there is disagreement on this point. However, the H-plane pattern of a quad is slightly broader than that of a Yagi at the half-power points. This means that the quad covers a wider area in the vertical plane.

The full-wave loop was discussed in Chapter 2. Two such loops, one as a driven element and one as a reflector, are shown in Fig 1. This is the original version of the quad; in subsequent development, loops tuned as directors have been added in front of the driven element. The square loops may be mounted either with the corners lying on horizontal and vertical lines, as shown at the left, or with two sides horizontal and two vertical (right). The feed points shown for these two cases will result in horizontal polarization, which is commonly used.

The parasitic element is tuned in much the same way as the parasitic element in a Yagi antenna. That is, the parasitic loop is tuned to a lower frequency than the driven element when the parasitic is to act as a reflector, and to a higher frequency when it is to act as a director. Fig 1 shows the parasitic element with an adjustable tuning stub, a convenient method of tuning since the resonant frequency can be changed simply by changing the position of the shorting bar on the stub. In practice, it has been found that the length around the loop should be approximately 3% greater than the self-resonant length if the element is a reflector, and about 3% shorter than the self-resonant length if the parasitic element is a director. Approximate formulas for the loop lengths in feet are

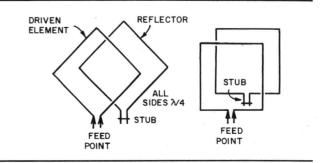

Fig 1—The basic two-element quad antenna, with driven loop and reflector loop. The driven loops are electrically one wavelength in circumference (¼ wavelength on a side); the reflectors are slightly longer. Both configurations shown give horizontal polarization; for vertical polarization, the driven element should be fed at one of the side corners in the arrangement at the left, or at the center of a vertical side in the "square" quad at the right.

$$\text{Driven element} = \frac{1005}{f(\text{MHz})}$$

$$\text{Reflector} = \frac{1030}{f(\text{MHz})}$$

$$\text{Director} = \frac{975}{f(\text{MHz})}$$

for quad antennas intended for operation below 30 MHz. At VHF, where the ratio of loop circumference to conductor diameter is usually relatively small, the circumference must be increased in comparison to the wavelength. For example, a one-wavelength loop constructed of ¼-inch tubing for 144 MHz should have a circumference about 2% greater than in the above equation for the driven element.

In any case, on-the-ground adjustment is required if optimum results are to be achieved, especially with respect to front-to-back ratio. The method of adjustment parallels that outlined in Chapter 11 for Yagi antennas.

Element spacings on the order of 0.14 to 0.2 wavelength are generally used. The smaller spacings are usually employed in antennas with more than two elements, where the structural support for elements with larger spacings tends to become difficult. The feed-point impedances of antennas having element spacings on this order have been found to be in the 40- to 60-ohm range, so the driven element can be fed directly with coaxial cable at only a small mismatch. For spacings on the order of 0.25 wavelength (physically feasible for two elements, or for several elements at 28 MHz) the impedance more closely approximates the impedance of a driven loop alone (see Chapter 2)—that is, 80 to 100 ohms. The feed methods described in Chapter 26 can be used, just as in the case of the Yagi.

Directive Patterns and Gain

The small gain of a one-wavelength loop over a half-wave dipole also appears in arrays of loop elements. That is, if a quad parasitic array and a Yagi with the same boom length are compared, the quad will have approximately 2 dB more gain than the Yagi, as mentioned earlier. This assumes that both antennas have the optimum number of elements for the antenna length; the number of elements is not necessarily the same in both when the antennas are long.

CONSTRUCTION OF QUADS

The sturdiness of a quad is directly proportional to the quality of the material used and the care with which it is constructed. The size and type of wire selected for use with a quad antenna is important because it will determine the capability of the spreaders to withstand high winds and ice.

Table 1

Quad Dimensions

Two-element quad (W7ZQ). Spacing given below; boom length given below.

	7 MHz	14 MHz	21 MHz	28 MHz
Reflector	144'11½"	72'4"	48'8"	35'7"
Driven Element	140'11½"	70'2"	47'4"	34'7"
Spacing	30'	13'	10'	6'6"
Boom length	30'	13'	10'	6'6"
Feed method	Directly with 23' RG-11, then any length of RG-8 coax.	Directly with 11'7" RG-11, then any length RG-8 coax.	Directly with 7'8½" RG-11, then any length RG-8 coax.	Directly with 5'8" RG-11, then any length RG-8 coax.

(Note that a spider or boomless quad arrangement could be used for the 14/21/28-MHz parts of the above dimensions, yielding a triband antenna.)

Four-element quad* (W0AIW (14 MHz)/W7ZQ K0KKU/K0EZH/W6FXB). Spacing: equal, 10 ft; Boom length: 30 ft.**

	14 MHz Phone	14 MHz CW	21 MHz	28 MHz
Reflector	72'1½"	72'5"	48'8"	35'8½"
Driven Element	70'1½"	70'5"	47'4"	34'8½"
Director 1	69'1"	69'1"	46'4"	33'7¼"
Director 2	69'1"	69'1"	46'4"	33'7¼"
Feed Method	Directly with 52-ohm coax.	Directly with 52-ohm coax.	Directly with 52-ohm coax.	Directly with 5'9" RG-11, then any length RG-8 coax.

*Common boom used to form a triband array.
**The two-element 7-MHz quad given above is added to form a four-band quad array.

Four-element quad (W7ZQ/K8DYZ*/K8YIB*/W7EPA*). Spacing: equal, 13'4"; Boom length: 40 ft.

	14 MHz	21 MHz	28 MHz
Reflector	72'5"	48'4"	35'8½"
Driven Element	70'5"	47'0"	34'8½"*
Director 1	69'1"	46'1"	(Directors 1-3 all 33'7")*
Director 2	69'1"	46'1"	
Feed method	Directly with 52-ohm coax.	Directly with 5'9" RG-11, then any length 52-ohm coax.	Directly with 52-ohm coax.

*For the 28-MHz band, the driven element is placed between the 14/21-MHz reflector and 14/21-MHz driven element. The 28-MHz reflector is placed on the same frame as the 14/21-MHz reflectors and the remaining 28-MHz directors are placed on the remaining 14/21-MHz frames. The 28-MHz portion is then a 5-element quad.

Six-element quad (W0YDM, W7UMJ). Spacing: equal, 12 ft; Boom length: 60 ft.

	14 MHz
Reflector	72'1½"
Driven Element	70'1½"
Directors 1, 2 and 3	69'1"
Director 4	69'4"
Feed Method	Directly with 52-ohm coax.

One of the more common problems confronting the quad owner is that of broken wires. A solid conductor is more apt to break than stranded wire under constant flexing conditions. For this reason, stranded copper wire is recommended. For 14, 21 or 28-MHz operation, no. 14 or no. 12 wire is a good choice. Soldering of the stranded wire at points where flexing is likely to occur should be avoided.

Connecting the wires to the spreader arms may be accomplished in many ways. The simplest method is to drill holes through the fiberglass at the approximate points on the arms and route the wires through the holes. Soldering a wire loop across the spreader, as shown later, is recommended. However, care should be taken to prevent solder from flowing to the corner point where flexing could break it.

Dimensions for quad elements and spacing have been given in texts and *QST* over the years. Quad tuning is not very critical, nor is element spacing. Table 1 is a collection of dimensions that will suit almost every amateur need for a quad system.

A boom diameter of 2 inches is recommended for systems having two or three elements for 14, 21 and 28 MHz. When the boom length reaches 20 feet or longer, as encountered in four- and five-element antennas, a 3-inch diameter boom is highly recommended. Wind creates two forces on the boom, vertical and horizontal. The vertical load on the boom can be reduced with a guy-wire truss cable. The horizontal forces on the boom are more difficult to relieve, so 3-inch diameter tubing is desirable.

Generally speaking, there are three grades of material which can be used for quad spreaders. The least expensive material is bamboo. Bamboo, however, is also the weakest material normally used for quad construction. It has a short life, typically only a few years, and will not withstand a harsh climate very well. Also, bamboo is heavy in contrast to fiberglass, which weighs only about a pound per 13-foot length. Fiberglass is the most popular type of spreader material, and will withstand normal winter climates. One step beyond the conventional fiberglass arm is the pole-vaulting arm. For quads designed to be used on 7 MHz, surplus "rejected" pole-vaulting poles are highly recommended. Their ability to withstand large amounts of bending is very desirable. The cost of these poles is high, and they are difficult to obtain. Table 2 contains information on availability of arms suitable for use as spreaders.

Table 2
Suppliers of Fiberglass Poles for Spreaders

Company	Material Size
Advanced Composites PO Box 15323 Salt Lake City, UT 84115	1¼ and 1½-in. OD, 12 and 20-foot lengths
dB+ Enterprises PO Box 24 Pine Valley, NY 14872	1-1/16-in. diameter, 13-ft lengths. Severe-duty cubical quads and accessories.
Dynaflex Manufacturing Corp. Rte 14, Box 370 Tallahassee, FL 32304	Three-section telescoping poles, 15-foot total length.
Sky-Pole Manufacturing, Inc 1922 Placentia Costa Mesa, CA 92627	Vaulting poles and tubing of various sizes and lengths. 1 to 1-5/8-in. tubing in odd lengths.
Moeller Instrument Co, Inc Kirk Electronics Div Main Street Ivoryton, CT 06442	Arms for quad antennas, 9- and 13-foot lengths. Hollow and tapered, but reinforced at the 14, 21 and 28-MHz drill points.

A Three-Band Quad Antenna System

Quads have been popular with amateurs during the past few decades because of their light weight, relatively small turning radius, and their unique ability to provide good DX performance even when mounted close to the ground. A two-element, three-band quad, for instance, with the elements mounted only 35 feet above ground, will give good performance in situations where a triband Yagi will not. Fig 2 shows a large quad antenna which can be used as a design basis for either smaller or larger arrays.

Five sets of element spreaders are used to support the three-element 14-MHz, four-element 21-MHz, and five-element 28-MHz wire-loop system. The spacing between elements has been chosen to provide optimum performance consistent with boom length and mechanical construction. Each of the parasitic loops is closed (ends soldered together) and requires no tuning. All of the loop sizes are listed in Table 3, and are designed for center frequencies of 14.1, 21.1 and 28.3 MHz. Because quads are rather broadband antennas, excellent performance is obtained in both the CW and SSB band segments of each band (with the possible exception of frequencies above 29 MHz). Changing the dimensions to favor a frequency 200 kHz higher in each band to create a "phone" antenna is not necessary.

The most obvious problem related to quad antennas is the ability to build a structurally sound system. If high winds or heavy ice are a normal part of the environment, special precautions are necessary if the antenna is to survive a winter season. Another stumbling block for would-be quad builders is the installation of a three-dimensional system (assuming a Yagi has only two important dimensions) on top of a tower—especially if the tower needs guy wires for support. With proper planning, however, many of these obstacles can be overcome. For example, a tram system may be used.

An X or a + Frame?

One question which comes up quite often is whether to mount the loops in a diamond or a square configuration. In other words, should one spreader be horizontal to the earth, or should the wire be horizontal to the ground (spreaders mounted in the fashion of an X)? From the electrical point of view, it is probably a trade-off. Some authorities indicate that separation of the current points in the diamond system

Table 3

Three-Band Quad Loop Dimensions

Band	Reflector	Driven Element	First Director	Second Director	Third Director
14 MHz	(A) 72'8"	(B) 71'3"	(C) 69'6"	—	—
21 MHz	(D) 48'6½"	(E) 47'7½"	(F) 46'5"	(G) 46'5"	—
28 MHz	(H) 36'2½"	(I) 35'6"	(J) 34'7"	(K) 34'7"	(L) 34'7"

Letters indicate loops identified in Fig 3.

gives slightly more gain than is possible with a square layout. It should be pointed out, however, that there has not been any substantial proof in favor of one or the other, electrically.

From the mechanical point of view there is no question which version is better. The diamond quad, with the associated horizontal and vertical spreader arms, is capable of holding an ice load much better than a system where no vertical support exists to hold the wire loops upright. Put another way, the vertical poles of a diamond array, if sufficiently strong, will hold the rest of the system erect. When water droplets are accumulating and forming into ice, it is very reassuring to see water running down the wires to a corner and dripping off, rather than just sitting there on the wires and freezing. The wires of a loop (or several loops, in the case of a multiband antenna) help support the horizontal spreaders under a load of ice. A square quad will droop severely under heavy ice conditions because there is nothing to hold it up straight.

Another consideration enters into the selection of a design for a quad. The support itself, if guyed, will require a diamond quad to be mounted a short distance higher on the mast or tower than an equivalent square array if the guy wires are not to interfere with rotation.

Fig 2—The three-band quad antenna.

The quad array shown in Figs 2 and 3 uses fiberglass spreaders available from Viking/Kirk Electronics of Chester, Connecticut (see Table 2). Bamboo is a suitable substitute (if economy is of great importance). However, the additional weight of the bamboo spreaders over fiberglass is an important consideration. A typical 12-foot bamboo pole weighs about 2 pounds; the fiberglass type weighs less than a pound. By multiplying the difference times 8 for a two-element array, times 12 for a three-element antenna, and

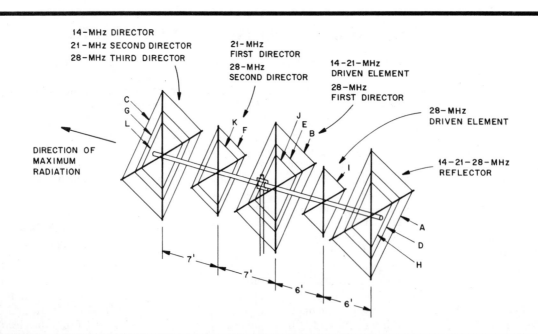

Fig 3—Dimensions of the three-band quad, not drawn to scale. See Table 3 for dimensions of lettered wires.

so on, it quickly becomes apparent that fiberglass is worth the investment if weight is an important factor. Properly treated, bamboo has a useful life of three or four years, while fiberglass life is probably 10 times longer.

Spreader supports (sometimes called spiders) are available from many different manufacturers. If the builder is keeping the cost at a minimum, he should consider building his own. The expense is about half that of a commercially manufactured equivalent and, according to some authorities, the homemade arm supports described below are less likely to rotate on the boom as a result of wind pressure.

A 3-foot length of steel angle stock, 1 inch per side, is used to interconnect the pairs of spreader arms. The steel is drilled at the center to accept a muffler clamp of sufficient size to clamp the assembly to the boom. The fiberglass is attached to the steel angle stock with automotive hose clamps, two per pole. Each quad-loop spreader frame consists of two assemblies of the type shown in Fig 4.

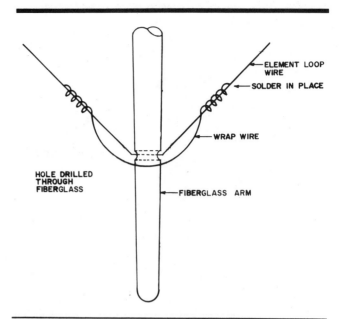

Fig 5—A method of assembling a corner of the wire loop of a quad element to the spreader arm.

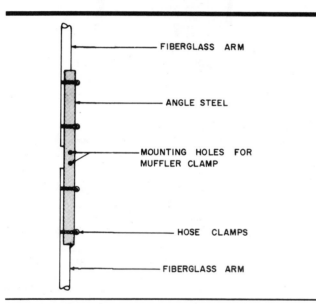

Fig 4—Details of one of two assemblies for a spreader frame. The two assemblies are jointed to form an X with a muffler clamp mounted at the position shown.

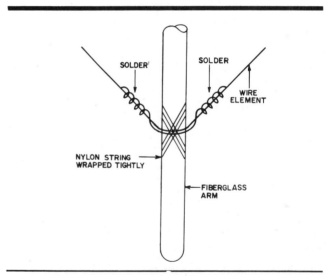

Fig 6—An alternative method of assembling the wire of a quad loop to the spreader arm.

Connecting the wires to the fiberglass can be done in a number of different ways. Holes can be drilled at the proper places on the spreader arms and the wires run through them. A separate wrap wire should be included at the entry/exit point to prevent the loop from slipping. Details are presented in Fig 5. Some amateurs have experienced cracking of the fiberglass, which might be a result of drilling holes through the material. However, this seems to be the exception rather than the rule. The model described here has no holes in the spreader arms; the wires are attached to each arm with a few layers of plastic electrical tape and then wrapped approximately 20 times in a criss-cross fashion with 1/8-inch diameter nylon string, as shown in Fig 6. The wire loops are left open at the bottom of each driven element where the coaxial cable is attached. See Fig 7. All of the parasitic elements are continuous loops of wire; the solder joint is at the base of the diamond.

A triband system requires that each driven element be fed separately. Two methods are possible. First, three individual sections of coaxial cable may be used. Quarter-wave transformers of 75-ohm line are recommended for this service. Second, a relay box may be installed at the center of the boom. A three-wire control system may be used to apply power to the proper relay for the purpose of changing bands. The circuit diagram of a typical configuration is presented in Fig 8 and its installation is shown in Fig 9. An alternative method of supplying a control signal to the remote switch is to make use of the feed line itself. Several articles on this subject have been published (see the bibliography at the end of this chapter).

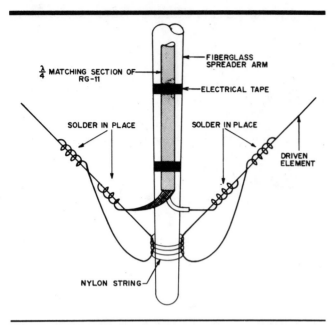

Fig 7—Assembly details of the driven element of a quad loop.

Fig 9—The relay box is mounted on the boom near the center. Each of the spreader-arm fiberglass poles is attached to steel angle stock with hose clamps.

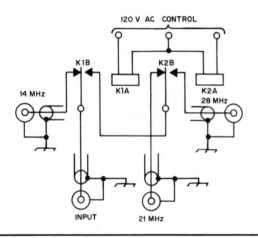

Fig 8—Suitable circuit for relay switching of bands for the three-band quad. A three-wire control cable is required. K1, K2—any type of relay suitable for RF switching, coaxial type not required (Potter and Brumfeld MR11A acceptable; although this type has double-pole contacts, mechanical arrangements of most single-pole relays make them unacceptable for switching of RF).

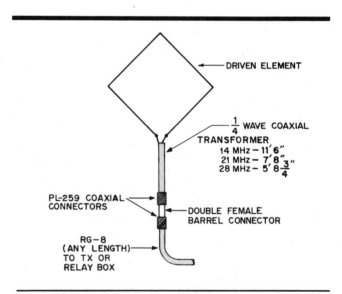

Fig 10—Showing installation of quarter-wave 75-Ω transformer section. The coax lengths indicated are based on a 66% velocity factor.

The quarter-wave transformers mentioned above are necessary to provide a match between the wire loop and a 52-ohm transmission line. This is simply a section of 75-ohm coax cable placed in series between the 52-ohm line and the antenna feed point, as shown in Fig 10. A pair of PL-259 connectors and a barrel connector may be used to splice the cables together. The connectors and the barrel should be wrapped well with plastic tape and then sprayed with acrylic for protection against the weather.

Every effort must be placed upon proper construction if freedom from mechanical problems is to be expected. Hardware must be secure or vibrations created by the wind may cause separation of assemblies. Solder joints should be clamped in place to keep them from flexing, which might fracture a connection point.

A 28-MHz Swiss Quad

The Swiss Quad is a two-element array with both elements driven. One element is longer than the other and is called the "reflector," while the shorter one is called the "director." Spacing between elements is usually 0.1 wavelength. The impedance of the antenna, using the 0.1-wavelength dimensions, is approximately 50 ohms.

Fig 11 is a drawing of the components of the beam. In its usual form, lengths of aluminum or copper tubing are bent to form the horizontal members. The element perimeters are completed with vertical wires. At the crossover points (X, Fig 11), which are connected together, voltage nodes occur.

The equations for the element sizes are based on the square (perimeter) and not the lengths of the wires. For the reflector the perimeter is equal to 1.148 × wavelength, and for the director 1.092 × wavelength, or

$$\text{Perimeter (inches)} = \frac{984}{f(\text{MHz})} \times 12 \times 1.148 \text{ (reflector)}$$

For example:

Perimeter for 28.1 MHz =

$$\frac{984}{28.1} \times 12 \times 1.148 = 482 \text{ in. (reflector)}$$

These equations apply only to the use of horizontal members of aluminum or copper tubing. Using PVC tubing and wire elements, the overall lengths of the perimeters are different and the correct lengths given later were determined experimentally.

One of the advantages of this antenna over the more conventional quad type is that plumber's delight type construction can be used. This means that both elements, at the top and bottom of the beam, can be grounded to the supporting mast. The structure is lightweight but strong, and an inexpensive TV rotator carries it nicely. Another feature is the small turning radius, which is less than half that of a three-element Yagi.

The antenna described here is made entirely of wire that is supported by two insulating frames constructed from rigid plastic water pipe. Rigid PVC water pipe is readily available from plumbing supply houses and from the large mail-order firms. The standard 10-foot lengths are just right for building the 28-MHz Swiss Quad. You can cut and drill PVC pipe with wood-working tools. PVC plastic sheds water, an advantage where winter icing is a problem. Heat from the intense summer sun has not softened or deformed the original quad structure.

To build the wire version of the Swiss Quad you will need the materials listed in Table 4 plus some wood screws and U bolts. Also required are a few scraps of wood dowel rod and some old toothbrushes.

Cut the PVC pipe to the lengths shown in Fig 12. Also cut several short lengths of dowel rod for reinforcement at the points indicated. These are held in place by means of epoxy cement. The bond is improved if the PVC surface is roughened with sandpaper and wiped clean before the cement is applied. A tack inserted through a tiny hole in the pipe will hold each dowel in place while the epoxy cures.

Reasonable care is required in forming the boom end joints so the two sections of ¾-inch pipe are parallel. The joining method used at WØERZ is illustrated in Fig 13. Parallel depressions were filed near each end of each boom

with a half-round rasp. These cradles are about 0.4 inch deep and their centers are 41.3 inches apart. Holes are drilled for the U bolts and the joints are completed with the U bolts and

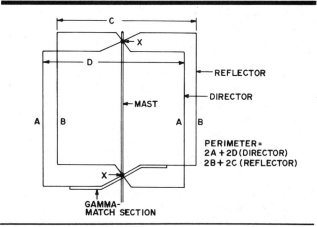

Fig 11—General arrangement for the Swiss Quad.

Table 4
Materials List, Swiss Quad

Four 10-ft lengths ½-in. rigid PVC pipe.
Two 10-ft lengths ¾-in. rigid PVC pipe.
One 10-ft length 1-in. rigid PVC pipe.
Twelve feet 1-1/8-in. or larger steel or aluminum tubing.
Epoxy cement (equal parts of resin and hardener).
100 ft hard-drawn copper wire, 14 or 16 gauge.

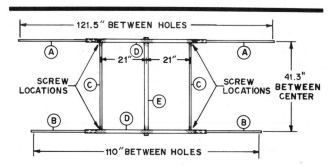

LOCATE DOWELS AT SHADED AREAS.

(A) ½"-PVC PIPE 40" LONG
(B) ½"-PVC PIPE 34" LONG
(C) ½"-PVC PIPE 40" LONG
(D) ¾"-PVC PIPE 54" LONG
(E) 1"-PVC PIPE 44" LONG

Fig 12—Dimensions and layout of the insulating frame.

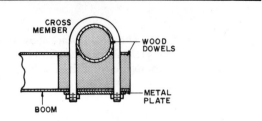

Fig 13—Boom end-joint detail.

the epoxy cement. Draw the bolts snug, but not so tight as to damage the PVC pipe. Final assembly of the insulating frames should be done on a level surface. Chalk an outline of the frame on the work surface so any misalignment will be easy to detect and correct. If the ½-inch pipe sections fit too loosely into the lateral members, shim them with two bands of masking tape before applying the epoxy cement.

Supports for the gamma-matching section can be made from old toothbrush handles or other scraps of plastic. Space the supports about 10 inches apart so that they support the gamma wire 2.5 inches on top of the lower PVC pipe. Attach the spacers with epoxy cement. Strips of masking tape can be used to hold the spacers in place while the epoxy is curing.

There are several ways to attach the frames to the vertical mast. The mounting hardware designed for the larger TV antennas should be quite satisfactory. Metal plates about 5 inches square can be drilled to accept four U bolts. Two U bolts should be used around the boom and two around the mast. A piece of wooden dowel inside the center of the boom prevents crushing the PVC pipe when the U bolts are tightened. The plates should not interfere with the element wires that must cross at the exact center of the frame. A 12-foot length of metal tubing serves as the vertical support. The galvanized steel tubing used as a top rail in chain link fences would be satisfactory.

When the epoxy resin has fully cured, you are ready to add the wire elements to produce the configuration shown in Fig 11. Start on the top side of the upper frame. Cut two pieces of copper wire (no. 14 or larger) at least 30.5 feet long and mark their centers. Thread the ends downward through holes spaced as shown in Fig 12 so that the wires cross at the top of the upper frame. Following the detail in Fig 14, drill pilot holes through the PVC pipe and drive four screws into the

dowels. The screws must be 41.3 inches apart and equidistant from the center of the frame. With the centers of the two wires together, bend the wires 45° around each screw and anchor with a short wrap of wire. Now pull the wires through the holes at the ends of the pipes until taut. A soldered wire wrap just below each hole prevents the element wires from sliding back through the holes.

Attach the wired upper frame about 2 feet below the top of the vertical mast. Make a bridle from stout nylon cord (or fiberglass-reinforced plastic clothesline), tying it from the top of the mast to each of four points on the upper frame to reduce sagging.

Now cut two 11.5-foot lengths of wire and attach them to the bottom of the lower frame. Also cut a 9-foot length for the gamma-matching section. If insulated wire is used, bare 6 inches at each end of the gamma wire. Details of the double gamma match are shown in Fig 15. Attach the wired lower frame to the mast about 9 feet below the upper frame and parallel to it. The ultimate spacing between the upper and lower frames, determined during the tuning process, will result in moderate tension in the vertical wires. Join the vertical wires to complete the elements of your Swiss Quad. All vertical wires must be of equal length. Do not solder the wire joints until you have tuned the elements.

Tune-Up

For tuning and impedance matching you will need a dip meter, an SWR indicator, and the station receiver and exciter. Stand the Swiss Quad vertically in a clear space with the lower frame at least 2 feet above ground. Using the dipper as a resonance indicator, prune a piece of 52-ohm coaxial cable to an integral multiple of a half-wavelength at the desired frequency. RG-8 and RG-58 with polyethylene insulation have a velocity factor of 0.66. At 28.6 MHz, a half-wavelength section (made from the above cables) is approximately 11.35 feet long. (Coaxial cable using polyfoam insulation has a velocity factor of approximately 0.80; consult the manufacturer's data.) Connect one end to the midpoint of the gamma section and the other to a 2-turn link. Couple the dipper to the link. You may observe several dips. Look for two pronounced dips, near 26 MHz and 31.4 MHz. Measure the frequencies at which these dips occur using your receiver to double-check the dip meter. Then multiply the frequencies and take the square root of this product; that is $\sqrt{f1 \times f2}$. If the result is less than 28.6, shorten the vertical wires equally and repeat the process until $\sqrt{f1 \times f2}$ lies between 28.6 MHz and 28.8 MHz. Your Swiss Quad is now tuned for the 28-MHz band.

Remove the link and connect the SWR bridge in its place.

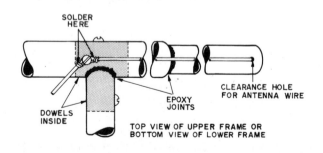

Fig 14—Details of the frame and wire assembly.

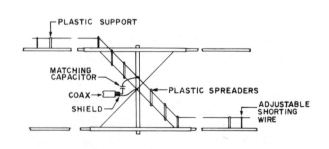

Fig 15—Details of the double gamma match.

Connect your exciter to the input terminals of the bridge, tune to 28.6 MHz, and apply just enough power to obtain a full-scale forward power indication. Measure the SWR. Now slide the two shorting wires of the matching section to new positions, equidistant from the center of the wire elements, and measure the SWR. Continue adjusting the shorting wires until minimum SWR is obtained. Insert a 100-pF variable capacitor between the center conductor of the coaxial cable feeder and the midpoint of the gamma wire. Adjust the capacitor for minimum SWR indication. It may be necessary to readjust both the shorting wires and the capacitor to obtain a satisfactory impedance match. With patience, a perfect match (SWR = 1:1) can be achieved. Solder the shorting wires.

The variable capacitor may be replaced with a short length of RG-59 coaxial cable. Each foot of this cable has a capacitance of approximately 20 pF. Measure or estimate the value to which the variable capacitor was finally set, add 10%, and cut a corresponding length of RG-59. Solder the shield braid to the midpoint of the gamma wire and the center wire to the center conductor of the 52-ohm transmission line, leaving the other end of the coaxial-cable capacitor open. You will probably observe that the SWR has increased. Snip short lengths from the open end of the capacitor until the original low SWR is obtained. When the antenna is raised to 40 feet the SWR should be less than 1.5:1 over the entire 28-MHz band.

Tape the capacitor to the PVC pipe boom, then wrap a few bands of tape around the sections where the wires run along the sides of the pipes. Check the solder joints and mechanical connections. Coat the solder joints and the cable ends with a weatherproof sealing compound (such as silicone bathtub caulk or RTV sealant) and hoist the Swiss Quad up the support.

Multiband Spider-Delta Loop

The following is a description of a no-compromise, full-wave loop antenna that can be constructed for operation at 7, 10, 14, 18, 21, 24 or 28 MHz. The 14, 18, 21, 24 or 28-MHz versions are manageable enough that they can be positioned on a tower by two people, one on the tower, and one on the ground. (The second person is required mainly for safety reasons.) The 4-band version (14 MHz and up, excluding 18 MHz) weighs about 50 pounds and is easily rotated with a Ham-M or equivalent rotator. This antenna was designed by Rich Guski, KC2MK, who has coined it the Spider-Delta Loop.

Measurements indicate that the gain and front-to-back ratio of this antenna are about the same as the conventional two-element quad. Depending on materials used and the number of bands covered, the cost of constructing this antenna should be far less than purchasing a comparable commercial antenna. The only complexity involved in building this antenna is the welding of steel angle stock for the spreaders.

The Spider-Delta Loop antenna is a hybrid of two familiar loop antenna designs, the two-element quad and the delta loop. Both antennas consist of two elements, one approximately 1-λ loop, used as a driven element, and another loop used as a reflector. The principal difference between the Spider-Delta Loop and a conventional quad is that the Spider uses triangular loops.

The traditional rotatable delta-loop antenna, which has a good reputation for DX performance, uses so-called plumber's delight construction. Two sides of the triangle loops consist of rigid material such as aluminum tubing. The apex formed by the two rigid sides is attached to a boom, which establishes the spacing between the loops. The third side of the triangle is made of wire. The triangles are normally oriented so the wire side is highest and parallel to the ground. The disadvantages of the delta loop configuration are that the antenna is top-heavy, and it can be built for only one band. The Spider-Delta Loop overcomes these difficulties.

The loops of a quad antenna are usually made of wire, suspended by two sets of four arms (spreaders) made of rigid nonconducting material. The spreaders of a conventional quad are attached to a boom that, like the delta loop, establishes the spacing between the loops.

Additional sets of loops can be added to the spreaders for multiband operation, but in the conventional quad all such loops must have the same spacing, resulting in optimum element spacing for only one band. The gain, front-to-back ratio and radiation resistance of a two-element loop antenna are largely dependent upon the spacing (in wavelengths) of the loops along the boom. The result, for the multiband conventional quad (with a boom) is a compromise for all but one of the bands covered by the antenna.

Another variation on the basic multielement loop antenna is the boomless quad, which offers an improvement over the conventional design. Instead of being supported by a boom, the spreaders are mounted at the center of the array and radiate outward. When viewed from the side of the array, the spreaders form two cones positioned point to point with the support mast between the points.

In a multiband boomless quad, the two longest elements (the elements for the lowest frequency of operation) are attached to the spreaders at the far ends. This positioning establishes the spacing of the two loops for that band. As the additional 1-λ loops for the other bands are attached to the spreaders, they will fit closer to the center of the array. The spacing of each of the shorter pairs of loops will be less than the spacing of the pair of longer loops. In this way it is possible to design a multiband two element wire loop antenna for which all pairs of loops have the optimum spacing (in wavelengths), and still share the same spreaders.

The Spider-Delta Loop is a boomless design similar to the one described above, so the weight and wind-loading problems associated with a conventional antenna are reduced. Three spreaders per loop array are used here, rather than the four used in a conventional design. Two fewer spreaders are needed for the entire antenna when compared to the conventional quad.

The two-element loop antenna system described here has approximately 0.12-λ spacing for all bands. This spacing provides good front-to-back ratio and gain, and a feed-point impedance close to 50 ohms at resonance.

Dimensions

The Spider-Delta Loop lengths and spacing are derived from the standard quad-loop length equations presented earlier in this chapter. Spacing between the loops is 0.12 times the free-space wavelength in use.

The spreader length is calculated by using the results of the above calculations as the starting point. The spreader length is the distance between the center of the array and the loop apexes when the loops are in the shape of an equilateral triangle, in parallel planes, and spaced 0.12 λ apart. The array is balanced, so the junction of the spreaders is the mechanical center of mass of the antenna. This is shown in Fig 16.

All spreaders are the same length. The actual spreader length required for this antenna is that which is required to support the longest set of loops. This is a function of the lowest frequency band on which the antenna is designed to operate.

Table 5 contains the results of the above calculations for selected design frequencies within the 7, 10, 14, 18, 21, 24 and 28-MHz bands. If you prefer to design your antenna for different center frequencies, the dimensions can be scaled easily based on the information given in Table 5. As mentioned earlier, however, quad antennas are inherently broad.

All the driven loops share the same three spreader poles. Similarly, all the reflector loops for the array share the same

Fig 16—The Spider-Delta Loop in place on the tower at KC2MK.

three spreader poles. These conditions hold true regardless of the number of bands covered. For example, using the data in Table 5, the construction of a 5-band Spider-Delta Loop covering 14 through 29.7 MHz requires six spreaders, each approximately 14½ feet long.

Feed System

The two-element Spider-Delta Loop has a feed-point impedance of about 55 ohms at resonance. This provides a good match to common coaxial cable such as RG-8, RG-8X or RG-58. The antenna may be fed directly by running separate cables to each driven element.

An alternative that offers a better directional pattern and improved front-to-back ratio is to use a separate balun for each of the driven-element feed points. The two-element triband version shown in Fig 16 uses this feed system. The baluns are homemade air-core transformers and are visible in the photographs. Refer to Chapter 26 for information on the construction and uses of balun transformers.

Materials

The mast used for the antenna shown in Fig 16 is a 2-inch steel pipe, 3 feet long, with the array attachments (described below) welded about 2 inches down from the top. Use the largest diameter steel mast that fits in your tower and rotator, to minimize the possibility of mast failure.

The two spider to mast attachments consist of steel angle stock 2 inches wide (on each side), and 7 inches long. These are welded directly to the mast as attachment points for the two spider halves. Two 3/8 × 1-inch steel bolts are also required for each of the two array attachments.

The two spider halves are each made of six pieces of steel angle stock. One of these, which forms the base of a spider half, is 2 inches wide and 17 inches long. The other five are all 1½ inches wide. Three of these, which will become the spreader mounts, are 20 inches long. Another piece, used to brace the two lower spreader mounts, is 17 inches long. The upper spreader mount brace is 5 inches long.

Two 3/8-inch diameter steel rods (or bolts), 5 inches long, are required to complete each of the spiders. They are needed to brace the lower spreaders.

The 14, 21 and 28-MHz Spider-Delta Loop shown in Fig 16 uses pole-vaulting poles for the spreaders. This antenna has survived years of ice and wind in the northeast. Although pole-vaulting poles or equivalent supports are required for a 7- or 10-MHz antenna, they are probably overkill for a 14-MHz or smaller antenna. Fiberglass poles suitable for use as spreaders are available from several companies (see Table 2).

Table 5
Loop and Spreader Lengths for Two-element Spider-Delta Loops for the 7- Through 28-MHz Amateur Bands. All Dimensions are in Feet.

Frequency (MHz)	7.175	10.125	14.175	18.100	21.250	24.930	28.600
Driven	140.07	99.26	70.90	55.52	47.29	40.31	35.14
Reflector	143.55	101.73	72.66	56.91	48.47	41.32	36.01
Spacing	16.50	11.70	8.36	6.54	5.58	4.75	4.14
Spreaders	28.83	20.43	14.59	11.43	9.74	8.30	7.23

[Note: At this writing the 17-meter band is not scheduled to be available to US amateurs until July 1989.—Ed.]

The spreaders are attached to the spreader mounts on the spider with adjustable stainless-steel hose clamps. Three hose clamps are used to attach each of the six spreaders.

The loops are no. 14 copper-clad steel wire. The lengths of the loops can be adjusted and locked using electrician's copper wire clamps. Table 6 lists the materials required to build a multiband Spider-Delta Loop antenna.

Construction

The spider attachments are welded to the mast as shown in the diagram of Fig 17. A 7-inch piece of steel angle stock as described in the materials list is used to construct each of the two spider attachments. Four 3/8-inch diameter steel bolts are permanently pinned between the angle stock and the mast with the bolt shafts facing outward through holes drilled in the angle stock. Carefully position the angle stock on the mast so the faces with the bolt holes are exactly opposite and parallel to each other. Be sure that you position the attachments high enough on the mast so the antenna will clear the tower when rotated. Weld each of the two pieces of angle stock to the mast along the entire length of the angle stock.

Center Spider Assembly

The center spider is constructed in two halves, one for the driven loop side and the other for the reflector side. This scheme permits raising the antenna one half at a time.

The two center spider halves are the structural heart of the antenna. Their construction is the most critical part of the project because they establish the shape and structural integrity of the antenna. They are made of steel angle stock and steel rods that are welded together to form the attachment points for the spreaders and the mast. Refer to Fig 18 for the layout of the spider halves.

A 17-inch long piece of 2-inch wide angle stock is used as the base of each of the spiders. Two holes are drilled in the base of this piece to receive the 3/8-inch bolts that are attached to the mast.

Refer to Fig 19. The upper spreader mount (as viewed from the favored direction of the antenna) is welded to the base immediately above the upper bolt hole. Be sure to leave enough room for the nut to clear the brace. The upper spreader mount is braced by a 5-inch piece of angle stock that is welded to the top of the spider base and to the spreader mount. The angle between the spider base and the spreader mount is 16.5°. This angle is important because it establishes the spacing for each of the multiband loops which will be

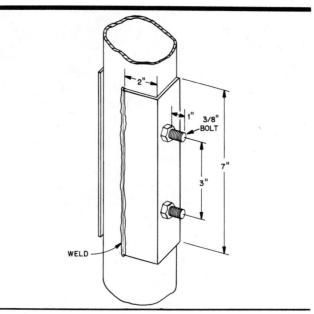

Fig 17—Diagram showing spider-mount attachments to the mast.

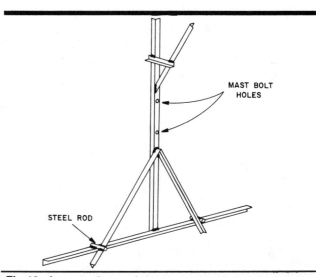

Fig 18—Layout of one of the two spider halves before attachment to the mast and spreader arms (not to scale).

Table 6

Materials Required for Construction of the Spider-Delta Loop

Quantity/Material	Application
48 in. of 2 in. × 2 in. steel angle stock	Two 7-in. lengths for spider to mast attachment, two 17-in. lengths for spider half-bases.
164 in. of 1½ × 1½ in. steel angle stock	Six 20-in. lengths for spreader mounts, two 17-in. lengths for lower spreader braces, two 5-in. lengths for upper spreader braces.
Four 3/8-in. diam × 1-in. steel bolts	Spider to mast attachments.
Four 3/8-in. diam × 5-in. steel bolts or rods	Lower spreader braces.
18 stainless steel hose clamps	Spreader to spreader-mount attachments.
Six fiberglass poles (see Table 5 for length)	Spreaders.
Copper-clad steel wire (see Table 5 for length)	Elements.
Several electrician's copper wire clamps	Element length adjustment. (Two per band required.)

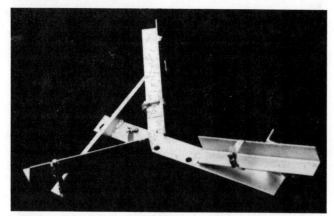

Fig 19—One of the spider halves.

attached to the spreaders.

The lower spreader mounts are welded to the spider at a point immediately below the bolt holes. They are positioned so that they form an angle of 120° with the upper spreader mount and each other (as viewed from the favored direction of the antenna). A 17-inch long piece of steel angle stock is used as a brace for the lower spreader mounts. The center of this piece is welded to the lower end of the spider base, parallel to the plane of the loops and at a 90° angle to the spider base. The 5-inch steel rods are welded to the ends of the brace, perpendicular to it and away from the spider. The other ends of the 5-inch rods are welded to the two lower spreader mounts. The lower spreader mounts, like the upper spreader mounts, must be angled out at 16.5° from a plane containing the mast and perpendicular to the favored direction of the antenna.

It is a good idea to spot-weld the parts together first. Take time to test fit everything together. Bolt the spider halves to the mast and check all angles to make sure the antenna will have the proper shape and dimensions before completing the welding.

Attaching the Spreaders

Now attach three spreaders to the spider. The poles rest in each of the three spreader mounts and are fastened with steel hose clamps. Short pieces of pipe with the same outside diameter as the inside diameter of the fiberglass poles should be slipped inside the poles where they meet the mounts. (This is to prevent crushing of the poles with the hose clamps, and to add strength.)

Should it become necessary to replace a wire loop or

access a feed point after the antenna has been installed, it is a simple matter to loosen the hose clamps holding the spreaders to their mounts, and pull the spreaders and wires close to the tower for service and adjustment. The antenna need be taken off the tower only for major servicing.

Cutting and Mounting the Wire Loops

With at least one half of the spiders and spreaders together, the wires for that side can be cut and fitted to the spreaders. The largest loop is attached to the poles first. Drill small holes through the poles to accept the wires. Depending on the poles used, you may want to use an alternative method of wire attachment, such as discussed earlier, especially if you are building a 7-MHz antenna and can make no compromises in structural strength.

Use the dimensions given in Table 5 to judge where on the spreaders to attach the wires. Cut each of the wires about 6 in. longer than the length shown in the table. This is to allow for tuning, which is done by adjusting the loop length where the wire ends meet and locking it with electrician's copper wire clamps. Refer to Fig 20. The wire clamps are located at the middle of the lower side of each of the loops. This makes the clamps accessible from the tower when the antenna is in place, as shown in Fig 21.

Feed Lines

Attach the feed lines to the driven loops where they attach to the upper spreader. The use of a balun at each feed point is optional. Each of the feed lines should be long enough to reach the center spider, with about 3 feet of excess. Terminate each cable with a PL-259.

The use of a remote antenna switch is optional. If used, it should be permanently attached to the spider of the driven half. The feed line switch box is visible in Fig 16.

Raising the Antenna

Place the mast with the spider attachments on the tower first. Insert the mast in the rotator and then align the rotator direction indicator. Raise the antenna one half at a time. If your antenna is for 14 MHz or higher, one person on the tower can pull one side up at a time and lift it into place on the mast

Fig 20—Close-up of one of the loop-length adjusting clamps used to tune the Spider-Delta Loop.

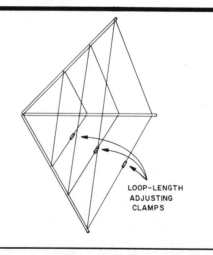

LOOP-LENGTH ADJUSTING CLAMPS

Fig 21—Diagram showing the placement of the loop-length adjusting clamps. Spider details are omitted for clarity. A triband version of the antenna is represented here.

to spider attachment points. Tighten the bolts and install the other half of the antenna the same way. If you are raising a larger version, you will need a gin pole or a heavy-duty pulley attached to the mast.

Tuning

The Spider-Delta Loop is tuned by lengthening or shortening the loops. Concentrating on one band at a time, adjust the driven loop for minimum SWR at the design frequency. Then adjust the reflector for the best SWR across the band. Alternatively, the reflector could be tuned for best gain or front-to-back ratio. The SWR curve of the Spider-Delta Loop is similar to that of the conventional quad antenna. Fig 22 shows typical SWR curves for the version described here.

Future Considerations

One modification to this design that may be valuable is the construction of a feed system that allows switching of the physical location of the feed point from the apex to one of the lower corners. Feeding the antenna at a lower corner changes the polarization of the antenna from horizontal to diagonal, almost vertical.

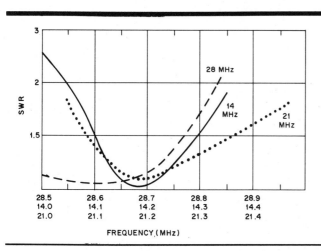

Fig 22—Typical SWR curves for the Spider-Delta Loop.

BIBLIOGRAPHY

Source material and more extended discussions of the topics covered in this chapter can be found in the references given below and in the textbooks listed at the end of Chapter 2.

P. S. Carter, C. W. Hansell and N. E. Lindenblad, "Development of Directive Transmitting Antennas by R.C.A. Communications," *Proc IRE*, Oct 1931.

C. Cleveland, "More'n One Way to Switch an Antenna," Technical Correspondence, *QST*, Nov 1986.

D. Cutter, "Simple Switcher," *73*, May 1980.

D. DeMaw, "A Remote Antenna Switcher for HF," *QST*, Jun 1986.

M. G. Knitter, Ed., *Loop Antennas—Design and Theory* (Cambridge, WI: National Radio Club, 1983).

J. Lindsay, "Quads and Yagis," *QST*, May 1968.

F. E. Terman, *Radio Engineering*, 3rd ed. (New York: McGraw-Hill Book Co, 1947).

E. M. Williams, "Radiating Characteristics of Short-Wave Loop Aerials," *Proc IRE*, Oct 1940.

Chapter 13

Long Wire and Traveling Wave Antennas

The power gain and directive characteristics of the harmonic wires (which are "long" in terms of wavelength) described in Chapter 2 make them useful for long-distance transmission and reception on the higher frequencies. In addition, long wires can be combined to form antennas of various shapes that will increase the gain and directivity over a single wire. The term "long wire," as used in this chapter, means any such configuration, not just a straight-wire antenna.

Long Wires Versus Multielement Arrays

In general, the gain obtained with long-wire antennas is not as great, when the space available for the antenna is limited, as can be obtained from the multielement arrays in Chapter 8. However, the long-wire antenna has advantages of its own that tends to compensate for this deficiency. The construction of long-wire antennas is simple both electrically and mechanically, and there are no especially critical dimensions or adjustments. The long-wire antenna will work well and give satisfactory gain and directivity over a 2-to-1 frequency range; in addition, it will accept power and radiate well on any frequency for which its overall length is not less than about a half wavelength. Since a wire is not "long," even at 28 MHz, unless its length is equal to at least a half wavelength on 3.5 MHz, any long-wire can be used on all amateur bands that are useful for long-distance communication.

Between two directive antennas having the same theoretical gain, one a multielement array and the other a long-wire antenna, many amateurs have found that the long-wire antenna seems more effective in reception. One possible explanation is that there is a diversity effect with a long-wire antenna because it is spread out over a large distance, rather than being concentrated in a small space. This may raise the average level of received energy for ionospheric-propagated signals. Another factor is that long-wire antennas have directive patterns that are sharp in both the horizontal and vertical planes, and tend to concentrate the radiation at the low vertical angles that are most useful at the higher frequencies. This is an advantage that some types of multielement arrays do not have.

General Characteristics of Long-Wire Antennas

Whether the long-wire antenna is a single wire running in one direction or is formed into a V, rhombic, or some other configuration, there are certain general principles that apply and some performance features that are common to all types. The first of these is that the power gain of a long-wire antenna as compared with a half-wave dipole is not considerable until the antenna is really long (its length measured in wavelengths rather than in a specific number of feet). The reason for this is that the fields radiated by elementary lengths of wire along the antenna do not combine, at a distance, in as simple a fashion as the fields from half-wave dipoles used as described in Chapter 8. There is no point in space, for example, where the distant fields from all points along the wire are exactly in phase (as they are, in the optimum direction, in the case of two or more collinear or broadside dipoles when fed with in-phase currents). Consequently, the field strength at a distance is always less than would be obtained if the same length of wire were cut up into properly phased and separately driven dipoles. As the wire is made longer, the fields combine to form an increasingly intense main lobe, but this lobe does not develop appreciably until the wire is several wavelengths long. This is indicated by the curve showing gain in Fig 1. The longer the antenna, the sharper the lobe becomes, and since it is really a hollow cone of radiation about the wire in free space, it becomes sharper in all planes. Also, the greater the length, the smaller the angle with the wire at which the maximum radiation occurs.

Directivity

Because many points along a long wire are carrying currents in different phase (usually with different current amplitude as well), the field pattern at a distance becomes more complex as the wire is made longer. This complexity is manifested in a series of minor lobes, the number of which increases with the wire length. The intensity of radiation from the minor lobes is frequently as great as, and sometimes greater than, the radiation from a half-wave dipole. The energy radiated in the minor lobes is not available to improve the gain in the major lobe, which is another reason why a long-wire antenna must be long to give appreciable gain in the desired direction.

Driven and parasitic arrays of the simple types described in Chapter 8 do not have minor lobes of any great consequence. For that reason they frequently seem to have much better directivity than long-wire antennas, because their responses in undesired directions are well down from their response in the desired direction. This is the case even if a multielement array and a long-wire antenna have the same actual gain in the favored direction. For amateur work, particularly with directive antennas that cannot be rotated, the minor lobes of a long-wire antenna have some advantages. In most directions the antenna will be as good as a half-wave dipole, and in addition will give high gain in the most favored

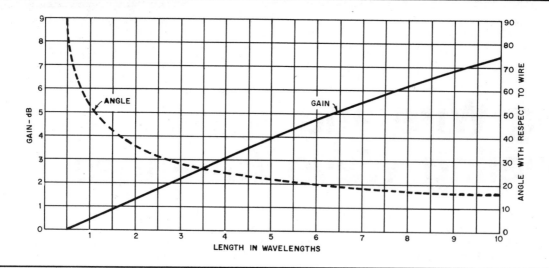

Fig 1—Theoretical gain of a long-wire antenna over a dipole as a function of wire length. The angle, with respect to the wire, at which the radiation intensity is maximum also is shown.

direction. Thus, a long-wire antenna (depending on the design) frequently is a good all-around radiator in addition to being a good directive antenna.

In the discussion of directive patterns of long-wire antennas in this chapter, keep in mind that the radiation patterns of resonant long wires are based on the assumption that each half-wave section of wire carries a current of the same amplitude. As pointed out in Chapter 2, this is not exactly true, since energy is radiated as it travels along the wire. For this reason it is to be anticipated that, although the theoretical pattern is bidirectional and identical in both directions, actually the radiation (and reception) will be best in one direction. This effect becomes more marked as the antenna is made longer.

Wave Angles

The wave angle at which maximum radiation takes place from a long-wire antenna depends largely on the same factors that determine the wave angles of simple dipoles and multielement antennas. That is, the directive pattern in the presence of ground is found by adding the free-space vertical-plane pattern of the antenna to the ground-reflection factors for the particular antenna height used. These factors are discussed in Chapter 2.

As mentioned earlier, the free-space radiation pattern of a long-wire antenna has a major lobe that forms a hollow cone around the wire. The angle at which maximum radiation takes place becomes smaller, with respect to the wire, as the wire length is increased. This is shown by the broken curve in Fig 1. For this reason a long-wire antenna is primarily a low-angle radiator when installed horizontally above the ground. Its performance in this respect is improved by selecting a height that also tends to concentrate the radiation at low wave angles (at least ½ λ for the lowest frequency). This is also discussed in Chapter 2.

Antenna systems formed from ordinary horizontal dipoles (that are not stacked) in most cases have a rather broad vertical pattern; the wave angle at which the radiation is maximum therefore depends chiefly on the antenna height. With a long-wire antenna, however, the wave angle at which

the major lobe is maximum can never be higher than the angle at which the first null occurs (see Fig 2). This is true even if the antenna height is very low. (The efficiency may be less at very low heights, partly because the pattern is affected in such a way as to put a greater proportion of the total power into the minor lobes.) The result is that when considering radiation at wave angles below 15 or 20 degrees, a long-wire antenna is less sensitive to height than are multielement arrays or simple dipoles. To assure good results, however, the antenna should have a height equivalent to at least a half wavelength at 14 MHz—that is, a minimum height of about 35 feet. Greater heights will give a worthwhile improvement at wave angles below 10°.

With an antenna of fixed physical length and height, both length and height (in terms of wavelength) increase as the frequency is increased. The overall effect is that both the antenna and the ground reflections tend to keep the system operating effectively throughout the frequency range. At low frequencies the wave angle is raised, but high wave angles are useful at 3.5 and 7 MHz. At high frequencies the inverse is true. Good all-around performance usually results on all bands when the antenna is designed for optimum performance in the 14-MHz band.

Calculating Length

In this chapter, lengths are always discussed in terms of wavelengths. There is nothing very critical about wire lengths in an antenna system that will work over a frequency range including several amateur bands. The antenna characteristics change very slowly with length, except when the wires are short (around one wavelength, for instance). There is no need to try to establish exact resonance at a particular frequency for proper antenna operation.

The formula for harmonic wires given in Chapter 2 is satisfactory for determining the lengths of any of the antenna systems to be described. For convenience, the formula is repeated here in slightly different form:

$$\text{Length (feet)} = \frac{984(N - 0.025)}{f(\text{MHz})}$$

where N is the antenna length in wavelengths. In cases where precise resonance is desired for some reason (for obtaining a resistive load for a transmission line at a particular frequency, for example) it is best established by trimming the wire length until the standing-wave ratio on the line is minimum.

LONG SINGLE WIRES

In Fig 1 the solid curve shows that the gain in decibels of a long wire increases almost linearly with the length of the antenna. The gain does not become appreciable until the antenna is about four wavelengths long, where it is equivalent to doubling the transmitter power (3 dB). The actual gain over a half-wave dipole when the antenna is at a practical height above ground will depend on the way in which the radiation resistance of the long-wire antenna and the comparison dipole are affected by the height. The exact way in which the radiation resistance of a long wire varies with height depends on its length. In general, the percentage change in resistance is not as great as in a half-wave antenna. This is particularly true at heights greater than one-half wavelength.

The nulls bounding the lobes in the directive pattern of a long wire are fairly sharp and are frequently somewhat obscured, in practice, by irregularities in the pattern. The locations of nulls and maxima for antennas up to eight wavelengths long are shown in Fig 2.

Orientation

The broken curve of Fig 1 shows the angle with the wire at which the radiation intensity is maximum. As shown in Chapter 2, there are two main lobes to the directive patterns of long-wire antennas; each makes the same angle with respect to the wire. The solid pattern, considered in free space, is the hollow cone formed by rotating the wire on its axis.

When the antenna is mounted horizontally above the ground, the situation depicted in Fig 3 exists. Only one of the two lobes is considered in this drawing, and its lower half

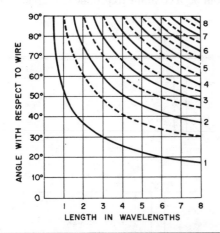

Fig 2—Angles at which radiation from long wires is maximum (solid curves) and zero (broken curves). The major lobe, no. 1, has the power gains given by Fig 1. Secondary lobes have smaller amplitude, but the maxima may exceed the radiation intensity from a half-wave dipole.

is cut off by the ground. The maximum intensity of radiation in the remaining half occurs through the broken-line semicircle; that is, the angle B (between the wire direction and the line marked wave direction) is the angle given by Fig 1 for the particular antenna length used.

In the practical case, there will be some wave angle (A) that is optimum for the frequency and the distance between the transmitter and receiver. Then, for that wave angle, the wire direction and the optimum geographical direction of transmission are related by the angle C. If the wave angle is very low, B and C will be practically equal. But as the wave angle becomes higher the angle C becomes smaller. In other words, the best direction of transmission and the direction of the wire more nearly coincide. They coincide exactly when C is zero; that is, when the wave angle is the same as the angle given by Fig 1.

The maximum radiation from the antenna can be aligned with a particular geographical direction at a given wave angle by means of the following formula:

$$\cos C = \frac{\cos B}{\cos A}$$

In most amateur work the chief requirement is that the wave angle should be as low as possible, particularly at 14 MHz and above. In this case it is usually satisfactory to make angle C the same as that given by Fig 1.

It should be borne in mind that only the maximum point of the lobe is represented in Fig 3. Radiation at higher and lower wave angles in any given direction will be proportional to the way in which the actual pattern shows the field strength to vary as compared with the maximum point of the lobe.

Tilted Wires

Fig 3 shows that when the wave angle is equal to the angle which the maximum intensity of the lobe makes with the wire, the best transmitting or receiving direction is that of the wire itself. If the wave angle is less than the lobe angle, the best direction can be made to coincide with the direction of the wire by tilting the wire enough to make the lobe and wave angle coincide. This is shown in Fig 4, for the case of a one-wavelength antenna tilted so that the maximum radiation from one lobe is horizontal to the left, and from the other is horizontal to the right (zero wave angle). The solid pattern can be visualized by imagining the plane diagram rotating about the antenna as an axis.

Since the antenna is neither vertical nor horizontal in this case, the radiation is part horizontally polarized and part vertically polarized. In computing the effect of the ground, the horizontal and vertical components must be handled separately. In general, the directive pattern at any given wave angle becomes unsymmetrical when the antenna is tilted. For small amounts of tilt (less than the amount that directs the lobe angle horizontally), and for low wave angles, the effect is to shift the optimum direction closer to the line of the antenna. This is true in the direction in which the antenna slopes downward. In the opposite direction the low-angle radiation is reduced.

Feeding Long Wires

It is pointed out in Chapter 26 that a harmonic antenna can be fed only at the end or at a current loop. Since a current loop changes to a node when the antenna is operated at any even multiple of the frequency for which it is designed, a long-

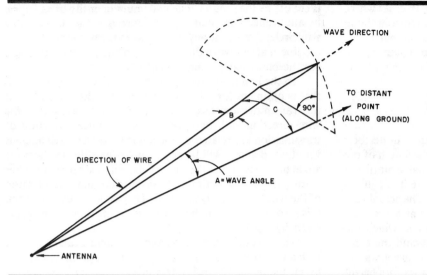

Fig 3—This drawing shows how the hollow cone of radiated energy from a long wire (broken-line arc) results in different wave angles (A) for various angles between the direction of the wire and the direction to the distant point (C).

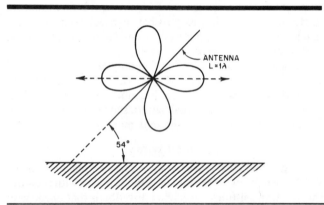

Fig 4—Alignment of lobes for horizontal transmission by tilting a long wire in the vertical plane.

wire antenna will operate as a true long wire on all bands only when it is fed at the end.

A common method of feeding is to use a resonant open-wire line, as described in Chapter 26. This system will work on all bands down to the one, if any, at which the antenna is only a half wave long. Any convenient line length can be used if the transmitter is matched to the line input impedance by the methods described in Chapter 25.

Two arrangements for using nonresonant lines are given in Fig 5. The one at A is useful for one band only since the matching section must be a quarter wave long, approximately, unless a different matching section is used for each band. In B, the Q-section impedance should be adjusted to match the antenna to the line as described in Chapter 26, using the value of radiation resistance given in Chapter 2. This method is best suited to working with a 600-ohm transmission line. Although it will work as designed on only one band, the antenna can be used on other bands by treating the line and matching transformer as a resonant line. In this case, as mentioned earlier, the antenna will not radiate as a true long wire on even multiples of the frequency for which the matching system is designed.

The end-fed arrangement, although the most convenient when tuned feeders are used, suffers the disadvantage that there is likely to be a considerable antenna current on the line, as described in Chapter 26. In addition, the antenna reactance changes rapidly with frequency for the reasons outlined in Chapter 2. Consequently, when the wire is several wavelengths long, a relatively small change in frequency—a fraction of the width of a band—may require major changes in the adjustment of the transmitter-to-line coupling apparatus. Also, the line becomes unbalanced at all frequencies between those at which the antenna is exactly resonant. This leads to a considerable amount of radiation from the line. The unbalance can be overcome by using two long wires in one of the arrangements described in succeeding sections.

COMBINATIONS OF RESONANT LONG WIRES

The directivity and gain of long wires may be increased by using two wires placed in relation to each other such that the fields from both combine to produce the greatest possible

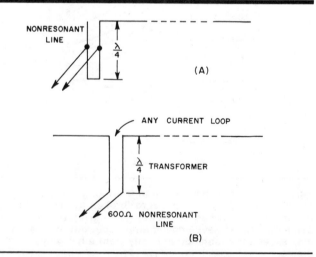

Fig 5—Methods of feeding long single-wire antennas.

field strength at a distant point. The principle is similar to that used in designing the multielement arrays described in Chapter 8. However, the maximum radiation from a long wire occurs at an angle of less than 90° with respect to the wire, so different physical relationships must be used.

Parallel Wires

One possible method of using two (or more) long wires is to place them in parallel, with a spacing of ½ wavelength or so, and feed the two in phase. In the direction of the wires the fields will add in phase. However, since the wave angle is greatest in the direction of the wire, as shown by Fig 3, this method will result in rather high-angle radiation unless the wires are several wavelengths long. The wave angle can be lowered, for a given antenna length, by tilting the wires as described earlier. With a parallel arrangement of this sort the gain should be about 3 dB over a single wire of the same length, at spacings in the vicinity of ½ wavelength.

THE V ANTENNA

Instead of using two long wires parallel to each other, they may be placed in the form of a horizontal V, with the angle between the wires equal to twice the angle given by Fig 1 for the particular length of wire used. The currents in the two wires should be *out of phase*. Under these conditions the plane directive patterns of the individual wires combine as shown in Fig 6. Along a line in the plane of the antenna and bisecting the V, the fields from the individual wires reinforce each other at a distant point. The other pair of lobes in the plane pattern is more or less eliminated, so the pattern becomes essentially bidirectional.

The directional pattern of an antenna of this type is sharper in both the horizontal and vertical planes than the patterns of the individual wires composing it. Maximum radiation in both planes is along the line bisecting the V. There are minor lobes in both the horizontal and vertical patterns, but if the legs are long in terms of wavelength the amplitude of the minor lobes is small. When the antenna is mounted horizontally above the ground, the wave angle at which the radiation from the major lobe is maximum is determined by the height, but cannot exceed the angle values shown in Fig 1 for the leg length used. Only the minor lobes give high-angle radiation.

The gain and directivity of a V depend on the length of the legs. An approximate idea of the gain for the V antenna may be obtained by adding 3 dB to the gain value from Fig 1 for the corresponding leg length. The actual gain will be affected by the mutual impedance between the sides of the V, and will be somewhat higher than indicated by the values determined as above, especially at longer leg lengths. With 8-wavelength legs, the gain is approximately 4 dB greater than that indicated for a single wire in Fig 1.

Lobe Alignment

It is possible to align the lobes from the individual wires with a particular wave angle by the method described in connection with Fig 3. At very low wave angles the required change in the apex angle is extremely small; for example, if the desired wave angle is 5° the apex angles of twice the value given in Fig 1 will not need to be reduced more than a degree or so, even at the longest leg lengths which might be used.

When the legs are long, alignment does not necessarily mean that the greatest signal strength will be obtained at the wave angle for which the apex angle is chosen. Keep in mind that the polarization of the radiated field is the same as that of a plane containing the wire. As illustrated by the diagram of Fig 3, at any wave angle other than zero, the plane containing the wire and passing through the desired wave angle is not horizontal. In the limiting case where the wave angle and the angle of maximum radiation from the wire are the same, the plane is vertical, and the radiation at that wave angle is vertically polarized. At in-between angles the polarization consists of both horizontal and vertical components.

When two wires are combined into a V, the polarization planes have opposite slopes. In the plane bisecting the V, this makes the horizontally polarized components of the two fields add together numerically, but the vertically polarized components are out of phase and cancel completely. As the wave angle is increased, the horizontally polarized components become smaller, so the intensity of horizontally polarized radiation decreases. On the other hand, the vertically polarized components become more intense but always cancel each other. The overall result is that although alignment for a given wave angle will increase the useful radiation at that angle, the wave angle at which maximum radiation occurs (in the direction of the line bisecting the V) is always below the wave angle for which the wires are aligned. As shown by Fig 7, the difference between the apex angles required for optimum alignment of the lobes at wave angles of 0° and 15° is rather small, even when the legs are many wavelengths long.

For long-distance transmission and reception, the lowest possible wave angle usually is the best. Consequently, it is good practice to choose an apex angle between the limits represented by the two curves in Fig 7. The actual wave angle

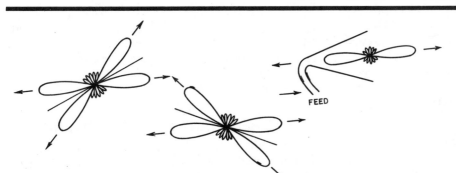

Fig 6—Two long wires and their respective patterns are shown at the left. If these two wires are combined to form a V with an angle that is twice that of the major lobes of the wires and with the wires excited out of phase, the radiation along the bisector of the V adds and the radiation in the other directions tends to cancel.

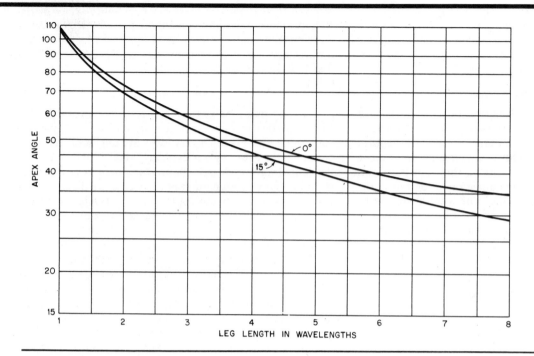

Fig 7—Apex angle of V antenna for alignment of main lobe at different wave angles, as a function of leg length in wavelengths.

at which the radiation is maximum will depend on the shape of the vertical pattern and the height of the antenna above ground.

When the leg length is small, there is some advantage in reducing the apex angle of the V because this changes the mutual impedance in such a way as to increase the gain of the antenna. For example, the optimum apex angle in the case of 1-λ legs is 90°.

Multiband Design

When a V antenna is used over a range of frequencies—such as 14 to 28 MHz—its characteristics over the frequency range will not change greatly if the legs are sufficiently long at the lowest frequency. The apex angle, at zero wave angle, for a 5-wavelength V (each leg approximately 350 feet long at 14 MHz) is 44°. At 21 MHz, where the legs are 7.5 wavelengths long, the optimum angle is 36°, and at 28 MHz where the leg length is 10 wavelengths it is 32 degrees. Such an antenna will operate well on all three frequencies if the apex angle is about 35°. From Fig 7, a 35-degree apex angle with a 5-wavelength V will align the lobes at a wave angle of something over 15°, but this is not too high when it is kept in mind that the maximum radiation actually will be at a lower angle. At 28 MHz the apex angle is a little large, but the chief effect will be a small reduction in gain and a slight broadening of the horizontal pattern, together with a tendency to reduce the wave angle at which the radiation is maximum. The same antenna can be used at 3.5 and 7 MHz, and on these bands the fact that the wave angle is raised is of less consequence, as high wave angles are useful. The gain will be small, however, because the legs are not very long at these frequencies.

Other V Combinations

A gain increase of about 3 dB can be had by stacking two Vs one above the other, a half wavelength apart, and feeding them with in-phase currents. This will result in a lowered angle of radiation. The bottom V should be at least a quarter wavelength above the ground, and preferably a half wavelength.

Two V antennas can be broadsided to form a W, giving an additional 3-dB gain. However, two transmission lines are required and this, plus the fact that five poles are needed to support the system, renders it impractical for the average amateur.

The V antenna can be made unidirectional by using a second V placed an odd multiple of a quarter wavelength in back of the first and exciting the two with a phase difference of 90°. The system will be unidirectional in the direction of the antenna with the lagging current. However, the V reflector is not normally employed by amateurs at low frequencies because it restricts the use to one band and requires a fairly elaborate supporting structure. Stacked Vs with driven reflectors could, however, be built for the 200- to 500-MHz region without much difficulty. The overall gain for such an antenna (two stacked Vs, each with a V reflector) is about 9 dB greater than the gains given in Fig 1.

Feeding the V

The V antenna is most conveniently fed with tuned feeders, since they permit multiband operation. Although the length of the wires in a V beam is not at all critical, it is important that both wires be of the same electrical length. If the use of a nonresonant line is desired, probably the most appropriate matching system is that using a stub or quarter-wave matching section. The adjustment of such a system is described in Chapter 26.

THE RESONANT RHOMBIC ANTENNA

The diamond-shaped or rhombic antenna shown in Fig 8 can be looked upon as two acute-angle Vs placed end-to-end. This arrangement is called a resonant rhombic, and has two advantages over the simple V that have caused it to be

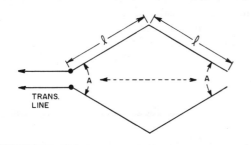

Fig 8—The resonant rhombic or diamond-shaped antenna. All legs are the same length, and opposite angles of the diamond are equal. Length ℓ is an integral number of half wavelengths.

favored by amateurs. For the same total wire length it gives somewhat greater gain than the V. A rhombic 4 wavelengths on a leg, for example, has better than 1 dB gain over a V antenna with 8 wavelengths on a leg. And the directional pattern of the rhombic is less frequency sensitive than the V when the antenna is used over a wide frequency range. This is because a change in frequency causes the major lobe from one leg to shift in one direction while the lobe from the opposite leg shifts the other way. This tends to make the optimum direction stay the same over a considerable frequency range. The leg lengths of the rhombic must be an integral number of half wavelengths in order to avoid reactance at its feed point. It is for this reason that the antenna bears the name *resonant* rhombic. The disadvantage of the rhombic as compared with the V is that one additional support is required.

The same factors that govern the design of the V antenna apply in the case of the resonant rhombic. The angle A in Fig 8 is the same as that for a V having a leg length equal to ℓ, Fig 8. If it is desired to align the lobes from individual wires with the wave angle, the curves of Fig 7 may be used, again using the length of one leg in taking the data from the curves. The diamond-shaped antenna also can be operated as a terminated antenna, as described later in this chapter, and much of the discussion in that section applies to the resonant rhombic as well.

The direction of maximum radiation with a resonant rhombic is given by the broken-line arrows in Fig 8; that is, the antenna is bidirectional. There are minor lobes in other directions, their number and intensity depending on the leg length. When used at frequencies below the VHF region, the rhombic antenna is always mounted with the plane containing the wires horizontal. The polarization in this plane, and also in the perpendicular plane that bisects the rhombic, is horizontal. At 144 MHz and above, the dimensions are such that the antenna can be mounted with the plane containing the wires vertical if vertical polarization is desired.

When the rhombic antenna is to be used on several HF amateur bands, it is advisable to choose the apex angle, A, on the basis of the leg length in wavelengths at 14 MHz. This point is covered in more detail in connection with both the V and the terminated rhombic. Although the gain on higher frequency bands will not be quite as favorable as if the antenna had been designed for the higher frequencies, the system will radiate well at the low angles that are necessary at such frequencies. At frequencies below the design fre-

quency, the greater apex angle of the rhombic (as compared with a V of the same total length) is more favorable to good radiation than in the case of the V.

The resonant rhombic antenna can be fed in the same way as the V. Resonant feeders are necessary if the antenna is to be used in several amateur bands.

TERMINATED LONG-WIRE ANTENNAS

All the antenna systems considered so far in this chapter have been based on operation with standing waves of current and voltage along the wire. Although most antenna designs are based on using resonant wires, resonance is by no means a necessary condition for the wire to radiate and intercept electromagnetic waves efficiently. The result of using nonresonant wires is reactance at the feed point, unless the antenna is terminated.

In Fig 9, let us suppose that the wire is parallel with the ground (horizontal) and is terminated by a load Z equal to its characteristic impedance, Z_0. The load Z can represent a receiver matched to the line. The resistor R is also equal to the Z_0 of the wire. A wave coming from direction X will strike the wire first at its far end and sweep across the wire at some angle until it reaches the end at which Z is connected. In so doing, it will induce voltages in the antenna, and currents will flow as a result. The current flowing toward Z is the useful output of the antenna, while the current flowing toward R will be absorbed in R. The same thing is true of a wave coming from the direction X'. In such an antenna there are no standing waves, because all received power is absorbed at either end.

The greatest possible power will be delivered to the load Z when the individual currents induced as the wave sweeps across the wire all combine properly on reaching the load. The currents will reach Z in optimum phase when the time required for a current to flow from the far end of the antenna to Z is exactly one-half cycle longer than the time taken by the wave to sweep over the antenna. A half cycle is equivalent to a half wavelength greater than the distance traversed by the wave from the instant it strikes the far end of the antenna to the instant that it reaches the near end. This is shown by the small drawing, where AC represents the antenna, BC is a line perpendicular to the wave direction, and AB is the

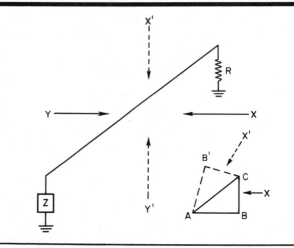

Fig 9—Terminated long-wire antenna.

distance traveled by the wave in sweeping past AC. AB must be one-half wavelength shorter than AC. Similarly, AB´ must be the same length as AB for a wave arriving from X´.

A wave arriving at the antenna from the opposite direction Y (or Y´), will similarly result in the largest possible current at the far end. However, since the far end is terminated in R, which is equal to Z, all the power delivered to R by the wave arriving from Y will be absorbed in R. The current traveling to Z will produce a signal in Z in proportion to its amplitude. If the antenna length is such that all the individual currents arrive at Z in such phase as to add up to zero, there will be no current through Z. At other lengths the resultant current may reach appreciable values. The lengths that give zero amplitude are those which are odd multiples of ¼ wavelength, beginning at ¾ wavelength. The response from the Y direction is greatest when the antenna is any even multiple of ½ wavelength long; the higher the multiple, the smaller the response.

Directional Characteristics

The explanation above considers the phase but not the relative amplitudes of the individual currents reaching the load. When the appropriate correction is made, the angle with the wire at which radiation or response is maximum is given by the curve of Fig 10. The response drops off gradually on either side of the maximum point, resulting in lobes in the directive pattern much like those for harmonic antennas, except that the system is essentially unidirectional. Typical patterns are shown in Fig 11. When the antenna length is 3/2 λ or greater, there are also angles at which secondary maxima (minor lobes) occur; these secondary maxima have peaks approximately at angles for which the length AB, Fig 9, is less than AC by any odd multiple of one-half wavelength. When AB is shorter than AC by an even multiple of a half wavelength, the induced currents cancel each other completely at Z, and in such cases there is a null for waves arriving in the direction perpendicular to BC.

The antenna of Fig 9 responds to horizontally polarized signals when mounted horizontally. If the wire lies in a plane that is vertical with respect to the earth, it responds to vertically polarized signals. By reciprocity, the directive characteristics for transmitting are the same as for receiving. For average conductor diameters and heights above ground, 20 or 30 feet, the Z_0 of the antenna is of the order of 500 to 600 ohms.

It is apparent that an antenna operating in this way has much the same characteristics as a transmission line. When it is properly terminated at both ends there are traveling waves—but no standing waves—on the wire. Consequently the current is essentially the same all along the wire over any given period of time. Actually, it decreases slightly in the direction in which the current is flowing because of energy loss by radiation as well as by ohmic loss in the wire and the ground. The antenna can be looked upon as a transmission line terminated in its characteristic impedance, but having such wide spacing between conductors (the second conductor in this case is the image of the antenna in the ground) that radiation losses are by no means inconsequential.

A wire terminated in its characteristic impedance will work on any frequency, but its directional characteristics change with frequency as shown by Fig 10. To give any appreciable gain over a dipole, the wire must be at least a few wavelengths along. The angle at which maximum response occurs can be in any plane that contains the wire axis, so in

free space the major lobe will be a hollow cone. In the presence of ground, the discussion given in connection with Fig 3 applies, with the modification that the angles of best radiation or response are those given in Fig 10, rather than by Figs 1 or 2. As comparison of the curves will show, the difference

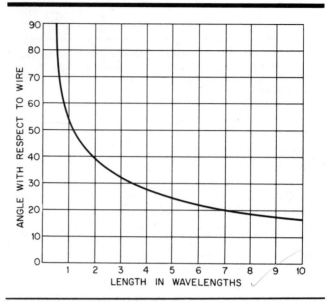

Fig 10—Angle with respect to wire axis at which the radiation from a terminated long-wire antenna is maximum.

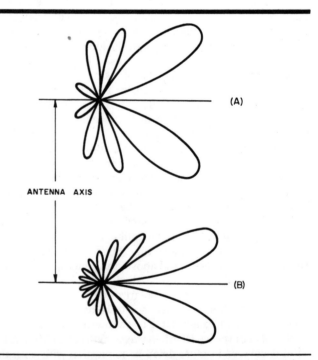

Fig 11—Typical radiation patterns (cross section of solid figure) for terminated long wires. At A, the length is two wavelengths, at B, four wavelengths, both for an idealized case in which there is no decrease of current along the wire. In practice, the pattern is somewhat distorted by wire attenuation.

in the optimum angle between resonant and terminated wires is quite small.

The Sloping V

The sloping V antenna, illustrated in Fig 12, is a terminated system. Even though it is simple to construct and offers multiband operation, it has not seen much use by amateurs. Only a single support is required, and the antenna should provide several decibels of gain over a frequency ratio of 3 to 1 or greater.

For satisfactory performance, the leg length, ℓ, should be a minimum of one wavelength at the lowest operating frequency. The height of the support may be ½ to ¾ of the leg length. The feed point impedance of the sloping V is on the order of 600 ohms. Therefore, open-wire line may be used for the feeder or, alternatively, a coaxial transmission line and a step-up transformer balun at the apex of the V may be used.

The terminating resistors should each be noninductive with a value of 300 ohms and a dissipation rating equal to one-half the transmitter output power. The grounded end of the resistors should be connected to a good RF ground, such as a radial system extending beneath the wires of the V. A single ground stake at each termination point will likely be insufficient; a pair of wires, one running from each termination point to the base of the support will probably prove superior.

By using the data presented earlier in this chapter, it should be possible to calculate the apex angle and support height for optimum lobe alignment from the two wires at a given frequency. Ground reflections will complicate the calculations, however, as both vertical and horizontal polarization components are present. Dimensions that have proved useful for point-to-point communications work on frequencies from 14 to 30 MHz are a support height of 60 feet, a leg length of 100 feet, and an apex angle of 36°.

THE TERMINATED RHOMBIC ANTENNA

The highest development of the long-wire antenna is the terminated rhombic, shown schematically in Fig 13. It consists of four conductors joined to form a diamond, or rhombus. All sides of the antenna have the same length and the opposite corner angles are equal. The antenna can be considered as being made up of two V antennas placed end to end and terminated by a noninductive resistor to produce a unidirectional pattern. The terminating resistor is connected between the far ends of the two sides, and is made approximately equal to the characteristic impedance of the antenna as a unit. The rhombic may be constructed either horizontally or vertically, but is practically always constructed horizontally at frequencies below 54 MHz, since the pole height required is considerably less. Also, horizontal polarization is equally, if not more, satisfactory at these frequencies.

The basic principle of combining lobes of maximum radiation from the four individual wires constituting the rhombus or diamond is the same in either the terminated type shown in Fig 13, or the resonant type described earlier in this chapter. The included angles should differ slightly because of the differences between resonant and terminated wires, as just described, but the differences are almost negligible.

Tilt Angle

In dealing with the terminated rhombic, it is a matter of custom to talk about the "tilt angle" (ϕ in Fig 13), rather than the angle of maximum radiation with respect to an individual wire. The tilt angle is simply 90° minus the angle of maximum radiation. In the case of a rhombic antenna designed for zero wave angle, the tilt angle is 90° minus the values given in Fig 10.

Fig 14 shows the tilt angle as a function of the antenna leg length. The curve marked "0°" is used for a wave angle of 0°; that is, maximum radiation in the plane of the antenna. The other curves show the proper tilt angles to use when aligning the major lobe with a desired wave angle. For a wave angle of 5° the difference in tilt angle is less than 1° for the range of lengths shown. Just as in the case of the resonant V and resonant rhombic, alignment of the wave angle and lobes always results in still greater radiation at a lower wave angle, and for the same reason, but also results in the greatest possible radiation at the desired wave angle.

The broken curve marked "optimum length" shows the leg length at which maximum gain is obtained at any given wave angle. Increasing the leg length beyond the optimum will result in lessened gain, and for that reason the curves do not extend beyond the optimum length. Note that the optimum length becomes greater as the desired wave angle decreases. Leg lengths over 6 λ are not recommended because the directive pattern becomes so sharp that the antenna performance is highly variable with small changes in the angle, both horizontal and vertical, at which an incoming wave reaches the antenna. Since these angles vary to some extent

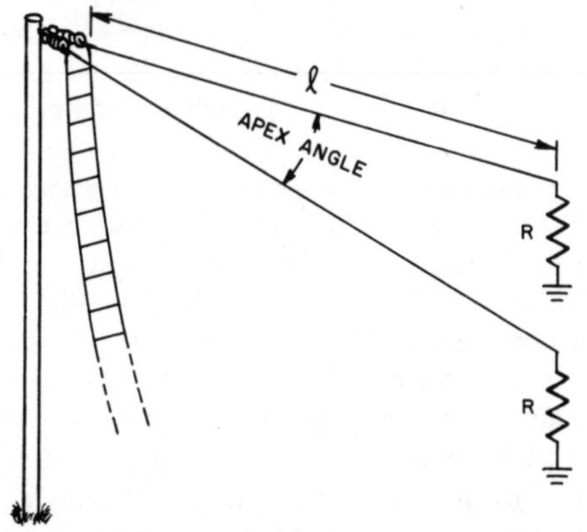

Fig 12—The sloping V antenna.

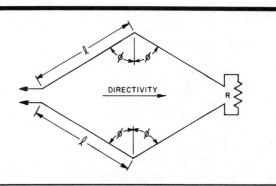

Fig 13—The terminated rhombic antenna.

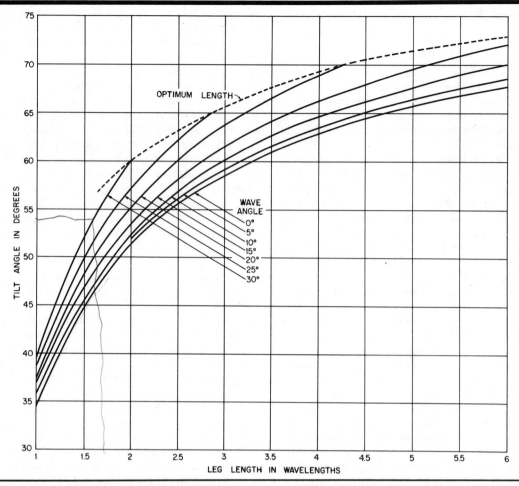

Fig 14—Rhombic-antenna design chart. For any given leg length, the curves show the proper tilt angle to give maximum radiation at the selected wave angle. The broken curve marked "optimum length" shows the leg length that gives the maximum possible output at the selected wave angle. The optimum length as given by the curves should be multiplied by 0.74 to obtain the leg length for which the wave angle and main lobe are aligned (see text, "Alignment of Lobes").

in ionospheric propagation, it does not pay to attempt to use too great a degree of directivity.

Multiband Design

When a rhombic antenna is to be used over a considerable frequency range, it is worth paying some attention to the effect of the tilt angle on the gain and directive pattern at various frequencies. For example, suppose the antenna is to be used at frequencies up to and including the 28-MHz band, and that the leg length is to be 6 λ on that band. For zero wave angle, the optimum tilt angle is 68°, and the calculated free-space directive pattern in the vertical plane bisecting the antenna is shown in Fig 15, at B. At 14 MHz, this same antenna has a leg length of three wavelengths, which calls for a tilt angle of 58.5° for maximum radiation at zero wave angle. The calculated patterns for tilt angles of 58.5 and 68° are shown at A in Fig 15. These show that if the optimum tilt for 28-MHz operation is used, the gain will be reduced and the wave angle raised at 14 MHz. In an attempt at a compromise, we might select a wave angle of 15°, rather than zero, for 14 MHz. As shown by Fig 14, the tilt angle here is larger and thus more nearly coincides with the tilt angle for zero wave angle on 28 MHz. From the chart, the tilt angle for 3 wavelengths on a leg and a 15-degree wave angle is 61.5°. The

patterns with this tilt angle are shown in Fig 15 for both the 14- and 28-MHz cases. The effect at 28 MHz is to decrease the gain at zero wave angle by more than 6 dB and to split the radiation in the vertical plane into two lobes, one of which is at a wave angle too high to be useful at this frequency.

Inasmuch as the gain increases with the leg length in wavelengths, it is probably better to favor the lower frequency in choosing the tilt angle. In the present example, the best compromise probably would be to split the difference between the optimum tilt angle for the 15° wave angle at 14 MHz and that for zero wave angle at 28 MHz; that is, use a tilt angle of about 64°. Design dimensions for such an antenna are given in Fig 16.

The patterns of Fig 15 are in the vertical plane through the center of the antenna only. In vertical planes making an angle with the antenna axis, the patterns may differ considerably. The effect of a tilt angle that is smaller than the optimum is to broaden the horizontal pattern, so at 28 MHz the antenna in the example would be less directive in the horizontal plane than would be the case if it were designed for optimum performance at that frequency. It should also be noted that the patterns given in Fig 15 are free-space patterns and must be multiplied by the ground-reflection factors for the actual antenna height used, if the actual vertical

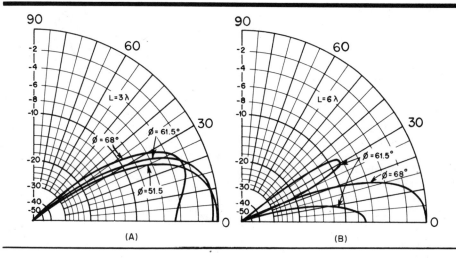

Fig 15—These drawings show the effect of tilt angle on the free-space vertical pattern of a terminated rhombic antenna having a leg length of 3 wavelengths at one frequency and 6 wavelengths at twice the frequency. These patterns apply only in the direction of the antenna axis. Minor lobes above 30° are not shown.

patterns are to be determined. (Also see later discussion on lobe alignment.)

Power Gain

The theoretical power gain of a terminated rhombic antenna over a dipole (both in free space) is given by the curve of Fig 17. This curve is for zero wave angle and includes an allowance of 3 dB for power dissipated in the terminating resistor. The actual gain of an antenna mounted horizontally above the ground, as compared with a dipole at the same height, can be expected to vary a bit either way from the figures given by the curve. The power lost in the terminating resistor is probably less than 3 dB in the average installation, since more than half of the input power is radiated before the end of the antenna is reached. However, there is also more power loss in the wire and in the ground under the antenna than in the case of a simple dipole, so the 3 dB figure is probably a representative estimate of overall loss.

Termination

Although there is no marked difference in the gain obtainable with resonant and terminated rhombics of comparable design, the terminated antenna has the advantage that over a wide frequency range it presents an essentially resistive and constant load to the transmitter. In addition, terminated operation makes the antenna essentially unidirectional, while the unterminated or resonant rhombic is always bidirectional (although not symmetrically so). In a sense, the power dissipated in the terminating resistor can be considered power that would have been radiated in the other direction had the resistor not been there. Therefore, the fact that some of the power (about one-third) is used up in heating the resistor does not mean an actual loss in the desired direction.

The characteristic impedance of an ordinary rhombic antenna, looking into the input end, is in the order of 700 to 800 ohms when properly terminated in a resistance at the far end. The terminating resistance required to bring about the matching condition usually is slightly higher than the input impedance because of the loss of energy through radiation by the time the far end is reached. The correct value usually will be found to be of the order of 800 ohms, and should be determined experimentally if the flattest possible antenna is desired. However, for average work a noninductive resistance of 800 ohms can be used with the assurance that the operation will not be far from optimum.

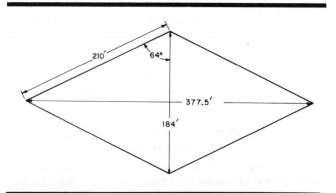

Fig 16—Rhombic antenna dimensions for a compromise design between 14- and 28-MHz requirements, as discussed in the text. The leg length is 6 λ at 28 MHz, 3 λ at 14 MHz.

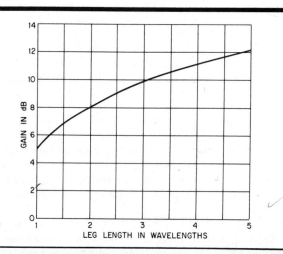

Fig 17—Theoretical gain of a terminated rhombic antenna over a half-wave dipole in free space. This curve includes an allowance of 3 dB for loss in the terminating resistor.

The terminating resistor must be practically a pure resistance at the operating frequencies; that is, its inductance and capacitance should be negligible. Ordinary wire-wound resistors are not suitable because they have far too much inductance and distributed capacitance. Small carbon resistors have satisfactory electrical characteristics but will not dissipate more than a few watts and so cannot be used, except when the transmitter power does not exceed 10 or 20 watts or when the antenna is to be used for reception only. The special resistors designed either for use as "dummy" antennas or for terminating rhombic antennas should be used in other cases. To allow a factor of safety, the total rated power dissipation of the resistor or resistors should be equal to half the power output of the transmitter.

To reduce the effects of stray capacitance it is desirable to use several units, say three, in series even when one alone will safely dissipate the power. The two end units should be identical and each should have one fourth to one third the total resistance, with the center unit making up the difference. The units should be installed in a weatherproof housing at the end of the antenna to protect them and to permit mounting without mechanical strain. The connecting leads should be short so that little extraneous inductance is introduced.

Alternatively, the terminating resistance may be placed at the end of an 800-ohm line connected to the end of the antenna. This will permit placing the resistors and their housing at a point convenient for adjustment rather than at the top of the pole. Resistance wire may be used for this line, so that a portion of the power will be dissipated before it reaches the resistive termination, thus permitting the use of lower wattage lumped resistors. The line length is not critical, since it operates without standing waves.

Multiwire Rhombics

The input impedance of a rhombic antenna constructed as in Fig 13 is not quite constant as the frequency is varied. This is because the varying separation between the wires causes the characteristic impedance of the antenna to vary along its length. The variation in Z_0 can be minimized by a conductor arrangement that increases the capacitance per unit length in proportion to the separation between the wires.

The method of accomplishing this is shown in Fig 18. Three conductors are used, joined together at the ends but with increasing separation as the junction between legs is approached. For HF work the spacing between the wires at the center is 3 to 4 feet, which is similar to that used in commercial installations using legs several wavelengths long. Since all three wires should have the same length, the top and bottom wires should be slightly farther from the support than the middle wire. Using three wires in this way reduces the Z_0 of the antenna to approximately 600 ohms, thus providing a better match for practical open-wire line, in addition to smoothing out the impedance variation over the frequency range.

A similar effect (although not quite as favorable) is obtained by using two wires instead of three. The 3-wire system has been found to increase the gain of the antenna by about 1 dB over that of a single-conductor version.

Front-to-Back Ratio

It is theoretically possible to obtain an infinite front-to-back ratio with a terminated rhombic antenna, and in practice very large values can be had. However, when the antenna is terminated in its characteristic impedance the infinite front-

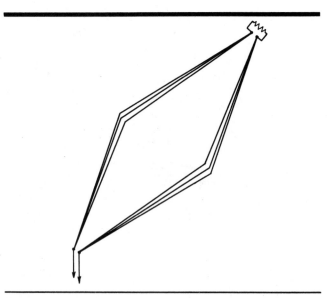

Fig 18—Three-wire rhombic antenna. Use of multiple wires improves the impedance characteristic of a terminated rhombic and increases the gain somewhat.

to-back ratio can be obtained only at frequencies for which the leg length is an odd multiple of a quarter wavelength, as described in the section on terminated long wires. The front-to-back ratio is smallest at frequencies for which the leg length is a multiple of a half wavelength.

When the leg length is not an odd multiple of a quarter wave at the frequency under consideration, the front-to-back ratio can be made very high by decreasing the value of terminating resistance slightly. This permits a small reflection from the far end of the antenna which cancels out the residual response at the input end. With large antennas, the front-to-back ratio may be made very large over the whole frequency range by experimental adjustment of the terminating resistance. Modification of the terminating resistance can result in a splitting of the back null into two nulls, one on either side of a small lobe in the back direction. Changes in the value of terminating resistance thus permit "steering" the back null over a small horizontal range so that signals coming from a particular spot not exactly to the rear of the antenna may be minimized.

Ground Effects

Reflections from the ground play exactly the same part in determining the vertical directive pattern of a horizontal rhombic antenna that they play with other horizontal antennas. Consequently, if a low wave angle is desired, it is necessary to make the height great enough to bring the wave angle into the desired range of values given by the charts in Chapter 2.

Alignment of Lobes, Wave Angle, and Ground Reflections

When maximum antenna response is desired at a particular wave angle (or maximum radiation is desired at that angle), the major lobe of the antenna can be aligned not only with the wave angle as previously described, but also with a maximum in the ground-reflection factor. When this is done it is no longer possible to consider the antenna height independently of other aspects of rhombic design. The wave angle, leg length, and height become mutually dependent.

This method of design is of particular value when the antenna is built to be used over fixed transmission distances for which the optimum wave angle is known. It has had wide application in commercial work with terminated rhombic antennas, but seems less desirable for amateur use where, for the long-distance work for which rhombic antennas are built, the lowest wave angle that can be obtained is the most desirable. Alignment of all three factors is limited in application because it leads to impractical heights and leg lengths for small wave angles. Consequently, when a fairly broad range of low wave angles is the objective, it is more satisfactory to design for a low wave angle and simply make the antenna as high as possible.

Fig 19 shows the lowest height at which ground reflections make the radiation maximum at a desired wave angle. It can be used in conjunction with Fig 14 for complete alignment of the antenna. For example, if the desired wave angle is 20°, Fig 19 shows that the height must be 0.75 λ. From Fig 14, the optimum leg length is 4.2 λ and the tilt angle is just under 70°. A rhombic antenna designed this way will have the maximum possible output that can be obtained at a wave angle of 20°; no other set of dimensions will be as good. However, it will have still greater output at some angle lower than 20°, for the reasons given earlier. When it is desired to make the maximum output of the antenna occur at the 20° wave angle, it may be accomplished by using the same height and tilt angle, but with the leg length reduced by 26%. Thus for such alignment, the leg length should be 4.2 × 0.74 = 3.1 λ. The output at the 20° wave angle will be smaller than with 4.2 λ legs, however, despite the fact that the smaller antenna has its maximum radiation at 20°. The reduction in gain is about 1.5 dB.

Methods of Feed

If the broad frequency characteristic of the rhombic antenna is to be utilized fully, the feeder system must be

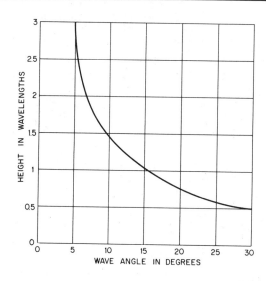

Fig 19—Antenna height to be used for maximum radiation at a desired wave angle. This curve applies to any type of horizontal antenna.

similarly broad. Open-wire transmission line of the same characteristic impedance as that shown at the antenna input terminals (approximately 700 to 800 ohms) may be used. Data for the construction of such lines is given in Chapter 24. While the usual matching stub can be used to provide an impedance transformation to more satisfactory line impedances, this limits the operation of the antenna to a comparatively narrow range of frequencies centering about that for which the stub is adjusted. Probably a more satisfactory arrangement would be to use a coaxial transmission line and a broadband transformer balun at the antenna feed point.

Wave Antennas

Perhaps the best known type of wave antenna is the Beverage. Many 160-meter enthusiasts have used Beverage antennas to enhance the signal-to-noise ratio while attempting to extract weak signals from the often high levels of atmospheric noise and interference on the low bands. Alternative antenna systems have been developed and used over the years, such as loops and long spans of unterminated wire on or slightly above the ground, but the Beverage antenna seems to be the best for 160-meter weak-signal reception. The information in this section was prepared by Rus Healy, NJ2L.

THE BEVERAGE ANTENNA

A Beverage is simply a wire antenna, at least one wavelength long, supported along its length at a fairly low height and terminated at the far end in its characteristic impedance. This antenna is shown in Fig 20A.

Improved HF reception with Beverage antennas may

result from propagation conditions at a given time. However, because the incoming sky waves above medium frequency arrive at moderate and high angles, and because their polarization changes at random during reflection from the ionosphere, these waves do not excite a Beverage in the same way as MF signals. The wave antenna is responsive mostly to very low angle incoming waves that maintain a constant (vertical) polarization. These conditions are nearly always satisfied on 160 meters, and most of the time on 80 meters. As the frequency is increased, however, the polarization and arrival angles are less and less constant and favorable, making Beverages less effective at these frequencies. Many amateurs have, however, reported consistently excellent performance from Beverage antennas at frequencies as high as 10 MHz.

Beverage Theory

The Beverage antenna acts like a long transmission line with one lossy conductor (the earth), and one good conductor

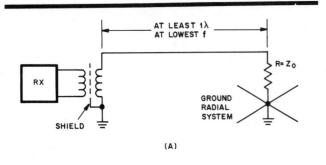

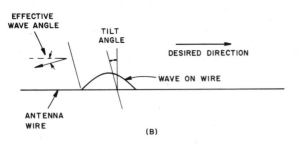

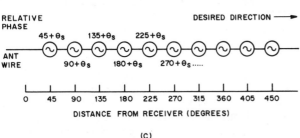

Fig 20—At A, a simple one-wire Beverage antenna with a variable termination impedance and a matching transformer for the receiver impedance. At B, a portion of a wave from the desired direction is shown traveling down the antenna wire. Its tilt angle and effective wave angle are also shown. At C, a situation analogous to the action of a Beverage on an incoming wave is shown. See text for discussion.

(the wire). Beverages have excellent directivity if erected properly, but they are quite inefficient. Therefore, they are not suitable for use as transmitting antennas.

Because the Beverage is a traveling wave antenna, it has no standing waves resulting from radio signals. After a wave strikes the end of the Beverage from the desired direction, the wave induces voltages along the antenna and continues in space as well. Fig 20B shows part of a wave on the antenna resulting from a desired signal. This diagram also shows the tilt of the wave. The signal induces equal voltages in both directions. The resulting currents are equal and travel in both directions; the component traveling toward the termination end moves *against* the wave and thus builds down to a very low level at the termination end. Any residual signal resulting from this direction of current flow will be absorbed in the termination (if the termination is equal to the antenna impedance). The component of the signal flowing in the other direction, as we will see, becomes a key part of the received signal.

As the wave travels along the wire, the wave in space travels at approximately the same velocity. (There is some

phase delay in the wire, as we shall see.) At any given point in time, the wave traveling along in space induces a voltage in the wire in addition to the wave already traveling on the wire (voltages already induced by the wave). Because these two waves are nearly in phase, the voltages add and build toward a maximum at the receiver end of the antenna.

This process can be likened to a series of signal generators lined up on the wire, with phase differences corresponding to their respective spacings on the wire (Fig 20C). At the receiver end, a maximum voltage is produced by these voltages adding in phase. For example, the wave component induced at the receiver end of the antenna will be in phase (at the receiver end) with a component of the same wave induced, say, 270° (or any other distance) down the antenna, after it travels to the receiver end.

In practice, there is some phase shift of the wave on the wire with respect to the wave in space. This phase shift results from the velocity factor of the antenna. (As with any transmission line, the signal velocity on the Beverage is somewhat less than in free space.) Velocity of propagation on a Beverage is typically between 85 and 98% of that in free space. As antenna height is increased to a certain optimum height (which is about 10 feet for 160 meters), the velocity factor increases. Beyond this height, only minimal improvement is afforded, as shown in Fig 21. These curves are the result of experimental work done in 1922 by RCA, and reported in a *QST* article (November 1922) entitled "The Wave Antenna for 200-Meter Reception," by H. H. Beverage. The curve for 160 meters was extrapolated from the other curves.

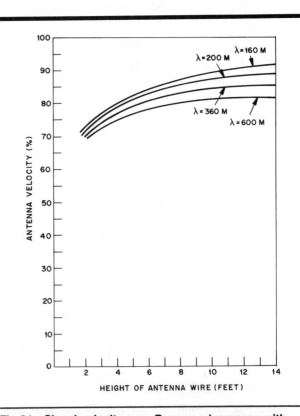

Fig 21—Signal velocity on a Beverage increases with height above ground, and reaches a practical maximum at about 10 feet. Improvement is minimal above this height. (The velocity of light is 100%.)

Phase shift (per wavelength) is shown as a function of velocity factor in Fig 22, and is given by

$$\theta = 360 \left(\frac{100}{k} - 1 \right)$$

where k = velocity factor of the antenna in percent.

The signals present on and around a Beverage antenna are shown graphically in A through D of Fig 23. These curves show relative voltage levels over a number of periods of the wave in space and their relative effects in terms of the total signal at the receiver end of the antenna.

Termination and Performance in Other Directions

The performance of a Beverage antenna in directions other than the favored one is quite different than previously discussed. Take, for instance, the case of a signal arriving perpendicular to the wire (90 degrees either side of the favored direction). In this case, the wave induces voltages along the wire that are essentially *in phase*, so they arrive at the receiver end more or less out of phase, and thus cancel. (This can be likened to a series of signal generators lined up along the antenna as before, but having no progressive phase differences.)

As a result of this cancellation, Beverages exhibit deep nulls off the sides. Some minor sidelobes will exist, as with other horizontal antennas, and will increase in number with the length of the antenna.

In the case of a signal arriving from the rear of the antenna, the behavior of the antenna is very similar to its performance in the favored direction. The major difference is that the signal from the rear adds in phase at the *termination* end and is absorbed by the termination impedance.

For proper operation, the Beverage must be terminated at both ends in an impedance equal to the Z_0 of the antenna. This consists of matching the receiver impedance to the antenna at one end and terminating the other end in a resistor of the correct value.

If the termination impedance is not equal to the characteristic impedance of the antenna, some part of the signal from the rear will be reflected back toward the receiver end of the antenna. If the termination impedance is merely an open circuit (no terminating resistor), total reflection will result and the antenna will exhibit a bidirectional pattern (still with very deep nulls off the sides). An unterminated Beverage will not have the same response to signals in the rearward direction as it exhibits to signals in the forward direction because of attenuation and reradiation of part of the reflected wave as it travels back toward the receiver end. The difference in response is typically on the order of 3 dB for a 1-λ single-wire Beverage (see Figs 24 and 25).

If the termination is between the extremes (open circuit

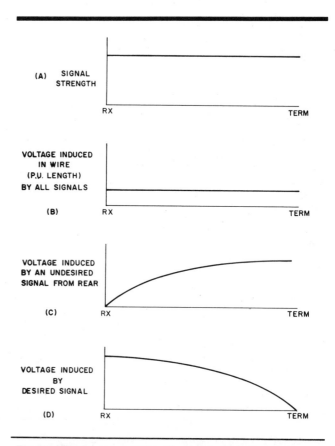

Fig 23—These curves show the voltages that appear in a Beverage antenna over a period of several cycles of the wave. Signal strength (at A) is constant over the length of the antenna during this period, as is voltage induced per unit length in the wire (at B). (The voltage induced in any section of the antenna is the same as the voltage induced in any other section of the same size, over the same period of time.) At C, the voltages induced by an undesired signal from the rearward direction add in phase and build to a maximum at the termination end, where they are dissipated in the termination (if $Z_{term} = Z_0$). The voltages resulting from a desired signal are shown at D. The wave on the wire travels closely with the wave in space, and the voltages resulting add in phase to a maximum at the receiver end of the antenna.

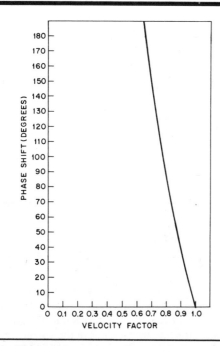

Fig 22—This curve shows phase shift (per wavelength) as a function of velocity factor on a Beverage antenna. Once the phase shift for the antenna goes beyond 90°, the gain drops off from its peak value, and any increase in antenna length will decrease gain.

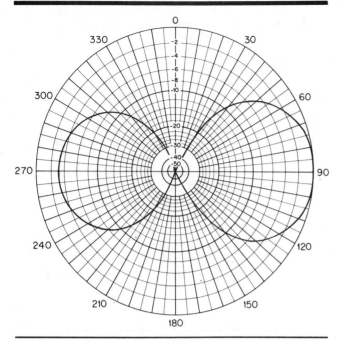

Fig 24—Directive pattern of a one-wavelength long unterminated Beverage antenna.

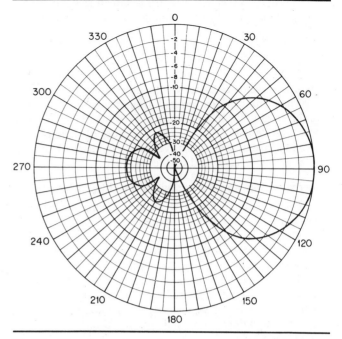

Fig 25—Directive pattern of a one-wavelength Beverage that is terminated in its characteristic impedance.

and perfect termination in Z_0), the peak direction and intensity of signals off the rear of the Beverage will change. As a result, an adjustable reactive termination can be employed to "steer" the nulls to the rear of the antenna (see Fig 26). This can be of great help in eliminating an interfering signal from a rearward direction (typically 30-40° either side of the back direction).

To determine the appropriate value for a terminating resistor, it is necessary to know the characteristic impedance (surge impedance), Z_0, of the Beverage. It is interesting to note that Z_0 is not a function of the *length* of the Beverage, but only the wire size and height above ground. Characteristic impedance can be found empirically by choosing some resistive value within the typical range of 400-600 ohms and then adjusting it for optimum rejection of rearward signals. This method takes time, patience, and (frequently) a second person to execute. It is far easier to start with a value that you know is close. The surge impedance of a single-wire Beverage is given by:

$$Z_0 = 138 \times \log\left(\frac{4h}{d}\right)$$

where

  Z_0 = characteristic impedance of the Beverage
  h = wire height above ground
  d = wire diameter (in the same units as h)

Another important aspect of terminating the Beverage is the assurance of a good RF ground for the termination. This is most easily accomplished by laying radial wires on the ground at the termination end. This is important, as the effective impedance of the termination will approach the Z_0 of the wire only if the RF ground at the termination is nearly ideal. This presents something of a problem for the Beverage builder, because, as mentioned earlier, the maximum signal will be induced into the antenna when the ground under

the antenna is poor. Some have quipped that the best location at which to erect a Beverage is in a desert with a salt marsh at the termination end.

As with many other antennas, improved directivity and gain can be achieved by lengthening the antenna and by arranging several antennas into an array. One item that must be kept in mind is that (as mentioned earlier) by virtue of the velocity factor of the antenna, there is some phase shift of the wave on the antenna with respect to the wave in space. Because of this phase shift, although the directivity will continue to sharpen with increased length, there will be some optimum length at which the gain of the antenna will peak. Beyond this length, the current increments arriving at the receiver end of the antenna will no longer be in phase, and will not add to produce a maximum signal at the receiver end. This optimum length is a function of velocity factor and frequency (it is also dependent on the number of wires—see later text), and is given by:

$$L = \frac{\lambda}{4\left(\frac{100}{k} - 1\right)}$$

where

  L = maximum effective length
  λ = signal wavelength in free space (same units as L)
  k = velocity factor of the antenna in percent

Because velocity factor increases with height (to a point, as mentioned earlier), optimum length is somewhat longer if the antenna height is increased. The maximum effective length also increases with the number of wires in the antenna system. For example, for a two-wire Beverage like the bidirectional version shown in Fig 26, the maximum effective length is about 20% longer than the single-wire version. A typical length for a single-wire 1.8 MHz Beverage (made of no. 16 wire and erected 10 feet above ground) is about 1200 feet.

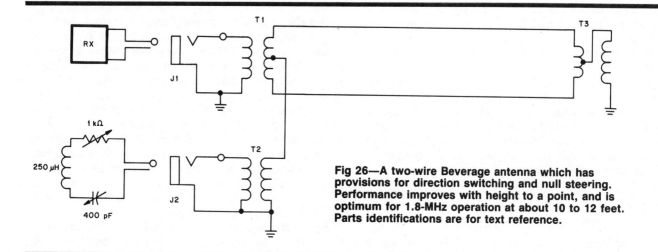

Fig 26—A two-wire Beverage antenna which has provisions for direction switching and null steering. Performance improves with height to a point, and is optimum for 1.8-MHz operation at about 10 to 12 feet. Parts identifications are for text reference.

The Two-Wire Beverage

The antenna shown in Fig 26 has the major advantage of having signals from both directions available at the receiver simultaneously. Also, because there are two wires in the system (equal amounts of signal voltage are induced in both wires), greater signal voltages will be produced.

Refer to Fig 26. A signal from the left direction induces equal voltages in both wires, and equal in-phase currents flow as a result. The reflection transformer (at the right-hand end of the antenna) then inverts the phase of these signals and reflects them back down the antenna toward the receiver, using the antenna wires as a balanced open-wire transmission line. This signal is transformed at T1 and is available at J1.

Signals traveling from right to left also induce equal voltages in each wire, and they travel in phase toward the receiver end, through T1, and into T2. Signals from this direction are available at J2.

Another convenient feature of the two-wire Beverage is the ability to steer the nulls off either end of the antenna while receiving in the opposite direction. For instance, if the series RLC network shown at J2 is adjusted while the receiver is connected to J1, signals can be received from the left direction while interference coming from the right can be partially or completely nulled. The nulls can be steered over a 60° (or more) area off the right-hand end of the antenna. The same null-steering capability exists in the opposite direction with the receiver connected at J2 and the termination connected at J1.

The two-wire Beverage is typically erected at the same height as a single-wire version. The two wires are at the same height and are spaced uniformly (12 to 18 inches apart). The impedance of the antenna depends on the wire size, spacing and height, and is given by

$$Z_0 = 69 \times \log\left[\frac{4h}{d} \sqrt{1 + \frac{(2h)^2}{S}}\right]$$

where

    Z_0 = Beverage impedance
    S = wire spacing
    h = height above ground
    d = wire diameter (in same units as S and h)

For proper operation, transformers T1, T2 and T3 must be carefully wound. Small toroidal ferrite cores are best for this application, with those of high permeability (μ_i = 125 to 5000) being the easiest to wind (fewest turns) and having the best high-frequency response. Trifilar-wound coils are most convenient. These principles also apply to single-wire Beverages. See Chapters 25 and 26 and *The ARRL Handbook* for information on winding toroidal transformers.

It should be mentioned that, even though Beverage antennas have excellent directive patterns if terminated properly, gain never exceeds about −3 dBi in most practical installations. However, the directivity that the Beverage provides results in a much higher signal-to-noise ratio for signals in the desired direction than almost any other antenna that can be used practically at low frequencies. The result of this is that instead of listening to an S9 signal with 20-dB over S9 noise and interference on a vertical, a Beverage will typically allow you to copy the same signal at S5 with only S1 (or lower) noise and interference, everything else being equal. This is certainly a worthwhile improvement!

Practical Considerations

There are a few basic principles that must be kept in mind when erecting Beverage antennas if optimum performance is to be realized.

1) Plan the installation thoroughly, including choosing an antenna length consistent with the optimum length values discussed earlier.

2) Keep the antenna as straight and as nearly level as possible over its entire run. Avoid following the terrain under the antenna too closely—keep the antenna level with the *average* terrain, avoiding changes in height over gullies, ditches, etc.

3) Use the largest wire practical and avoid joining multiple pieces of wire together to form the span if the antenna is to be permanent. The use of larger wire will keep losses, undesired phase shift, and fragility to a minimum. Joints in wire are subject to corrosion over time.

4) Minimize the lengths of vertical downleads at the ends of the antenna. Their effect is detrimental to the directive pattern of the antenna. It is best to slope the antenna wire from ground level to its final height (over a distance of 50

feet or so) at the feed-point end. Similar action should be taken at the termination end. Be sure to seal the transformers against weather.

5) Use a noninductive resistor for terminating a single-wire Beverage.

6) Use high-quality insulators for the Beverage wire where it comes into contact with the supports.

7) Keep the Beverage away from parallel conductors such as electric power and telephone lines for a distance of at least 200 feet. Perpendicular conductors may be crossed with relatively little interaction, but *do not cross any conductors that may pose a safety hazard*.

FISHBONE ANTENNAS

Another type of wave antenna is the fishbone, which, unlike the Beverage, is well suited to use at HF. A simple fishbone antenna is illustrated in Fig 27. Its impedance is approximately 400 ohms. The antenna is formed of closely spaced elements that are lightly coupled (capacitively) to a long, terminated transmission line. The capacitors are chosen to have a value that will keep the velocity of propagation of RF on the line more than 90% of that in air. The elements are usually spaced approximately 0.1 wavelength (or slightly more) so that an average of 7 or more elements are used for each full wavelength of transmission-line length. This antenna obtains low-angle response primarily as a function of its height, and therefore, is generally installed 60 to 120 feet above ground. If the antenna is to be used for transmission (for which it is well suited because of its excellent gain and broadband nature), transmitting-type capacitors must be used, since they will be required to handle substantial current.

The English HAD fishbone antenna, shown in its two-bay form in Fig 28, is less complicated than the one of Fig 27. It may be used singly, of course, and may be fed with 600-ohm open-wire line. Installation and operational characteristics are similar to the standard fishbone antenna.

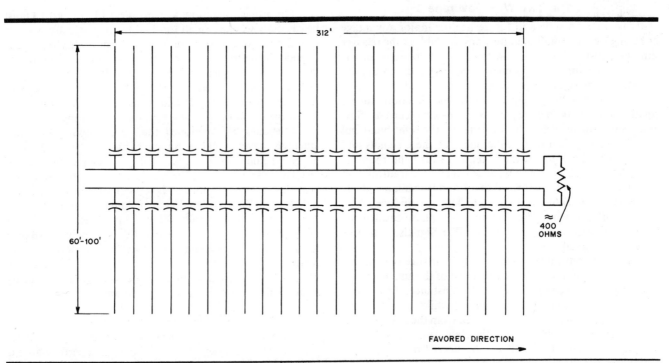

Fig 27—The fishbone antenna provides higher gain per acre than does a rhombic. It is essentially a wave antenna which evolved from the Beverage.

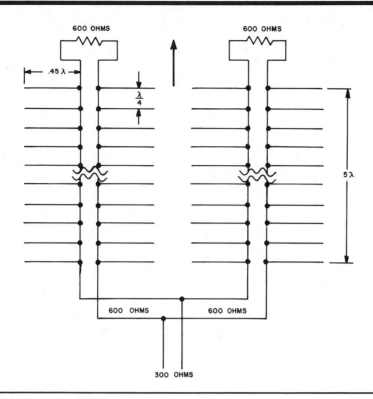

Fig 28—The English HAD fishbone antenna is a simplified version of the standard fishbone. It may be used as a single-bay antenna fed with 600-ohm open-wire line.

BIBLIOGRAPHY

Source material and more extended discussion of topics covered in this chapter can be found in the references given below.

A. Bailey, S. W. Dean and W. T. Wintringham, "The Receiving System for Long-Wave Transatlantic Radio Telephony," *The Bell System Technical Journal*, Apr 1929.

J. S. Belrose, "Beverage Antennas for Amateur Communications," Technical Correspondence, *QST*, Sep 1981.

H. H. Beverage, "Antennas," *RCA Review*, Jul 1939.

H. H. Beverage and D. DeMaw, "The Classic Beverage Antenna Revisited," *QST*, Jan 1982.

B. Boothe, "Weak-Signal Reception on 160—Some Antenna Notes," *QST*, Jun 1977.

E. Bruce, "Developments in Short-Wave Directive Antennas," *Proc IRE*, Aug 1931.

E. Bruce, A. C. Beck and L. R. Lowry, "Horizontal Rhombic Antennas," *Proc IRE*, Jan 1935.

P. S. Carter, C. W. Hansel and N. E. Lindenblad, "Development of Directive Transmitting Antennas by R.C.A. Communications," *Proc IRE*, Oct 1931.

J. Devoldere, *Low-Band DXing* (Newington: ARRL, 1987).

A. E. Harper, *Rhombic Antenna Design* (New York: D. Van Nostrand Co, Inc).

E. A. Laport, "Design Data for Horizontal Rhombic Antennas," *RCA Review*, Mar 1952.

G. M. Miller, *Modern Electronic Communication* (Englewood Cliffs, NJ: Prentice Hall, 1983).

V. A. Misek, *The Beverage Antenna Handbook* (Wason Rd, Hudson, NH: W1WCR, 1977).

F. E. Terman, *Radio Engineering*, Second Edition (New York: McGraw-Hill, 1937).

Chapter 14

Direction Finding Antennas

The use of radio for direction-finding purposes (RDF) is almost as old as its application for communications. Radio amateurs have learned RDF techniques and found much satisfaction by participating in hidden transmitter hunts. Other hams have discovered RDF through an interest in boating or aviation where radio direction finding is used for navigation and emergency location systems.

In many countries of the world, the hunting of hidden amateur transmitters takes on the atmosphere of a sport, as participants wearing jogging togs or track suits dash toward the area where they believe the transmitter is located. The sport is variously known as fox hunting, bunny hunting, ARDF (Amateur Radio direction finding) or simply transmitter hunting. In North America, most hunting of hidden transmitters is conducted from automobiles, although hunts on foot are gaining popularity.

There are less pleasant RDF applications as well, such as tracking down noise sources or illegal operators from unidentified stations. Jammers of repeaters, traffic nets and other amateur operations can be located with RDF equipment. Or sometimes a stolen amateur rig will be placed into operation by a person who is not familiar with Amateur Radio, and by being lured into making repeated transmissions, the operator unsuspectingly permits himself to be located with RDF equipment. The ability of certain RDF antennas to reject signals from selected directions has also been used to advantage in reducing noise and interference. Although not directly related to Amateur Radio, radio navigation is one application of RDF. The locating of downed aircraft is another, and one in which amateurs often lend their skills. Indeed, there are many useful applications for RDF.

Although sophisticated and complex equipment pushing the state of the art has been developed for use by governments and commercial enterprises, relatively simple equipment can be built at home to offer the radio amateur an opportunity to RDF. This chapter deals with antennas which are suited for that purpose.

RDF by Triangulation

It is impossible, using amateur techniques, to pinpoint the whereabouts of a transmitter from a single receiving location. With a directional antenna you can determine the direction of a signal source, but not how far away it is. To find the distance, you can then travel in the determined direction until you discover the transmitter location. However, that technique does not normally work very well.

A preferred technique is to take at least one additional direction measurement from a second receiving location. Then use a map of the area and plot the bearing or direction measurements as straight lines from points on the map representing the two locations. The approximate location of the transmitter will be indicated by the point where the two bearing lines cross. Even better results can be obtained by taking direction measurements from three locations and using the mapping technique just described. Because absolutely precise bearing measurements are difficult to obtain in practice, the three lines will almost always cross to form a triangle on the map, rather than at a single point. The transmitter will usually be located inside the area represented by the triangle. Additional information on the technique of triangulation may be found in recent editions of *The ARRL Handbook*.

DIRECTION FINDING SYSTEMS

Required for any RDF system are a directive antenna and a device for detecting the radio signal. In amateur applications the signal detector is usually a receiver; for convenience it will have a meter to indicate signal strength. Unmodified, commercially available portable or mobile receivers are generally quite satisfactory for signal detectors. At very close ranges a simple diode detector and dc microammeter may suffice for the detector.

On the other hand, antennas used for RDF techniques are not generally the types used for normal two-way communications. Directivity is a prime requirement, and here the word directivity takes on a somewhat different meaning than is commonly applied to antennas. Normally we associate directivity with gain, and we think of the ideal antenna pattern as one having a long, thin main lobe. Such a pattern may be of value for coarse measurements in RDF work, but precise bearing measurements are not possible. There is always a spread of a few (or perhaps many) degrees on the "nose" of the lobe, where a shift of antenna bearing produces no detectable change in signal strength. In RDF measurements, it is desirable to correlate an exact bearing or compass direction with the position of the antenna. In order to do this as accurately as possible, an antenna exhibiting a null in its pattern is used. A null can be very sharp in directivity, to within a half degree or less.

Loop Antennas

A simple antenna for RDF work is a small loop tuned to resonance with a capacitor. Several factors must be considered in the design of an RDF loop. The loop must be small compared with the wavelength. In a single-turn loop, the conductor should be less than 0.08 wavelength long. For 28 MHz, this represents a length of less than 34 inches (diameter of approximately 10 inches). Maximum response from the loop antenna is in the plane of the loop, with nulls exhibited at right angles to that plane.

To obtain the most accurate bearings, the loop must be balanced electrostatically with respect to ground. Otherwise, the loop will exhibit two modes of operation. One is the mode of a true loop, while the other is that of an essentially nondirectional vertical antenna of small dimensions. This second mode is called the "antenna effect." The voltages introduced by the two modes are seldom in phase and may add or subtract, depending upon the direction from which the wave is coming.

The theoretical true loop pattern is illustrated in Fig 1A. When properly balanced, the loop exhibits two nulls that are 180° apart. Thus, a single null reading with a small loop antenna will not indicate the exact direction toward the transmitter—only the line along which the transmitter lies. Ways to overcome this ambiguity are discussed later.

When the antenna effect is appreciable and the loop is tuned to resonance, the loop may exhibit little directivity, as shown in Fig 1B. However, by detuning the loop so as to shift the phasing, a pattern similar to 1C may be obtained. Although this pattern is not symmetrical, it does exhibit a null. Even so, the null may not be as sharp as that obtained with a loop that is well balanced, and it may not be at exact right angles to the plane of the loop.

By suitable detuning, the unidirectional cardioid pattern of Fig 1D may be approached. This adjustment is sometimes used in RDF work to obtain a unidirectional bearing, although there is no complete null in the pattern. A cardioid pattern can also be obtained with a small loop antenna by adding a sensing element. Sensing elements are discussed in a later section of this chapter.

An electrostatic balance can be obtained by shielding the loop, as shown in Fig 2. The shield is represented by the broken lines in the drawing, and eliminates the antenna effect. The response of a well constructed shielded loop is quite close to the ideal pattern of Fig 1A.

For the low-frequency amateur bands, single-turn loops of convenient physical size for portability are generally found to be unsatisfactory for RDF work. Therefore, multiturn loops are generally used instead. Such a loop is shown in Fig 3. This loop may also be shielded, and if the total conductor

length remains below 0.08 wavelength, the directional pattern is that of Fig 1A. A sensing element may also be used with a multiturn loop.

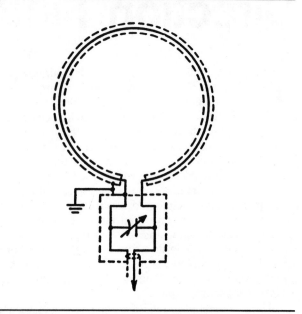

Fig 2—Shielded loop for direction finding. The ends of the shielding turn are not connected, to prevent shielding the loop from magnetic fields. The shield is effective against electric fields.

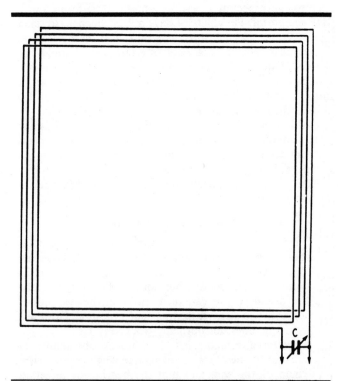

Fig 3—Small loop consisting of several turns of wire. The total conductor length is very much less than a wavelength. Maximum response is in the plane of the loop.

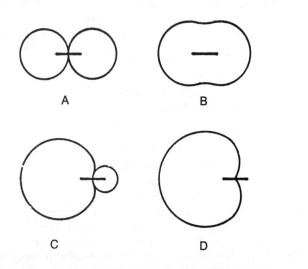

Fig 1—Small-loop field patterns with varying amounts of antenna effect—the undesired response of the loop acting merely as a mass of metal connected to the receiver antenna terminals. The heavy lines show the plane of the loop.

Loop Circuits and Criteria

No single word describes a direction-finding loop of high performance better than "symmetry." To obtain an undistorted response pattern from this type of antenna, it must be built in the most symmetrical manner possible. The next key word is "balance." The better the electrical balance, the deeper the loop null and the sharper the maxima.

The physical size of the loop for 7 MHz and below is not of major consequence. A 4-foot loop will exhibit the same electrical characteristics as one which is only an inch or two in diameter. The smaller the loop, however, the lower its efficiency. This is because its aperture samples a smaller section of the wave front. Thus, if loops that are very small in terms of a wavelength are used, preamplifiers are needed to compensate for the reduced efficiency.

An important point to keep in mind about a small loop antenna oriented in a vertical plane is that it is vertically polarized. It should be fed at the bottom for the best null response. Feeding it at one side, rather than at the bottom, will not alter the polarization and will only degrade performance. To obtain horizontal polarization from a small loop, it must be oriented in a horizontal plane, parallel to the earth. In this position the loop response is essentially omnidirectional.

The earliest loop antennas were of the "frame antenna" variety. These were unshielded antennas which were built on a wooden frame in a rectangular format. The loop conductor could be a single turn of wire (on the larger units) or several turns if the frame was small. Later, shielded versions of the frame antenna became popular, providing electrostatic shielding—an aid to noise reduction from such sources as precipitation static.

Ferrite Rod Antennas

With advances in technology, magnetic-core loop antennas later came into use. Their advantage was reduced size, and this appealed to the designers of aircraft and portable radios. Most of these antennas contain ferrite bars or cylinders, which provide high inductance and Q with a small number of coil turns.

Magnetic-core antennas consist essentially of many turns of wire around a ferrite rod. They are also known as loopstick antennas. Probably the best-known example of this type of antenna is that used in small portable AM broadcast receivers. Because of their reduced-size advantage, ferrite-rod antennas are used almost exclusively for portable work at frequencies below 150 MHz.

As implied in the earlier discussion of shielded loops in this chapter, the true loop antenna responds to the magnetic field of the radio wave, and not to the electrical field. The voltage delivered by the loop is proportional to the amount of magnetic flux passing through the coil, and to the number of turns in the coil. The action is much the same as in the secondary winding of a transformer. For a given size of loop, the output voltage can be increased by increasing the flux density, and this is done with a ferrite core of high permeability. A ½-inch diameter, 7-inch rod of Q2 ferrite ($\mu_i = 125$) is suitable for a loop core from the broadcast band through 10 MHz. For increased output, the turns may be wound on two rods that are taped together, as shown in Fig 4. Loopstick antennas for construction are described later in this chapter.

Maximum response of the loopstick antenna is broadside to the axis of the rod as shown in Fig 5, whereas

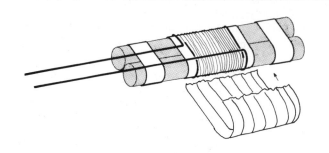

Fig 4—A ferrite-rod or loopstick antenna. Turns of wire may be wound on a single rod, or to increase the output from the loop, the core may be two rods taped together, as shown here. The type of core material must be selected for the intended frequency range of the loop. To avoid bulky windings, fine wire such as no. 28 or no. 30 is often used, with larger wire for the leads.

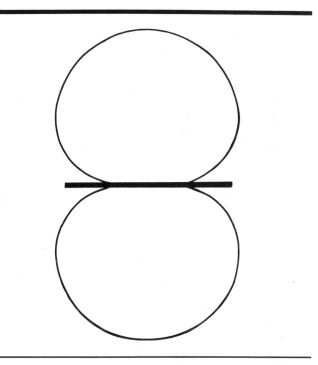

Fig 5—Field pattern for a ferrite rod antenna. The dark bar represents the rod on which the loop turns are wound.

maximum response of the ordinary loop is in a direction at right angles to the plane of the loop. Otherwise the performances of the ferrite-rod antenna and of the ordinary loop are similar. The loopstick may also be shielded to eliminate the antenna effect, such as with a U-shaped or C-shaped channel of aluminum or other form of "trough." The length of the shield should equal or slightly exceed the length of the rod.

Sensing Antennas

Because there are two nulls that are 180° apart in the directional pattern of a loop or a loopstick, an ambiguity exists as to which one indicates the true direction of the station being tracked. For example, assume you take a bearing measurement and the result indicates the transmitter is somewhere on a line running approximately east and west from your position. With this single reading, you have no way of knowing for sure if the transmitter is east of you or west of you.

If there is more than one receiving station taking bearings on a single transmitter, or if a single receiving station takes bearings from more than one position on the transmitter, the ambiguity may be worked out by triangulation, as described earlier. However, it is sometimes desirable to have a pattern with only one null, so there is no question about whether the transmitter in the above example would be east or west from your position.

A loop or loopstick antenna may be made to have a single null if a second antenna element is added. The element is called a sensing antenna, because it gives an added sense of direction to the loop pattern. The second element must be omnidirectional, such as a short vertical. When the signals from the loop and the vertical element are combined with a 90° phase shift between the two, a cardioid pattern results. The development of the pattern is shown in Fig 6A.

Fig 6B shows a circuit for adding a sensing antenna to a loop or loopstick. R1 is an internal adjustment and is used to set the level of the signal from the sensing antenna. For the best null in the composite pattern, the signals from the loop and the sensing antenna must be of equal amplitude, so R1 is adjusted experimentally during setup. In practice, the null of the cardioid is not as sharp as that of the loop, so the usual measurement procedure is to first use the loop alone to obtain a precise bearing reading, and then to add the sensing antenna and take another reading to resolve the ambiguity. (The null of the cardioid is 90° away from the nulls of the loop.) For this reason, provisions are usually made for switching the sensing element in and out of operation.

PHASED ARRAYS

Phased arrays are also used in amateur RDF work. Two general classifications of phased arrays are end-fire and broadside configurations. Depending on the spacing and phasing of the elements, end-fire patterns may exhibit a null in one direction along the axis of the elements. At the same time, the response is maximum off the other end of the axis, in the opposite direction from the null. A familiar arrangement is two elements spaced ¼ wavelength apart and fed 90° out of phase. The resultant pattern is a cardioid, with the null in the direction of the leading element. Other arrangements of spacing and phasing for an end-fire array are also suitable for RDF work. One of the best known is the Adcock array, discussed in the next section.

Broadside arrays are inherently bidirectional, which means there are always at least two nulls in the pattern. Ambiguity therefore exists in the true direction of the transmitter, but depending on the application, this may be no handicap. Broadside arrays are seldom used for amateur RDF applications.

The Adcock Antenna

Loops are adequate in RDF applications where only the ground wave is present. The performance of an RDF system for sky-wave reception can be improved by the use of an Adcock antenna, one of the most popular types of end-fire phased arrays. A basic version is shown in Fig 7.

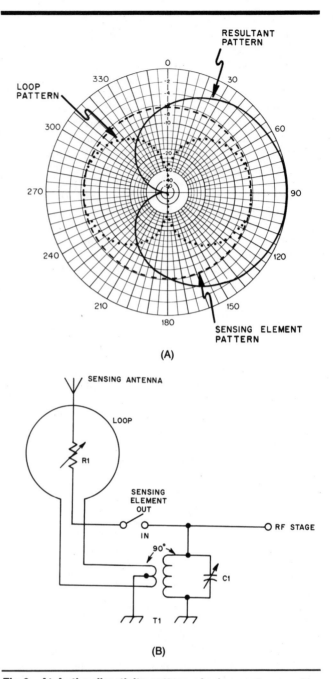

Fig 6—At A, the directivity pattern of a loop antenna with sensing element. At B is a circuit for combining the signals from the two elements. C1 is adjusted for resonance with T1 at the operating frequency.

This system was invented by F. Adcock and patented in 1919. The array consists of two vertical elements fed 180° apart, and mounted so the system may be rotated. Element spacing is not critical, and may be in the range from 1/10 to 3/4 wavelength. The two elements must be of identical lengths, but need not be self-resonant. Elements that are shorter than resonant are commonly used. Because neither the element spacing nor the length is critical in terms of wavelengths, an Adcock array may be operated over more than one amateur band.

The response of the Adcock array to vertically polarized

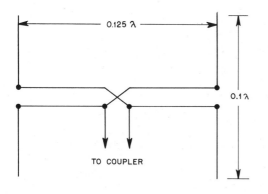

Fig 7—A simple Adcock antenna.

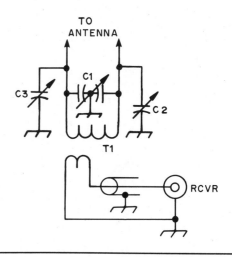

Fig 8—A suitable coupler for use with the Adcock antenna.

waves is similar to a conventional loop, and the directive pattern is essentially the same. Response of the array to a horizontally polarized wave is considerably different from that of a loop, however. The currents induced in the horizontal members tend to balance out regardless of the orientation of the antenna. This effect has been verified in practice when good nulls were obtained with an experimental Adcock under sky-wave conditions. The same circumstances produced poor nulls with small loops (both conventional and ferrite-loop models). Generally speaking, the Adcock antenna has attractive properties for amateur RDF applications. Unfortunately, its portability leaves something to be desired, making it more suitable to fixed or semi-portable applications. While a metal support for the mast and boom could be used, wood, PVC or fiberglass are preferable because they are nonconductors and would therefore cause less pattern distortion.

Since the array is balanced, a coupler is required to match the unbalanced input of a typical receiver. Fig 8 shows a suitable link-coupled network. C2 and C3 are null-clearing capacitors. A low-power signal source is placed some distance from the Adcock antenna and broadside to it. C2 and C3 are then adjusted until the deepest null is obtained. The coupler can be placed below the wiring-harness junction on the boom. Connection can be made by means of a short length of 300-ohm twin-lead.

The radiation pattern of the Adcock is shown in Fig 9A. The nulls are in directions broadside to the array, and become sharper with greater element spacings. However, with an element spacing greater than ¾ wavelength, the pattern begins to take on additional nulls in the directions off the ends of the array axis. At a spacing of 1 wavelength the pattern is that of Fig 9B, and the array is unsuitable for RDF applications.

Short vertical monopoles are often used in what is sometimes called the U-Adcock, so named because the elements with their feeders take on the shape of the letter U. In this arrangement the elements are worked against the earth as a ground or counterpoise. If the array is used only for reception, earth losses are of no great consequence. Short, elevated vertical dipoles are also used in what is sometimes called the H-Adcock.

The Adcock array, with two nulls in its pattern, has the same ambiguity as the loop and the loopstick. Adding a sensing element to the Adcock array has not met with great success. Difficulties arise from mutual coupling between the array elements and the sensing element, among other things. Because Adcock arrays are used primarily for fixed-station applications, the ambiguity presents no serious problem. The fixed station is usually one of a group of stations in an RDF network.

LOOPS VERSUS PHASED ARRAYS

Although loops can be made smaller than suitable phased arrays for the same frequency of operation, the phased arrays are preferred by some for a variety of reasons. In general, sharper nulls can be obtained with phased arrays, but this is also a function of the care used in constructing and feeding the individual antennas, as well as of the size of the phased array in terms of wavelengths. The primary constructional consideration is the shielding and balancing of the feed line against unwanted signal pickup, and the balancing of the antenna for a symmetrical pattern.

Loops are not as useful for skywave RDF work because of random polarization of the received signal. Phased arrays are somewhat less sensitive to propagation effects, probably because they are larger for the same frequency of operation and therefore offer some space diversity. In general, loops and loopsticks are used for mobile and portable operation, while phased arrays are used for fixed-station operation. However, phased arrays are used successfully above 144 MHz for portable and mobile RDF work. Practical examples of both types of antennas are presented later in this chapter.

THE GONIOMETER

Most fixed RDF stations for government and commercial work use antenna arrays of stationary elements, rather than mechanically rotatable arrays. This has been true since the earliest days of radio. The early-day device that permits finding directions without moving the elements is called a radiogoniometer, or simply a goniometer. Various types of goniometers are still used today in many installations, and offer the amateur many possibilities.

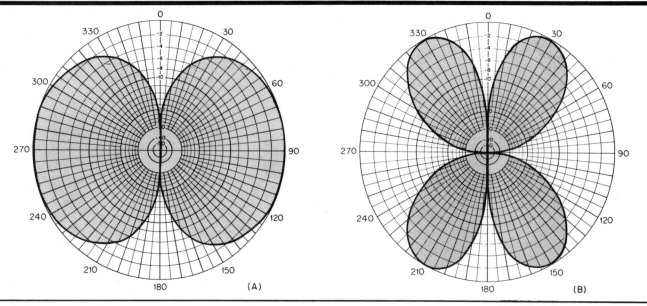

Fig 9—At A, the pattern of the Adcock array with an element spacing of ½ wavelength. In these plots the elements are aligned with the horizontal axis. As the element spacing is increased beyond ¾ wavelength, additional nulls develop off the ends of the array, and at a spacing of 1 wavelength the pattern at B exists. This pattern is unsuitable for RDF work.

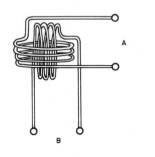

Fig 10—An early type of goniometer that is still used today in some RDF applications. This device is a special type of RF transformer that permits a movable coil in the center (not shown here) to be rotated and determine directions even though the elements are stationary.

The early style of goniometer is a special form of RF transformer, as shown in Fig 10. It consists of two fixed coils mounted at right angles to one another. Inside the fixed coils is a movable coil, not shown in Fig 10 to avoid cluttering the diagram. The pairs of connections marked A and B are connected respectively to two elements in an array, and the output to the detector or receiver is taken from the movable coil. As the inner coil is rotated, the coupling to one fixed coil increases while that to the other decreases. Both the amplitude and the phase of the signal coupled into the pickup winding are altered with rotation in a way that corresponds to actually rotating the array itself. Therefore, the rotation of the inner coil can be calibrated in degrees to correspond to bearing angles from the station location.

In the early days of radio, the type of goniometer just described saw frequent use with fixed Adcock arrays. A refinement of that system employed four Adcock elements, two arrays at right angles to each other. With a goniometer arrangement, RDF measurements could be taken in all compass directions, as opposed to none off the ends of a two-element fixed array. However, resolution of the four-element system was not as good as with a single pair of elements, probably because of mutual coupling among the elements. To overcome this difficulty a few systems of eight elements were installed.

Various other types of goniometers have been developed over the years, such as commutator switching to various elements in the array. A later development is the diode switching of capacitors to provide a commutator effect. As mechanical action has gradually been replaced with electronics to "rotate" stationary elements, the word *goniometer* is used less frequently these days. However, it still appears in many engineering reference texts. The more complex electronic systems of today are called beam-forming networks.

Electronic Antenna Rotation

With an array of many fixed elements, beam rotation can be performed electronically by sampling and combining signals from various individual elements in the array. Contingent upon the total number of elements in the system and their physical arrangement, almost any desired antenna pattern can be formed by summing the sampled signals in appropriate amplitude and phase relationships. Delay networks are used for some of the elements before the summation is performed. In addition, attenuators may be used for some elements to develop patterns such as from an array with binomial current distribution.

One system using these techniques is the Wullenweber antenna, employed primarily in government and military installations. The Wullenweber consists of a very large number of elements arranged in a circle, usually outside of (or in front of) a circular reflecting screen. Depending on the installation, the circle may be anywhere from a few hundred feet to more than a quarter of a mile in diameter. Although the Wullenweber is not one that would be constructed by an amateur, some of the techniques it uses may certainly be applied to Amateur Radio.

For the moment, consider just two elements of a Wullenweber antenna, shown as A and B in Fig 11. Also shown is the wavefront of a radio signal arriving from a distant transmitter. As drawn, the wavefront strikes element A first, and must travel somewhat farther before it strikes element B. There is a finite time delay before the wavefront reaches element B.

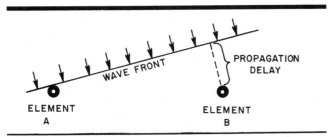

Fig 11—This diagram illustrates one technique used in electronic beam forming. By delaying the signal from element A by an amount equal to the propagation delay, the two signals may be summed precisely in phase, even though the signal is not in the broadside direction. Because this time delay is identical for all frequencies, the system is not frequency sensitive.

The propagation delay may be measured by delaying the signal received at element A before summing it with that from element B. If the two signals are combined directly, the amplitude of the resultant signal will be maximum when the delay for element A exactly equals the propagation delay. This results in an in-phase condition at the summation point. Or if one of the signals is inverted and the two are summed, a null will exist when the element-A delay equals the propagation delay; the signals will combine in a 180° out-of-phase relationship. Either way, once the time delay is known, it may be converted to distance. Then the direction from which the wave is arriving may be determined by trigonometry.

By altering the delay in small increments, the peak of the antenna lobe (or the null) can be steered in azimuth. This is true without regard to the frequency of the incoming wave. Thus, as long as the delay is less than the period of one RF cycle, the system is not frequency sensitive, other than for the frequency range that may be covered satisfactorily by the array elements themselves. Surface acoustic wave (SAW) devices or lumped-constant networks can be used for delay lines in such systems if the system is used only for receiving. Rolls of coaxial cable of various lengths are used in installations for transmitting. In this case, the lines are considered for the time delay they provide, rather than as simple phasing lines. The difference is that a phasing line is ordinarily designed for a single frequency (or for an amateur band), while a delay line offers essentially the same time delay at all frequencies.

By combining signals from other Wullenweber elements

appropriately, the broad beamwidth of the pattern from the two elements can be narrowed, and unwanted sidelobes can be suppressed. Then, by electronically switching the delays and attenuations to the various elements, the beam so formed can be rotated around the compass. The package of electronics designed to do this, including delay lines and electronically switched attenuators, is the beam-forming network. However, the Wullenweber system is not restricted to forming a single beam. With an isolation amplifier provided for each element of the array, several beam-forming networks can be operated independently. Imagine having an antenna system that offers a dipole pattern, a rhombic pattern, and a Yagi beam pattern, all simultaneously and without frequency sensitivity. One or more may be rotating while another is held in a particular direction. The Wullenweber was designed to fulfill this type of requirement.

One feature of the Wullenweber antenna is that it can operate at 360° around the compass. In many government installations, there is no need for such coverage, as the areas of interest lie in an azimuth sector. In such cases an in-line array of elements with a backscreen or curtain reflector may be installed broadside to the center of the sector. By using the same techniques as the Wullenweber, the beams formed from this array may be slewed left and right across the sector. The maximum sector width available will depend on the installation, but beyond 70 to 80° the patterns begin to deteriorate to the point that they are unsatisfactory for precise RDF work.

USING RDF ANTENNAS FOR COMMUNICATIONS

Because of their directional characteristics, RDF antennas would seem to be useful for two-way communications. It has not been mentioned earlier that the efficiency of receiving loops is poor. The radiation resistance is very low, on the order of 1 ohm, and the resistance of wire conductors by comparison is significant. For this reason it is common to use some type of preamplifier with receiving loops. Small receiving loops can often be used to advantage in a fixed station, to null out either a noise source or unwanted signals.

A loop that is small in terms of a wavelength may also be used for transmitting, but a different construction technique is necessary. A thick conductor is needed at HF, an inch or more in diameter. The reason for this is to decrease the ohmic losses in the loop. Special methods are also required to couple power into a small loop, such as links or a gamma match. A small loop is highly inductive, and the inductance may be canceled by inserting a capacitor in series with the loop itself. The capacitor must be able to withstand the high RF currents that flow during transmissions. Construction information for a small transmitting loop is contained in Chapter 5.

On the other hand, the Adcock antenna and other phased arrays have been used extensively for transmitting. In this application maximum response is off the ends of the Adcock, which is 90° away from the null direction used for RDF work.

RDF SYSTEM CALIBRATION AND USE

Once an RDF system is initially assembled, it should be "calibrated" or checked out before actually being put into use. Of primary concern is the balance or symmetry of the antenna pattern. A lopsided figure-8 pattern with a loop, for example, is undesirable; the nulls are not 180° apart nor are they at exact right angles to the plane of the loop. If this fact was not known in actual RDF work, measurement accuracy would suffer.

Initial checkout can be performed with a low-powered transmitter at a distance of a few hundred feet. It should be within visual range and must be operating into a vertical antenna. (A quarter-wave vertical or a loaded whip is quite suitable.) The site must be reasonably clear of obstructions, especially steel and concrete or brick buildings, large metal objects, nearby power lines, and so on. If the system operates above 30 MHz, trees and large bushes should also be avoided. An open field makes an excellent site.

The procedure is to "find" the transmitter with the RDF equipment as if its position were not known, and compare the RDF null indication with the visual path to the transmitter. For antennas having more than one null, each null should be checked.

If imbalance is found in the antenna system, there are two options available. One is to correct the imbalance. Toward this end, pay particular attention to the feed line. Using a coaxial feeder for a balanced antenna invites an asymmetrical pattern, unless an effective balun is used. A balun is not necessary if the loop is shielded, but an asymmetrical pattern can result with misplacement of the break in the shield itself. The builder may also find that the presence of a sensing antenna upsets the balance slightly. Experimenting with its position with respect to the main antenna may lead to correcting the error. You will also note that the position of the null shifts by 90° as the sensing element is switched in and out, and the null is not as deep. This is of little concern, however, as the intent of the sensing antenna is only to resolve ambiguities. The sensing element should be switched out when accuracy is desired.

The second option is to accept the imbalance of the antenna and use some kind of indicator to show the true directions of the nulls. Small pointers, painted marks on the mast, or an optical sighting system might be used. Sometimes the end result of the calibration procedure will be a compromise between these two options, as a perfect electrical balance may be difficult or impossible to attain.

The discussion above is oriented toward calibrating portable RDF systems. The same general suggestions apply if the RDF array is fixed, such as an Adcock. However, it won't be possible to move it to an open field. Instead, the array is calibrated in its intended operating position through the use of a portable or mobile transmitter. Because of nearby obstructions or reflecting objects, the null in the pattern may not appear to indicate the precise direction of the transmitter. Do not confuse this with imbalance in the RDF array. Check for imbalance by rotating the array 180° and comparing readings.

Once the balance is satisfactory, you should make a table of bearing errors noted in different compass directions. These error values should be applied as corrections when actual measurements are made. The mobile or portable transmitter should be at a distance of two or three miles for these measurements, and should be in as clear an area as possible during transmissions. The idea is to avoid conduction of the signal along power lines and other overhead wiring from the transmitter to the RDF site. Of course the position of the transmitter must be known accurately for each transmission.

FRAME LOOPS

It was mentioned earlier that the earliest style of receiving loops was the frame antenna. If carefully constructed, such an antenna performs well and can be built at low cost. Fig 12 illustrates the details of a practical frame type of loop antenna. This antenna was designed by Doug DeMaw, W1FB, and described in *QST* for July 1977. (See the bibliography at the end of this chapter.) The circuit at A is a 5-turn system which is tuned to resonance by C1. If the layout is symmetrical, good balance should be obtainable. L2 helps to achieve this objective by eliminating the need for direct coupling to the feed terminals of L1. If the loop feed was attached in parallel with C1, which is common practice, the chance for imbalance would be considerable.

L2 can be situated just inside or slightly outside of L1; a 1-inch separation works nicely. The receiver or preamplifier can be connected to terminals A and B of L2, as shown at B of Fig 12. C2 controls the amount of coupling between the loop and the preamplifier. The lighter the coupling, the higher is the loop Q, the narrower is the frequency response, and

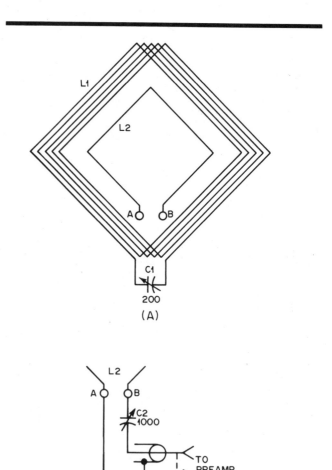

Fig 12—A multiturn frame antenna is shown at A. L2 is the coupling loop. The drawing at B shows how L2 is connected to a preamplifier.

the greater is the gain requirement from the preamplifier. It should be noted that no attempt is being made to match the loop impedance to the preamplifier. The characteristic impedance of small loops is very low—on the order of 1 ohm or less.

A supporting frame for the loop of Fig 12 can be constructed of wood, as shown in Fig 13. The dimensions given are for a 1.8-MHz frame antenna. For use on 75 or 40 meters, L1 of Fig 12A will require fewer turns, or the size of the wooden frame should be made somewhat smaller than that of Fig 13.

SHIELDED FRAME LOOPS

If electrostatic shielding is desired, the format shown in Fig 14 can be adopted. In this example, the loop conductor and the single-turn coupling loop are made from RG-58 coaxial cable. The number of loop turns should be sufficient to resonate with the tuning capacitor at the operating frequency. Antenna resonance can be checked by first connecting C1 (Fig 12A) and setting it at midrange. Then connect a small three-turn coil to the loop feed terminals, and couple to it with a dip meter. Just remember that the pickup coil will act to lower the frequency slightly from actual resonance.

In the antenna photographed for Fig 14, the 1-turn coupling loop was made of no. 22 plastic-insulated wire. However, electrostatic noise pickup occurs on such a coupling loop, noise of the same nature that the shield on the main loop prevents. This can be avoided by using RG-58 for the coupling loop. The shield of the coupling loop should be opened for about 1 inch at the top, and each end of the shield grounded to the shield of the main loop.

Larger single-turn frame loops can be fashioned from aluminum-jacketed Hardline, if that style of coax is available. In either case, the shield conductor must be opened at the electrical center of the loop, as shown in Fig 15 at A and B. The design example is based on 1.8-MHz operation.

Fig 14—An assembled table-top version of the electrostatically shielded loop. RG-58 cable is used in its construction.

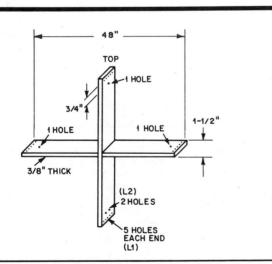

Fig 13—A wooden frame can be used to contain the wire of the loop shown in Fig 12.

In order to realize the best performance from an electrostatically shielded loop antenna, it must be operated near to and directly above an effective ground plane. An automobile roof (metal) qualifies nicely for small shielded loops. For fixed-station use, a chicken-wire ground screen can be placed below the antenna at a distance of 1 to 6 feet.

FERRITE-CORE LOOPS

Fig 16 contains a diagram for a rod loop (loopstick antenna). This antenna was also designed by Doug DeMaw, W1FB, and described in QST for July 1977. The winding (L1) has the appropriate number of turns to permit resonance with C1 at the operating frequency. L1 should be spread over approximately 1/3 of the core center. Litz wire will yield the best Q, but Formvar magnet wire can be used if desired. A layer of 3M Company glass tape (or Mylar tape) is recommended as a covering for the core before adding the wire. Masking tape can be used if nothing else is available.

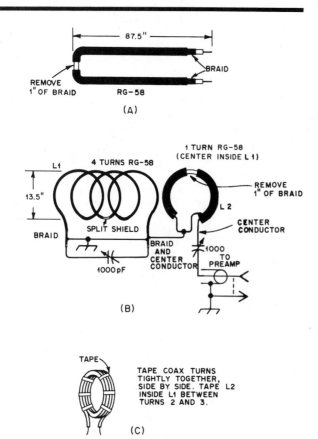

Fig 15—Components and assembly details of the shielded loop shown in Fig 14.

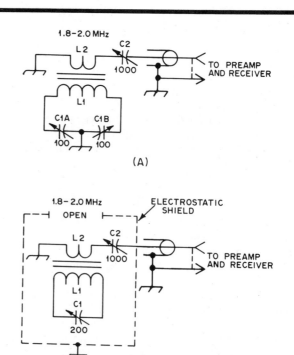

Fig 16—At A, the diagram of a ferrite loop. C1 is a dual-section air-variable capacitor. The circuit at B shows a rod loop contained in an electrostatic shield channel (see text). A suitable low-noise preamplifier is shown in Fig 19.

L2 functions as a coupling link over the exact center of L1. C1 is a dual-section variable capacitor, although a differential capacitor might be better toward obtaining optimum balance (not tried). The loop Q is controlled by means of C2, which is a mica compression trimmer.

Electrostatic shielding of rod loops can be effected by centering the rod in a U-shaped aluminum, brass or copper channel which extends slightly beyond the ends of the rod loop (1 inch is suitable). The open side (top) of the channel can't be closed, as that would constitute a shorted-turn condition and render the antenna useless. This can be proved by shorting across the center of the channel with a screwdriver blade when the loop is tuned to an incoming signal. The shield-braid gap in the coaxial loop of Fig 15 is maintained for the same reason.

Fig 17 shows the shielded rod loop assembly. This antenna was developed experimentally for 160 meters and uses two 7-inch ferrite rods which were glued end-to-end with epoxy cement. The longer core resulted in improved sensitivity during weak-signal reception. The other items in the photograph were used during the evaluation tests and are not pertinent to this discussion. This loop and the frame loop discussed in the previous section have bidirectional nulls, as shown in Fig 1A.

Obtaining a Cardioid Pattern

Although the bidirectional pattern of loop antennas can be used effectively in tracking down signal sources by means of triangulation, an essentially unidirectional loop response

Fig 17—The assembly at the top of the picture is a shielded ferrite-rod loop for 160 meters. Two rods have been glued end to end (see text). The other units in the picture are a low-pass filter (lower left), broadband preamplifier (lower center) and a Tektronix step attenuator (lower right). These were part of the test setup used when the antenna was evaluated.

will help to reduce the time spent when on a "hunting" trip. Adding a sensing antenna to the loop is simple to do, and it will provide the desired cardioid response. The theoretical pattern for this combination is shown in Fig 1D.

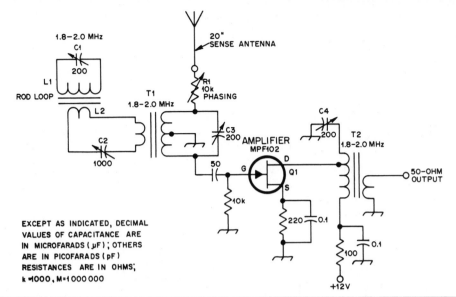

Fig 18—Schematic diagram of a rod-loop antenna with a cardioid response. The sensing antenna, phasing network and a pre-amplifier are shown also. The secondary of T1 and the primary of T2 are tuned to resonance at the operating frequency of the loop. T-68-2 to T-68-6 Amidon toroid cores are suitable for both transformers. Amidon also sells ferrite rods for this type of antenna.

EXCEPT AS INDICATED, DECIMAL VALUES OF CAPACITANCE ARE IN MICROFARADS (µF); OTHERS ARE IN PICOFARADS (pF) RESISTANCES ARE IN OHMS; k =1000, M=1 000 000

Fig 18 shows how a sensing element can be added to a loop or loopstick antenna. The link from the loop is connected via coaxial cable to the primary of T1, which is a tuned toroidal transformer with a split secondary winding. C3 is adjusted for peak signal response at the frequency of interest (as is C4), then R1 is adjusted for minimum back response of the loop. It will be necessary to readjust C3 and R1 several times to compensate for the interaction of these controls. The adjustments are repeated until no further null depth can be obtained. Tests at ARRL HQ showed that null depths as great as 40 dB could be obtained with the circuit of Fig 18 on 75 meters. A near-field weak-signal source was used during the tests.

The greater the null depth, the lower the signal output from the system, so plan to include a preamplifier with 25 to 40 dB of gain. Q1, as shown in Fig 18, will deliver approximately 15 dB of gain. The circuit of Fig 19 can be used following T2 to obtain an additional 24 dB of gain. In the interest of maintaining a good noise figure, even at 1.8 MHz, Q1 should be a low-noise device. A Siliconix U310 JFET would be ideal in this circuit, but a 2N4416, an MPF102 or a 40673 MOSFET would also be satisfactory. The sensing antenna can be mounted 6 to 15 inches from the loop. The vertical whip need not be more than 12 to 20 inches long. Some experimenting may be necessary in order to obtain the best results. Optimization will also change with the operating frequency of the antenna.

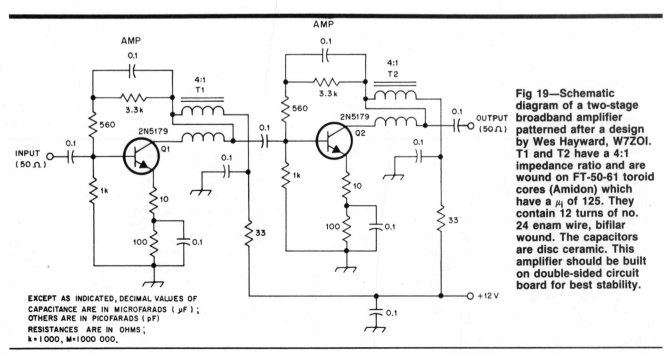

Fig 19—Schematic diagram of a two-stage broadband amplifier patterned after a design by Wes Hayward, W7ZOI. T1 and T2 have a 4:1 impedance ratio and are wound on FT-50-61 toroid cores (Amidon) which have a μ_i of 125. They contain 12 turns of no. 24 enam wire, bifilar wound. The capacitors are disc ceramic. This amplifier should be built on double-sided circuit board for best stability.

EXCEPT AS INDICATED, DECIMAL VALUES OF CAPACITANCE ARE IN MICROFARADS (µF); OTHERS ARE IN PICOFARADS (pF) RESISTANCES ARE IN OHMS; k = 1 000, M=1 000 000.

A SHIELDED LOOP WITH SENSING ANTENNA FOR 28 MHz

Fig 20 shows the construction and mounting of a simple shielded 10-meter loop. The loop was designed by Loren Norberg, W9PYG, and described in *QST* for April 1954. (See the bibliography at the end of this chapter.) It is made from an 18-inch length of RG-11 coax (solid or foam dielectric) secured to an aluminum box of any convenient size, with two coaxial cable hoods (Amphenol 83-1HP). The outer shield must be broken at the exact center. C1 is a 25-pF variable capacitor, and is connected in parallel with a 33-pF mica padder capacitor, C3. C1 must be tuned to the desired frequency while the loop is connected to the receiver in the same way as it will be used for RDF. C2 is a small differential capacitor used to provide electrical symmetry. The lead-in to the receiver is 67 inches of RG-59 cable (82 inches if the cable has foamed dielectric).

The loop can be mounted on the roof of the car with a rubber suction cup. The builder might also fabricate some kind of bracket assembly to mount the loop temporarily in the window opening of the automobile, allowing for loop rotation. Reasonably true bearings may be obtained through the windshield when the car is pointed in the direction of the hidden transmitter. More accurate bearings may be obtained with the loop held out the window and the signal coming toward that side of the car.

Sometimes the car broadcast antenna may interfere with accurate bearings. Disconnecting the antenna from the broadcast receiver may eliminate this trouble.

Sensing Antenna

A sensing antenna can be added to Norberg's loop to check on which of the two directions indicated by the loop is the correct one. Add a phono jack to the top of the aluminum case shown in Fig 20. The insulated center terminal of the jack should be connected to the side of the tuning capacitors that is common to the center conductor of the RG-59 coax feed line. The jack then takes a short vertical antenna rod of a diameter to fit the jack, or a piece of no. 12 or 14 solid wire may be soldered to the center pin of a phono plug for insertion in the jack. The sensing antenna can be plugged in as needed. Starting with a length of about four times the loop diameter, the length of the sensing antenna should be pruned until the pattern is similar to that of Fig 1D.

THE SNOOP LOOP—FOR CLOSE-RANGE RDF

Picture yourself on a hunt for a hidden 28-MHz transmitter. The night is dark, very dark. After you take off at the start of the hunt, heading in the right direction, the signal gets stronger and stronger. Your excitement increases with each additional S unit on the meter. You follow your loop closely, and it is working perfectly. You're getting out of town and into the countryside. The roads are unfamiliar. Now the null is beginning to swing rather rapidly, showing that you are getting close.

Suddenly the null shifts to give a direction at right angles to the car. With your flashlight you look carefully across the deep ditch beside the road and into the dark field where you know the transmitter is hidden. There are no roads into the field as far as you can see in either direction. You dare not waste miles driving up and down the road looking for an

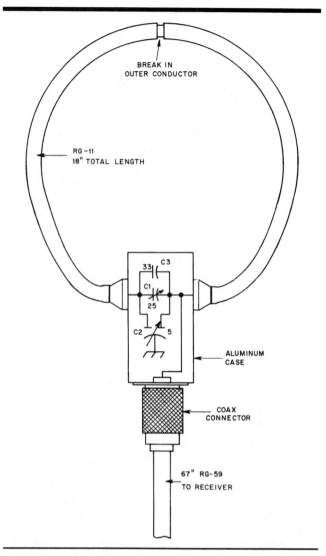

Fig 20—Sketch showing the constructional details of the 28-MHz RDF loop. The outer braid of the coax loop is broken at the center of the loop. The gap is covered with waterproof tape, and the entire assembly is given a coat of acrylic spray.

entrance, for each tenth of a mile counts. But what to do— your radio equipment is mobile, and requires power from the car battery.

In a brief moment your decision is made. You park beside the road, take your flashlight, and plunge into the veldt in the direction your loop null clearly indicated. But after taking a few steps, you're up to your armpits in brush and can't see anything forward or backward. You stumble on in hopes of running into the hidden transmitter—you're probably not more than a few hundred feet from it. But away from your car and radio equipment, it's like the proverbial hunt for the needle in the haystack. What you really need is a portable setup for hunting at close range, and you may prefer something that is inexpensive. The Snoop Loop was designed for just these requirements by Claude Maer, Jr, WØIC, and was described in *QST* for February 1957. (See the bibliography at the end of this chapter.)

The Snoop Loop is pictured in Fig 21. The loop itself is made from a length of RG-8 coax, with the shield broken at the top. A coax T connector is used for convenience and ease of mounting. One end of the coax loop is connected to a male plug in the conventional way, but the center conductor of the other end is shorted to the shield so the male connector at that end has no connection to the center prong. This results in an unbalanced circuit, but seems to give good bidirectional null readings as well as an easily detectable maximum reading when the grounded end of the loop is pointed in the direction of the transmitter. Careful tuning with C1 will improve this maximum reading. Don't forget to remove one inch of shielding from the top of the loop. You won't get much signal unless you do.

The detector and amplifier circuit for the Snoop Loop is shown in Fig 22. The model photographed does not include the meter, as it was built for use only with high-impedance headphones. The components are housed in an aluminum box. Almost any size box of sufficient size to contain the meter can be used. At very close ranges, reduction of sensitivity with R2 will prevent pegging the meter.

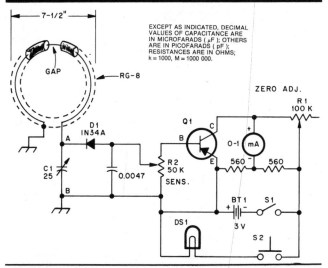

Fig 22—The Snoop Loop circuit for 28-MHz operation. The loop is a single turn of RG-8 inner conductor, the outer conductor being used as a shield. Note the gap in the shielding; about a 1-inch section of the outer conductor should be cut out. Refer to Fig 23 for alternative connection at points A and B for other frequencies of operation.

BT1—Two penlight cells.
C1—25-pF midget air padder.
D1—Small-signal germanium diode such as 1N34A or equiv.
DS1—Optional 2-cell penlight lamp for meter illumination, such as no. 222.
Q1—PNP transistor such as ECG102 or equiv.
R1—100-kΩ potentiometer, linear taper. May be PC-mount style.
R2—50-kΩ potentiometer, linear taper.
S1—SPST toggle.
S2—Optional momentary push for illuminating meter.

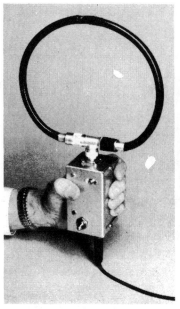

Fig 21—The box containing the detector and amplifier is also the "handle" for the Snoop Loop. The loop is mounted with a coax T as a support, a convenience but not an essential part of the loop assembly. The loop tuning capacitor is screwdriver adjusted. The on-off switch and the meter sensitivity control may be mounted on the bottom.

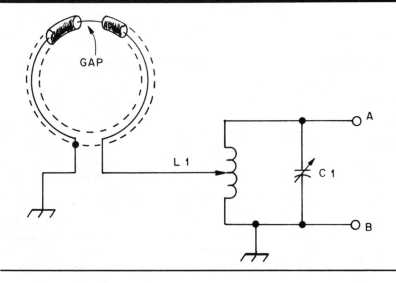

Fig 23—Input circuit for lower frequency bands. Points A and B are connected to corresponding points in the circuit of Fig 22, substituting for the loop and C1 in that circuit. L1-C1 should resonate within the desired amateur band, but the L/C ratio is not critical. After construction is completed, adjust the position of the tap on L1 for maximum signal strength. Instead of connecting the RDF loop directly to the tap on L1, a length of low impedance line may be used between the loop and the tuned circuit, L1-C1.

The Snoop Loop is not limited to the 10-meter band or to a built-in loop. Fig 23 shows an alternative circuit for other bands and for plugging in a separate loop connected by a low-impedance transmission line. Select coil and capacitor combinations that will tune to the desired frequencies. Plug-in coils could be used. It is a good idea to have the RF end of the unit fairly well shielded, to eliminate signal pickup except through the loop. This little unit should certainly help you on those dark nights in the country. (Tip to the hidden-transmitter operator—if you want to foul up some of your pals using these loops, just hide near the antenna of a 50-kW broadcast transmitter.)

A LOOPSTICK FOR 3.5 MHz

Figs 24 through 26 show an RDF loop suitable for the 3.5-MHz band. It uses a construction technique that has had considerable application in low-frequency marine direction finders. The loop is a coil wound on a ferrite rod from a broadcast-antenna loopstick. The loop was designed by John Isaacs, W6PZV, and described in *QST* for June 1958. Because it is possible to make a coil of high Q with the ferrite core, the sensitivity of such a loop is comparable to a conventional loop that is a foot or so in diameter. The output of the vertical-rod sensing antenna, when properly combined with that of the loop, gives the system the cardioid pattern shown in Fig 1D.

To make the loop, remove the original winding on the ferrite core and wind a new coil as shown in Fig 25. Other types of cores than the one specified may be substituted; use the largest coil available and adjust the winding so that the circuit resonates in the 3.5-MHz band within the range of C1. The tuning range of the loop may be checked with a dip meter.

The sensing system consists of a 15-inch whip and an adjustable inductance that will resonate the whip as a quarter-wave antenna. It also contains a potentiometer to control the output of the antenna. S1 is used to switch the sensing antenna in and out of the circuit.

The whip, the loopstick, the inductance L1, the capacitor C1, the potentiometer R1, and the switch S1 are all mounted on a 4 × 5 × 3-inch box chassis as shown in Fig 26. The loopstick may be mounted and protected inside a piece of ½-inch PVC pipe. A section of ½-inch electrical conduit is

attached to the bottom of the chassis box and this supports the instrument.

To produce an output having only one null there must be a 90-degree phase difference between the outputs of the loop and sensing antennas, and the signal strength from each must be the same. The phase shift is obtained by tuning the sensing antenna slightly off frequency by means of the slug in L1. Since the sensitivity of the whip antenna is greater than that of the loop, its output is reduced by adjusting R1.

Adjustment

To adjust the system, enlist the aid of a friend with a mobile transmitter and find a clear spot where the transmitter and RDF receiver can be separated by several hundred feet. Use as little power as possible at the transmitter. (Remove your own transmitter antenna before trying to make any loop adjustments and remember to leave it off during transmitter hunts.) With the test transmitter operating on the proper frequency, disconnect the sensing antenna with S1, and peak the loopstick using C1, while watching the S meter on the receiver. Once the loopstick is peaked, no further adjustment

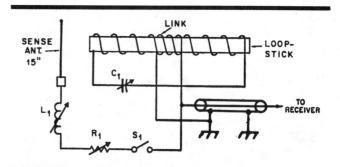

Fig 25—Circuit of the 3.5-MHz direction finder.

C1—140 pF variable (125-pF ceramic trimmer in parallel with
15-pF ceramic fixed).
L1—Approx. 140 μH adjustable (Miller No. 4512 or equivalent).
R1—1-kΩ carbon potentiometer.
S1—SPST toggle.
Loopstick—App. 15 μH (Miller 705-A, with original winding removed and wound with 20 turns of no. 22 enam.) Link is two turns at center. Winding ends secured with Scotch electrical tape.

Fig 24—Unidirectional 3.5-MHz RDF using ferrite-core loop with sensing antenna. Adjustable components of the circuit are mounted in the aluminum chassis supported by a short length of tubing.

of C1 will be necessary. Next, connect the sensing antenna and turn R1 to minimum resistance. Then vary the adjustable slug of L1 until a maximum reading of the S meter is again obtained. It may be necessary to turn the unit a bit during this adjustment to obtain a larger reading than with the loopstick alone. The last turn of the slug is quite critical, and some hand-capacitance effect may be noted.

Now turn the instrument so that one side (not an end) of the loopstick is pointed toward the test transmitter. Turn R1 a complete revolution and if the proper side was chosen a definite null should be observed on the S meter for one particular position of R1. If not, turn the RDF 180° and try again. This time leave R1 at the setting which produces the minimum reading. Now adjust L1 very slowly until the S-meter reading is reduced still further. Repeat this several times, first R1, and then L1, until the best minimum is obtained.

Finally, as a check, have the test transmitter move around the RDF and follow it by turning the RDF. If the tuning has been done properly the null will always be broadside to the loopstick. Make a note of the proper side of the RDF for the null, and the job is finished.

Fig 26—Components of the 3.5-MHz RDF are mounted on the top and sides of a Channel-lock type box. In this view R1 is on the left wall at the upper left and C1 is at the lower left. L1, S1 and the output connector are on the right wall. The loopstick and whip mount on the outside.

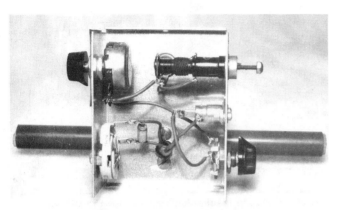

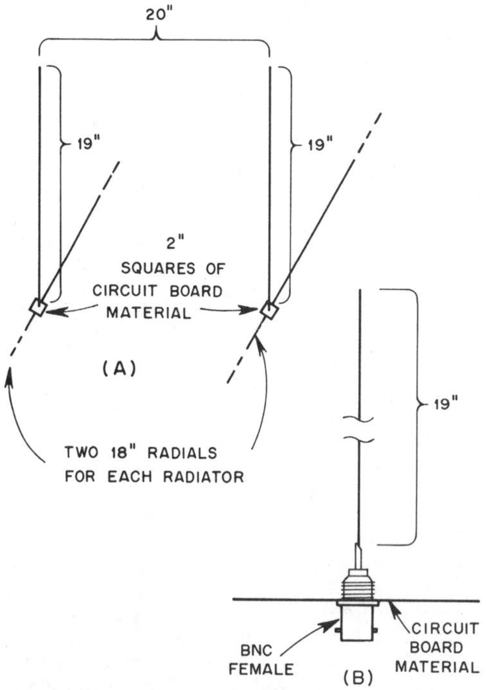

Fig 27—At A is a simple configuration that can produce a cardioid pattern. At B is a convenient way of fabricating a sturdy mount for the radiator using BNC connectors.

A 144-MHz ANTENNA FOR RDF

Although there may be any number of different antennas that will produce a cardioid pattern, the simplest design is depicted in Fig 27. Two ¼-wavelength vertical elements are spaced ¼-wavelength apart and are fed 90° out of phase. Each radiator is shown with two radials approximately 5 percent shorter than the radiators. This array was designed by Pete O'Dell, KB1N, and described in *QST* for March 1981.

During the design phase of this project a personal computer was used to predict the impact on the antenna pattern of slight alterations in its size, spacing and phasing of the elements. The results suggest that this system is a little touchy and that the most significant change comes at the null. Very slight alterations in the dimensions caused the notch to become much more shallow and, hence, less usable for RDF. Early experience in building a working model bore this out.

This means that if you build this antenna, you will find it advantageous to spend a few minutes to tune it carefully for the deepest null. If it is built using the techniques presented here, then this should prove to be a small task which is well worth the extra effort. Tuning is accomplished by adjusting the length of the vertical radiators, the spacing between them and, if necessary, the lengths of the phasing harness that connects them. Tune for the deepest null on your S meter when using a signal source such as a moderately strong repeater. This should be done outside, away from buildings and large metal objects. Initial indoor tuning on this project was tried in the kitchen, which revealed that reflections off the appliances were producing spurious readings. Beware too of distant water towers, radio towers, and large office or apartment buildings. They can reflect the signal and give false indications.

Construction is simple and straightforward. Fig 27B shows a female BNC connector (Radio Shack 278-105) that has been mounted on a small piece of PC-board material. The BNC connector is held "upside down," and the vertical radiator is soldered to the center solder lug. A 12-inch piece of brass tubing provides a snug fit over the solder lug. A second piece of tubing, slightly smaller in diameter, is telescoped inside the first. The outer tubing is crimped slightly at the top after the inner tubing is installed. This provides positive contact between the two tubes. For 146 MHz the length of the radiators is calculated to be about 19 inches. You should be able to find small brass tubing at a hobby store. If none is available in your area, consider brazing rods. These are often available in hardware sections of discount stores. It will probably be necessary to solder a short piece to the top since these come in 18-inch sections. Also, tuning will not be quite as convenient. Two 18-inch radials are added to each element by soldering them to the board. Two 36-inch pieces of heavy brazing rod were used in this project.

The Phasing Harness

As shown in Fig 28, a T connector is used with two different lengths of coaxial line to form the phasing harness. This method of feeding the antenna is superior over other simple systems toward obtaining equal currents in the two radiators. Unequal currents tend to reduce the depth of the null in the pattern, all other factors being equal.

The ½-wavelength section can be made from either RG-58 or RG-59 because it should act as a 1-to-1 transformer. With no radials or with two radials perpendicular to the vertical element, it was found that a ¼-wavelength section made of RG-59 75-Ω coax produced a deeper notch than a ¼-wavelength section made of RG-58 50-Ω line. However,

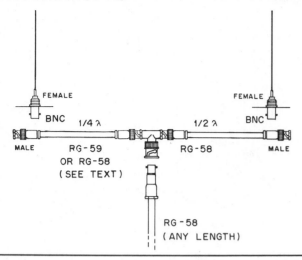

Fig 28—The phasing harness for the phased 144-MHz RDF array. The phasing sections must be measured from the center of the T connector to the point that the vertical radiator emerges from the shielded portion of the upside-down BNC female. Don't forget to take the length of the connectors into account when constructing the harness. If care is taken and coax with polyethylene dielectric is used, you should not have to prune the phasing line. With this phasing system, the null will be in a direction that runs along the boom, on the side of the ¼-wavelength section.

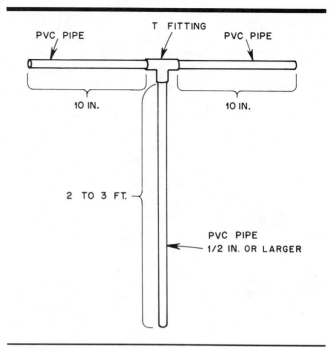

Fig 29—A simple mechanical support for the DF antenna, made of PVC pipe and fittings.

with the two radials bent downward somewhat, the RG-58 section seemed to outperform the RG-59. Because of minor differences in assembly techniques from one antenna to another, it will probably be worth your time and effort to try both types of coax and determine which works best for your antenna. You may also want to try bending the radials down at slightly different angles for the best null performance.

The most important thing about the coax for the harness is that it be of the highest quality (well shielded and with a polyethylene dielectric). The reason for avoiding foam dielectric is that the velocity factor can vary from one roll to the next—some say that it varies from one foot to the next. Of course, it can be used if you have test equipment available that will allow you to determine its electrical length. Assuming that you do not want to or cannot go to that trouble, stay with coax having a solid polyethylene dielectric. Avoid coax that is designed for the CB market or do-it-yourself cable-TV market. (A good choice is Belden 8240 for the RG-58 or Belden 8241 for the RG-59.)

Both RG-58 and RG-59 with polyethylene dielectric have a velocity factor of 0.66. Therefore, for 146 MHz a quarter wavelength of transmission line will be 20.2 inches × 0.66 = 13.3 inches. A half-wavelength section will be twice this length or 26.7 inches. One thing you must take into account is that the transmission line is the total length of the cable *and the connectors*. Depending on the type of construction and the type of connectors that you choose, the actual length of the coax by itself will vary somewhat. You will have to determine that for yourself.

Y connectors that mate with RCA phono plugs are widely available and the phono plugs are easy to work with. Avoid

the temptation to substitute these for the T and BNC connectors. Phono plugs and a Y connector were tried. The results with that system were not satisfactory. The performance seemed to change from day to day and the notch was never as deep as it should have been. Although they are more difficult to find, BNC T connectors will provide superior performance and are well worth the extra cost. If you must make substitutions, it would be preferable to use UHF connectors (type PL-259).

Fig 29 shows a simple support for the antenna. PVC tubing is used throughout. Additionally, you will need a T fitting, two end caps, and possibly some cement. (By not cementing the PVC fittings together, you will have the option of disassembly for transportation.) Cut the PVC for the dimensions shown, using a saw or a tubing cutter. A tubing cutter is preferred because it produces smooth, straight edges without making a mess. Drill a small hole through the PC board near the female BNC of each element assembly. Measure the 20-inch distance horizontally along the boom and mark the two end points. Drill a small hole vertically through the boom at each mark. Use a small nut and bolt to attach each element assembly to the boom.

Tuning

The dimensions given throughout this section are those for approximately 146 MHz. If the signal you will be hunting is above that frequency, then the measurements should be a bit shorter. If you are to operate below that frequency, then they will need to be somewhat longer. Once you have built the antenna to the rough size, the fun begins. You will need a signal source near the frequency that you will be using for your RDF work. Adjust the length of the radiators and the spacing between them for the deepest null on your S meter.

Make changes in increments of ¼ inch or less. If you must adjust the phasing line, make sure that the ¼-wavelength section is exactly one-half the length of the half-wavelength section. Keep tuning until you have a satisfactorily deep null on your S meter.

THE DOUBLE-DUCKY DIRECTION FINDER

For direction finding, most amateurs use antennas having pronounced directional effects, either a null or a peak in signal strength. FM receivers are designed to eliminate the effects of amplitude variations, and so they are difficult to use for direction finding without looking at an S meter. Most modern portable transceivers do not have S meters.

This "Double-Ducky" direction finder (DDDF) was designed by David Geiser, WA2ANU, and described in *QST* for July 1981. It works on the principle of switching between two nondirectional antennas, as shown in Fig 30. This creates phase modulation on the incoming signal that is heard easily on the FM receiver. When the two antennas are exactly the same distance (phase) from the transmitter, Fig 31, the tone disappears.

In theory the antennas may be very close to each other, but in practice the amount of phase modulation increases directly with the spacing, up to spacings of a half wavelength. While a half-wavelength separation on 2 meters (40 inches) is pretty large for a mobile array, a quarter wavelength gives entirely satisfactory results, and even an eighth wavelength (10 inches) is acceptable.

Think in terms of two antenna elements with fixed spacing. Mount them on a ground plane and rotate that ground plane. The ground plane held above the hiker's head or car roof reduces the needed height of the array and the directional-distorting effects of the searcher's body or other conducting objects.

The DDDF is bidirectional and, as described, its tone null points both toward and away from the signal origin. An L-shaped search path would be needed to resolve the ambiguity. Use the techniques of triangulation described earlier in this chapter.

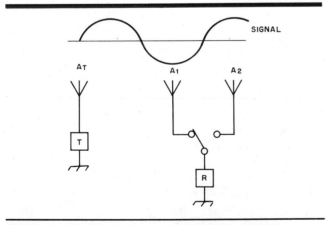

Fig 30—At the left, A$_T$ represents the antenna of the hidden transmitter, T. At the right, rapid switching between antennas A$_1$ and A$_2$ at the receiver samples the phase at each antenna, creating a pseudo-Doppler effect. An FM detector detects this as phase modulation.

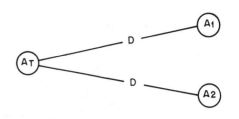

Fig 31—If both receiving antennas are an equal distance (D) from the transmitting antenna, there will be no difference in the phase angles of the signals in the receiving antennas. Therefore, the detector will not detect any phase modulation, and the audio tone will disappear from the output of the detector.

Specific Design

It is not possible to find a long-life mechanical switch operable at a fairly high audio rate, such as 1000 Hz. Yet we want an audible tone, and the 400- to 1000-Hz range is perhaps most suitable considering audio amplifiers and average hearing. Also, if we wish to use the transmit function of a transceiver, we need a switch that will carry perhaps 10 watts without much problem.

A solid-state switch, the PIN (positive-intrinsic-negative) diode, has been developed within the last several years. The intrinsic region of this type of diode is ordinarily bare of current carriers and, with a bit of reverse bias, looks like a low-capacitance open space. A bit of forward bias (20 to 50 mA) will load the intrinsic region with current carriers that are happy to dance back and forth at a 148-MHz rate, looking like a resistance of an ohm or so. In a 10-watt circuit, the diodes do not dissipate enough power to damage them.

Because only two antennas are used, the obvious approach is to connect one diode "forward" to one antenna, to connect the other "reverse" to the second antenna, and to drive the pair with square-wave audio-frequency ac. Fig 32 shows the necessary circuitry. RF chokes (Ohmite Z144, J. W. Miller RFC-144 or similar VHF units) are used to let the audio through to bias the diodes while blocking RF. Of course, the reverse bias on one diode is only equal to the forward bias on the other, but in practice this seems sufficient.

A number of PIN diodes were tried in the particular setup built. These were the Hewlett-Packard HP5082-3077, the Alpha LE-5407-4, the KSW KS-3542 and the Microwave Associates M/A-COM 47120. All worked well, but the HP diodes were used because they provided a slightly lower SWR (about 3:1).

A type 567 IC is used as the square-wave generator. The output does have a dc bias that is removed with a nonpolarized coupling capacitor. This minor inconvenience is more than rewarded by the ability of the IC to work well with between 7 and 15 volts (a nominal 9-V minimum is recommended).

The nonpolarized capacitor is also used for dc blocking when the function switch is set to XMIT. D3, a light-emitting diode (LED), is wired in series with the transmit bias to indicate selection of the XMIT mode. In that mode there is a high battery current drain (20 mA or so).

S1 should be a center-off locking type toggle switch. An ordinary center-off switch may be used but beware. If the

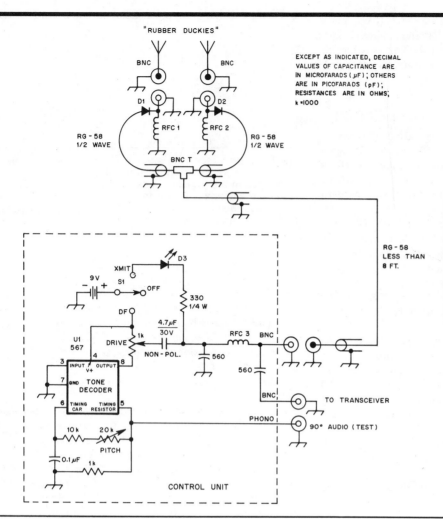

EXCEPT AS INDICATED, DECIMAL
VALUES OF CAPACITANCE ARE
IN MICROFARADS (μF); OTHERS
ARE IN PICOFARADS (pF);
RESISTANCES ARE IN OHMS;
k =1000

Fig 32—Schematic diagram of the DDDF circuit. Construction and layout are not critical. Components inside the broken lines should be housed inside a shielded enclosure. Most of the components are available from Radio Shack, except D1, D2, the antennas and RFC1-RFC3. These components are discussed in the text. S1—See text.

switch is left on XMIT you will soon have dead batteries.

Cables going from the antenna to the coaxial T connector were cut to an electrical ½ wavelength to help the open circuit, represented by the reverse-biased diode, look open at the coaxial T. (The length of the line within the T was included in the calculation.)

The length of the line from the T to the control unit is not particularly critical. If possible, keep the total of the cable length from the T to the control unit to the transceiver under 8 feet, because the capacitance of the cable does shunt the square-wave generator output.

Ground-plane dimensions are not critical. See Fig 33. Slightly better results may be obtained with a larger ground plane than shown. Increasing the spacing between the pickup antennas will give the greatest improvement. Every doubling (up to a half wavelength) will cut the width of the null in half. A 1° wide null can be obtained with 20-inch spacing.

DDDF Operation

Switch the control unit to DF and advance the drive potentiometer until a tone is heard on the desired signal. Do not advance the drive high enough to distort or "hash up" the voice. Rotate the antenna for a null in the fundamental tone. Note that a tone an octave higher may appear. The cause of the effect is shown in Fig 34. In Fig 34A, an oscilloscope

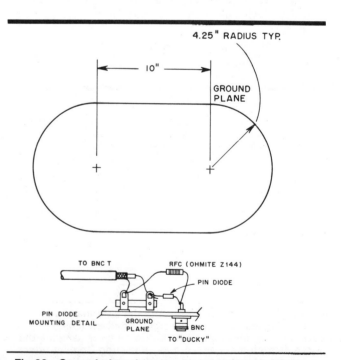

Fig 33—Ground-plane layout and detail of parts at the antenna connectors.

synchronized to the "90° audio" shows the receiver output with the antenna aimed to one side of the null (on a well-tuned receiver). Fig 34B shows the null condition and a twice-frequency (one octave higher) set of pips, while C shows the output with the antenna aimed to the other side of the null.

If the incoming signal is quite out of the receiver linear region (10 kHz or so off frequency), the off-null antenna aim may present a fairly symmetrical AF output to one side, Fig 35A. It may also show instability at a sharp null position, indicated by the broken line on the display in Fig 35B. Aimed to the other side of a null, it will give a greatly increased AF output, Fig 35C. This is caused by the different parts of the receiver FM detector curve used. The sudden tone change is the tip-off that the antenna null position is being passed.

The user should practice with the DDDF to become acquainted with how it behaves under known situations of signal direction, power and frequency. Even in difficult nulling situations where a lot of second-harmonic AF exists, rotating the antenna through the null position causes a very distinctive tone change. With the same frequencies and amplitudes present, the quality of the tone (timbre) changes. It is as if a note were first played by a violin, and then the same note played by a trumpet. (A good part of this is the change of phase of the fundamental and odd harmonics with respect to the even harmonics.) The listener can recognize differences (passing through the null) that would give an electronic analyzer indigestion.

DIRECTION FINDING WITH AN INTERFEROMETER

In New Mexico, an interferometer RDF system is used by the National ELT Location Team to aid in locating downed aircraft. The method can be used for other VHF RDF activities as well. With a little practice, you can take long-distance bearings that are accurate to within one degree. That's an error of less than 2000 feet from 20 miles away. The interferometer isn't complicated. It consists of a receiver, two antennas, and two lengths of coaxial cable. The system and techniques described here were developed by Robert E. Cowan, K5QIN, and Thomas A. Beery, WD5CAW, and were described in *QST* for November 1985.

Interferometer Basics

The theory of interferometer operation is simple. Signals from two antennas are combined out of phase to give a sharp null in signal strength when the antennas are located on a line of constant phase. Fig 36 shows that if you know the location of two points on a line of constant phase, you can get an accurate fix on the transmitter.

Most DF bearings are taken several miles from the transmitter. At these distances, the equal-phase circles appear as straight lines. As shown in Fig 37A, if you put the antennas at points A and B on a line of equal phase and connect them to a receiver with equal lengths of transmission line, the signals from the two antennas will add. By moving either one of the antennas back and forth across the equal-phase line, you will notice a broad peak in signal strength. Now if antenna B is moved halfway between two lines of equal phase, as shown in Fig 37B, the signals arriving at the receiver will be exactly out of phase; they will cancel each other completely. A sharp null in signal strength will be noted when either antenna is moved even slightly. This null is very easy to find just by listening to the receiver.

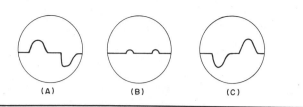

Fig 34—Typical on-channel responses. See text for discussion of the meaning of the patterns.

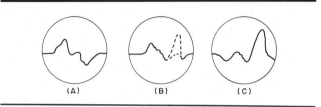

Fig 35—Representative off-channel responses. See text for discussion of the meaning of the patterns.

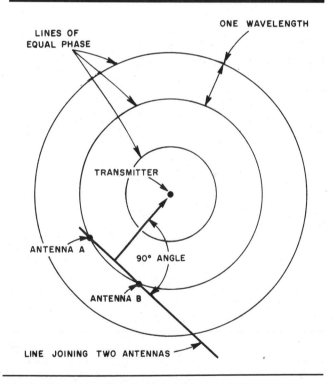

Fig 36—The transmitter is at a right angle to the center of the line that joins two points of equal phase.

It is this sharp null that you always look for when using the interferometer. However, the setup shown in Fig 37B doesn't put us on a line of equal phase. To do that, you must make the signals at the receiver 180° out of phase. This can be done by having one feed line a half wavelength shorter than

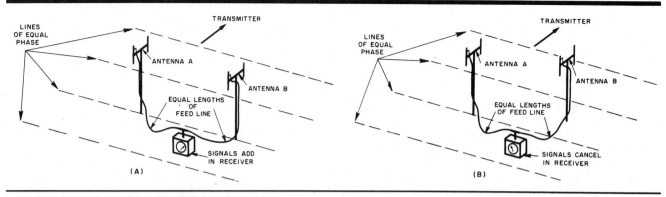

Fig 37—At A, placing two antennas on a line of equal phase and connecting them to a receiver with equal lengths of transmission line causes the signals to add. If the two antennas are placed a half wavelength apart in the direction of the transmitter, as shown at B, the signals cancel to form a null.

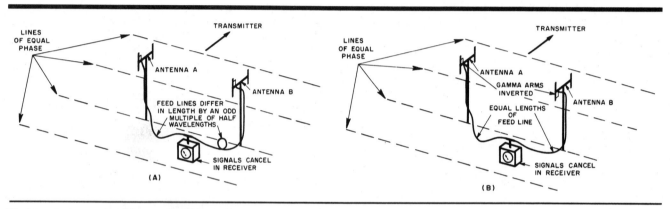

Fig 38—At A, placing two antennas on a line of equal phase and connecting them to a receiver with feed lines that differ in length by an odd multiple of half wavelengths causes the signals to cancel. Another way to obtain signal cancellation, shown at B, is to use equal-length feeders and invert the gamma arm on one of the antennas.

the other, as shown in Fig 38A. Now you will get a sharp null when the two antennas are on a line of equal phase. Another way of getting the phase reversal is shown in Fig 38B. If the gamma arms of the antennas are reversed (one pointing up, the other pointing down) and equal lengths of feed line are used, the signals will cancel.

After you have located two or more points of equal phase, you can draw a straight line through them. The transmitter will be 90° from this line. By locating an equal-phase line that is 30 to 50 feet long, you can take an accurate compass bearing down the line. This long base line is the secret of the interferometer's accuracy. Other DF systems have a much narrower aperture, and their accuracy is poorer than the interferometer. (Going much beyond 50 feet doesn't improve accuracy unless you have a transit for taking the bearing.) Fig 39 shows a typical interferometer setup in the field.

Using the Interferometer

To use the interferometer, first connect a receiver to one directional antenna. Rotate it for maximum signal strength to get an approximate bearing to the transmitter. Then add the second antenna to the system using a T connector. Its feed line must be a half wavelength shorter (or longer) than the one connected to the first, unless one of the gamma arms is inverted as shown in Fig 38B. Set up the second antenna about a wavelength away from the first one. The base of the second antenna mast should be at a right angle to the direction in which the first antenna is pointing. Now move the second antenna back and forth in the direction of the transmitter you are hunting, always keeping the mast vertical. As you do this, you will notice a sharp drop in signal strength from the receiver. Find the spot where the signal is weakest and mark its position on the ground.

Now move the second antenna a few steps farther from the first one, and on a line with the first null. Find the null again and mark its position. Continue "walking out the nulls" for 30 to 50 feet. Next take a compass bearing between the two antenna masts or down the line of nulls. The hidden transmitter will be on a line exactly 90° from the bearing. Now move to another location a few miles away and take another interferometer bearing. Plot your locations and bearings on

Fig 39—Typical field setup of an interferometer. One antenna is mounted on a stand while the second is moved to find the null locations. A compass bearing is taken down the line of nulls that is marked with surveyor's flags.

a map. The point where the bearings cross is the transmitter location.

Equipment for the Interferometer

Having an S meter on the interferometer receiver is handy, but by no means necessary. The nulls are a little easier to find if you have a meter to watch. You should have some way of adjusting the receiver sensitivity. An RF or IF gain control is convenient, but an RF attenuator in the feed line also works well. The sensitivity should be adjusted so the maximum incoming signal is about 20 to 30 dB above the noise. If the signal in the receiver is too strong, you may have trouble finding the first null. If it is too weak, the null will be broad and you will find it difficult to position the antenna precisely on the line of constant phase.

Almost any kind of antenna can be used to make an interferometer. Simple dipoles will give the correct null, but because you need to first get an approximate bearing to the transmitter, some sort of directional antenna is preferred. A 2- or 3-element Yagi has adequate directivity, yet is small enough to be carried to the field. It is important that both antennas be constructed alike. In this way their phase centers

are the same and you may take a compass bearing between any two similar features on the antennas—the masts, for example.

Reflected signals can cause problems when you are trying to find a good null. Having the interferometer antennas ten feet or more above the ground helps eliminate reflections from nearby objects such as rocks, cars and people.

Unless you use your antennas with one of the gamma arms inverted, the feed lines that connect the antennas to the receiver must differ in length by an odd multiple of half wavelengths. Don't forget to include the velocity factor of the cable in your calculations. You may want to set your receiver on the ground halfway between the two antennas, as shown in Fig 39. In this case the feeder lengths need be only a half wavelength different.

It is sometimes more convenient to mount your receiver on the mast of the second antenna. In this case one feed line will be only a few feet long and the other may be 40 feet long. This is fine as long as the difference in feed-line length is an odd multiple of half wavelengths. Having one feeder longer than the other creates some amplitude imbalance, but this will not affect your bearing. It is not necessary to be extremely precise when cutting your cables to give a length difference of an odd multiple of half wavelengths. At 146 MHz, a cutting error of a few inches will result in a bearing error of less than one degree.

The two feed lines are connected to the receiver input terminal with a coaxial T connector. Purists will argue that you need an impedance-matching device at this junction. Experiments with both resistive power combiners and Wilkinson hybrids indicate that neither works better than a T connector.

To take full advantage of the interferometer's accuracy, you must be able to take reliable compass bearings. Sighting compasses are the best compromise between hand-held transits and inexpensively priced lensatic compasses. With a moderately priced sighting compass you can take bearings that are accurate to within one degree. Don't forget to account for the magnetic declination of your area when plotting the bearings on a map. If you are unfamiliar with map and compass techniques, consult your local library or bookstore for references on the subject.

Some Fine Points on Interferometer Use

With a little practice, you will find that the interferometer is very easy to use. There are a few things to watch out for, though. Here are some that are based on experience.

Pick your DF site carefully. Although the interferometer works extremely well when reflected signals are present, you can get fooled. The best DF sites are in open terrain and well away from reflecting objects such as buildings, cars, fences and power lines.

Mark your null points on the ground as you find them. Use surveyor's flags, rocks, or other markers to indicate the null points. When you are done, look at this line of nulls. The markers should be in a straight line. Take your compass bearing down this line, and the accuracy will be better than sighting between the antenna masts. This is because you will be averaging several null readings when you take a bearing on the line, whereas a sighting between the masts gives you only a single null reading. Take compass bearings from both directions and average them for best accuracy. If the line has a periodic wiggle to it, this means that you have some reflections at your site. This is discussed later, but the correct sighting line will be down the center of the wiggles.

Beware of DFing pure reflections! Sometimes the only signal to be heard will be a reflection from a mountain or a building. Take two (or more) bearings, go to the area where they cross, and take more bearings to confirm the location. If you are in a critical situation such as locating a downed aircraft, don't commit all your resources until you know for sure that you are not DFing a pure reflection.

The most important thing to remember is that you must get a definite null at each point along the phase front. "Null all the way—okay," is a good rule. If you can't find a null, that means a strong reflection is entering the system. Inevitably it will give you bad information. The best recourse is to pack up the interferometer and move it to a new location. You may need to go a few hundred yards or perhaps a mile, but you will get a good bearing for your efforts. Remember— good nulls give good bearings, and good bearings will locate the transmitter.

Interferometer Radiation Patterns

At this point you have enough information to assemble and operate an interferometer, but some additional information may provide an insight into how it works. When you connect two antennas to a receiver (or transmitter), the two antennas and an out-of-phase feed line combine to form unique radiation patterns. These change dramatically as the two antennas are moved apart. There is always one null that faces the incoming signal. Other nulls are also present, and their location depends on the distance between the antennas. From page 2-16 of Jasik (see bibliography at the end of this chapter), the equation that is used to calculate the antenna pattern is as follows:

$$E = |\cos(180\ d \cos\theta + \phi/2)| \qquad \text{(Eq 1)}$$

where

E is the relative field strength
d is the spacing between antennas, wavelengths
θ is the angle at which the field strength is calculated, degrees
ϕ is the difference in phase between the two antennas, electrical degrees

Fig 40 shows the relationship of the terms in Eq 1.

Antenna patterns for an interferometer using vertical dipole antennas are shown in Fig 41. Note that one null always faces the incoming signal, indicated by 0° azimuth on the plots. This null becomes sharper as the antenna spacing is increased. The patterns also contain lobes and other nulls. In a field setup, the lobes and nulls change position as the antenna spacing is varied. Interfering signals that arrive at an angle from the main signal will be attenuated differently as these lobes and nulls change position with different antenna spacings. If directional antennas are used in the interferometer, the pattern will be modified by the pattern of the individual antennas.

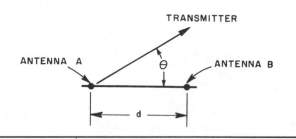

Fig 40—The relationship of the terms used in Eq 1 to calculate interferometer patterns. θ **is expressed in degrees, d in wavelengths.**

Effect of Reflected Signals on the Interferometer

All RDF systems are affected by multipath signals. The interferometer works better in multipath situations than any other system tried. Two effects are noticed when reflected signals are present. The first is a periodic curvature, or wiggle, of the apparent phase front. The second is a change in the depth of the nulls that are encountered. These two effects are caused by vector addition of the main signal and a reflected signal in the interferometer system.

If the amplitude of the reflected signal is low, you will be able to find the nulls and mark their positions on the ground. This is shown in Fig 42. The nulls are marked with surveyor's flags, and a compass bearing is being taken down the center of the wiggles. Surveyor's tape is used to mark the exact line of bearing. A drawing of the null locations obtained with multipath signals is shown in Fig 43. Notice, as indicated in Fig 43, that at some points you will obtain deep nulls, while only shallow nulls can be obtained at other points.

If you can successfully find all of the nulls in a multipath situation, you can easily determine the direction from which the reflected signal is arriving. To do this, first measure the period of the wiggle with a tape measure (the value of P as shown in Fig 43). The angle of the reflected signal with respect to the main signal can then be calculated from the equation

$$\theta = \arcsin P/\lambda \qquad \text{(Eq 2)}$$

where

θ is the angle of reflected signal with respect to main signal, degrees
P is the period of the wiggles, feet
λ is the length of 1 wavelength at the frequency you are using, feet, from the equation
$\lambda = 984/\text{frequency (MHz)}$

Eq 2 doesn't tell you whether the reflected signal is to the right or left of the main signal. Generally, though, you can resolve this because your directional antenna will point somewhere between the two signals.

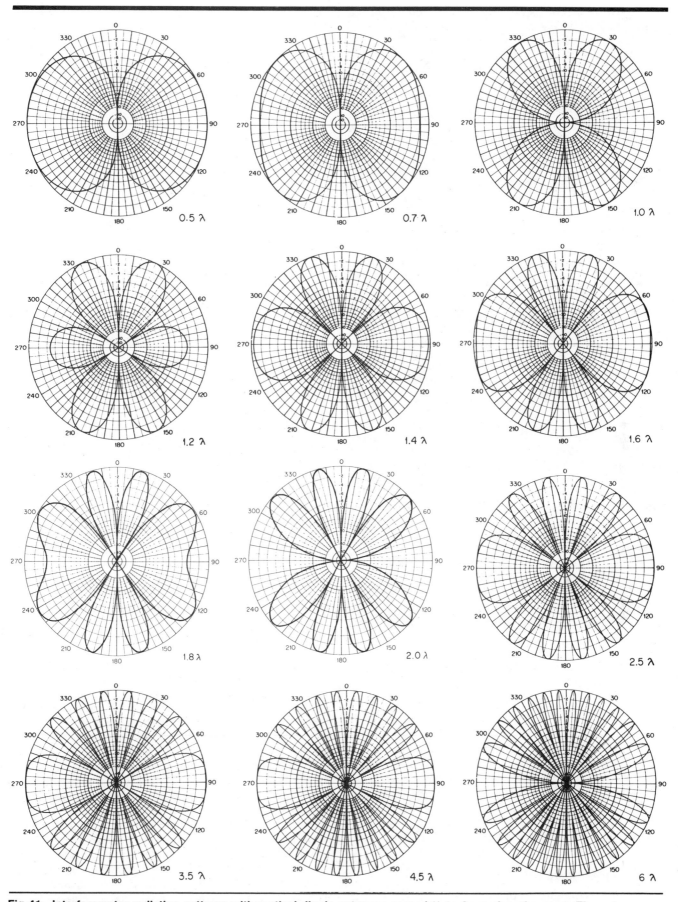

Fig 41—Interferometer radiation patterns with vertical dipole antennas spaced ½ to 6 wavelengths apart. The antennas are placed on the horizontal axis.

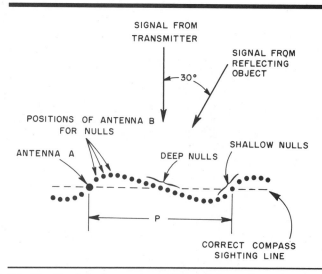

Fig 43—The interferometer pattern of the null markers, obtained in the presence of a direct signal and a reflected signal.

to the area where the bearings cross. This point will usually be within a half mile or less of the transmitter location. You can then home in on the transmitter using signal-strength techniques or hand-held DF units. The interferometer is a very useful tool to add to your collection of DF techniques. With a little practice, it can provide long-distance bearings that will quickly lead you to the hidden transmitter.

AN ADCOCK ANTENNA

Information in this section is condensed from an August 1975 *QST* article by Tony Dorbuck, K1FM, ex-W1YNC. Earlier in this chapter it was mentioned that loops are adequate in applications where only the ground wave is present. But the question arises, what can be done to improve the performance of an RDF system for sky-wave reception? One type of antenna that has been used successfully for this purpose is the Adcock antenna. There are many possible variations, but the basic configuration is shown in Fig 44.

The operation of the antenna when a vertically polarized wave is present is very similar to a conventional loop. As can be seen from Fig 44, currents I1 and I2 will be induced in the vertical members by the passing wave. The output current in the transmission line will be equal to their difference. Consequently, the directional pattern will be identical to the loop with a null broadside to the plane of the elements and with maximum gain occurring in end-fire fashion. The magnitude of the difference current will be proportional to the spacing, d, and the length of the elements. Spacing and length are not critical, but somewhat more gain will occur for larger dimensions than for smaller ones. In an experimental model, the spacing was 21 feet (approximately 0.15 wavelength at 7 MHz) and the element length was 12 feet.

Response of the Adcock antenna to a horizontally polarized wave is considerably different from that of a loop. The currents induced in the horizontal members (dotted arrows in Fig 44) tend to balance out regardless of the orientation of the antenna. This effect is borne out in practice, since good nulls can be obtained under sky-wave conditions

Fig 42—Null locations obtained with a reflected signal are marked with flags. A compass bearing is being taken on a line through the center of the wiggles, marked with a surveyor's tape.

If you try to use the interferometer in a location where the reflected signal is very strong, there will be certain antenna spacings where you just can't find a null, yet at other spacings the nulls will be quite deep. If you take lots of time, you might be able to figure out the proper bearing in the case of severe multipath, but generally it's not worth the effort. Moving the interferometer a short distance may allow you to find a "null all the way" and to take a good bearing.

DF Strategy Using the Interferometer

Locating transmitters with the interferometer is best done as a team effort. Several two-person teams can take bearings and send their data to one location where plotting is done. The person doing the map plotting can then direct the teams

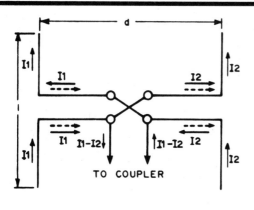

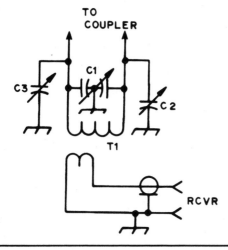

Fig 44—A simple Adcock antenna and suitable coupler (see text).

that produce only poor nulls with small loops, either conventional or ferrite-loop models. Generally speaking, the Adcock antenna has very attractive properties for fixed-station RDF work or for semi-portable applications. Wood, PVC tubing or pipe, or other nonconducting material is preferable for the mast and boom. Distortion of the pattern may result from metal supports.

Since a balanced feed system is used, a coupler is needed to match the unbalanced input of the receiver. It consists of T1, which is an air-wound coil with a two-turn link wrapped around the middle. The combination is then resonated to the operating frequency with C1. C2 and C3 are null-clearing capacitors. A low-power signal source is placed some distance from the Adcock antenna and broadside to it. C2 and C3 are then adjusted until the deepest null is obtained. The coupler can be placed on the ground below the wiring-harness junction on the boom and connected by means of a short length of 300-ohm twin-lead. A length of PVC tubing used as a mast

facilitates rotation and provides a means of attaching a compass card for obtaining bearings.

Tips on tuning and adjusting a fixed-location RDF array are presented earlier in this chapter. See the section, "RDF System Calibration and Use."

BIBLIOGRAPHY

Source material and more extended discussion of topics covered in this chapter can be found in the references given below and in the textbooks listed at the end of Chapter 2.

W. U. Amphar, "Unidirectional Loops for Transmitter Hunting," *QST*, Mar 1955.

G. Bonaguide, "HF DF—A Technique for Volunteer Monitoring," *QST*, Mar 1984.

D. S. Bond, *Radio Direction Finders*, 1st edition (New York: McGraw-Hill Book Co).

R. E. Cowan and T. A. Beery, "Direction Finding with the Interferometer," *QST*, Nov 1985.

D. DeMaw, "Beat the Noise with a Scoop Loop," *QST*, Jul 1977.

D. DeMaw, "Maverick Trackdown," *QST*, Jul 1980.

T. Dorbuck, "Radio Direction-Finding Techniques," *QST*, Aug 1975.

D. T. Geiser, "Double-Ducky Direction Finder," *QST*, Jul 1981.

N. K. Holter, "Radio Foxhunting in Europe," Parts 1 and 2, *QST*, Aug and Nov 1976.

J. Isaacs, "Transmitter Hunting on 75 Meters," *QST*, Jun 1958.

H. Jasik, *Antenna Engineering Handbook*, 1st edition (New York: McGraw-Hill, 1961).

R. Keen, *Wireless Direction Finding*, 3rd edition (London: Wireless World).

J. Kraus, *Antennas* (New York: McGraw-Hill Book Co, 1950).

J. Kraus, *Electromagnetics* (New York: McGraw-Hill Book Co).

C. M. Maer, Jr, "The Snoop-Loop," *QST*, Feb 1957.

L. R. Norberg, "Transmitter Hunting with the DF Loop," *QST*, Apr 1954.

P. O'Dell, "Simple Antenna and S-Meter Modification for 2-Meter FM Direction Finding," Basic Amateur Radio, *QST*, Mar 1981.

Ramo and Whinnery, *Fields and Waves in Modern Radio* (New York: John Wiley & Sons).

F. Terman, *Radio Engineering* (New York: McGraw-Hill Book Co.)

Radio Direction Finding, published by the Happy Flyers, 1811 Hillman Ave, Belmont, CA 94002.

Chapter 15

Portable Antennas

For many amateurs, the phrase "portable antennas" may conjure visions of antenna assemblies that can be broken down and carried in a backpack, suitcase, golf bag, or what-have-you, for transportation to some out-of-the way place where they will be used. Or the vision could be of larger arrays that can be disassembled and moved by pickup truck to a Field Day site, and then erected quickly on temporary supports. Portable antennas come in a wide variety of sizes and shapes, and can be used on any amateur frequency.

Strictly speaking, the phrase "portable antenna" really means *transportable antenna*—one that is moved to some (usually temporary) operating position for use. As such, portable antennas are not placed into service when they are being transported. This puts them in a different class from mobile antennas, which are intended to be used while in motion. Of course this does not mean that mobile antennas cannot be used during portable operation. Rather, true portable antennas are designed to be packed up and moved, usually with quick reassembly being one of the design requisites. This chapter describes antennas that are designed for portability. However, many of these antennas can also be used in more permanent installations.

Any of several schemes can be employed to support an antenna during portable operation. For HF antennas made of wire, probably the most common support is a conveniently located tree at the operating site. Temporary, lightweight masts are also used. An aluminum extension ladder, properly guyed, can serve as a mast for Field Day operation. Such supports are discussed in Chapter 22.

A SIMPLE TWIN-LEAD ANTENNA FOR HF PORTABLE OPERATION

The typical portable HF antenna is a random-length wire flung over a tree and end-fed through a Transmatch. Low power Transmatches can be made quite compact, but each additional piece of necessary equipment makes portable operation less attractive. The station can be simplified by using resonant impedance-matched antennas for the bands of interest. Perhaps the simplest antenna of this type is the half-wave dipole, center-fed with 50- or 75-ohm coax. Unfortunately, RG-58, RG-59 or RG-8 cable is quite heavy and bulky for backpacking, and the miniature cables such as RG-174 are too lossy.

A practical solution to the coax problem, developed by Jay Rusgrove, W1VD, and Jerry Hall, K1TD, is to use folded dipoles made from lightweight TV twin-lead. The characteristic impedance of this type of dipole is near 300 ohms, but this can easily be transformed to a 50-ohm impedance. The transformation is obtained by placing a lumped capacitive

reactance at a strategic distance from the input end of the line. Fig 1 illustrates the construction method and gives important dimensions for the twin-lead dipole. Shorting connections must be made a short distance inside the ends of the radiator, as shown in Fig 1. (This subject is covered more fully in Chapter 2.) The twin-lead may also be shorted at each end of the dipole.

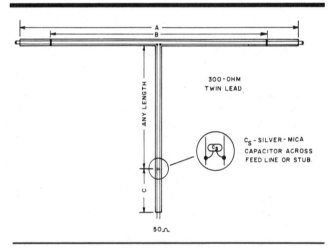

Fig 1—A twin-lead folded dipole makes an excellent portable antenna that is easily matched to 50-ohm equipment. See text and Table 1 for details.

A silver-mica capacitor is shown for the reactive element, but an open-end stub of twin-lead can serve as well, provided it is dressed at right angles to the transmission line for some distance. The stub method has the advantage of easy adjustment of the system resonant frequency.

The dimensions and capacitor values for twin-lead dipoles for the HF bands are given in Table 1. To preserve the balance of the feeder, a 1:1 balun must be used at the end of the feed line. In most applications the balance is not critical, and the twin-lead can be connected directly to a coaxial output jack—one lead to the center contact, and one lead to the shell.

Because of the transmission-line effect of the shorted radiator sections, a folded dipole exhibits a wider bandwidth than a single-conductor type. The antennas described here are not as broad as a standard folded dipole because the impedance-transformation mechanism is frequency selective. However, the bandwidth should be adequate. An antenna cut

Table 1
Twin-Lead Dipole Dimensions and Capacitor Values

Frequency	Length A	Length B	Length C	C_S	Stub Length
3.75 MHz	124' 9½"	104' 11½"	13' 0"	289 pF	37' 4"
7.15	65' 5½"	55' ½"	6' 10"	151 pF	19' 7"
10.125	46' 2½"	38' 10½"	4' 10"	107 pF	13' 10"
14.175	33' 0"	27' 9¼"	3' 5½"	76 pF	9' 10½"
18.118	25' 10"	21' 8¾"	2' 8½"	60 pF	7' 9"
21.225	22' ½"	18' 6½"	2' 3½"	51 pF	6' 7"
24.94	18' 9"	15' 9½"	1' 11½"	43 pF	5' 7½"
28.5	16' 5"	13' 9¾"	1' 8½"	38 pF	4' 11"

for 14.175 MHz, for example, will present an SWR of less than 2:1 over the entire 14-MHz band.

ZIP-CORD ANTENNAS

Zip cord is readily available at hardware and department stores, and it's not expensive. The nickname, zip cord, refers to that parallel-wire electrical cord with brown or white insulation used for lamps and many small appliances. The conductors are usually no. 18 stranded copper wire, although larger sizes may also be found. Zip cord is light in weight and easy to work with.

For these reasons, zip cord can be pressed into service as both the transmission line and the radiator section for an emergency dipole antenna system. This information by Jerry Hall, K1TD, appeared in *QST* for March 1979. The radiator section of a zip-cord antenna is obtained simply by "unzipping" or pulling the two conductors apart for the length needed to establish resonance for the operating frequency band. The initial dipole length can be determined from the equation $\ell = 468/f$, where ℓ is the length in feet and f is the frequency in megahertz. (It would be necessary to unzip only half the length found from the formula, since each of the two wires becomes half of the dipole.) The insulation left on the wire will have some loading effect, so a bit of length-trimming may be needed for exact resonance at the desired frequency.

For installation, you may want to use the electrician's knot shown in Fig 2 at the dipole feed point. This is a "balanced" knot that will keep the transmission-line part of the system from unzipping itself under the tension of dipole suspension. This way, if zip cord of sufficient length for both the radiator and the feed line is obtained, a solder-free installation can be made right down to the input end of the line. (Purists may argue that knots at the feed point will create an impedance mismatch or other complications, but as will become evident in the next section, this is not a major consideration.) Granny knots (or any other variety) can be used at the dipole ends with cotton cord to suspend the system. You end up with a lightweight, low-cost antenna system that can serve for portable or emergency use.

But just how efficient is a zip-cord antenna system? Since it is easy to locate the materials and simple to install, how about using such for a more permanent installation? Upon casual examination, zip cord looks about like 72-ohm balanced feed line. Does it work as well?

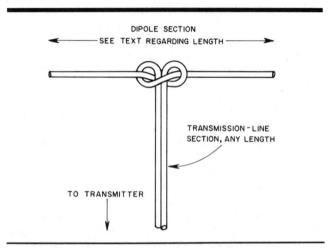

Fig 2—This electrician's knot, often used inside lamp bases and appliances in lieu of a plastic grip, can also serve to prevent the transmission-line section of a zip-cord antenna from unzipping itself under the tension of dipole suspension. To tie the knot, first use the right-hand conductor to form a loop, passing the wire behind the unseparated zip cord and off to the left. Then pass the left-hand wire of the pair behind the wire extending off to the left, in front of the unseparated pair, and thread it through the loop already formed. Adjust the knot for symmetry while pulling on the two dipole wires.

Zip Cord as a Transmission Line

In order to determine the electrical characteristics of zip cord as a radio-frequency transmission line, a 100-foot roll was subjected to tests in the ARRL laboratory with an RF impedance bridge. Zip cord is properly called parallel power cord. The variety tested was manufactured for GC Electronics, Rockford, IL, being 18-gauge, brown, plastic-insulated type SPT-1, GC cat. no. 14-118-2G42. Undoubtedly, minor variations in the electrical characteristics will occur among similar cords from different manufacturers, but the results presented here are probably typical.

The characteristic impedance was determined to be 107 ohms at 10 MHz, dropping in value to 105 ohms at 15 MHz and to a slightly lower value at 29 MHz. The nominal value is 105 ohms at HF. The velocity factor of the line was determined to be 69.5 percent.

Who needs a 105-ohm line, especially to feed a dipole? A dipole in free space exhibits a feed-point resistance of 73 ohms, and at heights above ground of less than ¼ wavelength the resistance can be even lower. An 80-meter dipole at 35 feet over average soil, for example, will exhibit a feed-point resistance of about 35 ohms. Thus, for a resonant antenna, the SWR in the zip-cord transmission line can be 105/35 or 3:1, and maybe even higher in some installations. Depending on the type of transmitter in use, the rig may not like working into the load presented by the zip-cord antenna system.

But the really bad news is still to come—line loss! Fig 3 is a plot of line attenuation in decibels per hundred feet of line versus frequency. Chart values are based on the assumption that the line is perfectly matched (sees a 105-ohm load as its terminating impedance).

In a feed line, losses up to about 1 decibel or so can be tolerated, because at the receiver a 1-dB difference in signal strength is just barely detectable. But for losses above about 1 dB, beware. Remember that if the total losses are 3 dB, half of your power will be used just to heat the transmission line. Additional losses over those charted in Fig 3 will occur when standing waves are present. (See Chapter 24.) The trouble is, you can't use a 50- or 75-ohm SWR instrument to measure the SWR in zip-cord line accurately.

Based on this information, we can see that a hundred feet or so of zip-cord transmission line on 80 meters might be acceptable, as might 50 feet on 40 meters. But for longer lengths and higher frequencies, the losses become appreciable.

Zip Cord Wire as the Radiator

For years, amateurs have been using ordinary copper house wire as the radiator section of an antenna, erecting it without bothering to strip the plastic insulation. Other than the loading effects of the insulation mentioned earlier, no noticeable change in performance has been noted with the insulation present. And the insulation does offer a measure of protection against the weather. These same statements can be applied to single conductors of zip cord.

The situation in a radiating wire covered with insulation is not quite the same as in two parallel conductors, where there may be a leaky dielectric path between the two conductors. In the parallel line, it is the current leakage that contributes to line losses. This leakage current is set up by the voltage potential that exists on the two adjacent wires. The current flowing through the insulation on a single radiating wire is quite small by comparison, and so as a radiator the efficiency is high.

In short, communications can certainly be established with a zip-cord antenna in a pinch on 160, 80, 40, 30 and perhaps 20 meters. For higher frequencies, especially with long line lengths for the feeder, the efficiency of the system is so low that its value becomes questionable.

A TREE-MOUNTED HF GROUNDPLANE ANTENNA

A tree-mounted, vertically polarized antenna may sound silly. But is it, really? Perhaps engineering references do not recommend it, but such an antenna does not cost much, is inconspicuous, and it works. This idea was described by Chuck Hutchinson, K8CH, in *QST* for September 1984.

The antenna itself is simple, as shown in Fig 4. A piece of RG-58 cable runs to the feed point of the antenna, and is attached to a porcelain insulator. Two radial wires are

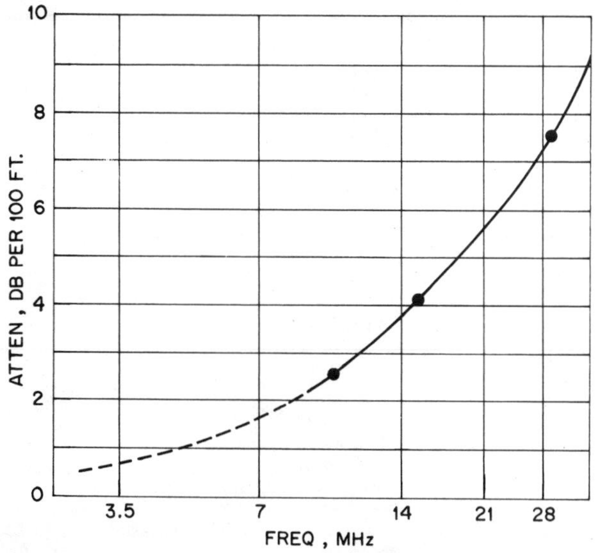

Fig 3—Attenuation of zip cord in decibels per hundred feet when used as a transmission line at radio frequencies. Measurements were made only at the three frequencies where plot points are shown, but the curve has been extrapolated to cover all high-frequency amateur bands.

Fig 4—The feed point of the tree-mounted groundplane antenna. The opposite ends of the two radial wires may be connected to stakes or other convenient anchor points.

soldered to the coax-line braid at this point. Another piece of wire forms the radiator. The top of the radiator section is suspended from a tree limb or other convenient support, and in turn supports the rest of the antenna.

The dimensions for the antenna are given in Fig 5. All three wires of the antenna are ¼ wavelength long. This generally limits the usefulness of the antenna for portable operation to 7 MHz and higher bands, as temporary supports higher than 35 or 40 feet are difficult to come by. Satisfactory operation might be had on 3.5 MHz with an inverted-L configuration of the radiator, if you can overcome the accompanying difficulty of "erecting" the antenna at the operating site.

The tree-mounted vertical idea can also be used for fixed-station installations to make an "invisible" antenna. Shallow trenches can be slit for burying the coax feeder and the radial wires. The radiator itself is difficult to see unless you are standing right next to the tree.

A TWO-BAND TRAP VERTICAL ANTENNA FOR THE TRAVELING HAM

This antenna can be built to cover two amateur bands in the following pairs: 10 and 14 MHz, 14 and 21 MHz, or 21 and 28 MHz. The original version was designed for 14 and 21 MHz by Doug DeMaw, W1FB, for operation from an RV camper or on a DXpedition. The antenna was described in *QST* for October 1980.

Short lengths of aluminum tubing that telescope into one another are used to fabricate the antenna. A 2-inch ID piece of aluminum tubing or a heavy-duty cardboard mailing tube will serve nicely as a container for shipping or carrying. Iron-pipe thread protectors can be used as plugs for the ends of the carrying tube. The antenna trap, mounting plate and coaxial feed line should fit easily into a suitcase with the operator's personal effects.

Six lengths of aluminum tubing are used in the construction of the antenna. The ends of these tubing sections are cut with a hacksaw to permit securing the joints by means of stainless steel hose clamps. The trap is constructed on a form of PVC tubing. It is held in place by two hose clamps that compress the PVC coil form and the ½-inch aluminum tubing sections onto ½-inch dowel-rod plugs. See Fig 6.

Strips of flashing copper (parts identified as G in Fig 6) slide inside sections B and C of the vertical. The opposite ends of the strips are placed under the hose clamps, which compress the PVC coil form. This provides an electrical contact between the trap coil and the tubing sections. The ends of the coil winding are soldered to the copper strips. Silicone grease should be put on the ends of strips G where they enter tubing sections B and C. This will retard corrosion. Grease can be applied to all mating surfaces of the telescoping sections for the same reason.

A suitable length of 50- or 75-ohm coaxial cable can be used as a trap capacitor. RG-58 or RG-59 cable is suggested for RF power levels below 150 watts. RG-8 or RG-11 will handle a few hundred watts without arcing or overheating. The advantage of using coaxial line as the trap capacitor is that the trap can be adjusted to resonance by selecting a length of cable that is too long, then trimming it until the trap is resonant. This is possible because each type of coax exhibits a specific amount of capacitance between the conductors. (See Chapter 24 for a table that lists coaxial cable characteristics.)

The trap (after final adjustment) should be protected against weather conditions. A plastic drinking glass can be inverted and mounted above the trap, or several coats of high-dielectric glue (Polystyrene Q-Dope) can be applied to the coil winding. If a coaxial-cable trap capacitor is used, it should be sealed at each end by applying noncorrosive RTV compound.

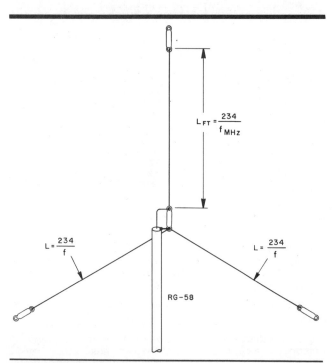

Fig 5—Dimensions and construction of the tree-mounted groundplane antenna.

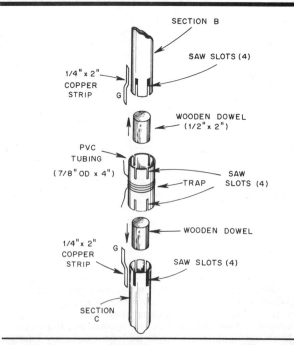

Fig 6—Breakdown view of the PVC trap for the 2-band vertical. The hose clamps used over the ends of the PVC coil form are not shown.

Tune the trap to resonance prior to installing it in the antenna. It should be resonant in the center of the desired operating range, that is, at 21.05 MHz if you prefer to operate from 21 to 21.1 MHz. Tuning can be done while using an accurately calibrated dip meter. If the dial isn't accurate, locate the dipper signal using a calibrated receiver *while the dipper is coupled to the coil and is set for the dip*.

A word of caution is in order here. Once the trap is installed in the antenna, it will not yield a dip at the same frequency as before. This is because it becomes absorbed in the overall antenna system and will appear to have shifted much lower in frequency. For this 2-band vertical, the apparent resonance will drop some 5 MHz. Ignore this condition and proceed with the installation.

The Tubing Sections

The assembled two-band vertical is shown in Fig 7, and dimensions for the tubing lengths appear in Table 2. The tubing diameters indicated in Fig 7 are suitable for 14- and 21-MHz use. The longer the overall antenna, the larger should be the tubing diameter adequate strength.

A short length of test-lead wire is used at the base of the antenna to join it to the coaxial connector on the mounting plate, as shown in Fig 7. A banana plug is attached to the end of the wire to permit connection to a UHF style of bulkhead connector. This method aids in easy breakdown of the antenna. A piece of PVC tubing slips over the bottom of section F to serve as an insulator between the antenna and the mounting plate.

If portable operation isn't planned, you may use fewer tubing sections. Only two sections are needed below the trap, and two sections will be sufficient above the trap. Two telescoping sections are necessary in each half of the antenna to permit resonating the system during final adjustment.

Other Bands

One additional band can be accommodated by using a top resonator, that is, a coil and a capacitance hat at the top of the antenna. This is equivalent to "top loading" the vertical antenna. Assume you have completed the antenna for two bands but also want to use the system on 7 MHz. One way to do this is to construct a 7-MHz loading coil which can be installed as shown in Fig 8. A number of commercial trap verticals use this technique. Another technique is to use a short rod of stiff, solid wire above the loading coil, rather than the capacitance hat, following the idea of commercial mobile-antenna resonators.

The loading assembly is called a "resonator" because it makes the complete antenna resonant at the lowest chosen operating frequency (7 MHz in this example). The coil turns must be adjusted while the antenna is assembled and installed in its final location. The remainder of the antenna must be adjusted for proper operation on all of the bands *before* the resonator is trimmed for 7-MHz resonance.

If you use capacitance-hat wires and they are short (approximately 12 inches), you can assume a capacitance of roughly 10 pF, which gives us an X_C of 2275 ohms. Therefore, the resonator will also have an X_L of 2275 ohms. This becomes 51 μH for operation at 7.1 MHz, since $L_{\mu H} = X_L/2\pi f$. The resonator coil should be wound for roughly 10% more inductance than needed, to allow some leeway for trimming it to resonance. Alternatively, the resonator can be wound for 51 μH and the capacitance-hat wires shortened or lengthened until resonance in the selected part of the 7-MHz

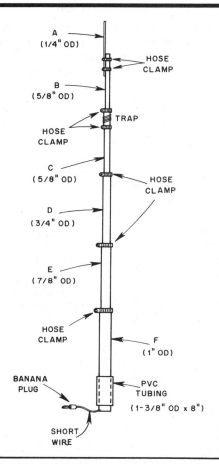

Fig 7—Assembly details for the two-band trap vertical. The coaxial-cable trap capacitor is taped to the lower end of section B. The aluminum tubing diameters shown here are suitable for 14- and 21-MHz use (see text).

Table 2

Tubing Length in Inches for Various Frequency Pairings, Two-Band Trap Vertical

Bands (MHz)	Tubing Section						C1 (pF)	L1 (Approx μH)
	A	B	C	D	E	F		
10 & 14	42	42	54	54	54	49	39	3.25
14 & 21	38	33	37	37	37	33	25	2.25
21 & 28	25	16	25	25	25	33	18	1.70

Note: Tubing sections are identified in Fig 7.

band is obtained. If you use a vertical rod above the coil instead of the capacitance hat, you can prune either the number of coil turns or the rod length (or both) for resonance.

As was true of the traps, the resonator coil should be wound on a low-loss form. The largest conductor size practical should be used to minimize losses and elevate the power-handling capability of the coil. Details of how a homemade resonator might be built are provided in Fig 9. The drawing in Fig 8 shows how the antenna would look with the resonator in place.

Ground System

There is nothing as rewarding as a big ground system. That is, the more radials the better, up to the point of diminishing returns. Some manufacturers of multiband trap verticals specify two radial wires for each band of operation.

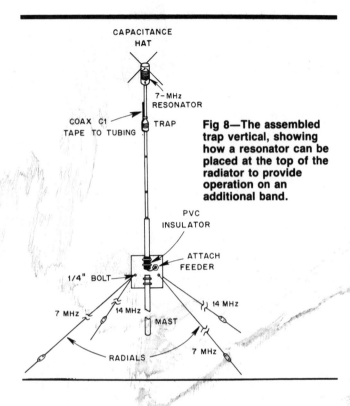

Fig 8—The assembled trap vertical, showing how a resonator can be placed at the top of the radiator to provide operation on an additional band.

CAPACITANCE HAT

7-MHz RESONATOR

COAX C1 TAPE TO TUBING — TRAP

PVC INSULATOR

ATTACH FEEDER

1/4" BOLT

7 MHz   14 MHz   14 MHz

MAST

7 MHz

RADIALS

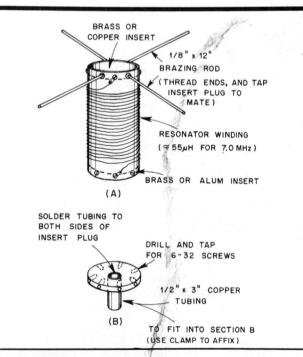

Fig 9—Details for building a resonator—a homemade top-loading coil and capacitance hat. The completed resonator should be protected against the weather to prevent detuning and deterioration.

BRASS OR COPPER INSERT

1/8" x 12" BRAZING ROD (THREAD ENDS, AND TAP INSERT PLUG TO MATE)

RESONATOR WINDING (≈ 55μH FOR 7.0 MHz)

BRASS OR ALUM INSERT

(A)

SOLDER TUBING TO BOTH SIDES OF INSERT PLUG

DRILL AND TAP FOR 6-32 SCREWS

1/2" x 3" COPPER TUBING

(B)

TO FIT INTO SECTION B (USE CLAMP TO AFFIX)

Admittedly, an impedance match can be had that way, and performance will be reasonably good. So during temporary operations where space for radial wires is at a premium, use two wires for each band. Four wires for each band will provide greatly improved operation. The slope of the wires will affect the feed-point impedance. The greater the downward slope, the higher the impedance. This can be used to advantage when adjusting for the lowest SWR.

A DELUXE RV 5-BAND ANTENNA

This antenna was designed to be mounted on a 31-foot Airstream travel trailer. With minor changes it can be used with any other recreational vehicle (RV). Perhaps the best feature of this antenna is that it requires no radials or ground system other than the RV itself. This section contains information by Charles Schecter, W8UCG, and was published in *QST* for October 1980.

The installation involves the use of a Hustler 4BTV vertical with the normal installation dimensions radically changed. See Figs 10 and 11. The modified antenna is mounted on a special mast that is hinged near the top to allow it to rest on the RV roof during travel.

The secret of the neat appearance of this installation is the unusual mast material used to support the antenna. Commonly known as Unistrut, it is often used by electrical contractors to build switching and control panels for industrial and commercial installations. The size selected is 1-5/8 inch

Fig 10—Ready to go! The W8UCG deluxe RV antenna is shown mounted at the rear of a 31-ft Airstream.

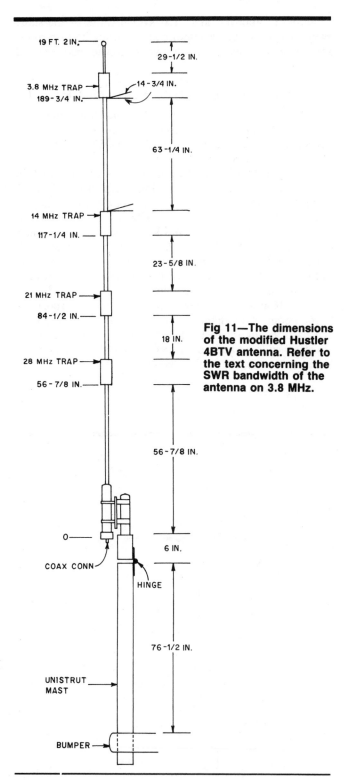

19 FT. 2 IN.

29-1/2 IN.

3.8 MHz TRAP
189-3/4 IN.   14-3/4 IN.

63-1/4 IN.

14 MHz TRAP
117-1/4 IN.

23-5/8 IN.

21 MHz TRAP
84-1/2 IN.

18 IN.

28 MHz TRAP
56-7/8 IN.

56-7/8 IN.

0

COAX CONN

6 IN.

HINGE

76-1/2 IN.

UNISTRUT MAST

BUMPER

Fig 11—The dimensions of the modified Hustler 4BTV antenna. Refer to the text concerning the SWR bandwidth of the antenna on 3.8 MHz.

line. Any brackets mounted higher detract from the neat appearance and will necessitate a complete change of dimensions for proper tuning. Install the antenna on the curb side of the vehicle to hide the lowered antenna behind the awning and provide greater safety to the person raising the antenna. This precaution is primarily for safety when stopping alongside a highway to meet a schedule. (Caution—beware of overhead power lines!)

Mounting the Hustler

The 4BTV base is U bolted to a 19-inch piece of 1-5/16 inch OD galvanized steel pipe (1-inch water pipe). This is inserted into and welded along the edges of a short piece of Unistrut, which is attached to the lower portion of the mast by means of a heavy duty, welded-on hinge. It is not feasible to bolt these pieces together, as the inside of the pipe must be completely clear to accept the end of a 54-inch piece of ½-inch water pipe (7/8 inch OD) that is used as a removable raising fixture and handle. This handle is wrapped with vinyl tape at the top end and also about 12 inches back. The tape forms a loose-fitting shim that provides a better fit inside the 1-inch water pipe. At 15 inches from the top end, a thicker wrap of tape acts as a stop to allow the handle to be inserted the same distance into the base support pipe in every instance. A short projecting bolt near the bottom end of the handle provides a means of lifting it off the lock pin (a bolt) which is mounted inside the Unistrut on an L-shaped bracket. See Figs 12 and 13.

The top-hat spider rods should be installed only on one side of the antenna so as not to poke holes in the top of the trailer. No effect on antenna performance will be noted. Bring the coaxial cable into the trailer at a point close to the antenna. This is preferable to running it beneath the trailer, where it can be more easily damaged and where ground-loop paths for RF current may be created. Be sure to use drip loops at both the antenna and at the point of entry into the RV. Silicone rubber sealant should be used at the outside connector end and at the RV entry hole. Clear acrylic spray will provide corrosion protection for any hardware used. Lock washers and locknuts should be used on all bolts; good workmanship will result in first-class appearance and long, trouble-free service.

To ensure optimum results, great care should be taken to obtain a good ground return from the antenna all the way back to the transceiver. Clean, tight connections, together with heavy duty tinned copper braid, should be used across the mast hinge, the mast-to-RV

Fig 12—The handle and raising fixture, seated on the lock pin.

square 12-gauge U channel. The open edges of the U are folded in for greater strength; the material is an extremely tough steel which resists bending (as well as drilling and cutting!). The U channel is available with a zinc-plated, galvanized or painted finish to prevent rust and corrosion; it may be repainted to match any RV color scheme.

The supporting mast is secured to the rear frame or bumper of the RV by means of 3/8-inch diameter bolts. A ¾-inch wrap-around strap was attached at the RV center trim

Fig 13—In this photo the open edges of the Unistrut channel are facing the viewer. The locking pin is visible in the middle of the channel.

frame and to bond the frame to the equipment chassis. This is absolutely necessary if the vertical quarter-wave antenna is to work properly.

Antenna Pruning and Tuning

The antenna must be carefully tuned to resonance on each band starting with 28 MHz. The most radical departure from the manufacturer's antenna dimensions (for home use) takes place with the 28-MHz section. There, a 30-inch length of tubing is cut off. Only 3 inches need be removed from the 21-MHz section. The 7-MHz (top) section must be lengthened, however, because of the radical shortening of the 28-MHz section. The easiest way to do this, short of buying a longer piece of 1¼-inch OD aluminum tubing, is merely to lengthen one of the top-hat spider rods. A 15½-inch length of ½-inch diameter aluminum tubing (with one end flattened and properly drilled) can be held in place under the RM-75S resonator. It is a good idea to start with a longer piece of tubing and trim as necessary to obtain resonance at 7.15 MHz.

Installing, grounding and tuning of the antenna as described here resulted in an SWR of 1.0:1 at resonance on the 3.8, 7, 14 and 21-MHz bands. At the lower end of the 28-MHz band, the SWR is 1.05:1. These low SWR values remain exactly the same regardless of whether or not the RV is grounded externally.

Exclusive of 10, 18 and 24 MHz, this system design also provides full band coverage on the 7- through 28-MHz bands with an SWR of less than 2:1. Band coverage on 3.8 MHz

is limited to approximately 100 kHz because of the short overall length of the resonator coil and whip. The tip rod is adjustable to enable you to select your favorite 100-kHz band segment.

A PORTABLE QUAD FOR 144 MHz

Figs 14 and 15 show a portable quad for home construction. This collapsible design was the product of Bob Decesari, WA9GDZ, and was described in *QST* for September 1980 and June 1981.

Both the driven and reflector elements of this array fold back on top of each other, resulting in a package about 17 inches long. When collapsed, the wire loop elements may be held in place around the boom with an elastic band. To support the antenna once it has been erected, the container is used as a stand. To provide more stability, four small removable struts slip into holes in the base of the container. Both the support rods and the struts fit inside the container when the antenna is disassembled.

Figs 16 and 17 show a method of attaching the spreaders to the boom. A mechanical stop is machined into the hub, and elastic bands are used to hold the spacers erect. The bands are attached to an additional strut to hold the spacers open. When not in use, the strut pulls out and sits across the hub,

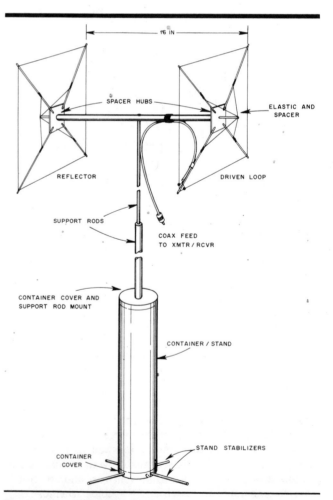

Fig 14—The basic portable quad assembly. An element spacing of 16 inches is used so the quad spacers will fold neatly between the hubs.

Fig 15—The portable quad stowed and ready for travel. Two long dowels are used as support rods. Four smaller dowels are used to stabilize the container when it is erected as a support stand.

Fig 16—This version of the portable quad uses mechanical stops machined into the hub; elastic bands hold the spacers open.

and the spacers can be folded back. Details are shown in Fig 17.

Building Materials

The portable quad antenna may be fabricated from any one of several plastic or wood materials. The most inexpensive method is to use wood doweling, available at most hardware stores. Wood is inexpensive and easily worked with hand tools; 1/4-inch doweling may be used for the quad spacers, and 3/8- or 1/2-inch doweling for the boom and support elements. A hardwood is recommended for the hub assembly, since a softwood may tend to crack along its grain if the hub is impacted or dropped. Plastics will also work well, but the cost will rise sharply if the material is purchased from a supplier. Plexiglas is an excellent choice for the hub. Fiberglass or phenolic rods are also excellent for the quad elements and support.

The loops are made with no. 18 AWG copper wire. If no insulation is used on the wire and wood doweling is used

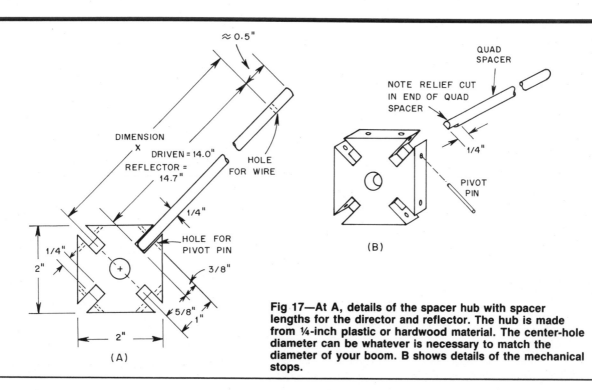

Fig 17—At A, details of the spacer hub with spacer lengths for the director and reflector. The hub is made from ¼-inch plastic or hardwood material. The center-hole diameter can be whatever is necessary to match the diameter of your boom. B shows details of the mechanical stops.

for the spacers, a coat of spar varnish in and around the spacer hole through which the wire runs is recommended. The loop wire is terminated at one element by attaching it to heavy gauge copper-wire posts inserted into tightly fitting holes in the element. For the driven element, three posts are used to allow the RG-58 feed-line braid, center conductor and matching capacitor to be attached. A single post is used on the reflector to complete the loop circuitry.

Fig 18 shows how to calculate the quad loop dimensions. The boom is 16 inches long. The feed-point matching system is detailed in Fig 19. The matching system uses a 3½-inch length of 300-Ω twin-lead as a shorted stub. Adjustment of the match is made at the 9- to 35-pF variable capacitor that is connected in series with the coaxial feed line.

The storage container shown in the photographs was made from a heavy cardboard tube originally used to store roll paper. Any rigid cylindrical housing of the proper dimensions may be used. Two wood end pieces were fabricated to cap the cardboard cylinder. The bottom end piece is cemented in place and has four holes drilled at 90° angles around the circumference. These holes hold the 4-inch struts that provide additional support when the antenna is erected. The top end piece is snug fitting and removable. It is of sufficient thickness (about 5/8 inch) to provide sufficient support for the antenna-supporting elements. A mounting hole for the supporting elements is drilled in the center of the top end piece. This hole is drilled only about three-quarters of the way through the end piece and should provide a snug fit for the antenna support. One or more antenna support elements may be used, depending on the height the builder wishes to have. Keep in mind, however, that the structure will be more prone to blowing over at greater heights above the ground! Doweling and snug-fitting holes are used to mate the support elements and the antenna boom.

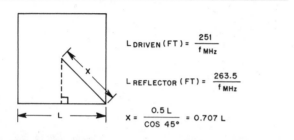

$$L_{DRIVEN} (FT) = \frac{251}{f_{MHz}}$$

$$L_{REFLECTOR} (FT) = \frac{263.5}{f_{MHz}}$$

$$X = \frac{0.5 L}{COS\ 45°} = 0.707\ L$$

Fig 18—Quad loop dimensions. Dimension X is the distance from the center of the hub to the hole drilled in each spacer for the loop wire. At 146 MHz, dimension X for the driven element is 1.216 feet (14.6 inches), and dimension X for the reflector is 1.276 feet (15.3 inches).

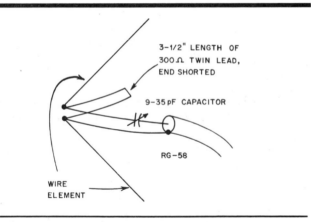

3-1/2" LENGTH OF 300 Ω TWIN LEAD, END SHORTED

9-35 pF CAPACITOR

RG-58

WIRE ELEMENT

Fig 19—Matching system for the portable quad. The stub may be taped to the element.

BIBLIOGRAPHY

Source material and more extended discussion of topics covered in this chapter can be found in the references given below.

R. J. Decesari, "A Portable Quad for 2 Meters," *QST*, Oct 1980, and "Portable Quad for 2 Meters, Part 2," Technical Correspondence, *QST*, Jun 1981.

D. DeMaw, "A Traveling Ham's Trap Vertical," *QST*, Oct 1980.

J. Hall, "Zip-Cord Antennas—Do They Work?" *QST*, Mar 1979.

C. L. Hutchinson, "A Tree-Mounted 30-Meter Ground-Plane Antenna," *QST*, Sep 1974.

C. W. Schecter, "A Deluxe RV 5-Band Antenna," *QST*, Oct 1980.

Chapter 16

Mobile and Maritime Antennas

Mobile antennas are those designed for use while they are in motion. At the mention of mobile antennas, most amateurs think of a whip mounted on an automobile or other highway vehicle, perhaps on a recreational vehicle (RV) or maybe on an off-road vehicle. While it is true that most mobile antennas are vertical whips, mobile antennas can also be found in other places. For example, antennas mounted aboard a boat or ship are mobile, and are usually called maritime antennas. Fig 1 shows yet another type of mobile antenna—those for use on handheld transceivers. Because they may be used while in motion, even these antennas are mobile by literal definition.

Pictured in Fig 1 is a telescoping full-size quarter-wave antenna for 144 MHz, and beside it a "stubby" antenna for the same band. The stubby is a helically wound radiator, made of stiff copper wire enclosed in a protective covering of rubber-like material. The inductance of the helical windings provides electrical loading for the antenna. For frequencies above 28 MHz, most mobile installations permit the use of a full-size antenna, but sometimes smaller, loaded antennas are used for convenience. The stubby, for example, is convenient for short-range communications, avoiding the problems of a lengthier, cumbersome antenna attached to a handheld radio.

Below 28 MHz, physical size becomes a problem with full-size whips, and some form of electrical loading (as with the stubby) is usually employed. Commonly used loading techniques are to place a coil at the base of the whip (base loading), or at the center of the whip (center loading). These and other techniques are discussed in this chapter.

Few amateurs construct their own antennas for HF mobile and maritime use, since safety reasons dictate very sound mechanical construction. However, construction projects are included in this chapter for those who may wish to build their own mobile antenna. Even if commercially made antennas are installed, most require some adjustment for the particular installation and type of operation desired, and the information given here may provide a better understanding of the optimization requirements.

HF-MOBILE FUNDAMENTALS

Fig 2 shows a typical bumper-mounted center-loaded whip suitable for operation in the HF range. The antenna could also be mounted on the car body itself (such as a fender), and mounts are available for this purpose. The base spring acts as a shock absorber for the bottom of the whip, as the continual flexing while in motion would otherwise weaken the antenna. A short heavy mast section is mounted between the base spring and loading coil. Some models have a mechanism which allows the antenna to be tipped over for adjustment or for fastening to the roof of the car when not in use. It is also advisable to extend a couple of guy lines from the base of the loading coil to clips or hooks fastened to the roof gutter on the car. Nylon fishing line (about 40-pound test) is suitable for this purpose. The guy lines act as safety cords and also reduce the swaying motion of the antenna considerably. The feed line to the transmitter is connected to

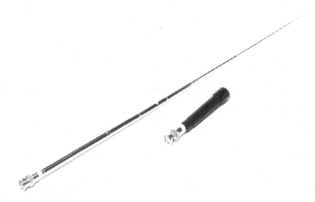

Fig 1—Two mobile antennas—mobile because they may be used while in motion. Shown here are a telescoping ¼-λ antenna and a "stubby" antenna, both designed for use at 144 MHz. The ¼-λ antenna is 19 in. long, while the stubby antenna is only 3½ in. long. (Both dimensions exclude the length of the BNC connectors. The stubby is a helically wound radiator.

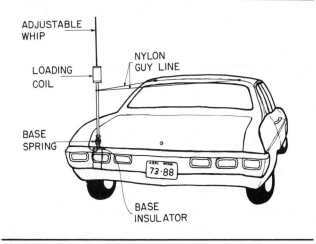

Fig 2—A typical bumper-mounted HF mobile antenna. Note the nylon guy lines.

the bumper and base of the antenna. Good low-resistance connections are important here.

Tune-up of the antenna is usually accomplished by changing the height of the adjustable whip section above the precut loading coil. First, tune the receiver and try to determine where the signals seem to peak up. Once this frequency is found, check the SWR with the transmitter on, and find the frequency where the SWR is lowest. Shortening the adjustable section will increase the resonant frequency, and making it longer will lower the frequency. It is important that the antenna be away from surrounding objects such as overhead wires by ten feet or more, as considerable detuning can occur. Once the setting is found where the SWR is lowest at the center of the desired frequency range, the length of the adjustable section should be recorded.

Propagation conditions and ignition noise are usually the limiting factors for mobile operation on 10 through 28 MHz. Antenna size restrictions affect operation somewhat on 7 MHz and much more on 3.5 and 1.8 MHz. From this standpoint, perhaps the optimum band for HF-mobile operation is 7 MHz. The popularity of the regional mobile nets on 7 MHz is perhaps the best indication of its suitability. For local work, 28 MHz is also useful, as antenna efficiency is high and relatively simple antennas without loading coils are easy to build.

As the frequency of operation is lowered, an antenna of fixed length looks (at its feed point) like a decreasing resistance in series with an increasing capacitive reactance. The capacitive reactance must be tuned out, which necessitates the use of an equivalent series inductive reactance or loading coil. The amount of inductance required will be determined by the placement of the coil in the antenna system.

Base loading requires the lowest value of inductance for a fixed-length antenna, and as the coil is placed farther up the whip, the necessary value increases. This is because the capacitance of the shorter antenna section (above the coil) to the car body is now lower (higher capacitive reactance), requiring more inductance to tune the antenna to resonance. The advantage is that the current distribution on the whip is improved, which increases the radiation resistance. The disadvantage is that requirement of a larger coil also means the coil size and losses increase. Center loading has been generally accepted as a good compromise with minimal construction problems. Placing the coil ⅔ the distance up the whip seems to be about the optimum position.

For typical antenna lengths used in mobile work, the difficulty in constructing suitable loading coils increases as the frequency of operation is lowered. Since the required resonating inductance gets larger and the radiation resistance decreases at lower frequencies, most of the power is dissipated in the coil resistance and in other ohmic losses. This is one reason why it is advisable to buy a commercially made loading coil with the highest power rating possible, even if only low-power operation is planned.

Coil losses in the higher-power loading coils are usually less (percentage-wise), with subsequent improvement in radiation efficiency, regardless of the power level used. Of course, the above philosophy also applies to homemade loading coils, and design considerations will be considered in a later section.

Once the antenna is tuned to resonance, the input impedance at the antenna terminals will look like a pure resistance. Neglecting losses, this value drops from nearly 15 ohms at 21 MHz to 0.1 ohm at 1.8 MHz for an 8-foot whip. When coil and other losses are included, the input resistance increases to approximately 20 ohms at 1.8 MHz and 16 ohms at 21 MHz. These values are for relatively high-efficiency systems. From this it can be seen that the radiation efficiency is much poorer at 1.8 MHz than at 21 MHz under typical conditions.

Since most modern gear is designed to operate with a 52-ohm transmission line, a matching network may be necessary with the high-efficiency antennas previously mentioned. This can take the form of either a broadband transformer, a tapped coil, or an LC matching network. With homemade or modified designs, the tapped-coil arrangement is perhaps the easiest one to build, while the broadband transformer requires no adjustment. As the losses go up, so does the input resistance, and in less efficient systems the matching network may be eliminated.

The Equivalent Circuit of a Typical Mobile Antenna

In the previous section, some of the general considerations were discussed, and these will now be taken up in more detail. It is customary in solving problems involving electric and magnetic fields (such as antenna systems) to try to find an equivalent network with which to replace the antenna for analysis reasons. In many cases, the network may be an accurate representation over only a limited frequency range. However, this is often a valuable method in matching the antenna to the transmission line.

Antenna resonance is defined as the frequency at which the input impedance at the antenna terminals is purely resistive. The shortest length at which this occurs for a vertical antenna over a ground plane is when the antenna is an electrical quarter wavelength at the operating frequency; the impedance value for this length (neglecting losses) is about 36 ohms. The idea of resonance can be extended to antennas shorter (or longer) than a quarter wave, and means only that the input impedance is purely resistive. As pointed out previously, when the frequency is lowered, the antenna looks like a series RC circuit, as shown in Fig 3. For the average 8-foot whip, the reactance of C_A may range from about 150 ohms at 21 MHz to as high as 8000 ohms at 1.8 MHz, while the radiation resistance R_R varies from about 15 ohms at 21 MHz to as low as 0.1 ohm at 1.8 MHz.

For an antenna less than 0.1 wavelength long, the approximate radiation resistance may be determined from the following:

$$R_R = 273 \times (\ell\, f)^2 \times 10^{-8}$$

where ℓ is the length of the whip in inches, and f is the frequency in megahertz.

Since the resistance is low, considerable current must flow in the circuit if any appreciable power is to be dissipated in the form of radiation in R_R. Yet it is apparent that little current can be made to flow in the circuit as long as the comparatively high series reactance remains.

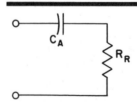

Fig 3—At frequencies below resonance, the whip antenna will show capacitive reactance as well as resistance. R_R is the radiation resistance, and C_A represents the capacitive reactance.

Antenna Capacitance

Capacitive reactance can be canceled by connecting an equivalent inductive reactance, (coil L_L) in series, as shown in Fig 4, thus tuning the system to resonance.

The capacitance of a vertical antenna shorter than a quarter wavelength is given by:

$$C_A = \frac{17\ell}{\left[\left(\ln \frac{24\ell}{D}\right) - 1\right]\left[1 - \left(\frac{f\ell}{234}\right)^2\right]}$$

where

C_A = capacitance of antenna in pF
ℓ = antenna height in feet
D = diameter of radiator in inches
f = operating frequency in MHz

$$\ln \frac{24\ell}{D} = 2.3 \log_{10} \frac{24\ell}{d}$$

Fig 5 shows the approximate capacitance of whip antennas of various average diameters and lengths. For 1.8, 4 and 7 MHz, the loading coil inductance required (when the loading coil is at the base) will be approximately the inductance required to resonate in the desired band (with the whip capacitance taken from the graph). For 10 through 21 MHz, this rough calculation will give more than the required inductance, but it will serve as a starting point for the final experimental adjustment that must always be made.

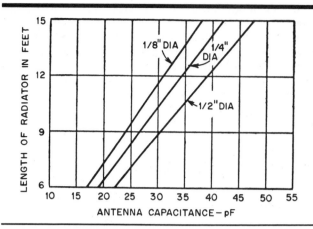

Fig 4—The capacitive reactance at frequencies below the resonant frequency of the whip can be canceled by adding an equivalent inductive reactance in the form of a loading coil in series with the antenna.

LOADING COIL DESIGN

To minimize loading coil loss, the coil should have a high ratio of reactance to resistance (that is, a high Q). A 4-MHz loading coil wound with small wire on a small-diameter solid form of poor quality, and enclosed in a metal protector, may have a Q as low as 50, with a resistance of 50 ohms or more. High-Q coils require a large conductor, air-wound construction, large spacing between turns, and the best insulating material available. A diameter not less than half the length of the coil (not always mechanically feasible) and a minimum of metal in the field of the coil are also necessities for optimum efficiency. Such a coil for 4 MHz may show a Q of 300 or more, with a resistance of 12 ohms or less.

The coil could then be placed in series with the feed line at the base of the antenna to tune out the unwanted capacitive

Fig 5—Graph showing the approximate capacitance of short vertical antennas for various diameters and lengths. These values should be approximately halved for a center-loaded antenna.

reactance, as shown in Fig 4. Such a method is often referred to as base loading, and many practical mobile antenna systems have been built using this scheme.

Over the years, the question has come up as to whether or not more efficient designs are possible compared with simple base loading. While many ideas have been tried with varying degrees of success, only a few have been generally accepted and incorporated into actual systems. These are center loading, continuous loading, and combinations of the latter with more conventional antennas.

Base Loading and Center Loading

If a whip antenna is short compared to a wavelength *and the current is uniform along the length ℓ*, the electric field strength E, at a distance d, away from the antenna is approximately:

$$E = \frac{120\pi I\ell}{d\lambda}$$

where

I is the antenna current in amperes
λ is the wavelength in the same units as d and ℓ

A uniform current flowing along the length of the whip is an idealized situation, however, since the current is greatest at the base of the antenna and goes to a minimum at the top. In practice, the field strength will be less than that given by the above equation, because it is a function of the current distribution on the whip.

The reason that the current is not uniform on a whip antenna can be seen from the circuit approximation shown in Fig 6. A whip antenna over a ground plane is similar in many respects to a tapered coaxial cable where the center conductor remains the same diameter along its length, but with an increasing diameter outer conductor. The inductance per unit length of such a cable would increase along the line, while the capacitance per unit length would decrease. In Fig 6 the antenna is represented by a series of LC circuits in which C1 is greater than C2, which is greater than C3, and so on. L1 is less than L2, which is less than succeeding inductances. The net result is that most of the antenna current returns to ground near the base of the antenna, and very little near the top.

Two things can be done to improve this distribution and make the current more uniform. One would be to increase the capacitance of the top of the antenna to ground through the use of top loading or a capacitance hat, as discussed in Chapter 2. Unfortunately, the wind resistance of the hat makes it somewhat unwieldy for mobile use. The other method is to place the loading coil farther up the whip, as shown in Fig 7, rather than at the base. If the coil is resonant (or nearly so) with the capacitance to ground of the section above the coil, the current distribution is improved as also shown in Fig 7. The result with both top loading and center loading is that the radiation resistance is increased, offsetting the effect of losses and making matching easier.

Table 1 shows the approximate loading coil inductance for the various amateur bands. Also shown in the table are approximate values of radiation resistance to be expected with an 8-foot whip, and the resistances of loading coils—one group having a Q of 50, the other a Q of 300. A comparison of radiation and coil resistances will show the importance of reducing the coil resistance to a minimum, especially on the three lower frequency bands. Table 2 shows suggested loading-coil dimensions for the inductance values given in Table 1.

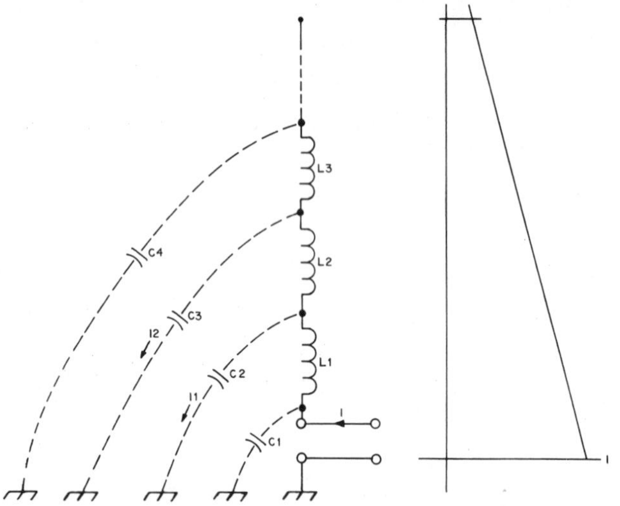

Fig 6—A circuit approximation of a simple whip over a perfectly conducting ground plane. The shunt capacitance per unit length gets smaller as the height increases, and the series inductance per unit length gets larger. Consequently, most of the antenna current returns to the ground plane near the base of the antenna, giving the current distribution shown at the right.

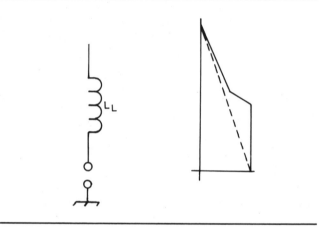

Fig 7—Improved current distribution resulting from center loading.

Table 1

Approximate Values for 8-ft Mobile Whip

f(MHz)	Loading L μH	R_C (Q50) Ohms	R_C (Q300) Ohms	R_R Ohms	Feed R* Ohms	Matching L μH
Base Loading						
1.8	345	77	13	0.1	23	3
3.8	77	37	6.1	0.35	16	1.2
7.2	20	18	3	1.35	15	0.6
10.1	9.5	12	2	2.8	12	0.4
14.2	4.5	7.7	1.3	5.7	12	0.28
18.1	3.0	5.0	1.0	10.0	14	0.28
21.25	1.25	3.4	0.5	14.8	16	0.28
24.9	0.9	2.6	—	20.0	22	0.25
29.0	—	—	—	—	36	0.23
Center Loading						
1.8	700	158	23	0.2	34	3.7
3.8	150	72	12	0.8	22	1.4
7.2	40	36	6	3.0	19	0.7
10.1	20	22	4.2	5.8	18	0.5
14.2	8.6	15	2.5	11.0	19	0.35
18.1	4.4	9.2	1.5	19.0	22	0.31
21.25	2.5	6.6	1.1	27.0	29	0.29

R_C = loading coil resistance; R_R = radiation resistance.
*Assuming loading coil Q = 300, and including estimated ground-loss resistance.

Table 2

Suggested Loading Coil Dimensions

Req'd L(μH)	Turns	Wire Size	Dia In.	Length In.
700	190	22	3	10
345	135	18	3	10
150	100	16	2-1/2	10
77	75	14	2-1/2	10
77	29	12	5	4-1/4
40	28	16	2-1/2	2
40	34	12	2-1/2	4-1/4
20	17	16	2-1/2	1-1/4
20	22	12	2-1/2	2-3/4
8.6	16	14	2	2
8.6	15	12	2-1/2	3
4.5	10	14	2	1-1/4
4.5	12	12	2-1/2	4
2.5	8	12	2	2
2.5	8	6	2-3/8	4-1/2
1.25	6	12	1-3/4	2
1.25	6	6	2-3/8	4-1/2

Table 3

Variables used in Eqs 1 through 17

A = area in degree-amperes
a = antenna radius in English or metric units
dB = signal loss in decibels
E = efficiency in percent
f(MHz) = frequency in megahertz
H = height in English or metric units
h = height in electrical degrees
h_1 = height of base section in electrical degrees
h_2 = height of top section in electrical degrees
I = I_{base} = 1 ampere base current
K = 0.0128
K_m = mean characteristic impedance
K_{m1} = mean characteristic impedance of base section
K_{m2} = mean characteristic impedance of top section
L = length or height of the antenna in feet
P_I = power fed to the antenna
P_R = power radiated
Q = coil figure of merit
R_C = coil loss resistance in ohms
R_G = ground-loss resistance in ohms
R_R = radiation resistance in ohms
X_L = loading-coil inductive reactance

OPTIMUM DESIGN OF SHORT COIL-LOADED HF MOBILE ANTENNAS

Optimum design of short HF mobile antennas results from a careful balance of the appropriate loading coil Q-factor, loading coil position in the antenna, ground loss resistance, and the length to diameter ratio of the antenna. The optimum balance of these parameters can be realized only through a thorough understanding of how they interact. This section presents a mathematical approach to designing mobile antennas for maximum radiation efficiency. This approach was first presented by Bruce Brown, W6TWW, in The *ARRL Antenna Compendium, Vol. 1.* (See the bibliography at the end of this chapter.)

The optimum location for a loading coil in an antenna can be found experimentally, but it requires many hours of designing and constructing models and making measurements to ensure the validity of the design. A faster and more reliable way of determining optimum coil location is through the use of a personal computer. This approach allows the variation of any single variable while observing the cumulative effects on the system. When plotted graphically, the data reveals that the placement of the loading coil is critical if maximum radiation efficiency is to be realized.

Radiation Resistance

The determination of radiation efficiency requires the knowledge of resistive power losses and radiation losses. Radiation loss is expressed in terms of radiation resistance. Radiation resistance is defined as the resistance that would dissipate the same amount of power that is radiated by the antenna. The variables used in the equations that follow are defined once in the text, and are summarized in Table 3. Radiation resistance of vertical antennas shorter than 45 electrical degrees (1/8 wavelength) is approximately

$$R_R = \frac{h^2}{312} \qquad \text{(Eq 1)}$$

where
  R_R = radiation resistance in ohms
  h = antenna length in electrical degrees

Antenna height in electrical degrees is expressed by

$$h = \frac{\ell}{984} \times f(MHz) \times 360 \qquad \text{(Eq 2)}$$

where
  ℓ = antenna length in feet
  f(MHz) = operating frequency in megahertz

End effect is purposely omitted to ensure that an antenna is electrically long. This is so that resonance at the design frequency can be obtained easily by removing a turn or two from the loading coil.

Eq 1 is valid only for antennas having a sinusoidal current distribution and no reactive loading. However, it can be used as a starting point for deriving an equation that is useful for shortened antennas with other than sinusoidal current distributions.

Refer to Fig 8. The current distribution on an antenna 90° long electrically (¼ wavelength) varies with the cosine of the length in electrical degrees. The current distribution of the top 30° of the antenna is essentially linear. It is this linearity that allows for derivation of a simpler, more useful equation for radiation resistance.

The radiation resistance of an electrically short base-loaded vertical antenna can be conveniently defined in terms of a geometric figure, a triangle, as shown in Fig 9. The radiation resistance is given by

$$R_R = KA^2 \qquad \text{(Eq 3)}$$

where
  K is a constant (to be derived shortly)
  A = area of the triangular current distribution in degree-amperes.

Degree-ampere area is expressed by

$$A = \frac{1}{2}h \times I_{base} \qquad \text{(Eq 4)}$$

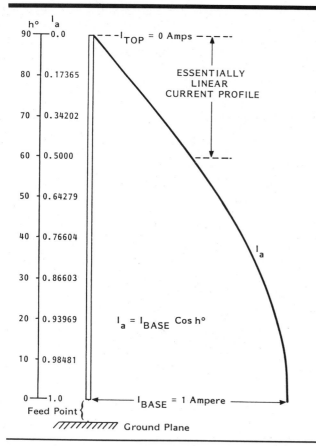

Fig 8—Relative current distribution on a vertical antenna of height H = 90 electrical degrees.

By combining Eqs 1 and 3 and solving for K, we get

$$K = \frac{h^2}{312 \times A^2} \qquad \text{(Eq 5)}$$

By substituting the values from Fig 9 into Eq 5 we get

$$K = \frac{30^2}{312 \times (0.5 \times 30 \times 1)^2} = 0.0128$$

and by substituting the derived value of K into Eq 3 we get

$$R_R = 0.0128 \times A^2 \qquad \text{(Eq 6)}$$

Eq 6 is useful for determining the radiation resistance of coil-loaded vertical antennas less than 30° in length. The derived constant differs slightly from that presented by Laport (see bibliography), as he used a different equation for radiation resistance (Eq 1).

When the loading coil is moved up an antenna (away from the feed point), the current distribution is modified as shown in Fig 10. The current varies with the cosine of the height in electrical degrees at any point in the base section. Therefore, the current flowing into the bottom of the loading coil is less than the current flowing at the base of the antenna.

But what about the current in the top section of the antenna? The loading coil acts as the lumped constant that it is, and disregarding losses and coil radiation, maintains the same current flow throughout. As a result, the current at the top of a high-Q coil is essentially the same as that at the bottom of the coil. This is easily verified by installing RF ammeters immediately above and below the loading coil in a test antenna. Thus, the coil "forces" much more current into the top section than would flow in the equivalent section

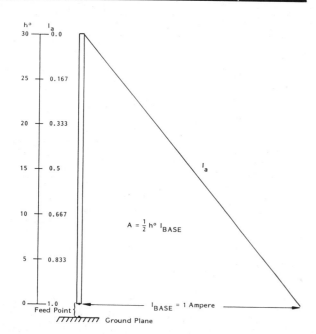

Fig 9—Relative current distribution on a base-loaded vertical antenna of height H = 30 electrical degrees (linearized). A base loading coil is omitted.

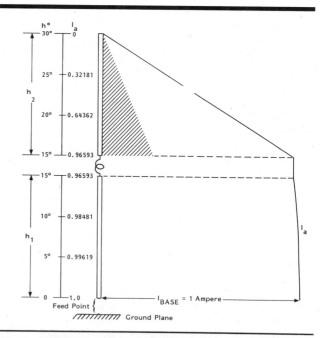

Fig 10—Relative current distribution on a center-loaded antenna with base and top sections each equal to 15 electrical degrees in length. The cross-hatched area shows the current distribution that would exist in the top 15° of a 90°-high vertical fed with 1 ampere at the base.

of a full 90° long antenna. This occurs as a result of the extremely high voltage that appears at the top of the loading coil. This higher current flow results in more radiation than would occur from the equivalent section of a quarter-wave antenna. (This is true for conventional coils. However, radiation from long thin coils allows coil current to decrease, as in helically wound antennas.)

The cross-hatched area in Fig 10 shows the current that would flow in the equivalent part of a 90° high antenna, and reveals that the degree-ampere area of the whip section of the short antenna is greatly increased as a result of the modified current distribution. The current flow in the top section decreases almost linearly to zero at the top. This can be seen in Fig 10.

The degree-ampere area of Fig 10 is the sum of the triangular area represented by the current distribution in the top section, and the nearly trapezoidal current distribution in the base section. Radiation from the coil is not included in the degree-ampere area because it is small and difficult to define. Any radiation from the coil can be considered a bonus.

The degree-ampere area is expressed by

$$A = \frac{1}{2}\left[h_1 \times \left(1 + \cos h_1\right) + h_2 \left(\cos h_1\right)\right] \qquad \text{(Eq 7)}$$

where
    h_1 = electrical length in degrees of the base section
    h_2 = electrical height in degrees of the top section.

The degree-ampere area (calculated by substituting Eq 7 into Eq 6) can be used to determine the radiation resistance when the loading coil is at any position other than the base of the antenna. Radiation resistance has been calculated with these equations and plotted against loading coil position at three different frequencies for 8- and 11-foot antennas, Fig 11. Eight feet is a typical length for commercial antennas, and 11-foot antennas are about the maximum practical length that can be installed on a vehicle.

In Fig 11, the curves reveal that the radiation resistance rises almost linearly as the loading coil is moved up the antenna. They also show that the radiation resistance rises rapidly as the frequency is increased. If the analysis were stopped at this point, one might conclude that the loading coil should be placed at the top of the antenna. This is not so, and will become apparent shortly.

Required Loading Inductance

Calculation of the loading-coil inductance needed to resonate a short antenna can be done easily and accurately by using the antenna transmission-line analog described by Boyer in *Ham Radio*. For a base-loaded antenna, Fig 9, the loading coil reactance required to resonate the antenna is given by

$$X_L = -j\, K_m \cot h \qquad \text{(Eq 8)}$$

where
    X_L = inductive reactance required
    K_m = mean characteristic impedance (defined in Eq 9)

The $-j$ term indicates that the antenna presents capacitive reactance at the feed point. This reactance must be canceled by a loading coil.

The mean characteristic impedance of an antenna is expressed by

$$K_m = 60 \times \left[\left(\ln \frac{2H}{a}\right) - 1\right] \qquad \text{(Eq 9)}$$

where
    H = physical antenna height (excluding the length of the loading coil)
    a = radius of the antenna in the same units as H.

From Eq 9 it can be seen that decreasing the height-to-diameter ratio of an antenna by increasing the radius results in a decrease in K_m. With reference to Eq 8, a decrease in K_m decreases the inductive reactance required to resonate an antenna. As will be shown later, this will increase radiation efficiency. In mobile applications, we quickly run into wind-loading problems if we attempt to use an antenna that is physically large in diameter.

If the loading coil is moved away from the base of the antenna, the antenna is divided into a base and top section, as depicted in Fig 10. The loading-coil reactance required to resonate the antenna when the coil is away from the base is given by

$$X_L = K_{m2} \times (\cotan h_2) - j\, K_{m1} (\tan h_1) \qquad \text{(Eq 10)}$$

In mobile-antenna design and construction, the top section is usually a whip with a much smaller diameter than the base section. Because of this, it is necessary to compute separate values of K_m for the top and base sections. K_{m1} and K_{m2} are the mean characteristic impedances of the base and top sections, respectively.

Loading coil reactance curves for the 3.8-MHz antennas of Fig 11 have been calculated and plotted in Fig 12. These curves show the influence of the loading coil position on the reactance required for resonance. The curves in Fig 12 show that the required reactance decreases with longer antennas. The curves also reveal that the required loading coil reactance grows at an increasingly rapid rate after the coil passes the center of the antenna. Because the highest possible loading

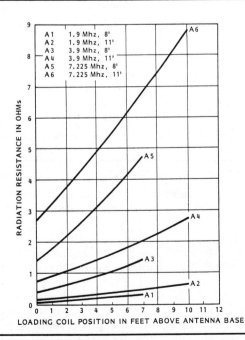

Fig 11—Radiation resistance plotted as a function of loading coil position.

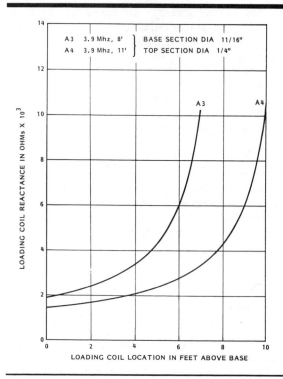

A3   3.9 Mhz, 8'   BASE SECTION DIA   11/16"
A4   3.9 Mhz, 11'   TOP SECTION DIA   1/4"

Fig 12—Loading coil reactance required for resonance, plotted as a function of coil height above the antenna base. The resonant frequency is 3.9 MHz.

coil Q factor is needed, and because optimum Q is attained when the loading coil diameter is twice the loading coil length, the coil would grow like a smoke ring above the center of the antenna, and would quickly reach an impractical size. It is for this reason that the highest loading coil position is limited to one foot from the top of the antenna in all computations.

Loading Coil Resistance

Loading coil resistance constitutes one of the losses that consumes power that could otherwise be radiated by the antenna. Heat loss in the loading coil is not of any benefit, so it should be minimized by using the highest possible loading coil Q. Loading coil loss resistance is a function of the coil Q and is given by

$$R_C = \frac{X_L}{Q} \qquad \text{(Eq 11)}$$

where

R_C = loading coil loss resistance in ohms
X_L = loading coil reactance
Q = coil figure of merit

Inspection of Eq 11 reveals that, for a given value of inductive reactance, loss resistance will be lower for higher Q coils. Measurements made with a Q meter show that typical, commercially manufactured coil stock produces a Q between 150 and 160 in the 3.8-MHz band.

Higher Q values can be obtained by using larger diameter coils having a diameter to length ratio of two, by using larger diameter wire, by using more spacing between turns, and by using low-loss polystyrene supporting and enclosure materials. Loading coil turns should not be shorted for tuning purposes

because shorted turns degrade Q. Pruning to resonance should be done only by removing turns from the coil.

Radiation Efficiency

The ratio of power radiated to power fed to an antenna determines the radiation efficiency. It is given by

$$E = \frac{P_R}{P_I} \times 100\% \qquad \text{(Eq 12)}$$

where

E = radiation efficiency in percent
P_R = power radiated
P_I = power fed to the antenna at the feed point

In a short, coil-loaded mobile antenna, a large portion of the power fed to the antenna is dissipated in ground and coil resistances. A relatively insignificant amount of power is also dissipated in the antenna conductor resistance and in the leakage resistance of the base insulator. Because these last two losses are both very small and difficult to estimate, they are neglected in calculating radiation efficiency.

Another loss worth noting is matching network loss. Because we are concerned only with power fed to the antenna in the determination of radiation efficiency, matching network loss is not considered in any of the equations. Suffice it to say that matching networks should be designed for minimum loss in order to maximize the transmitter power available at the antenna.

The radiation efficiency equation may be rewritten and expanded as follows:

$$E = \frac{I^2 \times R_R \times 100}{I^2 \times R_R + I^2 R_G + (I \cos h_1)^2 \times R_C} \qquad \text{(Eq 13)}$$

where

I = antenna base current in amperes
R_G = ground loss resistance in ohms
R_C = coil loss resistance in ohms

Each term of Eq 13 represents the power dissipated in its associated resistance. All the current terms cancel, simplifying this equation to

$$E = \frac{R_R}{R_R + R_G + R_C \times (\cos^2 h_1)} \qquad \text{(Eq 14)}$$

For base-loaded antennas the term $\cos^2 h_1$ drops to unity and may be omitted.

Ground Loss

Eq 14 shows that the total resistive losses in the antenna system are:

$$R_T = R_R + R_G + R_C \times \cos^2 h_1 \qquad \text{(Eq 15)}$$

where R_T is the total resistive loss. Ground loss resistance can be determined by rearranging Eq 15 as follows:

$$R_G = R_T + R_R - R_C \times \cos^2 h_1 \qquad \text{(Eq 16)}$$

R_T may be measured in a test antenna installation on a vehicle using an R-X noise bridge. R_R and R_C can then be calculated.

Ground loss is a function of vehicle size, placement of the antenna on the vehicle, and conductivity of the ground over which the vehicle is traveling. Only the first two variables can be feasibly controlled. Larger vehicles provide better ground planes than smaller ones. The vehicle ground plane is only partial, so the result is considerable RF current flow (and ground loss) in the ground around and under the vehicle.

By raising the antenna base as high as possible on the vehicle, the ground losses are decreased. This results from a decrease in antenna capacitance to ground, which increases the capacitive reactance to ground. This, in turn, reduces ground currents and ground losses.

This effect has been verified by installing the same antenna at three different locations on two different vehicles, and by determining the ground loss from Eq 16. In the first test, the antenna was mounted 6 inches below the top of a large station wagon, just behind the left rear window. This placed the antenna base 4 feet 2 inches above the ground, and resulted in a measured ground loss resistance of 2.5 ohms. The second test used the same antenna mounted on the left rear fender of a mid-sized sedan, just to the left of the trunk lid. In this test, the measured ground loss resistance was 4 ohms. The third test used the same mid-sized car, but the antenna was mounted on the rear bumper. In this last test, the measured ground loss resistance was 6 ohms.

The same antenna therefore sees three different ground loss resistances as a direct result of the antenna mounting location and size of the vehicle. It is important to note that the measured ground loss increases as the antenna base nears the ground. The importance of minimizing ground losses in mobile antenna installations cannot be overemphasized.

Efficiency Curves

With the equations defined previously, a computer was used to calculate the radiation-efficiency curves depicted in Figs 13 through 16. These curves were calculated for 3.8- and

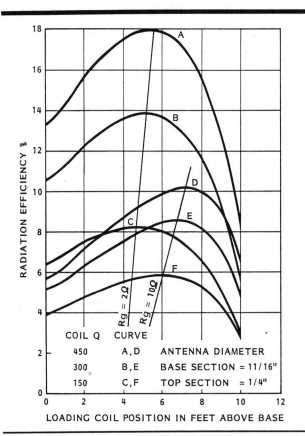

Fig 14—Radiation efficiency of 11-foot antennas at 3.9 MHz.

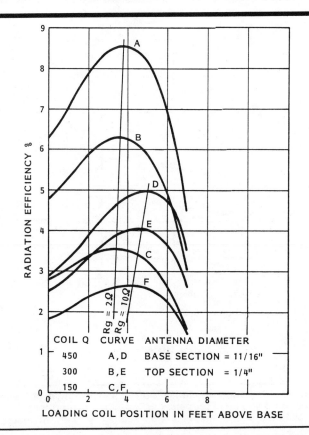

Fig 13—Radiation efficiency of 8-foot antennas at 3.9 MHz.

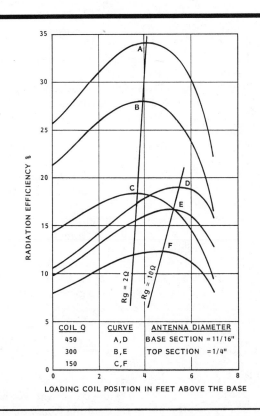

Fig 15—Radiation efficiency of 8-foot antennas at 7.225 MHz.

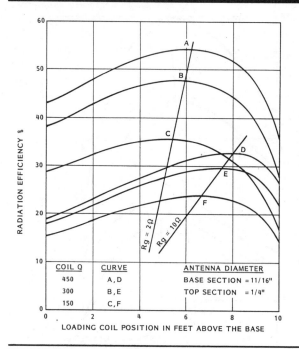

Fig 16—Radiation efficiency of 11-foot antennas at 7.225 MHz.

of a base section of larger diameter than the whip when the coil is above the base.

The curves in Figs 13 through 16 were calculated with constant (but not equal) diameter base and whip sections. Because of wind loading, it is not desirable to increase the diameter of the whip section. However, the base-section diameter can be increased within reason to further improve radiation efficiency. Fig 17 was calculated for base-section diameters ranging from 11/16 inch to 3 inches. The curves reveal that a small increase in radiation efficiency results from larger diameter base sections.

The curves in Figs 13 through 16 show that radiation efficiencies can be quite low in the 3.8-MHz band compared to the 7-MHz band. They are lower yet in the 1.8-MHz band. To gain some perspective on what these low efficiencies mean in terms of signal strength, Fig 18 was calculated using the following equation:

$$dB = \log \frac{100}{E} \qquad \text{(Eq 17)}$$

where

dB = signal loss in decibels
E = efficiency in percent

The curve in Fig 18 reveals that an antenna having 25% efficiency has a signal loss of 6 dB (approximately one S unit) below a quarter-wave vertical antenna over perfect ground. An antenna efficiency in the neighborhood of 6% will produce a signal strength on the order of two S units or about 12 dB below the same quarter-wave reference vertical. By careful optimization of mobile-antenna design, signal strengths from

7-MHz antennas of 8- and 11-foot lengths. Several values of loading coil Q were used, for both 2 and 10 ohms of ground loss resistance. For the calculations, the base section is ½-inch-diameter electrical EMT, which has an outside diameter of 11/16 inch. The top section is fiberglass bicycle-whip material covered with Belden braid. These are readily available materials which can be used by the average amateur to construct an inexpensive but rugged antenna.

Upon inspection, these radiation-efficiency curves reveal some significant information:

1) Higher coil Q produces higher radiation efficiencies,
2) longer antennas produce higher radiation efficiencies,
3) higher frequencies produce higher radiation efficiencies,
4) lower ground loss resistances produce higher radiation efficiencies,
5) higher ground loss resistances force the loading coil above the antenna center to reach a crest in the radiation-efficiency curve, and
6) higher coil Q sharpens the radiation-efficiency curves, resulting in the coil position being more critical for optimum radiation efficiency.

Note that the radiation-efficiency curves reach a peak and then begin to decline as the loading coil is raised farther up the antenna. This is because of the rapid increase in loading coil reactance required above the antenna center. Refer to Fig 12. The rapid increase in coil size required for resonance results in the coil loss resistance increasing much more rapidly than the radiation resistance. This results in decreased radiation efficiency, as shown in Fig 11.

A slight reverse curvature exists in the curves between the base-loaded position and the one-foot coil-height position. This is caused by a shift in the curve resulting from insertion

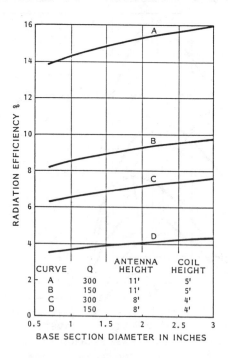

Fig 17—Radiation efficiency plotted as a function of base-section diameter. Frequency = 3.9 MHz, ground loss resistance = 2 ohms, and whip section = ¼-in. diameter.

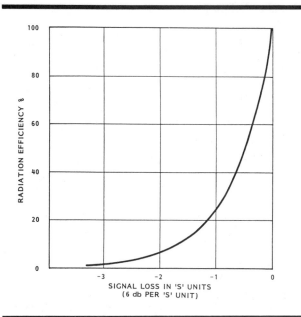

RADIATION EFFICIENCY %

SIGNAL LOSS IN 'S' UNITS
(6 db PER 'S' UNIT)

Fig 18—Mobile antenna signal loss as a function of radiation efficiency, compared to a quarter-wave vertical antenna over perfect ground.

mobiles can be made fairly competitive with those from fixed stations using comparable power.

Impedance Matching

The input impedance of short, high-Q coil-loaded antennas is quite low. For example, an 8-foot antenna optimized for 3.9 MHz with a coil Q of 300 and a ground loss resistance of two ohms has a base input impedance of about 13 ohms. This low impedance value causes a standing wave ratio of 4:1 on a 52-ohm coaxial line at resonance. This high SWR is not compatible with the requirements of solid-state transmitters. Also, the bandwidth of shortened vertical antennas is very narrow. This severely limits the capability to maintain transmitter loading over even a small frequency range.

Impedance matching can be accomplished by means of L networks or impedance-matching transformers, but the narrow bandwidth limitation remains. A more elegant solution to the impedance matching and narrow bandwidth problem is to install an automatic tuner at the antenna base. Such a device matches the antenna and coaxial line automatically, and permits operation over a wide frequency range. Another option is a device such as the KØYEH "Dollar Special" discussed later in this chapter.

In summary, mathematical modeling with a personal computer reveals that loading-coil Q factor and ground loss resistance greatly influence the optimum loading coil position in a short vertical antenna. It also shows that longer antennas, higher coil Q, and higher operating frequencies produce higher radiation efficiencies.

The tools are now available to tailor a mobile antenna design to produce maximum radiation efficiency. One of the missing elements is the availability of very high-Q commercial coils.

End effect has not been included in any of the equations

to assure that the loading coil will be slightly larger than necessary. Pruning the antenna to resonance should be done only by removing coil turns, rather than by shorting turns or shortening the whip section. Shortening the whip section reduces radiation efficiency, by both shortening the antenna and moving the optimum coil position. Shorting turns degrades the coil Q factor.

Shortened Dipoles

The mathematical modeling technique can be applied to shortened dipoles by using zero ground loss resistance and by doubling the computed values of radiation resistance and feed-point impedance. Radiation efficiency, however, does not double. Rather, it remains unchanged, because a second loading coil is required in the other leg of the dipole. The addition of the second coil offsets the gain in efficiency that occurs when the feed-point impedance and radiation resistance are doubled. There is a gain in radiation efficiency over a vertical antenna worked against ground, though, because the dipole configuration allows ground loss resistance to be eliminated from the calculations.

CONTINUOUSLY LOADED ANTENNAS

The design of high-Q air core inductors for RF work is complicated by the number of parameters which must be optimized simultaneously. One of these factors which affects coil Q adversely is radiation. Therefore, the possibility of cutting down the other losses while incorporating the coil radiation into that from the rest of the antenna system is an attractive one.

The general approach has been to use a coil made from heavy wire (no. 14 or larger), with length to diameter ratios as high as 21. British experimenters have reported good results with 8-foot overall lengths on the 1.8- and 3.5-MHz bands. The idea of making the entire antenna out of one section of coil has also been tried with some success. This technique is referred to as linear loading. Further information on linear-loaded antennas can be found in Chapter 6.

While going to extremes in trying to find a perfect loading arrangement may not improve antenna performance very much, a poor system with lossy coils and high-resistance connections must be avoided if a reasonable signal is to be radiated.

MATCHING TO THE TRANSMITTER

Most modern transmitters require a 52-ohm output load, and because the feed-point impedance of a mobile whip is quite low, a matching network is usually necessary. Although calculations are helpful in the initial design, considerable experimenting is often necessary in final tune-up. This is particularly true for the lower bands, where the antenna is electrically short compared with a quarter-wave whip. The reason is that the loading coil is required to tune out a very large capacitive reactance, and even small changes in component values result in large reactance variations. Since the feed-point resistance is low to begin with, the problem is even more aggravated. This is one reason why it is advisable to guy the antenna and to make sure that no conductors such as overhead wires are near the whip during tune-up.

Transforming the low resistance of the whip to a value

suitable for a 52-ohm system can be accomplished with an RF transformer or with a shunt-feed arrangement, such as an L network. The latter may only require one extra component at the base of the whip, since the circuit of the antenna itself may be used as part of the network. The following example illustrates the calculations involved.

Assume that a center-loaded whip antenna, 8.5 feet in overall length, is to be used on 7.2 MHz. From Table 1, earlier in this chapter, we see that the feed-point resistance of the antenna will be approximately 19 ohms, and from Fig 5 that the capacitance of the whip, as seen at its base, is approximately 24 pF. Since the antenna is to be center loaded, the capacitance value of the section above the coil will be cut approximately in half, to 12 pF. From this, it may be calculated that a center-loading inductor of 40.7 μH is required to resonate the antenna, that is, to cancel out the capacitive reactance. (This figure agrees with the approximate value of 40 μH shown in Table 1. The resulting feed-point impedance would then be 19 + j0 ohms—a good match, *if* one happens to have a supply of 19-ohm coax.

Solution: The antenna can be matched to a 52-ohm line by tuning it either above or below resonance and then canceling out the undesired component with an appropriate shunt element, inductive or capacitive. The way in which the impedance is transformed up can be seen by plotting the admittance of the series RLC circuit made up of the loading coil, antenna capacitance, and feed-point resistance. Such a plot is shown in Fig 19 for a constant feed-point resistance of 19 ohms. There are two points of interest, P1 and P2, where the input conductance is 19.2 millisiemens, which corresponds to 52 ohms. The undesired susceptance is shown as $1/X_p$ and $-1/X_p$, which must be canceled with a shunt element of the opposite sign, but with the same magnitude. The value of the canceling shunt reactance, X_p, may be found from the formula:

$$X_p = \frac{R_f Z_0}{\sqrt{R_f (Z_0 - R_f)}}$$

where X_p is the reactance in ohms, R_f is the feed-point resistance, and Z_0 is the feed-line impedance. For Z_0 = 52 ohms and R_f = 19 ohms, X_p = ±39.5 ohms. A coil or good quality mica capacitor may be used as the shunt element. With the tune-up procedure described later, the value is not critical, and a fixed-value component may be used.

To arrive at point P1, the value of the center loading-coil inductance would be less than that required for resonance. The feed-point impedance would then appear capacitive, and an inductive shunt-matching element would then be required. To arrive at point P2, the center loading coil should be more inductive than required for resonance, and the shunt element would need to be capacitive.

The value of the center loading coil required for the shunt-matched and resonated condition may be determined from the equation:

$$L = \frac{10^6}{4\pi^2 f^2 c} \pm \frac{X_s}{2\pi f}$$

where addition is performed if a capacitive shunt is to be used, or subtraction performed if the shunt is inductive, and where L is in μH, f is the frequency in MHz, C is the capacitance of the antenna section being matched in pF, and

$$X_s = \sqrt{R_f (Z_0 - R_f)}$$

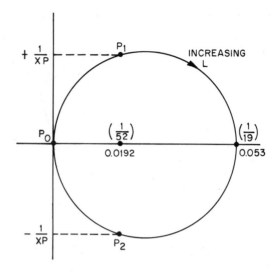

Fig 19—Admittance diagram of the RLC circuit consisting of the whip capacitance, radiation resistance and loading coil discussed in text. The horizontal axis represents conductance, and the vertical axis susceptance. The point P_0 is the input admittance with no whip loading inductance. Points P1 and P2 are described in the text. The conductance equals the reciprocal of the resistance, if no reactive components are present. For a series RX circuit, the conductance is given by

$$G = \frac{R}{R^2 + X^2}$$

and the susceptance is given by

$$B = \frac{-X}{R^2 + X^2}$$

Consequently, a parallel equivalent GB circuit of the series RX one can be found which makes computations easier. This is because conductances and susceptances add in parallel the same way resistances and reactances add in series.

For the example given, where Z_0 = 52 Ω, R_f = 19 Ω, f = 7.2 MHz, and C = 12 pF, X_s is found to be 25.0 ohms. The required loading inductance is either 40.2 μH or 41.3 μH, depending on the type of shunt. The various matching configurations for this example are shown in Fig 20. At A, the antenna is shown as tuned to resonance with L_L, a 40.7 μH coil, but with no provisions included for matching the resulting 19-ohm impedance to the 52-ohm line. At B, L_L has been reduced to 40.2 μH to make the antenna appear capacitive, and L_M, having a reactance of 39.5 ohms, is added in shunt to cancel the capacitive reactance and transform the feed-point impedance to 52 ohms. The arrangement at C is similar to that at B except that L_L has been increased to 41.3 μH, and C_M (a shunt capacitor having a reactance of 39.5 ohms) is added, which also results in a 52-ohm nonreactive termination for the feed line.

The values determined for the loading coil in the above example point out an important consideration concerning the matching of short antennas—that relatively small changes in values of the loading components will have a greatly magnified effect on the matching requirements. A change of less than 3% in the loading-coil inductance value necessitates a completely different matching network! Likewise, calculations show that a 3% change in antenna capacitance will give similar results, and the value of the precautions mentioned earlier becomes clear. The sensitivity of the circuit with regard to frequency variations is also quite critical, and an excursion around practically the entire circle in Fig 19 may represent only 600 kHz, centered around 7.2 MHz, for the above example. This is why tuning up a mobile antenna can be very frustrating unless a systematic procedure is followed.

Tune-Up

Assume that inductive shunt matching is to be used with the antenna in the previous example, Fig 20B, where 39.5 Ω is needed for L_M. This means that at 7.2 MHz, a coil of 0.87 μH will be needed across the whip terminals to ground. With a 40-μH loading coil in place, the adjustable whip section above the loading coil should be set for minimum height. Signals in the receiver will sound weak and the whip should be lengthened a bit at a time until signals start to peak. Turn the transmitter on and check the SWR at a few frequencies to find where a minimum occurs. If it is below the desired frequency, shorten the whip slightly and check again. It should be moved approximately ¼ inch at a time until the SWR is minimum at the center of the desired range. If the frequency where the minimum SWR occurs is above the desired frequency, repeat the above, but lengthen the whip only slightly.

If a shunt capacitance is to be used, as in Fig 20C, a value of 560 pF would correspond to the required 39.5 ohms of reactance at 7.2 MHz. With a capacitive shunt, start with the whip in its longest position and shorten it until signals peak up.

TOP-LOADING CAPACITANCE

Because the coil resistance varies with the inductance of the loading coil, the resistance can be reduced by removing turns from the coil. This can be compensated by adding capacitance to the portion of the mobile antenna that is *above* the loading coil (Fig 21). To achieve resonance, the inductance

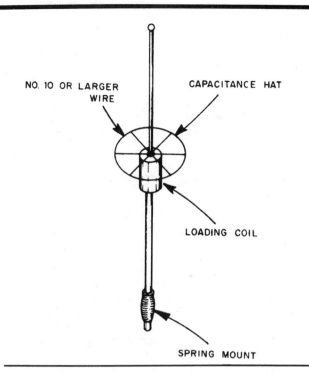

Fig 21—A capacitance hat can be used to improve the performance of base- or center-loaded whips. A solid metal disc can be used in place of the skeleton disc shown here.

of the coil is reduced proportionally. Capacitance hats, as they are called, can consist of a single stiff wire, two wires or more, or a disc made up of several wires like the spokes of a wheel. A solid metal disc can also be used, but is less practical for mobile work. The larger the capacitance hat (physically), the greater is the capacitance. The greater the capacitance, the less is the inductance required for resonance at a given frequency.

Capacitance-hat loading is applicable in either base-loaded or center-loaded systems. Since more inductance is required for center-loaded whips to make them resonant at a given frequency, capacitance hats are particularly useful in improving their efficiency.

TAPPED-COIL MATCHING NETWORK

Some of the drawbacks of the previous circuits can be eliminated by the use of the tapped-coil arrangement shown in Fig 22. While tune-up is still critical, a smaller loading coil is required, reducing the coil losses. Coil L2 can be inside the car body, at the base of the antenna, or at the base of the whip. As L2 helps determine the resonance of the antenna, L1 should be tuned to resonance in the desired part of the band with L2 in the circuit. The top section of the whip can be telescoped until a field-strength maximum is found. The tap on L2 is then adjusted for the lowest reflected power. Repeat these two adjustments until no further increase in field strength can be obtained; this point should coincide with the lowest SWR. The number of turns needed for L2 will have to be determined experimentally. The construction project that follows uses this technique.

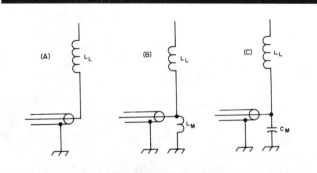

Fig 20—At A, a whip antenna which is resonated with a center loading coil. At B and C, the value of the loading coil has been altered slightly to make the feed-point impedance appear reactive, and a matching component is added in shunt to cancel the reactance. This provides an impedance transformation to match the Z_0 of the feed line. An equally acceptable procedure, rather than altering the loading coil inductance, is to adjust the length of the top section above the loading coil for the best match, as described in the tune-up section of the text.

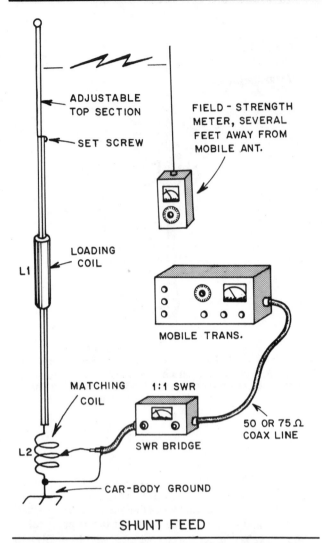

SHUNT FEED

Fig 22—A mobile antenna using shunt-feed matching. Overall antenna resonance is determined by the combination of L1 and L2. Antenna resonance is set by pruning the turns of L1, or adjusting the top section of the whip, while observing the field-strength meter or SWR indicator. Then adjust the tap on L2 for the lowest SWR.

MOBILE IMPEDANCE-MATCHING COIL

This shunt-feed impedance-matching arrangement for HF mobile antennas was designed by Bob Hawk, KØYEH, and has been dubbed the KØYEH Dollar Special. Its primary purpose is to provide a very efficient match to 52-Ω coax line, and not to base-load the antenna. The antenna itself should already be resonated for the band of operation, such as with base loading, center loading, or a resonator. See Fig 23. (A resonator provides a form of near-center loading, but uses a loading *assembly* that consists of a coil and a short radiator section.) Of these methods, base loading is the least efficient. Center loading and ⅔ loading (coil ⅔ the distance from the base to the tip) are more efficient, and the latter is recommended.

The Dollar Special is a great performer, fun to build, and costs only about a dollar for parts. If you have a junk box, you probably already have just about everything you'll need.

With the matcher properly installed and adjusted, you will be able to get on any of the HF bands (3.5 through 29.7 MHz) for which your antenna is designed, with a 1:1 SWR. You will be able to work stations with better mobile signal reports (both transmitted and received) than you've had before—especially if you have not used good impedance matching at the feed point in the past. KØYEH has consistently received flattering signal reports since he began using this matching method.

The matching unit is shown in Fig 24. It adapts easily to passenger cars, pickup trucks, vans, trailers and RVs, *as long as there is a metal body for a ground.* Body mounts are better than bumper mounts for a number of reasons. The matcher can, however, be bumper mounted, and will perform well in this configuration.

The coil is easy and efficient to use. After having made a prior tune-up and adjustment with the alligator clip for the best coil-turn position for each band, the turn position may be marked with fingernail polish. Band changing may then be done easily, usually within no more than three or four minutes (depending on the length of time you need to adjust your antenna or change resonators).

Design Philosophy

It has been well documented that the 52-ohm output of an HF transmitter (or transceiver) looks at a huge mismatch at the feed point of a mobile antenna. For example, a 3.5-MHz mobile whip, resonated as an electrical quarter-wave

Fig 23—Bob Hawk showing off his mobile antenna, which uses the KØYEH Dollar Special matching unit.

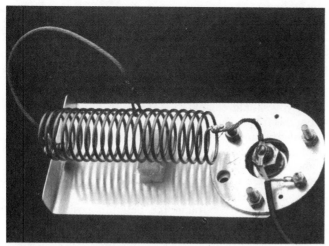

Fig 24—The assembled Dollar Special, ready for mounting.

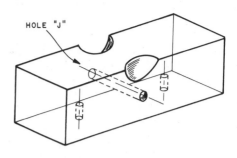

Fig 26—Insulated standoff (to support the coil). Mount on the base at holes E and F with two small (no. 5) sheet-metal screws. Trim the top center, as shown, to about 1/8 to 1/4 in. wide. The insulation block is about 1/2 in. square × 1-3/4 in. Drill a 1/8-in. hole at J (for fishing line) to tie to the bottom of one coil turn.

antenna, typically presents a load impedance of about 8 ohms, which represents a mismatch of more than 6 to 1! Similar mismatches (but of lesser magnitudes as frequency is increased) occur on the higher bands as well.

Partial (apparent or false) matches may be achieved by detuning the resonator from the true quarter-wave resonance position, but this results in a badly altered radiation pattern, with significant high-angle radiation and poor transmitted and received signals.

With the Dollar Special and resonator properly adjusted, a 1:1 match may be accomplished, resulting in the antenna being tuned for optimum performance as a true quarter wave. The same 1:1 match and the same optimum results are attainable on all the bands for which your antenna is designed to work.

Construction

Fig 25 shows a drilling template for the matching-coil assembly, and Fig 26 shows details of the insulation standoff block. Table 4 contains a list of materials needed for the Dollar Special.

Carefully lay out the reference lines on the base plate, using a needle-point scribe and a ruler. Mark and drill all holes. The large 1.34-inch dia hole may be cut out with a

Table 4

Materials Needed to Construct the Dollar Special

1) Aluminum or brass sheet 3-3/4 in. wide, 7-3/4 in. long, and about 0.040 to 0.050 in. thick.
2) One 9-1/2 foot length of no. 10 solid copper wire.
3) Flexible braid about 5/16 to 3/8 in. wide: one length 7-1/2 in. long with a terminal for a no. 10 metal bolt on one end and a no. 30c Mueller clip (small copper alligator clip) soldered to the other end. The second piece of braid should be 3-3/8 in. long with a terminal for a no. 8 screw soldered to one end.
4) One piece of dielectric (insulating) material about 1/2 to 5/8 in. square and about 2 in. long. This can be plastic such as nylon, Teflon, polyethylene or phenolic, or dry wood (if wood, preferably painted or boiled in paraffin).
5) One no. 10-32 x 3/4 in. bolt, three star washers, two flat washers, and one lock washer.
6) Two no. 5 sheet-metal screws, 3/8-in. long, to mount the dielectric standoff at points X and Y.

nibbling tool or a chassis punch, or you can drill several holes in the area and file them out. After drilling and cutting all holes, make a 90° bend at bend-line Z.

To form the loading coil, find a piece of 1¼-inch dia tubing or pipe that is at least a foot long. Use it as a core, and wind the no. 10 copper wire tightly around it. About 20½ turns make up the coil. After winding, carefully spread the coil turns as evenly as possible so that the coil is 5 inches long with 20 turns. On one end of the coil, fashion a loop to fit snugly around the 3/16-inch bolt. (This bolt will be attached at point D, shown at the left in Fig 25.)

Bend the extra ½ turn at the feed-point end of the coil at a 45° angle (about ½ inch from the end of the 20th turn) and cut off the excess. Attach the end of the 2-3/8-inch length of braid at this point and solder. Wrapping the joint with fine solid copper wire (about no. 24) before soldering makes the soldering job easier.

Fabricate the standoff insulator as shown in Fig 26. With a file or knife, remove material at the top center, as shown, to avoid sharp edges against the coil tie-down material. Next

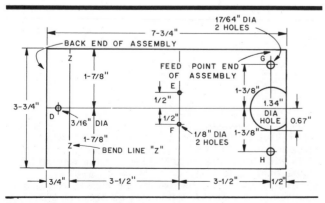

Fig 25—Drilling template for the base plate of the Dollar Special.

mount the dielectric standoff to the base at points E and F using two no. 5 screws. In mounting the dielectric piece, make sure that hole J is parallel to the base plate and to the axis of the mounted coil.

Secure the ground end of the coil and the terminal end of the 7-inch length of braid at hole D with the 3/16-inch bolt assembly. Connect the one bottom turn of the coil to the standoff with a 2- or 3-foot piece of cord or string (fishing line works well) through hole J.

Your Dollar Special should now be complete and ready for mounting. The secret of the outstanding performance of an antenna equipped with the Dollar Special is good grounding. Be sure to observe the precautions given in the next section about removing paint from the vehicle body.

Mounting the Matcher

The Dollar Special is easily mounted. If you have a standard (preferably heavy duty) swivel mount on your vehicle, remove two of the (usually 3) bolts from the mount and slip the base of the matcher underneath the heavy ring plate (approximately 4 inches dia). Connect the "hot" (feed-point) end of the coil, with attached terminal, to the same feed-point connector as the center conductor of the coax. Make sure the shield of the coax is grounded to the large mounting ring with a short length of the shield braid (two inches long or less).

Make sure the hole in the matcher base (about 1.34 inches) is properly centered so it does not touch and short out the center bolt assembly of the antenna. It is a good idea to make sure you have at least about 1/8 to 3/16 inch of clearance here.

It is a good idea to remove the antenna mount completely and remove all the paint and primer from at least a 1-inch dia area around each of the bolt holes on the inner side of the mount. It is important to obtain the best possible ground to the vehicle body. "No ground—no work!" is an axiom KØYEH has stressed to the dozens of amateurs he has helped install the Dollar Special. Star washers should be used between all contacting surfaces, and the hardware must be tightened well.

If you do not have a standard mount, make the appropriate connections to the antenna you are using based on these instructions. Mounting may take a little creativity, but the Dollar Special can be made to work with virtually any kind of mobile antenna.

Tune-Up

Place an SWR meter in the transmission line at the output of the transceiver. To avoid possible damage to the final amplifier and to prevent any unnecessary interference, tune-up should be done with the SWR meter at maximum sensitivity and the RF drive adjustments at no more than necessary to get an accurate SWR indication. Because 7 MHz is one of the most popular mobile bands, it is best to begin the tune-up procedure there. (Adjustment of each of the other bands is similar.)

First, move the alligator clip on the matching coil to the eighth turn from the feed-point end of the coil. Make a few spot SWR checks and determine where the SWR is lowest. If the SWR improves as you move toward the top of the band (7.3 MHz), you'll need to lengthen the resonator whip a small amount or use more inductance in your center loading coil. Conversely, if you find the SWR best at the bottom of the band, you will need to shorten the whip or use less

center-loading inductance. You will also need to move the alligator clip on the Dollar Special coil (check the SWR while you do this) a turn or half-turn at a time until you eventually find the coil-tap position that yields the best match.

After you have completed the tuning, the SWR should be at (or near) 1:1 at the desired frequency. On the 7-MHz band you should be able to move 10 to 15 kHz either way from this frequency with less than a 1.5:1 SWR. On 14 MHz and higher bands, you should be able to work the entire SSB subband with less than a 1.5:1 SWR. (These figures will vary somewhat depending on the antenna that you are using, but these numbers are typical for a Hustler antenna.) Measured SWR curves are shown in Fig 27.

Once you have found the best tap position on the matching coil for 7 MHz, mark it with red enamel or fingernail polish. This single tap position on the matching coil should be usable across the entire 7-MHz band. Frequency excursions of more than 15 or 20 kHz from the center of the desired frequency range will require changing the length of the whip top section accordingly.

The other bands are tuned in a similar manner. Approximate tap positions on the matching coil for the other bands (counting from the feed-point end of the coil) are as follows.

    3.5 MHz—15 turns
    7 MHz—8 turns
    14 MHz—4½ turns
    21 MHz—3 turns
    28 MHz—2 turns

Once you have the Dollar Special installed and tuned properly, you can expect good success in your HF mobile activities.

Most commercially made masts (Hustler, for example) are 4.5 feet long, and are made of approximately ½-inch OD tubing with 3/8-inch × 24 threaded fittings. If you are fortunate enough to find the material and have the capability, make a 1½ foot extension to add to the top of your mast, or else use a 6-foot mast. Your reward will be significantly improved operation on the 3.5- through 21-MHz bands. Fig 23 shows one of these masts ready for use.

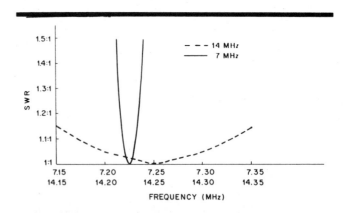

Fig 27—Typical SWR curves for the 7- and 14-MHz amateur bands. At 14 MHz, with adjustment centered on 14.25 MHz, the entire voice band is covered with an SWR of less than 1.2 to 1. Operation is similar at 21 and 28 MHz. At 7 MHz, the bandwidth is narrower, which is also true for 3.5 MHz. A match may be obtained after significant frequency shifts in these bands by adjusting the antenna resonator.

TWO-BAND HF ANTENNA WITH AUTOMATIC BAND SWITCHING

A popular HF mobile antenna is a center-loaded whip consisting of a loading coil mounted 2 to 4 feet from the base, with a whip atop the coil. A shorting-tap wire is provided to short out turns at the bottom end of the coil, bringing the antenna to resonance. Another popular scheme uses a resonator, consisting of a coil and a short top section, mounted on the short mast.

It is obvious that to change bands with these HF antennas, the operator must stop the car, get out, and change the coil tap or resonator. Further, if a matching arrangement is used in the trunk of the vehicle at the antenna mount (such as a shunt L or shunt C), the matching reactance must also be changed. The antenna described in this section was developed by William T. Schrader, K2TNO, as a means of providing instant band changing.

One approach to instant band changing is to install a pair of relays, one to switch the loading-coil tap and one to switch the matching reactance. (Of course, this is not practical with an antenna resonator.) In addition to the problem of running relay lines through the passenger compartment, this approach is a poor one because the coil-tap changing relay would need to be at the coil, adding weight and wind load to an already cumbersome antenna. Furthermore, that relay would need to be sealed, as it would be exposed to the weather.

The solution to be applied here allows automatic band changing, depending upon only the frequency of the signal applied to the antenna. The antenna described here provides gratifying results; it shows an SWR of less than 1.2:1 at both the 7- and 14-MHz design frequencies. The chosen method employs two resonant circuits, one that switches the matching capacitance in and out, and one that either shorts or opens turns of the coil, depending upon the excitation frequency. See Fig 28.

Coil-Tap Switching

A series LC circuit looks electrically like a dead short at its resonant frequency. Below that frequency it presents a capacitive reactance; above resonance it looks inductive. A series resonant network, L2-C1, is resonant at the 14-MHz design frequency. One end of C1 is connected to the 14-MHz tap point on the coil, and the other end is connected to the bottom of the coil. On 14 MHz, the network looks like a short circuit and shorts out the unwanted turns at the bottom end of the coil. At 7 MHz the network is not a short, and therefore opens the bottom turns (but adds some reactance to the antenna).

A coil-tapping clip is soldered to the stud at one end of C1. The other end of C1 is connected to L2. A dip meter is used to prune L2 until the L2-C1 network is resonant somewhere in the 14-MHz CW band. The design of the plastic supports on L2 limits pruning of the coil to ¼-turn increments. One lead of L2 should be cut close to the plastic and the short pigtail attached with a machine screw to the capacitor stud. The far end of L2 should have a long pigtail (about 5 to 6 inches) to secure the lower end of the network to the bottom of the antenna loading coil, L1. While resonating the network, the long pigtail can be bent around to clip to the top of the capacitor.

Any doorknob capacitor between about 25 and 100 pF could be used for C1. The lower the value of C, the larger the coil inductance will need to be. A 1000-V silver mica capacitor would also work, but the doorknob is preferred

because of the mechanical stability it provides.

The LC network should be mounted to the main coil, with the lower coil pigtail extended down roughly parallel to the main coil. Some turns adjustment will be required, so this pigtail should not be tight. The mounting details are visible in Fig 29.

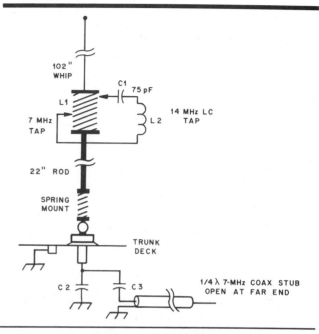

Fig 28—Details of the 7- and 14-MHz mobile antenna.
C1—Ceramic doorknob capacitor. See text.
C2—14 MHz matching. See text for determining value. May be made up of two or three parallel-connected 1000-V silver mica capacitors to obtain the required value.
C3—7 MHz matching, parallel-connected 1000-V silver mica capacitors or air variable. See text.
L1—Multiband center loading coil for mobile antenna.
L2—8½ turn coil (B & W no. 3046), 1¼ in. diameter, 6 turns per inch.

Fig 29—Close-up of the mounting arrangement of the 14-MHz LC network on the main tuning coil. (The antenna was pulled to a nearly horizontal position and the camera tilted slightly for this photograph.)

Tuning the Antenna

Once the LC network is attached, the antenna must be tuned for the 7- and 14-MHz bands. This job *requires* the use of an impedance-measuring device such as an R-X noise bridge (home-built or commercial, either is fine). As with many antenna projects, you're just wasting your time if you try to do the job with an SWR meter alone. Prepare a length of coax feed line which is an electrical half wavelength at the 7-MHz design frequency. Do not attempt to use the vehicle coax feed line unless you want to do a lot of Smith Chart calculations.

Once the special feed line is attached, install the impedance bridge and begin the tuning as follows. The antenna must first be resonated to each band by adjusting the taps on L1 first for 7 and then 14 MHz. Mark these two tap locations on the coil. Then using the steps that follow, perform tuning for the 14-MHz design frequency.

1) Move the 7-MHz tap wire up the coil to a new position that leaves about 60% of the original turns unshorted.

2) Listen at 14 MHz and adjust the impedance bridge for a null. The reactance dial should show capacitive reactance. Move the LC-network tap point down the coil about ¼ turn at a time until the bridge indicates pure resistance.

3) Switch to 7 MHz and follow the same procedure. On this band, move the shorting wire about ½ turn at a time. Do not be surprised if it takes some hunting to find resonance; tuning is very critical on 7 MHz.

4) The two adjustments interact; repeat steps 2 and 3 of this section for both bands until the measured impedance is purely resistive at both design frequencies.

5) Remove the impedance bridge and install an SWR meter. Determine the SWR on both bands. The minimum SWR should be about 1.5:1 on 14 MHz and about 2.2:1 on 7 MHz. Shift the VFO frequency about 10 kHz above and below the design frequencies on both bands to verify that the minimum SWR occurs at the design frequencies. Do not expect the minimum SWR to be 1:1, because the antenna is not yet matched to the line. Alternate bands and adjust the two taps slightly for minimum SWR at the desired frequencies for both bands.

6) Record the SWR and tap points for both bands. This completes the adjustments for the resonating work.

Designing the Matching Networks

Since the feed-point impedance is not 52 Ω on either band, a matching network is needed for each. The simple approach is to ignore the mismatch and let 'er rip with an amplifier! This strategy is known as the ''watts are simpler than brains'' approach.

Matching can be done easily with an L network, as described in Chapter 25. This network consists of a shunt capacitor from the antenna feed point to ground and a compensating increase in the coil inductance of L1, obtained by moving the tap slightly. The value of the matching capacitor is calculated by knowing R_A, the antenna feed-point resistance at resonance, Z_0, the impedance of the coax feed line, and f, the operating frequency in kHz.

1) Calculate the antenna feed-point resistance from the relationship SWR = Z_0/R_A. Do this calculation for both bands. For the antenna Schrader constructed, values of R_A were 33.3 Ω on 14 MHz and 21.4 Ω on 7 MHz.

2) Calculate the value for C2, the 14-MHz matching capacitor. This is the value obtained for C_M from

$$C_M = \frac{\sqrt{R_A (Z_0 - R_A)}}{2\pi f Z_0 R_A} \times 10^6 \qquad \text{(Eq 1)}$$

where

C_M is the matching capacitance in pF
R_A and Z_0 are in ohms
f is in MHz

Using Schrader's value of R_A as an example, the capacitance is calculated as follows.

$$C_M = \frac{\sqrt{33.3 (52 - 33.3)}}{2\pi \times 14.06 \times 52 \times 33.3} \times 10^6 = 163 \text{ pF}$$

This is the value for C2. A practical value is 160 pF.

3) From Eq 1, calculate the total matching capacitance required for 7 MHz. Again using Schrader's value,

$$C_M = \frac{\sqrt{21.4 (52 - 21.4)}}{2\pi \times 7.06 \times 52 \times 21.4} \times 10^6 = 518 \text{ pF}$$

4) Because C2 is present in the matching circuit at both 7 and 14 MHz, the value of C3 is not the C_M value just calculated. Calculate the value of C3 from

C3 = C_M − C2

where C_M is the value calculated in step 3 of this section. In this example, C3 = 518 − 163 = 355 pF.

Final Tuning of the Antenna

Install C2 from the antenna feed point to ground. Now readjust the tap point of the 14-MHz LC network to add just enough additional inductance to give a 52-Ω feed-point resistance. The tap point will be moved down (more turns in use) as the match is approached.

1) Attach the SWR meter and apply RF at 14 MHz (10 watts or so). Note that the SWR is higher than it was before C2 was added.

2) Move the tap point down the coil about 1/8 turn at a time. Eventually the SWR will begin to fall, and there will be a point where it approaches 1:1. For the antenna in the photos, almost a full additional coil turn was necessary on 14 MHz.

3) Verify (by shifting the VFO) that the minimum SWR occurs at the 14-MHz design frequency. Adjust the tap point until this condition is met. Note: If the SWR never falls to nearly 1:1, either C_M was miscalculated, the SWR was not measured correctly, the antenna was not resonant, or the coax feed line was not actually ½ wavelength long on 7 MHz.

4) Add C3 in parallel with C2. Repeat steps 2 and 3 of this section at 7 MHz, moving the 7-MHz tap wire.

5) Recheck 14 and 7 MHz. Both bands should now show a low SWR (less than 1.2:1) at the design frequencies. Note: The grounded end of C3 must be lifted when you recheck 14 MHz and reconnected for 7 MHz.

Now the antenna is resonant and properly matched on both bands, but C3 must be manually grounded and ungrounded to change bands. This problem may be solved as described below.

Matching Capacitor Switching

A length of coaxial cable (any impedance) that is exactly one-quarter wavelength long at a given frequency and is open-circuited at its far end will be resonant at that frequency. At this frequency, the input end of the coax appears as a dead short. If a signal of twice the frequency is applied, the line is ½ λ long at that frequency, and the input terminals of the line are not shorted, but rather present a very high impedance (an open circuit, in theory). This property of quarter- and half-wavelength transmission lines can be used as a switch in this antenna, because the two frequencies in use are harmonically related.

Cut a length of RG-58 to resonate at the 7-MHz design frequency (about 22 feet), and leave the far end open. High RF voltages exist at this end, so it is a good idea to insulate it. Strip back the braid about 3/16 inch and tape the end of the cable. This length of coax acts as a switch to either ground or lift the low side of C3.

Connect one lead of C3 to the antenna feed point, and the other end to the center conductor of the coax stub, as shown in the diagram of Fig 28. Ground the braid of the coax at the base of the antenna. This circuit grounds the low end of the capacitor on 7.060 MHz, but opens it on 14.060 MHz automatically, depending on the frequency of the signal applied to it. Details of the matching network are shown in Fig 30.

Coil the coax stub and place it out of the way (in the trunk or wherever is convenient). Coiling does not affect stub tuning at all.

Operation of the Antenna

With antenna adjustments completed, remove the ½-wavelength feed line and reinstall the regular feed line. The antenna should now be operable on either band with a very low SWR. Because of the high Q of the open-wire coil and the antenna, bandwidth is limited on 7 MHz. A Transmatch can be used to allow wide frequency excursions. If only a small segment of the 7-MHz band is to be used, no Transmatch is necessary.

The L2-C1 network should be positioned behind the main coil for minimum wind buffeting. As its attachment point is dictated by the electrical requirements, the network can be rotated behind the coil by installing a washer on the 3/8-inch × 24 stud where the bottom of the coil is attached to the lower mast. The antenna is shown installed on a vehicle in Fig 31.

Orientation of the tap wire and the LC bottom tap wire have a large effect on tuning. Be sure to orient these leads during tuning in the same way that you will when using the antenna.

SWR measurements have been made with various tap positions of both the 14-MHz LC tap and the 7-MHz tap wire. The results are summarized in Table 5. With the matching and switching system installed as described, the antenna showed an SWR of 1:1 at the transmitter on both bands. The 2:1 SWR bandwidth was about 40 kHz on 7 MHz, and over 350 kHz on 14 MHz.

Table 5 includes typical coil-tap settings for changing the resonant frequency on both bands. Exact tap positions will depend upon the geometry of the antenna, its position on the vehicle and the arrangement of the leads themselves. The table also shows how the two band adjustments interact. For

Fig 31—This photo shows the antenna mounted on the trunk of a car. The structure is somewhat cumbersome, so it is guyed appropriately.

Fig 30—Details of the matching network located at the base of the antenna inside the vehicle. The mica capacitors are visible at the center. The coaxial stub used to switch them in and out of the circuit comes in from the left, and the feed line exits toward the bottom of the photo.

Table 5

**Coil Tap Positions for the
Two-Band Mobile Antenna**

Unshorted Turns[1]		Resonant Frequency (MHz)[2]	
14-MHz LC	7-MHz Tap	14-MHz Band	7-MHz Band
6¼	11½	14.190	7.267
	11¾	14.170	7.144
	12	14.160	7.104
	12¼	—	7.034
6½	12	14.085	7.207
	12¼	14.070	7.080
	12½	14.020	7.005

[1]Turns in use (measured from the top of the coil).
[2]Frequency at which SWR is 1:1.

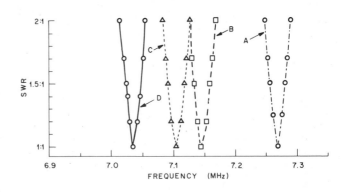

Fig 33—SWR curves for the antenna in the 7-MHz band. The 14-MHz LC tap was 6¼ turns from the top. The 7-MHz tap for curve A was set at 11½ turns, 11¾ turns for B, 12 turns for C, and 12¼ turns for D.

example, with the 14-MHz LC tap at 6¼ turns, changing the 7-MHz tap from 11½ to 12 turns moved the 7-MHz resonance point from 7.267 to 7.104 MHz. There was also a 30-kHz change in the 14-MHz resonance point, from 14.190 to 14.160 MHz. The inverse effect was even more pronounced. With 12 turns in use for the 7-MHz tap, moving the 14-MHz LC tap from 6¼ to 6½ turns altered the 14-MHz frequency from 14.160 to 14.085 MHz. Simultaneously the 7-MHz resonant frequency shifted from 7.104 to 7.207 MHz. Thus, both settings interact strongly.

Since the bandwidth on 14 MHz is nearly sufficient to cover the entire amateur band without adjustment, the settings of the 14-MHz LC network are not very critical. However, as Table 5 shows, slight readjustments of either tap will have marked effects upon 7-MHz performance.

Typical SWR curves for the two bands are shown in Figs 32 and 33. Fig 32 shows that moving the 14-MHz LC tap point from 6½ to 6¼ turns raised the resonant frequency from 14.040 to 14.168 MHz. The 7-MHz tap was set at 12 turns for these measurements. When the 7-MHz tap was moved to 11½ turns, the 14-MHz resonant frequency was raised to 14.190 MHz. The 14-MHz LC tap was kept constant for the measurements shown in Fig 33, and the difference in resonant frequency that results from moving the 7-MHz tap is shown.

The matching network, L2-C1, is quite broadbanded. Once the feed-point matching capacitors (C_M) and the retuned coil were adjusted, the minimum SWR was 1:1 at all tap settings on both bands. Thus, the matching arrangement does not require adjustment. If a compromise setting is chosen for the 14-MHz LC tap position to allow both CW and SSB operation on that band, only adjustment of the 7-MHz tap will be required during routine operation. To this end, the plots shown in Fig 34 were obtained. The curves show the 7-MHz resonant frequency as a function of tap position. Also included is a plot showing the effect at 7 MHz of altering the 14-MHz LC tap point.

Other Considerations

There is no reason why the strategy described here could

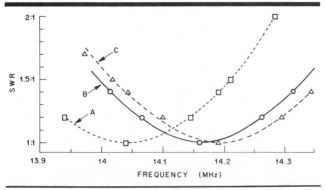

Fig 32—SWR curves for the antenna in the 14-MHz band. The 7-MHz tap was 12 turns from the top. Curves are shown for the 14-MHz LC tap positioned at 6½ turns (A) and at 6¼ turns (B). In the last case, moving the 7-MHz tap to 11½ turns altered the resonant frequency as shown at C.

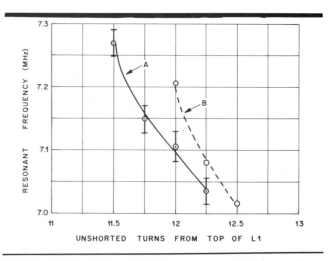

Fig 34—Effect of tap positions on resonant frequency in the 7 MHz band. The 14-MHz LC tap was set at 6¼ turns from the top. At A, each circled dot shows the resonant frequency at which the SWR is 1:1. Bars about each point show the frequency limits at which the SWR is 2:1. The measurements were repeated with the 14-MHz LC tap set at 6½ turns, yielding the circled points on curve B.

not be applied to any two bands, as long as the desired operating frequencies are harmonically related. Other likely candidates would be 3.5-MHz CW/7-MHz CW using a 3.5-MHz coil, 14 MHz/28 MHz using a 14-MHz coil, and 7-MHz SSB/14-MHz SSB. A combination that would probably not work is 3.8-MHz SSB/7-MHz SSB, but it might be worth a try.

The antenna performs very well on the design frequencies. It is too big for routine city use, but it sure makes a great open-highway antenna.

A MOBILE J ANTENNA FOR 144 MHz

The J antenna is a mechanically modified version of the Zepp (Zeppelin) antenna. It consists of a half-wavelength radiator fed by a quarter-wave matching stub. This antenna exhibits an omnidirectional pattern with little high angle radiation, but does not require the ground plane that ¼-wave and 5/8-wave antennas do to work properly. The material in this section was prepared by Domenic Mallozzi, N1DM, and Allan White, W1EYI.

Fig 35 shows two common configurations of the J antenna. Fig 35A shows the shorted-stub version that is usually fed with 200- to 600-ohm open-wire line. Some have attempted to feed this antenna directly with coax, which leads to less than optimum results. Among the problems with this configuration are a lack of reproducibility and heavy coupling with nearby objects. To eliminate these problems, many amateurs have used a 4:1 half-wave balun between the feed point and a coaxial feed line. This simple addition results in an antenna that can be easily reproduced and that does not interact so heavily with surrounding objects. The bottom of the stub may be grounded (for mechanical or other reasons) without impairing the performance of the antenna.

The open-stub-fed J antenna shown in Fig 35B can be connected directly to low-impedance coax lines with good results. The lack of a movable balun (which allows some impedance adjustment) may make this antenna a bit more difficult to adjust for minimum SWR, however.

The Length Factor

Dr. John S. Belrose, VE2CV, noted in *The Canadian Amateur* that the diameter of the radiating element is important to two characteristics of the antenna—its bandwidth and its physical length. (See bibliography at the end of this chapter.) As the element diameter is increased, the usable bandwidth increases, while the physical length of the

radiating element decreases with respect to the free space half-wavelength. The increased diameter makes the end effect more pronounced, and also slows the velocity of propagation on the element. These two effects are related to resonant antenna lengths by a factor, "k." This factor is expressed as a decimal fraction giving the equivalent velocity of propagation on the antenna wire as a function of the ratio of the element diameter to a wavelength. The k factor is discussed at length in Chapter 2.

The length of the radiating element is given by

$$l = \frac{5904 \ k}{f}$$

where

　　l = length in inches
　　f = frequency in MHz
　　k = k factor

The k factor can have a significant effect. For example, if you use a 5/8-inch diameter piece of tubing for the radiator at 144 MHz, the k value is 0.907 (9.3% shorter than a free-space half wavelength).

The J antenna gives excellent results for both mobile and portable work. The mobile described here is similar to an antenna described by W. B. Freely, K6HMS, in April 1977 *QST*. This design uses mechanical components that are easier to obtain. As necessary with all mobile antennas, significant attention has been paid to a strong, reliable, mechanical design. It has survived not only three New England winters, but also two summers of 370-mile weekend commutes. During this time, it has maintained consistent electrical performance with no noticeable deterioration.

The mechanical mount to the bumper is a 2 × 2-inch stainless steel angle iron, ten inches long. It is secured to the bumper with stainless steel hardware, as shown in Fig 36. A stainless steel ½-inch pipe coupling is welded to the left side of the bracket, and an SO-239 connector is mounted at the right side of the bracket. The bracket is mounted to the bumper so a vertical pipe inserted in the coupling will allow

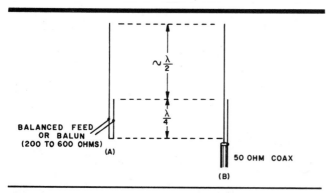

Fig 35—Two configurations of the J antenna.

Fig 36—The mount for the mobile J is made from stainless steel angle stock and secured to the bumper with stainless steel hardware. Note the ½-in. pipe plug and a PL-259 (with a copper disc soldered in its unthreaded end). These protect the mount and connector threads when the antenna is not in use.

the hatchback of the vehicle to be opened with the antenna installed, Fig 37.

A ½-inch galvanized iron pipe supports the antenna so the radiating portion of the J is above the vehicle roof line. This pipe goes into a bakelite insulator block, visible in Fig 37. The insulator block also holds the bottom of the stub. This block was first drilled and then split with a band saw, as shown in Fig 38. After splitting, the two portions are weatherproofed with varnish and rejoined with 10-32 stainless hardware. The corners of the insulator are cut to clear the L sections at the shorted end of the stub.

The quarter-wave matching section is made of ¼-inch type L copper tubing (5/16-inch ID, 3/8-inch OD). The short at the bottom of the stub is made from two copper L-shaped sections and a short length of ¼-inch tubing. Drill a 1/8-inch hole in the bottom of this piece of tubing to drain any water that may enter or condense in the stub.

A 5/16-inch diameter brass rod, 1½ to 2 inches long, is partially threaded with a 5/16 × 24 thread to accept a Larsen whip connector. This rod is then sweated into one of the legs of the quarter-wave matching section. A 40-inch whip is then inserted into the Larsen connector.

The antenna is fed with 52-ohm coaxial line and a coaxial 4:1 half-wave balun. This balun is described in Chapter 26. As with any VHF antenna, use high quality coax for the

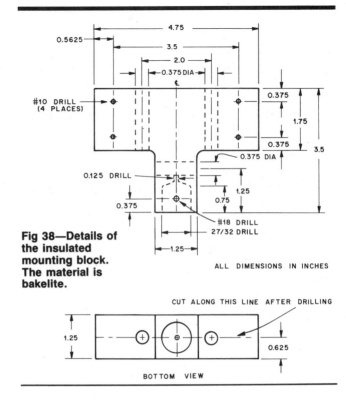

Fig 38—Details of the insulated mounting block. The material is bakelite.

ALL DIMENSIONS IN INCHES

Fig 37—The J antenna, ready for use. Note the bakelite insulator and the method of feed. Tie wraps are used to attach the balun to the mounting block and to hold the coax to the support pipe. Clamps made of flashing copper are used to connect the balun to the J antenna just above the insulating block. The ends of the balun should be weatherproofed.

balun. Seal all open cable ends and the rear of the SO-239 connector on the mount with RTV sealant.

Adjustment is not complicated. Set the whip so that its tip is 41 inches above the open end of the stub, and adjust the balun position for lowest SWR. Then adjust the height of the whip for the lowest SWR at the center frequency you desire. Fig 39 shows the measured SWR of the antenna after adjustments are completed.

THE SUPER-J MARITIME ANTENNA

This 144-MHz vertical antenna doesn't have stringent grounding requirements and can be made from easy to find parts. The material in this section was prepared by Steve

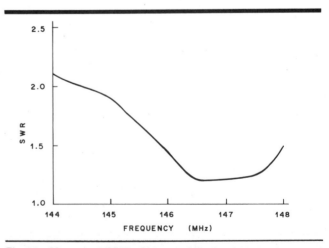

Fig 39—Measured SWR of the mobile J antenna.

Cerwin, WA5FRF, who developed the Super-J for use on his boat.

Antennas for maritime use must overcome difficulties that other kinds of mobile antennas normally do not encounter. For instance, the transom of a boat is the logical place to mount an antenna. But the transoms of many boats are composed mostly of fiberglass, and they ride some distance out of the water—from several inches to a few feet, depending on the size of the vessel. Because the next best thing to a ground plane (the water surface) is more than an appreciable fraction of a wavelength away at 144 MHz, none of the popular gain-producing antenna designs requiring a counterpoise are suitable. Also, since a water surface does a good job of assuming the earth's lowest mean elevation (at least on a calm day), anything that can be done to get the radiating part of the antenna up in the air is helpful.

One answer is the venerable J-pole, with an extra in-phase half-wave section added on top... the Super-J antenna. The two vertical half waves fed in phase give outstanding omnidirectional performance for a portable antenna. Also, the "J" feed arrangement provides the desired insensitivity to height above ground (or water) plus added overall antenna height. Best of all, a ¼-wave CB whip provides enough material to build the whole driven element of the antenna, with a few inches to spare. The antenna has enough bandwidth to cover the entire 144-MHz band, and affords a measure of lightning protection by being a "grounded" design.

Antenna Operation

The antenna is represented schematically in Fig 40. The classic J-pole antenna is the lower portion shown between points A and C. The half-wave section between points B and C does most of the radiating. The added half-wave section of the Super-J version is shown between points C and E. The side-by-side quarter-wave elements between points A and B comprise the J feed arrangement.

At first glance, counterproductive currents in the J section between points A and B may seem a waste of element material, but it is through this arrangement that the antenna is able to perform well in the absence of a good ground. The two halves of the J feed arrangement, side by side, provide a loading mechanism regardless of whether or not a ground plane is present.

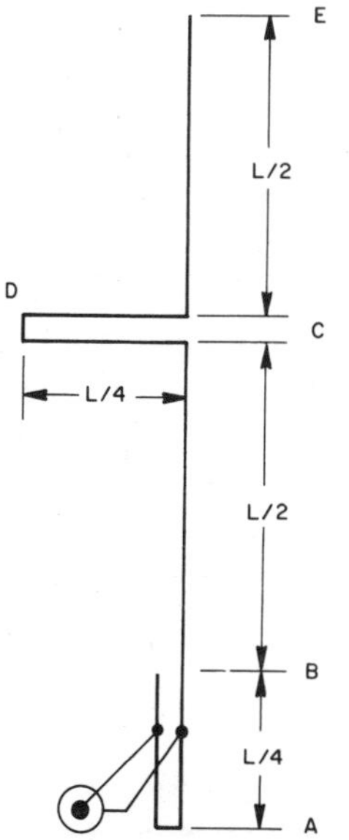

The radiation resistance of any antenna fluctuates as a function of height above ground, but the magnitude of this effect is small compared to the wildly changing impedance encountered when the distance from a ground plane element to its counterpoise is varied. Also, the J section adds ¼ wavelength of antenna height, reducing the effect of ground-height variations even further. Reducing ground-height sensitivity is particularly useful in maritime operation on those days when the water is rough.

The gain afforded by doubling the aperture of a J-pole with the extra half-wave section can be realized only if the added section is excited in phase with the half-wave element B-C. This is accomplished in the Super-J in a conventional manner, through the use of the quarter-wave phasing stub shown between C and D.

Construction and Adjustment

The completed Super-J is shown in Fig 41. Details of the individual parts are given in Fig 42. The driven element can be liberated from a quarter-wave CB whip antenna and cut to the dimensions shown. All other metal stock can be obtained from metal supply houses or machine shops. Metal may even be scrounged for little or nothing as scraps or remnants, as were the parts for the antenna shown here.

The center insulator and the two J stub spacers are made of ½-inch fiberglass and stainless steel stock, and the end caps are bonded to the insulator sections with epoxy. If you don't have access to a lathe to make the end caps, a simpler one-piece insulator design of wood or fiberglass could be used. However, keep in mind that good electrical connections must be maintained at all joints, and strength is a consideration for the center insulator.

The quarter-wave phasing stub is made of 1/8-inch stainless steel tubing, Fig 43. The line comprising this stub is bent in a semicircular arc to narrow the vertical profile and to keep the weight distribution balanced. This makes for an attractive appearance and keeps the antenna from leaning to one side.

The bottom shorting bar and base mounting plate are of ¼-inch stainless steel plate, shown in Fig 44. The J stub is made of 3/16-inch stainless steel rod stock. The RF connector may be mounted on the shorting bar as shown, and connected to the adjustable slider with a short section of coaxial cable. RTV sealant

Fig 40—Schematic representation of the Super-J maritime antenna. The radiating section is two half waves in phase.

Fig 41—Andy and the assembled Super-J antenna.

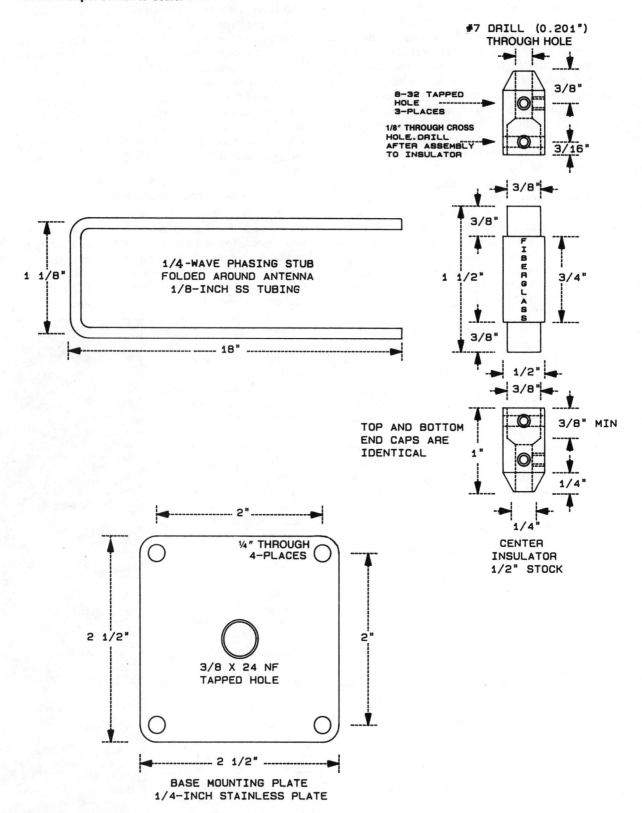

#7 DRILL (0.201")
THROUGH HOLE

8-32 TAPPED
HOLE
3-PLACES

1/8" THROUGH CROSS
HOLE. DRILL
AFTER ASSEMBLY
TO INSULATOR

3/8"

3/16"

1/4-WAVE PHASING STUB
FOLDED AROUND ANTENNA
1/8-INCH SS TUBING

1 1/8"

18"

3/8"

3/8"

FIBERGLASS

1 1/2"

3/4"

3/8"

1/2"

3/8"

TOP AND BOTTOM
END CAPS ARE
IDENTICAL

3/8" MIN

1"

1/4"

1/4"

CENTER
INSULATOR
1/2" STOCK

2"

¼" THROUGH
4-PLACES

2 1/2"

2"

3/8 X 24 NF
TAPPED HOLE

2 1/2"

BASE MOUNTING PLATE
1/4-INCH STAINLESS PLATE

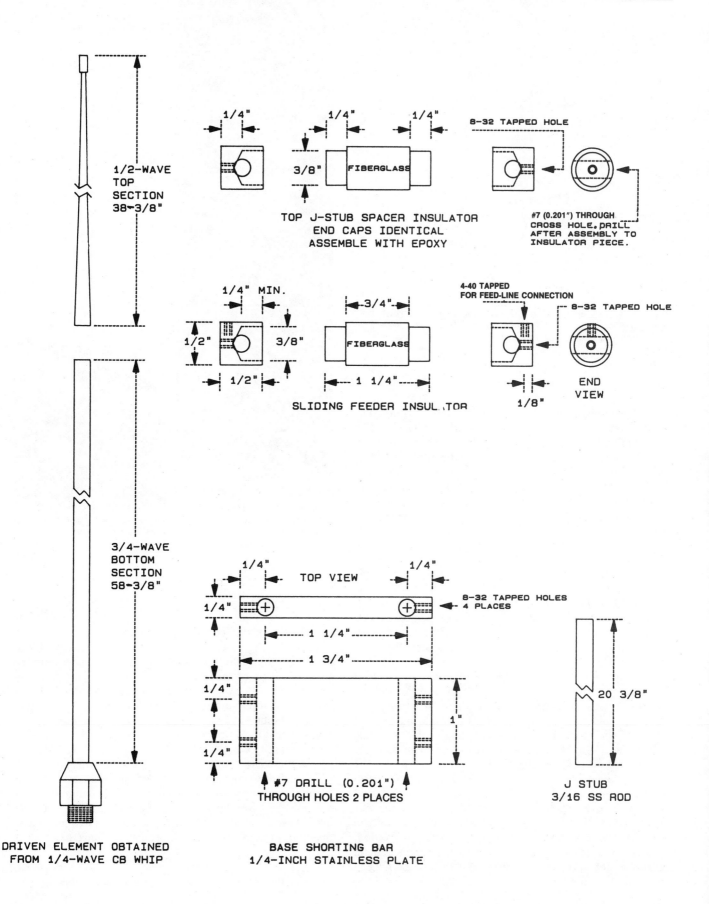

1/2-WAVE TOP SECTION 38-3/8"

3/4-WAVE BOTTOM SECTION 58-3/8"

DRIVEN ELEMENT OBTAINED FROM 1/4-WAVE CB WHIP

1/4"   1/4"   1/4"

3/8"   FIBERGLASS

TOP J-STUB SPACER INSULATOR
END CAPS IDENTICAL
ASSEMBLE WITH EPOXY

8-32 TAPPED HOLE

#7 (0.201") THROUGH
CROSS HOLE, DRILL
AFTER ASSEMBLY TO
INSULATOR PIECE.

1/4" MIN.

1/2"   3/8"

1/2"

FIBERGLASS

1 1/4"

3/4"

SLIDING FEEDER INSULATOR

4-40 TAPPED
FOR FEED-LINE CONNECTION

8-32 TAPPED HOLE

END VIEW

1/8"

TOP VIEW

1/4"   1/4"

1/4"

8-32 TAPPED HOLES
4 PLACES

1 1/4"

1 3/4"

1/4"

1/4"

1/4"

1"

#7 DRILL (0.201")
THROUGH HOLES 2 PLACES

BASE SHORTING BAR
1/4-INCH STAINLESS PLATE

20 3/8"

J STUB
3/16 SS ROD

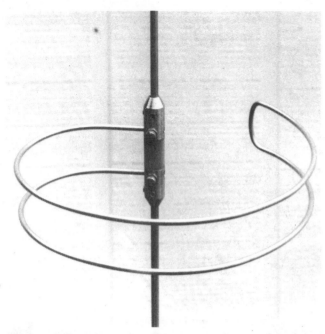

Fig 43—A close-up look at the ¼-λ phasing section of the Super-J. The insulator fitting is made of stainless steel end caps and fiberglass rod.

Fig 44—The bottom shorting bar and base mounting plate assembly.

Fig 45—The Super-J in portable use at a field site.

should be used at the cable ends to keep out moisture. The all-stainless construction looks nice and weathers well in maritime mobile applications.

The antenna should work well over the whole 144-MHz band if cut to the dimensions shown. The only tuning required is adjustment of the sliding feed point for minimum SWR in the center of the band segment you use most. Setting the slider 2-13/16 inches above the top of the shorting bar gave the best match for this antenna and may be used for a starting point.

Performance

Initial tests of the Super-J were performed in portable use and were satisfactory, if not exciting. Fig 45 shows the Super-J mounted on a wooden mast at a portable site. Simplex communication with a station 40 miles away with a 10-watt mobile rig was full quieting both ways. Stations were worked through distant repeaters that were thought inaccessible from this location.

Comparative tests between the Super-J and a commercial 5/8-wave antenna mounted on the car showed the Super-J to give superior performance, even when the Super-J was lowered to the same height as the car roof. The mast shown in Fig 45 was made from two 8-foot lengths of 1 × 2-inch pine. (The two mast sections and the Super-J can be easily transported in most vehicles.)

The Super-J offers a gain of about 6 dB over a quarter-wave whip and around 3 dB over a 5/8-wave antenna. Actual performance, especially under less-than-ideal or variable ground conditions, is substantially better than other vertical antennas operated under the same conditions. The freedom from ground-plane radials proves to be a real benefit in maritime mobile operation, especially for those passengers in the back of the boat with sensitive ribs!

A TOP-LOADED 144-MHz MOBILE ANTENNA

Earlier in this chapter, the merits of various loading schemes for shortened whip antennas were discussed. Quite naturally, one might be considering HF mobile operation for the application of those techniques. But the principles may be applied at any frequency. Fig 46 shows a 144-MHz antenna that is both top and center loaded. This antenna is suitable for both mobile and portable operation, being intended for use on a handheld transceiver. This antenna was devised by Don Johnson, W6AAQ, and Bruce Brown, W6TWW.

A combination of top and center loading offers improved efficiency over continuously loaded antennas such as the "stubby" pictured at the beginning of this chapter. This antenna also offers low construction cost. The only materials needed are a length of stiff wire and a scrap of circuit-board material, in addition to the appropriate connector.

Construction

The entire whip section with above-center loading coil

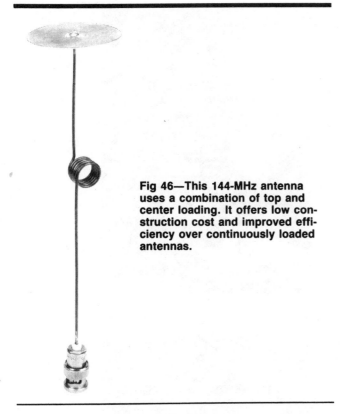

Fig 46—This 144-MHz antenna uses a combination of top and center loading. It offers low construction cost and improved efficiency over continuously loaded antennas.

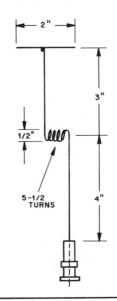

Fig 47—Dimensions for the top-loaded 144-MHz antenna. See text regarding coil length.

is made of one continuous length of material. An 18-inch length of brazing rod or no. 14 Copperweld wire is suitable.

In the antenna pictured in Fig 46, the top loading disk was cut from a scrap of circuit-board material, but flashing copper or sheet brass stock could be used instead. Aluminum is not recommended.

The dimensions of the antenna are given in Fig 47. First wind the center loading coil. Use a ½-inch bolt, wood dowel, or other cylindrical object for a coil form. Begin winding at a point 3 inches from one end of the wire, and wrap the wire tightly around the coil form. Wind 5½ turns, with just enough space between turns so they don't touch.

Remove the coil from the form. Next, determine the length necessary to insert the wire into the connector you'll be using. Cut the long end of the wire to this length plus 4 inches, measured from the center of the coil. Solder the wire to the center pin and assemble the connector. A tight-fitting sleeve made of Teflon or Plexiglas rod may be used to support and insulate the antenna wire inside the shell. An alternative is to fill the shell with epoxy cement, and allow the cement to set while the wire is held centered in the shell.

The top loading disk may be circular, cut with a hole saw. A circular disk is not required, however—it may be of any shape. Just remember that with a larger disk, less coil inductance will be required, and vice versa. Drill a hole at the center of the disk for mounting it to the wire. For a more rugged antenna, reinforce the hole with a brass eyelet. Solder the disk in place at the top of the antenna, and construction is completed.

Tune-Up

Adjustment consists of spreading the coil turns for the correct amount of inductance. Do this at the center frequency of the range you'll normally be using. Optimum inductance is determined with the aid of a field-strength meter at a distance of 10 or 15 feet.

Attach the antenna to a handheld transceiver operating on low power, and take a field-strength reading. With the transmitter turned off, spread the coil turns slightly, and then take another reading. By experiment, spread or compress the coil turns for the maximum field-strength reading. Very little adjustment should be required. There is one precaution, however. You must keep your body, arms, legs, and head in the same relative position for each field-strength measurement. It is suggested that the transceiver be placed on a nonmetal table and operated at arm's length for these checks.

Once the maximum field-strength reading is obtained, adjustments are completed. With this antenna in operation, you'll likely find it possible to access repeaters that are difficult to reach with other shortened antennas. W6AAQ reports that in distant areas his antenna even outperforms a 5/8-λ vertical.

VHF QUARTER-WAVELENGTH VERTICAL

Ideally, a VHF vertical antenna should be installed over a perfectly flat reflector to assure uniform omnidirectional radiation. This suggests that the center of the automobile roof is the best place to mount it for mobile use. Alternatively, the flat portion of the trunk deck can be used, but will result in a directional pattern because of car-body obstruction.

Fig 48 illustrates how a Millen high-voltage connector can be used as a roof mount for a VHF whip. The hole in the roof can be made over the dome light, thus providing accessibility through the upholstery. RG-59 and the ¼-wave matching section, L (Fig 48C), can be routed between the car roof and the ceiling upholstery and brought into the trunk compartment, or down to the dashboard of the car. Instead of a Millen connector, some operators install an SO-239 coax connector on the roof for mounting the whip. The method is similar to that shown in Fig 48.

It has been established that in general, ¼-λ vertical antennas for mobile repeater work are not as effective as 5/8-λ

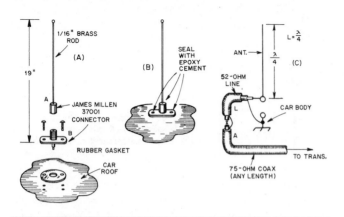

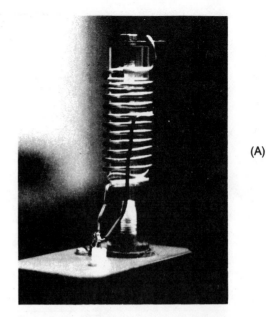

(A)

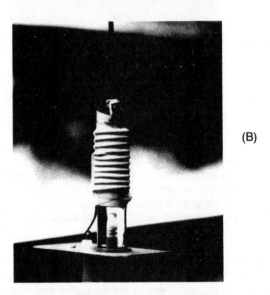

(B)

Fig 48—At A and B, an illustration of how a quarter-wavelength vertical antenna can be mounted on a car roof. The whip section should be soldered into the cap portion of the connector and then screwed into the base socket. This arrangement allows for the removal of the antenna when desired. Epoxy cement should be used at the two mounting screws to prevent the entry of moisture through the screw holes. Diagram C is discussed in the text.

Fig 49—At A, a photograph of the 5/8-wavelength vertical base section. The matching coil is affixed to an aluminum bracket that screws onto the inner lip of the car trunk. At B, the completed assembly. The coil has been wrapped with vinyl electrical tape to keep out dirt and moisture.

verticals are. With a 5/8-λ antenna, more of the transmitted signal is directed at a low wave angle, toward the horizon, offering a gain of about 3 dB over the ¼-λ vertical. However, in areas where the repeater is located nearby on a very high hill or a mountain top, the ¼-λ antenna will usually offer more reliable performance than a 5/8-λ antenna. This is because there is more power in the lobe of the ¼-λ vertical at higher angles.

144-MHz 5/8-WAVELENGTH VERTICAL

Perhaps the most popular antenna for 144-MHz FM mobile and fixed-station use is the 5/8-wavelength vertical. As compared to a ¼-wavelength vertical, it has 3 dB gain.

This antenna is suitable for mobile or fixed-station use because it is small, omnidirectional, and can be used with radials or a solid-plane ground (such as a car body). If radials are used, they need be only ¼ wavelength long.

Construction

The antenna shown here is made from low-cost materials. Fig 49 shows the base coil and aluminum mounting plate. The coil form is a piece of low-loss solid rod, such as Plexiglas or phenolic. The dimensions for this and other parts of the antenna are given in Fig 50. A length of brazing rod is used as the whip section.

The whip should be 47 inches long. However, brazing rod comes in standard 36-inch lengths, so if used, it is necessary to solder an 11-inch extension to the top of the whip. A piece of no. 10 copper wire will suffice. Alternatively, a stainless-steel rod can be purchased to make a 47-inch whip. Shops that sell CB antennas should have such rods for replacement purposes on base-loaded antennas. The limitation one can expect with brazing rod is the relative fragility of the material, especially when the threads are cut for screwing the rod into the base coil form. Excessive stress can cause the rod

to break where it enters the form. The problem is complicated somewhat in this design because a spring is not used at the antenna mounting point. Builders of this antenna can find all kinds of solutions to the problems just outlined by changing the physical design and using different materials when constructing the antenna. The main purpose of this description is to provide dimensions and tune-up information.

The aluminum mounting bracket must be shaped to fit the car with which it will be used. The bracket can be used to effect a no-holes mount with respect to the exterior portion of the car body. The inner lip of the vehicle trunk (or hood) can be the point where the bracket is attached by means of no. 6 or no. 8 sheet-metal screws. The remainder of the bracket is bent so that when the trunk lid or car hood is raised

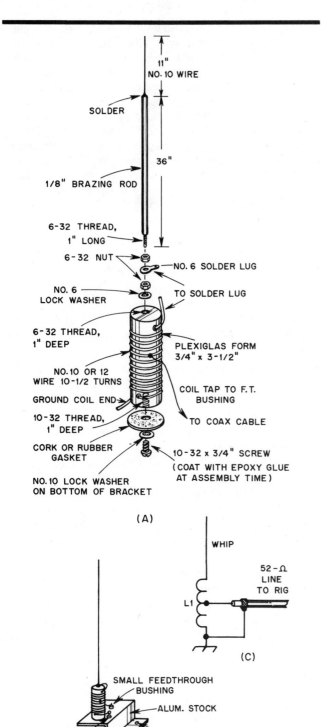

(A)

(B)

(C)

Fig 50—Structural details for the 2-meter 5/8-λ antenna are provided at A. The mounting bracket is shown at B and the equivalent circuit is given at C.

and lowered, there is no contact between the bracket and the moving part. Details of the mounting unit are given in Fig 50B. A 14-gauge metal (or thicker) is recommended for rigidity.

Wind 10½ turns of no. 10 or no. 12 copper wire on the ¾-inch diameter coil form. The tap on L1 is placed approximately four turns below the whip end. A secure solder joint is imperative.

Tune-Up

After the antenna has been mounted on the vehicle, connect an SWR indicator in the 52-Ω transmission line. Key the 144-MHz transmitter and experiment with the coil tap placement. If the whip section is 47 inches long, an SWR of 1:1 can be obtained when the tap is at the right location. As an alternative method of adjustment, place the tap at four turns from the top of L1, make the whip 50 inches long, and trim the whip length until an SWR of 1:1 occurs. Keep the antenna well away from other objects during tune-up, as they may detune the antenna and yield false adjustments for a match.

A 5/8-WAVELENGTH 220-MHz MOBILE ANTENNA

The antenna shown in Figs 51 and 52 was developed to fill the gap between a homemade ¼-λ mobile antenna and a commercially made 5/8-λ model. While antennas can be made by modifying CB models, that presents the problem of cost in acquiring the original antenna. The major cost in this setup is the whip portion. This can be any tempered rod that will spring easily.

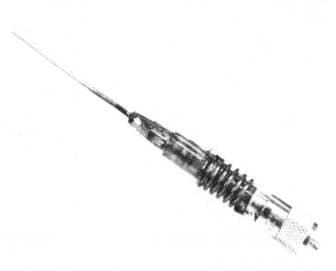

Fig 51—The 220-MHz 5/8-λ mobile antenna. The coil turns are spaced over a distance of 1 in., and the bottom end of the coil is soldered to the coax connector.

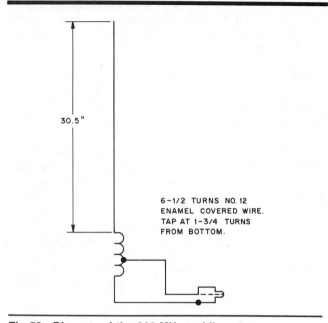

30.5"

6-1/2 TURNS NO. 12
ENAMEL COVERED WIRE.
TAP AT 1-3/4 TURNS
FROM BOTTOM.

Fig 52—Diagram of the 220-MHz mobile antenna.

Construction

The base insulator portion is made of ½-inch Plexiglas rod. A few minutes' work on a lathe is sufficient to shape and drill the rod. (The innovative builder can use an electric drill and a file for the "lathe" work.) The bottom ½ inch of the rod is turned down to a diameter of 3/8 inch. This portion will now fit into a PL-259 UHF connector. A 1/8-inch diameter hole is drilled through the center of the rod. This hole will hold the wires that make the connections between the center conductor of the connector and the coil tap. The connection between the whip and the top of the coil is also run through this opening. A stud is force-fitted into the top of the Plexiglas rod. This allows for removal of the whip from the insulator.

The coil should be initially wound on a form slightly smaller than the base insulator. When the coil is transferred to the Plexiglas rod, it will keep its shape and will not readily move. After the tap point has been determined, a longitudinal hole is drilled into the center of the rod. A no. 22 wire can then be inserted through the center of the insulator into the connector. This method is also used to attach the whip to the top of the coil. After the whip has been fully assembled, a coating of epoxy cement is applied. This seals the entire assembly and provides some additional strength. During a full winter's use there was no sign of cracking or other mechanical failure. The adjustment procedure is the same as for the 144-MHz version described previously.

BIBLIOGRAPHY

Source material and more extended discussions of topics covered in this chapter can be found in the references given below and in the textbooks listed at the end of Chapter 2.

J. S. Belrose, "Short Antennas for Mobile Operation," *QST*, Sep 1953.

J. S. Belrose, "Vertical J Antenna for 2 Meters," *The Canadian Amateur*, Jul/Aug 1979, pp 23-26.

J. M. Boyer, "Antenna-Transmission Line Analog," *Ham Radio*, May 1977.

B. F. Brown, "Tennamatic: An Auto-Tuning Mobile Antenna System," *73*, Jul 1979.

B. F. Brown, "Optimum Design of Short Coil-Loaded High-Frequency Mobile Antennas," *ARRL Antenna Compendium, Vol 1* (Newington: ARRL, 1985), p. 108.

W. B. Freely, "A Two-Meter J Antenna," *QST*, Apr 1977, pp 35-36.

E. A. Laport, *Radio Antenna Engineering* (New York: McGraw-Hill Book Co., 1952), p 23.

C. E. Smith and E. M. Johnson, "Performance of Short Antennas," Proceedings of the IRE, Oct 1947.

F. E. Terman, *Radio Engineering Handbook*, 3rd edition (New York: McGraw-Hill Book Co., 1947), p 74.

Chapter 17

Repeater Antenna Systems

There is an old adage in Amateur Radio that goes "If your antenna did not fall down last winter, it wasn't big enough." This adage might apply to antennas for MF and HF work, but at VHF things are a bit different, at least as far as antenna size is concerned. VHF antennas are smaller than their HF counterparts, but yet the theory is the same; a dipole is a dipole, and a Yagi is a Yagi, regardless of frequency. A 144-MHz Yagi may pass as a TV antenna, but most neighbors can easily detect a radio hobbyist if a 14-MHz Yagi looms over his property.

Repeater antennas are discussed in this chapter. Because the fundamental operation of these antennas is no different than presented in Chapter 2, there is no need to to delve into any exotic theory. Certain considerations must be made and certain precautions must be observed, however, as most repeater operations—amateur and commercial—take place at VHF and UHF.

Basic Concepts

The antenna is a vital part of any repeater installation. Because the function of a repeater is to extend the range of communications between mobile and portable stations, the repeater antenna should be installed in the best possible location to provide the desired coverage. This usually means getting the antenna as high above average terrain as possible. In some instances, a repeater may need to have coverage only in a limited area or direction. When this is the case, antenna installation requirements will be completely different, with certain limits being set on height, gain and power.

Horizontal and Vertical Polarization

Until the upsurge in FM repeater activity several years ago, most antennas used in amateur VHF work were horizontally polarized. These days, very few repeater groups use horizontal polarization. (One of the major reasons for using horizontal polarization is to allow separate repeaters to share the same input and/or output frequencies with closer than normal geographical spacing.) The vast majority of VHF and UHF repeaters use vertically polarized antennas, and all the antennas discussed in this chapter are of that type.

Transmission Lines

Repeaters provide the first venture into VHF and UHF work for many amateurs. The uninitiated may not be aware that the transmission lines used at VHF become very important because feed-line losses increase with frequency.

The characteristics of feed lines commonly used at VHF are discussed in Chapter 24. Although information is provided for RG-58 and RG-59, these should not be used except for very short feed lines (25 feet or less). These cables are very lossy at VHF. In addition, the losses can be much higher if the fittings and connections are not carefully installed.

The differences in loss between solid polyethylene dielectric types (RG-8 and RG-11) and those using foam polyethylene (FM8 and FM11) are significant. If you can afford the line with the least loss, buy it.

If coaxial cable must be buried, check with the manufacturer before doing so. Many popular varieties of coaxial cable should not be buried, as the dielectric can become contaminated from moisture and soil chemicals. Some coaxial cables are labeled as noncontaminating. Such a label is the best way to be sure your cable can be buried without damage.

Matching

Losses are lowest in transmission lines that are matched to their characteristic impedances. If there is a mismatch at the end of the line, the losses increase.

The *only way* to reduce the SWR on a transmission line is by matching the line *at* the antenna. Changing the length of a transmission line does not reduce the SWR. The SWR is established by the impedance of the line and the impedance of the antenna, so matching must be done at the antenna end of the line.

The importance of matching, as far as feed-line losses are concerned, is sometimes overstressed. But under some conditions, it is necessary to minimize feed-line losses related to SWR if repeater performance is to be consistent. It is important to keep in mind that most VHF/UHF equipment is designed to operate into a 52-Ω load. The output circuitry will not be loaded properly if connected to a mismatched line. This leads to a loss of power, and in some cases, damage to the transmitter.

Repeater Antenna System Design

Choosing a repeater or remote-base antenna system is as close as most amateurs come to designing a commercial-grade antenna system. The term *system* is used because most repeaters utilize not only an antenna and a transmission line, but also include duplexers, cavity filters, circulators or isolators in some configuration. Assembling the proper combination of these items in constructing a reliable system is both an art and a science. In this section prepared by Domenic Mallozzi, N1DM, the functions of each component in a repeater antenna system and their successful integration are discussed. While every possible complication in constructing a repeater is not foreseeable at the outset, this discussion should serve to steer you along the right lines in solving any problems encountered.

The Repeater Antenna

The most important part of the system is the antenna itself. As with any antenna, it must radiate and collect RF energy as efficiently as possible. Many repeaters use omnidirectional antennas, but this is not always the best choice. For example, suppose a group wishes to set up a repeater to cover towns A and B and the interconnecting state highway shown in Fig 1. The X shows the available repeater site on the map. No coverage is required to the west or south, or over the ocean. If an omnidirectional antenna is used in

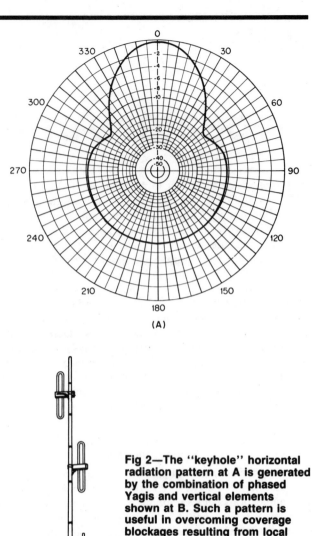

(A)

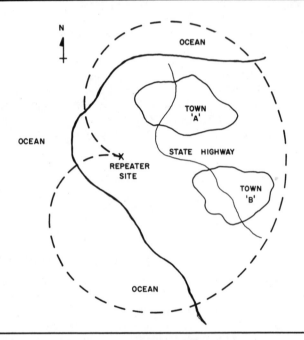

Fig 1—There are many situations where equal repeater coverage is not desired in all directions from the "machine." One such situation is shown here, where the repeater is needed to cover only towns A and B and the interconnecting highway. An omnidirectional antenna would provide coverage in undesired directions, such as over the ocean. The broken line shows the radiation pattern of an antenna that is better suited to this circumstance.

Fig 2—The "keyhole" horizontal radiation pattern at A is generated by the combination of phased Yagis and vertical elements shown at B. Such a pattern is useful in overcoming coverage blockages resulting from local terrain features. (Based on a design by Decibel Products, Inc)

(B)

this case, a significant amount of the radiated signal goes in undesired directions. By using an antenna with a cardioid pattern, as shown in Fig 1, the coverage is concentrated in the desired directions. The repeater will be more effective in these locations, and signals from low power portables and mobiles will be more reliable.

In many cases, antennas with special patterns are more expensive than omnidirectional models. This is an obvious consideration in designing a repeater antenna system.

Over terrain where coverage may be difficult in some direction from the repeater site, it may be desirable to skew the antenna pattern in that direction. This can be accomplished by using a phased-vertical array or a combination of a Yagi and a phased vertical to produce a "keyhole" pattern. See Fig 2.

As repeaters are established on 440 MHz and above, many groups are investing in high-gain omnidirectional antennas. A consequence of getting high gain from an omnidirectional antenna is vertical beamwidth reduction. In most cases, these antennas are designed to radiate their peak gain at the horizon, resulting in optimum coverage when the antenna is located at a moderate height over normal terrain. Unfortunately, in cases where the antenna is located at a very high site (overlooking the coverage area) this is not the most desirable pattern. In a case like this, the vertical pattern of the antenna can be tilted downward to facilitate coverage of the desired area. This is called *vertical-beam downtilt*.

An example of such a situation is shown in Fig 3. The repeater site overlooks a town in a valley. A 450-MHz repeater is needed to serve low power portable and mobile stations.

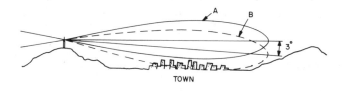

Fig 3—Vertical-beam downtilt is another form of radiation-pattern distortion useful for improving repeater coverage. This technique can be employed in situations where the repeater station is at a greater elevation than the desired coverage area, when a high-gain omnidirectional antenna is used. Pattern A shows the normal vertical-plane radiation pattern of a high-gain omnidirectional antenna with respect to the desired coverage area (the town). Pattern B shows the pattern tilted down, and the coverage improvement is evident.

Constraints on the repeater dictate the use of an antenna with a gain of 11 dBi. (An omnidirectional antenna with this gain has a vertical beamwidth of approximately 6 degrees.) If the repeater antenna has its peak gain at the horizon, a major portion of the transmitted signal and the best area from which to access the repeater exists *above* the town. By tilting the pattern down 3°, the peak radiation will occur in the town.

Vertical-beam downtilt is generally produced by feeding the elements of a collinear vertical array slightly out of phase with each other. Lee Barrett, K7NM, showed such an array in *Ham Radio*. (See the bibliography at the end of this chapter.) Barrett gives the geometry and design of a four-pole array with progressive phase delay, and a computer program to model it. The technique is shown in Fig 4.

Commercial antennas are sometimes available (at extra cost) with built-in downtilt characteristics. Before ordering such a commercial antenna, make sure that you really require it; they generally are special order items and are not returnable.

There are disadvantages to improving coverage by means of vertical-beam downtilt. When compared to a standard collinear array, an antenna using vertical-beam downtilt will have somewhat greater extraneous lobes in the vertical pattern, resulting in reduced gain (usually less than 1 dB). Bandwidth is also slightly reduced. The reduction in gain, when combined with the downtilt characteristic, results in a reduction in total coverage area. These trade-offs, as well as the increased cost of a commercial antenna with downtilt, must be compared to the improvement in total performance in a situation where vertical-beam downtilt is required.

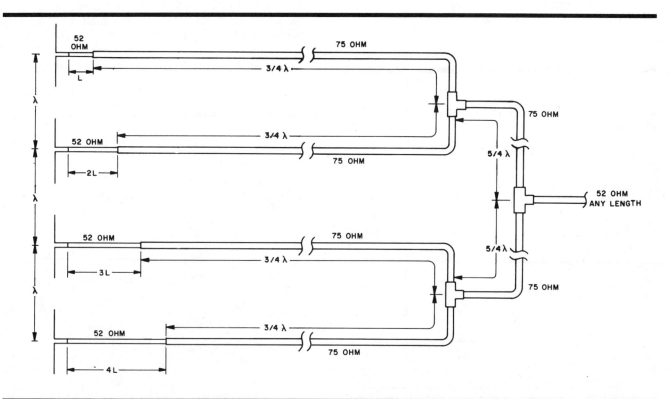

Fig 4—Vertical-beam downtilt can be facilitated by inserting 52-Ω delay lines in series with the 75-Ω feed lines to the collinear elements of an omnidirectional antenna. The delay lines to each element are progressively longer so the phase shift between elements is uniform. Odd ¼-λ coaxial transformers are used in the main (75-Ω) feed system to match the dipole impedances to the driving point. Tilting the vertical beam in this way often produces minor lobes in the vertical pattern that do not exist when the elements are fed in phase.

Top Mounting and Side Mounting

Amateur repeaters often share towers with commercial and public service users. In many of these cases, other antennas are at the top of the tower, so the amateur antenna must be side mounted. A consequence of this arrangement is that the free-space pattern of the repeater antenna is distorted by the tower. This effect is especially noticeable when an omnidirectional antenna is side mounted on a structure.

The effects of supporting structures are most pronounced at close antenna spacings to the tower and with large support dimensions. The result is a measurable increase in gain in one direction and a partial null in the other direction (sometimes 15 dB deep). The shape of the supporting structure also influences pattern distortion. Many antenna manufacturers publish radiation patterns showing the effect of side mounting antennas in their catalogs.

Side mounting is not always a disadvantage. In cases where more (or less) coverage is desired in one direction, the supporting structure can be used to advantage. If pattern distortion is not acceptable, a solution is to mount antennas around the perimeter of the structure and feed them with the proper phasing to synthesize an omnidirectional pattern. Many manufacturers make antennas to accommodate such situations.

The effects of different mounting locations and arrangements can be illustrated with an array of exposed dipoles, Fig 5. Such an array is a very versatile antenna because, with simple rearrangement of the elements, it can develop either an omnidirectional pattern or an offset pattern. Drawing A of Fig 5 shows a basic collinear array of 4 vertical ½-λ elements. The vertical spacing between adjacent elements is 1 λ. All elements are fed in phase. If this array is placed in the clear and supported by a nonconducting mast, the calculated radiation resistance of each dipole element is on the order of 63 Ω. If the feed line is completely decoupled, the resulting azimuth pattern is omnidirectional. The vertical-plane pattern is shown in Fig 6.

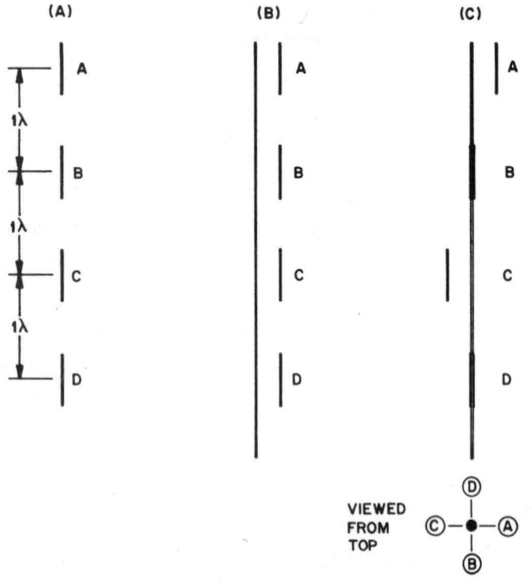

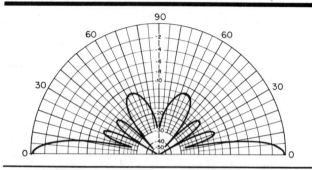

Fig 6—Calculated vertical-plane pattern of the array of Fig 5A, assuming a nonconducting mast support and complete decoupling of the feeder. In azimuth the array is omnidirectional. The calculated gain of the array is 8.6 dBi at 0° elevation; the −3 dB point is at 6.5°.

Fig 5B shows the same array in a side mounting arrangement, at a spacing of ¼ λ from a conducting mast. In this mounting arrangement, the mast takes on the role of a reflector, producing a F/B ratio on the order of 5.7 dB. The azimuth pattern is shown in Fig 7. The vertical pattern is not significantly different from that of Fig 6, except the four small minor lobes (two on either side of the vertical axis) tend to become distorted. They are not as "clean," tending to merge into one minor lobe at some mast heights. This apparently is a function of currents in the supporting mast. The proximity of the mast also alters the feed-point impedance. For elements that are resonant in the con-

Fig 5—Various arrangements of exposed dipole elements. At A is the basic collinear array of four elements. B shows the same elements mounted on the side of a mast, and C shows the elements in a side-mounted arrangement around the mast for omnidirectional coverage. See text and Figs 6 through 8 for radiation pattern information.

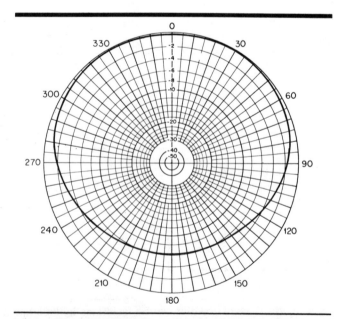

Fig 7—Calculated azimuth pattern of the side-mounted array of Fig 5B, assuming ¼-λ spacing from a 4-inch mast. The calculated gain in the favored direction, away from the mast and through the elements, is 10.6 dBi.

figuration of Fig 5A, the calculated impedance in the arrangement of Fig 5B is in the order of 72 + j10 Ω.

If side mounting is the only possibility and an omnidirectional pattern is required, the arrangement of Fig 5C may be used. The calculated azimuth pattern takes on a slight cloverleaf shape, but is within 1.5 dB of being circular. However, gain performance suffers, and the idealized vertical pattern of Fig 6 is not achieved. See Fig 8. Spacings other than ¼ λ from the mast were not investigated.

One very important consideration in side mounting an antenna is mechanical integrity. As with all repeater components, reliability is of great importance. An antenna hanging by the feed line and banging against the tower provides far from optimum performance and reliability. Use a mount that is appropriately secured to the tower and the antenna. Also use good hardware, preferably stainless steel (or bronze). If your local hardware store does not carry stainless steel hardware, try a boating supplier.

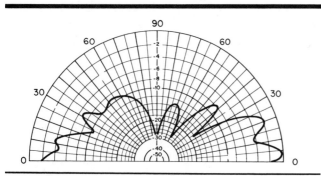

Fig 8—Calculated vertical pattern of the array of Fig 5C, assuming ¼-λ element spacing from a 4-inch mast. The azimuth pattern is circular within 1.5 dB, and the calculated gain is 4.4 dBi.

Be certain that the feed line is properly supported along its length. Long lengths of cable are subject to contraction and expansion with temperature from season to season, so it is important that the cable not be so tight that contraction causes it to stress the connection at the antenna. This can cause the connection to become intermittent (and noisy) or, at worst, an open circuit. This is far from a pleasant situation if the antenna connection is 300 feet up a tower, and it happens to be the middle of the winter!

Effects of Other Conductors

Feed-line proximity and tower-access ladders or cages also have an effect on the radiation patterns of side-mounted antennas. This subject was studied by Connolly and Blevins, and their findings are given in *IEEE Conference Proceedings* (see the bibliography at the end of this chapter). Those considering mounting antennas on air conditioning evaporators or maintenance penthouses on commercial buildings should consult this article. It gives considerable information on the effects of these structures on both unidirectional and omnidirectional antennas.

Metallic guy wires also affect antenna radiation patterns. Yang and Willis studied this and reported the results in *IRE Transactions on Vehicular Communications*. As expected, the closer the antenna is to the guy wires, the greater the effect on the radiation patterns. If the antennas are near the point where the guy wires meet the tower, the effect of the guy wires can be minimized by breaking them up with insulators every 0.75 λ for 2.25 λ to 3.0 λ.

ISOLATION REQUIREMENTS IN REPEATER ANTENNA SYSTEMS

Because repeaters generally operate in full duplex (the transmitter and receiver operate simultaneously), the antenna system must act as a filter to keep the transmitter from blocking the receiver. The degree to which the transmitter and receiver must be isolated is a complex problem. It is quite dependent on the equipment used and the difference in transmitter and receiver frequencies (offset). Instead of going into great detail, a simplified example can be used for illustration.

Consider the design of a 144-MHz repeater with a 600-kHz offset. The transmitter has an RF output power of 10 watts, and the receiver has a squelch sensitivity of 0.1 μV. This means there must be at least 1.9×10^{-16} watts at the 52-ohm receiver-antenna terminals to detect a signal. If both the transmitter and receiver were on the same frequency, the isolation (attenuation) required between the transmitter and receiver antenna jacks to keep the transmitter from activating the receiver would be

$$\text{Isolation} = 10 \log \frac{10 \text{ watts}}{1.9 \times 10^{-16} \text{ watts}} = 167 \text{ dB}$$

Obviously there is no need for this much attenuation, because the repeater does not transmit and receive on the same frequency.

If the 10-watt transmitter has noise 600 kHz away from the carrier frequency that is 45 dB below the carrier power, that 45 dB can be subtracted from the isolation requirement. Similarly, if the receiver can detect a 0.1-μV on-frequency signal in the presence of a signal 600 kHz away that is 40 dB greater than 0.1 μV, this 40 dB can also be subtracted from the isolation requirement. Therefore, the isolation requirement is

167 dB − 45 dB − 40 dB = 82 dB

Other factors enter into the isolation requirements as well. For example, if the transmitter power is increased by 10 dB (from 10 to 100 watts), this 10 dB must be added to the isolation requirement. Typical requirements for 144- and 440-MHz repeaters are shown in Fig 9.

Obtaining the required isolation is the first problem to be considered in constructing a repeater antenna system. There are three common ways to obtain this isolation:

1) Physically separate the receiving and transmitting antennas so the combination of path loss for the spacing and the antenna radiation patterns results in the required isolation.

2) Use a combination of separate antennas and high-Q filters to develop the required isolation. (The high-Q filters serve to reduce the physical distance required between antennas.)

3) Use a combination filter and combiner system to allow the transmitter and receiver to share one antenna. Such a filter and combiner is called a *duplexer*.

Repeaters operating on 28 and 50 MHz generally use separate antennas to obtain the required isolation. This is largely because duplexers in this frequency range are both large and very expensive. It is generally less expensive to buy two antennas and link the sites by a committed phone line or an RF link than to purchase a duplexer. At 144 MHz and higher, duplexers are more commonly used. Duplexers are discussed in greater detail in a later section.

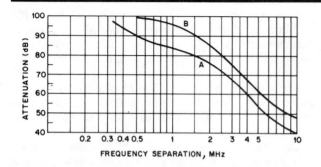

Fig 9—Typical isolation requirements for repeater transmitters and receivers operating in the 132-174 MHz band (Curve A), and the 400-512 MHz band (Curve B). Required isolation in dB is plotted against frequency separation in MHz. These curves were developed for a 100-W transmitter. For other power levels, the isolation requirements will differ by the change in decibels relative to 100 W. Isolation requirements will vary with receiver sensitivity. (The values plotted were calculated for transmitter-carrier and receiver-noise suppression necessary to prevent more than 1 dB degradation in receiver 12-dB SINAD sensitivity.)

Separate Antennas

Receiver desensing (gain limiting caused by the presence of a strong off-frequency signal) can be reduced, and often eliminated, by separation of the transmitting and receiving antennas. Obtaining the 55 to 90 dB of isolation required for a repeater antenna system requires separate antennas to be spaced a considerable distance apart (in wavelengths).

Fig 10 shows the distances required to obtain specific values of isolation for vertical dipoles having horizontal separation (at A) and vertical separation (at B). The isolation gained by using separate antennas is subtracted from the total isolation requirement of the system. For example, if the transmitter and receiver antennas for a 450-MHz repeater are separated horizontally by 400 feet, the total isolation requirement in the system is reduced by about 64 dB.

Note from Fig 10B that a vertical separation of only about 25 feet also provides 64 dB of isolation. Vertical separation yields much more isolation than does horizontal separation. Vertical separation is also more practical than horizontal, as only a single support is required.

An explanation of the significant difference between the two graphs is in order. The vertical spacing requirement for 60 dB attenuation (isolation) at 155 MHz is about 43 feet. The horizontal spacing for the same isolation level is on the order of 700 feet. Fig 11 shows why this difference exists. The radiation patterns of the antennas at A overlap; each antenna

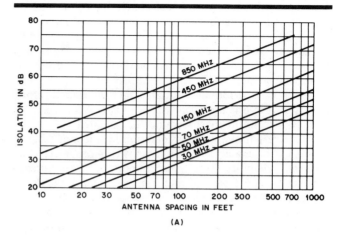

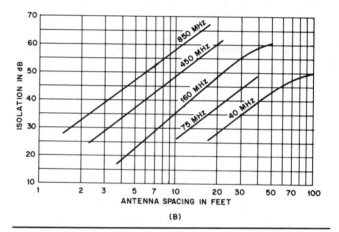

Fig 10—At A, the amount of attenuation (isolation) provided by horizontal separation of vertical dipole antennas. At B, isolation afforded by vertical separation of vertical dipoles. Spacing is that between antenna centers.

has gain in the direction of the other. The path loss between the antennas is given by

$$\text{Path loss (dB)} = 20 \log \frac{4\pi d}{\lambda}$$

where

d = distance between antennas
λ = wavelength, in the same units as d

The isolation between the antennas in Fig 11A is the path loss less the antenna gains. Conversely, the antennas at B share pattern nulls, so the isolation is the path loss added to the depth of these nulls. This significantly reduces the spacing requirement for vertical separation. Because the depth of the pattern nulls is not infinite, some spacing is required. Combined horizontal and vertical spacing is much more difficult to quantify because the results are dependent on both radiation patterns and the positions of the antennas relative to each other.

Separate antennas have one major disadvantage: They create disparity in transmitter and receiver coverage. For

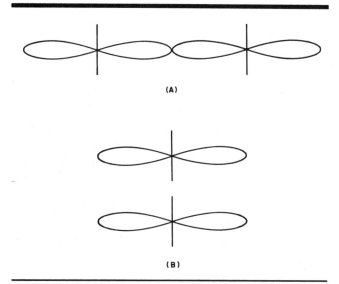

(A)

(B)

Fig 11—A relative representation of the isolation advantage afforded by separating antennas horizontally (A) and vertically (B) is shown. A great deal of isolation is provided by vertical separation, but horizontal separation requires two supports and much greater distance to be as effective. Separate-site repeaters (those with transmitter and receiver at different locations) benefit much more from horizontal separation than do single-site installations.

example, say a 50-MHz repeater is installed over average terrain with the transmitter and repeater separated by 2 miles. If both antennas had perfect omnidirectional coverage, the situation depicted in Fig 12 would exist. In this case, stations able to hear the repeater may not be able to access it, and vice versa. In practice, the situation can be considerably worse. This is especially true if the patterns of both antennas are not omnidirectional. If this disparity in coverage cannot be tolerated, the solution involves skewing the patterns of the antennas until their coverage areas are essentially the same.

Cavity Resonators

As just discussed, receiver desensing can be reduced by

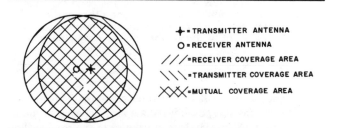

+ = TRANSMITTER ANTENNA
O = RECEIVER ANTENNA
/// = RECEIVER COVERAGE AREA
\\\ = TRANSMITTER COVERAGE AREA
XXX = MUTUAL COVERAGE AREA

Fig 12—Coverage disparity is a major problem for separate-site repeater antennas. The transmitter and receiver coverage areas overlap, but are not entirely mutually inclusive. Solving this problem requires a great deal of experimentation, as many factors are involved. Among these factors are terrain features and distortion of the antenna radiation patterns from supports.

separating the transmitter and receiver antennas. But the amount of transmitted energy that reaches the receiver input must often be decreased even farther. Other nearby transmitters can cause desensing as well. A *cavity resonator* (cavity filter) can be helpful in solving these problems. When properly designed and constructed, this type of resonator has very high Q. A commercially made cavity is shown in Fig 13.

A cavity resonator placed in series with a transmission line acts as a band-pass filter. For a resonator to operate in series, it must have input and output coupling loops (or probes).

A cavity resonator can also be connected across (in parallel with) a transmission line. The cavity then acts as a band-reject (notch) filter, greatly attenuating energy at the frequency to which it is tuned. Only one coupling loop or probe is required for this method of filtering. This type of cavity could be used in the receiver line to "notch" the transmitter signal. Several cavities can be connected in series or parallel to increase the attenuation in a given configuration. The graphs of Fig 14 show the attenuation of a single cavity (A) and a pair of cavities (B).

The only situation in which cavity filters would not help is the case where the off-frequency noise of the transmitter was right on the receiver frequency. With cavity resonators, an important point to remember is that addition of a cavity across a transmission line may change the impedance of the system. This change can be compensated by adding tuning stubs along the transmission line.

Duplexers

The material in this section was prepared by Domenic Mallozzi, N1DM. Most amateur repeaters in the 144, 220 and 440-MHz bands use duplexers to obtain the necessary transmitter to receiver isolation. Duplexers have been commonly used in commercial repeaters for many years. The duplexer

Fig 13—A coaxial cavity filter of the type used in many amateur and commercial repeater installations. Center-conductor length (and thus resonant frequency) is varied by adjustment of the knob (top).

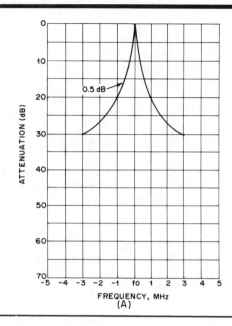

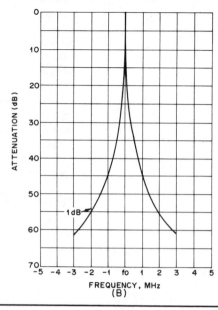

Fig 14—Frequency response curves for a single cavity (A) and two cavities cascaded (B). These curves are for cavities with coupling loops, each having an insertion loss of 0.5 dB. (The total insertion loss is indicated in the body of each graph.) Selectivity will be greater if lighter coupling (greater insertion loss) can be tolerated.

consists of two high-Q filters. One filter is used in the feed line from the transmitter to the antenna, and another between the antenna and the receiver. These filters must have low loss at the frequency to which they are tuned while having very high attenuation at the surrounding frequencies. To meet the high attenuation requirements at frequencies within as little as 0.4% of the frequency to which they are tuned, the filters usually take the form of cascaded transmission-line cavity filters. These are either band-pass filters, or band-pass filters with a rejection notch. (The rejection notch is tuned to the center frequency of the other filter.) The number of cascaded filter sections is determined by the frequency separation and the ultimate attenuation requirements.

Duplexers for the amateur bands represent a significant technical challenge, because in most cases amateur repeaters operate with significantly less frequency separation than their commercial counterparts. Information on home construction of duplexers is presented in a later section of this chapter. Many manufacturers market high quality duplexers for the amateur frequencies.

Duplexers consist of very high Q cavities whose resonant frequencies are determined by mechanical components, in particular the tuning rod. Fig 15 shows the cutaway view of a typical duplexer cavity. The rod is usually made of a material which has a limited thermal expansion coefficient (such as Invar). Detuning of the cavity by environmental changes introduces unwanted losses in the antenna system. An article by Arnold in *Mobile Radio Technology* considered the causes of drift in the cavity (see the bibliography at the end of this chapter). These can be broken into four major categories.

1) Ambient temperature variation (which leads to mechanical variations related to the thermal expansion coefficients of the materials used in the cavity).

2) Humidity (dielectric constant) variation.

3) Localized heating from the power dissipated in the cavity (resulting from its insertion loss).

4) Mechanical variations resulting from other factors (vibration, etc).

In addition, because of the high-Q nature of these cavities, the insertion loss of the duplexer increases when the

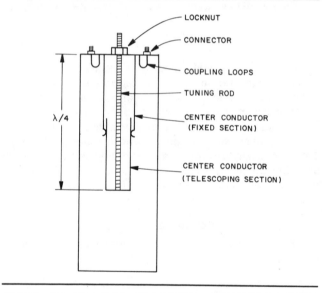

Fig 15—Cutaway view of a typical cavity. Note the relative locations of the coupling loops to each other and to the center conductor of the cavity. A locknut is used to prevent movement of the tuning rod after adjustment.

signal is not at the peak of the filter response. This means, in practical terms, less power is radiated for a given transmitter output power. Also, the drift in cavities in the receiver line results in increased system noise figure, reducing the sensitivity of the repeater.

As the frequency separation between the receiver and the transmitter decreases, the insertion loss of the duplexer reaches certain practical limits. At 144 MHz, the minimum insertion loss for 600-kHz spacing is 1.5 dB per filter.

Testing and using duplexers requires some special considerations (especially as frequency increases). Because duplexers are very high-Q devices, they are very sensitive to the termination impedances at their ports. A high SWR on any port is a serious problem, because the apparent insertion loss of the duplexer will increase, and the isolation may appear to decrease. Some have found that, when duplexers are used at the limits of their isolation capabilities, a small change in antenna SWR is enough to cause receiver desensitization. This occurs most often under ice-loading conditions on antennas with open-wire phasing sections.

The choice of connectors in the duplexer system is important. BNC connectors are good for use below 300 MHz. Above 300 MHz, their use is discouraged because even though many types of BNC connectors work well up to 1 GHz, older style standard BNC connectors are inadequate at UHF and above. Type N connectors should be used above 300 MHz. It is false economy to use marginal quality connectors. Some commercial users have reported deteriorated isolation in commercial UHF repeaters when using such connectors. The location of a bad connector in a system is a complicated and frustrating process. Despite all these considerations, the duplexer is still the best method for obtaining isolation in the 144- to 925-MHz range.

ADVANCED TECHNIQUES

As the number of available antenna sites decreases and the cost of various peripheral items (such as coaxial cable) increases, amateur repeater groups are required to devise advanced techniques if repeaters are to remain effective. Some of the techniques discussed here have been applied in commercial services for many years, but until recently have not been economically justified for amateur use.

One technique worth consideration is the use of *cross-band couplers*. To illustrate a situation where a cross-band coupler would be useful, consider the following example. A repeater group plans to install 144- and 902-MHz repeaters on the same tower. The group intends to erect both antennas on a horizontal cross-arm at the 325-foot level. A 325-foot run of 7/8-inch Heliax® costs approximately $1400. If both antennas are to be mounted at the top of the tower, the logical approach would require two separate feed lines. A better solution involves the use of a single feed line for both repeaters, along with a cross-band coupler at each end of the line.

The use of the cross-band coupler is shown in Fig 16. As the term implies, the coupler allows two signals on different bands to share a common transmission line. Such couplers cost approximately $200 each. In our hypothetical example, this represents a saving of $1000 over the cost of using separate feed lines. But, as with all compromises, there are disadvantages. Cross-band couplers have a loss of about 0.5 dB per unit. Therefore, the pair required represents a loss of 1.0 dB in *each transmission path*. If this loss can be tolerated, the cross-band coupler is a good solution.

Cross-band couplers do not allow two repeaters *on the same band* to share a single antenna and feed line. As repeater sites and tower space become more scarce, it may be desirable to have two repeaters on the same band share the same antenna. The solution to this problem is the use of a *transmitter multicoupler*. The multicoupler is related to the duplexers discussed earlier. It is a cavity filter and combiner which allows multiple transmitters and receivers to share the same antenna. This is a common commercial practice. A block diagram of a multicoupler system is shown in Fig 17.

The multicoupler, however, is a very expensive device, and has the disadvantage of even greater loss per transmission path than the standard duplexer. For example, a well-designed

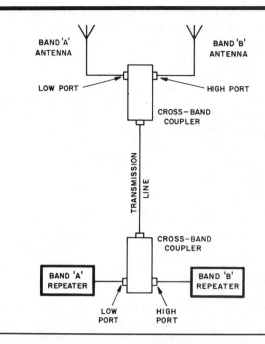

Fig 16—Block diagram of a system using cross-band couplers to allow the use of a single feed line for two repeaters. If the feeder to the antenna location is long (more than 200 ft or so), cross-band couplers may provide a significant saving over separate feed lines, especially at the higher amateur repeater frequencies. Cross-band couplers cannot be used with two repeaters on the same band.

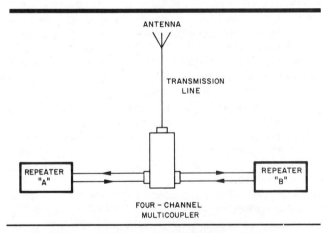

Fig 17—Block diagram of a system using a transmitter multicoupler to allow a single feed line and antenna to be used by two repeaters on one band. The antenna must be designed to operate at all frequencies that the repeaters utilize. More than two repeaters can be operated this way by using a multicoupler with the appropriate number of input ports.

duplexer for 600-kHz spacing at 146 MHz has a loss per transmission path of approximately 1.5 dB. A four-channel multicoupler (the requirement for two repeaters) has an insertion loss per transmission path on the order of 2.5 dB or more. Another constraint of such a system is that the antenna must present a good match to the transmission line at all frequencies on which it will be used (both transmitting and receiving). This becomes difficult for the system with two repeaters operating at opposite ends of a band.

If you elect to purchase a commercial base station antenna that requires you to specify a frequency to which the antenna must be tuned, be sure to indicate to the manufacturer the intended use of the antenna and the frequency extremes. In some cases, the only way the manufacturer can accommodate your request is to provide an antenna with some vertical-beam uptilt at one end of the band and some downtilt at the other end of the band. In the case of antennas with very high gain, this in itself may become a serious problem. Careful analysis of the situation is necessary before assembling such a system.

Diversity Techniques for Repeaters

Mobile flutter, "dead spots" and similar problems are a real problem for the mobile operator. The popularity of hand-held transceivers using low power and mediocre antennas causes similar problems. A solution to these difficulties is the use of some form of diversity reception. Diversity reception works because signals do not fade at the same rate when received by antennas at different locations (space diversity) or of different polarizations (polarization diversity).

Repeaters with large transmitter coverage areas often have difficulty "hearing" low power stations in peripheral areas or in dead spots. Space diversity is especially useful in such a situation. Space diversity utilizes separate receivers at different locations that are linked to the repeater. The repeater uses a circuit called a *voter* that determines which receiver has the best signal, and then selects the appropriate receiver from which to feed the repeater transmitter. This technique is helpful in urban areas where shadowing from large buildings and bridges causes problems. Space-diversity receiving, when properly executed, can give excellent results. But with the improvement come some disadvantages: added initial cost, maintenance costs, and the possibility of failure created by the extra equipment required. If installed and maintained carefully, problems are generally minimal.

A second improvement technique is the use of circularly polarized repeater antennas. This technique has been used in the FM broadcast field for many years, and has been considered for use in the mobile telephone service as well. Some experiments by amateurs have proved very promising, as discussed by Pasternak and Morris (see the bibliography at the end of this chapter).

The improvement afforded by circular polarization is primarily a reduction in mobile flutter. The flutter on a mobile signal is caused by reflections from large buildings (in urban settings) or other terrain features. These reflections cause measurable polarization shifts, sometimes to the point where a vertically polarized signal at the transmitting site may appear to be primarily horizontally polarized after reflection.

A similar situation results from multipath propagation, where one or more reflected signals combine with the direct signal at the repeater, having varying effects on the signal. The multipath signal is subjected to large amplitude and phase variations at a relatively rapid rate.

In both of the situations described here, circular polarization can offer considerable improvement. This is because circularly polarized antennas respond equally to all linearly polarized signals, regardless of the plane of polarization. At this writing, there are no known sources of commercial circularly polarized omnidirectional antennas for the amateur bands. Pasternak and Morris describe a circularly polarized antenna made by modifying two commercial four-pole arrays.

EFFECTIVE ISOTROPIC RADIATED POWER (EIRP)

It is useful to know effective isotropic radiated power (EIRP) in calculating the coverage area of a repeater. The FCC formerly required EIRP to be entered in the log of every amateur repeater station. Although logging EIRP is no longer required, it is still useful to have this information on hand for repeater coordination purposes and so system performance can be monitored periodically.

Calculation of EIRP is straightforward. The PEP output of the transmitter is simply multiplied by the gains and losses in the transmitting antenna system. (These gains and losses are best added or subtracted in decibels and then converted to a multiplying factor.) The following worksheet and example illustrates the calculations.

Feed-line loss	_____	dB
Duplexer loss	_____	dB
Isolator loss	_____	dB
Cross-band coupler loss	_____	dB
Cavity filter loss	_____	dB
	_____	
Total losses (L)	_____	dB

G (dB) = antenna gain (dBi) − L

where G = antenna system gain. (If antenna gain is specified in dBd, add 2.14 dB to obtain the gain in dBi.)

$M = 10^{G/10}$

where M = multiplying factor

EIRP (watts) = transmitter output (PEP) × M

Example

A repeater transmitter has a power output of 50 W PEP (50-W FM transmitter). The transmission line has 1.8 dB loss.

The duplexer used has a loss of 1.5 dB, and a circulator on the transmitter port has a loss of 0.3 dB. There are no cavity filters or cross-band couplers in the system. Antenna gain is 5.6 dBi.

Feed-line loss	1.8 dB
Duplexer loss	1.5 dB
Isolator loss	0.3 dB
Cross-band coupler loss	0 dB
Cavity filter loss	0 dB
Total losses (L)	3.6 dB

Antenna system gain in dB = G = antenna gain (dBi) − L

$$G = 5.6 \text{ dBi} - 3.6 \text{ dB} = 2 \text{ dB}$$

$$\text{Multiplying factor} = M = 10^{G/10}$$

$$M = 10^{2/10} = 1.585$$

$$\text{EIRP (watts)} = \text{transmitter output (PEP)} \times M$$

$$\text{EIRP} = 50 \text{ W} \times 1.585 = 79.25 \text{ W}$$

If the antenna system is lossier than this example, G may be *negative,* resulting in a multiplying factor less than 1. The result is an EIRP that is less than the transmitter output power. This situation can occur in practice, but for obvious reasons is not desirable.

Assembling a Repeater Antenna System

The charts and tables in this section are included to aid you in planning and assembling your repeater antenna system. The material was prepared by Domenic Mallozzi, N1DM. Consult Chapter 23 for information on propagation for the band of your interest.

First, a repeater antenna selection checklist such as this will help you in evaluating the antenna system for your needs.

Gain Needed _____ dBi

Pattern Required _____ Omnidirectional

_____ Offset

_____ Cardioidal

_____ Bidirectional

_____ Special Pattern

(specify)_____

Mounting _____ Top of Tower

_____ Side of Tower (determine effects of tower on pattern. Is the result consistent with the pattern required?).

Is Downtilt Required? _____ Yes

_____ No

Type of RF Connector _____ UHF

_____ N

_____ BNC

_____ other (specify) _____

Size (length) _____

Weight _____

Maximum Cost $_____

Tables 1 and 2 have been compiled to provide general information on commercial components available for repeater and remote-base antenna systems. Table 1 lists the various components in a matrix format by manufacturer, for equipment designed to operate in the various amateur bands. Table 2 provides addresses for the manufacturers listed in Table 1.

Although every effort has been made to make this data complete, the ARRL is not responsible for omissions or errors. The listing of a product in this chart does not constitute an endorsement by ARRL. Manufacturers are urged to contact the editors with updating information.

Even though almost any antenna can be used for a repeater, the companies indicated in the *antenna* column in Table 1 are known to have produced heavy-duty antennas to commercial standards for repeater service. Many of these companies offer their antennas with special features for repeater service (such as vertical-beam downtilt). It is best to obtain catalogs of current products from the manufacturers listed, both for general information and to determine which special options are available on their products.

Table 1
Product Matrix Showing Repeater Equipment and Manufacturer by Frequency Band

Source	Antennas							Duplexers					Cavity Filters				
	28	50	144	220	450	902	1296	50	144	220	450	902	50	144	220	450	902
Ant Spec	C	C	C		C												
Austin			C	C	C	S	S										
Celwave	C	C	C	C	C	C			C	C	C	C		C	C	C	C
Ckt Bd Spc								K	K				K	K			
Cushcraft		C	C	C	C												
Dec Prod		C	C	C	C	C			C	C				C		C	
Encomm			C		C		C										
MA/COM														C		C	
NCG			C		C	C	C										
Sinclair	C	C	C						C		C			C		C	C
Telwave									C	C	C			C		C	C
TX/RX									C	C	C			C		C	C

Note: Coaxial cable is not listed, because most manufacturers sell only to dealers.

Key to codes used:
C = catalog (standard) item
S = special-order item
K = kit

Table 2
Manufacturers of Repeater Antenna System Equipment and Addresses

Antenna Specialists Inc
P O Box 12370
Cleveland, OH 44112-0370

Austin Custom Antennas
P O Box 357
Sandown, NH 03873

Celwave RF Inc
Route 79
Marlboro, NJ 07746

Circuit Board Specialists
P O Box 951
Pueblo, Co 81002

Cushcraft Corp
P O Box 4680
48 Perimeter Road
Manchester, NH 03108

Decibel Products Inc
P O Box 47128
3184 Quebec Street
Dallas, TX 75247

Encomm (Diamond) Inc
1506 Capital Avenue, Suite 300
Plano, TX 75074

MA/COM Land Mobile Communications
21 Continental Blvd
Merrimack, NH 03054

NCG, Inc
1275 North Grove Street
Anaheim, CA 92806

Sinclair Radio Laboratories, Inc
675 Ensminger Rd
Tonawanda, NY 14150

Telwave, Inc
1155 Terra Bella
Mountain View, CA 94043

TX/RX Systems Inc
8625 Industrial Parkway
Angola, NY 14006

A 144-MHz Duplexer

Obtaining sufficient isolation between the transmitter and receiver of a repeater can be difficult. Many of the solutions to this problem compromise receiver sensitivity or transmitter power output. Other solutions create an imbalance between receiver and transmitter coverage areas. When a duplexer is used, insertion loss is the compromise. But a small amount of insertion loss is more than offset by the use of one antenna for both the transmitter and receiver. Using one antenna assures equal antenna patterns for both transmitting and receiving, and reduces cost, maintenance and mechanical complexity.

As mentioned earlier in this chapter, duplexers may be built in the home workshop. Bob Shriner, WA0UZO, presented a small, mechanically simple duplexer for low power

| Isolators/Circulators | | | | | | Transmitter Multicouplers | | | | Cross-Band Couplers | | | | Source |
28	50	144	220	450	902	144	220	450	902	0-174 and 450-512	0-512 and 800-960	50-174 and 806-960	406-512 and 806-960	
														Ant Spec
														Austin
			C	C		C		C	C					Celwave
														Ckt Bd Spc
														Cushcraft
		C		C	C		C	C	C					Dec Prod
														Encomm
C	C			C	C					C	C			MA/COM
														NCG
		C		C	C									Sinclair
		C		C	C	C		C	C			C	C	Telwave
									C					TX/RX

applications in April 1979 *QST*. Shriner's design is unique, as the duplexer cavities are constructed of circuit board material. Low cost and simplicity are the result, but with a trade-off in performance. A silver-plated version of Shriner's design has an insertion loss of approximately 5 dB at 146 MHz. The loss is greater if the copper is not plated, and increases as the inner walls of the cavities tarnish.

This duplexer construction project by John Bilodeau, W1GAN, represents an effective duplexer. The information originally appeared in July 1972 *QST*. It is a time proven project used by many repeater groups, and can be duplicated relatively easily. Its insertion loss is just 1.5 dB.

Fig 18 will help you visualize the requirements for a duplexer, which can be summed up as follows. The duplexer must attenuate the transmitter carrier to avoid overloading the receiver and thereby reducing its sensitivity. It must also attenuate any noise or spurious frequencies from the transmitter on or near the receiver frequency. In addition, a duplexer must provide a proper impedance match between transmitter, antenna, and receiver.

As shown in Fig 18, transmitter output on 146.94 MHz going from point C to D should not be attenuated. However, the transmitter energy should be greatly attenuated between points B and A. Duplexer section 2 should attenuate any noise or signals that are on or near the receiver input frequency of 146.34 MHz. For good reception the noise and spurious signal level must be less than -130 dBm (0 dBm = 1 milliwatt into 50 ohms). Typical transmitter noise 600 kHz away from the carrier frequency is 80 dB below the transmitter power output. For 60 watts of output ($+48$ dBm), the noise level is -32 dBm. The duplexer must make up the difference between -32 dBm and -130 dBm, or 98 dB.

The received signal must go from point B to A with a minimum of attenuation. Section 1 of the duplexer must also provide enough attenuation of the transmitter energy to prevent receiver overload. For an average receiver, the transmitter signal must be less than -30 dBm to meet this requirement. The difference between the transmitter output of $+48$ dBm and the receiver overload point of -30 dBm, 78 dB, must be made up by duplexer section 1.

THE CIRCUIT

Fig 19 shows the completed 6-cavity duplexer, and Fig 20 shows the assembly of an individual cavity. A ¼-λ resonator was selected for this duplexer design. The length of the center conductor is adjusted by turning a threaded rod, which changes the resonant frequency of the cavity. Energy

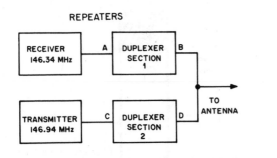

Fig 18—Duplexers permit using one antenna for both transmitting and receiving in a repeater system. Section 1 prevents energy at the transmitter frequency from interfering with the receiver, while section 2 attenuates any off-frequency transmitter energy that is at or near the receiver frequency.

Fig 19—A six-cavity duplexer for use with a 144-MHz repeater. The cavities are fastened to a plywood base for mechanical stability. Short lengths of double-shielded cable are used for connections between individual cavities. An insertion loss of less than 1.5 dB is possible with this design.

is coupled into and out of the tuned circuit by the coupling loops extending through the top plate.

The cavity functions as a series resonant circuit. When a reactance is connected across a series resonant circuit, an anti-resonant notch is produced, and the resonant frequency is shifted. If a capacitor is added, the notch appears below the resonant frequency. Adding inductance instead of capacitance makes the notch appear above the resonant frequency. The value of the added component determines the spacing between the notch and the resonant frequency of the cavity.

Fig 21 shows the measured band-pass characteristics of the cavity with shunt elements. With the cavity tuned to 146.94 MHz and a shunt capacitor connected from input to output, a 146.34-MHz signal is attenuated by 35 dB. If an inductance is placed across the cavity and the cavity is tuned to 146.34 MHz, the attenuation at 146.94 MHz is 35 dB. Insertion loss in both cases is 0.4 dB. Three cavities with shunt capacitors are tuned to 146.94 MHz and connected together

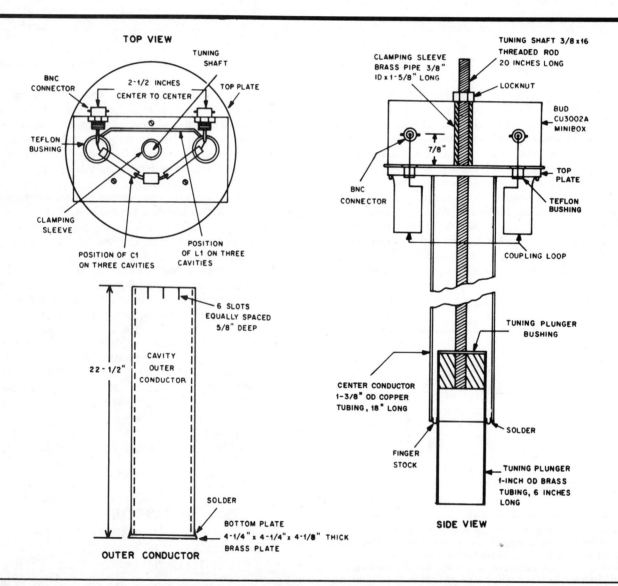

Fig 20—The assembly of an individual cavity. A Bud Minibox is mounted on the top plate with three screws. A clamping sleeve made of brass pipe is used to prevent crushing the box when the locknut is tightened on the tuning shaft. Note that the positions of both C1 and L1 are shown, but that three cavities will have C1 installed and three will have L1 in place.

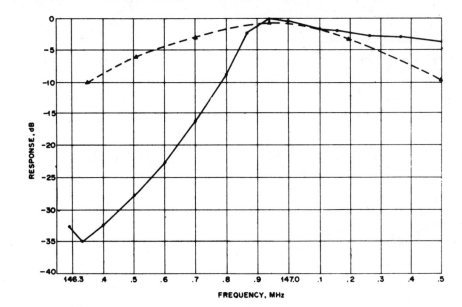

Fig 21—Typical frequency response of a single cavity of the type used in the duplexer. The dotted line represents the passband characteristics of the cavity alone; the solid line for the cavity with a shunt capacitor connected between input and output. An inductance connected in the same manner will cause the rejection notch to be above the frequency to which the cavity is tuned.

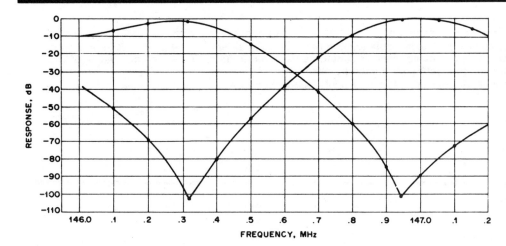

Fig 22—Frequency response of the six-cavity duplexer. One set of three cavities is tuned to pass 146.34 MHz and notch 146.94 MHz (the receiver leg). The remaining set of three cavities is tuned to pass 146.94 MHz and notch 146.34 MHz. This duplexer provides approximately 100 dB of isolation between the transmitter and receiver when properly tuned.

in cascade with short lengths of coaxial cable. The attenuation at 146.34 MHz is more than 100 dB, and insertion loss at 146.94 MHz is 1.5 dB. Response curves for a six-cavity duplexer are given in Fig 22.

Construction

The schematic diagram for the duplexer is shown in Fig 23. Three parts for the duplexer must be machined; all others can be made with hand tools. A small lathe can be used to machine the brass top plate, the threaded tuning plunger bushing, and the Teflon insulator bushing. The dimensions of these parts are given in Fig 24.

Type DWV copper tubing is used for the outer conductor of the cavities. The wall thickness is 0.058 inch, with an outside diameter of 4-1/8 inches. You will need a tubing cutter large enough to handle this size (perhaps borrowed or rented). The wheel of the cutter should be tight and sharp. Make slow, careful cuts so the ends will be square. The outer conductor is 22½ inches long.

The inner conductor is made from type M copper tubing having an outside diameter of 1-3/8 inches. A 6-inch length of 1-inch OD brass tubing is used to make the tuning plunger.

The tubing types mentioned above are designations used in the plumbing and steam-fitting industry. Other types may be used in the construction of a duplexer, but you should check the sizes carefully to assure that the parts will fit each

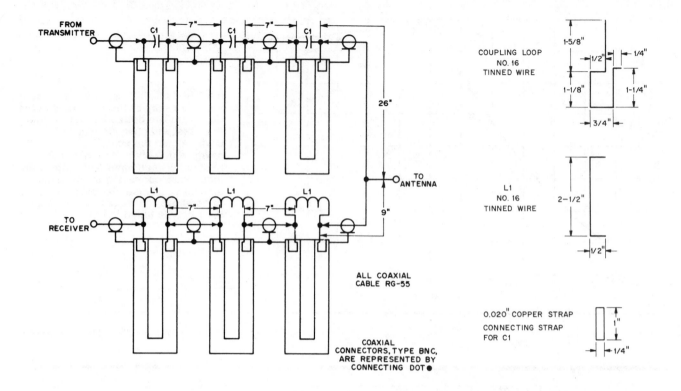

Fig 23—Diagram of the six-cavity duplexer. Coaxial cable lengths between cavities are critical and must be followed closely. Double-shielded cable and high-quality connectors should be used throughout. The sizes and shapes of the coupling loops, L1, and the straps for connecting C1 should be observed.
C1—1.7-11 pF circuit-board mount, E. F. Johnson 189-5-5 or equiv. Set at ¾ closed for initial alignment.

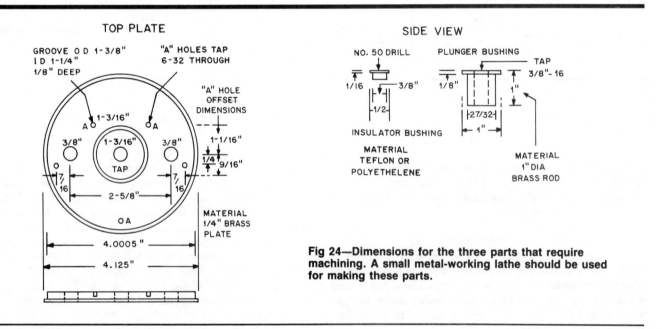

Fig 24—Dimensions for the three parts that require machining. A small metal-working lathe should be used for making these parts.

other. Tubing with a greater wall thickness will make the assembly heavier, and the expense will increase accordingly.

Soft solder is used throughout the assembly. Unless you have experience with silver solder, do not use it. Eutectic type 157 solder with paste or acid flux makes very good joints. This type has a slightly higher melting temperature than ordinary tin-lead alloy, but has considerably greater strength.

First solder the inner conductor to the top plate (Fig 25). The finger stock can then be soldered inside the lower end of the inner conductor, while temporarily held in place with a plug made of aluminum or stainless steel. While soldering, do not allow the flame from the torch to overheat the finger stock. The plunger bushing is soldered into the tuning plunger and a 20-inch length of threaded rod is soldered into the bushing.

Fig 25—Two of the center conductor and top plate assemblies. In the assembly at the left, C1 is visible just below the tuning shaft, mounted by short straps made from sheet copper. The assembly on the right has L1 in place between the BNC connectors. The Miniboxes are fastened to the top plate by a single large nut in these units. Using screws through the Minibox into the top plate, as described in the text, is preferred.

Cut six slots in the top of the outer conductor. They should be 5/8 inch deep and equally spaced around the tubing. The bottom end of the 4-inch tubing is soldered to the square bottom plate. The bottom plates have holes in the corners so they can be fastened to a plywood base by means of wood screws. Because the center conductor has no support at one end, the cavities must be mounted vertically.

The size and position of the coupling loops are critical. Follow the given dimensions closely. Both loops should be 1/8 inch away from the center conductor on opposite sides. Connect a solder lug to the ground end of the loop, then fasten the lug to the top plate with a screw. The free end of the loop is insulated by Teflon bushings where it passes through the top plate for connection to the BNC fittings.

Before final assembly of the parts, clean them thoroughly. Soap-filled steel wool pads and hot water work well for

this. Be sure the finger stock makes firm contact with the tuning plunger. The top plate should fit snugly in the top of the outer conductor—a large hose clamp tightened around the outer conductor will keep the top plate in place.

ADJUSTMENT

After the cavities have been checked for band-pass characteristics and insertion loss, install the anti-resonant elements, C1 and L1. (See Fig 21.) It is preferable to use laboratory test equipment when tuning the duplexer. An option is to use a low power transmitter with an RF probe and an electronic voltmeter. Both methods are shown in Fig 26.

With the test equipment connected as shown in Fig 26A, adjust the signal generator frequency to the desired repeater input frequency. Connect a calibrated step attenuator between points X and Y. With no attenuation, adjust the HP-415 for 0 on the 20-dB scale. You can check the calibration of the

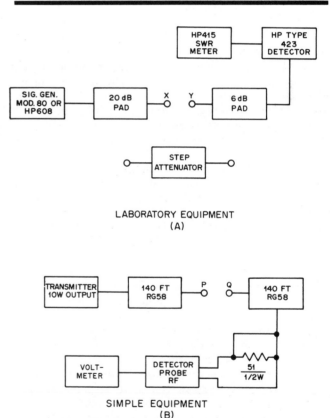

Fig 26—The duplexer can be tuned by either of the two methods shown here, although the method depicted at A is preferred. The signal generator should be modulated by a 1-kHz tone. If the setup shown at B is used, the transmitter should not be modulated, and should have a minimum of noise and spurious signals. The cavities to be aligned are inserted between X and Y in the setup at A, and between P and Q in B.

415 by switching in different amounts of attenuation and noting the meter reading. You may note a small error at either high or very low signal levels.

Next, remove the step attenuator and replace it with a cavity that has the shunt inductor, L1, in place. Adjust the tuning screw for maximum reading on the 415 meter. Remove the cavity and connect points X and Y. Set the signal generator to the repeater output frequency and adjust the 415 for a 0 reading on the 20-dB scale.

Reinsert the cavity between X and Y and adjust the cavity tuning for minimum reading on the 415. The notch should be sharp and have a depth of at least 35 dB. It is important to maintain the minimum reading on the meter while tightening the locknut on the tuning shaft.

To check the insertion loss of the cavity, the output from the signal generator should be reduced, and the calibration of the 415 meter checked on the 50-dB expanded scale. Use a fixed 1-dB attenuator to make certain the error is less than 0.1 dB. Replace the attenuator with the cavity and read the loss. The insertion loss should be 0.5 dB or less. The procedure is the same for tuning all six cavities, except that the frequencies are reversed for those having the shunt capacitor installed.

Adjustment with Minimum Equipment

A transmitter with a minimum of spurious output should be used for this method of adjustment. Most modern transmitters meet this requirement. The voltmeter in use should be capable of reading 0.5 volt (or less), full scale. The RF probe used should be rated to 100 MHz or higher. Sections of RG-58 cable are used as attenuators, as shown in Fig 26B. The loss in these 140-foot lengths is nearly 10 dB, and helps to isolate the transmitter in case of mismatch during tuning.

Set the transmitter to the repeater input frequency and connect P and Q. Obtain a reading between 1 and 3 volts on the voltmeter. Insert a cavity with shunt capacitors in place between P and Q and adjust the cavity tuning for a minimum reading on the voltmeter. (This reading should be between 0.01 and 0.05 volt.) The rejection in dB can be calculated by

$$dB = 20 \log (V1/V2)$$

This should be at least 35 dB. Check the insertion loss by putting the receiver on the repeater output frequency and noting the voltmeter reading with the cavity out of the circuit.

A 0.5-dB attenuator can be made from a 7-foot length of RG-58. This 7-foot cable can be used to check the calibration of the detector probe and the voltmeter.

Cavities with shunt inductance can be tuned the same way, but with the frequencies reversed. If two or more cavities are tuned while connected together, transmitter noise can cause the rejection readings to be low. In other words, there will be less attenuation.

Results

The duplexer is conservatively rated at 150 watts input, but, if constructed carefully, should be able to handle as much as 300 watts. Silver plating the interior surfaces of the cavities

is recommended if input power is to be greater than 150 watts. A duplexer of this type with silver-plated cavities has an insertion loss of less than 1 dB, and a rejection of more than 100 dB. Unplated cavities should be disassembled at least every two years, cleaned thoroughly, and then retuned.

Miscellaneous Notes

1) Double-shielded cable and high-quality connectors are *required* throughout the system.

2) The SWR of the antenna should not exceed 1.2:1 for proper duplexer performance.

3) Good shielding of the transmitter and receiver at the repeater is essential.

4) The antenna should have four or more wavelengths of vertical separation from the repeater.

5) Conductors in the near field of the antenna should be well bonded and grounded to eliminate noise.

6) The feed line should be electrically bonded and mechanically secured to the tower or mast.

7) Feed lines and other antennas in the near field of the repeater antenna should be well bonded and as far from the repeater antenna as possible.

8) Individual cavities can be used to improve the performance of separate antenna or separate site repeaters.

9) Individual cavities can be used to help solve intermodulation problems.

BIBLIOGRAPHY

Source material and more extended discussions of the topics covered in this chapter can be found in the references below.

P. Arnold, "Controlling Cavity Drift in Low-Loss Combiners," *Mobile Radio Technology*, Apr 1986, pp 36-44.

L. Barrett, "Repeater Antenna Beam Tilting," *Ham Radio*, May 1983, pp 29-35. (See correction, *Ham Radio*, Jul 1983, p 80.)

W. F. Biggerstaff, "Operation of Close Spaced Antennas in Radio Relay Systems," *IRE Transactions on Vehicular Communications*, Sep 1959, pp 11-15.

J. J. Bilodeau, "A Homemade Duplexer for 2-Meter Repeaters," *QST*, Jul 1972, pp 22-26, 47.

W. B. Bryson, "Design of High Isolation Duplexers and a New Antenna for Duplex Systems," *IEEE Transactions on Vehicular Communications*, Mar 1965, pp 134-140.

K. Connolly and P. Blevins, "A Comparison of Horizontal Patterns of Skeletal and Complete Support Structures," *IEEE 1986 Vehicular Technology Conference Proceedings*, pp 1-7.

S. Kozono, T. Tsuruhara and M. Sakamoto, "Base Station Polarization Diversity Reception for Mobile Radio," *IEEE Transactions on Vehicular Technology*, Nov 1984, pp 301-306.

J. Kraus, *Antennas* (New York: McGraw-Hill Book Co, 1950).

W. Pasternak and M. Morris, *The Practical Handbook of Amateur Radio FM & Repeaters* (Blue Ridge Summit, PA: Tab Books Inc, 1980), pp 355-363.

M. W. Scheldorf, "Antenna-To-Mast Coupling in Communications," *IRE Transactions on Vehicular Communications*, Apr 1959, pp 5-12.

R. D. Shriner, "A Low Cost PC Board Duplexer," *QST*, Apr 1979, pp 11-14.

W. V. Tilston, "Simultaneous Transmission and Reception with a Common Antenna," *IRE Transactions on Vehicular Communications*, Aug 1962, pp 56-64.

E. P. Tilton, "A Trap-Filter Duplexer for 2-Meter Repeaters," *QST*, Mar 1970, pp 42-46.

R. Wheeler, "Fred's Advice solves Receiver Desense Problem," *Mobile Radio Technology*, Feb 1986, pp 42-44.

R. Yang and F. Willis, "Effects of Tower and Guys on Performance of Side Mounted Vertical Antennas," *IRE Transactions on Vehicular Communications*, Dec 1960, pp 24-31.

Chapter 18

VHF and UHF Antenna Systems

A good antenna system is one of the most valuable assets available to the VHF/UHF enthusiast. Compared to an antenna of lesser quality, an antenna that is well designed, is built of good quality materials, and is well maintained, will increase transmitting range, enhance reception of weak signals, and reduce interference problems. The work itself is by no means the least attractive part of the job. Even with high gain antennas, experimentation is greatly simplified at VHF and UHF because the antennas are a physically manageable size. Setting up a home antenna range is within the means of most amateurs, and much can be learned about the nature and adjustment of antennas. No large investment in test equipment is necessary.

Selecting an Antenna

Selecting the best VHF or UHF antenna for a given installation involves much more than scanning gain figures and prices in a manufacturer's catalog. There is no one "best" VHF or UHF antenna design for all purposes. The first step in choosing an antenna is figuring out what you want it to do.

GAIN

At VHF and UHF, it is possible to build Yagi antennas with very high gain—15 to 20 dB—on a physically manageable boom. Such antennas can be combined in arrays of two, four, six, eight, or more antennas. These arrays are attractive for EME, tropospheric scatter or other weak-signal communications modes.

FREQUENCY RESPONSE

The ability to work over an entire VHF band may be important in some types of work. The response of an antenna element can be broadened somewhat by increasing the conductor diameter, and by tapering it to something approximating a cigar shape, but this is done mainly with simple antennas. More practically, wide frequency coverage may be a reason to select a collinear array, rather than a Yagi. On the other hand, the growing tendency to channelize operations in small segments of our bands tends to place broad frequency coverage low on the priority list of most VHF operators.

RADIATION PATTERNS

Antenna radiation can be made omnidirectional, bidirectional, practically unidirectional, or anything between these conditions. A VHF net operator may find an omnidirectional system almost a necessity, but it may be a poor choice otherwise. Noise pickup and other interference problems tend to be greater with such antennas, and such antennas having some gain are especially bad in these respects. Maximum gain and low radiation angle are usually prime interests of the weak signal DX aspirant. A clean pattern, with lowest possible pickup and radiation off the sides and back, may be important in high activity areas, or where the noise level is high.

HEIGHT GAIN

In general, the higher the better in VHF and UHF antenna installations. If raising the antenna clears its view over nearby obstructions, it may make dramatic improvements in coverage. Within reason, greater height is almost always worth its cost, but height gain (see Chapter 23) must be balanced against increased transmission-line loss. This loss can be considerable, and it increases with frequency. The best available line may not be very good if the run is long in terms of wavelengths. Line loss considerations (shown in table form in Chapter 24) are important in antenna planning.

PHYSICAL SIZE

A given antenna design for 432 MHz has the same gain as the same design for 144 MHz, but being only one-third as large intercepts only one-third as much energy in receiving. In other words, the antenna has less pickup efficiency at 432 MHz (see Chapter 2). To be equal in communication effectiveness, the 432-MHz array should be at least equal in *size* to the 144-MHz antenna, which requires roughly three times as many elements. With all the extra difficulties involved in using the higher frequencies effectively, it is best to keep antennas as large as possible for these bands.

DESIGN FACTORS

With the objectives sorted out in a general way, decisions on specifics, such as polarization, type of transmission line, matching methods and mechanical design must be made.

POLARIZATION

Whether to position antenna elements vertically or horizontally has been widely questioned since early VHF pioneering. Tests have shown little evidence as to which polarization sense is most desirable. On long paths, there is no consistent advantage either way. Shorter paths tend to yield higher signal levels with horizontally polarized antennas over some kinds of terrain. Man-made noise, especially ignition interference, also tends to be lower with horizontal antennas. These factors make horizontal polarization somewhat more desirable for weak-signal communications. On the other hand, vertically polarized antennas are much simpler to use in omnidirectional systems and in mobile work.

Vertical polarization was widely used in early VHF work, but horizontal polarization gained favor when directional arrays started to become widely used. The major trend to FM and repeaters, particularly in the 144-MHz band, has tipped the balance in favor of vertical antennas in mobile and repeater use. Horizontal polarization predominates in other communication on 50 MHz and higher frequencies. Additional loss of 20 dB or more can be expected when cross-polarized antennas are used.

TRANSMISSION LINES

Transmission line principles are covered in detail in Chapter 24. Techniques that apply to VHF and UHF operation are dealt with in greater detail here. The principles of carrying RF from one location to another via a feed line are the same for all radio frequencies. As at HF, RF is carried principally via open wire lines and coaxial cables at VHF/UHF. Certain aspects of these lines characterize them as good or bad for use above 50 MHz.

Properly built open wire line can operate with very low loss in VHF and UHF installations. A total line loss under 2 dB per 100 feet at 432 MHz can easily be obtained. A line made of no. 12 wire, spaced ¾ inch or more with Teflon spreaders and run essentially straight from antenna to station, can be better than anything but the most expensive coax. Such line can be made at a fraction of the cost of coaxial cables, with comparable loss characteristics. Careful attention must be paid to efficient impedance matching if the benefits of this system are to be realized. A similar system for 144 MHz can easily provide a line loss under 1 dB.

Small coax such as RG-58 or RG-59 should never be used in VHF work if the run is more than a few feet. Lines of ½-inch diameter (RG-8 or RG-11) work fairly well at 50 MHz, and are acceptable for 144-MHz runs of 50 feet or less. These lines are somewhat better if they employ foam instead of ordinary PE dielectric material. Aluminum jacketed lines with large inner conductors and foam insulation are well worth their cost, as they are easily waterproofed and can last almost indefinitely.

Beware of any "bargains" in coax for VHF or UHF use. Feed-line loss can be compensated to some extent by increasing transmitter power, but once lost, a weak signal can never be recovered in the receiver.

Effects of weather on transmission lines should not be ignored. Well constructed open wire line works optimally in nearly any weather, and it stands up well. Twin-lead is almost useless in heavy rain, wet snow or icing. The best grades of coax are completely impervious to weather; they can be run underground, fastened to metal towers without insulation, and bent into any convenient position with no adverse effects on performance.

G-LINE

Conventional two-conductor transmission lines and most coaxial cables are quite lossy in the upper UHF and microwave ranges. If the station and antenna are separated by more than 100 feet, common coaxial cables (such as RG-8) are almost useless for serious work. Unless the very best rigid coax with the proper fittings can be obtained, it is worthwhile to explore alternative methods of carrying RF energy between the station and antenna.

There is a single conductor transmission line, invented by Georg Goubau (called "G-line" in his honor), that can be effectively used in this frequency range. Papers by the inventor appeared some years ago, in which seemingly fantastic claims for line loss were made—under 1 dB per 100 feet in the microwave region, for example. (See the bibliography at the end of this chapter.) Especially attractive was the statement that the matching device was broadband in nature, making it appear that a single G-Line installation might be made to serve on, say, 432, 903 and 1296 MHz.

The basic idea is that a single conductor can be an almost lossless transmission line at UHF, if a suitable "launching device" is used. A similar "launcher" is placed at the other end. Basically, the launcher is a cone-shaped device that is a flared extension of the coaxial cable shield. In effect, the cone begins to carry the RF as the outer conductor is gradually "removed." These launch cones should be at least 3 λ long. The line should be large and heavily insulated, such as no. 14, vinyl covered.

Propagation along a G-Line is similar to "ground wave," or "surface wave" propagation over perfectly conducting earth. The dielectric material confines the energy to the vicinity of the wire, preventing radiation. The major drawback of G-Line is that it is very sensitive to deviation from straight lines. If any bends must be made, they should be in the form of a large radius arc. This is preferable to even an obtuse angle change in the direction of the run. The line must be kept several inches away from metal objects and should be supported with as few insulators as possible.

WAVEGUIDES

Above 2 GHz, coaxial cable is a losing proposition for communications work. Fortunately, at this frequency the wavelength is short enough to allow practical, efficient energy transfer by an entirely different means. A *waveguide* is a conducting tube through which energy is transmitted in the form of electromagnetic waves. The tube is not considered as carrying a current in the same sense that the wires of a two-conductor line do, but rather as a *boundary* that confines the waves in the enclosed space. Skin effect prevents any electromagnetic effects from being evident outside the guide. The energy is injected at one end, either through capacitive or inductive coupling or by radiation, and is removed from the other end in a like manner. Waveguide merely confines the energy of the fields, which are propagated through it to the receiving end by means of reflections against its inner walls.

Analysis of waveguide operation is based on the assumption that the guide material is a perfect conductor of electricity. Typical distributions of electric and magnetic fields in a rectangular guide are shown in Fig 1. The intensity of the electric field is greatest (as indicated by closer spacing of the lines of force) at the center along the X dimension (Fig 1C), diminishing to zero at the end walls. The fields must diminish in this manner, because the existence of any electric field parallel to the walls at the surface would cause an infinite

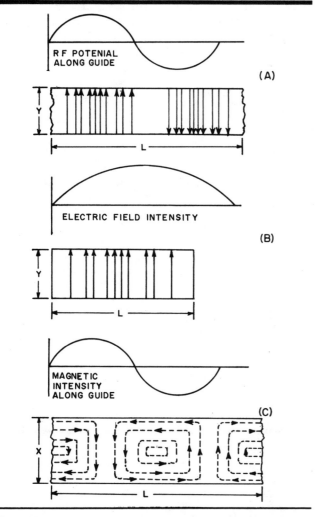

R F POTENTIAL ALONG GUIDE

(A)

ELECTRIC FIELD INTENSITY

(B)

MAGNETIC INTENSITY ALONG GUIDE

(C)

Fig 1—Field distribution in a rectangular waveguide. The TE$_{10}$ mode of propagation is depicted.

current to flow in a perfect conductor. Waveguides, of course, cannot carry RF in this fashion.

Modes of Propagation

Fig 1 represents the most basic distribution of the electric and magnetic fields in a waveguide. There are an infinite number of ways in which the fields can arrange themselves in a waveguide (for frequencies above the low cutoff frequency of the guide in use). Each of these field configurations is called a *mode*.

The modes may be separated into two general groups. One group, designated *TM* (transverse magnetic), has the magnetic field entirely transverse to the direction of propagation, but has a component of the electric field in that direction. The other type, designated *TE* (transverse electric) has the electric field entirely transverse, but has a component of magnetic field in the direction of propagation. TM waves are sometimes called E waves, and TE waves are sometimes called H waves, but the TM and TE designations are preferred.

The mode of propagation is identified by the group letters followed by two subscript numerals. For example, TE$_{10}$, TM$_{11}$, etc. The number of possible modes increases with frequency for a given size of guide, and there is only one possible mode (called the *dominant* mode) for the lowest frequency that can be transmitted. The dominant mode is the one generally used in amateur work.

Waveguide Dimensions

In rectangular guide the critical dimension is X in Fig 1. This dimension must be more than ½ λ at the lowest frequency to be transmitted. In practice, the Y dimension usually is made about equal to ½ X to avoid the possibility of operation in other than the dominant mode.

Cross-sectional shapes other than the rectangle can be used, the most important being the circular pipe. Much the same considerations apply as in the rectangular case.

Wavelength dimensions for rectangular and circular guides are given in Table 1, where X is the width of a rectangular guide and r is the radius of a circular guide. All figures apply to the dominant mode.

Coupling to Waveguides

Energy may be introduced into or extracted from a waveguide or resonator by means of either the electric or magnetic field. The energy transfer frequently is through a coaxial line. Two methods for coupling to coaxial line are shown in Fig 2. The probe shown at A is simply a short extension of the inner conductor of the coaxial line, oriented so that it is parallel to the electric lines of force. The loop shown at B is arranged so that it encloses some of the magnetic lines of force. The point at which maximum coupling is obtained depends upon the mode of propagation in the guide or cavity. Coupling is maximum when the coupling device is in the most intense field.

Coupling can be varied by turning the probe or loop through a 90° angle. When the probe is perpendicular to the electric lines the coupling is minimum; similarly, when the plane of the loop is parallel to the magnetic lines the coupling is minimum.

If a waveguide is left open at one end it will radiate energy. This radiation can be greatly enhanced by flaring the waveguide to form a pyramidal horn antenna. The horn acts as a transition between the confines of the waveguide and free space. To effect the proper impedance transformation the horn must be at least ½ λ on a side. A horn of this dimension (cutoff) has a unidirectional radiation pattern with a null toward the waveguide transition. The gain at the cutoff frequency is 3 dB, increasing 6 dB with each doubling of frequency. Horns are used extensively in microwave work, both as primary radiators and as feed elements for elaborate focusing systems. Details for constructing 10-GHz horn antennas are given later in this chapter.

Evolution of a Waveguide

Suppose an open wire line is used to carry RF energy from a generator to a load. If the line has any appreciable length it must be mechanically supported. The line must be well insulated from the supports if high losses are to be avoided. Because high quality insulators are difficult to construct at microwave frequencies, the logical alternative is to support the transmission line with ¼-λ stubs, shorted at the end opposite the feed line. The open end of such a stub presents an infinite impedance to the transmission line, provided the shorted stub is nonreactive. However, the shorting link has a finite length, and therefore some inductance. The effect of this inductance can be removed by making the RF current flow on the surface of a plate rather than a thin wire. If the plate is large enough, it will prevent the magnetic lines of force from encircling the RF current.

An infinite number of these ¼-λ stubs may be connected in parallel without affecting the standing waves of voltage and current. The transmission line may be supported from the top as well as the bottom, and when an infinite number of supports are added, they form the walls of a waveguide at its cutoff frequency. Fig 3 illustrates how a rectangular waveguide evolves from a two-wire parallel transmission line as described. This simplified analysis also shows why the cutoff dimension is ½ λ.

While the operation of waveguides is usually described in terms of fields, current does flow on the inside walls, just as on the conductors of a two-wire transmission line. At the waveguide cutoff frequency, the current is concentrated in the center of the walls, and disperses toward the floor and ceiling as the frequency increases.

IMPEDANCE MATCHING

Impedance matching is covered in detail in Chapters 25 and 26, and the theory is the same for frequencies above 50 MHz. Practical aspects are similar, but physical size can be a major factor in the choice of methods. Only the matching devices used in practical construction examples later in this chapter are discussed in detail here. This should not rule out consideration of other methods, however, and a reading of relevant portions of Chapter 26 is recommended.

Table 1
Waveguide Dimensions

	Rectangular	Circular
Cutoff wavelength	2X	3.41r
Longest wavelength transmitted with little attenuation	1.6X	3.2r
Shortest wavelength before next mode becomes possible	1.1X	2.8r

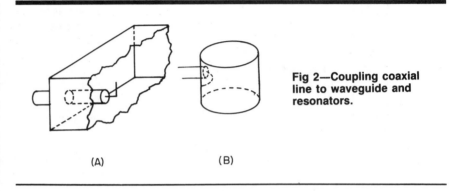

Fig 2—Coupling coaxial line to waveguide and resonators.

(A)   (B)

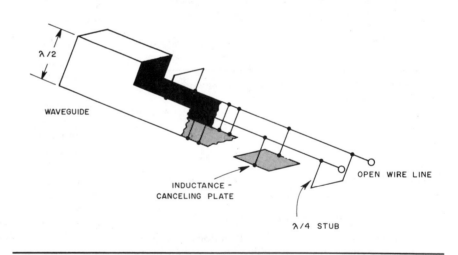

Fig 3—At its cutoff frequency a rectangular waveguide can be thought of as a parallel two-conductor transmission line supported from top and bottom by an infinite number of ¼-λ stubs.

UNIVERSAL STUB

As its name implies, the double adjustment stub of Fig 4A is useful for many matching purposes. The stub length is varied to resonate the system, and the transmission line attachment point is varied until the transmission line and stub impedances are equal. In practice this involves moving both the sliding short and the point of line connection for zero reflected power, as indicated on an SWR bridge connected in the line.

The universal stub allows for tuning out any small reactance present in the driven part of the system. It permits matching the antenna to the line without knowledge of the actual impedances involved. The position of the short yielding the best match gives some indication of the amount of reactance present. With little or no reactive component to be tuned out, the stub must be approximately ½ λ from load toward the short.

The stub should be made of stiff bare wire or rod, spaced

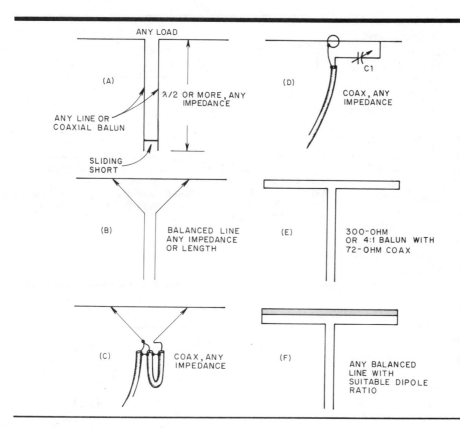

Fig 4—Matching methods commonly used at VHF. The universal stub, A, combines tuning and matching. The adjustable short on the stub and the points of connection of the transmission line are adjusted for minimum reflected power on the line. In the delta match, B and C, the line is fanned out and connected to the dipole at the point of optimum impedance match. Impedances need not be known in A, B or C. The gamma match, D, is for direct connection of coax. C1 tunes out inductance in the arm. A folded dipole of uniform conductor size, E, steps up antenna impedance by a factor of four. Using a larger conductor in the unbroken portion of the folded dipole, F, gives higher orders of impedance transformation.

no more than 1/20 λ apart. Preferably it should be mounted rigidly, on insulators. Once the position of the short is determined, the center of the short can be grounded, if desired, and the portion of the stub no longer needed can be removed.

It is not necessary that the stub be connected directly to the driven element. It can be made part of an open wire line, as a device to match coaxial cable to the line. The stub can be connected to the lower end of a delta match or placed at the feed point of a phased array. Examples of these uses are given later.

DELTA MATCH

Probably the most basic impedance matching device is the delta match, fanned ends of an open wire line tapped onto a ½-λ antenna at the point of most efficient power transfer. This is shown in Fig 4B. Both the side length and the points of connection either side of the center of the element must be adjusted for minimum reflected power on the line, but as with the universal stub, the impedances need not be known. The delta match makes no provision for tuning out reactance,

so the universal stub is often used as a termination for it, to this end.

At one time, the delta match was thought to be inferior for VHF applications because of its tendency to radiate if improperly adjusted. The delta has come back into favor now that accurate methods are available for measuring the effects of matching. It is very handy for phasing multiple bay arrays with open wire lines, and its dimensions in this use are not particularly critical. It should be checked out carefully in applications like that of Fig 4C, where no tuning device is used.

GAMMA MATCH

An application of the same principle allowing direct connection of coax is the gamma match, Fig 4D. Because the RF voltage at the center of a ½-λ dipole is zero, the outer conductor of the coax is connected to the element at this point. This may also be the junction with a metallic or wooden boom. The inner conductor, carrying the RF current, is tapped out on the element at the matching point. Inductance of the arm is tuned out by means of C1, resulting in electrical balance. Both the point of contact with the element and the setting of the capacitor are adjusted for zero reflected power, with a bridge connected in the coaxial line.

The capacitance can be varied until the required value is found, and the variable capacitor replaced with a fixed unit of that value. C1 can be mounted in a waterproof box. The maximum required value should be about 100 pF for 50 MHz and 35 to 50 pF for 144 MHz.

The capacitor and arm can be combined in one coaxial assembly with the arm connected to the driven element by means of a sliding clamp, and the inner end of the arm sliding inside a sleeve connected to the center conductor of the coax. An assembly of this type can be constructed from concentric pieces of tubing, insulated by plastic sleeving. RF voltage across the capacitor is low when the match is adjusted properly, so with a good dielectric, insulation presents no great problem. The initial adjustment should be made with low power. A clean, permanent high conductivity bond between arm and element is important, as the RF current is high at this point.

FOLDED DIPOLE

The impedance of a ½-λ antenna broken at its center is about 70 ohms. If a single conductor of uniform size is folded to make a ½-λ dipole as shown in Fig 4E, the impedance is stepped up four times. Such a folded dipole can be fed directly with 300-ohm line with no appreciable mismatch. If a 4:1 balun is used, the antenna can be fed with 75-Ω coaxial cable. (See balun information presented below.) Higher step-up impedance transformation can be obtained if the unbroken portion is made larger in cross-section than the fed portion, as shown in Fig 4F. Chapter 2 contains information on determining step-up ratios.

HAIRPIN MATCH

The feed-point resistance of most multielement Yagi arrays is less than 52 ohms. If the driven element is split and fed at the center, it may be shortened from its resonant length to add capacitive reactance at the feed point. Then, shunting the feed point with a wire loop resembling a hairpin causes a step-up of the feed-point resistance. The hairpin match is used in two of the 50-MHz arrays described later in this chapter.

BALUNS AND TRANSMATCHES

Conversion from balanced loads to unbalanced lines (or vice versa) can be performed with electrical circuits, or their equivalents made of coaxial cable. A balun made from flexible coax is shown in Fig 5A. The looped portion is an electrical ½ λ. The physical length depends on the velocity factor of the line used, so it is important to check its resonant frequency as shown in Fig 5B. The two ends are shorted, and the loop

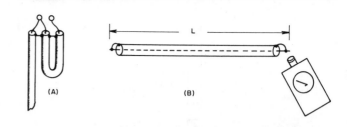

Fig 5—Conversion from unbalanced coax to a balanced load can be done with a ½-λ coaxial balun at A. Electrical length of the looped section should be checked with a dip meter, with the ends shorted, as at B. The ½-λ balun gives a 4:1 impedance step-up.

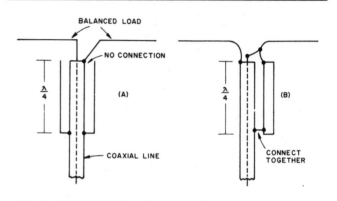

Fig 6—The balun conversion function, with no impedance transformation, can be accomplished with ¼-λ lines, open at the top and connected to the coax outer conductor at the bottom. The coaxial sleeve at A is preferred.

at one end is coupled to a dip meter coil. This type of balun gives an impedance step-up of 4:1 (typically 50 to 200 ohms, or 75 to 300 ohms).

Coaxial baluns that yield 1:1 impedance transformations are shown in Fig 6. The coaxial sleeve, open at the top and connected to the outer conductor of the line at the lower end (A) is the preferred type. At B, a conductor of approximately the same size as the line is used with the outer conductor to form a ¼-λ stub. Another piece of coax, using only the outer conductor, will serve this purpose. Both baluns are intended to present an infinite impedance to any RF current that might otherwise flow on the outer conductor of the coax.

The functions of the balun and the impedance transformer can be handled by various tuned circuits. Such a device, commonly called an antenna tuner or Transmatch, can provide a wide range of impedance transformations. Additional selectivity inherent in the Transmatch can reduce RFI problems.

THE YAGI AT VHF AND UHF

The relatively small size of VHF and UHF arrays opens a wide range of construction possibilities. Finding components is becoming difficult for home constructors of ham gear in general, but this should not limit the antenna constructor. Radio and TV parts distributors have many useful antenna parts and materials. Hardware stores, metals suppliers, lumber yards, welding supply and plumbing supply houses, and even junkyards, should not be overlooked. A little imagination makes parts procurement much less difficult than it might otherwise be.

Element and Boom Dimensions

Tables 2 through 5 list element and boom dimensions for several Yagi configurations for 50, 144, 220 and 432 MHz. Boom materials and sizes recommended in Chapter 21 for VHF and UHF arrays are usable with dimensions given in Table 2. Larger diameters broaden the frequency response; smaller ones sharpen it. Much smaller diameters than those recommended require longer elements (see element diameter scaling in Chapter 2).

The four tables are based on information contained in "Yagi Antenna Design" by Peter Viezbicke at the National

Table 2

NBS 50.1-MHz Yagi Dimensions

Electrical Boom Length	0.4 λ		0.8 λ		1.2 λ		2.2 λ	
Boom Length	7′10″		15′8-1/2″		23′6-7/8″		39′3-3/8″	
Boom Diameter	1-1/4″		2″		2″		2″	
Element Diameter	1/2″		3/4″		3/4″		3/4″	
Insulated Elements	Yes	No	Yes	No	Yes	No	Yes	No
Reflector	9′7″	9′7-3/4″	9′6-1/2″	9′7-3/4″	9′6-1/2″	9′7-3/4″	9′6-1/2″	9′7-3/4″
Driven	9′1-3/4″	9′1-3/4″	9′1-3/4″	9′1-3/4″	9′1-3/4″	9′1-3/4″	9′1-3/4″	9′1-3/4″
Dir. 1	9′5/8″	9′1-3/8″	8′9-1/8″	8′10-1/4″	8′9-1/8″	8′10-1/4″	8′9-7/8″	8′11″
Dir. 2	—	—	8′8-3/8″	8′9-5/8″	8′7-3/4″	8′8-7/8″	8′7″	8′8-1/8″
Dir. 3	—	—	8′9-1/8″	8′10-1/4″	8′7-3/4″	8′8-7/8″	8′5-3/8″	8′6-1/2″
Dir. 4	—	—	—	—	8′9-1/8″	8′10-1/4″	8′3-1/2″	8′4-5/8″
Dir. 5	—	—	—	—	—	—	8′1-3/4″	8′3″
Dir. 6	—	—	—	—	—	—	8′1-3/4″	8′3″
Dir. 7	—	—	—	—	—	—	8′1-3/4″	8′3″
Dir. 8	—	—	—	—	—	—	8′1-3/4″	8′3″
Dir. 9	—	—	—	—	—	—	8′3-1/2″	8′4-5/8″
Dir. 10	—	—	—	—	—	—	8′5-3/8″	8′6-1/2″

Table 3

NBS 144.1-MHz Yagi Dimensions

	0.8 λ		1.2 λ		2.2 λ	
Electrical Boom Length	0.8 λ		1.2 λ		2.2 λ	
Boom Length	5'5-9/16"		8'2-5/16"		15'-1/4"	
Boom Diameter	1"		1"		1-1/4"	
Element Diameter	3/16"		3/16"		3/16"	
Insulated Elements	Yes	No	Yes	No	Yes	No
Reflector	3'4"	3'4-5/8"	3'4"	3'4-5/8"	3'4"	3'4-13/16"
Driven	3'2-3/16"	3'2-3/16"	3'2-3/16"	3'2-3/16"	3'2-3/16"	3'2-3/16"
Dir. 1	3'7/8"	3'1-1/2"	3'7/8"	3'1-1/2"	3'1-1/8"	3'1-15/16"
Dir. 2	3'11/16"	3'1-3/8"	3'7/16"	3'1-1/8"	3'5/16"	3'1-1/8"
Dir. 3	3'7/8"	3'1-1/2"	3'7/16"	3'1-1/8"	2'11-13/16"	3'5/8"
Dir. 4	—	—	3'7/8"	3'1-1/2"	2'11-1/4"	3'0"
Dir. 5	—	—	—	—	2'10-9/16"	2'11-3/8"
Dir. 6	—	—	—	—	2'10-9/16"	2'11-3/8"
Dir. 7	—	—	—	—	2'10-9/16"	2'11-3/8"
Dir. 8	—	—	—	—	2'10-9/16"	2'11-3/8"
Dir. 9	—	—	—	—	2'11-1/4"	3'0"
Dir.10	—	—	—	—	2'11-13/16"	3'5/8"
Dir.11	—	—	—	—	—	—
Dir.12	—	—	—	—	—	—
Dir.13	—	—	—	—	—	—
Dir.14	—	—	—	—	—	—
Dir.15	—	—	—	—	—	—

Table 4

NBS 220.1-MHz Yagi Dimensions

	0.8 λ		1.2 λ		2.2 λ	
Electrical Boom Length	0.8 λ		1.2 λ		2.2 λ	
Boom Length	3'6-15/16"		5'4-3/8"		9'10"	
Boom Diameter	1"		1"		1"	
Element Diameter	3/16"		3/16"		3/16"	
Insulated Elements	Yes	No	Yes	No	Yes	No
Reflector	2'2-1/6"	2'2-3/4"	2'2-1/16"	2'2-3/4"	2'2-1/6"	2'2-3/4"
Driven	2'1"	2'1"	2'1"	2'1"	2'1"	2'1"
Dir. 1	1'11-13/16"	2'1/2"	1'11-13/16"	2'1/2"	2'1/16"	2'3/4"
Dir. 2	1'11-11/16"	2'3/8"	1'11-9/16"	2'1/4"	1'11-5/16"	2'1/16"
Dir. 3	1'11-13/16"	2'1/2"	1'11-9/16"	2'1/4"	1'10-15/16"	1'11-5/8"
Dir. 4	—	—	1'11-3/16"	2'1/2"	1'10-1/2"	1'11-1/4"
Dir. 5	—	—	—	—	1'10-1/8"	1'10-7/8"
Dir. 6	—	—	—	—	1'10-1/8"	1'10-7/8"
Dir. 7	—	—	—	—	1'10-1/8"	1'10-7/8"
Dir. 8	—	—	—	—	1'10-1/8"	1'10-7/8"
Dir. 9	—	—	—	—	1'10-1/2"	1'11-1/4"
Dir. 10	—	—	—	—	1'10-15/16"	1'11-5/8"
Dir. 11	—	—	—	—	—	—
Dir. 12	—	—	—	—	—	—
Dir. 13	—	—	—	—	—	—
Dir. 14	—	—	—	—	—	—
Dir. 15	—	—	—	—	—	—

	3.2 λ		4.2 λ	
	21'10-1/16"		28'8-1/8"	
	1-1/2"		1-1/2"	
	3/16"		3/16"	
	Yes	No	Yes	No
	3'4"	3'5-1/16"	3'3-3/8"	3'4-1/2"
	3'2-3/16"	3'2-3/16"	3'2-3/16"	3'2-3/16"
	3'7/8"	3'1-15/16"	3'9/16"	3'1-5/8"
	3'9/16"	3'1-3/8"	3'9/16"	3'1-5/8"
	2'11-3/4"	3'13/16"	3'3/8"	3'1-7/16"
	2'11-1/8"	3'3/16"	2'11-5/8"	3'11/16"
	2'10-7/8"	3'0"	2'11-1/2"	3'9/16"
	2'10-9/16"	2'11-5/8"	2'11-1/8"	3'3/16"
	2'10-5/16"	2'11-3/8"	2'10-13/16"	2'11-7/8"
	2'10-5/16"	2'11-3/8"	2'10-9/16"	2'11-5/8"
	2'10-5/16"	2'11-3/8"	2'10-9/16"	2'11-5/8"
	2'10-5/16"	2'11-3/8"	2'10-9/16"	2'11-5/8"
	2'10-5/16"	2'11-3/8"	2'10-9/16"	2'11-5/8"
	2'10-5/16"	2'11-3/8"	2'10-9/16"	2'11-5/8"
	2'10-5/16"	2'11-3/8"	2'10-9/16"	2'11-5/8"
	2'10-5/16"	2'11-3/8"	—	—
	2'10-5/16"	2'11-3/8"	—	—

	3.2 λ		4.2 λ	
	14'3-11/16"		18'9-5/16"	
	1-1/4"		1-1/2"	
	3/16"		3/16"	
	Yes	No	Yes	No
	2'2-1/16"	2'3"	2'1-11/16"	2'2-3/4"
	2'1"	2'1"	2'1"	2'1"
	1'11-13/16"	2'3/4"	1'11-5/8"	2'11/16"
	1'11-9/16"	2'7/16"	1'11-5/8"	2'11/16"
	1'10-15/16"	1'11-7/8"	1'11-7/16"	2'1/2"
	1'10-1/2"	1'11-7/16"	1'10-7/8"	2'0"
	1'10-5/16"	1'11-1/4"	1'10-3/4"	1'11-13/16"
	1'10-1/8"	1'11"	1'10-1/2"	1'11-9/16"
	1'9-7/8"	1'10-13/16"	1'10-5/16"	1'11-3/8"
	1'9-7/8"	1'10-13/16"	1'10-1/8"	1'11-3/16"
	1'9-7/8"	1'10-13/16"	1'10-1/8"	1'11-3/16"
	1'9-7/8"	1'10-13/16"	1'10-1/8"	1'11-3/16"
	1'9-7/8"	1'10-13/16"	1'10-1/8"	1'11-3/16"
	1'9-7/8"	1'10-13/16"	1'10-1/8"	1'11-3/16"
	1'9-7/8"	1'10-13/16"	1'10-1/8"	1'11-3/16"
	1'9-7/8"	1'10-13/16"	—	—
	1'9-7/8"	1'10-13/16"	—	—

Table 5
NBS 432.1-MHz Yagi Dimensions

Electrical Boom Length	1.2 λ		2.2 λ		3.2 λ		4.2 λ	
Boom Length	2'8-13/16"		5'1/8"		7'3-15/32"		9'6-25/32"	
Boom Diameter	1"		1"		1"		1"	
Element Diameter	3/16"		3/16"		3/16"		3/16"	
Insulated Elements	Yes	No	Yes	No	Yes	No	Yes	No
Reflector	1'1-3/16"	1'1-15/16"	1'1-3/16"	1'1-15/16"	1'1-3/16"	1'1-15/16"	1'1"	1'1-3/4"
Driven	1'23/32"	1'23/32"	1'23/32"	1'23/32"	1'23/32"	1'23/32"	1'23/32"	1'23/32"
Dir. 1	11-13/16"	1'17/32"	11-29/32"	1'21/32"	11-27/32"	1'9/16"	11-11/16"	1'7/16"
Dir. 2	11-5/8"	1'11/32"	11-7/16"	1'3/16"	11-5/8"	1'11/32"	11-11/16"	1'7/16"
Dir. 3	11-5/8"	1'11/32"	11-1/4"	1'0"	11-1/4"	1'0"	11-19/32"	1'11/32"
Dir. 4	11-13/16"	1'17/32"	11"	11-3/4"	11"	11-3/4"	11-1/4"	1'0"
Dir. 5	—	—	10-13/16"	11-17/32"	10-29/32"	11-5/8"	11-5/32"	11-7/8"
Dir. 6	—	—	10-13/16"	11-17/32"	10-13/16"	11-17/32"	11"	11-3/4"
Dir. 7	—	—	10-13/16"	11-17/32"	10-11/16"	11-13/32"	10-29/32"	11-5/8"
Dir. 8	—	—	10-13/16"	11-17/32"	10-11/16"	11-13/32"	10-13/16"	11-17/32"
Dir. 9	—	—	11"	11-3/4"	10-11/16"	11-13/32"	10-13/16"	11-17/32"
Dir. 10	—	—	11-1/4"	1'0"	10-11/16"	11-13/32"	10-13/16"	11-17/32"
Dir. 11	—	—	—	—	10-11/16"	11-13/32"	10-13/16"	11-17/32"
Dir. 12	—	—	—	—	10-11/16"	11-13/32"	10-13/16"	11-17/32"
Dir. 13	—	—	—	—	10-11/16"	11-13/32"	10-13/16"	11-17/32"
Dir. 14	—	—	—	—	10-11/16"	11-13/32"	—	—
Dir. 15	—	—	—	—	10-11/16"	11-13/32"	—	—

Bureau of Standards (see bibliography at the end of this chapter). Viezbicke offers element dimensions for maximum gain Yagi arrays, as well as for other types of antennas. The original information provides various element and boom diameters. The information shown in Tables 2 through 5 represents a highly condensed set of antenna designs, making use of standard and readily available material. Element and boom diameters have been chosen to produce lightweight, yet very rugged antennas.

Because these antennas are designed for maximum forward gain, the front to back (F/B) ratios are somewhat lower than those for some other designs. F/B ratios on the order of 15 to 25 dB are common for these antennas and should be more than adequate for most installations. The patterns are quite clean, with the side lobes well suppressed. The driven element lengths for the antennas represent good starting point dimensions. The type of feed system used on the array may require longer or shorter driven element lengths, as appropriate. Generally speaking, a balanced feed system is preferred in order to prevent pattern skewing and the possibility of unwanted side lobes that result from unbalanced feed systems.

Element spacings for the various arrays are presented in Fig 7 in terms of the wavelength of the boom, as noted in the first lines of Tables 2 through 5. The 0.4, 0.8, 2.2 and 3.2-λ boom antennas have equally spaced elements for both reflector and directors; 1.2- and 4.2-λ boom antennas have different reflector and director spacings. As all of the antenna parameters are interrelated, changes in element diameter, boom diameter and element spacing require design changes.

As the information presented by Viezbicke is straightforward, the serious antenna experimenter should have no difficulty designing antennas with different dimensions from those presented in the tables. For antennas for different frequencies within the band, standard scaling techniques may be applied. See Chapter 2.

Trigonal Reflectors

One of the experiments Viezbicke documents concerns reflector arrangements other than single cylindrical elements. The report claims that a 4.2-λ Yagi gains 0.75 dB when the reflector configuration of Fig 8 is used in place of the conventional single-element reflector. This modification was not applied to the shorter arrays, but the report suggests a similar advantage would be had. In cases where one is restricted to a short boom, a trigonal reflector may produce a significant improvement—possibly equivalent to adding a director to a conventional design.

STACKING YAGIS

Where suitable provision can be made for supporting them, two Yagis mounted one above the other and fed in phase can provide better performance than one long Yagi with the same theoretical or measured gain. The pair occupies a much smaller turning space for the same gain, and their lower radiation angle can provide excellent results. On long ionospheric paths, a stacked pair occasionally may show an

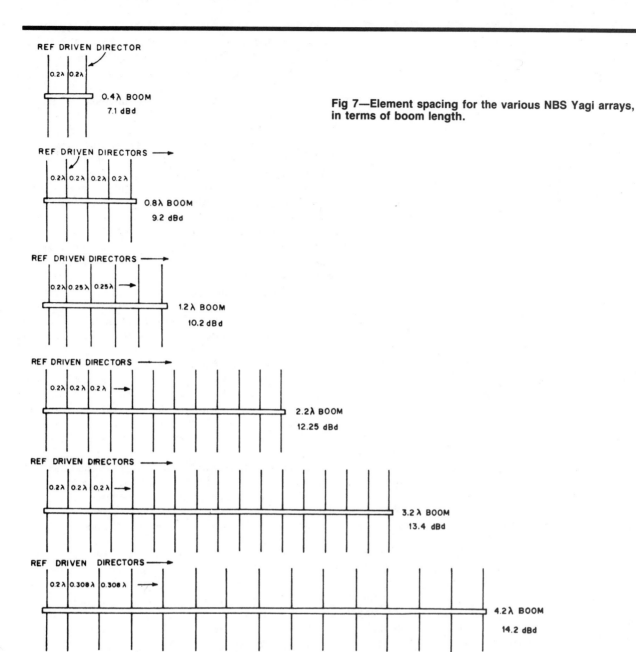

REF DRIVEN DIRECTOR

0.2λ 0.2λ

0.4λ BOOM
7.1 dBd

REF DRIVEN DIRECTORS →

0.2λ 0.2λ 0.2λ 0.2λ

0.8λ BOOM
9.2 dBd

REF DRIVEN DIRECTORS →

0.2λ 0.25λ 0.25λ →

1.2λ BOOM
10.2 dBd

REF DRIVEN DIRECTORS →

0.2λ 0.2λ 0.2λ →

2.2λ BOOM
12.25 dBd

REF DRIVEN DIRECTORS →

0.2λ 0.2λ 0.2λ →

3.2λ BOOM
13.4 dBd

REF DRIVEN DIRECTORS →

0.2λ 0.308λ 0.308λ →

4.2λ BOOM
14.2 dBd

Fig 7—Element spacing for the various NBS Yagi arrays, in terms of boom length.

apparent gain much greater than the measured 2 to 3 dB of stacking gain.

Optimum spacing for Yagis of five elements or more is 1 λ, but this may be too much for many builders of 50-MHz antennas to handle. Worthwhile results can be obtained with as little as ½ λ (10 feet), but 5/8 λ (12 feet) is markedly better. The difference between 12 and 20 feet may not be worth the added structural problems involved in the wider spacing, at least at 50 MHz. The closer spacings give lower measured gain, but the antenna patterns are cleaner than with 1 λ spacing.

The extra gain with wider spacings is usually the objective on 144 MHz and the higher frequency bands, where the structural problems are not as severe.

Yagis can also be stacked in the same plane (collinear elements) for sharp azimuth directivity. A spacing of 5/8 λ between the ends of the inner elements yields the maximum gain within the main lobe of the array.

If individual bays of a stacked array are properly designed, they "look like" noninductive resistors to the phasing system that connects them. The impedances involved

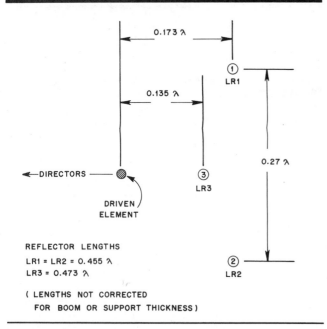

REFLECTOR LENGTHS

LR1 = LR2 = 0.455 λ
LR3 = 0.473 λ

(LENGTHS NOT CORRECTED
FOR BOOM OR SUPPORT THICKNESS)

Fig 8—A trigonal reflector (end view shown) can increase the gain of a Yagi by as much as 0.75 dB.

can thus be treated the same as resistances in parallel, if the phasing lines are ½ λ long or a multiple thereof. This is true because, disregarding losses, the impedance at the end of a transmission line is repeated at every electrical ½ λ along the length.

Three sets of stacked dipoles are shown in Fig 9. Whether these are merely dipoles or the driven elements of Yagi arrays makes no difference for the purpose of these examples. Two 300-Ω antennas at A are 1 λ apart, resulting in a feed-point impedance of approximately 150 Ω at the center. (Actually it is slightly less than 150 Ω because of coupling between bays, but this can be neglected for illustrative purposes.) This value remains the same regardless of the impedance of the phasing line. Thus, any convenient line can be used for phasing, as long as the *electrical* length is correct.

The velocity factor of the line must be taken into account as well. As with coax, this is subject to so much variation that it is important to make a resonance check on the actual line

used. The method for doing this is shown in Fig 5B. A ½-λ line is resonant both open and shorted, but the shorted condition (both ends) is usually the more convenient test condition.

The impedance transforming property of a ¼-λ line section can be used in combination matching and phasing lines, as shown in Fig 9B and C. At B, two bays spaced ½ λ apart are phased and matched by a 400-Ω line, acting as a double Q section, so that a 300 Ω main transmission line is matched to two 300-Ω bays. The two halves of this phasing line could also be 3/4 λ or 5/4 λ long, if such lengths serve a useful mechanical purpose. (An example is the stacking of two Yagis where the desirable spacing is more than ½ λ.)

A double Q section of coaxial line is illustrated in Fig 9C. This is useful for feeding stacked bays that were designed for 52-Ω feed. A spacing of 5/8 λ is optimum for small Yagis, and this is the equivalent of a full electrical wavelength of solid-dielectric coax such as RG-11. If the phasing line is electrically ¼ λ on one side of the feed and ¾ λ on the other, the connection to one driven element should be reversed with respect to the other to keep the RF currents in the elements in phase. If the number of ¼-λ is the same on either side of the feed point, the two connections should be in the same position, and not reversed as shown in Fig 9C.

One marked advantage of coaxial phasing lines is that they can be wrapped around the vertical support, taped or grounded to it, or arranged in any way that is mechanically convenient. The spacing between bays can be set at the most desirable value, and the phasing line placed anywhere necessary.

In stacking horizontal Yagis one above the other on a single support, certain considerations apply whether the bays are for different bands or for the same band. As a rule of thumb, the minimum desirable spacing is half the boom length for two bays on the same band, or half the boom length of the higher frequency array where two bands are involved.

Assume the stacked two-band array of Fig 10 is for 50 and 144 MHz. The 50-MHz, 4-element Yagi is going to "look like ground" to the 7-element 144-MHz Yagi above it, if it has any effect at all. It is well known that the impedance of an antenna varies with height above ground, passing through the free-space value at ¼ λ and multiples thereof. At ¼ λ and at the *odd* multiples thereof, ground also acts like a reflector, causing considerable radiation straight up. This effect is least at the ½-λ points, where the impedance also passes through the free-space value. Preferably, then, the

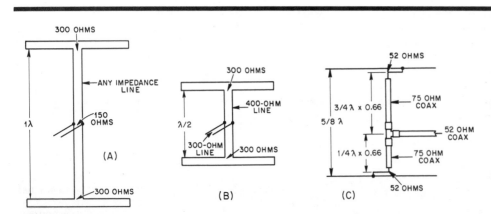

Fig 9—Three methods of feeding stacked VHF arrays. A and B are for bays having balanced driven elements, where a balanced phasing line is desired. Array C has an all-coaxial matching and phasing system. If the lower section is also ¾ λ, no transposition of line connections is needed.

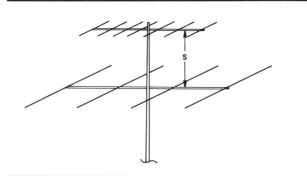

Fig 10—In stacking Yagi arrays one above the other, the minimum spacing between bays (S) should be about half the boom length of the smaller array. Wider spacing is desirable, in which case it should be ½ λ, or some multiple thereof, at the frequency of the smaller array. If the beams shown are for 50 and 144 MHz, S should be at least 40 in., but 80 in. is preferred. Similar conditions apply for stacking antennas for a single band.

spacing S should be ½ λ, or multiple thereof, at the frequency of the smaller antenna. The "half the boom length" rule gives about the same answer in this example. For this size 144-MHz antenna, 40 inches is the minimum desirable spacing, but 80 inches would be better.

The effect of spacing on the larger array is usually negligible. If spacing closer than half the boom length or ½ λ must be used, the principal concern is variation in feed impedance of the higher frequency antenna. If this antenna has an adjustable matching device, closer spacings can be used in a pinch, if the matching is adjusted for best SWR. Very close spacing and interlacing of elements should be avoided unless the builder is prepared to go through an extensive program of adjustments of both matching and element lengths.

QUADS FOR VHF

The quad antenna can be built of very inexpensive materials, yet its performance is comparable to other arrays of its size. Adjustment for resonance and impedance matching can be accomplished readily. Quads can be stacked horizontally and vertically to provide high gain, without sharply limiting frequency response. Construction of quad antennas for VHF use is covered later in this chapter.

Stacking Quads

Quads can be mounted side by side or one above the other, or both, in the same general way as other antennas. Sets of driven elements can also be mounted in front of a screen reflector. The recommended spacing between adjacent element sides is ½ λ. Phasing and feed methods are similar to those employed with other antennas described in this chapter.

Adding Directors

Parasitic elements ahead of the driven element work in a manner similar to those in a Yagi array. Closed loops can be used for directors by making them 5% shorter than the driven element. Spacings are similar to those for conventional Yagis. In an experimental model the reflector was spaced

0.25 λ and the director 0.15 λ. A square array using four 3-element bays works extremely well.

VHF AND UHF QUAGIS

At higher frequencies, especially 420 MHz and above, Yagi arrays using dipole-driven elements are difficult to feed and match. The cubical quad described earlier overcomes the feed problems to a large extent. When many parasitic elements are used, however, the loops are not nearly as convenient to assemble and tune as are straight cylindrical ones. The Quagi, designed and popularized by Wayne Overbeck, N6NB, is an antenna having a full-wave loop driven element and reflector, and Yagi type of straight rod directors. Construction details and examples are given in the projects later in this chapter.

COLLINEAR ANTENNAS

The information given earlier in this chapter pertains mainly to parasitic arrays, but the collinear array is worthy of consideration in VHF/UHF operations. Because it is inherently broad in frequency response, the collinear array is a logical choice where coverage of an entire band is desired (see Fig 11). This tolerance also makes collinear arrays easy to build and adjust for any VHF application. The use of many driven elements is popular in very large phased arrays, such as those required in moonbounce (EME) communications.

Large Collinear Arrays

Bidirectional curtain arrays of four, six, and eight half waves in phase are shown in Fig 12. Usually reflector elements are added, normally at about 0.2 λ behind each driven element, for more gain and a unidirectional pattern. Such parasitic elements are omitted from the sketch in the interest of clarity.

The feed-point impedance of two half waves in phase is high, typically 1000 Ω or more. When they are combined in parallel and parasitic elements are added, the feed impedance is low enough for direct connection to open wire line or twin-lead, connected at the points indicated by black dots. With coaxial line and a balun, it is suggested that the universal stub match, Fig 4A, be used at the feed point. All elements should be mounted at their electrical centers, as indicated by open

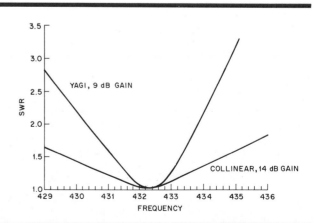

Fig 11—Comparison of the frequency responses of a small Yagi antenna and a large collinear array. A Yagi of comparable gain has an even sharper frequency response.

circles in Fig 12. The framework can be metal or insulating material. The metal supporting structure is entirely behind the plane of the reflector elements. Sheet-metal clamps can be cut from scraps of aluminum for this kind of assembly.

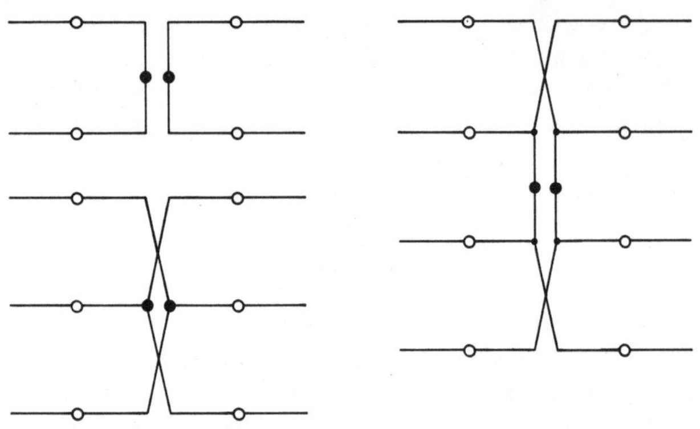

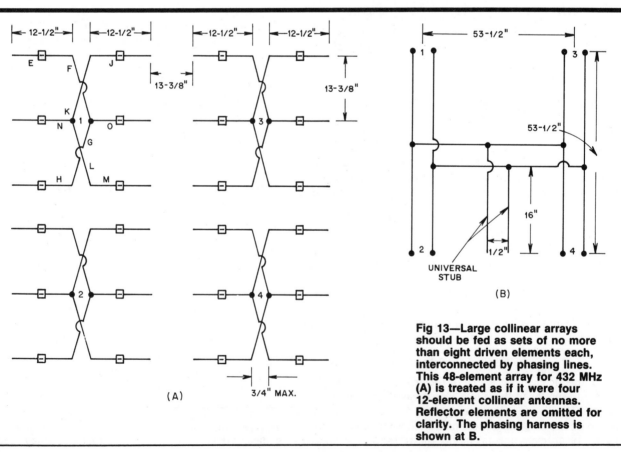

Fig 12—Element arrangements for 8, 12 and 16-element collinear arrays. Elements are ½ λ long and spaced ½ λ. Parasitic reflectors, omitted here for clarity, are 5% longer and 0.2 λ behind the driven elements. Feed points are indicated by black dots. Open circles show recommended support points. The elements can run through wood or metal booms, without insulation, if supported at their centers in this way. Insulators at the element ends (points of high RF voltage) detune and unbalance the system.

Collinear elements of this type should always be mounted at their centers (where the RF voltage is zero), never at their ends, where the voltage is high and insulation losses and detuning can be very harmful.

Collinear arrays of 32, 48, 64 and even 128 elements can give outstanding performance. Any collinear array should be fed at the center of the system, to ensure balanced current distribution. This is very important in large arrays, where sets of six or eight driven elements are treated as "sub arrays," and are fed through a balanced harness. The sections of the harness are resonant lengths, usually of open wire line. The 48-element collinear array for 432 MHz in Fig 13 illustrates this principle.

A reflecting plane, which may be sheet metal, wire mesh, or even closely spaced elements of tubing or wire, can be used in place of parasitic reflectors. To be effective, the plane reflector must extend on all sides to at least ¼ λ beyond the area occupied by the driven elements. The plane reflector provides high F/B ratio, a clean pattern, and somewhat more gain than parasitic elements, but large physical size limits it to use above 420 MHz. An interesting space-saving possibility lies in using a single plane reflector with elements for two different bands mounted on opposite sides. Reflector spacing from the driven element is not critical. About 0.2 λ is common.

THE CORNER REFLECTOR

When a single driven element is used, the reflector screen may be bent to form an angle, giving an improvement in the radiation pattern and gain. At 220 and 420 MHz its size

Fig 13—Large collinear arrays should be fed as sets of no more than eight driven elements each, interconnected by phasing lines. This 48-element array for 432 MHz (A) is treated as if it were four 12-element collinear antennas. Reflector elements are omitted for clarity. The phasing harness is shown at B.

assumes practical proportions, and at 902 MHz and higher, practical reflectors can approach ideal dimensions (very large in terms of wavelengths), resulting in more gain and sharper patterns. The corner reflector can be used at 144 MHz, though usually at much less than optimum size. For a given aperture, the corner reflector does not equal a parabola in gain, but it is simple to construct, broadbanded, and offers gains from about 10 to 15 dB, depending on the angle and size. This section was written by Paul M. Wilson, W4HHK.

The corner angle can be 90, 60 or 45°, but the side length must be increased as the angle is narrowed. For a 90° corner, the driven element spacing can be anything from 0.25 to 0.7 λ, 0.35 to 0.75 λ for 60°, and 0.5 to 0.8 λ for 45°. In each case the gain variation over the range of spacings given is about 1.5 dB. Because the spacing is not very critical to gain, it may be varied for impedance matching purposes. Closer spacings yield lower feed-point impedances, but a folded dipole radiator could be used to raise this to a more convenient level.

Radiation resistance is shown as a function of spacing in Fig 14. The maximum gain obtained with minimum spacing is the primary mode (the one generally used at 144, 220 and 432 MHz to maintain reasonable side lengths). A 90° corner, for example, should have a minimum side length (S, Fig 15) equal to twice the dipole spacing, or 1 λ long for 0.5-λ spacing. A side length greater than 2 λ is ideal. Gain with a 60° or 90° corner reflector with 1-λ sides is about 10 dB. A 60° corner with 2-λ sides has about 12 dB gain, and a 45° corner with 3-λ sides has about 13 dB gain.

Reflector length (L, Fig 15) should be a minimum of 0.6 λ. Less than that spacing causes radiation to increase to the sides and rear, and decreases gain.

Spacing between reflector rods (G, Fig 15) should not exceed 0.06 λ for best results. A spacing of 0.06 λ results in a rear lobe that is about 6% of the forward lobe (down 12 dB). A small mesh screen or solid sheet is preferable at the higher frequencies to obtain maximum efficiency and highest F/B ratio, and to simplify construction. A spacing of 0.06 λ at 1296 MHz, for example, requires mounting

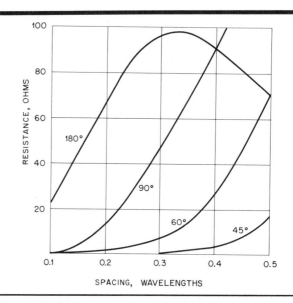

Fig 14—Radiation resistance of the driven element in a corner reflector array for corner angles of 180° (flat sheet), 90°, 60° and 45° as a function of spacing D, as shown in Fig 15.

reflector rods about every ½ inch along the sides. Rods or spines may be used to reduce wind loading. The support used for mounting the reflector rods may be of insulating or conductive material. Rods or mesh weave should be parallel to the radiator.

A suggested arrangement for a corner reflector is shown in Fig 15. The frame may be made of wood or metal, with a hinge at the corner to facilitate portable work or assembly atop a tower. A hinged corner is also useful in experimenting with different angles. Table 6 gives the principal dimensions for corner reflector arrays for 144 to 2300 MHz. The arrays for 144, 220 and 420 MHz have side lengths of twice to four times the driven element spacing. The 915-MHz corner

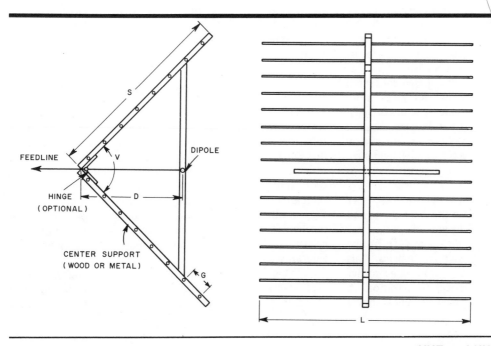

Fig 15—Construction of a corner reflector array. The frame can be wood or metal. Reflector elements are stiff wire or tubing. Dimensions for several bands are given in Table 6. Reflector element spacing, G, is the maximum that should be used for the frequency; closer spacings are optional. The hinge permits folding for portable use.

Table 6

Dimensions of Corner Reflector Arrays for VHF and UHF

Freq, MHz	Side Length S, in.	Dipole to Vertex D, in.	Reflector Length L, in.	Reflector Spacing G, in.	Corner Angle, V°	Feed Impedance, Ohms
144*	65	27½	48	7¾	90	70
144	80	40	48	4	90	150
220*	42	18	30	5	90	70
220	52	25	30	3	90	150
220	100	25	30	Screen	60	70
420	27	8¾	16¼	2-5/8	90	70
420	54	13½	16¼	Screen	60	70
915	20	6½	25¾	0.65	90	70
915	51	16¾	25¾	Screen	60	65
915	78	25¾	25¾	Screen	45	70
1296	18	4½	27½	½	90	70
1296	48	11¾	27½	Screen	60	65
1296	72	18¼	27½	Screen	45	70
2304	15½	2½	20½	¼	90	70
2304	40	6¾	20½	Screen	60	65
2304	61	10¼	20½	Screen	45	70

*Side length and number of reflector elements somewhat below optimum—slight reduction in gain.

Notes

915 MHz	*1296 MHz*	*2304 MHz*
Wavelength is 12.9 in.	Wavelength is 9.11 in.	Wavelength is 5.12 in.
Side length S is 3 × D, dipole to vertex distance	Side length S is 4 × D, dipole to vertex distance	Side length S is 6 × D, dipole to vertex distance
Reflector length L is 2.0 λ	Reflector length L is 3.0 λ	Reflector length L is 4.0 λ
Reflector spacing G is 0.05 λ	Reflector spacing G is 0.05 λ	Reflector spacing G is 0.05 λ

reflectors use side lengths of three times the element spacing, 1296-MHz corners use side lengths of four times the spacing, and 2304-MHz corners employ side lengths of six times the spacing. Reflector lengths of 2, 3 and 4 wavelengths are used on the 915, 1296 and 2304 MHz reflectors, respectively. A 4 × 6 λ reflector closely approximates a sheet of infinite dimensions.

A corner reflector may be used for several bands, or for UHF television reception, as well as amateur UHF work. For operation on more than one frequency, side length and reflector length should be selected for the lowest frequency, and reflector spacing for the highest frequency. The type of driven element plays a part in determining bandwidth, as does the spacing to the corner. A fat cylindrical element (small λ/dia ratio) or triangular dipole (bow tie) gives more bandwidth than a thin driven element. Wider spacings between driven element and corner give greater bandwidths. A small increase in gain can be obtained for any corner reflector by mounting collinear elements in a reflector of sufficient size, but the simple feed of a dipole is lost if more than two elements are used.

A dipole radiator is usually employed with a corner reflector. This requires a balun between the coaxial line and the balanced feed-point impedance of the antenna. Baluns are easily constructed of coaxial line on the lower VHF bands, but become more difficult at the higher frequencies. This problem may be overcome by using a ground-plane corner reflector, which can be used for vertical polarization. A ground-plane corner with monopole driven element is shown in Fig 16. The corner reflector and a ¼-λ radiator are mounted on the ground plane, permitting direct connection to a coaxial line if the proper spacing is used. The effective aperture is reduced, but at the higher frequencies, second- or third-mode radiator spacing and larger reflectors can be employed to obtain more gain and offset the loss in effective aperture. A J antenna could be used to maintain the aperture area and provide a match to a coaxial line.

For vertical polarization work, four 90° corner reflectors built back-to-back (with common reflectors) could be used for scanning 360° of horizon with modest gain. Feed-line switching could be used to select the desired sector.

TROUGH REFLECTORS

To reduce the overall dimensions of a large corner reflector the vertex can be cut off and replaced with a plane reflector. Such an arrangement is known as a *trough reflector*. See Fig 17. Performance similar to that of the large corner reflector can thereby be had, provided that the dimensions of S and T as shown in Fig 17 do not exceed the limits indicated in the figure. This antenna provides performance very similar to the corner reflector, and presents fewer mechanical problems because the plane center portion is relatively easy to mount on the mast. The sides are considerably shorter, as well.

The gain of both corner reflectors and trough reflectors may be increased by stacking two or more and arranging them to radiate in phase, or alternatively by adding further collinear

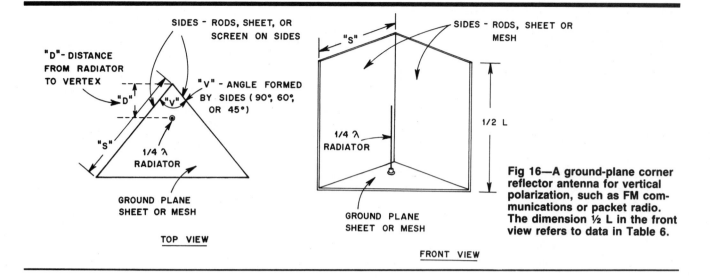

Fig 16—A ground-plane corner reflector antenna for vertical polarization, such as FM communications or packet radio. The dimension ½ L in the front view refers to data in Table 6.

TOP VIEW

FRONT VIEW

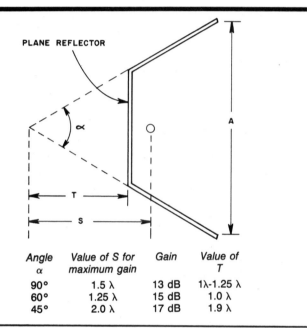

Angle α	Value of S for maximum gain	Gain	Value of T
90°	1.5 λ	13 dB	1λ-1.25 λ
60°	1.25 λ	15 dB	1.0 λ
45°	2.0 λ	17 dB	1.9 λ

Fig 17—The trough reflector. This is a useful modification of the corner reflector. The vertex has been cut off and replaced by a simple plane section. The tabulated data shows the gain obtainable for greater values of S than those covered in Table 6, assuming that the reflector is of adequate size.

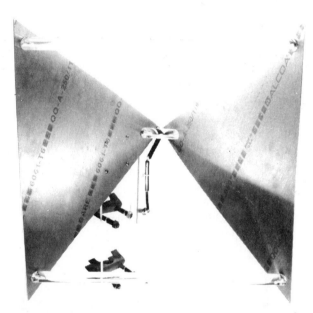

Fig 18—An experimental two-sided pyramidal horn constructed in the ARRL laboratory. A pair of muffler clamps allows mounting the antenna on a mast. This model has sheet-aluminum sides, although window screen would work as well. Temporary elements could be made from cardboard covered with aluminum foil. The horizontal spreaders are Plexiglas rod. Oriented as shown here, the antenna radiates horizontally polarized waves.

dipoles (fed in phase) within a wider reflector. Not more than two or three radiating units should be used, because the great virtue of the simple feeder arrangement would then be lost.

HORN ANTENNAS FOR THE MICROWAVE BANDS

Horn antennas were briefly introduced in the section on coupling energy into and out of waveguides. For amateur

purposes, horns begin to show usable gain with practical dimensions in the 902-MHz band.

It isn't necessary to feed a horn with waveguide. If only two sides of a pyramidal horn are constructed, the antenna may be fed at the apex with a two-conductor transmission line. The impedance of this arrangement is on the order of 300 to 400 Ω. A 60° two-sided pyramidal horn with 18-inch sides is shown in Fig 18. This antenna has a theoretical gain of 15 dBi at 1296 MHz, although the feed system detailed in

Fig 19 probably degrades this value somewhat. A ¼-λ, 150 Ω matching section made from two parallel lengths of twin-lead connects to a bazooka balun made from RG-58 cable and a brass tube. This matching system was assembled strictly for the purpose of demonstrating the two-sided horn in a 52-Ω system. In a practical installation the horn would be fed with open wire line and matched to 52 Ω at the station equipment.

PARABOLIC ANTENNAS

When an antenna is located at the focus of a parabolic reflector (dish), it is possible to obtain considerable gain. Furthermore, the beamwidth of the radiated energy will be very narrow, provided all the energy from the driven element is directed toward the reflector. This section was written by Paul M. Wilson, W4HHK.

Gain is a function of parabolic reflector diameter, surface accuracy and proper illumination of the reflector by the feed. Gain may be found from

$$G = 10 \log k \left(\frac{\pi D}{\lambda}\right)^2 \qquad \text{(Eq 1)}$$

where

G = gain over an isotropic antenna, dB (subtract 2.14 dB for gain over a dipole)
k = efficiency factor, usually about 55%
D = dish diameter in feet
λ = wavelength in feet

See Table 7 for parabolic antenna gain for the bands 420 MHz through 10 GHz and diameters of 2 to 30 feet.

A close approximation of beamwidth may be found from

$$\psi = \frac{70 \lambda}{D} \qquad \text{(Eq 2)}$$

where

ψ = beamwidth in degrees at half-power points (3 dB down)
D = dish diameter in feet
λ = wavelength in feet

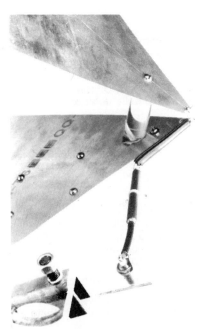

Fig 19—Matching system used to test the horn. Better performance would be realized with open wire line. See text.

Table 7
Gain, Parabolic Antennas*

Frequency	2	4	6	10	15	20	30
420 MHz	6.0	12.0	15.5	20.0	23.5	26.0	29.5
902	12.5	18.5	22.0	26.5	30.0	32.5	36.0
1215	15.0	21.0	24.5	29.0	32.5	35.0	38.5
2300	20.5	26.5	30.0	34.5	38.0	40.5	44.0
3300	24.0	30.0	33.5	37.5	41.5	43.5	47.5
5650	28.5	34.5	38.0	42.5	46.0	48.5	52.0
10 GHz	33.5	39.5	43.0	47.5	51.0	53.5	57.0

Dish Diameter (Feet) across top columns.

*Gain over an isotropic antenna (subtract 2.1 dB for gain over a dipole antenna). Reflector efficiency of 55% assumed.

At 420 MHz and higher, the parabolic dish becomes a practical antenna. A simple, single feed point eliminates phasing harnesses and balun requirements. Gain is dependent on good surface accuracy, which is more difficult to achieve with increasing frequency. Surface errors should not exceed 1/8 λ in amateur work. At 430 MHz 1/8 λ is 3.4 inches, but at 10 GHz it is 0.1476 inch! Mesh can be used for the reflector surface to reduce weight and wind loading, but hole size should be less than 1/12 λ. At 430 MHz the use of 2-inch hole diameter poultry netting (chicken wire) is acceptable. Fine mesh aluminum screening works well as high as 10 GHz.

A support form may be fashioned to provide the proper parabolic shape by plotting a curve (Fig 20) from

$$Y^2 = 4SX$$

as shown in the figure.

Optimum illumination occurs when power at the reflector edge is 10 dB less than that at the center. A circular waveguide

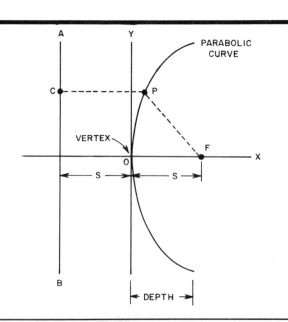

Fig 20—Details of the parabolic curve, Y² = 4SX. This curve is the locus of points which are equidistant from a fixed point, the focus (F), and a fixed line (AB) which is called the *directrix*. Hence, FP = PC. The focus (F) is located at coordinates S,0.

feed of correct diameter and length for the frequency and correct beamwidth for the dish focal length to diameter (f/D) ratio provides optimum illumination at 902 MHz and higher. This, however, is impractical at 432 MHz, where a dipole and plane reflector are often used. An f/D ratio between 0.4 and 0.6 is considered ideal for maximum gain and simple feeds.

The focal length of a dish may be found from

$$f = \frac{D^2}{16d}$$ (Eq 3)

where

  f = focal length
  D = diameter
  d = depth distance from plane at mouth of dish to vertex (see Fig 20)

The units of f are the same as those used to measure the depth and diameter.

Table 8 gives the subtended angle at focus for dish f/D ratios from 0.2 to 1.0. A dish, for example, with a typical f/D of 0.4 requires a 10-dB beamwidth of 130°. A circular waveguide feed with a diameter of approximately 0.7 λ provides nearly optimum illumination, but does not uniformly illuminate the reflector in both the magnetic (TM) and electric

(TE) planes. Fig 21 shows data for plotting radiation patterns from circular guides. The waveguide feed aperture can be modified to change the beamwidth.

Table 8
f/D Versus Subtended Angle at Focus of a Parabolic Reflector Antenna

f/D	Subtended Angle (Deg.)	f/D	Subtended Angle (Deg.)
0.20	203	0.65	80
0.25	181	0.70	75
0.30	161	0.75	69
0.35	145	0.80	64
0.40	130	0.85	60
0.45	117	0.90	57
0.50	106	0.95	55
0.55	97	1.00	52
0.60	88		

Taken from graph "f/D vs Subtended Angle at Focus," page 170 of the 1966 *Microwave Engineers' Handbook and Buyers Guide*. Graph courtesy of K. S. Kelleher, Aero Geo Astro Corp, Alexandria, Virginia.

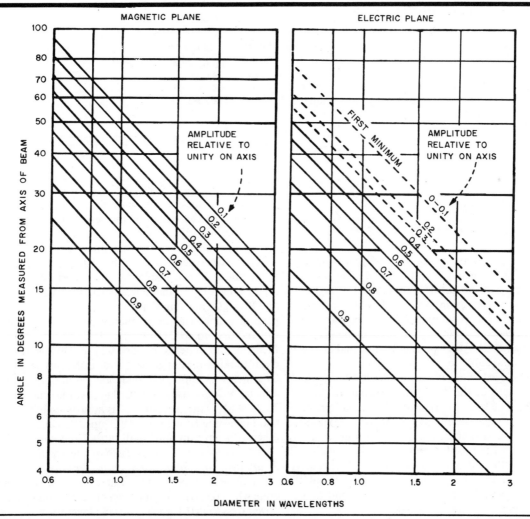

Fig 21—This graph can be used in conjunction with Table 8 for selecting the proper diameter waveguide to illuminate a parabolic reflector.

One approach used successfully by some experimenters is the use of a disc at a short distance behind the aperture as shown in Fig 22. As the distance between the aperture and disc is changed, the TM plane patterns become alternately broader and narrower than with an unmodified aperture. A disc about 2λ in diameter appears to be as effective as a much larger one. Some experimenters have noted a 1 to 2 dB increase in dish gain with this modified feed. Rectangular waveguide feeds can also be used, but dish illumination is not as uniform as with round guide feeds.

The circular feed can be made of copper, brass, aluminum or even tin in the form of a coffee or juice can, but the latter must be painted on the outside to prevent rust or corrosion. The circular feed must be within a proper size (diameter) range for the frequency being used. This feed operates in the dominant circular waveguide mode known as the TE_{11} mode. The guide must be large enough to pass the TE_{11} mode with no attenuation, but smaller than the diameter that permits the next higher TM_{01} mode to propagate. To support the desirable TE_{11} mode in circular waveguide, the cutoff frequency, F_C, is given by

$$F_C\,(TE_{11}) = \frac{6917.26}{d\,(inches)} \qquad (Eq\ 4)$$

where

$\quad f_C$ = cutoff frequency in MHz for TE_{11} mode
$\quad d$ = waveguide inner diameter

Circular waveguide will support the TM_{01} mode having a cutoff frequency

$$F_C\,(TM_{01}) = \frac{9034.85}{d\,(inches)} \qquad (Eq\ 5)$$

The wavelength in a waveguide always exceeds the free-space wavelength and is called guide wavelength, λ_g. It is related to the cutoff frequency and operating frequency by the equation

$$\lambda_g = \frac{11802.85}{\sqrt{f_0{}^2 - f_C{}^2}} \qquad (Eq\ 6)$$

where

$\quad \lambda_g$ = guide wavelength, inches
$\quad f_0$ = operating frequency, MHz
$\quad f_C$ = TE_{11} waveguide cutoff frequency, MHz

An inside diameter range of about 0.66 to 0.76 λ is suggested. The lower frequency limit (longer dimension) is dictated by proximity to the cutoff frequency. The higher frequency limit (shorter dimension) is dictated by higher order waves. See Table 9 for recommended inside diameter dimensions for the 902- to 10,000-MHz amateur bands.

The probe that excites the waveguide and makes the transition from coaxial cable to waveguide is $\frac{1}{4}\,\lambda$ long and spaced from the closed end of the guide by $\frac{1}{4}$ guide wavelength. The length of the feed should be two to three guide wavelengths. The latter is preferred if a second probe is to be mounted for polarization change or for polaplexer work where duplex communication (simultaneous transmission and reception) is possible because of the isolation between two properly located and oriented probes. The second probe for polarization switching or polaplexer work should be spaced $\frac{3}{4}$ guide wavelength from the closed end and mounted at right angles to the first probe.

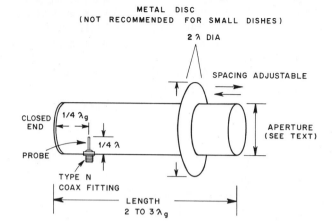

Probe spacing to closed end and probe length adjusted for minimum SWR and maximum signal. Vertical-polarization probe shown. Note: Drawing not to scale.

Fig 22—Details of a circular waveguide feed.

Table 9
Circular Waveguide Dish Feeds

Freq. (MHz)	Inside Diameter Circular Waveguide Range (in.)
915	8.52-9.84
1296	6.02-6.94
2304	3.39-3.91
3400	2.29-2.65
5800	1.34-1.55
10,250	0.76-0.88

The feed aperture is located at the focal point of the dish and aimed at the center of the reflector. The feed mounts should permit adjustment of the aperture either side of the focal point and should present a minimum of blockage to the reflector. Correct distance to the dish center places the focal point about 1 inch inside the feed aperture. The use of a nonmetallic support minimizes blockage. PVC pipe, fiberglass and Plexiglas are commonly used materials. A simple test by placing a material in a microwave oven reveals if it is satisfactory up to 2450 MHz. PVC pipe has tested satisfactorily and appears to work well at 2300 MHz. A simple, clean-looking mount for a 4-foot dish with 18 inches focal length, for example, can be made by mounting a length of 4-inch PVC pipe using a PVC flange at the center of the dish. At 2304 MHz the circular feed is approximately 4 inches ID, making a snug fit with the PVC pipe. Precautions should be taken to keep rain and small birds from entering the feed.

Never look into the open end of a waveguide when power is applied, or stand directly in front of a dish while transmitting. Tests and adjustments in these areas should be done while receiving or at extremely low levels of transmitter power (less than 0.1 watt). The US Government has set a limit of 10 mW/cm² averaged over a 6-minute period as the safe maximum. Other authorities believe even lower levels should be used. Destructive thermal heating of body tissue results from excessive exposure. This heating effect is especially dangerous to the eyes. The accepted safe level of 10 mW/cm² is reached in the near field of a parabolic antenna if the level at 2D²/λ is 0.242 mW/cm². The equation for power density is

$$\text{Power density} = \frac{3\lambda P}{64D^2} = \frac{158.4\ P}{D^2}\ \text{mW/cm}^2 \qquad \text{(Eq 7)}$$

where

P = average power in kilowatts
D = antenna diameter in feet
λ = wavelength in feet

New commercial dishes are expensive, but surplus ones can often be purchased at low cost. Some amateurs build theirs, while others modify UHF TV dishes or circular metal snow sleds for the amateur bands. Fig 23 shows a dish using the homemade feed just described. Photos showing a highly ambitious dish project under construction by ZL1BJQ appear in Figs 24 and 25. Practical details for constructing this type of antenna are given in Chapter 19. Dick Knadle, K2RIW, described modern UHF antenna test procedures in February 1976 *QST* (see bibliography).

Fig 23—Coffee-can 2304-MHz feed described in text and Fig 22 mounted on a 4-ft dish.

Omnidirectional Antennas for VHF and UHF

Local work with mobile stations requires an antenna with wide coverage capabilities. Most mobile work is on FM, and the polarization used with this mode is generally vertical. Some simple vertical systems are described later in this chapter. Additional material on antennas of this type is presented in Chapter 16.

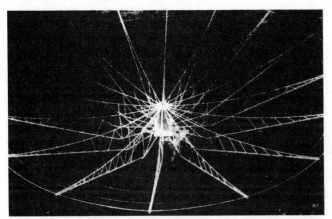

Fig 24—Aluminum framework for a 23-foot dish under construction by ZL1BJQ.

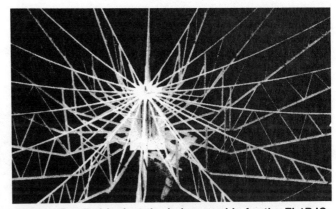

Fig 25—Detailed look at the hub assembly for the ZL1BJQ dish. Most of the structural members are made from ¾-inch T section.

A Low-Cost Yagi for 50 MHz

The antenna described here can be considered a "beginner" antenna in that it permits the newcomer to explore the 50-MHz band with a minimum of labor and expense. However, it can also be a permanent fixture for serious efforts if space is limited. It is optimized for gain and represents the best that can be done with three elements on an 8-foot boom. In its first week on the air (November 1981), this model produced contacts with American Samoa and The Gambia from Connecticut, with only 10 watts of transmitter power. Dennis Lusis, W1LJ, built this antenna in the ARRL laboratory.

Figs 26, 27 and 28 show the important construction details. The element lengths depart slightly from those specified in the NBS chart (Table 2) because of the larger tubing used. A balanced feed system to a split driven element was used on the unit shown, but an all-metal (plumber's delight) construction using a gamma match would be a justifiable simplification. A 3-element Yagi has a fairly broad major lobe, so any pattern skewing caused by feed imbalance should not be objectionable, if noticeable.

This antenna can be adjusted near the ground if it is pointed straight up at the sky away from utility lines or other antennas. The reflector can be supported by two wooden sawhorses or chairs, and the boom can be steadied with a pair of broomsticks. Adjust the driven element length and the matching system for minimum reflected power, which should be near zero. The dimensions given in Fig 26 are optimized for 50.1 MHz, but the antenna performs well up to 51 MHz. If this antenna is stacked above an HF beam, the separation should be at least 4 feet for a good impedance match, or 10 feet for low-angle radiation. (See the section on stacking.)

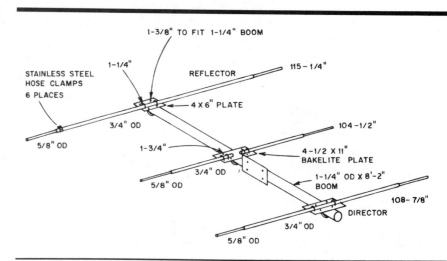

Fig 26—Dimensional drawing of a 3-element maximum-gain Yagi for 50-MHz. The feed system is similar to that shown in Fig 30, except that the hairpin loop is 1 in. shorter.

Fig 27—Dennis Lusis, W1LJ, and the 3-element 50-MHz beam ready for installation. One person can easily handle this array.

Fig 28—This photo shows how the driven element and feed system are attached to the boom. The center of the hairpin loop may be connected to the boom electrically and mechanically if desired.

A 5-Element Yagi for 50 MHz

The antenna described here was designed from information contained in Table 2 and Fig 7. This antenna has a theoretical gain of 9.2 dB over a ½-λ dipole and should exhibit an F/B ratio of roughly 18 dB. The pattern is quite clean. A hairpin matching system is used, and if the dimensions are followed closely, no adjustment should be necessary. The antenna is rugged, yet lightweight, and should be easy to install on any tower or mast. This antenna was built by Jay Rusgrove, W1VD, in the ARRL laboratory.

Mechanical Details

Constructional details of the antenna are given in Figs 29 and 30. The boom of the antenna is 17 feet long and is made from a single piece of 2-inch aluminum irrigation tubing that has a wall thickness of 0.047 inch. Irrigation tubing is normally supplied in 20-foot sections, so a few feet may be removed from the length.

Elements are constructed from ¾-inch OD aluminum tubing of the 6061-T6 variety, with a wall thickness of 0.058 inch. Each element, with the exception of the driven element, is made from a single length of tubing. The driven element is split in the center and insulated from the boom to provide a balanced feed system. The reflector and directors have short lengths of 7/8-inch aluminum tubing telescoped over the center of the elements for reinforcement purposes. Four boom to element clamps were fashioned from aluminum

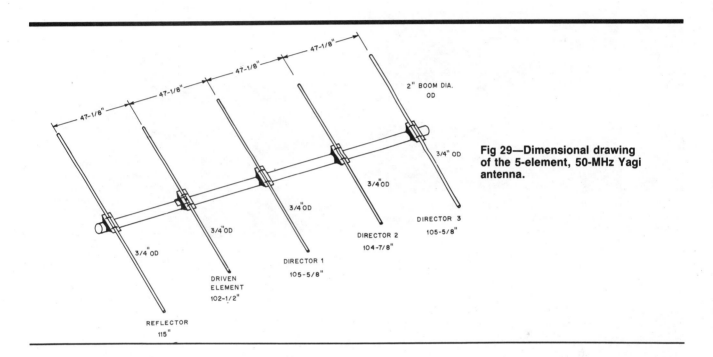

Fig 29—Dimensional drawing of the 5-element, 50-MHz Yagi antenna.

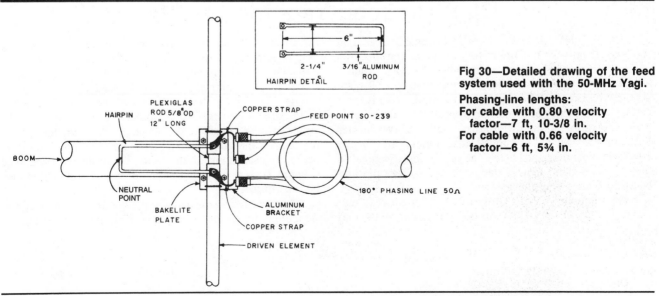

Fig 30—Detailed drawing of the feed system used with the 50-MHz Yagi.

Phasing-line lengths:
For cable with 0.80 velocity factor—7 ft, 10-3/8 in.
For cable with 0.66 velocity factor—6 ft, 5¾ in.

plate stock, 3/16 inch thick, as shown in Figs 31 and 32.

Two muffler clamps hold each plate to the boom, and two U bolts affix each element to the plate. Exact dimensions of the plates are not critical, but they should be large enough to accommodate the two muffler clamps and two U bolts. The element and clamp structure may seem to be a bit over-engineered, but the antenna is designed to withstand the severe weather conditions common in New England. This antenna has withstood many windstorms and several ice storms, and no maintenance has been required.

The driven element is mounted to the boom on a Bakelite plate of similar dimension to the reflector and director boom plates. A piece of 5/8-inch Plexiglas rod, 12 inches long, is inserted into each half of the driven element. The Plexiglas

piece allows the use of a single clamp on each side of the element and also seals the center of the elements against moisture. Self-tapping screws are used for connection to the driven element. A length of ¼-inch polypropylene rope is inserted into each element, and end caps are placed on the elements. The ropes dampen element vibrations that could lead to element and hardware fatigue.

Feed System

Details of the feed system are shown in Figs 30 and 31. A bracket made from a piece of scrap aluminum is used to mount the three SO-239 connectors to the driven element plate. A ½-λ phasing line connects the two element halves, providing the necessary 180° phase difference between them. The hairpin is connected directly across the element halves. The exact center of the hairpin is electrically neutral and may be fastened to the boom or allowed to hang. The driven element is the shortest element in this array. While this may seem a bit unusual, it is necessitated by the hairpin matching system.

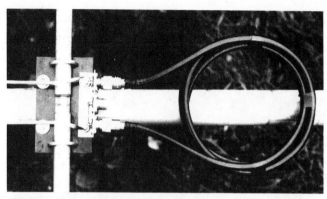

Fig 31—Close-up photo of the driven element and feed system for the 5-element 50-MHz Yagi. The phasing line is coiled and taped to the boom.

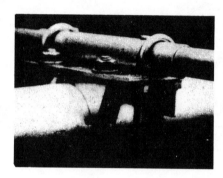

Fig 32—The element to boom clamp. U bolts are used to hold the element to the plate, and 2-in. plated muffler clamps hold the plates to the boom.

DL Style Long Yagi for 50 MHz

The previously described 50-MHz antennas are based on the NBS designs for maximum forward gain. Optimization of the F/B ratio or SWR bandwidth, rather than gain, requires adjustment of the element lengths and spacing. All three characteristics cannot be simultaneously optimized. A large array can, however, be assembled to produce the maximum gain, F/B ratio and bandwidth of a smaller array. Guenter Schwarzbeck, DL1BU, applied that philosophy to the 7-element Yagi described here. The gain is roughly equal to that of the 5-element NBS design, but the F/B ratio and bandwidth are somewhat greater.

An aluminum boom just over 22 feet long is required, and it should have an OD of 2 inches. Depending on the wall

thickness of the material selected, some additional bracing may be required. The elements are made from 3/4-inch and 5/8-inch OD aluminum tubing with a wall thickness of 0.058 inch. A thinner wall may be used for the 5/8-inch tubing. Fig 33 gives the dimensions of the array. The length of the driven element is not especially critical, and may be varied over a considerable range to effect an impedance match. A standard gamma match, such as pictured in Fig 34 or 35, is used. One of the more elaborate balanced feed systems, such as a hairpin match or beta match, could also be substituted.

Each element is fastened to the boom by a single U bolt extending through holes drilled at the center of the element.

This is detailed in Fig 36. It may be desirable to reinforce the element where the holes are drilled. One way to do this is to put a short piece of 7/8-inch OD tubing around the center of the ¾-inch tubing so the U bolt goes through a double thickness of tubing.

The antenna is dimensioned for 50.1 MHz, but works well up to 52 MHz. The 3-dB beamwidth is on the order of 40° in the azimuth plane. This is a sharp pattern, so a good rotator and an accurate direction indicator are required for the most effective operation.

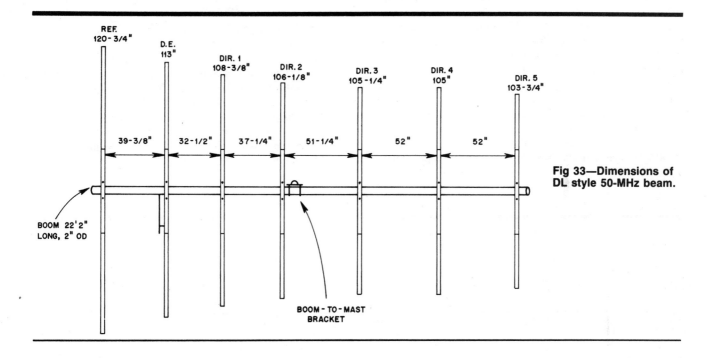

Fig 33—Dimensions of DL style 50-MHz beam.

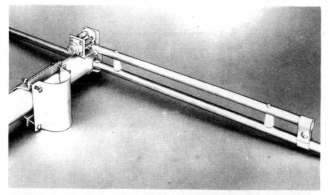

Fig 34—Typical gamma match construction. The variable capacitor, 50 pF, should be mounted in an inverted plastic cup or other device to protect it from the weather. The gamma arm is about 12 inches long for 50 MHz. The same construction technique may be used for 144 MHz, with an arm length of about 5 in.

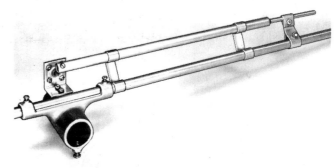

Fig 35—Gamma matching section using tubular capacitor. The sheet-aluminum clip at the right is moved along the driven element for matching. The small rod can be slid in and out of the 15-in. tube for adjustment of series capacitance. The rod should be about 14 in. long for 50 MHz.

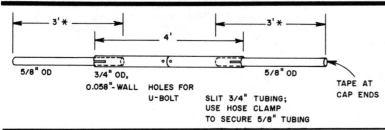

Fig 36—Construction of a typical element for the DL style beam. The telescoping lengths marked with an asterisk (*) are approximate. This length for the reflector should be 3 ft 3 in., and may be shorter for directors. See Fig 33.

A 15-Element Yagi for 432 MHz

At 144 MHz and above, the mechanical problems of antenna construction and installation are not as complex as those found at and below 50 MHz; the wavelength is shorter. This is a fortunate situation, because more elements are required to maintain good pickup efficiency as the frequency increases, as mentioned previously.

Another term for pickup efficiency is *aperture*, and this is proportional to the physical size of the array. For a 220-MHz antenna to occupy the same physical volume as a similar one designed for 50 MHz, it must have four to five times as many elements. The array described in this section illustrates the construction of Yagis for the UHF range.

Dimensions for a 432-MHz Yagi antenna are given in Figs 37 and 38. This antenna was built by Jay Rusgrove, W1VD, in the ARRL laboratory. The theoretical gain of this antenna is 14.2 dB over a dipole, with an F/B ratio of approximately 22 dB. The pattern is very sharp and quite clean, as would be expected from a well tuned array of this size. Four of these antennas in a "box" or "H" array serve well for terrestrial work, while eight make a respectable EME system.

Mechanical Details

The boom is made from a length of 1-inch aluminum tubing. Each of the elements is mounted through the boom, and only the driven element is insulated. Auveco 8715 external retaining rings secure each of the parasitic elements in place. These rings are available at most hardware supply houses.

The driven element is insulated from the boom by a 1¾-inch length of ½-inch dia Teflon rod. This rod is drilled lengthwise to accept the ¼-inch dia driven element. A press fit secures the Teflon piece in the boom of the antenna. An exact fit can be obtained by drilling the hole slightly undersize

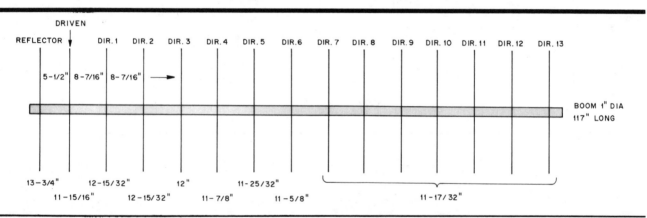

Fig 37—Dimensional drawing of the 15-element Yagi for 432 MHz. Elements are aluminum tubing, ¼ in. OD. The balance point of the antenna is between directors 5 and 6. Dimensions are based on an NBS design.

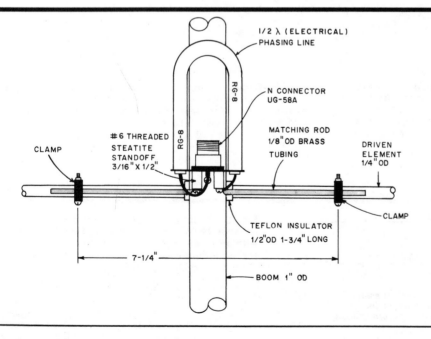

Fig 38—Detailed drawing of the feed system used with the Yagi. Center to center spacing between the driven element and matching rods is 5/8 in. A small copper plate is attached to the coaxial connector plate assembly for connection of the phasing line braid. As indicated, the braid is soldered to the plate. A coat of clear lacquer or enamel is recommended for waterproofing the feed system.

and enlarging the hole with a hand reamer, a small amount at a time. If the hole is too large, RTV (silicone sealant) can be used to secure the Teflon in place.

It should be possible to use a driven element that is not insulated from the boom. Small changes in the position of the matching rods or clamps might be necessary. The performance of such a configuration was not tested while this antenna was being built.

Details of the feed system are shown in Fig 38. This is a form of the T match, where the driven element is shortened from its resonant length to provide the necessary capacitance to tune out the reactance of the matching rods. With this system no variable capacitors are required, as in the more conventional T-match systems used at HF. This is a definite advantage in terms of antenna endurance in harsh weather environments.

The center pin of the UG-58 type N connector is attached to one of the matching rods. A half wavelength of 52-Ω, foam dielectric cable is used to provide the 180° phase shift from one half of the element to the other. An alternative to the large and cumbersome cable used here is miniature copper Hardline with Teflon dielectric material, such as RG-401.

Each of the matching rods is secured to two threaded steatite standoffs at the center of the antenna. These standoffs provide tie points for the ends of the phasing lines, the center pin of the coaxial connector, and the ends of the matching rods. Solder lugs are used for each of the connections for easy assembly and disassembly. The clamps that connect the matching rods to the driven element are constructed from pieces of aluminum measuring 1/4 × 1/2 × 1-5/16 inches. These pieces are drilled and slotted so that when the screws are tightened the pieces are compressed slightly to provide a snug fit. Alternatively, simple clamps could be fashioned from strips of aluminum.

Adjustment

If the dimensions given in the drawings are followed closely, little adjustment should be necessary. If adjustment is necessary, as indicated by an SWR greater than 1.5:1, move the clamps a short distance along the driven element. Keep in mind that the two clamps should be located equidistant from the center of the boom.

A 144-MHz 2-Element Quad

The basic 2-element quad array for 144 MHz is shown in Fig 39. The supporting frame is 1 × 1-inch wood, of any kind suitable for outdoor use. Elements are no. 8 aluminum wire. The driven element is 1 λ (83 inches) long, and the reflector five percent longer (87 inches). Dimensions are not critical, as the quad is relatively broad in frequency response.

The driven element is open at the bottom, its ends fastened to a plastic block. The block is mounted at the bottom of the forward vertical support. The top portion of the element runs through the support and is held firmly by a screw running into the wood and then bearing on the aluminum wire. Feed is by means of 52-Ω coax, connected to the driven-element loop.

The reflector is a closed loop, its top and bottom portions running through the rear vertical support. It is held in position with screws at the top and bottom. The loop can be closed by fitting a length of tubing over the element ends, or by hammering them flat and bolting them together as shown in the sketch.

The elements in this model are not adjustable, though this can easily be done by the use of stubs. It would then be desirable to make the loops slightly smaller to compensate for the wire in the adjusting stubs. The driven element stub would be trimmed for length and the point of connection for the coax would be adjustable for best match. The reflector stub can be adjusted for maximum gain or maximum F/B ratio, depending on the builder's requirements.

In the model shown only the spacing is adjusted, and this is not particularly critical. If the wooden supports are made as shown, the spacing between the elements can be adjusted for best match, as indicated by an SWR meter connected in the coaxial line. The spacing has little effect on the gain (from 0.15 to 0.25 λ), so the variation in impedance with spacing can be used for matching. This also permits use of either 52 or 75-Ω coax for the transmission line.

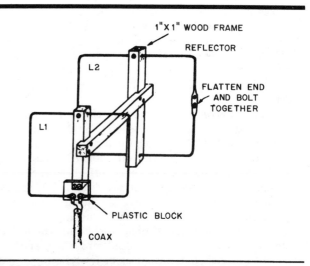

Fig 39—Mechanical details of a 2-element quad for 144 MHz. The driven element, L1, is one wavelength long; reflector L2 is 5% longer. With the transmission line connected as shown here, the resulting radiation is horizontally polarized. Sets of elements of this type can be stacked horizontally and vertically for high gain with broad frequency response. Recommended bay spacing is ½ λ between adjacent element sides. The example shown may be fed directly with 52-Ω coax.

A Portable 144-MHz 4-Element Quad

Element spacing for quad antennas found in the literature ranges from 0.14 λ to 0.25 λ. Factors such as the number of elements in the array and the parameters to be optimized (F/B ratio, forward gain, bandwidth, etc), determine the optimum element spacing within this range. The 4-element quad antenna described here was designed for portable use, so a compromise between these factors was chosen. This antenna, pictured in Fig 40, was designed and built by Philip D'Agostino, W1KSC.

Based on several experimentally determined correction factors related to the frequency of operation and the wire size, optimum design dimensions were found to be as follows.

$$\text{Reflector length (ft)} = \frac{1046.8}{f_{MHz}}$$

$$\text{Driven element (ft)} = \frac{985.5}{f_{MHz}}$$

$$\text{Directors (ft)} = \frac{937.3}{f_{MHz}}$$

Cutting the loops for 146 MHz provides satisfactory performance across the entire 144-MHz band.

Materials

The quad was designed for quick and easy assembly and disassembly, as illustrated in Fig 41. Wood (clear trim pine) was chosen as the principal building material because of its light weight, low cost, and ready availability. Pine is used for the boom and element supporting arms. Round wood clothes closet poles comprise the mast material. Strips connecting the mast sections are made of heavier pine trim. Elements are made of no. 8 aluminum wire. Plexiglas is used to support the feed point. Table 10 lists the hardware and other parts needed to duplicate the quad.

Construction

The elements of the quad are assembled first. The mounting holes in the boom should be drilled to accommodate 1½ inch no. 8 hardware. Measure and mark the locations where the holes are to be drilled in the element spreaders, Fig 42. Drill the holes in the spreaders just large enough to accept the no. 8 wire elements. It is important to drill all the holes straight so the elements line up when the antenna is assembled.

Construction of the wire elements is easiest if the directors are made first. A handy jig for bending the elements can be made from a piece of 2 × 3-inch wood cut to the side length of the directors. It is best to start with about 82 inches of wire for each director. The excess can be cut off when the elements are completed. (The total length of each director is 77 inches.) Two bends should initially be made so the directors can be slipped into the spreaders before the remaining corners are bent. See Fig 43. Electrician's copper-wire clamps can be used to join the wires after the final bends are made, and they facilitate adjustment of element length. The reflector is made the same way as the directors, but the total length is 86 inches.

The driven element, total length 81 inches, requires special attention, as the feed attachment point needs to be adequately supported. An extra hole is drilled in the driven element spreader to support the feed-point strut, as shown

Fig 40—The 4-element 144-MHz portable quad, assembled and ready for operation. Sections of clothes closet poles joined with pine strips make up the mast. *(Photo by Adwin Rusczek, W1MPO)*

Fig 41—The complete portable quad, broken down for travel. Visible in the foreground is the driven element. The pine box in the background is a carrying case for equipment and accessories. A hole in the lid accepts the mast, so the box doubles as a base for a short mast during portable operation. *(W1MPO photo)*

in Fig 44. A Plexiglas plate is used at the feed point to support the feed-point hardware and the feed line. The feed-point support strut should be epoxied to the spreader, and a wood screw used for extra mechanical strength.

For vertical polarization, locate the feed point in the center of one side of the driven element, as shown in Fig 44. Although this arrangement places the spreader supports at voltage maxima points on the four loop conductors, D'Agostino reports no adverse effects during operation. However, if the antenna is to be left exposed to the weather, the builder may wish to modify the design to provide support for the loops at current maxima points, such as shown in Fig 39. (The elements of Fig 39 should be rotated 90° for vertical polarization.)

Orient the driven element spreader so that it mounts properly on the boom when the antenna is assembled. Bend the driven element the same way as the reflector and directors, but do not leave any overlap at the feed point. The ends of the wires should be ¾ inch apart where they mount on the Plexiglas plate. Leave enough excess that small loops can be bent in the wire for attachment to the coaxial feed line with stainless steel hardware.

Drill the boom as shown in Fig 45. It is a good idea to use hardware with wing nuts to secure the element spreaders to the boom. After the boom is drilled, clean all the wood parts with denatured alcohol, sand them, and give them two coats of glossy polyurethane. After the polyurethane dries, wax all the wooden parts.

The boom to mast attachment is made next. Square the ends of a 6-foot section of clothes closet pole (a miter box is useful for this). Drill the center holes in both the boom attachment piece and one end of the mast section (Fig 46). Make certain that the mast hole is smaller than the flat-head screw to be used to ensure a snug fit. Accurately drill the holes for attachment to the boom as shown in Fig 46.

Countersink the hole for the flat-head screw to provide a smooth surface for attachment to the boom. Apply epoxy cement to the surfaces and screw the boom attachment piece securely to the mast section. One 6-foot mast is used for attachment to the other mast sections.

Two additional 6-foot mast sections are prepared next. This brings the total mast height to 18 feet. It is important to square the ends of each pole so the mast

Table 10
Parts List for the 144-MHz 4-Element Quad

Boom: ¾ × ¾ × 48-in. pine
Driven element support (spreader): ½ × ¾ × 21¼ in. pine
Driven element feed point strut: ½ × ¾ × 7½ in. pine
Reflector support (spreader): ½ × ¾ × 22½ in. pine
Director supports (spreaders): ½ × ¾ × 20¼ in. pine, 2 req'd
Mast brackets: ¾ × 1½ × 12 in. heavy pine trim, 4 req'd
Boom to mast bracket: ½ × 1-5/8 × 5 in. pine
Element wire: Aluminum ground wire (Radio Shack no. 15-035)
Wire clamps: ¼-in. electrician's copper or zinc plated steel clamps, 3 req'd
Boom hardware:
  6 no. 8-32 × 1½ in. stainless steel machine screws
  6 no. 8-32 stainless steel wing nuts
  12 no. 8 stainless steel washers
Mast hardware:
  8 hex bolts, ¼-20 × 3½ in.
  8 hex nuts, ¼-20
  16 flat washers
Mast material: 1-5/16 in. × 6 ft wood clothes closet poles, 3 req'd
Feed point support plate: 3½ × 2½ in. Plexiglas sheet
Wood preparation materials: Sandpaper, clear polyurethane, wax
Feed line: 52 Ω RG-8 or RG-58 cable
Feed line terminals: Solder lugs for no. 8 or larger hardware, 2 req'd
Miscellaneous hardware:
  4 small machine screws, nuts, washers; 2 flat-head wood screws

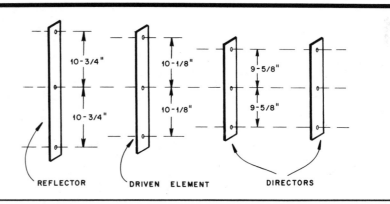

Fig 42—Dimensions for the pine element spreaders for the 144-MHz 4-element quad.

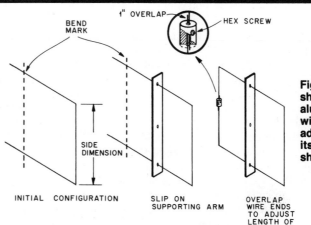

Fig 43—Illustration showing how the aluminum element wires are bent. The adjustment clamp and its location are also shown.

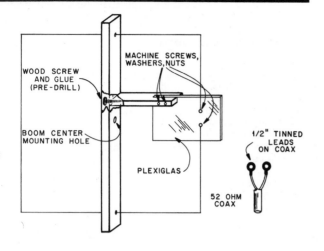

NOTE: SIDEARM STRUT RESTS ON BOOM

Fig 44—Layout of the driven element of the 144-MHz quad. The leads of the coaxial cable should be stripped to ½ in. and solder lugs attached for easy connection and disconnection. See text regarding impedance at loop support points.

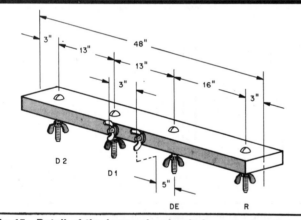

Fig 45—Detail of the boom showing hole center locations and boom to mast connection points.

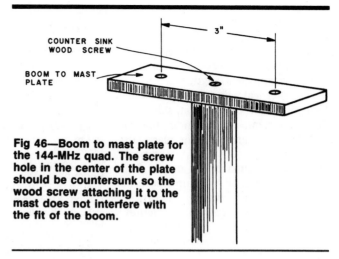

Fig 46—Boom to mast plate for the 144-MHz quad. The screw hole in the center of the plate should be countersunk so the wood screw attaching it to the mast does not interfere with the fit of the boom.

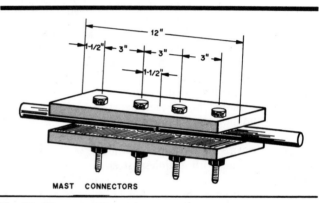

Fig 47—Mast coupling connector details for the portable quad. The plates should be drilled two at a time to ensure the holes line up.

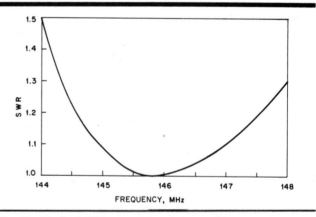

Fig 48—Typical SWR curve for the 144-MHz portable quad. The large wire diameter and the quad design provide excellent bandwidth.

stands straight when assembled. Mast-section connectors are made of pine as shown in Fig 47. Using 3½ × ¼-inch hex bolts, washers, and nuts, sections may be attached as needed, for a total height of 6, 12 or 18 feet. Drill the holes in two connectors at a time. This ensures good alignment of the holes. A drill press is ideal for this job, but with care a hand drill can be used if necessary.

Line up two mast sections end to end, being careful that they are perfectly straight. Use the predrilled connectors to maintain pole straightness, and drill through the poles, one at a time. If good alignment is maintained, a straight 18-foot mast section can be made. Label the connectors and poles immediately so they are always assembled in the same order.

When assembling the antenna, install all the elements on the boom before attaching the feed line. Connect the coax to the screw connections on the driven element support plate and run the cable along the strut to the boom. From there, the cable should be routed directly to the mast and down. Assemble the mast sections to the desired height. The antenna provides good performance, and has a reasonable SWR curve over the entire 144-MHz band (Fig 48).

Building Quagi Antennas

The Quagi antenna was designed by Wayne Overbeck, N6NB. He first published information on this antenna in 1977 (see bibliography). There are a few tricks to Quagi building, but nothing very difficult or complicated is involved. In fact, Overbeck mass produced as many as 16 in one day. Tables 11 and 12 give the dimensions for Quagis for various frequencies up to 446 MHz.

For the designs of Tables 11 and 12, the boom is *wood* or any other nonconductor (such as, fiberglass or Plexiglas). If a metal boom is used, a new design and new element lengths will be required. Many VHF antenna builders go wrong by failing to follow this rule: If the original uses a metal boom, use the same size and shape metal boom when you duplicate it. If it calls for a wood boom, use a nonconductor. Many amateurs dislike wood booms, but in a salt air environment they outlast aluminum (and surely cost less). Varnish the boom for added protection.

The 144-MHz version is usually built on a 14-foot, 1 × 3 inch boom, with the boom tapered to 1 inch at both ends. Clear pine is best because of its light weight, but construction grade Douglas fir works well. At 220 MHz the boom is under 10 feet long, and most builders use 1 × 2 or (preferably) ¾ × 1¼ inch pine molding stock. At 432 MHz, except for long-boom versions, the boom should be ½ inch thick or less. Most builders use strips of ½-inch exterior plywood for 432 MHz.

Table 12

432-MHz, 15-Element, Long Boom Quagi Construction Data

Element Lengths, Inches	Interelement Spacing, Inches
R—28	R-DE—7
DE—26-5/8	DE-D1—5-1/4
D1—11-3/4	D1-D2—11
D2—11-11/16	D2-D3—5-7/8
D3—11-5/8	D3-D4—8-3/4
D4—11-9/16	D4-D5—8-3/4
D5—11-1/2	D5-D6—8-3/4
D6—11-7/16	D6-D7—12
D7—11-3/8	D7-D8—12
D8—11-5/16	D8-D9—11-1/4
D9—11-5/16	D9-D10—11-1/2
D10—11-1/4	D10-D11—9-3/16
D11—11-3/16	D11-D12—12-3/8
D12—11-1/8	D12-D13—13-3/4
D13—11-1/16	

Boom: 1 × 2-in. × 12-ft Douglas fir, tapered to 5/8 in. at both ends.
Driven element: No. 12 TW copper wire loop in square configuration, fed at bottom center with type N connector and 52-Ω coax.
Reflector: No. 12 TW copper wire loop, closed at bottom.
Directors: 1/8-in. rod passing through boom.

Table 11

Dimensions, 8-Element Quagi

Element Lengths	Frequency 144.5 MHz	147 MHz	222 MHz	432 MHz	446 MHz
Reflector[1]	86-5/8"	85"	56-3/8"	28"	27-1/8"
Driven[2]	82"	80"	53-1/2"	26-5/8"	25-7/8"
Directors	35-15/16" to 35" in 3/16" steps	35-5/16" to 34-3/8" in 3/16" steps	23-3/8" to 23-3/4" in 1/8" steps	11-3/4" to 11-7/16" in 1/16" steps	11-3/8" to 11-1/16" in 1/16" steps
Spacing					
R-DE	21"	20-1/2"	13-5/8"	7"	6.8"
DE-D1	15-3/4"	15-3/8"	10-1/4"	5-1/4"	5.1"
D1-D2	33"	32-1/2"	21-1/2"	11"	10.7"
D2-D3	17-1/2"	17-1/8"	11-3/8"	5.85"	5.68"
D3-D4	26.1"	25-5/8"	17"	8.73"	8.46"
D4-D5	26.1"	25-5/8"	17"	8.73"	8.46"
D5-D6	26.1"	25-5/8"	17"	8.73"	8.46"
Stacking Distance Between Bays					
	11"	10'10"	7'1-1/2"	3'7"	3'5-5/8"

[1] All no. 12 TW (electrical) wire, closed loops.
[2] All no. 12 TW wire loops, fed at bottom.

Fig 49—A close-up view of the feed method used on a 432-MHz Quagi. This arrangement produces a low SWR and gain in excess of 13 dBi with a 4-ft 10-in. boom! The same basic arrangement is used on lower frequencies, but wood may be substituted for the Plexiglas spreaders. The boom is ½-in. exterior plywood.

Fig 50—A view of the 10-element version of the 1296-MHz Quagi. It is mounted on a 30-in. Plexiglas boom with a 3 × 3-in. square of Plexiglas to support the driven element and reflector. Note how the driven element is attached to a standard UG-290 BNC connector. The elements are held in place with silicone sealing compound.

The quad elements are supported at the current maxima (the top and bottom, the latter beside the feed point) with Plexiglas or small strips of wood. See Fig 49. The quad elements are made of no. 12 copper wire, commonly used in house wiring. Some builders may elect to use no. 10 wire on 144 MHz and no. 14 on 432 MHz, although this changes the resonant frequency slightly. Solder a type N connector (an SO-239 is often used at 144 MHz) at the midpoint of the driven element bottom side, and close the reflector loop.

The directors are mounted through the boom. They can be made of almost any metal rod or wire of about 1/8 inch diameter. Welding rod or aluminum clothesline wire works well if straight. (The designer uses 1/8-inch stainless steel rod obtained from an aircraft surplus store.)

A TV type U bolt mounts the antenna on a mast. A single machine screw, washers and a nut are used to secure the spreaders to the boom so the antenna can be quickly "flattened" for travel. In permanent installations two screws are recommended.

Construction Reminders

Based on the experiences of Quagi builders, the following hints are offered. First, remember that at 432 MHz even a 1/8-inch measurement error results in performance deterioration. Cut the loops and elements as carefully as possible. No precision tools are needed, but accuracy is necessary. Also make sure to get the elements in the right order. The longest director goes closest to the driven element.

Finally, remember that a balanced antenna is being fed with an unbalanced line. Every balun the designer tried introduced more trouble in terms of losses than the feed imbalance caused. Some builders have tightly coiled several turns of the feed line near the feed point to limit line radiation. In any case, the feed line should be kept at right angles to the antenna. Run it from the driven element directly to the supporting mast and then up or down perpendicularly for best results.

QUAGIS FOR 1296 MHz

The Quagi principle has recently been extended to the 1296-MHz band, where good performance is extremely difficult to obtain from homemade conventional Yagis. Fig 50 shows the construction and Table 13 gives the design

Table 13

Dimensions, 1296-MHz Quagi Antennas

Note: All lengths are gross lengths. See text and photos for construction technique and recommended overlap at loop junctions. All loops are made of no. 18 AWG solid-covered copper bell wire. The Yagi type directors are 1/16-in. brass brazing rod. See text for a discussion of director taper.

Feed: Direct with 52-Ω coaxial cable to UG-290 connector at driven element; run coax symmetrically to mast at rear of antenna.

Boom: ¼-in. thick Plexiglas, 30 in. long for 10-element quad or Quagi and 48 in. long for 15-element Quagi; 84 in. for 25-element Quagi.

10-Element Quagi for 1296 MHz

Element	Length, Inches	Construction	Element	Interelement Spacing, In.
Reflector	9.5625	Loop	R-DE	2.375
Driven	9.25	Loop	DE-D1	2.0
Director 1	3.91	Brass rod	D1-D2	3.67
Director 2	3.88	Brass rod	D2-D3	1.96
Director 3	3.86	Brass rod	D3-D4	2.92
Director 4	3.83	Brass rod	D4-D5	2.92
Director 5	3.80	Brass rod	D5-D6	2.92
Director 6	3.78	Brass rod	D6-D7	4.75
Director 7	3.75	Brass rod	D7-D8	3.94
Director 8	3.72	Brass rod		

15-Element Quagi for 1296 MHz

The first 10 elements are the same lengths as above, but the spacing from D6 to D7 is 4.0 in.; D7 to D8 is also 4.0 in.

Director 9	3.70	D8-D9	3.75
Director 10	3.67	D9-D10	3.83
Director 11	3.64	D10-D11	3.06
Director 12	3.62	D11-D12	4.125
Director 13	3.59	D12-D13	4.58

25-Element Quagi for 1296 MHz

The first 15 elements use the same element lengths and spacings as the 15-element model. The additional directors are evenly spaced at 3.0-in. intervals and taper in length successively by 0.02 in. per element. Thus, D23 is 3.39 in.

information for antennas with 10, 15 and 25 elements.

At 1296 MHz, even slight variations in design or building materials can cause substantial changes in performance. The 1296-MHz antennas described here work every time—but only if the same materials are used and the antennas are built *exactly* as described. This is not to discourage experimentation, but if modifications to these 1296-MHz antenna designs are contemplated, consider building one antenna as described here, so a reference is available against which variations can be compared.

The Quagis (and the cubical quad) are built on ¼ inch thick Plexiglas booms. The driven element and reflector (and also the directors in the case of the cubical quad) are made of insulated no. 18 AWG solid copper bell wire, available at hardware and electrical supply stores. Other types and sizes of wire work equally well, but the dimensions vary with the wire diameter. Even removing the insulation usually necessitates changing the loop lengths.

Quad loops are approximately square (Fig 51), although the shape is relatively noncritical. The element lengths, however, *are* critical. At 1296 MHz, variations of 1/16 inch alter the performance measurably, and a 1/8 inch departure can cost several decibels of gain. The loop lengths given are *gross* lengths. Cut the wire to these lengths and then solder the two ends together. There is a 1/8 inch overlap where the two ends of the reflector (and director) loops are joined, as shown in Fig 51.

The driven element is the most important of all. The no. 18 wire loop is soldered to a standard UG-290 chassis-mount BNC connector as shown in the photographs. This exact type of connector must be used to ensure uniformity in construction. Any substitution may alter the driven element electrical length. One end of the 9¼-inch driven loop is pushed as far as it can go into the center pin, and is soldered in that position. The loop is then shaped and threaded through small holes drilled in the Plexiglas support. Finally, the other end is fed into one of the four mounting holes on the BNC connector and soldered. In most cases, the best SWR is obtained if the end of the wire just passes through the hole so it is flush with the opposite side of the connector flange.

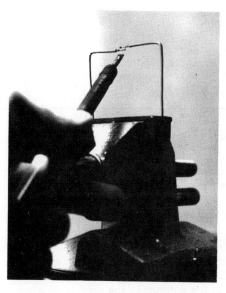
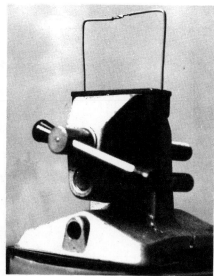

Fig 51—These photos show the construction method used for the 1296-MHz quad type parasitic elements. The two ends of the no. 18 bell wire are brought together with an overlap of 1/8 in. and soldered.

Loop Yagis for 1296 MHz

Described here are loop Yagis for the 1296-MHz band. The loop Yagi fits into the quad family of antennas, as each element is a closed loop with a length of approximately 1 λ. Several versions are described, so the builder can choose the boom length and frequency coverage desired for the task at hand. Mike Walters, G3JVL, brought the original loop Yagi design to the amateur community in the 1970s. Since then, many versions have been developed with different loop and boom dimensions. Chip Angle, N6CA, developed the antennas shown here.

Three sets of dimensions are given. Good performance can be expected if the dimensions are carefully followed. Check all dimensions before cutting or drilling anything. The 1296-MHz version is intended for weak-signal operation, while the 1270-MHz version is optimized for FM and mode L satellite work. The 1283-MHz antenna provides acceptable performance from 1280 to 1300 MHz.

These antennas have been built on 6- and 12-foot booms. Results of gain tests at VHF conferences and by individuals around the country show the gain of the 6-foot model to be about 18 dBi, while the 12-foot version provides about 20.5 dBi. Swept measurements indicate that gain is about 2 dB down from maximum gain at ± 30 MHz from the design frequency. The SWR, however, deteriorates within a few megahertz on the low side of the design center frequency.

The Boom

The dimensions given here apply only to a ¾-inch OD boom. If a different boom size is used, the dimensions must be scaled accordingly. Many hardware stores carry aluminum tubing in 6- and 8-foot lengths, and that tubing is suitable for a short Yagi. If a 12-foot antenna is planned, find a piece of more rugged boom material, such as 6061-T6 grade aluminum. Do not use anodized tubing. The 12-foot antenna must have additional boom support to minimize boom sag. The 6-foot version can be rear mounted. For rear mounting, allow 4½ inches of boom behind the last reflector to eliminate SWR effects from the support.

The antenna is attached to the mast with a gusset plate. This plate mounts at the boom center. See Fig 52. Drill the plate mounting holes perpendicular to the element mounting holes (assuming the antenna polarization is to be horizontal).

Elements are mounted to the boom with no. 4-40 machine screws, so a series of no. 33 (0.113-inch) holes must be drilled along the center of the boom to accommodate this hardware. Fig 53 shows the element spacings for different parts of the band. Dimensions should be followed as closely as possible.

Parasitic Elements

The reflectors and directors are cut from 0.032-inch thick aluminum sheet and are ¼ inch wide. Fig 54 indicates the lengths for the various elements. These lengths apply only to elements cut from the specified material. For best results, the element strips should be cut with a shear. If the edges are left sharp, birds won't sit on the elements.

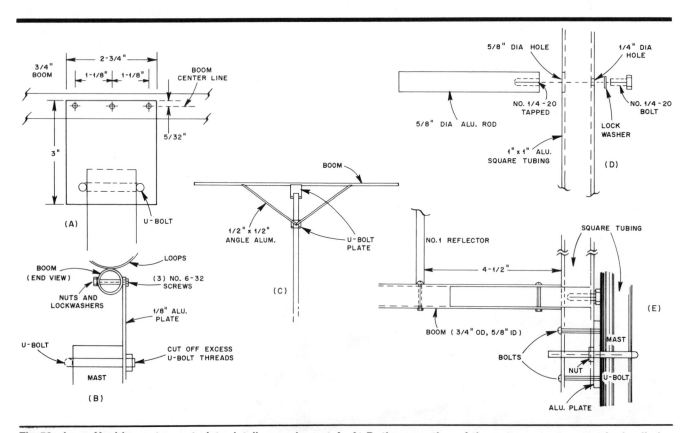

Fig 52—Loop Yagi boom to mast plate details are given at A. At B, the mounting of the antenna to the mast is detailed. A boom support for long antennas is shown at C. The arrangement shown in D and E may be used to rear-mount antennas up to 6 or 7 ft long.

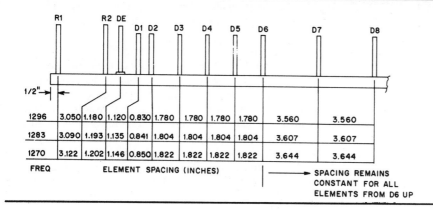

Fig 53—Boom drilling dimensions. These dimensions must be carefully followed and the same materials used if performance is to be optimum. Element spacings are the same for all directors after D6—use as many as necessary to fill the boom.

FREQ	ELEMENT SPACING (INCHES)								SPACING REMAINS CONSTANT FOR ALL ELEMENTS FROM D6 UP	
1296	3.050	1.180	1.120	0.830	1.780	1.780	1.780	1.780	3.560	3.560
1283	3.090	1.193	1.135	0.841	1.804	1.804	1.804	1.804	3.607	3.607
1270	3.122	1.202	1.146	0.850	1.822	1.822	1.822	1.822	3.644	3.644

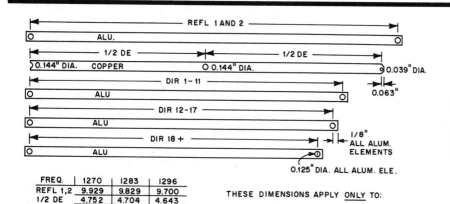

Fig 54—Parasitic elements for the loop Yagi are made from aluminum sheet, the driven element from copper sheet. The dimensions given are for ¼-in. wide by 0.0325-in. thick elements only. Lengths specified are hole to hole distances; the holes are located 1/8 in. from each element end.

FREQ.	1270	1283	1296
REFL 1,2	9.929	9.829	9.700
1/2 DE	4.752	4.704	4.643
DIR 1-11	8.445	8.359	8.250
DIR 12-17	8.189	8.106	8.000
DIR 18+	7.882	7.802	7.700

ELEMENT LENGTHS (INCHES)
(HOLE TO HOLE)

THESE DIMENSIONS APPLY ONLY TO:

0.250" ELEMENT WIDTH
0.0325" ELEMENT THICKNESS
0.750" DIAMETER BOOM

NOTE: ALL DIMENSIONS ARE IN INCHES (").

Drill the mounting holes as shown in Fig 54 after carefully marking their locations. After the holes are drilled, form each strap into a circle. This is easily done by wrapping the element around a round form. (A small juice can works well.)

Mount the loops to the boom with no. 4-40 × 1-inch machine screws, lock washers and nuts. See Fig 55. It is best to use only stainless steel or plated brass hardware. Although the initial cost is higher than for ordinary plated steel hardware, stainless or brass hardware will not rust and need replacement after a few years. Unless the antenna is painted, the hardware will definitely deteriorate.

Driven Element

The driven element is cut from 0.032-inch copper sheet and is ¼ inch wide. Drill three holes in the strap as detailed in Fig 54. Trim the ends as shown and form the strap into a loop similar to the other elements. This antenna is like a quad; if the loop is fed at the top or bottom, it is horizontally polarized.

Driven element mounting details are shown in Fig 56. A mounting fixture is made from a ¼-20 × 1¼-inch brass bolt. File the bolt head to a thickness of 1/8 inch. Bore a 0.144-inch (no. 27 drill) hole lengthwise through the center

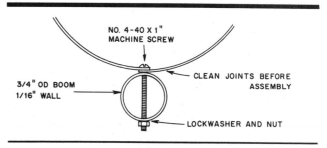

Fig 55—Element to boom mounting details.

of the bolt. A piece of 0.141-inch semi-rigid Hardline (UT-141 or equivalent) mounts through this hole and is soldered to the driven loop feed point. The point at which the UT-141 passes through the copper loop and brass mounting fixture should be left unsoldered at this time to allow for matching adjustments when the antenna is completed, although the range of adjustment is not very large.

The UT-141 can be any convenient length. Attach the

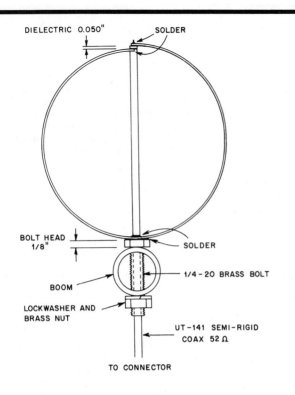

DIELECTRIC 0.050"   **SOLDER**

BOLT HEAD 1/8"

SOLDER

1/4 – 20 BRASS BOLT

BOOM

LOCKWASHER AND BRASS NUT

UT – 141 SEMI-RIGID COAX 52 Ω

TO CONNECTOR

Fig 56—Driven element details. See Fig 54 and the text for additional information.

connector of your choice (preferably type N). Use a short piece of low-loss RG-8 size cable (or ½-inch Hardline) for the run down the boom and mast to the main feed line. For best results, the main feed line should be the lowest loss 52-Ω cable obtainable. Good 7/8-inch Hardline has 1.5 dB of loss per

100 feet and virtually eliminates the need for remote mounting of the transmit converter or amplifier.

Tuning the Driven Element

If the antenna is built carefully to the dimensions given, the SWR should be close to 1:1. Just to be sure, check the SWR if you have access to test equipment. Be sure the signal source is clean, however; wattmeters respond to "dirty" signals and can give erroneous readings. If problems are encountered, recheck all dimensions. If they look good, a minor improvement may be realized by changing the shape of the driven element. Slight bending of reflector 2 may also improve the SWR. When the desired match has been obtained, solder the point where the UT-141 jacket passes through the loop and brass bolt.

Tips for 1296-MHz Antenna Installations

Construction practices that are common on lower frequencies cannot be used on 1296 MHz. This is the most important reason why all who venture to these frequencies are not equally successful. First, when a proven design is used, copy it exactly—don't change *anything*. This is especially true for antennas.

Use the best feed line you can get. Here are some realistic measurements of common coaxial cables at 1296 MHz (loss per 100 feet).

RG-8, 213, 214: 11 dB
1/2-inch foam/copper Hardline: 4 dB
7/8-inch foam/copper Hardline: 1.5 dB

Mount the antennas to keep feed line losses to an absolute minimum. Antenna height is less important than keeping the line losses low. *Do not* allow the mast to pass through the elements, as is common on antennas for lower frequencies. Cut all U bolts to the minimum length needed; ¼ λ at 1296 MHz is only a little over 2 inches. Avoid any unnecessary metal around the antenna.

Groundplane Antennas for 144, 220 and 440 MHz

For the FM operator living in the primary coverage area of a repeater, the ease of construction and low cost of a ¼-λ groundplane antenna make it an ideal choice. Three different types of construction are detailed in Figs 57 through 60; the choice of construction method depends upon the materials at hand and the desired style of antenna mounting.

The 144-MHz model shown in Fig 57 uses a flat piece of sheet aluminum, to which radials are connected with machine screws. A 45° bend is made in each of the radials. This bend can be made with an ordinary bench vise. An SO-239 chassis connector is mounted at the center of the aluminum plate with the threaded part of the connector facing down. The vertical portion of the antenna is made of no. 12 copper wire soldered directly to the center pin of the SO-239 connector.

The 220-MHz version, Fig 58, uses a slightly different technique for mounting and sloping the radials. In this case the corners of the aluminum plate are bent down at a 45° angle with respect to the remainder of the plate. The four radials are held to the plate with machine screws, lock washers and nuts. A mounting tab is included in the design of this antenna as part of the aluminum base. A compression type of hose clamp could be used to secure the antenna to a mast. As with the 144-MHz version, the vertical portion of the antenna is soldered directly to the SO-239 connector.

A very simple method of construction, shown in Figs 59 and 60, requires nothing more than an SO-239 connector and some no. 4-40 hardware. A small loop formed at the inside end of each radial is used to attach the radial directly to the mounting holes of the coaxial connector. After the radial is fastened to the SO-239 with no. 4-40 hardware, a large soldering iron or propane torch is used to solder the radial

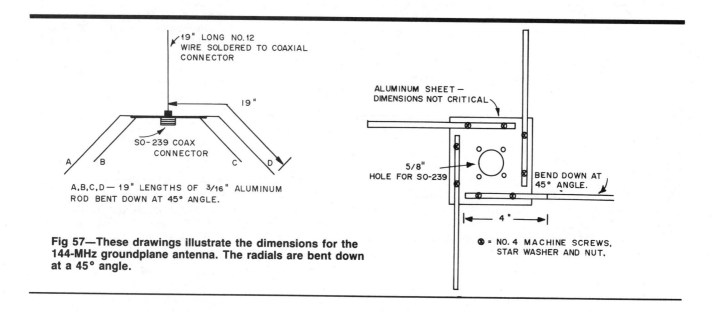

Fig 57—These drawings illustrate the dimensions for the 144-MHz groundplane antenna. The radials are bent down at a 45° angle.

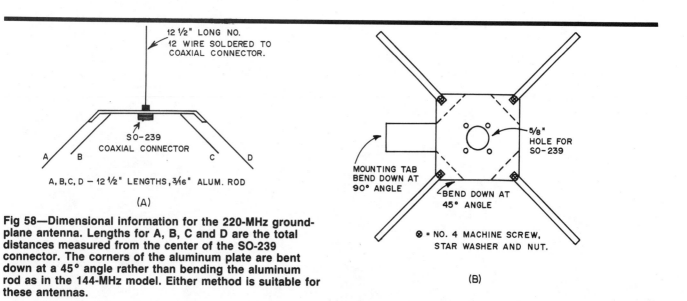

Fig 58—Dimensional information for the 220-MHz ground-plane antenna. Lengths for A, B, C and D are the total distances measured from the center of the SO-239 connector. The corners of the aluminum plate are bent down at a 45° angle rather than bending the aluminum rod as in the 144-MHz model. Either method is suitable for these antennas.

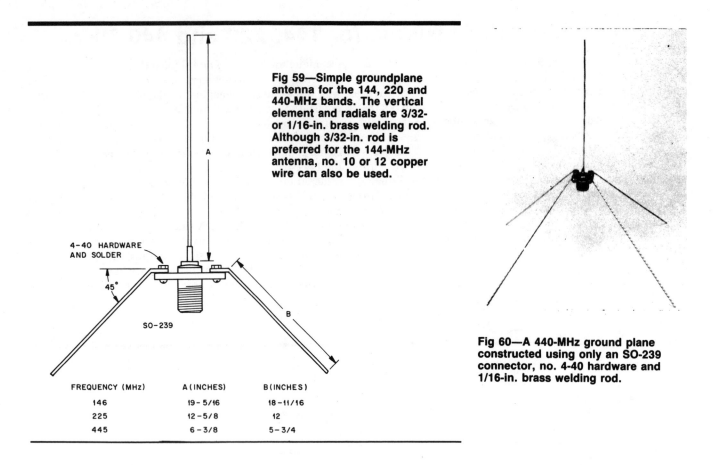

Fig 59—Simple groundplane antenna for the 144, 220 and 440-MHz bands. The vertical element and radials are 3/32- or 1/16-in. brass welding rod. Although 3/32-in. rod is preferred for the 144-MHz antenna, no. 10 or 12 copper wire can also be used.

4-40 HARDWARE AND SOLDER

45°

SO-239

A

B

FREQUENCY (MHz)	A (INCHES)	B (INCHES)
146	19 – 5/16	18 – 11/16
225	12 – 5/8	12
445	6 – 3/8	5 – 3/4

Fig 60—A 440-MHz ground plane constructed using only an SO-239 connector, no. 4-40 hardware and 1/16-in. brass welding rod.

and the mounting hardware to the coaxial connector. The radials are bent to a 45° angle and the vertical portion is soldered to the center pin to complete the antenna. The antenna can be mounted by passing the feed line through a mast of ¾ inch ID plastic or aluminum tubing. A compression hose clamp can be used to secure the PL-259 connector, attached to the feed line, in the end of the mast. Dimensions for the 144, 220 and 440-MHz bands are given in Fig 59.

If these antennas are to be mounted outside it is wise to apply a small amount of RTV sealant or similar material around the areas of the center pin of the connector to prevent the entry of water into the connector and coax line.

VHF and UHF Parabeams

Fig 61 shows two types of Parabeam Yagis for use at VHF and UHF. This style of gain antenna was developed in the United Kingdom by J-Beam, Ltd to offer high gain and wide bandwidth. The design is suitable for use at 144, 220, 432, 903 and 1296 MHz. The Parabeam utilizes a skeleton slot radiator and reflector (original J-Beam format), but employs conventional Yagi parasitic directors.

The 144-MHz Parabeam of Fig 61A has a claimed gain of 15 dBd and a half-power horizontal beamwidth of 24°, according to information contained in the RSGB *Radio Communication Handbook*. Fig 61B shows a 432-MHz Parabeam with a claimed gain of 17 dBd and a half-power beamwidth of 28°. The reflector of the 432-MHz version consists of a pair of ½-λ elements joined to form a full-wavelength loop. The 2:1 SWR bandwidth for this and the 144-MHz version is about 20% of the center frequency.

Details of the skeleton slot driven element are shown in Fig 62. A forward pitch of 11° is provided. According to the designers, this is necessary to obtain optimum launching into the parasitic directors. The feed impedance of these Yagis is on the order of 280 to 300 Ω. A 4:1 balun can be used to provide a match to 75-Ω transmission line.

Dimensional details for a 432-MHz version of the Parabeam are given in Fig 63. There are 18 elements used in the system, and the antennas can be stacked in the conventional manner if desired. Good results should be possible by scaling this design to 144, 220, 903 or 1296 MHz. Experimental adjustment of the element spacing should be done, however, to ensure optimum forward gain from the scaled versions.

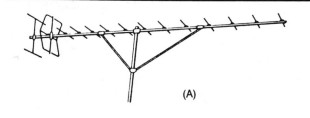

Fig 61—Illustration of a 144-MHz parabeam (A) and a 432-MHz version (B).

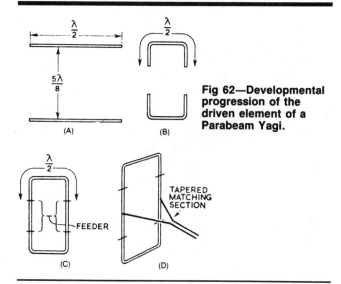

Fig 62—Developmental progression of the driven element of a Parabeam Yagi.

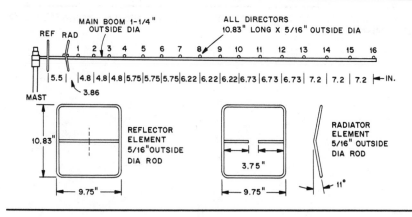

Fig 63—RSGB version of an 18-element Parabeam for 432 MHz.

Trough Reflectors for 432 and 1296 MHz

Dimensions are given in Fig 64 for 432- and 1296-MHz trough reflectors. The gain to be expected is 15 dB and 17 dB, respectively. A very convenient arrangement, especially for portable work, is to use a metal hinge at each angle of the reflector. This permits the reflector to be folded flat for transit. It also permits experiments to be carried out with different apex angles.

A housing is required at the dipole center to prevent the entry of moisture and, in the case of the 432-MHz antenna, to support the dipole elements. The dipole may be moved in and out of the reflector to get either minimum SWR or, if this cannot be measured, maximum gain. If a two-stub tuner or other matching device is used, the dipole may be placed to give optimum gain and the matching device adjusted to give optimum match. In the case of the 1296-MHz antenna, the dipole length can be adjusted by means of the brass screws at the ends of the elements. Locking nuts are essential.

The reflector should be made of sheet aluminum for 1296 MHz, but can be constructed of wire mesh (with twists parallel to the dipole) for 432 MHz. To increase the gain by 3 dB, a pair of these arrays can be stacked so the reflectors are barely separated (to prevent the formation of a slot radiator by the edges). The radiating dipoles must then be fed in phase, and suitable feeding and matching must be arranged. A two-stub tuner can be used for matching either a single- or double-reflector system.

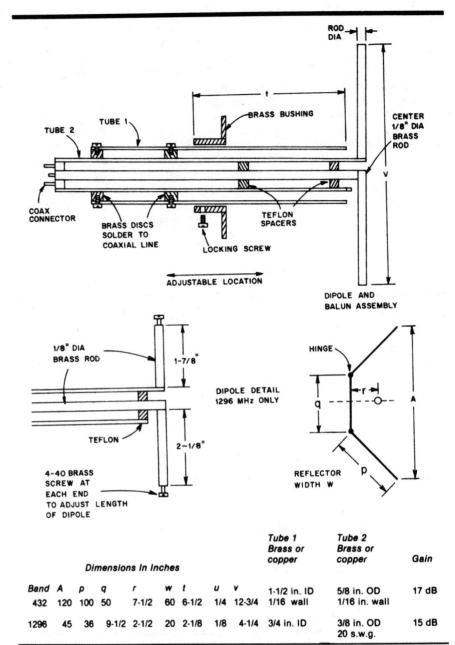

Dimensions In Inches								Tube 1 Brass or copper	Tube 2 Brass or copper	Gain
Band	A	p	q	r	w	t	u	v			
432	120	100	50	7-1/2	60	6-1/2	1/4	12-3/4	1-1/2 in. ID 1/16 wall	5/8 in. OD 1/16 in. wall	17 dB
1296	45	36	9-1/2	2-1/2	20	2-1/8	1/8	4-1/4	3/4 in. ID	3/8 in. OD 20 s.w.g.	15 dB

Fig 64—Practical construction information for trough reflector antennas for 432 and 1296 MHz.

A Horn Antenna for 10 GHz

The horn antenna is the easiest antenna for the beginner on 10 GHz to construct. It can be made out of readily available flat sheet brass. Because it is inherently a broadband structure, minor constructional errors can be tolerated. The one drawback is that horn antennas become physically cumbersome at gains over about 25 dB, but for most line-of-sight work this much gain is rarely necessary. This antenna was designed by Bob Atkins, KA1GT, and appeared in *QST* for April and May 1987.

Horn antennas are usually fed by waveguide. When operating in its normal frequency range, waveguide propagation is in the TE_{10} mode. This means that the electric (E) field is across the short dimension of the guide and the magnetic (H) field is across the wide dimension. This is the reason for the E-plane and H-plane terminology shown in Fig 65.

There are many varieties of horn antennas. If the waveguide is flared out only in the H-plane, the horn is called an H-plane sectoral horn. Similarly, if the flare is only in the E-plane, an E-plane sectoral horn results. If the flare is in both planes, the antenna is called a pyramidal horn.

For a horn of any given aperture, directivity (gain along the axis) is maximum when the field distribution across the aperture is uniform in magnitude and phase. When the fields are not uniform, side lobes that reduce the directivity of the antenna are formed. To obtain a uniform distribution, the horn should be as long as possible with minimum flare angle. From a practical point of view, however, the horn should be as short as possible, so there is an obvious conflict between performance and convenience.

Fig 66 illustrates this problem. For a given flare angle and a given side length, there is a path-length difference from the apex of the horn to the center of the aperture (L), and from the apex of the horn to the edge of the aperture (L ′). This causes a phase difference in the field across the aperture, which in turn causes formation of side lobes, degrading directivity (gain along the axis) of the antenna. If L is large this difference is small, and the field is almost uniform. As L decreases however, the phase difference increases and directivity suffers. An optimum (shortest possible) horn is constructed so that this phase difference is the maximum allowable before side lobes become excessive and axial gain markedly decreases.

The magnitude of this permissible phase difference is different for E-plane and H-plane horns. For the E-plane horn, the field intensity is quite constant across the aperture. For the H-plane horn, the field tapers to zero at the edge. Consequently, the phase difference at the edge of the aperture in the E-plane horn is more critical and should be held to less than 90° (¼ λ). In an H-plane horn, the allowable phase difference is 144° (0.4 λ). If the aperture of a pyramidal horn exceeds one wavelength in both planes, the E-plane and H-plane patterns are essentially independent and can be analyzed separately.

The usual direction for orienting the waveguide feed is

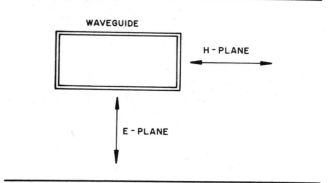

Fig 65—10-GHz antennas are usually fed with waveguide. See text for a discussion of waveguide propagation characteristics.

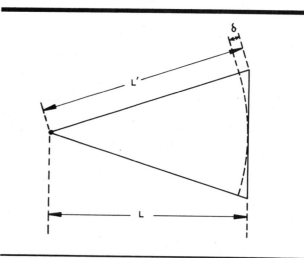

Fig 66—The path-length (phase) difference between the center and edge of a horn antenna is δ.

with the broad face horizontal, giving vertical polarization. If this is the case, the H-plane sectoral horn has a narrow horizontal beamwidth and a very wide vertical beamwidth. This is not a very useful beam pattern for most amateur applications. The E-plane sectoral horn has a narrow vertical beamwidth and a wide horizontal beamwidth. Such a radiation pattern could be useful in a beacon system where wide coverage is desired.

The most useful form of the horn for general applications is the optimum pyramidal horn. In this configuration the two beamwidths are almost the same. The E-plane (vertical)

Fig 67—This pyramidal horn has 18.5 dBi gain at 10 GHz. Construction details are given in the text.

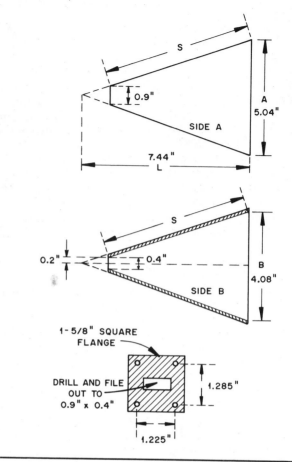

Fig 68—Dimensions of the brass pieces used to make the 10-GHz horn antenna. Construction requires two of each of the triangular pieces (side A and side B).

beamwidth is slightly less than the H-plane (horizontal), and also has greater side lobe intensity.

Building the Antenna

A 10-GHz pyramidal horn with 18.5 dBi gain is shown in Fig 67. The first design parameter is usually the required gain, or the maximum antenna size. These are of course related, and the relationships can be approximated by the following:

L = H-plane length (λ) = 0.0654 × gain   (Eq 1)
A = H-plane aperture (λ) = 0.0443 × gain   (Eq 2)
B = E-plane aperture (λ) = 0.81 A   (Eq 3)
where
    gain is expressed as a *ratio*; 20 dB gain = 100
    L, A and B are dimensions shown in Fig 68

From these equations, the dimensions for a 20 dB gain horn for 10.368 GHz can be determined. One wavelength at 10.368 GHz is 1.138 inches. The length (L) of such a horn is 0.0654 × 100 = 6.54 λ. At 10.368 GHz, this is 7.44 inches. The corresponding H-plane aperture (A) is 4.43 λ (5.04 inches), and the E-plane aperture (B), 4.08 inches.

The easiest way to make such a horn is to cut pieces from brass sheet stock and solder them together. Fig 68 shows the dimensions of the triangular pieces for the sides and a square piece for the waveguide flange. (A standard commercial waveguide flange could also be used.) Because the E-plane and H-plane apertures are different, the horn opening is not square. Sheet thickness is unimportant; 0.02 to 0.03 inch works well. Brass sheet is often available from hardware or hobby shops.

Note that the triangular pieces are trimmed at the apex to fit the waveguide aperture (0.9 × 0.4 inch). This necessitates that the length, from base to apex, of the smaller triangle (side B) is shorter than that of the larger (side A). Note that the length, S, of the two different sides of the horn must be the same if the horn is to fit together! For such a simple looking object, getting the parts to fit together properly requires careful fabrication.

The dimensions of the sides can be calculated with simple geometry, but it is easier to draw out templates on a sheet of cardboard first. The templates can be used to build a mock antenna to make sure everything fits together properly before cutting the sheet brass.

First, mark out the larger triangle (side A) on cardboard. Determine at what point its width is 0.9 inch and draw a line parallel to the base as shown in Fig 68. Measure the length of the side S; this is also the length of the sides of the smaller (side B) pieces.

Mark out the shape of the smaller pieces by first drawing

a line of length B and then constructing a second line of length S. One end of line S is an end of line B, and the other is 0.2 inch above a line perpendicular to the center of line B as shown in Fig 68. (This procedure is much more easily followed than described.) These smaller pieces are made slightly oversize (shaded area in Fig 68) so you can construct the horn with solder seams on the outside of the horn during assembly.

Cut out two cardboard pieces for side A and two for side B and tape them together in the shape of the horn. The aperture at the waveguide end should measure 0.9 × 0.4 inch and the aperture at the other end should measure 5.04 × 4.08 inches.

If these dimensions are correct, use the cardboard templates to mark out pieces of brass sheet. The brass sheet should be cut with a bench shear if one is available, because scissors type shears tend to bend the metal. Jig the pieces together and solder them on the *outside* of the seams. It is important to keep both solder and rosin from contaminating the inside of the horn; they can absorb RF and reduce gain at these frequencies.

Assembly is shown in Fig 69. When the horn is completed, it can be soldered to a standard waveguide flange, or one cut out of sheet metal as shown in Fig 68. The transition between the flange and the horn must be smooth. This antenna provides an excellent performance to cost ratio (about 20 dB gain for about five dollars in parts).

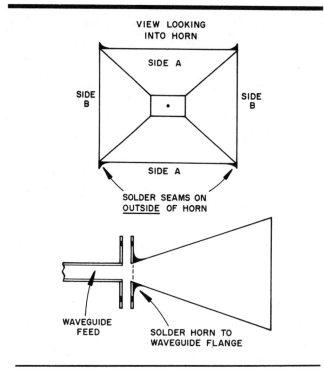

Fig 69—Assembly of the 10-GHz horn antenna.

Periscope Antenna Systems

One problem common to all who use microwaves is that of mounting an antenna at the maximum possible height while trying to minimize feed-line losses. The higher the frequency, the more severe this problem becomes, as feeder losses increase with frequency. Because parabolic dish reflectors are most often used on the higher bands, there is also the difficulty of waterproofing feeds (particularly waveguide feeds). Inaccessibility of the dish is also a problem when changing bands. Unless the tower is climbed every time and the feed changed, there must be a feed for each band mounted on the dish. One way around these problems is to use a periscope antenna system (sometimes called a "flyswatter antenna").

The material in this section was prepared by Bob Atkins, KA1GT, and appeared in *QST* for January and February 1984. Fig 70 shows a schematic representation of a periscope antenna system. A plane reflector is mounted at the top of a rotating tower at an angle of 45°. This reflector can be elliptical with a major to minor axis ratio of 1.41, or rectangular. At the base of the tower is mounted a dish or other type of antenna such as a Yagi, pointing straight up. The advantage of such a system is that the feed antenna can be changed and worked on easily. Additionally, with a correct choice of reflector size, dish size, and dish to reflector spacing, feed losses can be made small, increasing the effective system gain. In fact, for some particular system configurations, the gain of the overall system can be greater than that of the feed antenna alone.

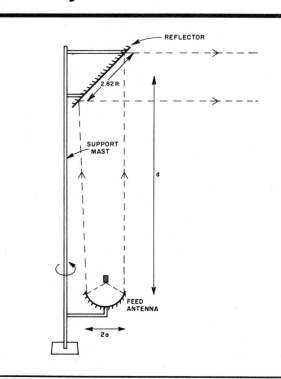

Fig 70—The basic periscope antenna. This design makes it easy to adjust the feed antenna.

Gain of a Periscope System

Fig 71 shows the relationship between the effective gain of the antenna system and the distance between the reflector and feed antenna for an elliptical reflector. At first sight, it is not at all obvious how the antenna system can have a higher gain than the feed alone. The reason lies in the fact that, depending on the feed to reflector spacing, the reflector may be in the near field (Fresnel) region of the antenna, the far field (Fraunhoffer) region, or the transition region between the two.

In the far field region, the gain is proportional to the reflector area and inversely proportional to the distance between the feed and reflector. In the near field region, seemingly strange things can happen, such as decreasing gain with decreasing feed to reflector separation. The reason for this gain decrease is that, although the reflector is intercepting more of the energy radiated by the feed, it does not all contribute in phase at a distant point, and so the gain decreases.

In practice, rectangular reflectors are more common than elliptical. A rectangular reflector with sides equal in length to the major and minor axes of the ellipse will, in fact, normally give a slight gain increase. In the far field region, the gain will be proportional to the area of the reflector. To use Fig 71 with a rectangular reflector, R^2 may be replaced by A/π, where A is the projected area of the reflector. The antenna pattern depends in a complicated way on the system parameters (spacing and size of the elements), but Table 14 gives an approximation of what to expect. R is the radius of

Table 14

Radiation Patterns of Periscope Antenna Systems

	Elliptical Reflector	Rectangular Reflector
3-dB beamwidth, degr	60 λ/2R	52 λ/b
6-dB beamwidth, degr	82 λ/2R	68 λ/b
First minimum, degr from axis	73 λ/2R	58 λ/b
First maximum, degr from axis	95 λ/2R	84 λ/b
Second minimum, degr from axis	130 λ/2R	116 λ/b
Second maximum, degr from axis	156 λ/2R	142 λ/b
Third minimum, degr from axis	185 λ/2R	174 λ/b

the projected circular area of the elliptical reflector (equal to the minor axis radius), and b is the length of the side of the projected square area of the rectangular reflector (equal to the length of the short side of the rectangle).

For those wishing a rigorous mathematical analysis of this type of antenna system, several references are given in the bibliography at the end of this chapter.

Mechanical Considerations

There are some problems with the physical construction of a periscope antenna system. Since the antenna gain of a microwave system is high and, hence, its beamwidth narrow, the reflector must be accurately aligned. If the reflector does

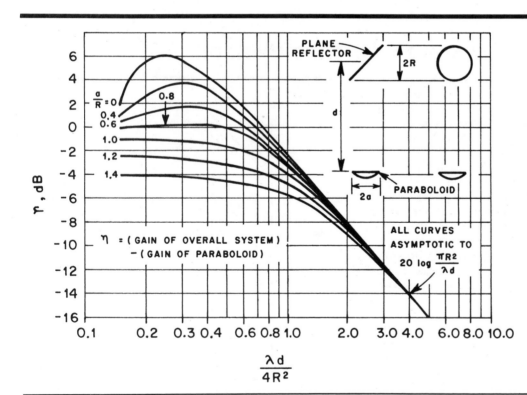

Fig 71—Gain of a periscope antenna using a plane elliptical reflector (after Jasik—see bibliography).

not produce a beam that is horizontal, the useful gain of the system will be reduced. From the geometry of the system, an angular misalignment of the reflector of X degrees in the vertical plane will result in an angular misalignment of 2X degrees in the vertical alignment of the antenna system pattern. Thus, for a dish pointing straight up (the usual case), the reflector must be at an angle of 45° to the vertical and should not fluctuate from factors such as wind loading.

The reflector itself should be flat to better than 1/10 λ for the frequency in use. It may be made of mesh, provided that the holes in the mesh are also less than 1/10 λ in diameter. A second problem is getting the support mast to rotate about a truly vertical axis. If the mast is not vertical, the resulting beam will swing up and down from the horizontal as the system is rotated, and the effective gain at the horizon will fluctuate. Despite these problems, amateurs have used periscope antennas successfully on the bands through 10 GHz. Periscope antennas are used frequently in commercial service, though usually for point-to-point transmission. Such a commercial system is shown in Fig 72.

Fig 72—Commercial periscope antennas, such as this one, are often used for point-to-point communication.

Circular polarization is not often used for terrestrial work, but if it is used with a periscope system there is an important point to remember. The circularity sense changes when the signal is reflected. Thus, for right hand circularity with a periscope antenna system, the feed arrangement on the ground should produce left hand circularity. It should also be mentioned that it is possible (though more difficult for amateurs) to construct a periscope antenna system using a parabolically curved reflector. The antenna system can then be regarded as an offset fed parabola. More gain is available from such a system at the added complexity of constructing a parabolically curved reflector, accurate to 1/10 λ.

BIBLIOGRAPHY

Source material and more extended discussion of topics covered in this chapter can be found in the references given below and in the textbooks listed at the end of Chapter 2.

B. Atkins, "Periscope Antenna Systems," The New Frontier, *QST*, Jan and Feb 1984.

B. Atkins, "Horn Antennas for 10 GHz," The New Frontier, *QST*, Apr and May 1987.

J. Drexler, "An Experimental Study of a Microwave Periscope," *Proc. IRE*, Correspondence, Vol 42, Jun 1954, p 1022.

D. Evans and G. Jessop, *VHF-UHF Manual*, 3rd ed. (London: RSGB), 1976.

N. Foot, "WA9HUV 12-Foot Dish for 432 and 1296 MHz," The World Above 50 Mc., *QST*, Jun 1971, pp 98-101, 107.

N. Foot, "Cylindrical Feed Horn for Parabolic Reflectors," *Ham Radio*, May 1976.

G. Gobau, "Single-Conductor Surface-Wave Transmission Lines," *Proc IRE*, Vol 39, Jun 1951, pp 619-624; also see *Journal of Applied Physics*, Vol 21 (1950), pp 1119-1128.

R. E. Greenquist and A. J. Orlando, "An Analysis of Passive Reflector Antenna Systems," *Proc IRE*, Vol 42, Jul 1954, pp 1173-1178.

G. A. Hatherell, "Putting the G Line to Work," *QST*, Jun 1974, pp 11-15, 152, 154, 156.

D. L. Hilliard, "A 902-MHz Loop Yagi Antenna," *QST*, Nov 1985, pp 30-32.

W. C. Jakes, Jr, "A Theoretical Study of an Antenna-Reflector Problem," *Proc IRE*, Vol 41, Feb 1953, pp 272-274.

H. Jasik, *Antenna Engineering Handbook*, 1st ed. (New York: McGraw-Hill, 1961).

R. T. Knadle, "UHF Antenna Ratiometry," *QST*, Feb 1976, pp 22-25.

T. Moreno, *Microwave Transmission Design Data* (New York: McGraw-Hill, 1948).

W. Overbeck, "The VHF Quagi," *QST*, Apr 1977, pp 11-14.

W. Overbeck, "The Long-Boom Quagi," *QST*, Feb 1978, pp 20-21.

W. Overbeck, "Reproducible Quagi Antennas for 1296 MHz," *QST*, Aug 1981, pp 11-15.

G. Southworth, *Principles and Applications of Waveguide Transmission* (New York: D. Van Nostrand Co, 1950).

P. P. Viezbicke, "Yagi Antenna Design," *NBS Technical Note 688* (US Dept of Commerce/National Bureau of Standards, Boulder, CO), Dec 1976.

D. Vilardi, "Easily Constructed Antennas for 1296 MHz," *QST*, Jun 1969.

D. Vilardi, "Simple and Efficient Feed for Parabolic Antennas," *QST*, Mar 1973.

Radio Communication Handbook, 5th ed. (London: RSGB, 1976).

Chapter 19

Antenna Systems for Space Communications

There are two basic modes of space communications: satellite and earth-moon-earth (EME—also referred to as moonbounce). Both require consideration of the effects of polarization and elevation angle, along with the azimuth directions of transmitted and received signals.

Signal polarization is generally of little concern on the HF bands, as the original polarization direction is lost after the signal passes through the ionosphere. Vertical antennas receive sky-wave signals emanating from horizontal antennas, and vice versa. It is not beneficial to provide a means of varying the elevation angle in this case, because at HF the takeoff angle is not significantly affected. With satellite communications, however, because of polarization changes, a signal that would disappear into the noise on one antenna may be S9 on one that is not sensitive to polarization direction. Elevation angle is also important from the standpoint of tracking and avoiding indiscriminate ground reflections that may cause nulls in signal strength.

These are the characteristics common to both satellite and EME communications. There are also characteristics unique to each mode, and these cause the antenna requirements to differ in several ways—some subtle, others profound. Each mode is dealt with separately in this chapter after some basic information pertaining to all space communications is presented.

Antenna Positioning

Where high-gain antennas are required in space communications, precise and accurate azimuth and elevation control and indication are necessary. High gain implies narrow beamwidth in at least one plane. Low orbit satellites such as FO-12 move through the window very quickly, so azimuth and elevation tracking are essential if high-gain antennas are used.

These satellites are fairly easy to access with moderate power and broad coverage antennas. The low power, high-gain approach is more sophisticated, but the high power, low-gain solution may be more practical and economical.

Some EME arrays are fixed, but these are limited to narrow time windows for communication. The az-el positioning systems described in the following sections are adaptable to either satellite or modest EME arrays. Figs 1 and 2 illustrate one of the more ambitious ventures in positioning a large EME array.

AN AZ-EL MOUNT FOR CROSSED YAGIS

The mounting system of Figs 3, 4 and 5 was originally described by Katashi Nose, KH6IJ, in June 1973 *QST*. (See the bibliography at the end of this chapter.) The basic criteria in the design of this system were low cost and ease of assembly. The choice of a crossed Yagi system was influenced by the ready availability of Yagi antennas from dealers. Methods of feeding such arrays are discussed later in this chapter.

Fig 1—An aggressive approach to steering a giant EME antenna—a 5-in. gun turret from a destroyer.

Fig 3 shows the assembled array. The antennas are eight-element Yagis. Fig 4 is a head-on view of the array, showing the antennas mounted at 90° with respect to each other and 45° with respect to the cross arm. Coupling between the two Yagis is minimal at 90°. By setting the angle at 45° with respect to the cross arm, coupling is reduced (but not eliminated).

Determine length d in Fig 4 by pointing the array straight up and rotating it. Length d should be the minimum distance necessary for the elements to clear the tripod base when this is done. In the array shown in Fig 3, a 5-foot section of TV mast serves the purpose.

The Mounting Tripod

A mounting tripod can be made of aluminum railing, called "NuRail," of which all manner of swivels, crosses and T fittings are available. The least expensive method, however, is to purchase a TV-antenna tripod. These tripods sell for such low prices that there is little point in constructing your own. Spread the legs of the tripod more than usual to assure greater support, and be sure that the elements of the antenna clear the base in the straight-up position.

Elevation and Azimuth Rotators

Any medium-duty rotator can be used for azimuth rotation in this system. The elevation rotator should be one that allows the cross arm of the array to be rotated on its axis when supported at the center.

Fig 5 shows the mounting of the two rotators. The flat

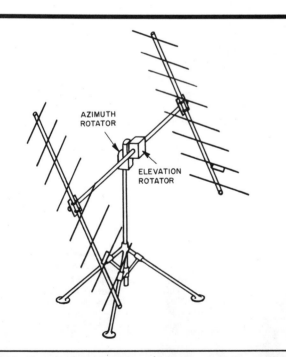

Fig 2—The gun mount of Fig 1 with its "warhead" attached—a homemade 42-ft parabolic dish. This is part of the arsenal of Ken Kucera, KAØY.

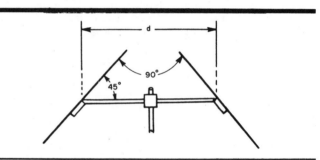

Fig 4—An end-on view of the crossed Yagi antennas shows that they are mounted at 90° to each other, and at 45° to the cross boom.

Fig 3—A crossed Yagi antenna system can be assembled using off-the-shelf components such as Hy-Gain Yagis, Cornell-Dubilier or Blonder-Tongue rotators and a commercially made tripod.

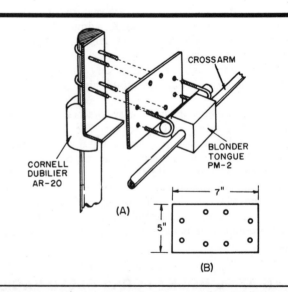

Fig 5—The method of mounting two rotators together. A pair of PM-2 rotators may also be used. The adapter plate (B) may be fabricated from ¼-in. thick aluminum stock, or a ready made plate is available from Blonder-Tongue.

portion of the Cornell-Dubilier AR-20 rotator makes an ideal mounting surface for the elevation rotator. If commercially fabricated components are to be used throughout, a mounting plate similar to that shown in Fig 5B can be purchased. The adapter plate may be used to fasten two rotators together.

ELEVATION CONTROL USING SYNCHROS

Many amateurs have adapted TV rotators such as the Alliance U-100 and U-110 for use as elevation rotators. For small OSCAR antennas with wide beamwidths, these rotators perform satisfactorily. Unfortunately, however, the elevation of antennas with the stock U-100 and U-110 rotators is limited to increments of 10°. This limitation, combined with the possibility of the control box losing synchronization with the motor, can cause the actual antenna elevation to differ from that desired by as much as 30° or 40° at times. With high gain, narrow beamwidth arrays, such as those needed for EME work and for high altitude satellites (Phase III), this large a discrepancy is unsuitable. (Rotators designed specifically for use in the horizontal position should be used for EME antennas. The elevation readout system described here will provide superior accuracy when used with most rotators.)

This indication system uses a pair of *synchro transformers* to provide an accurate, continuous readout of the elevation angle of the antenna array. The Alliance rotator control unit is modified so that the motor can be operated to provide a continuously variable angle of antenna elevation. Jim Bartlett, K1TX, described this system in June 1979 *QST*.

The synchro or Selsyn™ is a specialized transformer. See Fig 6A. It can be best described as a transformer having three secondary windings and a single *rotating* primary winding. Synchros are sometimes called "one-by-threes" for this reason. When two synchros are connected together as in Fig 6B and power is applied to their primary windings, the shaft attached to the rotating primary in one synchro will track the position of the shaft and winding in the other. When two synchros are used together in such an arrangement, the system is called a *synchro repeater loop*.

In repeater loops, one synchro transformer is usually designated as the one where motion is initiated, and the other repeats this motion. When two synchro transformers are used in such a repeater loop, the individual units can be thought of as "transmitter" and "receiver," or *synchro generator* and *synchro motor*, respectively. In this application (where one unit is located at the antenna array and another is used as an indicator), the antenna unit is referred to as the generator and the indicator unit the motor.

The synchro generator is so named because it electrically transmits a rotational force to the synchro motor. The motor, also sometimes called the *receiver, follower* or *repeater*, receives this energy from the generator, and its shaft turns accordingly.

Physical Characteristics

Synchro transformers, both generator and motor types, resemble small electric motors, with only minor differences. Generator and motor synchros are identical in design for all practical purposes. The only difference between them is the

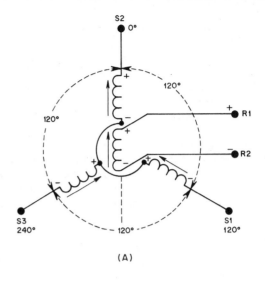

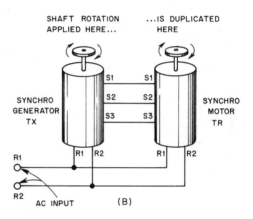

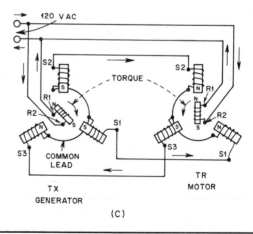

Fig 6—At A, a schematic diagram of the synchro or Selsyn transformer. Connection of two synchros in a repeater loop is shown at B. The drawing at C shows the instantaneous forces in the repeater loop with the rotor shafts at different positions. The "TX" and "TR" notations stand for *torque transmitter and torque receiver*, respectively. Synchros are sometimes listed in catalogs by these "type" symbols.

presence of an *inertia damper*—a special flywheel—on units specifically designated as synchro motors. For antenna use, the inertia damper is not a necessity.

Fig 6A shows the synchro transformer schematically. In each synchro, there are two elements: the fixed secondary windings, called the *stator*, and the rotatable primary, called the *rotor*. The rotor winding is connected to a source of alternating current, and the shaft is coupled to a controlling shaft or load—in this case, the antenna array or elevation readout pointer. An alternating field is set up by the rotor winding as a result of the ac voltage applied to it. This causes voltages to be induced in the stator windings. These voltages are representative of the angular position of the rotor.

The stator consists of many coils of wire placed in slots around the inside of a laminated field structure, much like that in an electric motor. The stator coils are divided into three groups spaced 120° around the inside of the field with some overlap to provide a uniform magnitude of attractive force on the rotor. The leads from the rotor and stator windings are attached to insulated terminal strips, usually located at the rear of the motor or generator housing. The rotor connections are labeled R1 and R2, and the stator connections S1, S2 and S3. These are shown in Fig 6A. These rotor and stator designations are standard identifications.

Synchro Transformer Action

Synchros operate much like transformers. The main difference between them is that in a synchro, the primary winding (rotor) can be rotated through 360°.

The ac applied to the synchro rotor coil varies, but the most common ratings are 115 V/60 Hz, 115 V/400 Hz and 26 V/400 Hz. The 400-Hz varieties are easier to find on the surplus market, but are more difficult to use, as a 400-Hz supply must be built. Bartlett, K1TX, used 90-V/60-Hz synchros for this project, and the 90 volts required was obtained by using two surplus transformers back to back (one 6.3 volt and one 5 volt). Regardless of the voltage or line frequency used, synchros should be fused, and *isolated from the ac mains by a transformer*. This is important to ensure a safe installation.

The voltages induced in the stator windings are determined by the position of the rotor. As the rotor changes position and different values are induced, the direction of the resultant fields changes.

When a second synchro transformer is connected to the first, forming a generator/motor pair or repeater loop, the voltages induced in the three generator stator coils are also induced into the respective motor stator coils. As long as the two rotor shafts are in the same position, the voltages induced in the stator windings of the generator and motor units are equal. These voltages are of opposite polarity, however, because of the way the two units are connected together. This results in a zero potential difference between the stators in the two synchro units, and no current flows in either set of stator coils.

With the absence of current flow, no magnetic field is set up by the stator windings, and the system is in mechanical equilibrium. (There are no unbalanced forces acting on either rotor.) This situation exists whenever the two rotors are aligned in identical angular positions, regardless of the specific angle of displacement from the zero point (S2).

The repeater action of the two-synchro system occurs when one rotor is moved, causing the voltages in the system to become unbalanced. When this happens, current flows through the stator coils, setting up magnetic fields that tend to pull the rotors together so that the static (equilibrium) condition again exists. A torque results from the magnetic fields set up in both units, causing the two rotors to turn in opposite directions until they align themselves.

The generator shaft, however, is usually attached to a control shaft or large load (relative to that attached to the motor shaft) so that it cannot freely rotate. Thus, as long as the motor rotor is free to move, it will remain in alignment with the generator rotor. Fig 6C shows the instantaneous forces present in a repeater loop when one rotor is turned.

Selecting the Synchros

Synchro operating voltages are not critical. Most units will function with voltages as much as 20% above or 30% below their nominal ratings. Make sure the transformer(s) you use will handle the necessary current. Fig 7A shows how to connect two transformers to obtain 90 V for the units used in this project.

Synchro transformers normally found in surplus catalogs and at flea markets may not be suitable for this application. Some of the types you should *not* buy are ones marked *differential generator, differential synchro* or *resolver synchro*. These synchros are designed for different uses.

Most catalogs list synchro transformers with their ratings and prices. Look for the least expensive set of synchros that will operate at the required voltage and line frequency. When comparing specifications, look for synchros that have a high *torque gradient* (accuracy). It is possible to obtain accuracy as good as ±1° with a properly installed synchro readout system.

When the synchros have been obtained and a power supply designed, begin construction of the elevation system. Check the synchros by connecting the two units as shown in Fig 6B. Verify proper operation. Set the synchros aside and begin modification of the Alliance rotator-control unit (if you have decided to use an Alliance rotator).

The Alliance Rotator-Control Unit

Remove the transformer, capacitor and pilot light from the control unit and discard the rest. Mount the transformer and capacitor in a small, shallow enclosure, like the one shown in Fig 8. The synchro power supply will also be mounted in this box.

Wire the rotator control circuit as shown in Fig 7A. The transformer, pilot light and capacitor shown are the ones removed from the Alliance control unit. Add a fuse at the point shown. The 120-V input to this circuit can be tied to that of the power supply circuit if desired. This allows for a common fuse and power switch. The rating of the fuse depends upon the current drain of the synchros used, but a 1-A fuse should be ample to handle the control and power supply circuits. Note that there are four wires in the Alliance control system. Only three are needed here; the fourth wire is not used.

Test the control unit before mounting the rotator on the mast. Connect the motor to the modified control unit and check to see that it rotates properly in both directions when S1 is activated. This switch should be a DPDT, momentary on, center off toggle switch. Next install the synchro power supply inside the rotator-control enclosure. Some type of multiconnector plug and jack combination should be used at

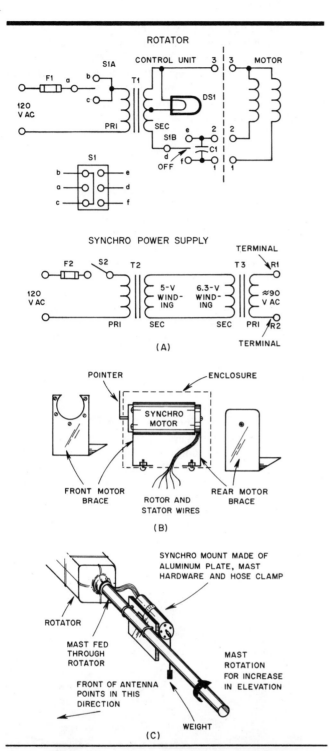

Fig 7—Shown at A are the circuits for the modified control unit and the synchro power supply. T1, DS1 and C1 are from the stock U-100 or U-110 control box. See text. At B, the mounting method used to secure the synchro motor is shown. Details of the synchro generator mounting are shown at C. See text for description of materials.

F1, F2—1 A, 250 V fuse.
S1—DPDT momentary contact, center off toggle switch.
S2—SPST toggle switch.
T2, T3—Transformers selected for proper voltage to synchro rotor.

Fig 8—The completed control/readout unit for antenna elevation. The dial face was made from a plastic protractor.

the rear of the cabinet so the rotator and synchro control wires can be easily disconnected from the control box. Eight wires are used between the control unit and the synchro and rotator mounted at the antenna array. An 8-pin, octal connector set and standard 8-wire rotator cable were used in this project. A suitable alternative connector set is Calectro F3-248 (male cord) and F3-268 (female chassis).

Mechanical Details

The synchro motor providing the elevation readout is mounted inside a cube-shaped chassis. (Any suitable chassis will do.) Two aluminum brackets support the motor inside the box, as shown in Fig 7B. The motor is positioned to allow the shaft to protrude through the front panel of the enclosure. The pointer is fashioned from a scrap of copper sheet, and soldered to the edge of a washer. This is secured to the shaft between two nuts. A large protractor that fits the front of the enclosure serves as the dial face.

Mounting and Calibration

The synchro generator mounting is shown in Fig 7C. An aluminum plate is drilled and fitted with standard hardware. Cut two slots between the clamps, and insert a large stainless steel hose clamp through the slots and around the generator casing. After positioning the synchro, tighten the clamp. The generator is mounted close to the rotator and directly behind the elevation mast when the antennas are pointed at the horizon.

The elevation and azimuth rotators are mounted in the normal fashion, as shown in Fig 9. Elevation of the antennas causes generator shaft rotation through a weighted rod fastened to the synchro shaft, as shown in Fig 7C.

As the antenna array is elevated, the synchro generator moves through an arc starting behind the elevation mast, through a position directly below the mast, to one in front of it. During the swing through this arc, gravity keeps the weighted rod perpendicular to the ground, and the synchro shaft turns in proportion to the elevation angle. (If high winds are common in your area, keep the "plumb-line" swing arm short so gusts of wind won't cause fluctuations in the elevation readout.)

The easiest way to calibrate the system is to attach the antennas and synchro to the mast when the elevation rotator is at the end of rotation (at a stop). Do this so any movement must be in the direction that will elevate the array with respect to the horizon. With the antennas pointing at the horizon, set the synchro motor pointer to 0° at one end of the protractor scale. The proper "zero" end depends upon the specific mounting scheme used at the antenna.

If the generator is mounted as shown in Fig 7C and all connections are properly made, the elevation needle should swing from right to left as the antennas move from zero through 90 to 180°. If not, remove power from the system and interchange the S1 and S3 wires at the indicator motor.

A RADIO-COMPASS ELEVATION READOUT SYSTEM

As described by Jim Bartlett, K1TX, in September 1979 *QST*, an MN-98 Canadian radio compass and a Sperry R5663642 synchro transmitter combine to make a highly precise elevation indicator. These components, displayed in Fig 10, may be available from Fair Radio Sales Co, PO Box 1105, Lima, OH 45802. The AY-201 transmitter is *not* suitable for this project.

Place the MN-98 indicator face down on a soft cloth on a flat surface and remove the rear cover of the indicator unit. Disconnect the four wires that go to the glass-metal feedthrough located on the back panel. This frees the rear cover. Remove the rear cover and put it aside. Drill a small hole in the rear of the case, next to the edge of the feedthrough. (See Fig 11A.) Do this *carefully*, making sure that the drill bit doesn't push through into the inside of the indicator shell and get tangled in the wiring. When the bit breaks through the metal casing, the pressurized seal will be broken.

Using a small screwdriver and a hammer, tap each of the individual glass feedthrough inserts, cracking them. Try to keep the screwdriver from pushing the broken pieces of glass down into the enclosure where they could get lodged in the dial mechanism. Attempt to shake all the pieces of glass out of the case. The remaining part of the feedthrough can be removed by heating with a soldering iron and prying with a screwdriver or needle-nosed pliers.

After the feedthrough has been removed, gently pull the ends of the wires out through the hole left by the feedthrough. Clip off the feedthrough terminal pins. There are five wires—a group of three and two others. The group of three will probably be blue, yellow and black. The other two wires twisted together should be red and black. Fig 11B shows how these are connected to the terminals on the synchro transmitter in a five-wire system.

Construction of the System

Fig 11C shows the schematic diagram of a simple 6.3-V ac power supply for the indicator system. Because the synchro and indicator were originally designed to operate from 26 V at 400 Hz, a 6.3-V transformer is acceptable for use at 60 Hz. A 22-Ω resistor is wired in series with the synchros to limit

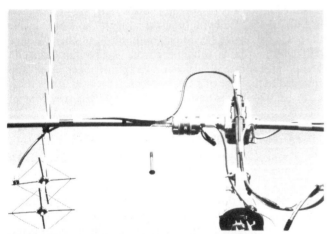

Fig 9—Close-up photo of the synchro transformer mounting method. The weighted arm is kept short to minimize wind effects on elevation readout.

Fig 10—The MN-98 Canadian radio compass and Sperry R5663642 synchro transmitter. Note the small knob at the upper right-hand corner of the indicator face. This can be used to calibrate the system without making any changes at the antenna end. By turning this knob, you can rotate the degree markings around the outside of the dial face so that any desired heading can be placed in line with the pointer.

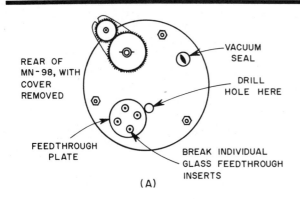

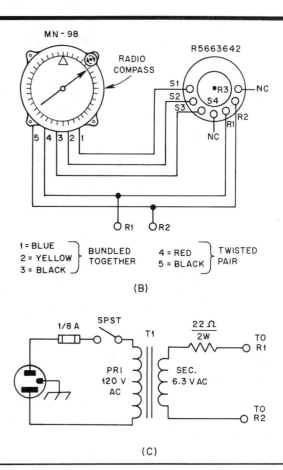

Fig 11—The rear of the MN-98 Canadian radio compass is shown at A. The drawing at B shows the interconnecting method used between the MN-98 and the Sperry synchro transmitter. The schematic diagram at C shows the power supply used with this indicator system. T1 can be Radio Shack 273-1384 or any junk-box 6.3-V transformer.

current and thus eliminate an annoying buzzing sound in the indicator unit at certain pointer positions.

The indicator, along with the power supply, can be mounted in a small metal enclosure. Include a fuse, ON-OFF switch, and three-wire line cord. At the synchro transmitter end (at the antenna), provide some kind of shield to keep weather from affecting the system.

A small weight, cut in the shape of a large pie section and drilled to fit the synchro transmitter shaft, can be mounted on the shaft and shielded with a small margarine tub which is taped or glued to the outside of the synchro casing. This arrangement should allow free movement of the weight, yet keep high winds or heavy icing from affecting the indicator. The synchro transmitter should be mounted to the mast in such a way that it will rotate with the antennas, causing the weight to turn the shaft.

Antennas for Satellite Work

This section contains a number of antenna systems that are practical for satellite communications. Some of the simpler ones bring space communications into the range of any amateur's budget.

RECEIVING ANTENNAS FOR 29.4 MHz

Fig 12 shows three antennas suitable for satellite downlink reception at 29 MHz. At A is a turnstile, an antenna that is omnidirectional in the azimuth plane. The vertical pattern depends on the height above ground. (This subject is treated in detail in Chapter 3.) The circular polarization of the turnstile at high elevation angles reduces signal fading from satellite rotation and ionospheric effects.

The antenna at B is a simple rotatable dipole for use when a satellite is near the horizon and some directivity is helpful. When horizontally mounted, the full-wave loop at C gives

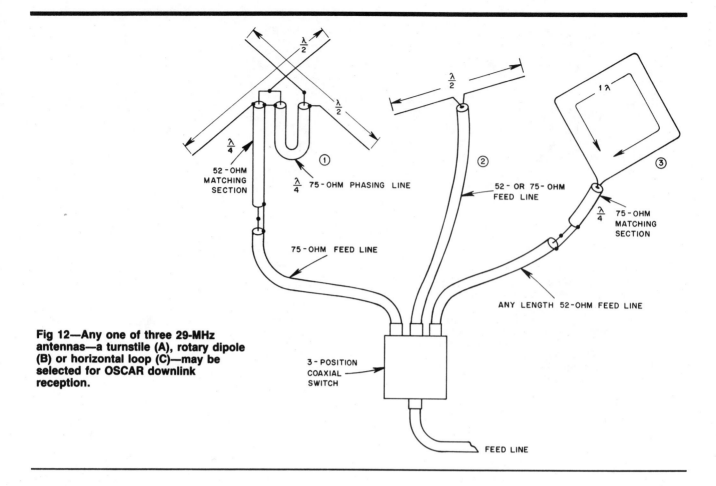

Fig 12—Any one of three 29-MHz antennas—a turnstile (A), rotary dipole (B) or horizontal loop (C)—may be selected for OSCAR downlink reception.

good omnidirectional reception for elevation angles above 30°. It should be mounted at least 1/8 λ above ground. It is difficult to predict which antenna will deliver the best signal under any circumstances. All are inexpensive, and the most effective amateur satellite stations have all three, with a means of selecting the best one for the existing conditions. For low-altitude satellites, conditions should be expected to change in the matter of a few minutes.

A 146-MHz TURNSTILE ANTENNA

The 146-MHz antenna of Fig 13 is simple and effective for use with OSCAR Modes A, B and J. The antenna, called a turnstile-reflector array, can be built very inexpensively and put into operation without the need for test equipment. The information contained here is based on a September 1974 *QST* article by Martin Davidoff, K2UBC.

Experience with several amateur satellites has shown that rapid fading is a severe problem in satellite work. Fortunately, the ground station has control over two important parameters affecting fading: cross polarization between the ground-

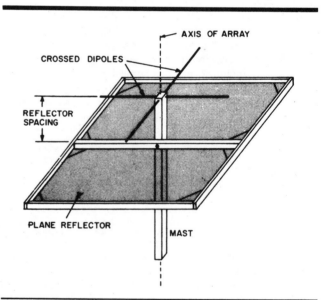

Fig 13—The turnstile-reflector (TR) array consists of crossed dipoles above a 4-ft square screen reflector.

station antenna and OSCAR antenna, and nulls in the ground-station antenna pattern. Fading that results from cross polarization can be reduced by using a circularly polarized ground-station antenna. Fading caused by radiation pattern nulls can be overcome by (1) using a rotatable, tiltable array and continuously tracking the satellite or (2) using an antenna with a broad, null-free pattern. The turnstile-reflector array solves these problems, as it is circularly polarized at high elevation angles and has a balloon-like high-angle directivity pattern. At lower elevation angles the polarization is elliptical. (Circular and elliptical polarizations are discussed later in this chapter.)

Construction

The mast used to support the two dipoles is made of wood, being 2 inches square and 8 feet long. The dipoles may be made of no. 12 copper wire, aluminum rod or tubing. The reflecting screen is 20-gauge hexagonal poultry netting, 1-inch mesh, stapled to a 4-foot square frame made of furring strips. Hardware cloth can be used in place of the poultry netting. Corner bracing of the reflector screen provides increased mechanical stability. Spar varnish applied to the wooden members will extend the service life of the assembly.

Dimensions for the two dipole antennas and the phasing network are shown in Fig 14. Spacing between the dipole antennas and the reflecting screen affects the antenna radiation pattern. Choose the spacing for the pattern that best suits your needs from data in Chapter 3, and construct the antenna accordingly. A spacing of $3/8 \lambda$ (30 inches) is suggested. This distance provides a theoretical pattern response of ± 1.5 dB at all angles above $15°$. Spacings greater than 30 inches will increase the response at elevation angles lower than $15°$, but at the expense of nulls in the pattern at

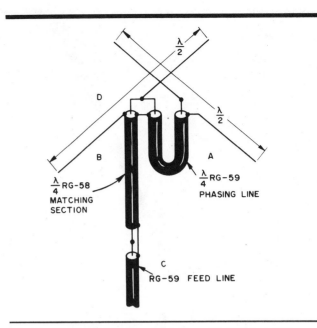

Fig 14—Dimensions and connections for the turnstile antenna. The phasing line is 13.3 in. of RG-59 coax (velocity factor = 0.66). A similar length of RG-58 cable is used as a matching section between the turnstile and the feed line. The phasing line length should be corrected for lines with other velocity factors.

higher angles. The feed-point impedance of the array will vary somewhat, depending on the spacing between the dipole elements and the reflecting screen.

Circular Polarization

The ideal antenna for random polarization is one with a circularly polarized radiation pattern. There are two commonly used methods for obtaining circular polarization. One is with crossed linear elements such as dipoles or Yagis. An array of crossed Yagis is shown in Fig 15. The second common method is with the helical antenna, described later in this chapter. Other methods also exist, such as with the quadrifilar helix (see Chapter 20).

Polarization *sense* is a critical factor, especially in EME work or if the satellite uses a circularly polarized antenna. In physics, clockwise rotation of an *approaching* wave is called "right circular polarization," but the IEEE standard uses the term "clockwise circular polarization" for a *receding* wave.

Amateur terminology follows the IEEE standard, calling clockwise polarization for a receding wave as right-hand. Either clockwise or a counter-clockwise sense can be selected by reversing the phasing harness of a crossed Yagi antenna. The sense of a helical antenna is fixed, determined by its physical construction.

In working through a satellite with a circularly polarized antenna, it is necessary to have the capability of switching the polarization sense. This is because the sense of the received signal reverses when the satellite passes its nearest point to you. If the received signal has right hand circular polarization as the satellite approaches, it will have left hand circularity as the satellite recedes. There is a sense reversal in EME work,

Fig 15—This VHF crossed Yagi antenna design by KH6IJ was presented in January 1973 *QST*. Placement of the phasing harness and T connector is shown in the lower half of the photograph. Note that the gamma match is mounted somewhat off element center for better balance of RF voltages on elements.

as well, because of a phase reversal of the signal as it is reflected from the surface of the moon. A signal transmitted with right hand circularity will be returned to the earth with left hand circularity.

Mathematically, linear and circular polarization are special cases of elliptical polarization. Consider two electric-field vectors at right angles to each other. The frequencies are the same, but the magnitudes and phase angles vary. If either one or the other of the magnitudes is zero, linear polarization results. If the magnitudes are the same and the phase angle between the two vectors (in time) is 90°, circular polarization results. Any combination between these two limits gives elliptical polarization.

Crossed Linear Antennas

Dipoles radiate linearly polarized signals, and the polarization direction depends upon the orientation of the antenna. Fig 16 shows the electric-field or E-plane patterns of horizontal and vertical dipoles at A and B. If the two outputs are combined with the correct phase difference (90°), a circularly polarized wave results, and the resulting electric field pattern is shown in Fig 16C. Note that because the electric fields are identical in magnitude, the power from the transmitter must be equally divided between the two fields. Another way of looking at this is to consider the power as being divided between the two antennas; hence the gain of each is decreased by 3 dB when taken alone in the plane of its orientation.

As previously mentioned, a 90° phase shift must exist between the two antennas. The simplest way to obtain the shift

is to use two feed lines with one section that is ¼ λ longer than the other, as shown in Fig 17A. These separate feed lines are then paralleled to a common transmission line to the transmitter or receiver. Therein lies one of the headaches of this system—assuming negligible coupling between the crossed antennas, the impedance presented to the common transmission line by the parallel combination is one half that of either section alone. (This is not true when there is mutual coupling between the antennas, as in phased arrays). A practical construction method for implementing the system of Fig 17A is given in Fig 18.

Another factor to consider is the attenuation of the cables used in the harness, along with the connectors. Good low loss coaxial line should be used. Type N or BNC connectors are preferable to the UHF variety.

Another method of obtaining circular polarization is to use equal length feed lines and place one antenna ¼ λ ahead of the other. This method is shown at B of Fig 17. The advantage of equal-length feed lines is that identical load impedances will be presented to the common feeder. With the phasing-line method, any mismatch at one antenna will be magnified by the extra ¼ λ of transmission line. This upsets the current balance between the two antennas, resulting in a loss of polarization circularity.

Fig 17C shows a popular method of mounting off-the-shelf Yagi arrays—at right angles to each other. The two arrays may be physically offset by ¼ λ and fed in parallel, as shown, or they may be mounted with no offset and fed 90° out of phase. Neither of these arrangements produces true circular polarization. Instead, polarization diversity is obtained with elliptical polarization from such a system.

ELLIPTICALLY POLARIZED ANTENNAS FOR 144- AND 432-MHz SATELLITE WORK

The antenna system described here offers polarization diversity, with switchable right hand or left hand elliptical polarization. The array can be positioned in both azimuth and elevation. This system makes use of commercially available antennas (KLM 9-element 145-MHz and KLM 14-element 435-MHz antennas), rotators (Alliance U-110 and Telex/Hy-

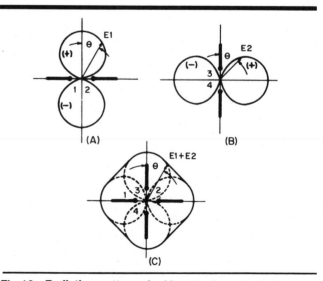

Fig 16—Radiation patterns looking head-on at dipoles.

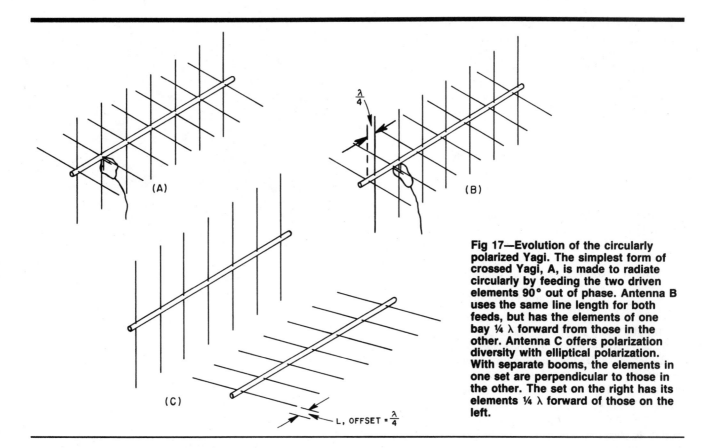

Fig 17—Evolution of the circularly polarized Yagi. The simplest form of crossed Yagi, A, is made to radiate circularly by feeding the two driven elements 90° out of phase. Antenna B uses the same line length for both feeds, but has the elements of one bay ¼ λ forward from those in the other. Antenna C offers polarization diversity with elliptical polarization. With separate booms, the elements in one set are perpendicular to those in the other. The set on the right has its elements ¼ λ forward of those on the left.

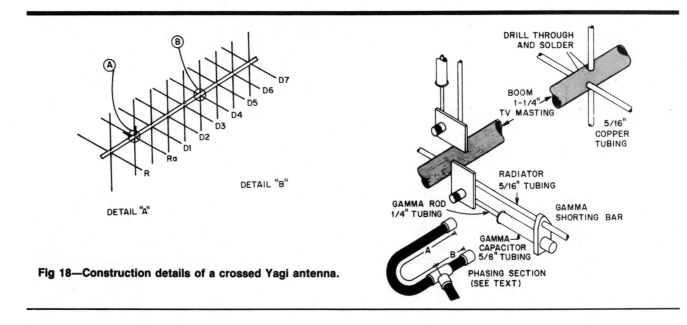

Fig 18—Construction details of a crossed Yagi antenna.

Gain Ham series or Tailtwister) and coaxial relays which are combined in a way that offers total flexibility.

This setup is suited for Mode B or Mode J satellite operation. As shown in Figs 19 and 20, the whole assembly is built on a heavy-duty TV tripod so that it can be roof-mounted. The idea for this system came from Clarke Greene, K1JX.

System Outline

The antennas shown in the photographs are actually two totally separate systems sharing the same azimuth and elevation positioning systems. Each system is identical in the way it performs—one system for 145 MHz and one for 435 MHz. Individual control lines allow independent control of the polarization sense for each system. This is mandatory, as often a different polarization sense is required for the uplink and downlink. Also, throughout any given pass of a satellite, the required sense may change.

Mechanical Details

The azimuth rotator is mounted inside the tripod by means of a Rohn 25 type of rotator plate. See Fig 20. U bolts around the tripod legs secure the plate to the tripod. A length of 1-inch galvanized water pipe (the mast) extends from the top of the rotator through a homemade aluminum bearing at the top of the tripod. Because a relatively small diameter mast is used, several pieces of shim material are required between it and the body of the rotator to assure that the mast will be aligned in the bearing through 360° of rotation. This is covered in detail in the Telex/Hy-Gain rotator instruction sheets.

The Alliance U-110 elevation rotator is mounted to the 1-inch water pipe mast by means of a 1/8-inch aluminum plate. TV U-bolt hardware provides a good fit for this mast material. The cross arm that supports the antennas is a piece of 1¼ inch thick fiberglass rod, 6 feet in length. Other materials can be used, but the strength of fiberglass makes it desirable as a cross arm. This should be a consideration if you live in an area that is frequented by ice storms. Although it is relatively expensive (about $3 to $4 per foot), one piece should last a lifetime.

Electrical Details

As the antenna systems are identical, this description applies to both. As mentioned earlier, it is possible to obtain polarization diversity with two separate antennas mounted apart from each other as shown in the photographs. One advantage of this system is that the weight distribution on each side of the elevation rotator is equal. As long as the separation between antennas is small, performance should be nearly as good as when both sets of elements are on a single boom. There is no operational difference between true circular polarization and the polarization diversity provided by this antenna system.

Because of mutual coupling between the arrays, the two feed-point impedances will not be identical, but from a practical standpoint the differences are almost insignificant. One antenna must be fed 90° out of phase with respect to the other. For switchable right hand and left hand polarization, some means must be included to shift a 90° phasing line

Fig 19—An elliptically polarized antenna system for satellite communications on 146 and 435 MHz. The array is assembled from KLM log periodic Yagis.

Fig 20—The polarization sense of the antenna is controlled by coaxial relays and phasing lines. The 146- and 435-MHz systems are controlled independently.

in series with either antenna. Such a scheme is shown in Fig 21. Since two antennas are essentially connected in parallel, the feed impedance will be half that of either antenna alone. The antennas used in this system have a 52-Ω feed-point impedance. RG-133 (95-Ω coax) proves difficult to locate. RG-63 (125-Ω impedance) may be used with a slightly higher mismatch. As can be seen in the drawing, the phasing line is always in series with the system feed point and one of the antennas. As shown, the antenna on the left receives energy 90° ahead of the one on the right. When the relay is switched, the opposite is true.

It is not necessary to use single quarter wavelengths of line. For example, the 75-ohm impedance-transforming lines between each antenna and the relay can be any odd multiple of 1/4 λ, such as 3/4, 5/4, 7/4 λ, etc. The same is true for the 95- or 125-Ω phasing line.

Keep track of phasing-line lengths. This is especially important when determining which position of the relay will yield right or left hand polarization. You will probably find it necessary to use a number of quarter wavelengths, because a single quarter wavelength of line is extremely short (when the velocity factor is taken into consideration). The lengths used in this system are shown in Fig 22. Try to use the shortest practical lengths, because the SWR bandwidth of the array decreases as the number of quarter wavelengths of line is increased.

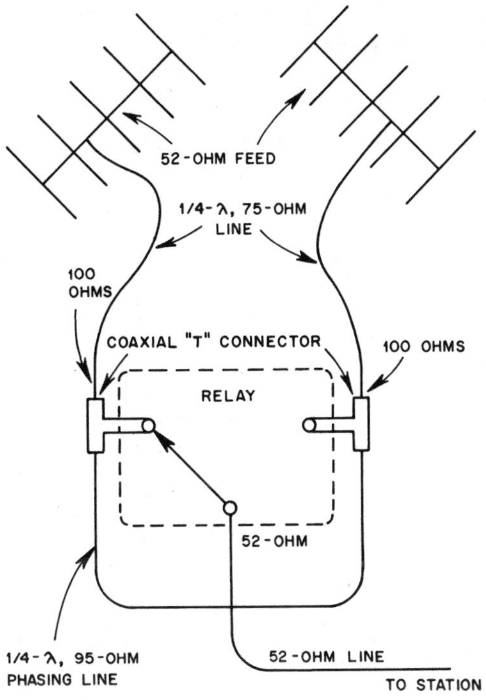

Fig 21—Electrical diagram of the switchable polarization antenna system, complete with cable specifications. When calculating the lengths of individual cables, be sure to include the proper velocity factor of the cable used.

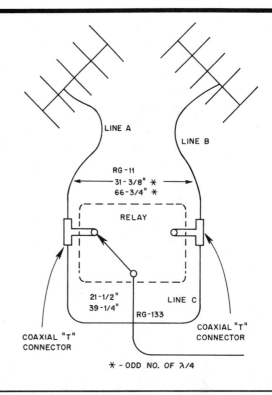

Fig 22—The basic antenna system for switchable right or left hand elliptical polarization. Lines A and B step the 52-ohm antenna impedance up to 100 ohms. The phasing line is made from 95-ohm coaxial cable to provide a good match to the 100-ohm system. See text for a detailed description of the system. The shorter lengths are for 435.15 MHz and the longer lengths are for 145.925 MHz. The line lengths shown are for a 66% velocity factor.

Antenna Systems for EME Communications

The tremendous path loss incurred over an EME circuit places stringent requirements on earth station performance. Low-noise receiving equipment, maximum legal power and large antenna arrays are required for successful EME operation. Although it is possible to copy some of the better-equipped stations with a single Yagi antenna, it is unlikely that such an antenna can provide reliable two-way communication. Antenna gain of at least 20 dB is required for reasonable success. Generally speaking, more antenna gain yields the most noticeable improvement in station performance, as the increased gain improves both the received and transmitted signals.

Several types of antennas have become popular among EME enthusiasts. Perhaps the most popular antenna for 144-MHz work is an array of either 4 or 8 long boom (14 to 15 dB gain) Yagis. The 4-Yagi array provides approximately 20 dB gain, and the 8-antenna system gives an approximate 3 dB increase over the 4-antenna array. At 432 MHz, 8 or 16 long boom Yagis are often used. Yagi antennas are commercially available, and can be constructed from readily available materials. Information on maximum gain Yagi antennas is presented in Chapter 18.

A moderately sized Yagi array has the advantage that it is relatively easy to construct, and can be positioned in azimuth and elevation with commercially available equipment. Matching and phasing lines present few problems. The main disadvantage of Yagi arrays is that the polarization plane of the individual Yagis cannot be conveniently changed. One way around this is to use cross polarized Yagis and a relay switching system to select the desired polarization, as described in the previous section. This represents a considerable increase in system complexity to select the desired polarization. Some amateurs have gone as far as building complicated chain-driven systems to allow constant polarization adjustment of all the Yagis in a large array. Polarization shift of EME signals at 144 MHz is fairly slow, and the added complexity of the cross polarized antenna system or a sophisticated chain driven polarity adjustment scheme may not be worth the effort. At 432 MHz, where the polarization shifts at a somewhat faster rate, an adjustable polarization system offers a definite advantage over a fixed one.

The Yagi antenna system used by Ed Stallman, N5BLZ, is shown in Fig 23. The system is comprised of twelve 144-MHz long boom 17-element Yagi antennas. The Yagi arrays of Timo Korhonen, OH6NU, and Steve Powlishen, K1FO, are shown in Figs 24 and 25, respectively.

Quagi antennas (made from both quad and Yagi elements) are also popular for EME work. Slightly more gain per unit boom length is possible as compared to the conventional Yagi. Additional information on the Quagi is presented in Chapter 18.

The collinear array is another popular type of antenna for EME work. A 40-element collinear array has approximately the same frontal area as an array of four Yagis, but produces approximately 1 to 2 dB less gain. One attraction to a collinear array is that the depth dimension is considerably less than the long boom Yagis. An 80-element collinear is marginal for EME communications, providing approximately 19 dB gain. Many operators using collinear arrays use 160-element or larger systems.

Fig 23—The EME antenna system used at N5BLZ consists of twelve 17-element, long boom 144-MHz Yagis. The tractor, lower left, really puts this array into perspective!

Fig 24—The Yagi array used for EME at OH6NU/OH6NM.

Fig 25—K1FO uses this system for serious moonbounce work.

As with Yagi and Quagi antennas, the collinear cannot be adjusted easily for polarity changes. From a constructional standpoint, there is little difference in complexity and material costs between the collinear and Yagi arrays.

The parabolic dish is another antenna that is used extensively for EME work. Unlike the other antennas described, the major problems associated with dish antennas are mechanical ones. Dishes approaching 20 feet in diameter are required for successful EME operation on 432 MHz. Structures of this size with wind and ice loading place a severe strain on the mounting and positioning system. Extremely rugged mounts are required for large dish antennas, especially when used in windy locations.

Several aspects of parabolic dish antennas make the extra mechanical problems worth the trouble, however. For example, the dish antenna is inherently broadbanded, and may be used on several different bands by simply changing the feed. An antenna that is suitable for 432-MHz work is also usable for each of the higher amateur bands. Increased gain is available as the frequency of operation is increased.

Another advantage of this antenna is in the feed system. The polarization of the feed, and therefore the polarization of the antenna, can be adjusted with little difficulty. It is a relatively easy matter to devise a system whereby the feed can be rotated remotely from the shack. Changes in polarization of the signal can thereby be compensated for at the operating position.

Because polarization changes can account for as much as 30 dB of signal attenuation, the rotatable feed can make the difference between consistent communications and no

communications at all. A parabolic dish antenna under construction by Dave Wardley, ZL1BJQ is shown in Fig 26. The 20-foot stressed parabolic dish used at F2TU is shown in Fig 27. More information on parabolic dish antennas is given later in this chapter and in Chapter 18.

Antennas suitable for EME work are by no means limited to the types described thus far. Rhombics, quad arrays, helicals and others can also be used. These antennas have not gained the popularity of the Yagi, Quagi, collinear and parabolic dish, however.

Fig 26—The ½-in. wire mesh is about all that is needed to complete this 7-meter dia dish at ZL1BJQ.

Fig 27—This 20-foot stressed parabolic dish is used for EME work at F2TU on 432 and 1296 MHz.

A 12-Foot Stressed Parabolic Dish

Very few antennas evoke as much interest among UHF amateurs as the parabolic dish, and for good reason. First, the parabola and its cousins—Cassegrain, hog horn and Gregorian—are probably the ultimate in high gain antennas. One of the highest gain antennas in the world (148 dB) is a parabola. This is the 200-inch Mt. Palomar telescope. (The very short wavelength of light rays causes such a high gain to be realizable.) Second, the efficiency of the parabola does not change as size increases. With collinear arrays, the loss of the phasing harness increases as the size increases. The corresponding component of the parabola is lossless air between the feed horn and the reflecting surface. If there are few surface errors, the efficiency of the system stays constant regardless of antenna size. This project was presented by Richard Knadle, K2RIW, in August 1972 *QST*.

Some amateurs reject parabolic antennas because of the belief that they are all heavy, hard to construct, have large wind-loading surfaces, and require precise surface accuracy. However, with modern construction techniques, a prudent choice of materials, and an understanding of accuracy requirements, these disadvantages can be largely overcome. A parabola may be constructed with a 0.6 f/d (focal length/diameter) ratio, producing a rather flat dish, which makes it easy to surface and allows the use of recent advances in high efficiency feed horns. This results in greater gain for a given dish size over conventional designs.

Such an antenna is shown in Fig 28. This parabolic dish is lightweight, portable, easy to build, and can be used for 432- and 1296-MHz mountaintopping, as well as on 2300, 3450 and 5760 MHz. Disassembled, it fits into the trunk of a car, and can be assembled in 45 minutes.

The usually heavy structure that supports the surface of

most parabolic dish antennas has been replaced in this design by aluminum spokes bent into a near parabolic shape by string. These strings serve the triple function of guying the focal point, bending the spokes, and reducing the error at the dish perimeter (as well as at the center) to nearly zero. By contrast, in conventional designs, the dish perimeter (which has a greater surface area than the center) is farthest from the supporting center hub. For these reasons, it often has the greatest error. This error becomes more severe when the wind blows. Here, each of the spokes is basically a cantilevered beam with end loading. The equations of beam bending predict a near perfect parabolic curve for extremely small deflections. Unfortunately the deflections in this dish are not that small, and the loading is not perpendicular. For these reasons, mathematical prediction of the resultant curve is quite difficult. A much better solution is to measure the surface error with a template and make the necessary correction by bending each of the spokes to fit. This procedure is discussed in a later section.

The uncorrected surface is accurate enough for 432- and 1296-MHz use. Trophies taken by this parabola in antenna gain contests were won using a completely natural surface with no error correction.

By placing the transmission line inside the central pipe that supports the feed horn, the area of the shadows or blockages on the reflector surface is much smaller than in other feeding and supporting systems, thus increasing gain. For 1296 MHz, a backfire feed horn may be constructed to take full advantage of this feature. At 432 MHz, a dipole and reflector assembly produces 1.5 dB additional gain over a corner reflector feed system. Because the preamplifier is located right at the horn on 2300 MHz, a conventional feed horn may be used.

Construction

Table 1 is a list of materials required for construction. Care must be exercised when drilling holes in the connecting center plates so assembly problems will not be experienced later. See Fig 29. A notch in each plate allows them to be assembled in the same relative positions. The two plates should be clamped together and drilled at the same time. Each of the eighteen ½-inch dia aluminum spokes has two no. 28

Fig 28—A 12-foot stressed parabolic dish set up for reception of Apollo or Skylab signals near 2280 MHz. A preamplifier is shown taped below the feed horn. The dish was designed by K2RIW, standing at the right. From *QST*, August 1972.

Table 1

Materials List for the 12-Foot Stressed Parabolic Dish

1) Aluminum tubing, 12-ft × ½-in. OD × 0.049-in. wall, 6061-T6 alloy, 9 required to make 18 spokes.
2) Octagonal mounting plates 12 × 12 × 1/8 in., 2024-T3 alloy, 2 required.
3) 1¼ in. ID pipe flange with setscrews.
4) 1¼ in. × 8 ft TV mast tubing, 2 required.
5) Aluminum window screening, 4 × 50 ft.
6) 130-pound test Dacron trolling line (available from Finney Sports, 2910 Glansman Rd, Toledo, OH 43614.)
7) 38 ft no. 9 galvanized fence wire (perimeter), Montgomery Ward Farm and Garden Catalog.
8) Two hose clamps, 1½ in.; two U bolts; ½ × 14-in. Bakelite rod or dowel; water-pipe grounding clamp; 18 eye bolts; 18 S hooks.

holes drilled at the base to accept no. 6-32 machine screws that go through the center plates. The 6-foot long spokes are cut from standard 12-foot lengths of tubing. A fixture built from a block of aluminum assures that the holes are drilled in exactly the same position in each spoke. The front and back center plates constitute an I-beam type of structure that gives the dish center considerable rigidity.

A side view of the complete antenna is shown in Fig 30. Aluminum alloy (6061-T6) is used for the spokes, while 2024-T3 aluminum alloy sheet, 1/8 inch thick, is used for the center plates. (Aluminum has approximately three times the strength to weight ratio of wood, and aluminum cannot warp or become water logged.) The end of each of the 18 spokes has an eyebolt facing the dish focal point, which serves a dual purpose:

1) To accept the no. 9 galvanized fence wire that is routed through the screw eyes to define the dish perimeter, and

2) To facilitate rapid assembly by accepting the S hooks which are tied to the end of each of the lengths of 130-pound test Dacron fishing string.

The string bends the spokes into a parabolic curve; the dish may be adapted for many focal lengths by tightening or slackening the strings. Dacron was chosen because it has the same chemical formula as Mylar. This is a low-stretch material that keeps the dish from changing shape. The galvanized perimeter wire has a 5-inch overlap area that is bound together with baling wire after the spokes have been hooked to the strings.

The aluminum window screening is bent over the perimeter wire to hold it in place on the back of the spokes. Originally, there was concern that the surface perturbations (the spokes) in front of the screening might decrease the gain. The total spoke area is so small, however, that this fear proved unfounded.

Placing the aluminum screening in front of the spokes requires the use of 200 pieces of baling wire to hold the screening in place. This procedure increases the assembly time by at least an hour. For contest and mountaintop operation

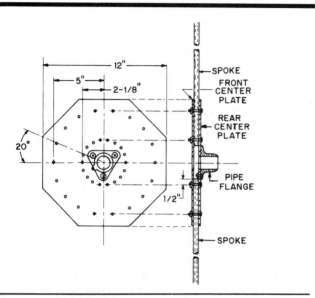

Fig 29—Center plate details. Two center plates are bolted together to hold the spokes in place.

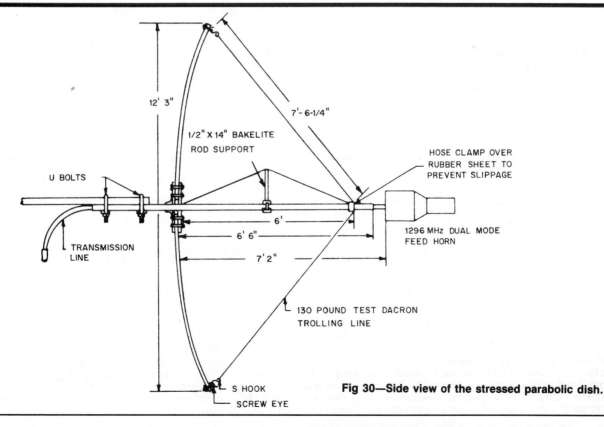

Fig 30—Side view of the stressed parabolic dish.

(when the screening is on the back of the spokes) no fastening technique is required other than bending the screen to overlap the wire perimeter.

The Parabolic Surface

A 4-foot wide roll of aluminum screening 50 feet long is cut into appropriate lengths and laid parallel with a 3-inch overlap between the top of the unbent spokes and hub assembly. The overlap seams are sewn together on one half of the dish using heavy Dacron thread and a sailmaker's curved needle. Every seam is sewn twice; once on each edge of the overlapped area. The seams on the other half are left open to accommodate the increased overlap that occurs when the spokes are bent into a parabola. The perimeter of the screening is then trimmed. Notches are cut in the 3-inch overlap to accept the screw eyes and S hooks.

The first time the dish is assembled, the screening strips are anchored to the inside surface of the dish and the seams sewn in this position. It is easier to fabricate the surface by placing the screen on the back of the dish frame with the structure inverted. The spokes are sufficiently strong to support the complete weight of the dish when the perimeter is resting on the ground.

The 4-foot wide strips of aluminum screening conform to the compound bend of the parabolic shape very easily. If the seams are placed parallel to the E field polarization of the feed horn, minimum feedthrough will occur. This feedthrough, even if the seams are placed perpendicular to the E field, is so small that it is negligible. Some constructors may be tempted to cut the screening into pie-shaped sections. This procedure will increase the seam area and construction time considerably. The dish surface appears most pleasing from the front when the screening perimeter is slipped between the spokes and the perimeter wire, and is then folded back over the perimeter wire. In disassembly, the screening is removed in one piece, folded in half and rolled.

The Horn and Support Structure

The feed horn is supported by 1¼-inch aluminum television mast. The Hardline that is inserted into this tubing is connected first to the front of the feed horn, which then slides back into the tubing for support. A setscrew assures that no further movement of the feed horn occurs. During antenna gain competition the setscrew is omitted, allowing the ½-inch semirigid CATV transmission line to move in or out while adjusting the focal length for maximum gain. The TV mast is held firmly at the center plates by two setscrews in the pipe flange that is mounted on the rear plate. At 2300 MHz, the dish is focused for best gain by loosening these setscrews on the pipe flange and sliding the dish along the TV mast tubing. (The dish is moved instead of the feed horn.)

The fishing strings are held in place by attaching them to a hose clamp that is permanently connected to the TV tubing. A piece of rubber sheet under the hose clamp prevents slippage and keeps the hose clamp from cutting the fishing string. A second hose clamp is mounted below the first as extra protection against slippage.

The high efficiency 1296-MHz dual mode feed horn, detailed in Fig 31, weighs 5¾ pounds. This weight causes some bending of the mast tubing, but this is corrected by a ½-inch diameter Bakelite support, as shown in Fig 30. This support is mounted to a pipe grounding clamp with a no. 8-32 screw inserted in the end of the rod. The Bakelite rod and grounding clamp are mounted midway between the hose clamp and the center plates on the mast. A double run of fishing string slipped over the notched upper end of the Bakelite rod counteracts bending.

The success of high efficiency parabolic antennas is primarily determined by feed horn effectiveness. The multiple diameter of this feed horn may seem unusual. This patented dual mode feed, designed by Dick Turrin, W2IMU, achieves efficiency by launching two different kinds of waveguide modes simultaneously. This causes the dish illumination to be more constant than conventional designs.

Illumination drops off rapidly at the perimeter, reducing spillover. The feedback lobes are reduced by at least 35 dB because the current at the feed perimeter is almost zero; the phase center of the feed system stays constant across the angles of the dish reflector. The larger diameter section is a phase corrector and should not be changed in length. In theory, almost no increase in dish efficiency can be achieved without increasing the feed size in a way that would increase complexity, as well as blockage.

The feed is optimized for a 0.6 f/d dish. The dimensions of the feeds are slightly modified from the original design in order to accommodate the cans. Either feed type can be

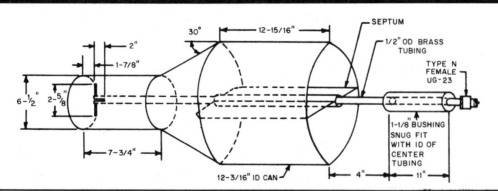

Fig 31—Backfire type 1296-MHz feed horn, linear polarization only. The small can is a Quaker State oil container; the large can is a 50-pound shortening container (obtained from a restaurant, Gold Crisp brand). Brass tubing, ½-in. OD, extends from UG-23 connector to dipole. Center conductor and dielectric are obtained from 3/8-in. Alumafoam coaxial cable. The dipole is made from 3/32-in. copper rod. The septum and 30° section are made from galvanized sheet metal. Styrofoam is used to hold the septum in position. The primary gain is 12.2 dBi.

constructed for other frequencies by changing the scale of all dimensions.

Multiband Use

Many amateurs construct multiband antenna arrays by putting two dishes back to back on the same tower. This is cost inefficient. The parabolic reflector is a completely frequency independent surface, and studies have shown that a 0.6 f/d surface can be steered seven beamwidths by moving the feed horn from side to side before the gain diminishes by 1 dB. Therefore, the best dual band antenna can be built by mounting separate horns side by side. At worst, the antenna may have to be moved a few degrees (usually less than a beamwidth) when switching between horns, and the unused horn increases the shadow area slightly. In fact, the same surface can function simultaneously on multiple frequencies, making crossband duplex operation possible with the same dish.

Order of Assembly

1) A single spoke is held upright behind the rear center plate with the screw eye facing forward. Two 6-32 machine screws are pushed through the holes in the rear center plate, through the two holes of the spoke, and into the corresponding holes of the front center plate. Lock washers and nuts are placed on the machine screws and hand tightened.

2) The remaining spokes are placed between the machine screw holes. Make sure that each screw eye faces forward. Machine screws, lock washers, and nuts are used to mount all 18 spokes.

3) The no. 6-32 nuts are tightened using a nut driver.

4) The mast tubing is attached to the spoke assembly, positioned properly, and locked down with the setscrews on the pipe flange at the rear center plate. The S hooks of the 18 Dacron strings are attached to the screw eyes of the spokes.

5) The ends of two pieces of fishing string (which go over the Bakelite rod support) are tied to a screw eye at the forward center plate.

6) The dish is laid on the ground in an upright position and no. 9 galvanized wire is threaded through the eyebolts. The overlapping ends are lashed together with baling wire.

7) The dish is placed on the ground in an inverted position with the focus downward. The screening is placed on the back of the dish and the screening perimeter is fastened as previously described.

8) The extension mast tubing (with counterweight) is connected to the center plate with U bolts.

9) The dish is mounted on a support and the transmission line is routed through the tubing and attached to the horn.

Parabola Gain Versus Errors

How accurate must a parabolic surface be? This is a frequently asked question. According to the Rayleigh limit for telescopes, little gain increase is realized by making the mirror accuracy greater than $\pm 1/8 \lambda$ peak error. John Ruze of the MIT Lincoln Laboratory, among others, has derived an equation for parabolic antennas and built models to verify it. The tests show that the tolerance loss can be predicted within a fraction of a decibel, and less than 1 dB of gain is sacrificed with a surface error of $\pm 1/8 \lambda$. (An eighth λ is 3.4 inches at 432 MHz, 1.1 inches at 1296 MHz and 0.64 inch at 2300 MHz.)

Some confusion about requirements of greater than $1/8$-λ accuracy may be the result of technical literature describing highly accurate surfaces. Low sidelobe levels are the primary interest in such designs. Forward gain is a much greater concern than low sidelobe levels in amateur work; therefore, these stringent requirements do not apply.

When a template is held up against a surface, positive and negative ($\pm$) peak errors can be measured. The graphs of dish accuracy requirements are frequently plotted in terms of RMS error, which is a mathematically derived function much smaller than $\pm$ peak error (typically $\frac{1}{3}$). These small RMS accuracy requirements have discouraged many constructors who confuse them with $\pm$ peak errors.

Fig 32 may be used to predict the resultant gain of various dish sizes with typical errors. There are a couple of surprises, as shown in Fig 33. As the frequency is increased for a given dish, the gain increases 6 dB per octave until the tolerance errors become significant. Gain deterioration then increases rapidly. Maximum gain is realized at the frequency where the tolerance loss is 4.3 dB. Notice that at 2304 MHz, a 24-foot dish with ± 2-inch peak errors has the same gain as a 6-foot dish with ± 1-inch peak errors. Quite startling, when it is realized that a 24-foot dish has 16 times the area of a 6-foot dish. Each time the diameter or frequency is doubled or halved, the gain changes by 6 dB. Each time all the errors are halved, the frequency of maximum gain is doubled. With this information, the gain of other dish sizes with other tolerances can be predicted.

These curves are adequate for predicting gain, assuming a high efficiency feed horn is used (as described earlier) which realizes 60% aperture efficiency. At frequencies below 1296 MHz where the horn is large and causes considerable blockage, the curves are somewhat optimistic. A properly built dipole and splasher feed will have about 1.5 dB less gain when used with a 0.6 f/d dish than the dual mode feed system described.

The worst kind of surface distortion is where the surface curve in the radial direction is not parabolic but gradually departs in a smooth manner from a perfect parabola. The decrease in gain can be severe, because a large area is involved. If the surface is checked with a template, and if reasonable construction techniques are employed, deviations are controlled and the curves represent an upper limit to the gain that can be realized.

If a 24-foot dish with ± 2-inch peak errors is being used with 432-MHz and 1296-MHz multiple feed horns, the constructor might be discouraged from trying a 2300-MHz feed because there is 15 dB of gain degradation. The dish will still have 29 dB of gain on 2300 MHz, however, making it worthy of consideration.

The near-field range of this 12-foot stressed dish (actually 12 feet 3 inches) is 703 feet at 2300 MHz. By using the sun as a noise source and observing receiver noise power, it was found that the antenna had two main lobes about 4° apart. The template showed a surface error (insufficient spoke bending at ¾ radius), and a correction was made. A recheck showed one main lobe, and the solar noise was almost 3 dB stronger.

Other Surfacing Materials

The choice of surface materials is a compromise between RF reflecting properties and wind loading. Aluminum screening, with its very fine mesh (and weight of 4.3 pounds per

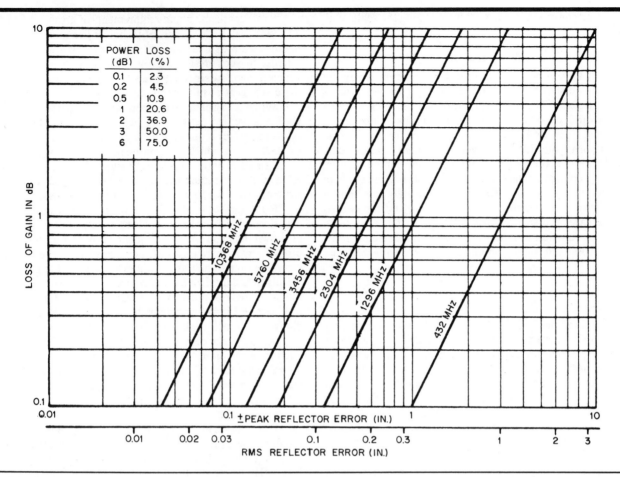

Fig 32—Gain deterioration versus reflector error. Basic information obtained from J. Ruze, British *IEE*.

100 square feet) is useful beyond 10 GHz because of its very close spacing. This screening is easy to roll up and is therefore ideal for a portable dish. This close spacing causes the screen to be a 34% filled aperture, bringing the wind force at 60 miles per hour to more than *400 pounds* on this 12-foot dish. Those considering a permanent installation of this dish should investigate other surfacing materials.

Hexagonal 1-inch poultry netting (chicken wire), which is an 8% filled aperture, is nearly ideal for 432-MHz operation. It weighs 10 pounds per 100 square feet, and exhibits only 81 pounds of force with 60 mile per hour winds. Measurement on a large piece reveals 6 dB of feedthrough at 1296 MHz, however. Therefore, on 1296 MHz, one fourth of the power will feed through the surface material. This will cause a loss of only 1.3 dB of forward gain. Since the low wind loading material will provide a 30-dB gain potential, it is a very good trade-off.

Poultry netting is very poor material for 2300 MHz and above, because the hole dimensions approach ½ λ. As with all surfacing materials, minimum feedthrough occurs when the E-field polarization is parallel to the longest dimension of the surfacing holes.

Hardware cloth with ½-inch mesh weighs 20 pounds per 100 square feet and has a wind loading characteristic of 162 pounds with 60 mile per hour winds. The filled aperture is 16%, and this material is useful to 2300 MHz.

A rather interesting material worthy of investigation is ¼-inch reinforced plastic (described in Montgomery Ward Farm and Garden Catalog). It weighs only 4 pounds per 100 square feet. The plastic melts with many universal solvents such as lacquer thinner. If a careful plastic-melting job is done, what remains is the ¼-inch spaced aluminum wires with a small blob of plastic at each junction to hold the matrix together.

There are some general considerations to be made in selecting surface materials:

1) Joints of screening do not have to make electrical contact. The horizontal wires reflect the horizontal wave.

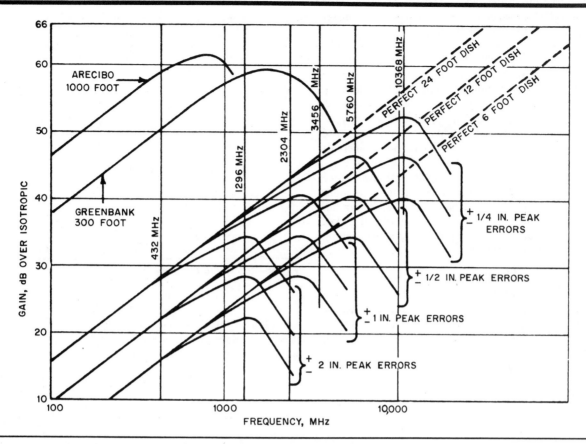

Fig 33—Parabolic-antenna gain versus size, frequency, and surface errors. All curves assume 60% aperture efficiency and 10-dB power taper. Reference: J. Ruze, British *IEE*.

Skew polarizations are merely a combination of horizontal and vertical components which are thus reflected by the corresponding wires of the screening. To a horizontally polarized wave, the spacing and diameter of only the horizontal wires determine the reflection coefficient (see Fig 34). Many amateurs have the mistaken impression that screening materials that do not make electrical contact at their junctions are poor reflectors.

2) By measuring wire diameter and spacings between the wires, a calculation of percentage of aperture that is filled can be made. This will be one of the major determining factors of wind pressure when the surfacing material is dry. Under ice and snow conditions, smaller aperture materials may become clogged, causing the surfacing material to act as a solid "sail." Ice and snow have a rather minor effect on the reflecting properties of the surface, however.

3) Amateurs who live in areas where ice and snow are prevalent should consider a deicing scheme such as weaving enameled wire through the screening and passing a current

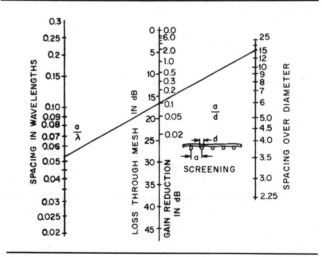

Fig 34—Surfacing material quality.

through it, fastening water-pipe heating tape behind the screening, or soldering heavy leads to the screening perimeter and passing current through the screening itself.

A Parabolic Template

At and above 2300 MHz (where high surface accuracy is required), a parabolic template should be constructed to measure surface errors. A simple template may be constructed (see Fig 35) by taking a 12-foot 3-inch length of 4-foot wide tar paper and drawing a parabolic shape on it with chalk. The points for the parabolic shape are calculated at 6-inch intervals and these points are connected with a smooth curve.

For those who wish to use the template with the surface material installed, the template should be cut along the chalk line and stiffened by cardboard or a wood lattice frame. Surface error measurements should take place with all spokes installed and deflected by the fishing strings, as some bending of the center plates does take place.

Variations

All the possibilities of the stressed parabolic antenna have not been explored. For instance, a set of fishing strings or guy wires can be set up behind the dish for error correction, as long as this does not cause permanent bending of the aluminum spokes. This technique also protects the dish against

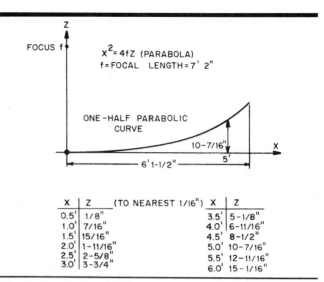

$x^2 = 4fZ$ (PARABOLA)
f = FOCAL LENGTH = 7' 2"

ONE-HALF PARABOLIC CURVE

X	Z	(TO NEAREST 1/16")	X	Z
0.5'	1/8"		3.5'	5-1/8"
1.0'	7/16"		4.0'	6-11/16"
1.5'	15/16"		4.5'	8-1/2"
2.0'	1-11/16"		5.0'	10-7/16"
2.5'	2-5/8"		5.5'	12-11/16"
3.0'	3-3/4"		6.0'	15-1/16"

Fig 35—Parabolic template for 12-ft., 3-in. dish.

wind loading from the rear. An extended piece of TV mast is an ideal place to hang a counterweight and attach the rear guys. This strengthens the structure considerably.

The Helical Antenna

The axial-mode helical antenna was introduced by Dr John Kraus, W8JK, in the 1940s. The material in this section was prepared by Domenic Mallozzi, N1DM.

This antenna has two characteristics that make it especially interesting and useful in many applications. First, the helix is circularly polarized. As discussed earlier, circular polarization is simply linear polarization that continually rotates as it travels through space. In the case of a helical array, the rotation is about the axis of the antenna. This can be pictured as the second hand of a watch moving at the same rate as the applied frequency, where the position of the second hand can be thought of as the instantaneous polarization of the signal.

The second interesting property of the helical antenna is its predictable pattern, gain and impedance characteristics over a wide frequency range. This is one of the few antennas that has both broad bandwidth and high gain. The benefit of this property is that, when used for narrow-band applications, the helical antenna is very forgiving of mechanical inaccuracies.

Probably the most common amateur use of the helical antenna is in satellite communications, where the spinning of the satellite antenna system (relative to the earth) and the effects of Faraday rotation cause the polarization of the satellite signal to be unpredictable. Using a linearly polarized antenna in this situation results in deep fading, but with the helical antenna (which responds equally to linearly polarized signals), fading is essentially eliminated.

This same characteristic makes helical antennas useful in polarization diversity systems. The advantages of circular polarization have been demonstrated by Bill Sykes, G2HCG, on VHF voice schedules over nonoptical paths, in cases where linearly polarized beams did not perform satisfactorily. See bibliography. An array of linear antennas was used to develop a circularly polarized radiation pattern in this case. The helix is also a good antenna for long-haul commercial TV reception.

Another use for the helical antenna is the transmission of color ATV signals. Many beam antennas (when adjusted for maximum gain) have far less bandwidth than the required 6 MHz, or have nonuniform gain over this frequency range. The result is significant distortion of the transmitted and received signals, affecting color reproduction and other features. This problem becomes more aggravated over nonoptical paths. The helix exhibits maximum gain (within 1 dB) over at least 6 MHz anywhere above 420 MHz.

The helical antenna can be used to advantage with multimode rigs, especially above 420 MHz. Not only does the helix give high gain over an entire amateur band, but it also allows operation on FM, SSB and CW without the need for separate vertically and horizontally polarized antennas.

HELICAL ANTENNA BASICS

The helical antenna is an unusual specimen in the antenna world, in that its physical configuration gives a hint to its electrical performance. A helix looks like a large air-wound coil with one of its ends fed against a ground plane, as shown in Fig 36. The ground plane is a screen of 0.8 λ to 1.1 λ

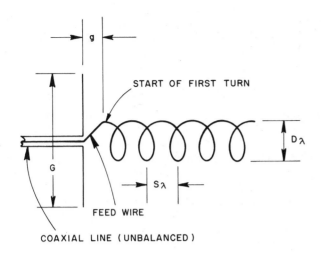

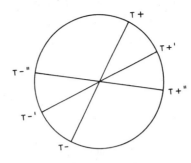

Fig 37—A helical antenna with an axial ratio of 1.0 produces pure circular polarization. See text.

C_λ = 0.75 to 1.33 λ
S_λ = 0.2126 C_λ to 0.2867 C_λ
G = 0.8 to 1.1 λ
g = 0.12 to 0.13 λ

$$AR \text{ (axial ratio)} = \frac{2n + 1}{2n}$$

S_λ = axial length of one turn
D_λ = diameter of winding
G = ground plane diameter (or side length)
g = ground plane to first turn distance
$C_\lambda = \pi D_\lambda$ = circumference of winding
n = number of turns
Gain (dBi) = 11.8 + 10 log ($C_\lambda^2 n S_\lambda$)

$$\text{Half power beamwidth (HPBW)} = \frac{52}{C_\lambda \sqrt{n S_\lambda}} \text{ degrees}$$

$$\text{Beamwidth to first nulls} = \frac{115}{C_\lambda \sqrt{n S_\lambda}} \text{ degrees}$$

Input impedance = 140 C_λ ohms
L_λ = length of conductor in one turn = $\sqrt{(\pi D_\lambda)^2 + S_\lambda^2}$

Fig 36—The basic helical antenna and design equations.

diameter (or on a side for a square ground plane). The circumference (C_λ) of the coil form must be between 0.75 λ and 1.33 λ for the antenna to radiate in the axial mode. The coil should have at least three turns to radiate in this mode. (It is possible, through special techniques, to make axial-mode helicals with as little as one turn.) The ratio of the spacing between turns (in wavelengths), S_λ to C_λ, should be in the range of 0.2126 to 0.2867. This ratio range results from the requirement that the pitch angle, α, of the helix be between 12° and 16°, where

$$\alpha = \arctan \frac{S_\lambda}{C_\lambda}$$

These constraints result in a single main lobe along the axis of the coil. This is easily visualized from Fig 37. Assume

the winding of the helix comes out of the page with a clockwise winding direction. (The winding can also be a counterclockwise—this results in the opposite polarization sense.)

A helix with a C_λ of 1 λ has a wave propagating from one end of the coil (at the ground plane). The "peak" (+) of the wave appears opposite the "valley" (−) of the wave. This corresponds to a dipole—"across" the helix—with the same polarization as the instantaneous polarization of the helix at time T.

At a later time (T '), the "peak" and "valley" of the wave are at a slightly different angle relative to the original dipole. The polarization of the dipole antenna at this instant is slightly different. At an instant of time later yet (T "), the dipole has again "moved," changing the polarization slightly again.

The electrical rotation of this dipole produces circularly polarized radiation. Because the wave is moving along the helix conductor at nearly the speed of light, the rotation of the electrical dipole is at a very high rate. True circular polarization results.

Physicists and engineers formerly had opposite terms for the same sense of polarization. Recently, the definition of polarization sense used by the Institute of Electrical and Electronic Engineers (IEEE) has become the standard. The IEEE definition, in simple terms, is that when viewing the antenna from the feed-point end, a clockwise wind results in right-hand circular polarization, and a counterclockwise wind results in left-hand circular polarization. This is important, because when two stations use helical antennas *over a nonreflective path*, both must use antennas with the same polarization sense. If antennas of opposite sense are used, a signal loss of at least 30 dB results from the cross polarization alone.

As mentioned previously, circularly polarized antennas can be used in communications with any linearly polarized antenna (horizontal or vertical), because circularly polarized antennas respond equally to all linearly polarized signals. The gain of a helix is 3 dB less than the theoretical gain in this case, because the linearly polarized antenna does not respond to linear signal components that are orthogonally polarized relative to it.

The response of a helix to all polarizations is indicated by a term called *axial ratio*, also known as circularity. Axial ratio is the ratio of amplitude of the polarization that gives *maximum* response to the amplitude of the polarization that gives *minimum* response. An ideal circularly polarized antenna has an axial ratio of 1.0. A well designed practical helix exhibits an axial ratio of 1.0 to 1.1. The axial ratio of a helix is

$$AR = \frac{2n + 1}{2n}$$

where

AR = axial ratio
n = the number of turns in the helix

Axial ratio can be measured in two ways. The first is to excite the helix and use a linearly polarized antenna with an amplitude detector to measure the axial ratio directly. This is done by rotating the linearly polarized antenna in a plane perpendicular to the axis of the helix and comparing the maximum and minimum amplitude values. The ratio of maximum to minimum is the axial ratio.

Another method of measuring axial ratio was presented in *73* by A. Bridges, WB4VXP. (See the bibliography at the end of this chapter.) The linear antenna is replaced by two circularly polarized antennas of equal gain but opposite polarization sense. Taking the amplitude measurement with first one and then the other, the following equation is used to calculate axial ratio:

$$AR = \frac{E_{rcp} + E_{lcp}}{E_{rcp} - E_{lcp}}$$

where

E_{rcp} is the voltage measured with the right-hand circularly polarized test antenna
E_{lcp} is the voltage measured with the left-hand circularly polarized test antenna

This equation gives not only the axial ratio, but also indicates the polarization sense. If the result is greater than zero, the antenna being excited is right-hand circularly polarized, and left-hand if negative. This method is useful to those measuring other types of elliptically polarized antennas with polarization senses that are not easily determined.

The impedance of the helix is easily predictable. The terminal impedance of a helix is unbalanced, and is defined by

$$Z = 140 \times C_\lambda$$

where Z is the impedance of the helix in ohms.

The gain of a helical antenna is determined by its physical characteristics. Gain can be calculated from

Gain (dBi) $= 11.8 + 10 \log (C_\lambda^2 nS_\lambda)$

The beamwidth of the helical antenna (in degrees) at the half-power points is

$$BW = \frac{52}{C_\lambda \sqrt{nS_\lambda}}$$

The diameter of the helical antenna conductor should be between 0.006 λ and 0.05 λ, but smaller diameters have been used successfully at 144 MHz. The previously noted diameter of the ground plane (0.8 to 1.1 λ) should not be exceeded if a clean radiation pattern is desired. As the ground plane size is increased, the sidelobe levels also increase. (The ground plane need not be solid; it can be in the form of a spoked wheel or a frame covered with hardware cloth or poultry netting.)

MATCHING SYSTEMS

Because helical antennas present impedances on the order of 110 to 180 Ω, the antenna must be matched for use with a 52-Ω transmission line. Matching systems for helical antennas are classified two ways: narrow band and wide band. Narrow band is generally recognized to represent bandwidths less than 25%. Narrow band matching techniques are relatively straightforward; matching systems useful over the full frequency range of a helix are a bit more involved.

Many matching techniques are available. Some of the proven methods are discussed here. For narrow band use, the simplest impedance matching technique is the use of a ¼-λ series transformer. A 1-λ circumference helix has a feed point impedance of approximately 140 Ω, so the transformer must be ¼ λ of 84-Ω transmission line. This line can be fabricated in microstrip form, or a piece of air-dielectric coax can be built, as shown by Doug DeMaw in November 1965 *QST*. (See bibliography.)

Another solution is to design the helix so that its feed-point impedance allows the use of a standard impedance line for the matching transformer. This method was shown by D. Mallozzi in the March 1978 *AMSAT Newsletter*. This helix was designed with a circumference of 0.8 λ, resulting in an input impedance of 112 Ω. Standard 75-Ω coaxial cable can be used for the ¼-λ matching transformer in this case, as shown in Fig 38. Yet another matching method is to use a series section. For example, a 125-Ω helix may be matched to 52-Ω line by inserting 0.125 λ of RG-133 (95-Ω impedance) in the 52-Ω line at a distance of 0.0556 λ from the antenna feed point. Series section matching is discussed in Chapter 26.

The physical construction of a ¼-λ transformer or a series section at UHF is a project requiring careful measurement and assembly. For narrow bandwidths at relatively low frequencies (below 148 MHz), the familiar pi network can be used for impedance matching to helical antennas. Other matching methods are discussed in the references listed in the bibliography at the end of this chapter.

Two series transformers can be used to allow operation of a helical antenna over its entire bandwidth (see Fig 39). This method is in use in a number of helical antenna installations and provides good performance.

SPECIAL CONFIGURATIONS

Many special helical antenna configurations have been

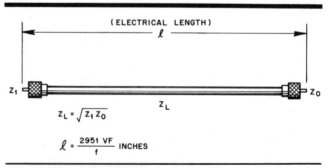

Fig 38—Narrow-band matching technique using a ¼-λ series transformer. In the length equation, VF = velocity factor of the cable; f = frequency, MHz.

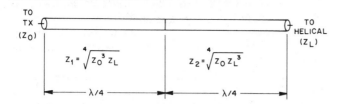

Fig 39—The dual series quarter-wave transformer is another means of matching 52-Ω coaxial cable to a helical antenna.

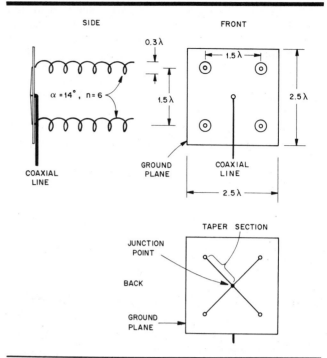

Fig 40—An array of four helicals on a common ground screen can provide as much gain as a single helix with four times as many turns as the individual helicals in the array. The benefits of this design include a cleaner radiation pattern and much smaller turning radius than a single long helix. See Fig 41 for detail of taper section.

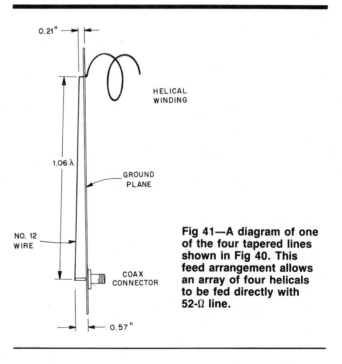

Fig 41—A diagram of one of the four tapered lines shown in Fig 40. This feed arrangement allows an array of four helicals to be fed directly with 52-Ω line.

developed. These special configurations usually address improvements in one or more of four areas:

1) Easing mechanical construction.
2) "Cleaning up" the radiation pattern (reducing side-lobes and backlobes), and increasing gain.
3) Maximizing bandwidth.
4) Improving terminal characteristics.

Increasing the bandwidth of a helical antenna is not usually required in amateur applications. Many of the professional journals listed in the bibliography have published articles discussing this subject, however. The other improvements listed all have applications in amateur work.

For example, the mechanical difficulty of making a helical antenna of the required diameter with the required conductor diameter is formidable at and below 148 MHz. Square or triangular winding forms are simpler than round forms at these frequencies. These configurations offer more mechanical stability under icing and wind conditions. Measurements indicate that if the perimeter of the form remains constant, it makes little difference in helical antenna characteristics if the cross sectional shape is circular, square or triangular.

Helical Antenna Variations

"Cleaning up" the radiation pattern (minimizing extraneous minor lobes) and increasing the gain of the helix can be done in a number of ways. The most common method of doing this is to mount three or four helical radiators on a single reflector and feed them in phase. This results in high gain with a cleaner radiation pattern than can be obtained with a single helical radiator having enough turns to obtain the same gain. Four 6-turn helicals mounted as shown in Fig 40 exhibit essentially the same gain as a single 24-turn helix. The turning radius of the quad array of helicals is smaller than the turning radius of a single helix of equal gain. The four elements are fed in parallel, using the feed method shown in Fig 41.

Another method of reducing extraneous lobes is to use the helix to excite a conical horn. This method is somewhat cumbersome mechanically, but is useful in situations where very clean radiation patterns are required.

Combining two "good" antennas can sometimes result in a single "better" antenna. This is the case when a helix is used to feed a parabolic dish. The high gain inherent in the dish and the circular polarization afforded by the helix combine to make an excellent antenna for satellite and EME communications.

Such an antenna was built and tested at 465 MHz. The bandwidth was measured at more than 60 MHz. The antenna produces circular polarization with a sense *opposite* that of the feed helix. (The sense is reversed in reflection of the wave front from the parabolic surface.) This antenna is much easier to build than other types of circularly polarized dishes, because the mechanical construction of the feed is simpler.

This feed system is attractive to those who wish to use any of the common TVRO dishes that are available at reasonable cost. The gain of this combination at a given frequency is based on the illumination efficiency and size of the dish used.

When using short helical antennas in a quad array for reception, wiring a series resistor of 155 Ω ¼ λ from the open end of each helix improves the array performance. The sidelobe levels decrease, and matching and circularity (axial ratio) increase as a result of this modification. This performance improvement has been attributed to resistor dissipation of the unradiated energy reflected back toward the feed from the open end of the windings.

Publications such as *IEEE Transactions on Propagation and Antennas* and similar professional journals are a good source of information on the uses of helical antennas. University libraries often have these publications available for reference.

A Switchable Sense Helical Antenna

Constructing a pair of helix antennas for the 435-MHz band is quite simple. One antenna is wound for RHCP, and the other for LHCP, as shown in Fig 42. A good UHF relay and some Hardline are all that is needed to complete the system. Inexpensive, readily available materials are used for construction, and the dimensions of the helicals are not critical. Fig 36 shows the helix formulas and dimensions.

This antenna has a 70% bandwidth, and is ideal for a high gain, broad beamwidth satellite tracking antenna. This switchable antenna system and 50 to 100 watts of RF output yield respectable signals on the Phase III satellites.

A detail of the complicated portion of the helix is shown in Fig 43. Table 2 contains a keyed list of parts for the array. A good starting point for construction is the reflector, which is made of heavy wire mesh. This wire mesh is used in most UHF TV "bow tie" antennas. Wire companies and many hardware stores supply this material in 4-foot widths. It is 14-gauge galvanized steel, and sells for approximately $1.60 to $2 per lineal foot. A piece of mesh 2 × 4 feet is required to build two antennas. Trim the mesh so that no sharp ends stick out.

The next step is to make the reflector mounting plates and boom brackets. Follow the dimensions shown in Fig 44. Heavy aluminum material is recommended; 0.060 inch is the minimum recommended thickness. Thicker material is more difficult to bend, but two bends of 45° spaced about ¼ inch apart will work fine for the brackets in this case. The measurements shown are for TV type 1¾-inch U bolts. If you use another size, change the dimensions appropriately. Drill the four holes in the reflector mounting plate and mount the coax receptacle, using pop rivets or stainless steel hardware.

Check the clearance between the coax receptacle and the elevation boom before final assembly. The thickness of the U bolt spacers will affect this clearance. Mount a short piece of pipe (the same size as the elevation boom you will be using) to the U bolts, wire mesh reflector, reflector mounting plate and boom brackets. The elevation boom is shown in Fig 43. Position the plate in the center of the wire mesh reflector. (It may be necessary to bend some of the mesh to clear the U bolts.) Finger tighten the U bolts so the plate can be adjusted to fit the mesh.

The wood boom assembly shown in Fig 43 consists of two 6-foot wooden tomato stakes joined by spacers in three places. Mount one spacer in the center and the other spacers 1 foot from each end. Notch the ends of the boom to fit into the mesh. When the correct alignment is obtained, clamp the assembly together and drill holes for rivets or bolts through the reflector mounting plate, brackets and wood boom assembly. When drilling the boom holes, place the reflector flat on the floor and use a square so the boom is perpendicular to the reflector. Mark the boom through the holes in the boom bracket. When the assembly is complete, coat the wood boom with marine varnish.

The most unusual aspect of this antenna is its use of

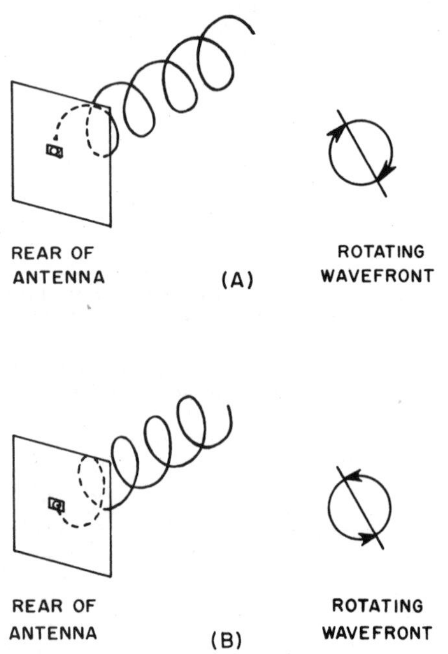

Fig 42—Right-hand circular polarization, A. Left-hand circular polarization, B.

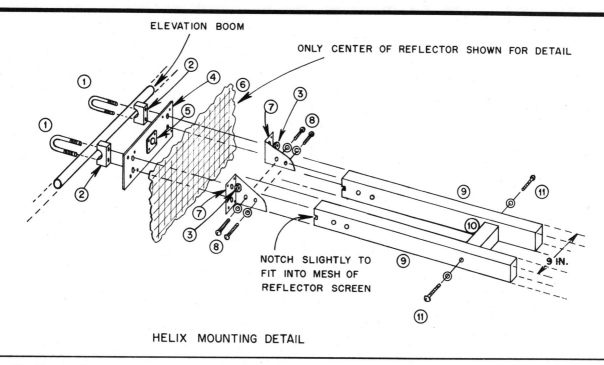

ELEVATION BOOM

ONLY CENTER OF REFLECTOR SHOWN FOR DETAIL

NOTCH SLIGHTLY TO
FIT INTO MESH OF
REFLECTOR SCREEN

9 IN.

HELIX MOUNTING DETAIL

Fig 43—The details of the helix mounting arrangement. See Table 2 for a number-keyed parts list.

Table 2
Parts List for the Helix Mounting Detail Shown in Fig 43

Piece No.	Description	Comments
1	U bolt, TV type	Use to bolt antenna to elevation boom
2	U bolt spacer	As above
3	U bolt nut with lock washer	As above
4	Reflector mounting plate (see Fig 44)	Rivet through reflector to boom brackets
5	Type N coaxial receptacle	Rivet to mounting plate
6	1 × 2-in. heavy gauge wire mesh	Reflector, cut approx. 22 in. square
7	Helix boom to reflector brackets	Rivet through reflector to mounting plate
8	No. 8-32 bolts with nuts and washers	Bolt boom brackets to boom
9	Boom, approx. 1 × 1-in. tomato stake	2 pieces, 6 ft long
10	Boom spacer, 1 × 1-in. tomato stake	Boom to bolt; cut to give 9-in. spacing
11	No. 8 wood screws with washers	Attach spacers to boom (three places)

Notes
1) Mount reflector mounting plate to boom brackets, leaving 9-in. clearance for boom.
2) Wire mesh may be bent to provide clearance for U bolts.
3) When positioning the reflector mounting plate, try to center the coaxial receptacle in the wire mesh screen.

coaxial cable for the helix conductor. Coax is readily available, inexpensive, lightweight, and easy to shape into the coil required for the helix. Nine turns requires about 22 feet of cable, but start with 25 feet and trim off any excess. The antenna of Fig 45 uses FM-8 coaxial cable, but any coax that is near the ½-inch diameter required can be used. (The cable used must have a center conductor and shield that can be soldered together.)

Strip about 4 inches off one end of the cable down to the center conductor, but leave enough braid to solder to the center conductor. Solder the braid to the center conductor at this point. Measure the exposed center conductor 3.3 inches from the short and cut off the excess. (This is dimension g in Fig 36.)

Wind the 25-foot length of coax in a coil about 10 inches in diameter. Fig 42 shows which way to wind the coil for RHCP or LHCP. Slip the coil over the boom and move the stripped end of the cable toward the coax receptacle, which is the starting point of the nine turns. Solder the center conductor to the coax receptacle, and start the first turn 3.3 inches from the point of connection at the coax receptacle.

Use tie wraps to fasten the coax to the wood boom. Mark the boom using dimension S_λ in Fig 36. The first tie wrap is only half this distance when it first comes in contact with the boom; each successive turn on that side of the boom will be spaced by dimension S_λ. Use two tie wraps so they form an X around the boom and coax. Once the first wrap is secure, wind each turn and fasten the cable one point at a time. Before each turn is tightened, make sure the dimensions are correct.

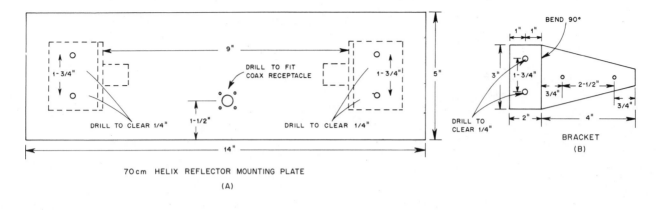

70cm HELIX REFLECTOR MOUNTING PLATE
(A)

BRACKET
(B)

Fig 44—At A, the helix reflector mounting plate (part no. 4 in Table 2). At B, the boom brackets (part no. 7 in Table 2).

When all nine turns are wound, check all dimensions again. Cut the coax at the ninth turn, strip the end, and solder the braid to the center conductor. The exposed solder connections at each end of the coax conductor should be sealed to weatherproof them.

A coaxial 75-Ω ¼-λ matching section as shown in Fig 38 is connected in series with the feed line at the antenna feed point. The length of this cable (including connectors) is 4.5 inches if the cable used has a velocity factor of 0.66. Lengths for other types of cable can be calculated from the equation in the drawing.

The impedance of the helix is approximately 140 Ω. To match the 52-Ω transmission line, a transformer of 85.3 Ω is required. The 75-Ω cable used here is close enough to this value for a good match. The transformer should be connected directly to the female connector mounted on the reflector mounting plate. Use a double female adapter to connect the feed line to the matching transformer. Weatherproof the connectors appropriately.

To mount these antennas on an elevation boom, a counterbalance is required. The best way to do this is to mount an arm about 2 feet long to the elevation boom, at some point that is clear of the rotator, mast and other antennas. Point the arm away from the direction the helicals are pointing, and add weight to the end of the arm until balance is obtained. The completed antenna is shown in Fig 45.

Do not run long lengths of coax to this antenna, unless you use Hardline. Even short runs of good RG-8 coax are quite lossy; 50 feet of foam dielectric RG-8 has a loss of 2 dB at 430 MHz. There are other options if you must make long runs and can't use Hardline. Some amateurs mount the converters, transverters, amplifiers and filters at the antenna.

This can be easily done with the helix antenna; the units can be mounted behind the reflector. (This also adds counterweight.) If this approach is used, check local electrical codes before running any power lines to the antenna.

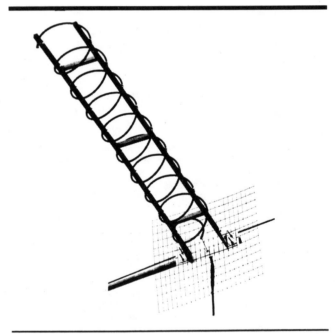

Fig 45—A close-up view of the 435-MHz helical antenna, designed and built by Bernie Glassmeyer, W9KDR.

52-OHM HELIX FEED

Joe Cadwallader, K6ZMW, presented this feed method in June 1981 *QST*. Terminate the helix in an N connector mounted on the ground screen *at the periphery of the helix* (Fig 46). Connect the helix conductor to the N connector as close to the ground screen as possible (Fig 47). Then adjust the first turn of the helix to maintain uniform spacing of the turns.

This modification goes a long way toward curing a deficiency of the helix—the 140-Ω nominal feed-point impedance. The traditional ¼-λ matching section has proved difficult to fabricate and maintain. But if the helix is fed at the periphery, the first half turn of the helix conductor (leaving the N connector) acts much like a transmission line—a single conductor over a perfectly conducting ground plane. The impedance of such a transmission line is

$$Z_0 = 138 \log \frac{4h}{d}$$

where

Z_0 = line impedance in ohms
h = height of the center of the conductor above the ground plane
d = conductor diameter (in the same units as h).

The impedance of the helix is 140 Ω a turn or two away from the feed point. But as the helix conductor swoops down toward the feed connector (and the ground plane), h gets smaller, so the impedance decreases. The 140-Ω nominal impedance of the helix is transformed to a lower value. For any particular conductor diameter, an optimum height can be found that will produce a feed-point impedance equal to 52 Ω. The height should be kept very small, and the diameter should be large. Apply power to the helix and measure the SWR at the operating frequency. Adjust the height for an optimum match.

Typically, the conductor diameter may not be large enough to yield a 52-Ω match at practical (small) values of h. In this case, a strip of thin brass shim stock or flashing copper can be soldered to the first quarter turn of the helix conductor (Fig 48). This effectively increases the conductor diameter, which causes the impedance to decrease further yet. The edges of this strip can be slit every ½ inch or so, and the strip bent up or down (toward or away from the ground plane) to tune the line for an optimum match.

This approach yields a perfect match to nearly any coax. The usually wide bandwidth of the helix (70% for less than 2:1 SWR) will be reduced slightly (to about 40%) for the same conditions. This reduction is not enough to be of any consequence for most amateur work. The improvements in performance, ease of assembly and adjustment are well worth the effort in making the helix more practical to build and tune.

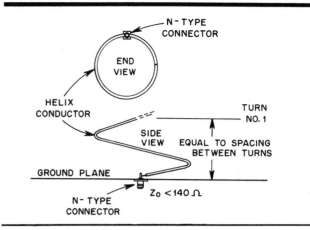

Fig 46—End view and side view of peripherally fed helix.

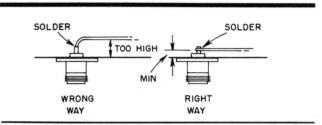

Fig 47—Wrong and right ways to attach helix to a type N connector for 52-Ω feed.

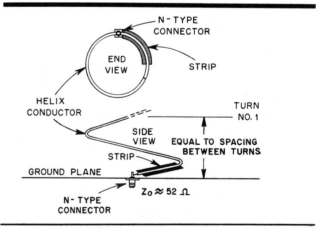

Fig 48—End view and side view of peripherally fed helix with metal strip added to improve transformer action.

Portable Helix for 435 MHz

Helicals for 435 MHz are excellent uplink antennas for Mode B satellite communications. The true circular polarization afforded by the helix minimizes signal "spin fading" that is so predominant in these operations. The antenna shown in Fig 49 fills the need for an effective portable uplink antenna for OSCAR operation. Speedy assembly and disassembly and light weight are among the benefits of this array. This antenna was designed by Jim McKim, WØCY.

As mentioned previously, the helix is about the most tolerant of any antenna in terms of dimensions. The dimensions given here should be followed as closely as possible, however. Most of the materials specified are available in any well supplied "do it yourself" hardware or building supply store. The materials required to construct the portable helix are listed in Table 3.

The portable helix consists of eight turns of ¼-inch soft copper tubing spaced around a 1-inch fiberglass tube or maple dowel rod 4 feet 7 inches long. Surplus aluminum jacket Hardline can be used in lieu of the copper tubing if necessary. The turns of the helix are supported by 5-inch lengths of ¼-inch maple dowel that are mounted through the 1-inch rod in the center of the antenna. Fig 50A shows the overall dimensions of the antenna. Each of these support dowels has a V shaped notch in the end to locate the tubing (see Fig 50B).

The rod in the center of the antenna terminates at the feed-point end in a 4-foot piece of 1-inch-ID galvanized steel pipe. The pipe serves as a counterweight for the heavier end of the antenna (that with the helical winding). The 1-inch rod material that is inside the helix must be nonconductive. Near the point where the nonconductive rod and the steel pipe are joined, a piece of aluminum screen or hardware cloth is used as a reflector screen.

If you have trouble locating the ¼-inch soft copper tubing, try a refrigeration supply house. The perforated aluminum screening can be cut easily with tin snips. This material is usually supplied in 30 × 30-inch sheets, making this size convenient for a reflector screen. Galvanized ¼-inch hardware cloth or copper screen could also be used for the screen, but aluminum is lighter and easier to work with.

A 1/8-inch thick aluminum sheet is used as the support plate for the helix and the reflector screen. Surplus rack panels provide a good source of this material. Fig 51 shows the layout of this plate.

Fig 52 shows how aluminum channel stock is used to support the reflector screen. (Aluminum tubing also works well for this. Discarded TV antennas provide plenty of this material if the channel stock is not available.) The screen is mounted on the bottom of the 10-inch aluminum center plate. The center plate, reflector screen and channel stock are connected together with plated hardware or pop rivets. This support structure is very sturdy.

Fiberglass tubing is the best choice for the center rod material. Maple dowel can be used, but is generally not available in lengths over 3 feet. If maple must be used, the dowels can be spliced together by drilling holes in the center of each end and inserting a short length of smaller dowel into one of them. One of the large dowel ends should be notched, and the end of the other cut in a chisel shape so that they fit together. The small dowel can then be epoxied into both ends

Fig 49—The portable 435-MHz helix, assembled and ready for operation. (WØCY photo)

Table 3
Parts List for the Portable 435-MHz Helix

Qty	Item
1	Type N female chassis mount connector
18 ft	¼-in. soft copper tubing
4 ft	1-in. ID galvanized steel pipe
1	5 ft × 1-in. fiberglass tube or maple dowel
14	5-in. pieces of ¼-in. maple dowel (6 ft total)
1	1/8-in. aluminum plate, 10 in. diameter
3	2 × ¾-in. steel angle brackets
1	30 × 30-in. (round or square) aluminum screen or hardware cloth
8 ft	½ × ½ × ½-in. aluminum channel stock or old TV antenna element stock
3	Small scraps of Teflon or polystyrene rod (spacers for first half turn of helix)
1	1/8 × 5 × 5-in. aluminum plate (boom to mast plate)
4	1½-in. U bolts (boom to mast mounting)
3 ft	No. 22 bare copper wire (helix turns to maple spacers)

Assorted hardware for mounting connector, aluminum plate and screen, etc.

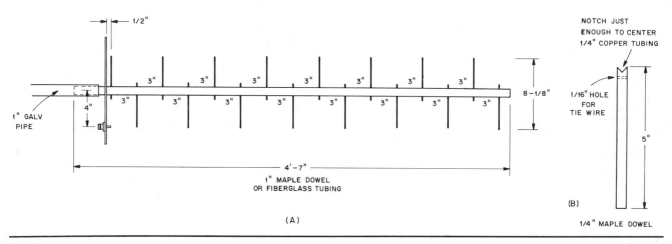

Fig 50—At A, the layout of the portable 435-MHz helix is shown. Spacing between the first 5-in. winding-support dowel and the ground plane is ½ in.; all other dowels are spaced 3 in. apart. At B, the detail of notching the winding-support dowels to accept the tubing. As indicated, drill a 1/16-in. hole below the notch for a piece of small wire to hold the tubing in place.

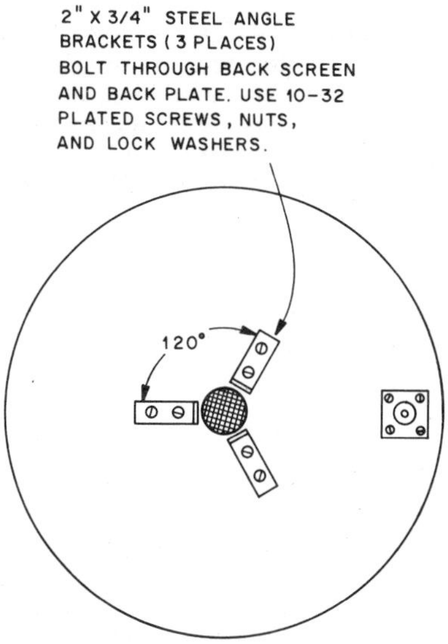

Fig 51—The ground plane and feed-point support assembly. The circular piece is a 10-in. dia, 1/8-in. thick piece of aluminum sheet. (A square plate may be used instead.) Three 2 × ¾-in. angle brackets are bolted through this plate to the back side of the reflector screen to support the screen on the pipe. The type N female chassis connector is mounted in the plate four inches from the 1-in. diameter center hole.

Fig 52—The method of reinforcing the reflector screen with aluminum channel stock. In this version of the antenna, the three angle brackets of Fig 51 have been replaced with a surplus aluminum flange assembly. (WØCY photo)

when they are fitted together. Fig 53 illustrates this method of splicing dowels. The splice in the dowels should be placed as far from the center plate as possible to minimize stress on the connection.

Mount the type N connector on the bottom of the center plate with the appropriate hardware. The center pin should be exposed enough to allow a flattened end of the copper tubing to be soldered to it. Tin the end of the tubing after it is flattened so that no moisture can enter it. If the helix

is to be removable from the ground-plane screen, do not solder the copper tubing to the connector. Instead, prepare a small block of brass, drilled and tapped at one side for a no. 6-32 screw. Drill another hole in the brass block to accept the center pin of the type N connector, and solder this connection. Now the connection to the copper tubing helix can be made in the field with a no. 6-32 screw instead of with a soldering iron.

Refer to Fig 50A. Drill the fiberglass or maple rod at the positions indicated to accept the 5-inch lengths of ½-inch dowel. (If maple doweling is used, the wood must be weatherproofed as described below before drilling.) Drill a 1/16-inch hole near the notch of each 5-inch dowel to accept a piece of no. 22 bare copper wire. (The wire is used to keep the copper tubing in place in the notch.) Sand the ends of the 5-inch dowels so the glue will adhere properly, and epoxy them

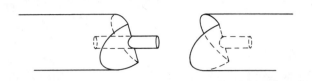

Fig 53—Close-up view of the dowel-splicing method. One dowel is notched, and the other is cut in a wedge shape to fit into the notch. Before this is done, both ends are drilled to accept a small piece of dowel (¼ inch), which is glued into one of the ends. The large dowels should both be weatherproofed before splicing.

into the main support rod.

Begin winding the tubing in a clockwise direction from the reflector screen end. First drill a hole in the flattened end of the tubing to fit over the center pin of the type N connector. Solder it to the connector, or put the screw into the brass block described earlier. Carefully proceed to bend the tubing in a circular winding from one support to the next.

Fig 54 shows how the first half turn of the helix tubing must be positioned about ¼ inch above the reflector assembly. It is important to maintain this spacing, as extra capacitance between the tubing and ground is required for impedance-matching purposes.

Insert a piece of no. 22 copper wire in the hole in each support as you go. Twist the wire around the tubing and the support dowel. Solder the wire to the tubing and to itself to keep the tubing in the notches. Continue in this way until all eight turns have been wound. After winding the helix, pinch the far end of the tubing together and solder it closed.

Weatherproofing the Wood

A word about preparing the maple doweling is in order. Wood parts must be protected against the weather to ensure long service life. A good way to protect wood is to boil it in paraffin for about half an hour. Any holes to be drilled in the wooden parts should be drilled *after* the paraffin is applied, as epoxy does not adhere well to wood after it has been coated with paraffin. The small dowels can be boiled in a saucepan. Caution must be exercised here—the wood can be scorched if the paraffin is too hot. Paraffin is sold for canning purposes at most grocery stores.

The center maple dowel is too long to put in a pan for boiling. A hair drier can be used to heat the long dowel, and paraffin can then be rubbed onto it. Heat the wood again to impregnate the surface with paraffin. This process should be repeated several times to ensure proper weatherproofing. Wood parts can also be protected with three or four coats of spar varnish. Each coat must be allowed to dry fully before another coat is applied.

The fiberglass tube or wood dowel must fit snugly with the steel pipe. The dowel can be sanded or turned down to the appropriate diameter on a lathe. If fiberglass is used, it can be coupled to the pipe with a piece of wood dowel that fits snugly inside the pipe and the tubing. Epoxy the dowel splice into the pipe for a permanent connection.

Drill two holes through the pipe and dowel and bolt them together. The pipe provides a solid mount to the boom of the rotator, as well as most of the weight needed to counter-

balance the antenna. More weight can be added to the pipe if the assembly is "front-heavy." (Cut off some of the pipe if the balance is off in the other direction.)

The helix has a nominal impedance of about 105 Ω in this configuration. By varying the spacing of the first half turn of tubing, a good match to 52-Ω coax should be

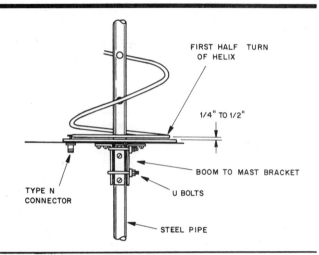

Fig 54—Side view of the helix feed-point assembly. The first half turn of the helix should be kept between ¼ and ½ inch above the ground screen during winding. The height above the screen is adjusted for optimum match to a 52-Ω transmission line after the antenna is completed.

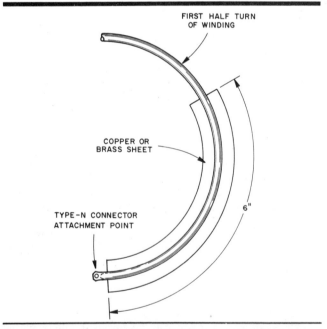

Fig 55—If the match to the antenna cannot be obtained with the tubing alone, a 6-inch piece of copper flashing material can be soldered to the bottom of the first turn of the helix, starting very close to the feed point. The spacing can then be adjusted for best match as described in the text. When the appropriate spacing has been found, affix Teflon or polystyrene blocks between the screen and the winding with silicone sealant to maintain the spacing.

obtainable. If the SWR cannot be brought below about 1.5:1, a 6-inch length of copper flashing material can be added to the first half turn of the helix, as shown in Fig 55. The flashing material should be added as close to the coaxial cable connector as possible.

When the spacing has been established for the first half turn to provide a good match, add pieces of polystyrene or Teflon rod stock between the tubing and the reflector assembly to maintain the spacing. These can be held in place on the reflector assembly with silicone sealant. Be sure to seal the type N connector with the same material.

BIBLIOGRAPHY

Source material and more extended discussion of the topics covered in this chapter can be found in the references given below and in the textbooks listed at the end of Chapter 2.

W. Allen, "A Mode J Helix," *AMSAT Newsletter*, Jun 1979, pp 30-31.

D. J. Angelakos and D. Kajfez, "Modifications on the Axial-Mode Helical Antenna," *Proc IEEE*, Apr 1967, pp 558-559.

J. Bartlett, "An Accurate, Low-Cost Antenna Elevation System," *QST*, Jun 1979, pp 19-22.

J. Bartlett, "A Radio-Compass Antenna-Elevation Indicator," *QST*, Sep 1979, pp 24-25.

A. L. Bridges, "Really Zap OSCAR With this Helical Antenna," *73*, in three parts, Jul, Aug and Sep 1975.

K. R. Carver, "The Helicone—A Circularly Polarized Antenna with Low Sidelobe Level," *Proc IEEE*, Apr 1967, p 559.

M. R. Davidoff, "A Simple 146-MHz Antenna for Oscar Ground Stations," *QST*, Sep 1974, pp 11-13.

M. R. Davidoff, *The Satellite Experimenter's Handbook* (Newington, CT: ARRL, 1984).

D. DeMaw, "The Basic Helical Beam," *QST*, Nov 1965, pp 20-25, 170.

D. Evans and G. Jessop, *VHF-UHF Manual*, 3rd ed. (London: RSGB, 1976).

N. Foot, "WA9HUV 12-Foot Dish for 432 and 1296 MHz," The World Above 50 Mc., *QST*, Jun 1971, pp 98-101, 107.

N. Foot, "Cylindrical Feed Horn for Parabolic Reflectors," *Ham Radio*, May 1976.

O. J. Glasser and J. D. Kraus, "Measured Impedances of Helical Beam Antennas," *Journal of Applied Physics*, Feb 1948, pp 193-197.

H. E. Green, "Paraboidal Reflector Antenna with a Helical Feed," *Proc IRE* of Australia, Feb 1960, pp 71-83.

G. A. Hatherell, "Putting the G Line to Work," *QST*, Jun 1974, pp 11-15, 152, 154, 156.

G. R. Isely and W. G. Smith, "A Helical Antenna for Space Shuttle Communications," *QST*, Dec 1984, pp 14-18.

H. Jasik, *Antenna Engineering Handbook*, 1st ed. (New York: McGraw-Hill, 1961).

R. T. Knadle, "A Twelve-Foot Stressed Parabolic Dish," *QST*, Aug 1972, pp 16-22.

J. D. Kraus, "Helical Beam Antenna," *Electronics*, Apr 1947, pp 109-111.

J. D. Kraus and J. C. Williamson, "Characteristics of Helical Antennas Radiating in the Axial Mode," *Journal of Applied Physics*, Jan 1948, pp 87-96.

J. D. Kraus, "Helical Beam Antenna for Wide Band Applications," *Proc of the IRE*, Oct 1948, pp 1236-1242.

J. D. Kraus, *Antennas* (New York: McGraw-Hill Book Co., 1950).

J. D. Kraus, *Big Ear* (Powell, OH: Cygnus-Quasar Books, 1976).

J. D. Kraus, "A 50-Ohm Input Impedance for Helical Beam Antenna," *IEEE Transactions on Antennas and Propagation*, Nov 1977, p 913.

T. S. M. MacLean, "Measurements on High-Gain Helical Aerial and on Helicals of Triangular Section," *Proc of IEE* (London), Jul 1964, pp 1267-1270.

D. M. Mallozzi, "The Tailored Helical," *AMSAT Newsletter*, Mar 1978, pp 8-9.

T. Moreno, *Microwave Transmission Design Data* (New York: McGraw-Hill, 1948).

K. Nose, "Crossed Yagi Antennas for Circular Polarization," *QST*, Jan 1973, pp 21-24.

K. Nose, "A Simple Az-El Antenna System for Oscar," *QST*, Jun 1973, pp 11-12.

C. Richards, "The 10 Turn Chopstick Helical (Mk2) for OSCAR 10 432 MHz Uplink," *Radio Communication*, Oct 1984, pp 844-845.

S. Sander and D. K. Cheng, "Phase Center of Helical Beam Antennas," *IRE National Convention Record Part 6*, 1958, pp 152-157.

E. A. Scott and H. E. Banta, "Using the Helical Antenna at 1215 Mc.," *QST*, Jul 1962, pp 14-16.

G. Southworth, *Principles and Applications of Waveguide Transmission* (New York: D. Van Nostrand Co, 1950).

B. Sykes, "Circular Polarization and Crossed-Yagi Antennas," Technical Topics, *Radio Communication*, Feb 1985, p 114.

H. E. Taylor and D. Fowler, "A V-H-F Helical Beam Antenna," *CQ*, Apr 1949, pp 13-16.

G. Tillitson, "The Polarization Diplexer—A Polaplexer," *Ham Radio*, Mar 1977.

P. P. Viezbicke, "Yagi Antenna Design," *NBS Technical Note 688* (US Dept of Commerce/National Bureau of Standards, Boulder, CO), Dec 1976.

D. Vilardi, "Easily Constructed Antennas for 1296 MHz," *QST*, Jun 1969.

D. Vilardi, "Simple and Efficient Feed for Parabolic Antennas," *QST*, Mar 1973.

J. L. Wong and H. E. King, "Broadband Quasi-Tapered Helical Antenna," *IEEE Transactions on Antennas and Propagation*, Jan 1979, pp 72-78.

Radio Communication Handbook, 5th ed. (London: RSGB, 1976).

Chapter 20

Spacecraft Antennas

Amateur Radio entered the space age on December 12, 1961. On that date, after two years of effort on the part of many amateurs, OSCAR 1 was launched from Vandenberg Air Force Base, California. Traveling as a hitchhiker, OSCAR 1 was piggybacked aboard a Thor-Agena rocket to become the first non-government satellite ever to be placed in orbit.

One might imagine that OSCAR 1 was bristling with antennas for transponders, beacons, and other transmitting and receiving equipment. But no. A single spring-loaded ¼-λ monopole served the need. OSCAR 1 carried only a 140-milliwatt, 145-MHz beacon transmitter.

As technology in aerospace communication has advanced since the launch of OSCAR 1, so have the antennas used aboard amateur spacecraft. This chapter is not intended to provide do-it-yourself construction data for those who are building spacecraft antennas. Rather, it is intended to provide a source of general information to the radio amateur on unusual or unique antennas that are employed aboard spacecraft.

THE QUADRIFILAR HELIX ANTENNA

The quadrifilar helix antenna is seeing use on commercial as well as amateur spacecraft. For amateur applications it first saw use on AMSAT-OSCAR 7. The information in this section is based on a paper prepared by M. Walter Maxwell, W2DU, at RCA's Astro-Electronics Division. (See the bibliography at the end of this chapter.)

Radiation from the quadrifilar helix antenna is circularly polarized and of the same screw sense everywhere throughout the radiation sphere. The antenna embodies a unique configuration and method of feeding loop elements to produce radiation having a controllable pattern shape. Shown in Fig 1, the quadrifilar antenna comprises two bifilar helical loops oriented in a mutually orthogonal relation on a common axis. The terminals of each loop are fed 180° out of phase, and the currents in the two loops are in phase quadrature (90° out of phase).

By selecting the appropriate configuration of the loops, a wide range of radiation pattern shapes is available, with excellent axial ratio appearing over a large volume of the pattern. The basic form of the quadrifilar antenna was developed by Dr. C. C. Kilgus of the Applied Physics Laboratory, Johns Hopkins University, who has published several articles that establish the theoretical basis for its operation. (See bibliography.) This version of the quadrifilar helix uses a novel feed system which includes the infinite balun. The feed system also includes a means for attaining the 90° current phase relationship without requiring additional components to achieve separate differential-phase excitations.

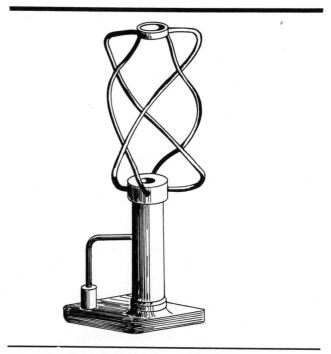

Fig 1—Artist's drawing of a quadrifilar helix antenna. Radiation from the antenna is circularly polarized and of the same screw sense everywhere throughout its radiation sphere.

The quadrifilar antenna is applicable for general use in the frequency range above 30 MHz. It is especially attractive for certain spacecraft applications because it can provide omnidirectional radiation in a single hemisphere (a cardioid volume of revolution) without requiring a ground plane. This no-ground-plane feature affords a dramatic saving in weight. In addition, the inherent cardioid radiation characteristic which eliminates the need for the ground plane also affords freedom in the choice of a mounting position on a spacecraft. The profile of the volume beneath the antenna has relatively little effect on the radiation pattern, provided that the quadrifilar is mounted at least a quarter-wavelength above conducting surfaces which would form a reflective plane.

Quadrifilars designed to operate in the frequency range of 1800 and 2200 MHz can be extremely light, weighing only about 0.7 ounce. Developed by RCA Astro-Electronics, several of these antennas are flying on Air Force satellites of the Defense Meteorological Support Program. A slightly smaller version of similar design (donated by RCA Astro-Electronics) is the 2304.1-MHz beacon antenna, which flew on AMSAT-OSCAR 7.

Quadrifilars having a somewhat different configuration from that shown in Fig 1 have also been developed by RCA for use on NOAA TIROS-N weather satellites. The design differs because the radiation pattern requirements are different. The design of the quadrifilars for the TIROS-N spacecraft is noteworthy because the radiation pattern has been shaped to maintain a nearly constant signal level versus slant-range distance between the spacecraft and the ground station during the in-view portion of the spacecraft orbit.

Physical Characteristics

There is a tendency to confuse the quadrifilar with the conventional helical antenna, probably because portions of the quadrifilar are helically shaped. As a result, the radiation characteristics of the quadrifilar antenna are sometimes misunderstood. Although the term quadrifilar is not incorrect, the term itself often fails to conjure up a true picture of the physical characteristics. Therefore, since several characteristics of the quadrifilar antenna differ radically from those of the conventional helical antenna (for example, opposite screw sense of the circular polarization), the following paragraphs examine some aspects of both the physical and the radiation characteristics of the quadrifilar to clarify the confusion.

In describing the physical characteristics of the quadrifilar, we will concentrate on the simple half-turn, half-wavelength ($\lambda/2$) configuration. As stated earlier, the quadrifilar helix is a combination of two bifilar helicals arranged in a mutually orthogonal relationship along a common axis, and enveloping a common volume.

What is a bifilar helix? One way of visualizing the half-turn $\lambda/2$ bifilar helix is to develop it from a continuous square loop of wire one wavelength in perimeter, as shown in Fig 2. As in the driven element of the conventional cubical-quad antenna, each side of this square loop is a quarter-wavelength, and the feed terminals are formed by opening the loop at the midpoint of the bottom side. We call this square-loop configuration a zero-turn, $\lambda/2$ bifilar loop. There is a half-wavelength of wire radiator extending away from each feed terminal around the loop to the antipodal point on the opposite side of the square (loop half-length $L_e = \lambda/2$). This loop is a balanced-input device, requiring a balanced, two-wire feed line with push-pull currents.

To visualize the development of the quadrifilar helix, insert an imaginary cylinder of diameter $D = \lambda/4$ inside the loop. And then, while holding the bottom side of the loop fixed, grasp the top side and give it one half turn of rotation with respect to the bottom. As a result, Fig 3, each of the two vertical sides of the square loop becomes a half-turn helix as it curves around the surface of the imaginary cylinder. However, because of the curved paths of the once-straight vertical sides, the distance between the top and bottom has shrunk, and the axial length, L_p, is less than the $\lambda/4$ diameter.

With these particular proportions of the half-turn $\lambda/2$ bifilar, some of the radiation characteristics are not particularly attractive. However, some physical parameters of the loop may be selected to obtain characteristics that make this antenna especially attractive. Such parameters include the electrical length of the conductors, the number of turns, the cylindrical diameter D, and length L_p. In addition, for a given conductor length, the length to diameter ratio, L/D, of the cylinder is also an important variable, controlled by the diameter and number of turns. For example, the antenna shown in Fig 1 is a half-turn $\lambda/2$ quadrifilar. But by reducing

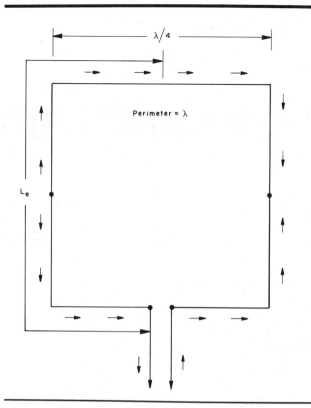

Fig 2—Horizontally polarized square loop radiator. The small arrows represent the direction of current flow.

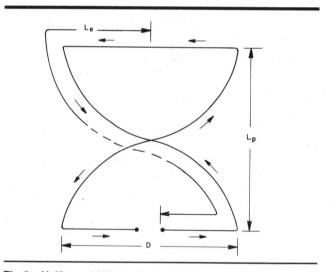

Fig 3—Half-turn bifilar helix loop. L_e = loop half-length.

the diameter from 0.25 λ to 0.18 λ for this model, the axial length is inherently increased to 0.27 λ. This results in vastly improved radiation characteristics.

Electrical Characteristics

Let's take a short qualitative look at some basic changes in the radiation pattern that result from the half-turn twist of the square loop. First, recall that the square loop having a perimeter P = 1 λ is basically a broadside array of two dipole elements, with the top element being voltage fed from the bottom element. In this loop the currents in the top and bottom sides flow in the same direction, as shown in Fig 2. The horizontally polarized fields produced by the currents in both top and bottom sides are therefore in phase. The two fields add to form the conventional figure-8 broadside radiation pattern. The bidirectional lobes in the pattern are at right angles to the plane of the loop. The null in the pattern appears bidirectionally on a horizontal line that is in the plane of the loop midway between the top and bottom sides.

In the vertical sides of the loop, the current in the top half of each side flows in the direction opposite to that in the corresponding bottom half. Consequently, the vertically polarized fields produced by both halves of each vertical side are mutually out of phase, and add to zero in all directions. This cancellation of the vertical fields results in zero net vertically polarized radiation.

On the other hand, in the bifilar helix loop having the half-turn twist, the currents in the top and bottom sides flow in opposite directions because of the physical half-turn rotation of the top side. This current relationship is shown in Fig 3. Thus, the fields produced by the currents in the top and bottom sides are now out of phase with each other, forming an end-fire array relationship.

In the direction broadside to the plane formed by the top and bottom sides, the fields now cancel completely. As in the ordinary end-fire array, this results in zero radiation in the broadside direction, where maximum radiation appeared with the square loop. The conventional lobes of end-fire radiation contributed by the top and bottom sides now occur toward the top and bottom of the antenna as drawn in Fig. 3. This radiation is horizontally polarized.

The currents flowing in the helical portions of the loop retain the same current-flow pattern as in Fig 2. However, the physical positions of each current segment in the twisted vertical wires has now been shifted to a new position and to a new orientation in the helical paths. This results in a corresponding different position and vector direction for each elemental field produced by the helical current elements. As would be expected, the addition of all these elemental fields now results in a composite field consisting of both horizontally and vertically polarized fields.

Difference Between Quadrifilar and Conventional Helix

Before considering the radiation characteristics of the bifilar helix further, it may be helpful to mention some critical distinctions between the Kilgus bifilar helix and the conventional helical antenna configurations. In the conventional helical antenna having more than one radiating element, the elements are generally fed in phase. However, Kilgus found that interesting results were obtained with a helical antenna which has two elements spaced radially at 180° when the two elements are fed 180° out of phase. This finding led Kilgus to his further investigation of the bifilar helical antenna having out-of-phase feed, and still further to the quadrifilar configuration. Thus, an alternative way of visualizing the bifilar helix is as a conventional helical antenna with two elements radially spaced at 180°, but with the outer ends radially connected to each other, and with the elements fed out of phase.

Dr. Kilgus' analyses provide a more complete theoretical basis for the functioning of both the bifilar and the quadrifilar helix. He shows the current distribution and radiation characteristics of the half-turn λ/2 bifilar helix to be similar to those of a loop-dipole combination described by Brown and Woodward (see bibliography). Brown and Woodward analyzed a combination horizontal loop and vertical dipole sharing a common axis. They show that while both loop and dipole produce identical toroid-shaped radiation patterns, the electric field produced by the dipole current in this arrangement is vertical, and the electric field produced by the loop current is horizontal. When the currents in the loop and dipole are in phase, their fields are in phase quadrature, a requirement for circularly polarized radiation. When the vertical and horizontal fields are also equal in magnitude, the resulting radiation is circularly polarized, with the same screw sense everywhere throughout the radiation sphere.

Experimental data obtained by Kilgus supports his theoretical analysis, and confirms the similarity of radiation characteristics between the half-turn λ/2 bifilar helix and the loop-dipole combination. For the purpose of obtaining additional data pertinent to the development of quadrifilar antennas for use on the TIROS-N spacecraft, M. W. Maxwell performed extensive measurements of pattern shape, polarization, and axial ratio on many different bifilar and quadrifilar helix configurations. The data obtained from Maxwell's measurements agree with the Kilgus data, adding further validity to the earlier findings as well as extending them.

Development of the Quadrifilar Helix Antenna

In this section we will examine the intrinsic cardioid, or omnidirectional hemispherical radiation characteristic of the quadrifilar. In doing so, the toroidal radiation pattern of the bifilar helix becomes of primary interest. Recall that in the toroidal radiation pattern of the loop-dipole, maximum radiation occurs broadside to the axis, with the null on the axis. On the other hand, as mentioned earlier, maximum radiation from the bifilar loop occurs bidirectionally along the axis of the loop, while the null appears in the direction perpendicular to the plane formed by the top and bottom sides, or radials of the loop. Except for this 90° rotation of the axes, the radiation characteristics of the two antennas are similar.

We may develop a quadrifilar helix from the bifilar (which we will call bifilar A) by adding a second bifilar, B, on the same axis and enclosing the same space, but rotated 90° relative to bifilar A. This arrangement is shown in Fig 4. The fields radiated by bifilar B are identical to those of bifilar A, except that the entire radiation pattern of bifilar B is rotated 90° relative to that of bifilar A. Consequently, the null of bifilar A and the maximum radiation of bifilar B appear at the same point in space, and vice versa. On the other hand, in the axial directions the radiations from both bifilars are equal.

This field relationship is the key to the cardioid radiation characteristic of the quadrifilar. When both bifilars are fed in a 0-180 and 90-270° phase relationship (excited simultaneously with a mutual 90° current-phase relationship) the cardioid radiation pattern appears in the far field. This is because the fields of both bifilars are in phase in one axial direction. In this direction, the two fields add, while in the opposite direction, the fields are out of phase and cancel. Since the cancellation is not perfect off axis, the result is a cardioid-shaped pattern of revolution about the axis. In the broadside direction normal to the axis, the respective nulls and maxima of the two individual bifilars compensate each other to

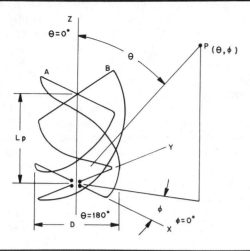

Fig 4—Half-turn quadrifilar (two half-turn bifilar loops).

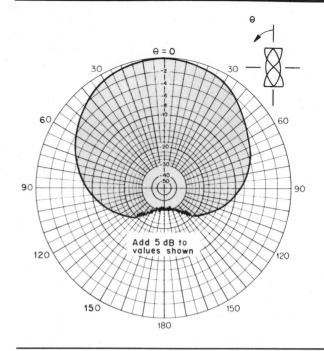

Add 5 dB to values shown

Fig 5—Radiation pattern of the half-turn λ/2 quadrifilar antenna.

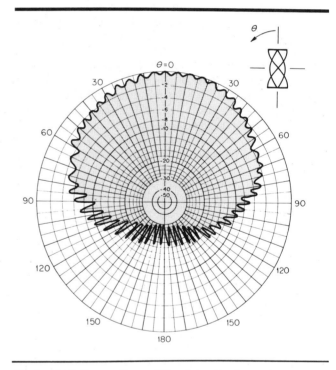

Fig 6—Radiation pattern with axial ratio data added.

produce a uniform omnidirectional radiation pattern around the axis.

The Quadrifilar Shape Factor

As stated earlier, the radiation pattern shape and the axial ratio of the polarization can be controlled. This is done by tailoring both the length of wire and number of turns in the loop, and the length to diameter ratio, L/D, of the formed cylinder. Kilgus' data shows these relationships graphically for certain ranges of the parameter values. However, let us now examine how the Air Force 5D, the amateur OSCAR 7, and the TIROS-N quadrifilar designs were tailored to obtain the radiation characteristics required for their particular missions.

When using bifilars having a diameter D = 0.25 λ formed by placing a half twist in the square loop of Fig 2, the L/D ratio is less than one, which yields poor front-to-back and axial ratios. By decreasing the diameter to D = 0.18 λ while retaining the half-turn 0.5-λ half loop, the axial length increases to L = 0.27 λ. The resulting ratio L/D = 1.5 yields a vast improvement in both front-to-back and axial ratios. These are the design values used in the Air Force 5D and OSCAR 7 quadrifilars. The radiation patterns obtained with this design are presented in Figs 5 and 6. The radiation patterns shown here are presented in the standard IEEE spherical-coordinate system of notation, and the orientation of the antenna relative to the coordinate axes is shown in Fig 4.

The θ pattern shown in Fig 5 was measured while using circularly polarized illumination of the quadrifilar. This pattern shows an on-axis gain of 5 dBic (decibels relative to isotropic, circular), and a front-to-back ratio greater than 20 dB. The radiation at any angle θ is uniform for all values of φ around the Z axis. Thus, the single θ pattern (at any value of φ) represents the shape of the envelope of the volume of revolution about the Z axis, which defines the solid radiation pattern in all directions.

The pattern shown in Fig 6 was obtained by using a spinning dipole to illuminate the quadrifilar, to obtain axial-ratio information. The periodic ripple appearing in the pattern results from the rotation of the dipole. The maxima and minima, respectively, correspond to the times when the dipole is parallel to the major and minor axes of the polarization ellipse. The axial ratio may be determined from the difference between adjacent maximum and minimum values. Thus, the axial ratio of this design is seen to be less than 2 dB over a beamwidth of ±30°.

As mentioned earlier, the shape-factor parameters for the TIROS-N configuration were determined empirically from measurements taken by Maxwell. The measurements were made on many combinations of quadrifilar configurations in which each physical parameter was separately varied in small increments while holding the other variables constant. Hundreds of shape-factor combinations were measured, which provided families of patterns from which to select the parameters for the desired pattern shape. The TIROS-N mission required a radiation pattern shape which provides a nearly constant signal level to the ground station during in-view time. This pattern shape is shown in Fig 7. The bifilar parameters which yield this pattern shape were found in the acquired data families to be 1½ turns, 1.25 λ half loop, D = 0.1 λ, L_p = 1.0 λ, and L/D = 10.0.

inherent, balun, combined with a novel method of self-phasing the two bifilar loops to achieve the 90° differential current relationship between the loops. The constructional simplicity is apparent in Fig 1, which illustrates the quadrifilar configuration used on both the Air Force 5D satellites and the amateur satellite OSCAR 7.

The Infinite or Inherent Balun

In the infinite balun, shown in Fig 8, we see the coaxial feed line extending into the loop and shaped to form the first half of a bifilar loop. At the end of the coax, the inner conductor is connected to the opposite or second half of the bifilar loop to form the feed point. The other end of the second half of the loop is connected to the outside surface of the coax feed line. Thus, the loop is closed at the antipodal point of the loop, where the feed line enters.

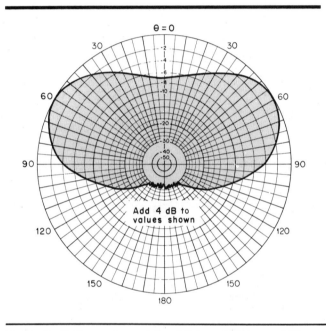

Fig 7—Radiation pattern of the 1½-turn 1.25-λ quadrifilar antenna.

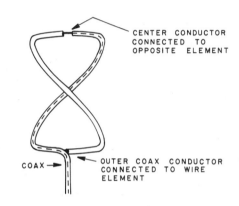

Fig 8—Half-turn bifilar loop with infinite balun feed.

Methods of Feeding the Quadrifilar

Feeding the quadrifilar with a single, unbalanced coaxial line requires special attention. Because the individual bifilar loops are balanced-input devices, some form of balun is required to provide balanced push-pull currents to the terminals of each bifilar. In addition, to obtain the unidirectional radiation characteristic of the quadrifilar, the two bifilar loops require separate excitations having a relative phase difference of 90°. Further, the sense of the 90° phase relationship determines from which end the quadrifilar radiates.

Several different balun and quadrature-phase circuit arrangements are available for feeding the quadrifilar, such as the folded balun, the split-sheath balun, or a combination of 90° and 180° hybrids, and so forth, as described by Bricker and Rickert (see bibliography). However, to save weight and to effect simplicity in the construction, a unique feeding arrangement is used. This technique features the infinite, or

In operation, current flowing on the inner conductor emerges at the feed point to flow onto the second half of the loop. For current flowing on the inside of the outer conductor of the coax, on its arrival at the end of the coax, the only path for current flow is around the end and onto the outside of the outer conductor. Now such external feed-line current is the desired antenna current, because the outside portion of coax extending from the feed point to the antipodal point is the radiator. Externally, the antipodal point demarks the end of the feed line and the beginning of the loop radiator. Because of skin effect, the transmission-line currents flowing inside the coax portion of the loop are completely divorced from the currents flowing externally on the loop. Their only relationship is that the internal currents emerge at the feed point, where they become the external currents.

Since the feed line is dressed away from the antipodal point symmetrically relative to the loop, currents induced on the feed line because of coupling from each half of the loop are equal and flow in opposite directions. The opposing currents on the line thus cancel each other, decoupling the feed line from the loop. In other words, from the external viewpoint of the loop radiator, the source generator can be considered to exist directly between the two input terminals of the loop at the feed point, and the feed line effectively disappears. Thus, the current-mode transition in the coaxial-line portion of the loop—from an internal unbalanced mode entering at the antipodal point, to an external balanced mode emerging at the feed point—constitutes an inherent balun device. Such a device is called an "infinite balun."

Self-Phased Quadrature Feed

As stated, a quadrature-phase current relationship is required between the two bifilar loops of the quadrifilar array. This requirement is met by using the self-phasing method. The orthogonal bifilar loops are designed such that one loop is larger relative to the desired resonant frequency length and therefore inductive, while the other loop is smaller and therefore capacitive. Using this method, the two loops are fed in parallel by connecting the terminals of both loops together at the feed point, as shown in Fig 9. This self-phasing method requires only one coaxial feed line, and any of the balun arrangements mentioned earlier (shown by Bricker and Rickert) may be used.

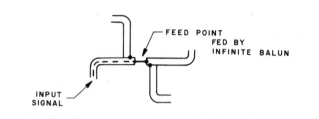

Fig 9—Feed arrangement for 90° self-phasing of loops.

The larger inductive loop is designed such that, at the operating frequency, the reactive component X_L of the loop terminal impedance is equal to the resistive component, R. Similarly, the smaller capacitive loop is designed so its reactive component $X_C = R$ at the operating frequency. The $\pm X = R$ relationship is important, because to obtain a relative current phase of 90° between the two loops, the larger loop current must lag by 45° and the smaller must lead by 45°.

For the current phase of the larger loop to lag or the smaller loop to lead by 45°, their phase angles must be 45°, or have an arc tangent of ± 1. This occurs only when $\pm X = R$. When the two loops are added in parallel, the relative currents in the loops differ in phase by 90° without requiring any additional components to obtain separate differential-phase excitations. Dimensions that yield the correct phase relationship with a loop wire diameter of 0.0088 λ are as follows.

Smaller loop:
    D = 0.156 λ
    L = 0.238 λ
    Perimeter = 1.016 λ

Larger loop:
    D = 0.173 λ
    L = 0.260 λ
    Perimeter = 1.120 λ

where

    D = diameter of imaginary cylinder on which bifilar is wound
    L = distance L_p shown in Fig 3

The resultant impedance versus frequency response of the parallel-loop combination is shown in the Smith Chart, Fig 10. The formation of the cusp in the impedance locus is the design goal that signifies that the 90° phase relationship exists between the loops.

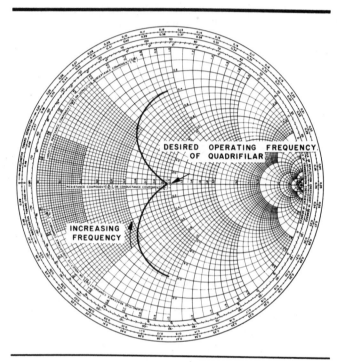

Fig 10—Impedance versus frequency plot of self-phased quadrifilar antenna.

In summary, the quadrifilar helix antenna differs significantly from the conventional helical antenna. Control of pattern shape and other radiation characteristics are available by selection of appropriate dimensional parameters. A novel balun for feeding this balanced-input device from coaxial feed line is used, as is a method of self-phasing the loop radiating elements to obtain a quadrature current relation between the loops. The quadrifilar helix is a durable antenna, one that is seeing use on both amateur and commercial satellites.

BIBLIOGRAPHY

Source material and more extended discussion of topics covered in this chapter can be found in the references given below.

R. W. Bricker and H. H. Rickert, "S-Band Resonant Quadrifilar Antenna for Satellite Communications," *RCA Engineer*, Vol 20, No. 5, Feb-Mar 1975. (Reprint from original published in 1974 *International IEEE AP-S Symposium Digest*, Atlanta.)

R. W. Bricker, "A Shaped-Beam Antenna for Satellite Data Communication," 1976 *International IEEE/AP-S Symposium Digest* (Amherst).

G. H. Brown and O. M. Woodward, Jr, "Circularly Polarized Omni-Directional Antenna," *RCA Review*, Vol 8, Jun 1947, pp 259- 260.

C. C. Kilgus, "Multi-element Fractional Turn Helices," *IEEE Trans*, Vol AP-16, Jul 1968, pp 499-500.

C. C. Kilgus, "Resonant Quadrifilar Helix," *IEEE Trans*, Vol AP-17, May 1969, pp 349-351.

C. C. Kilgus, "Resonant Quadrifilar Helix Design," *Microwave Journal*, Dec 1970, pp 49-54.

M. W. Maxwell, "Some Aspects of the Quadrifilar Helix Antenna," RCA Astro-Electronics, Princeton, NJ (private distribution).

M. W. Maxwell, "Cover Story: 2304 MHz Quadrifilar Antenna for AMSAT-OSCAR 7," *AMSAT Newsletter*, Vol 7, No. 1, Mar 1975.

M. W. Maxwell, "Some Aspects of the Quadrifilar Helix Antenna," *Trends in HF and VHF/UHF Antenna Design*, IEEE Electro 77 Professional Program, New York, April 19-21, 1977.

Chapter 21

Antenna Materials and Accessories

This chapter contains information on materials amateurs use to construct antennas—what types of material to look for in a particular application, tips on working with and using various materials, and information on where to purchase those materials.

Basically, antennas for MF, HF, VHF and the lower UHF range consist simply of one or more conductors that radiate (or receive) electromagnetic waves. However, an antenna system must also include some means to support those conductors and maintain their relative positions—the boom for a Yagi antenna and the halyards for a wire dipole, for example. In this chapter we'll look at materials for those applications, too. Structural supports such as towers, masts, poles, etc. are discussed in another chapter.

There are two main types of material used for antenna conductors, wire and tubing. Wire antennas are generally simple and therefore easier to construct, although some arrays of wire elements can become rather complex. When tubing is required, aluminum tubing is used most often because of its light weight. Aluminum tubing is discussed in a subsequent section of this chapter.

Wire Antennas

Although wire antennas are relatively simple, they can constitute a potential hazard unless properly constructed. Antennas should *never* be run under or over public utility (telephone or power) lines. Several amateurs have lost their lives by failing to observe this precaution.

The National Electric Code® of the National Fire Protection Association contains a section on amateur stations in which a number of recommendations are made concerning minimum size of antenna wire and the manner of bringing the transmission line into the station. Chapter 1 contains more information about this code. The code in itself does not have the force of law, but it is frequently made a part of local building regulations, which are enforceable. The provisions of the code may also be written into, or referred to, in fire and liability insurance documents.

The RF resistance of copper wire increases as the size of the wire decreases. However, in most types of antennas that are commonly constructed of wire (even quite small wire), the radiation resistance will be much higher than the RF resistance, and the efficiency of the antenna will still be adequate. Wire sizes as small as no. 30, or even smaller, have been used quite successfully in the construction of "invisible" antennas in areas where more conventional antennas cannot be erected. In most cases, the selection of wire for an antenna will be based primarily on the physical properties of the wire, since the suspension of wire from elevated supports places a strain on the wire.

WIRE TYPES

Wire having an enamel coating is preferable to bare wire, since the coating resists oxidation and corrosion. Several types of wire having this type of coating are available, depending on the strength needed. "Soft-drawn" or annealed copper wire is easiest to handle; unfortunately, it stretches considerably under stress. Soft-drawn wire should be avoided, except for applications where the wire will be under little or no tension, or where some change in length can be tolerated. (For example, the length of a horizontal antenna fed at the center with open-wire line is not critical, although a change in length may require some readjustment of coupling to the transmitter.)

"Hard-drawn" copper wire or copper-clad steel wire (also known as Copperweld™) is harder to handle, because it has a tendency to spiral when it is unrolled. These types of wire are ideal for applications where significant stretch cannot be tolerated. Care should be exercised in using this wire to make sure that kinks do not develop—the wire will have a far greater tendency to break at a kink. After the coil has been unwound, suspend the wire a few feet above ground for a day or two before using it. The wire should not be recoiled before it is installed.

Several factors influence the choice of wire type and size. Most important to consider are the length of the unsupported span, the amount of sag that can be tolerated, the stability of the supports under wind pressure, and whether or not an unsupported transmission line is to be suspended from the span. Table 1 shows the wire dia, current-carrying capacity and resistance of various sizes of copper wire. Table 2 shows the maximum rated working tensions of hard-drawn and copper-clad steel wire of various sizes. These two tables can be used to select the appropriate wire size for an antenna.

Table 1
Copper-Wire Table

Wire Size AWG (B&S)	Dia in Mils[1]	Dia in mm	Turns per Linear Inch Enamel	Feet per Pound Bare	Ohms per 1000 ft 25°C	Cont.-duty current[2] Single Wire in Open Air	Wire Size AWG (B&S)	Dia in Mils[1]	Dia in mm	Turns per Linear Inch Enamel	Feet per Pound Bare	Ohms per 1000 ft 25°C	Cont.-duty current[2] Single Wire in Open Air
1	289.3	7.348	—	3.947	0.1264	—	21	28.5	0.723	33.1	407.8	13.05	—
2	257.6	6.544	—	4.977	0.1593	—	22	25.3	0.644	37.0	514.2	16.46	—
3	229.4	5.827	—	6.276	0.2009	—	23	22.6	0.573	41.3	648.4	20.76	—
4	204.3	5.189	—	7.914	0.2533	—	24	20.1	0.511	46.3	817.7	26.17	—
5	181.9	4.621	—	9.980	0.3195	—	25	17.9	0.455	51.7	1031	33.00	—
6	162.0	4.115	—	12.58	0.4028	—	26	15.9	0.405	58.0	1300	41.62	—
7	144.3	3.665	—	15.87	0.5080	—	27	14.2	0.361	64.9	1639	52.48	—
8	128.5	3.264	7.6	20.01	0.6405	73	28	12.6	0.321	72.7	2067	66.17	—
9	114.4	2.906	8.6	25.23	0.8077	—	29	11.3	0.286	81.6	2607	83.44	—
10	101.9	2.588	9.6	31.82	1.018	55	30	10.0	0.255	90.5	3287	105.2	—
11	90.7	2.305	10.7	40.12	1.284	—	31	8.9	0.227	101	4145	132.7	—
12	80.8	2.053	12.0	50.59	1.619	41	32	8.0	0.202	113	5227	167.3	—
13	72.0	1.828	13.5	63.80	2.042	—	33	7.1	0.180	127	6591	211	—
14	64.1	1.628	15.0	80.44	2.575	32	34	6.3	0.160	143	8310	266	—
15	57.1	1.450	16.8	101.4	3.247	—	35	5.6	0.143	158	10480	335	—
16	50.8	1.291	18.9	127.9	4.094	22	36	5.0	0.127	175	13210	423	—
17	45.3	1.150	21.2	161.3	5.163	—	37	4.5	0.113	198	16660	533	—
18	40.3	1.024	23.6	203.4	6.510	16	38	4.0	0.101	224	21010	673	—
19	35.9	0.912	26.4	256.5	8.210	—	39	3.5	0.090	248	26500	848	—
20	32.0	0.812	29.4	323.4	10.35	11	40	3.1	0.080	282	33410	1070	—

[1]A mil is 0.001 inch.
[2]Max wire temp of 212° F and max ambient temp of 135° F.

Table 2
Stressed Antenna Wire

American Wire Gauge	Recommended Tension[1] (pounds)		Weight (pounds per 1000 feet)	
	Copper-clad steel[2]	Hard-drawn copper	Copper-clad steel[2]	Hard-drawn copper
4	495	214	115.8	126.0
6	310	130	72.9	79.5
8	195	84	45.5	50.0
10	120	52	28.8	31.4
12	75	32	18.1	19.8
14	50	20	11.4	12.4
16	31	13	7.1	7.8
18	19	8	4.5	4.9
20	12	5	2.8	3.1

[1]Approximately one-tenth the breaking load. Might be increased 50% if end supports are firm and there is no danger of ice loading.
[2]Copperweld,™ 40% copper.

Wire Tension

If the tension on a wire can be adjusted to a known value, the expected sag of the wire (Fig 1) may be determined before installation using Table 2 and the nomograph of Fig 2. Even though there may be no convenient method to determine the tension in pounds, calculation of the expected sag for practicable working tensions is often desirable. If the calculated sag is greater than allowable it may be reduced by any one or a combination of the following:

1) Providing additional supports, thereby decreasing the span

2) Increasing the tension in the wire if less than recommended

3) Decreasing the size of the wire

Instructions for Using the Nomograph

1) From Table 2, find the weight (pounds/1000 feet) for the particular wire size and material to be used.

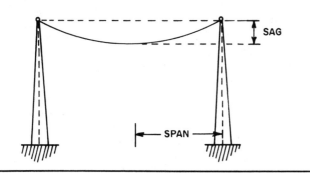

Fig 1—The span and sag of a long-wire antenna.

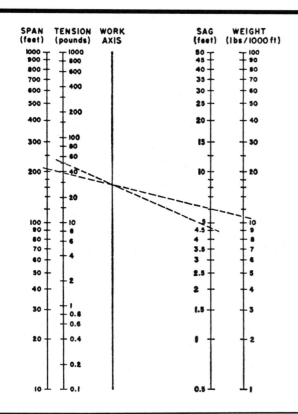

Fig 2—Nomograph for determining wire sag. *(John Elengo, Jr, K1AFR)*

2) Draw a line from the value obtained above, plotted on the weight axis, to the desired span (feet) on the span axis, Fig 2. Note in Fig 1 that the span is one half the distance between the supports.

3) Choose an operating tension level (in pounds) consistent with the values presented in Table 2 (preferably less than the recommended wire tension).

4) Draw a line from the tension value chosen (plotted on the tension axis) through the point where the work axis crosses the original line constructed in step 2, and continue this new line to the sag axis.

5) Read the sag in feet on the sag axis.
Example:
   Weight = 11 pounds/1000 feet
   Span = 210 feet
   Tension = 50 pounds
   Answer: Sag = 4.7 feet

These calculations do not take into account the weight of a feed line supported by the antenna wire.

Wire Splicing

Wire antennas should preferably be made with unbroken lengths of wire. In instances where this is not feasible, wire sections should be spliced as shown in Fig 3. The enamel insulation should be removed for a distance of about 6 inches from the end of each section by scraping with a knife or rubbing with sandpaper until the copper underneath is bright. The turns of wire should be brought up tight around the standing part of the wire by twisting with broad-nose pliers.

The crevices formed by the wire should be completely filled with solder. An ordinary soldering iron or gun may not provide sufficient heat to melt solder outdoors; a propane torch is desirable. The joint should be heated sufficiently so the solder flows freely into the joint when the source of heat is removed momentarily. After the joint has cooled completely, it should be wiped clean with a cloth, and then sprayed generously with acrylic to prevent corrosion.

ANTENNA INSULATION

To prevent loss of RF power, the antenna should be well insulated from ground, unless of course it is a shunt-fed system. This is particularly important at the outer end or ends of wire antennas, since these points are always at a comparatively high RF potential. If an antenna is to be installed indoors (in an attic, for instance) the antenna may be suspended directly from the wood rafters without additional insulation, if the wood is permanently dry. Much greater care should be given to the selection of proper insulators when the antenna is located outside where it is exposed to wet weather.

Insulator Leakage

Antenna insulators should be made of material that will not absorb moisture. Most insulators designed specifically for

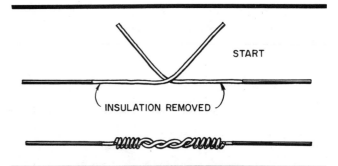

Fig 3—Correct method of splicing antenna wire. Solder should be flowed into the wraps after the connection is completed. After cooling, the joint should be sprayed with acrylic to prevent oxidation and corrosion.

antenna use are made of glass or glazed porcelain. The length of an insulator relative to its surface area is indicative of its comparative insulating ability. A long thin insulator will have less leakage than a short thick insulator. Some antenna insulators are deeply ribbed to increase the surface leakage path without increasing the physical length of the insulator. Shorter insulators can be used at low-potential points, such as at the center of a dipole. If such an antenna is to be fed with open-wire line and used on several bands, however, the center insulator should be the same as those used at the ends, because high RF potential may exist across the center insulator on some bands.

Insulator Stress

As with the antenna wire, the insulator must have sufficient physical strength to support the stress of the antenna without danger of breakage. Long elastic bands or lengths of nylon fishing line provide long leakage paths and make satisfactory insulators within their limits to resist mechanical strain. They are often used in antennas of the "invisible" type mentioned earlier.

For low-power work with short antennas not subject to appreciable stress, almost any small glass or glazed-porcelain insulator will do. Homemade insulators of Lucite rod or sheet will also be satisfactory. More care is required in the selection of insulators for longer spans and higher transmitter power.

For a given material, the breaking tension of an insulator will be proportional to its cross-sectional area. It should be remembered, however, that the wire hole at the end of the insulator decreases the effective cross-sectional area. For this reason, insulators designed to carry heavy strains are fitted with heavy metal end caps, the eyes being formed in the metal cap, rather than in the insulating material itself. The following stress ratings of antenna insulators are typical:

5/8 in. square by 4 in. long—400 lb
1 in. diameter by 7 or 12 in. long—800 lb
1½ in. diameter by 8, 12 or 20 in. long, with special metal
end caps—5000 lb

These are rated breaking tensions. The actual working tensions should be limited to not more than 25% of the breaking rating.

The antenna wire should be attached to the insulators as shown in Fig 4. Care should be taken to avoid sharp angular bends in the wire when it is looped through the insulator eye. The loop should be generous enough in size that it will not bind the end of the insulator tightly. If the length of the antenna is critical, the length should be measured to the outward end of the loop, where it passes through the eye of the insulator. The soldering should be done as described earlier for the wire splice.

Strain Insulators

Strain insulators have their holes at right angles, since they are designed to be connected as shown in Fig 5. It can be seen that this arrangement places the insulating material under compression, rather than tension. An insulator connected this way can withstand much greater stress. Furthermore, the wire will not collapse if the insulator breaks, since the two wire loops are interlocked. Because the wire is wrapped around the insulator, however, the leakage path is reduced drastically, and the capacitance between the wire loops provides an additional leakage path. For this reason, the use of the stain insulator is usually confined to such applications as breaking up resonances in guy wires, where high levels of stress prevail, and where the RF insulation is of less importance. Such insulators might be suitable for use at low-potential points on an antenna, such as at the center of a dipole. These insulators may also be fastened in the conventional manner if the wire will not be under sufficient tension to break the eyes out.

Insulators for Ribbon-Line Antennas

Fig 6A shows the sketch of an insulator designed to be used at the ends of a folded dipole or a multiple dipole made of ribbon line. It should be made approximately as shown, out of Lucite or bakelite material about ¼ inch thick. The advantage of this arrangement is that the strain of the antenna is shared by the conductors and the plastic webbing of the ribbon, which adds considerable strength. After soldering, the screw should be sprayed with acrylic.

Fig 6B shows a similar arrangement for suspending one dipole from another in a multiple-dipole system. If better insulation is desired, these insulators can be wired to a conventional insulator.

PULLEYS AND HALYARDS
Pulleys and halyards commonly used to raise and lower

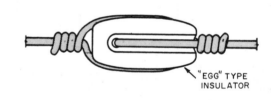

Fig 5—Conventional manner of fastening wire to a strain insulator. This method decreases the leakage path and increases capacitance, as discussed in the text.

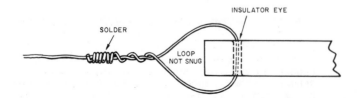

Fig 4—When fastening antenna wire to an insulator, do not make the wire loop too snug. After the connection is complete, flow solder into the turns. Then when the joint has cooled completely, spray it with acrylic.

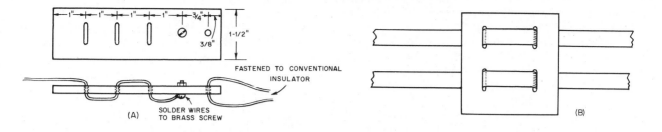

Fig 6—At A, an insulator for the ends of folded dipoles, or multiple dipoles made of 300-ohm ribbon. At B, a method of suspending one ribbon dipole from another in a multiband dipole system.

a wire antenna must also be capable of taking the same strain as the antenna wire and insulators. Unfortunately little specific information on the stress ratings of most pulleys is available. Several types of pulleys are readily available at almost any hardware store. Among these are small galvanized pulleys designed for awnings and several styles and sizes of clothesline pulleys. Heavier and stronger pulleys are those used in marine work. The factors that determine how much stress a pulley will handle include the diameter of the shaft, how securely the shaft is fitted into the sheath and the size and material that the frame is made of.

Another important factor to be considered in the selection of a pulley is its ability to resist corrosion. Galvanized awning pulleys are probably the most susceptible to corrosion. While the frame or sheath usually stands up well, these pulleys usually fail at the shaft. The shaft rusts out, allowing the grooved wheel to break away under tension.

Most good-quality clothesline pulleys are made of alloys which do not corrode readily. Since they are designed to carry at least 50 feet of line loaded with wet clothing in stiff winds, they should be adequate for normal spans of 100 to 150 feet between stable supports. One type of clothesline pulley has a 4-inch diameter plastic wheel with a ¼-inch shaft running in bronze bearings. The sheath is made of cast or forged corrosion-proof alloy. Such pulleys sell for about four dollars in hardware stores. Some look-alike low-cost pulleys of this type have an aluminum shaft with no bearings. For antenna work, these cheap pulleys are of little long-term value.

Marine pulleys have good weather-resisting qualities, since they are usually made of bronze, but they are comparatively expensive and are not designed to carry heavy loads. For extremely long spans, the wood-sheathed pulleys used in "block and tackle" devices and for sail hoisting should work well.

Halyards

Table 3 shows the recommended maximum tensions for various sizes and types of line and rope suitable for hoisting halyards. Probably the best type for general amateur use for spans up to 150 or 200 feet is ¼-inch nylon rope. Nylon is somewhat more expensive than ordinary rope of the same size, but it weathers much better. Nylon also has a certain amount of elasticity to accommodate gusts of wind, and is particularly recommended for antennas using trees as supports. A disadvantage of new nylon rope is that it stretches by a significant percentage. After an installation with new rope, it will be necessary to repeatedly take up the slack created by stretching. This process will continue over a period of several

Table 3

Approximate Safe Working Tension for Various Halyard Materials

Material	Dia, In.	Tension, Lb
Manila hemp rope	1/4	120
	3/8	270
	1/2	530
	5/8	800
Polypropylene rope	1/4	270
	3/8	530
	1/2	840
Nylon rope	1/4	300
	3/8	660
	1/2	1140
7 × 11 galvanized sash cord	1/16	30
	1/8	125
	3/16	250
	1/4	450
High-strength stranded galvanized steel guy wire	1/8	400
	3/16	700
	1/4	1200
Rayon-filled plastic clothesline	7/32	60 to 70

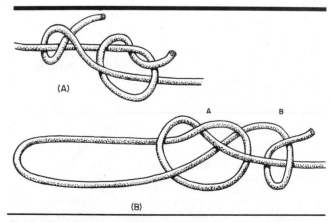

Fig 7—This is one type of knot that will hold with smooth rope, such as nylon. Shown at A, the knot for splicing two ends. B shows the use of a similar knot in forming a loop, as might be needed for attaching an insulator to a halyard. Knot A is first formed loosely 10 or 12 in. from the end of the rope; then the end is passed through the eye of the insulator and knot A. Knot B is then formed and both knots pulled tight. (Richard Carruthers, K7HDB)

weeks, at which time most of the stretching will have taken place. Even a year after installation, however, some slack may still arise from stretching.

Most types of synthetic rope are slippery, and some types of knots ordinarily used for rope will not hold well. Fig 7 shows a knot that should hold well, even in nylon rope or plastic line.

For exceptionally long spans, stranded galvanized steel sash cord makes a suitable support. Cable advertised as "wire rope" usually does not weather well. A boat winch, sold at marinas and at Sears, is a great convenience in antenna hoisting (and usually a necessity with metal halyards).

Antennas of Aluminum Tubing

Aluminum is a malleable, ductile metal with a mass density of 2.70 grams per cubic centimeter. The density of aluminum is approximately 35% that of iron and 30% that of copper. Aluminum can be polished to a high brightness, and it will retain this polish in dry air. In the presence of moisture, aluminum forms an oxide coating (Al_2O_3) that protects the metal from further corrosion. Direct contact with certain metals, however (especially ferrous metals such as iron or steel), in an outdoor environment can bring about galvanic corrosion of aluminum and its alloys. Some protective coating should be applied to any point of contact between two dissimilar metals. Much of this information about aluminum and aluminum tubing was prepared by Ralph Shaw, K5CAV.

Aluminum is non-toxic; it is used in cooking utensils and to hold and cover "TV dinners" and other frozen foods, so it is certainly safe to work with. The ease with which it can be drilled or sawed makes it a pleasure to work with. Aluminum products lend themselves to many and varied applications.

Aluminum alloys can be used to build amateur antennas, as well as for towers and supports. Light weight and high conductivity make aluminum ideal for these applications. Alloying lowers the conductivity ratings, but the tensile strength can be increased by alloying aluminum with one or more metals such as manganese, silicon, copper, magnesium or zinc. Cold rolling can be employed to further increase the strength.

A four-digit system is used to identify aluminum alloys, such as 6061. Aluminum alloys starting with a 6 contain di-magnesium silicide (Mg_2Si). The second digit indicates modifications of the original alloy or impurity limits. The last two digits designate different aluminum alloys within the category indicated by the first digit.

In the 6000 series, the 6061 alloy is commonly used for antenna applications. Type 6061 has good resistance to corrosion and has medium strength. A further designation like T-6 denotes thermal treatment (heat tempering). More information on the available aluminum alloys can be found in Table 4.

SELECTING ALUMINUM TUBING

Table 5 shows the standard sizes of aluminum tubing that are stocked by most aluminum suppliers or distributors in the United States and Canada. Note that all tubing comes in 12-foot lengths (local hardware stores sometimes stock 6- and 8-foot lengths). Note also that any diameter tubing will fit snugly into the next larger size, if the larger size has a 0.058-inch wall thickness. For example, 5/8-inch tubing has an outside diameter of 0.625 inch. This will fit into ¾-inch tubing with a 0.058-inch wall, which has an inside diameter of 0.634 inch. A clearance of 0.009 inch is just right for a slip fit or for slotting the tubing and then using hose clamps. Always get the next larger size and specify a 0.058-inch wall to obtain the 0.009-inch clearance.

A little figuring with Table 5 will give you all the information you need to build a beam, including what the antenna will weigh. The 6061-T6 type of aluminum has a relatively high strength and has good workability. It is highly resistant to corrosion and will bend without taking a "set."

SOURCES FOR ALUMINUM

Aluminum can be purchased new, and suppliers are listed later in this chapter. But don't overlook the local metal scrap yard. The price varies, but between 35 and 60 cents per pound is typical for scrap aluminum. Some aluminum items to look

Table 4

Aluminum Numbers for Amateur Use

Common Alloy Numbers

Type	Characteristic
2024	Good formability, high strength
5052	Excellent surface finish, excellent corrosion resistance, normally not heat treatable for high strength
6061	Good machinability, good weldability, can be brittle at high tempers
7075	Good formability, high strength

Common Tempers

Type	Characteristics
T0	Special soft condition
T3	Hard
T6	Very hard, possibly brittle
TXXX	Three digit tempers—usually specialized high strength heat treatments, similar to T6

General Uses

Type	Uses
2024-T3 7075-T3	Chassis boxes, antennas, anything that will be bent or flexed repeatedly
6061-T6	Mounting plates, welded assemblies or machined parts

Table 5
Aluminum Tubing Sizes

6061-T6 (61S-T6) Round Aluminum Tube In
12-Foot Lengths

Tubing Diameter	Wall Thickness Inches	Wall Thickness Stubs Ga.	ID, Inches	Approximate Weight Pounds Per Foot	Approximate Weight Pounds Per Length
3/16 in.	0.035	(No. 20)	0.117	0.019	0.228
	0.049	(No. 18)	0.089	0.025	0.330
1/4 in.	0.035	(No. 20)	0.180	0.027	0.324
	0.049	(No. 18)	0.152	0.036	0.432
	0.058	(No. 17)	0.134	0.041	0.492
5/16 in.	0.035	(No. 20)	0.242	0.036	0.432
	0.049	(No. 18)	0.214	0.047	0.564
	0.058	(No. 17)	0.196	0.055	0.660
3/8 in.	0.035	(No. 20)	0.305	0.043	0.516
	0.049	(No. 18)	0.277	0.060	0.720
	0.058	(No. 17)	0.259	0.068	0.816
	0.065	(No. 16)	0.245	0.074	0.888
7/16 in.	0.035	(No. 20)	0.367	0.051	0.612
	0.049	(No. 18)	0.339	0.070	0.840
	0.065	(No. 16)	0.307	0.089	1.068
1/2 in.	0.028	(No. 22)	0.444	0.049	0.588
	0.035	(No. 20)	0.430	0.059	0.708
	0.049	(No. 18)	0.402	0.082	0.984
	0.058	(No. 17)	0.384	0.095	1.040
	0.065	(No. 16)	0.370	0.107	1.284
5/8 in.	0.028	(No. 22)	0.569	0.061	0.732
	0.035	(No. 20)	0.555	0.075	0.900
	0.049	(No. 18)	0.527	0.106	1.272
	0.058	(No. 17)	0.509	0.121	1.452
	0.065	(No. 16)	0.495	0.137	1.644
3/4 in.	0.035	(No. 20)	0.680	0.091	1.092
	0.049	(No. 18)	0.652	0.125	1.500
	0.058	(No. 17)	0.634	0.148	1.776
	0.065	(No. 16)	0.620	0.160	1.920
	0.083	(No. 14)	0.584	0.204	2.448
7/8 in.	0.035	(No. 20)	0.805	0.108	1.308
	0.049	(No. 18)	0.777	0.151	1.810
	0.058	(No. 17)	0.759	0.175	2.100
	0.065	(No. 16)	0.745	0.199	2.399
1 in.	0.035	(No. 20)	0.930	0.123	1.476
	0.049	(No. 18)	0.902	0.170	2.040
	0.058	(No. 17)	0.884	0.202	2.424
	0.065	(No. 16)	0.870	0.220	2.640
	0.083	(No. 14)	0.834	0.281	3.372
1-1/8 in.	0.035	(No. 20)	1.055	0.139	1.668
	0.058	(No. 17)	1.009	0.228	2.736
1-1/4 in.	0.035	(No. 20)	1.180	0.155	1.860
	0.049	(No. 18)	1.152	0.210	2.520
	0.058	(No. 17)	1.134	0.256	3.072
	0.065	(No. 16)	1.120	0.284	3.408
	0.083	(No. 14)	1.084	0.357	4.284
1-3/8 in.	0.035	(No. 20)	1.305	0.173	2.076
	0.058	(No. 17)	1.259	0.282	3.384
1-1/2 in.	0.035	(No. 20)	1.430	0.180	2.160
	0.049	(No. 18)	1.402	0.260	3.120
	0.058	(No. 17)	1.384	0.309	3.708
	0.065	(No. 16)	1.370	0.344	4.128
	0.083	(No. 14)	1.334	0.434	5.208
	*0.125	1/8 in.	1.250	0.630	7.416
	*0.250	1/4 in.	1.000	1.150	14.832
1-5/8 in.	0.035	(No. 20)	1.555	0.206	2.472
	0.058	(No. 17)	1.509	0.336	4.032
1-3/4 in.	0.058	(No. 17)	1.634	0.363	4.356
	0.083	(No. 14)	1.584	0.510	6.120
1-7/8 in.	0.058	(No. 17)	1.759	0.389	4.668
2 in.	0.049	(No. 18)	1.902	0.350	4.200
	0.065	(No. 16)	1.870	0.450	5.400
	0.083	(No. 14)	1.834	0.590	7.080
	*0.125	1/8 in.	1.750	0.870	9.960
	*0.250	1/4 in.	1.500	1.620	19.920
2-1/4 in.	0.049	(No. 18)	2.152	0.398	4.776
	0.065	(No. 16)	2.120	0.520	6.240
	0.083	(No. 14)	2.084	0.660	7.920
2-1/2 in.	0.065	(No. 16)	2.370	0.587	7.044
	0.083	(No. 14)	2.334	0.740	8.880
	*0.125	1/8 in.	2.250	1.100	12.720
	*0.250	1/4 in.	2.000	2.080	25.440
3 in.	0.065	(No. 16)	2.870	0.710	8.520
	*0.125	1/8 in.	2.700	1.330	15.600
	*0.250	1/4 in.	2.500	2.540	31.200

*These sizes are extruded. All other sizes are drawn tubes.

for include aluminum vaulting poles, tent poles, tubing and fittings from scrapped citizen's band antennas, and aluminum angle stock. The scrap yard may even have a section or two of triangular aluminum tower.

Aluminum vaulting poles are 12 or 14 feet long and range in diameter from 1½ to 1¾ inches. These poles are suitable for the center-element sections of large 14-MHz beams or as booms for smaller antennas. Tent poles range in length from 2½ to 4 feet. The tent poles are usually tapered; they can be split on the larger end and then mated with the smaller end of another pole of the same diameter. A small stainless steel hose clamp (sometimes also available at scrap yards!) can be used to fasten the poles at this junction. A 14- or 21-MHz element can be constructed from several tent poles in this fashion. If a longer continuous piece of tubing is available, it can be used for the center section to decrease the number of junctions and clamps.

Other aluminum scrap is sometimes available, such as US Army aluminum mast sections designated AB-85/GRA-4 (J&H Smith Mfg). These are 3 foot sections with a 1-5/8 inch diameter. The ends are swaged so they can be assembled one into another. These are ideal for making a portable mast for a 144-MHz beam or for field day applications.

CONSTRUCTION WITH ALUMINUM TUBING

Most antennas built for frequencies of 14 MHz and above are made to be rotated. Constructing a rotatable antenna requires materials that are strong, lightweight and easy to obtain. The materials required to build a suitable antenna will vary, depending on many factors. Perhaps the most important factor that determines the type of hardware needed is the weather conditions normally encountered. High winds usually don't cause as much damage to an antenna as does ice, especially ice along with high winds. Aluminum element and boom sizes should be selected so the various sections of tubing will telescope to provide the necessary total length.

The boom size for a rotatable Yagi or quad should be selected to provide stability to the entire system. The best diameter for the boom depends on several factors; most important are the element weight, number of elements and overall length. Tubing diameters of 1¼ inches can easily support three-element 28-MHz arrays and perhaps a two-element 21-MHz system. A 2-inch diameter boom will be adequate for larger 28-MHz antennas or for harsh weather conditions, and for antennas up to three elements on 14 MHz or four elements on 21 MHz. It is not recommended that 2-inch diameter booms be made any longer than 24 feet unless additional support is given to reduce both vertical and horizontal bending forces. Suitable reinforcement for a long 2-inch boom can consist of a truss or a truss and lateral support, as shown in Fig 8.

A boom length of 24 feet is about the point where a 3-inch diameter begins to be very worthwhile. This dimension provides a considerable improvement in overall mechanical stability as well as increased clamping surface area for element hardware. Clamping surface area is extremely important if heavy icing is common and rotation of elements around the boom is to be avoided. Pinning an element to the boom with a large bolt helps in this regard. On smaller diameter booms, however, the elements sometimes work loose and tend to elongate the pinning holes in both the element and the boom. After some time the elements shift their positions slightly (sometimes from day to day!) and give a rather ragged appearance to the system, even though this doesn't generally harm the electrical performance.

A 3-inch diameter boom·with a wall thickness of 0.065 inch is satisfactory for antennas up to about a five-element, 14-MHz array that is spaced on a 40 foot long boom. A truss is recommended for any boom longer than 24 feet.

There is no RF voltage at the center of a parasitic element, so no insulation is required in mounting elements that are centered on the boom (driven elements excepted). This is true whether the boom is metal or a nonconducting material. Metal booms have a small "shortening effect" on elements that run through them. With materials sizes commonly employed, this is not more than one percent of the element length, and may not be noticeable in many applications. It is just perceptible with ½-inch tubing booms used on

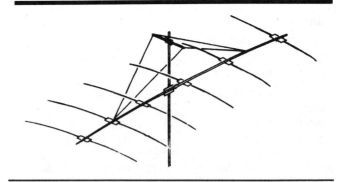

Fig 8—A long boom needs both vertical and horizontal support. The cross bar mounted above the boom can support a double truss to help keep the antenna in position.

432 MHz, for example. Design-formula lengths can be used as given, if the matching is adjusted in the frequency range one expects to use. The center frequency of an all-metal array will tend to be 0.5 to 1 percent higher than a similar system built of wooden supporting members.

Element Assembly

While the maximum safe length of an antenna element depends to some extent on its diameter, the only laws that specify the minimum diameter of an element are the laws of nature. That is, the element must be rugged enough to survive whatever weather conditions it will encounter.

Fig 9 shows designs for tapered Yagi elements that should survive all but the most extreme weather conditions. Designs that are more rugged and will therefore withstand more ice loading are shown in Fig 10. (These designs also have more wind area.) No guarantees are made as to their ability to handle a heavy load of ice in high winds, but if you lose an antenna made with elements like these you'll have plenty of company among your neighbors with commercially made antennas!

Fig 11 shows several methods of fastening antenna element sections together. The slot and hose clamp method shown in Fig 11A is probably the best for joints where adjustments are required. Generally, one adjustable joint per

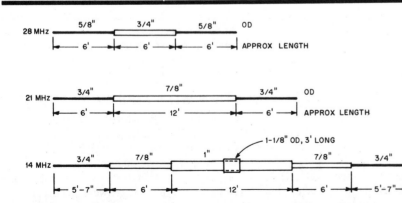

Fig 9—Element designs for Yagi antennas. (Use 0.058-in.-wall aluminum tubing except for the end pieces, where thinner-wall tubing may be used.)

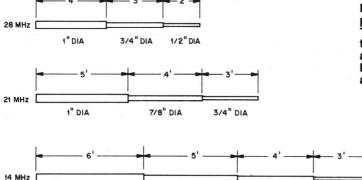

Fig 10—A more rugged schedule of taper proportions for Yagi half-elements than Fig 9. The other side of the element is identical, and the center section can be a single piece twice as long as the length shown here for the largest diameter section. See Table 5 for aluminum tubing details.

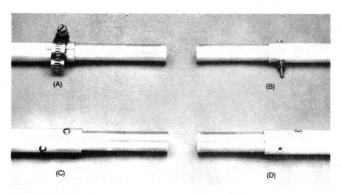

Fig 11—Methods of connecting telescoping tubing sections to build beam elements. See text for a discussion of each method.

Table 6
Hose-Clamp Diameters

| | Clamp Diameter (In.) | |
Size No.	Min	Max
06	7/16	7/8
08	7/16	1
10	1/2	1-1/8
12	5/8	1-1/4
16	3/4	1-1/2
20	7/8	1-3/4
24	1-1/8	2
28	1-3/8	2-1/4
32	1-5/8	2-1/2
36	1-7/8	2-3/4
40	2-1/8	3
44	2-5/16	3-1/4
48	2-5/8	3-1/2
52	2-7/8	3-3/4
56	3-1/8	4
64	3-1/2	4-1/2
72	4	5
80	4-1/2	5-1/2
88	5-1/8	6
96	5-5/8	6-1/2
104	6-1/8	7

element half is sufficient to tune the antenna. Stainless-steel hose clamps (beware—some "stainless steel" models do not have a stainless screw and will rust) are recommended for longest antenna life. Table 6 shows available hose-clamp sizes.

Figs 11B, 11C and 11D show possible fastening methods for joints that do not require adjustment. At B, machine screws and nuts hold the elements in place. At C, sheet metal screws are used. At D, rivets secure the tubing. If the antenna is to be assembled permanently, rivets are the best choice. Once in place, they are permanent. They will never work free, regardless of vibration or wind. If aluminum rivets with aluminum mandrels are used, they will never rust. In addition, there is no danger of dissimilar-metal corrosion with aluminum rivets and aluminum antenna elements. If the antenna is to be disassembled and moved periodically, either B or C will work. If machine screws are used, however, take all possible precautions to keep the nuts from vibrating free. Use lock washers, lock nuts and flexible sealant such as silicone bathtub sealant to keep the hardware in place.

Very strong elements can be made by using a double thickness of tubing, made by telescoping one size inside another for the total length. This is usually done at the center of an element where more element strength is desired at the boom support point. Other materials can be used as well, such as wood dowels, fiberglass rods, and so forth.

In each case where a smaller diameter length of tubing is telescoped inside a larger diameter one, it's a good idea to coat the inside of the joint with Penetrox or a similar substance to ensure a good electrical bond. Antenna elements have a tendency to vibrate when they are mounted on a tower, and one way to dampen the vibrations is by running a piece of clothesline rope through the length of the element. Cap or tape the end of the element to secure the clothesline. If mechanical requirements dictate (a U-bolt going through the center of the element, for instance), the clothesline may be cut into two pieces.

Antennas for 50 MHz need not have elements larger than ½-inch diameter, although up to 1 inch is used occasionally.

At 144 and 220 MHz the elements are usually 1/8 to 1/4 inch in diameter. For 420 MHz, elements as small as 1/16 inch diameter work well, if made of stiff rod. Aluminum welding rod of 3/32 to 1/8 inch diameter is fine for 420-MHz arrays, and 1/8 inch or larger is good for the 220-MHz band. Aluminum rod or hard-drawn wire works well at 144 MHz.

Tubing sizes recommended in the paragraph above are usable with most formula dimensions for VHF/UHF antennas. Larger diameters broaden the frequency response; smaller ones sharpen it. Much smaller diameters than those recommended will require longer elements, especially in 50-MHz arrays.

Element Taper and Electrical Length

The builder should be aware of one important aspect of telescoping or tapered elements. When the element diameters are tapered, as shown in Figs 9 and 10, the electrical length is not the same as it would be for a cylindrical element of the same total length. Length corrections for tapered elements are discussed in Chapter 2.

Other Materials for Antenna Construction

Wood is very useful in antenna work. It is available in a great variety of shapes and sizes. Rug poles of wood or bamboo make fine booms. Bamboo is quite satisfactory for spreaders in quad antennas.

Round wood stock (doweling) is found in many hardware stores in sizes suitable for small arrays. Wood is good for the framework of multibay arrays for the higher bands, as it keeps down the amount of metal in the active area of the array. Square or rectangular boom and frame materials can be cut to order in most lumber yards if they are not available from the racks in suitable sizes.

Wood used for antenna construction should be well seasoned and free of knots or damage. Available materials vary, depending on local sources. Your lumber dealer can help you better than anyone else in choosing suitable materials. Joining wood members at right angles can be done with gusset plates, as shown in Fig 12. These can be made of thin outdoor-grade plywood or Masonite. Round materials can be handled in ways similar to those used with metal components, with U clamps and with other hardware.

In the early days of Amateur Radio, hardwood was used as insulating material for antennas, such as at the center and ends of dipoles, or for the center insulator of a driven element made of tubing. Wood dowels cut to length were the most common source. To drive out moisture and prevent the subsequent absorption of moisture into the wood, it was treated before use by boiling it in paraffin. Of course today's technology has produced superior materials for insulators in terms of both strength and insulating qualities. However, the technique is worth consideration in an emergency situation or if low cost is a prime requirement. "Baking" the wood in an oven for a short period at 200° F should drive out any moisture. Then treatment as described in the next paragraph should prevent moisture absorption. The use of wood insulators should be avoided at high-voltage points if high power is being used.

All wood used in outdoor installations should be protected from the weather with varnish or paint. A good grade of marine spar varnish or polyurethane varnish will offer protection for years in mild climates, and one or more seasons in harsh climates. Epoxy-based paints also offer good protection.

Plastics

Plastic tubing and rods of various sizes are available from many building-supplies stores. The uses for the available plastic materials are limited only by your imagination. Some amateurs have built beam antennas for VHF using wire elements run inside thin PVC plumbing pipe. The pipe gives the elements a certain amount of physical strength. Other hams have built temporary antennas by wrapping plastic pipe with aluminum foil or other conductive material. Plastic plumbing pipe fittings can also be used to enclose baluns and as the center insulator or end insulators of a dipole, as shown in Fig 13. Plastic or Teflon rod can be used as the core of a loading coil for a mobile antenna (Fig 14) but the material for this use should be selected carefully. Some plastics become quite warm in the presence of a strong RF field, and the loading-coil core might melt or catch fire!

Fiberglass

Fiberglass poles are the preferred material for spreaders for quad antennas. They are lightweight, they withstand harsh

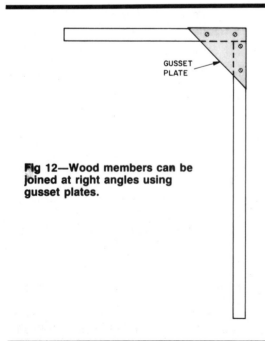

Fig 12—Wood members can be joined at right angles using gusset plates.

GUSSET PLATE

Fig 13—Plastic plumbing parts can be used as antenna center and end insulators.

Fig 14—A mobile-antenna loading coil wound on a polystyrene rod.

weather well, and their insulating qualities are excellent. One disadvantage of fiberglass poles is that they may be crushed rather easily. Fracturing occurs at the point where the pole is crushed, causing it to lose its strength. A crushed pole is next to worthless. Some amateurs have repaired crushed poles with fiberglass cloth and epoxy, but the original strength is nearly impossible to regain.

Fiberglass poles can also be used to construct other types of antennas. Examples are helically wound Yagi elements or verticals, where a wire is wound around the pole.

CONCLUSION

The antenna should be put together with good quality hardware. Stainless steel is best for long life. Rust will quickly attack plated steel hardware, making nuts difficult, if not impossible, to remove. If stainless steel muffler clamps and hose clamps are not available, the next best thing is to have them plated. If you can't have them plated, at least paint them with a good zinc-chromate primer and a finish coat or two.

Galvanized steel generally has a longer life than plated steel, but this depends on the thickness of the galvanizing coat. Even so, in harsh climates rust will usually develop on galvanized fittings in a few years. For the ultimate in long-term protection, galvanized steel should be further protected with zinc-chromate primer and then paint or enamel before exposing it to the weather.

Good quality hardware is expensive initially, but if you do it right the first time, you won't have to take the antenna down in a few years and replace the hardware. When the time does come to repair or modify the antenna, nothing is more frustrating than fighting rusty hardware at the top of the tower.

Basically any conductive material can be used as the radiating element of an antenna. Almost any insulating material can be used as an antenna insulator. The materials used for antenna construction are limited mainly by physical considerations (required strength and resistance to outdoor exposure) and by the availability of materials. Don't be afraid to experiment with radiating materials and insulators.

Antenna Manufacturers Product List

Parts procurement is often the most difficult aspect of an antenna project. Suppliers of aluminum exist in most major metropolitan areas, and can be found in the Yellow Pages section of the phone book. Some careful searching of the Yellow Pages may also reveal sources of other materials and accessories. If you live away from a metropolitan area, these telephone books for the nearest large metropolitan area may

be available in the reference section of your local library.

Many dealers and distributors will ship their products by freight or by mail. The listings of Tables 7 and 8 have been compiled by Domenic Mallozzi, N1DM, to allow you to quickly locate the names of manufacturers of antenna products. All attempts have been made to make this list complete and accurate. The American Radio Relay League takes no responsibility for errors or omissions. Similarly, a listing here does not represent an endorsement of a manufacturer or his products by the ARRL. We recommend that you refer to the product reviews in *QST* for the particular products of interest to you. (See Chapter 29.)

Because manufacturers' product lines change periodically, we recommend that you obtain a catalog from the manufacturers that interest you before contacting a distributor. In addition, all indications of sales policies and prices for catalogs are given for general information only, and are subject to change without notice. In cases where overseas manufacturers do not have US offices, we have given the name of a known importing agent in square brackets, [agent].

Antenna products for repeaters, remotely controlled stations and wide-coverage digipeaters are included in a separate listing of products at the end of the chapter on repeater antenna systems.

Table 7
Antenna Manufacturers' Product List

Addresses of suppliers are contained in Table 8. To the best of our knowledge the suppliers listed are willing to sell products to amateurs by mail unless indicated in Table 8. A listing here does not necessarily indicate endorsement by the ARRL. This listing will be updated with each edition of *The Antenna Book*. Suppliers wishing to be listed are urged to contact the editors.

Key to symbols used:

N = No direct sales, sells through distributors only
D = Direct sales only
K = Product available in kit form only
    (most beam antennas require some assembly but are not listed as kits)
U = Sells used parts/equipment
[name] = importing agent

ACTIVE ANTENNAS (RECEIVING)
(Unless otherwise specified the antennas shown are for HF receiving)
    Ameco
    Datong [Gilfer]
    Dressler (HF and VHF/UHF versions available) [Gilfer]
    Grove
    Hamtronics (VHF/UHF version only) (K)
    Heath (K)
    MFJ
    Palomar
    Stoner

ANTENNA TUNERS AND MATCHING SYSTEMS—HF
    Ameritron
    AMP Supply
    Barker & Williamson
    Daiwa
    Heath (K)
    ICOM (mobile & fixed)
    MFJ
    Nye
    Texas Radio (mobile only)
    THL [Encomm]

ANTENNA TUNERS AND MATCHING SYSTEMS—VHF
    Barker & Williamson (144-MHz mobile, low power only)

BALUNS
    Antennas Etc
    Barker & Williamson
    Caddell
    Cushcraft
    Heath (K)
    MFJ
    Mirage
    Moeller
    Palomar
    Radiokit (K)
    Telex/Hy-Gain
    Telrex
    The Radio Works
    Van Gorden

CLIMBING AND SAFETY EQUIPMENT
    Avatar
    UPI

COLLINEAR ARRAYS—144, 220 and 432 MHz
    Cushcraft

COMBINERS, POWER DIVIDERS AND PHASING HARNESSES
(These devices are usually made by a manufacturer for use when stacking his antenna in pairs or quads)
    Byers (Not specific to particular antennas, kits only for 144-1250 MHz)
    Cushcraft
    Down East
    Mirage
    Spectrum International
    Tonna [PX Shack]

CONNECTORS—RF
(Major manufacturers of connectors will not sell direct to amateurs. Almost all ham distributors sell connectors. The companies listed below specialize in selling RF connectors and transmission lines.)
    Nemal
    RF Industries
    The RF Connection

DDRR ANTENNAS
    Kilo-Tec (144-MHz mobile)

DELTA LOOP BEAMS
    Delta Loop (14, 21, 24 & 28-MHz single-band versions)

DIPOLES, INVERTED V DIPOLES AND SLOPERS—HF
    Alpha Delta
    Antennas, Etc
    Barker & Williamson
    Mosley
    Nye
    Spi-Ro (50 MHz also available)
    Telrex
    The Radio Works
    W6TIK
    W9INN

DIRECTIONAL COUPLERS (COAXIAL AND WAVEGUIDE)
    Lectronic (U)

FERRITE CORES AND RODS
    Amidon
    Palomar
    Radiokit

FILTERS—TVI (LOW PASS AND HIGH PASS)
    Ameco
    Antennas Etc

HARDWARE
    Antennas Etc
    DC Sales (trailer hitch antenna mount)
    Electro Sonic
    Elwick
    Kilo-Tec (weatherboots for connectors)
    Moeller (quad spreaders and hubs)
    N'Tenna
    Spi-Ro
    The Radio Works
    United Ropeworks (nonconductive guy cable)
HELICAL ANTENNAS
    G3RUH (432 and 1250 MHz)
    Sommer [Theiler]
HAND-HELD TRANSCEIVER ANTENNAS (RUBBER
DUCKS)
    AEA
    Antenna Specialists
    Cushcraft
    Diamond [Encomm]
    Hustler
    ICOM
    The Radio Works
HAND-HELD TRANSCEIVER ANTENNAS—EXPANDABLE
(TELESCOPIC)
    AEA
    Austin
    Cushcraft
    ICOM
    Orion
    The Radio Works
INSULATORS
    Antennas Etc
    Barker & Williamson
    Kilo-Tec
    Radiokit
    Spi-Ro
    Telex/Hy-Gain
    The Radio Works
    W1JC
LIGHTNING ARRESTERS
    Alpha Delta
    Cushcraft
    RF Industries
    Telex/Hy-Gain
LOG PERIODIC ANTENNAS
    Creative Design [Orion]
    Mirage
    Telex/Hy-Gain
LOOP YAGIS—VHF/UHF
    Down East (902 & 1250 MHz, 2.4 & 3.4 GHz)
    Spectrum International (902 and 1250 MHz)
LOOPS (HF FULL SIZE)
    W6TIK
MASTS (OTHER THAN TOWERS)
    Radio Shack
MICROWAVE ANTENNAS
    Down East
    Mirage
MOBILE ANTENNAS—HF
    Anteck
    Hustler
    ICOM
    Mobile
    Mosley
    Ten Tec
    Texas Radio
    The Radio Works
    Valor

MOBILE ANTENNAS—VHF (BELOW 200 MHz)
    Antenna Specialists
    Austin (both vertical and horizontal polarized available)
    Cushcraft
    Diamond [Encomm]
    Hustler
    ICOM
    Kilo-Tec (144-MHz DDRR)
    Mission (horizontal polarized)
    Mobile
    Mosley
    NCG
    Telex/Hy-Gain
    Texas Radio Products
    The Radio Works
    Valor
MOBILE ANTENNAS—UHF
    Antenna Specialists
    Austin
    Cushcraft
    Diamond [Encomm]
    Hustler
    ICOM
    Mobile
    NCG
    The Radio Works
    Valor
MULTIBAND DIPOLES—HF
    Alpha Delta
    Barker & Williamson
    Kilo-Tec
    Maxcom
    Mosley
    Spi-Ro
    Telex/Hy-Gain
    Telrex
    Texas Radio
    The Radio Works
    W6TIK
    W9INN
QUADS—7 MHz
    Moeller
QUADS—14 MHz
    Moeller
    N'Tenna
QUADS—14/21/28 MHz
    Cubex
    Gem Quad (144-MHz option available)
    Moeller
    N'Tenna
QUADS—21 MHz
    Moeller
    N'Tenna (including 28 MHz)
QUADS—28 MHz
    Moeller
    N'Tenna
QUADS—50 MHz
    Moeller
QUADS—144 MHz
    Antenna Company of America
    Moeller
QUADS—220 MHz
    Antenna Company of America
QUADS—432 MHz
    Antenna Company of America

REDUCED-SIZE DIPOLES—HF (INCLUDING MULTIBAND)
Alpha Delta
AMP Supply
Barker & Williamson
Creative Design [Orion]
Kilo-Tec
Mirage (rotatable)
Mor-Gain
Mosley
Spi-Ro
Telex/Hy-Gain
Texas Radio
The Radio Works
W1JC
W6TIK
W9INN

ROTATORS
Craig's Antenna & Tower Service (rotator repair)
Creative Design [Orion]
Daiwa
KenPro (including az-el and computer control) [Encomm]
Mirage (elevation rotators)
ProSearch
Radio Shack
Silicon Solutions (satellite tracking software only)
Telex/Hy-Gain
Telrex

SATELLITE ANTENNAS (ALL BANDS)
Austin (144 and 432 MHz)
Cushcraft (144 and 432 MHz)
Down East (1250 MHz)
G3RUH (432 and 1250 MHz)
Mirage (144, 432 and 1250 MHz)
Sommer [Theiler] (432 MHz)
Spectrum International (144, 432 and 1250 MHz)
Telex/Hy-Gain (144 and 432 MHz)

SMALL HF BEAMS
Butternut
Creative Design [Orion]
Moeller
Stewart

SMALL HF TRANSMITTING ANTENNAS (OTHER THAN BEAMS)
Barker & Williamson
Bilal Company
Kilo-Tec
MFJ
Mor-Gain
Mosley
Stewart
W1JC
W6TIK

SPREADERS, FIBERGLASS
Advanced Composites (12 & 20 ft lengths)
dB + (13 ft lengths)
Dynaflex (3-section telescoping poles, 15 ft total length)
Sky-Pole (vaulting poles, various length tubing)
Moeller/Kirk (9 & 13 ft lengths, tapered)

STACKING FRAMES
(Unless otherwise noted these frames are for use in stacking the manufacturer's own antennas in pairs or quads. These stacking kits are for VHF or UHF antennas only)
Cushcraft
Down East
Mirage
Mosley
Spectrum International

SWITCHES (MANUAL COAX)
Alpha Delta
Barker & Williamson
Daiwa
Heath (K)
MFJ

SWITCHES (REMOTE COAX)
Ameritron
Antennas Etc
Heath (K)

SWR & WATTMETERS (HF)
Bird
Coaxial Dynamics
Daiwa
Dielectric Communications
Heath (K)
MFJ
Mirage
Nye (including audible version for the visually impaired)
Palomar
Texas Radio
The Radio Works
Welz [Encomm]

SWR & WATTMETERS (VHF AND ABOVE)
Bird
Coaxial Dynamics
Daiwa
Dielectric Communications
MFJ
Mirage
Nye
Texas Radio
Welz [Encomm]

TOWERS
Aluma
Creative Design [Orion] (rooftop towers only)
Delhi [Electro Sonic]
Martin
Rohn
Telex/Hy-Gain
Triangle International
Trylon [BJX]
US Tower

TRANSMISSION LINES—COAXIAL
(Major manufacturers of cable will not sell direct to amateurs. Almost all ham distributors sell coax cables. The companies listed below specialize in selling RF connectors and transmission lines.)
Nemal
The RF Connection
The Radio Works

TRANSMISSION LINES—HARDLINE
Major manufacturers of cable will not sell direct to amateurs. Almost all ham distributors sell coax cables. The companies listed below specialize in selling RF connectors and transmission lines.)
AGW
Nemal

TRANSMISSION LINES—TRANSMITTING TWIN-LEAD (LADDER LINE)
(Major manufacturers of cable will not sell direct to amateurs. Almost all ham distributors sell coax cables. The companies listed below specialize in selling RF connectors and transmission lines.)
Radiokit
The RF Connection
The Radio Works
W1JC
W9INN

TRAPS
Antennas Etc
Barker & Williamson
Spi-Ro

TUBING, ALUMINUM
Metal & Cable Corporation, Inc.
(Also check the Yellow Pages)

VERTICALS—HF
- Barker & Williamson
- Butternut
- Creative Design [Orion]
- Cushcraft
- Hustler
- Mirage
- Mosley
- Telex/Hy-Gain

VERTICALS—VHF
- AEA
- Antenna Specialists
- Austin
- Butternut
- Cushcraft
- Diamond [Encomm]
- Hustler
- Kilo-Tec (144 MHz)
- Mirage
- Mobile
- Mosley
- NCG
- Telex/Hy-Gain

VERTICALS—UHF
- AEA
- Antenna Specialists
- Austin
- Butternut
- Cushcraft
- Diamond [Encomm]
- Hustler
- Mirage
- NCG
- Telex/Hy-Gain

VERTICALS—WIDEBAND VHF & UHF
- Austin
- Diamond [Encomm]
- Heath (K)
- ICOM

WIRE
- Kilo-Tec
- Spi-Ro
- The Radio Works
- W1JC
- W9INN

YAGIS—14/21/28 MHz
- Creative Design [Orion]
- Cushcraft
- Mirage
- Mosley
- Sommer [Theiler]
- Telex/Hy-Gain

YAGIS—14/18/21/24/28 MHz
- Cushcraft
- Mosley
- Sommer [Theiler]

YAGIS—3.5 MHz
(Rotatable 3.5-MHz Yagis are all physically smaller than the full electrical length with maximum element lengths of 90 to 100 feet)
- Creative Design [Orion]
- Mirage

YAGIS—7 MHz
- Creative Design [Orion]
- Cushcraft
- Mirage
- Mosley
- Sommer [Theiler] (part of multiband structure)
- Telex/Hy-Gain
- Telrex

YAGIS—10 MHz
- Creative Design [Orion]
- Telex/Hy-Gain

YAGIS—14, 21 and 28 MHz
- Creative Design [Orion]
- Cushcraft
- Mirage
- Mosley
- Telex/Hy-Gain
- Telrex

YAGIS—50 MHz
- Creative Design [Orion]
- Cushcraft
- Hustler
- Mirage
- Mosley
- Telex/Hy-Gain
- Telrex
- Tonna [PX Shack]

YAGIS—144 MHz
- Austin
- Creative Design [Orion]
- Cushcraft
- Hustler
- Mirage
- Mosley
- Sommer [Theiler]
- Spectrum International
- Telex/Hy-Gain
- Telrex
- Tonna [PX Shack]

YAGIS—220 MHz
- Austin
- Cushcraft
- Mirage
- Mosley
- Telrex

YAGIS—432 MHz
- Austin
- Creative Design [Orion]
- Cushcraft
- Mirage
- Mosley
- Sommer [Theiler]
- Spectrum International
- Telrex
- Tonna [PX Shack]

YAGIS—902 MHz
- Austin
- Tonna [PX Shack]

YAGIS—1250 MHz
- Mirage
- Tonna [PX Shack]

Table 8
Addresses of Suppliers

Key to symbols used:
N = No direct sales, sells through distributors only
D = Direct sales only
K = Product available in kit form only (most beam antennas require some assembly but are not listed as kits)
U = Sells used parts/equipment

Advanced Composites
PO Box 15323
Salt Lake City, UT 84115

AEA - Advanced Electronic Applications, Inc
PO Box C-2160
Lynnwood, WA 98036-0918

AGW Enterprises, Inc (D)
RD #10, Route 206
Vincentown, NJ 08088

Alpha Delta Communications
PO Box 571
Centerville, OH 45459

Aluma Tower Company
1639 Old Dixie Highway, Box 2806
Vero Beach, FL 32961-2806

Ameco Equipment Company
220 East Jericho Turnpike
Mineola, NY 11501

Ameritron Division
Prime Instruments Inc
9805 Walford Avenue
Cleveland, OH 44102

Amidon Associates
12033 Otsego Street
North Hollywood, CA 91607

AMP Supply Company
208 Snow Avenue
PO Box 147
Raleigh, NC 27602

Anteck, Inc
Box 415, Route 1
Hansen, ID 83334

Antenna Company of America (D)
PO Box 2308
Santa Clara, CA 95051

Antennas Etc (Unadilla/Reyco/Inline Division)
PO Box 215BV
Andover, MA 01810-0814

The Antenna Specialists Company (N)
12345 Euclid Avenue
Cleveland, OH 44106-4386

Austin Custom Antennas
PO Box 357
Tenney Road
Sandown, NH 03873

Avatar Magnetics (D)
Ronald C. Williams, W9JVF
1408 W Edgewood Ave
Indianapolis, IN 46217-3618

Barker & Williamson Company
10 Canal Street
Bristol, PA 19007

Bilal Company
Eucha, OK 74342

Bird Electronic Corporation
30303 Aurora Road
Solon, OH 44139

BJX Supply Company
PO Box 388
Corfu, NY 14036

Butternut Electronics (N)
405 East Market Street
Lockhart, TX 78644

Byers Chassis Kit
Charles Byers, K3IWK
5120 Harmony Grove Road
Dover, PA 17315

Caddell Coil Corporation
35 Main Street
Poultney, VT 05764

Craig's Antenna & Tower Service
7368 SR 105
Pemberville, OH 43450

Coaxial Dynamics Inc
15210 Industrial Parkway
Cleveland, OH 44135

Cubex Corporation
PO Box 732
Altadena, CA 91001

Cushcraft Corporation
48 Perimeter Road
PO Box 4680
Manchester, NH 03108

Daiwa Electronics Corporation
1980A Del Amo Boulevard
Torrance, CA 90501

Datong Electronics Ltd
Spence Mills
Mill Lane
Bramley, Leeds LS13 3HE
England (UK)

dB + Enterprises
PO Box 24
Pine Valley, NY 14872

DC Sales
1602 Chestnut Ridge Road
Kingwood, TX 77339

Delta Loop Antennas (D)
44 Old State Road, Unit 1-B
New Milford, CT 06776

Dielectric Communications
Raymond, ME 04071

Down East Microwave
Box 2310, RR 1
Troy, ME 04987

Dressler Hochfrequenztechnik GMBH
Wuerselener 73-75
D-5190 Stolberg
Federal Republic of Germany

Dynaflex Manufacturing Corp
Route 14, Box 370
Tallahassee, FL 32304

Electro Sonic Inc (D)
1100 Gordon Baker Road
Willowdale, Ontario, Canada M2H 3B3
Catalog $25.00, minimum order $25.00

Elwick Supply Company (D)
230 Woods Lane
Somerdale, NJ 08083

ENCOMM, Inc (N)
1506 Capital
Plano, TX 75074

G3RUH, J. R. Miller (D)
3 Benny's Way
Coton, Cambridge CV3 7PS
England (UK)

Gem Quad Products (1987) Ltd (D)
PO Box 291
Boissevain, Manitoba R0K 0E0
Canada

Gilfer Associates, Inc (D)
52 Park Avenue
PO Box 239
Park Ridge, NJ 07656

Grove Enterprises
PO Box 98
Brasstown, NC 28902

Hamtronics, Inc (D)
65 Moul Road
Hilton, NY 14468

Heath Company (D or through local stores)
Benton Harbor, MI 49022

Hustler/Newtronics Antenna Corporation (N)
One Newtronics Place
Mineral Wells, TX 76067

ICOM America Inc (N)
2380 116th Avenue NE
Bellevue, WA 98004

ICOM Canada (N)
3071-#5 Road, Unit 9
Richmond, BC V6X 2T4
Canada

Kilo-Tec (D)
PO Box 1001
Oakview, CA 93022

Lectronic Research Labs (D)
Atlantic & Ferry Avenues
Camden, NJ 08104

Glen Martin Engineering
PO Box 7253
Boonville, MO 65233

Metal & Cable Corporation, Inc (D)
2170 East Aurora Road
PO Box 117
Twinsburg, OH 44087
Minimum order $50.00

MFJ Enterprises, Inc
Box 494
Mississippi State, MS 39762

Mirage Communications Equipment, Inc
PO Box 1000
Morgan Hill, CA 95037

Mission Communications
11903 Alief Clodine Road #500
Houston, TX 77082

Mobile Antennas and Accessories
Division of AC & DC Electronics
Route 1, Box 406-C, Pond Road
Burlington, NC 27215

Moeller Instrument Company, Inc
Kirk Electronics Division
Main Street
Ivoryton, CT 06442
Minimum order $25.00

Mor-Gain
2200 South 4th Street
PO Box 329
Leavenworth, KS 66048

Mosley Electronics, Inc
1344 Baur Boulevard
Saint Louis, MO 63132

NCG Companies
1275 North Grove Street
Anaheim, CA 92806

N'Tenna Quad Kits (D)
Box 5332 Viewpoint
Hickory, NC 28603

Nemal Electronics International, Inc (D)
12240 NE 14th Avenue
North Miami, FL 33161
Catalog $4.00

William M. Nye Company
1614 130th Avenue NE
Bellevue, WA 98005

Orion Hi-Tech (N)
PO Box 8771
Calabasas, CA 91302-8771

Palomar Engineers
PO Box 455
Escondido, CA 92025

Pro-Search Electronics Company, Inc
1344 Baur Boulevard
Saint Louis, MO 63132

The PX Shack
53 Stonywyck Drive
Belle Mead, NJ 08502

Radiokit (D)
PO Box 973
Pelham, NH 03076

Radio Shack
(Contact your local store)

RF Industries
690 West 28th Street
Hialeah, FL 33010-1293

Rohn
Division of Unarco
6718 West Plank Road
PO Box 2000
Peoria, IL 61656

Silicon Solutions, Inc (D)
PO Box 742546
Houston, TX 77274-2546

Sky-Pole Manufacturing, Inc
1922 Placentia
Costa Mesa, CA 92627

Sommer GMBH
Kandelstrasse 35
D-7819 Denzlingen
Federal Republic of Germany

Spectrum International, Inc (D)
PO Box 1084
Concord, MA 01742
Catalog 3 first class stamps

Spi-Ro Manufacturing, Inc
PO Box 1538
Hendersonville, NC 28793

Stoner Communications, Inc
McKay Dymek Division
9119 Milliken Avenue
Rancho Cucamonga, CA 91730

H. Stewart Designs
PO Box 643
Oregon City, OR 97045

Telex Communications, Inc (N)
Hy-Gain Division
9600 Aldrich Avenue South
Minneapolis, MN 55420

Telrex Labs, Inc (D)
PO Box 879
Asbury Park, NJ 07712

Ten Tec, Inc
Highway 411, E
Sevierville, TN 37862

Texas Radio Products
5 East Upshaw
Temple, TX 76501

The RF Connection (D)
213 North Frederick Avenue, Suite 11-F
Gaithersburg, MD 20877

The Radio Works (D)
Box 6159
Portsmouth, VA 23703

H. J. Theiler Corporation, (D)
c/o KI4KN
PO Box 5369, Highway I-85
Spartansburg, SC 29304

Tonna F9FT
Antennes Tonna
132, BD Dauphinot
51100 Reims
France

Triangle International Towers (D)
PO Box 1056
Mandeville, LA 70448

Trylon Manufacturing Company, Ltd
(See BJX for US orders)
21 Howard Avenue
Elmira, Ontario
Canada

United Ropeworks (USA), Inc (R, N)
(Phillystran)
20 Commerce Drive
PO Box 306
Montgomeryville, PA 18936

UPI Communications Systems, Inc
PO Box 886
Saddle Brook, NJ 07662

US Tower Corporation
8975 West Goshen Avenue
Visalia, CA 93291

Valor Enterprises, Inc (N)
185 West Hamilton Street
West Milton, OH 45383

Van Gorden Engineering
Box 21305
South Euclid, OH 44121

W1JC-TV Evans (D)
113 Stratton Brook Road
Simsbury, CT 06070

W6TIK-Rudy Plak
PO Box 966
San Marcos, CA 92069

W9INN Antennas
PO Box 393
Mt. Prospect, IL 60056

Length Conversions

Throughout this book, equations may be found for determining the design length and spacing of antenna elements. For convenience, the equations are written to yield a result in feet. (The answer may be converted to meters simply by multiplying the result by 0.3048.) If the result in feet is not an integral number, however, it is necessary to make a conversion from a decimal fraction of a foot to inches and fractions before the physical distance can be determined with a conventional tape measure. Table 9 may be used for this conversion, showing inches and fractions for increments of

Table 9

Conversion, Decimal Feet to Inches (Nearest 16th)

					Decimal Increments					
	0.00	0.01	0.02	0.03	0.04	0.05	0.06	0.07	0.08	0.09
0.0	0-0	0-1/8	0-1/4	0-3/8	0-1/2	0-5/8	0-3/4	0-13/16	0-15/16	1-1/16
0.1	1-3/16	1-5/16	1-7/16	1-9/16	1-11/16	1-13/16	1-15/16	2-1/16	2-3/16	2-1/4
0.2	2-3/8	2-1/2	2-5/8	2-3/4	2-7/8	3-0	3-1/8	3-1/4	3-3/8	3-1/2
0.3	3-5/8	3-3/4	3-13/16	3-15/16	4-1/16	4-3/16	4-5/16	4-7/16	4-9/16	4-11/16
0.4	4-13/16	4-15/16	5-1/16	5-3/16	5-1/4	5-3/8	5-1/2	5-5/8	5-3/4	5-7/8
0.5	6-0	6-1/8	6-1/4	6-3/8	6-1/2	6-5/8	6-3/4	6-13/16	6-15/16	7-1/16
0.6	7-3/16	7-5/16	7-7/16	7-9/16	7-11/16	7-13/16	7-15/16	8-1/16	8-3/16	8-1/4
0.7	8-3/8	8-1/2	8-5/8	8-3/4	8-7/8	9-0	9-1/8	9-1/4	9-3/8	9-1/2
0.8	9-5/8	9-3/4	9-13/16	9-15/16	10-1/16	10-3/16	10-5/16	10-7/16	10-9/16	10-11/16
0.9	10-13/16	10-15/16	11-1/16	11-3/16	11-1/4	11-3/8	11-1/2	11-5/8	11-3/4	11-7/8

0.01 foot. The table deals with only the fractional portion of a foot. The integral number of feet remains the same. For example, if a calculation yields a result of 11.63 feet, Table 9 indicates the equivalent fraction for 0.63 feet is 7-9/16 inches. The total length is thus 11 feet 7-9/16 inches.

Similarly, Table 10 may be used to make the conversion from inches and fractions to decimal fractions of a foot. This table is convenient for using measured distances in equations. For example, a length of 19 feet 7-3/4 inches converts to 19 + 0.646 = 19.646 feet.

BIBLIOGRAPHY

Source material and more extended discussion of topics covered in this chapter can be found in the reference given below.

J. J. Elengo, Jr., "Predicting Sag in Long Wire Antennas," *QST*, Jan 1966, pp 57-58.

Table 10
Conversion, Inches and Fractions to Decimal Feet

| | *Fractional Increments* | | | | | | | |
	0	1/8	1/4	3/8	1/2	5/8	3/4	7/8
0-	0.000	0.010	0.021	0.031	0.042	0.052	0.063	0.073
1-	0.083	0.094	0.104	0.115	0.125	0.135	0.146	0.156
2-	0.167	0.177	0.188	0.198	0.208	0.219	0.229	0.240
3-	0.250	0.260	0.271	0.281	0.292	0.302	0.313	0.323
4-	0.333	0.344	0.354	0.365	0.375	0.385	0.396	0.406
5-	0.417	0.427	0.438	0.448	0.458	0.469	0.479	0.490
6-	0.500	0.510	0.521	0.531	0.542	0.552	0.563	0.573
7-	0.583	0.594	0.604	0.615	0.625	0.635	0.646	0.656
8-	0.667	0.677	0.688	0.698	0.708	0.719	0.729	0.740
9-	0.750	0.760	0.771	0.781	0.792	0.802	0.813	0.823
10-	0.833	0.844	0.854	0.865	0.875	0.885	0.896	0.906
11-	0.917	0.927	0.938	0.948	0.958	0.969	0.979	0.990

Chapter 22

Antenna Supports

A prime consideration in the selection of a support for an antenna is that of structural safety. Building regulations in many localities require that a permit be obtained in advance of the erection of certain structures, often including antenna poles or towers. In general, localities having such requirements also have building safety codes that must be observed. Such regulations may govern the method and materials used in construction of, for example, a self-supporting tower. Checking with your local government building department before putting up a tower may save a good deal of difficulty later, because a tower would have to be taken down or modified if not approved by the building inspector on safety grounds.

Municipalities have the right and duty to enforce any reasonable regulations having to do with the safety of life or property. The courts generally have recognized, however, that municipal authority does not extend to esthetic questions. The fact that someone may object to the mere presence of a pole, tower or other antenna structure because in his opinion it detracts from the beauty of the neighborhood, is not grounds for refusing to issue a permit for a safe structure to be erected. Since the introduction of PRB-1 (federal preemption of unnecessarily restrictive antenna ordinances), this principle has been borne out in many courts. Permission for erecting amateur towers is more easily obtained than in the recent past because of this legislation.

Even where local regulations do not exist or are not enforced, the amateur should be careful to select a location and a type of support that contribute as much safety as possible to the installation. If collapse occurs, the chances of personal injury or property damage should be minimized by careful choice of design and erection methods. A single injury can be far more costly than the price of a more rugged support, in terms of both monetary loss and damage to the respect with which Amateur Radio is viewed by the public.

TREES AS ANTENNA SUPPORTS

From the beginning of Amateur Radio, trees have been used widely for supporting wire antennas. Trees cost nothing to use, and often provide a means of supporting a wire antenna at considerable height. As antenna supports, trees are unstable in the presence of wind, except in the case of very large trees used to support antennas well down from the top branches. As a result, tree supported antennas must be constructed much more sturdily than is necessary with stable supports. Even with rugged construction, it is unlikely that an antenna suspended from a tree, or between trees, will stand up indefinitely. Occasional repair or replacement usually must be expected.

There are two general methods of securing a pulley to a tree. If the tree can be climbed safely to the desired level,

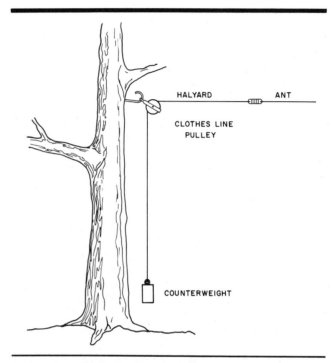

Fig 1—A method of counterweighting to minimize antenna movement and avoid its breaking from tree movement in the wind. The antenna may be lowered without climbing the tree by removing the counterweight and tying additional rope at the bottom end of the halyard. Excess rope may be left at the counterweight for this purpose, as the knot at the lower end of the halyard will not pass through the pulley.

a pulley can be attached to the trunk of the tree, as shown in Fig 1. To clear the branches of the tree, the antenna end of the halyard can be tied temporarily to the tree at the pulley level. Then the remainder of the halyard is coiled up, and the coil thrown out horizontally from this level, in the direction in which the antenna runs. It may help to have the antenna end of the halyard weighted.

After attaching the antenna to the halyard, the other end is untied from the tree, passed through the pulley, and brought to ground along the tree trunk in as straight a line as possible. The halyard need only be long enough to reach the ground after the antenna has been hauled up. (Additional rope can be tied to the halyard when it becomes necessary to lower the antenna.)

The other method consists of passing a line over the tree from ground level, and using this line to haul a pulley up into the tree and hold it there. Several ingenious methods have been used to accomplish this. The simplest method employs a weighted pilot line, such as fishing line or mason's chalk line. By grasping the line about two feet from the weight, the weight is swung back and forth, pendulum style, and then heaved with an underhand motion in the direction of the tree top.

Several trials may be necessary to determine the optimum size of the weight for the line selected, the distance between the weight and the hand before throwing, and the point in the arc of the swing where the line released. The weight, however, must be sufficiently large to carry the pilot line back to ground after passing over the tree. Flipping the end of the line up and down so as to put a traveling wave on the line often helps to induce the weight to drop down if the weight is marginal. The higher the tree, the lighter the weight and the pilot line must be. A glove should be worn on the throwing hand, because a line running swiftly through the bare hand can cause a severe burn.

If there is a clear line of sight between ground and a particularly desirable crotch in the tree, it may eventually be possible to hit the crotch after a sufficient number of tries. Otherwise, it is best to try to heave the pilot line completely over the tree, as close to the center line of the tree as possible. If it is necessary to retrieve the line and start over again, the line should be drawn back very slowly; otherwise the swinging weight may wrap the line around a small limb, making retrieval impossible.

Stretching the line out straight on the ground before throwing may help to keep the line from snarling, but it places extra drag on the line, and the line may snag on obstructions overhanging the line when it is thrown. Another method is to make a stationary reel by driving eight nails, arranged in a circle, through a 1-inch board. After winding the line around the circle formed by the nails, the line should reel off readily when the weighted end of the line is thrown. The board should be tilted at approximately right angles to the path of the throw.

Other devices that have been used successfully to pass a pilot line over a tree are the bow and arrow with heavy thread tied to the arrow, and the short casting rod and spinning reel used by fishermen. The Wrist Rocket slingshot made from surgical rubber tubing and a metal frame has proved highly effective as an antenna launching device. Still another method that has been used where sufficient space is available is flying a kite to sufficient altitude, walking around the tree until the kite string lines up with the center of the tree, and paying out string until the kite falls to the earth. This method can be used to pass a line over a patch of woods between two higher supports, which may be impossible using any other method.

The pilot line can be used to pull successively heavier lines over the tree until one of adequate size to take the strain of the antenna has been reached. This line is then used to haul a pulley up into the tree after the antenna halyard has been threaded through the pulley. The line that holds the pulley must be capable of withstanding considerable chafing where it passes through the crotch, and at points where lower branches may rub against the standing part. For this reason, it may be advisable to use galvanized sash cord or stranded guy wire for raising the pulley.

Larger lines or cables require special attention when they must be spliced to smaller lines. A splice that minimizes the chances of coming undone when coaxed through the tree crotch must be used. One type of splice is shown in Fig 2.

The crotch in which the line first comes to rest may not be sufficiently strong to stand up under the tension of the

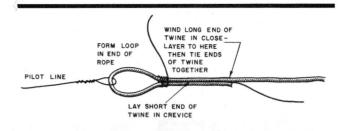

Fig 2—In connecting the halyard to the pilot line, a large knot that might snag in the crotch of a tree should be avoided, as shown.

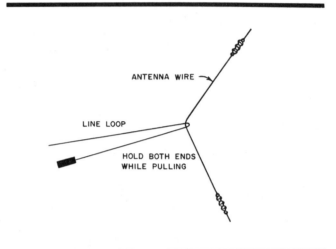

Fig 3—A weighted line thrown over the antenna can be used to pull the antenna to one side of overhanging obstructions, such as tree branches, as the antenna is pulled up. When the obstruction has been cleared, the line can be removed by releasing one end.

antenna. If, however, the line has been passed over (or close to) the center line of the tree, it will usually break through the lighter crotches and come to rest in a stronger one lower in the tree.

Needless to say, any of the suggested methods should be used with due respect to persons or property in the immediate vicinity. A child's sponge-rubber ball (baseball size) makes a safe weight for heaving a heavy thread line or fishing line.

If the antenna wire snags in the lower branches of the tree when the wire is pulled up, or if other trees interfere with raising the antenna, a weighted line thrown over the antenna and slid to the appropriate point is often helpful in pulling the antenna wire to one side to clear the interference as the antenna is being raised. This is shown in Fig 3.

Wind Compensation

The movement of an antenna suspended between supports that are not stable in the wind can be reduced by the use of heavy springs, such as screen-door springs under tension, or by a counterweight at the end of one halyard. This is shown in Fig 1. The weight, which may be made up of junk-yard metal, window sash weights, or a galvanized pail filled with sand or stone, should be adjusted experimentally for best results under existing conditions. Fig 4 shows as convenient way of fastening the counterweight to the halyard. It eliminates the necessity for untying a knot in the halyard which may have hardened under tension and exposure to the weather.

TREES AS SUPPORTS FOR VERTICAL WIRE ANTENNAS

Trees can often be used to support vertical as well as horizontal antennas. If the tree is tall and has overhanging branches, the scheme of Fig 5 may be used. The top end of the antenna is secured to a halyard passed over the limb, brought back to ground level, and fastened to the trunk of the tree.

MAST MATERIALS

Where suitable trees are not available, or a more stable support is desired, masts are suitable for wire antennas of reasonable span length. At one time, most amateur masts were constructed of lumber, but the TV industry has brought out metal masts that are inexpensive and much more durable than wood. However, there are some applications where wood is necessary or desirable.

A Ladder Mast

A temporary antenna support is sometimes needed for an antenna system for antenna testing, site selection, emergency exercises or Field Day. Ordinary aluminum extension ladders are ideal candidates for this service. They are strong, light, extendable, weatherproof and easily transported. Additionally, they are readily available and can be returned to normal use once the project is concluded. A ladder tower will support a lightweight triband beam and rotator.

With patience and ingenuity one person can erect this assembly. One of the biggest problems is holding the base down while "walking" the ladder to a vertical position. The ladder can be guyed with ¼-inch polypropylene rope. Rope guys are arranged in the standard fashion with three at each level. If help is available, the ladder can be walked up in its retracted position and extended after the antenna and rotator are attached. The lightweight pulley system on most extension ladders is not strong enough to lift the ladder extension. This mechanism must be replaced (or augmented) with a heavy duty pulley and rope. Make sure when attaching the guy ropes that they do not foul the operation of the sliding upper section of the ladder.

There is one hazard in this system that must be avoided: Do not climb or stand on the ladder when it is being extended—even as much as one rung. *Never* stand on the ladder and attempt to raise or lower the upper section. Do all the extending and retracting with the heavy-duty rope and pulley!

If the ladder is to be raised by one person, use the following guidelines. First, make sure the rung-latching mechanism operates properly before beginning. The base must be hinged so that it does not slip along the ground during erection. The guy ropes should be tied and positioned in such a way that they serve as safety constraints in the event that control of the assembly is lost. Have available a device (such as another ladder) for supporting the ladder during rest periods. (See Fig 6.)

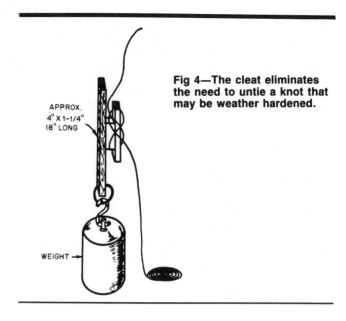

APPROX. 4" X 1-1/4" 18" LONG

WEIGHT

Fig 4—The cleat eliminates the need to untie a knot that may be weather hardened.

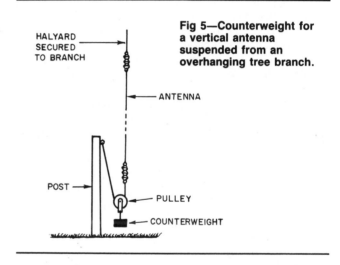

HALYARD SECURED TO BRANCH

ANTENNA

POST

PULLEY

COUNTERWEIGHT

Fig 5—Counterweight for a vertical antenna suspended from an overhanging tree branch.

After the ladder is erect and the lower section guys tied and tightened, raise the upper portion one rung at a time. *Do not raise the upper section higher than it is designed to go;* safety is far more important than a few extra feet of height.

For a temporary installation, finding suitable guy anchors can be an exercise in creativity. Fence posts, trees, and heavy pipes are all possibilities. If nothing of sufficient strength is available, anchor posts or pipes can be driven into the soil. Sandy soil is the most difficult to work with because it does a very poor job of holding anchors. A discarded car axle can be driven into the ground as an anchor, if its mass and strength are substantial. A chain and car-bumper jack can be used to remove the axle when the operation is done.

Above all else, keep the tower and antenna away from power lines. Make sure that nothing can touch the lines if the assembly falls. Disassemble by reversing the process. Ladder towers are handy for "quickie" antenna supports, but as with any improvisation of support materials, care must be taken to ensure safe construction.

Fig 6—Walking the ladder up to its vertical position. Keith, VE2AQU, supports the mast with a second ladder while Chris, VE2FRJ, checks the ropes. *(photo by Keith Baker, VE2XL)*

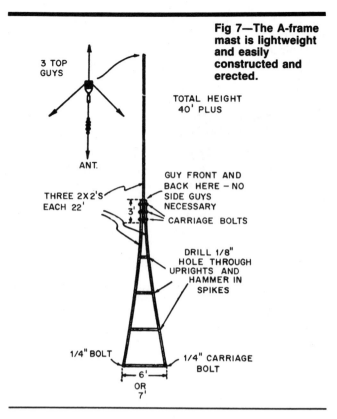

Fig 7—The A-frame mast is lightweight and easily constructed and erected.

3 TOP GUYS

ANT.

TOTAL HEIGHT 40' PLUS

THREE 2X2'S EACH 22'

GUY FRONT AND BACK HERE – NO SIDE GUYS NECESSARY

CARRIAGE BOLTS

3'

DRILL 1/8" HOLE THROUGH UPRIGHTS AND HAMMER IN SPIKES

1/4" BOLT

1/4" CARRIAGE BOLT

6' OR 7'

The A-Frame Mast

A light and relatively inexpensive mast is shown in Fig 7. In lengths up to 40 feet it is very easy to erect and will stand the pull of ordinary wire antenna systems. The lumber used is 2 × 2-inch straight-grained pine (which many lumber yards know as hemlock) or even fir stock. The uprights can be as long as 22 feet each (for a mast slightly over 40 feet high) and the cross pieces are cut to fit. Four pieces of 2 × 2 lumber, each 22 feet long, provides more than enough. The only other materials required are five ¼-inch carriage bolts 5½ inches long, a few spikes, about 300 feet of stranded or solid galvanized wire for guying, enough glazed porcelain compression ("egg") insulators to break up the guys into sections, and the usual pulley and halyard rope. If the strain insulators are put in every 20 feet, approximately 15 of them will be enough.

After selecting and purchasing the lumber—which should be straight-grained and knot-free—sawhorses or boxes should be set up and the mast assembled as shown in Fig 8. At this stage it is wise to give the mast a coat of primer and a coat of outside white latex paint.

After the coat of paint is dry, attach the guys and rig the pulley for the antenna halyard. The pulley anchor should be at the point where the top stays are attached so the back stay will assume the greater part of the load tension. It is better to use wire wrapped around the mast with a small through-bolt to prevent sliding down than to use eye bolts.

If the mast is to stand on the ground, a couple of stakes should be driven to keep the bottom from slipping. At this point the mast may be "walked up" by a helper. If it is to go on a roof, first stand it up against the side of the building and then hoist it, from the roof, keeping it vertical. The whole assembly is light enough for two men to perform the complete operation—lifting the mast, carrying it to its permanent berth, and fastening the guys with the mast vertical all the while. It is therefore entirely practicable to put up this kind of mast on a small flat area of roof that would prohibit the erection of one that had to be raised to the vertical in its final location.

TV Mast Material

TV mast is available in 5- and 10-foot lengths, 1¼ inches

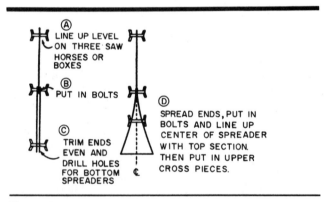

Ⓐ LINE UP LEVEL ON THREE SAW HORSES OR BOXES

Ⓑ PUT IN BOLTS

Ⓒ TRIM ENDS EVEN AND DRILL HOLES FOR BOTTOM SPREADERS

Ⓓ SPREAD ENDS, PUT IN BOLTS AND LINE UP CENTER OF SPREADER WITH TOP SECTION. THEN PUT IN UPPER CROSS PIECES.

Fig 8—Method of assembling the A-frame mast on sawhorses.

diameter, in both steel and aluminum. These sections are crimped at one end to permit sections to be joined together. A form that is usually more convenient is the telescoping mast available from many electronic supply houses. The masts may be obtained with three, four or five 10-foot sections, and come complete with guying rings and a means of locking the sections in place after they have been extended. These masts are stronger than the nontelescoping type because the diameters of the sections increase toward the bottom of the mast. For instance, the top section of a 50-foot mast is 1¼ inches diameter, and the bottom section is 2½ inches diameter.

Guy rings are provided at 10-foot intervals, but guys may not be required at every point. Guying is essential at the top and at least one other place near the center of the mast. If the mast has any tendency to whip in the wind, or to bow under the stress of the antenna, additional guys should be added at the appropriate points.

MAST GUYING

Three guy wires in each set are usually adequate for a mast. These should be spaced equally around the mast. The required number of sets of guys depends on the height of the mast, its natural sturdiness, and the required antenna tension. A 30-foot mast usually requires two sets of guys, and a 50-foot mast needs at least three sets. One guy of the top set should be anchored to a point directly opposite the direction in which the antenna runs. The other two guys of the same set should be spaced 120° with respect to the first, as shown in Fig 7.

Generally, the top guys should be anchored at distances from the base of the mast of at least 60% of the mast height. At the 60% distance, the stress on the guy wire opposite the antenna is approximately twice the tension on the antenna. As the distance between the guy anchor and the base of the mast is decreased, the tension on the rear guy in proportion to the tension on the antenna rises rapidly. The extra tension results in additional compression on the mast, increasing the tendency for the mast to buckle.

Additional sets of guys serve to correct for any tendency that the mast may have to buckle under the compression imposed by the top guys. To eliminate possible mechanical resonance in the mast that might cause the mast to vibrate, the sets of guys should not be spaced equally on the mast. A second set of guys should be placed at approximately 60% of the distance between the ground and the top of the mast. A third set should be placed at about 60% of the distance between the ground and the second set of guys.

The additional set of guys should be anchored at distances from the base of the mast equal to at least 60% of the distance between the ground and the points of attachment on the mast. In practice, the same anchors are usually used for all sets of guys, automatically meeting this requirement if the top set has been anchored at the correct distance.

Electrical resonances that might cause distortion of the radiation pattern of the antenna can be eliminated by breaking each guy into nonresonant lengths by the insertion of strain insulators (see Figs 9 and 10). This subject is covered in detail later in this chapter.

Guy Material

Within their stress ratings, any of the halyard materials listed in Chapter 21 may be used for the construction of guys. Nonmetallic materials have the advantage that there is no need to break them up into sections to avoid resonances. All of these materials are subject to stretching, however, which causes mechanical problems in permanent installations. At rated working load tension, dry manila rope stretches about 5%, while nylon rope stretches about 20%.

Antenna wire is also suitable for guys, particularly the copper-clad steel types. Solid galvanized steel wire is also widely used for guying. This wire has approximately twice the tension ratings of similar sizes of copper-clad wire, but it is more susceptible to corrosion. Stranded galvanized wire sold for guying TV masts is also suitable for light-duty applications, but is also susceptible to corrosion.

Guy Anchors

Figs 11 and 12 show two different kinds of guy anchors. In Fig 11, one or more pipes are driven into the ground at

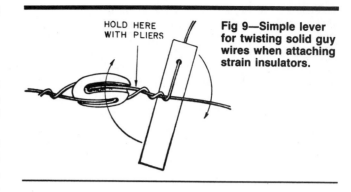

Fig 9—Simple lever for twisting solid guy wires when attaching strain insulators.

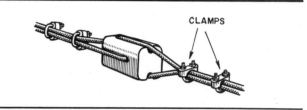

Fig 10—Stranded guy wire should be attached to strain insulators by means of standard cable clamps made to fit the size of wire used.

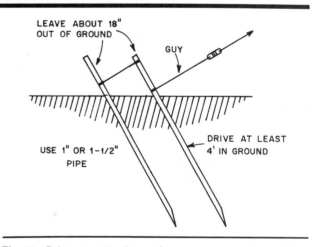

Fig 11—Driven guy anchors. One pipe is usually sufficient for a small mast. For added strength, a second pipe may be added, as shown.

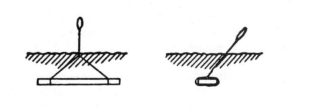

Fig 12—Buried "dead-man" guy anchor (see text).

right angles to the guy wire. If a single pipe proves to be inadequate, another pipe can be added in tandem, as shown. Steel fence posts may be used in the same manner. Fig 12 shows a "dead-man" type of anchor. The buried anchor may consist of one or more pipes 5 or 6 feet long, or scrap automobile parts, such as bumpers or wheels. The anchors should be buried 3 or 4 feet in the ground. Some tower manufacturers make heavy auger-type anchors that screw into the earth. These anchors are usually heavier than required for guying a mast, although they may be more convenient to install. Trees and buildings may also be used as guy anchors if they are located appropriately. Care should be exercised, however, to make sure that the tree is of adequate size, and that the fastening to a building can be made sufficiently secure.

Guy Tension

Most troubles encountered in mast guying are a result of pulling the guy wires too tight. Guy-wire tension should never be more than necessary to correct for obvious bowing or movement under wind pressure. In most cases, the tension needed does not require the use of turnbuckles, with the possible exception of the guy opposite the antenna. If any great difficulty is experienced in eliminating bowing from the mast, the guy tension should be reduced.

ERECTING A MAST OR OTHER SUPPORT

Masts less than 30 feet high usually can be simply "walked" up after blocking the bottom end securely. Blocking must be done so that the base can neither slip along the ground or upend when the mast is raised. An assistant should be stationed at each guy wire, and may help by pulling the proper guy wire as the mast nears the vertical position. Halyards can be used in the same manner.

As the mast is raised, it may be helpful to follow the underside of the mast with a scissors rest (Fig 13), should a pause in the hoisting become necessary. The rest may also be used to assist in the raising, if each leg is manned by an assistant.

As the mast nears the vertical position, those holding the guy wires should be ready to temporarily fasten the guys to

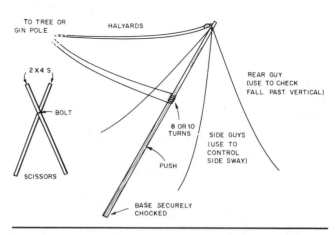

Fig 13—Pulling on a gin line fastened slightly above the center point of the mast and on the halyards can assist in erecting a tall mast. The tensions should be just enough to keep the mast in as straight a line as possible. The "scissors" may be used to push on the under side and to serve as a rest if a pause in raising becomes necessary.

prevent the mast from falling. The guys can then be adjusted until the mast is perfectly straight.

For masts over 30 feet long, a "gin" of some form may be required, as shown in Fig 13. Several turns of rope are wound around a point on the mast above center. The ends of the rope are then brought together and passed over a tree limb. The rope should be pulled as the mast is walked up to keep the mast from bending at the center. If a tree is not available, a post, such as a 2 × 4, temporarily erected and guyed, can be used. After the mast has been erected, the assisting rope can be removed by walking one end around the mast (inside the guy wires).

Telephone poles and towers are much sturdier supports. Such supports may require no guying, but they are not often used solely for the support of wire antennas because of their relatively high cost. For antenna heights in excess of 50 feet, however, they are usually the most practical form of support.

Tower Selection and Installation

The selection of a tower, its height, and the type of antenna and rotator to be used may seem like a complicated matter, particularly for the newcomer. These aspects of an antenna system are interrelated, and one should consider the overall system before making any decisions as to a specific component. Perhaps the most important consideration for many amateurs is the effect of the antenna system on the surrounding environment. If plenty of space is available for a tower installation and there is little chance of the antenna causing esthetic distress on the part of other family members or the neighbors, the amateur is indeed fortunate. The limitations in this case are mostly financial. For most, however, the size of the property, the effect of the system on others, local ordinances, and the proximity of power lines and

poles influence the selection of antenna components considerably.

The amateur must consider the practical limitations for installation. Some points for consideration are given below:

1) A tower should not be installed in a position where it could fall onto a neighbor's property.

2) The antenna must be located in such a position that *it cannot possibly tangle with power lines, either during normal operation or if the structure fell.*

3) Sufficient yard space must be available to position a guyed tower *properly*. The guy anchors should be between 60% and 80% of the tower height in distance from the base of the tower.

4) Provisions must be made to keep children from

climbing the support. (Poultry netting around the tower base will serve this need.)

5) Local ordinances should be checked to determine if any legal restrictions affect the proposed installation.

Other important considerations are (1) the total dollar amount to be invested, (2) the size and weight of the antenna desired, (3) the climate, and (4) the ability of the owner to climb a fixed tower.

The selection of a tower support usually is dictated more by circumstances than by desire. The most economical system, in terms of feet per dollar investment, is a guyed tower.

Once a decision has been tentatively made, the next step is to write to the manufacturer (several are listed in Chapter 21) and request specifications for the equipment that may be needed. Locate and mark guy anchor points to ensure that they fit on the available property. The specification sheet for the tower should give a wind-load capability; antennas can then be chosen based on the ratings of the structure.

It is often very helpful to the novice tower installer to visit other local amateurs who have installed towers. Look over their hardware and ask questions. If possible, have a few local experienced amateurs look over your plans—before you commit yourself. They may be able to offer a great deal of help. If someone in your area is planning to install a tower and antenna system, be sure to offer your assistance. There is no substitute for experience when it comes to tower work, and your experience there may prove invaluable to you later.

THE TOWER

The most common variety of tower is the guyed tower made of stacked identical sections. The information in Fig 14 is based on data taken from the Unarco-Rohn catalog. Rohn calls for a maximum vertical separation of 35 feet between sets of guy wires. At A, the tower is 70 feet high, and there are two sets of evenly spaced guy wires. At B, the tower is 80 feet high, and there are three sets of evenly spaced guy wires. Exceeding the vertical spacing requirements (under-guying) could result in the tower buckling.

This may not seem to be a likely occurrence unless the function of guy wires is well understood. Guy wires restrain the tower against the force of the wind. They translate the lateral force of the wind into a downward compression that forces the tower down onto the base. Manufacturers usually specify the initial tension in the guy wires. This is another force that is translated into the downward compression on the tower. If there are not enough guys and if they are not properly spaced, a heavy gust of wind may over-stress the structure, causing the tower to buckle at a weak point.

An overhead view of a guyed tower is given in Fig 14C. Manufacturers usually call for equal angular spacing between guy wires. If it is necessary to deviate from this spacing, the engineering staff of the tower manufacturer or a civil engineer should be contacted for advice.

Unguyed Towers

Some towers are not normally guyed—these are usually referred to as free-standing or self-supporting towers. The principles involved are the same regardless of the term the manufacturers use to describe them. The wind blowing against the side of the tower creates an overturning movement that would topple the tower if it were not for the anchoring at the base. Fig 15 details the action and reaction involved. The tower is restrained by the base. As the wind blows against

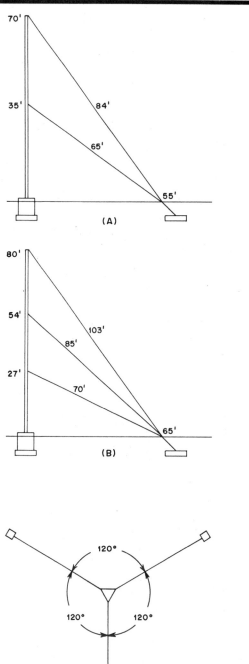

Fig 14—The proper method of installation of a guyed tower.

one side of the tower, the opposite side is compressed downward much as in the guyed installation.

Because there are no guys to restrain the top, the side on which the wind is blowing is simultaneously pulled up (uplift). The force of the wind creates a moment that tends to pivot about a point at the base of the tower. The base of the guyed tower simply must hold the tower up, but the base

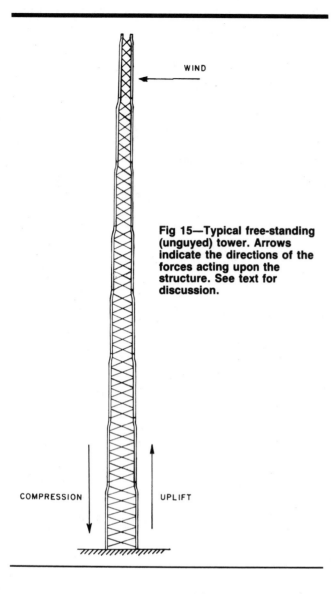

Fig 15—Typical free-standing (unguyed) tower. Arrows indicate the directions of the forces acting upon the structure. See text for discussion.

WIND

COMPRESSION

UPLIFT

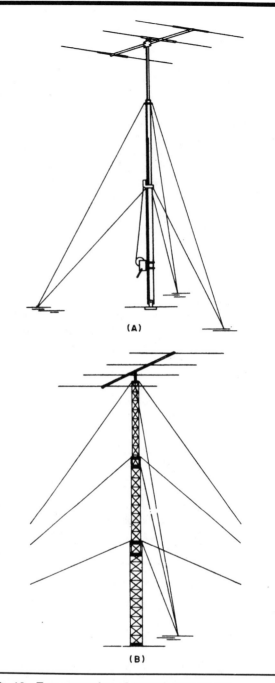

(A)

(B)

Fig 16—Two examples of "crank-up" towers.

of the free-standing tower must simultaneously hold one side of the tower up and the other side *down!* For this reason, manufacturers often call for a great deal more concrete in the base of free-standing towers than they do in the base of guyed towers.

Fig 16 shows two variations of another popular type of tower, the crank-up. In regular guyed or free-standing towers, each section is bolted atop the next lower section. The height of the tower is the sum of the heights of the sections (minus any overlap). Crank-up towers use a different system. The outer diameter of each section is smaller than the inner diameter of the next lower section. Instead of bolting together, the sections are attached with a set of cables and pulleys. The overall height of the tower is adjusted by using the pulleys and cables to "telescope" the sections together or apart.

Depending on the design, the manufacturer may or may not require guy wires. The primary advantage of the crank-up tower is that antenna work can be done near the ground. A second advantage is that the tower can be kept retracted except during use, which reduces the guying needs. (Presumably, the tower would not be extended during periods of high wind.) The disadvantages include mechanical complexity and (usually) cost. *NEVER* climb on an extended crank-up tower,

even if it is extended only a small amount. Serious injury could result if the hoisting system fails.

Some towers have another convenience feature—a hinged section that permits the owner to fold over all or a portion of the tower. The primary benefit is in allowing antenna work to be done close to ground level, without the necessity of removing the antenna and lowering it for service. Fig 17 shows a hinged base; of course, the hinged section can be designed for portions of the tower other than the base. Also, a hinge feature can be added to some crank-up towers.

Misuse of hinged sections during tower erection is a dangerously common practice among radio amateurs. Unfortunately, these episodes often end in accidents. If you do not have a good grasp of the fundamentals of physics, it might

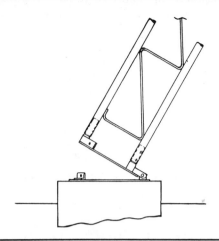

Fig 17—Fold-over or tilting base. There are several different kinds of hinged sections permitting different types of installation. Great care should be exercised when raising or lowering a tilting tower.

be wise to avoid hinged towers or to consult an expert if there are any questions about safely installing and using such a tower. It is often far easier (and safer) to erect a regular guyed tower or self-supporting tower with gin pole and climbing belt than it is to try to "walk up" an unwieldy hinged tower.

TOWER BASES

Tower manufacturers can provide customers with detailed plans for properly constructing tower bases. Fig 18 is an example of one such plan. This plan calls for a hole that is 3½ × 3½ × 6 feet. Steel reinforcement bars are lashed together and placed in the hole. The bars are positioned so that they will be completely embedded in the concrete, yet will not contact any metallic object in the base itself. This is done to minimize the possibility of a direct discharge path for lightning through the base. Should such a discharge occur, the concrete base would likely explode and bring about the collapse of the tower.

A strong wooden form is constructed around the top of the hole. The hole and the wooden form are filled with concrete so that the resultant block will be 4 inches above grade. The anchor bolts are embedded in the concrete before it hardens. Usually it is easier to ensure that the base is level and properly aligned by attaching the mounting base and the first section of the tower to the concrete anchor bolts. Manufacturers can provide specific, detailed instructions for the proper mounting procedure. Fig 19 shows a slightly different design for a tower base.

The one assumption so far is that "normal" soil is predominant in the area in which the tower is to be installed. "Normal soil" is a mixture of clay, loam, sand and small rocks. More conservative design parameters for the tower base should be adopted (usually, more concrete) if the soil is sandy, swampy or extremely rocky. If there are any doubts about the soil, the local agricultural extension office can usually provide specific technical information about the soil in a given area. When this information is in hand, contact the engineering department of the tower manufacturer or a civil engineer for specific recommendations with regard to compensating

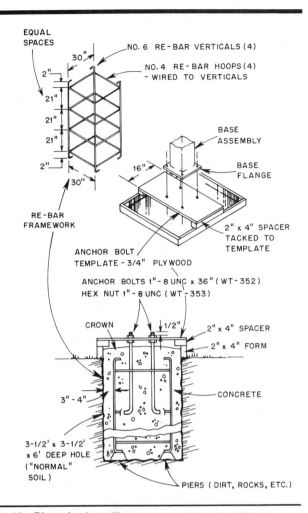

Fig 18—Plans for installing concrete base for Wilson ST-77B tower. Although the instructions and dimensions vary from tower to tower, this is representative of the type of concrete base specified by most manufacturers.

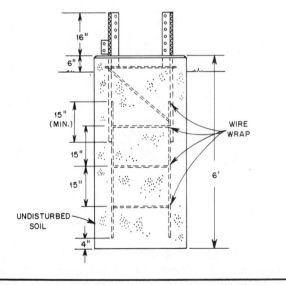

Fig 19—Another example of a concrete base (Tri-Ex LM-470).

for any special soil characteristics.

TOWER INSTALLATION

The installation of a tower is not difficult when the proper techniques are used. A guyed tower, in particular, is not hard to erect, because each of the individual sections are relatively lightweight and can be handled with only a few helpers and some good quality rope. A gin pole is a handy device for working with tower sections. The gin pole shown in Fig 20 is designed to fit around the leg of Rohn 25 tower and clamp in place. The tubing, which is about 12 feet long, has a pulley on one end. A rope is routed through the tubing and over the pulley. When the gin pole is attached to the tower and the tubing is extended into place and locked, the rope can be used to haul tower sections and the antenna into place.

One of the most important aspects of any tower installation project is the safety of all persons involved. Chapter 1 details several safety points to be observed. Basically, the use of hard hats is highly recommended for all assistants helping from the ground. Helpers should always stand clear of the tower base to prevent being hit by a dropped tool or hardware. A good climber's safety belt should be used by each person working on the tower. When climbing the tower, if more than one person is involved, one should climb into position before the other begins climbing. The same procedure is required for climbing down a tower after the job is completed. The purpose is to have the nonclimbing person stand relatively still so as not to drop any tools or objects on the climbing person, or unintentionally obstruct his movements. When two persons are working on top of a tower, only one should change position (unbelt and move) at a time.

For most installations, a good-quality ½-inch dia manila hemp rope can adequately handle the work load for the hoisting tasks. The rope must be periodically inspected to assure that no tearing or chafing has developed, and if the rope should get wet from rain, it should be hung out to dry at the first opportunity. Safety knots should be used to assure that the rope stays tied during the hoisting of a tower section or antenna.

ATTACHING GUY WIRES

In typical Amateur Radio installations, guy wires may experience loads in excess of 1000 pounds. Under such circumstances, the wires cannot merely be twisted together and expected to hold. Figs 21, 22 and 23 depict the traditional method for fixing the end of a guy wire. A thimble is used to prevent the wire from breaking because of a sharp bend at the point of intersection. Three cable clamps follow to hold the wire securely. As a final backup measure, the individual strands of the free end are unraveled and wrapped around

Fig 21—Proper tension can be placed on the guy wires with the aid of a block-and-tackle system. *(photo by K1WA)*

Fig 20—A gin pole is helpful in positioning antennas and tower sections. The weight of the assembly is held by the ground crew via a heavy rope, making tower work safer and less tiring. *(photo by Dave Pietraszewski, K1WA)*

Fig 22—A length of guy cable is used to assure that the turnbuckles remain in place after they are tightened. This procedure is an absolute requirement in guyed tower systems. *(photo by K1WA)*

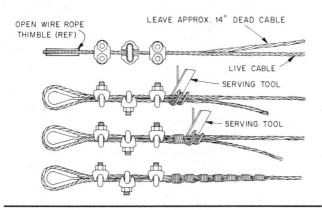

Fig 23—Traditional method for securing the end of a guy wire.

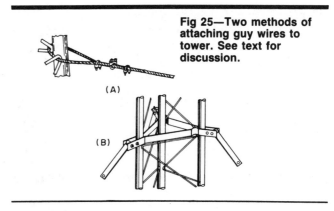

Fig 25—Two methods of attaching guy wires to tower. See text for discussion.

the guy wire. It is a lot of work, but it is necessary to ensure a safe and permanent connection.

Fig 24 shows the use of a device that replaces the clamps and twisted strands of wire. These devices are known as dead ends. They are far more convenient to use than are clamps. The guy wires must be cut to the proper length. The dead end of each wire is installed into the object to which the guy wire is being attached (use a thimble, if needed). One side of the dead end is then wrapped around the guy wire. The other side of the dead end follows. The savings in time and trouble more than make up for the slightly higher cost.

As indicated in Chapter 21, guy wire comes in different sizes, strengths and types. Typically, 3/16-inch EHS guy wire is adequate for moderate tower installations at most amateur stations. Some amateurs prefer to use 5/32-inch "aircraft" cable. Although this cable is somewhat more flexible than 3/16-inch EHS, it is only about 70% as strong. Standard guy wire at least 3/16-inch EHS is the safest bet in tower guying.

Fig 25 shows two different methods for attaching guy wires to towers. At A, the guy wire is simply looped around the tower leg and terminated in the usual manner. At B, a "torque bracket" has been added. There is not much difference in performance for wind forces that tend to "push the tower over." If more loading area (antennas, feed lines, etc) is present on one side of the tower than the other, the

force of the wind causes the tower to "twist" into the ground. The torque bracket is far more effective in resisting this twisting motion than the simpler installation. The trade-off is in terms of initial cost.

There are two types of commonly used guy anchors. Fig 26A depicts an earth screw. These are usually 4 to 6 feet long. The screw blade at the bottom typically measures 6 to 8 inches diameter. Fig 26B illustrates two people installing the anchor. The shaft is tilted so that it will be in line with the guy wires. Earth screws are suitable for use in "normal" soil where permitted by local building codes.

The alternative to earth screws is the concrete block anchor. Fig 26C shows the installation of this type of anchor; it is suitable for any soil condition, with the possible exception of a bed of lava rock or coral. Consult the instructions from

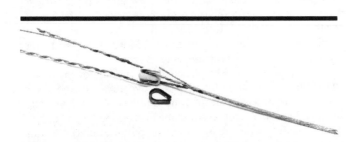

Fig 24—Alternative method for attaching guy wires using dead ends. The dead end on the right is completely assembled (the end of the guy wire extends beyond the grip for illustrative purposes). On the left, one side of the dead end is partially attached to the guy wire. In front, a thimble is used where a sharp bend might cause the guy wire or dead end to break.

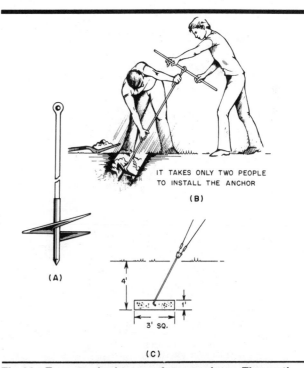

Fig 26—Two standard types of guy anchors. The earth screw shown at A is easy to install and widely available, but may not be suitable for use in certain soils. The concrete anchor is more difficult to install properly, but it is suitable for use with a wide variety of soil conditions and will satisfy most building code requirements.

the manufacturer for the precise method of installation.

Turnbuckles and associated hardware are used to attach guy wires to anchors and to provide a convenient method of adjusting tension on the guy wires. Fig 27A shows a turnbuckle of a single guy wire attached to the eye of the anchor. Turnbuckles are usually fitted with either two eyes, or one eye and one jaw. The eyes are the oval ends, while the jaws are U-shaped with a bolt through each tip. Fig 27B depicts two turnbuckles attached to the eye of an anchor. The procedure for installation is to remove the bolt from the jaw, pass the jaw over the eye of the anchor and reinstall the bolt through the jaw, through the eye of the anchor, and through the other side of the jaw.

If two or more guy wires are attached to one anchor, equalizer plates should be installed (Fig 27C). In addition to providing a convenient point to attach the turnbuckles, the plates pivot slightly and equalize the tension on the guy wires. Once the installation is complete, a safety wire should be passed through the turnbuckles in a "figure-eight" fashion to prevent the turnbuckles from turning under load.

Resonance in Guy Wires

If guy wires are resonant at or near the operating frequency, they can receive and reradiate RF energy. By behaving as parasitic elements, the guy wires may alter and thereby distort the radiation pattern of a nearby antenna. For low frequencies where a dipole or other simple antenna is used, this is generally of little or no consequence. But at the higher frequencies where a unidirectional antenna is installed, it is desirable to avoid pattern distortion if at all possible. The symptoms of reradiating guy wires are usually a lower front to back ratio and a lower front to side ratio than the antenna is capable of producing. The gain of the antenna and the feed-point impedance will usually not be significantly affected, although sometimes changes in SWR can be noted as the antenna is rotated. (Of course other conductors in the vicinity of the antenna can also produce these same symptoms.)

The amount of reradiation from a guy wire depends on two factors—its resonant frequency, and the degree of coupling to the antenna. Resonant guy wires near the antenna will have a greater effect on performance than those which are farther away. Therefore, the upper portion of the top level of guy wires should warrant the most attention with horizontally polarized arrays. The lower guy wires are usually closer to horizontal than the top level, but by virtue of their increased distance from the antenna, are not coupled as tightly to the antenna.

To avoid resonance, the guys should be broken up by means of egg or strain insulators. Fig 28 shows wire lengths that fall within 10% of ½-λ resonance (or a multiple of ½-λ) for all the HF amateur bands. Unfortunately, no single length greater than about 14 feet avoids resonance in all bands. If you operate just a few bands, you can locate greater lengths from Fig 28 that will avoid resonance. For example, if you operate only the 14, 21 and 24-MHz bands, guy wire lengths of 27 feet or 51 feet would be suitable, along with any length less than 16 feet.

ANTENNA INSTALLATION

All antenna installations are different in some respects. Therefore, thorough planning is the most important first step in installing any antenna. Before anyone climbs the tower, the whole process should be discussed to be sure each crew member understands what is to be done. Consider what tools and parts must be assembled and what items must be taken up the tower, and plan alternative actions for possible trouble spots. Extra trips up and down the tower can be avoided by careful planning.

Raising a beam antenna requires planning. If done properly, the actual work of getting the antenna into position can be executed quite easily with only one person at the top of the tower. The ground crew should do all the heavy work and leave the person on the tower free to guide the antenna into position. Because the ground crew does all the lifting, a large pulley, preferably on a gin pole placed at the top of the tower, is essential. Local radio clubs often have gin poles available for use by their members. Stores that sell tower materials frequently rent gin poles as well.

A gin pole should be placed along the side of the tower so the pulley is no more than 2 feet above the top of the tower (or the point at which the antenna is to be placed). Normally this height is sufficient to allow the antenna to be positioned easily. An important reason that the pulley is placed at this level, however, is that there can be considerable strain on the pole when the antenna is maneuvered past the guy wires.

The rope (halyard) through the pulley must be somewhat longer than twice the tower height so the ground crew can raise the antenna from ground level. The rope should be 1/2 or 5/8 inch diameter for both strength and ease of handling.

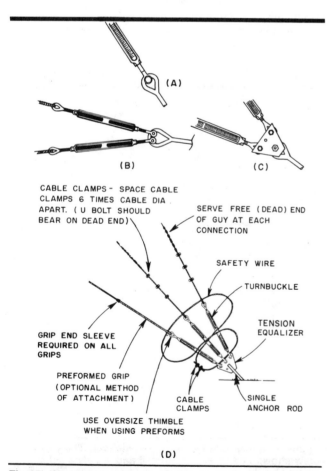

CABLE CLAMPS - SPACE CABLE CLAMPS 6 TIMES CABLE DIA APART. (U BOLT SHOULD BEAR ON DEAD END)

SERVE FREE (DEAD) END OF GUY AT EACH CONNECTION

SAFETY WIRE

TURNBUCKLE

TENSION EQUALIZER

GRIP END SLEEVE REQUIRED ON ALL GRIPS

PREFORMED GRIP (OPTIONAL METHOD OF ATTACHMENT)

CABLE CLAMPS

SINGLE ANCHOR ROD

USE OVERSIZE THIMBLE WHEN USING PREFORMS

Fig 27—Variety of means available for attaching guy wires and turnbuckles to anchors.

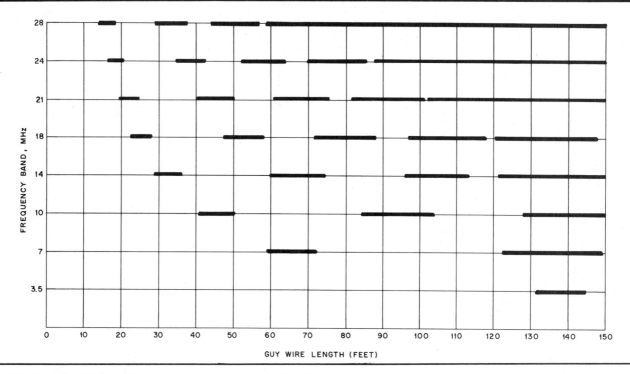

Fig 28—The black bars indicate ungrounded guy wire lengths to avoid for the eight HF amateur bands. This chart is based on resonance within 10% of any frequency in the band. Grounded wires will exhibit resonance at odd multiples of a quarter wavelength. (by Jerry Hall, K1TD)

Smaller diameter rope is less easily manipulated; it has a tendency to jump out of the pulley track and foul pulley operation.

The first person to climb the tower should carry an end of the halyard so that the gin pole can be lifted and secured to the tower. Those climbing the tower *must* have safety belts. Belts provide safety and convenience; it is simply impossible to work effectively while hanging onto the tower with one hand.

Once positioned, the gin pole and pulley allow parts and tools to be sent quickly up the tower. A useful trick for sending up small items like bolts and pliers is for a ground crew member to slide them through the rope strands where they are held by the rope for the trip to the top of the tower. Items that might be dislodged by contact with the tower should either be taped or tied to the halyard.

Remember, once someone is on the tower, no one should be allowed to stand near the base of the tower! Ever present is the hazard of falling tools or hardware. It is foolish to stand near a tower when someone is working above. Hard-hats should be worn by ground-crew members as extra insurance.

Raising the Antenna

A technique that can save much effort in raising the antenna is outlined here. First, the halyard is passed through the gin-pole pulley, and the leading end of the rope is returned to the ground crew where it is tied to the antenna. The assembled antenna should be placed in a clear area of the yard (or the roof) so the boom points toward the tower. The halyard is then passed *under* the front elements of the beam to a position past the midpoint of the antenna, where it is securely tied to the boom (Fig 29A).

Note that once the antenna is installed, the tower worker must be able to reach and untie the halyard from the boom; the rope must be tied less than an arm's length along the boom from the mounting point. If necessary, a large loop may be placed around the first element located beyond the midpoint of the boom, with the knot tied near the center of the antenna. The rope may then be untied easily after completion of the installation. The halyard should be tied to the boom at the front of the antenna by means of a short piece of light rope or twine.

While the antenna is being raised, the ground crew does all the pulling. As soon as the front of the antenna reaches the top of the mast, the person atop the tower unties the light rope and prevents the front of the antenna from falling, as the ground crew continues to lift the antenna (Fig 29B). When the center of the antenna is even with the top of the tower, the tower worker puts one bolt through the mast and the antenna mounting bracket on the boom. The single bolt acts as a pivot point and the ground crew continues to lift the back of the antenna with the halyard (Fig 29C). After the antenna is horizontal, the tower worker secures the rest of the mounting bolts and unties the halyard. By using this technique, the tower worker performs no heavy lifting.

Avoiding Guy Wires

Although the same basic methods of installing a Yagi apply to any tower, guyed towers pose a special problem. Steps must be taken to avoid snagging the antenna on the guy wires. With proper precautions, however, even large antennas can be pulled to the top of a tower, even if the mast is guyed at several levels.

Sometimes one of the top guys can provide a track to

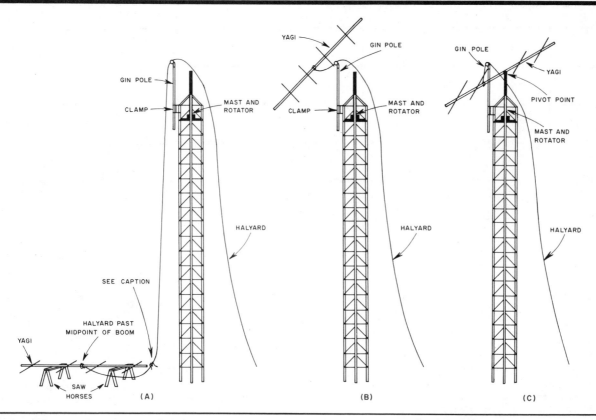

Fig 29—Raising a Yagi antenna to the top of a tower. At A the Yagi is placed in a clear area, with the boom pointing toward the tower. The halyard is passed under the elements, then secured to the boom beyond the midpoint. B shows the antenna approaching the top of the mast. The person on the tower guides it after the lifting rope has been untied from the front of the antenna. At C the antenna is pulled into a horizontal position by the ground crew. The tower worker inserts the pivot bolt and secures it. Note: A short piece of rope is tied around the halyard and the boom at the front of the antenna. It is removed by the tower worker when the antenna reaches the top.

support the antenna as it is pulled upward. Insulators in the guys, however, may obstruct the movement of the antenna. A better track made with rope is an alternative. One end of the rope is secured outside the guy anchors. The other end is passed over the top of the tower and back down to an anchor near the first anchor. So arranged, the rope forms a narrow V track strung outside the guy wires. Once the V track is secured, the antenna may simply be pulled up the track.

Another method is to tie a rope to the back of the antenna (but within reach of the center). The ground crews then pull the antenna out away from the guys as the antenna is raised. With this method, some crew members are pulling up the antenna to raise it while others are pulling down and out to keep the beam clear of the guys. Obviously, the opposing crews must act in coordination to avoid damaging the antenna. The beam is especially vulnerable when it begins to tip into the horizontal position. If the crew continues to pull out and down against the antenna, the boom can be broken. Another problem with this approach is that the antenna may rotate on the axis of the boom as it is raised. To prevent such rotation, long lengths of twine may be tied to outer elements, one piece on each side of the boom. Ground personnel may then use these "tag lines" to stabilize the antenna. Where this is done, provisions must be made for untying the twine once the antenna is in place.

A third method is to tie the halyard to the center of the antenna. A crew member, wearing a safety belt, walks the

antenna up the tower as the crew on the ground raises it. Because the halyard is tied at the balance point, the tower worker can rotate the elements around the guys. A tag line can be tied to the bottom end of the boom so that a ground worker can help move the antenna around the guys. The tag line must be removed while the antenna is still vertical.

THE PVRC MOUNT

The methods described earlier in this chapter for hoisting antennas are sometimes not satisfactory for large arrays. The best way to handle large Yagis is to assemble them on top of the tower. One way to do this easily is by using the "PVRC mount." Many members of the Potomac Valley Radio Club have successfully used this method to install large antennas. Simple and ingenious, the idea involves offsetting the boom from the mast to permit the boom to tilt 360° and rotate *axially* 360°. This permits the entire length of the boom to be brought alongside the tower, allowing the elements to be attached one by one. (It also allows *any part* of the antenna to be brought alongside the tower for antenna maintenance.)

See Figs 30 through 34. The mount itself consists of a short length of pipe of the same diameter as the rotating mast (or greater), a steel plate, eight U bolts and four pinning bolts. The steel plate is the larger, horizontal one shown in Fig 31. Four U bolts attach the plate to the rotating mast, and four

Fig 30—The PVRC mount, boom plate, mast and rotator ready to go. The mast and rotator are installed on the tower first.

Fig 32—Working at the 70-foot level. A gin pole makes pulling up and mounting the boom to the boom plate a safe and easy procedure.

Fig 31—Close-up of the PVRC mount. The long pipe (horizontal in this photo) is the rotating mast. The U bolts in the vertical plate at the left are ready to accept the antenna boom. The heads of two locking pins (bolts) are visible at the midline of the boom plate. The other two pins help secure the horizontal pipe to the large steel mast plate. (The head of the bolt nearest the camera blends in with the right hand leg of the U bolt behind it.)

attach the horizontal pipe to the plate. The horizontal pipe provides the offset between the antenna boom and the tower. The antenna boom-to-mast plate is mounted at the outer end of the short pipe. Four bolts are used to ensure that the antenna ends up parallel to the ground, two pinning each plate to the short pipe. When the mast plate pinning bolts are removed and the four U bolts loosened, the short pipe and boom plate can be rotated through 360°, allowing either half of the boom to come alongside the tower.

First assemble the antenna on the ground. Carefully mark all critical dimensions, and then remove the antenna elements from the boom. Once the rotator and mast have been installed on the tower, a gin pole is used to bring the mast plate and short pipe to the top of the tower. There, the "top crew" unpins the horizontal pipe and tilts the antenna boom plate to place it in the vertical plane. The boom is attached to the boom plate at the balance point *of the assembled antenna*. It is important that the boom be rotated axially so the bottom side of the boom is closest to the tower. This will allow the boom to be tilted without the elements striking the tower.

During installation it may be necessary to remove one guy wire temporarily to allow for tilting of the boom. As a safety precaution, a temporary guy should be attached to the same leg of the tower just low enough so the assembled antenna will clear it.

The elements are assembled on the boom, starting with those closest to the center of the boom, working out alternately to the farthest director and reflector. This procedure *must* be followed. If all the elements are put first on one half of the boom, it will be dangerous (if not impossible) to put on the remaining elements. By starting at the middle and working outward, the balance point of the partly assembled antenna will never be so far removed from the tower that tilting of the boom becomes impossible.

When the last element is attached, the boom is brought parallel to the ground, the horizontal pipe is pinned to the

Fig 33—Mounting the last element prior to positioning the boom in a horizontal plane.

Fig 34—The U bolts securing the short pipe to the mast plate are loosened and the boom is turned to a horizontal position. This puts the elements in a vertical plane. Then the pipe U bolts are tightened and pinning bolts secured. The boom U bolts are then loosened and the boom turned axially 90°.

mast plate, and the mast plate U bolts tightened. At this point, all the antenna elements will be positioned vertically. Next, loosen the U bolts that hold the boom and rotate the boom axially 90°, bringing the elements parallel to the ground. Tighten the boom bolts and double check all the hardware.

Many long boom Yagis employ a truss to prevent boom sag. With the PVRC mount, the truss must be attached to a pipe that is independent of the rotating mast. A short length of pipe is attached to the boom as close as possible to the balance point. The truss then moves with the boom whenever the boom is tilted or twisted.

THE TOWER ALTERNATIVE

A cost saving alternative to the ground mounted tower is the roof mounted tripod. Units suitable for small HF or VHF antennas are commercially available. Perhaps the biggest problem with a tripod is determining how to fasten it securely to the roof.

One method of mounting a tripod on a roof is to nail 2 × 6 boards to the undersides of the rafters. Bolts can be extended from the leg mounts through the roof and the 2 × 6s. To avoid exerting too much pressure on the area of the roof between rafters, place another set of 2 × 6s on top of the roof (a mirror image of the ones in the attic). Installation details are shown in Figs 35 through 38.

The 2 × 6s are cut 4 inches longer than the outside distance between two rafters. Bolts are cut from a length of ¼-inch threaded rod. Nails are used to hold the boards in place during installation, and roofing tar is used to seal the area to prevent leaks.

Find a location on the roof that will allow the antenna to turn without obstruction from such things as trees, TV antennas and chimneys. Determine the rafter locations. (Chimneys and vent pipes make good reference points.) Now the tower is set in place atop three 2 × 6s. A plumb line run from the top center of the tower can be used to center it on the peak of the roof. Holes for the mounting bolts can now be drilled through the roof.

Before proceeding, the bottom of the 2 × 6s and the area of the roof under them should be given a coat of roofing tar. Leave about 1/8 inch of clear area around the holes to ensure easy passage of the bolts. Put the tower back in place and insert the bolts and tighten them. Apply tar to the bottom of the legs and the wooden supports, including the bolts. For added security the tripod can be guyed. Guys should be anchored to the frame of the house.

If a rotator is to be mounted above the tripod, pressure will be applied to the bearings. Wind load on the antenna will be translated into a "pinching" of one side of the bearings. Make sure that the rotator is capable of handling this additional stress.

ROTATOR SYSTEMS

There are not many choices when it comes to antenna rotators for the amateur antenna system. However, making the correct decision as to how much capacity the rotator must have is very important if trouble free operation is desired. There are basically four grades of rotators available to the amateur. The lightest duty rotator is the type typically used

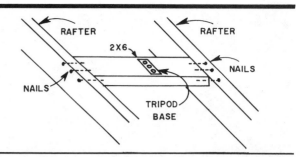

Fig 36—This cutaway view illustrates how the tripod tower is secured to the roof rafters. The leg to be secured to the cross piece is placed on the outside of the roof. Another cross member is fastened to the underside of the rafters. Bolts, inserted through the roof and the two cross pieces, hold the inner cross member in place because of pressure applied. The inner cross piece can be nailed to the rafter for added strength.

Fig 35—This tripod tower supports a rotary beam antenna. In addition to saving yard space, a roof-mounted tower can be more economical than a ground-mounted tower. A ground lead fastened to the lower part of the frame is for lightning protection. The rotator control cable and the coaxial line are dressed along two of the legs. *(photo courtesy of Jane Wolfert)*

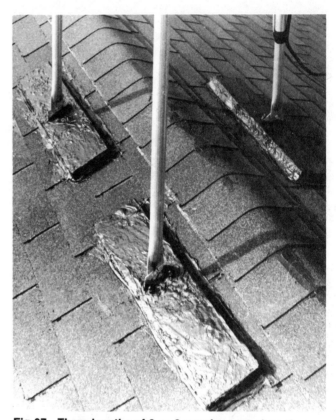

Fig 37—Three lengths of 2 × 6 wood mounted on the outside of the roof and reinforced under the roof by three identical lengths provide a durable means for anchoring the tripod. A thick coat of roofing tar guards against weathering and leaks.

to turn TV antennas. Without much difficulty, these rotators will handle a small three-element tribander array (14, 21 and 28 MHz) or a single 21- or 28-MHz monoband three-element antenna. The important consideration with a TV rotator is that it lacks braking or holding capability. High winds turn the rotator motor via the gear train in a reverse fashion. Broken gears sometimes result.

The next grade up from the TV class of rotator usually includes a braking arrangement whereby the antenna is held in place when power is not applied to the rotator. Generally speaking, the brake prevents gear damage on windy days. If adequate precautions are taken, this group of rotators is capable of holding and turning stacked monoband arrays, or up to a five-element 14-MHz system. The next step up in rotator strength is more expensive. This class of rotator will turn just about anything the most demanding amateur might want to install.

Fig 38—The strengthened anchoring for the tripod. Bolts are placed through two 2 × 6s on the underside of the roof and through the 2 × 6 on the top of the roof, as shown in Fig 37.

and as far below the top of the tower as possible. The mast absorbs the torsion developed by the antenna during high winds, as well as during starting and stopping.

Some amateurs have used a long mast from the top to the base of the tower. Rotator installation and service can be accomplished at ground level. A mast length of 10 feet or more between the rotator and the antenna will add greatly to the longevity of the entire system. Another benefit of mounting the rotator 10 feet or more below the antenna is that any misalignment among the rotator, mast and the top of the tower is less significant. A tube at the top of the tower (a sleeve bearing) through which the mast protrudes almost completely eliminates any lateral forces on the rotator casing. All the rotator must do is support the downward weight of the antenna system and turn the array.

While the normal weight of the antenna and the mast is usually not more than a couple of hundred pounds, even with a large system, one can ease this strain on the rotator by installing a thrust bearing at the top of the tower. The bearing is then the component that holds the weight of the antenna system, and the rotator need perform only the rotating task.

Indicator Alignment

A problem often encountered in amateur installations is that of misalignment between the direction indicator in the rotator control box and the heading of the antenna. With a light duty rotator, this happens frequently when the wind blows the antenna to a different heading. With no brake, the gear train and motor of the rotator are moved by the wind force, while the indicator remains fixed. Such rotator systems have a mechanical stop to prevent continuous rotation during operation, and provision is usually included to realign the indicator against the mechanical stop from inside the shack. During installation, the antenna must be oriented correctly for the mechanical stop position, which is usually north.

In larger rotator systems with an adequate brake, indicator misalignment is caused by mechanical slippage in the antenna boom-to-mast hardware. Many texts suggest that the boom be pinned to the mast with a heavy duty bolt and the rotator be similarly pinned to the mast. There is a trade-off here. If there is sufficient wind to cause slippage in the couplings without pins, with pins the wind could break a rotator casting. The slippage will act as a clutch release, which *may* prevent serious damage to the rotator. On the other hand, the amateur might not like to climb the tower and realign the system after each heavy windstorm.

A description of antenna rotators would not be complete without the mention of the prop pitch class. The prop pitch rotator system consists of a surplus aircraft propeller blade pitch motor coupled to an indicator system and a power supply. There are mechanical problems of installation, however, resulting mostly from the size and weight of these motors. It has been said that a prop pitch rotator system, properly installed, is capable of turning a house. Perhaps in the same class as the prop pitch motor (but with somewhat less capability) is the electric motor of the type used for opening garage doors. These have been used successfully in turning large arrays.

Proper installation of the antenna rotator can provide many years of trouble free service; sloppy installation can cause problems such as a burned out motor, slippage, binding and casting breakage. Most rotators are capable of accepting mast sizes of different diameters, and suitable precautions must be taken to shim an undersized mast to assure dead center rotation. It is very desirable to mount the rotator inside

Delayed-Action Braking for the Ham-M Rotator

On most rotators equipped with braking capabilities, the brake is applied almost instantly after power is removed from the rotator motor to stop the array from rotating and hold it at a chosen bearing. Because of inertia, however, the array itself does not stop rotating instantly. The larger and heavier the antenna, the more it tends to continue its travel, in which case the mast may absorb the torsion, the entire tower may twist back and forth, or the brake of the rotator may shear or jam. A more suitable system involves removing power from

the rotator motor during rotation before the desired bearing is reached, allowing the beam to coast to a slower speed or to a complete stop before the brake is applied. Delayed action braking may be added to the Ham-M rotator system by adding a couple of components inside the control head case. Fig 39 is a partial schematic diagram showing the necessary changes.

Circuit Operation

The 5-kΩ relay is energized by the operating switch and

held closed after release for approximately 1¾ seconds by means of the 500-μF capacitor. The relay contacts supply 120 V to the primary of the main transformer, which continues to hold the brake off after rotation power is removed.

Note the addition of the 200-μF capacitor in parallel with the original 30-μF filter. This is required because the 500-μF capacitor across the relay coil increases the control voltage, thereby causing approximately a 15° error between readings.

The additional 200-μF capacitor increases the control voltage such that identical readings are obtained during rotation or at rest. This modification also causes the unit to read position whenever it is plugged in. An ON-OFF switch is easily added.

To increase the indicator lamp life, change the lamps to 28-V types. The relay is approximately 1/2 × 5/8 inch and fits nicely near the left front just above the screwdriver-adjust calibration control. The capacitors are fitted easily near the rear of the meter.

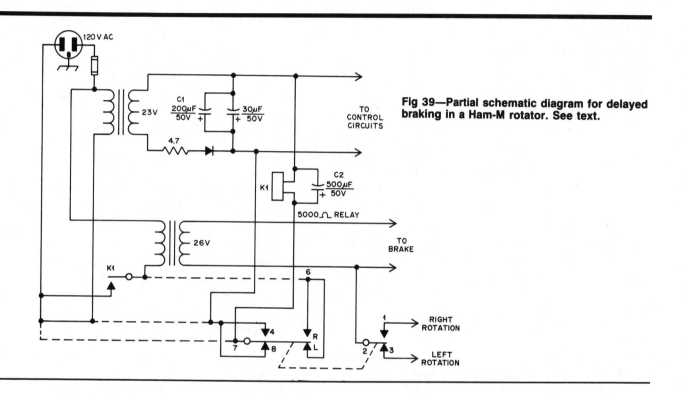

Fig 39—Partial schematic diagram for delayed braking in a Ham-M rotator. See text.

A Delayed Brake Release for the Ham-II

Not only is it wise to delay braking in a rotator system, but it is even more important that rotation in the opposite direction is not initiated until the system is at rest. The circuit presented in Fig 40 offers the protection of delayed braking, and it also disables the direction selector switches. In this manner, the antenna system coasts to a stop before rotation may begin in the opposite direction. The automatic delay prevents damage to the antenna system and rotator, even during a contest when the operator's attention is not on the rotator control.

In the circuit of Fig 40, S3, S4 and S5 are the existing Ham-II control unit brake release and direction switches. S4 selects clockwise (cw) rotation and S5 selects counter-clockwise (ccw) rotation. These switches are replaced by K3, K4 and K5, respectively, in the modified control unit.

A pair of NAND gates in U1 form a debouncing circuit for each direction switch to prevent false triggering of the

brake from contact bounce. Pressing S4 causes pin 3 of U2 to go high ($+V_{DD}$), or to a logical 1, which forces pin 3 of U3 low (near 0 V), pin 11 of U5 high, and energizes both the brake-release relay K3 and the BRAKE RELEASED LED, D1. In addition, pressing only S4 forces pin 10 of U2 low and pin 11 of U3 high, energizing K4, the cw rotation control relay. When S4 is released, a short pulse appears at pin 2 of U4, triggering the monostable multivibrator. While pin 3 of U4 is high, the brake remains released, and the selection switches are disabled by the logical 1 on pins 9 and 13 of U3. In a similar fashion, pressing S5 energizes the brake-release relay K3, LED D1, and the cw rotation control relay, K5. Whenever one of the direction control relays is energized, the ROTATE LED, D2, illuminates to indicate the rotator is turning.

The circuit has been designed to detect the simultaneous selection of both rotation directions using a NAND gate in U2. If both are pressed, a transition to 0 at pin 4 of U2 triggers

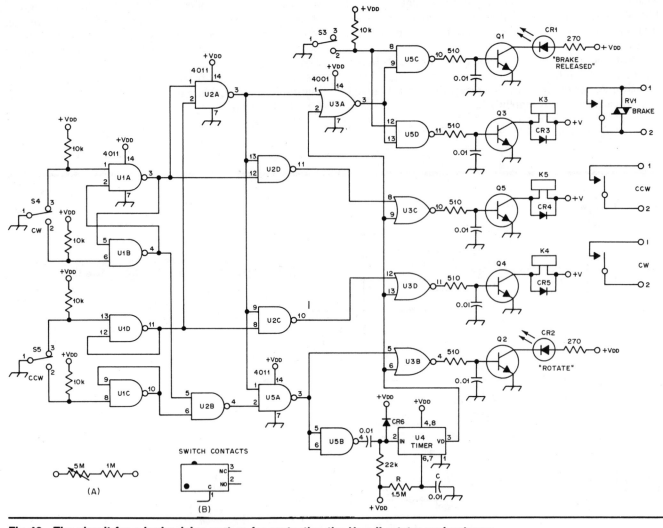

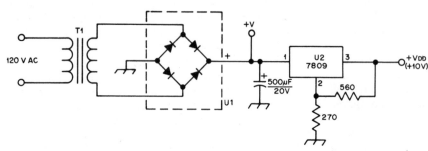

Fig 40—The circuit for a brake-delay system for protection the Ham-II rotator and antenna.

D1, D2—Light-emitting diode, Motorola type MLED600 or equiv.
D3-D6, incl—Silicon signal diode, 1N914 or equiv.
K3-K5, incl—Switching relay, 12 V dc, 1200 ohms, 10 mA; contact rating 1 A; 125 V ac; Radio Shack 275-003 or equiv.

Q1-Q5, incl—Silicon NPN transistor, 2N3904 or equiv.
RV1—Varistor, GE 750 or equiv.
U1, U2, U5—CMOS quad NAND-gate IC RCA CD-4011A or equiv.
U3—CMOS quad NOR-gate IC, RCA CD-4001A, or equiv.
U4—Timer IC, 555 or equiv.

Fig 41—Regulated power supply for the delayed brake release system.

T1—Power transformer; pri 120 V; sec 12 V, 300 mA; Radio Shack 273-1385 or equiv.
U1—Bridge rectifier, 50 PIV, 1.5 A; Radio Shack 267-1151 or equiv.
U2—Monolithic three-terminal positive-voltage regulator, 9 V, 500 mA; Fairchild 7809 or equiv.

the monostable multivibrator, forcing a brake delay period. In this way, the rapid rocking of the antenna back and forth is eliminated. After the end of the delay cycle, if both direction switches are still pressed, neither control relay is energized, because both pins 8 and 12 of U3 are high, keeping Q4 and Q5 off.

If a longer delay is desired, the brake can be released manually with S3. D1 signals when the brake-release is energized, but no delay cycle is initiated.

U4, the delay timer (NE555) is connected in a monostable multivibrator configuration. The components R and C at pins 6 and 7 determine the length of the delay. The values shown provide a delay period of about 3 seconds. An alternative is to use a potentiometer for R as shown in Fig 40A to yield a variable delay of 2 to 8 seconds.

Construction

CMOS integrated circuits were used in this design because of their high noise margin, low power dissipation, and tolerance of varying supply voltage. CMOS units operate with a V_{DD} ranging from 3 to 15 volts, although the 10-volt regulator shown in Fig 41 is used in this unit. TTL circuits may be substituted, but some RF immunity is sacrificed and, of course, the pin connections of the devices are different.

The transistor drivers Q1 through Q5 are necessary, as the CMOS devices cannot sink enough current to energize either the relays or the LEDs. The 0.01-μF capacitor on the base of each transistor is included to eliminate false keying of the relays by stray RF. An added precaution is the transient suppressor shown across the contacts of K3. The brake-release relay connects the line voltage to the primary of the brake and rotation power transformer. Without the suppressor, the contacts of K3 would pit badly because of arcing when the relay contacts open.

Fig 42—Modification of the Ham-II control unit showing the Vector circuit board and components.

The circuit as shown in Fig 42 is constructed on a Vector IC circuit breadboard using IC sockets and standard wire-wrap techniques. Homemade printed-circuit boards or other fabrication techniques could also be used, as the layout is not critical.

Fig 43 illustrates the Ham-II circuit modifications. Relays K3, K4 and K5 replace S3, S4 and S5 in the original diagram,

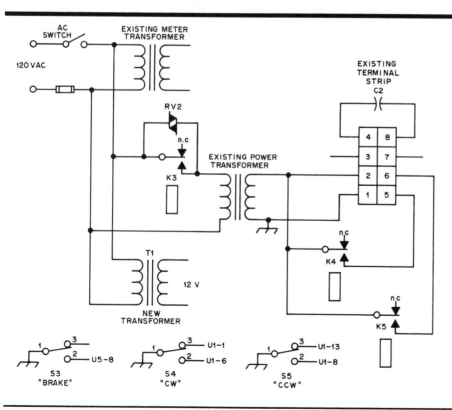

Fig 43—The Ham-II circuit modifications. T1 is the power supply transformer shown in Fig 41.

and the primary of a small 12-V power transformer is connected to the control unit ac power switch.

There is more than enough room beneath the Ham-II chassis to mount the delay-circuit card. It may be necessary to relocate the phasing capacitor, C2, above the chassis. The wires that were originally connected to S3, S4 and S5 are relocated, connecting them to the corresponding relay contacts. The switches are connected to the delay circuit inputs. These are single-pole double-throw microswitches with the contact configuration shown in Fig 40B. The LEDs are mounted below the switches in the front panel, as pictured in Fig 44.

Operation

The modified rotator control unit is used in the same manner as always, except that the operation of S3 (the brake release) is now automatic. Both LEDs, D1 and D2, are illuminated during rotation and D1 (BRAKE RELEASED) remains on through the brake delay cycle after rotation. Because an average size antenna coasts approximately 10°, the operator must release the rotation switch about 10° before the antenna reaches the desired heading. With practice, the early release becomes natural.

BIBLIOGRAPHY

Source material and more extended discussions of the topics covered in this chapter can be found in the references listed below and in the texts listed at the end of Chapter 2.

L. H. Abraham, "Guys for Guys Who Have To Guy,", *QST*, Jun 1955.

K. Baker, "A Ladder Mast," *QST*, Jun 1981, p 24.

W. R. Gary, "Toward Safer Tower Installations," *QST*, Jan 1980.

C. L. Hutchinson, "A Tree-Mounted 30-Meter Ground-Plane Antenna," *QST*, Sep 1984, pp 16-18.

M. P. Keown and L. L. Lamb, "A Simple Technique for Tower-Section Separation," *QST*, Sep 1979, pp 37-38.

R. A. Lodwig, "Wind Force on A Yagi Antenna," *QST*, Jul 1974.

P. O'Dell, "The Ups and Downs of Towers," *QST*, Jul 1981, p 35.

S. Phillabaum, "Installation Techniques for Medium and Large Yagis," *QST*, Jun 1979, p 23.

A. B. White, "A Delayed Brake Release for the Ham-II," *QST*, Aug 1977, p 14.

B. White, E. White and J. White, "Assembling Big Antennas on Fixed Towers," *QST*, Mar 1982, pp 28-29

L. Wolfert, "The Tower Alternative," *QST*, Nov 1980, p 36.

Structural Standards for Steel Antenna Towers and Antenna Supporting Structures, EIA Standard EIA-222-D, Electronic Industries Association, Oct 1986. May be purchased from EIA, 2001 Eye St, NW, Washington, DC 20006.

Fig 44—A view of the control panel of the Ham-II rotator.

Chapter 23

Radio Wave Propagation

Because radio communication is carried on by means of electromagnetic waves traveling through the earth's atmosphere, it is important to understand the nature of these waves and their behavior in the propagation medium. Most antennas will radiate the power applied to them efficiently, but if they are not erected in a manner that allows the energy they radiate to reach desired destinations, the time and money that they represent will not have been well spent. No antenna can do all things well under all circumstances. Whether you design and build your own antennas, or buy them and have them put up by a professional, you'll need propagation know-how for best results.

THE NATURE OF RADIO WAVES

You probably have some familiarity with the concept of electric and magnetic fields. A radio wave is a combination of both, with the energy divided equally between them. If the wave could originate at a point source in free space, it would spread out in an ever-growing sphere, with the source at the center. No antenna can be designed to do this, but the theoretical *isotropic antenna* is useful in explaining and measuring the performance of practical antennas we *can* build. It is, in fact, the basis for any discussion or evaluation of antenna performance.

Our theoretical spheres of radiated energy would expand very rapidly—at the same speed as the propagation of light—approximately 186,000 miles or 300,000,000 meters per second. These values are close enough for practical purposes, and are used throughout this book. If one wishes to be more precise, light propagates in a vacuum at the speed of 299.793077 meters per microsecond, and slightly slower in air.

The path of a ray from the source to any point on the spherical surface is considered to be a straight line—a radius of the sphere. An observer on the surface of the sphere would think of it as flat, just as the earth seems flat to us. A radio wave far enough from its source to appear flat is called a *plane wave*. We will be discussing primarily plane waves from here on.

It helps to understand the radiation of electromagnetic energy if we visualize a plane wave as being made up of electric and magnetic forces as shown in Fig 1. The nature of wave propagation is such that the electric and magnetic lines of force are always perpendicular. The plane containing the sets of crossed lines represents the wave front. The direction of travel is always perpendicular to the wave front; "forward" or "backward" is determined by the relative directions of the electric and magnetic forces.

The speed of travel of a wave through anything but a vacuum is always less than 300,000,000 meters per second. How much less depends on the medium. If it is air, the reduction in propagation speed can be ignored in most

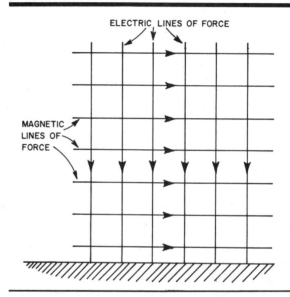

Fig 1—Representation of the magnetic and electric fields of a vertically polarized plane wave traveling along the ground. The arrows indicate instantaneous directions of the fields for a wave traveling perpendicularly out of the page toward the reader. Reversal of the direction of one set of lines reverses the direction of travel. There is no change in direction when both sets are reversed. Such a dual reversal occurs, in fact, once each half cycle.

discussions of propagation at frequencies below 30 MHz. In the VHF range and higher, temperature and moisture content of the medium have increasing effects on the communications range, as will be discussed later. In solid insulating materials the speed is considerably less. In distilled water (a good insulator) the speed is 1/9 that in free space. In good conductors the speed is so low that the opposing fields set up by the wave front occupy practically the same space as the wave itself, and thus cancel it out. This is the reason for "skin effect" in conductors at high frequencies, making metal enclosures good shields for electrical circuits working at radio frequencies.

Phase and Wavelength

Because the velocity of wave propagation is so great, we tend to ignore it. Only 1/7 of a second is needed for a radio wave to travel around the world—but in working with antennas the *time* factor is extremely important. The wave concept evolved because an alternating current flowing in a wire (antenna) sets up moving electric and magnetic fields. We can hardly discuss antenna theory or performance at all without involving travel time, consciously or otherwise.

Waves used in radio communication may have frequencies from about 10,000 to several billion hertz (Hz). Suppose the frequency is 30,000,000 Hz, more commonly called 30 megahertz (MHz). One cycle, or period, is completed in 1/30,000,000 second. The wave is traveling at 300,000,000 meters per second, so it will move only 10 meters during the time that the current is going through one complete period of alternation. The electromagnetic field 10 meters away from the antenna is caused by the current that was flowing one period earlier in time. The field 20 meters away is caused by the current that was flowing two periods earlier, and so on.

If each period of the current is simply a repetition of the one before it, the currents at corresponding instants in each period will be identical, and the fields caused by those currents will also be identical. As the fields move outward from the antenna they become more thinly spread over larger and larger surfaces; their amplitudes decrease with distance from the antenna, but they do not lose their identity with respect to the instant of the period at which they were generated. They are, and they remain, *in phase*. In the example above, at intervals of 10 meters, measured outward from the antenna, the phase of the waves at any given instant is identical.

From this information we can define both "wave front" and "wavelength." Consider the wave front as an imaginary surface. On every part of this surface, the wave is in the same phase. The wavelength is the distance between two wave fronts having the same phase at any given instant. This distance must be measured perpendicular to the wave fronts—along the line that represents the direction of travel. The abbreviation for wavelength is the Greek letter lambda, λ, which is used throughout this book.

Expressed in an equation, the length of a wave is

$$\lambda = \frac{v}{f}$$

where

λ = wavelength in distance
v = velocity of wave in distance per unit time
f = frequency of current causing the wave, in cycles per unit time

The wavelength will be in the same length units as the velocity when the frequency is expressed in the same time units as the velocity. For waves traveling in free space (and near enough for waves traveling through air) the wavelength is

$$\lambda \ (\text{meters}) = \frac{300}{f(\text{MHz})}$$

There will be few pages in this book where phase, wavelength and frequency do not come into the discussion. It is essential to have a clear understanding of their meaning in order to understand the design, installation, adjustment or use of antennas, matching systems or transmission lines in detail. In essence, "phase" means "time." When something goes through periodic variations, as an alternating current does, corresponding instants in succeeding periods are in phase.

The points A, B and C in Fig 2 are all in phase. They are corresponding instants in the current flow, at 1-λ intervals. This is a conventional view of a sine-wave alternating current, with time progressing to the right. It also represents a "snapshot" of the intensity of the traveling fields, if distance is substituted for time in the horizontal axis. The distance between A and B or between B and C is one wavelength. The

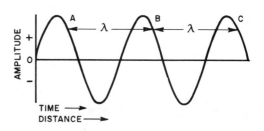

Fig 2—The instantaneous amplitude of both fields (electric and magnetic) varies sinusoidally with time as shown in this graph. Since the fields travel at constant velocity, the graph also represents the instantaneous distribution of field intensity along the wave path. The distance between two points of equal phase, such as A-B and B-C, is the length of the wave.

field-intensity distribution follows the sine curve, in both amplitude and polarity, corresponding exactly to the time variations in the current that produced the fields. Remember that this is an *instantaneous* picture—the wave moves outward much as a wave in water does.

Polarization

A wave like that in Fig 1 is said to be polarized in the direction of the electric lines of force. The polarization here is vertical, because the electric lines are perpendicular to the earth. It is one of the laws of electromagnetic action that electric lines touching the surface of a good conductor must do so perpendicularly. Most ground is a rather good conductor at frequencies below about 10 MHz, so waves of these frequencies, over good ground, are mainly vertically polarized. Over partially conducting ground there may be a forward tilt to the wave front, the tilt in the electric lines of force increasing as the energy loss in the ground becomes greater.

Waves traveling in contact with the surface of the earth are of little use in amateur communication because, as the frequency is raised, the distance over which they will travel without excessive energy loss becomes smaller and smaller. The surface wave is most useful at low frequencies and through the standard broadcast band. At high frequencies the wave reaching the receiving antenna has had little contact with the ground, and its polarization is not necessarily vertical.

If the electric lines of force are horizontal, the wave is said to be horizontally polarized. Horizontally and vertically polarized waves may be classified generally under linear polarization. Linear polarization can be anything between horizontal and vertical. In free space, "horizontal" and "vertical" have no meaning, the reference of the seemingly horizontal surface of the earth having been lost.

In many cases the polarization of waves is not fixed, but rotates continually, somewhat at random. When this occurs the wave is said to be elliptically polarized. A gradual shift in polarization in a medium is known as Faraday rotation. For space communications, circular polarization is commonly used to overcome the effects of Faraday rotation. A circularly polarized wave rotates its polarization through 360° as it travels a distance of one wavelength in the propagation medium. The direction of rotation as viewed from the transmitting antenna defines the direction of circularity, right-hand (clockwise) or left-hand (counterclockwise). Linear and circular polarization may be considered as special cases of elliptical polarization.

Field Intensity

The energy from a propagated wave decreases with distance from the source. This decrease in strength is caused by the spreading of the wave energy over ever larger spheres as the distance from the source increases.

A measurement of the strength of the wave at a distance from the transmitting antenna is its field intensity, which is synonymous with field strength. The strength of a wave is measured as the voltage between two points lying on an electric line of force in the plane of the wave front. The standard of measure for field intensity is the voltage developed in a wire that is one meter long, expressed as volts per meter. (If the wire were 2 meters long, the voltage developed therein would be divided by 2 to determine the field strength in volts per meter.)

The voltage in a wave is usually quite low, so the measurement is made in millivolts or microvolts per meter. The voltage goes through time variations like those of the current that caused the wave. It is measured like any other ac voltage—in terms of the effective value or, sometimes, the peak value.

It is fortunate that in amateur work it is not necessary to measure actual field strength, as the equipment required is elaborate. We need to know only if an adjustment has been beneficial, so relative measurements are quite satisfactory. These can be made quite easily, with homebuilt equipment.

Wave Attenuation

In free space the field intensity of the wave varies inversely with the distance from the source. If the field strength at one mile from the source is 100 millivolts per meter, it will be 50 millivolts per meter at two miles, and so on.

The relationship between field intensity and power density is similar to that for voltage and power in ordinary circuits. They are related by the impedance of free space, which has been determined to be 377 ohms. A field intensity of one volt per meter is therefore equivalent to a power density of

$$P = \frac{E^2}{Z} = \frac{(1 \text{ volt/m})^2}{377}$$

$$= \frac{1}{377} \text{ W/square meter} = 2.65 \text{ mW per square meter}$$

Because of the relationship between voltage and power, the power density therefore varies with the square root of the field intensity, or inversely with the *square* of the distance. If the power density at one mile is 4 mW per square meter, then at a distance of two miles it will be 1 mW per square meter.

It is important to remember this spreading loss or attenuation when antenna performance is being considered. Gain can come only from narrowing the radiation pattern of an antenna, to concentrate the radiated energy in the desired direction. There is no "antenna magic" by which the total energy radiated can be increased.

In practice, attenuation of the wave energy may be much greater than the "inverse-distance" law would indicate. The wave does not travel in a vacuum, and the receiving antenna seldom is situated so there is a clear line of sight. The earth is spherical and the waves do not penetrate its surface appreciably, so communication beyond visual distances must be by some means that will bend the waves around the curvature of the earth. These means involve additional energy losses that increase the path attenuation with distance, above that for the theoretical spreading loss in a vacuum.

Bending of Radio Waves

Radio waves and light waves are both propagated as electromagnetic energy; their major difference is in wavelength, though radio-reflecting surfaces are usually much smaller in terms of wavelength than those for light. In material of a given electrical conductivity, long waves penetrate farther than short ones, and so require a thicker mass for good reflection. Thin metal is a good reflector of radio waves of even quite long wavelength. With poorer conductors, such as the earth's crust, long waves may penetrate quite a few feet below the surface.

Reflection occurs at any boundary between materials of differing dielectric constant. Familiar examples with light are reflections from water surfaces and window panes. Both water and glass are transparent for light, but their dielectric constants are very different from that of air. Light waves, being very short, seem to "bounce off" both surfaces. Radio waves, being much longer, are practically unaffected by glass, but their behavior upon encountering water may vary, depending on the purity of that medium. Distilled water is a good insulator; salt water is a relatively good conductor.

Depending on their length (or frequency), radio waves may be reflected by buildings, trees, vehicles, the ground, water, ionized layers in the upper atmosphere, or at boundaries between air masses having different temperatures and moisture content. Most of these factors can affect antenna performance. Ionospheric and atmospheric conditions are important in practically all communication beyond purely local ranges.

Refraction is the bending of a ray as it passes from one medium to another at an angle. The appearance of bending of a straight stick, where it is made to enter water at an angle, is an example of light refraction known to us all. The degree of bending of radio waves at boundaries between air masses increases with the radio frequency. There is some slight atmospheric effect in our HF bands. It becomes noticeable at 28 MHz, more so at 50 MHz, and it is much more of a factor in the higher VHF range and in UHF and microwave propagation.

Diffraction of light over a solid wall prevents total darkness on the far side from the light source. This is caused largely by the spreading of waves around the top of the wall, from the interference of one part of the beam with another. The dielectric constant of the surface of the obstruction may affect what happens to our radio waves when they encounter terrestrial obstructions—but the radio "shadow area" is never totally "dark."

The three terms, reflection, refraction and diffraction, were in use long before the radio age began. Radio propagation is nearly always a mix of these phenomena, and it may not be easy to identify or separate them in practical communications experience. This book tends to rely on the words *bending* and *scattering* in its discussions, with appropriate modifiers as needed. The important thing to remember is that any alteration of the path taken by energy as it is radiated from an antenna is almost certain to affect on-the-air results—which is why this chapter on *propagation* is included in an *antenna* book.

The Ground Wave

As we have already seen, radio waves are affected in many ways by the media through which they travel. This has led to some confusion of terms in literature concerning wave propagation. Waves travel close to the ground in several ways, some of which involve relatively little contact with the ground itself. The term *ground wave* has had several meanings in antenna literature, but more or less by common consent it has come to be applied to any wave that stays close to the earth, reaching the receiving point without leaving the earth's lower atmosphere. It could be traveling in actual contact with the ground, as in Fig 1. It could travel directly between the transmitting and receiving antennas, when they are high enough so they can "see" each other. In this text we include waves that are made to follow the earth's curvature by bending in the earth's lower atmosphere, or troposphere, usually no more than a few miles above the ground. Often called *tropospheric bending*, this mode is a major factor in amateur communication above 50 MHz.

The Surface Wave

The surface wave travels in contact with the earth's surface. It can provide coverage up to about 100 miles in the standard broadcast band in daytime, but attenuation is quite high. As can be seen from Fig 3, the attenuation increases with frequency. The surface wave is of little value in amateur communication, except possibly at 1.8 MHz. Vertical antennas must be used, which tends to limit amateur surface-wave communication except where large vertical systems can be erected.

ground, the angle at which the wave strikes the ground will be rather small. For a horizontally polarized signal, such a reflection reverses the phase of the wave. If the distances traveled by both parts of the wave were the same, the two parts would arrive out of phase, and would therefore cancel each other.

The ground-reflected ray in Fig 4 must travel a little farther, so the phase difference between the two depends on the lengths of the paths, measured in wavelengths. The wavelength in use is important in determining the useful signal strength in this type of communication.

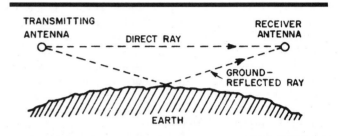

Fig 4—The ray traveling directly from the transmitting antenna to the receiving antenna combines with a ray reflected from the ground to form the space wave. For a horizontally polarized signal, a reflection as shown here reverses the phase of the ground-reflected ray.

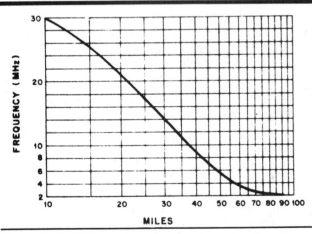

Fig 3—Typical HF ground-wave range as a function of frequency.

The Space Wave

Propagation between two antennas situated within line of sight of each other is shown in Fig 4. Energy traveling directly between the antennas is attenuated to about the same degree as in free space. Unless the antennas are very high or quite close together, an appreciable portion of the energy is reflected from the ground. This combines with direct radiation to produce the actual signal received.

In most communication between two stations on the

If the difference in path length is 3 meters, the phase difference with 160-meter waves would be only 6.8° (360° × 3/160). This is a negligible difference from the 180° shift caused by the reflection, so the effective signal strength over the path would still be very small because of cancellation of the two waves. But with 6-meter radio waves the phase length would be 360° × 3/6 = 180°. With the additional 180° shift on reflection, the two rays would add. Thus, the space wave is a negligible factor at low frequencies, but it can be increasingly useful as the frequency is raised. It is a dominant factor in local amateur communication at 50 MHz and higher.

Interaction between the direct and reflected waves is the principle cause of "mobile flutter" observed in local VHF communication between fixed and mobile stations. The flutter effect decreases once the stations are separated enough that the reflected ray becomes inconsequential. The reflected energy can also confuse the results of field-strength measurements during tests on antennas for VHF use.

As with most propagation explanations, the space-wave picture presented here is simplified, and practical considerations dictate modifications. There is energy loss when the wave is reflected from the ground, and the phase of the ground-reflected wave is not shifted exactly 180°, so the waves never cancel completely. At UHF, ground-reflection losses can be greatly reduced or eliminated by using highly directive antennas. By confining the antenna pattern to something approaching a flashlight beam, nearly all the energy is in the direct wave. The resulting energy loss is low enough that microwave relays, for example, can operate with moderate power levels over hundreds and even thousands of miles. Thus we see that, while the space wave is inconsequential below about 20 MHz, it can be a prime asset in the VHF realm and higher.

VHF Propagation Beyond Line of Sight

From Fig 4 it appears that use of the space wave depends on direct line of sight between the antennas of the communicating stations. This is not literally true, though that belief was common in the early days of amateur communication on frequencies above 30 MHz. When equipment was built that operated efficiently and antenna techniques improved, it soon became clear that VHF waves were bent or scattered in several ways, permitting reliable communication somewhat beyond visual distances between the two stations. This was found true with low power and simple antennas. The average communications range can be approximated by assuming the waves travel in straight lines, but that the earth's radius is increased by one-third. The distance to the "radio horizon" is then given as

$$D \text{ (mi)} = 1.415 \sqrt{H \text{ (ft)}}$$

or

$$D \text{ (km)} = 4.124 \sqrt{H \text{ (m)}}$$

where H is the height of the transmitting antenna, as shown in Fig 5. The formula assumes that the earth is smooth out to the horizon, so any obstructions along the path must be taken into consideration. For an elevated receiving antenna the communications distance is equal to D + D1, that is, the sum of the distances to the horizon of both antennas. Radio horizon distances are given in graphic form in Fig 6. Two stations on a flat plain, one with its antenna 60 feet above ground

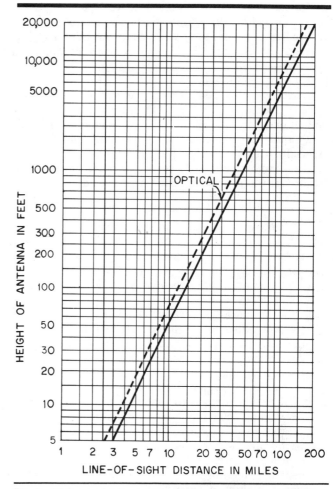

Fig 6—Distance to the horizon from an antenna of given height. The solid curve includes the effect of atmospheric refraction. The optical line-of-sight distance is given by the broken curve.

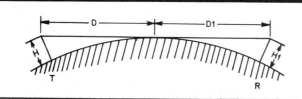

Fig 5—The distance, D, to the horizon from an antenna of height H is given by equations in the text. The maximum line of sight distance between two elevated antennas is equal to the sum of their distances to the horizon, as indicated here.

and the other 40 feet, could be up to about 20 miles apart for strong-signal line-of-sight communication (11 + 9 mi). The terrain is almost never completely flat, and variations along the way may add to or subtract from the distance for reliable communication. Remember that energy is absorbed, reflected or scattered in many ways, in nearly all communications situations. The formula or the chart will be a good guide for estimating the potential radius of coverage for a VHF FM repeater, assuming the users are mobile or portable with simple, omnidirectional antennas. Coverage with optimum home-station equipment, high-gain directional arrays, and SSB or CW is quite a different manner. A much more detailed method for estimating coverage on frequencies above 50 MHz is given later in this chapter.

For maximum use of the ordinary space wave it is important to have the antenna as high as possible above nearby buildings, trees, wires and surrounding terrain. A hill that rises above the rest of the countryside is a good location for an amateur station of any kind, and particularly so for extensive coverage on the frequencies above 50 MHz. The highest point on such an eminence is not necessarily the best location for the antenna. In the example shown in Fig 7, the hilltop would be a good site in all directions. But if maximum

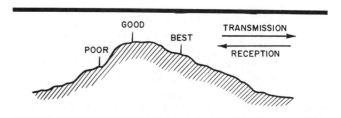

Fig 7—Propagation conditions are generally best when the antenna is located slightly below the top of a hill on the side that faces the distant station. Communication is poor when there is a sharp rise immediately in front of the antenna in the direction of communication.

performance to the right was the objective, a point just below the crest might do better. This would involve a trade-off with reduced coverage in the opposite direction. Conversely, an antenna situated on the left side, lower down the hill, might do well to the left, but almost certainly would be inferior in performance to the right.

Selection of a home site for its radio potential is a complex business, at best. A VHF enthusiast dreams of the highest hill. The DX-minded ham may be more attracted by a dry spot near a salt marsh. A wide saltwater horizon, especially from a high cliff, just *smells* of DX. In shopping for ham radio real estate, a mobile or portable rig for the frequencies you're most interested in can provide useful clues.

Antenna Height and Polarization

If effective communication over long distances were the only consideration, we would be concerned mainly with radiation of energy at the lowest possible angle above the horizon. However, being engaged in a residential avocation often imposes practical restrictions on our antenna projects. As an example, our 1.8- and 3.5-MHz bands are used primarily for short-distance communication because they serve that purpose with antennas that are not difficult or expensive to put up. Out to a few hundred miles, simple wire antennas for these bands do quite well, even though their radiation is mostly at high angles above the horizon. Vertical systems might be better for long-distance use, but they require extensive ground systems for optimum performance.

Horizontal antennas that radiate well at low angles are more easily erected for 7 MHz and higher frequencies, so horizontal wires and arrays are almost standard practice for work on 7 through 29.7 MHz. Vertical antennas are also used in this frequency range, such as a single omnidirectional antenna of multiband design. An antenna of this type may be a good solution to the space problem for a city dweller on a small lot, or even for the resident of an apartment building.

High-gain antennas are almost always used at 50 MHz and higher frequencies, and most of them are horizontal. The principal exception is mobile communication with FM, through repeaters, discussed in Chapter 17. The height question is answered easily for VHF enthusiasts—the higher the better.

The theoretical and practical effects of height above ground are treated in detail in Chapter 3. Note that it is the height in *wavelengths* that is important—a good reason to think in the metric system, rather than in feet and inches.

In working locally on any amateur frequency band, best results will be obtained with the same polarization at both stations, except on rather rare occasions when polarization shift is caused by terrain obstructions or reflections from buildings. Where such shift is observed, mostly above 100 MHz or so, horizontal polarization tends to work better than vertical. This condition is found primarily on short paths, so it is not too important. Polarization shift may occur on long paths where tropospheric bending is a factor, but here the effect tends to be random. Long-distance communication by way of the ionosphere produces random polarization effects routinely, so polarization matching is of little or no importance. This is fortunate for the HF mobile enthusiast, who will find that even his short, inductively loaded whips work very well at all distances other than local.

Because it responds to all plane polarizations equally, circular polarization may pay off on circuits where the arriving polarization is random, but it exacts a 3-dB penalty when used with a single-plane polarization of any kind. Circular systems find greatest use in work with orbiting satellites. It should be remembered that "horizontal" and "vertical" are meaningless terms in space, where the plane-earth reference is lost.

Polarization Factors Above 50 MHz

In most VHF communication over short distances, the polarization of the space wave tends to remain constant. Polarization discrimination is quite high, usually in excess of 20 dB, so the same polarization should be used at both ends of the circuit. Horizontal, vertical and circular polarization all have certain advantages, above 50 MHz, so there has never been complete standardization on any one of them.

Horizontal systems are popular, in part because they tend to reject man-made noise, much of which is vertically polarized. There is some evidence that vertical polarization shifts to horizontal in hilly terrain, more readily than horizontal shifts to vertical. With large arrays, horizontal systems may be easier to erect, and they tend to give higher signal strengths over irregular terrain, if any difference is observed.

Practically all work with VHF mobiles is now handled with vertical systems. For use in a VHF repeater system, the vertical antenna can be designed to have gain without losing the desired omnidirectional quality. In the mobile station a small vertical whip has obvious esthetic advantages. Often a telescoping whip used for broadcast reception can be pressed into service for the 144-MHz FM rig. A car-top mount is preferable, but the broadcast whip is a practical compromise. Tests with at least one experimental repeater have shown that horizontal polarization can give a slightly larger service area, but mechanical advantages of vertical systems have made them the almost unanimous choice in VHF FM communication. Except for the repeater field, horizontal is the standard VHF system almost everywhere.

In communication over the earth-moon-earth (EME) route the polarization picture is blurred, as might be expected with such a diverse medium. If the moon were a flat target we could expect a 180-degree phase shift from the moon reflection process. But it is not flat, and the moon's libration and wave travel both ways through the earth's entire atmosphere and magnetic field provides other variables that confuse the phase and polarization issue. Building a huge array that will track the moon, and give gains in excess of 20 dB, is enough of a task that most EME enthusiasts tend to take their chances with phase and polarization problems. Where rotation of the element plane has been tried it has helped to stabilize signal levels, but it is not widely employed.

Tropospheric Propagation of VHF Waves

The effects of changes in the dielectric constant of the propagation medium were discussed earlier. Varied weather patterns over most of the earth's surface can give rise to boundaries between air masses of very different temperature and humidity characteristics. These boundaries can be anything from local anomalies to air-circulation patterns of continental proportions.

Under stable weather conditions, large air masses can retain their characteristics for hours or even days at a time. See Fig 8. Stratified warm dry air over cool moist air, flowing slowly across the Great Lakes region to the Atlantic Seaboard, can provide the medium for east-west communication on 144 MHz and higher amateur frequencies over as much as

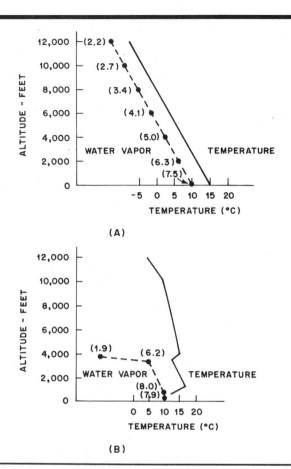

Fig 8—Upper air conditions that produce extended range communication on the VHF bands. At the top is shown the US Standard Atmosphere temperature curve. The humidity curve (dotted) is what would result if the relative humidity were 70% from the ground level to 12,000 feet elevation. There is only slight refraction under this standard condition. At the bottom is shown a sounding that is typical of marked refraction of VHF waves. Figures in parentheses are the "mixing ratio"—grams of water vapor per kilogram of dry air. Note the sharp break in both curves at about 3500 feet.

1200 miles. More common, however, are communications distances of 400 to 600 miles under such conditions.

A similar *inversion* along the Atlantic Seaboard as a result of a tropical storm air-circulation pattern may bring VHF and UHF openings extending from the Maritime Provinces of Canada to the Carolinas. Propagation across the Gulf of Mexico, sometimes with very high signal levels, enlivens the VHF scene in coastal areas from Florida to Texas. The California coast, from below the Bay Area to Mexico, is blessed with a similar propagation aid during the warmer months. Tropical storms moving west, across the Pacific below the Hawaiian Islands, may provide a transpacific long-distance VHF medium. This was first exploited by amateurs on 144, 220 and 432 MHz, in 1957. It has been used fairly often in the summer months since, although not yearly.

The examples of long-haul work cited above may occur infrequently, but lesser extensions of the minimum operating range are available almost daily. Under minimum conditions there may be little more than increased signal strength over paths that are workable at any time.

There is a diurnal effect in temperate climates. At sunrise the air aloft is warmed more rapidly than that near the earth's surface, and as the sun goes lower late in the day the upper air is kept warm, while the ground cools. In fair, calm weather the sunrise and sunset temperature inversions can improve signal strength over paths beyond line of sight as much as 20 dB over levels prevailing during the hours of high sun. The diurnal inversion may also extend the operating range for a given strength by some 20 to 50%. If you would be happy with a new antenna, try it first around sunrise!

There are other short-range effects of local atmospheric and topographical conditions. Known as *subsidence*, the flow of cool air down into the bottom of a valley, leaving warm air aloft, is a familiar summer-evening pleasure. The daily inshore-offshore wind shift along a seacoast in summer sets up daily inversions that make coastal areas highly favored as VHF sites. Ask any jealous 144-MHz operator who lives more than a few miles inland!

Tropospheric effects can show up at any time, in any season. Late spring and early fall are the most favored periods, although a winter warming trend can produce strong and stable inversions that work VHF magic almost equal to that of the more familiar spring and fall events.

Regions where the climate is influenced by large bodies of water enjoy the greatest degree of tropospheric bending. Hot, dry desert areas see little of it, at least in the forms described above.

Tropospheric Ducting

Tropospheric propagation of VHF and UHF waves can influence signal levels at all distances from purely local to something beyond 4000 km (2500 mi). The outer limits are not well known. At the risk of over-simplification we will divide the modes into two classes—extended local and long distance. This concept must be modified depending on the frequency under consideration, but in the VHF range the extended-local effect gives way to a form of propagation much like that of microwaves in a waveguide, called *ducting*. The transition distance is ordinarily somewhere around 200 miles. The basic difference lies in whether the atmospheric condition producing the bending is localized or continental in scope. Remember, we're concerned here with frequencies in the VHF range, and perhaps up to 500 MHz. At 10 GHz, for example, the scale is much smaller.

In VHF propagation beyond a few hundred miles, more than one weather front is probably involved, but the wave is propagated between the inversion layers and ground, in the main. On long paths over the ocean (two notable examples are California to Hawaii and Ascension Island to Brazil), propagation is likely to be between two atmospheric layers. On such circuits the communicating station antennas must be in the duct, or capable of propagating strongly into it. Here again, we see that the positions and radiation angles of the antennas are important. As with microwaves in a waveguide, the low-frequency limit for the duct is quite critical. In long-distance ducting it is also very variable. Airborne equipment has shown that duct capability exists well down into the HF region in the stable atmosphere west of Ascension Island. Some contacts between Hawaii and Southern California on 50 MHz are believed to have been by way of tropospheric ducts. Probably all work over these paths on 144 MHz and higher bands is because of duct propagation.

Amateurs have played a major part in the discovery and eventual explanation of tropospheric propagation. In recent years they have shown that, contrary to beliefs widely held in earlier times, long-distance communication using tropospheric modes is possible to some degree on all amateur frequencies from 50 to at least 10,000 MHz.

RELIABLE VHF COVERAGE

In the preceding sections we discussed means by which our bands above 50 MHz may be used intermittently for communication far beyond the visual horizon. In emphasizing distance we should not neglect a prime asset of the VHF bands: reliable communication over relatively short distances. The VHF region is far less subject to disruption of local communication than are frequencies below about 30 MHz. Since much amateur communication is essentially local in nature, our VHF assignments can carry a great load, and such use of the VHF bands helps solve interference problems on lower frequencies.

Because of age-old ideas, misconceptions about the coverage obtainable in our VHF bands persist. This reflects the thoughts that VHF waves travel only in straight lines, except when the DX modes described above happen to be present. However, let us survey the picture in the light of modern wave-propagation knowledge and see what the bands above 50 MHz are good for on a day-to-day basis, ignoring the anomalies that may result in extensions of normal coverage.

It is possible to predict with fair accuracy how far you should be able to work consistently on any VHF or UHF band, provided a few simple facts are known. The factors affecting operating range can be reduced to graph form, as described in this section. The information was originally published in November 1961 *QST* by D. W. Bray, K2LMG (see the bibliography at the end of this chapter).

To estimate your station's capabilities, two basic numbers must be determined: station gain and path loss. Station gain is made up of eight factors: receiver sensitivity, transmitted power, receiving antenna gain, receiving antenna height gain, transmitting antenna gain, transmitting antenna height gain and required signal-to-noise ratio. This looks complicated but it really boils down to an easily made evaluation of receiver, transmitter, and antenna performance. The other number, path loss, is readily determined from the nomogram, Fig 9. This gives path loss over smooth earth, for 99% reliability.

For 50 MHz, lay a straightedge from the distance between stations (left side) to the appropriate distance at the right side.

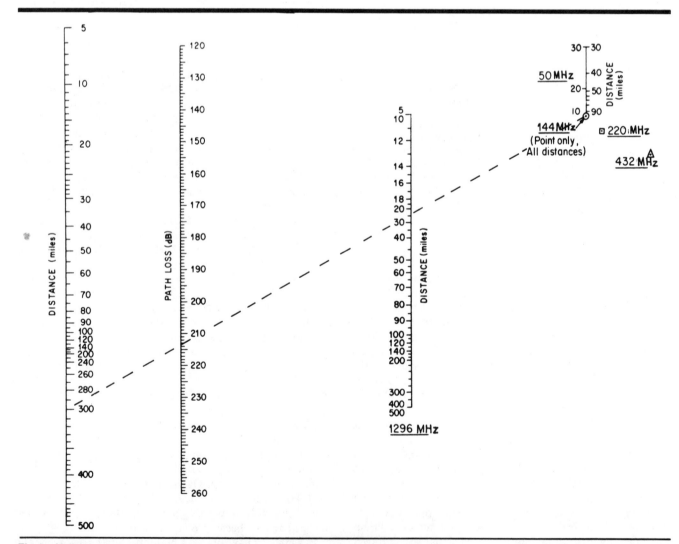

Fig 9—Nomogram for finding the capabilities of stations on amateur bands from 50 to 1300 MHz. Either the path loss for a given distance or vice versa may be found if one of the two factors is known.

For 1296 MHz, use the full scale, right center. For 144, 220 and 432, use the dot in the circle, square or triangle, respectively. Example: At 300 miles the path loss for 144 MHz is 214 dB.

To be meaningful, the losses determined from this nomograph are necessarily greater than simple free-space path losses. As described in an earlier section, communication beyond line-of-sight distances involves propagation modes that increase the path attenuation with distance.

Station Gain

The largest of the eight factors involved in station gain is receiver sensitivity. This is obtainable from Fig 10, if you know the approximate receiver noise figure and transmission-line loss. If you can't measure noise figure, assume 3 dB for 50 MHz, 5 for 144 or 220, 8 for 432 and 10 for 1296 MHz, if you know your equipment is working moderately well. These noise figures are well on the conservative side for modern solid-state receivers.

Line loss can be taken from information in Chapter 24 for the line in use, if the antenna system is fed properly. Lay a straightedge between the appropriate points at either side of Fig 10, to find effective receiver sensitivity in decibels below 1 watt (dBW). Use the narrowest bandwidth that is practical for the emission intended, with the receiver you will be using. For CW, an average value for effective work is about 500 Hz.

Phone bandwidth can be taken from the receiver instruction manual.

Antenna gain is next in importance. Gains of amateur antennas are often exaggerated. For well designed Yagis they run close to 10 times the boom length in wavelengths. (Example: A 24-foot Yagi on 144 MHz is 3.6 wavelengths long; 3.6 × 10 = 36, or about 15½ dB.) Add 3 dB for stacking, where used properly. Add 4 dB more for ground-reflection gain. This varies in amateur work, but averages out near this figure.

We have one more plus factor—antenna height gain, obtainable from Fig 11. Note that this is greatest for short distances. The left edge of the horizontal center scale is for 0 to 10 miles, the right edge for 100 to 500 miles. Height gain for 10 to 30 feet is assumed to be zero. It will be seen that for 50 feet the height gain is 4 dB at 10 miles, 3 dB at 50 miles, and 2 dB at 100 miles. At 80 feet the height gains are roughly 8, 6 and 4 dB for these distances. Beyond 100 miles the height gain is nearly uniform for a given height, regardless of distance.

Transmitter power output must be stated in decibels above 1 watt. If you have 500 watts output, add 10 log (500/1), or 27 dB, to your station gain. The transmission line loss must be subtracted from the station gain. So must the required signal-to-noise ratio. The information is based on CW work, so the additional signal needed for other modes

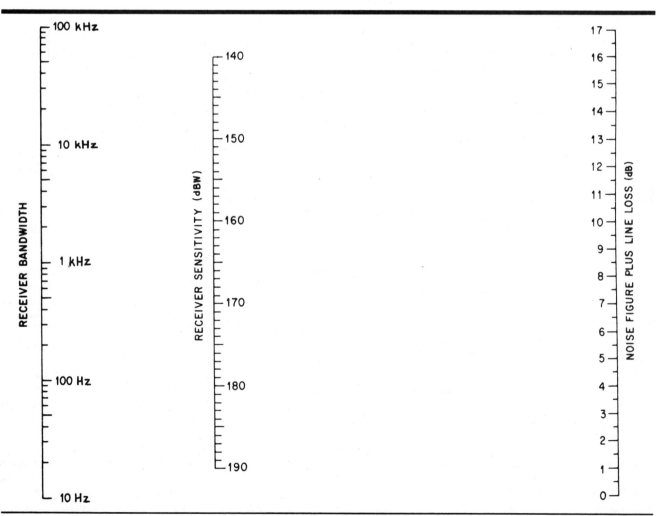

Fig 10—Nomogram for finding effective receiver sensitivity.

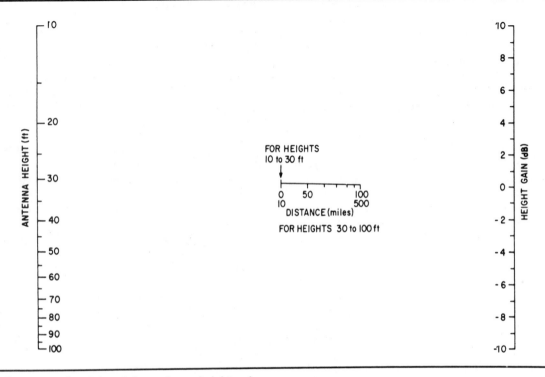

Fig 11—Nomogram for determining antenna-height gain.

must be subtracted. Use 3 dB for SSB. Fading losses must be accounted for. It has been shown that for distances beyond 100 miles the signal will vary plus or minus about 7 dB from the average level, so 7 dB must be subtracted from the station gain for high reliability. For distances under 100 miles, fading diminishes almost linearly with distance. For 50 miles, use − 3.5 dB for fading.

What It All Means

Add all the plus and minus factors to get the station gain. Use the final value to find the distance over which you can expect to work reliably, from the nomogram, Fig 9. Or work it the other way around: Find the path loss for the distance you want to cover from the nomogram and then figure out what station changes will be needed to overcome it.

The significance of all this becomes more obvious when we see path loss plotted against frequency for the various bands, as in Fig 12. At the left this is done for 50% reliability. At the right is the same information for 99% reliability. For near-perfect reliability, a path loss of 195 dB (easily encountered at 50 or 144 MHz) is involved in 100-mile communication. But look at the 50% reliability curve: The same path loss takes us out to well over 250 miles. Few amateurs demand near-perfect reliability. By choosing our times, and by accepting the necessity for some repeats or occasional loss of signal, we can maintain communication out to distances far beyond those usually covered by VHF stations.

Working out a few typical amateur VHF station setups with these curves will show why an understanding of these factors is important to any user of the VHF spectrum. Note that path loss rises very steeply in the first 100 miles or so. This is no news to VHF operators; locals are very strong, but stations 50 or 75 miles away are much weaker. What happens beyond 100 miles is not so well known to some of us.

From the curves of Fig 12, we see that path loss levels off markedly at what is the approximate limit of working range for average VHF stations using wideband modulation modes. Work out the station gain for a 50-watt station with an average receiver and moderate-sized antenna, and you'll find that it comes out around 180 dB. This means you'd have about a 100-mile working radius in average terrain, for good but no perfect reliability. Another 10 dB may extend the range to as much as 250 miles. Just changing from AM phone to SSB and CW makes a major improvement in daily coverage on the VHF bands.

A bigger antenna, a higher one if your present beam is not at least 50 feet up, an increase in power to 500 watts from 50, an improvement in receiver noise figure if it is presently poor—any of these things can make a big improvement in reliable coverage. Achieve all of them, and you will have very likely tripled your sphere of influence, thanks to that hump in the path-loss curves. This goes a long way toward explaining why using a 10-watt packaged station with a small antenna, fun though it may be, does not begin to show what the VHF bands are *really* good for.

About Terrain

The coverage figures derived from the above procedure are for average terrain. What of stations in mountainous country? Although an open horizon is generally desirable for the VHF station site, mountain country should not be considered hopeless. Help for the valley dweller often lies in the optical phenomenon known as knife-edge refraction. A flashlight beam pointed at the edge of a partition does not cut off sharply at the partition edge, but is refracted around it, partially illuminating the shadow area. A similar effect is observed with VHF waves passing over ridges; there is a shadow effect, but not a complete blackout. If the signal is

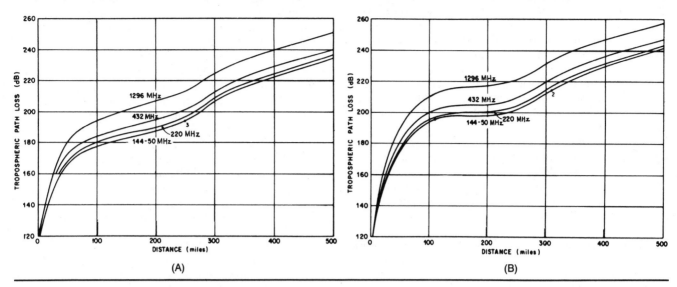

Fig 12—Path loss versus distance for amateur frequencies above 50 MHz. At A are curves for 50% of the time; at B, for 99%. The curves at A are more representative of Amateur Radio requirements.

strong where it strikes the mountain range, it will be heard well in the bottom of a valley on the far side.

This is familiar to all users of VHF communications equipment who operate in hilly terrain. Where only one ridge lies in the way, signals on the far side may be almost as good as on the near. Under ideal conditions (a very high and sharp-edged obstruction near the midpoint of a long enough path so that signals would be weak over average terrain), knife-edge refraction may yield signals even stronger than would be possible with an open path.

The obstruction must project into the radiation patterns of the antennas used. Often mountains that look formidable to the viewer are not high enough to have an appreciable effect, one way or the other. Since the normal radiation from a VHF array is several degrees above the horizontal,

mountains that are less than about three degrees above the horizon, as seen from the antenna, are missed by the radiation from the array. Moving the mountains out of the way would have substantially no effect on VHF signal strength in such cases.

Rolling terrain, where obstructions are not sharp enough to produce knife-edge refraction, still does not exhibit a complete shadow effect. There is no complete barrier to VHF propagation—only attenuation, which varies widely as the result of many factors. Thus, even valley locations are usable for VHF communication. Good antenna systems, preferably as high as possible, the best available equipment, and above all, the willingness and ability to work with weak signals may make outstanding VHF work possible, even in sites that show little promise by casual inspection.

Sky-Wave Propagation

From earliest times we've had trouble with radio propagation language. We were never sure how long or short a "short wave" or a "long wave" was. Changing to *frequency* as a measurement term in place of *wavelength* was a useful innovation, but with technology outstripping terminology, adding modifiers such as *low, medium, high, very high, ultra high* and *super high*—all relative terms—didn't clarify much.

There is similar confusion in the accepted names for propagation modes and media. The term "ground wave" is commonly applied to propagation that is confined to the earth's lower atmosphere, but that definition allows use of the name for a mode that has been shown to work up to at least 4000 km, on occasion. Now we are about to use "sky

wave" to describe a mode that covers the same distances, but by a different medium, and at different frequencies, with different degrees of reliability.

THE IONOSPHERE

There will be inevitable "gray areas" in our discussion of the earth's atmosphere and the changes wrought in it by the sun and by the associated changes in the earth's magnetic field. This is not a story that can be told in neat equations, or values carried out to a satisfying number of decimal places. But the story must be told, and understood—with its well-known limitations—if we are to put up good antennas and make them serve us well.

Thus far in this chapter we have been concerned with what might be called our above-ground living space—that portion of the total atmosphere wherein we can survive without artificial breathing aids, or up to about 6 km (4 miles). The boundary area is a broad one, but life (and radio propagation) undergo basic changes beyond this life and weather zone. Somewhat farther out, but still technically within the earth's atmosphere, the role of the sun in the wave propagation picture is a dominant one.

This is the *ionosphere*—a region where the air pressure is so low that free electrons and ions can move about for some time without getting close enough to recombine into neutral atoms. A radio wave entering this rarefied atmosphere, a region of many free electrons, is affected in the same way as in entering a medium of different dielectric constant—its direction of travel is altered.

Ultraviolet radiation from the sun is the primary cause of ionization in the outer atmosphere. The degree of ionization does not increase uniformly with distance from the earth's surface. Instead there are relatively dense regions (layers) of ionization, each quite thick and more or less parallel to the earth's surface, at fairly well-defined intervals outward from about 40 to 300 km (25 to 200 miles).

Ionization is not constant within each layer, but tapers off gradually on either side of the maximum at the center of the layer. The total ionizing energy from the sun reaching a given point, at a given time, is never constant, so the height and intensity of the ionization in the various regions will also vary. Thus, the practical effect on long-distance communication is an almost continuous variation in signal level, related to the time of day, the season of the year, the distance between the earth and the sun, and both short-term and long-term variations in solar activity. It would seem from all this that only the very wise or the very foolish would attempt to predict radio propagation conditions, but it is now possible to do so with a fair chance of success.

Layer Characteristics

The lowest known ionized region, called the D layer, lies between 60 and 92 km (37 to 57 miles) above the earth. In this relatively low and dense part of the atmosphere, atoms broken up into ions by sunlight recombine quickly, so the ionization level is directly related to sunlight. It begins at sunrise, peaks at local noon and disappears at sundown. When electrons in this dense medium are set in motion by a passing wave, collisions between particles are so frequent that a major portion of their energy may be used up as heat.

The probability of collisions depends on the distance an electron travels under the influence of the wave—in other words, on the wavelength. Thus, our 1.8- and 3.5-MHz bands, having the longest wavelengths, suffer the highest daytime absorption loss, particularly for waves that enter the medium at the lowest angles. But at times of high solar activity (peak years of the solar cycle) even waves entering the D layer vertically suffer almost total energy absorption around midday, making these bands almost useless for communication over appreciable distances during the hours of high sun. They "go dead" quickly in the morning, but come alive again the same way in late afternoon. The diurnal D-region effect is less at 7 MHz (though still quite marked), slight at 14 MHz and inconsequential on higher amateur frequencies.

The D layer is ineffective in bending HF waves back to earth, so its role in long-distance communication by amateurs

is largely a negative one. It is the principal reason why our frequencies up through the 7-MHz band are useful mainly for short-distance communication, in the high-sun hours.

The lowest portion of the ionosphere that is useful for long-distance communication by amateurs is the E region, about 100 to 115 km (62 to 71 miles) above the earth. In the E layer, at intermediate atmospheric density, ionization varies with the sun angle above the horizon, but solar ultraviolet radiation is not the sole ionizing agent. Solar X-rays and meteors entering this portion of the earth's atmosphere also play a part. Ionization increases rapidly after sunrise, reaches maximum around noon local time, and drops off quickly after sundown. The minimum is after midnight, local time. As with the D region, the E layer absorbs wave energy in the lower frequency amateur bands when the sun angle is high, around mid-day. The other varied effects of E-region ionization will be discussed later.

Most of our long-distance communications capability stems from the tenuous outer reaches of the earth's atmosphere known as the F region. At heights above 100 miles, ions and electrons recombine more slowly, so the observable effects of the sun angle develop more slowly. Also, the region holds its ability to reflect wave energy back to earth well into the night. The maximum usable frequency (MUF) for F-layer propagation on east-west paths thus peaks just after noon at the midpoint, and the minimum occurs after midnight.

Using the F region effectively is by no means that simple, however. The layer height may be from 160 to more than 500 km (100 to over 310 miles), depending on the season of the year, the latitudes, the time of day and, most capricious of all, what the sun has been doing in the last few minutes and in perhaps the last three days before the attempt is made. The MUF between Eastern US and Europe, for example, has been anything from 7 to 70 MHz, depending on the conditions mentioned above, plus the point in the long-term solar activity cycle at which the check is made. Nevertheless, the MUF for a given circuit can be estimated with fair accuracy after one "gets the feel" for it.

Easy-to-use prediction charts appear in *The ARRL Operating Manual*. Propagation information tailored to amateur needs is transmitted in all information bulletin periods by the ARRL Headquarters station, W1AW. Finally, solar and geomagnetic field data, transmitted hourly and updated eight times daily, is given in brief bulletins carried by the National Bureau of Standards station, WWV. But more on these services later.

During the day the F region may split into two layers. The lower and weaker F1 layer, about 160 km (100 miles) up, has only a minor role, acting more like the E than the F2 region. At night the F1 layer disappears and the F2 layer height drops somewhat.

Bending in the Ionosphere

The degree of bending of a wave path in an ionized layer depends on the density of the ionization and the length of the wave (or its frequency). The bending at any given frequency or wavelength will increase with increased ionization density. With a given ionization density, bending increases with wavelength (decreases with frequency). Two extremes are thus possible. If the intensity of the ionization is sufficient and the frequency low enough, even a wave entering the layer perpendicularly will be reflected back to earth. Conversely, if the frequency is high enough or the ionization decreases to a low enough density, a condition is reached where the wave angle is not affected enough by the ionosphere to cause a useful portion of the wave energy to return to the earth. This basic principle has been used for many years to "sound" the ionosphere for determining its communication potential at various wave angles and frequencies.

A simplified example, showing only one layer, is given in Fig 13. The effects of additional layers are shown in Fig 14. The simple case of Fig 13 illustrates several important facts about antenna design for long-distance communication. At the left we see three waves that will do us no good—they all take off at angles high enough that they pass through the layer and are lost in space. Note that as the angle of radiation decreases (that is, the wave is closer to the horizon), the amount of bending needed for sky-wave communication also decreases. The fourth wave from the left takes off at what is called the *critical angle*—the highest that will return the wave to earth at a given density of ionization in the layer for the frequency under consideration.

We can communicate with point A at this frequency, but not much closer to our transmitter site. Neither, under this set of conditions as to layer height, layer density and *wave angle*, can we communicate much farther than point A. But suppose we install an antenna that radiates at a lower angle, as with the fifth wave from the left. This will bring our signal

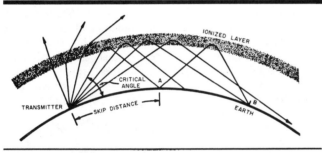

Fig 13—Behavior of waves encountering the ionosphere. Rays entering the ionized region at angles above the critical angle are not bent enough to be returned to earth, and are lost to space. Waves entering at angles below the critical angle reach the earth at increasingly greater distances as the angle approaches the horizontal. The maximum distance that may normally be covered in a single hop is 4000 km. Greater distances may be covered with multiple hops.

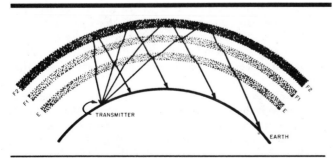

Fig 14—Typical daytime propagation of high frequencies (14 to 28 MHz). The waves are partially bent in going through the E and F1 layers, but not enough to be returned to earth. The actual reflection is from the F2 layer.

down to earth appreciably farther away than the higher (critical) angle did. Perhaps we can accomplish even more if we can achieve a very low radiation angle. Our sixth wave, with its radiation angle lower still, comes back to earth much farther away, at point B.

The lowest wave drawn in Fig 13 reaches the earth at a still greater distance, beyond point B. If the radio wave leaves the earth at a radiation angle of zero degrees, just toward the horizon, the maximum distance that may be reached under usual ionospheric conditions is about 4000 kilometers (2500 miles).

The earth itself acts as a reflector of radio waves. Quite often a radio signal will be reflected from the reception point on the earth into the ionosphere again, reaching the earth a second time at a still more distant point. This effect is also illustrated in Fig 13, where the critical-angle wave travels from the transmitter via the ionosphere to point A, in the center of the drawing. The signal reflected from point A travels via the ionosphere again to point B, at the right. Signal travel from the earth through the ionosphere and back to the earth is called a hop. Signal hopping is covered in more detail in a subsequent section.

In each case in Fig 13, the distance at which a ray reaches the earth in a single hop depends on the angle at which it left the transmitting antenna. This radiation angle comes into play throughout this book.

Skip Distance

When the critical angle is less than 90° there will always be a region around the transmitting site where the ionospherically propagated signal cannot be heard, or is heard weakly. This area lies between the outer limit of the ground-wave range and the inner edge of energy return from the ionosphere. It is called the *skip zone*, and the distance between the originating site and the beginning of the ionospheric return is called the *skip distance*. The signal may often be heard to some extent within the skip zone, through various forms of scattering, but it will ordinarily be marginal. When the skip distance is short, both ground-wave and sky-wave signals may be received at distances not far from the transmitter. In such instances the sky wave frequently is stronger than the ground wave, even as close as a few miles from the transmitter. The ionosphere is an efficient communications medium under favorable conditions. Comparatively, the ground wave is not.

MULTIHOP PROPAGATION

The information in Fig 13 is greatly simplified in the interest of explanation and example. In actual communication the picture is complicated by many factors. One is that the transmitted energy spreads over a considerable area after it leaves the antenna. Even with an antenna array having the sharpest practical beam pattern, there is what might be described as a cone of radiation centered on the wave lines shown in the drawing. The "reflection" in the ionosphere is also varied, and is the cause of considerable spreading and scattering.

As already mentioned, a radio signal will often be reflected from the reception point on the earth into the ionosphere again, reaching the earth a second time at a still more distant point. As in the case of light waves, the angle of reflection is the same as the angle of incidence, so a wave striking the surface of the earth at an angle of, say, 15° is reflected upward from the surface at approximately the same angle. Thus, the distance to the second point of reception will be about twice the distance of the first, that is, the distance from the transmitter to point A versus to point B in Fig 13. Under some conditions it is possible for as many as four or five signal hops to occur over a radio path, but no more than two or three hops is the norm. In this way, HF communications can be conducted over thousands of miles.

An important point should be recognized with regard to signal hopping. A significant loss of signal occurs with each hop. Lower layers of the ionosphere absorb energy from the signals as they pass through, and the ionosphere tends to scatter the radio energy in various directions, rather than confining it in a tight bundle. The earth also scatters the energy at a reflection point.

Assuming that both waves do reach point B in Fig 13, the *low-angle* wave will contain more energy at point B. This wave passes through the lower layers just twice, compared to four passes through these layers plus encountering an earth reflection for the higher angle route. Measurements indicate that although there can be great variation in the relative strengths of the two signals, the 1-hop signal will generally be from 7 to 10 dB stronger. The terrain at the mid-path reflection point for the 2-hop wave, the angle at which the wave is reflected from the earth, and the condition of the ionosphere in the vicinity of all the refraction points are the primary factors in determining the signal-strength ratio.

The loss per hop becomes significant at greater distances. It is because of these losses that no more than four or five propagation hops are useful; the received signal becomes too weak to be usable over more hops. Although modes other than signal hopping also account for the propagation of radio waves over thousands of miles, backscatter studies of actual radio propagation have displayed signals with as many as 5 hops. So the hopping mode is one distinct possibility for long-distance communications.

Present propagation theory holds that for communications distances of many thousands of kilometers, signals do not always hop in relatively short increments along the entire path. Instead, the wave is thought to propagate inside the ionosphere throughout some portion of the path length, tending to be ducted in the ionized layer. This theory is supported by the results of propagation studies which show that a medium-angle ray sometimes reaches the earth at a greater distance from the transmitter than a low-angle ray, as shown in Fig 15. This higher-angle ray, named the Pederson ray, is believed to penetrate the layer farther than lower-angle rays. In the less densely ionized upper edge of the layer, the amount of refraction is less, nearly equaling the curvature of the layer itself as it encircles the earth. This nonhopping theory is further supported by studies of propagation times for signals that travel completely around the world. The time required is significantly less than would be necessary to hop between the earth and the ionosphere 10 or more times while circling the earth.

Propagation between two points thousands of kilometers apart may consist of a combination of ducting and hopping. Whatever the exact mechanics of long-distance wave propagation may be, the signal must first enter the ionosphere at some point. The amateur wanting to work great distances should have the lowest possible wave angle, for years of amateur experience have shown this to be a decided advantage under all usual conditions.

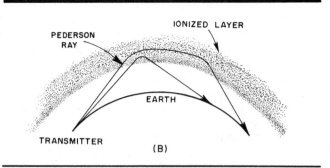

Fig 15—Studies have shown that under some conditions, rays entering the layer at intermediate angles will propagate farther than those entering at lower angles. The higher-angle wave is known as the Pederson ray.

Despite all the complex factors involved, most long-distance propagation can be seen to follow certain rules. Thus, much commercial and military point-to-point communication over long distances employs antennas designed to make maximum use of known radiation angles and layer heights, even on paths where multihop propagation is assumed.

In amateur work we try for the lowest practical radiation angle, hoping to keep reflection losses to a minimum. The geometry of propagation via the F2 layer limits our maximum distance along the earth's surface to about 4000 km (2500 miles) for a single hop. With higher radiation angles this same distance may require two or more hops (with higher reflection loss), so the fewer hops the better, in most cases. If you have a near neighbor who consistently outperforms you on the longer paths, a radiation angle difference in his favor may be the reason.

Virtual Height and Critical Frequency

Ionospheric sounding devices have been in service at enough points over the world's surface that a continuous record of ionospheric propagation conditions going back many years is available for current use, or for study. The sounding principle is similar to that of radar, making use of travel time to measure distance. The sounding is made at vertical incidence, to measure the useful heights of the ionospheric layers. This can be done at any one frequency, but the sounding usually is done over a frequency range wider than the expected return-frequency spread, so information related to the maximum usable frequency (MUF) is also obtained.

The distance so measured, called the *virtual height*, is that from which a pure reflection would have the same effect as the rather diffused refraction that actually happens. The method is illustrated in Fig 16. Some time is consumed in the refraction process, so the virtual height is slightly more than the actual.

The sounding procedure involves pulses of energy at progressively higher frequencies, or transmitters with the output frequency swept at many kilohertz per second. As the frequency rises, the returns show an area where the virtual height seems to increase rapidly, and then cease. The highest frequency returned is known as the *vertical incidence critical frequency*. The critical frequency can be used to determine

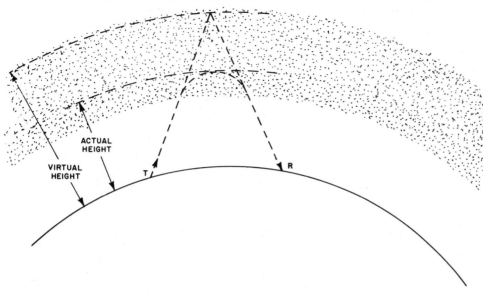

Fig 16—The virtual height of the refracting layer is measured by sending a wave vertically to the layer and measuring the time it takes to come back to the receiver. The actual height is somewhat less because of the time required for the wave to "turn around" in the ionized region.

the maximum usable frequency for long-distance communication by way of the layer, at that time. As shown in Fig 13, the amount of bending required decreases as the radiation angle decreases. At the lowest practical angle the range for a single hop reaches the 4000-km limit.

MAXIMUM USABLE FREQUENCY

The vertical incidence critical frequency is the maximum usable frequency for local sky-wave communications. It is also useful in the selection of optimum working frequencies and the determination of the maximum usable frequency for distant points at a given time. The abbreviation *MUF* will be used hereafter.

The critical frequency ranges between about 1 and 4 MHz for the E layer, and between 2 and 13 MHz for the F2 layer. The lowest figures are for nighttime conditions in the lowest years of the solar cycle. The highest are for the daytime hours in the years of high solar activity. These are average figures. Critical frequencies have reached something around 20 MHz briefly during exceptionally high solar activity.

The MUF for a 4000-km distance is about 3.5 times the critical frequency existing at the path midpoint. For 1-hop signals, if a uniform ionosphere is assumed, the MUF decreases with shorter distances along the path. This is true because the higher-frequency waves are not bent sufficiently to reach the earth at closer ranges, and so a lower frequency (where more bending occurs) must be used.

Precisely speaking, a maximum usable frequency or MUF is defined for communication between *two specific points* on the earth's surface, for the conditions existing at the time. At the same time and with the same conditions, the MUF from either of these two points to a third point may be quite different. Therefore, the MUF cannot be expressed broadly as a single frequency, even for any given location at a particular time. The ionosphere is never uniform, and in fact at a given time and for a fixed distance, the MUF changes

significantly with changes in compass direction for almost any point on the earth. Under usual conditions, the MUF will always be highest in the direction toward the sun—to the east in the morning, to the south at noon (from northern latitudes), and to the west in the afternoon and evening.

For the strongest signals at the greatest distance, especially where the limited power levels of the Amateur Radio Service are concerned, it is important to work fairly near the MUF. It is at these frequencies where signals suffer the least loss. The MUFs can be estimated with sufficient accuracy by using the prediction charts that appear in *The ARRL Operating Manual*. They can also be *observed*, with the use of a continuous coverage communications receiver. Frequencies up to the MUFs are in round-the-clock use today. When you "run out of signals" while tuning upward in frequency from your favorite ham band, you have a pretty good clue as to which band is going to work well, right then. Of course it helps to know the direction to the transmitters whose signals you are hearing. Shortwave broadcasters know what frequencies to use, and you can hear them anywhere, if conditions are good. Time-and-frequency stations are excellent indicators, since they operate around the clock. See Table 1. WWV is also a reliable source of propagation data, *hourly*, as discussed in more detail later in this chapter.

The value of working near the MUF is two-fold. Under average conditions, the absorption loss decreases with higher frequency. Perhaps more important, the hop distance is considerably greater as the MUF is approached. A transcontinental contact is much more likely to be made on a single hop on 28 MHz than on 14 MHz, so the higher frequency will give the stronger signal most of the time. The strong-signal reputation of the 28-MHz band is founded on this fact.

LOWEST USABLE FREQUENCY

There is also a lower limit to the range of frequencies that provide useful communication between two given points

Call	Frequency (MHz)	Location
WWV	2.5, 5, 10, 15, 20	Ft Collins, Colorado
WWVH	Same as WWV, but no 20	Kekaha, Kauai, Hawaii
CHU	3.330, 7.335, 14.670	Ottawa, Ontario, Canada
RID	5.004, 10.004, 15.004	Irkutsk, USSR*
RWM	4.996, 9.996, 14.996	Novosibirsk, USSR*
ZUO	2.5, 5	Pretoria, South Africa
VNG	7.5	Lyndhurst, Australia
BPV	5, 10, 15	Shanghai, China
JJY	2.5, 5, 10, 15	Tokyo, Japan
LOL	5, 10, 15	Buenos Aires, Argentina

*The call, taken from an international table, may not be that used during actual transmission. Locations and frequencies appear to be as given.

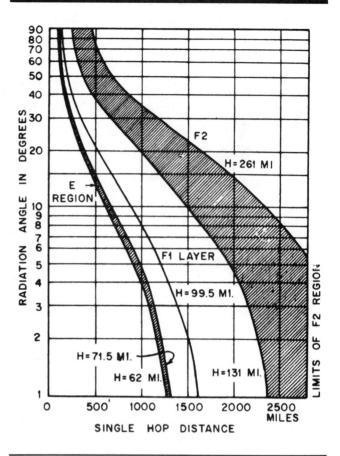

Fig 17—Distance plotted against wave angle (one-hop transmission) for the nominal range of virtual heights for the E and F2 layers, and for the F1 layer.

by way of the ionosphere. Lowest usable frequency is abbreviated *LUF*. If it were possible to start near the MUF and work gradually lower in frequency, the signal would decrease in strength and eventually disappears into the ever-present "background noise." This happens because the absorption increases at lower frequencies. The frequency nearest the point where reception became unusable would be the LUF. It is not likely that you would want to work at the LUF, although reception could be improved if the station could increase power by a considerable amount. By contrast, if there is a wide range between the LUF and the MUF, operation near the MUF can be quite satisfactory, even with a transmitter power of 1 or 2 watts.

Frequently, the "window" between the LUF and the MUF for two fixed points is very narrow, and there may be no amateur frequencies available inside the window. On occasion the LUF may be *higher* than the MUF between two points. This means that, for the highest possible frequency that will propagate through the ionosphere for that path, the absorption is so great as to make even that frequency unusable. Under these conditions it is not possible to establish amateur sky-wave communications between those two points, no matter what frequency is used. (It would normally be possible, however, to communicate between either point and other points on *some* frequency under the existing conditions.) Conditions when amateur sky-wave communications are impossible between two fixed points occur commonly for long distances where the total path is in darkness, and for very great distances in the daytime during periods of low solar activity.

TRANSMISSION DISTANCE AND LAYER HEIGHT

It was shown in connection with Fig 13 that the distance at which a ray returns to earth depends on the angle at which it left the earth (the wave angle). Although it is not shown specifically in that drawing, distance also depends on the layer height at the time.

There is a large difference in the distance covered in a

single hop, depending on whether the E or the F2 layer is used. The maximum distance via the E layer is about 2000 km (1250 miles)—or about half the maximum distance via the F2 layer. Practical communicating distances for single-hop E or F layer work at various wave angles are shown in graphic form in Fig 17.

For various reasons, actual communications experience will not fit these patterns exactly on all occasions. The particular amateur band in use will complicate the picture, in part because the E layer will be a factor much more often on our lower frequencies. Application of the chart to multihop paths is difficult, in part because both E and F layers may be involved. Also, the lowest wave angles are more practical at the higher frequencies.

Data on angle-of-arrival of signals in six amateur bands for a 3500-mile trans-atlantic communications circuit are shown in Table 2 (in part, from information in A. E. Harper's *Rhombic Design*). If signal hopping is considered to be the propagation mode, this distance would be covered in two hops of 1750 miles. Fig 17 indicates that for F2-layer propagation, the wave angles should fall in the range from about 7° to 19°, depending on the layer height. Note that for the lower frequencies, the angles run somewhat above this range, while

Table 2

Measured Vertical Angles of Arrival of Signals from England at Receiving Location in New Jersey

Freq, MHz	Angle below which signals arrived 99% of the time	Angle above which signals arrived 50% of the time	Angle above which signals arrived 99% of the time
7	35°	22°	10°
10	26°	17°	8°
14	17°	11°	6°
18	12°	8°	5°
21	11°	7°	4°
28	9°	5°	3°

the opposite is true for the higher frequencies. The conclusion from this information is that even where the same antennas and frequency are used over a test period, the angle of arrival will change over a wide range, possibly with accompanying large variations in signal level.

ONE-WAY PROPAGATION

On occasion a signal may be started on the way back toward the earth by reflection from the F layer, only to come down into the *top* of the E region and be reflected back up again. This set of conditions is one explanation for the often-reported phenomenon called *one-way skip*. The reverse path may not necessarily have the same multilayer characteristic, and the effect is more often a difference in the signal strengths, rather than a complete lack of signal in one direction. It is important to remember this possibility, when a long-path test with a new antenna system yields apparently conflicting evidence. Even many tests, on paths of different lengths and headings, may provide data that are difficult to understand. Communication by way of the ionosphere is not always a source of consistent answers to antenna questions.

SHORT OR LONG PATH?

Propagation between any two points on the earth's surface is usually by the shortest direct route—the great-circle path found by stretching a string tightly between the two points on a globe. If an elastic band going completely around the globe in a straight line is substituted for the string, it will show another great-circle path, going "the long way around." The long path may serve for communication over the desired circuit when conditions are favorable along the longer route. There may be times when communication is possible over the long path but not possible at all over the short path. Especially if there is knowledge of this potential at the other end, long-path communication may work very well. Cooperation is almost essential, because both the antenna aiming and the timing of the attempts must be right for any worthwhile result.

Sunlight is a required element in long-distance communication via the F layer. This fact tends to define long-path timing and antenna aiming. Both are essentially the reverse of the "normal" for a given circuit. We know also that salt-water paths work better than overland ones. This can be significant in long-path work.

We can better understand several aspects of long-path propagation if we become accustomed to thinking of the earth as a ball. This is easy if we use a globe frequently. A flat map of the world, of the azimuthal-equidistant projection type, is a useful substitute. The ARRL World Map is one, centered on Wichita, Kansas. A somewhat similar world map prepared by N5KR and centered on Newington, Connecticut, is shown in Fig 18. These help to clarify paths involving those areas of the world.

Long-Path Examples

There are numerous long-path routes well known to DX-minded amateurs, those who continually seek foreign countries. Two long paths that work frequently and well on 28 MHz from northeastern US are New England to Perth, Australia, and New England to Tokyo. Although they represent different beam headings and distances, they share some favorable conditions. By long path, Perth is close to halfway around the world; Tokyo is about three-quarters of the way. On 28 MHz, both areas come through in the early daylight hours, Eastern Time, but not necessarily on the same days. Both paths are at their best around the equinoxes. (The sunlight is more uniformly distributed over transequatorial paths at these times.) Probably the factor that most favors both is the nature of the first part of the trip at the US end. To work Perth via long path, northeastern US antennas are aimed southeast, out over salt water for thousands of miles— the best low-loss start a signal could have. It is salt water essentially all the way, and the distance, about 13,000 miles, is not too much greater than the "short" path.

The long path to Japan is more toward the south, but still with no major land mass at the early reflection points. It is much longer, however, than that to western Australia. Japanese signals are more limited in number on the long path than on the short, and signals average somewhat weaker, probably because of the greater distance.

On the short path an amateur in the Perth area is looking at the worst conditions—away from the ocean, and out across a huge land mass unlikely to provide strong ground reflections. The short paths to both Japan and western Australia, from most of the eastern half of North America, are hardly favorable. The first hop comes down in various western areas likely to be desert or mountains, or both, and not favored as reflection points.

A word of caution: Don't count on the long-path signals

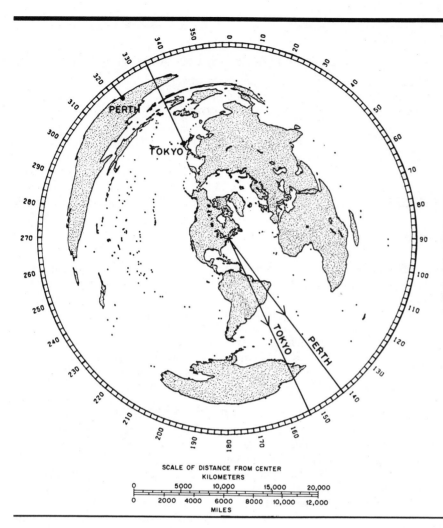

Fig 18—N5KR's computer-generated azimuthal-equidistant projection centered on Newington, Connecticut. (See bibliography for ordering information.) Shading of the land masses and information showing long paths to Perth and Tokyo have been added. Notice that the paths in both cases lie almost entirely over water, rather than over land masses.

SCALE OF DISTANCE FROM CENTER
KILOMETERS

0	5000	10,000	15,000	20,000

0	2000	4000	6000	8000	10,000	12,000
MILES

always coming in on the same beam heading. There can be noticeable differences in the line of propagation via the ionosphere on even relatively short distances. There can be more variations on long path, especially on circuits close to halfway around the world. Remember, for a point exactly halfway around, all directions of the compass represent great circle paths.

FADING

Taking into account all the variable factors in long-distance communication, it is not surprising that signals vary in strength during almost every contact beyond the local range. In VHF communication we can encounter some fading, at any distances greater than just to the visible horizon. These are mainly the result of changes in the temperature and moisture content of the air in the first few thousand feet above the ground.

On paths covered by ionospheric modes, the causes of fading are very complex—constantly changing layer height and density, random polarization shift, portions of the signal arriving out of phase, and so on. The energy arriving at the receiving antenna has components which have been acted upon differently by the ionosphere. Often the fading is very different for small changes in frequency. With a signal of a wideband nature, such as high-quality FM, or even double-

sideband AM, the sidebands may have different fading rates from each other, or from the carrier. This causes severe distortion, resulting in what is termed *selective fading*. The effects are greatly reduced when single-sideband (SSB) is used. Some immunity from fading during reception (but not to the distortion induced by selective fading) can be had by using two or more receivers on separate antennas, preferably with different polarizations, and combining the receiver outputs in what is known as a *diversity receiving system*.

MINOR PROPAGATION MODES

In propagation literature there is a tendency to treat the various propagation modes as if they were separate and distinct phenomena. This they may be at times, but often there is a shifting from one to another, or a mixture of two or more kinds of propagation affecting communication at one time. In the upper part of the usual frequency range for F-layer work, for example, there may be enough tropospheric bending at one end (or both ends) to have an appreciable effect on the usable path length. There is the frequent combination of E- and F-layer propagation in long-distance work. And in the case of the E layer, there are various causes of ionization that have very different effects on communication. Finally, there are weak-signal variations of both tropospheric and ionospheric modes, lumped under the term "scatter." We look at these phenomena separately here, but in practice we may have to deal with them in combination, more often than not.

Sporadic-E (E_s)

First, note that this is E-subscript-s, a usefully descriptive term, wrongly written "Es" so often that it is sometimes called "ease"—which is certainly *not* descriptive. Sporadic E is ionization at E-layer height, but of quite different origin and communications potential from the E layer that affects mainly our lower amateur frequencies.

The formative mechanism for sporadic E is believed to be wind shear. This explains ambient ionization being redistributed and compressed into a ledge of high density, without the need for production of extra ionization. Neutral winds of high velocity, flowing in opposite directions at slightly different altitudes, produce shears. In the presence of the magnetic field the ions are collected at a particular altitude, forming a thin, overdense layer. Data from rockets entering E_s regions confirm the electron density, wind velocities and height parameters.

The ionization is formed in clouds of high density, lasting only a few hours at a time and distributed randomly. They vary in density and, in the middle latitudes in the northern hemisphere, move rapidly from southeast to northwest. Although E_s can develop at any time, it is most prevalent in the northern hemisphere between May and August, with a minor season about half as long beginning in December (the summer and winter solstices). The seasons and distribution in the southern hemisphere are not so well known. Australia and New Zealand seem to have conditions much like those in the US, but with the length of the seasons reversed, of course. Much of what is known about E_s came as the result of amateur pioneering in the VHF range.

Correlation of E_s openings with observed natural phenomena, including sunspot activity, is not readily apparent, although there is a meteorological tie-in with high-altitude winds. There is also a form of E_s, mainly in the northern part of the north temperate zone, that is associated with auroral phenomena.

At the peak of the long E_s season, most commonly in late June and early July, ionization becomes extremely dense and widespread. This extends the usable range from the more common "single-hop" maximum of about 1400 miles to "double-hop" distances, mostly 1400 to 2500 miles. With 50-MHz techniques and interest improving in recent years, it has been shown that distances considerably beyond 2500 miles can be covered. There is also an E_s "link-up" possibility with other modes, believed to be involved in some 50-MHz work between antipodal points, or even long-path communication beyond 12,500 miles.

The MUF for E_s is not known precisely. It was long thought to be around 100 MHz, but in the last 25 years or so there have been thousands of 144-MHz contacts during the summer E_s season. Presumably, the possibility also exists at 220 MHz. The skip distance at 144 MHz does average much longer than at 50 MHz, and the openings are usually brief and extremely variable.

The terms "single" and "double" hop may not be accurate technically, since it is likely that cloud-to-cloud paths are involved. There may also be "no-hop" E_s. At times the very high ionization density produces critical frequencies up to the 50-MHz region, with no skip distance at all. It is often said that the E_s mode is a great equalizer. With the reflecting region practically overhead, even a simple dipole close to the ground may do as well over a few hundred miles as a large stacked antenna array designed for low-angle radiation. It's a great mode for low power and simple antennas on 28 and 50 MHz.

Scatter Modes

The term "skip zone" should not be taken literally. Two stations communicating over a single ionospheric hop can be heard to some degree at almost any point along the way, unless they are running quite low power and using simple antennas. Some of the wave energy is scattered in all directions, including back to the starting point and farther. The wave energy of VHF stations is not gone after it reaches the radio horizon, described early in this chapter. It is scattered, but it can be heard to some degree for hundreds of miles. Everything on earth, and in the regions of space up to at least 100 miles, is a potential scattering agent.

Tropospheric scatter is always with us. Its effects are often hidden, masked by more effective propagation modes on the lower frequencies. But beginning in the VHF range, scatter from the lower atmosphere extends the reliable range quite markedly if we make use of it. Called "tropo scatter," this is what produces that nearly flat portion of the curves given in an earlier section on reliable VHF coverage. We are not out of business at somewhere between 50 and 100 miles, on the VHF and even UHF bands, especially if we don't mind weak signals and something less than 99% reliability. As long ago as the early 1950s, VHF enthusiasts found that VHF contests could be won with high power, big antennas and a good ear for signals deep in the noise. They still can.

Ionospheric scatter works much the same as the tropo version, except that the scattering medium is the E region of the ionosphere, with some help from the D and F. Ionospheric scatter is useful mainly *above* the MUF, so its useful frequency range depends on geography, time of day, season, and the state of the sun. With near maximum legal power, good antennas, and quiet locations, ionospheric scatter can fill in the skip zone with marginally readable signals scattered from ionized trails of meteors, small areas of random ionization, cosmic dust, satellites and whatever may come into the antenna patterns at 50 to 150 miles or so above the earth. It's mostly an E-layer business, so it works at E-layer distances. Good antennas and keen ears help.

Backscatter is a sort of ionospheric radar. Because it is mainly scattering from the earth at the point where the strong ionospherically propagated signal comes down, it is, in fact, a part of long-distance radar techniques. It is also a great "filler-inner" of the skip zone, particularly in work near the MUF where propagation is best. It was proved by amateurs using sounding techniques that you can tell to what part of the world a band is usable (single-hop F) by probing the backscatter with a directive antenna, even when the earth-contact point is open ocean. In fact, that's where the mode is at its best.

Backscatter is very useful on 28 MHz, particularly when that band *seems* dead simply because nobody is active in the right places. The mode keeps the 10-meter band lively in the low years of the solar cycle, thanks to the never-say-die attitude of some users. The mode is also an invaluable tool of 50-MHz DX aspirants, in the high years of the sunspot cycle, for the same reasons. On a high-MUF morning, hundreds of 6-meter beams may zero in on a hot spot somewhere in the Caribbean or South Atlantic, where there is no land, let alone other 6-meter stations—keeping in contact while they wait for the band to open to a place where there *is* somebody.

Sidescatter is similar to backscatter, except the ground scatter zone is merely somewhat off the direct line between the participants. A typical example, often observed during the lowest years of the solar cycle, is communication on 28 MHz between eastern US (and adjacent areas of Canada) and much of the European continent. Often, this may start as "backscatter chatter" between stations on the east side of the Atlantic, with antennas beamed toward the Azores. Then suddenly the North Americans join the fun, perhaps for only a few minutes, but sometimes much longer, with beams also pointed toward the Azores. Duration of the game can be extended, at times, by careful reorientation of antennas at both ends, as with backscatter. The secret, of course, is to keep hitting the highest-MUF area of the ionosphere and the most favorable ground-reflection points.

The favorable route is usually, but not always, south of the direct great-circle heading (for stations in the northern hemisphere). There can also be sidescatter from the auroral regions. Sidescatter signals are stronger than backscatter signals using the same general area of ground scattering.

Sidescatter signals have been observed frequently on the 14-MHz band, and can take place on any band where there is a large window between the MUF and the LUF. For sidescatter communications to occur, the thing to look for is a common area to which the band is open from both ends of the path (the Azores, in the above example), when there is no direct-path opening. It helps if the common area is in the open ocean, where there is less scattering loss than over land. A study of the *Operating Manual* propagation charts will indicate areas where backscatter and sidescatter are likely to occur, when a direct path may not work. In the above example, the charts would show an opening between North America and South Africa at the same time as an opening between Europe and South America.

Transequatorial scatter (TE) was an amateur 50-MHz discovery in the years 1946-1947. It was turned up almost simultaneously on three separate north-south paths, by amateurs of all continents. These amateurs tried to communicate on 50 MHz even though the predicted MUF was around 40 MHz for the favorable daylight hours. The first success came at night, when the MUF was thought to be even lower. A remarkable research program inaugurated by amateurs in Europe, Cyprus, Southern Rhodesia (now Zimbabwe) and South Africa eventually provided technically sound theories to explain the then-unknown mode.

It had been known for years that the MUF is higher and less seasonally variable on transequatorial circuits, but the full extent of the difference was not learned until amateur work brought it to light. Briefly, the ionosphere over equatorial regions is higher, thicker and more dense than elsewhere. Because of its more constant exposure to solar radiation, the equatorial belt has high nighttime-MUF possibilities. It is now known that the TE mode can often work marginally at 144 MHz, and even at 432 MHz on occasion. The potential MUF varies with solar activity, but not to the extent that conventional F-layer propagation does. It is a late-in-the-day mode, taking over about when normal F-layer propagation goes out.

The TE range is usually within about 4000 km (2500 miles) either side of the geomagnetic equator. The earth's magnetic axis is tilted with respect to the geographical axis

so the TE belt appears as a curving band on conventional flat maps of the world. See Fig 19. As a result, TE has a different latitude coverage in the Americas from that shown in the drawing. The TE belt just reaches into the southern US. Puerto Rico, Mexico and even the northern parts of South America encounter the mode more often than do most US areas. It is no accident that TE was discovered as a result of 50-MHz work in Mexico City and Buenos Aires.

Within its optimum regions of the world, the TE mode extends the usefulness of the 50-MHz band far beyond that of conventional F-layer propagation, since the practical TE MUF runs around 1.5 times that of normal F2. Both its seasonal and diurnal characteristics also are extensions of what is considered normal for 50-MHz propagation. In that part of the Americas south of about 20° N latitude, the existence of TE affects the whole character of band usage, especially in years of high solar activity.

Auroral Propagation

Sudden bursts of solar activity are accompanied by the ejection of charged particles from the sun. These particles travel in various directions, and some enter the earth's atmosphere, usually 24 to 36 hours after the event. Here they may react with the magnetic field to produce a visible or radio aurora, visible if their time of entry is after dark. Some information on major solar outbursts is obtainable from WWV propagation bulletins, discussed later in this chapter. (From WWV information, the possibility of auroral activity can be known in advance.)

The visible aurora is, in effect, fluorescence at E-layer height—a curtain of ions capable of refracting radio waves in the frequency range above about 20 MHz. D-region absorption increases on lower frequencies during auroras. The exact frequency ranges depend on many factors: time, season, position with relation to the earth's auroral regions, and the level of solar activity at the time, to name a few.

The auroral effect on VHF waves is another amateur discovery, this one dating back to the 1930s. The discovery

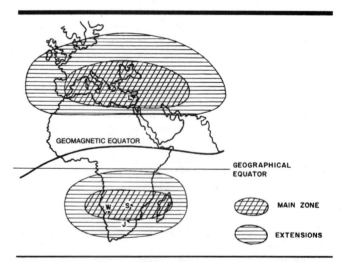

Fig 19—Main and occasional zones of transequatorial 50-MHz propagation show Limassol, Cyprus, and Salisbury, Zimbabwe, to be almost ideally positioned with respect to the curving geomagnetic equator. Windhoek, Namibia, is also in a favorable spot; Johannesburg somewhat less so.

came coincidentally with improved transmitting and receiving techniques. The returning signal is diffused in frequency by the diversity of the auroral curtain as a refracting (scattering) medium. The result is a modulation of a CW signal, from just a slight burbling sound to what is best described as a keyed roar. Before SSB took over in VHF work, voice was all but useless for auroral paths. A sideband signal suffers, too, but its narrower bandwidth helps to retain some degree of understandability. Distortion induced by a given set of auroral conditions increases with the frequency in use. In general, 50-MHz signals are much more intelligible than those on 144 MHz on the same path at the same time. On 144 MHz, CW is almost mandatory for effective auroral communication.

The number of auroras that can be expected per year varies with the geomagnetic latitude. Drawn with respect to the earth's magnetic poles instead of the geographical ones, these latitude lines in the US tilt upward to the northwest. For example, Portland, Oregon, is 2 degrees farther north (geographic latitude) than Portland, Maine. But the Maine city's geomagnetic latitude line crosses the Canadian border before it gets as far west as its Oregon namesake. In terms of auroras intense enough to produce VHF propagation results, Portland, Maine, is likely to see about 10 times as many per year. Oregon's auroral prospects are more like those of southern New Jersey or central Pennsylvania.

The antenna requirements for auroral work are mixed. High gain helps, but the area of the aurora yielding the best returns sometimes varies quite rapidly, so very high directivity can be a disadvantage. So could a very low radiation angle, or a beam pattern very sharp in the vertical plane. Experience indicates that few amateur antennas are sharp enough in either plane to present a real handicap. The beam heading for maximum signal can change, however, so a bit of scanning in azimuth may turn up some interesting results. A very large array such as is commonly used for moonbounce (with az-el control) should be worthwhile.

The incidence of auroras, their average intensity, and their geographical distribution as to visual sightings and VHF propagation effects all vary to some extent with solar activity. There is some indication that the peak period for auroras lags the sunspot-cycle peak by a year or two. But like sporadic E, an unusual auroral opening can come at any season. There is a marked diurnal swing in the number of auroras. Favored times are late afternoon and early evening, late evening through early morning, and early afternoon, in about that order. Major auroras often start in early afternoon and carry through to early morning the next day.

GRAY-LINE PROPAGATION

The *gray line*, sometimes called the twilight zone, is a band around the earth between the sunlit portion and darkness. Astronomers call this the terminator. The termi-

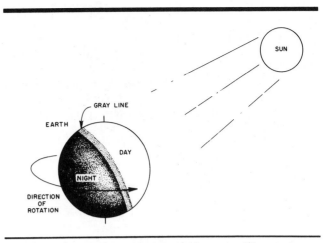

Fig 20—The gray line or terminator is a transition region between daylight and darkness. One side of the earth is coming into sunrise, and the other is just past sunset.

nator is a somewhat diffused region because the earth's atmosphere tends to scatter the light into the darkness. Fig 20 illustrates the gray line. Notice that on one side of the earth, the gray line is coming into daylight (sunrise), and on the other side it is coming into darkness (sunset).

Propagation along the gray line is very efficient, so greater distances can be covered than might be expected for the frequency in use. One major reason for this is that the D layer, which absorbs HF signals, disappears rapidly on the sunset side of the gray line, and has not yet built up on the sunrise side.

The gray line runs generally north and south, but varies as much as 23° either side of the north-south line. This variation is caused by the tilt of the earth's axis relative to its orbital plane around the sun. The gray line will be exactly north and south at the equinoxes (March 21 and September 21). On the first day of northern hemisphere summer, June 21, it is tilted to the maximum of 23° one way, and on December 21, the first day of winter, it is tilted 23° the other way.

To an observer on the earth, the direction of the terminator is always at right angles to the direction of the sun at sunrise or sunset. It is important to note that, except at the equinoxes, the gray-line direction will be different at sunrise from that at sunset. This means you can work different areas of the world in the evening than you worked in the morning.

It isn't necessary to be located inside the twilight zone in order to take advantage of gray-line propagation. The effects can be used to advantage before sunrise and after sunset. This is because the sun "rises" earlier and "sets" later on the ionospheric layers than it does on the earth below.

The Role of the Sun

Everything that happens in radio propagation, as with all life on earth, is the result of radiation from the sun. The variable nature of radio propagation reflects the ever-changing intensity of ultraviolet and X-ray radiation, the ionizing agents in solar energy. Every day, solar dynamics are turning hydrogen into helium, releasing an unimaginable blast of energy into space in the process.

Why the intensity of this release changes, and what the sun is going to do at any time in the future, are known only in a general way. Still, tremendous progress has been made within our lifetimes in understanding the nature of the sun and predicting its future course.

SUNSPOTS

The most readily observed characteristic of the sun, other than its blinding brilliance, is its tendency to have grayish-black blemishes, seemingly at random times and at random places, on its fiery surface. (See Fig 21.) There are written records of naked-eye sightings of sunspots in the Orient back to more than 2000 years ago. As far as is known, the first indication that sunspots were recognized as part of the sun was the result of observations by Galileo in the early 1600s, not long after he developed one of the first practical telescopes.

Galileo also developed the projection method for observing the sun safely, but probably not before he had suffered severe eye damage by trying to look at the sun directly. (He was afflicted with blindness in his last years.) His drawings of sunspots, indicating their variable nature and position, are the first such record known to have been made. His reward for this brilliant work was immediate condemnation by church authorities of the time, which probably set back progress in learning more about the sun for generations. But one way or another, observation of the sun continued, mostly in secret.

The systematic study of solar activity began about 1750, so a fairly reliable record of sunspot numbers goes back that far. (There are some gaps in the early data.) The record shows clearly that the sun is always in a state of change. It never looks exactly the same from one day to the next. The most obvious daily change is the movement of visible activity centers (sunspots or groups thereof) across the solar disc, from east to west, at a constant rate. This movement was soon found to be the result of the rotation of the sun, at a rate of approximately four weeks for a complete round. The average is about 27.5 days.

Sunspot Numbers

Since the earliest days of systematic observation, our traditional measure of solar activity has been based on a count of sunspots. In these hundreds of years we have learned that the average number of spots goes up and down in cycles very roughly approximating a sine wave. In 1848, a method was introduced for the daily measurement of sunspot numbers. That method, which is still used today, was devised by the Swiss astronomer, Johann Rudolph Wolf. The observer counts the total number of spots visible on the face of the sun and the number of groups into which they are clustered, because neither quantity alone provides a satisfactory measure of sunspot activity. The observer's sunspot number for that day is computed by multiplying the number of groups he sees by 10, and then adding to this value the number of individual

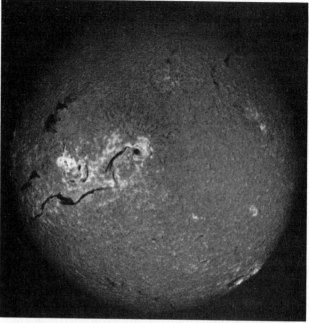

Fig 21—Much more than sunspots can be seen when the sun is viewed through selective optical filters. This photo was taken through a hydrogen-alpha filter that passes a narrow light segment at 6562 angstroms. The bright patches are active areas around and often between sunspots. Dark irregular lines are filaments of activity having no central core. Faint magnetic field lines are visible around a large sunspot group near the disc center. *(Photo courtesy of Sacramento Peak Observatory, Sunspot, New Mexico)*

spots. Where possible, sunspot data collected prior to 1848 have been converted to this system.

As can readily be understood, results from one observer to another can vary greatly, since measurement depends on the capability of the equipment in use and on the stability of the earth's atmosphere at the time of observation, as well as on the experience of the observer. A number of observatories around the world cooperate in measuring solar activity. A weighted average of the data is used to determine the International Sunspot Number or ISN for each day. (An amateur astronomer can approximate the determination of ISN values by multiplying his or her values by a correction factor determined empirically.)

A major step forward was made with the development of various methods for observing narrow portions of the sun's spectrum. Narrowband light filters that can be used with any good telescope perform a visual function very similar to the aural function of a sharp filter added to a communications receiver. This enables the observer to see the actual area of the sun doing the radiating of the ionizing energy, in addition to the sunspots, which are more a by-product than a cause. The photo of Fig 21 was made through such a filter. Studies of the ionosphere with instrumented probes, and later with satellites, manned and unmanned, have added greatly to our knowledge of the effects of the sun on radio communication.

Daily sunspot counts are recorded, and monthly and yearly averages determined. The averages are used to see trends and observe patterns. Sunspot records were formerly

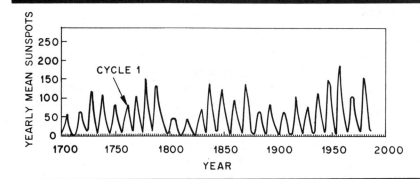

Fig 22—Yearly means of sunspot numbers, from data for 1700 through 1986. This plot clearly shows that sunspot activity takes place in cycles of approximately 11 years duration. Cycle 1, the first complete cycle to be examined by systematic observation, began in 1755.

kept in Zurich, Switzerland, and the values were known as Zurich sunspot numbers. They were also known as Wolf sunspot numbers. Since 1981, international records have been compiled and kept at the Sunspot Index Data Center, 3 avenue Circulaire, B 1180 Bruxelles, Belgium. In the United States, solar data records are maintained at the National Geophysical Data Center, Solar-Terrestrial Physics Division (E/GC2), 325 Broadway, Boulder, CO 80303. Publications related to sunspot data are available from both Brussels and Boulder.

The yearly means (averages) of sunspot numbers from 1700 through 1986 are plotted in Fig 22. The cyclic nature of solar activity becomes readily apparent from this graph. The duration of the cycles varies from 9.0 to 12.7 years, but averages approximately 11.1 years. The first complete cycle to be observed systematically began in 1755, and is numbered cycle 1. Solar cycle numbers thereafter are consecutive. Cycle 22 began in 1987.

THE MAUNDER MINIMUM

In the relatively brief span of recorded sunspot activity, we have no way of knowing for sure if the cyclic nature of solar activity is continuous, or perhaps a recurring phenomena of some grander, overall cycle. How nice it would be if we had thousands of years of records to examine!

Although reliable solar activity records go back only a few hundred years, there are other ways to derive information about the history of the sun's behavior. Records of aurora sightings are one. Historical accounts of the appearance of the solar corona during an eclipse are another, since the shape of the corona varies with solar activity. A third is by analysis of the carbon-14 isotope in tree rings. The isotope is formed continuously in the atmosphere by action of cosmic rays, and the level of cosmic rays entering the atmosphere is modulated by solar activity. When the sun is quiet the carbon-14 content rises, and it is lower in times of high solar activity. The carbon-14 isotope is assimilated by the trees along with carbon dioxide, and shows up in the annual growth rings.

Based on strong evidence from various sources, including some circumstantial but convincing, it appears that there have been long periods with a lack of any solar activity. The most recent of these was a 70-year period beginning about 1645. This period has been named the Maunder Minimum, after E. W. Maunder, who was the superintendent of the Solar Department of Greenwich Observatory in the late 1800s.

During those 70 years, there appears to have been very little or no solar activity at all. Another long period of little or no activity apparently began about 1460 and extended for 90 years.

On the other side of the coin, carbon-14 data indicates a long period of exceptionally high solar activity in the 12th and early 13th centuries. This maximum also appears, though more vaguely, in natural and historic records. From this information, we can conclude that the 11-year solar cycle may not be a regular feature in the life of the sun.

SMOOTHED SUNSPOT NUMBERS

For more than 50 years it has been well known that radio propagation phenomena vary with the number and size of sunspots. The daily and seasonal variations in the ionized layers resulting from changes in the amount of ultraviolet light received from the sun have already been mentioned. The so-called 11-year sunspot cycle affects propagation conditions because there is a direct correlation between sunspot activity and ionization.

As mentioned earlier, activity on the surface of the sun is changing continually. Individual sunspots may vary in size and appearance, or even disappear totally, within a single day. In general, larger active areas persist through several rotations of the sun. Some active areas have been identified over periods up to about two years. Because of these continual changes in solar activity, there are continual changes in the state of the ionosphere and resulting changes in propagation conditions. A short-term burst of solar activity may trigger unusual propagation conditions which last for less than an hour.

Sunspot data are averaged or smoothed to remove the effects of short-term changes. The sunspot values used most often for correlating propagation conditions are "smoothed sunspot numbers," often called 12-month running average values. Data for 13 consecutive months are required to determine a smoothed sunspot number.

Each smoothed number is an average of 13 monthly means centered on the month of concern. The 1st and 13th months are given a weight of 0.5. A monthly mean is simply the sum of the daily ISN values for a calendar month divided by the number of days in that month. We would commonly call this value a monthly average.

This may all sound very complicated, but an example should clarify the procedure. Suppose we wished to calculate the smoothed sunspot number for June 1986. We would require monthly mean values for 6 months prior and 6 months after this month, or from December 1985 through December 1986. The monthly mean ISN values for these months are

Dec 85, 17.3	Jul 86, 18.1
Jan 86, 2.5	Aug 86, 7.4
Feb 86, 23.2	Sep 86, 3.8
Mar 86, 15.1	Oct 86, 35.4
Apr 86, 18.5	Nov 86, 15.2
May 86, 13.7	Dec 86, 6.8
Jun 86, 1.1	

First we find the sum of the values, but using only ½ the amounts indicated for the 1st and 13th months in the listing. This value is 166.05. Then we determine the smoothed value by dividing the sum by 12; 166.05/12 = 13.8. (Values beyond the first decimal place are not warranted.) Thus, 13.8 is the smoothed sunspot number for June 1986. From this example, you can see that the smoothed sunspot number for any month cannot be determined until six months afterwards.

Generally the plots we see of sunspot numbers are averaged data. As already mentioned, smoothed numbers make it easier to observe trends and see patterns, but sometimes this data can be misleading. The plots tend to imply that solar activity varies smoothly, indicating, for example, that at the onset of a new cycle the activity just gradually increases. But this is definitely not so! Significant changes in solar activity can take place within hours, causing sudden band openings at frequencies well above the MUF values predicted from smoothed sunspot number curves. The duration of such openings may be brief, or they may recur for several days running, depending on the nature of the solar activity.

SOLAR FLUX

In more recent years an additional method of determining solar activity has been put to use—the measurement of solar radio flux. The quiet sun emits radio energy across a broad frequency spectrum, with a slowly varying intensity. Solar flux is a measure of energy received per unit time, per unit area, per unit frequency interval. These radio fluxes, which originate from atmospheric layers high in the sun's chromosphere and low in its corona, change gradually from day to day, in response to the activity causing sunspots. Thus, there is a degree of correlation between solar flux values and sunspot numbers.

One solar flux unit equals the factor 10 to the power -22 Joules per second per square meter per hertz. Solar flux values are measured daily at 2800 MHz (10.7 cm) at the Algonquin Radio Observatory near Ottawa, Ontario, where daily data have been collected since February 1947. Measurements are also made at other observatories around the world, at several frequencies. With some variation, the measured flux values increase with increasing frequency of measurement, at least up to 15.4 GHz. The daily 2800-MHz Ottawa value (taken at 1700 UT) is sent to Boulder, Colorado, where it is incorporated into WWV propagation bulletins (see later section). Daily solar flux information is of value in determining current propagation conditions, as sunspot numbers on a given day do not relate directly to maximum

usable frequency. Solar flux values are much more reliable for this purpose.

Correlating Sunspot Numbers and Solar Flux Values

Based on historical data, an exact mathematical relationship does not exist to correlate sunspot data and solar flux values. Comparing daily values yields almost no correlation. Comparing monthly mean values (often called monthly averages) produces a degree of correlation, but the spread in data is quite significant. This is indicated in Fig 23, a scatter-diagram plot of monthly mean sunspot numbers versus the monthly means of solar flux values adjusted to one astronomical unit. (This adjustment applies a correction for differences in distance between the sun and the earth at different times of the year.)

A closer correlation exists when smoothed (12-month running average) sunspot numbers are compared with smoothed (12-month running average) solar flux values adjusted to one astronomical unit. A scatter diagram for smoothed data appears in Fig 24. Note how the plot points establish a better defined pattern in Fig 24. But yet the correlation is still no better than a few percent, for records indicate a given smoothed sunspot number does not always correspond with the same smoothed solar flux value, and vice versa. Table 3 illustrates some of the inconsistencies that exist in the historical data. Smoothed or 12-month running average values are shown.

Even though there is no precise mathematical relationship between sunspot numbers and solar flux values, it is helpful to have some way to convert from one to the other. The primary reason is that sunspot numbers are valuable as a long-term link with the past, but the great usefulness of solar flux values are their immediacy, and their direct bearing on

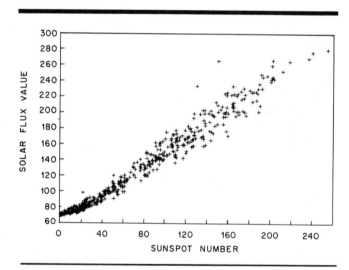

Fig 23—Scatter diagram or X-Y plot of monthly mean sunspot numbers and monthly mean 2800-MHz solar flux values. Data values are from February 1947 through February 1987. Each "+" mark represents the intersection of data for a given month. If the correlation between sunspot number and flux values were consistent, all the marks would align to form a smooth curve.

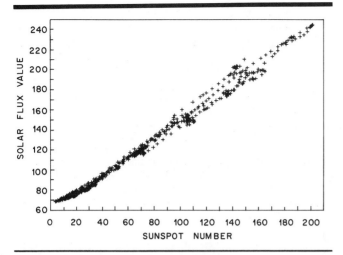

Fig 24—Scatter diagram of smoothed or 12-month running averages, sunspot numbers versus 2800-MHz solar flux values. The correlation of smoothed values is better than for monthly means, shown in Fig 23.

Table 3

Selected Historical Data Showing Inconsistent Correlation Between Sunspot Number and Solar Flux

Month	Smoothed Sunspot Number	Smoothed Solar Flux Value
May 1953	17.4	75.6
Sep 1965	17.4	78.5
Jul 1985	17.4	74.7
Jun 1969	106.1	151.4
Jul 1969	105.9	151.4
Dec 1982	94.6	151.4
Aug 1948	141.1	180.5
Oct 1959	141.1	192.3
Apr 1979	141.1	180.4
Aug 1981	141.1	203.3

our field of interest. (Remember, a smoothed sunspot number cannot be calculated until six months after the fact.)

The following mathematical approximation has been derived to convert a smoothed sunspot number to a solar flux value.

$$F = 63.75 + 0.728\,S + 0.00089\,S^2 \qquad \text{(Eq 1)}$$

where

F = solar flux value
S = sunspot number

This equation has been found to yield errors as great as 10% when historical data was examined. (Look at the August 1981 data in Table 3.) The reason for this is the lack of exact correlation in the data, as discussed earlier. Therefore, conversions should be rounded to the nearest whole number, as decimal places are unwarranted.

To make conversions from flux to sunspot number, the following approximation may be used. It was derived from Eq 1.

$$S = 33.52\,\sqrt{85.12 + F}\ -\ 408.99 \qquad \text{(Eq 2)}$$

A graphic representation of Eq 1 is given in Fig 25. Use this chart to make conversions graphically, rather than by calculations. With the graph, solar flux and sunspot number conversions can be made either way.

PROPAGATION PREDICTIONS

Quite reliable methods of determining the MUF for any given radio path have been developed over the last 50 years. These methods are all based on the smoothed sunspot number as the measure of solar activity. (The smoothed sunspot number was calculated for June 1986 in an earlier example.) It is for this reason that smoothed sunspot numbers hold so much meaning for radio amateurs and others concerned with radio-wave propagation—they are the link to past (and future) propagation conditions.

Early on, the prediction of propagation conditions required rather tedious work with numerous graphs, along with charts of frequency contours overlaid or overprinted on world maps. The basic materials were available from an agency of the US government. Monthly publications provided the frequency-contour data a few months in advance.

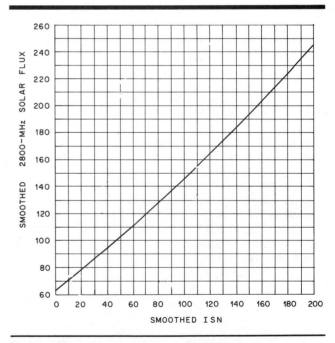

Fig 25—Chart for conversions between smoothed International Sunspot Numbers and smoothed 2800-MHz solar flux. This curve is based on the mathematical approximation given in the text.

Amateur predictions for a distant future date were unheard of.

Today, serious calculations are done exclusively by computer. Large main-frame systems can take many variables into account in determining the MUF and LUF, such as transmitted power, receiver sensitivity, the radiated wave angle (related to antenna height), and even the earth's magnetic field. Further, sunspot numbers can now be estimated years in advance, with a known degree of certainty, so making predictions for the distant future is as simple as making predictions for tomorrow.

Excellent programs have also been written for personal computers. The basic information required by these programs is the smoothed sunspot number, the date (month and day), and the latitudes and longitudes at the two ends of the radio path. Some programs accept a solar flux value and convert it to a sunspot-number equivalent. The latitude and longitude information, of course, is used to determine the great-circle radio path. The date is used to determine the latitude of the sun, and this, with the sunspot number, is used to determine the properties of the ionosphere at critical points on the path.

One such computer program is MINIMUF, written in BASIC and presented in December 1982 *QST* by Bob Rose, K6GKU. Rose reports that many modified versions of MINIMUF have evolved with "enhancements," but several of these modifications degrade the accuracy of the MUF calculations.

Another program written with more "user friendliness" and many features for the amateur is MINIPROP, by Sheldon Shallon, W6EL. (See bibliography.) In addition to making both short-path and long-path MUF computations between two points, this program takes absorption into account in determining band openings, giving an indication of signal strengths to be expected on a band-by-band basis. MINIPROP also calculates and displays bearings, distances, radiation angles, the number of hops, and sunrise/sunset times. It further calculates a "DX compass," showing low-radiation-angle MUF values for a single hop at 30° intervals around the compass for a particular date, time, and smoothed sunspot number (or solar flux value).

Obtaining Sunspot Number/Solar Flux Data

In essence, an amateur wishing to make reliable estimations of propagation conditions for past, present, or future periods may do so readily if he or she is equipped with a personal computer and a good propagation-prediction program. But there is still one more necessary ingredient—a knowledge (or an estimation) of the sunspot number or solar flux value for the period in question.

Fig 26 shows a graph of smoothed sunspot numbers for solar cycles 17 through 21 and predictions for cycles 22 through 26. The graph covers a period of 100 years, from 1940 to 2040, and may be used for making long-term or historical calculations. Just remember that the graph shows *smoothed* numbers. The solar activity at any given time can be

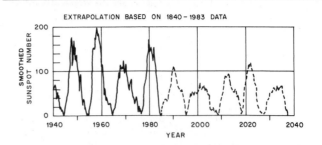

Fig 26—Smoothed sunspot number prediction for the next five sunspot cycles. *(Courtesy of Naval Ocean Systems Center, San Diego.)*

significantly lower or significantly higher than the graph indicates.

WWV PROPAGATION DATA

For the most current data on what the sun is doing, National Bureau of Standards station WWV broadcasts information on solar activity. (See the frequency listing in Table 1.) At 18 minutes past each hour, propagation bulletins give the solar flux, geomagnetic A-Index, Boulder K-Index, and a brief statement of solar and geomagnetic activity in the past and coming 24-hour periods, in that order. The solar flux and A-Index are changed daily with the 1818 UT bulletin, the rest every three hours—0018, 0318, 0618 UT and so on.

Radio amateurs are probably the largest audience for this service. Their use of the data runs from an occasional check on "the numbers" to round-the-clock recording and detailed record keeping and charting of the data. Charts of the data are useful for correlating propagation effects and predicting recurring activity with each 27½ day rotational period of the sun.

The A-Index

The WWV A-Index is a daily figure for the state of activity of the earth's magnetic field. It is updated with the 1818 UT bulletin. The A-Index tells you mainly how yesterday was, but it is very revealing when charted regularly, because geomagnetic disturbances nearly always recur at four-week intervals.

The K-Index

The K-Index (new every three hours) reflects Boulder readings on the geomagnetic field in the hours just preceding the bulletin data changes. It is the nearest thing to current data on radio propagation available. With new data every three hours, K-Index *trend* is important. Rising is bad news; falling is good, especially related to propagation on paths involving latitudes above 30° north. Because this is a *Boulder* reading of geomagnetic activity, it may not correlate closely with conditions in other areas.

The K-Index is also a timely clue to aurora possibilities. Values of 3, and rising, warn that conditions associated with auroras and degraded HF propagation are present in the Boulder area at the time of the bulletin's preparation.

Other Data

The timing of major flares is announced, usually in the next bulletin period, and the flare is given a rating. Some hours later, if the particles are intercepted by our geostationary satellites that are designed for that purpose, a *proton event* is announced. Around 12 to 24 hours later the solar particles enter the earth's atmosphere.

Since auroras are related to solar flares and the periods of increased geomagnetic activity that follows them, we have an aurora-alert system of sorts in the WWV propagation bulletins. Major flares don't always produce major geomagnetic disturbances and major auroras, but they are a good warning that an aurora is likely, usually 24 to 36 hours after the flare.

W1AW PROPAGATION BULLETINS

Propagation bulletins are transmitted from W1AW several times a day. These bulletins contain information on solar and geomagnetic activity. This information helps to make interpretation of the charts in *The ARRL Operating Manual* more accurate for a given time. New bulletins are normally prepared each Tuesday, but they can be revised at any time if their content does not match current conditions. The propagation information is part of the regular W1AW bulletin service, on all suitable modes. See April and October *QST*, yearly, for W1AW summer and winter schedules.

Weekly sunspot information is also included in W1AW propagation bulletins. This is the freshest information available.

BIBLIOGRAPHY

Source material and more extended discussion of topics covered in this chapter can be found in the references given below.

D. Bray, "Method of Determining VHF Station Capabilities," *QST*, Nov 1961.

K. Davies, *Ionospheric Radio Propagation*, National Bureau of Standards. Also known as NBS Monograph 80. Out of print, but may be found in technical libraries. Excellent technical reference.

A. E. Harper, *Rhombic Antenna Design* (New York: D. Van Nostrand Co, Inc, 1941).

W. D. Johnston, Computer-calculated and computer-drawn great-circle maps are offered. An 11 × 14-in. map is custom made for your location. Write to N5KR, PO Box 370, White Sands, NM 88002.

J. L. Lynch, "The Maunder Minimum," *QST*, Jul 1976, pp 24-26.

R. B. Rose, "MINIMUF: A Simplified MUF-Prediction Program for Microcomputers," *QST*, Dec 1982, pp 36-38.

S. C. Shallon: MINIPROP, a user-supported program written for 16-bit PC/MS-DOS and for 8-bit CP/M systems. The source code is proprietary. Available from some bulletin boards around the country, or write W6EL at 11058 Queensland St, Los Angeles, CA 90034-3029.

E. P. Tilton, "The DXer's Crystal Ball," *QST*, Jun, Aug and Sep 1975.

E. P. Tilton, "Propagation—Past and Prospects," *QST*, Aug 1979, pp 24-27.

Chapter 24

Transmission Lines

The desirability of installing an antenna in a clear space, not too near buildings or power and telephone lines, cannot be stressed too strongly. On the other hand, the transmitter that generates the RF power for driving the antenna is usually, as a matter of necessity, located some distance from the antenna terminals. The connecting link between the two is the RF transmission line, feeder or feed line. Its sole purpose is to carry RF power from one place to another, and to do it as efficiently as possible. That is, the ratio of the power *transferred* by the line to the power *lost* in it should be as large as the circumstances permit.

At radio frequencies, every conductor that has appreciable length compared with the wavelength in use *radiates* power. That is, every conductor is an antenna. Special care must be used, therefore, to minimize radiation from the conductors used in RF transmission lines. Without such care, the power radiated by the line may be much larger than that which is lost in the resistance of conductors and dielectrics (insulating materials). Power loss in resistance is inescapable, at least to a degree, but loss by radiation is largely avoidable.

Preventing Radiation

Radiation loss from transmission lines can be prevented by using two conductors arranged and operated so the electromagnetic field from one is balanced everywhere by an equal and opposite field from the other. In such a case, the resultant field is zero everywhere in space; in other words, there is no radiation from the line.

For example, Fig 1A shows two parallel conductors having currents I1 and I2 flowing in opposite directions. If the current I1 at point Y on the upper conductor has the same amplitude as the current I2 at the corresponding point X on the lower conductor, the fields set up by the two currents are equal in magnitude. Because the two currents are flowing in opposite directions, the field from I1 at Y is 180° out of phase with the field from I2 at X. However, it takes a measurable interval of time for the field from X to travel to Y. If I1 and I2 are alternating currents, the phase of the field from I1 at Y changes in such a time interval, so at the instant the field from X reaches Y, the two fields at Y are not exactly 180° out of phase. The two fields are exactly 180° out of phase at every point in space only when the two conductors occupy the same space—an obviously impossible condition if they are to remain separate conductors.

The best that can be done is to make the two fields cancel each other as completely as possible. This can be achieved by keeping the distance d between the two conductors small enough so the time interval during which the field from X is moving to Y is a very small part of a cycle. When this is the case, the phase difference between the two fields at any given point is so close to 180° that cancellation is nearly complete.

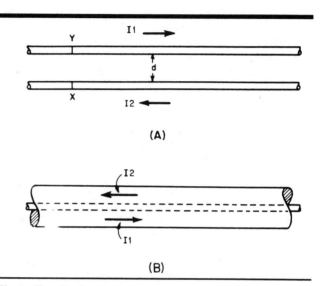

Fig 1—Two basic types of transmission lines.

Practical values of d (the separation between the two conductors) are determined by the physical limitations of line construction. A separation that meets the condition of being "very small" at one frequency may be quite large at another. For example, if d is 6 inches, the phase difference between the two fields at Y is only a fraction of a degree if the frequency is 3.5 MHz. This is because a distance of 6 inches is such a small fraction of a wavelength (1 λ = 281 feet) at 3.5 MHz. But at 144 MHz, the phase difference is 26°, and at 420 MHz, it is 77°. In neither of these cases could the two fields be considered to "cancel" each other. Conductor separation must be very small in comparison with the wavelength used; it should never exceed 1% of the wavelength, and smaller separations are desirable. Transmission lines consisting of two parallel conductors as in Fig 1A are called parallel-conductor lines, open-wire lines or two-wire lines.

A second general type of line construction is shown in Fig 1B. In this case, one of the conductors is tube-shaped and encloses the other conductor. This is called a coaxial line ("coax") or concentric line. The current flowing on the inner conductor is balanced by an equal current flowing in the opposite direction on the *inside surface* of the outer conductor. Because of skin effect, the current on the inner surface of the outer conductor does not penetrate far enough to appear on the outside surface. In fact, the total electromagnetic field outside the coaxial line (as a result of currents flowing on the conductors inside) is always zero, because the outer conductor acts as a shield at radio frequencies. The separation between the inner conductor and the outer conductor is therefore unimportant from the standpoint of reducing radiation.

A third general type of transmission line is the *waveguide*. Waveguides are discussed in detail in Chapter 18.

CURRENT FLOW IN LONG LINES

In Fig 2, imagine that the connection between the battery and the two wires is made instantaneously and then broken. During the time the wires are in contact with the battery terminals, electrons in wire no. 1 will be attracted to the positive battery terminal and an equal number of electrons in wire no. 2 will be repelled from the negative terminal. This happens only near the battery terminals at first, because electromagnetic waves do not travel at infinite speed. Some time does elapse before the currents flow at the more extreme parts of the wires. By ordinary standards, the elapsed time is very short. Because the speed of wave travel along the wires may be almost 300,000,000 meters per second, it becomes necessary to measure time in millionths of a second (microseconds), rather than in more familiar time units.

For example, suppose that the contact with the battery is so short that it can be measured in a very small fraction of a microsecond. Then the "pulse" of current that flows at the battery terminals during this time can be represented by the vertical line in Fig 3. At the speed of light this pulse travels 30 meters along the line in 0.1 microsecond, 60 meters in 0.2 microsecond, 90 meters in 0.3 microsecond, and so on, as far as the line reaches.

The current does not exist all along the wires; it is only present at the point that the pulse has reached in its travel. At this point it is present in both wires, with the electrons moving in one direction in one wire and in the other direction in the other wire. If the line is infinitely long and has no resistance (or other cause of energy loss), the pulse will travel undiminished forever.

By extending the example of Fig 3, it is not hard to see that if, instead of one pulse, a whole series of them were started on the line at equal time intervals, the pulses would travel along the line with the same time and distance spacing between them, each pulse independent of the others. In fact, each pulse could have a different amplitude if the battery voltage were varied. Furthermore, the pulses could be so closely spaced that they touched each other, in which case current would be present everywhere along the line simultaneously.

Wavelength

It follows from this that an alternating voltage applied to the line would give rise to the sort of current flow shown in Fig 4. If the frequency of the ac voltage is 10,000,000 hertz (cycles per second) or 10 MHz, each cycle occupies 0.1 microsecond, so a complete cycle of current will be present along each 30 meters of line. This is a distance of one wavelength. Any currents at points B and D on the two conductors occur one cycle later in time than the currents at A and C. Put another way, the currents initiated at A and C do not appear at B and D, one wavelength away, until the applied voltage has gone through a complete cycle.

Because the applied voltage is always changing, the currents at A and C change in proportion. The current a short distance away from A and C—for instance, at X and Y—is not the same as the current at A and C. This is because the current at X and Y was caused by a value of voltage that occurred slightly earlier in the cycle. This situation holds true all along the line; at any instant the current anywhere along the line from A to B and C to D is different from the current at any other point on that section of the line.

The remaining series of drawings in Fig 4 shows how the instantaneous currents might be distributed if we could take snapshots of them at intervals of ¼ cycle. The current travels out from the input end of the line in waves. At any given point on the line, the current goes through its complete range of ac values in one cycle, just as it does at the input end. Therefore (if there are no losses) an ammeter inserted in either conductor reads exactly the same current at any point along the line, because the ammeter averages the current over a whole cycle. (The phases of the currents at any two separate points is different, but the ammeter cannot show phase.)

VELOCITY OF PROPAGATION

In the example above it was assumed that energy travels

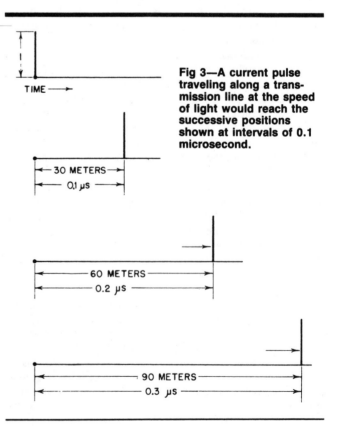

Fig 3—A current pulse traveling along a transmission line at the speed of light would reach the successive positions shown at intervals of 0.1 microsecond.

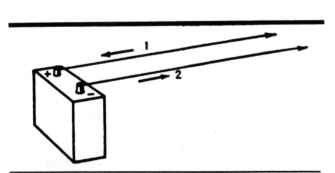

Fig 2—A representation of current flow on a long transmission line.

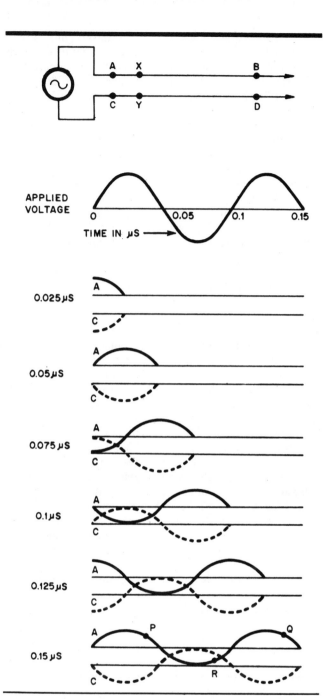

Fig 4—Instantaneous current along a transmission line at successive time intervals. The frequency is 10 MHz; the time of each cycle is 0.1 microsecond.

along the line at the velocity of light. The actual velocity is very close to that of light only in lines in which the insulation between conductors is air. The presence of dielectrics other than air reduces the velocity.

Electrical Length

Current does not flow at the speed of light in any medium other than a vacuum. Therefore, the time required for a signal of a given frequency to travel down a length of transmission line is *longer* than the time required for the same signal to travel the same distance in free space. Because of this propagation delay, 360° of a given wave exists in a physically shorter distance on a given transmission line than in free space.

The exact delay for a given transmission line is a function of the properties of the line, including the dielectric constant of the insulating material between the conductors. This delay is expressed in terms of the speed of light (either as a percentage or a decimal fraction), and is referred to as velocity factor (VF). The velocity factor is related to the dielectric constant (ϵ) by

$$VF = \frac{1}{\sqrt{\epsilon}}$$

The wavelength in a practical line is always shorter than the wavelength in free space. Whenever reference is made to a line as being a "half wavelength" or "quarter wavelength" long (½ λ or ¼ λ), it is understood that what is meant by this is the *electrical* length of the line. The physical length corresponding to an electrical wavelength on a given line is given by

$$1 \, \lambda \, (ft) = \frac{983.6}{f} \times VF \qquad (Eq \ 1)$$

where
f = frequency in MHz
VF = velocity factor

Values of VF for several common types of lines are given later in this chapter. The actual VF of a given cable varies slightly from one production run or manufacturer to another, even though the cables may have exactly the same specifications.

Because a quarter-wavelength line is frequently used as an impedance transformer, it is convenient to calculate the length of a quarter-wave line directly by

$$\frac{1}{4} \, \lambda = \frac{245.9}{f} \times VF \qquad (Eq \ 1A)$$

CHARACTERISTIC IMPEDANCE

If the line is "perfect"—having no resistance—a question immediately comes up: What is the amplitude of the current in the pulse? Will a larger voltage result in a larger current, or is the current theoretically infinite for any applied voltage, as we would expect from applying Ohm's Law to a circuit without resistance? The answer is that the current does depend directly on the voltage, just as though resistance were present.

The reason for this is that the current flowing in the line is something like the charging current that flows when a battery is connected to a capacitor. That is, the line has capacitance. However, it also has inductance. Both of these are "distributed" properties. We may think of the line as being composed of a whole series of small inductors and capacitors, connected as in Fig 5, where each coil is the

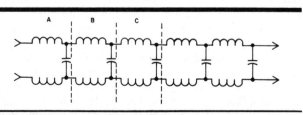

Fig 5—Equivalent of a transmission line in terms of ordinary circuit elements (lumped constants). The values of inductance and capacitance depend on the line construction.

inductance of an extremely small section of wire, and the capacitance is that existing between the same two sections. Each inductance limits the rate at which each immediately following capacitor can be charged, and the effect of the chain is to establish a definite relationship between current and voltage. In this way the line has an apparent "resistance," called its characteristic impedance or surge impedance. The conventional symbol for characteristic impedance is Z_0.

TERMINATED LINES

The value of the characteristic impedance is equal to L/C in a perfect line—that is, one in which the conductors have no resistance and there is no leakage between them—where L and C are the inductance and capacitance, respectively, per unit length of line. The inductance decreases with increasing conductor diameter, and the capacitance decreases with increasing spacing between the conductors. Hence a line with closely spaced large conductors has a relatively low characteristic impedance, while one with widely spaced thin conductors has a high impedance. Practical values of Z_0 for parallel-conductor lines range from about 200 to 800 ohms. Typical coaxial lines have characteristic impedances from 30 to 100 ohms. Physical constraints on the practical wire diameters and spacings limit Z_0 values to these ranges.

In the earlier discussion of current traveling along a transmission line, we assumed that the line was infinitely long. Practical lines have a definite length, and they are terminated in a load at the "output" end (the end to which the power is delivered). If the load is a pure resistance of a value equal to the characteristic impedance of the line, Fig 6, the current traveling along the line to the load finds that the load "looks like" more transmission line of the same characteristic impedance.

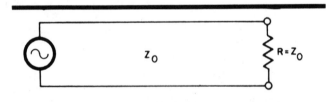

Fig 6—A transmission line terminated in a resistive load equal to the characteristic impedance of the line.

The reason for this can be more easily understood by considering it from another viewpoint. Along a transmission line, power is transferred successively from one elementary section in Fig 5 to the next. When the line is infinitely long, this power transfer goes on in one direction—away from the source of power.

From the standpoint of Section B, Fig 5, for instance, the power transferred to section C has simply disappeared in C. As far as section B is concerned, it makes no difference whether C has absorbed the power itself or has transferred it along to more transmission line. Consequently, if we substitute a load for section C that has the same electrical characteristics as the transmission line, section B will transfer power into it just as if it were more transmission line. A pure resistance equal to the characteristic impedance of C, which is also the characteristic impedance of the line, meets this condition. It absorbs all the power just as the infinitely long line absorbs all the power transferred by section B.

Matched Lines

A line terminated in a purely resistive load equal to the characteristic line impedance is said to be *matched*. In a matched transmission line, power is transferred outward along the line from the source until it reaches the load, where it is completely absorbed. Thus, with either the infinitely long line or its matched counterpart, the impedance presented to the source of power (the line-input impedance) is the same *regardless of the line length*. It is simply equal to the characteristic impedance of the line. The current in such a line is equal to the applied voltage divided by the characteristic impedance, and the power put into it is E^2/Z_0 or I^2Z_0, by Ohm's Law.

Mismatched Lines

Now take the case where the terminating resistance, R, is *not* equal to Z_0, as in Fig 7. The load R no longer "looks like" more line to the section of line immediately adjacent. Such a line is said to be *mismatched*. The more that R differs from Z_0, the greater the mismatch. The power reaching R is not totally absorbed, as it was when R was equal to Z_0, because R requires a voltage to current ratio that is different from the one traveling along the line. The result is that R absorbs only part of the power reaching it (the *incident* power); the remainder acts as though it had bounced off a wall and starts back along the line toward the source. This is *reflected power*, and the greater the mismatch, the larger is the percentage of the incident power that is reflected. In the extreme case where R is zero (a short circuit) or infinity (an open circuit), *all* of the power reaching the end of the line is reflected.

Whenever there is a mismatch, power is transferred in both directions along the line. The voltage to current ratio is the same for the reflected power as for the incident power,

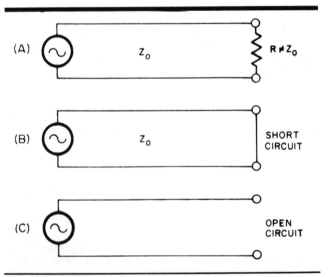

Fig 7—Mismatched lines. A—termination not equal to Z_0; B—short-circuited line; C—open-circuited line.

because this ratio is determined by the Z_0 of the line. The voltage and current travel along the line in both directions in the same wave motion shown in Fig 4. If the source of power is an ac generator, the incident (outgoing) voltage and the reflected (returning) voltage are simultaneously present all along the line. The actual voltage at any point along the line is the algebraic sum of the two components. The same is true of the current.

The effect of the incident and reflected components on the behavior of the line can be understood more readily by considering first the two limiting cases—the short-circuited line and the open-circuited line. If the line is short-circuited as in Fig 7B, the voltage at the end must be zero. Thus the incident voltage must disappear suddenly at the short. It can do this only if the reflected voltage is opposite in phase and of the same amplitude. This is shown by the vectors in Fig 8. The current, however, does not disappear in the short circuit; in fact, the incident current flows through the short and there is in addition the reflected component in phase with it and of the same amplitude.

The reflected voltage and current must have the same amplitudes as the incident voltage and current, because no power is dissipated in the short circuit; all the power starts back toward the source. Reversing the phase of *either* the current or voltage (but not both) reverses the direction of power flow; in the short-circuited case the phase of the voltage is reversed on reflection, but the phase of the current is not.

If the line is open-circuited (Fig 7C) the current must be zero at the end of the line. In this case the reflected current is 180° out of phase with the incident current and has the same amplitude. By reasoning similar to that used in the short-circuited case, the reflected voltage must be in phase with the incident voltage, and must have the same amplitude. Vectors for the open-circuited case are shown in Fig 9.

Where there is a finite value of resistance at the end of the line, Fig 7A, only part of the power reaching the end of the line is reflected. That is, the reflected voltage and current are smaller than the incident voltage and current. If R is less than Z_0, the reflected and incident voltage are 180° out of phase, just as in the case of the short-circuited line, but the amplitudes are not equal because all of the voltage does not

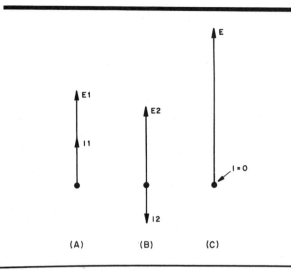

Fig 9—Voltage and current at the end of an open-circuited line. A—outgoing voltage and current; B—reflected voltage and current; C—resultant.

disappear at R. Similarly, if R is greater than Z_0, the reflected and incident currents are 180° out of phase (as they were in the open-circuited line), but all of the current does not appear in R. The amplitudes of the two components are therefore not equal. These two cases are shown in Fig 10. Note that the resultant current and voltage are in phase in R, because R is a pure resistance.

Reflection Coefficient

The ratio of the reflected voltage to the incident voltage is called the *reflection* coefficient. Thus

$$\rho = \frac{E_r}{E_f} \qquad \text{(Eq 2)}$$

where

ρ = reflection coefficient
E_r = reflected voltage
E_f = forward (incident) voltage

(In some professional literature, Γ is used in place of ρ to represent the reflection coefficient.)

The reflection coefficient is determined by the relationship between the line Z_0 and the actual load at the terminated end of the line. The coefficient can never be larger than 1 (which indicates that all the incident power is reflected) nor smaller than zero (indicating that the line is perfectly matched to the load).

If the load is purely resistive, the reflection coefficient can be found from

$$\rho = \frac{R - Z_0}{R + Z_0} \qquad \text{(Eq 2A)}$$

where R is the resistance of the load terminating the line. In this expression ρ is positive if R is larger than Z_0, and negative if R is smaller than Z_0. The change in signs accompanies the change in phase of the reflected voltage described above.

STANDING WAVES

As might be expected, reflection cannot occur at the load without some effect on the voltages and currents all along the line. A detailed description tends to become complicated, and

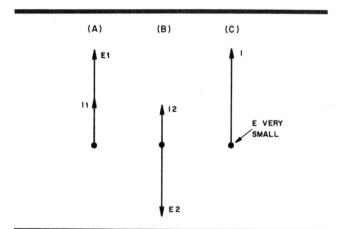

Fig 8—Voltage and current at the short circuit on a short-circuited line. These vectors show how the outgoing voltage and current (A) combine with the reflected voltage and current (B) to result in high current and very low voltage in the short circuit (C).

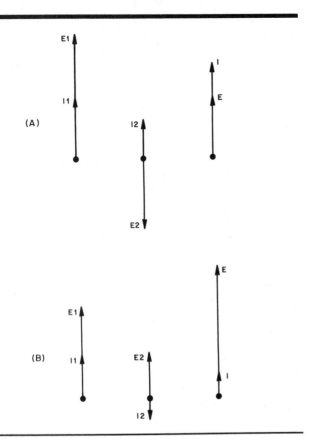

Fig 10—Incident and reflected components of voltage and current when the line is terminated in a pure resistance not equal to Z_0. In the case shown, the reflected components have half the amplitude of the incident components. A—R less than Z_0; B—R greater than Z_0.

the effects are most simply shown by vector diagrams. Fig 11 is an example in the case where R is less than Z_0. The voltage and current vectors at the load, R, are shown in the reference position; they correspond with the vectors in Fig 10A.

Back along the line from R toward the power source, the incident vectors, E1 and I1, lead the vectors at the load according to their position along the line measured in electrical degrees. (The corresponding distances in fractions of a wavelength are also shown.) The vectors representing reflected voltage and current, E2 and I2, successively lag the same vectors at the load. This lag is the natural consequence of the direction in which the incident and reflected components are traveling, together with the fact that it takes time for power to be transferred along the line. The resultant voltage, E, and current, I, at each of these positions are shown dotted. Although the incident and reflected components maintain their respective amplitudes (the reflected component is shown at half the incident-component amplitude in this drawing), their phase relationships vary with position along the line. The phase shifts cause both the amplitude and phase of the *resultants* to vary with position on the line.

If the amplitude variations (disregarding phase) of the resultant voltage and current are plotted against position along the line, graphs like those of Fig 12A will result. If we could go along the line with a voltmeter and ammeter measuring the current and voltage at each point, plotting the collected data would give curves like these. In contrast, if the load matched the Z_0 of the line, similar measurements along the line would show that the voltage is the same everywhere (and similarly for the current). The mismatch between load and line is responsible for the variations in amplitude which, because of their wave-like appearance, are called *standing waves*.

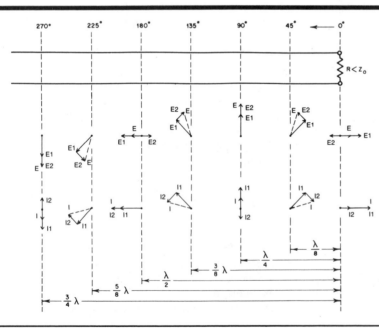

Fig 11—Incident and reflected components at various positions along the transmission line, together with resultant voltages and currents at the same positions. The case shown is for R less than Z_0.

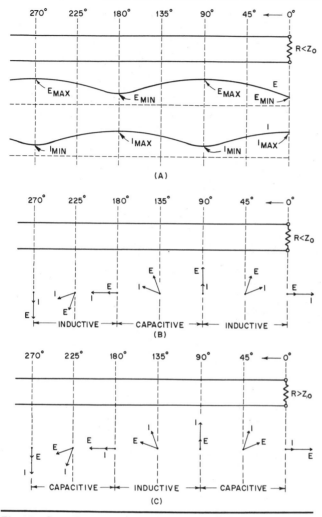

Fig 12—Standing waves of current and voltage along the line for R less than Z₀ (A). Resultant voltages and currents along a mismatched line are shown at B and C. B—R less than Z₀; C—R greater than Z₀.

From the earlier discussion it should be clear that when R is greater than Z_0 the voltage will be largest and the current smallest at the load. This is just the reverse of the case shown in Fig 12A. In such a case, the curve labeled E (voltage) would become the I (current) curve, while the current curve would become the voltage curve.

Some general conclusions can be drawn from inspection of the standing-wave curves: At a position 180° (½ λ) from the load, the voltage and current have the same values they do at the load. At a position 90° from the load, the voltage and current are "inverted." That is, if the voltage is lowest and current highest at the load (when R is less than Z_0), then 90° from the load the voltage reaches its highest value. The current reaches its lowest value at the same point. In the case where R is greater than Z_0, so the voltage is highest and the current lowest at the load, the voltage is lowest and the current is highest 90° from the load.

Note that the conditions at the 90° point also exist at the 270° point (¾ λ). If the graph were continued on toward the source of power it would be found that this duplication occurs

at every point that is an odd multiple of 90° (odd multiple of ¼ λ) from the load. Similarly, the voltage and current are the same at every point that is a multiple of 180° (any multiple of ½ λ) away from the load.

Standing-Wave Ratio

The ratio of the maximum voltage along the line to the minimum voltage—that is, the ratio of E_{max} to E_{min} in Fig 12A, is defined as the *voltage standing-wave ratio* (VSWR) or simply *standing-wave ratio* (SWR).

$$SWR = \frac{E_{max}}{E_{min}} \qquad \text{(Eq 3)}$$

The ratio of the maximum current to the minimum current (I_{max}/I_{min}) is the same as the VSWR, so either current or voltage can be measured to determine the standing-wave ratio. Fig 13 contains a convenient nomograph from which SWR can be determined in accordance with forward and reflected readings on an RF wattmeter.

The standing-wave ratio is an index of many of the properties of a mismatched line. It can be measured with fairly simple equipment, so is a convenient quantity to use in making calculations on line performance. *If the load contains no reactance*, the SWR is numerically equal to the ratio between the load resistance, R, and the characteristic impedance of the line. When R is greater than Z_0,

$$SWR = \frac{R}{Z_0} \qquad \text{(Eq 4)}$$

When R is less than Z_0,

$$SWR = \frac{Z_0}{R} \qquad \text{(Eq 4A)}$$

(The smaller quantity is always used in the denominator of the fraction so the ratio will be a number greater than 1).

This relationship shows that the greater the mismatch—that is, the greater the difference between Z_0 and R—the larger the SWR. In the case of open- and short-circuited lines the SWR is infinite. On such lines the voltage and current are zero at the minimum points (E_{min} and I_{min}), because total reflection occurs at the end of the line, and the incident and reflected components have equal amplitudes.

Flat Lines

As discussed earlier, all the power that is transferred along a transmission line is absorbed in the load if that load is a resistance value equal to the Z_0 of the line. In this case, the line is said to be perfectly matched. None of the power is reflected back toward the source. As a result, no standing waves of current or voltage will be developed along the line. For a line operating in this condition, the waveforms drawn in Fig 12A become straight lines, representing the voltage and current delivered by the source. The voltage along the line is constant, so the minimum value is the same as the maximum value. The voltage standing-wave ratio is therefore 1:1. Because a plot of the voltage standing wave is a straight line, the transmission line is also said to be "flat."

INPUT IMPEDANCE

The relationship between voltage and current at any point along a transmission line (including the effects of the incident and reflected components) becomes clear when only the

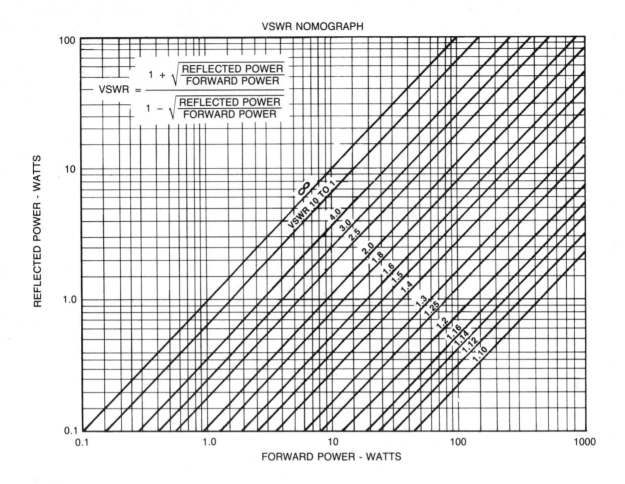

Fig 13—SWR as a function of forward and reflected power.

resultant voltage and current are shown, as in Fig 12B. Note that the voltage and current are in phase not only at the load, but also at every point that is a multiple of 90° from the load.

Suppose the line were cut at one of these points and the generator or source of power connected to the portion terminated in R. The generator would then "see" a pure resistance, just as it would if it were connected directly to R. However, the value of the resistance it sees would depend on the line length. If the length is 90°, or an odd multiple of 90° (where the voltage is high and the current low), the resistance seen by the generator would be greater than R. If the length is 180° or a multiple of 180°, the voltage and current relationships are the same as in R, and therefore the generator "sees" a resistance equal to the actual load resistance.

With a resistive load, the current and voltage are exactly in phase only at points that are multiples of 90° from the load. At all other points the current either leads or lags the voltage, so the load seen by the generator when the line length is not an exact multiple of 90° is not a pure resistance. The input impedance of the line (the impedance seen by the generator connected to the line) in such a case has both resistive and reactive components. When the current lags the voltage the reactance is inductive; when it leads the voltage the reactance is capacitive. Fig 12B shows that when the line is terminated

in a resistance smaller than Z_0, the reactance is inductive in each odd 90° of line from the load toward the generator (first, third, etc), and capacitive in the even 90° sections toward the generator.

Fig 12C illustrates the case where R is greater than Z_0. The voltage and current vectors are merely interchanged, because, as explained in connection with Fig 11, in this case the vector for the reflected current is the one that is reversed in phase on reflection. The reactance becomes capacitive in the first 90°, inductive in the second, and so on.

Factors Determining the Input Impedance

The magnitude and phase angle of the input impedance depend on the SWR, the line length, and the Z_0 of the line. If the SWR is low, the input impedance is principally resistive at all line lengths. If the SWR is high, however, the reactive component may be relatively large.

The input impedance of the line can be represented by a series circuit of resistance and reactance, as shown in Fig 14, where R_s is the resistive component and X_s is the reactive component. The series-equivalent impedance is usually denoted as $R + jX$, with the "s" subscripts omitted. The j is an operator function, used to indicate that the values for R and X cannot be added directly. Vector addition must

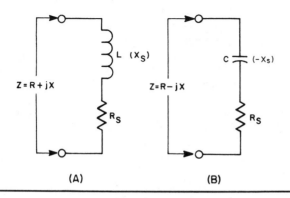

Fig 14—Series circuits that may be used to represent the input impedance of a length of transmission line.

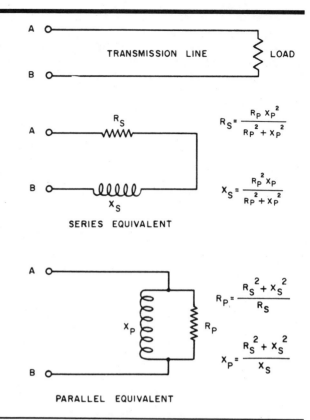

Fig 15—Input impedance of a line terminated in a pure resistance. This impedance can be represented by either a resistance and reactance in series, or a resistance and reactance in parallel. The relationships between the R and X values in the series and parallel equivalents are given by the equations. X may be either inductive or capacitive, depending on the line length, Z_0 and the load impedance.

be used if the polar impedance (magnitude and phase angle) is to be determined. (This is analogous to solving a right triangle for the length of its hypotenuse, where R and X represent the length of its two sides. The length of the hypotenuse represents Z, the overall impedance.) By convention, a plus sign is assigned to j when the reactance is inductive (R + jX), and a minus sign is used when the reactance is capacitive (R − jX).

Equivalent Circuits for the Input Impedance

The series circuits shown in Fig 14 are equivalent to the actual input impedance of the line, because they have the same total impedance and the same phase angle. It is also possible to form a circuit with resistance and reactance in parallel that has the same total impedance and phase angle as the line. This equivalence is shown in Fig 15. The individual values in the parallel circuit are not the same as those in the series circuit (although the overall result is the same), but are related to the series-circuit values by the equations shown in the drawing.

Either of the two equivalent circuits may be used, depending on which is more convenient for the particular purpose. These circuits are important from the standpoint of designing coupling networks to take the desired amount of power from the source.

REACTIVE TERMINATIONS

So far, the only type of load considered has been a pure resistance. In most instances, the load will be fairly close to being a pure resistance. This is because most transmission lines used by amateurs are connected to resonant antenna systems, which are principally resistive in nature. Consequently, the resistive load is an important practical case.

An antenna system is purely resistive only at one fundamental frequency, however, and when it is operated over a band of frequencies without readjustment—the usual condition—its impedance will contain a certain amount of reactance along with resistance. The effect of such a combination is to increase the SWR. For instance, a purely resistive load of, say, 100 Ω has a lower SWR on any given line than a reactive load having a 100-Ω total impedance. Also, a purely resistive load has a lower SWR on any given line than another load with the same resistive component but also a reactive component. If the resistive and reactive components of the load are known, the SWR may be determined from

$$SWR = \frac{A + B}{A - B} \qquad (Eq\ 5)$$

where

$$A = \sqrt{(R + Z_0)^2 + X^2}$$

$$B = \sqrt{(R - Z_0)^2 + X^2}$$

R = resistive component, ohms
X = reactive component, ohms
Z_0 = characteristic impedance of the line

The effect of reactance in the load is to shift the phase of the current with respect to the voltage both in the load itself and in the reflected components of voltage and current. This causes a shift in the phase of the resultant current with respect to the resultant voltage. The net result is to shift the points along the line at which the various effects already described will occur. With a load having inductive reactance, the point of maximum voltage and minimum current is shifted toward the load. The opposite is true when the load reactance is capacitive.

ATTENUATION

The discussion so far in this chapter applies to all types

of transmission lines, regardless of their physical construction. It is, however, based on the assumption that there is no power loss in the line. Every practical line will have some inherent loss, partly because of the resistance of the conductors, partly because power is consumed in every dielectric used for insulating the conductors, and partly because in many cases a small amount of power escapes from the line by radiation.

Losses in a line modify its characteristic impedance slightly, but usually not significantly. They will also affect the input impedance; in this case the theoretical values will be modified only slightly if the line is short and has only a small loss. But the input impedance may be changed considerably if an appreciable proportion of the power input to the line is dissipated by the line itself. A large loss may exist because the line is long, because it has inherently high loss per unit length, because the SWR is high, or because of a combination of these factors.

The reflected power returning to the input terminals of the line is lower when the line has losses than it would be if there were no losses. The overall effect is that *the SWR changes* along the line, being highest at the load and smallest at the input terminals. As far as its input impedance is concerned, a long, high-loss line therefore tends to act as though the impedance match at the load end were better than is actually the case.

Line Losses

The power lost in a transmission line is not directly proportional to the line length, but varies logarithmically with the length. That is, if 10% of the input power is lost in a section of line of certain length, 10% of the remaining power will be lost in the next section of the same length, and so on. For this reason it is customary to express line losses in terms of decibels per unit length, since the decibel is a logarithmic unit. Calculations are very simple because the total loss in a line is found by multiplying the decibel loss per unit length by the total length of the line. Line loss is usually expressed in decibels per 100 feet. It is necessary to specify the frequency for which the loss applies, because the loss does vary with frequency.

Conductor loss and dielectric loss both increase as the operating frequency is increased, but not in the same way. This, together with the fact that the relative amount of each type of loss depends on the actual construction of the line, makes it impossible to give a specific relationship between loss and frequency that will apply to all types of lines. Each line must be considered individually. Actual loss values are given in a later section of this chapter.

Effect of SWR

The power lost in a given line is least when the line is terminated in a resistance equal to its characteristic impedance. The loss increases with an increase in the SWR. This is because the effective values of both current and voltage become greater. The increase in effective current raises the ohmic losses (I^2R) in the conductors, and the increase in effective voltage increases the losses in the dielectric (E^2/R).

The increased loss caused by an SWR greater than 1 may or may not be serious. If the SWR at the load is not greater than 2:1, the additional loss caused by the standing waves, as compared with the loss when the line is perfectly matched, does not amount to more than about ½ dB even on very long

lines. One-half dB is an undetectable change in signal strength. Therefore, it can be said that, from a practical standpoint, an SWR of 2 or less is every bit as good as a perfect match, so far as losses are concerned.

The above statement generally applies at frequencies below 30 MHz. However, above these frequencies, in the VHF and especially the UHF range, matched-line losses for commonly available types of coax can be relatively high. This means that even a slight mismatch may become a concern regarding additional transmission line losses.

The additional loss caused by the standing waves may be calculated from the following equation.

$$\text{Additional loss (dB)} = 10 \log \left[\frac{\alpha^2 - \rho^2}{\alpha(1 - \rho^2)} \right] - \text{dB} \quad \text{(Eq 6)}$$

where

$\alpha = 10^{\text{dB}/10}$

$\rho = \dfrac{\text{SWR} - 1}{\text{SWR} + 1}$

dB = the line loss in dB when perfectly matched

SWR = SWR at the load

It is of interest to note that when the line loss is high with perfect matching, the additional loss in decibels caused by the SWR tends to be constant regardless of the matched line loss. The reason for this is that the amount of power available to be reflected from the load is reduced, because relatively little power reaches the load in the first place. For example, if the line loss with perfect matching is 6 dB, only 25% of the power originally put into the line reaches the load. If the mismatch at the load (the SWR at the load) is 4 to 1, 36% of the power reaching the load will be reflected (from Eq 10 in a later section). Of the power originally put into the line, then, $0.25 \times 0.36 = 0.09$ or 9% will be reflected. This in turn will be attenuated 6 dB in traveling back to the input end of the line, so that only $0.09 \times 0.25 = 0.0225$ or slightly over 2% of the original power actually gets back to the input terminals. With such a small proportion of power returning to the input terminals, the SWR measured at the input end of the line would be only about 1.35 to 1—although it is 4 to 1 at the load. In the presence of line losses the SWR always decreases along the line going from the load to the input end.

On lines having low losses when perfectly matched, a high SWR may increase the power loss by a large factor. However, in this case the total loss may still be inconsequential in comparison with the power delivered to the load. An SWR of 10 on a line having only 0.3 dB loss when perfectly matched will cause an additional loss of 1 dB. This loss would produce a just-detectable difference in signal strength.

NONRESISTIVE TERMINATIONS

In most of the preceding discussions, we considered loads containing only resistance or resistance and reactance. Such a resistive load will consume some, if not all, of the power that has been transferred along the line. However, a nonresistive load such as a pure reactance can also terminate a length of line. Such terminations, of course, will consume no power, but will reflect all of the energy arriving at the end of the line. In this case the theoretical SWR in the line will be infinite, but in practice, losses in the line will limit the SWR to some finite value at line positions back toward the source.

At first you might think there is little or no point in terminating a line with a nonresistive load. But consider the effect on the input impedance of the line. From an earlier section we learned that the value of input impedance depends on the value of the load impedance, on the length of the line and on the characteristic impedance of the line. There are times when a line terminated in a nonresistive load can be used to advantage, such as in phasing or matching applications. Remote switching of reactive terminations on sections of line can be used to reverse the beam heading of an antenna array, for example. More information on such applications is presented later in this chapter and in other chapters. The point of this discussion is that a line need not always be terminated in a load that will consume power.

Short- and Open-Circuit Terminations

Two types of nonresistive line terminations are quite useful—short and open circuits. The impedance of the short-circuit termination is $0 + j0$, and the impedance of the open-circuit termination is infinite. Such terminations are used in stub matching. (See Chapters 26 and 28.) These terminations are also useful in evaluating the quality of a given piece of transmission line, as described later.

SPECIAL CASES

Beside the primary purpose of transporting power from one point to another, transmission lines have properties that are useful in a variety of ways. One such special case is a line an exact multiple of ¼-λ (90°) long. As shown earlier, such a line will have a purely resistive input impedance when the termination is a pure resistance. Also, short-circuited or open-circuited lines can be used in place of conventional inductors and capacitors since such lines have an input impedance that is substantially a pure reactance when the line losses are low.

The Half-Wavelength Line

When the line length is an even multiple of 90° (that is, a multiple of ½ λ), the input resistance is equal to the load resistance, regardless of the line Z_0. As a matter of fact, a line an exact multiple of ½ λ in length (disregarding line losses) simply repeats, at its input or sending end, whatever impedance exists at its output or receiving end. It does not matter whether the impedance at the receiving end is resistive, reactive, or a combination of both. Sections of line having such length can be cut in or out without changing any of the operating conditions, at least when the losses in the line itself are negligible.

Impedance Transformation with Quarter-Wave Lines

The input impedance of a line an odd multiple of ¼ λ long is

$$Z_i = \frac{Z_0{}^2}{Z_L} \qquad \text{(Eq 7)}$$

where Z_i is the input impedance and Z_L is the load impedance. If Z_L is a pure resistance, Z_i will also be a pure resistance. Rearranging this equation gives

$$Z_0 = \sqrt{Z_i Z_L} \qquad \text{(Eq 7A)}$$

This means that if we have two values of impedance that we wish to "match," we can do so if we connect them together

by a ¼-λ transmission line having a characteristic impedance equal to the square root of their product.

A ¼-λ line is, in effect, a transformer. It is frequently used as such in antenna work when it is desired, for example, to transform the impedance of an antenna to a new value that will match a given transmission line. This subject is considered in greater detail in a later chapter.

Lines as Circuit Elements

An open- or short-circuited line does not deliver any power to a load, and for that reason is not, strictly speaking a "transmission" line. However, the fact that a line of the proper length has inductive reactance makes it possible to substitute the line for a coil in an ordinary circuit. Likewise, another line of appropriate length having capacitive reactance can be substituted for a capacitor.

Sections of lines used as circuit elements are usually ¼ λ or less long. The desired type of reactance (inductive or capacitive) or the desired type of resonance (series or parallel) is obtained by shorting or opening the far end of the line. The circuit equivalents of various types of line sections are shown in Fig 16.

When a line section is used as a reactance, the amount of reactance is determined by the characteristic impedance and the electrical length of the line. The type of reactance exhibited at the input terminals of a line of given length depends on whether it is open- or short-circuited at the far end.

The equivalent "lumped" value for any "inductor" or "capacitor" may be determined with the aid of the Smith Chart. Line losses may be taken into account if desired, as explained in Chapter 28.

In the case of a line having no losses, and to a close approximation when the losses are small, the inductive reactance of a short-circuited line less than ¼ λ in length is

$$X_L \text{ (ohms)} = Z_0 \tan \ell \qquad \text{(Eq 8)}$$

where ℓ is the length of the line in electrical degrees and Z_0 is the characteristic impedance of the line.

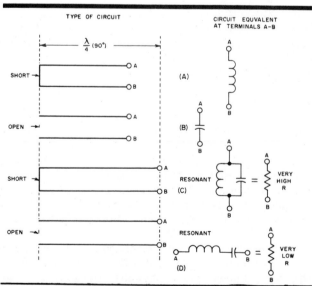

Fig 16—Lumped-constant circuit equivalents of open- and short-circuited transmission lines.

The capacitive reactance of an open-circuited line less than ¼ λ in length is

$$X_C \text{ (ohms)} = Z_o \cot \ell \qquad \text{(Eq 9)}$$

Lengths of line that are exact multiples of ¼ λ have the properties of resonant circuits. With an open-circuit termination, the input impedance of the line acts like a series-resonant circuit. With a short-circuit termination, the line input simulates a parallel-resonant circuit. The effective Q of such linear resonant circuits is very high if the line losses, both in resistance and by radiation, are kept down. This can be done without much difficulty, particularly in coaxial lines, if air insulation is used between the conductors. Air insulated open wire lines are likewise very good at frequencies for which the conductor spacing is very small in terms of wavelength.

Applications of line sections as circuit elements in connection with antenna and transmission-line systems are discussed in later chapters.

Voltages, Currents and Line Losses

The power reflected from a mismatched load does not represent an actual loss, except as it is attenuated in traveling back to the input end of the line. It merely represents power returned, and with a conjugate match at the line input (see Chapter 25), the actual effect is to reduce the power taken from the source. That is, the returned power reduces the coupling between the power source and the line. This is easily overcome by readjusting the coupling until the actual power put into the line is the same as it would be with a matched load. In doing this, of course, the voltages and currents at loops along the line are increased.

As an example, suppose that a line having a characteristic impedance of 600 Ω is matched by a resistive load of 600 Ω and that 100 watts of power goes into the input terminals. The line simply looks like a 600-Ω resistance to the source of power. By Ohm's Law the current and voltage in such a matched line are

$$I = \sqrt{P/R} \text{ and } E = \sqrt{PR}$$

Substituting 100 watts for P and 600 Ω for R, the current is 0.408 ampere and the potential is 245 volts. Assume for the moment that the line has no losses. All the power will reach the load, so the voltage and current at the load will be the same as at the input terminals.

Now suppose that the load is 60 Ω instead of 600 Ω. From Eq 4A, the SWR is therefore 600/60 = 10. From Eq 2A, the reflection coefficient, or ratio of the reflected voltage or current to the voltage or current arriving at the load, is

$$\rho = \frac{SWR - 1}{SWR + 1} \qquad \text{(Eq 10)}$$

In this case the reflection coefficient is $(10 - 1)/(10 + 1)$ = 9/11 = 0.818, so the reflected voltage and current are both equal to 81.8% of the incident voltage and current. The reflected power is proportional *to the square* of either the current or voltage, and so is equal to $0.818^2 = 0.67$ times the incident power, or 67 watts. Since we have assumed that the line has no losses, this amount of power arrives back at the input terminals and subtracts from the original 100 watts, leaving only 33 watts as the amount of power actually taken from the source.

In order to put 100 watts into the 60-Ω load, the coupling to the source must be increased so the incident power minus the reflected power equals 100 watts. And since the power absorbed by the load is only 33% of that reaching it, the incident power must equal 100/0.33 = 303 watts. In a perfectly matched line, the current and voltage with 303 watts input would be 0.71 ampere and 426 volts, respectively. The reflected current and voltage are 0.818 times these values, or 0.581 ampere and 348 volts. At current maxima or loops the current will therefore be 0.71 + 0.58 = 1.29 A, and at a minimum point will be 0.71 − 0.58 = 0.13 A. The voltage maxima and minima will be 426 + 348 = 774 volts and 426 − 348 = 78 volts. (Because of rounding values in the calculation process, the SWR does not work out to be exactly 10 in either the voltage or current case, but the error is very small.)

In the interests of simplicity this example has been based on a line with no losses, but the effect of line attenuation could be included without much difficulty. If the matched-line loss were 3 dB, for instance, only half the input power would reach the load, so new values of current and voltage at the load would be computed accordingly. The reflected power would then be based on the attenuated figure, and then itself attenuated 3 dB to find the power arriving back at the input terminals. The overall result would be, as stated before, a reduction in the SWR at the input terminals as compared with that at the load, along with less actual power delivered to the load for the same power input to the line.

LINE VOLTAGES AND CURRENTS

It is often desirable to know the voltages and currents that are developed in a line operating with standing waves. The voltage maximum may be calculated from Eq 11, and the other values determined from the result.

$$E_{max} = \sqrt{P \times Z_0 \times SWR} \qquad \text{(Eq 11)}$$

where

E_{max} = voltage maximum along the line in the presence of standing waves

P = power delivered by the source to the line input, watts

Z_0 = characteristic impedance of the line, ohms

SWR = SWR at the load

Using the values from the previous example, E_{max} = $\sqrt{100 \times 600 \times 10}$ = 774.6 V. Based on Eq 3, E_{min}, the minimum voltage along the line equals E_{max}/SWR = 774.6/10 = 77.5 V. The maximum current may be found by using Ohm's Law; $I_{max} = E_{max}/Z_0 = 774.6/600 = 1.29$ A. The minimum current equals I_{max}/SWR = 1.29/10 = 0.129 A.

The voltage determined from Eq 11 is the *RMS* value—that is, the voltage that would be measured with an ordinary RF voltmeter. If voltage breakdown is a consideration, the value from Eq 11 should be converted to an *instantaneous peak voltage.* Do this by multiplying times $\sqrt{2}$ (assuming the RF waveform is a sine wave). Thus, the maximum instantaneous peak voltage in the above example is $774.6 \times \sqrt{2} = 1095.4$ V.

Strictly speaking, the values obtained as above apply only near the load in the case of lines with appreciable losses. However, the resultant values are the maximum possible that can exist along the line, whether there are line losses or not. For this reason they are useful in determining whether or not a particular line can operate safely with a given SWR. Voltage ratings for various cable types are given in a later section.

Fig 17 shows the ratio of current or voltage at a loop, in the presence of standing waves, to the current or voltage that would exist with the same power in a perfectly matched line. As with Eq 11 and related calculations, the curve literally applies only near the load.

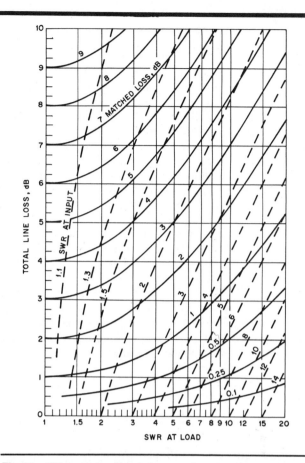

Fig 18—This graph relates the matched-line attenuation loss in dB to the total line loss and to SWR values at the line input and load. The horizontal line at the bottom of the graph represents 0 dB of matched-line loss. *(Graph design by K1TD)*

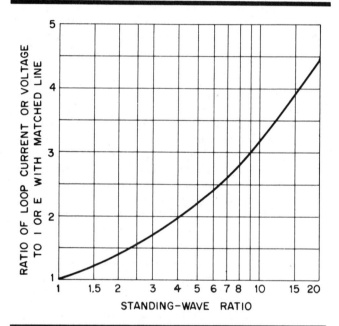

Fig 17—Increase in maximum value of current or voltage on a line with standing waves, as referred to the current or voltage on a perfectly matched line, for the same power delivered to the load. Voltage and current at minimum points are given by the reciprocals of the values along the vertical axis. The curve is plotted from the relationship, current (or voltage) ratio = the square root of SWR.

SWR AND LINE LOSSES

As previously stated, because of line losses the standing-wave ratio changes along the line, being highest at the load end and smallest at the input end. The interrelated effects of SWR and line loss are shown in Fig 18. The chart is a plot of four related variables: (1) SWR at the line input, (2) SWR at the load, (3) line loss when perfectly matched, and (4) total line loss taking the SWR into account. If any two of these variables are known, the other two may be determined from Fig 18.

The horizontal axis of the chart represents the SWR at the load, and the vertical axis represents the total attenuation in the line, taking the SWR into account. The dashed (diagonal) lines represent the SWR at the line input, and the curved lines represent the loss in the line when perfectly matched (flat-line loss). Interpolation of values may be made between the curves.

A pair of examples best illustrate the use of the chart. Let's assume we have 100 feet of RG-8 coax connected between a 28-MHz transmitter and an antenna. The SWR as measured *at the antenna* is 3:1. We desire to know the total line loss and the SWR value at the line input. First we must determine the matched loss of the length of coaxial line. Data given later in this chapter indicates that RG-8 has 1.0 dB of loss per 100 feet at 30 MHz, a value close enough for our needs. Now we proceed by looking along the horizontal axis at the bottom of the chart, locating the value of 3 for the SWR at the load. Then we follow the vertical "3" line up until it intersects with the "1" solid-line curve that represents 1 dB of matched-line loss. At this intersection we read (from the calibration scale at the left) that the total line loss is 1.5 dB. At this same intersection, by interpolating between the dashed-line curves, we also see that the SWR at the line input is approximately 2.3 to 1.

As another example, assume we are using a length of line

at VHF with a matched loss of 3 dB, and we measure the SWR at the *line input* to be 2:1. We wish to know the total line loss and the SWR at the load. First we locate the diagonal dashed line representing a 2:1 SWR at the line input. Then we find the point where this line intersects the "3" solid-line curve, representing the matched line loss in dB. At this intersection we read (scale at left) that the total line loss is just a bit more than 5 dB, and (scale at bottom) that the SWR at the load is 5:1. From this result we realize that a 5:1 load mismatch is costing us 2 dB in additional line losses.

By measuring the SWR at the line input and again at the load, we may determine the matched-line loss from Fig 18, by locating the intersection of the appropriate vertical SWR AT LOAD line with the appropriate diagonal SWR AT INPUT line. At the intersecting point we read a value from the MATCHED LOSS curves. For example, if the load SWR is 1.7:1 and the input SWR is 1.3:1, Fig 18 shows the matched-line loss to be 3 dB.

SWR and Line Losses by Equation

For greater resolution than Fig 18 offers in determining line losses, you may calculate the information. Equations relating load SWR, input SWR, matched-line loss and total transmission line loss are given below.

$$\text{SWR at load} = S_L = \frac{A + B}{A - B} \qquad \text{(Eq 12)}$$

$$\text{SWR at input} = S_i = \frac{B + C}{B - C} \qquad \text{(Eq 13)}$$

$$\text{Total loss (dB)} = L_t = 10 \log \left[\frac{B^2 - C^2}{B(1 - C^2)} \right]$$

$$= 10 \log \left[\frac{C(A^2 - 1)}{A(1 - C^2)} \right] = 10 \log \left[\frac{B(A^2 - 1)}{A^2 - B^2} \right] \text{(Eq 14)}$$

$$\text{Matched-line loss (dB)} = L_m = 10 \log AC \qquad \text{(Eq 15)}$$

where

$$A = \frac{S_i + 1}{S_i - 1}$$

$$B = 10^{L_m/10}$$

$$C = \frac{S_L - 1}{S_L + 1}$$

TESTING TRANSMISSION LINES

Coaxial cable loss should be checked at least every two years if the cable is installed outdoors or buried. (See later section on losses and deterioration.) Testing of any type of line can be done using the technique illustrated in Fig 19. If the measured loss in watts equates to more than 1 dB over the rated loss per 100 feet, the line should be replaced. The matched-line loss in dB can be determined from

$$dB = 10 \log \frac{P1}{P2} \qquad \text{(Eq 16)}$$

where
    P1 is the power at the transmitter output
    P2 is the power measured at R_L of Fig 19

Another method of determining the losses in a line was presented in the preceding section. With any convenient load

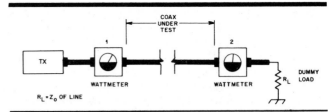

Fig 19—Method for determining losses in transmission lines. The impedance of the dummy load must equal the Z_0 of the line for accurate results.

terminating the line, take two SWR measurements, one at the load and one at the line input. The matched-line loss may then be determined from Fig 18 or from Eq 15. An advantage of using this method is that the line need not be terminated in its characteristic impedance. This method is valid even in the presence of standing waves on the line.

Yet other methods of determining line losses may be used. If the line input impedances can be measured accurately with a short- and then an open-circuit termination, the electrical line length and the matched-line loss may be calculated for the frequency of measurement. The procedure is described in Chapter 28.

Determining line characteristics as just mentioned requires the use of a laboratory style of impedance bridge, or at least an impedance or noise bridge calibrated to a high degree of accuracy. But useful information about a transmission line can also be learned with just an SWR indicator, if it offers reliable readings at high SWR values.

A lossless line theoretically exhibits an infinite SWR when terminated in an open or a short circuit. A practical line will contain losses, and therefore limit the SWR at the line input to some finite value. Provided the signal source can operate

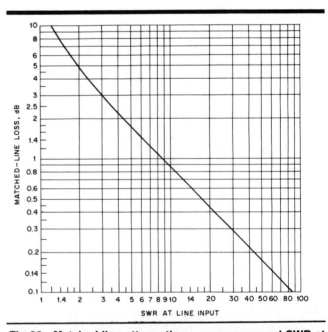

Fig 20—Matched-line attenuation versus measured SWR at the line input when the line is terminated with either a short or an open circuit. Use this graph to determine the loss in a given length of transmission line with just a single SWR measurement.

safely into a severe mismatch, an SWR indicator can be used to determine the line loss. The instruments available to most amateurs lose accuracy at SWR values greater than about 5:1, so this method is useful principally as a go/no-go check on lines that are fairly long. For short, low-loss cables, only significant deterioration can be detected by the open-circuit SWR test.

First, either open or short circuit one end of the line. It makes no difference which termination is used, as the terminating SWR is theoretically infinite in either case. Then measure the SWR at the other end of the line. The matched-line loss for the frequency of measurement may then be determined from

$$L_m \text{ (dB)} = 10 \log \frac{SWR + 1}{SWR - 1} \qquad \text{(Eq 17)}$$

where SWR = the SWR value measured at the line input

Instead of using Eq 17, you may use Fig 20 to obtain a graphic solution for the matched-line loss. The graph is based on Eq 17.

Line Construction and Operating Characteristics

The two basic types of transmission lines, parallel conductor and coaxial, can be constructed in a variety of forms. Both types can be divided into two classes, (1) those in which the majority of the insulation between the conductors is air, where only the minimum of solid dielectric necessary for mechanical support is used, and (2) those in which the conductors are embedded in and separated by a solid dielectric. The first variety (air insulated) has the lowest loss per unit length, because there is no power loss in dry air if the voltage between conductors is below the value at which corona forms. At the maximum power permitted in amateur transmitters, it is seldom necessary to consider corona unless the SWR on the line is very high.

AIR INSULATED LINES

A typical construction technique used for parallel conductor or "two wire" air insulated transmission lines is shown in Fig 21. The two wires are supported a fixed distance apart by means of insulating rods called spacers. Spacers may be made from material such as Teflon, Plexiglas, phenolic, polystyrene, plastic clothespins or plastic hair curlers. Materials commonly used in high quality spacers are isolantite, Lucite and polystyrene. (Teflon is generally not used because of its higher cost.) The spacer length varies from 2 to 6 inches. The smaller spacings are desirable at the higher frequencies (28 MHz) so radiation from the transmission line is minimized.

Spacers must be used at small enough intervals along the line to keep the two wires from moving appreciably with respect to each other. For amateur purposes, lines using this construction ordinarily have no. 12 or no. 14 conductors, and the characteristic impedance is from 500 to 600 Ω. Although once used nearly exclusively, such lines have now been largely superseded by prefabricated lines.

Prefabricated open-wire lines (sold principally for television receiving applications) are available in nominal characteristic impedances of 300 and 450 Ω. The spacers, of low-loss material such as polystyrene, are molded on the conductors at relatively small intervals (typically 6 inches apart) so there is no tendency for the conductors to move with respect to each other. A conductor spacing of 1 inch is used in the 450-Ω line and ½ inch in the 300-Ω line. The conductor size is usually about no. 18. The impedances of such lines are somewhat lower than given by Fig 22 for the same conductor size and spacing, because of the effect of the dielectric constant of the spacers used. The attenuation is quite low and

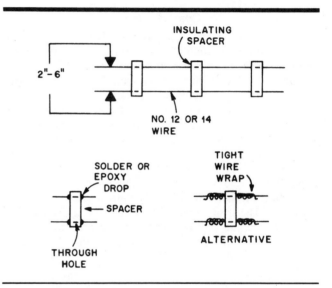

Fig 21—Typical open-wire line construction. The spacers may be held in place by beads of solder or epoxy cement. Wire wraps can also be used, as shown.

lines of this type are entirely satisfactory for transmitting applications at amateur power levels.

Where an air insulated line with still lower characteristic impedance is needed, metal tubing from ¼ to ½ inch diameter is frequently used. With the larger conductor diameter and relatively close spacing, it is possible to build a line having a characteristic impedance as low as about 200 Ω. This construction technique is principally used for ¼-λ matching transformers at the higher frequencies.

Characteristic Impedance

The characteristic impedance of an air insulated parallel conductor line, neglecting the effect of the spacers, is given by

$$Z_0 = 276 \log \frac{2S}{d} \qquad \text{(Eq 18)}$$

where

Z_0 = characteristic impedance in ohms

S = center-to-center distance between conductors

d = outer diameter of conductor (in the same units as S)

Impedances for common sizes of conductors over a range of spacings are given in Fig 22. Equations for determining the characteristic impedances of common lines are given in Fig 23.

Four-Wire Lines

Another parallel conductor line that is useful in some applications is the four-wire line (Fig 23B). In cross section, the conductors of the four-wire line are at the corners of a square. Spacings are on the same order as those used in two-wire lines. The conductors at opposite corners of the square are connected to operate in parallel. This type of line has a lower characteristic impedance than the simple two-wire type. Also, because of the more symmetrical construction, it has better electrical balance to ground and other objects that are close to the line. The spacers for a four-wire line may be discs of insulating material, X-shaped members, etc.

Coaxial Lines

In air insulated coaxial lines (Fig 23C), a considerable proportion of the insulation between conductors may actually be a solid dielectric, because the separation between the inner and outer conductors must be constant. This is particularly likely to be true in small diameter lines. The inner conductor, usually a solid copper wire, is supported at the center of the copper tubing outer conductor by insulating beads or a helically wound strip of insulating material. The beads usually are isolantite, and the wire is generally crimped on each side of each bead to prevent the beads from sliding. The material of which the beads are made, and the number of beads per unit length of line, will affect the characteristic impedance of the line. The greater the number of beads in a given length, the lower the characteristic impedance compared with the value obtained with air insulation only. Teflon is ordinarily

used as a helically wound support for the center conductor. A tighter helical winding lowers the characteristic impedance.

The presence of the solid dielectric also increases the losses in the line. On the whole, however, a coaxial line of this type tends to have lower actual loss, at frequencies up to about 100 MHz, than any other line construction, provided the air inside the line can be kept dry. This usually means that air-tight seals must be used at the ends of the line and at every joint.

The characteristic impedance of an air insulated coaxial line is given by

$$Z_0 = 138 \log \frac{D}{d} \qquad \text{(Eq 19)}$$

where

Z_0 = characteristic impedance in ohms
D = inside diameter of outer conductor
d = Outside diameter of inner conductor (in same units as D)

Values for typical conductor sizes are graphed in Fig 24.

The equation and the graph for coaxial lines are approximately correct for lines in which bead spacers are used, provided the beads are not too closely spaced.

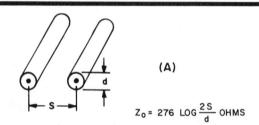

(A)

$$Z_0 = 276 \ \text{LOG} \ \frac{2S}{d} \ \text{OHMS}$$

TWO-WIRE LINE

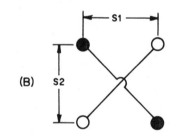

(B)

FOUR-WIRE LINE

$$Z_0 = 138 \ \text{LOG} \ \frac{2 \ S2}{d \sqrt{1+(S2/S1)^2}}$$

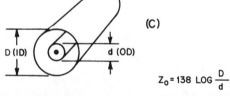

(C)

$$Z_0 = 138 \ \text{LOG} \ \frac{D}{d}$$

COAXIAL LINE

Fig 23—Construction of air insulated transmission lines.

CHARACTERISTIC IMPEDANCE

WIRE SIZE

NO. 22
NO. 20
NO. 18
NO. 16
NO. 14
NO. 12
NO. 10
NO. 8

TUBING DIA

1/4"
3/8"
1/2"

SPACING (S) INCHES, CENTER TO CENTER

Fig 22—Characteristic impedance as a function of conductor spacing and size for parallel conductor lines.

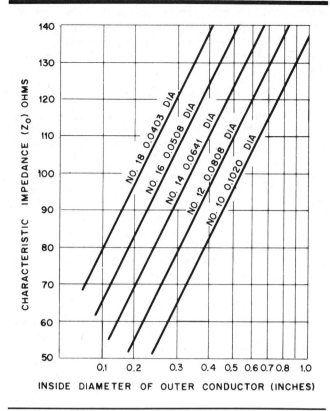

Fig 24—Characteristic impedance of typical air insulated coaxial lines.

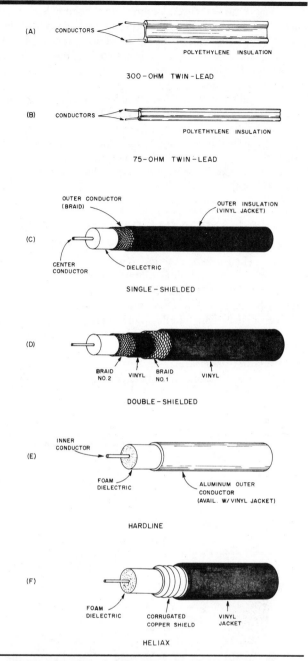

Fig 25—Construction of flexible parallel conductor and coaxial lines with solid dielectric. A common variation of the double shielded design at D has the braids in continuous electrical contact.

FLEXIBLE LINES

Transmission lines in which the conductors are separated by a flexible dielectric have a number of advantages over the air insulated type. They are less bulky, weigh less in comparable types and maintain more uniform spacing between conductors. They are also generally easier to install, and are neater in appearance. Both parallel conductor and coaxial lines are available with flexible insulation.

The chief disadvantage of such lines is that the power loss per unit length is greater than in air insulated lines. Power is lost in heating of the dielectric, and if the heating is great enough (as it may be with high power and a high SWR), the line may break down mechanically and electrically.

Parallel Conductor Lines

The construction of a number of types of flexible line is shown in Fig 25. In the most common 300-Ω type (twin-lead), the conductors are stranded wire equivalent to no. 20 in cross sectional area, and are molded in the edges of a polyethylene ribbon about ½ inch wide. The effective dielectric is partly solid and partly air. The presence of the solid dielectric lowers the characteristic impedance of the line as compared with the same conductors in air. The resulting impedance is approximately 300 Ω.

Because part of the field between the conductors exists outside the solid dielectric, dirt and moisture on the surface of the ribbon tend to change the characteristic impedance of the line. The operation of the line is therefore affected by weather conditions. The effect will not be very serious in a line terminated in its characteristic impedance, but if there is a considerable mismatch, a small change in Z_0 may cause wide fluctuations of the input impedance. Weather effects can be minimized by cleaning the line occasionally and giving it a thin coating of a water repellent material such as silicone grease or car wax.

To overcome the effects of weather on the characteristic impedance and attenuation of ribbon type line, another type of twin-lead is made using a polyethylene tube with an air core or a foamed dielectric core. The conductors are molded diametrically opposite each other in the walls. This increases the leakage path across the dielectric surface. Also, much of the electric field between the conductors is in the hollow (or foam-filled) center of the tube. This type of line is fairly impervious to weather effects. Care should be used when installing it, however, so any moisture that condenses on the inside with changes in temperature and humidity can drain out at the bottom end of the tube and not be trapped in one section. This type of line is made in two conductor sizes (with different tube diameters), one for receiving applications and the other for transmitting.

Transmitting type 75-Ω twin lead uses stranded conductors nearly equivalent to solid no. 12 wire, with quite close spacing between conductors. Because of the close spacing, most of the field is confined to the solid dielectric, with very little existing in the surrounding air. This makes the 75-Ω line much less susceptible to weather effects than the 300-Ω ribbon type.

COAXIAL CABLES

Coaxial cable is available in flexible and semiflexible varieties. The fundamental design is the same in all types, as shown in Fig 25. The outer diameter varies from 0.06 inch to over 5 inches. Power handling capability and cable size are directly proportional, as larger dielectric thickness and larger conductor sizes can handle higher voltages and currents. Generally, losses decrease as cable diameter increases. The extent to which this is true is dependent on the properties of the insulating material.

Some coaxial cables have stranded wire center conductors

while others use a solid copper conductor. Similarly, the outer conductor (shield) may be a single layer of copper braid, a double layer of braid (more effective shielding), solid aluminum (Hardline), aluminum foil, or a combination of these.

Losses and Deterioration

The power handling capability and loss characteristics of coaxial cable depend largely on the dielectric material between the conductors. The commonly used cables and some of their properties are listed in Table 1. Fig 26 is a graph of the attenuation characteristics versus frequency for the most popular lines. The outer insulating jacket of the cable (usually PVC) is used solely as protection from dirt, moisture and chemicals. It has no electrical function. Exposure of the inner insulating material to moisture and chemicals over time contaminates the dielectric and increases cable losses. Foam dielectric cables are less prone to contamination than are solid-polyethylene insulated cables.

Impregnated cables, such as Decibel Products *VB-8* and Times Wire & Cable Co. *Imperveon*, are immune to water and chemical damage, and may be buried if desired. They also have a self-healing property that is valuable when rodents chew into the line. Cable loss should be checked at least every two years if the cable has been outdoors or buried. See the earlier section on testing transmission lines.

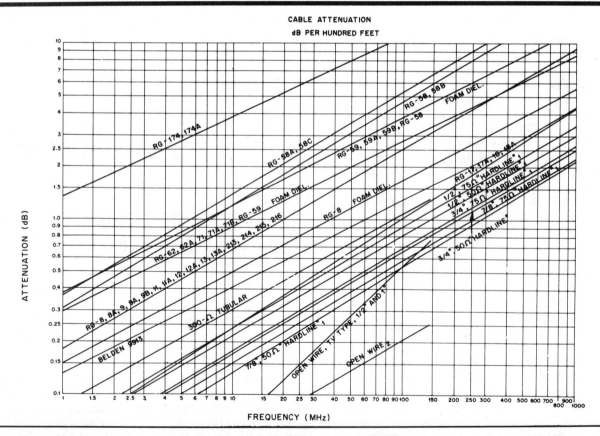

CABLE ATTENUATION
dB PER HUNDRED FEET

Fig 26—Nominal attenuation in decibels per 100 ft of various common transmission lines. Total attenuation is directly proportional to length. Attenuation will vary somewhat in actual cable samples, and generally increases with age in coaxial cables having a type I jacket. Cables grouped together in the above chart have approximately the same attenuation. Types having foam polyethylene dielectric have slightly lower loss than equivalent solid types, when not specifically shown above.

Table 1

Characteristics of Commonly Used Transmission Lines

Type of line	Z_0 Ohms	VF %	pF per foot	OD	Dielectric Material	Max Operating Volts (RMS)
RG-6	75.0	75	18.6	0.266	Foam PE	—
RG-8X	52.0	75	26.0	0.242	Foam PE	—
RG-8	52.0	66	29.5	0.405	PE	4000
RG-8 foam	50.0	80	25.4	0.405	Foam PE	1500
RG-8A	52.0	66	29.5	0.405	PE	5000
RG-9	51.0	66	30.0	0.420	PE	4000
RG-9A	51.0	66	30.0	0.420	PE	4000
RG-9B	50.0	66	30.8	0.420	PE	5000
RG-11	75.0	66	20.6	0.405	PE	4000
RG-11 foam	75.0	80	16.9	0.405	Foam PE	1600
RG-11A	75.0	66	20.6	0.405	PE	5000
RG-12	75.0	66	20.6	0.475	PE	4000
RG-12A	75.0	66	20.6	0.475	PE	5000
RG-17	52.0	66	29.5	0.870	PE	11000
RG-17A	52.0	66	29.5	0.870	PE	11000
RG-55	53.5	66	28.5	0.216	PE	1900
RG-55A	50.0	66	30.8	0.216	PE	1900
RG-55B	53.5	66	28.5	0.216	PE	1900
RG-58	53.5	66	28.5	0.195	PE	1900
RG-58 foam	53.5	79	28.5	0.195	Foam PE	600
RG-58A	53.5	66	28.5	0.195	PE	1900
RG-58B	53.5	66	28.5	0.195	PE	1900
RG-58C	50.0	66	30.8	0.195	PE	1900
RG-59	73.0	66	21.0	0.242	PE	2300
RG-59 foam	75.0	79	16.9	0.242	Foam PE	800
RG-59A	73.0	66	21.0	0.242	PE	2300
RG-62	93.0	86	13.5	0.242	Air space PE	750
RG-62 foam	95.0	79	13.4	0.242	Foam PE	700
RG-62A	93.0	86	13.5	0.242	Air space PE	750
RG-62B	93.0	86	13.5	0.242	Air space PE	750
RG-133A	95.0	66	16.2	0.405	PE	4000
RG-141	50.0	70	29.4	0.190	PTFE	1900
RG-141A	50.0	70	29.4	0.190	PTFE	1900
RG-142	50.0	70	29.4	0.206	PTFE	1900
RG-142A	50.0	70	29.4	0.206	PTFE	1900
RG-142B	50.0	70	29.4	0.195	PTFE	1900
RG-174	50.0	66	30.8	0.100	PE	1500
RG-213	50.0	66	30.8	0.405	PE	5000
RG-214*	50.0	66	30.8	0.425	PE	5000
RG-215	50.0	66	30.8	0.475	PE	5000
RG-216	75.0	66	20.6	0.425	PE	5000
RG-223*	50.0	66	30.8	0.212	PE	1900
9913 (Belden)*	50.0	84	24.0	0.405	Air space PE	—
9914 (Belden)*	50.0	78	26.0	0.405	Foam PE	—

Aluminum Jacket, Foam Dielectric

Type of line	Z_0 Ohms	VF %	pF per foot	OD	Dielectric Material	Max Operating Volts (RMS)
1/2 inch	50.0	81	25.0	0.500		2500
3/4 inch	50.0	81	25.0	0.750		4000
7/8 inch	50.0	81	25.0	0.875		4500
1/2 inch	75.0	81	16.7	0.500		2500
3/4 inch	75.0	81	16.7	0.750		3500
7/8 inch	75.0	81	16.7	0.875		4000
Open wire	—	97	—	—		—
75-ohm transmitting twin lead	75.0	67	19.0	—		—
300-ohm twin lead	300.0	82	5.8	—		—
300-ohm tubular	300.0	80	4.6	—		—

Open Wire, TV Type

Type of line	Z_0 Ohms	VF %	pF per foot	OD	Dielectric Material	Max Operating Volts (RMS)
1/2 inch	300.0	95	—	—		—
1 inch	450.0	95	—	—		—

Dielectric Designation	Name	Temperature Limits
PE	Polyethylene	−65° to +80°C
Foam PE	Foamed polyethylene	−65° to +80°C
PTFE	Polytetrafluoroethylene (Teflon)	−250° to +250°C

*Double shield

The pertinent characteristics of unmarked coaxial cables can be determined from the equations in Table 2. The most common impedance values are 52, 75 and 95 ohms. However, impedances from 25 to 125 ohms are available in special types of manufactured line. The 25-ohm cable (miniature) is used extensively in magnetic-core broadband transformers.

Cable Capacitance

The capacitance between the conductors of coaxial cable varies with the impedance and dielectric constant of the line. Therefore, the lower the impedance, the higher the capacitance per foot, because the conductor spacing is decreased. Capacitance also increases with dielectric constant.

Voltage and Power Ratings

Selection of the correct coaxial cable for a particular application is not a casual matter. Not only is the attenuation loss of significance, but breakdown and heating (voltage and power) also need to be considered. If a cable were lossless, the power handling capability would be limited only by the breakdown voltage. RG-58, for example, can withstand an operating potential of 1900 V RMS. In a 52-Ω system this equates to more than 69 kW, but the current corresponding to this power level is 36.5 amperes, which would obviously melt the conductors in RG-58. In practical coaxial cables, the copper and dielectric losses, rather than breakdown voltage, limit the maximum power that can be accommodated. If 1000 W is applied to a cable having a loss of 3 dB, only 500 W is delivered to the load. The remaining 500 W must be dissipated in the cable. The dielectric and outer jacket are good thermal insulators, which prevent the conductors from efficiently transferring the heat to free air.

As the operating frequency increases, the power handling capability of a cable decreases because of increasing conductor loss (skin effect) and dielectric loss. RG-58 with foam dielectric has a breakdown rating of only 600 V, yet it can handle substantially more power than its ordinary solid dielectric counterpart because of the lower losses. Normally, the loss is inconsequential (except as it affects power handling capability) below 10 MHz in amateur applications. This is true unless extremely long runs of cable are used. In general, full legal amateur power can be safely applied to inexpensive RG-58 coax in the bands below 10 MHz. Cables of the RG-8 family can withstand full amateur power through the VHF spectrum, but connectors must be carefully chosen in these applications. Connector choice is discussed in a later section.

Excessive RF operating voltage in a coaxial cable can cause noise generation, dielectric damage and eventual breakdown between the conductors.

Coaxial Fittings

There is a wide variety of fittings and connectors designed to go with various sizes and types of solid-dielectric coaxial line. The "UHF" series of fittings is by far the most widely used type in the amateur field, largely because they are widely available and are inexpensive. These fittings, typified by the PL-259 plug and SO-239 chassis fitting (military designations) are quite adequate for VHF and lower frequency applications, but are not weatherproof. Neither do they exhibit a 52-Ω impedance.

Type N series fittings are designed to maintain constant impedance at cable joints. They are a bit harder to assemble than the "UHF" type, but are better for frequencies above 300 MHz or so. These fittings are weatherproof.

The BNC fittings are for small cable such as RG-58, RG-59 and RG-62. They feature a bayonet-locking arrangement for quick connect and disconnect, and are weatherproof. They exhibit a constant impedance.

Methods of assembling connectors on the cable are shown in Figs 27 through 31. The most common or longest established connector in each series is illustrated. Several variations of each type exist. Assembly instructions for coaxial fittings not shown here are available from the manufacturers.

PL-259 Assembly

Fig 27 shows how to install the solder type of PL-259

Table 2
Coaxial Cable Equations

$$C \text{ (pf/ft)} = \frac{7.36\epsilon}{\log (D/d)} \qquad \text{(Eq 1)}$$

$$L \text{ } (\mu\text{H/ft}) = 0.14 \log \frac{D}{d} \qquad \text{(Eq 2)}$$

$$Z_0 \text{ (ohms)} = \sqrt{\frac{L}{C}} = \left(\frac{138}{\sqrt{\epsilon}}\right)\left(\log \frac{D}{d}\right) \qquad \text{(Eq 3)}$$

$$VF \text{ \% (velocity factor, ref. speed of light)} = \frac{100}{\sqrt{\epsilon}} \qquad \text{(Eq 4)}$$

$$\text{Time delay (ns/ft)} = 1.016 \sqrt{\epsilon} \qquad \text{(Eq 5)}$$

$$f \text{ (cutoff/GHz)} = \frac{7.50}{\sqrt{\epsilon} (D + d)} \qquad \text{(Eq 6)}$$

$$\text{Refl coef} = \rho = \frac{Z_L - Z_0}{Z_L + Z_0} = \frac{SWR - 1}{SWR + 1} \qquad \text{(Eq 7)}$$

$$SWR = \frac{1 + \rho}{1 - \rho} \qquad \text{(Eq 8)}$$

$$V \text{ peak} = \frac{(1.15 \text{ } Sd) (\log D/d)}{K} \qquad \text{(Eq 9)}$$

$$A = \frac{0.435}{Z_0 D}\left[\frac{D}{d} (K1 + K2)\right]\sqrt{f} + 2.78 \sqrt{\epsilon} (PF)(f) \qquad \text{(Eq 10)}$$

where
- A = atten in dB/100 ft
- d = OD of inner conductor
- D = ID of outer conductor
- S = max voltage gradient of insulation in volts/mil
- ϵ = dielectric constant
- K = safety factor
- K1 = strand factor
- K2 = braid factor
- f = freq in MHz
- PF = power factor

Note: Obtain K1 and K2 data from manufacturer.

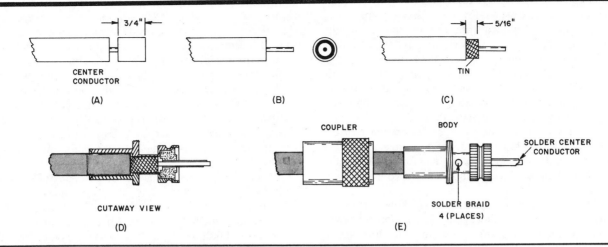

Fig 27—The PL-259 or UHF connector is almost universal for amateur HF work and is popular for equipment operating in the VHF range. Steps for assembly are given in detail in the text.

83-58FCP

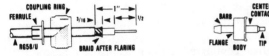

1. Strip cable — *don't nick braid, dielectric or conductor.* Slide ferrule, then coupling ring on cable. Flare braid slightly by rotating conductor and dielectric in circular motion.

2. Slide body on dielectric, barb going under braid until flange is against outer jacket. Braid will fan out against body flange.

3. Slide nut over body. Grasp cable with hand and push ferrule over barb until braid is captured between ferrule and body flange. Squeeze crimp tip only of center contact with pliers; alternate-solder tip.

83-1SP (PL-259) PLUG WITH ADAPTERS (UG-176/U OR UG-175/U)

1) Cut end of cable even. Remove vinyl jacket 3/4″ — don't nick braid. Slide coupling ring and adapter on cable.

2) Fan braid slightly and fold back over cable.

3) Position adapter to dimension shown. Press braid down over body of adapter and trim to 3/8″. Bare 5/8″ of conductor. Tin exposed center conductor.

4) Screw the plug assembly on adapter. Solder braid to shell through solder holes. Solder conductor to contact sleeve.

5) Screw coupling ring on plug assembly.

Fig 28—Crimp-on connectors and adapters for use with standard PL-259 connectors are popular for connecting to RG-58 and RG-59 coax. *(This material courtesy of Amphenol Electronic Components, RF Division, Bunker Ramo Corp)* connector on RG-8 type cable. Proper preparation of the cable end is the key to success. Follow these simple steps.

1) Measure back ¾ inch from the cable end and slightly score the outer jacket around its circumference.

2) With a sharp knife, cut along the score line through the outer jacket, through the braid, and through the dielectric material, right down to the center conductor. Be careful not to score the center conductor. Cutting through all outer layers at once keeps the braid from separating.

3) Pull the severed outer jacket, braid and dielectric off the end of the cable as one piece. Inspect the area around the cut, looking for any strands of braid hanging loose. If there are any, snip them off. There won't be any if your knife was sharp enough.

4) Next, score the outer jacket 5/16 inch back from the first cut. Cut through the jacket lightly; do not score the braid. This step takes practice. If you score the braid, start again.

5) Remove the outer jacket. Tin the exposed braid and center conductor, but apply the solder sparingly. Avoid melting the dielectric.

6) Slide the coupling ring onto the cable. *(Don't forget this important step!)*

7) Screw the connector body onto the cable. If you prepared the cable to the right dimensions, the center conductor will protrude through the center pin, the braid will show through the solder holes, and the body will actually thread itself onto the outer cable jacket.

8) With a large soldering iron, solder the braid through each of the four solder holes. Use enough heat to flow the solder onto the connector body, but not so much as to melt the dielectric. Poor connection to the braid is the most common form of PL-259 failure. This connection is just as important as that between the center conductor and the

connector. With some practice you'll learn how much heat to use.

9) Allow the connector body to cool somewhat, and then solder the center connector to the center pin. The solder should flow on the inside, not the outside of the pin. Trim the center conductor to be even with the end of the center pin. Use a small file to round the end, removing any solder that may have built up on the outer surface of the center pin. Use a sharp knife, very fine sandpaper, or steel wool to remove any solder flux from the outer surface of the center pin.

10) Screw the coupling onto the body, and the job is finished.

Fig 28 shows two options for using RG-58 or RG-59 cable with PL-259 connectors. The crimp-on connectors manufactured for the smaller cable work well if installed correctly. The alternative method involves using adapters for the smaller cable with standard PL-259 connectors made for RG-8. Prepare the cable as shown in Fig 28. Once the braid is prepared, screw the adapter into the PL-259 shell and finish the job as you would with RG-8 cable.

Fig 29 shows how to assemble female SO-239 connectors onto coaxial cable. Figs 30 and 31 respectively show the assembly of BNC and type N connectors.

SINGLE WIRE LINE

There is one type of line, in addition to those already described, that deserves mention because it is still used to a limited extent. This is the *single wire line*, consisting simply of a single conductor running from the transmitter to the antenna. The "return" circuit for such a line is the earth; in fact, the second conductor of the line can be considered to be the image of the actual conductor in the same way that

1) Cut end of cable even: Remove vinyl jacket to dimension appropriate for type of hood. Tin exposed braid.

2) Remove braid and dielectric to expose center conductor. Do not nick conductor.

3) Remove braid to expose dielectric to appropriate dimension. Tin center conductor. Soldering assembly depends on the hood used, as illustrated.

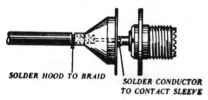

4) Slide hood over braid. Solder conductor to contact. Slide hood flush against receptacle and bolt both to chassis. Solder hood to braid as illustrated. Tape this junction if necessary. (For UG-177/U.)

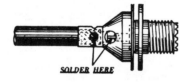

5) Slide hood over braid. Bring receptacle flush against hood. Solder hood to braid and conductor to contact sleeve through solder holes as illustrated. Tape junction if necessary. (For UG-372/U.)

6) Slide hood over braid and force under vinyl. Place inner conductor in contact sleeve and solder. Push hood flush against receptacle. Solder hood to braid through solder holes. Tape junction if necessary. (For UG-106/U.)

Fig 29—Assembly of the 83 series (SO-239) with hoods. Complete electrical shield integrity in the UHF female connector requires that the shield be attached to the connector flange by means of a hood.

an antenna strung above the earth has an image (see Chapter 3). The characteristic impedance of the single wire line depends on the conductor size and the height of the wire above ground, ranging from 500 to 600 ohms for no. 12 or no. 14 conductors at heights of 10 to 30 feet. The characteristic impedance may be calculated from

$$Z_0 = 138 \log \frac{4h}{d} \qquad \text{(Eq 20)}$$

where

  Z_0 = characteristic impedance of the single wire line
  h = antenna height
  d = wire diameter, in same units as h

By connecting the line to the antenna at a point that represents a resistive impedance of 500 to 600 ohms, the line can be matched and operated without standing waves.

Although the single wire line is very simple to install, it has at least two outstanding disadvantages. First, because the return circuit is through the earth, the behavior of the system depends on the kind of ground over which the antenna and transmission lines are erected. In practice, it may not be possible to get the necessary good connection to actual ground that is required at the transmitter. Second, the line always radiates, because there is no nearby second conductor to cancel the fields. Radiation is minimum when the line is properly terminated, because the line current is lowest under

BNC CONNECTORS

Standard Clamp

1. Cut cable and even. Strip jacket. Fray braid and strip dielectric. *Don't nick braid or center conductor.* Tin center conductor.

2. Taper braid. Slide nut, washer, gasket and clamp over braid. Clamp inner shoulder should fit squarely against end of jacket.

3. With clamp in place, comb out braid, fold back smooth as shown. Trim center conductor.

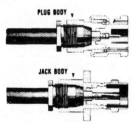

4. Solder contact on conductor through solder hole. Contact should butt against dielectric. Remover excess solder from outside of contact. Avoid excess heat to prevent swollen dielectric which would interfere with connector body.

5. Push assembly into body. Screw nut into body with wrench until tight. *Don't rotate body on cable to tighten.*

Improved Clamp

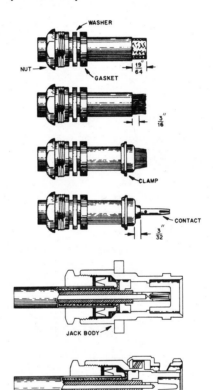

Follow 1, 2, 3 and 4 in BNC connectors (standard clamp) except as noted. Strip cable as shown. Slide gasket on cable *with groove facing clamp.* Slide clamp on cable *with sharp edge facing gasket.* Clamp *should* cut gasket to seal properly.

C. C. Clamp

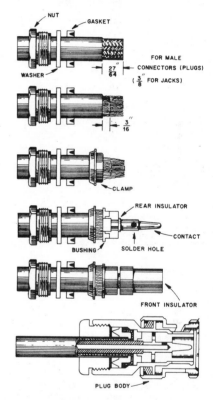

1) Follow steps 1, 2 and 3 as outlined for the standard-clamp BNC connector.

2) Slide on the bushing, rear insulator and contact. The parts must butt securely against each other, as shown.

3) Solder the center conductor to the contact. Remove flux and excess solder.

4) Slide the front insulator over the contact, making sure it butts against the contact shoulder.

5) Insert the prepared cable end into the connector body and tighten the nut. Make sure that the sharp edge of the clamp seats properly in the gasket.

Fig 30—BNC connectors are common on VHF and UHF equipment at low power levels. (*Courtesy of Amphenol Electronic Components, RF Division, Bunker Ramo Corp*)

TYPE N CONNECTORS

Standard Clamp

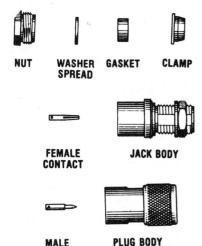

NUT **WASHER SPREAD** **GASKET** **CLAMP**

FEMALE CONTACT **JACK BODY**

MALE CONTACT **PLUG BODY**

1) Cut cable and even. Remove 9/16″ of vinyl jacket. When using double-shielded cable remove 5/8″.

2) Comb out copper braid as shown. Cut off dielectric 7/32″ from end. Tin center conductor.

3) Taper braid as shown. Slide nut, washer and gasket over vinyl jacket. Slide clamp over braid with internal shoulder of clamp flush against end of vinyl jacket. When assembling connectors with gland, be sure knife-edge is toward end of cable and groove in gasket is toward the gland.

4) Smooth braid back over clamp and trim. Soft-solder contact to center conductor. Avoid use of excessive heat and solder. See that end of dielectric is clean. Contact must be flush against dielectric. Outside of contact must be free of solder. Female contact is shown; procedure is similar for male contact.

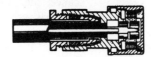

5) Slide body into place carefully so that contact enters hole in insulator (male contact shown). Face of dielectric must be flush against insulator. Slide completed assembly into body by pushing nut. When nut is in place, tighten with wrenches. In connectors with gland, knife edge should cut gasket in half by tightening sufficiently.

Improved Clamp

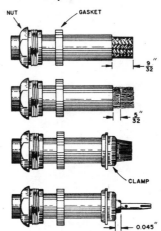

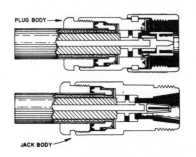

1) Follow instructions 1 through 4 as detailed in the standard clamp (be sure to use the correct dimensions).

2) Slide the body over the prepared cable end. Make sure the sharp edges of the clamp seat properly in the gasket. Tighten the nut.

C. C. Clamp

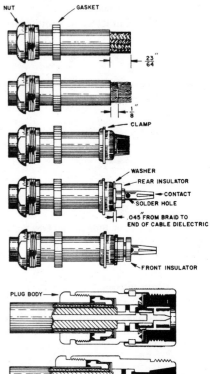

1) Follow instructions 1 through 3 as outlined for the standard-clamp Type N connector.

2) Slide on the washer, rear insulator and contact. The parts must butt securely against each other.

3) Solder the center conductor to the contact. Remove flux and excess solder.

4) Slide the front insulator over the contact, making sure it butts against the contact shoulder.

5) Insert the prepared cable end into the connector body and tighten the nut. Make sure the sharp edge of the clamp seats properly in the gasket.

Fig 31—Type N connectors are required for high-power operation at VHF and UHF. *(Courtesy of Amphenol Electronic Components, RF Division, Bunker Ramo Corp)* these conditions. The line is, however, always a part of the radiating antenna system, to some extent.

LINE INSTALLATION

One great advantage of coaxial line, particularly the flexible dielectric type, is that it can be installed with almost no regard for its surroundings. It requires no insulation, can be run on or in the ground or in piping, can be bent around corners with a reasonable radius, and can be "snaked" through places such as the space between walls where it would be impractical to use other types of lines. However, coaxial lines should always be operated in systems that permit a low SWR, and precautions must be taken to prevent RF currents from flowing on the *outside* of the line. This is discussed in Chapter 26. Additional information on line installation is given in Chapter 4.

Parallel Wire Lines

In installing a parallel wire line, care must be used to prevent it from being affected by moisture, snow and ice. In home construction, only spacers that are impervious to moisture and are unaffected by sunlight and weather should be used on air insulated lines. Steatite spacers meet this requirement adequately, although they are somewhat heavy. The wider the line spacing, the longer the leakage path across the spacers, but this cannot be carried too far without running into line radiation, particularly at the higher frequencies. Where an open wire line must be anchored to a building or other structure, standoff insulators of a height comparable with the line spacing should be used if mounted in a spot that is open to the weather. Lead-in bushings for bringing the line into a building also should have a long leakage path.

The line should be kept away from other conductors, including downspouts, metal window frames, flashing, etc, by a distance of two or three times the line spacing. Conductors that are very close to the line will be coupled to it to some degree, and the effect is that of placing an additional load across the line at the point where the coupling occurs. Reflections take place from this coupled "load," raising the SWR. The effect is at its worst when one wire is closer than the other to the external conductor. In such a case one wire carries a heavier load than the other, with the result that the line currents are no longer equal. The line then becomes unbalanced.

Solid dielectric, two-wire lines have a relatively small external field because of the small spacing, and can be mounted within a few inches of other conductors without much danger of coupling between the line and such conductors. Standoff insulators are available for supporting lines of this type when run along walls or similar structures.

Sharp bends should be avoided in any type of transmission line, because such bends cause a change in the characteristic impedance. The result is that reflections take place from each bend. This is of less importance when the SWR is high than when an attempt is being made to match the load to the line Z_0. It may be impossible to get the SWR to the desired figure until bends in the line are made very gradual.

BIBLIOGRAPHY

Source material and more extended discussion of topics covered in this chapter can be found in the references given below and in the textbooks listed at the end of Chapter 2.

C. Brainard and K. Smith, "Coaxial Cable—The Neglected Link," *QST*, Apr 1981, pp 28-31.

D. DeMaw, "In-Line RF Power Metering," *QST*, Dec 1969.

D. Geiser, "Resistive Impedance Matching with Quarter-Wave Lines," *QST*, Feb 1963, pp 56-57.

H. Jasik, *Antenna Engineering Handbook*, 1st ed (New York: McGraw-Hill, 1961).

R. C. Johnson and H. Jasik, *Antenna Engineering Handbook*, 2nd ed (New York: McGraw-Hill, 1984), pp 43-27 to 43-31.

R. W. P. King, H. R. Mimno and A. H. Wing, *Transmission Lines, Antennas and Waveguides* (New York: Dover Publications, Inc, 1965).

J. D. Kraus, *Antennas* (New York: McGraw-Hill Book Co, 1950).

M. W. Maxwell, "Another Look at Reflections," *QST*, Apr, Jun, Aug and Oct 1973, Apr and Dec 1974, and Aug 1976.

T. McMullen, "The Line Sampler, an RF Power Monitor for VHF and UHF," *QST*, April 1972.

H. Weinstein, "RF Transmission Cable for Microwave Applications," *Ham Radio*, May 1985, p 106.

Chapter 25

Coupling the Transmitter to the Line

In any system using a transmission line to feed the antenna, the load that the transmitter "sees" is the input impedance of the line. As shown in the previous chapter, this impedance is completely determined by the line length, the Z_0 of the line, and the impedance of the load (the antenna) at the output end of the line. The line length and Z_0 are generally matters of choice regardless of the type of antenna used. The antenna impedance, which may or may not be known accurately, is (with Z_0) the factor that determines the standing-wave ratio.

The SWR can be measured with relative ease, and from it the limits of variation in the line input impedance can be determined with little difficulty. It may be said, therefore, that the problem of transferring power from the transmitter to the line can be approached purely on the basis of the known Z_0 of the line and the maximum SWR that may be encountered.

Coupling systems that will deliver power into a flat line are readily designed. For all practical purposes the line can be considered to be flat if the SWR is no greater than about 1.5:1. That is, a coupling system designed to work into a pure resistance equal to the line Z_0 should have enough leeway to take care of the small variations in input impedance that will occur when the line length is changed, if the SWR is higher than 1:1 but no greater than 1.5:1.

So far as the transmitter itself is concerned, the requirements of present-day amateur operation almost invariably include complete shielding and provision for the use of low-pass filters to prevent harmonic interference with television reception. In almost all cases this means that the use of coaxial cable at the output of the transmitter is mandatory because it is inherently shielded. Current practice in transmitter design is to provide an output circuit that will work into a coaxial line of 52 to 75 ohms characteristic impedance. This does not mean that a coaxial line must be used to feed the *antenna*; any type of line can be used.

If the input impedance of the transmission line that is to be connected to the transmitter differs appreciably from the impedance value that the transmitter output circuit is designed to operate, an impedance-matching network must be inserted between the transmitter and the line input terminals.

THE CONJUGATE MATCH AND THE Z_0 MATCH

One purpose of matching impedances is to enable a power source or generator, having an optimum load impedance, to deliver its maximum available power to a load having a different impedance. In antenna systems, that power is usually delivered through a transmission line having still a third impedance. Chapter 24 discusses matched and mismatched lines, standing waves, and standing-wave ratios (SWR). Various schemes may be used for matching a transmitter to the input of an antenna system. The same principles apply, and the same wave actions and reflections occur, whether the matching device is a stub on a line, a ¼-λ line transformer, or a network of lumped reactances such as a pi-network tank circuit or a Transmatch. (The various types of matching just mentioned are discussed later in this chapter or in other chapters.) To aid in understanding the principles of matching, we shall consider transmission lines and reactance components of the Transmatch as lossless elements, and then treat the effects of loss later.

The conjugate match is defined by the National Bureau of Standards as the condition for maximum power absorption by a load, in which the impedance seen looking toward the load at a point in the circuit is the complex conjugate of that seen looking toward the source. The Z_0 match is the condition in which the impedance seen looking into a transmission line is equal to the characteristic impedance of the line.

If a generator, a load, and an ideal lossless line connecting them all have the same Z_0 impedance, both a Z_0 match and a conjugate match exist. (A Z_0 match is also a special condition of the conjugate match.) Under these conditions the generator delivers its maximum available power into the line, which transfers it to the load where it is completely absorbed.

If the Z_0 load is now replaced by another, such as an antenna having an impedance $Z_L = R + jX \neq Z_0$, we have two mismatches at the load, a Z_0 mismatch and a conjugate mismatch. The Z_0 mismatch creates a reflection having a magnitude

$$\rho = \frac{Z_L - Z_0}{Z_L + Z_0}$$

causing a reflection loss ρ^2 that is referred back along the line to the generator. This in turn causes the generator to see the same magnitude of Z_0 mismatch at the line input. This referred Z_0 mismatch causes the generator to deliver less than its maximum available power by the amount equal to the reflection loss. There is no power lost in the line—only a reduction of power delivered. And all of the reduced power delivered is absorbed by the load.

If a matching device is now inserted between the line and the load, the device provides a conjugate match at the load by supplying a reflection gain which cancels the reflection loss, so the line sees a Z_0 match at the input of the matching device.

Since the line is terminated in a Z_0 match, the generator now sees a Z_0 match at the line input, and will again deliver its maximum available power into the line, to be completely absorbed by the load.

However, if a Transmatch is inserted between the generator and the line instead of a matching device at the load, the Transmatch provides the conjugate match at the line input, and the generator sees a Z_0 match at the Transmatch input. At the load, the line now sees a conjugate match, but also sees a Z_0 mismatch. The Z_0 mismatch again causes a reflection loss that is referred back to the line input as before. With the Transmatch at the line input, the reflection gain, which again cancels the reflection loss, shields the generator from seeing the referred Z_0 mismatch again appearing at the line input.

The reflection gain of the Transmatch also creates the conjugate match at the load, and thus enables the load to absorb the maximum available power delivered by the generator. Since the generator sees a Z_0 match at the Transmatch input, it continues delivering its maximum available power into the line through the Transmatch, and the power is still completely absorbed by the load because of the conjugate match.

The amount of forward power flowing on a lossless line is determined by the expression

$$\frac{1}{1 - \rho^2}$$

where ρ = voltage coefficient of reflection

On a real line with attenuation, forward power at the conjugate matching point is

$$\frac{1}{1 - \rho^2 \epsilon^{-4\alpha}}$$

where

α = line attenuation in nepers = dB/8.686
ϵ = 2.71828..., the base of natural logarithms

The power arriving at the mismatched load with attenuation in the line is

$$\frac{\epsilon^{-2\alpha}}{1 - \rho^2 \epsilon^{-4\alpha}}$$

and the power absorbed in the load is

$$\frac{(1 - \rho^2)\ \epsilon^{-2\alpha}}{1 - \rho^2\ \epsilon^{-4\alpha}}$$

The power reflected is

$$\frac{\rho^2 \epsilon^{-2\alpha}}{1 - \rho^2 \epsilon^{-4\alpha}}$$

The Conjugate Matching Theorem

To understand how a conjugate match is created at the load with the Transmatch at the line input, examine the Conjugate Matching Theorem and the wave actions which produce the conjugate match. It is the conjugate match that enables the load to absorb the maximum available power from the generator despite the Z_0 load mismatch.

"If a group of four-terminal networks containing only reactances (or lossless lines) are arranged in tandem to connect a generator to its load, then if at any junction there is a conjugate match of impedances, there will be a conjugate match of impedances at every other junction in the system."

To paraphrase from the NBS definition, "conjugate match" means that if in one direction from a junction the impedance has the dimensions $R + jX$, then in the opposite direction the impedance will have the dimensions $R - jX$. And according to the theorem, when a conjugate match is accomplished at any of the junctions in the system, any reactance appearing at any junction is canceled by an equal and opposite reactance, which also includes any reactance appearing in the load. This reactance cancellation results in a net system reactance of zero, establishing resonance in the entire system. In this resonant condition the generator delivers its maximum available power to the load. This is why an antenna operated away from its natural resonant frequency is tuned to resonance by a Transmatch connected at the input to the transmission line.

Conjugate matching is obtained through a controlled wave interference between two sets of reflected waves. One set of waves is reflected by the load mismatch, and the other by a complementary mismatch introduced by the reactances in the Transmatch. The mismatch introduced by the reactances of the Transmatch creates reflected waves of voltage and current at the matching point (the Transmatch input). These waves are equal in magnitude but opposite in phase relative to those arriving from the load mismatch. The two sets of reflected voltage and current waves created by the two mismatches combine at the matching point to produce resultant waves of voltage and current that are respectively at 0° and 180° in phase relative to the source wave.

The phase relationship of 180° between the resultant voltage and current waves creates a virtual open circuit to waves traveling toward the source, which totally re-reflects both sets of reflected waves back into the line. This re-reflection prevents the reflected waves from traveling beyond the matching point, which is the reason an SWR indicator connected at the input of the Transmatch shows zero reflected power at the conclusion of the tune-up procedure. On re-reflection the voltage and current components of the reflected waves emerge in phase with the corresponding components of the source wave, so by superposition, all of the power contained in the reflected waves is added to the power in the source wave. Consequently, the forward power in the line is greater than the source power whenever the line is terminated in a Z_0 mismatch. This is why the forward-power wattmeter connected at the output of the Transmatch indicates a value higher than the source power, by the amount equal to the reflected power.

This enlargement of the forward power is called "reflection gain." Reflection gain cancels the reflection loss, creating the conjugate match that enables the Z_0-mismatched load to absorb the maximum available power delivered by the generator. When the source power enlarged by the reflected power reaches the Z_0-mismatched load, Z_L, the power previously reflected and added to the source power is again subtracted by the Z_0-mismatch reflection, leaving the source power to be completely absorbed by the load. This is why all power entering the line is absorbed in the load, regardless of the mismatch or SWR.

In lines having attenuation, all power entering the line is absorbed in the load, except for that lost because of attenuation. When the matching is performed at the line input, the attenuation increases as the Z_0 load mismatch increases. This happens because, in addition to the attenuation of the forward power, the reflected power is also attenuated in the same proportion. However, when the matched-line attenuation is low, as it is in typical amateur installations, the additional loss because of Z_0 mismatch is small. The additional losses are so low that even with moderate to high SWR values, the difference in power radiated compared to that with a 1:1 SWR is too small to be discerned by the receiving station. Information for calculating the additional power lost versus SWR is given in Chapter 24.

THE MATCHING SYSTEM

The basic system is as shown in Fig 1. Assuming that the transmitter is capable of delivering its rated power into a load on the order of 52 to 75 ohms, the coupling problem is one of designing a matching circuit that will transform the actual line input impedance into a resistance of 52 or 75 ohms. This resistance will be unbalanced; that is, one side will be grounded. The line to the antenna may be unbalanced (coaxial cable) or balanced (parallel-conductor line).

Many factors will influence the choice of line to be used with a given antenna. The shielding of coaxial cable offers advantages in incidental radiation and routing flexibility. Low loss and low cost should ensure that open wire line is around for a long time. Coaxial cable can perform acceptably even with significant SWR. (Refer to information in Chapter 24.) A length of 100 feet of RG-8 line has 1 dB loss at 30 MHz. If this line were used with an SWR at the load of 5:1, the total line loss would be 2.1 dB. Open-wire line, in the same circumstances, has a matched loss of less than 0.1 dB. If this line was used with an SWR of 20:1 the total loss would be less than 1 dB.

Several types of matching circuits are available. With some, such as an L network using fixed-value components, it is necessary to know the actual line-input impedance with fair accuracy in order to arrive at a proper design. This information is not essential with the circuits to be described,

since they are capable of adjustment over a wide range.

MATCHING WITH INDUCTIVE COUPLING

Inductively coupled matching circuits are shown in basic form in Fig 2. R1 is the actual load resistance to which the power is to be delivered, and R2 is the resistance seen by the power source. R2 depends on the circuit design and adjustment; in general, the objective is to make it equal to 52 or 75 ohms. L1 and C1 form a resonant circuit capable of being tuned to the operating frequency. The coupling between L1 and L2 is adjustable.

The circuit formed by C1, L1 and L2 is equivalent to a transformer having a primary-to-secondary impedance ratio adjustable over wide limits. The resistance "coupled into" L2 from L1 depends on the effective Q of the circuit L1-C1-R1, the reactance of L2 at the operating frequency, and the coefficient of coupling, k, between the two coils. The approximate relationship is (assuming C1 is properly tuned)

$$R2 = k^2 X_{L2} Q$$

where X_{L2} is the reactance of L2 at the operating frequency. The value of L2 is optimum when $X_{L2} = R2$, in which case the desired value of R2 is obtained when

$$k = \frac{1}{\sqrt{Q}}$$

This means that the desired value of R2 may be obtained by adjusting either the coupling, k, between the two coils, or by changing the Q of the circuit L1-C1-R1—or, if necessary, by doing both. If the coupling is fixed, as is often the case, Q must be adjusted to attain a match. Note that increasing the value of Q is equivalent to tightening the coupling, and vice versa.

If L2 does not have the optimum value, the match may still be obtained by adjusting k and Q, but one or the other—

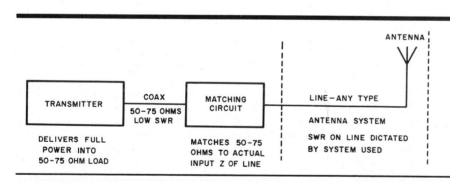

Fig 1—Essentials of a coupling system between transmitter and transmission line.

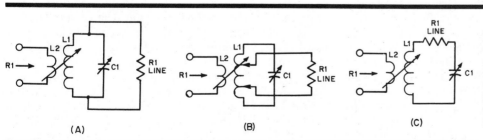

Fig 2—Circuit arrangements for inductively coupled impedance-matching circuit. A and B use a parallel-tuned coupling tank; B is equivalent to A when the taps are at the ends of L1. The series-tuned circuit at C is useful for very low values of load resistance, R1.

or both—must have a larger value than is needed when X_{L2} is equal to R2. In general, it is desirable to use as low a value of Q as is practicable, since low Q values mean that the circuit requires little or no readjustment when shifting frequency within a band (provided R1 does not vary appreciably with frequency).

Circuit Q

In Fig 2A, Q is equal to R1 in ohms divided by the reactance of C1 in ohms, assuming L1-C1 is tuned to the operating frequency. This circuit is suitable for comparatively high values of R1—from several hundred to several thousand ohms. In Fig 2C, Q is equal to the reactance of C1 divided by the resistance of R1, L1-C1 again being tuned to the operating frequency. This circuit is suitable for low values of R1—from a few ohms up to a hundred or so ohms. In Fig 2B the Q depends on the placement of the taps on L1 as well as on the reactance of C1. This circuit is suitable for matching all values of R1 likely to be encountered in practice.

Note that to change Q in either A or C, Fig 2, it is necessary to change the reactance of C1. Since the circuit is tuned essentially to resonance at the operating frequency, this means that the L/C ratio must be varied in order to change Q. In Fig 2B a fixed L/C ratio may be used, since Q can be varied by changing the tap positions. The Q will increase as the taps are moved closer together, and will decrease as they are moved farther apart on L1.

Reactive Loads—Series and Parallel Coupling

More often than not, the load represented by the input impedance of the transmission line is reactive as well as resistive. In such a case the load cannot be represented by a simple resistance, such as R1 in Fig 2. As stated in an earlier chapter, we have the option of considering the load to be a resistance in parallel with a reactance, or as a resistance in series with a reactance. In Fig 2, at A and B, it is convenient to use the parallel equivalent of the line input impedance. The series equivalent is more suitable for Fig 2C.

Thus, in A and B of Fig 3 the load might be represented by R1 in parallel with the capacitive reactance C, and in Fig 3C by R1 in series with a capacitive reactance C. In A, the capacitance C is in parallel with C1 and so the total capacitance is the sum of the two. This is the effective capacitance that, with L1, tunes to the operating frequency. Obviously the setting of C1 will be at a lower value of capacitance with such a load than it would with a purely resistive load such as in Fig 2A.

In Fig 3B the capacitance of C also increases the total capacitance effective in tuning the circuit. However, in this case the increase in effective tuning capacitance depends on the positions of the taps; if the taps are close together the effect of C on the tuning is relatively small, but it increases as the taps are moved farther apart.

In Fig 3C, the capacitance C is in series with C1 and so the total capacitance is less than either. Hence the capacitance of C1 must be increased in order to resonate the circuit, as compared with the purely resistive load shown in Fig 2C.

If the reactive component of the load impedance is inductive, similar considerations apply. In such case an inductance would be substituted for the capacitance C shown in Fig 3. The effect in Fig 3A and 3B would be to decrease the effective inductance in the circuit, so C1 would require a larger value of capacitance in order to resonate the circuit at the operating frequency. In Fig 3C the effective inductance

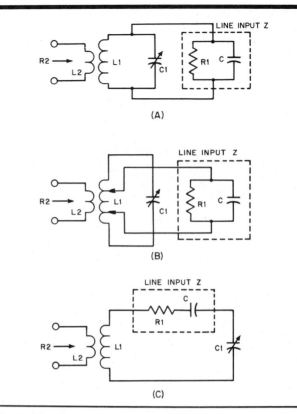

Fig 3—Line input impedances containing both resistance and reactance can be represented as shown enclosed in dotted lines, for capacitive reactance. If the reactance is inductive, a coil is substituted for the capacitance C.

would be increased, thus making it necessary to set C1 at a lower value of capacitance for resonating the circuit.

Effect of Line Reactance on Circuit Q

The presence of reactance in the line input impedance can affect the Q of the matching circuit. If the reactance is capacitive, the Q will not change if resonance can be maintained by adjustment of C1 without changing either the value of L1 or the position of the taps in Fig 3B (as compared with the Q when the load is purely resistive and has the same value of resistance, R1). If the load reactance is inductive the L/C ratio changes because the effective inductance in the circuit is changed and, in the ordinary case, L1 is not adjustable. This increases the Q in all three circuits of Fig 3.

When the load has appreciable reactance it is not always possible to adjust the circuit to resonance by readjusting C1, as compared with the setting it would have with a purely resistive load. Such a situation may occur when the load reactance is low compared with the resistance in the parallel-equivalent circuit, or when the reactance is high compared with the resistance in the series-equivalent circuit. The very considerable detuning of the circuit that results is often accompanied by an increase in Q, sometimes to values that lead to excessively high circulating currents in the circuit. This causes the efficiency to suffer. (Ordinarily the power loss in matching circuits of this type is inconsequential, if the Q is below 10 and a good coil is used.) An unfavorable ratio of reactance to resistance in the input impedance of the line can exist if the SWR is high and the line length is near an odd multiple of one-eighth wavelength (45°).

Q of Line Input Impedance

The ratio between reactance and resistance in the equivalent input circuit—that is, the Q of the line input impedance—is a function of line length and SWR. There is no specific value of this Q of which it can be said that lower values are satisfactory while higher values are not. In part, the maximum tolerable value depends on the tuning range available in the matching circuit. If the tuning range is restricted (as it will be if the variable capacitor has relatively low maximum capacitance), compensating for the line input reactance by absorbing it in the matching circuit—that is, by retuning C1 in Fig 3—may not be possible. Also, if the Q of the matching circuit is low, the effect of the line input reactance will be greater than it will when the matching-circuit Q is high.

As stated earlier, the optimum matching-circuit design is one in which the Q is low, that is, a low reactance-to-resistance ratio.

Compensating for Input Reactance

When the reactance/resistance ratio in the line input impedance is unfavorable it is advisable to take special steps to compensate for it. This can be done as shown in Fig 4. Compensation consists of supplying external reactance of the same numerical value as the line reactance, but of the opposite kind. Thus in Fig 4A, where the line input impedance is represented by resistance and capacitance in parallel, an inductance L having the same numerical value of reactance as C can be connected across the line terminals to "cancel out" the line reactance. (This is actually the same thing as tuning the line to resonance at the operating frequency.) Since the parallel combination of L and C is equivalent to an extremely high resistance at resonance, the input impedance of the line becomes a pure resistance having essentially the same resistance as R1 alone.

The case of an inductive line is shown in Fig 4B. In this case the external reactance required is capacitive, of the same numerical value as the reactance of L. Where the series equivalent of the line input impedance is used, the external reactance is connected in series, as shown at C and D in Fig 4.

In general, these methods need not be used unless the matching circuit does not have sufficient range of adjustment to provide compensation for the line reactance as described earlier, or when such a large readjustment is required that the matching-circuit Q becomes undesirably high. The latter condition usually is accompanied by heating of the coil used in the matching network.

Methods for Variable Coupling

The coupling between L1 and L2, Figs 2 and 3, preferably should be adjustable. If the coupling is fixed, such as with a fixed-position link, the placement of the taps on L1 for proper matching becomes rather critical. The additional matching adjustment afforded by adjustable coupling between the coils facilitates the matching procedure considerably. L2 should be coupled to the center of L1 for the sake of maintaining balance, since the circuit is used with balanced lines.

If adjustable inductive coupling such as a swinging link is not feasible for mechanical reasons, an alternative is to use a variable capacitor in series with L2. This is shown in Fig 5. Varying C2 changes the total reactance of the circuit formed by L2-C2, with much the same effect as varying the actual

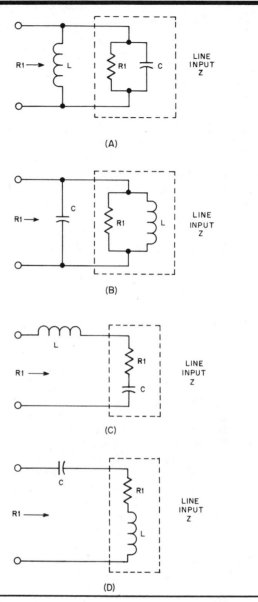

Fig 4—Compensating for reactance present in the line input impedance.

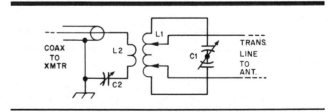

Fig 5—Using a variable capacitance, C2, as an alternative to variable mutual inductance between L1 and L2.

mutual inductance between L1 and L2. The capacitance of C2 should be such as to resonate with L2 at the lowest frequency in the band of operation. This calls for a fairly large value of capacitance at low frequencies (about 100 pF at

3.5 MHz for 52-ohm line) if the reactance of L2 is equal to the line Z_0. To utilize a capacitor of more convenient size—maximum capacitance of perhaps 250-300 pF—a value of inductance may be used for L2 that will resonate at the lowest frequency with the maximum capacitance available.

On the higher frequency bands the problem of variable capacitors does not arise since a reactance of 52 to 75 ohms is within the range of conventional components.

Circuit Balance

Fig 5 shows C1 as a balanced or split-stator capacitor. This type of capacitor is desirable in a practical matching circuit to be used with a balanced line, since the two sections are symmetrical. The rotor assembly of the balanced capacitor may be grounded, if desired, or it may be left "floating" and the center of L1 may be grounded; or both may "float." Which method to use depends on considerations discussed later in connection with antenna currents on transmission lines. As an alternative to using a split-stator type of capacitor, a single-section capacitor may be used.

Measurement of Line Input Current

The RF ammeters shown in Fig 6 are not essential to the adjustment procedure but they, or some other form of output indicator, are useful accessories. In most cases the circuit adjustments that lead to a match as shown by the SWR indicator will also result in the most efficient power transfer to the transmission line. However, it is possible that a good match will be accompanied by excessive loss in the matching circuit. This is unlikely to happen if the steps described for obtaining a low Q are taken. If the settings are highly critical or it is impossible to obtain a match, the use of additional reactance compensation as described earlier is indicated.

RF ammeters are useful for showing the comparative output obtained with various matching-network settings, and also for showing the improvement in output resulting from the use of reactance compensation when it seems to be required. Providing no basic circuit changes (such as grounding or ungrounding some part of the matching circuit) are made during such comparisons, the current shown by the ammeters will increase whenever the power put into the line is increased. Thus, the highest reading indicates the greatest transfer efficiency, assuming that the power input to the transmitter is kept constant.

If the line Z_0 is matched by the antenna, the current can be used to determine the actual power input to the line. The power at the input terminals is then equal to I^2Z_0, where I is the current and Z_0 is the characteristic impedance of the line. If there are standing waves on the line this relationship does not hold. In such a case the current that will flow into the line is determined by the line length, the SWR, and whether the antenna impedance is higher or lower than the line impedance. Information in Chapter 24 shows how the maximum current to be expected will vary with the standing-wave ratio. This information can be used in selecting the proper ammeter range.

Two ammeters, one in each line conductor, are shown in Fig 6. The use of two instruments gives a check on the line balance, since the currents should be the same. However, a single meter can be switched from one conductor to the other. If only one instrument is used it is preferably left out of the circuit except when adjustments are being made, since it will add capacitance to the side in which it is inserted and thus cause some unbalance. This is particularly important when the instrument is mounted on a metal panel.

Since the resistive component of the input impedance of a line operating with an appreciable SWR is seldom known accurately, the RF current is of little value as a check on power input to such a line. However, it shows in a relative way the efficiency of the system as a whole. The set of coupling adjustments that results in the largest line current with the least final-amplifier input power is the most desirable—and most efficient.

For adjustment purposes, it is possible to substitute small flashlight lamps, shunted across a few inches of the line wires, for the RF ammeters. Their relative brightness shows when the current increases or decreases. They have the advantage of being inexpensive and of such small physical size that they do not unbalance the circuit.

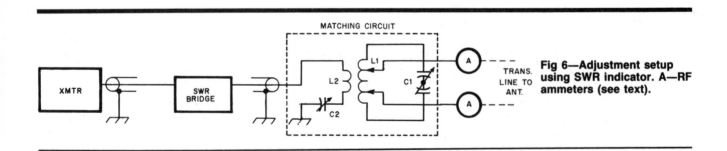

Fig 6—Adjustment setup using SWR indicator. A—RF ammeters (see text).

A Link-Coupled Matching Network

Link coupling offers many advantages over other types of systems where a direct connection between the equipment and antenna is required. This is particularly true at 3.5 MHz, where commercial broadcast stations often induce sufficient voltage to cause either rectification or front-end overload. Transceivers and receivers that show this tendency can usually be cured by using only magnetic coupling between the transceiver and antenna system. There is no direct connection, and better isolation results, along with the inherent band-pass characteristics of magnetically coupled tuned circuits.

Although link coupling can be used with either single-ended or balanced antenna systems, its most common application is with balanced feed. The model shown here is designed for 3.5- through 28-MHz operation.

The Circuit

The Transmatch or matching networks shown in Figs 7 through 9 is a band-switched link coupler. L2 is the link and C1 is used to adjust the coupling. S1B selects the proper amount of link inductance for each band. L1 and L3 are located on each side of the link and are the coils to which the antenna is connected. Alligator clips are used to connect the antenna to the coil because antennas of different impedances must be connected at different points (taps) along the coil. Also, with most antennas it will be necessary to change taps for different bands of operation. C2 tunes L1 and L3 to resonance at the operating frequency.

Switch sections S1A and S1C select the amount of inductance necessary for each of the HF bands. The inductance of each of the coils has been optimized for antennas in the impedance range of roughly 20 to 600 Ω. Antennas that exhibit impedances well outside this range may require that some of the fixed connections to L1 and L3 be changed. Should this be necessary, remember that the L1 and L3 sections must be kept symmetrical—the same number of turns on each coil.

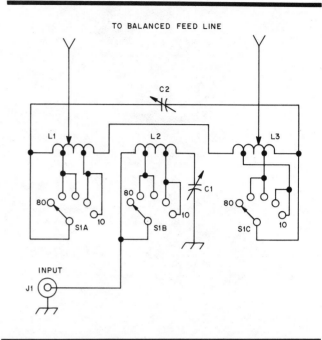

Fig 8—Schematic diagram of the link coupler. The connections marked as "to balanced feed line" are steatite feedthrough insulators. The arrows on the other ends of these connections are alligator clips.
C1—350 pF maximum, 0.0435-in. plate spacing or greater.
C2—100 pF maximum, 0.0435-in. plate spacing or greater.
J1—Coaxial connector.
L1, L2, L3—B&W 3026 Miniductor stock, 2-in. diameter, 8 turns per inch, no. 14 wire. Coils assembly consists of 48 turns, L1 and L3 are each 17 turns tapped at 8 and 11 turns from outside ends. L2 is 14 turns tapped at 8 and 12 turns from C1 end. See text for additional details.
S1—3-pole, 5-position ceramic rotary switch.

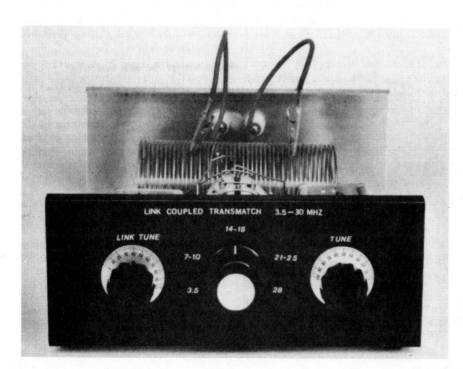

Fig 7—Exterior view of the band-switched link coupler. Alligator clips are used to select the proper tap positions of the coil.

Construction

The unit is housed in a homemade aluminum enclosure that measures 9 × 8 × 3½ inches. Any cabinet with similar dimensions that will accommodate the components may be used. L1, L2 and L3 are a one-piece assembly of B&W 3026 Miniductor stock. The individual coils are separated from each other by cutting two of the turns at the appropriate spots along the length of the coil. Then the inner ends of the outer sections are joined by a short wire that is run through the center of L2. Position the wire so it will not come into contact with L2. Locate each of the fixed tap points on L1, L2 and L3, and attach lengths of hookup wire. The coil is mounted in the enclosure, and the connections between the coil and the band switch are made. Every second turn of L1 and L3 are pressed in toward the center of the coil to facilitate connection of the alligator clips.

As can be seen from the schematic, C2 must be isolated from ground. This can be accomplished by mounting the capacitor on steatite cones or other suitable insulating material. Make sure that the hole through the front panel for the shaft of C2 is large enough so the shaft does not come into contact with the chassis.

Tune-Up

The transmitter should be connected to the input of the Transmatch through some sort of instrument that will indicate SWR. Set S1 to the band of operation, and connect the balanced line to the insulators on the rear panel of the coupler. Attach alligator clips to the mid points of coils L1 and L3, and apply power. Adjust C1 and C2 for minimum reflected power. If a good match is not obtained, move the antenna tap points either closer to the ends or center of the coils. Again apply power and tune C1 and C2 until the best possible match is obtained. Continue moving the antenna taps until a 1-to-1 match is obtained.

The circuit described here is intended for power levels up to roughly 200 watts. Balance was checked by means of two RF ammeters, one in each leg of the feed line. Results showed the balance to be well within 1 dB.

A Transmatch for Balanced or Unbalanced Lines

Most modern transmitters are designed to operate into loads of approximately 52 Ω. Solid-state transmitters produce progressively lower output power as the SWR on the transmission line increases, because of built-in SWR protection circuits. Therefore, it is useful to employ a matching network between the transmitter and the antenna feeder when antennas with complex impedances are used.

One example of this need can be seen in the case of a

3.5-MHz, coax-fed dipole antenna which has been cut for resonance at, say, 3.6 MHz. If this antenna were used in the 3.8-MHz phone band, the SWR would be fairly high. A Transmatch could be used to give the transmitter a 52-Ω load, even though a significant mismatch was present at the antenna feed point. It is important to remember that the Transmatch will not correct the actual SWR condition on the line, but it will resonate the *antenna system* and allow the transmitter to deliver full power to the load. There will be some additional loss caused by SWR on the line; that topic is covered in Chapter 24.

A Transmatch is useful also when employing a single-wire antenna for multiband use. By means of a balun at the Transmatch output it is possible to operate the transmitter into a balanced transmission line, such as a 300- or 600-Ω feed system of the type that would be used with a multiband tuned dipole, V beam or rhombic antenna.

A secondary benefit can be realized from a Transmatch. Under most conditions of normal application it will attenuate harmonics from the transmitter. The amount of attenuation depends on the circuit design and the loaded Q (Q_L) of the network after the impedance has been matched. The higher the Q_L, the greater the attenuation. The SPC Transmatch pictured in Fig 10 was designed to provide high harmonic attenuation. See Fig 11. The SPC (series-parallel capacitance) circuit maintains a band-pass response under load conditions of less than 25 Ω to more than 1000 Ω (from a 52-Ω transmitter). This is because a substantial amount of capacitance is always in parallel with the rotary inductor (C2B and L1 of Fig 12).

The circuit of Fig 12 operates from 1.8 to 30 MHz with the values shown. The SPC exhibits somewhat sharper tuning than other designs. This arises from the relatively high network Q, and is especially prominent at 1.8, 3.5 and 7 MHz. For this reason there are reduction drives with vernier dials on C1 and C2. They are also useful in logging the dial settings for changing bands or antennas.

Construction

Figs 10 and 13 show the structural details of the Transmatch. The cabinet is homemade from 16-gauge aluminum sheeting. L brackets are affixed to the right and left sides of the lower part of the cabinet to permit attachment of the U-shaped cover.

The conductors which join the components should be of heavy gauge material to minimize stray inductance and heating. Wide strips of flashing copper are suitable for the conductor straps. The center conductor and insulation from RG-59 polyfoam coaxial cable is used in this model for the wiring between the switch and the related components. The insulation is sufficient to prevent breakdown and arcing at 2 kW PEP input to the transmitter.

All leads should be kept as short as possible to help prevent degradation of the circuit Q. The stators of C1 and C2 should face toward the cabinet cover to minimize the stray capacitance between the capacitor plates and the bottom of the cabinet (important at the upper end of the Transmatch frequency range). Insulated ceramic shaft couplings are used between the reduction drives and C1 and C2, since the rotors of both capacitors are "floating" in this circuit. C1 and C2 are supported above the bottom plate on steatite cone insulators. S1 is attached to the rear apron of the cabinet by means of two metal standoff posts.

Fig 10—Exterior view of the SPC Transmatch. Radio Shack vernier drives are used for adjusting the tuning capacitors. A James Millen turns-counter drive is coupled to the rotary inductor. Green paint and green Dymo tape labels are used for panel decor. The cover is plain aluminum with a lightly grooved finish (sandpapered) which has been coated with clear lacquer. An aluminum foot holds the Transmatch at an easy access angle.

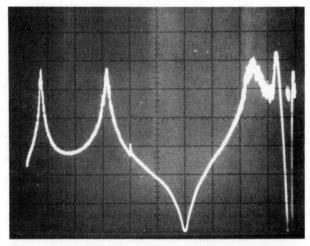

Fig 11—Spectrograph of the response characteristics of the SPC Transmatch looking into a 1000-Ω termination from a 52-Ω signal source. The scale divisions are 2 MHz horizontal and 10 dB vertical. The fundamental frequency is 8 MHz; the second harmonic is attenuated by 28 dB with these matching conditions.

Operation

The SPC Transmatch shown here is designed to handle the output from transmitters that operate up to 2 kW PEP. L2 improves the circuit Q at 10 and 15 meters. However, it may be omitted from the circuit if the rotary inductor (L1) has a tapered pitch at the minimum-inductance end. It may be necessary to omit L2 if the stray wiring inductance of the builder's version is high. Otherwise, it may be impossible to obtain a matched condition at 28 MHz with certain loads.

An SWR indicator is used between the transmitter and

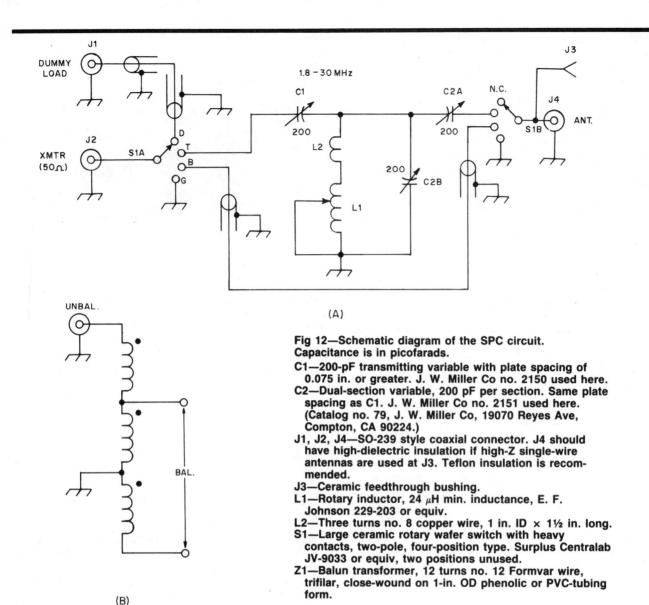

(A)

(B)

Fig 12—Schematic diagram of the SPC circuit. Capacitance is in picofarads.

C1—200-pF transmitting variable with plate spacing of 0.075 in. or greater. J. W. Miller Co no. 2150 used here.

C2—Dual-section variable, 200 pF per section. Same plate spacing as C1. J. W. Miller Co no. 2151 used here. (Catalog no. 79, J. W. Miller Co, 19070 Reyes Ave, Compton, CA 90224.)

J1, J2, J4—SO-239 style coaxial connector. J4 should have high-dielectric insulation if high-Z single-wire antennas are used at J3. Teflon insulation is recommended.

J3—Ceramic feedthrough bushing.

L1—Rotary inductor, 24 μH min. inductance, E. F. Johnson 229-203 or equiv.

L2—Three turns no. 8 copper wire, 1 in. ID × 1½ in. long.

S1—Large ceramic rotary wafer switch with heavy contacts, two-pole, four-position type. Surplus Centralab JV-9033 or equiv, two positions unused.

Z1—Balun transformer, 12 turns no. 12 Formvar wire, trifilar, close-wound on 1-in. OD phenolic or PVC-tubing form.

Fig 13—Interior view of the SPC Transmatch. L2 is mounted on the rear wall by means of two ceramic standoff insulators. C1 is on the left and C2 is at the right. The coaxial connectors, ground post and J3 are on the lower part of the rear panel.

the Transmatch to show when a matched condition is attained. The builder may want to integrate an SWR meter in the Transmatch circuit between J2 and the arm of S1A (Fig 12A). If this is done there should be room for an edgewise panel meter above the reduction drive for C2.

Initial transmitter tuning should be done with a dummy antenna connected to J1, and with S1 in the D position. This will prevent interference that could otherwise occur if tuning is done on the air. After the transmitter is properly tuned into the dummy antenna, unkey the transmitter and switch S1 to T (Transmatch). Never "hot switch" a Transmatch, as this can damage both transmitter and Transmatch. Set C1 and C2 at midrange. With a few watts of RF, adjust L1 for a decrease in reflected power. Then adjust C1 and C2 alternately for the lowest possible SWR. If the SWR cannot be reduced to 1:1, adjust L1 slightly and repeat the procedure. Finally, increase the transmitter power to maximum and touch up the Transmatch controls if necessary. When tuning, keep your transmissions brief and identify your station.

The air-wound balun of Fig 12B can be used outboard from the Transmatch if a low-impedance balanced feeder is contemplated. Ferrite or powdered-iron core material is not used in the interest of avoiding TVI and harmonics which can result from core saturation. A detailed discussion of baluns can be found later in this chapter.

The B position of S1 permits switched-through operation when the Transmatch is not needed. The G position is used for grounding the antenna system, as necessary; a good quality earth ground should be attached at all times to the Transmatch chassis.

Final Comments

Surplus coils and capacitors are suitable for use in this circuit. L1 should have at least 25 μH of inductance, and the tuning capacitors need to have 150 pF or more of capacitance per section. Insertion loss through this Transmatch was measured at less than 0.5 dB at 600 watts of RF power on 7 MHz.

L NETWORKS

A comparatively simple but very useful matching circuit for unbalanced loads is the L network, Fig 14A. L-network Transmatches are normally used for only a single band of operation, although multiband versions with switched or variable coil taps exist. To determine the range of circuit values for a matched condition, the input and load impedance values must be known or assumed. Otherwise a match may be found by trial.

In Fig 14A, L1 is shown as the series reactance, X_s, and C1 as the shunt or parallel reactance, X_p. However, a capacitor may be used for the series reactance and an inductor for the shunt reactance, to satisfy mechanical or other considerations.

The ratio of the series reactance to the series resistance, X_s/R_s, is defined as the network Q. The four variables, R_s, R_p, S_x and S_p, are related as given in the equations below. When any two values are known, the other two may be calculated.

$$Q = \sqrt{\frac{R_p}{R_s} - 1} = \frac{X_s}{R_s} = \frac{R_p}{X_p}$$

$$X_s = QR_s = \frac{QR_p}{1 + Q^2}$$

$$X_p = \frac{R_p}{Q} = \frac{R_pR_s}{X_s} = \frac{R_s^2 + X_s^2}{X_s}$$

$$R_s = \frac{R_p}{1 + Q^2} = \frac{X_sX_p}{R_p}$$

$$R_p = R_s(1 + Q^2) = QX_p = \frac{R_s^2 + X_s^2}{R_s}$$

The reactance of loads that are not purely resistive may be taken into account and absorbed or compensated for in the reactances of the matching network. Inductive and capacitive reactance values may be converted to inductor and capacitor values for the operating frequency with standard reactance equations.

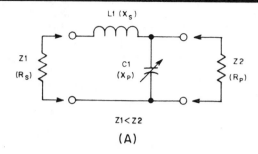

(A)

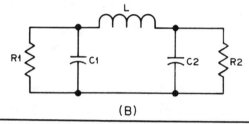

(B)

Fig 14—At A, the L matching network, consisting of L1 and C1, matches Z1 and Z2. The lower of the two impedances to be matched, Z1, must always be connected to the series-arm of the network and the higher impedance, Z2, to the shunt-arm side. The positions of the inductor and capacitor may be interchanged in the network. At B, the pi network, matching R1 to R2. The pi provides more flexibility than the L as a Transmatch circuit. See equations in the text for calculating component values.

PI NETWORKS

The pi network, shown in Fig 14B, offers more flexibility than the L since the operating Q may be chosen practically at will. The only limitation on the circuit values that may be used is that the reactance of the series arm, the inductor L in the figure, must not be greater than the square root of the product of the two values of resistive impedance to be matched. As the circuit is applied to amateur use, this limiting value of reactance would represent a network with an undesirably low operating Q, and the circuit values ordinarily used are well on the safe side of the limiting values. If R1 and R2 are known or assumed, these equations may be used to determine the component values required for a match. The value of Q may be arbitrarily chosen, usually between 5 and 15 for most applications.

$R1 > R2$

$$X_{C1} = R1/Q$$

$$X_{C2} = R2 \sqrt{\frac{R1/R2}{Q^2 + 1 - R1/R2}}$$

$$X_L = \frac{Q \cdot R1 + R1 \cdot R2/X_{C2}}{Q^2 + 1}$$

The pi network may be used to match a low impedance to a rather high one, such as 52 Ω to several thousand ohms. Conversely, it may be used to match 52 Ω to a quite low value, such as 1 Ω or less. For Transmatch applications, C1 and C2 may be independently variable. L may be a roller inductor or a coil with switchable taps. Alternatively, a lead fitted with a suitable clip may be used to short out turns of a fixed inductor. In this way, a match may be obtained through trial. It will be possible to match two values of impedances with several different settings of L, C1 and C2. This results because the Q of the network is being changed. If a match is maintained with other adjustments, the Q of the circuit rises with increased capacitance at C1.

Quite often the load and source have reactive components along with resistance, but in most instances the pi network can still be used. The effect of these reactive components can be compensated for by changing one of the reactive elements in the matching network. For example, if some reactance was shunted across R2, the setting of C2 could be changed to compensate, whether that shunt reactance be inductive or capacitive.

The Conjugate Match and the Pi Network

In transceivers having tubes in the RF output stage, the pi network serves as a conjugate matching network as well as a harmonic-controlling tank circuit. When properly adjusted, the pi network matches the optimum load resistance of the tubes to the impedance seen at the input of the coaxial line connected to the output of the network. In the ever-present quest for the perfect 1:1 SWR, many amateurs are unaware that the pi network is usually capable of matching the optimum tube load resistance to a wider range of output load impedances than defined by a 2:1 SWR; some will match to as high as 3:1, and sometimes even 4:1. During tune-up, if a resonance dip in plate current can be obtained at the proper loaded value of current within the range of the tuning and loading capacitors in the pi network, the network has correctly matched the tubes to whatever impedance is seen at the output. This is true no matter how high the SWR. When correctly tuned to the plate-current dip, the tubes see the desired resistive (nonreactive) load, the network has supplied a conjugate match, and any reflected power as shown by an SWR indicator has been re-reflected in the same manner as described above for the Transmatch. In other words, the pi network can serve as the Transmatch for whatever SWR it can load properly.

S-P NETWORKS

The S-P (series-parallel) network is an LC impedance transformation circuit with many applications. This section is based on a June 1986 *QST* article by Warren Bruene, W5OLY. A diagram of the S-P network is shown in Fig 15. It uses four elements, in contrast to the two found in an L network or three in a T or pi network. It has the properties of a series resonant circuit on the low resistance side and a parallel resonant circuit on the high resistance side with the characteristics of a perfect transformer in between. It has a geometrically symmetrical band-pass response and zero phase delay at the center frequency. An equivalent circuit is shown in Fig 16.

This impedance transformation circuit has been used in filter design, but it is very appropriate for antenna-coupler design. The network is discussed here as a separate circuit but, of course, other circuits can be added at either end.

Component Relationships

The circuit designer may choose any desired impedance

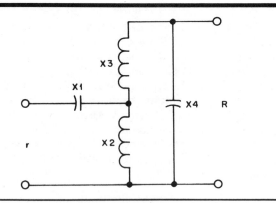

Fig 15—Basic S-P network configuration.

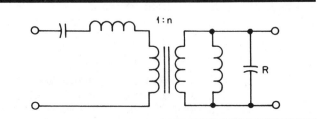

Fig 16—Equivalent circuit for the S-P network.

step-up (or step-down by reversing the input and output connections). One may also choose the value of any single element. This choice must be made with care because it determines the frequency response, as explained later. The signs of all reactances can be reversed. That means that in Fig 15 the capacitors can be replaced with inductors, and the inductors with capacitors. The resulting circuit is shown in Fig 17, where all circuit elements have the same magnitudes of reactance as their counterparts in Fig 15. The frequency response remains the same, however.

Component value relationships are easily grasped: X2 and X3 in series resonate with X4; also, X2 and X3 in parallel resonate with X1. The ratio $-X4/X3$ is equal to the voltage step-up ratio, n.

Frequency Response Shape

The S-P network is the equivalent of a classical 2-pole band-pass filter as shown by the equivalent circuit in Fig 16. Each pole (resonant circuit) has a Q defined as:

$$Q1 = \frac{X1}{r} \qquad \qquad \text{(Eq 1)}$$

$$Q2 = \frac{R}{X4} \qquad \qquad \text{(Eq 2)}$$

where r, R, X1 and X4 are as illustrated in Fig 15. (Omit the sign of the reactance when computing Q.)

Two factors affect the shape of the frequency response curve. One is the ratio Q1/Q2, and the other is whether one or both ends of the network are terminated with the design values of r and R. The effective internal resistance of the source must be the same as the design value of the input end

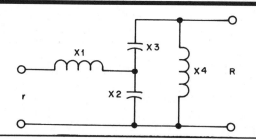

Fig 17—Alternative S-P network configuration.

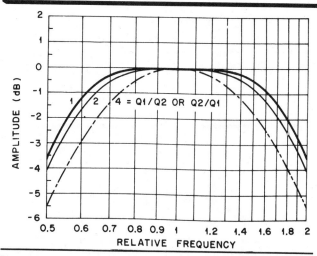

Fig 18—Network response is affected by Q1/Q2 ratio.

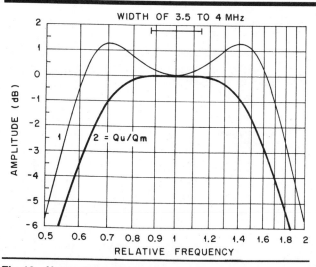

Fig 19—Network response where only one end is matched for two Q_u/Q_m cases for R/r = 4.

of the network to provide a "terminated" input. A signal generator with a 52-Ω output is designed to have an effective internal resistance of 52 Ω. A receiving antenna matched to a 52-Ω load also has a source resistance of 52 Ω.

Mismatching one end of the network changes the frequency response. For this discussion we examine four cases.

Case 1: Both ends of the network are matched. An example is an input from an antenna which is matched to 52 Ω and a network load resistance of 208 Ω.

Case 2: The input is matched as in Case 1 to the low resistance end, but the high resistance end is open-circuited. An example is an antenna coupled to an FET.

Case 3: A matched load is connected to the high resistance end of the network, but the input is driven by a zero-impedance source. A class-D transistor amplifier approximates this condition.

Case 4: A matched load is connected to the low resistance end and the high resistance end is fed by a very high impedance source such as a tetrode linear amplifier.

When both ends of the network are terminated as for Case 1, the input and output Q values are equal for a maximally flat response. Fig 18 shows the response for this case and for two widely different Q ratios when R/r = 4. A network design with nominally equal Q values will be most tolerant of moderate mismatches on either or both ends. The frequency responses of Cases 2, 3 and 4 are the same when the ratio of the network Q on the unmatched end (Q_u) to the Q on the matched end (Q_m) is the same. For a maximally flat response, the ratio Q_u/Q_m equals two.

When equal Q values are used for these cases there will be a double-humped response with a rise of a little over 1 dB, as shown in Fig 19. The bandwidth is widened by the double-humped response, however.

Bandwidth

The bandwidth of this network for a given response shape is determined principally by the impedance transformation ratio R/r. It is interesting that the product QP = Q1 × Q2 = n − 1. Fig 20 shows the maximally flat responses for several R/r ratios when the network is matched on both ends.

Calculating Element Values

First choose two of the values in one of the following set of three equations and compute the third.

$$n = \sqrt{\frac{R}{r}} \text{ or } R = rn^2 \text{ or } r = \frac{R}{n^2} \qquad \text{(Eq 3)}$$

Then compute the Q product:

$$QP = n - 1 \qquad \text{(Eq 4)}$$

From a selected ratio of Q1/Q2, compute Q1 and Q2:

$$Q1 = \sqrt{QP \frac{Q1}{Q2}} \qquad \text{(Eq 5)}$$

$$Q2 = \frac{QP}{Q1} \qquad \text{(Eq 6)}$$

Then

$$X4 = \frac{R}{Q2} \qquad \text{(Eq 7)}$$

$$X2 = \frac{-X4}{n} \qquad \text{(Eq 8)}$$

$$X3 = -X4 - X2 \qquad \text{(Eq 9)}$$

$$X1 = r\,Q1 = \frac{1}{\dfrac{1}{X2} + \dfrac{1}{X3}} \qquad \text{(Eq 10)}$$

Now convert the reactance values, minus (capacitance) to pF and plus (inductance) to μH. Capacitance (in pF) is calculated by

$$C = \frac{1,000,000}{2\pi fX} \qquad \text{(Eq 11)}$$

where

f is the geometrical center frequency in MHz
X is the (negative) reactance value

Inductance (in μH) is calculated from

$$L = \frac{X}{2\pi f} \qquad \text{(Eq 12)}$$

where X is the (positive) reactance value.

Using a Tapped Coil

One coil with a tap can be used in place of the two series inductors in the middle. In fact the bandwidth will be even wider because of the mutual coupling between the two parts of the coil. The tap must be located on the coil to give the desired voltage step-up. The voltage along the coil is not proportional to the number of turns (except at the middle) because the end turns of a solenoid do not contribute as much to the total inductance as turns in the center. There are ways to calculate the correct tap position, but it is probably easiest to find it by experiment.

Tuning

A fixed-tuned S-P network can be tuned with a grid-dip meter. Disconnect both ends of the network and adjust X4 to resonate at the desired center frequency. Then short-circuit both ends and adjust X1 to resonate at the same frequency.

The amount of resistance transformation can be varied by adjusting X2 (or the coil tap). X3 or X4 and also X1 should be readjusted after a change in X2. The network behaves as a double-tuned circuit when used as an antenna coupler.

Comparison to L, T and Pi Networks

In Fig 21, you can see the frequency responses of L, T, pi and S-P networks when the source resistance is matched for the case of a 4-to-1 impedance step-up. Low-pass 90° (equal capacitive and inductive reactance) T and pi networks are shown because they give wider bandwidths than networks with larger Q values. The responses of the T and pi networks are identical. Fig 22 shows the comparison when each is designed for a source resistance of zero (constant voltage). The low-pass pi network response is not shown because the input shunt element across a voltage source does not affect the network response. The S-P network response can be widened a little by choosing the Q ratio to give a double-humped type of response. Two examples of the input phase angle variation with frequency are shown in Fig 23. In each case the network is terminated in a matched resistive load.

BALUNS

A center-fed antenna with open ends, of which the half-wave dipole is an example, is inherently a balanced radiator. When opened at the center and fed with a parallel-conductor line, this balance is maintained throughout the system, so long as the causes of feeder unbalance discussed in Chapter 26 are avoided.

If the antenna is fed at the center through a coaxial line, this balance is upset because one side of the radiator is

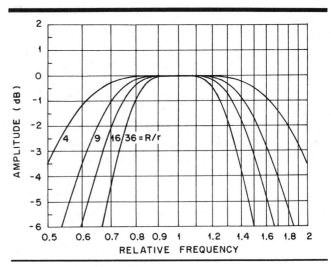

Fig 20—The bandwidth is related to the R/r step-up ratio.

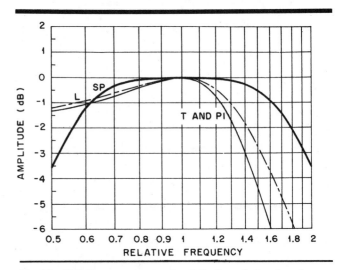

Fig 21—Relative responses for S-P, T and pi networks where both ends are matched.

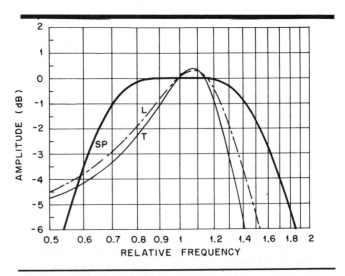

Fig 22—Relative responses for S-P, T and pi networks connected to a voltage source.

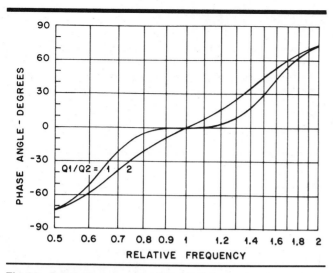

Fig 23—Input phase angle of the S-P network with frequency for resistive loads and R/r = 4.

matching with balanced-to-unbalanced operation, but has the disadvantage that it uses resonant circuits and thus can work over only a limited band of frequencies without readjustment. However, if a fixed impedance ratio in the balun can be tolerated, the coil balun described below can be used without adjustment over a frequency range of about 10:1, or 3 to 30 MHz, for example.

Coil Baluns

The type of balun known as the "coil balun" is based on the principles of linear-transmission-line balun as shown in the upper drawing of Fig 24. Two transmission lines of equal length having a characteristic impedance (Z_0) are connected in series at one end and in parallel at the other. At the series-connected end the lines are balanced to ground and will match an impedance equal to $2Z_0$. At the parallel-connected end the lines will be matched by an impedance equal to $Z_0/2$. One side may be connected to ground at the parallel-connected end, provided the two lines have a length such that, considering each line as a single wire, the balanced end is effectively decoupled from the parallel-connected end. This requires a length that is an odd multiple of a quarter wavelength.

A definite line length is required only for decoupling purposes, and as long as there is adequate decoupling, the system will act as a 4:1 impedance transformer regardless of line length. If each line is wound into a coil, as in the lower drawing of Fig 24, the inductances so formed will act as choke coils and will tend to isolate the series-connected end from any ground connection that may be placed on the parallel-connected end. Balun coils made in this way will operate over a wide frequency range, as the choke inductance is not critical. The lower frequency limit is where the coils are no longer effective in isolating one end from the other; the length of line in each coil should be about equal to a quarter wavelength at the lowest frequency to be used.

The principal application of such coils is in going from a 300-Ω balanced line to a 75-Ω coaxial line. This requires that the Z_0 of the lines forming the coils be 150 Ω.

A balun of this type is simply a fixed-ratio transformer, when matched. It cannot compensate for inaccurate matching elsewhere in the system. With a "300-Ω" line on the balanced end, for example, a 75-Ω coax cable will not be matched unless the 300-Ω line actually is terminated in a 300-Ω load.

connected to the shield while the other is connected to the inner conductor. On the side connected to the shield, a current can flow down over the *outside* of the coaxial line. The fields thus set up cannot be canceled by the fields from the inner conductor because the fields *inside* the line cannot escape through the shielding afforded by the outer conductor. Therefore, these "antenna" currents flowing on the outside of the line will be responsible for radiation.

Line radiation can be prevented by a number of devices intended to detune or decouple the line for "antenna" currents and thus greatly reduce their amplitude. Such devices generally are known as *baluns* (a contraction for "balanced to unbalanced").

The need for baluns arises in coupling a transmitter to a balanced transmission line, since the output circuits of most transmitters have one side grounded. (This type of output circuit is desirable for a number of reasons including TVI reduction.) The most flexible type of balun for this purpose is the inductively coupled matching network described in a previous section in this chapter. This combines impedance

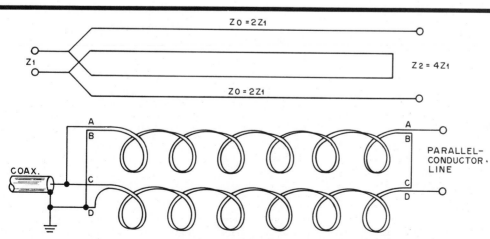

Fig 24—Baluns for matching between coaxial line and parallel-conductor line. The impedance ratio is 4:1 from the balanced side to the unbalanced side. Coiling the lines (lower drawing) increases the frequency range over which satisfactory operation is obtained.

Broadband Toroidal Baluns

Air-wound balun transformers are somewhat bulky when designed for operation in the 1.8- to 30-MHz range. A more compact broadband transformer can be realized by using toroidal ferrite core material as the foundation for bifilar-wound coil balun transformers. Two such baluns are described here.

In Fig 25A, a 1:1 ratio balanced to unbalanced line transformer is shown. This transformer is useful in converting

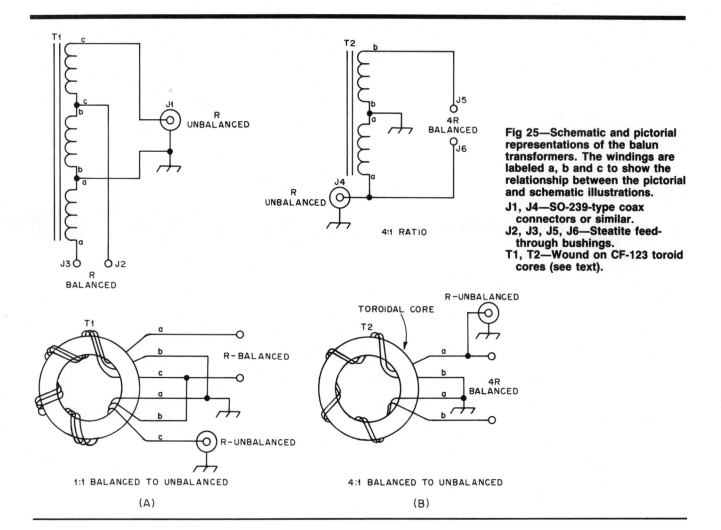

Fig 25—Schematic and pictorial representations of the balun transformers. The windings are labeled a, b and c to show the relationship between the pictorial and schematic illustrations.

J1, J4—SO-239-type coax connectors or similar.
J2, J3, J5, J6—Steatite feed-through bushings.
T1, T2—Wound on CF-123 toroid cores (see text).

a 52-Ω balanced line condition to one that is 52 Ω, unbalanced. Similarly, the transformer will work between balanced and unbalanced 75-Ω impedances. A 4:1 ratio transformer is illustrated in Fig 25B. This balun is useful for converting a 208-Ω balanced condition to one that is 52 Ω, unbalanced. In a like manner, the transformer can be used between a balanced 300-Ω point and a 75-Ω unbalanced line. Both balun transformers will handle 1000 watts of RF power and are designed to operate from 1.8 through 60 MHz.

Low-loss high-frequency ferrite core material is used for T1 and T2. The cores are made from Q2 material and are 0.5 inch thick, have an OD of 2.4 inches, and the ID is 1.4 inches. The permeability rating of the cores is 40. Packaged 1-kilowatt balun kits, with winding instructions for 1:1 or 4:1 impedance transformation ratios, are available, but use a core of slightly different dimensions. Ferrite cores are available from several sources. See Chapter 21.

Winding Information

The transformer shown in Fig 25A has a trifilar winding consisting of 10 turns of no. 14 Formvar-insulated copper wire. A 10-turn bifilar winding of the same type of wire is used for the balun of Fig 25B. If the cores have rough edges, they should be carefully sanded until smooth enough to prevent damage to the Formvar wire insulation. The windings should be spaced around the entire core as shown in Fig 26. Insulation can be used between the core material and the windings to increase the breakdown voltage of the balun.

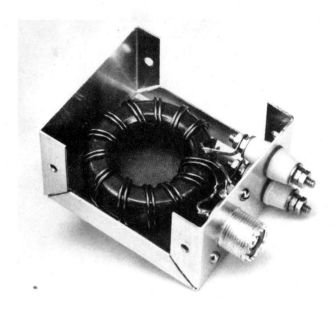

Fig 26—Layout of a kilowatt 4:1 toroidal balun transformer. Phenolic insulating board is mounted between the transformer and the Minibox wall to prevent short-circuiting. The board is held in place with epoxy cement. Cement is also used to secure the transformer to the board. For outdoor use, the Minibox cover can be installed, then sealed against the weather by applying epoxy cement along the seams of the box.

A 52- to 75-Ohm Broadband Transformer

Shown in Figs 27 through 29 is a simple 52- to 75-Ω or 75- to 52-Ω transformer that is suitable for operation in the 2- to 30-MHz frequency range. A pair of these transformers is ideal for using 75-Ω CATV Hardline in a 52-Ω system. In this application, one transformer is used at each end of the cable run. At the antenna, one transformer steps the 52-Ω impedance of the antenna up to 75 Ω, thereby presenting a match to the 75-Ω cable. At the station end, a transformer is used to step the 75-Ω line impedance down to 52 Ω.

The schematic diagram of the transformer is shown in Fig 27, and the winding details are given in Fig 28. C1 and C2 are compensating capacitors; the values shown were determined through swept return-loss measurements using a spectrum analyzer and a tracking generator. The transformer consists of a trifilar winding of no. 14 enameled copper wire wound over an FT-200-61 (Q1 material) or equivalent core. As shown in Fig 28, one winding has only half the number of turns of the other two. Care must be taken when connecting the loose ends so the proper phasing of the turns is maintained. Improper phasing will become apparent when power is applied to the transformer.

If the core has sharp edges it is a good idea to either sand

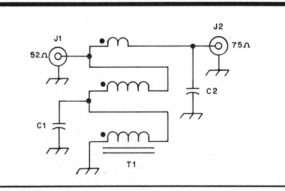

Fig 27—Schematic diagram of the 52- to 75-Ω transformer described in the text. C1 and C2 are compensating capacitors.
C1—100 pF silver mica.
C2—10 pF, silver mica.
J1, J2—Coaxial connectors, builder's choice.
T1—Transformer, 6 trifilar turns, no. 14 enameled copper wire on an FT-200-61 (Q1 material, μ_i = 125) core.
The upper winding has one-half the number of turns of the other two.

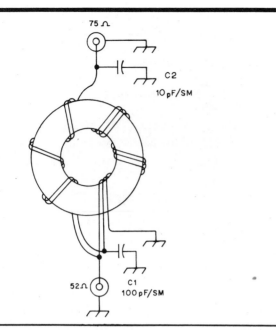

Fig 28— Pictorial drawing of the 52- to 75-Ω transformer showing details of the windings.

Fig 29—The 52- to 75-Ω transformers. The units are identical.

the edges until they are relatively smooth or wrap the core with tape. The one shown in the photograph was wrapped with ordinary vinyl electrical tape, although glass-cloth insulating tape would be better. The idea is to prevent chafing of the wire insulation.

Construction

The easiest way to construct the transformer is to wind the three lengths of wire on the core at the same time. Different color wires will aid in identifying the ends of the windings. After all three windings are securely in place, the appropriate winding may be unwound three turns as shown in the diagram. This wire is the 75-Ω connection point. Connections at the 52-Ω end are a bit tricky, but if the information in Fig 28 is followed carefully no problems should be encountered. Use the shortest connections possible, as long leads will degrade the high-frequency performance.

The balun is housed in a homemade aluminum enclosure measuring 3½ × 3¾ × 1¼ inches. Any commercial cabinet of similar dimensions will work fine. In the unit shown in the photograph, several "blobs" of silicone seal (RTV) were used to hold the core in position. Alternatively, a piece of phenolic insulating material may be used between the core and the aluminum enclosure. Silicone seal is used to protect the inside of the unit from moisture. All joints and screw heads should receive a generous coating of RTV.

Checkout

Checkout of the completed transformer or transformers is quite simple. If a 75-Ω dummy antenna is available connect it to the 75-Ω terminal of the transformer. Connect a transmitter and SWR indicator (52 Ω) to the 52-Ω terminal of the transformer. Apply power (on each of the HF bands) and measure the SWR looking into the transformer. Readings should be well under 1.3 to 1 on each of the bands. If a 75-Ω load is not available and two transformers have been

constructed, they may be checked out simultaneously as follows. Connect the 75-Ω terminals of both transformers together, either directly through a coaxial adapter or through a length of 75-Ω cable. Attach a 52-Ω load to one of the 52-Ω terminals and connect a transmitter and SWR indicator (52 Ω) to the remaining 52-Ω terminal. Apply power as outlined above and record the measurements. Readings should be under 1.3 to 1.

The transformers in the photo were checked in the ARRL laboratory under various mismatched conditions at the 1500-watt power level. No spurious signals (indicative of core saturation) could be found while viewing the MF, HF and VHF frequency range with a spectrum analyzer. A key-down, 1500-watt signal produced no noticeable core heating and only a slight increase in the temperature of the windings.

Using the Baluns

For indoor use, the transformers can be assembled open style, without benefit of a protective enclosure. For outdoor installations, such as at the antenna feed point, the balun should be encapsulated in epoxy resin or mounted in a suitable weatherproof enclosure. A Minibox, sealed against moisture, works nicely.

Balun Terminations

A word about baluns in Transmatches may be in order. Broadband transformers of the type found in many Transmatches are not suitable for use at high impedances. Disastrous results can be had when using these transformers with loads higher than, say, 300 Ω during high-power operation. The effectiveness of the transformer is questionable as well. At high peak RF voltages (high-Z load conditions such as 600-Ω feeders or an end-fed random-length antenna), the core can saturate and the RF voltage can cause arcs between turns or between the winding and the core material. If a balanced-to-unbalanced transformation must be effected, try to keep the load impedance at 300 Ω or less. An airwound 1:1 balun with a trifilar winding is recommended over a transformer with ferrite or powdered-iron core material.

The principles on which baluns operate should make it obvious that the termination must be essentially a pure resistance in order for the proper impedance transformation to take place. If the termination is not resistive, the input impedance of each bifilar winding will depend on its electrical characteristics and the input impedance of the main transmission line; in other words, the impedance will vary just as it does with any transmission line, and the transformation ratio likewise will vary over wide limits.

Baluns alone are convenient as matching devices when the above condition can be met, since they require no adjustment. When used with a matching network as described earlier, however, the impedance-transformation ratio of a balun becomes of only secondary importance, and loads containing reactance may be tolerated so long as the losses in the balun itself do not become excessive.

BIBLIOGRAPHY

Source material and more extended discussion of topics covered in this chapter can be found in the references given below and in the textbooks listed at the end of Chapter 2.

D. K. Belcher, "RF Matching Techniques, Design and Example," *QST*, Oct 1972, p 24.

W. Bruene, "Introducing the Series-Parallel Network," *QST*, Jun 1986, p 21.

T. Dorbuck, "Matching-Network Design," *QST*, Mar 1979, p 26.

G. Grammer, "Simplified Design of Impedance-Matching Networks," in three parts, *QST*, Mar, Apr and May 1957.

M. W. Maxwell, "Another Look at Reflections," *QST*, Apr, Jun, Aug, and Oct 1973; Apr and Dec 1974, and Aug 1976.

B. Pattison, "A Graphical Look at the L Network," *QST*, Mar 1979, p 24.

E. Wingfield, "New and Improved Formulas for the Design of Pi and Pi-L Networks," *QST*, Aug 1983, p 23.

Chapter 26

Coupling the Line to the Antenna

Throughout the discussion of transmission-line principles in Chapter 24, the operation of the line was described in terms of an abstract "load." This load had the electrical properties of resistance and, sometimes, reactance. It did not, however, have any physical attributes that associated it with a particular electrical device. That is, it could be anything at all that exhibits electrical resistance and reactance. The fact is that so far as the line is concerned, it does not matter what the load *is*, just as long as it will accept power.

Many amateurs make the mistake of confusing transmission lines with antennas, believing that because two identical antennas have different kinds of lines feeding them, or the same kind of line with different methods of coupling to the antenna, the "antennas" are different. There may be practical reasons why one system (including antenna, transmission line, and coupling method) may be preferred over another in a particular application. But to the transmission line, an antenna is just a load that terminates it, and the important thing is what that load looks like to the line in terms of resistance and reactance. *Any* kind of transmission line can be used with *any* kind of antenna, if the proper measures are taken to couple the two together.

Frequency Range and SWR

Probably the principal factor that determines the way a transmission line is operated is the frequency range over which the antenna is to work. Very few types of antennas will present essentially the same load impedance to the line on harmonically related frequencies. As a result, the builder often is faced with choosing between (1) an antenna system that will permit operating the transmission line with a low SWR, but is confined to one operating frequency or a narrow band of frequencies, and (2) a system that will permit operation in several harmonically related bands but with a large SWR on the line. (There are "multiband" systems which, in principle, make one antenna act as though it were a ½-λ dipole on each of several amateur bands, by using "trap" circuits or multiple wires. Information on these is given in other chapters. Such an antenna can be assumed to be equivalent to a resonant ½-λ dipole on each of the bands for which it is designed, and may be fed through a line, coaxial or otherwise, that has a Z_0 matching the antenna impedance, as described later in this chapter for simple dipoles.)

From a practical standpoint, therefore, methods of coupling the line to the antenna may be divided into two classes. In the first, operation on several amateur bands is the prime consideration and the SWR is secondary. The SWR is normally rather large and the input impedance of the line depends on the line length and the operating frequency.

In the second class, a conscious attempt is made, when necessary, to transform the antenna impedance to a value that matches the characteristic impedance of the line. When this is done the line is operated with a very low SWR and its input impedance is essentially a pure resistance, regardless of the line length. A transmission line can be considered to be "flat," within practical limits, if the SWR is not more than about 1.5 to 1.

Losses

A principal reason for matching the antenna to the line impedance is that a flat line operates with the least power loss. While it is always desirable to reduce losses and thus increase efficiency, the effect of standing waves in this connection is often overemphasized. This is particularly true at the lower amateur frequencies, where the inherent loss in most types of lines is quite low even for runs that, in the average amateur installation, are rather long.

For example, 100 feet of 300-ohm receiving type twin-lead has a loss of only 0.18 dB at 3.5 MHz, as shown in Chapter 24. Even with an SWR as high as 10 to 1 the additional loss caused by standing waves is less than 0.7 dB. Since 1 dB represents the minimum detectable change in signal strength, it does not matter from this standpoint whether the line is flat or not. But at 144 MHz the loss in the same length of line when perfectly matched is 2.8 dB, and an SWR of 10 to 1 would result in an *additional* loss of 3.9 dB. At the higher frequency, then, it *is* worthwhile to match the antenna and line as closely as possible.

Power Limitations

Another reason for matching is that certain types of lines, particularly those with solid dielectric, have definite voltage and current limitations. At the lower frequencies this is a far more compelling reason than power loss for at least approximate matching. Where the voltage and current must not exceed definite maximum values, the amount of power that the line can handle is inversely proportional to the SWR. If the safe rating on the 300-ohm line in the example above is 500 watts when perfectly matched, the line can handle only 50 watts with equal safety when the SWR is 10 to 1. Thus, despite the fact that the line losses are low enough to make no appreciable difference in the signal strength, the high SWR could be tolerated only with low-power transmitters.

Line Radiation

Aside from power considerations, there is a more-or-less common belief that a flat line "does not radiate" while one with a high SWR does radiate. This impression is unjustified. It is true that the radiation from a parallel-conductor line increases with the current in the line, and that the effective line current increases with the SWR. However, the loss by radiation from a properly balanced line is so small (and is, furthermore, independent of the line length) that multiplying it several times still does not bring it out of the "negligible" classification.

Whenever a line radiates it is because of faulty installation (resulting in unbalance with parallel-conductor lines) or "antenna currents" on the line. Radiation from "antenna currents" can take place from either resonant or nonresonant line, parallel-conductor or coaxial.

UNMATCHED SYSTEMS

In many multiband systems or simple antennas where no attempt is made to match the antenna impedance to the characteristic impedance of the line, the customary practice is to connect the line either to the center of the antenna (center feed) as indicated in Fig 1A, or to one end (end feed) as shown in Fig 1B.

Because the line may operate at a rather high SWR, the best type to use is the open-wire line. Solid twin-lead of the 300-ohm receiving variety can also be used, but the power limitations discussed in the preceding section should be kept in mind. Although the manufacturers have placed no power rating on receiving-type 300-ohm line, it seems reasonable to make the assumption, based on the conductor size, that a current of 2 A can readily be carried by the line installed so that there is free air circulating about it. This corresponds to a power of 1200 watts in a matched 300-ohm line. When there are standing waves, the safe power can be found by dividing 1200 by the SWR. In a center-fed ½-λ antenna, as in Fig 1A, the SWR should not exceed about 5 to 1 (at the fundamental

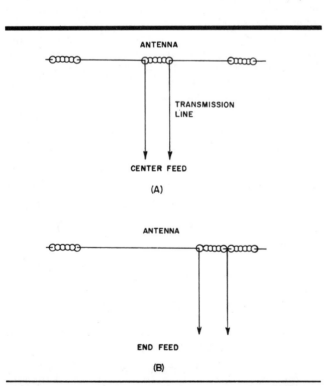

Fig 1—Center and end feed as used in simple antenna systems.

frequency), so receiving type 300-ohm twin-lead would appear to be safe for power outputs up to 250 watts or so.

Since there is little point in using a mismatched line to feed an antenna that is to operate on one amateur band only, the discussion to follow will be based on the assumption that the antenna is to be operated on its harmonics for multiband work.

"Current" and "Voltage" Feed

Usual practice is to connect the transmission line to the antenna at a point where either a current or voltage loop (minimum) occurs. If the feed point is at a current loop the antenna is said to be current fed; if at a voltage loop the antenna is voltage fed.

These terms should not be confused with center feed and end feed, because they do not necessarily have corresponding meanings. Typical cases of current feed are shown at A, B and C in Fig 2. The feed point is at a current loop, which always occurs at the midpoint of a ½-λ section of the antenna. In order to feed at a current loop, the transmission line must be connected at a point that is an odd multiple of ¼ λ from either end of the resonant antenna. A center-fed antenna is also current fed *only* when the antenna length is an *odd* multiple of ½ λ. Thus, the antenna in Fig 2B is both center fed and current fed since it is 3/2 λ long. It would also be center fed and current fed if it were five, seven, etc, half wavelengths long.

There is always a voltage loop at the end of a resonant antenna, no matter what the number of ½ λ, so a resonant end-fed antenna is always voltage fed. This is illustrated at D and E in Fig 2 for end-fed antennas that are ½ λ long (antenna fundamental frequency) and 1 λ long (second harmonic). It would continue to be true for an end-fed antenna operated on any harmonic. However, Fig 2F shows voltage feed at the *center* of the antenna; in this case the antenna has a total length of two ½ λ, each of which is voltage fed. Voltage feed is determined not by the physical position of the transmission line connection to the antenna, but by the fact that a voltage loop occurs on the antenna at the feed point. Since voltage loops always occur at integral multiples of ½ λ from either end of a resonant antenna, feeding the antenna at any ½-λ point constitutes voltage feed.

To current feed a 1-λ antenna, or any resonant antenna having a length that is an *even* multiple of ½ λ, it is necessary to shift the feed point from the center of the antenna (where a voltage loop always occurs in such a case) to the middle of one of the ½-λ sections. This is indicated in Fig 2C in the case of a 1-λ antenna; current feed can be used if the line is connected to the antenna at a point ¼ λ from either end.

Operation on Harmonics

In the usual case of an antenna operated on several bands, the point at which the transmission line is attached is, of course, fixed. The antenna length is usually such that it is resonant at some frequency in the lowest frequency band to be used, and the transmission line is connected either to the center or the end. The current and voltage distribution along antennas fed at both points is shown in Fig 3. With end feed, A to F inclusive, there is always a voltage loop at the feed point. Also, the current distribution is such that in every case the antenna operates as a true harmonic radiator of the type described in Chapter 2.

With center feed, the feed point is always at a current loop on the fundamental frequency and all *odd* multiples of

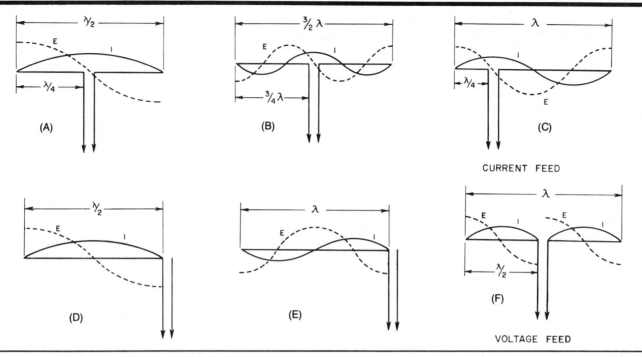

Fig 2—Current and voltage feed in antennas operated at the fundamental frequency, 2 times the fundamental, and 3 times the fundamental. The current and voltage distribution on the antenna are identical with both methods only at the fundamental frequency.

the fundamental. In these cases the current and voltage distribution are identical with the distribution on an end-fed antenna. This can be seen by comparing A and G, C and I, and E and K, Fig 3. (In I, the phase is reversed as compared with C, but this is merely for convenience in drawing; the actual phases of the currents in each ½-λ section reverse each half cycle so it does not matter whether the current curve is drawn above or below the line, so long as the *relative* phases are properly shown in the same antenna.) On odd multiples of the fundamental frequency, therefore, the antenna operates as a true harmonic antenna.

On *even* multiples of the fundamental frequency the feed point with center feed is always at a voltage loop. This is shown at H, J and L in Fig 3. Compare B and H, which show that the current distribution is different with center feed than with end feed. With center feed the currents in both ½-λ sections of the antenna are in the same phase, but with end feed the current in one ½-λ section is in reverse phase to the current in the other. This does not mean that one antenna is a better radiator than the other, but simply that the two will have different directional characteristics. The center-fed arrangement is commonly known as "two half-waves in phase," while the end-fed system is a "1-λ antenna" or "second-harmonic" antenna.

Similarly, the system at J has a different current and voltage distribution than the system at D, although both resonate at four times the fundamental frequency. A similar comparison can be made between F and L. The center-fed arrangement at J really consists of two 1-λ antennas, while the arrangement at L has two 2-λ antennas. These have different directional characteristics than the 2-λ and 4-λ antennas (D and F) that resonate at the same multiple, respectively, of the fundamental frequency.

The reason for this difference between odd and even

multiples of the fundamental frequency in the case of the center-fed antenna can be explained with the aid of Fig 4. Recall from Chapter 2 that the direction of antenna current flow reverses in each half wavelength of wire. Also, in any transmission line the currents in the two wires must always be equal and flowing in opposite directions at any point along the line. Starting from the end of the antenna, the current must be flowing in one direction throughout the first ½-λ section, whether this section is entirely antenna or partly antenna and partly one wire of the transmission line. Thus, in A, Fig 4, the current flows in the same direction from P to Q, since this is all the same conductor. However, one quarter wave is in the antenna and one in the transmission line. The current in the other line wire, starting from R, must flow in the opposite direction in order to balance the current in the first wire, as shown by the arrow. And since the distance from R to S is ½ λ, the current continues to flow in the same direction all the way to S. The currents in the two halves of the *antenna* are therefore flowing in the same direction. Furthermore, the current maximum is ¼ λ from the ends of the antenna, as previously explained, and so both the currents are maximum at the junction of the antenna and transmission line. This makes the current distribution along the length of the antenna exactly the same as with an end fed antenna.

Fig 4B shows the case where the overall length of the antenna is 1 λ, making ½ λ on each side. A half wavelength along the transmission line also is shown. If we assume that the current is flowing downward in the line conductor from Q to R, it must be flowing upward from S to T if the line currents are to balance. However, the distance from Q to P is ½ λ, and so the current in this section of the antenna must flow in the opposite direction to the current flowing in the section from Q to R. The current in section PQ is therefore flowing *away* from Q. Also, the current in section TU must

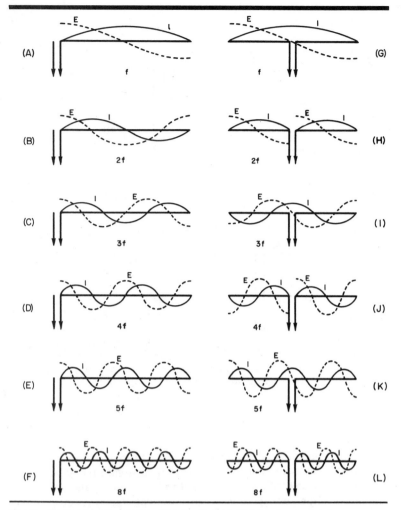

Fig 3—Current and voltage distribution at the fundamental frequency and various multiples, with both end feed and center feed. The distributions are the same with both types of feed only when the frequency is an odd multiple of the fundamental.

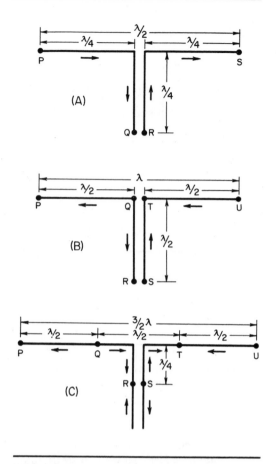

Fig 4—Showing how the type of feed changes from current to voltage, with a center-fed antenna, on twice the fundamental frequency, and back to current feed on three times the fundamental. The same change occurs between all even and odd frequency multiples.

be flowing in the opposite direction to the current in ST, and so is flowing *toward* T. The currents in the two ½-λ sections of the antenna are therefore flowing in the same direction. That is, they are in the same phase.

With the above in mind, the direction of current flow in a 1½ λ antenna, Fig 4C, should be easy to follow. The center ½-λ section QT corresponds to the ½-λ antenna in A. The currents in the end sections, PQ and TU simply flow in the opposite direction to the current in QT. Thus the currents are out of phase in alternate ½-λ sections.

The shift in voltage distribution between odd and even multiples of the fundamental frequency can be demonstrated by a similar method, making allowance for the fact that the voltage is maximum where the current is minimum, and vice versa. On all even multiples of the fundamental frequency there is a current minimum at the junction of the line and antenna, with center feed, because there is an integral number of ½ λ in each side of the antenna. The voltage is maximum at the junction in such a case, and we have voltage feed. Where the multiple of the fundamental is odd, there is always a current maximum at the junction of the transmission line and antenna, as demonstrated by A and C in Fig 4. At these points the voltage is minimum and we therefore have current feed.

"Zepp" or End Feed

In the early days of short-wave communication an antenna consisting of a ½-λ radiator, end-fed through a ¼-λ transmission line, was developed as a trailing antenna for Zeppelin airships. In its use by amateurs, over the years, it has become popularly known as the "Zeppelin" or "Zepp" antenna. The term is now applied to practically any resonant antenna fed at the end by a two-wire transmission line. The term is sometimes also applied loosely to any horizontal antenna, no matter how it is fed.

The mechanism of end feed is perhaps somewhat difficult to visualize, since only one of the two wires of the transmission line is connected to the antenna while the other is simply left free. The difficulty lies in the natural tendency to think in terms of current flow in ordinary electrical circuits, where it is necessary to have a complete loop between both terminals of the power source before any current can flow at all. But as explained in Chapter 2, this limitation applies only to circuits in which the electromagnetic fields reach the most distant part of the circuit in a time interval that is negligible in comparison with the time of one cycle. When the circuit dimensions are comparable with the wavelength, no such complete loop is necessary. *The antenna itself is an example of an "open" circuit in which large currents can flow.*

One way of looking at end feed is to consider the entire length of wire, including both antenna and feeder, as a single unit. For example, suppose we have a wire 1 λ long, as in Fig 5A, fed at a current loop by a source of RF power. The current distribution will be as shown by the curves, with the assumed directions indicated by the arrows. If we now fold back the ¼-λ section to the left of the power source, as shown at B, the overall current distribution will be similar, but the currents in the two wires of the folded section will be flowing in opposite directions. The amplitudes of the currents at any point along the folded-back portion will be equal in the two wires. The folded section, therefore, has become a ¼-λ transmission line, since the fields from the equal and opposite currents cancel. There is, however, nothing to prevent current from continuing to flow in the right-hand ½-λ section, since there was current there before the left-hand section was folded.

This picture, although showing how power can flow from the transmission line to an antenna through end feed, lacks completeness. It does not take into account the fact that the current I1 in the transmission line is greatly different from the current I in the antenna. A more basic viewpoint is the one already mentioned in Chapter 2. The current is caused by electromagnetic fields traveling along the wire and simply constitutes a measurable manifestation of those fields; the current does not cause the fields. From this standpoint the transmission-line conductors merely serve as "guides" for the fields so the electromagnetic energy will go where we want it to go. When the energy reaches the end of the transmission line it meets another guide, in the form of the antenna, and continues along it. However, the antenna is a different form of guide; it has a single conductor while the line has two; it has no provision for preventing radiation while the line is designed for that very purpose. This is simply another way of saying that the impedance of the antenna differs from that of the transmission line, so there will be reflection when the energy traveling along the line arrives at the antenna. We are then back on familiar ground, in that we have a transmission line terminated in an impedance different from its characteristic impedance.

Feeder Unbalance

With end feed, the currents in the two line wires do not balance exactly and there is therefore some radiation from the line. The reason for this is that the current at the end of the free wire is zero (neglecting a small charging current in the insulator at the end) while the current does not go to zero at the junction of the "active" line wire and the antenna. This is because not all the energy going into the antenna is reflected back from the far end, some being radiated; hence the incident and reflected currents cannot completely cancel at a node.

In addition to this unavoidable line radiation a further unbalance will occur if the antenna is not exactly resonant at the operating frequency. If the frequency is too high (antenna too long) the current node does not occur at the junction of the antenna and "live" feeder, but moves out on the antenna. When the frequency is too low the node moves down the active feeder. Since the node on the free feeder has to occur at the end, either case is equivalent to shifting the position of the standing wave along one feeder wire but not the other. The further off resonance the antenna is operating, the greater the unbalance and the greater the line radiation. With center feed this unbalance does not occur, because the system is symmetrical with respect to the line.

To avoid line radiation it is always best to feed the antenna at its center of symmetry. In the case of simple antennas for operation in several bands, this means that center feed should be used. End feed is required only when the antenna is operated on an even harmonic to obtain a desired directional characteristic, and then only when it must be used on more than one band. For single-band operation it is always possible to feed an even-harmonic antenna at a current loop in one of the ½-λ sections nearest the center.

SWR with Wire Antennas

When a line is connected to a single-wire antenna at a current loop the SWR can be estimated with good-enough accuracy from information found in Chapter 2. Although the actual value of the radiation resistance, as measured at a current loop, will vary with the height of the antenna above ground, the theoretical values will at least serve to establish whether the SWR will be high or low.

With center feed the line will connect to the antenna at a current loop on the fundamental frequency and all odd multiples, as shown by Fig 3. At the fundamental frequency and usual antenna heights, the antenna resistance should lie between 50 and 100 ohms, so with a line having a characteristic impedance of 450 ohms the SWR will be $Z_0/R_L = 450/50 = 9$ to 1 as one limit and $450/100 = 4.5$ to 1 as the other. On the third harmonic the theoretical resistance is near 100 ohms, so the SWR should be about 4.5 to 1. For 300-ohm line the SWR can be expected to be between 3 and 6 on the antenna fundamental and about 3 to 1 on the third harmonic.

The impedances to be expected at voltage loops are less readily determined. Theoretical values are in the neighborhood of 5000 to 8000 ohms, depending on the antenna conductor size and the number of half wavelengths along the wire. Such experimental figures as are available indicate a lower order of resistance, with measurements and estimates running from 1000 to 5000 ohms. In any event, there will be some difference between end feed and center feed, since the current distribution on the antenna is different in these two cases at any given even multiple of the fundamental frequency. Also, the higher the multiple the lower the resistance at a voltage loop, so the SWR can be expected to decrease when an antenna is operated at a high multiple of its fundamental frequency. Assuming 4000 ohms for a wire antenna two ½ λ long, the SWR would be about 6 or 7 with a 600-ohm line

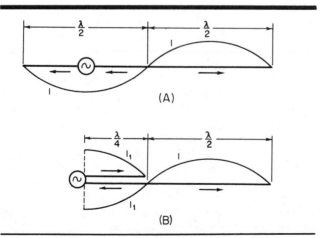

Fig 5—Folded-antenna analogy of transmission line for an end-fed antenna.

and around 12 with a 300-ohm line. However, considerable variation is to be expected.

ANTENNA CURRENTS ON TRANSMISSION LINES

In any discussion of transmission-line operation it is always assumed that the two conductors carry equal and opposite currents throughout their length. This is an ideal condition that may or may not be realized in practice. In the average case the chances are rather good that the currents will *not* be balanced unless special precautions are taken. Whether the line is matched or not has little to do with the situation.

Consider the ½-λ antenna shown in Fig 6 and assume that it is somehow fed by a source of power at its center, and that the instantaneous direction of current flow is as indicated by the arrows. In the neighborhood of the antenna is a group of conductors disposed in various ways with respect to the antenna itself. All of these conductors are in the field of the antenna and are therefore coupled to it. Consequently, when current flows in the antenna a voltage will be induced in each conductor. This causes a current flow determined by the induced voltage and the impedance of the conductor.

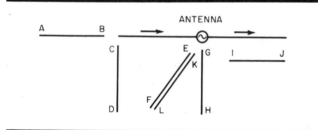

Fig 6—Coupling between antenna and conductors in the antenna field.

The degree of coupling depends on the position of the conductor with respect to the antenna, assuming that all the conductors in the figure are the same length. The coupling between the antenna and conductor IJ is greater than in any other case, because IJ is close to and parallel with the antenna. Ideally, the coupling between conductor GH and the antenna is zero, because the voltage induced by current flowing in the left-hand side of the antenna is exactly balanced by a voltage of opposite polarity induced by the current flowing in the right-hand side. This is because the two currents are flowing in opposite directions with respect to GH. Complete cancellation of the induced voltages can occur, of course, only if the currents in the two halves of the antenna are symmetrically distributed with respect to the center of the antenna, and also only if every point along GH is equidistant from any two points along the antenna that are likewise equidistant from the center. This cannot be true of any of the other conductors shown, so a finite voltage will be induced in any conductor in the vicinity of the antenna except one perpendicular to the antenna at its center.

Transmission Line in the Antenna Field

Now consider the two conductors EF and KL, which are parallel and very close together. Except for the negligible spacing between them, the two conductors lie in the same position with respect to the antenna. Therefore, identical voltages will be induced in both, and the resulting currents will be flowing *in the same direction in both conductors*. It is only a short step to visualizing conductors EF and KL as the two conductors of a section of transmission line in the vicinity of the antenna. Because of coupling to the antenna, it is not only possible but *certain* that a voltage will be induced in the two conductors of the transmission line in parallel. The resulting current flow is in the same direction in both conductors, whereas the true transmission line currents are always flowing in opposite directions at each point along the line. These "parallel" currents are of the same nature as the current in the antenna itself, and hence are called "antenna" currents on the line. They are responsible for most of the radiation that takes place from transmission lines.

When there is an antenna current of appreciable amplitude on the line it will be found that not only are the line currents unbalanced but the apparent SWR is different in each conductor, and that the loops and nodes of current in one wire do not occur at corresponding points in the other wire. Under these conditions it is impossible to measure the true SWR.

It should be obvious from Fig 6 that only in the case of a center-fed antenna can the coupling between the line and antenna be reduced to zero. There is always some such coupling when the antenna is end fed, so there is always the possibility that antenna currents of appreciable amplitude will exist on the line, contributing further to the inherent line unbalance in the end-fed arrangement. But the center-fed system will also have appreciable antenna-to-line coupling if the line is not brought off at right angles to the antenna for a distance of at least a ½ λ.

Antenna currents will be induced on lines of any type of construction. If the line is coax, the antenna current flows only on the *outside* of the outer conductor; no current is induced *inside* the line. However, an antenna current on the outside of coax is just as effective in causing radiation as a similar current induced in the two wires of a parallel-conductor line.

Detuning the Line for Antenna Currents

The antenna current flowing on the line as a result of voltage induced from the antenna will be small if the overall circuit, considering the line simply as a single conductor, is not resonant at the operating frequency. The frequency (or frequencies) at which the system is resonant depends on the total length and whether the transmission line is grounded or not at the transmitter end.

If the line is connected to a coupling circuit that is not grounded, either directly or through a capacitance of more than a few picofarads, it is necessary to consider only the length of the antenna and line. In the end-fed arrangement, shown at A in Fig 7, the line length, L, should not be an integral multiple or close to such a multiple of ½ λ. In the center-fed system, Fig 7B, the length of the line plus *one side* of the antenna should not be a multiple of ½ λ. In this case the two halves of the antenna are simply in parallel so far as resonance for the induced "antenna" current on the line is concerned, because the line conductors themselves act in parallel. When the antenna is to be used in several bands, resonance of this type should be avoided at all frequencies to be used.

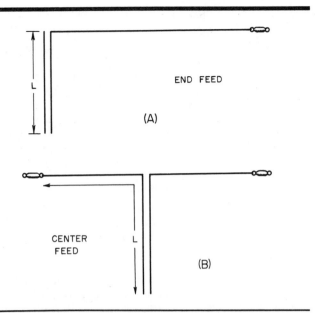

Fig 7—L indicates the important length for resonance to antenna currents coupled from the antenna to the line. In the center-fed system at B, one side of the antenna is part of the "parallel"-resonant system.

Transmission lines usually have bends, are at varying heights above ground, and so on, all of which will modify the resonant frequency. It is advisable to check the system for resonance at and near all operating frequencies before assuming that the line is safely detuned for antenna currents. This can be done by temporarily connecting the ends of the line together and coupling them through a small capacitance (not more than a few picofarads) to a resonance indicator such as a dip meter. Very short leads should be used between the meter and antenna. Fig 8 shows the method. Once the resonance points are known it is a simple matter to prune the feeders to get as far away as possible from resonance at any frequency to be used.

Resonances in systems in which the coupling apparatus is grounded at the transmitter are not so easily predicted. The "ground" in such a case is usually the metal chassis of the transmitter itself, and not actual ground. In the average amateur station it is not possible to get a connection to real

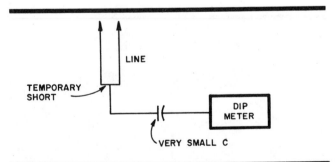

Fig 8—Using a dip meter to check resonance of the antenna system for antenna currents on the transmission line. The small capacitance may be a short length of wire connected to the feed line, coupled lightly to the dip-oscillator coil with a 1-turn loop.

ground without having a lead that is an appreciable fraction of wavelength long. At the higher frequencies, and particularly in the VHF region, the distance from the transmitter to ground may be a wavelength or more. Probably the best plan in such cases is to make the length L in Fig 7 equal to a multiple of $\frac{1}{2}\lambda$. If the transmitter has fairly large capacitance to ground, a system of this length will be effectively detuned for the fundamental and all even harmonics when grounded to the transmitter at the coupling apparatus. However, the resonance frequencies will depend on the arrangement and constants of the coupling system even in such a case, and preferably should be checked by means of the dip meter. If this test shows resonance at or near the operating frequency, alternative grounds (to a heating radiator, for example) should be tried until a combination is found that detunes the whole system.

It should be quite clear, from the mechanism that produces antenna currents on a transmission line, that such currents are entirely independent of the normal operation as a true transmission line. It does not matter whether the line is perfectly matched or is operated with a high SWR. Nor does it matter what kind of line is used, air insulated or solid dielectric, parallel conductor or coax. In every case, the antenna currents should be minimized by detuning the line if the line is to fulfill only its primary purpose of transferring power to the antenna.

Other Causes of Unbalance

Unbalance in center-fed systems can arise even when the line is brought away at right angles to the antenna for a considerable distance. If both halves of the antenna are not symmetrically placed with respect to nearby conductors (such as power and telephone wires and downspouting) the antenna itself becomes unbalanced and the current distribution is different in the two halves. Because of this unbalance a voltage will be induced in the line even if the line is symmetrical with respect to the antenna.

MATCHED LINES

Operating the transmission line at a low SWR requires that the line be terminated, at its output end, in a resistive load matching the characteristic impedance of the line as closely as possible. The problem can be approached from two standpoints: (1) selecting a transmission line having a characteristic impedance that matches the antenna resistance at the point of connection; or (2) transforming the antenna resistance to a value that matches the Z_0 of the line selected.

The first approach is simple and direct, but its application is limited because the antenna impedance and line impedance are alike only in a few special cases. The second approach provides a good deal of freedom in that the antenna and line can be selected independently. Its disadvantage is that it is more complicated constructionally. Also, it sometimes calls for a somewhat tedious routine of measurement and adjustment before the desired match is achieved.

Operating Considerations

As pointed out earlier in this chapter, most antenna systems show a marked change in resistance when going from the fundamental to multiples of the fundamental frequency. For this reason it is usually possible to match the line impedance only on one frequency. A matched antenna system is consequently a one-band affair, in most cases. It can, however, usually be operated over a fair frequency range in a given band. The frequency range over which the SWR is low is determined by how rapidly the impedance changes as the frequency is changed. If the change in impedance is small for a given change in frequency, the SWR will be low over a fairly wide band of frequencies. However, if the impedance change is rapid (a sharply resonant or high-Q antenna—see discussion of Q later in this chapter) the SWR will also rise rapidly as the operating frequency is shifted away from antenna resonance, where the line is matched.

Antenna Resonance

A point that needs emphasis in connection with matching the antenna to the line is that, with the exception of a few special cases discussed later in this chapter, the impedance at the point where the line is connected must be a *pure resistance*. This means that the antenna system must be resonant at the frequency for which the line is to be matched. (Some types of long-wire antennas are exceptions, in that their input impedances are resistive over a wide band of frequencies. Such systems are essentially nonresonant.) The higher the Q of the antenna system, the more essential it is that exact resonance be established before an attempt is made to match the line. This is particularly true of close-spaced parasitic arrays. With simple dipole and harmonic antennas, the tuning is not so critical and it is usually sufficient to cut the antenna to the length given by the appropriate equation in Chapter 2. The frequency should be selected to be at the center of the range of frequencies (which may be the entire width of an amateur band) over which the antenna is to be used.

DIRECT MATCHING

As discussed in Chapter 2, the impedance at the center of a resonant ½-λ antenna at heights of the order of ¼ λ and more is resistive and is in the neighborhood of 70 ohms. This is fairly well matched by transmitting-type twin-lead having a characteristic impedance of 75 ohms. It is possible, therefore, to operate with a low SWR using the arrangement shown in Fig 9. No precautions are necessary beyond those already described in connection with antenna-to-line coupling.

This system is badly mismatched on *even* multiples of the fundamental frequency, since the feed in such cases is at a high-impedance point. However, it is reasonably well matched at *odd* multiples of the fundamental. For example, an antenna that is resonant near the low-frequency end of the 7-MHz band will operate with a low SWR over the 21-MHz band (three times the fundamental).

The same method may be used to feed a harmonic antenna at any current loop along the wire. For lengths up to 3 or 4 λ the SWR should not exceed 2 to 1 if the antenna is ¼ or ½ λ above ground.

At the fundamental frequency the SWR should not exceed about 2 to 1 within a frequency range ±2% from the frequency of exact resonance. Such a variation corresponds approximately to the entire width of the 7-MHz band, if the antenna is resonant at the center of the band. A wire antenna is assumed. Antennas having a greater ratio of diameter to length will have a lower change in SWR with frequency.

Coaxial Cable

Instead of using twin-lead as just described, the center of a ½-λ dipole may be fed through 75-ohm coaxial cable such as RG-11, as shown in Fig 10. Cable having an impedance of approximately 52 ohms, such as RG-8, also may be used, particularly in those cases where the antenna height is such as to lower the radiation resistance of the antenna, below ¼ λ. (See Chapter 3.) The principle is exactly the same as with twin-lead, and the same remarks as to SWR apply. However, there is a considerable practical difference between the two types of line. With the parallel-conductor line the system is symmetrical, but with coaxial line it is inherently unbalanced.

Stated broadly, the unbalance with coaxial line is caused by the fact that the outside of the outer conductor is not coupled to the antenna in the same way as the inner conductor and the inside of the outer conductor. The overall result is that current will flow on the outside of the outer conductor in the simple arrangement shown in Fig 10. The unbalance is rather small if the line diameter is very small compared with the length of the antenna, a condition that is met fairly well at the lower amateur frequencies. It is not negligible in the VHF and UHF range, however, nor should it be ignored at 28 MHz. The current that flows on the outside of the line because of this unbalance, it should be noted, does not arise from the same type of coupling as the "antenna" current previously discussed. The coupling pictured in Fig 6 can still occur, *in addition*. However, the remedy is the same in both cases—the system must be detuned for currents on the outside of the line. This can be done by an actual resonance check using the method shown in Fig 8.

BALANCING DEVICES

The unbalanced coupling described in the preceding

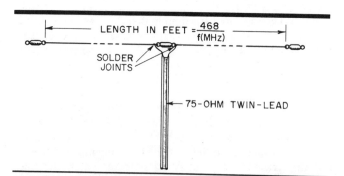

Fig 9—A ½-λ dipole fed with 75-ohm twin-lead, giving a close match between antenna and line impedance. The leads in the "Y" from the end of the line to the ends of the center insulator should be as short as possible.

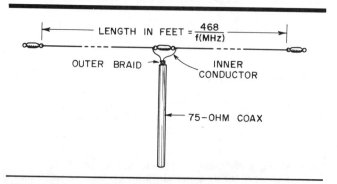

Fig 10—A ½-λ antenna fed with 75-ohm coaxial cable. The outside of the outer conductor of the line may be grounded for lightning protection.

paragraph can be nullified by the use of devices that prevent the unwanted current from flowing on the outside of the coaxial line. This may be done either by making the current cancel itself out or by choking it off. Devices of this type fall in a class of circuits usually termed *baluns*, a contraction for "balanced to unbalanced." The baluns described in Chapter 25 perform the same function, but the techniques described here are generally more suitable for mechanical reasons in coupling the line to the antenna.

The voltages at the antenna terminals in Fig 10 are equal in amplitude with respect to ground but opposite in phase. Both these voltages act to cause a current to flow on the outside of the coax, and if the currents produced by both voltages were equal, the resultant current on the outside of the line would be zero since the currents are out of phase and would cancel each other. But since one antenna terminal is directly connected to the cable shield while the other is only weakly coupled to it, the voltage at the directly connected terminal produces a much larger current, and so there is relatively little cancellation.

The two currents could be made equal in amplitude by making a direct connection between the outside of the line and the antenna terminal that is connected to the inner conductor, but if it were done right at the antenna terminals the line and antenna would be short-circuited. If the connection is made through a conductor parallel to the line and ¼ λ long, as shown in Fig 11A, the second conductor and the outside of the line act as a ¼-λ "insulator" for the normal voltage and current at the antenna terminals. (This is because a line that is ¼ λ long and short-circuited at the far end exhibits a very high resistive impedance, as explained in Chapter 24. On the other hand, any unbalanced current flowing on the outside of the line because of the direct connection between it and the antenna has a counterpart in an equal current flowing on the second conductor, because the latter is directly connected to the *other* antenna terminal. Where the two conductors are joined together at the bottom, the resultant of the two currents is zero, since they are of opposite phase. Thus, no current flows on the remainder of the transmission line.

Note that the length of the extra conductor has no particular bearing on its operation in balancing out the undesired current. The length is critical only with respect to preventing the normal operation of the antenna from being upset.

Combined Balun and Matching Stub

In certain antenna systems the balun length can be considerably shorter than ¼ λ; the balun is, in fact, used as part of the matching system. This requires that the radiation resistance be fairly low as compared with the line Z_0 so that a match can be brought about by first shortening the antenna to make it have a capacitive reactance, and then using a shunt inductor across the antenna terminals to resonate the antenna and simultaneously raise the impedance to a value equal to the line Z_0. (See later section on matching stubs.) The balun is then made the proper length to exhibit the desired value of inductive reactance.

The basic matching method is shown at A in Fig 12, and the balun adaptation to coaxial feed is shown at B. The matching stub in B is a parallel-line section, one conductor of which is the outside of the coax between point X and the antenna; the other stub conductor is an equal length of wire. (A piece of coax may be used instead, as in the balun in Fig 11A.) The spacing between the stub conductors can be 2 to 3 inches. The stub of Fig 12 is ordinarily much shorter than ¼ λ, and the impedance match can be adjusted by altering the stub length along with the antenna length. With simple coax feed, even with a ¼-λ balun as in Fig 11, the match depends entirely on the actual antenna impedance and the Z_0 of the cable; no adjustment is possible.

Adjustment

When a ¼-λ balun is used it is advisable to resonate it before connecting the antenna. This can be done without much difficulty if a dip meter is available. In the system shown in Fig 11A, the section formed by the two parallel pieces of line should first be made slightly longer than the length given

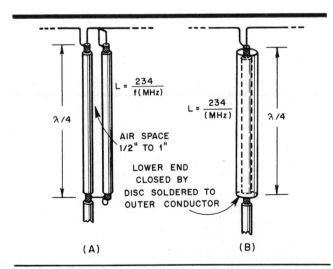

Fig 11—Methods of balancing the termination when a coaxial cable is connected to a balanced antenna.

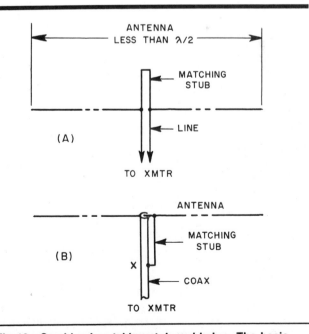

Fig 12—Combined matching stub and balun. The basic arrangement is shown at A. B shows the balun arrangement achieved by using a section of the outside of the coax feed line as one conductor of a matching stub.

by the equation. The shorting connection at the bottom may be installed permanently. With the dip meter coupled to the shorted end, check the frequency and cut off small lengths of the shield braid (cutting both lines equally) at the open ends until the stub is resonant at the desired frequency. In each case leave just enough inner conductor remaining to make a short connection to the antenna. After resonance has been established, solder the inner and outer conductors of the second piece of coax together and complete the connections indicated in Fig 11A.

Another method is to first adjust the antenna length to the desired frequency, with the line and stub disconnected, then connect the balun and recheck the frequency. Its length may then be adjusted so that the overall system is again resonant at the desired frequency.

Construction

In constructing a balun of the type shown in Fig 11A, the additional conductor and the line should be maintained parallel by suitable spacers. It is convenient to use a piece of coax for the second conductor; the inner conductor can simply be soldered to the outer conductor at both ends since it does not enter into the operation of the device. The two cables should be separated sufficiently so that the vinyl covering represents only a small proportion of the dielectric between them. Since the principal dielectric is air, the length of the ¼-λ section is based on a velocity factor of 0.95, approximately.

Detuning Sleeves

The detuning sleeve shown in Fig 11B also is essentially an air-insulated ¼-λ line, but of the coaxial type, with the sleeve constituting the outer conductor and the outside of the coax line being the inner conductor. Because the impedance at the open end is very high, the unbalanced voltage on the coax line cannot cause much current to flow on the outside of the sleeve. Thus the sleeve acts like a choke coil in isolating the remainder of the line from the antenna. (The same viewpoint can be used in explaining the action of the ¼-λ arrangement shown at A, but is less easy to understand in the case of baluns less than ¼ λ long.)

A sleeve of this type may be resonated by cutting a small longitudinal slot near the bottom, just large enough to take a single-turn loop which is, in turn, link-coupled to the dip meter. If the sleeve is a little long to start with, a bit at a time can be cut off the top until the stub is resonant.

The diameter of the coaxial detuning sleeve in B should be fairly large compared with the diameter of the cable it surrounds. A diameter of two inches or so is satisfactory with half-inch cable. The sleeve should be symmetrically placed with respect to the center of the antenna so that it will be equally coupled to both sides. Otherwise a current will be induced from the antenna to the outside of the sleeve. This is particularly important at VHF and UHF.

In both the balancing methods shown in Fig 11 the ¼-λ section should be cut to be resonant at exactly the same frequency as the antenna itself. These sections tend to have a beneficial effect on the impedance-frequency characteristic of the system, because their reactance varies in the opposite direction to that of the antenna. For instance, if the operating frequency is slightly below resonance the antenna has capacitive reactance, but the shorted ¼-λ sections or stubs have inductive reactance. Thus the reactances tend to cancel, which prevents the impedance from changing rapidly and helps maintain a low SWR on the line over a band of frequencies.

Impedance Step-up Balun

A coax-line balun may also be constructed to give an impedance step-up ratio of 4:1. This form of balun is shown in Fig 13. If 75-Ω line is used, as indicated, the balun will provide a match for a 300-Ω terminating impedance. The U-shaped section of line must be an electrical ½ λ long, taking the velocity factor of the line into account. In most installations using this type of balun, it is customary to roll up the length of line represented by the U-shaped section into a coil of several inches in diameter. The coil turns may be bound together with electrical tape. Because of the bulk and weight of the balun, this type is seldom used with wire-line antennas suspended by insulators at the antenna ends. More commonly it is used with multielement antennas, where its weight may be supported by the boom of the antenna system.

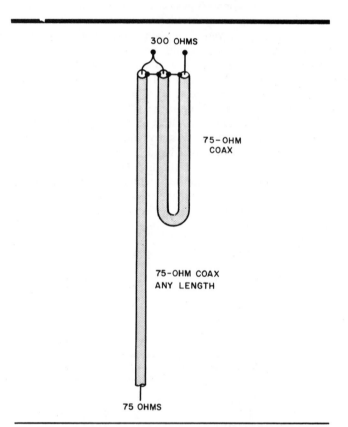

Fig 13—A balun that provides an impedance step-up ratio of 4:1. The electrical length of the U-shaped section of line is ½ λ.

Coax-Line RF Choke

As was discussed earlier in this section, the unbalanced coupling that results from connecting coaxial line to a balanced antenna may be nullified by choking off the current from flowing on the outside of the feed line. A direct approach to this objective is shown in Fig 14; where the line itself is formed into a coil at the antenna feed point. Ten turns of coax line coiled at a diameter of 6 to 8 inches has been found

Fig 14—An RF choke formed by coiling the feed line at the point of connection to the antenna. Electrical tape may be used to hold the coils in place.

effective for the HF bands. The turns may be secured in a tight coil with electrical tape. This approach offers the advantage of requiring no pruning adjustments and no balun losses, other than in the few additional feet of coaxial line. The effectiveness of a choke of this sort decreases at the higher frequencies, however, because of the distributed capacitance among the turns.

Ferrite-Core Baluns

The frequency range of the coax-line RF choke, or choke balun, can be extended to well below 2.0 MHz by using a core of high-permeability ferrite instead of air. Looping the turns of coaxial line through a large toroidal form is a simple means of making a balun of this type. With higher core permeability, the choke inductance increases dramatically, thereby retaining the high reactance needed to minimize current flowing on the outside of the coax braid at the lower frequencies. Of great importance, no core saturation occurs at high power levels in the choke balun (a serious problem in transformer-type baluns), because the core excitation is low level, produced only by current on the outside, and not by the high internal current that feeds the antenna.

Bead Baluns

Greatly improved choke balun performance can be realized by placing several ferrite beads or sleeves of even higher permeability around the coaxial feed line. This concept was introduced by Walter Maxwell, W2DU, in *QST* for March 1983 (see bibliography at the end of this chapter.)

Bead materials of various size and RF characteristics are available that dramatically increase both the reactance and resistance of a conductor. (Adding resistance to the reactance in this circuit improves the operational bandwidth of the balun with no increase in loss.) In general, the impedance of the outer coaxial braid surface increases almost proportionately with the number of beads placed over it. A combination of 52-Ω teflon-dielectric RG-303 cable (RG-141, with the fabric covering removed) and ferrite beads having an ID of 0.197 inch and a length of 0.190 inch, form a superb, compact, wide-band balun. While the inner conductor and shield of the coaxial cable remain unaffected, the beads introduce a high impedance in series with the braid outer surface. This configuration effectively isolates the external output terminal of the feed line from that at the input end.

Maxwell made a test balun by slipping 300 no. 73 beads (μ = 2500 to 4000) over a piece of RG-303 coaxial cable. The impedance of the outer conductor of the cable measured 4500 + *j*3800 Ω at 4.0 MHz. With a single bead the impedance measured 15.6 + *j*13.1 Ω. For practical baluns for 1.8 to 30 MHz (less than 12 inches long, including connector), use 50 no. 73 beads (Amidon no. FB-73-2401 or Fair-Rite no. 2673002401-0). For 30 to 250 MHz, use 25 no. 43 beads (μ = 950 to 3000, Amidon no. FB-43-2401 or Fair-Rite no. 2643002401). No. 64 beads (μ = 250 to 375) are recommended for use above 200 MHz, but Maxwell did not experiment with them. The coaxial cable need be only long enough to hold the beads, and to access the end connectors.

The graphs in Fig 15 show the frequency versus measured values of series resistance (R), reactance (X) and impedance (Z) of the outer braid surface of a choke balun, for bead baluns containing both 25 and 50 beads. With either balun, external currents will be negligible.

At full legal input levels, no power-handling problems will arise using these baluns, because the CW power-handling capability of the cable is 3.5 kW at 50 MHz, and 9 kW at 10 MHz. Any suitable connector that will mate with the load end of your feed line can be used at the input of the balun, and the balanced-output terminals may simply be pigtails formed by the inner and outer conductors of the feed line.

PATTERN DISTORTION

Fig 16 shows the classic "figure-eight" radiation pattern of a ½-λ dipole in free space. This is actually an idealized

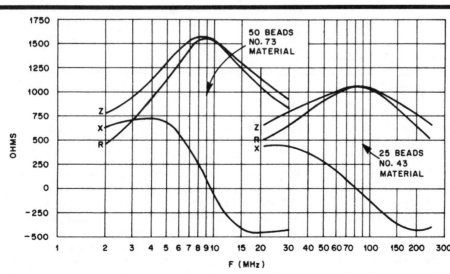

Fig 15—Graph of frequency versus series impedance of coaxial-balun shield outer surface.

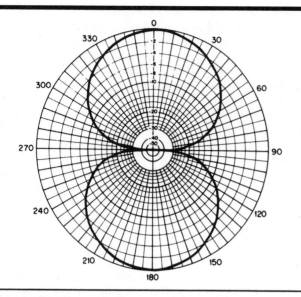

Fig 16—Classic response pattern of a ½-λ dipole in free space. The concentric-circle scale is indicated in decibels down, relative to the response in a broadside direction from the axis of the dipole. The outer scale shows degrees of departure from one broadside direction. The axis of the conductor is common with the line between the 90° and 270° outer scale markings.

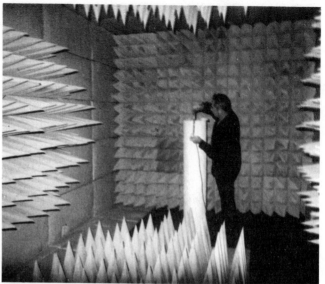

Fig 17—Positioning one of the test antennas on the rotatable Styrofoam support in the RF anechoic chamber.

pattern based on current flowing only in the antenna. As was stated earlier, current may flow on the outside of coaxial cable used to feed a dipole directly. That current will cause radiation and hence a distortion of the theoretical pattern.

Bruce Eggers, WA9NEW, conducted tests in the radio-frequency anechoic chamber at North Carolina State University to determine dipole patterns. The results, published in April 1980 *QST*, are summarized here. An RF anechoic chamber is simply a room in which the walls, floor and ceiling are covered with a material that is designed to break up an electromagnetic wave and absorb its energy. An antenna placed in such a chamber can not "see," or be influenced by, any surface or objects that can reflect or reradiate electromagnetic energy. It is a simulation of "free space" right here on earth.

Two antennas were used for the tests. The source of RF was a ½-λ balun-fed dipole, mounted horizontally at one end of the chamber. The type of balun used is shown in Fig 11B. It was mounted at the same height as the receiving antenna and fed a few milliwatts of power at 1.6 GHz. The test antennas were then mounted, one at a time, horizontally, at the other end of the chamber, on a rotating support. The supports for both antennas were made of Styrofoam. The test antennas were then rotated a full 360°. The received signal was carried to the receiver outside of the chamber on a coaxial feed line. The feed line dropped away from the antenna perpendicularly for a distance of about 9 λ. The chamber is shown in Fig 17.

Fig 18 shows the pattern of the balun-fed antenna. The signal level in the nulls off the ends of the antenna is about 32 dB below the "broadside" signal level. Noise precluded indentifying nulls significantly deeper than that level with the setup used.

Fig 19 shows the pattern of a dipole without the benefit of the balun. The peak amplitude of the signal is about 5 dB

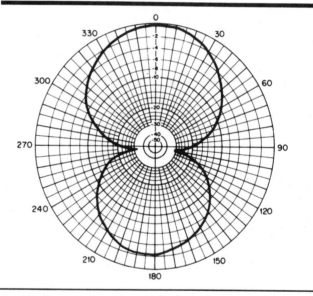

Fig 18—Response pattern of the balun-fed ½-λ dipole in the RF anechoic chamber. The apparent front-to-back ratio exists in part because the antenna was not located at the exact center of the rotating support. This response and that of Fig 19 are drawn to the same relative scale.

below that of the balun-fed antenna and one of the nulls, 30° from broadside, is just as deep as was the null off the end of the balun-fed antenna. A couple of points about this trace need to be considered. First, the exact location of peaks and nulls is highly dependent on the relative location of the feed line as the antenna is rotated. In repeating the experiment with a different relative position of either, the pattern changed. Fig 19 can be considered as only representative of how a ½-λ dipole performs as a receiving antenna when used without a balun and when used with a long feed line. Second, the overall drop in signal level is not necessarily representative of what

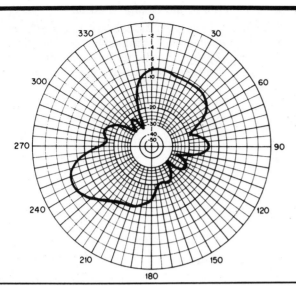

Fig 19—Response pattern of the ½-λ dipole without a balun. The pattern changed significantly during tests if the coaxial feed line was relocated, no doubt caused by changes in the amplitude and phase of currents flowing on the outside of the line.

you should expect from the antenna in a transmitting application. Reciprocity notwithstanding, antenna currents flowing on the outside of the coax are, in general, lost to the receiver. These same currents, in the transmitting mode, can radiate energy that effectively fills in the nulls noted here. The pattern of Fig 18 is fully predictable and can be easily reproduced in a repeated experiment. That of Fig 19 cannot.

A balun gives a predictable pattern. The biggest benefit that accrues from this feature is applicable to using a balanced element in a directional array.

The above should not necessarily be interpreted to mean that installing a balun on a 3.5-MHz dipole is going to result in any detectable differences. Antennas interact with all kinds of reflecting and reradiating objects. Every piece of material in the vicinity of an antenna has an effect. The pattern of a 3.5-MHz dipole might not look as bad as Fig 19 does, but it probably doesn't look like Fig 18, either. The majority of the variations between the pattern of a practical antenna and an idealized pattern, at least with regard to simple antennas on the lower frequencies, will result from objects in the near field of the antenna. The additional variations introduced as a result of not using a balun in an application of a coaxial-fed balanced antenna will become most significant at higher frequencies with multielement antennas.

QUARTER-WAVE TRANSFORMERS

The impedance-transforming properties of a ¼-λ transmission line can be used to good advantage in matching the antenna impedances to the characteristic impedance of the line. As described in Chapter 24, the input impedance of a ¼-λ line that is terminated in a resistive impedance Z_R is

$$Z_i = \frac{Z_0^2}{Z_L}$$

where

Z_i = the impedance at the input end of the line
Z_0 = the characteristic impedance of the line
Z_L = the impedance at the load end of the line

Rearranging this equation gives

$$Z_0 = \sqrt{Z_i Z_L}$$

This means that any value of load impedance Z_L can be transformed into any desired value of impedance Z_i at the input terminals of a ¼-λ line, provided the line can be constructed to have a characteristic impedance Z_0 equal to the square root of the product of the other two impedances. The factor that limits the range of impedances that can be matched by this method is the range of values for Z_0 that is physically realizable. The latter range is approximately 52 to 600 Ω. Practically any type of line can be used for the matching section, including both air-insulated and solid-dielectric lines.

One application of this type of matching section is in feeding a ½-λ antenna with a 600-Ω line, as shown in Fig 20. Assuming that the antenna has a resistive impedance in the vicinity of 65 Ω, the required Z_0 of the matching section is $\sqrt{65 \times 600}$ or approximately 197 Ω. A section of this type of line can be constructed of parallel tubing, from the data in Chapter 24.

The ¼-λ transformer may be adjusted to resonance before being connected to the antenna by short-circuiting one end and coupling that end inductively to a dip meter. The length of the short-circuiting conductor lowers the frequency slightly, but this can be compensated for by adding half the length of the shorting bar to each conductor after resonating, measuring the shorting-bar length between the centers of the conductors.

Driven Beam Elements

Another application for the ¼-λ "linear transformer" is in matching the very low antenna impedances encountered in close-spaced directional arrays to a transmission line having

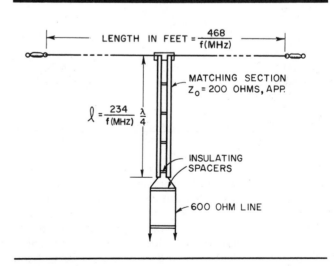

Fig 20—Matching a ½-λ antenna to a 600-Ω line through a ¼-λ linear transformer.

a characteristic impedance of 300 to 600 Ω. The observed impedances at the antenna feed point in such cases range from about 8 to 20 Ω. A matching section having a Z_0 of 75 Ω is useful with such arrays. The impedance at its input terminals will vary from approximately 700 Ω with an 8-Ω load to 280 Ω with a 20-Ω load.

Transmitting twin-lead is suitable for this application; such a short length is required that the loss in the matching section should not exceed about 0.6 dB even though the SWR in the matching section may be almost 10 to 1 in the extreme case.

SERIES-SECTION TRANSFORMERS

The series-section transformer has advantages over either stub tuning or the ¼-λ transformer. Illustrated in Fig 21, the series-section transformer bears considerable resemblance to the ¼-λ transformer. (Actually, the ¼-λ transformer is a special case of the series-section transformer.) The important differences are (1) that the matching section need not be located exactly at the load, (2) the matching section may be less than a quarter wavelength long, and (3) there is great freedom in the choice of the characteristic impedance of the matching section.

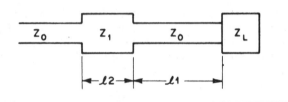

Fig 21—Series-section transformer Z_1 for matching transmission-line Z_0 to load, Z_L.

In fact, the matching section can have *any* characteristic impedance that is not too close to that of the main line. Because of this freedom, it is almost always possible to find a length of commercially available line that will be suitable as a matching section. As an example, consider a 75-Ω line, a 300-Ω matching section, and a pure-resistance load. It can be shown that a series-section transformer of 300-Ω line may be used to match *any* resistance between 5 Ω and 1200 Ω to the main line.

Frank Regier, OD5CG, described series-section transformers in July 1978 *QST*. This information is based on that article. The design of a series-section transformer consists of determining the length $\ell 2$ of the series or matching section and the distance $\ell 1$ from the load to the point where the section should be inserted into the main line. Three quantities must be known. These are the characteristic impedances of the main line and of the matching section, both assumed purely resistive, and the complex-load impedance. Either of two design methods may be used. One is a graphic method using the Smith Chart, and the other is algebraic. You can take your choice. (Of course the algebraic method may be adapted to obtaining a computer solution.) The Smith Chart graphic method is described in Chapter 28.

Algebraic Design Method

The two lengths $\ell 1$ and $\ell 2$ are to be determined from the characteristic impedances of the main line and the matching section, Z_0 and Z_1, respectively, and the load impedance $Z_L = R_L + jX_L$. The first step is to determine the normalized impedances.

$$n = \frac{Z_1}{Z_0}$$

$$r = \frac{R_L}{Z_0}$$

$$x = \frac{X_L}{Z_0}$$

Next, $\ell 2$ and $\ell 1$ are determined from

$$\ell 2 = \arctan B \text{ where}$$

$$B = \pm\sqrt{\frac{(r-1)^2 + x^2}{r\left(n - \frac{1}{n}\right)^2 - (r-1)^2 - x^2}}$$

$$\ell 1 = \arctan A \text{ where}$$

$$A = \frac{\left(n - \frac{r}{n}\right)B + x}{r + xnB - 1}$$

Lengths $\ell 2$ and $\ell 1$ as thus determined are electrical lengths in degrees (or radians). The electrical lengths in wavelengths are obtained by dividing by 360° (or by 2π radians). The physical lengths (main line or matching section, as the case may be), are then determined from multiplying by the free-space wavelength and by the velocity factor of the line.

The sign of B may be chosen either positive or negative, but the positive sign is preferred because it results in a shorter matching section. The sign of A may not be chosen but can turn out to be either positive or negative. If a negative sign occurs and a computer or electronic calculator is then used to determine $\ell 1$, a negative electric length will result for $\ell 1$. If this happens, add 180°. The resultant electrical length will be correct both physically and mathematically.

In calculating B, if the quantity under the radical is negative, an imaginary value for B results. This would mean that Z_1, the impedance of the matching section, is too close to Z_0 and should be changed.

Limits on the characteristic impedance of Z_1 may be calculated in terms of the SWR produced by the load on the main line without matching. For matching to occur, Z_1 should either be greater than $Z_0\sqrt{SWR}$ or less than $Z_0/\sqrt{SWR}$.

An Example

As an example, suppose we want to feed a 29-MHz ground plane vertical antenna with RG-58 type foam-dielectric coax (Fig 22). We'll assume the antenna impedance to be 36 Ω, pure resistance, and use a length of RG-59 foam-dielectric coax as the series section.

Z_0 is 52 Ω, Z_1 is 75 Ω, and both cables have a velocity factor of 0.79. Because the load is a pure resistance we may determine the SWR to be 52/36 = 1.444. From the above, Z_1 must have an impedance greater than 52 × $\sqrt{1.444}$ or

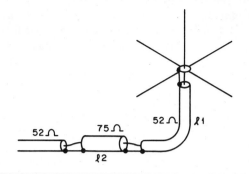

Fig 22—Example of series-section matching. A 36-Ω antenna is matched to 52-Ω coax by means of a length of 75-Ω cable.

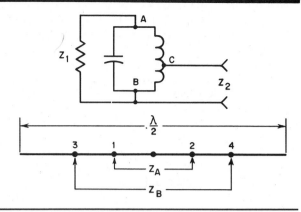

Fig 23—Impedance transformation with a resonant circuit, together with antenna analogy.

62.5 Ω, or less than $52/\sqrt{1.444}$ or 43.3 Ω.) The design steps are as follows.

From the earlier equations, n = 75/52 = 1.4423, r = 36/52 = 0.6923, and x = 0.

Further, B = 0.5678 (positive sign chosen), and $\ell 2$ = 29.6° or 0.0822 wavelength. The value of A is −1.7757. Calculating $\ell 1$ yields −60.6°. Adding 180° to obtain a positive result gives $\ell 1$ = 119.4°, or 0.3316 wavelength.

To find the physical lengths $\ell 1$ and $\ell 2$ we first find the free-space wavelength.

$$\lambda = \frac{984}{f(MHz)} = 33.93 \text{ ft}$$

Multiply this value by 0.79 (the velocity factor for both types of line), and we obtain the electrical wavelength in coax as 26.81 feet. From this, $\ell 1$ = 0.3316 × 26.81 = 8.89 feet, and $\ell 2$ = 0.0822 × 26.81 = 2.20 feet.

This completes the calculations. Construction consists of cutting the main coax at a point 8.89 feet from the antenna and inserting a 2.2-foot length of the 75-Ω cable.

The Quarter-Wave Transformer

The antenna in the preceding example could have been matched by a ¼-λ transformer at the load. Such a transformer would have a characteristic impedance of 43.27 Ω. It is interesting to see what happens in the design of a series-section transformer if this value is chosen as the characteristic impedance of the series section.

Following the same steps as before, we find n = 0.8321, r = 0.6923, and x = 0.

From these values we find B = ∞ and $\ell 2$ = 90°. Further, A = 0 and $\ell 1$ = 0°. These results represent a ¼-λ section at the load, and indicate that, as stated earlier, the ¼-λ transformer is indeed a special case of the series-section transformer.

DELTA MATCHING

Among the properties of a coil and capacitor resonant circuit is that of transforming impedances. If a resistive impedance, Z_1 in Fig 23, is connected across the outer terminals AB of a resonant LC circuit, the impedance Z_2 as viewed looking into another pair of terminals such as BC will

also be resistive, but will have a different value depending on the mutual coupling between the parts of the coil associated with each pair of terminals. Z_2 will be less than Z_1 in the circuit shown. Of course this relationship will be reversed if Z_1 is connected across terminals BC and Z_2 is viewed from terminals AB.

A resonant antenna has properties similar to those of a tuned circuit. The impedance presented between any two points symmetrically placed with respect to the center of a ½-λ antenna will depend on the distance between the points. The greater the separation, the higher the value of impedance, up to the limiting value that exists between the open ends of the antenna. This is also suggested in Fig 23, in the lower drawing. The impedance Z_A between terminals 1 and 2 is lower than the impedance Z_B between terminals 3 and 4. Both impedances, however, are purely resistive if the antenna is resonant.

This principle is used in the *delta* matching system shown in Fig 24. The center impedance of a ½-λ dipole is too low to be matched directly by any practicable type of air-insulated parallel-conductor line. However, it is possible to find, between two points, a value of impedance that can be matched to such a line when a "fanned" section or delta is used to couple the line and antenna. The antenna length ℓ should be based on the equation in Chapter 2, using the appropriate

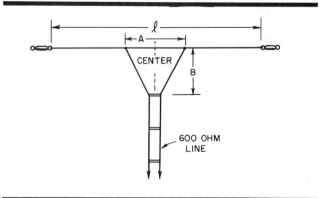

Fig 24—The delta matching system.

factor for the length/diameter ratio. The ends of the delta or "Y" should be attached at points equidistant from the center of the antenna. When so connected, the terminating impedance for the line will be essentially purely resistive.

Based on experimental data for the case of a simple ½-λ antenna coupled to a 600-Ω line, the total distance, A, between the ends of the delta should be 0.120 λ for frequencies below 30 MHz, and 0.115 λ for frequencies above 30 MHz. The length of the delta, distance B, should be 0.150 λ. These values are based on a wavelength in air, and on the assumption that the center impedance of the antenna is approximately 70 Ω. The dimensions will require modifications if the actual impedance is very much different.

The delta match can be used for matching the driven element of a directive array to a transmission line, but if the impedance of the element is low—as is frequently the case—the proper dimensions for A and B must be found by experimentation.

The delta match is somewhat awkward to adjust when the proper dimensions are unknown, because both the length and width of the delta must be varied. An additional disadvantage is that there is always some radiation from the delta. This is because the conductor spacing does not meet the requirement for negligible radiation: The spacing should be very small in comparison with the wavelength.

FOLDED DIPOLES

Basic information on the folded dipole antenna appears in Chapter 2. The two-wire system of Chapter 2 is an especially useful one. The input impedance is so close to 300 Ω that it can be fed directly with 300-Ω twin-lead or with open line without any other matching arrangement, and the line will operate with a low SWR. The antenna itself can be built like an open-wire line; that is, the two conductors can be held apart by regular feeder spreaders. TV "ladder" line is quite suitable. It is also possible to use 300-Ω line for the antenna, in addition to using it for the transmission line. Additional construction information is contained in Chapter 15. Since the antenna section does not operate as a transmission line, but simply as two wires in parallel, the velocity factor of twin-lead can be ignored in computing the antenna length. The reactance of the folded-dipole antenna varies less rapidly with frequency changes away from resonance than a single-wire antenna. Therefore it is possible to operate over a wider range of frequencies, while maintaining a low SWR on the line, than with a simple dipole. This is partly explained by the fact that the two conductors in parallel form a single conductor of greater effective diameter.

For reasons described in Chapter 2, a folded dipole will not accept power at twice the fundamental frequency. However, the current distribution is correct for harmonic operation on odd multiples of the fundamental. Because the radiation resistance is not greatly different for a 3/2-λ antenna and one that is ½ λ, a folded dipole can be operated on its third harmonic with a low SWR in a 300-Ω line. A 7-MHz folded dipole, consequently, can be used for the 21-MHz band as well.

Spacing Adjustment of Multi- and Unequal-Conductor Dipoles

Chapter 2 shows how a wide range of impedance step-up ratios is available by using conductors of different size or by using more than two. Because the relatively large effective

thickness of the antenna reduces the rate of change of reactance with frequency, the tuning becomes relatively broad. It is a good idea, however, to check the resonant frequency with a dip meter in making length adjustments. The transmission line should be disconnected and the antenna terminals temporarily short-circuited when this check is being made.

As shown by the charts in Chapter 2, there are two special cases where the impedance ratio of the folded dipole is independent of the spacing between conductors. These are for a ratio of 4:1 with the two-conductor dipole and a ratio of 9:1 in the three-conductor case. In all other cases the impedance ratio can be varied by adjustment of the spacing. The adjustment range is quite limited when ratios near 4 and 9, respectively, are used, but increases with the departure in either direction from these "fixed" values. This offers a means for final adjustment of the match to the transmission line when the antenna resistance is known approximately but not exactly.

If a suitable match cannot be obtained by adjustment of spacing, there is no alternative but to change the ratio of conductor diameters. The impedance ratio decreases with an increase in spacing, and vice versa. Hence, if a match cannot be brought about by changing the spacing, such a change will at least indicate whether the ratio of d_2/d_1 (using the notation of Chapter 2) should be increased or decreased.

THE T AND GAMMA

The T matching system shown in Fig 25 has a considerable resemblance to the folded dipole; in fact, if the distance A is extended to the full length of the antenna, the system becomes an ordinary folded dipole. The T has considerable flexibility in impedance ratio and is more convenient, constructionally, than the folded dipole when used with the driven element of a rotatable parasitic array. Since it is a symmetrical system it is inherently balanced, and so is well suited to use with parallel-conductor transmission lines. If coaxial line is used, some form of balun, as described earlier, should be installed. Alternatively, the gamma form described below can be used with unbalanced lines.

The current flowing at the input terminals of the T consists of the normal antenna current divided between the radiator and the T conductors in a way that depends on their relative diameters and the spacing between them, with a superimposed transmission-line current flowing in each half of the T and its associated section of the antenna. Each such T conductor and the associated antenna conductor can be

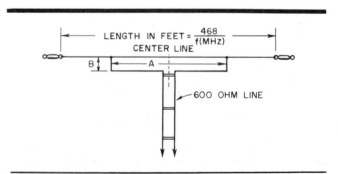

Fig 25—The T matching system, applied to a ½-λ antenna and 600-Ω line.

looked upon as a section of transmission line shorted at the end. Since it is shorter than ¼ λ it has inductive reactance; as a consequence, if the antenna itself is exactly resonant at the operating frequency, the input impedance of the T will show inductive reactance as well as resistance. The reactance must be tuned out if a good match to the transmission line is to be obtained. This can be done either by shortening the antenna to obtain a value of capacitive reactance that will reflect through the matching system to cancel the inductive reactance at the input terminals, or by inserting a capacitance of the proper value in series at the input terminals as shown in Fig 26, upper drawing.

A theoretical analysis has shown that the part of the impedance step-up arising from the spacing and ratio of conductor diameters is approximately the same as given for the folded dipole in Chapter 2. The actual impedance ratio is, however, considerably modified by the length A of the matching section (Fig 25). The trends can be stated as follows:

1) The input impedance increases as the distance A is made larger, but not indefinitely. There is in general a distance A that will give a maximum value of input impedance, after which further increase in A will cause the impedance to decrease.

2) The distance A at which the input impedance reaches a maximum is smaller as d_2/d_1 (using the notation of Chapter 2) is made larger, and becomes smaller as the spacing between the conductors is increased.

3) The maximum impedance values occur in the region where A is 40 to 60 percent of the antenna length in the average case.

4) Higher values of input impedance can be realized when the antenna is shortened to cancel the inductive reactance of the matching section.

Simple Dipole Matching

For a dipole having an approximate impedance of 70 Ω, the T matching section dimensions for matching a 600-Ω line are given by the following formulas:

$$A \text{ (feet)} = \frac{180.5}{f(MHz)}$$
$$B \text{ (inches)} = \frac{114}{f(MHz)}$$

These equations apply for wire antennas with the matching section made of the same size wire. With an antenna element of different impedance, or for matching a line having a Z_0 other than 600 Ω, the matching section dimensions can be determined experimentally.

The Gamma

The gamma arrangement shown in Fig 27 is an unbalanced version of the T, suitable for use with coaxial lines. Except for the fact that the matching section is connected between the center and one side of the antenna, the remarks above about the behavior of the T apply equally well. The inherent reactance of the matching section can be canceled either by shortening the antenna appropriately or by using the resonant length and installing a capacitor C, as shown in the lower drawing of Fig 26.

The gamma match has been widely used for matching coaxial cable to all-metal parasitic beams for a number of years. Because it is well suited to "plumber's delight" construction, where all the metal parts are electrically and

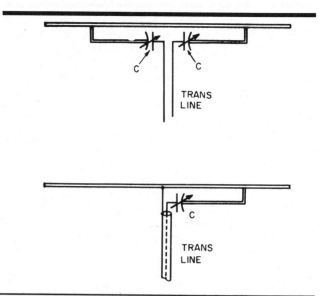

Fig 26—Series capacitors for tuning out residual reactance with the T and gamma matching systems. A maximum capacitance of 150 pF in each capacitor should provide sufficient adjustment range, in the average case, for 14-MHz operation. Proportionately smaller capacitance values can be used on higher frequency bands. Receiving type plate spacing will be satisfactory for power levels up to a few hundred watts.

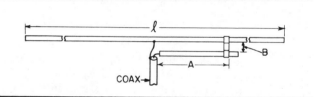

Fig 27—The gamma match, as used with tubing elements. The transmission line may be either 52-Ω or 75-Ω coax.

mechanically connected, it has become quite popular for amateur arrays. Construction details for a gamma-matching section are given in Fig 28.

Because of the many variable factors—driven-element length, gamma rod length, rod diameter, spacing between rod and driven element, and value of series capacitors—a number of combinations will provide the desired match. The task of finding a proper combination can be a tedious one, however, as the settings are interrelated. A few "rules of thumb" have evolved that provide a starting point for the various factors. For matching a multielement array made of aluminum tubing to 52-Ω line, the length of the rod should be 0.04 to 0.05 λ, its diameter ⅓ to ½ that of the driven element, and its spacing (center to center from the driven element), approximately 0.007 λ. The capacitance value should be approximately 7 pF per meter of wavelength. This translates to about 140 pF for 20-meter operation. The exact gamma dimensions and value for the capacitor will depend on the radiation resistance of the driven element, and whether or not it is resonant. These starting-point dimensions are for an array having a feed-point impedance of about 25 Ω, with the driven element shortened

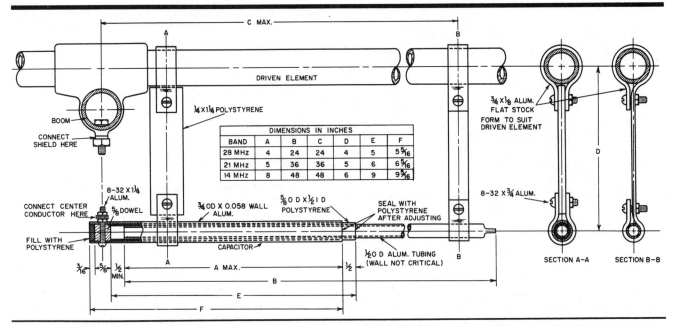

DIMENSIONS IN INCHES						
BAND	A	B	C	D	E	F
28 MHz	4	24	24	4	5	5 5/16
21 MHz	5	36	36	5	6	6 5/16
14 MHz	8	48	48	6	9	9 5/16

Fig 28—Constructional details of a gamma matching section for 52-Ω coax line. The reactance compensating capacitor is in tubular form. It is made by dividing the gamma rod or bar into two telescoping sections separated by a length of polystyrene tubing, which serves as the dielectric.

approximately 3% from resonance. A graphical approximation for determining the settings for a gamma match may be obtained with the aid of a Smith Chart. The technique is described in Chapter 28.

Adjustment

After installation of the antenna, the proper constants for the T and gamma must be determined experimentally. The use of the variable series capacitors, as shown in Fig 26, is recommended for ease of adjustment. With a trial position of the tap or taps on the antenna, measure the SWR on the transmission line and adjust C (both capacitors simultaneously in the case of the T) for minimum SWR. If it is not close to 1:1, try another tap position and repeat. It may be necessary to try another size of conductor for the matching section if satisfactory results cannot be brought about. Changing the spacing will show which direction to go in this respect, just as in the case of the folded dipole discussed in the preceding section.

THE OMEGA MATCH

The *omega match* is a slightly modified form of the gamma match. In addition to the series capacitor, a shunt capacitor is used to aid in canceling a portion of the inductive reactance introduced by the gamma section. This is shown in Fig 29. C1 is the usual series capacitor. The addition of C2 makes it possible to use a shorter gamma rod, or makes it easier to obtain the desired match when the driven element is resonant. During adjustment, C2 will serve primarily to determine the resistive component of the load as seen by the coax line, and C1 serves to cancel any reactance.

THE HAIRPIN AND BETA MATCHES

The usual form of the *hairpin match* is shown in Fig 30. Basically, the hairpin is a form of an L-matching network.

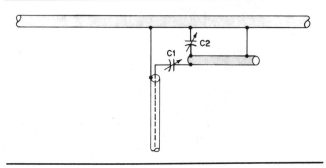

Fig 29—The omega match.

Because it is somewhat easier to adjust for the desired terminating impedance than the gamma match, it is preferred by many amateurs. Its disadvantages, compared with the gamma, are that it must be fed with a balanced line (a balun may be used with a coax feeder, as shown in Fig 30), and the driven element must be split at the center. This latter requirement complicates the mechanical mounting arrangement for the element, by ruling out "plumber's delight" construction.

As indicated in Fig 30, the center point of the hairpin is electrically neutral. As such, it may be grounded or connected to the remainder of the antenna structure. The hairpin itself is usually secured by attaching this neutral point to the boom of the antenna array. The *beta match* is electrically identical to the hairpin match, the difference being in the mechanical construction of the matching section. With the beta match, the conductors of the matching section straddle the boom, one conductor being located on either side, and the electrically neutral point consists of a sliding or adjustable shorting clamp placed around the boom and the

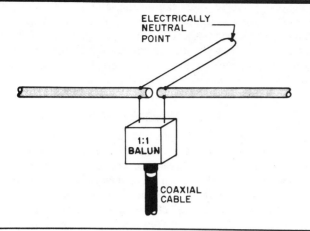

Fig 30—The hairpin match.

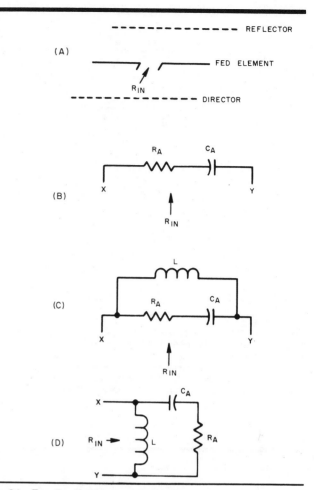

Fig 31—For the Yagi antenna shown at A, the driven element is shorter than its resonant length. The input impedance at resonance is represented at B. By adding an inductor, as shown at C, a low value of R_A is made to appear as a higher impedance at terminals XY. At D, the diagram of C is redrawn in the usual L-network configuration.

two matching-section conductors.

The electrical operation of the hairpin match has been treated extensively by Gooch, Gardner and Roberts (see bibliography at the end of this chapter). The driven element is shortened so it presents a capacitive reactance. Then the antenna is matched to the transmission line by forming an equivalent parallel-resonant circuit in which the antenna resistance appears in series with the capacitance. The impedance of this type parallel-resonant circuit varies almost inversely with the series antenna resistance, and therefore can cause a very small antenna resistance to appear as a very large resistance at the terminals of the resonant circuit. The values of inductance and capacitance are chosen to transform the antenna resistance to a resistance value equal to the characteristic impedance of the transmission line.

As mentioned above, the capacitive portion of this circuit is produced by slightly shortening the antenna driven element, shown in Fig 31A. For a given frequency the impedance of a shortened ½-λ element appears as the antenna resistance and a capacitance in series, as indicated schematically in Fig 31B. The inductive portion of the resonant circuit at C is a hairpin of heavy wire or small tubing which is connected across the driven-element center terminals. The diagram of C is redrawn in D to show the circuit in conventional L-network form. R_A, the radiation resistance, is a smaller value than R_{IN}, the impedance of the feed line. (In L-network matching, the higher of the two impedances to be matched is connected to the shunt-arm side of the network, and the lower impedance to the series-arm side.)

If the approximate radiation resistance of the antenna system is known, Figs 32 and 33 may be used to gain an idea of the hairpin dimensions necessary for the desired match. The curves of Fig 32 were obtained from design equations for L-network matching. Fig 33 is based on the equation, $X_p = j \tan \theta$, which gives the inductive reactance as normalized to the Z_0 of the hairpin, looking at it as a length of transmission line terminated in a short circuit. For example, if an antenna-system impedance of 20 ohms is to be matched to 52-ohm line, Fig 32 indicates that the inductive reactance required for the hairpin is 41 ohms. If the hairpin is constructed of ¼-inch tubing spaced 1½ inches, its characteristic impedance is 300 ohms (from Chapter 24.) Normalizing the required 41-ohm reactance to this impedance, 41/300 = 0.137.

By entering the graph of Fig 33 with this value, 0.137, on the scale at the bottom, it may be seen that the hairpin length should be 7.8 electrical degrees, or 7.8/360 λ. For purposes of these calculations, taking a 97.5% velocity factor into account, the wavelength in inches is $11,508/f_{MHz}$. If the antenna is to be used on 14 MHz, the required hairpin length is 7.8/360 × 11,508/14 = 17.8 inches. The length of the hairpin affects primarily the resistive component of the terminating impedance as seen by the feed line. Greater resistances are obtained with longer hairpin sections, and smaller resistances with shorter sections. Reactance at the feed point terminals is tuned out by adjusting the length of the driven element, as necessary. If a fixed-length hairpin section is in use, a small range of adjustment may be made in the effective value of the inductance by spreading or squeezing together the conductors of the hairpin. Spreading the conductors apart will have the same effect as lengthening the hairpin, while placing them closer together will effectively shorten it.

Instead of using a hairpin of stiff wire or tubing, this same matching technique may be used with a lumped-constant inductor connected across the antenna terminals. Such a method of matching has been dubbed the *helical hairpin*. The

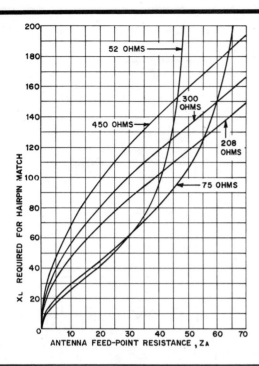

Fig 32—Reactance required for a hairpin to match various antenna resistances to common line or balun impedance.

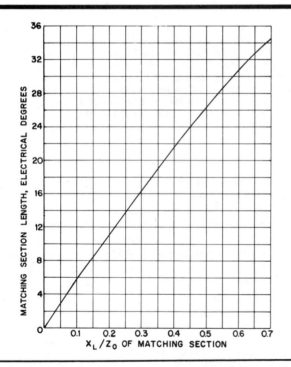

Fig 33—Inductive reactance (normalized to Z_0 of matching section), scale at bottom, versus required hairpin matching section length, scale at left. To determine the length in wavelengths divide the number of electrical degrees by 360. For open-wire line, a velocity factor of 97.5% should be taken into account when determining the electrical length.

inductor, of course, must exhibit the same reactance at the operating frequency as the hairpin which it replaces. A cursory examination with computer calculations indicates that a helical hairpin may offer a slightly improved SWR bandwidth over a true hairpin, but the effects of different length/diameter ratios of the driven element were not investigated.

MATCHING STUBS

As explained in Chapter 24, a mismatch-terminated transmission line less than ¼ λ long has an input impedance that is both resistive and reactive. The equivalent circuit of the line input impedance can be formed either of resistance and reactance in series or resistance and reactance is parallel. Depending on the line length, the series resistance component, R_s, can have any value between the terminating resistance, Z_R (when the line has zero length) and Z_0^2/Z_R (when the line is exactly ¼ wave long). The same thing is true of R_p, the parallel-resistance component. R_s and R_p do not have the same values at the same line length, however, other than zero and ¼ λ. With either equivalent there is some line length that will give a value of R_s or R_p equal to the characteristic impedance of the line. However, there will always be reactance along with the resistance. But if provision is made for canceling or "tuning out" this reactive part of the input impedance, only the resistance will remain. Since this resistance is equal to the Z_0 of the transmission line, the section from the reactance-cancellation point back to the generator will be properly matched.

Tuning out the reactance in the equivalent series circuit requires that a reactance of the same value as X_s (but of opposite kind) be inserted in series with the line. Tuning out the reactance in the equivalent parallel circuit requires that a reactance of the same value as X_p (but of opposite kind) be connected across the line. In practice it is more convenient to use the parallel-equivalent circuit. The transmission line is simply connected to the load (which of course is usually a resonant antenna) and then a reactance of the proper value is connected across the line at the proper distance from the load. From this point back to the transmitter there are no standing waves on the line.

A convenient type of reactance to use is a section of transmission line less than ¼ λ long, terminated with either an open circuit or a short circuit, depending on whether capacitive reactance or inductive reactance is called for. Reactances formed from sections of transmission line are called *matching stubs*, and are designated as *open* or *closed* depending on whether the free end is open or short circuited. The two types of matching stubs are shown in the sketches of Fig 34.

The distance from the load to the stub (dimension A in Fig 34) and the length of the stub, B, depend on the characteristic impedances of the line and stub and on the ratio of Z_R to Z_0. Since the ratio of Z_R to Z_0 is also the standing-wave ratio in the absence of matching (and with a resonant antenna), the dimensions are a function of the SWR. If the line and stub have the same Z_0, dimensions A and B are dependent on the SWR only. Consequently, if the SWR can be measured before the stub is installed, the stub can be properly located and its length determined even though the actual value of load impedance is not known.

Typical applications of matching stubs are shown in Fig 35, where open-wire line is being used. From inspection of these drawings it will be recognized that when an antenna is fed at a current loop, as in Fig 35A, Z_R is less than Z_0 (in the average case) and therefore an open stub is called for, installed within the first ¼ λ of line measured from the antenna. Voltage feed, as at B, corresponds to Z_R greater than Z_0 and therefore requires a closed stub.

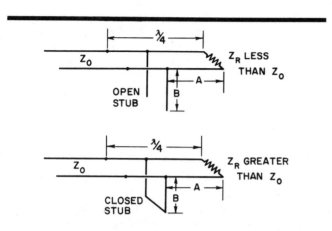

Fig 34—Use of open or closed stubs for canceling the parallel reactive component of input impedance.

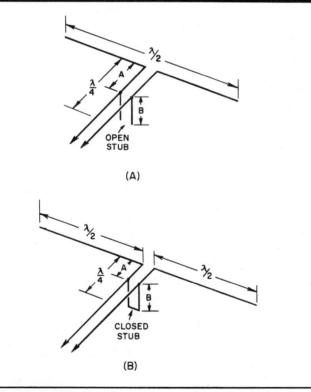

(A)

(B)

Fig 35—Application of matching stubs to common types of antennas.

The Smith Chart may be used to determine the length of the stub and its distance from the load (see Chapter 28). If the load is a pure resistance and the characteristic impedances of the line and stub are identical, the lengths may be determined by equations. For the closed stub when Z_R is greater than Z_0, they are

$$\arctan A = \sqrt{SWR} \quad \text{and} \quad \text{arccot } B = \frac{SWR - 1}{\sqrt{SWR}}$$

For the open stub when Z_R is less than Z_0

$$\text{arccot } A = \sqrt{SWR} \quad \text{and} \quad \arctan B = \frac{SWR - 1}{\sqrt{SWR}}$$

In these equations the lengths A and B are the distance from the stub to the load and the length of the stub, respectively, as shown in Fig 35. These lengths are expressed in electrical degrees, equal to 360 times the lengths in wavelengths.

In using the above equations it must be remembered that the wavelength along the line is not the same as in free space. If an open-wire line is used the velocity factor of 0.975 will apply. When solid-dielectric line is used, the free-space wavelength as determined above must be multiplied by the appropriate velocity factor to obtain the actual lengths of A and B (see Chapter 24.)

Although the equations above do not apply when the characteristic impedances of the line and stub are not the same, this does not mean that the line cannot be matched under such conditions. The stub can have any desired characteristic impedance if its length is chosen so that it has the proper value of reactance. By using the Smith Chart, the correct lengths can be determined without difficulty for dissimilar types of line.

In using matching stubs it should be noted that the length and location of the stub should be based on the SWR *at the load*. If the line is long and has fairly high losses, measuring the SWR at the input end will not give the true value at the load. This point is discussed in Chapter 24 in the section on attenuation.

Reactive Loads

In this discussion of matching stubs it has been assumed that the load is a pure resistance. This is the most desirable condition, since the antenna that represents the load preferably should be tuned to resonance before any attempt is made to match the line. Nevertheless, matching stubs can be used even when the load is considerably reactive. A reactive load simply means that the loops and nodes of the standing waves of voltage and current along the line do not occur at integral multiples of ¼ λ from the load. If the reactance at the load is known, the Smith Chart may be used to determine the correct dimensions for a stub match.

Stubs on Coaxial Lines

The principles outlined in the preceding section apply also to coaxial lines. The coaxial cases corresponding to the open-wire cases shown in Fig 34 are given in Fig 36. The equations given earlier may be used to determine dimensions A and B. In a practical installation the junction of the transmission line and stub would be a T connector.

A special case of the use of a coaxial matching stub in

which the stub is associated with the transmission line in such a way as to form a balun has been described earlier in this chapter (Fig 12). The principles used are those just described. The antenna is shortened to introduce just enough reactance at its feed point to permit the matching stub to be connected there, rather than at some other point along the transmission line as in the general cases discussed here. To use this method the antenna resistance must be lower than the Z_0 of the main transmission line, since the resistance is transformed to a higher value. In beam antennas this will nearly always be the case.

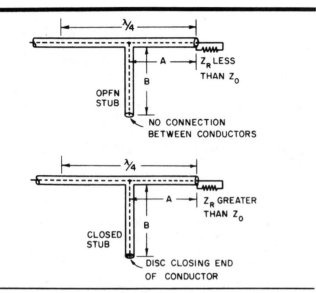

Fig 36—Open and closed stubs on coaxial lines.

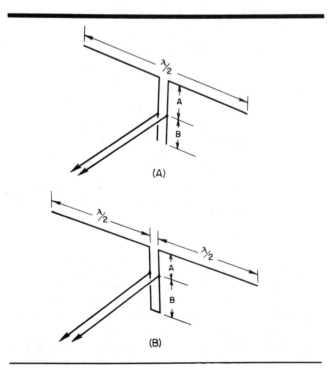

Fig 37—Application of matching sections to common antenna types.

Matching Sections

If the two antenna systems in Fig 35 are redrawn in somewhat different fashion, as shown in Fig 37, a system results that differs in no consequential way from the matching stubs described previously, but in which the stub formed by A and B together is called a "quarter-wave matching section." The justification for this is that a ¼-λ section of line is similar to a resonant circuit, as described earlier in this chapter. It is therefore possible to use the ¼-λ section to transform impedances by tapping at the appropriate point along the line.

Earlier equations give design data for matching sections, A being the distance from the antenna to the point at which the line is connected, and A + B being the total length of the matching section. The equations apply only in the case where the characteristic impedance of the matching section and transmission line are the same. Equations are available for the case where the matching section has a different Z_0 than the line, but are somewhat complicated. A graphic solution for different line impedances may be obtained with the Smith Chart (Chapter 28).

Adjustment

In the experimental adjustment of any type of matched line it is necessary to measure the SWR with fair accuracy in order to tell when the adjustments are being made in the proper direction. In the case of matching stubs, experience has shown that experimental adjustment is unnecessary, from a practical standpoint, if the SWR is first measured with the stub not connected to the transmission line, and the stub is then installed according to the design data.

FLEXIBLE SECTIONS FOR ROTATABLE ARRAYS

When open-wire transmission line is used there is likely to be trouble with shorting or grounding of feeders in rotatable arrays unless some special precautions are taken. Usually some form of insulated flexible line is connected between the antenna and a stationary support at the top of the tower or mast on which the antenna is mounted.

Such a flexible section can take several forms, and it can be made to do double duty. Probably the most satisfactory system for arrays that are not designed to be fed with coaxial line, is to use a flexible section of coax with coaxial baluns at both ends. The outer conductor of the coax may be grounded to the tower or to the beam antenna framework, wherever it is advantageous to do so. Such a flexible section is shown in Fig 38. If the coaxial section is made any multiple of ½ λ in electrical length, the impedance of the array will be repeated at the bottom of the flexible section.

Another method is to use twin-lead for the flexible section. The 300-ohm type designed for transmitting applications is recommended. Here, again, ½-λ sections repeat the antenna feed impedance at the bottom end. The twin-lead section may also be made an odd multiple of ¼ λ, in which case it will act as a transformer section, giving an impedance step-down between a 450-ohm line and an antenna impedance of 200 ohms.

GROUNDPLANE ANTENNAS

The same principles discussed earlier also apply to an unsymmetrical system such as the grounded antenna or the groundplane antenna. In the case of the ¼-λ groundplane

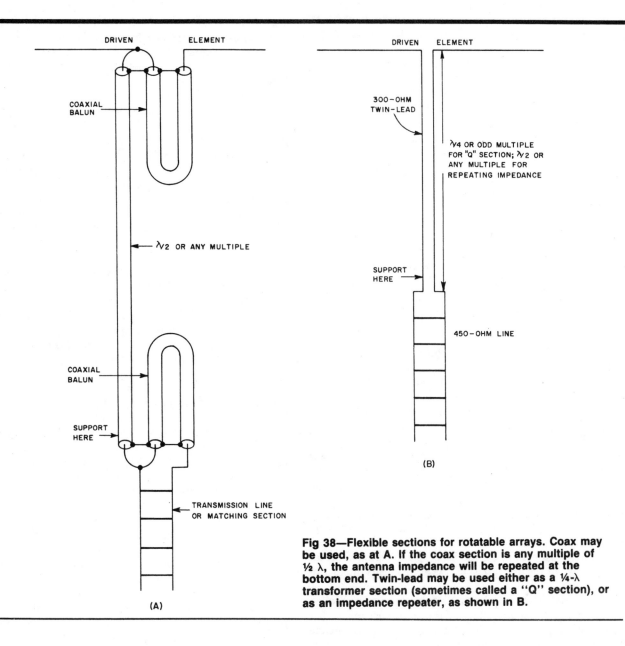

Fig 38—Flexible sections for rotatable arrays. Coax may be used, as at A. If the coax section is any multiple of ½ λ, the antenna impedance will be repeated at the bottom end. Twin-lead may be used either as a ¼-λ transformer section (sometimes called a "Q" section), or as an impedance repeater, as shown in B.

antenna, a straightforward design procedure for matching is possible because the radiation resistance is essentially independent of the physical height of the system (provided the radiator is reasonably clear of other conductors in the vicinity) and there is no ground-connection resistance to be included in the total resistance to be matched.

The groundplane antenna lends itself well to direct connection to coaxial line, so this type of line is nearly always used. Several matching methods are available. If the antenna length can be adjusted to resonance, the stub matching system previously described is convenient.

Probably the most convenient and therefore the most popular method of "matching" an elevated groundplane antenna to 52-Ω line is to "droop" the radials. If the radials all lie in a single horizontal plane, the feed-point resistance of the antenna is approximately 36 Ω. Lowering the radials so they form an angle greater than 90° with the vertical radiator raises the feed-point resistance. In the limiting case, when the radials form a 180° angle with the radiator (radials

surround the feeder to form a sleeve), the feed-point resistance is approximately 72 Ω. Angles between 90° and 180° will present a resistance between 36 and 72 Ω. Angles on the order of 135° provide a good match to 52-Ω coax.

Another method of matching, particularly convenient for small antennas (28 MHz and higher frequencies) mounted on top of a supporting mast or pole, requires shortening the antenna to the high frequency side of resonance so it shows a particular value of capacitive reactance at its base. The antenna terminals are then shunted by an inductive reactance, which may have the physical form either of a coil or a closed stub, to restore resonance and simultaneously transform the radiation resistance to the proper value for matching the transmission line. This concept is the same as for the hairpin match, described in detail earlier.

Tapped-Coil Matching

The matching arrangement shown in Fig 39 is a more general form of the method just mentioned, in that it does

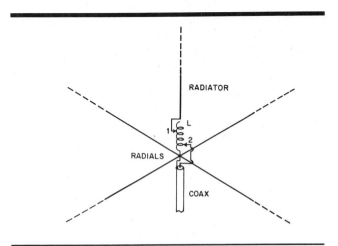

Fig 39—Matching to groundplane antenna by tapped coil. This requires that the antenna (but not radials) be shorter than the resonant length.

not require adjusting the radiator height to an exact value. The radiator must be shortened so that the system will show capacitive reactance, but any convenient amount of shortening can be used. This system is particularly useful on lower frequencies where it may not be possible to obtain a radiator height approximating ¼ λ.

The antenna impedance is matched to the characteristic impedance of the line by adjusting the taps on L. As a preliminary adjustment, before attaching tap 2 for the line, tap 1 for the radiator may be set for resonance at the operating frequency as indicated by a dip meter coupled to L. Tap 2 is then moved along the coil to find the point that gives minimum SWR, as indicated by an SWR indicator. To bring the SWR down to 1 to 1, it will usually be necessary to make a small readjustment of the radiator tap and perhaps further "touch up" of the line tap, since the adjustments interact to some extent.

This method is equivalent to tapping down on a parallel-resonant circuit to match a low value of resistance to a higher value connected across the whole circuit. The antenna impedance can be represented by a capacitance in parallel with a resistance which is much higher than the actual radiation resistance. The transformation of resistance comes about by using the parallel equivalent of the radiation resistance and capacitive reactance in series, using the relationship given in Chapter 24.

Matching by Length Adjustment

Still another method of matching may be used when the antenna length is not fixed by other considerations. As shown in Chapter 2 under "Grounded Antennas," the radiation resistance as measured at the base of a groundplane antenna increases with the antenna height. It is possible to choose a height such that the base radiation resistance will equal the Z_0 of the transmission line to be used. The heights of most interest are a little over 100° (0.28 λ), where the resistance is approximately 52 Ω, and about 113° (0.32 λ), where the resistance is 75 Ω, to match the two common types of coaxial

line. These heights are quite practicable for groundplane antennas for 14 MHz and higher frequencies. The lengths (heights) in degrees as given above do not require any correction for length/diameter ratio; they are free-space lengths.

Since the antenna is not resonant at these lengths, its input impedance will be reactive as well as resistive. The reactance must be tuned out in order to make the line see a purely resistive load equal to its characteristic impedance. This can be done with a series capacitor of the proper value. The approximate value of capacitive reactance required, for antennas of typical length/diameter ratio, is about 100 Ω for the 52-Ω case and about 200 Ω for the 75-Ω case. The corresponding capacitance values for the frequency in question can be determined from appropriate charts or by equation. Variable capacitors of sufficient range should be used.

In an analysis by Robert Stephens, W3MIR, the impedance and reactance versus height data of Chapter 2 for groundplane antennas has been converted to conductance and susceptance information. For parallel-equivalent matching, an input conductance of 1/52 siemen is needed. For an antenna length to diameter ratio of approximately 1000, there are two heights for which the conductance is this value—one at 0.234 λ and the other at 0.255 λ. At the shorter height, the susceptance is positive (capacitive), 0.0178 siemen, and at the longer it is negative, 0.0126 siemen. As far as radiation is concerned, one is as good as the other and the choice becomes the one of the simpler mechanical approach for the particular antenna. If the antenna is made 0.234 λ high, its capacitive reactance may be canceled with a shunt inductor having a reactance of 56 Ω at the operating frequency, resulting in a 52-Ω termination for the feed line. If the antenna is 0.255 λ, a 52-Ω match may be obtained with a shunt capacitor of 79 Ω. Similarly, a match may be obtained for 75-Ω coax with a height of 0.23 λ and a shunt inductor of 56 Ω, or with a height of 0.26 λ and a shunt capacitor of 78 Ω. These reactance values may be obtained with lumped constants or with stubs, as described earlier. As mentioned above, these heights do not require any correction factor; they are free-space lengths.

The adjustment of systems like these requires only that the capacitance or inductance be varied until the lowest possible SWR is obtained. If the lengths mentioned above are used, the SWR should be close enough to 1 to 1 to make a fine adjustment of the length unnecessary.

Broadband Matching Transformers

Broadband transformers have been used widely because of their inherent bandwidth ratios (as high as 20,000:1) from a few tens of kilohertz to over a thousand megahertz. This is possible because of the transmission-line nature of the windings. The interwinding capacitance is a component of the characteristic impedance and therefore, unlike the conventional transformer, forms no resonances which seriously limit the bandwidth. At low frequencies, where interwinding capacitances can be neglected, these transformers are similar in operation to the conventional transformer. The main difference (and a very important one from a power standpoint) is that the windings tend to cancel out the induced flux in the core. Thus, high permeability ferrite cores, which are not only highly nonlinear but also suffer serious damage even at flux levels as low as 200 to 500 gauss, can be used. This greatly extends the low frequency range of performance. Since higher permeability also permits fewer turns at the lower frequencies, HF performance is also improved since the upper cutoff is determined mainly from transmission line considerations. At the high frequency cutoff, the effect of the core is negligible.

Bifilar matching transformers lend themselves to unbalanced operation. That is, both input and output terminals can have a common ground connection. This eliminates the third magnetizing winding required in balanced to unbalanced (balun) operation. (See Chapter 25 for a discussion of baluns.) By adding third and fourth windings, as well as by tapping windings at appropriate points, various combinations of broadband matching can be obtained. Fig 40 shows a 4:1 unbalanced to unbalanced configuration. No. 14 wire can be used and it will easily handle 1000 watts of power. By tapping at points ¼, ½ and ¾ of the way along the top winding, ratios of approximately 1.5:1, 2:1 and 3:1 can also be obtained. It should be noted that one of the wires should be covered with vinyl electrical tape in order to prevent voltage breakdown between the windings. This is necessary when a step-up ratio is used at high power to match antennas with impedances greater than 52 Ω.

Fig 41 shows a transformer with four windings, permitting wide-band matching ratios as high as 16:1. Fig 42 shows a four-winding transformer with taps at 4:1, 6:1, 9:1 and 16:1. In tracing the current flow in the windings when using the 16:1 tap, one sees that the top three windings carry the same current. The bottom winding, in order to maintain the proper potentials, sustains a current three times greater. The bottom current cancels out the core flux caused by the other three windings. If this transformer is used to match into low impedances, such as 3 to 4 Ω, the current in the bottom winding can be as high as 15 amperes. This value is based on the high side of the transformer being fed with 52-Ω cable handling a kilowatt of power. If one needs a 16:1 match like this at high power, then cascading two 4:1 transformers is recommended. In this case, the transformer at the lowest impedance side requires each winding to handle only 7.5 A. Thus, even no. 14 wire would suffice in this application.

The popular cores used in these applications are 2.5 inch OD ferrites of Q1 and Q2 material, and powdered iron cores of 2 inches OD. The permeabilities of these cores, μ, are nominally 125, 40 and 10 respectively. Powdered iron cores of permeabilities 8 and 25 are also available.

In all cases these cores can be made to operate over the 1.8- to 28-MHz bands with full power capability and very low

loss. The main difference in their design is that lower permeability cores require more turns at the lower frequencies. For example, Q1 material requires 10 turns to cover the 1.8-MHz band. Q2 requires 12 turns, and powdered iron ($\mu = 10$) requires 14 turns. Since the more common powdered iron core is generally smaller in diameter and requires more turns because of lower permeability, higher ratios are sometimes difficult to obtain because of physical limitations. When you are working with low impedance levels, unwanted parasitic inductances come into play, particularly on 14 MHz and above. In this case lead lengths should be kept to a minimum.

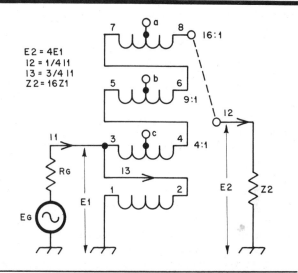

Fig 41—Four-winding, broadband, variable impedance transformer. Connections a, b, and c can be placed at appropriate points to yield various ratios from 1.5:1 to 16:1.

Fig 42—A 4-winding, wide-band transformer (with front cover removed) with connections made for matching ratios of 4:1, 6:1, 9:1 and 16:1. The 6:1 ratio is the top coaxial connector and, from left to right, 16:1, 9:1 and 4:1 are the others. There are 10 quadrifilar turns of no. 14 enameled wire on a Q1, 2.5-in. OD ferrite core.

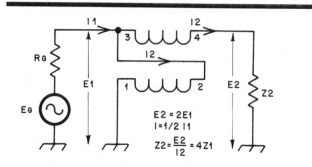

Fig 40—Broadband bifilar transformer with a 4:1 impedance ratio. The upper winding can be tapped at appropriate points to obtain other ratios such as 1.5:1, 2:1, and 3:1.

BANDWIDTH AND ANTENNA Q

Although more properly a subject for discussion in connection with antenna fundamentals, the bandwidth of the antenna is considered here because as a practical matter the change in antenna impedance with frequency is reflected as a change in the SWR on the transmission line. Thus, when an antenna is matched to the line at one frequency—usually in the center of the band of frequencies over which the antenna is to be used—a shift in the operating frequency will be accompanied by a change in the SWR. This would not occur if the antenna impedance were purely resistive and constant regardless of frequency, but unfortunately no practical antennas are that "flat."

In the frequency region around resonance the resistance change is fairly small and, by itself, would not affect the SWR enough to matter, practically. The principal cause of the change in SWR is the change in the reactive component of the antenna impedance when the frequency is varied. If the reactance changes rapidly with frequency the SWR will rise rapidly off resonance, but if the rate of reactance change is small the shift in SWR likewise will be small. Therefore, an antenna that has a relatively slow rate of reactance change will cover a wider frequency band, for a given value of SWR at the band limits (such as 2 to 1 or 3 to 1), than one having a relatively rapid rate of reactance change.

In the region around the resonant frequency of the antenna, the impedance as measured at a current loop varies with frequency in essentially the same way as the impedance of a series-resonant circuit using lumped constants. It is therefore possible to define a quantity Q for the antenna in the same way as Q is defined in a series-resonant circuit. The Q of the antenna is a measure of the antenna selectivity, just as the Q of an ordinary circuit is a measure of its selectivity.

The Q of the antenna can be found by measuring its input resistance and reactance at some frequency close to the resonant frequency (at exact resonance the antenna is purely resistive and there is no reactive component). Then, for frequency changes of less than 5% from the exact resonant frequency, the antenna Q is given with sufficient accuracy by the following formula:

$$Q = \frac{X}{R} \times \frac{1}{2n}$$

where

   X = the measured reactance
   R = the measured resistance
   n is the percentage difference, expressed as a decimal, between the antenna resonant frequency and the frequency at which X and R were measured. Example: If the frequency used for the measurement differs from the resonant frequency by 2%, n = 0.02.

For an ordinary $\frac{1}{2}$-λ dipole, the approximate Q values vary from about 14 for a length/diameter ratio of 25,000 to about 8 for a ratio of 1250. In parasitic arrays with close spacing between elements the input Q may be well over 50, depending on the spacing and tuning (see Chapter 11).

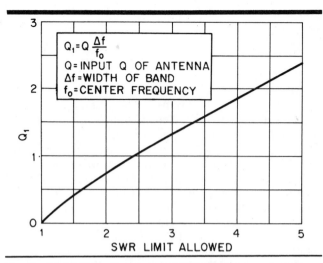

Fig 43—Bandwidth, in terms of SWR limits, as a function of antenna Q. The inset equation gives an effective Q. Q1 is determined by the fractional band ($\Delta f/f_0$) and the actual antenna Q as defined in the text.

SWR Versus Q

If the Q of the antenna is known, the variation in SWR over the operating band can be determined from Fig 43. It is assumed that the antenna is matched to the line at the center frequency of the band. Conversely, if a limit is set on the SWR, the SWR bandwidth can be found from Fig 43. As an example, suppose that a dipole having a Q of 15 (more or less typical of a wire antenna) is to be used over the 3.5-4 MHz band and that it is matched with a 1:1 SWR at the band center. Then $\Delta f/f_0 = 0.5/3.75 = 0.133$ and $Q1 = 15 \times 0.133 = 2$. The SWR that can be expected at the band edges, 3.5 and 4 MHz, is shown by the chart to be a bit over 4 to 1. If it should be decided arbitrarily that no more than a 2:1 SWR is allowable, Q1 is 0.75 and from the equation in Fig 43 the total bandwidth is found to be 187.5 kHz.

Effect of Matching Network

The measurement of resistance and reactance to determine Q should be made at the input terminals of the matching network, if one is required. The selectivity of the matching network has just as much effect on the bandwidth, in terms of SWR on the line, as the selectivity of the antenna itself. When the greatest possible bandwidth is wanted, a low-Q matching network must be used. This is not always controllable, particularly when the antenna resistance differs considerably from the Z_0 of the line to which it is to be matched. A large impedance ratio usually means that large values of reactance must be used in the matching section. In other words, the Q of the matching section in such cases tends to be higher than desirable. Simple systems having direct matching, such as a dipole fed with 75-Ω line or a folded dipole matched to the line, will have the greatest bandwidth, other things being equal, because no matching network is required.

BIBLIOGRAPHY

Source material and more extended discussions of topics covered in this chapter can be found in the references given below and in the textbooks listed at the end of Chapter 2.

B. A. Eggers, "An Analysis of the Balun," *QST*, Apr 1980, pp 19-21.

D. Geiser, "Resistive Impedance Matching with Quarter-Wave Lines," *QST*, Feb 1963, pp 56-57.

J. D. Gooch, O. E. Gardner, and G. L. Roberts, "The Hairpin Match," *QST*, Apr 1962, pp 11-14, 146, 156.

G. Grammer, "Simplified Design of Impedance-Matching Networks," in three parts, *QST*, Mar, Apr and May 1957.

D. J. Healey, III, "An Examination of the Gamma Match," *QST*, Apr 1969, pp 11-15, 57.

J. D. Kraus and S. S. Sturgeon, "The T-Matched Antenna," *QST*, Sep 1940, pp 24-25.

M. W. Maxwell, "Another Look at Reflections," *QST*, Apr, Jun, Aug and Oct 1973, Apr and Dec 1974, and Aug 1976.

M. W. Maxwell, "Some Aspects of the Balun Problem," *QST*, Mar 1983, pp 38-40.

F. A. Reiger, "Series-Section Transmission-Line Impedance Matching," *QST*, Jul 1978, pp 14-16.

J. Sevick, "Simple Broadband Matching Networks," *QST*, Jan 1976, pp 20-23.

R. E. Stephens, "Admittance Matching the Ground-Plane Antenna to Coaxial Transmission Line," Technical Correspondence, *QST*, Apr 1973, pp 55-57.

Chapter 27

Antenna and Transmission-Line Measurements

The principal quantities to be measured on transmission lines are line current or voltage, and standing-wave ratio. Measurements of current or voltage are made for the purpose of determining the power input to the line. SWR measurements are useful in connection with the design of coupling circuits and the adjustment of the match between the antenna and transmission line, as well as in the adjustment of matching circuits.

For most practical purposes a *relative* measurement is sufficient. An uncalibrated indicator that shows when the largest possible amount of power is being put into the line is just as useful, in most cases, as an instrument that measures the power accurately. It is seldom necessary to know the actual number of watts going into the line unless the overall efficiency of the system is being investigated. An instrument that shows when the SWR is close to 1 to 1 is all that is needed for most impedance-matching adjustments. Accurate measurement of SWR is necessary only in studies of antenna characteristics such as bandwidth, or for the design of some types of matching systems, such as a stub match.

Quantitative measurements of reasonable accuracy demand good design and careful construction in the measuring instruments. They also require intelligent use of the equipment, including a knowledge not only of its limitations but also of stray effects that often lead to false results. Until the complete conditions of the measurements are known, a certain amount of skepticism regarding numerical data resulting from amateur measurements with simple equipment is justified. On the other hand, purely qualitative or relative measurements are easy to make and are reliable for the purposes mentioned above.

LINE CURRENT AND VOLTAGE

A current or voltage indicator that can be used with coaxial line is a useful piece of equipment. It need not be elaborate or expensive. Its principal function is to show when the maximum power is being taken from the transmitter; for any given set of line conditions (length, SWR, etc) this will occur when the transmitter coupling is adjusted for maximum current or voltage at the input end of the line. Although the final-amplifier plate or collector current meter is frequently used for this purpose, it is not always a reliable indicator. In many cases, particularly with a screen-grid tube in the final stage, minimum loaded plate current does not occur simultaneously with maximum power output.

RF VOLTMETER

A germanium diode in conjunction with a low-range milliammeter and a few resistors can be assembled to form an RF voltmeter suitable for connecting across the two conductors of a coaxial line, as shown in Fig 1. It consists of a voltage divider, R1-R2, having a total resistance about 100 times the Z_0 of the line (so the power consumed will be negligible) with a diode rectifier and milliammeter connected across part of the divider to read relative RF voltage. The purpose of R3 is to make the meter readings directly proportional to the applied voltage, as nearly as possible, by "swamping" the resistance of D1, since the diode resistance will vary with the amplitude of the current through the diode.

The voltmeter may be constructed in a small metal box, indicated by the dashed line in the drawing, and fitted with coax receptacles. R1 and R2 should be composition resistors. The power rating for R1 should be 1 watt for each 100 watts

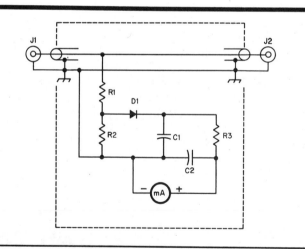

Fig 1—RF voltmeter for coaxial line.
C1, C2—0.005- or 0.01-µF ceramic.
D1—Germanium diode, 1N34A.
J1, J2—Coaxial fittings, chassis-mounting type.
M1—0-1 milliammeter (more sensitive meter may be used if desired; see text).
R1—6.8 kΩ, composition, 1 watt for each 100 watts of RF power.
R2—680 Ω, ½- or 1-watt composition.
R3—10 kΩ, ½ watt (see text).

of carrier power in the matched line; separate 1- or 2-watt resistors should be used to make up the total power rating required, to a total resistance as given. Any type of resistor can be used for R3; the total resistance should be such that about 10 volts dc will be developed across it at full scale. For example, a 0-1 milliammeter would require 10 kΩ, a 0-500 microammeter would take 20 kΩ, and so on. For comparative measurements only, R3 may be a variable resistor so the sensitivity can be adjusted for various power levels.

In constructing such a voltmeter, care should be used to prevent inductive coupling between R1 and the loop formed by R2, D1 and C1, and between the same loop and the line conductors in the assembly. With the lower end of R1 disconnected from R2 and grounded to the enclosure, but without changing its position with respect to the loop, there should be no meter indication when full power is going through the line.

If more than one resistor is used for R1, the units should be arranged end to end with very short leads. R1 and R2 should be kept ½ inch or more from metal surfaces parallel to the body of the resistor. If these precautions are observed the voltmeter will give consistent readings at frequencies up to 30 MHz. Stray capacitance and stray coupling limit the accuracy at higher frequencies but do not affect the utility of the instrument for comparative measurements.

Calibration

The meter may be calibrated in RF voltage by comparison with a standard such as an RF ammeter. This requires that the line be well matched so the impedance at the point of measurement is equal to the actual Z_0 of the line. Since in that case $P = I^2 Z_0$, the power can be calculated from the current. Then $E = \sqrt{P Z_0}$. By making current and voltage measurements at a number of different power levels, enough points may be obtained to permit drawing a calibration curve for the voltmeter.

RF AMMETERS

An RF ammeter can be mounted in any convenient location at the input end of the transmission line, the principal precaution in its mounting being that the capacitance to ground, chassis, and nearby conductors should be low. A bakelite-case instrument can be mounted on a metal panel without introducing enough shunt capacitance to ground to cause serious error up to 30 MHz. When a metal-case instrument is installed on a metal panel it should be mounted on a separate sheet of insulating material in such a way that there is 1/8 inch or more separation between the edge of the case and the metal.

A 2-inch instrument can be mounted in a 2 × 4 × 4-inch metal box as shown in Fig 2. This is a convenient arrangement for use with coaxial line.

Installed this way, a good quality RF ammeter will measure current with an accuracy that is entirely adequate for calculating power in the line. As discussed above in connection with calibrating RF voltmeters, the line must be closely matched by its load so the actual impedance will be resistive and equal to Z_0. The scales of such instruments are cramped at the low end, however, which limits the range of power that can be measured by a single meter. The useful current range is about 3 to 1, corresponding to a power range of about 9 to 1.

Fig 2—A convenient method of mounting an RF ammeter for use in a coaxial line. This is a metal-case instrument mounted on a thin bakelite panel. The cutout in the metal clears the edge of the meter by about 1/8 in.

SWR Measurements

On parallel-conductor lines it is possible to measure the standing-wave ratio by moving a current (or voltage) indicator along the line, noting the maximum and minimum values of current (or voltage) and then computing the SWR from these measured values. This cannot be done with coaxial line since it is not possible to make measurements of this type inside the cable. The technique is, in fact, seldom used with open lines, because it is not only inconvenient but sometimes impossible to reach all parts of the line conductors. Also, the method is subject to considerable error from antenna currents flowing on the line.

Present-day SWR measurements made by amateurs practically always use some form of "directional coupler" or RF bridge circuit. The indicator circuits themselves are fundamentally simple, but considerable care is required in their construction if the measurements are to be accurate. The requirements for indicators used only for the adjustment of impedance-matching circuits, rather than actual SWR measurement, are not so stringent and an instrument for this purpose can be made easily.

BRIDGE CIRCUITS

Two commonly used bridge circuits are shown in Fig 3. The bridges consist essentially of two voltage dividers in parallel, with a voltmeter connected between the junctions of each pair of "arms," as the individual elements are called. When the equations shown to the right of each circuit are satisfied there is no potential difference between the two junctions, and the voltmeter indicates zero voltage. The bridge is then said to be in "balance."

Taking Fig 3A as an illustration, if $R1 = R2$, half the applied voltage, E, will appear across each resistor. Then if $R_s = R_x$, $\frac{1}{2} E$ will appear across each of these resistors and the voltmeter reading will be zero. Remember that a matched transmission line has a purely resistive input impedance, and suppose that the input terminals of such a line are substituted for R_x. Then if R_s is a resistor equal to the Z_0 of the line, the bridge will be balanced. If the line is not perfectly matched, its input impedance will not equal Z_0 and hence will not equal R_s, since the latter is chosen to equal Z_0. There will then be a difference in potential between points X and Y, and the voltmeter will show a reading. Such a bridge therefore can be used to show the presence of standing waves on the line, because the line input impedance will be equal to Z_0 only when there are no standing waves.

Considering the nature of the incident and reflected components of voltage that make up the actual voltage at the input terminals of the line, as discussed in Chapter 24, it should be clear that when $R_s = Z_0$, the bridge is always in balance for the incident component. Thus the voltmeter does not respond to the incident component at any time but reads only the reflected component (assuming that R2 is very small compared with the voltmeter impedance). The incident component can be measured across either R1 or R2, if they are equal resistances. The standing-wave ratio is then

$$SWR = \frac{E1 + E2}{E1 - E2} \qquad \text{(Eq 1)}$$

where E1 is the incident voltage and E2 is the reflected voltage. It is often simpler to normalize the voltages by expressing E2 as a fraction of E1, in which case the formula becomes

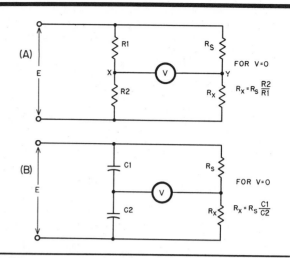

Fig 3—Bridge circuits suitable for SWR measurement. A—Wheatstone type using resistance arms. B—Capacitance-resistance bridge ("Micromatch"). Conditions for balance are independent of frequency in both types.

$$SWR = \frac{1 + k}{1 - k} \qquad \text{(Eq 2)}$$

where $k = E2/E1$.

The operation of the circuit in Fig 3B is essentially the same, although this circuit has arms containing reactance as well as resistance.

It is not necessary that $R1 = R2$ in Fig 3A; the bridge can be balanced, in theory, with any ratio of these two resistances provided R_s is changed accordingly. In practice, however, the accuracy is highest when the two are equal; this circuit is generally so used.

A number of types of bridge circuits appear in Fig 4, many of which have been used in amateur products or amateur construction projects. All except that at G can have the generator and load at a common potential. At G, the generator and detector are at a common potential. The positions of the detector and transmitter (or generator) may be interchanged in the bridge, and this may be an advantage in some applications.

Bridges shown at D, E, F and H may have one terminal of the generator, detector and load common. Bridges at A, B, E, F, G and H have constant sensitivity over a wide frequency range. Bridges at B, C, D and H may be designed to show no discontinuity (impedance lump) with a matched line, as shown in the drawing. Discontinuities with A, E and F may be small.

Bridges are usually most sensitive when the detector bridges the midpoint of the generator voltage, as in G or H, or in B when each resistor equals the load impedance. Sensitivity also increases when the currents in each leg are equal.

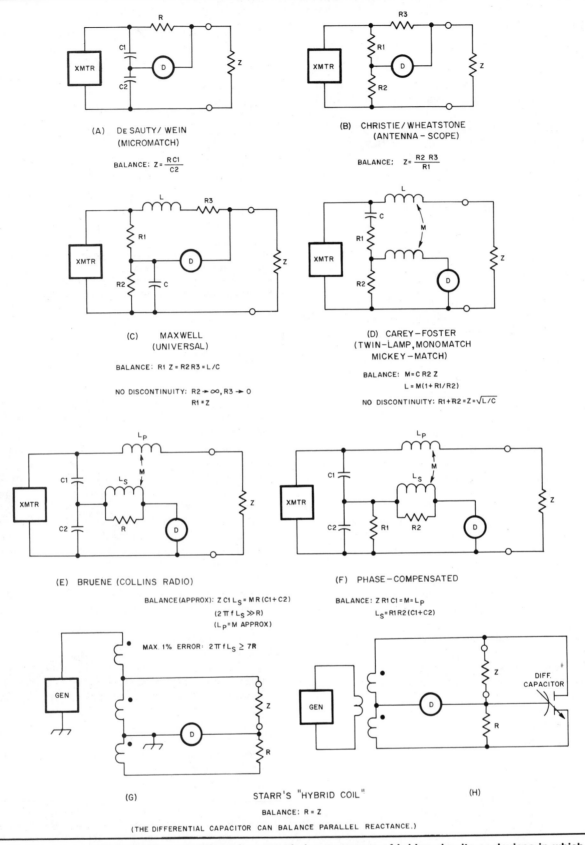

(A) DE SAUTY/ WEIN
(MICROMATCH)

BALANCE: $Z = \dfrac{R C1}{C2}$

(B) CHRISTIE/ WHEATSTONE
(ANTENNA – SCOPE)

BALANCE: $Z = \dfrac{R2\ R3}{R1}$

(C) MAXWELL
(UNIVERSAL)

BALANCE: $R1\ Z = R2\ R3 = L/C$

NO DISCONTINUITY: $R2 \to \infty,\ R3 \to 0$
$R1 = Z$

(D) CAREY – FOSTER
(TWIN – LAMP, MONOMATCH
MICKEY – MATCH)

BALANCE: $M = C\ R2\ Z$
$L = M(1 + R1/R2)$

NO DISCONTINUITY: $R1 + R2 = Z = \sqrt{L/C}$

(E) BRUENE (COLLINS RADIO)

BALANCE (APPROX): $Z\ C1\ L_S = M R (C1 + C2)$
$(2\pi f\ L_S \gg R)$
$(L_P = M\ \text{APPROX})$

MAX 1% ERROR: $2\pi f\ L_S \geq 7R$

(F) PHASE – COMPENSATED

BALANCE: $Z\ R1\ C1 = M = L_P$
$L_S = R1\ R2\ (C1 + C2)$

(G)

STARR'S "HYBRID COIL"

(H)

BALANCE: $R = Z$

(THE DIFFERENTIAL CAPACITOR CAN BALANCE PARALLEL REACTANCE.)

Fig 4—Various types of SWR indicator circuits and commonly known names of bridge circuits or devices in which they have been used. Detectors (D) are usually semiconductor diodes with meters, isolated with RF chokes and capacitors. However, the detector may be a radio receiver. In each circuit, Z represents the load being measured. *(This information provided by David Geiser, WA2ANU)*

Resistance Bridge

The basic bridge type shown in Fig 3A may be home constructed and is reasonably accurate for SWR measurement. A practical circuit for such a bridge is given in Fig 5 and a representative layout is shown in Fig 6. Properly built, a bridge of this design can be used for measurement of standing-wave ratios up to about 15 to 1 with good accuracy.

Important constructional points to be observed are:

1) Keep leads in the RF circuit short, to reduce stray inductance.

2) Mount resistors two or three times their body diameter away from metal parts, to reduce stray capacitance.

3) Place the RF components so there is as little inductive and capacitive coupling as possible between the bridge arms.

In the instrument shown in Fig 6, the input and line connectors, J1 and J2, are mounted fairly close together so the standard resistor, R_s, can be supported with short leads directly between the center terminals of the connectors. R2 is mounted at right angles to R_s, and a shield partition is used between these two components and the others.

The two 47-kΩ resistors, R5 and R6 in Fig 5, are voltmeter multipliers for the 0-100 microammeter used as an indicator. This is sufficient resistance to make the voltmeter linear (that is, the meter reading is directly proportional to the RF voltage) and no voltage calibration curve is needed. D1 is the rectifier for the reflected voltage and D2 is for the incident voltage. Because of manufacturing variations in resistors and diodes, the readings may differ slightly with two multipliers of the same nominal resistance value, so a correction resistor, R3, is included in the circuit. Its value should be selected so that the meter reading is the same with S1 in either position, when RF is applied to the bridge with the line connection open. In the instrument shown, a value of 1000 ohms was required in series with the multiplier for reflected voltage; in other cases different values probably would be needed and R3 might have to be put in series with the multiplier for the incident voltage. This can be determined by experiment.

The value used for R1 and R2 is not critical, but the two resistors should be matched to within 1 or 2 percent if possible. The resistance of R_s should be as close as possible to the actual Z_0 of the line to be used (generally 52 or 75 ohms). The resistor should be selected by actual measurement with an accurate resistance bridge, if one is available.

R4 is for adjusting the incident-voltage reading to full scale in the measurement procedure described below. Its use is not essential, but it offers a convenient alternative to exact adjustment of the RF input voltage.

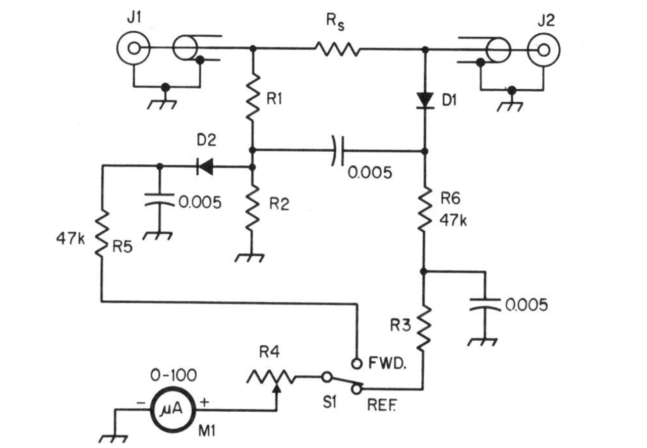

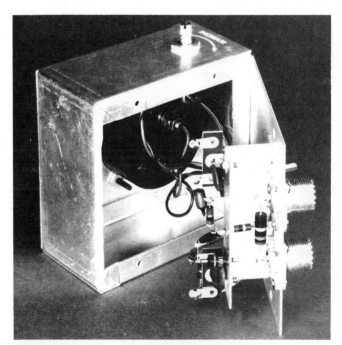

Fig 5—Resistance bridge for SWR measurement. Capacitors are disc ceramic. Resistors are ½-watt composition except as noted below.

D1, D2—Germanium diode, high back resistance type (1N34A, 1N270, etc)
J1, J2—Coaxial connectors, chassis-mounting type.
M1—0-100 dc microammeter.
R1, R2—47 Ω, ½-watt composition (see text).
R3—See text.
R4—50-kΩ volume control.
R_s—Resistance equal to line Z_0 (½ or 1 watt composition).
S1—SPDT toggle.

Fig 6—A 2 × 4 × 4-in. aluminum box is used to house this SWR bridge, which uses the circuit of Fig 5. The variable resistor, R4, is mounted on the side. The bridge components are mounted on one side plate of the box and a subchassis formed from a piece of aluminum. The input connector is at the top in this view. R_s is connected directly between the two center posts of the connectors. R2 is visible behind it and perpendicular to it. One terminal of D1 projects through a hole in the chassis so the lead can be connected to J2. R1 is mounted vertically to the left of the chassis in this view, with D2 connected between the junction of R1-R2 and a tie point.

Testing

R1, R2 and R_s should be measured with a reliable ohmmeter or resistance bridge after wiring is completed, in order to make sure their values have not changed from the heat of soldering. Disconnect one side of the microammeter and leave the input and output terminals of the unit open during such measurements, in order to avoid stray shunt paths through the rectifiers.

Check the two voltmeter circuits as described above, applying enough RF (about 10 volts) to the input terminals to give a full-scale reading with the line terminals open. If necessary, try different values for R3 until the reading is the same with S1 in either position.

With J2 open, adjust the RF input voltage and R4 for full-scale reading with S1 in the incident-voltage position. Then switch S1 to the reflected-voltage position. The reading should remain at full scale. Next, short-circuit J2 by touching a screwdriver between the center terminal and the frame of the connector to make a low-inductance short. Switch S1 to the incident-voltage position and readjust R4 for full scale, if necessary. Then throw S1 to the reflected-voltage position, keeping J2 shorted, and the reading should be full scale as before. If the readings differ, R1 and R2 are not the same value, or there is stray coupling between the arms of the bridge. It is necessary that the reflected voltage read full scale with J2 either open or shorted, when the incident voltage is set to full scale in each case, in order to make accurate SWR measurements.

The circuit should pass these tests at all frequencies at which it is to be used. It is sufficient to test at the lowest and highest frequencies, usually 1.8 or 3.5 and 28 or 50 MHz. If R1 and R2 are poorly matched but the bridge construction is otherwise good, discrepancies in the readings will be substantially the same at all frequencies. A difference in behavior at the low and high ends of the frequency range can be attributed to stray coupling between bridge arms, or stray inductance or capacitance in the arms.

To check the bridge for balance, apply RF and adjust R4 for full scale with J2 open. Then connect a resistor identical with R_s (the resistance should match within 1 or 2 percent) to the line terminals, using the shortest possible leads. It is convenient to mount the test resistor inside a cable connector (PL-259), a method of mounting that also minimizes lead inductance. When the test resistor is connected, the reflected-voltage reading should drop to zero. The incident voltage should be reset to full scale by means of R4, if necessary. The reflected reading should be zero at any frequency in the range to be used. If a good null is obtained at low frequencies but some residual current shows at the high end, the trouble may be the inductance of the test resistor leads, although it may also be caused by stray coupling between the arms of the bridge itself. If there is a constant low (but not zero) reading at all frequencies the cause is poor matching of the resistance values. Both effects can be present simultaneously. A good null must be obtained at all frequencies before the bridge is ready for use.

Bridge Operation

The RF power input to a bridge of this type must be limited to a few watts at most, because of the power-dissipation ratings of the resistors. If the transmitter has no

provision for reducing power output to a very low value—less than 5 watts—a simple "power absorber" circuit can be made up as shown in Fig 7. The lamp DS1 tends to maintain constant current through the resistor over a fairly wide power range, so the voltage drop across the resistor also tends to be constant. This voltage is applied to the bridge, and with the constants given is in the right range for resistance-type bridges.

To make the measurement, connect the unknown to J2 and apply sufficient RF voltage to J1 to give a full-scale incident-voltage reading. Use R4 to set the indicator to exactly full scale. Then throw S1 to the reflected-voltage position and note the meter reading. The SWR is then found by substituting the readings in Eq 1.

For example, if the full-scale calibration of the dc instrument is 100 μA and the reading with S2 in the reflected-voltage position is 40 μA, the SWR is

$$SWR = \frac{100 + 40}{100 - 40} = \frac{140}{60} = 2.33 \text{ to } 1$$

Instead of determining the SWR value by calculations, the *voltage* curve of Fig 8 may be used. In this example the ratio of reflected to forward voltage is $40/100 = 0.4$, and from Fig 8 the SWR value is seen to be about 2.3 to 1.

The meter scale may be calibrated in any arbitrary units so long as the scale has equal divisions, since it is the ratios of the voltages, and not the actual values, that determine the SWR.

AVOIDING ERRORS IN SWR MEASUREMENTS

The principal causes of inaccuracies within the bridge are differences in the resistances of R1 and R2, stray inductance and capacitance in the bridge arms, and stray coupling between arms. If the checking procedure described above is followed carefully, the bridge of Fig 5 should be amply

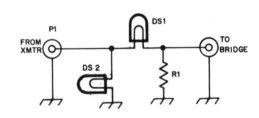

Fig 7—"Power absorber" circuit for use with resistance-type SWR bridges when the transmitter has no special provisions for power reduction. For RF powers up to 50 watts, DS1 is a 117-volt 40-watt incandescent lamp and DS2 is not used. For higher powers, use sufficient additional lamp capacity at DS2 to load the transmitter to about normal output; for example, for 250 watts output DS2 may consist of two 100-watt lamps in parallel. R1 is made from three 1-watt 68-Ω resistors connected in parallel. P1 and P2 are cable-mounting coaxial connectors. Leads in the circuit formed by the lamps and R1 should be kept short, but convenient lengths of cable may be used between this assembly and the connectors.

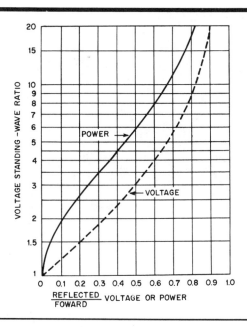

Fig 8—Chart for finding voltage standing-wave ratio when the ratio of reflected-to-forward voltage or reflected-to-forward power is known.

accurate for practical use. The accuracy is highest for low standing-wave ratios because of the nature of the SWR calculation; at high ratios the divisor in the equation above represents the difference between two nearly equal quantities, so a small error in voltage measurement may mean a considerable difference in the calculated SWR.

The standard resistor R_s must equal the actual Z_0 of the line. The actual Z_0 of a sample of line may differ by a few percent from the nominal figure because of manufacturing variations, but this has to be tolerated. In the 52- to 75-ohm range, the RF resistance of a composition resistor of ½- or 1-watt rating is essentially identical with its dc resistance.

"Antenna" Currents

As explained in Chapter 26, there are two ways in which "parallel" or "antenna" currents can be caused to flow on the *outside* of a coaxial line—currents induced on the line because of its spatial relationship to the antenna, and currents that result from the direct connection between the coax outer conductor and (usually) one side of the antenna. The induced current usually will not be troublesome if the bridge and the transmitter (or other source of RF power for operating the bridge) are shielded so that any RF currents flowing on the outside of the line cannot find their way into the bridge. This point can be checked by "cutting in" an additional section of line (1/8 to 1/4 electrical wavelength, preferably) of the same Z_0. The SWR indicated by the bridge should not change except possibly for a slight decrease because of the additional line loss. If there is a marked change, better shielding may be required.

Parallel-type currents caused by the connection to the antenna will cause a change in SWR with line length, even though the bridge and transmitter are well shielded and the shielding is maintained throughout the system by the use of coaxial fittings. Often, merely moving the transmission line around will cause the indicated SWR to change. This is because the outside of the coax tends to become part of the antenna system, being connected to the antenna at the feed point, and so constitutes a load on the line, along with the desired load represented by the antenna itself. The SWR on the line then is determined by the composite load of the antenna and the outside of the coax, and since changing the line length (or position) changes one component of this composite load, the SWR changes too.

The remedy for such a situation is to use a good balun or to detune the outside of the line by proper choice of length. It is well to note that this is not a measurement error, since what the instrument reads is the actual SWR on the line. However, it is an undesirable condition since the line is operating at a higher SWR than it should—and would—if the parallel-type current on the outside of the coax were eliminated.

Spurious Frequencies

Off-frequency components in the RF voltage applied to the bridge may cause considerable error. The principal components of this type are harmonics and low-frequency subharmonics that may be fed through the final stage of the transmitter driving the bridge. The antenna is almost always a fairly selective circuit, and even though the system may be operating with a very low SWR at the desired frequency, it is practically always mismatched at harmonic and sub-harmonic frequencies. If such spurious frequencies are applied to the bridge in appreciable amplitude, the SWR indication will be very much in error. In particular, it may not be possible to obtain a null on the bridge with any set of adjustments of the matching circuit. The only remedy is to filter out the unwanted components by increasing the selectivity of the circuits between the transmitter final amplifier and the bridge.

REFLECTOMETERS

Low-cost reflectometers that do not have a guaranteed wattmeter calibration are not ordinarily reliable for accurate numerical measurement of standing-wave ratio. They are, however, very useful as aids in the adjustment of matching networks, since the objective in such adjustment is to reduce the reflected voltage or power to zero. Relatively inexpensive devices can be used for this, since only good bridge balance is required, not actual calibration. Bridges of this type are usually "frequency-sensitive"—that is, the meter response becomes greater with increasing frequency, for the same applied voltage. When matching and line monitoring, rather than SWR measurement, is the principal use of the device, this is not a serious handicap.

Various simple reflectometers, useful for matching and monitoring, have been described from time to time in *QST* and in *The ARRL Handbook*. Because most of these are frequency sensitive, it is difficult to calibrate them accurately for power measurement, but their low cost and suitability for use at moderate power levels, combined with the ability to show accurately when a matching circuit has been properly adjusted, make them a worthwhile addition to the amateur station.

An In-Line RF Wattmeter

Considerable attention was devoted to the resistance-type SWR bridge in the preceding section because it is the simplest type that is capable of adequate accuracy in measuring standing-wave ratio. Its disadvantage is that it must be operated at a very low power level, and thus is not suitable for continuous monitoring of the SWR in actual transmission. To do this the instrument must be capable of carrying the entire power output of the transmitter, and should do it with negligible loss. An RF wattmeter meets this requirement.

It is neither costly nor difficult to build an RF wattmeter. And, if the instrument is equipped with a few additional components, it can be switched to read reflected power as well as forward power. With this feature the instrument can be used as an SWR meter for antenna matching and Transmatch adjustments. The wattmeter shown in Figs 9 through 12 meets these requirements. The instrument uses a directional type of coupler for sampling the energy on the transmission line. The indicator sensitivity of this instrument is not related to frequency, as is the case with some types of directional couplers. This unit may be calibrated for power levels as low as 1 watt, full scale, in any part of the HF spectrum. With suitable calibration, it has good accuracy over the 3-30 MHz range. It is built in two parts, an RF head for inserting in the coaxial transmission line, and a control-meter box which can be placed in any location where it can be operated conveniently. Only direct current flows in the cable connecting the two pieces.

Design Philosophy

See the circuit of Fig 10. The transmission line center conductor, W1, passes through the center of a toroid core and becomes the primary of T1. The multiturn winding on the core functions as the transformer secondary. Current flowing on W1 induces a voltage in the secondary which causes a current to flow through resistors R1 and R2. The voltage drops across these resistors are equal in amplitude, but 180° out of phase with respect to common or ground. They are thus, for practical purposes, respectively in and out of phase with the line current. Capacitive voltage dividers, C1-C3 and C2-C4, are connected across the line to obtain equal-amplitude voltages *in phase* with the line voltage, the division ratio being adjusted so that these voltages match the voltage drops across R1 and R2 in amplitude. (As the current/voltage ratio in the line depends on the load, this can be done only for a particular value of load impedance. Load values chosen for this standardization are pure resistances that match the characteristic impedance of the transmission line with which the bridge is to be used, 52 or 75 ohms usually.) Under these conditions, the voltages rectified by D1 and D2 represent, in the one case, the vector *sum* of the voltages caused by the line current and voltage, and in the other, the vector *difference*. With respect to the resistance for which the circuit has been set up, the sum is proportional to the forward component of a traveling wave such as occurs on a transmission line, and the difference is proportional to the reflected component.

Component Selection

R1 and R2 should be selected for the best null reading when adjusting the bridge into a resistive 52- or 75-ohm load. Normally, the value will be somewhere between 10 and 47 ohms. The 10-ohm value worked well with the instruments shown here. Half-watt composition resistors are suitable to

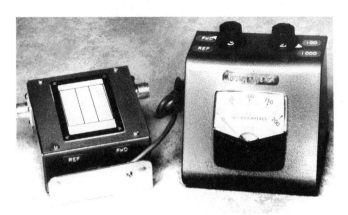

Fig 9—The RF wattmeter consists of two parts, the RF head (left), and the control-meter box (right). The paper scale affixed to the RF head contains the calibration information which appears in Fig 10.

30 MHz. R1 and R2 should be as closely matched in resistance as possible. Their exact value is not critical, so an ohmmeter may be used to match them.

Ideally, C3 and C4 should be matched in value. Silver-mica capacitors are usually close enough in tolerance that special selection is not required, providing there is enough leeway in the ranges of C1 and C2 to compensate for any difference in the values of C3 and C4.

Diodes D1 and D2 should also be matched for best results. An ohmmeter can be used to select a pair of diodes having forward dc resistances within a few ohms of being the same. Similarly, the back resistances of the diodes can be matched. The matched diodes will help to assure equal meter readings when the bridge is reversed. (The bridge should be perfectly bilateral in its performance characteristics.) Germanium diodes should be used to avoid misleading results when low values of reflected power are present during antenna adjustments. The SWR can appear to be perfect when actually it isn't. The germanium diodes conduct at approximately 0.3 volt, making them more suitable for low-power readings than silicon diodes (conduction at 0.7 V).

Any meter having a full-scale reading between 50 μA and 1 mA can be used at M1. The more sensitive the meter, the more difficult it will be to get an absolute reflected-power reading of zero. Some residual current will flow in the bridge circuit no matter how carefully the circuit is balanced, and a sensitive instrument will indicate this current flow. Also, the more sensitive the meter, the larger will have to be the calibrating resistances, R3 through R6, to provide high-power readings. A 0-200 μA meter represents a good compromise for power ranges between 100 and 2000 watts.

Construction

It is important that the layout of any RF bridge be as symmetrical as possible if good balance is to be had. The circuit-board layout of Fig 12 meets the requirement for this instrument. The input and output ports of the equipment should be isolated from the remainder of the circuit so that only the sampling circuits feed voltage to the bridge. A shield across the end of the box which contains the input and output jacks and W1 is necessary. If stray RF gets into the bridge

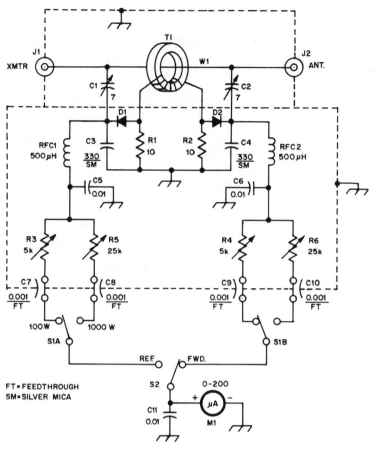

Fig 10—Schematic diagram of the RF wattmeter. A calibration scale for M1 is shown also. Fixed-value resistors are ½-watt composition. Fixed-value capacitors are disc ceramic unless otherwise noted. Decimal-value capacitances are in microfarads. Others are picofarads. Resistance is in ohms; k = 1000.

C1, C2—1.3- to 6.7-pF miniature trimmer (E. F. Johnson 189-502-4, available from Newark Electronics, Chicago, Illinois)
C3-C11, incl—Numbered for circuit-board identification.
D1, D2—Matched small-signal germanium diodes, 1N34A, etc (see text).
J1, J2—Chassis-mount coax connector of builder's choice. Type SO-239 used here.
M1—0-200 μA meter (Triplett type 330-M used here).
R1, R2—Matched 10-Ω resistors (see text).
R3, R4—5-kΩ printed-circuit carbon control (IRC R502-B).
R5, R6—25-kΩ printed-circuit carbon control (IRC R242-B).
RFC1, RFC2—500-μH RF choke (Millen 34300-500 or similar).
S1—DPDT single-section phenolic wafer switch (Mallory 3222J).
S2—SPDT phenolic wafer switch (Centralab 1460).
T1—Toroidal transformer; 35 turns of no. 26 enam wire to cover entire core of Amidon T-68-2 toroid (available from Amidon Assoc or Radiokit).
W1—Numbered for text discussion.

WATTS	M1	WATTS
100	200	1000
90	180	900
80	170	800
70	155	700
60	145	600
50	125	500
40	105	400
30	85	300
20	65	200
10	40	100
5	20	50

circuit, it will be impossible to obtain a complete zero reflected-power reading on M1 even though a 1:1 SWR exists.

All of the RF head components except J1, J2 and the feedthrough capacitors are assembled on the board. The board is held in place by means of a homemade aluminum L bracket at the end nearest T1. The circuit board end nearest the feedthrough capacitors is secured with a single no. 6 spade bolt. Its hex nut is outside the box, and is used to secure a solder lug which serves as a connection point for the ground braid in the cable which joins the control box to the RF head.

T1 fits into a cutout area of the circuit board. A 1-inch long piece of RG-8 coax is stripped of its vinyl jacket and shield braid, and is snug-fit into the center hole of T1. The inner conductor is soldered to the circuit board to complete the W1 connection between J1 and J2.

The upper dashed lines of Fig 10 represent the shield partition mentioned above. It can be made from flashing copper or thin brass.

The control box, a sloping-panel utility cabinet measuring 4 × 5 inches, houses S1, S2 and the meter, M1. Four-conductor shielded cable—the shield serving as the common lead—is used to join the two pieces. There is no reason the entire instrument cannot be housed in one container, but it is sometimes awkward to have coaxial cables attached to a unit that occupies a prominent place in the operating position. When built as shown, the two-piece instrument permits the RF pickup head to be concealed behind the transmitter, while the control head can be mounted where it is accessible to the operator.

Adjustment

Perhaps the most difficult task faced by the constructor

Fig 11—Top view of the RF head for the circuit of Fig 10. A flashing-copper shield isolates the primary RF line and T1 from the rest of the circuit. The second shield (thicker) is not required and can be eliminated from the circuit. If a 2000-watt scale is desired, fixed-value resistors of approximately 22 kΩ can be connected in series with high-range printed-circuit controls. Instead, the 25-kΩ controls shown here can be replaced by 50-kΩ units.

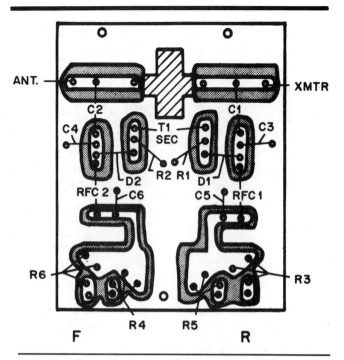

Fig 12—Etching pattern and parts layout for the RF wattmeter, as viewed from the foil side of the board. The etched-away portions of the foil are shown as darkened areas in this drawing. The area with diagonal lines is to be cut out for the mounting of T1.

is that of calibrating the power meter for a desired wattage range. The least involved method is to use a commercial wattmeter as a standard. If one is not available, the power output of the test transmitter can be computed by means of an RF ammeter in series with a 52-ohm dummy load, using the standard formula, $P = I^2R$. The calibration chart of Fig 10 is representative, but the actual calibration of a particular instrument will depend on the diodes used at D1 and D2. Frequently, individual scales are required for the two power ranges.

Connect a noninductive 52-ohm dummy load to J2. A Heath Cantenna or similar load will serve nicely for adjustment purposes. Place S2 in the FORWARD position, and set S1 for the 100-watt range. An RF ammeter or calibrated power meter should be connected between J2 and the dummy load during the tests, to provide power calibration points against which to plot the scale of M1. Apply transmitter output power to J1, gradually, until M1 begins to deflect upward. Increase the transmitter power and adjust R4 so that a full-scale meter reading occurs when 100 watts is indicated on the RF ammeter or other standard in use. Next, switch S2 to REFLECTED and turn the transmitter off. Temporarily short across R3, turn the transmitter on, and gradually increase power until a meter reading is noted. With an insulated screwdriver, adjust C2 for a null in the meter reading.

The next step is to reverse the coax-connections to J1 and J2. Place S2 in the REFLECTED position and apply transmitter power until the meter reads full scale at 100 watts output. In

this mode the REFLECTED position actually reads forward power because the bridge is reversed. Calibrating resistance R3 is set to obtain 100 watts full scale during this adjustment. Now, switch S2 to FORWARD and temporarily place a short across R4. Adjust C1 for a null reading on M1. Repeat the foregoing steps until no further improvement can be obtained. It will not be necessary to repeat the nulling adjustments on the 1000-watt range, but R5 and R6 will have to be adjusted to provide a full-scale meter reading at 1000 watts. If insufficient meter deflection is available for nulling adjustments on the 100-watt range, it may be necessary to adjust C1 and C2 at some power level higher than 100 watts. If the capacitors tune through a null, but the meter will not drop all the way to zero, chances are that some RF is leaking into the bridge circuit through stray coupling. If this is the case, it may be necessary to experiment with the shielding of the through-line section of the RF head. If only a small residual reading is noted it will be of minor importance and can be ignored.

With the component values given in Fig 10, the meter readings track for both power ranges. That is, the 10-watt level on the 100-watt range and the 100-watt point on the 1000-watt range fall at the same place on the meter scale, and so on. This no doubt results from the fact that the diodes are conducting in the most linear portion of their curve. Ordinarily, this desirable condition does not exist, making it necessary to plot separate scales for the different power ranges.

Tests indicate that the SWR caused by insertion of the

power meter in the transmission line is negligible. It was checked at 28 MHz and no reflected power could be noted on a commercially built RF wattmeter. Similarly, the insertion loss was so low that it could not be measured with ordinary instruments.

Operation

It should be remembered that when the bridge is used in a mismatched feed line that has not been properly matched at the antenna, a reflected-power reading will result. The reflected power must be subtracted from the forward power to obtain the actual power output. If the instrument is calibrated for, say, a 52-ohm line, the calibration will not hold for other values of line Z_0.

If the instrument is to be used for determining SWR, the reflected-to-forward power ratio can easily be converted into the corresponding voltage ratio for use in Eq 2 given earlier. Since power is proportional to voltage squared, the normalized formula becomes

$$SWR = \frac{1 + \sqrt{k}}{1 - \sqrt{k}}$$

where k is the ratio of reflected power to forward power. The *power* curve of Fig 8 is based on the above relationship, and may be used in place of the equation to determine the SWR.

An Inexpensive VHF Directional Coupler

Precision in-line metering devices capable of reading forward and reflected power over a wide range of frequencies are very useful in amateur VHF and UHF work, but their rather high cost puts them out of the reach of many VHF enthusiasts. The device shown in Figs 14 through 16 is an inexpensive adaptation of their basic principles. It can be made for the cost of a meter, a few small parts, and bits of copper pipe and fittings that can be found in the plumbing stocks at many hardware stores.

Construction

The sampler consists of a short section of handmade coaxial line, in this instance, of 52-ohms impedance, with a reversible probe coupled to it. A small pickup loop built into the probe is terminated with a resistor at one end and a diode at the other. The resistor matches the impedance of the loop, not the impedance of the line section. Energy picked up by the loop is rectified by the diode, and the resultant current is fed to a meter equipped with a calibration control.

The principal metal parts of the device are a brass plumbing T, a pipe cap, short pieces of 3/4-inch-ID and

Fig 14—Major components of the line sampler. The brass T and two end sections are at the upper left in this picture. A completed probe assembly is at the right. The N connectors have their center pins removed. The pins are shown with one inserted in the left end of the inner conductor and the other lying in the right foreground.

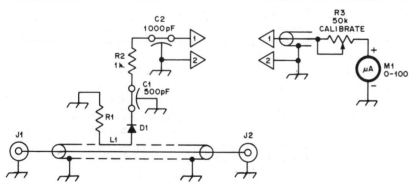

Fig 13—Circuit diagram for the line sampler.
C1—500-pF feedthrough capacitor, solder-in type.
C2—1000-pF feedthrough capacitor, threaded type.
D1—Germanium diode 1N34, 1N60, 1N270, 1N295, or similar.
J1, J2—Coaxial connector, type N (UG-58A).
L1—Pickup loop, copper strap 1 in. long × 3/16 in. wide. Bend into "C" shape with flat portion 5/8 in. long.
M1—0-100 μA meter.
R1—Composition resistor, 82 to 100 ohms. See text.
R3—50-kΩ composition control, linear taper.

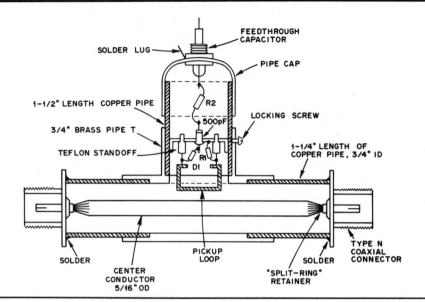

Fig 15—Cross-section view of the line sampler. The pickup loop is supported by two Teflon standoff insulators. The probe body is secured in place with one or more locking screws through holes in the brass T.

5/16-inch-OD copper pipe, and two coaxial fittings. Other available tubing combinations for 52-ohm line may be usable. The ratio of outer conductor ID to inner conductor OD should be 2.4/1. For a sampler to be used with other impedances of transmission line, see Chapter 24 for suitable ratios of conductor sizes. The photographs and Fig 15 show construction details.

Soldering of the large parts can be done with a 300-watt iron or a small torch. A neat job can be done if the inside of the T and the outside of the pipe are tinned before assembling. When the pieces are reheated and pushed together, a good mechanical and electrical bond will result. If a torch is used, go easy with the heat, as an over-heated and discolored fitting will not accept solder well.

Coaxial connectors with Teflon or other heat-resistant insulation are recommended. Type N, with split-ring retainers for the center conductors, are preferred. Pry the split-ring washers out with a knife point or small screwdriver. Don't lose them, as they'll be needed in the final assembly.

The inner conductor is prepared by making eight radial cuts in one end, using a coping saw with a fine-toothed blade, to a depth of ½ inch. The fingers so made are then bent together, forming a tapered end, as shown in Figs 14 and 15. Solder the center pin of a coaxial fitting into this, again being careful not to overheat the work.

In preparation for soldering the body of the coax connector to the copper pipe, it is convenient to use a similar fitting clamped into a vise as a holding fixture. Rest the T assembly on top, held in place by its own weight. Use the partially prepared center conductor to assure that the coax connector is concentric with the outer conductor. After being sure that the ends of the pipe are cut exactly perpendicular to the axis, apply heat to the coax fitting, using just enough so a smooth fillet of solder can be formed where the flange and pipe meet.

Before completing the center conductor, check its length. It should clear the inner surface of the connector by the thickness of the split ring on the center pin. File to length;

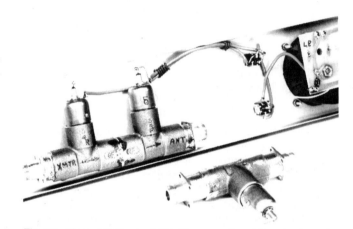

Fig 16—Two versions of the line sampler. The single unit described in detail here is in the foreground. Two sections in a single assembly provide for monitoring forward and reflected power without probe reversal.

if necessary, slot as with the other end, and solder the center pin in place. The fitting can now be soldered onto the pipe, to complete the 52-ohm line section.

The probe assembly is made from a 1½ inch length of the copper pipe, with a pipe cap on the top to support the upper feedthrough capacitor, C2. The coupling loop is mounted by means of small Teflon standoffs on a copper disc, cut to fit inside the pipe. The disc has four small tabs around the edge for soldering inside the pipe. The diode, D1, is connected between one end of the loop and a 500-pF feedthrough capacitor, C1, soldered into the disc. The terminating resistor, R1, is connected between the other end of the loop and ground, as directly as possible.

When the disc assembly is completed, insert it into the pipe, apply heat to the outside, and solder the tabs in place by melting solder into the assembly at the tabs. The position of the loop with respect to the end of the pipe will determine the sensitivity of a given probe. For power levels up to 200 watts the loop should extend beyond the face of the pipe about 5/32 inch. For use at higher power levels the loop should protrude only 3/32 inch. For operation with very low power levels the best probe position can be determined by experiment.

The decoupling resistor, R2, and feedthrough capacitor, C2, can be connected, and the pipe cap put in place. The threaded portion of the capacitor extends through the cap. Put a solder lug over it before tightening its nut in place. Fasten the cap with two small screws that go into threaded holes in the pipe.

Calibration

The sampler is very useful for many jobs even if it is not accurately calibrated, although it is desirable to calibrate it against a wattmeter of known accuracy. A good 52-ohm dummy load is required.

The first step is to adjust the inductance of the loop, or the value of the terminating resistor, for lowest reflected power reading. The loop is the easier to change. Filing it to reduce its width will increase its impedance. Increasing the cross-section of the loop will lower the impedance, and this can be done by coating it with solder. When the reflected power reading is reduced as far as possible, reverse the probe and calibrate for forward power by increasing the transmitter power output in steps and making a graph of the meter readings obtained. Use the calibration control, R3, to set the maximum reading.

Variations

Rather than to use one sampler for monitoring both forward and reflected power by repeatedly reversing the probe, it is better to make two assemblies by mounting two T fittings end-to-end, using one for forward and one for reflected power. The meter can be switched between the probes, or two meters can be used.

The sampler described was calibrated at 146 MHz, as it was intended for repeater use. On higher bands the meter reading will be higher for a given power level, and it will be lower for lower frequency bands. Calibration for two or three adjacent bands can be achieved by making the probe depth adjustable, with stops or marks to aid in resetting for a given band. Of course more probes can be made, with each probe calibrated for a given band, as is done in some of the commercially available units.

Other sizes of pipe and fittings can be used by making use of information given in Chapter 24 to select conductor sizes required for the desired impedances. (Since it is occasionally possible to pick up good bargains in 75-ohm line, a sampler for this impedance might be desirable.)

Type N fittings were used because of their constant impedance and their ease of assembly. Most have the splitting-ring retainer, which is simple to use in this application. Some have a crimping method, as do apparently all BNC connectors. If a fitting must be used and cannot be taken apart, drill a hole large enough to clear a soldering-iron tip in the copper-pipe outer conductor. A hole of up to 3/8-inch diameter will have very little effect on the operation of the sampler.

A Calorimeter For VHF And UHF Power Measurements

A quart of water in a Styrofoam ice bucket, a roll of small coaxial cable and a thermometer are all the necessary ingredients for an accurate RF wattmeter. Its calibration is independent of frequency. The wattmeter works on the calorimeter principle: A given amount of RF energy is equivalent to an amount of heat, which can be determined by measuring the temperature rise of a known quantity of thermally insulated material. This principle is used in many of the more accurate high-power wattmeters. This procedure was developed by James Bowen, WA4ZRP, and was first described in December 1975 QST.

The roll of coaxial cable serves as a dummy load to convert the RF power into heat. RG-174 cable was chosen for use as the dummy load in this calorimeter because of its high loss factor, small size, and low cost. It is a standard 52-ohm cable of approximately 0.11 inch diameter. A pre-packaged roll marked as 60 feet long, but measured to be 68 feet, was purchased at a local electronics store. A plot of measured RG-174 loss factor as a function of frequency is shown in Fig 17.

In use, the end of the cable not connected to the transmitter is left open circuited. Thus, at 50 MHz, the reflected wave returning to the transmitter (after making a round trip of 136 feet through the cable) is 6.7 dB × 1.36 = 9.11 dB below the forward wave. A reflected wave 9.11 dB down represents an SWR to the transmitter of 2.08:1. While this

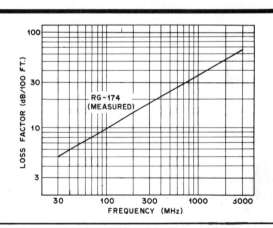

Fig 17—Loss factor of RG-174 coax used in the calorimeter.

value seems larger than would be desired, keep in mind that most 50-MHz transmitters can be tuned to match into an SWR of this magnitude efficiently. To assure accurate results, merely tune the transmitter for maximum power into the load before making the measurement. At higher frequencies the cable loss increases so the SWR goes down. Table 1 presents

Table 1

Calculated Input SWR for 68 Feet of Unterminated RG-174 Cable

Freq (MHz)	SWR
50	2.08
144	1.35
220	1.20
432	1.06
1296	1.003
2304	1.0003

Fig 18—The calorimeter ready for use. The roll of coaxial cable is immersed in one quart of water in the left-hand compartment of the Styrofoam container. Also shown is the thermometer, which doubles as a stirring rod.

the calculated input SWR values at several frequencies for 68 feet of RG-174. At 1000 MHz and above, the SWR caused by the cable connector will undoubtedly exceed the very low cable SWR listed for these frequencies.

In operation, the cable is submerged in a quart of water and dissipated heat energy flows from the cable into the water, raising the water temperature. See Fig 18. The calibration of the wattmeter is based on the physical fact that one calorie of heat energy will raise one gram of liquid water 1° Celsius. Since one quart of water contains 946.3 grams, the transmitter must deliver 946.3 calories of heat energy to the water to raise its temperature 1°C. One calorie of energy is equivalent to 4.186 joules and a joule is equal to 1 watt for 1 second. Thus, the heat capacitance of 1 quart of water expressed in joules is $946.3 \times 4.186 = 3961$ joules/°C.

The heat capacitance of the cable is small with respect to that of the water, but nevertheless its effect should be included for best accuracy. The heat capacitance of the cable was determined in the manner described below.

The 68-foot roll of RG-174 cable was raised to a uniform temperature of 100°C by immersing it in a pan of boiling water for several minutes. A quart of tap water was poured into the Styrofoam ice bucket and its temperature was measured at 28.7°C. The cable was then transferred quickly from the boiling water to the water in the ice bucket. After the water temperature in the ice bucket had ceased to rise, it measured 33.0°C. Since the total heat gained by the quart of water was equal to the total heat lost from the cable, we can write the following equation:

$$(\Delta T_{WATER})(C_{WATER}) = -(\Delta T_{CABLE})(C_{CABLE})$$

where

ΔT_{WATER} = the change in water temperature
C_{WATER} = the water heat capacitance
ΔT_{CABLE} = the change in cable temperature
C_{CABLE} = the cable heat capacitance

Substituting and solving:

$$(33.0 - 28.7)(3961) = -(33.0 - 100)(C_{CABLE})$$

$$\frac{(4.3)(3961)}{67} = C_{CABLE}$$

$$254 \text{ joules/}° = C_{CABLE}$$

Thus, the total heat capacitance of the water and cable in the calorimeter is $3961 + 254 = 4215$ joules/°C. Since 1°F = 5/9°C, the total heat capacitance can also be expressed as $4215 \times 5/9 = 2342$ joules/°F.

Materials and Construction

The quart of water and cable must be thermally insulated to assure that no heat is gained from or lost to the surroundings. A Styrofoam container is ideal for this purpose since Styrofoam has a very low thermal conductivity and a very low thermal capacitance. A local variety store was the source of a small Styrofoam cold chest with compartments for carrying sandwiches and drink cans. The rectangular compartment for sandwiches was found to be just the right size for holding the quart of water and coax.

The thermometer can be either a Celsius or Fahrenheit type, but try to choose one which has divisions for each degree spaced wide enough so that the temperature can be estimated readily to one-tenth degree. Photographic supply stores carry darkroom thermometers, which are ideal for this purpose. In general, glass bulb thermometers are more accurate than mechanical dial-pointer types.

The RF connector on the end of the cable should be a constant-impedance type. A BNC type connector especially designed for use on 0.11-inch diameter cable was located through surplus channels. If you cannot locate one of these, wrap plastic electrical tape around the cable near its end until the diameter of the tape wrap is the same as that of RG-58. Then connect a standard BNC connector for RG-58 in the normal fashion.

Carefully seal the opposite open end of the cable with plastic tape or silicone caulking compound so no water can leak into the cable at this point.

Procedure for Use

Pour 1 quart of water (4 measuring cups) into the

Styrofoam container. As long as the water temperature is not very hot or very cold, it is unnecessary to cover the top of the Styrofoam container during measurements. Since the transmitter will eventually heat the water several degrees, water initially a few degrees cooler than air temperature is ideal because the average water temperature will very nearly equal the air temperature and heat transfer to the air will be minimized.

Connect the RG-174 dummy load to the transmitter through the shortest possible length of lower loss cable such as RG-8. Tape the connectors and adapter at the RG-8 to the RG-174 joint carefully with plastic tape to prevent water from leaking into the connectors and cable at this point. Roll the RG-174 into a loose coil and submerge it in the water. Do not bind the turns of the coil together in any way, as the water must be able to freely circulate among the coaxial cable turns. All the RG-174 cable must be submerged in the water to ensure sufficient cooling. Also submerge part of the taped connector attached to the RG-174 as an added precaution.

Upon completing the above steps, quickly tune up the transmitter for maximum power output into the load. Cease transmitting and stir the water slowly for a minute or so until its temperature has stabilized. Then measure the water temperature as precisely as possible. After the initial temperature has been determined, begin the test "transmission," measuring the total number of seconds of key-down time accurately. Stir the water slowly with the thermometer and continue transmitting until there is a significant rise in the water temperature, say 5 to 10 degrees. The test may be broken up into a series of short periods, as long as you keep track of the total key-down time. When the test is completed, continue to stir the water slowly and monitor its temperature. When the temperature ceases to rise, note the final indication as precisely as possible.

To compute the transmitter power output, multiply the calorimeter heat capacitance (4215 for C or 2342 for F) by the difference in initial and final water temperature. Then divide by the total number of seconds of key-down time. The resultant is the transmitter power in watts. A nomogram which can also be used to find transmitter power output is given in Fig 19. With a straight line, connect the total number of key-down seconds in the time column to the number of degrees change (F or C) in the temperature rise column, and read off the transmitter power output at the point where the straight line crosses the power-output column.

Power Limitation

The maximum power-handling capability of the calorimeter is limited by the following. At very high powers

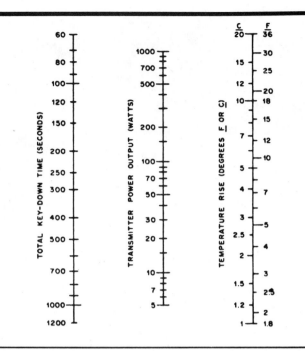

Fig 19—Nomogram for finding transmitter power output for the calorimeter.

the dielectric material in the coaxial line will melt because of excessive heating or the cable will arc over from excessive voltage. As the transmitter frequency gets higher, the excessive-heating problem is accentuated, as more of the power is dissipated in the first several feet of cable. For instance, at 1296 MHz, approximately 10 percent of the transmitting power is dissipated in the first foot of cable. Overheating can be prevented when working with high power by using a low duty cycle to reduce the average dissipated power. Use a series of short transmissions, such as two seconds on, ten seconds off. Keep count of the total key-down time for power calculation purposes. If the cable arcs over, use a larger-diameter cable, such as RG-58, in place of the RG-174. The cable should be long enough to assure that the reflected wave will be down 10 dB or more at the input. It may be necessary to use more than one quart of water in order to submerge all the cable conveniently. If so, be sure to calculate the new value of heat capacitance for the larger quantity of water. Also you should measure the new coaxial cable heat capacitance using the method previously described.

A Noise Bridge For 1.8 Through 30 MHz

The noise bridge, sometimes referred to as an antenna (RX) noise bridge, is an instrument that will allow the user to measure the impedance of an antenna or other electrical circuits. The unit shown in Figs 20 through 24, designed for use in the 1.8- through 30-MHz range, provides adequate accuracy for most measurements. Battery operation and small physical size make this unit ideal for remote-location use. Tone modulation is applied to the wide-band noise generator

as an aid for obtaining a null indication. A detector, such as the station receiver, is required for operation.

The noise bridge consists of two parts—the noise generator and the bridge circuitry. See Fig 21. A 6.8-volt Zener diode serves as the noise source. U1 generates an approximate 50% duty cycle, 1000-Hz square wave signal which is applied to the cathode of the Zener diode. The 1000-Hz modulation appears on the noise signal and provides a useful null detection

Fig 20—Exterior and interior views of the noise bridge. The unit is finished in red enamel. Press-on lettering is used for the calibration marks. Note that the potentiometer must be isolated from ground.

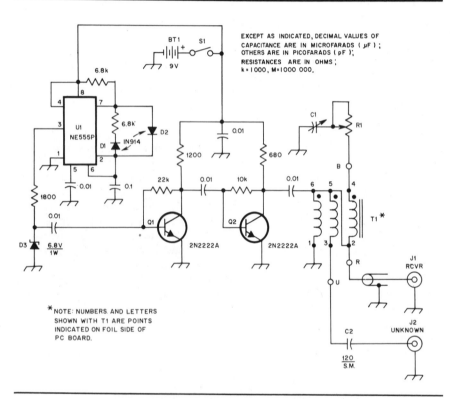

Fig 21—Schematic diagram of the noise bridge. Resistors are ¼-watt composition types. Capacitors are miniature ceramic units unless indicated otherwise. Component designations indicated in the schematic but not called out in the parts list are for text and parts-placement reference only.

BT1—9-volt battery, NEDA 1604A or equiv.
C1—Variable, 250 pF maximum. Use a good grade of capacitor.
C2—Approximately ½ of C1 value. Selection may be necessary—see text.
J1, J2—Coaxial connector, BNC type.
R1—Linear, 250 ohm, AB type. Use a good grade of resistor.
S1—Toggle, SPST.
T1—Broadband transformer, 8 trifilar turns of no. 26 enameled wire on an Amidon FT-37-43 toroid core.
U1—Timer, NE555 or equiv.

enhancement effect. The broadband-noise signal is amplified by Q1, Q2 and associated components to a level that produces an approximate S9 signal in the receiver. Slightly more noise is available at the lower end of the frequency range, as no frequency compensation is applied to the amplifier. Roughly 20 mA of current is drawn from the 9-volt battery, thus ensuring long battery life—providing the power is switched off after use!

The bridge portion of the circuit consists of T1, C1, C2 and R1. T1 is a trifilar wound transformer with one of the windings used to couple noise energy into the bridge circuit. The remaining two windings are arranged so that each one is in an arm of the bridge. C1 and R1 complete one arm and the UNKNOWN circuit, along with C2, comprise the remainder of the bridge. The terminal labeled RCVR is for connection to the detector.

Construction

The noise bridge is contained in a homemade aluminum enclosure that measures 5 × 2-3/8 × 3-3/4 inches. Many of the circuit components are mounted on a circuit board that is fastened to the rear wall of the cabinet. The circuit-board layout is such that the lead lengths to the board from the bridge and coaxial connectors are at a minimum. An etching pattern and a parts-placement guide for the circuit board are shown in Figs 23 and 24.

Care must be taken when mounting the potentiometer, R1. For accurate readings the potentiometer must be well insulated from ground. In the unit shown this was accomplished by mounting the control on a piece of Plexiglas, which in turn was fastened to the chassis with a piece of aluminum angle stock. Additionally, a ¼-inch control-shaft coupling and a length of phenolic rod were used to further isolate the control from ground where the shaft passes through the front panel. A high quality potentiometer is required if good measurement results are to be obtained.

Mounting the variable capacitor is not a problem since the rotor is grounded. As with the potentiometer, a good grade of capacitor is important. If you must cut corners to save money, look

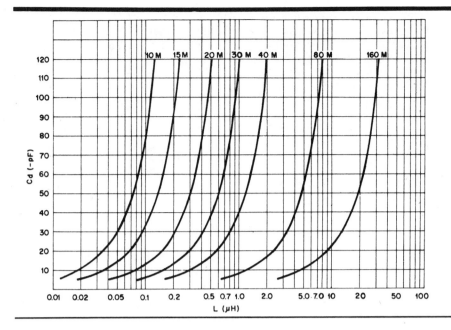

Fig 22—Graph for determining actual inductance from the calibration marks on the negative portion of the dial. These curves are accurate only for bridges having 120 pF at C2.

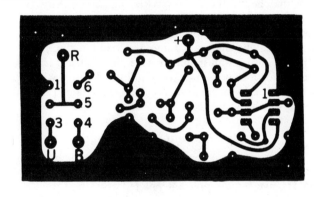

Fig 23—Etching pattern for the noise bridge pc board, at actual size. Black represents copper. This is the pattern for the bottom side of the board. The top side of the board is a complete ground plane with a small amount of copper removed from around the component holes.

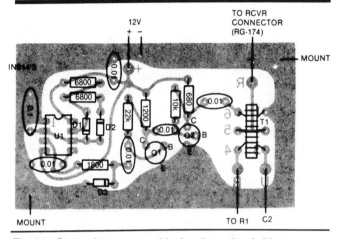

Fig 24—Parts-placement guide for the noise bridge as viewed from the component or top side of the board. Mounting holes are located in two corners of the board, as shown.

elsewhere in the circuit. Two BNC-type female coaxial fittings are provided on the rear panel for connection to a detector (receiver) and to the UNKNOWN circuit. There is no reason why other types of connectors can't be used. One should avoid the use of plastic-insulated phono connectors, however, as these might influence the accuracy at the higher frequencies. A length of miniature coaxial cable (RG-174) is used between the RCVR connector and the appropriate circuit board foils. Also, C2 has one lead attached to the circuit board

and the other connected directly to the UNKNOWN circuit connector.

Calibration and Use

Calibration of the bridge is straightforward and requires no special instruments. A receiver tuned to any portion of the 21-MHz band is connected to the RCVR terminal of the bridge. The power is switched on and a broadband noise with a 1000-Hz note should be heard in the receiver. Calibration of the resistance dial should be performed first. This is accomplished by inserting small composition resistors of appropriate values across the UNKNOWN connector of the bridge. The resistors should have the shortest lead lengths possible in order to mate with the connector. Start with 25 Ω of resistance (this may be made up of series or parallel connected units). Adjust the capacitance and resistance dials for a null of the signal as heard in the receiver. Place a calibration mark on the front panel at that location of the resistance dial. Remove the 25-Ω resistor and insert a 50-Ω resistor, 100-Ω unit and so on until the dial is completely calibrated.

The capacitance dial is calibrated in a similar manner. Initially, this dial is set so that the plates of C1 are exactly half meshed. If a capacitor having no stops is used, orient the knob so as to unmesh the plates when the knob is rotated into the positive capacitance region of the dial. A 50-Ω resistor is connected to the UNKNOWN terminal and the resistance control is adjusted for a null. Next, the reactance dial is adjusted for a null and its position is noted. If this setting is significantly different than the half-meshed position, the value of C2 will need to be changed. Manufacturing value variations of 120-pF capacitors may be sufficient to provide a suitable unit. Other values can be connected in series or parallel and tried in place of the 120-pF capacitor. The idea is to have the capacitance dial null as close as possible to the half-meshed position of C1.

Once the final value of C2 has been determined and the appropriate component installed in the circuit, the bridge should be adjusted for a null. The zero reactance point can be marked on the face of the unit. The next step is to place a 20-pF capacitor in series with the 50-Ω load resistor. Use a good grade of capacitor, such as a silver-mica type, and keep the leads as short as possible. Null with the capacitance dial and make a calibration mark at that point. Remove the 20-pF capacitor and insert a 40-pF unit in series with the 50-Ω resistor. Again null the bridge and make a calibration mark for 40-pF. Continue on in a similar manner until that half of the dial is completely calibrated.

To calibrate the negative half of the scale, the same capacitors may be used. This time they must be placed temporarily in parallel with C2. Connect the 50-Ω resistor to the UNKNOWN terminal and the 20-pF capacitor in parallel with C2. Null the bridge and place a calibration mark on the panel. Remove the 20-pF unit and temporarily install the 40-pF capacitor. Again null the bridge and make a calibration mark at that point. Continue this procedure until the capacitance dial is completely calibrated. It should be pointed out that the exact resistance and capacitance values used for calibration can be determined by the builder. If resistance values of 20, 40, 60, 80, 100 ohms and so on are more in line with the builder's needs, the scale may be calibrated in those terms. The same is true for the capacitance dial. The accuracy of the bridge is determined by the components that are used in the calibration process.

Many amateurs use a noise bridge simply to find the resonant (nonreactive) impedance of an antenna system. For this service it is necessary only to calibrate the zero-reactance point of the capacitance dial. This simplification relaxes the stringent quality requirement for the bridge capacitors, C1 and C2.

Operation

The resistance dial is calibrated directly in ohms, but the capacitance dial is calibrated in picofarads of capacitance. The +C half of the dial indicates that the load is capacitive and the −C portion is for inductive loads. To find the reactance of the load when the capacitance reading is positive, the dial setting must be applied to the standard capacitive reactance formula:

$$X = \frac{1}{2\pi fC}$$

The result will be a capacitive reactance.

Inductance values corresponding to negative capacitance dial readings may be taken from the graph of Fig 22. The reactance is then found from the formula:

$$X = 2\pi fL$$

When using the bridge remember that the instrument measures the impedance of loads as connected at the UNKNOWN terminal. This means that the actual load to be measured must be directly at the connector rather than being attached to the bridge by a length of coaxial cable. Even a short length of cable will transform the load impedance to some other value. Unless the electrical length of line is known and taken into account, it is necessary to place the bridge at the load. An exception to this would be if the antenna were to be matched to the characteristic impedance of the cable. In this case the bridge controls may be preset for 52-ohms resistance and 0-pF capacitance. With the bridge placed at any point along the coaxial line, the load (antenna) may be adjusted until a null is obtained. If the electrical length of line is known to be an integral multiple of half-wavelengths at the frequency of interest, the readings obtained from the bridge will be accurate.

Interpreting the Readings

A few words on how to interpret the measurements may be in order. For example, assume that the impedance of a 7-MHz inverted-V dipole antenna fed with a half-wavelength of cable was measured. The antenna had been cut for roughly the center of the band (7.150 MHz) and the bridge was nulled with the aid of a receiver tuned to that frequency. The results were 45-Ω resistive and 600 picofarads of capacitance. The 45-Ω resistance reading is about what would be expected for this type of antenna. The capacitive reactance calculates to be 37 Ω from the equation:

$$X = \frac{1}{2\pi \ (7.15 \times 10^6) \ (600 \times 10^{-12})} = 37 \ \Omega$$

When an antenna is adjusted for resonance, reactance will be zero. Since this antenna looks capacitive, it is too short, and wire should be added to each side of the antenna. An approximation of how much wire to add can be made by tuning the receiver higher in frequency until a point is reached where the bridge nulls with the capacitance dial at zero. The percentage difference between this new frequency and the desired frequency indicates the approximate amount that the antenna should be lengthened. The same system will work if the antenna has been cut too long. In this case the capacitance dial would have nulled in the −C region, indicating an inductive reactance. This procedure will work for any directly fed single-element antenna.

An Accurate RF Impedance Bridge

This antenna impedance bridge is designed and described by Wilfred N. Caron. Some of the unique features and the simplicity of this bridge can be realized after a brief study of the simplified bridge circuit shown in Fig 25A. The circuit diagram shows that there is no potentiometer for balancing the bridge, only a variable capacitor (C_R).

This bridge circuit provides a number of important advantages over conventional impedance bridges. The use of the capacitor as the variable element provides excellent resolution control, especially at the low resistance end of the dial, and eliminates problems of noise, wear, variable reactance and frequency limitations commonly associated with even the best of potentiometers. Another desirable characteristic of this type of bridge is that the interaction between the arms of the bridge and between the resistive and reactive components of the impedance of the test item are eliminated, thus providing a better null.

The bridge allows measurement of the resistance and reactance of the measured impedance in a parallel arrangement. Measurement range is between 5- and 500-Ω resistance. The maximum reactance value is limited by the size of the variable capacitor, C_X. (See Fig 25B.) A 365-pF capacitor is used in the practical bridge circuit. The reactance range, however, can be extended by placing fixed capacitors in parallel with C_X. The L-C switch (S1) permits the measurement of inductive and capacitive reactances.

Still another important feature of the bridge is the use of a "candelabra" or "binocular" balun as the input transformer. A ferrite candelabra balun along with a Type I balun provides the best balance between the two arms of the bridge. This, in turn, increases measurement accuracy by stabilizing the null across the frequency range of the bridge. The usable frequency range is between 1 MHz and at least 30 MHz.

Resistance Measurement

In the arrangement shown in Fig 25A, the bridge is balanced independently of frequency when the following relationship is satisfied:

$$R_{unknown} = \frac{R}{\dfrac{C1}{C_R} - 1} \qquad \text{(Eq 1)}$$

This shows that $R_{unknown}$ can be measured in a range from infinity (when $C_R = C1$) down to a fraction of an ohm, depending upon the maximum ratio of $C1/C_R$.

Reactance Measurement

The circuit of Fig 25A is not capable of measuring the reactive component (inductive or capacitive) of $R_{unknown}$. To remedy this limitation, one can connect a calibrated variable capacitor (C_X) to one of two arms of the bridge and adjust this capacitor to obtain resonance at the test frequency. This is shown in Fig 25B. At resonance the bridge sees only the effective parallel resistance (R_p) of $R_{unknown}$.

Series and Parallel Load Impedance Components

Many impedance bridges give the user the option of measuring the unknown load impedance in terms of its series or parallel resistance and reactance equivalents. However, this bridge can measure only parallel resistance and reactance components of the unknown impedance. If the equivalent

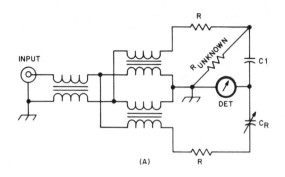

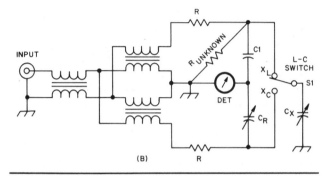

Fig 25—Simplified schematic of the impedance bridge.

series arrangement is required, some mathematical manipulation is necessary.

Every impedance can be expressed, for any frequency, as either a series or a parallel combination of resistance and reactance as shown in Fig 26. The relations between the elements of Fig 26 are:

$$R_p = \frac{R_s^2 + X_s^2}{R_s} \qquad \text{(Eq 2)}$$

$$X_p = \frac{R_s^2 + X_s^2}{X_s} \qquad \text{(Eq 3)}$$

and

$$R_s = \frac{R_p X_p^2}{R_p^2 + X_p^2} \qquad \text{(Eq 4)}$$

$$X_s = \frac{R_p^2 X_p}{R_p^2 + X_p^2} \qquad \text{(Eq 5)}$$

The Ferrite Candelabra Balun

Balance with respect to ground is essential in any bridge circuit. Some circuits handle this requirement with double-shielded transformers for coupling to a grounded, unbalanced signal source. The intershield capacitance across the arms of the bridge limits the operating frequency range. This limitation was overcome to a great extent by the use of a ferrite candelabra balun. The candelabra balun transforms the unbalanced input to a balanced output, and in the process steps up the impedance by a factor of four (see Fig 27).

For proper operation in a bridge circuit of this type, the

connection marked number 2 of the balun must be grounded. This helps the bridge balance problem, but it also presents a difficulty when one side of the input is also grounded. To eliminate this difficulty a second balun, a simple Type I balun, is used as indicated in Fig 25. The bifilar windings on a ferrite core provide a 1:1 impedance ratio. The twisted bifilar winding closely approximates a transmission line having a characteristic impedance of 50 to 100 Ω. Published data indicates that both the candelabra and Type I baluns can work satisfactorily to 1 GHz.

Construction Details

Figs 28 through 32 show the circuitry and construction details of the bridge. Be sure to make the wiring as short and as direct as possible. R1 and R2, the two 47-ohm resistors that constitute the two arms of the bridge, must be nonreactive and have a tolerance of 1% or better. (It is critical that the two resistors be of the same value to maintain bridge balance.) The C_R variable capacitor must be isolated from the metal case. A plastic case may be used but the shaft must be isolated to avoid hand-capacitance effects. Mounting the C_X capacitor is no problem because the rotor is grounded. Both the C_R and C_X variable capacitors are broadcast-band-receiver replacement units with semicircular plates. The semicircular plates provide a straight line capacitance curve. Measurements on a capacitance meter indicate a minimum capacitance of 14 pF and a maximum of 390 pF for a Calectro AI-227. These variable air capacitors have become scarce and may be obtained from Antique Electronic Supply, 688 W 1st St, Tempe, AZ 85281.

The entire bridge assembly is contained in a Bud CU234, $4\frac{1}{2} \times 3\frac{1}{2} \times 2$ inch aluminum case. The signal source is

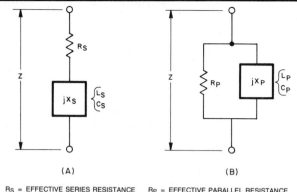

RS = EFFECTIVE SERIES RESISTANCE   RP = EFFECTIVE PARALLEL RESISTANCE
XS = EFFECTIVE SERIES REACTANCE   XP = EFFECTIVE PARALLEL REACTANCE
LS = EFFECTIVE SERIES INDUCTANCE   LP = EFFECTIVE PARALLEL INDUCTANCE
CS = EFFECTIVE SERIES CAPACITANCE   CP = EFFECTIVE PARALLEL CAPACITANCE

Fig 26—Series (A) and parallel (B) components of impedance.

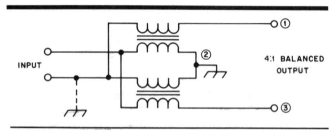

Fig 27—The ferrite candelabra balun.

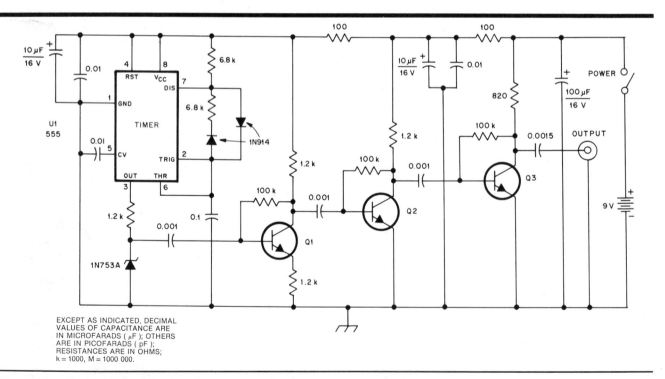

Fig 28—Schematic of the noise source.
Q1, Q2—NPN silicon transistor, general purpose, 250 mW, 2N2222A or equiv.
Q3—NPN silicon transistor, RF amplifier, 150 mW, ECG161 or equiv.
U1—Timer IC, NE555.

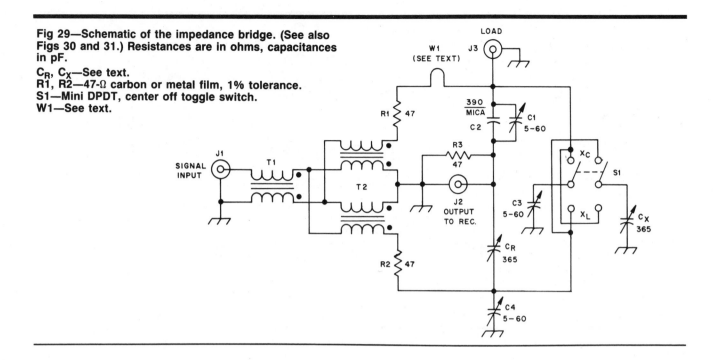

Fig 29—Schematic of the impedance bridge. (See also Figs 30 and 31.) Resistances are in ohms, capacitances in pF.

C_R, C_X—See text.
R1, R2—47-Ω carbon or metal film, 1% tolerance.
S1—Mini DPDT, center off toggle switch.
W1—See text.

external to the bridge. Any suitable connectors may be used on the rear panel for connecting to a receiver and the signal source. A good quality RF connector must be used for the load connector, J3, such as a type N, UHF or BNC.

The Signal Source

Any low-level signal generator can be used with the bridge. Perhaps the simplest is a wide band noise source and amplifier as shown in Fig 28. The circuitry provides high gain—sufficient gain to peg the S meter at the high-frequency limit. Although not essential, the noise is modulated at an audio rate in order to distinguish the noise of the generator from receiver noise. The null of the bridge is so sharp and deep that it is often obscured by receiver noise. Construction is straightforward.

Calibration Procedure

Calibration of the bridge is not complicated and requires no special instruments. It is convenient to have a precision 52-ohm dummy load but this is not essential. A communications receiver that tunes between about 2 MHz and 30 MHz is connected to the receiver terminal, J2, of the bridge (see Fig 29). The noise generator is connected to the input terminal, J1, and its power switch turned on. A broadband noise with a 1000-Hz note should be heard in the receiver.

Calibration of the Resistance Dial

Calibration of the resistance dial C_R should be performed first by following these steps.

1) Connect the signal source to J1, a receiver to J2 and a dummy load to J3.

2) Set C4 trimmer to minimum position.

3) Place S1 to center-off position.

4) Set the receiver to 2 MHz.

5) Null the bridge by adjusting C_R while watching the S meter of the receiver.

6) Note the position of the dial pointer.

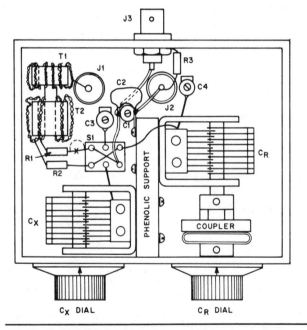

Fig 30—Layout of the impedance bridge.

7) Set the receiver to 30 MHz and recheck the null. If the null has not shifted, proceed to step 10. If the null position is not the same for the 2- and 30-MHz frequencies, disconnect one end of one of the matched resistors (R1 and R2) and insert W1, a short length of wire approximately ½ inch long, as shown in Fig 29.

8) Recheck the positions of the nulls at both frequencies. Null positions may either diverge or converge. If the nulls diverge, remove W1, the short wire, and reconnect the resistor.

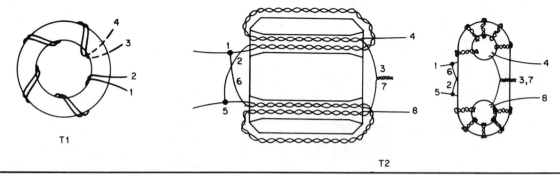

Fig 31—Construction details of the balun transformers. Transformer cores are available from Amidon Associates.

T1—10 turns no. 28 enam twisted 5 turns per inch on Amidon FT-50A-43 or equiv.

T2—5 turns each side no. 30 enam twisted 5 turns per inch on Amidon BLN-43-302 binocular core or equiv.

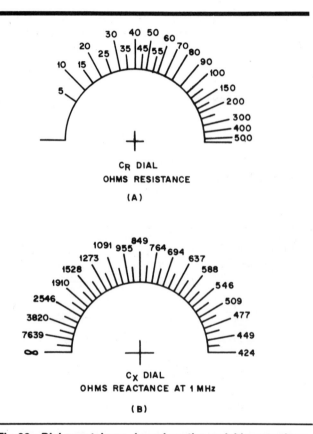

Fig 32—Dial escutcheon, based on the variable capacitors described in the text.

9) Place the short wire in the other leg of the bridge by disconnecting one end of the other matched resistor. The nulls should now start to converge. Continue replacing the wire with a longer piece, using increments of 1/8 inch, until the null position is the same for both the high and low frequencies. (In the designer's bridge a final wire length of 1¼ inches was required.)

10) With a 500-Ω precision resistor or a potentiometer connected as a load, set C1 so that the 500-Ω calibration point is at the far end of the dial (capacitor C_R will be fully closed).

11) With the potentiometer or an assortment of precision resistors, calibrate the C_R dial from 500 Ω to 5 Ω. Fig 32A illustrates the various calibration points involved. This completes the resistance dial calibration.

Calibrating the C_X Dial

Calibration of the C_X dial is somewhat different. If a 365-pF air variable capacitor is used, it will have a minimum capacitance of about 14 pF and a maximum of about 390 pF. The 14-pF minimum capacitance is undesirable and must be neutralized. That is the purpose of the trimmer, C3. Neutralizing the minimum capacitance has the effect of modifying the tuning range of C_X. For calibration purposes its minimum capacitance is zero and the maximum capacitance is now 390 − 14 = 376 pF. From this information, one can construct the C_X dial. Since C_X is a straight line capacitor, the dial can be marked off in equal divisions as shown in Fig 32B. The dial can be calibrated in pF or in reactance at 1 MHz from the following equation:

$$X_{C(1\ MHz)} = \frac{159,000}{C\ (pF)}$$

To obtain the reactance at other frequencies it is simply a matter of dividing $X_{C(1\ MHz)}$ by the test frequency. Positive (inductive) reactance (X_L) and negative (capacitive) reactance (X_C) measurements are selected by the position of S1.

The bridge is resonated (zero residual reactance) when a 52-ohm nonreactive load is measured and toggling S1 doesn't disturb the null reading. Resonating the bridge is accomplished as follows:

1) A zero reactance null must occur when C_X is set to its minimum position and S1 is toggled without disturbing the null. This is done by adjusting trimmer capacitors C3 and C4. This calibration is performed at 30 MHz. This completes the reactance dial calibration.

Operating the Bridge

When using the bridge, remember that the instrument measures the impedance of the load as connected at J3. This means that if the load being measured is connected to the bridge with a length of coaxial line, the line will act as a transformer and will transform the actual load impedance to some other value. Unless the actual electrical length of the line is known and the operator is capable of transforming the measured impedance to the actual load impedance, it will be necessary to place the bridge directly at the load terminals.

As an example for use of the bridge, assume we measure the impedance of a dipole antenna at 7.2 MHz. This measurement is made through an electrical ½ λ of line, and the resultant dial readings are 50-Ω resistance and 1400-Ω reactance. S1 is in the X_L position for a balanced condition, so the antenna reactance is inductive. Because the measurement is made at a frequency other than 1 MHz, we must apply a correction to the reactance dial reading. The corrected value is obtained simply by dividing the dial reading by the frequency in megahertz: 1400/7.2 = 194 ohms. Thus, the antenna impedance is 50-Ω resistance in shunt with 194 Ω of inductive reactance.

If equivalent series values of resistance and reactance are desired, we may use Eqs 4 and 5. The denominator term is common to both equations, so let's calculate that value first.

$$R_p{}^2 + X_p{}^2 = 50^2 + 194^2 = 40136$$

Then from Eq 4,

$$R_s = \frac{50 \times 194^2}{40136} = 46.9 \ \Omega$$

And from Eq 5,

$$X_s = \frac{50^2 \times 194}{40136} = 12.1 \ \Omega$$

The equivalent series impedance of the antenna is therefore 46.9 + j12.1 Ω.

A balanced line may be measured using the bridge by the use of a balun. When a balun is used, the bridge is operated exactly as if the line were a quarter of its actual impedance if the balun is a 4:1 type. Thus a 208-ohm transmission line properly terminated will read 52 ohms on the bridge, and a 600-ohm line will read 150 ohms. Thus, a bridge dial of 5 to 500 ohms becomes 20 to 2000 ohms.

Measuring Soil Conductivity

An important parameter for both vertical and horizontal antennas is soil conductivity. For horizontal antennas, the energy reflected from the earth beneath it affects the antenna impedance, thereby affecting the SWR and the current flowing in the antenna elements, which in turn affects the distant signal strength. (This is discussed in more detail in Chapter 3.) The conductivity of the ground within several wavelengths of the antenna also affects the ground reflection factors discussed in Chapter 3.

The conductivity of the soil under and in the near vicinity of a vertical antenna is most important in determining the extent of the radial system required and the overall performance. Short verticals with very small radial systems can be surprisingly effective—in the right location. The material in this section was prepared by Jerry Sevick, W2FMI.

Most soils are nonconductors of electricity when completely dry. Conduction through the soil results from conduction through the water held in the soil. Thus, conduction is electrolytic. Dc techniques for measuring conductivity are impractical because they tend to deplete the carriers of electricity in the vicinity of the electrodes. The main factors contributing to the conductivity of soil are

1) Type of soil.
2) Type of salts contained in the water.
3) Concentration of salts dissolved in the contained water.
4) Moisture content.
5) Grain size and distribution of material.
6) Temperature.
7) Packing density and pressure.

Although the type of soil is an important factor in determining its conductivity, rather large variations can take place between locations because of the other factors involved. Generally, loams and garden soils have the highest conductivities. These are followed in order by clays, sand and

Table 2

General Classification of Conductivity

Material	Conductivity (millisiemens per meter)
Poor Soil	1-5
Average Soil	10-15
Very Good Soil	100
Salt Water	5000
Fresh Water	10-15

gravel. Soils have been classified according to conductivity, as shown in Table 2. Although some differences are noted in the reporting of this mode of classification because of the many variables involved, the classification generally follows the values shown in the table. Approximate soil conductivities for the continental US are shown in Fig 33. Soil conductivity is also discussed in Chapter 3.

Making Conductivity Measurements

Since conduction through the soil is almost entirely electrolytic, ac measurement techniques are preferable. Many commercial instruments using ac techniques are available and described in the literature. But rather simple ac measurement techniques can be used that provide accuracies on the order of 25% and are quite adequate for the radio amateur. Such a setup was developed by Jerry Sevick, W2FMI, and M. C. Waltz, W2FNQ, and was published by Sevick in April 1978 and March 1981 *QST*. It is shown in Figs 34 through 36.

Four probes are used. Each is 9/16 inch in diameter, and

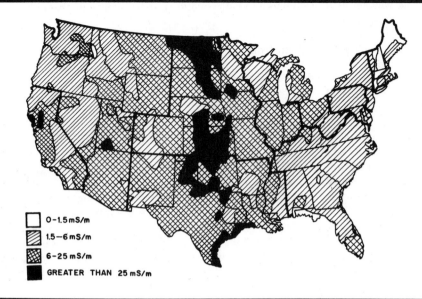

Fig 33—Approximate soil conductivity in the continental US. This information was adapted from US government publications.

0–1.5 mS/m
1.5–6 mS/m
6–25 mS/m
GREATER THAN 25 mS/m

Fig 34—The complete soil conductivity measuring setup. The four probes are cut to 18 in. lengths from an 8-foot copper-coated steel ground rod. (This length provides a measuring stick for spacing the probes when driving them into the soil.) The tip of each probe is ground to a point, and black electricians' tape indicates the depth to which it is to be driven for measurements. Two ground clamps provide for connections to the driven probes.

Fig 35—A standard 3½-in. electrical outlet box and a porcelain ceiling fixture may be used to construct the soil conductivity test set. The resistors comprising R1 are mounted on a tie-point strip inside the box, and test-point jacks provide for measuring the voltage drop across the resistor combination. Leads exiting the box through the cable clamp are protected with several layers of electricians' tape. These leads run approximately 4 ft to the power plug and to small alligator clips for attachment to the ground clamps shown in Fig 34. Large clips such as for connecting to automotive battery posts may be used instead of ground clamps.

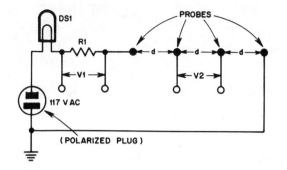

Fig 36—Schematic diagram, four-point probe method for measuring earth conductivity.
DS1—100-W electric light bulb.
R1—14.6 ohms, 5 W. A suitable resistance can be made by paralleling five 1-W resistors, three of 68 Ω and two of 82 Ω. (The dissipation rating of this combination will be 4.7 W.)
Probes—See text and Fig 34.

may be made of either iron or copper. The probes are inserted in a straight line at a spacing of 18 inches (dimension *d* in Fig 36). The penetration depth is 12 inches. *Caution*: Do not insert the probes with power applied! A shock hazard exists! After applying power, measure the voltage drops V1 and V2, as shown in the diagram. Depending on soil conditions, readings should fall in the range from 2 to 10 volts.

Earth conductivity, c, may be determined from

$$c = 21 \times \frac{V1}{V2} \text{ millisiemens per meter}$$

For example, assume the reading across the resistor (V1) is 4.9 V, and the reading between the two center probes (V2) is 7.2 V. The conductivity is calculated as 21 × 4.9/7.2 = 14 mS/m.

Soil conditions may not be uniform in different parts of your yard. A few quick measurements will reveal whether this is the case or not.

Fig 37 shows the conductivity readings taken in one location over a period of three months. It is interesting to note the general drop in conductivity over the three months, as well as the short-term changes from periods of rain.

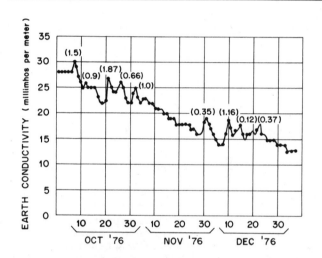

Fig 37—Earth conductivity at a central New Jersey location during a three-month period. Numbers in parentheses indicate inches of rainfall.

A Switchable RF Attenuator

A switchable RF attenuator is helpful in making antenna gain comparisons or plotting antenna radiation patterns; attenuation may be switched in or out of the line leading to the receiver to obtain an initial or reference reading on a signal strength meter. Some form of attenuator is also helpful for locating hidden transmitters, where the real trick is pin-

pointing the signal source from within a few hundred feet. At such a close distance, strong signals may overload the front end of the receiver, making it impossible to obtain any indication of a bearing.

The attenuator of Figs 38 and 39 is designed for low power levels, not exceeding ¼ watt. If for some reason the

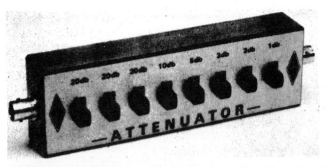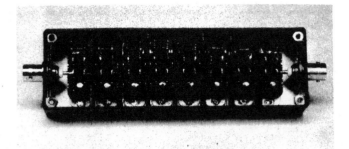

Fig 38—A construction method for a step attenuator. Double-sided circuit-board material, unetched (except for panel identification), is cut to the desired size and soldered in place. Flashing copper may also be used, although it is not as sturdy. Shielding partitions between sections are necessary to reduce signal leakage. Brass nuts soldered at each of the four corners allow machine screws to secure the bottom cover. The practical limit for total attenuation is 80 or 90 dB, as signal leakage around the outside of the attenuator will defeat attempts to obtain much greater amounts. A kit of parts is available from Circuit Board Specialists, PO Box 951, Pueblo, CO 81002.

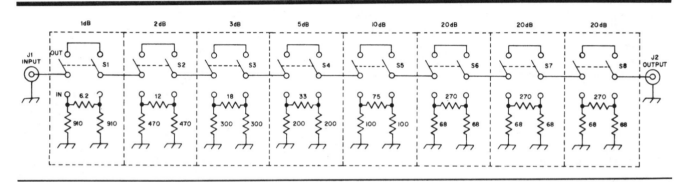

Fig 39—Schematic diagram of the step attenuator, designed for a nominal impedance of 52 ohms. Resistance values are in ohms. Resistors are ¼-watt, carbon-composition types, 5% tolerance. Broken lines indicate walls of circuit-board material. A small hole is drilled through each partition wall to route bus wire. Keep all leads as short as possible. The attenuator is bilateral; that is, the input and output ends may be reversed.

J1, J2—Female BNC connectors, Radio Shack 278-105 or equiv.
S1-S8, incl.—DPDT slide switches, standard size. (Avoid subminiature or toggle switches.) Stackpole S-5022CD03-0 switches are used here.

attenuator will be connected to a transceiver, a means of bypassing the unit during transmit periods must be devised. An attenuator of this type is commonly called a step attenuator, because any amount of attenuation from 0 dB to the maximum available (81 dB for this particular instrument) may be obtained in steps of 1 dB. As each switch is successively thrown from the OUT to the IN position, the attenuation sections add in cascade to yield the total of the attenuator steps switched in. The maximum attenuation of any single section is limited to 20 dB because leak-through would probably degrade the accuracy of higher values. The tolerance of resistor values also becomes more significant regarding accuracy at higher attenuation values.

A good quality commercially made attenuator will cost upwards from $150, but for less than $25 in parts and a few hours of work, an attenuator may be built at home. it will be suitable for frequencies up to 450 MHz. Double-sided pc board is used for the enclosure. The version of the attenuator shown in Fig 38 has identification lettering etched into the top surface (or front panel) of the unit. This adds a nice touch and is a permanent means of labeling. Of course rub-on transfers or Dymo tape labels could be used as well.

Female BNC single-hole, chassis-mount connectors are used at each end of the enclosure. These connectors provide a means of easily connecting and disconnecting the attenuator.

Construction

After all the box parts are cut to size and the necessary holes made, scribe light lines to locate the inner partitions. Carefully tack-solder all partitions in position. A 25-watt pencil type of iron should provide sufficient heat. Dress any pc board parts that do not fit squarely. Once everything is in proper position, run a solder bead all the way around the joints. Caution! Do not use excessive amounts of solder, as the switches must later be fit flat inside the sections. The top, sides, ends and partitions can be completed. Dress the outside of the box to suit your taste. For instance, you might wish to bevel the box edges. Buff the copper with steel wool, add lettering, and finish off the work with a coat of clear lacquer or polyurethane varnish.

Using a little lacquer thinner or acetone (and a lot of caution), soak the switches to remove the grease that was added during their manufacture. When they dry, spray the inside of the switches lightly with a TV tuner cleaner/lubricant. Use a sharp drill bit (about 3/16 inch will do), and countersink the mounting holes on the actuator side of the switch mounting plate. This ensures that the switches will fit flush against the top plate. At one end of each switch, bend the two lugs over and solder them together. Cut off the upper halves of the remaining switch lugs. (A close look at Fig 38 will help clarify these steps.)

Solder the series-arm resistors between the appropriate switch lugs. Keep the lead lengths as short as possible and do not overheat the resistors. Now solder the switches in place to the top section of the enclosure by flowing solder through the mounting holes and onto the circuit-board material. Be certain that you place the switches in their proper positions; correlate the resistor values with the degree of attenuation. Otherwise, you may wind up with the 1-dB step at the wrong end of the box—how embarrassing!

Once the switches are installed, thread a piece of no. 18 bare copper wire through the center lugs of all the switches,

passing it through the holes in the partitions. Solder the wire at each switch terminal. Cut the wire between the poles of each individual switch, leaving the wire connecting one switch pole to that of the neighboring one on the other side of the partition, as shown in Fig 38. At each of the two end switch terminals, leave a wire length of approximately 1/8 inch. Install the BNC connectors and solder the wire pieces to the connector center conductors.

Now install the shunt-arm resistors of each section. Use short lead lengths. Do not use excessive amounts of heat when soldering. Solder a no. 4-40 brass nut at each inside corner of the enclosure. Recess the nuts approximately 1/16 inch from the bottom edge of the box to allow sufficient room for the bottom panel to fit flush. Secure the bottom panel with four no. 4-40, ¼-inch machine screws and the project is completed. Remember to use caution, always, when your test setup provides the possibility of transmitting power into the attenuator.

A Portable Field-Strength Meter

Few amateur stations, fixed or mobile, are without need of a field-strength meter. An instrument of this type serves many useful purposes during antenna experiments and adjustments. When work is to be done from many wavelengths away, a simple wavemeter lacks the necessary sensitivity. Further, such a device has a serious fault because its linearity leaves much to be desired. The information in this section is based on a January 1973 *QST* article by Lew McCoy, W1ICP.

The field-strength meter described here takes care of these problems. Additionally, it is small, measuring only 4 × 5 × 8 inches. The power supply consists of two 9-volt batteries. Sensitivity can be set for practically any amount desired. However, from a usefulness standpoint, the circuit should not be too sensitive or it will respond to unwanted signals. This unit also has excellent linearity with regard to field strength. (The field strength of a received signal varies inversely with the distance from the source, all other things being equal.) The frequency range includes all amateur bands from 3.5 through 148 MHz, with band-switched circuits, thus avoiding the use of plug-in inductors. All in all, it is a quite useful instrument.

The unit is pictured in Figs 40 and 41, and the schematic diagram is shown in Fig 42. A type 741 op-amp IC is the heart of the unit. The antenna is connected to J1, and a tuned circuit is used ahead of a diode detector. The rectified signal is coupled as dc and amplified in the op amp. Sensitivity of the op amp is controlled by inserting resistors R3 through R6 in the circuit by means of S2.

With the circuit shown, and in its most sensitive setting, M1 will detect a signal from the antenna on the order of 100 μV. Linearity is poor for approximately the first 1/5 of the meter range, but then is almost straight-line from there to full-scale deflection. The reason for the poor linearity at the start of the readings is because of nonlinearity of the diodes at the point of first conduction. However, if gain measurements are being made this is of no real importance,

Fig 40—The linear field-strength meter. The control at the upper left is for C1 and the one to the right for C2. At the lower left is the band switch, and to its right the sensitivity switch. The zero-set control for M1 is located directly below the meter.

as accurate gain measurements can be made in the linear portion of the readings.

The 741 op amp requires both a positive and a negative voltage source. This is obtained by connecting two 9-volt batteries in series and grounding the center. One other feature of the instrument is that it can be used remotely by connecting an external meter at J2. This is handy if you want to adjust an antenna and observe the results without having to leave the antenna site.

L1 is the 3.5/7-MHz coil and is tuned by C1. The coil is wound on a toroid form. For 14, 21 or 28 MHz, L2 is switched in parallel with L1 to cover the three bands. L5 and

Fig 41—Inside view of the field-strength meter. At the upper right is C1 and to the left, C2. The dark leads from the circuit board to the front panel are the shielded leads described in the text.

C2 cover approximately 40 to 60 MHz, and L7 and C2 from 130 MHz to approximately 180 MHz. The two VHF coils are also wound on toroid forms.

Construction Notes

The majority of the components may be mounted on an etched circuit board. A shielded lead should be used between pin 4 of the IC and S2. The same is true for the leads from R3 through R6 to the switch. Otherwise, parasitic oscillations may occur in the IC because of its very high gain.

In order for the unit to cover the 144-MHz band, L6 and L7 should be mounted directly across the appropriate terminals of S1, rather than on a circuit board. The extra lead length adds too much stray capacitance to the circuit. It isn't necessary to use toroid forms for the 50- and 144-MHz coils. They were used in the version described here simply because they were available. Air-wound coils of the appropriate inductance can be substituted.

Calibration

The field-strength meter can be used "as is" for a

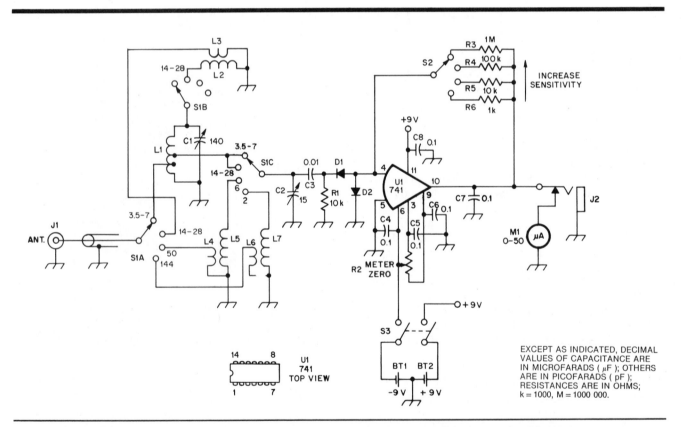

Fig 42—Circuit diagram of the linear field-strength meter. All resistors are ¼- or ½-watt composition types.

C1—140-pF variable.
C2—15-pF variable
D1, D2—1N914 or equiv.
L1—34 turns no. 24 enam. wire wound on an Amidon T-68-2 core, tapped 4 turns from ground end.
L2—12 turns no. 24 enam. wire wound on T-68-2 core.
L3—2 turns no. 24 enam. wire wound at ground end of L2.
L4—1 turn no. 26 enam. wire wound at ground end of L5.
L5—12 turns no. 26 enam. wire wound on T-25-12 core.

L6—1 turn no. 26 enam. wire wound at ground end of L7.
L7—1 turn no. 18 enam. wire wound on T-25-12 core.
M1—50 or 100 μA dc.
R2—10-kΩ control, linear taper.
S1—Rotary switch, 3 poles, 5 positions, 3 sections.
S2—Rotary switch, 1 pole, 4 positions.
S3—DPST toggle.
U1—Type 741 op amp. Pin nos. shown are for a 14-pin package.

relative-reading device. A linear indicator scale will serve admirably. However, it will be a much more useful instrument for antenna work if it is calibrated in decibels, enabling the user to check relative gain and front-to-back ratios. If one has access to a calibrated signal generator, it can be connected to the field-strength meter and different signal levels can be fed to the device for making a calibration chart. Signal-generator voltage ratios can be converted to decibels by using the equation,

$$dB = 20 \log (V1/V2)$$

where

V1/V2 is the ratio of the two voltages
log is the common logarithm (base 10)

Let's assume that M1 is calibrated evenly from 0 to 10. Next, assume we set the signal generator to provide a reading of 1 on M1, and that the generator is feeding a 100-μV signal into the instrument. Now we increase the generator output to 200 μV, giving us a voltage ratio of 2 to 1. Also let's assume M1 reads 5 with the 200-μV input. From the equation above, we find that the voltage ratio of 2 equals 6.02 dB between 1 and 5 on the meter scale. M1 can be calibrated more accurately between 1 and 5 on its scale by adjusting the generator and figuring the ratio. For example, a ratio of 126 μV to 100 μV is 1.26, corresponding to 2.0 dB. By using this method, all of the settings of S2 can be calibrated. In the instrument shown here, the most sensitive setting of S2 with

R3, 1 MΩ, provides a range of approximately 6 dB for M1. Keep in mind that the meter scale for each setting of S1 must be calibrated similarly for each band. The degree of coupling of the tuned circuits for the different bands will vary, so each band must be calibrated separately.

Another method for calibrating the instrument is using a transmitter and measuring its output power with an RF wattmeter. In this case we are dealing with power rather than voltage ratios, so this equation applies:

$$dB = 10 \log (P1/P2)$$

where P1/P2 is the power ratio.

With most transmitters the power output can be varied, so calibration of the test instrument is rather easy. Attach a pickup antenna to the field-strength meter (a short wire a foot or so long will do) and position the device in the transmitter antenna field. Let's assume we set the transmitter output for 10 watts and get a reading on M1. We note the reading and then increase the output to 20 watts, a power ratio of 2. Note the reading on M1 and then use Eq 2. A power ratio of 2 is 3.01 dB. By using this method the instrument can be calibrated on all bands and ranges.

With the tuned circuits and coupling links specified in Fig 42, this instrument has an average range on the various bands of 6 dB for the two most sensitive positions of S2, and 15 dB and 30 dB for the next two successive ranges. The 30-dB scale is handy for making front-to-back antenna measurements without having to switch S2.

An RF Current Probe

The RF current probe of Figs 43 through 45 operates on the magnetic component of the electromagnetic field, rather than the electric field. Since the two fields are precisely related, as discussed in Chapter 23, the relative field strength measurements are completely equivalent. The use of the magnetic field offers certain advantages, however. The instrument may be made more compact for the same sensitivity, but its principal advantage is that it may be used near a conductor to measure the current flow without cutting the conductor.

In the average amateur location there may be substantial currents flowing in guy wires, masts and towers, coaxial-cable braids, gutters and leaders, water and gas pipes, and perhaps even drainage pipes. Current may be flowing in telephone and power lines as well. All of these RF currents may have an influence on antenna patterns or be of significance in the case of RFI.

The circuit diagram of the current probe appears in Fig 44, and construction is shown in the photo, Fig 45. The winding data given here apply only to a ferrite rod of the particular dimensions and material specified. Almost any microammeter can be used, but it is usually convenient to use a rather sensitive meter and provide a series resistor to "swamp out" nonlinearity arising from diode conduction characteristics. A control is also used to adjust instrument

Fig 43—The RF current probe. The sensitivity control is mounted at the top of the instrument, with the tuning and band switches on the lower portion of the front panel. Frequency calibration of the tuning control was not considered necessary for the intended use of this particular instrument, but marks identifying the various amateur bands would be helpful. If the unit is provided with a calibrated dial, it can also be used as an absorption wavemeter.

sensitivity as required during operation. The tuning capacitor may be almost anything that will cover the desired range.

As shown in the photos, the circuit is constructed in a metal box. This enclosure shields the detector circuit from

Fig 44—Schematic diagram of the RF current probe. Resistances are in ohms; k = 1000. Capacitances are in picofarads; fixed capacitors are silver mica. Be sure to ground the rotor of C1, rather than the stator, to avoid hand capacitance. L1, L2 and L3 are each close-wound with no. 22 enam. wire on a single ferrite rod, 4 inches long and ½ inch dia, with μ = 125 (Amidon R61-50-400). Windings are spaced approximately ¼ inch apart. The ferrite rod, the variable capacitor, and other components may be obtained from Radiokit (see Chapter 21).

C1—Air variable, 6-140 pF; Hammarlund HF140 or equiv.
D1—Germanium diode; 1N34A, 1N270 or equiv.
L1—1.6-5 MHz; 30 turns, tapped at 3 turns from grounded end.
L2—5-20 MHz; 8 turns, tapped at 2 turns from grounded end.
L3—17-39 MHz; 2 turns, tapped at 1 turn.
M1—Any microammeter may be used. The one pictured is a Micronta

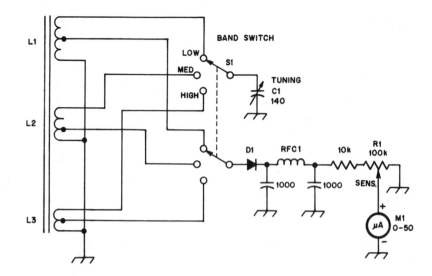

meter, Radio Shack no. 270-1751.
R1—Linear taper.
RFC1—1 mH; Miller no. 4642 or equiv. Value is not critical.

S1—Ceramic rotary switch, 1 section, 2 poles, 2 to 6 positions; Centralab PA2002 or PA2003 or equiv.

Fig 45—The current probe just before final assembly. Note that all parts except the ferrite rod are mounted on a single half of the 3 × 4 × 5-in. Minibox (Bud CU-2105B or equiv). Rubber grommets are fitted in holes at the ends of the slot to accept the rod during assembly of the enclosure. Leads in the RF section should be kept as short as possible, although those from the rod windings must necessarily be left somewhat long to facilitate final assembly.

the electric field of the radio wave. A slot must be cut with a hacksaw across the back of the box, and a thin file may be used to smooth the cut. This slot is necessary to prevent the box from acting as a shorted turn.

Using the Probe

In measuring the current in a conductor, the ferrite rod should be kept at right angles to the conductor, and at a constant distance from it. In its upright or vertical position, this instrument is oriented for taking measurements in vertical conductors. It must be laid horizontal to measure current in horizontal conductors.

Numerous uses for the instrument are suggested in an earlier paragraph. In addition, the probe is an ideal instrument for checking the current distribution in antenna elements. It is also useful for measuring RF ground currents in radial systems. A buried radial may be located easily by sweeping the ground. Current division at junctions may be investigated. "Hot spots" usually indicate areas where additional radials would be effective.

Stray currents in conductors not intended to be part of the antenna system may often be eliminated by bonding or by changing the physical lengths involved. Guy wires and other unwanted "parasitic" elements will often give a tilt to the plane of polarization and make a marked difference in front-to-back ratios. When the ferrite rod is oriented parallel to the electric field lines, there will be a sharp null reading that may be used to locate the plane of polarization quite accurately. When using the meter, remember that the magnetic field is at right angles to the electric field.

The current probe may also be used as a relative signal-strength meter. In making measurements on a vertical antenna, the meter should be located at least two wavelengths away, with the rod in a horizontal position. For horizontal antennas, the instrument should be at approximately the same height as the antenna, with the rod vertical.

Antenna Measurements

Of all the measurements made in Amateur Radio systems, perhaps the most difficult and least understood are various measurements of antennas. For example, it is relatively easy to measure the frequency and CW power output of a transmitter, the response of a filter, or the gain of an amplifier. These are all what might be called bench measurements because, when performed properly, all the factors that influence the accuracy and success of the measurement are under control. In making antenna measurements, however, the "bench" is now perhaps the backyard. In other words, the environment surrounding the antenna can affect the results of the measurement. Control of the environment is not at all as simple as it was for the bench measurement, because now the work area may be rather spacious. This section describes antenna measurement techniques which are closely allied to those used in an antenna-measuring event or contest. With these procedures the measurements can be made successfully and with meaningful results. These techniques should provide a better understanding of the measurement problems, resulting in a more accurate and less difficult task. The information in this section was provided by Dick Turrin, W2IMU, and originally published in November 1974 *QST*.

SOME BASIC IDEAS

An antenna is simply a transducer or coupler between a suitable feed line and the environment surrounding it. In addition to efficient transfer of power from feed line to environment, an antenna at VHF or UHF is most frequently required to concentrate the radiated power into a particular region of the environment.

To be consistent in comparing different antennas, it is necessary that the environment surrounding the antenna be standardized. Ideally, measurements should be made with the measured antenna so far removed from any objects causing environmental effects that it is literally in outer space—a very impractical situation. The purpose of the measurement techniques is therefore to simulate, under practical conditions, a controlled environment. At VHF and UHF, and with practical-size antennas, the environment *can* be controlled so that successful and accurate measurements can be made in a reasonable amount of space.

The electrical characteristics of an antenna that are most desirable to obtain by direct measurement are: (1) gain (relative to an isotropic source, which by definition has a gain of unity); (2) space-radiation pattern; (3) feed-point impedance (mismatch) and (4) polarization.

Polarization

In general the polarization can be assumed from the geometry of the radiating elements. That is to say, if the antenna is made up of a number of linear elements (straight lengths of rod or wire which are resonant and connected to the feed point) the polarization of the electric field will be linear and polarized parallel to the elements. If the elements are not consistently parallel with each other, then the polarization cannot easily be assumed. The following techniques are directed to antennas having polarization that is essentially linear (in one plane), although the method can be extended to include all forms of elliptic (or mixed) polarization.

Feed-Point Mismatch

The feed-point mismatch, although affected to some degree by the immediate environment of the antenna, does *not* affect the gain or radiation characteristics of an antenna. If the immediate environment of the antenna does not affect the feed-point impedance, then any mismatch intrinsic to the antenna tuning reflects a portion of the incident power back to the source. In a receiving antenna this reflected power is reradiated back into the environment, "free space," and can be lost entirely. In a transmitting antenna, the reflected power goes back to the final amplifier of the transmitter if it is not matched.

In general an amplifier by itself is *not* a matched source to the feed line, and, if the feed line has very low loss, the amplifier output controls are customarily altered during the normal tuning procedure to obtain maximum power transfer to the antenna. The power which has been reflected from the antenna combines with the source power to travel again to the antenna. This procedure is called conjugate matching, and the feed line is now part of a resonant system consisting of the mismatched antenna, feed line, and amplifier tuning circuits. It is therefore possible to use a mismatched antenna to its full gain potential, provided the mismatch is not so severe as to cause heating losses in the system, especially the feed line and matching devices. (See also the discussion of additional loss caused by SWR in Chapter 24.) Similarly, a mismatched receiving antenna may be conjugately matched into the receiver front end for maximum power transfer. In any case it should be clearly kept in mind that the feed-point mismatch does *not* affect the radiation characteristics of an antenna. It can only affect the system efficiency wherein heating losses are concerned.

Why then do we include feed-point mismatch as part of the antenna characteristics? The reason is that for efficient system performance, most antennas are resonant transducers and present a reasonable match over a relatively narrow frequency range. It is therefore desirable to design an antenna, whether it be a simple dipole or an array of Yagis, such that the final single feed-point impedance be essentially resistive and of magnitude consistent with the impedance of the feed line which is to be used. Furthermore, in order to make accurate, absolute gain measurements, it is vital that the antenna under test accept all the power from a matched-source generator, or that the reflected power caused by the mismatch be measured and a suitable error correction for heating losses be included in the gain calculations. Heating losses may be determined from information contained in Chapter 24.

While on the subject of feed-point impedance, mention should be made of the use of baluns in antennas. A balun is simply a device which permits a lossless transition between a balanced system—feed line or antenna—and an unbalanced feed line or system. If the feed point of an antenna is symmetric such as with a dipole and it is desired to feed this antenna with an unbalanced feed line such as coax, it is necessary to provide a balun between the line and the feed point. Without the balun, current will be allowed to flow on the outside of the coax. The current on the outside of the feed line will cause radiation and thus the feed line becomes part of the antenna radiation system. In the case of beam antennas where it is desired to concentrate the radiated energy is a specific direction, this extra radiation from the feed line will be detrimental, causing distortion of the expected antenna pattern.

ANTENNA TEST SITE SET-UP AND EVALUATION

Since an antenna is a reciprocal device, measurements of gain and radiation patterns can be made with the test antenna used either as a transmitting or as a receiving antenna. In general and for practical reasons, the test antenna is used in the receiving mode, and the source or transmitting antenna is located at a specified fixed remote site and unattended. In other words the source antenna, energized by a suitable transmitter, is simply required to illuminate or flood the receiving site in a controlled and constant manner.

As mentioned earlier, antenna measurements ideally should be made under "free-space" conditions. A further restriction is that the illumination from the source antenna be a plane wave over the effective aperture (capture area) of the test antenna. A plane wave by definition is one in which the magnitude and phase of the fields are uniform, and in the test-antenna situation, *uniform over the effective area plane of the test antenna*. Since it is the nature of all radiation to expand in a spherical manner at great distance from the source, it would seem to be most desirable to locate the source antenna as far from the test site as possible. However, since for practical reasons the test site and source location will have to be near the earth and not in outer space, the environment must include the effects of the ground surface and other obstacles in the vicinity of both antennas. These effects almost always dictate that the test range (spacing between source and test antennas) be as short as possible consistent with maintaining a nearly error-free plane wave illuminating the test *aperture*.

A nearly error-free plane wave can be specified as one in which the phase and amplitude, from center to edge of the illuminating field over the test aperture, do not deviate by more than about 30 degrees and 1 decibel, respectively. These conditions will result in a gain-measurement error of no more than a few percent less that the true gain. Based on the 30° phase error alone, it can be shown that the minimum range distance is approximately

$$S_{min} = 2 \frac{D^2}{\lambda}$$

where D is the largest aperture dimension and λ is the free-space wavelength in the same units as D. The phase error

over the aperture D for this condition is 1/16 wavelength.

Since aperture size and gain are related by

$$Gain = \frac{4\pi A_e}{\lambda^2}$$

where A_e is the effective aperture area, the dimension D may be obtained for simple aperture configurations. For a square aperture

$$D^2 = G \frac{\lambda^2}{4\pi}$$

which results in a minimum range distance for a square aperture of

$$S_{min} = G \frac{\lambda}{2\pi}$$

and for a circular aperture of

$$S_{min} = G \frac{2\lambda}{\pi^2}$$

For apertures with a physical area that is not well defined or is much larger in one dimension that in other directions, such as a long thin array for maximum directivity in one plane, it is advisable to use the maximum estimate of D from either the expected gain or physical aperture dimensions.

Up to this point in the range development, only the conditions for minimum range length, S_{min}, have been established, as though the ground surface were not present. This minimum S is therefore a necessary condition even under "free-space" environment. The presence of the ground further complicates the range selection, not in the determination of S but in the exact location of the source and test antennas above the earth.

It is always advisable to select a range whose intervening terrain is essentially flat, clear of obstructions, and of uniform surface conditions, such as all grass or all pavement. The extent of the range is determined by the illumination of the source antenna, usually a beam, whose gain is no greater than the highest gain antenna to be measured. For gain measurements the range consists essentially of the region in the beam of the test antenna. For radiation-pattern measurements, the range is considerably larger and consists of all that area illuminated by the source antenna, especially around and behind the test site. Ideally a site should be chosen where the test-antenna location is near the center of a large open area and the source antenna located near the edge where most of the obstacles (trees, poles, fences, etc) lie.

The primary effect of the range surface is that some of the energy from the source antenna will be reflected into the test antenna while other energy will arrive on a direct line-of-sight path. This is illustrated in Fig 46. The use of a flat, uniform ground surface assures that there will be essentially a mirror reflection even though the reflected energy may be slightly weakened (absorbed) by the surface material (ground). In order to perform an analysis it is necessary to realize that horizontally polarized waves undergo a 180° phase reversal upon reflection from the earth. The resulting illumination amplitude at any point in the test aperture is the vector sum of the electric fields arriving from the two directions, the direct path and the reflected path. If a perfect mirror reflection is assumed from the ground (it is nearly that for practical ground conditions at VHF/UHF) and the source antenna is isotropic, radiating equally in all directions, then a simple geometric analysis of the two path lengths will show

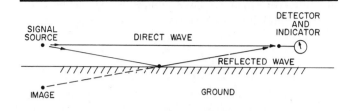

Fig 46—On an antenna test range, energy reaching the receiving equipment may arrive after being reflected from the surface of the ground, as well as by the direct path. The two waves may tend to cancel each other, or may reinforce one another, depending on their phase relationship at the receiving point.

that at various point in the vertical plane at the test-antenna site the waves will combine in different phase relationships. At some points the arriving waves will be in phase, and at other points they will be 180° out of phase. Since the field amplitudes are nearly equal, the resulting phase change caused by path length difference will produce an amplitude variation in the vertical test site direction similar to a standing wave, as shown in Fig 47.

The simplified formula relating the location of h2 for maximum and minimum values of the two-path summation in terms of h1 and S is

$$h2 = n \frac{\lambda}{4} \cdot \frac{S}{h1}$$

with n = 0, 2, 4, . . . for minimums and
  n = 1, 3, 5, . . . for maximums, and S is much larger than either h1 or h2.

The significance of this simple ground-reflection formula is that it permits the approximate location of the source antenna to be determined to achieve a nearly plane-wave amplitude distribution *in the vertical direction* over a particular test *aperture size*. It should be clear from exami-

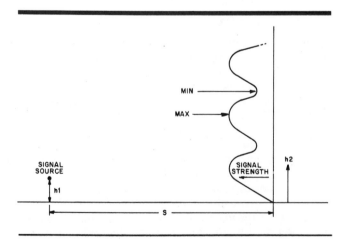

Fig 47—The vertical profile, or plot of signal strength versus test-antenna height, for a fixed height of the signal source above ground and at a fixed distance. See text for definitions of symbols.

nation of the height formula that as h1 is decreased, the vertical distribution pattern of signal at the test site, h2, expands. Also note that the signal level for h2 equal to zero is always zero on the ground regardless of the height of h1.

The objective in using the height formula then is, given an effective antenna aperture to be illuminated from which a minimum S (range length) is determined and a suitable range site chosen, to find a value for h1 (source antenna height). The required value is such that the *first* maximum of vertical distribution at the test site, h2, is at a practical distance above the ground and at the same time the signal amplitude over the aperture in the vertical direction does not vary more than about 1 dB. This last condition is not sacred but is closely related to the particular antenna under test. In practice these formulas are useful only to initialize the range set-up. A final check of the vertical distribution at the test site must be made by direct measurement. This measurement should be conducted with a small low-gain but unidirectional probe antenna such as a corner reflector or 2-element Yagi that is moved along a vertical line over the intended aperture site. Care should be exercised to minimize the effects of local environment around the probe antenna and that the beam of the probe be directed at the source antenna at all times for maximum signal. A simple dipole is undesirable as a probe antenna because it is susceptible to local environmental effects.

The most practical way to instrument the vertical distribution measurement is to construct some kind of vertical track, preferably of wood, with a sliding carriage or platform which may be used to support and move the probe antenna. It is assumed of course that a stable source transmitter and calibrated receiver or detector are available so variations of the order of ½ dB can be clearly distinguished.

Once these initial range measurements are completed successfully, the range is now ready to accommodate any aperture size less in vertical extent than the largest for which S_{min} and the vertical field distribution were selected. The test antenna is placed with the center of its aperture at the height h2 where maximum signal was found. The test antenna should be tilted so that its main beam is pointed in the direction of the source antenna. The final tilt is found by observing the receiver output for maximum signal. This last process must be done empirically since the apparent location of the source is somewhere between the actual source and its image, below the ground.

An example will illustrate the procedure. Assume that we wish to measure a 7-foot diameter parabolic reflector antenna at 1296 MHz (λ = 0.75 foot). The minimum range distance, S_{min}, can be readily computed from the formula for a circular aperture.

$$S_{min} = 2 \frac{D^2}{\lambda} = 2 \times \frac{49}{0.75} = 131 \text{ ft}$$

Now a suitable site is selected based on the qualitative discussion given before.

Next determine the source height, h1. The procedure is to choose a height h1 such that the first minimum above ground (n = 2 in formula) is at least two or three times the aperture size, or about 20 feet.

$$h1 = n \frac{\lambda}{4} \frac{S}{h2} = 2 \times \frac{0.75}{4} \times \frac{131}{20} = 2.5 \text{ ft}$$

Place the source antenna at this height and probe the vertical distribution over the 7-foot aperture location, which will be about 10 feet off the ground.

$$h2 = n \; \frac{\lambda}{4} \; \frac{S}{h1} = 1 \times \frac{0.75}{4} \times \frac{131}{2.5} = 9.8 \text{ ft}$$

The measured profile of vertical signal level versus height should be plotted. From this plot, empirically determine whether the 7-foot aperture can be fitted in this profile such that the 1-dB variation is not exceeded. If the variation exceeds 1 dB over the 7-foot aperture, the source antenna should be lowered and h2 raised. Small changes in h1 can quickly alter the distribution at the test site. Fig 48 illustrates the points of the previous discussion.

The same set-up procedure applies for either horizontal or vertical linear polarization. However, it is advisable to check by direct measurement at the site for each polarization to be sure that the vertical distribution is satisfactory. Distribution probing in the horizontal plane is unnecessary as little or no variation in amplitude should be found, since the reflection geometry is constant. Because of this, antennas with apertures which are long and thin, such as a stacked collinear vertical, should be measured with the long dimension parallel to the ground.

A particularly difficult range problem occurs in measurements of antennas which have depth as well as cross-sectional aperture area. Long end-fire antennas such as long Yagis, rhombics, V-beams, or arrays of these antennas, radiate as volumetric arrays and it is therefore even more essential that the illuminating field from the source antenna be reasonably uniform in depth as well as plane wave in cross section. For measuring these types of antennas it is advisable to make several vertical profile measurements which cover the depth of the array. A qualitative check on the integrity of the illumination for long end-fire antennas can be made by moving the array or antenna axially (forward and backward) and noting the change in received signal level. If the signal level varies less than 1 or 2 dB for an axial movement of several wavelengths then the field can be considered satisfactory for most demands on accuracy. Large variations indicate that the illuminating field is badly distorted over the array depth and subsequent measurements are questionable.

It is interesting to note in connection with gain measurements that any illuminating field distortion will always result in measurements that are lower than true values.

ABSOLUTE GAIN MEASUREMENT

Having established a suitable range, the measurement of gain relative to an isotropic (point source) radiator is almost always accomplished by direct comparison with a calibrated standard-gain antenna. That is, the signal level with the test antenna in its optimum location is noted. Then the test antenna is removed and the standard-gain antenna is placed with its aperture at the center of location where the test antenna was located. The difference in signal level between the standard and the test antennas is measured and appropriately added to or subtracted from the gain of the standard-gain antenna to obtain the absolute gain of the test antenna, absolute here meaning with respect to a point source which has a gain of unity by definition. The reason for using this reference rather than a dipole, for instance, is that it is more useful and convenient for system engineering. It is assumed that both standard and test antennas have been carefully matched to the appropriate impedance and an accurately calibrated and matched detecting device is being used.

A standard-gain antenna may be any type of unidirectional, preferably planar-aperture, antenna, which has been calibrated either by direct measurement or in special cases by accurate construction according to computed dimensions. A standard-gain antenna has been suggested by Richard F. H. Yang (see bibliography). Shown in Fig 49, it consists of two in-phase dipoles ½ λ apart and backed up with a ground plane 1 λ square.

In Yang's original design, the stub at the center is a balun formed by cutting two longitudinal slots of 1/8-inch width, diametrically opposite, on a 1/4-λ section of 7/8-inch rigid 52-Ω coax. An alternative method of feeding is to feed RG-8 or RG-213 coax through slotted 7/8-inch copper tubing. Be sure to leave the outer jacket on the coax to insulate it from the copper-tubing balun section. When constructed accurately to scale for the frequency of interest, this type of standard will have an absolute gain of 7.7 dBd (dB gain over a dipole) with an accuracy of plus or minus 0.25 dB.

RADIATION-PATTERN MEASUREMENTS

Of all antenna measurements, the radiation pattern is the most demanding in measurement and most difficult to interpret. Any antenna radiates to some degree in all directions into the space surrounding it. Therefore, the radiation pattern of an antenna is a three-dimensional representation of the magnitude, phase and polarization. In general, and in practical cases for Amateur Radio communications, the polarization is well defined and only the magnitude of radiation is important. Furthermore, in many of these cases the radiation in one particular plane is of primary interest, usually the plane corresponding to that of the earth's surface, regardless of polarization.

Because of the nature of the range setup, measurement of radiation pattern can be successfully made only in a plane nearly parallel to the earth's surface. With beam antennas it is advisable and usually sufficient to take two radiation pattern measurements, one in the polarization plane and one at right angles to the plane of polarization. These radiation patterns are referred to in antenna literature as the principal E-plane

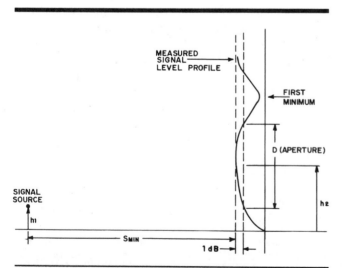

Fig 48—Sample plot of a measured vertical profile.

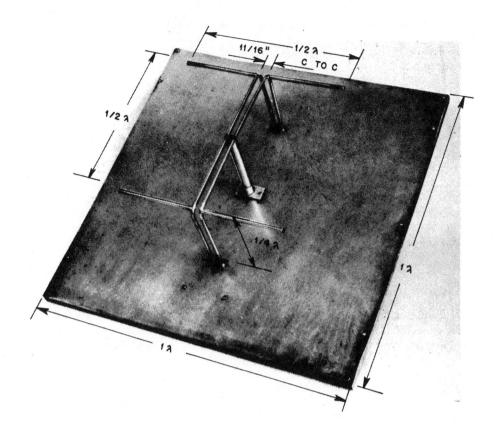

Fig 49—Standard-gain antenna. When accurately constructed for the desired frequency, this antenna will exhibit a gain of 7.7 dB over a dipole radiator, plus or minus 0.25 dB. In this model, constructed for 432 MHz, the elements are 3/8-in. dia tubing. The phasing and support lines are of 5/16-in. dia tubing or rod.

and H-plane patterns, respectively, E plane meaning parallel to the electric field which is the polarization plane and H plane meaning parallel to the magnetic field. The electric field and magnetic field are always perpendicular to each other in a plane wave as it propagates through space.

The technique in obtaining these patterns is simple in procedure but requires more equipment and patience than does making a gain measurement. First, a suitable mount is required which can be rotated in the azimuth plane (horizontal) with some degree of accuracy in terms of azimuth angle positioning. Second, a signal-level indicator calibrated over at least a 20-dB dynamic range with a readout resolution of at least 2 dB is required. A dynamic range of up to about 40 dB would be desirable but does not add greatly to the measurement significance.

With this much equipment, the procedure is to locate first the area of maximum radiation from the beam antenna by carefully adjusting the azimuth and elevation positioning. These settings are then arbitrarily assigned an azimuth angle of zero degrees and a signal level of zero decibels. Next, without changing the elevation setting (tilt of the rotating axis), the antenna is carefully rotated in azimuth in small steps which permit signal-level readout of 2 or 3 dB per step. These points of signal level corresponding with an azimuth angle are recorded and plotted on polar coordinate paper. A sample of the results is shown on ARRL coordinate paper in Fig 50.

On the sample radiation pattern the measured points are marked with an X and a continuous line is drawn in, since the pattern is a continuous curve. Radiation patterns should preferably be plotted on a logarithmic radial scale, rather than a voltage or power scale. The reason is that the log scale approximates the response of the ear to signals in the audio

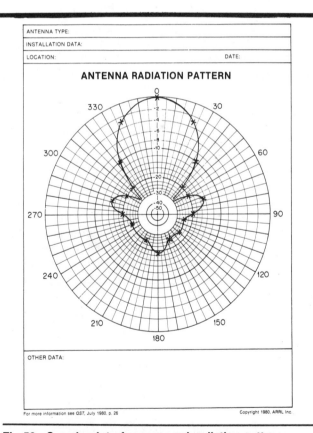

Fig 50—Sample plot of a measured radiation pattern, using techniques described in the text. The plot is on coordinate paper available from ARRL HQ. The form provides space for recording significant data and remarks.

range. Also many receivers have AGC systems that are somewhat logarithmic in response; therefore the log scale is more representative of actual system operation.

Having completed a set of radiation-pattern measurements, one is prompted to ask, "Of what use are they?" The primary answer is as a diagnostic tool to determine if the antenna is functioning as it was intended to. A second answer is to know how the antenna will discriminate against interfering signals from various directions.

Consider now the diagnostic use of the radiation patterns. If the radiation beam is well defined, then there is an approximate formula relating the antenna gain to the measured half-power beamwidth of the E- and H-plane radiation patterns. The half-power beamwidth is indicated on the polar plot where the radiation level falls to 3 dB below the main beam 0-dB reference on either side. The formula is

$$\text{Gain} \cong \frac{41,253}{\theta_E \phi_H}$$

where θ_E and ϕ_H are the half-power beamwidths in degrees of the E- and H-plane patterns, respectively. This equation assumes a lossless antenna system.

To illustrate the use of this equation, assume that we have a Yagi antenna with a boom length of 2 wavelengths. From known relations (described in Chapter 11) the expected gain of a Yagi with a boom length of 2 wavelengths is about 12 dB; its gain, G, equals 15.8. Using the above relationship, the product of $\theta_E \times \phi_H \cong 2600$ square degrees. Since a Yagi produces a nearly symmetric beam shape in cross section, $\theta_E \cong \phi_H = 51°$. Now if the measured values of θ_E and ϕ_H are much larger than 51°, then the gain will be much lower than the expected 12 dB.

As another example, suppose that the same antenna (a 2-wavelength-boom Yagi) gives a measured gain of 9 dB but the radiation pattern half power beamwidths are approximately 51°. This situation indicates that although the radiation patterns seem to be correct, the low gain shows inefficiency somewhere in the antenna, such as lossy materials or poor connections.

Large broadside collinear antennas can be checked for excessive phasing-line losses by comparing the gain computed from the radiation patterns with the direct-measured gain. It seems paradoxical, but it is indeed possible to build a large array with a very narrow beamwidth indicating high gain, but actually having very low gain because of losses in the feed distribution system.

In general, and for most VHF/UHF Amateur Radio communications, gain is the primary attribute of an antenna. However, radiation in other directions than the main beam, called sidelobe radiation, should be examined by measurement of radiation patterns for effects such as nonsymmetry on either side of the main beam or excessive magnitude of sidelobes. (Any sidelobe which is less than 10 dB below the main beam reference level of 0 dB should be considered excessive.) These effects are usually attributable to incorrect phasing of the radiating elements or radiation from other parts of the antenna which was not intended, such as the support structure or feed line.

The interpretation of radiation patterns is intimately related to the particular type of antenna under measurement. Reference data should be consulted for the antenna type of interest, to verify that the measured results are in agreement with expected results.

To summarize the use of pattern measurements, if a beam antenna is first checked for gain (the easier measurement to make) and it is as expected, then pattern measurements may be academic. However, if the gain is lower than expected it is advisable to make the pattern measurements as an aid in determining the possible cause of low gain.

Regarding radiation-pattern measurements, remember that the results measured under proper range facilities will not necessarily be the same as observed for the same antenna at a home-station installation. The reasons may be obvious now in view of the preceding information on the range setup, ground reflections, and the vertical-field distribution profiles. For long paths over rough terrain where many large obstacles may exist, the effects of ground reflection tend to become diffused, although they still can cause unexpected results. For these reasons it is usually unjust to compare VHF/UHF antennas over long paths.

BIBLIOGRAPHY

Source material and more extended discussion of topics covered in this chapter can be found in the references given below.

J. H. Bowen, "A Calorimeter for VHF and UHF Power Measurements," *QST*, Dec 1975, pp 11-13.

W. Bruene, "An Inside Picture of Directional Wattmeters," *QST*, Apr 1959.

D. DeMaw, "In-Line RF Power Metering," *QST*, Dec 1969.

L. McCoy, "A Linear Field-Strength Meter," *QST*, Jan 1973.

T. McMullen, "The Line Sampler, an RF Power Monitor for VHF and UHF," *QST*, Apr 1972.

J. Sevick, "Short Ground-Radial Systems for Short Verticals," *QST*, Apr 1978, pp 30-33.

J. Sevick, "Measuring Soil Conductivity," *QST*, Mar 1981, pp 38-39.

D. Turrin, "Antenna Performance Measurements," *QST*, Nov 1974, pp 35-41.

R. F. H. Yang, "A Proposed Gain Standard for VHF Antennas," *IEEE Transactions on Antennas and Propagation*, Nov 1966.

Chapter 28

Smith Chart Calculations

The Smith Chart is a sophisticated graphic tool for solving transmission line problems. One of the simpler applications is to determine the feed-point impedance of an antenna, based on an impedance measurement at the input of a random length of transmission line. By using the Smith Chart, the impedance measurement can be made with the antenna in place atop a tower or mast, and there is no need to cut the line to an exact multiple of half wavelengths. The Smith Chart may be used for other purposes, too, such as the design of impedance-matching networks. These matching networks can take on any of several forms, such as L and pi networks, a stub matching system, a series-section match, gamma match, and more. With a knowledge of the Smith Chart, the amateur can eliminate much "cut and try" work.

Named after its inventor, Phillip H. Smith, the Smith Chart was originally described in *Electronics* for January 1939. Smith Charts may be obtained at most university book stores. They may be ordered in quantities of 100 from Analog Instruments Co, PO Box 808, New Providence, NJ 07974. (For 8½- × 11-inch paper charts with normalized coordinates, request Form 82-BSPR.) Smith Charts with 50-Ω coordinates (Form 5301-7569) are available. Smith charts also are available from ARRL HQ. (See the caption for Fig 3.)

It is stated in Chapter 24 that the input impedance, or the impedance seen when "looking into" a length of line, is dependent upon the SWR, the length of the line, and the Z_0 of the line. The SWR, in turn, is dependent upon the load which terminates the line. There are complex mathematical relationships which may be used to calculate the various values of impedances, voltages, currents, and SWR values that exist in the operation of a particular transmission line. These equations can be solved with a personal computer and suitable software, or the parameters may be determined with the Smith Chart. Even if a computer is used, a fundamental knowledge of the Smith Chart will promote a better understanding of the problem being solved. And such an understanding might lead to a quicker or simpler solution than otherwise. If the terminating impedance is known, it is a simple matter to determine the input impedance of the line for any length by means of the chart. Conversely, as indicated above, with a given line length and a known (or measured) input impedance, the load impedance may be determined by means of the chart—a convenient method of remotely determining an antenna impedance, for example.

Although its appearance may at first seem somewhat formidable, the Smith Chart is really nothing more than a specialized type of graph. Consider it as having curved, rather than rectangular, coordinate lines. The coordinate system consists simply of two families of circles—the resistance family, and the reactance family. The resistance circles, Fig 1, are centered on the resistance axis (the only straight line on the chart), and are tangent to the outer circle at the

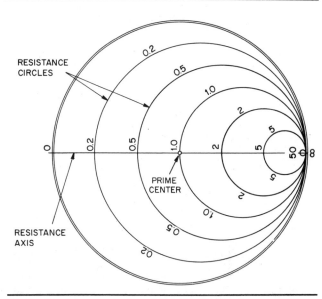

Fig 1—Resistance circles of the Smith Chart coordinate system.

right of the chart. Each circle is assigned a value of resistance, which is indicated at the point where the circle crosses the resistance axis. All points along any one circle have the same resistance value.

The values assigned to these circles vary from zero at the left of the chart to infinity at the right, and actually represent a *ratio* with respect to the impedance value assigned to the center point of the chart, indicated 1.0. This center point is called prime center. If prime center is assigned a value of 100 Ω, then 200 Ω resistance is represented by the 2.0 circle, 50 Ω by the 0.5 circle, 20 Ω by the 0.2 circle, and so on. If, instead, a value of 50 is assigned to prime center, the 2.0 circle now represents 100 Ω, the 0.5 circle 25 Ω, and the 0.2 circle 10 Ω. In each case, it may be seen that the value on the chart is determined by dividing the actual resistance by the number assigned to prime center. This process is called normalizing.

Conversely, values from the chart are converted back to actual resistance values by multiplying the chart value times the value assigned to prime center. This feature permits the use of the Smith Chart for any impedance values, and therefore with any type of uniform transmission line, whatever its impedance may be. As mentioned above, specialized versions of the Smith Chart may be obtained with a value of 50 Ω at prime center. These are intended for use with 50-Ω lines.

Now consider the reactance circles, Fig 2, which appear

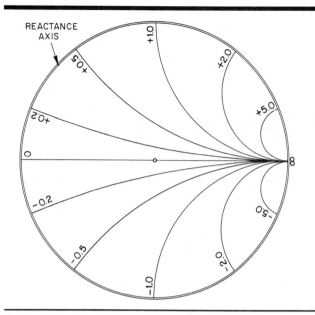

Fig 2—Reactance circles (segments) of the Smith Chart coordinate system.

the same impedance value, 50 + *j*100 Ω. How can this be?

These examples show that the same impedance may be plotted at different points on the chart, depending upon the value assigned to prime center. But two plotted points cannot represent the same impedance *at the same time!* It is customary when solving transmission-line problems to assign to prime center a value equal to the characteristic impedance, or Z_0, of the line being used. This value should always be recorded at the start of calculations, to avoid possible confusion later. (In using the specialized charts with the value of 50 at prime center, it is, of course, not necessary to normalize impedances when working with 50-Ω line. The resistance and reactance values may be read directly from the chart coordinate system.)

Prime center is a point of special significance. As just mentioned, it is customary when solving problems to assign the Z_0 value of the line to this point on the chart—50 Ω for a 50-Ω line, for example. What this means is that the center point of the chart now represents 50 + *j*0 ohms—a pure resistance equal to the characteristic impedance of the line. If this were a load on the line, we recognize from transmission-line theory that it represents a perfect match, with no reflected power and with a 1.0 to 1 SWR. Thus, prime center also represents the 1.0 SWR circle (with a radius of zero).

Short and Open Circuits

On the subject of plotting impedances, two special cases deserve consideration. These are short circuits and open circuits. A true short circuit has zero resistance and zero reactance, or 0 + *j*0. This impedance is plotted at the left of the chart, at the intersection of the resistance and the reactance axes. By contrast, an open circuit has infinite resistance, and therefore is plotted at the right of the chart, at the intersection of the resistance and reactance axes. These two special cases are sometimes used in matching stubs, described later.

as curved lines on the chart because only segments of the complete circles are drawn. These circles are tangent to the resistance axis, which itself is a member of the reactance family (with a radius of infinity). The centers are displaced to the top or bottom on a line tangent to the right of the chart. The large outer circle bounding the coordinate portion of the chart is the reactance axis.

Each reactance circle segment is assigned a value of reactance, indicated near the point where the circle touches the reactance axis. All points along any one segment have the same reactance value. As with the resistance circles, the values assigned to each reactance circle are normalized with respect to the value assigned to prime center. Values to the top of the resistance axis are positive (inductive), and those to the bottom of the resistance axis are negative (capacitive).

When the resistance family and the reactance family of circles are combined, the coordinate system of the Smith Chart results, as shown in Fig 3. Complex impedances (R + *j*X) can be plotted on this coordinate system.

IMPEDANCE PLOTTING

Suppose we have an impedance consisting of 50 Ω resistance and 100 Ω inductive reactance (Z = 50 + *j*100). If we assign a value of 100 Ω to prime center, we normalize the above impedance by dividing each component of the impedance by 100. The normalized impedance is then 50/100 + *j*(100/100) = 0.5 + *j*1.0. This impedance is plotted on the Smith Chart at the intersection of the 0.5 resistance circle and the +1.0 reactance circle, as indicated in Fig 3. Calculations may now be made from this plotted value.

Now say that instead of assigning 100 Ω to prime center, we assign a value of 50 Ω. With this assignment, the 50 + *j*100 Ω impedance is plotted at the intersection of the 50/50 = 1.0 resistance circle, and the 100/50 = 2.0 positive reactance circle. This value, 1 + *j*2, is also indicated in Fig 3. But now we have *two* points plotted in Fig 3 to represent

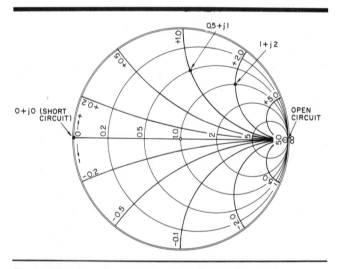

Fig 3—The complete coordinate system of the Smith Chart. For simplicity, only a few divisions are shown for the resistance and reactance values. Smith Chart forms are available from ARRL HQ in two types, standard and expanded. The standard forms are of the type shown here, while the expanded forms are designed for work with SWR values below 1.6:1. At the time of this writing, five 8½- × 11-inch Smith Chart forms are available for $1.

Standing-Wave-Ratio Circles

Members of a third family of circles, which are not printed on the chart but which are added during the process of solving problems, are standing-wave-ratio or SWR circles. See Fig 4. This family is centered on prime center, and appears as concentric circles inside the reactance axis. During calculations, one or more of these circles may be added with a drawing compass. Each circle represents a value of SWR, with every point on a given circle representing the same SWR. The SWR value for a given circle may be determined directly from the chart coordinate system, by reading the resistance value where the SWR circle crosses the resistance axis to the right of prime center. (The reading where the circle crosses the resistance axis to the left of prime center indicates the inverse ratio.)

Consider the situation where a load mismatch in a length of line causes a 3-to-1 SWR ratio to exist. If we temporarily disregard line losses, we may state that the SWR remains constant throughout the entire length of this line. This is represented on the Smith Chart by drawing a 3:1 constant SWR circle (a circle with a radius of 3 on the resistance axis), as in Fig 5. The design of the chart is such that any impedance encountered *anywhere* along the length of this mismatched line will fall on the SWR circle. The impedances may be read from the coordinate system merely by progressing around the SWR circle by an amount corresponding to the length of the line involved.

This brings into use the wavelength scales, which appear in Fig 5 near the perimeter of the Smith Chart. These scales are calibrated in terms of portions of an electrical wavelength along a transmission line. Both scales start from 0 at the left of the chart. One scale, running counterclockwise, starts at the generator or input end of the line and progresses toward the load. The other scale starts at the load and proceeds toward the generator in a clockwise direction. The complete

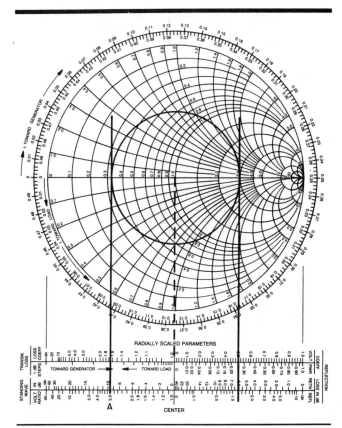

Fig 5—Example discussed in text.

circle around the edge of the chart represents ½ λ. Progressing once around the perimeter of these scales corresponds to progressing along a transmission line for ½ λ. Because impedances repeat themselves every ½ λ along a piece of line, the chart may be used for any length of line by disregarding or subtracting from the line's total length an integral, or whole number, of half wavelengths.

Also shown in Fig 5 is a means of transferring the radius of the SWR circle to the external scales of the chart, by drawing lines tangent to the circle. Another simple way to obtain information from these external scales is to transfer the radius of the SWR circle to the external scale with a drawing compass. Place the point of a drawing compass at the center or 0 line, and inscribe a short arc across the appropriate scale. It will be noted that when this is done in Fig 5, the external STANDING-WAVE VOLTAGE-RATIO scale indicates the SWR to be 3.0 (at A)—our condition for initially drawing the circle on the chart (and the same as the SWR reading on the resistance axis).

SOLVING PROBLEMS WITH THE SMITH CHART

Suppose we have a transmission line with a characteristic impedance of 50 Ω and an electrical length of 0.3 λ. Also, suppose we terminate this line with an impedance having a resistive component of 25 Ω and an inductive reactance of 25 Ω ($Z = 25 + j25$). What is the input impedance to the line?

The characteristic impedance of the line is 50 Ω, so we begin by assigning this value to prime center. Because the line is not terminated in its characteristic impedance, we know that standing waves will exist on the line, and that, therefore, the

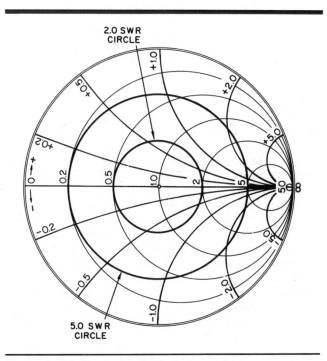

Fig 4—Smith Chart with SWR circles added.

input impedance to the line will not be exactly 50 Ω. We proceed as follows. First, normalize the load impedance by dividing both the resistive and reactive components by 50 (Z_0 of the line being used). The normalized impedance in this case is $0.5 + j0.5$. This is plotted on the chart at the intersection of the 0.5 resistance and the $+0.5$ reactance circles, as in Fig 6. Then draw a constant SWR circle passing through this point. Transfer the radius of this circle to the external scales with the drawing compass. From the external STANDING-WAVE VOLTAGE-RATIO scale, it may be seen (at A) that the voltage ratio of 2.62 exists for this radius, indicating that our line is operating with an SWR of 2.62 to 1. This figure is converted to decibels in the adjacent scale, where 8.4 dB may be read (at B), indicating that the ratio of the voltage maximum to the voltage minimum along the line is 8.4 dB. (This is mathematically equivalent to 20 times the log of the SWR value.)

Next, with a straightedge, draw a radial line from prime center through the plotted point to intersect the wavelengths scale. At this intersection, point C in Fig 6, read a value from the wavelengths scale. Because we are starting from the load, we use the TOWARD GENERATOR or outermost calibration, and read 0.088 λ.

To obtain the line input impedance, we merely find the point on the SWR circle that is 0.3 λ toward the generator from the plotted load impedance. This is accomplished by adding 0.3 (the length of the line in wavelengths) to the reference or starting point, 0.088; $0.3 + 0.088 = 0.388$. Locate 0.388 on the TOWARD GENERATOR scale (at D). Draw a second radial line from this point to prime center. The intersection of the new radial line with the SWR circle represents the normalized line input impedance, in this case $0.6 - j0.66$.

To find the unnormalized line impedance, multiply by 50, the value assigned to prime center. The resulting value is $30 - j33$, or 30 Ω resistance and 33 Ω capacitive reactance. *This* is the impedance that a transmitter must match if such a system were a combination of antenna and transmission line. This is also the impedance that would be measured on an impedance bridge if the measurement were taken at the line input.

In addition to the line input impedance and the SWR, the chart reveals several other operating characteristics of the above system of line and load, if a closer look is desired. For example, the voltage reflection coefficient, both magnitude and phase angle, for this particular load is given. The phase angle is read under the radial line drawn through the plot of the load impedance, where the line intersects the ANGLE OF REFLECTION COEFFICIENT scale. This scale is not included in Fig 6, but will be found on the Smith Chart just inside the wavelengths scales. In this example, the reading is 116.6 degrees. This indicates the angle by which the reflected voltage wave leads the incident wave at the load. It will be noted that angles on the bottom half, or capacitive-reactance half, of the chart are negative angles, a "negative" lead indicating that the reflected voltage wave actually lags the incident wave.

The magnitude of the voltage-reflection-coefficient may be read from the external REFLECTION COEFFICIENT VOLTAGE scale, and is seen to be approximately 0.45 (at E) for this example. This means that 45 percent of the incident voltage is reflected. Adjacent to this scale on the POWER calibration,

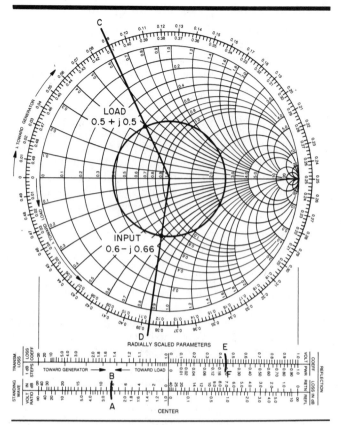

Fig 6—Example discussed in text.

it is noted (at F) that the power reflection coefficient is 0.20, indicating that 20 percent of the incident power is reflected. (The amount of reflected power is proportional to the square of the reflected voltage.)

ADMITTANCE COORDINATES

Quite often it is desirable to convert impedance information to admittance data—conductance and susceptance. Working with admittances greatly simplifies determining the resultant when two complex impedances are connected in parallel, as in stub matching. The conductance values may be added directly, as may be the susceptance values, to arrive at the overall admittance for the parallel combination. This admittance may then be converted back to impedance data, if desired.

On the Smith Chart, the necessary conversion may be made very simply. The equivalent admittance of a plotted impedance value lies diametrically opposite the impedance point on the chart. In other words, an impedance plot and its corresponding admittance plot will lie on a straight line that passes through prime center, and each point will be the same distance from prime center (on the same SWR circle). In the above example, where the normalized line input impedance is $0.6 - j0.66$, the equivalent admittance lies at the intersection of the SWR circle and the extension of the straight line passing from point D though prime center. Although not shown in Fig 6, the normalized admittance value may be read as $0.76 + j0.84$ if the line starting at D is extended.

In making impedance-admittance conversions, remember that capacitance is considered to be a positive susceptance and inductance a negative susceptance. This corresponds to the scale identification printed on the chart. The admittance in siemens is determined by dividing the normalized values by the Z_0 of the line. For this example the admittance is $0.76/50 + j0.84/50 = 0.0152 + j0.0168$ siemen.

Of course admittance coordinates may be converted to impedance coordinates just as easily—by locating the point on the Smith Chart that is diametrically opposite that representing the admittance coordinates, on the same SWR circle.

DETERMINING ANTENNA IMPEDANCES

To determine an antenna impedance from the Smith Chart, the procedure is similar to the previous example. The electrical length of the feed line must be known and the impedance value at the input end of the line must be determined through measurement, such as with an impedance-measuring or a good quality noise bridge. In this case, the antenna is connected to the far end of the line and becomes the load for the line. Whether the antenna is intended purely for transmission of energy, or purely for reception makes no difference; the antenna is still the terminating or load impedance on the line as far as these measurements are concerned. The input or generator end of the line is that end connected to the device for measurement of the impedance. In this type of problem, the measured impedance is plotted on the chart, and the TOWARD LOAD wavelengths scale is used in conjunction with the electrical line length to determine the actual antenna impedance.

For example, assume we have a measured input impedance to a 50-Ω line of $70 - j25$ Ω. The line is 2.35 λ long, and is terminated in an antenna. What is the antenna feed impedance? Normalize the input impedance with respect to 50 Ω, which comes out $1.4 - j0.5$, and plot this value on the chart. See Fig 7. Draw a constant SWR circle through the point, and transfer the radius to the external scales. The SWR of 1.7 may be read from the VOLTAGE RATIO scale (at A). Now draw a radial line from prime center through this plotted point to the wavelengths scale, and read a reference value (at B). For this case the value is 0.195, on the TOWARD LOAD scale. Remember, we are starting at the generator end of the transmission line.

To locate the load impedance on the SWR circle, add the line length, 2.35 λ, to the reference value from the wavelengths scale; $2.35 + 0.195 = 2.545$. Locate the new value on the TOWARD LOAD scale. But because the calibrations extend only from 0 to 0.5, we must first subtract a number of half wavelengths from this value and use only the remaining value. In this situation, the largest integral number of half wavelengths that can be subtracted with a positive result is 5, or 2.5 λ. Thus, $2.545 - 2.5 = 0.045$. Locate the 0.045 value on the TOWARD LOAD scale (at C). Draw a radial line from this value to prime center Now, the coordinates at the intersection of the second radial line and the SWR circle represent the load impedance. To read this value closely, some interpolation between the printed coordinate lines must be

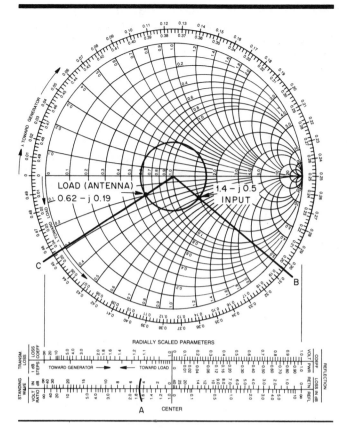

Fig 7—Example discussed in text.

made, and the value of $0.62 - j0.19$ is read. Multiplying by 50, we get the actual load or antenna impedance as $31 - j9.5$ Ω, or 31 Ω resistance with 9.5 Ω capacitive reactance.

Problems may be entered on the chart in yet another manner. Suppose we have a length of 50-Ω line feeding a base-loaded resonant vertical groundplane antenna which is shorter than ¼ λ. Further, suppose we have an SWR monitor in the line, and that it indicates an SWR of 1.7 to 1. The line is known to be 0.95 λ long. We want to know both the input and the antenna impedances.

From the information available, we have no impedances to enter into the chart. We may, however, draw a circle representing the 1.7 SWR. We also know, from the definition of resonance, that the antenna presents a purely resistive load to the line, that is, no reactive component. Thus, the antenna impedance must lie on the resistance axis. If we were to draw such an SWR circle and observe the chart with only the circle drawn, we would see two points which satisfy the resonance requirement for the load. These points are $0.59 + j0$ and $1.7 + j0$. Multiplying by 50, we see that these values represent 29.5 and 85 Ω resistance. This may sound familiar, because, as was discussed in Chapter 24, when a line is terminated in a pure resistance, the SWR in the line equals Z_R/Z_0 or Z_0/Z_R, where Z_R = load resistance and Z_0 = line impedance.

If we consider antenna fundamentals described in Chapter 2, we know that the theoretical impedance of a ¼-λ groundplane antenna is approximately 36 Ω. We therefore can quite logically discard the 85-Ω impedance figure in favor of the 29.5-Ω value. This is then taken as the load impedance value for the Smith Chart calculations. To find the line input impedance, we subtract 0.5 λ from the line length, 0.95, and find 0.45 λ on the TOWARD GENERATOR scale. (The wavelength-scale starting point in this case is 0.) The line input impedance is found to be 0.63 − j0.20, or 31.5 − j10 Ω.

DETERMINATION OF LINE LENGTH

In the example problems given so far in this chapter, the line length has conveniently been stated in wavelengths. The electrical length of a piece of line depends upon its physical length, the radio frequency under consideration, and the velocity of propagation in the line. If an impedance-measurement bridge is capable of quite reliable readings at high SWR values, the line length may be determined through line input-impedance measurements with short- or open-circuit line terminations. Information on the procedure is given later in this chapter. A more direct method is to measure the physical length of the line and calculate its electrical length from

$$N = \frac{Lf}{984\ VF} \tag{Eq 1}$$

where

N = number of electrical wavelengths in the line
L = line length in feet
f = frequency, MHz
VF = velocity or propagation factor of the line

The velocity factor may be obtained from transmission-line data tables in Chapter 24.

Line-Loss Considerations with the Smith Chart

The example Smith Chart problems presented in the previous section ignored attenuation, or line losses. Quite frequently it is not even necessary to consider losses when making calculations; any difference in readings obtained are often imperceptible on the chart. However, when the line losses become appreciable, such as for high-loss lines, long lines, or at VHF and UHF, loss considerations may become significant in making Smith Chart calculations. This involves only one simple step, in addition to the procedures previously presented.

Because of line losses, as discussed in Chapter 24 the SWR does not remain constant throughout the length of the line. As a result, there is a decrease in SWR as one progresses away from the load. To truly present this situation on the Smith Chart, instead of drawing a constant SWR circle, it would be necessary to draw a spiral inward and clockwise from the load impedance toward the generator, as shown in Fig 8. The rate at which the curve spirals toward prime center is related to the attenuation in the line. Rather than drawing spiral curves, a simpler method is used in solving line-loss problems, by means of the external scale TRANSMISSION LOSS 1-DB STEPS. This scale may be seen in Fig 9. Because this is only a relative scale, the decibel steps are not numbered.

If we start at the left end of this external scale and proceed in the direction indicated TOWARD GENERATOR, the first dB step is seen to occur at a radius from center corresponding to an SWR of about 9 (at A); the second dB step falls at an SWR of about 4.5 (at B), the third at 3.0 (at C), and so forth, until the 15th dB step falls at an SWR of about 1.05 to 1. This means that a line terminated in a short or open circuit (infinite SWR), and having an attenuation of 15 dB, would exhibit an SWR of only 1.05 at its input. It will be noted that the dB steps near the lower end of the scale are

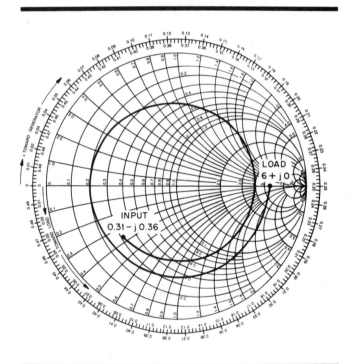

Fig 8—This spiral is the actual "SWR circle" when line losses are taken into account. It is based on calculations for a 16-ft length of RG-174 coax feeding a resonant 28-MHz 300-ohm antenna (50-Ω coax, velocity factor = 66%, attenuation = 6.2 dB per 100 ft). The SWR at the load is 6:1, while it is 3.6:1 at the line input. When solving problems involving attenuation, two constant SWR circles are drawn instead of a spiral, one for the line input SWR and one for the load SWR.

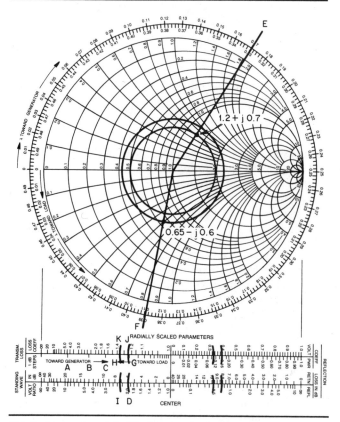

Fig 9—Example of Smith Chart calculations taking line losses into account.

very close together, and a line attenuation of 1 or 2 dB in this area will have only slight effect on the SWR. But near the upper end of the scale, corresponding to high SWR values, a 1 or 2 dB loss has considerable effect on the SWR.

Using a Second SWR Circle

In solving a problem using line-loss information, it is necessary only to modify the radius of the SWR circle by an amount indicated on the TRANSMISSION-LOSS 1-DB STEPS scale. This is accomplished by drawing a second SWR circle, either smaller or larger than the first, depending on whether you are working toward the load or toward the generator.

For example, assume that we have a 50-ohm line that is 0.282 λ long, with 1-dB inherent attenuation. The line input impedance is measured as 60 + j35 ohms. We desire to know the SWR at the input and at the load, and the load impedance. As before, we normalize the 60 + j35-ohm impedance, plot it on the chart, and draw a constant SWR circle and a radial line through the point. In this case, the normalized impedance is 1.2 + j0.7. From Fig 9, the SWR at the line input is seen to be 1.9 (at D), and the radial line is seen to cross the TOWARD LOAD scale at 0.328 (at E). To the 0.328 we add the line length, 0.282, and arrive at a value of 0.610. To locate this point on the TOWARD LOAD scale, first subtract 0.500, and locate 0.110 (at F); then draw a radial line from this point to prime center.

To account for line losses, transfer the radius of the SWR circle to the external 1-DB STEPS scale. This radius crosses the external scale at G, the fifth decibel mark from the left. Since the line loss was given as 1 dB, we strike a new radius (at H), one "tick mark" to the left (toward load) on the same scale. (This will be the fourth decibel tick mark from the left of the scale.) Now transfer this new radius back to the main chart, and scribe a new SWR circle of this radius. This new radius represents the SWR at the load, and is read as 2.3 on the external VOLTAGE RATIO scale. At the intersection of the new circle and the load radial line, we read 0.65 − j0.6. This is the normalized load impedance. Multiplying by 50, we obtain the actual load impedance as 32.5 − j30 ohms. The SWR in this problem was seen to increase from 1.9 at the line input to 2.3 (at I) at the load, with the 1-dB line loss taken into consideration.

In the example above, values were chosen to fall conveniently on or very near the "tick marks" on the 1-DB scale. Actually, it is a simple matter to interpolate between these marks when making a radius correction. When this is necessary, the relative distance between marks for each decibel step should be maintained while counting off the proper number of steps.

Adjacent to the 1-DB STEPS scale lies a LOSS COEFFICIENT scale. This scale provides a factor by which the matched-line loss in decibels should be multiplied to account for the increased losses in the line when standing waves are present. These added losses do not affect the SWR or impedance calculations; they are merely the additional dielectric and copper losses of the line caused by the fact that the line conducts more average current and must withstand more average voltage in the presence of standing waves. For the above example, from Fig 9, the loss coefficient at the input end is seen to be 1.21 (at J), and 1.39 (at K) at the load. As a good approximation, the loss coefficient may be averaged over the length of line under consideration; in this case, the average is 1.3. This means that the total losses in the line are 1.3 times the matched loss of the line (1 dB), or 1.3 dB. This is the same result that may be obtained from procedures given in Chapter 24 for this data.

Smith Chart Procedure Summary

To summarize briefly, any calculations made on the Smith Chart are performed in four basic steps, although not necessarily in the order listed.

1) Normalize and plot a line input (or load) impedance, and construct a constant SWR circle.

2) Apply the line length to the wavelengths scales.

3) Determine attenuation or loss, if required, by means of a second SWR circle.

4) Read normalized load (or input) impedance, and convert to impedance in ohms.

The Smith Chart may be used for many types of problems other than those presented as examples here. The transformer action of a length of line—to transform a high impedance (with perhaps high reactance) to a purely resistive impedance of low value—was not mentioned. This is known as "tuning the line," for which the chart is very helpful, eliminating the need for "cut and try" procedures. The chart may also be used to calculate lengths for shorted or open matching stubs in a system, described later in this chapter.

In fact, in any application where a transmission line is not perfectly matched, the Smith Chart can be of value.

ATTENUATION AND Z_0 FROM IMPEDANCE MEASUREMENTS

If an impedance bridge is available to make accurate measurements in the presence of very high SWR values, the attenuation, characteristic impedance and velocity factor of any random length of coaxial transmission line can be determined. This section was written by Jerry Hall, K1TD.

Homemade impedance bridges and noise bridges will seldom offer the degree of accuracy required to use this technique, but sometimes laboratory bridges can be found as industrial surplus at a reasonable price. It may also be possible for an amateur to borrow a laboratory type of bridge for the purpose of making some weekend measurements. Making these determinations is not difficult, but the procedure is not commonly known among amateurs. One equation treating complex numbers is used, but the math can be handled with a calculator supporting trig functions. Full details are given in the paragraphs that follow.

For each frequency of interest, two measurements are required to determine the line impedance. Just one measurement is used to determine the line attenuation and velocity factor. As an example, assume we have a 100-foot length of unidentified line with foamed dielectric, and wish to know its characteristics. We make our measurements at 7.15 MHz. The procedure is as follows.

1) Terminate the line in an open circuit. The best "open circuit" is one that minimizes the capacitance between the center conductor and the shield. If the cable has a PL-259 connector, unscrew the shell and slide it back down the coax for a few inches. If the jacket and insulation have been removed from the end, fold the braid back along the outside of the line, away from the center conductor.

2) Measure and record the impedance at the input end of the line. If the bridge measures admittance, convert the measured values to resistance and reactance. Label the values as $R_{oc} + jX_{oc}$. For our example, assume we measure $85 + j179$ Ω. (If the reactance term is capacitive, record it as negative.)

3) Now terminate the line in a short circuit. If a connector exists at the far end of the line, a simple short is a mating connector with a very short piece of heavy wire soldered between the center pin and the body. If the coax has no connector, removing the jacket and center insulation from a half inch or so at the end will allow you to tightly twist the braid around the center conductor. A small clamp or alligator clip around the outer braid at the twist will keep it tight.

4) Again measure and record the impedance at the input end of the line. This time label the values as $R_{sc} \pm jX$. Assume the measured value now is $4.8 - j11.2$ Ω.

This completes the measurements. Now we reach for the calculator.

As amateurs we normally assume that the characteristic impedance of a line is purely resistive, but it can (and does) have a small capacitive reactance component. Thus, the Z_0 of a line actually consists of $R_0 + jX_0$. The basic equation for calculating the characteristic impedance is

$$Z_0 = \sqrt{Z_{oc}Z_{sc}} \qquad \text{(Eq 2)}$$

where
$$Z_{oc} = R_{oc} + jX_{oc}$$
$$Z_{sc} = R_{sc} + jX_{sc}$$

From Eq 2 the following working equation may be derived.

$$Z_0 = \sqrt{(R_{oc}R_{sc} - X_{oc}X_{sc}) + j(R_{oc}X_{sc} + R_{sc}X_{oc})} \qquad \text{(Eq 3)}$$

The expression under the radical sign in Eq 3 is in the form of $R + jX$. By substituting the values from our example into Eq 3, the R term becomes $85 \times 4.8 - 179 \times (-11.2) = 2412.8$, and the X term becomes $85 \times (-11.2) + 4.8 \times 179 = -92.8$. So far we have determined that

$$Z_0 = \sqrt{2412.8 - j92.8}$$

The quantity under the radical sign is in rectangular form. Extracting the square root of a complex term is handled easily if it is in polar form, a vector value and its angle. The vector value is simply the square root of the sum of the squares, which in this case is

$$\sqrt{2412.8^2 + 92.8^2} = \sqrt{2414.58}$$

The tangent of the vector angle we are seeking is the value of the reactance term divided by the value of the resistance term. For our example this is arctan $-92.8/2412.8 = $ arctan -0.03846. The angle is thus found to be $-2.20°$. From all of this we have determined that

$$Z_0 = \sqrt{2414.58 \underline{/-2.20°}}$$

Extracting the square root is now simply a matter of finding the square root of the vector value, and taking half the angle. (The angle is treated mathematically as an exponent.)

Our result for this example is $Z_0 = 49.1 \underline{/-1.1°}$. The small negative angle may be ignored, and we now know that we have coax with a nominal 50-Ω impedance. (Departures of as much as 6 to 8% from the nominal value are not uncommon.) If the negative angle is large, or if the angle is positive, you should recheck your calculations and perhaps even recheck the original measurements. You can get an idea of the validity of the measurements by normalizing the measured values to the calculated impedance and plotting them on a Smith Chart as shown in Fig 10 for this example. Ideally, the two points should be diametrically opposite, but in practice they will be not quite 180° apart and not quite the same distance from prime center. Careful measurements will yield plotted points that are close to ideal. Significant departures from the ideal indicates sloppy measurements, or perhaps an impedance bridge that is not up to the task.

Determining Line Attenuation

The short circuit measurement may be used to determine the line attenuation. This reading is more reliable than the open circuit measurement because a good short circuit is a short, while a good open circuit is hard to find. (It is impossible to escape some amount of capacitance between conductors with an "open" circuit, and that capacitance presents a path for current to flow at the RF measurement frequency.)

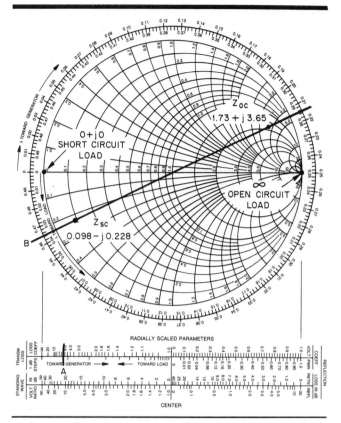

Fig 10—Determining the line loss and velocity factor with the Smith Chart from input measurements taken with open circuit and short circuit terminations.

Use the Smith Chart and the 1-DB STEPS external scale to find line attenuation. First normalize the short circuit impedance reading to the calculated Z_0, and plot this point on the chart. See Fig 10. For our example, the normalized impedance is $4.8/49.1 - j11.2/49.1$ or $0.098 - j0.228$. After plotting the point, transfer the radius to the 1-DB STEPS scale. This is shown at A of Fig 10.

Remember from discussions earlier in this chapter that the impedance for plotting a short circuit is $0 + j0$, at the left edge of the chart on the resistance axis. On the 1-DB STEPS scale this is also at the left edge. The total attenuation in the line is represented by the number of dB steps from the left edge to the radius mark we have just transferred. For this example it is 0.8 dB. Some estimation may be required in interpolating between the 1-dB step marks.

Determining Velocity Factor

The velocity factor is determined by using the TOWARD GENERATOR wavelength scale of the Smith Chart. With a straightedge, draw a line from prime center through the point representing the short-circuit reading, until it intersects the wavelengths scale. In Fig 10 this point is labeled B. Consider that during our measurement, the short circuit was the load at the end of the line. Imagine a spiral curve progressing from $0 + j0$ clockwise and inward to our plotted measurement

point. The wavelength scale, at D, indicates this line length is 0.464 λ. By rearranging the terms of Eq 1 given early in this chapter, we arrive at an equation for calculating the velocity factor.

$$VF = \frac{Lf}{984\,N} \qquad (Eq\ 4)$$

where

  VF = velocity factor
  L = line length, feet
  f = frequency, MHz
  N = number of electrical wavelengths in the line

Inserting the example values into Eq 4 yields VF = 100 × 7.15/(984 × 0.464) = 1.566, or 156.6%. Of course, this value is an impossible number—the velocity factor in coax cannot be greater than 100%. But remember, the Smith Chart can be used for lengths greater than ½ λ. Therefore, that 0.464 value could rightly be 0.964, 1.464, 1.964, and so on. When using 0.964 λ, Eq 4 yields a velocity factor of 0.753, or 75.3%. Trying successively greater values for the wavelength results in velocity factors of 49.6 and 37.0%. Because the cable we measured had foamed dielectric, 75.3% is the probable velocity factor. This corresponds to an electrical length of 0.964 λ. Therefore, we have determined from the measurements and calculations that our unmarked coax has a nominal 50-Ω impedance, an attenuation of 0.8 dB per hundred feet at 7.15 MHz, and a velocity factor of 75.3%.

It is difficult to use this procedure with short lengths of coax, just a few feet. The reason is that the SWR at the line input is too high to permit accurate measurements with most impedance bridges. In the example above, the SWR at the line input is approximately 12:1.

The procedure described above may also be used for determining the characteristics of balanced lines. However, impedance bridges are generally unbalanced devices, and the procedure for measuring a balanced impedance accurately with an unbalanced bridge is complicated. Using a balun device introduces impedance shifts, so their use is invalid.

LINES AS CIRCUIT ELEMENTS

Information is presented in Chapter 24 on the use of transmission-line sections as circuit elements. For example, it is possible to substitute transmission lines of the proper length and termination for coils or capacitors in ordinary circuits. While there is seldom a practical need for that application, lines are frequently used in antenna systems in place of lumped components to tune or resonate elements. Probably the most common use of such a line is in the hairpin match, where a short section of stiff open-wire line acts as a lumped inductor.

The equivalent "lumped" value for any "inductor" or "capacitor" may be determined with the aid of the Smith Chart. Line losses may be taken into account if desired, as explained earlier. See Fig 11. Remember that the top half of the Smith Chart coordinate system is used for impedances containing inductive reactances, and the bottom half for capacitive reactances. For example, a section of 600-Ω line 3/16-λ long (0.1875 λ) and short-circuited at the far end is

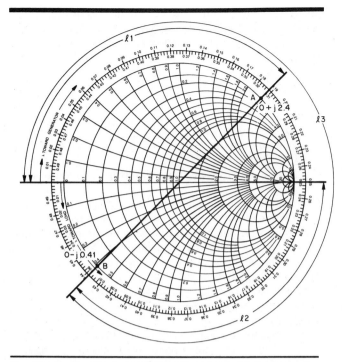

Fig 11—Smith Chart determination of input impedances for short- and open-circuited line sections, disregarding line losses.

At lengths of line that are exact multiples of ¼ λ, such lines have the properties of resonant circuits. At lengths where the input reactance passes through zero at the left of the Smith Chart, the line acts as a series-resonant circuit. At lengths for which the reactances theoretically pass from "positive" to "negative" infinity at the right of the Smith Chart, the line simulates a parallel-resonant circuit.

Designing Stub Matches with the Smith Chart

The design of stub matches is covered in detail in Chapter 26. Equations are presented there to calculate the electrical lengths of the main line and the stub, based on a purely resistive load and on the stub being the same type of line as the main line. The Smith Chart may also be used to determine these lengths, without the requirements that the load be purely resistive and that the line types be identical.

Fig 12 shows the stub matching arrangement in coaxial line. As an example, suppose that the load is an antenna, a close-spaced array fed with a 52-Ω line. Further suppose that the SWR has been measured as 3.1:1. From this information, a constant SWR circle may be drawn on the Smith Chart. Its radius is such that it intersects the right portion of the resistance axis at the SWR value, 3.1, as shown at point B in Fig 13.

Since the stub of Fig 12 is connected in parallel with the transmission line, determining the design of the matching arrangement is simplified if Smith Chart values are dealt with as admittances, rather than impedances. (An admittance is simply the reciprocal of the associated impedance. Plotted on the Smith Chart, the two associated points are on the same SWR circle, but diametrically opposite each other.) Using admittances leaves less chance for errors in making calculations, by eliminating the need for making series-equivalent to parallel-equivalent circuit conversions and back, or else for using complicated equations for determining the resultant value of two complex impedances connected in parallel.

A complex impedance, Z, is equal to R + jX, as described in Chapter 24. The equivalent admittance, Y, is equal to G − jB, where G is the conductance component and B the susceptance. (Inductance is taken as negative susceptance, and capacitance as positive.) Conductance and susceptance values are plotted and handled on the Smith Chart in the same manner as are resistance and reactance.

represented by ℓ1, drawn around a portion of the perimeter of the chart. The "load" is a short-circuit, $0 + j0$ Ω, and the TOWARD GENERATOR wavelengths scale is used for marking off the line length. At A in Fig 11 may be read the normalized impedance as seen looking into the length of line, $0 + j2.4$. The reactance is therefore inductive, equal to $600 \times 2.4 = 1440$ Ω. The same line open-circuited (termination impedance $= \infty$, the point at the right of the chart) is represented by ℓ2 in Fig 11. At B the normalized line-input impedance may be read as $0 - j0.41$; the reactance in this case is capacitive, $600 \times 0.41 = 246$ Ω. (Line losses are disregarded in these examples.) From Fig 11 it is easy to visualize that if ℓ1 were to be extended by ¼ λ, the total length represented by ℓ3, the line-input impedance would be identical to that obtained in the case represented by ℓ2 alone. In the case of ℓ2, the line is open-circuited at the far end, but in the case of ℓ3 the line is terminated in a short. The added section of line for ℓ3 provides the "transformer action" for which the ¼-λ line is noted.

The equivalent inductance and capacitance as determined above can be found by substituting these values in the equations relating inductance and capacitance to reactance, or by using the various charts and calculators available. The frequency corresponding to the line length in degrees must be used, of course. In this example, if the frequency is 14 MHz the equivalent inductance and capacitance in the two cases are 16.4 μH and 46.2 pF, respectively. Note that when the line length is 45° (0.125 λ), the reactance in either case is numerically equal to the characteristic impedance of the line. In using the Smith Chart it should be kept in mind that the electrical length of a line section depends on the frequency and velocity of propagation, as well as on the actual physical length.

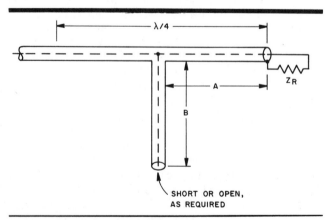

Fig 12—The method of stub matching a mismatched load on coaxial lines.

Assuming that the close-spaced array of our example has been resonated at the operating frequency, it will present a purely resistive termination for the load end of the 52-Ω line. From information in Chapter 24, it is known that the impedance of the antenna equals $Z_0/SWR = 52/3.1 = 16.8$ Ω. (We can logically discard the possibility that the antenna impedance is SWR/Z_0, or 0.06 Ω.) If this 16.8-Ω value were to be plotted as an impedance on the Smith Chart, it would first be normalized (16.8/52 = 0.32) and then plotted as $0.32 + j0$. Although not necessary for the solution of this example, this value is plotted at point A in Fig 13. What is necessary is a plot of the admittance for the antenna as a load. This is the reciprocal of the impedance; 1/16.8 Ω equals 0.060 siemen. To plot this point it is first normalized by multiplying the conductance and susceptance values by the Z_0 of the line. Thus, $(0.060 + j0) \times 52 = 3.1 + j0$. This admittance value is shown plotted at point B in Fig 13. It may be seen that points A and B are diametrically opposite each other on the chart. Actually, for the solution of this example, it wasn't necessary to compute the values for either point A or point B as in the above paragraph, for they were both determined from the known SWR value of 3.1. As may be seen in Fig 13, the points are located on the constant SWR circle which was already drawn, at the two places where it intersects the resistance axis. The plotted value for point A, 0.32, is simply the reciprocal of the value for point B, 3.1. However, an understanding of the relationship between impedance and admittance is easier to gain with simple examples such as this.

In stub matching, the stub is to be connected at a point in the line where the conductive component equals the Z_0 of the line. Point B represents the admittance of the load, which is the antenna. Various admittances will be encountered along the line, when moving in a direction indicated by the TOWARD GENERATOR wavelengths scale, but all admittance plots must fall on the constant SWR circle. Moving clockwise around the SWR circle from point B, it is seen that the line input conductance will be 1.0 (normalized Z_0 of the line) at point C, 0.082 λ toward the transmitter from the antenna. Thus, the stub should be connected at this location on the line.

The normalized admittance at point C, the point representing the location of the stub, is $1 - j1.2$ siemens, having an inductive susceptance component. A capacitive susceptance having a normalized value of $+ j1.2$ siemens is required across the line at the point of stub connection, to cancel the inductance. This capacitance is to be obtained from the stub section itself; the problem now is to determine its type of termination (open or shorted), and how long the stub should be. This is done by first plotting the susceptance required for cancellation, $0 + j1.2$, on the chart (point D in Fig 13). This point represents the input admittance as seen looking into the stub. The "load" or termination for the stub section is found by moving in the TOWARD LOAD direction around the chart, and will appear at the closest point on the resistance/conductance axis, either at the left or the right of the chart. Moving counterclockwise from point D, this is located at E, at the left of the chart, 0.139 λ away. From this we know the required stub length. The "load" at the far end of the stub, as represented on the Smith Chart, has a normalized admittance of $0 + j0$ siemen, which is equivalent to an open circuit.

When the stub, having an input admittance of $0 + j1.2$ siemens, is connected in parallel with the line at a point 0.082 λ from the load, where the line input admittance is $1.0 - j1.2$, the resultant admittance is the sum of the individual admittances. The conductance components are added directly, as are the susceptance components. In this case, $1.0 - j1.2 + j1.2 = 1.0 + j0$ siemen. Thus, the line from the point of stub connection to the transmitter will be terminated in a load which offers a perfect match. When determining the physical line lengths for stub matching, it is important to remember that the velocity factor for the type of line in use must be considered.

MATCHING WITH LUMPED CONSTANTS

It was pointed out earlier that the purpose of a matching stub is to cancel the reactive component of line impedance at the point of connection. In other words, the stub is simply a reactance of the proper kind and value shunted across the line. It does not matter what physical shape this reactance takes. It can be a section of transmission line or a "lumped" inductance or capacitance, as desired. In the above example with the Smith Chart solution, a capacitive reactance was required. A capacitor having the same value of reactance can be used just as well. There are cases where, from an installation standpoint, it may be considerably more convenient to connect a capacitor in place of a stub. This is particularly true when open-wire feeders are used. If a variable capacitor is used, it becomes possible to adjust the capacitance to the exact value required.

The proper value of reactance may be determined from Smith Chart information. In the previous example, the required susceptance, normalized, was $+ j1.2$ siemens. This is converted into actual siemens by dividing by the line Z_0; 1.2/52 = 0.023 siemen, capacitance. The required capacitive reactance is the reciprocal of this latter value, 1/0.023 = 43.5 Ω. If the frequency is 14.2 MHz, for instance, 43.5 Ω

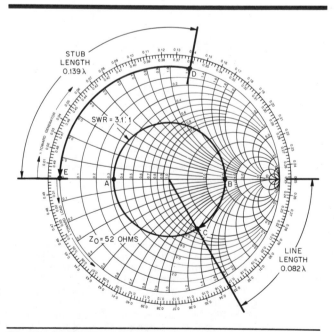

Fig 13—Smith Chart method of determining the dimensions for stub matching.

corresponds to a capacitance of 258 pF. A 325-pF variable capacitor connected across the line 0.082 λ from the antenna terminals would provide ample adjustment range. The RMS voltage across the capacitor is

$$E = \sqrt{P \times Z_0}$$

For 500 watts, for example, E = the square root of 500 × 52 = 161 volts. The peak voltage is 1.41 times the RMS value, or 227 volts.

The Series-Section Transformer

The series-section transformer is described in Chapter 26, and equations are given there for its design. The transformer can be designed graphically with the aid of a Smith Chart. This information is based on a *QST* article by Frank A. Regier, OD5CG. Using the Smith Chart to design a series-section match requires the use of the chart in its less familiar off-center mode. This mode is described in the next two paragraphs.

Fig 14 shows the Smith Chart used in its familiar centered mode, with all impedances normalized to that of the transmission line, in this case 75 ohms, and all constant SWR circles concentric with the normalized value r = 1 at the chart center. An actual impedance is recovered by multiplying a chart reading by the normalizing impedance of 75 ohms. If the actual (unnormalized) impedances represented by a constant SWR circle in Fig 14 are instead divided by a normalizing impedance of 300 ohms, a different picture results. A Smith Chart shows all possible impedances, and so a closed path such as a constant SWR circle in Fig 14 must

again be represented by a closed path. In fact, it can be shown that the path remains a circle, but that the constant SWR circles are no longer concentric. Fig 15 shows the circles that result when the impedances along a mismatched 75-Ω line are normalized by dividing by 300 ohms instead of 75. The constant SWR circles still surround the point corresponding to the characteristic impedance of the line (r = 0.25) but are no longer concentric with it. Note that the normalized impedances read from corresponding points on Figs 14 and 15 are different but that the actual, unnormalized, impedances are exactly the same.

An Example

Now turn to the example shown in Fig 16. A complex

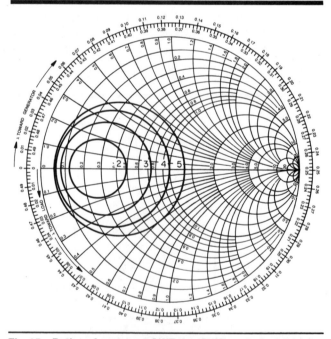

Fig 15—Paths of constant SWR for SWR = 2, 3, 4 and 5, showing impedance variation along 75-Ω line, normalized to 300 Ω. Normalized impedances differ from those in Fig 14, but actual impedances are obtained by multiplying chart readings by 300 Ω and are the same as those corresponding in Fig 14. Paths remain circles but are no longer concentric. One, the matching circle, SWR = 4 in this case, passes through the chart center and is thus the locus of all impedances which can be matched to a 300-Ω line.

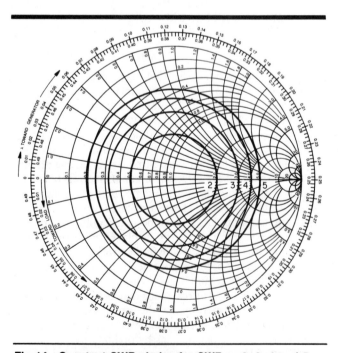

Fig 14—Constant SWR circles for SWR = 2, 3, 4 and 5, showing impedance variation along 75-Ω line, normalized to 75 Ω. The actual impedance is obtained by multiplying the chart reading by 75 Ω.

load of $Z_L = 600 + j900$ ohms is to be fed with 300-Ω line, and a 75-Ω series section is to be used. These characteristic impedances agree with those used in Fig 15, and thus Fig 15 can be used to find the impedance variation along the 75-Ω series section. In particular, the constant SWR circle which passes through the Fig 15 chart center, SWR = 4 in this case, passes through all the impedances (normalized to 300 ohms) which the 75-Ω series section is able to match to the 300-Ω main line. The length $\ell1$ of 300-Ω line has the job of transforming the load impedance to some impedance on this matching circle.

Fig 17 shows the whole process more clearly, with all impedances normalized to 300 Ω. Here the normalized load impedance $Z_L = 2 + j3$ is shown at R, and the matching circle appears centered on the resistance axis and passing through the points $r = 1$ and $r = n^2 = (75/300)^2 = 0.0625$. A constant SWR circle is drawn from R to an intersection with the matching circle at Q or Q′ and the corresponding length $\ell1$ (or $\ell1$ ′) can be read directly from the Smith Chart. The clockwise distance around the matching circle represents the length of the matching line, from either Q′ to P or from Q to P. Because in this example the distance QP is the shorter of the two for the matching section, we choose the length $\ell1$ as shown. By using values from the TOWARD GENERATOR scale, this length is found as 0.045 − 0.213, and adding 0.5 to obtain a positive result yields a value of 0.332 λ.

Although the impedance locus from Q to P is shown in Fig 17, the length $\ell2$ cannot be determined directly from this chart. This is because the matching circle is not concentric with the chart center, so the wavelength scales do not apply to this circle. This problem is overcome by forming Fig 18, which is the same as Fig 17 except that all normalized impedances have been divided by n = 0.25, resulting in a Smith Chart normalized to 75 ohms instead of 300. The matching circle and the chart center are now concentric, and the series-section length $\ell2$, the distance between Q and P, can be taken directly from the chart. By again using the TOWARD GENERATOR scale, this length is found as 0.250 − 0.148 = 0.102 λ.

In fact it is not necessary to construct the entire impedance locus shown in Fig 18. It is sufficient to plot Z_Q/n (Z_Q is read from Fig 17) and $Z_p/n = 1/n$, connect them by a circular arc centered on the chart center, and to determine the arc length $\ell2$ from the Smith Chart.

Procedure Summary

The steps necessary to design a series-section transformer by means of the Smith Chart can now be listed:

1) Normalize all impedances by dividing by the characteristic impedance of the main line.

2) On a Smith Chart, plot the normalized load im-

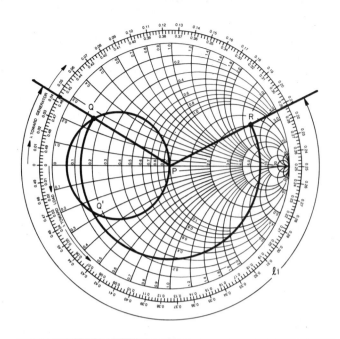

Fig 17—Smith Chart representation of the example shown in Fig 16. The impedance locus always takes a clockwise direction from the load to the generator. This path is first along the constant SWR circle from the load at R to an intersection with the matching circle at Q or Q′, and then along the matching circle to the chart center at P. Length $\ell1$ can be determined directly from the chart, and in this example is 0.332 λ.

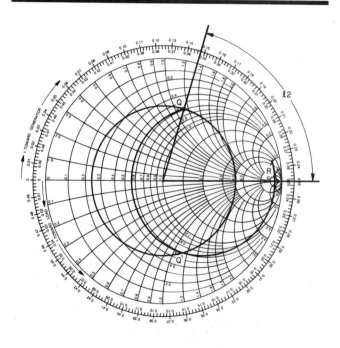

Fig 18—The same impedance locus as shown in Fig 17 except normalized to 75 Ω instead of 300. The matching circle is now concentric with the chart center, and $\ell2$ can be determined directly from the chart, 0.102 λ in this case.

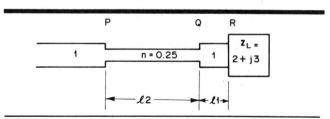

Fig 16—Example for solution by Smith Chart. All impedances are normalized to 300 Ω.

pedance Z_L at R and construct the matching circle so that its center is on the resistance axis and it passes through the points r = 1 and r = n².

3) Construct a constant SWR circle centered on the chart center through point R. This circle should intersect the matching circle at two points. One of these points, normally the one resulting in the shorter clockwise distance along the matching circle to the chart center, is chosen as point Q, and the clockwise distance from R to Q is read from the chart and taken to be ℓ_1.

4) Read the impedance Z_Q from the chart, calculate Z_Q/n and plot it as point Q on a second Smith Chart. Also plot r = 1/n as point P.

5) On this second chart construct a circular arc, centered on the chart center, clockwise from Q to P. The length of this arc, read from the chart, represents ℓ_2. The design of the

transformer is now complete, and the necessary physical line lengths may be determined.

The Smith Chart construction shows that two design solutions are usually possible, corresponding to the two intersections of the constant SWR circle (for the load) and the matching circle. These two values correspond to positive and negative values of the square-root radical in the equation given in Chapter 24 for a mathematical solution of the problem. It may happen, however, that the load circle misses the matching circle completely, in which case no solution is possible. The cure is to enlarge the matching circle by choosing a series section whose impedance departs more from that of the main line.

A final possibility is that, rather than intersecting the matching circle, the load circle is tangent to it. There is then but one solution—that of the ¼-λ transformer.

Calculating Gamma Dimensions

D. H. Healey, W3PG, has developed a method of determining by calculations whether or not a particular set of parameters for a gamma match will be suitable for obtaining the desired impedance transformation. The impedance of the antenna must be known or assumed for this procedure. The results of calculations are in close agreement with values calculated by other means. If the antenna impedance is not accurately known, this procedure provides a very good starting point for initial adjustments of a gamma match. This method was initially published in April 1969 *QST*. (See bibliography at the end of this chapter.) The procedure uses mathematical equations and the Smith Chart, and consists of the following basic steps.

1) Find the impedance step-up ratio for the gamma rod and element diameters and spacing from

$$r = \left(1 + \frac{\log \frac{2S}{d_1}}{\log \frac{2S}{d_2}}\right)^2 \qquad \text{(Eq 1)}$$

where

r = impedance step-up ratio
S = center-to-center spacing of conductors
d_1 = rod diameter
d_2 = element diameter

2) Determine the Z_0 of the "transmission line" formed by the gamma rod and the element, considering them as two parallel conductors, from

$$Z_0 = 276 \log_{10} \frac{2S}{\sqrt{d_1 d_2}} \text{ ohms} \qquad \text{(Eq 2)}$$

where the terms are as above

3) Assign (or assume) a length for the gamma rod, expressed in electrical degrees. Call this angle θ.

4) Determine the increased impedance of the driven

element over its center-point impedance, caused by its being fed off center. Use the equation

$$Z_2 = \frac{Z_1}{\cos^2\theta} \qquad \text{(Eq 3)}$$

where Z_2 is the impedance at the tap point and Z_1 is the complex impedance at the center of the element.

5) Determine the "load" impedance at the antenna end of the gamma "transmission line." This is the resultant value of step 1 above multiplied by the value for Z_2, taken as R + jX from step 4. Normalize this impedance value to the Z_0 of the gamma "transmission line" determined in step 2. Plot this normalized impedance on the Smith Chart.

6) Using the TOWARD GENERATOR wavelengths scale of the Smith Chart, take the "transmission line" length (rod length) into account and determine the normalized input impedance to this line. This impedance represents the portion of the total impedance at the gamma feed point which arises from the antenna alone.

7) In shunt with the impedance from step 6 is an inductive reactance caused by the short-circuit termination on the gamma "transmission line" itself. Determine the normalized value of this inductance either from the Smith Chart (taking 0 + $j0$) as the load and the rod length into account on the TOWARD GENERATOR wavelengths scale) or from the equation

$$X_p = j \tan \theta \text{ ohms} \qquad \text{(Eq 4)}$$

8) Invert the line input impedance (from step 6) to obtain the equivalent admittance, G + jB. This may be done by locating the point on the chart which is diametrically opposite that for the plot of the impedance. (Remember that inductance is considered to be a negative susceptance, and capacitance a positive susceptance.) Similarly, invert the inductance value from step 7. (This susceptance will simply be the reciprocal of the reactance.)

9) Add the two parallel susceptance components from

step 8, taking algebraic signs into account. Plot the new admittance on the Smith Chart, G (from step 8) + jB (from this step).

10) Invert the admittance of step 9 to impedance by locating the diametrically opposite point on the Smith Chart. Convert the normalized resistance and reactance components to ohms by multiplying each by the line Z_0 (from step 2). This impedance is that which terminates the transmission line with no gamma capacitor. A capacitor having the reactance of the X component of the impedance should be used to cancel the inductance, leaving a purely resistive line termination. If the dimensions were properly chosen, this value will be near the Z_0 of the coaxial feed line.

An Example

As an example, assume a 14.3-MHz Yagi beam is to be matched to 50-ohm line. The driven element is 1½ inches in diameter, and the gamma rod is a length of ½-inch tubing, spaced 6 inches from the element (center to center). Initially, the rod length is adjusted to 34 inches, which is 0.041 λ or 14.83°. The driven element has been shortened by 3% from its resonant length.

Following step 1, the impedance step-up ratio from Eq 1 is

$$r = \left(1 + \frac{\log 12/0.5}{\log 12/1.5}\right)^2 = \left(1 + \frac{1.3802}{0.9031}\right)^2 = 6.4$$

From Eq 2 in step 2, the Z_0 of the transmission line is

$$Z_0 = 276 \log \frac{12}{\sqrt{0.5 \times 1.5}} = 315 \text{ ohms}$$

From step 3, $\theta = 14.83°$.

For step 4, assume the antenna has a radiation resistance of 25 ohms and a capacitive reactance component of 25 ohms (about the reactance which would result from the 3% shortening). The overall impedance of the driven element is therefore $25 - j25$ ohms. From step 4, using this value for Z_1 in Eq 3, Z_2 is determined to be $26.75 - j26.75$.

In step 5, the value obtained above for Z_2, 26.75, $- j26.75$, is multiplied by the step-up ratio (step 1), 6.4. The resultant impedance is $171.2 - j171.2$ ohms. Normalized to the Z_0 of the gamma "transmission line," 315 ohms, this impedance is $0.54 - j0.54$. This value is plotted on the Smith Chart, shown at point A of Fig 19.

Taking the line length into account (0.041 λ) as indicated in step 6, the normalized line input impedance is found to be $0.44 - j0.30$, as shown at point B of Fig 19.

From step 7 and Eq 4, X_p is found to be $j0.265$. This same value may be determined from the Smith Chart, as shown at point C, for a matching-section length of 14.83°.

Point D is found on the Smith Chart as directed in step 8, and represents a normalized admittance of $1.56 + j1.06$ siemens. The inductance from step 7, above, inverted to susceptance, is $-j1/0.265 = -j3.89$. (This same value may be read diametrically opposite point C in Fig 19, at point C'.)

Proceeding as indicated in step 9, the admittance components of the parallel combination are $1.56 + j2.71$. This

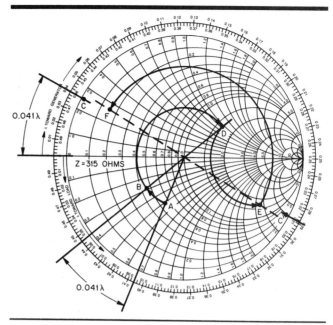

Fig 19—Smith Chart calculation of gamma dimensions. See text.

admittance is plotted as shown at point E in Fig 19.

Inverting the above admittance to its equivalent impedance, point F of Fig 19, the normalized value of $0.16 + j0.28$ is read. Multiplying each value by 315 (from step 2), the input impedance to the gamma section is found to be $50.4 + j88.2$ ohms.

Thus, a series capacitor having a reactance of 88.2 ohms is required to cancel the inductance in the gamma section (from the standard reactance equation the required capacitance is 126 pF), and a very good match is provided for 50-ohm line.

BIBLIOGRAPHY

Source material and more extended discussion of topics covered in this chapter can be found in the references given below and in the textbooks listed at the end of Chapter 2.

D. J. Healey, "An Examination of the Gamma Match," *QST*, Apr 1969, pp 11-15, 57.

C. MacKeand, "The Smith Chart in BASIC," *QST*, Nov 1984, pp 28-31.

R. A. Nelson, "Basic Gamma Matching," *Ham Radio*, Jan 1985, pp 29-31, 33.

F. A. Regier, "Series-Section Transmission-Line Impedance Matching," *QST*, Jul 78, pp 14-16.

P. H. Smith, *Electronic Applications of the Smith Chart*, reprint ed. (Malabar, FL: Krieger Pub Co, Inc, 1983).

H. F. Tolles, "How to Design Gamma-Matching Networks," *Ham Radio*, May 1983, pp 46-55.

Chapter 29

Topical Bibliography on Antennas

This chapter consists of a topical bibliography of articles published for the Radio Amateur, cross indexed, on the subjects of antennas, transmission lines and associated subjects. The information here was compiled by Domenic Mallozzi, N1DM, and John Almeida, KA1AIR. The indexed material has appeared in the following publications: *QST*, January 1963 through April 1988, *QEX*, issues no. 1 through no. 73 (December 1981 through March 1988), as well as other special references. Unless otherwise indicated, all references in the body of the bibliography are to *QST*.

A Note From the Authors

For quite a few years, one of the hottest topics in ham radio has been antennas. This has led to a large number of antenna articles being published in *QST*. In the past, locating articles on a particular antenna has meant sitting down with a bunch of annual indexes and searching. To help you, we have indexed over twenty years of antenna articles from *QST* by topic. We have also indexed all antenna related articles that have appeared in *QEX* from the very first issue. You will also find a few references to other publications we feel might be especially helpful to the active amateur.

Under each topic, the most recent articles are listed first, and the remaining articles are listed in reverse chronological order. If a particular article can be classified into more than one topic, we have listed it under all appropriate topics. At the beginning of some topics you will find a list of other topic headings to consult. These topics may contain material related to the topic at hand.

In cases where Technical Correspondence or corrections to a particular article have been published, you will find the date and page of these items in the remarks column, indexed under the original article. For example, "See 79/03 p 195" means see March 1979 QST, page 195, for further information. When the Technical Correspondence or correction refers only to a particular article, it will be listed only in the remarks column; it is not listed again separately.

Another feature of this bibliography is the "Commercial Products—Reviews and Modifications" topic. It lists not only reviews and product announcements, but it also gives reference to repair and modification articles on commercial products (even if they were not reviewed). These modification and repair articles may be listed not only under this topic, but also under other appropriate topics. To illustrate, an article on "A New Brake Control for the Acme Super Rotator" will also appear in the "Rotators and Direction Indicators" topic.

The reader should be aware of the following when using this bibliography:

1) Titles are not always exact citations of the original titles. Some have been abbreviated to meet space constraints.

2) Dates are listed in the form YY/MM (that is, 81/05 refers to *QST* for May 1981).

3) Dates and pages of other parts of multipart articles are listed in the remarks column.

4) At the end of each topic are references to other books and magazines. They are indicated by the date being listed as 00/00. The page number of these citations refers to a numbered reference in the "Special References" list at the end of this chapter.

We hope you will find this bibliography useful. Your comments and suggestions about how we could improve future editions of this bibliography are welcomed.

INDEX TOPICS AND ABBREVIATIONS USED IN THIS BIBLIOGRAPHY

The main topics used in this bibliography index are listed in Table 1. Commonly used abbreviations are listed below.

Note: The first letter of the abbreviation may or may not be capitalized in the index, depending on its context.

Acc.	Accessories	LDEs	Long delayed echoes
Adapt.	Adapter(s)	Mag.	Magnetic
Amat.	Amateur	Mech.	Mechanical
Ant.	Antenna(s)	Mod.	Modification(s)
Arrest.	Arrestor	Mtd.	Mounted
Atten.	Attenuation, attenuator(s)	Omni	Omnidirectional
Comm.	Communications	Pol.	Polarization
COCO	Color computer	Prog.	Program(s)
Condx.	Conditions	Prop.	Propagation
Configurat.	Configuration	Pwr.	Power
Conn.	Connector(s)	PX	Program Exchange (ARRL)
Construct.	Construction	Rot.	Rotator(s)
Dir.	Direction, director(s)	RS	Radio Shack
Ele.	Element(s)	Sat.	Satellite
FD	Field Day	Sw.	Switch(es)
GDO	Grid (gate) dip oscillator	Sys.	System
Gnd.	Ground(s)	TC	Technical Correspondence
H&K	Hints and Kinks	Temp.	Temperature
Hdwe.	Hardware	Thru	Through
Hor.	Horizontal	Vert.	Vertical
Induct.	Inductance	w/	With
Ind.	Indicator(s)	Wx.	Weather
Info.	Information	X	Reactance
Insul.	Insulation, insulator(s)	Z	Impedance

Table 1
Main Topics Used In This Bibliography

Active antennas
Antenna hardware and switches
Antenna theory
Apartment and hidden antennas—HF
Apartment and hidden antennas—VHF
Baluns
Beverage (wave) antennas
Collinear antennas
Commercial products—reviews and modifications
Comparisons between various antennas
Conical monopole and discone antennas
Curtain array antennas
DDRR antennas
Delta loop antennas
Diversity reception
Double bazooka antennas
Dummy loads
Filters/power dividers and duplexers
Gamma matches
Graphic design aids and computer programs
Grounds, grounding and lightning protection
Hairpin matches
Helical antennas
Inverted V dipoles, slopers and half-slopers
L and T antennas
Log periodic antennas
Long wire antennas
Loop antennas—receiving
Loop antennas—transmitting
Loop Yagis
Matching techniques (except Transmatches)
Misc./special antennas
Mobile antennas and mounts—HF
Mobile antennas and mounts—VHF
Parabolic (dish) antennas
Phased arrays
Phasing systems and networks

Portable/emergency antennas—HF
Portable/emergency antennas—VHF
Propagation—HF
Propagation—VHF/UHF/satellite
Quad antennas
Quagi antennas
Repeater (base) antennas
Rhombic antennas
Rotators and direction indicators
Satellite (OSCAR) antennas
Slot antennas
Small and short antennas (except beams)
Small beam antennas
Smith Chart
Standing wave ratio (SWR) theory
T matches
Test equipment—dip meters
Test equipment—misc. and measurement methods
Test equipment—noise bridges
Test equipment—RX bridges
Test equipment—SWR/power meters
Towers and antenna supports
Transmatches—HF
Transmatches—VHF
Transmission lines and waveguides
Trap antennas and trap design
Turnstile antennas
V beams
Vertical and groundplane antennas
VHF/UHF antennas (general articles)
W8JK beams
Windom antennas
Yagi antennas (general articles)
Zepp, double Zepp and J antennas
ZL Special beams
1.8-MHz antennas and topics
900 MHz and above antennas

ACTIVE ANTENNAS

	Date	Page	Remarks
(See also Loop Antennas—Receiving)			
Ant. Hint for DXers	80/04	60	
Active Ant. Covers 0.5 to 30 MHz	00/00	45	
Incredible Capacitive (Active) Antenna	00/00	49	
VLF-HF Active Ant.	00/00	62	

ANTENNA HARDWARE AND SWITCHES

	Date	Page	Remarks
Wire Gauge and Diameter in BASIC	87/09	42	
Connections for ½-In. Hardline	87/08	37	
Alpha Delta Comm. 4-Position Coax Switch	87/07	19	
Larsen Electronics AD-2/70 Coupler	87/05	29	Dual-band coupler
Radial Materials and Connections	87/05	38	
Hardline Coaxial Connectors You Can Make	87/04	32	
Antenna Hardware You Can Build	87/04	36	
Strain Relief for Coaxial Cables	87/04	58	
More on Coax Switch and Lightning Protection	86/10	50	
Inexpensive RF Switches for Shack	86/08	25	
Alternate Mount for Large Size Antenna	86/08	28	
Plastic Film Spools as Wire Spreader	86/08	38	
Source of Open Wire Line Spacers	86/08	38	
Coaxial Switches and Lightning Protection	86/07	43	
Ant. Insulators from the Golf Course	86/07	44	
Inexpensive Source of Light Wire	86/07	44	
Wx. Guard for Base Loaded Vertical Antennas	86/07	44	
Remote Antenna Switch for HF	86/06	24	See 86/09 p 51 & 86/11 p 45
Stress Calculator for Stacked Antennas	86/04	59	PX prog. announcement
Antenna Switching	84/06	42	
Connectors for 1½ Inch Hardline	84/03	44	
Flexible Mobile Antenna Mount	83/09	37	
Low Cost Antenna Wire	83/06	39	
End Caps for Antenna Elements	83/03	41	
Fiberglass Poles for Antenna Construction	83/03	41	
Yagi Element Mounting Advice	83/03	43	Tech. Correspondence
RG-8 and the PL259	82/09	43	
RF Connectors	82/09	67	
Cable Entrance for the Shack	82/04	52	
Enclosing Cables in Dryer Tubing	82/04	52	Stops TVI
Lightning Arrestor Antenna End Insulator	82/03	52	
Strain Relief for Wire Antennas	82/03	52	
Coaxial Cable Wire Stripper	82/01	48	
Countering Salt Water Corrosion	81/10	44	
Pin Diode Switching	81/10	51	
Tip for 2-M Antennas	81/07	45	
Julie's Custom Antenna Switch	81/06	30	
Hardline Connectors and Corrosion	81/05	43	
Preventing Wind Damage to Match Stubs	81/05	45	
Half Sloper Hardware	81/04	55	
Enclosure for Coax Lightning Arrestor	81/04	56	
TR Switching with PIN Diodes	81/03	19	See 81/04 p 53
Tempo S1 Transmission Line Adapter	81/03	52	
Weatherproofing Compounds	81/02	48	
Teflon for Preventing Ice Buildup	81/01	44	
Making Husky Ground Terminals	80/12	54	
Connectors for CATV Hardline and Heliax	80/09	43	
NOR-Gate Break-In Circuit	80/05	22	See 80/06 p 43
Grounding Guy Wires Eliminates QRM	80/04	57	
Attaching Coax Connectors	80/04	58	
Burndy Conn. Aid Adjusting Dipole Length	80/04	58	
Nonconductive Guy Lines	80/04	58	
Preventing Ice Buildup on Antennas	80/03	46	
Coffee Can Counterweight	80/02	43	
Ant. Corrosion Remover	80/02	45	
Hardline Coax Antenna for 2 M	80/01	51	
Soldering Tarnished Copper Wire	80/01	53	
Toward Safer Antenna Installations	80/01	56	
Switching Open Wire Fed Antennas	79/12	59	Balanced feed ant. type
Beam Ant. Color Codes (Mosley)	79/08	51	Mosley trap color code
Lightning Protection and Weatherproofing	79/06	42	
Novel Way to Mount a Rotary Beam Antenna	79/05	32	
Easy Dipole Center Insulator	79/04	45	
Ant. Accessories for the Beginner	79/02	15	See 79/11 p 49
Salvaging Coaxial Fittings	78/12	39	
Coaxial Fittings on Hardline	78/11	39	
Antennas—Keeping Them Up	78/08	26	

Antenna Hardware and Switches, continued

	Date	Page	Remarks
Ant. Leads into the House	78/06	43	
Ideas for Beam Antennas (Cable Mounting)	78/04	39	
2-M Antenna Mount	78/03	42	For mobile use
Off Center Loaded Antenna—Coil Hint	78/01	40	See 74/09 p 28
Copper Clad Wire for Antennas	77/12	44	
Novel Antenna Installation for Sailboat	77/08	17	
Magnetic Mount Antenna Inside	77/06	47	
Dielectric No-No (Coil Forms)	77/04	56	Materials not to use
Type F to BNC Adapters	77/04	57	
Inexpensive Traps for Wire Antennas	77/02	18	
Another Ant. Sealant	77/02	44	
Dipole Center Insulator	77/01	45	
Easier Installation of Coax Connectors	77/01	45	
Ant. Feedthrough Method	76/11	41	
RF Sensed Antenna Changeover Relay	76/08	21	
Open Wire Spreader Source	76/08	43	
Feed Line Feedthrough	76/07	41	
Ant. Feedthrough Panel	76/04	34	
Easy Side Mount Fixture for Antennas	76/03	73	For VHF antennas on tower
Coaxial Cable Straps	76/02	45	From bleach bottle
Bonding Aluminum with Solder and Flux	75/10	49	
Custom Made Weatherproof Enclosures	75/05	45	
Aluminum Wire Connections	75/03	51	
Low Cost Cable Hangers	75/03	51	
Source for Insulators	75/01	44	
Inexpensive Mobile Antenna Mount	74/10	38	
VHF Yagi Construction	74/10	39	
Remote Antenna Switch	74/08	41	
Element Mount (for VHF Yagis)	74/08	44	
Protecting Antenna Hardware from Corrosion	74/08	44	
Wind Force on Yagi Antenna	74/07	46	
Raising Large Antennas—A Simple Method	74/03	42	
Long Helical Coils for Antenna Loading	74/02	23	
Ant. Sealing Compound	73/09	44	
Supporting Open Wire Line	73/08	50	
Precision Alignment Jig (Element Holes)	73/07	54	
Source of Feeder Spreaders	73/07	54	
Putting Up Wire Antennas	73/06	58	
Ant. Changeover Sys. and Power Output Ind.	73/05	17	
Getting Curl Out of Steel Core Wire	73/03	56	
Wiring Coax Cable to a Wafer Switch	73/03	57	
Forms for Ant. Loading Coils	73/02	46	
Ant. Insulators from Clorox Bottles	73/02	47	
UHF Boom to Element Clamps	72/11	48	
Downspouts House Feed Lines	72/11	49	
Center Insulator from Plastic Plumbing	72/10	57	Tees
Plain Facts About Multiband Vert. Antennas	72/09	14	
Ant. Insulators from a Six Pack	72/09	53	
Care of Wire Rope	72/09	56	
Quad Saver—To Reduce Metal Fatigue	72/08	52	
Cheap Boom to Mast Fitting	72/07	45	
Coax Shield Separator	72/05	50	
Noninductive Guy Lines	72/02	48	
Making Connections to Aluminum Wire	71/12	40	
Quick Antenna Anchors	71/10	57	
Simplified Antenna Switching	71/04	30	
Wire for Antennas & Ground Radials	71/04	50	
Ant. Adjustment from the Drivers Seat	71/04	51	
T Brace for Large Arrays	71/04	51	
Plastic Tubing for Quad Arms	71/03	49	
Bamboo Poles (for Quad Arms)	71/02	46	
Center Insulator for Rotatable Dipole	71/02	46	
Raising Antennas	69/12	51	
Ant. Relay Box	69/07	46	TR relay
Moedras (Digital Readout for Antenna Switch)	69/07	56	
Ant. to Mast Bracket	69/03	44	
Silicon Rubber Seals	69/02	47	
Raising Portable Antennas	68/09	50	
Base Insulator (for Vertical)	68/07	40	
Simple Mounting of Beam to Utility Pole	68/05	52	
Method for Wiring 83-1SP (PL259) Plug	68/04	44	
Stepping Up TR Switch Performance	67/12	28	
Ant. Switching for Beginners	67/10	36	See 67/12 p 81
New High-Power Keyed Antenna Relay	67/08	32	
Nylon Line Insulators	67/08	40	
Ice-Breaking Insulators	67/03	50	

Antenna Hardware and Switches, continued

	Date	Page	Remarks
Lightweight Insulators	67/02	49	
A Really Rugged Coax Switch	67/01	40	
Plastic Quad Frame	66/12	45	
Keeping Feed Lines Untangled	66/07	66	
Boom Drilling Aid	66/06	75	
Nonconductive Guys	66/05	28	See 66/11 p 99
Neat Coaxial Shield Connectors	66/04	66	
Connection Weatherproofing	66/04	67	
Open Wire Line Spacers	66/03	45	
Predicting the Sag in Long Wire Antennas	66/01	57	
Dipole Center Insulator	65/10	94	
Store Bought Hardware for Quads	65/07	20	
Window Feedthrough	65/03	71	
Dipole Center Insulator	64/11	59	
Keyed Antenna Relay	64/07	29	
Ant. Relay for the Beginner	64/04	59	
Plastic Tubing Spreaders	63/11	64	
Beam Hoist for a Wood Pole	63/08	48	
Another Dipole Connector	63/05	90	
Coax Connector Removal	63/05	90	
Tune Into a Load—For Sure	00/00	26	
UHF-Type Connectors for ½-In. Heliax	00/00	79	

ANTENNA THEORY

	Date	Page	Remarks
Some Reflections on Vertical Antennas	87/07	15	See also 87/10 p 39
Broadband Dipoles—Some New Insights	86/10	27	See 87/01 p 37
Antennas—From the Ground Up	86/06	37	
Gaining on the Decibel—Part 3	86/04	29	See 86/03 p 28
Gaining on the Decibel—Part 2	86/03	28	See 86/04 p 29
Antenna Current	85/06	42	
Vertical Antenna Gain	84/12	51	
Antennas—Effect of Real Gnd. On—Part 5	84/11	35	5-part series
Current Distribution on Dipole Ant.	84/10	42	
Antennas and How They Operate	84/09	30	
Antennas—Effect of Real Gnd. On—Part 4	84/08	31	5-part series
Noise Temp.—Antenna Temp. and Sun Noise	84/07	69	
Antennas—Effect of Real Gnd. On—Part 3	84/06	30	5-part series
Antennas—Effect of Real Gnd. On—Part 2	84/04	34	5-part series
Equations for Impedance of Coup. Ant.	84/04	50	See 87/07 p 43
Antenna System for the New Novice	84/03	37	
Antennas—Effect of Real Gnd. On—Part 1	84/02	15	5-part series
Some Practical Antenna Considerations	84/01	30	
Antenna Pruning for 30 M	83/11	60	
True Antenna Height	83/11	60	
My Dipole Does Not Work Right	83/10	47	
Getting the Most Out of Your Antenna	83/07	34	
Effects of Supports on Simple Antennas	83/07	41	See 82/12 p 32
Scaling Antennas	83/05	75	
A Simple Approach to Antenna Impedances	83/03	16	
Antenna Gain Measurements—Part 2 Feedback	83/02	53	See 82/12 p 27
Antenna Gain Measurements—Part 2	82/12	27	See 82/11 p 35 & 83/02 p 53
Effects of Supports on Simple Antennas	82/12	32	See 83/07 p 41
Antenna Gain Measurements—Part 1	82/11	35	See 82/12 p 27 & 83/02 p 53
Gain of Vertical Collinear Antenna	82/10	40	
Go for the Gain NBS Style	82/08	34	See 83/03 p 43
W8JK Antenna—Recap and Update	82/06	11	
300-Ohm Ribbon J Antenna for 2—Analysis	82/04	43	
Power Versus Energy	82/04	47	
Apparent SWR—What is it	82/02	51	
Simple Gain Antennas for Beginner	81/08	32	See 81/10 p 50
Combined Vertical Directivity	81/02	19	
Ant. and Transmission Line Quiz	81/01	43	Answers in 81/02 p 46
Ant. Pruning Shortcut	80/04	46	
Imperfect Antenna System and How it Works	79/07	24	
Balanced Dipoles Fed by Coax Cable	79/05	43	
Radiation Resistance of Vertical Antennas	79/04	37	
Corvette Antenna Mount	79/03	45	
Aerial Performers of Radio Circuits	78/12	44	See 78/11 p 42
Aerial Performers of Radio Circuits	78/11	42	See 78/12 p 44
Basic Antenna Concepts	78/06	18	
Some Basic Antenna Information	77/04	15	
My Feed Line Tunes My Antenna	77/04	40	
Another Look at Reflections—Part 7	76/08	15	Multipart series
Impedance of Short Horizontal Dipole	76/01	32	
Pattern Factor Hor. Ant. Over Earth	75/11	19	Real earth effects

Antenna Theory, continued

Baluns, continued

	Date	Page	Remarks
Broadband Balun Benefits	80/01	56	
Ant. Accessories for the Beginner	79/02	15	See 79/11 p 49
Faulty Balun Equals TVI	74/07	37	
Switchable Impedance Balun	71/01	44	
Broadband Balun Transformers	69/04	42	See 69/11 p 73
A Neglected Form of Balun	69/04	49	
Is a Balun Required	68/12	28	
Matching with Homemade Baluns	68/10	46	
Beer Can Baluns for 144, 220 and 432 MHz	65/02	48	See 65/03 p 43
Broadband Balun Transformers	64/08	33	
Finding VHF Balun Lengths w/a GDO	64/04	56	
Twin-Lead Balun	63/04	47	
Transmission Line Transformers	00/00	75	Classic textbook

BEVERAGE (WAVE) ANTENNAS

	Date	Page	Remarks
On-Ground Low-Noise Receiving Antennas	88/04	30	"Snake" antennas
Beverage Ant. for Amateur Communications	83/01	22	See 84/05 p 46
Classic Beverage Antenna Revisited	82/01	11	
Mr. Beverage on the Beverage Antenna	81/12	55	See 82/03 p 51
Beverage Ant. for Amateur Communications	81/09	51	
Weak Signal Reception on 160 M	77/06	35	Antenna notes
The Beverage Antenna Handbook	00/00	31	Book
Beverage Wave Ant. for Broadcast Reception	00/00	56	
Beverage & Long Wire Ant.—Design and Theory	00/00	65	

COLLINEAR ANTENNAS

(See also VHF/UHF Antennas; Repeater Antennas)

	Date	Page	Remarks
12-M Broadside Loop Array	85/10	45	
Meet the Curtain-Quad Antenna	84/11	48	
Franklin Broadside	84/02	48	
Noise and the Cushcraft 4-Pole Array	84/01	46	
Quad J-Collinear Antenna	83/09	45	
Other Bands for the JF Array	83/04	39	
The JF Array	82/11	26	80/40/15 M
Gain of Vertical Collinear Antenna	82/10	40	
The Lowbanders One-Antenna Farm	82/02	23	
Extended Expanded Collinear Array	81/12	32	Construction article
Collinear Yagi Sextet Antenna	80/06	52	See 80/09 p 56
The Cornwall Collinear	79/07	28	Fixed 48 ele. for 6 M
VHF Antenna Arrays for High Performance	74/12	38	32 to 128 ele. collinear
10-M Collinear Yagi Quartet	69/11	11	
K6MYC Collinear (2 M)	67/04	87	Modified Cushcraft
Trap Collinear Antenna	63/08	30	
Beam Tilt	00/00	71	For vertical collinears

COMMERCIAL PRODUCTS—REVIEWS AND MODIFICATIONS

	Date	Page	Remarks
MFJ-931 Artificial RF Ground	88/04	40	
Hustler Mag. Mounts	88/03	20	
Archer Coaxial Cable Improved	88/03	32	
Antenna Specialists RF Power Dividers	88/03	39	
Hustler 10-M and WARC-Band Mobile Ant.	88/02	22	
Spi-Ro Manufacturing Multiband Antennas	87/12	19	
Austin Triband VHF/UHF Antennas	87/11	45	
Alpha Delta Comm. 4-Position Coax Switch	87/07	19	
Snyder Full-Band Wide-Band Antennas	87/07	29	
Larsen Electronics AD-2/70 Coupler	87/05	29	Dual-band coupler
Alpha Delta Comm. EMP Series Transi-Trap	87/04	59	See 85/04 p 17
Advanced Design Networks Microloop Ant.	87/01	33	
Cure for Arcing in Hy-Gain 402BA	86/10	49	
KLM 220-22 LBX 220-MHz Yagi	86/09	48	
Add 2 Ele. for 40 to Cushcraft A3 and A4	86/09	49	
Spider® HF Mobile Ant.	86/07	40	
Adapting the Hy-Gain Hy-Tower for 30 M	86/07	45	
Antennas—New From Larson	85/12	17	
Bird 4240-400 RF Interseries Adapters	85/12	45	
KLM 2M-22C and 435-40CX Yagis	85/10	43	
AR-200XL Antenna Rot.	85/10	44	
Down East Microwave Ant. and Accessories	85/10	44	
QSK1500 High Power RF Switch	85/09	39	
Coaxial Cable Weather Boots	85/08	40	
Beam Antenna Handbook	85/08	50	
World-Wide Sunrise/Sunset Tables	85/05	39	
Alpha Delta Comm. AC Transi-Trap	85/04	17	See 82/02 p 39
KLM 2M16LBX 2-M Beam	85/03	41	

	Date	Page	Remarks
KLM 144-148 13LBA 2-M Yagi	85/02	41	See 81/10 p 49
Maxcom Antenna Matcher and Dipole Kit	84/11	53	
Ameritron ATR-15 Antenna Tuner	84/10	20	
KLM 7.2-2 40-M Monoband Yagi	84/07	39	
Heath HFT-9 Antenna Tuner	84/07	41	QRP Tuner
Snyder Antenna Corp. Wire	84/07	41	
Toroid Corp. Toroid Power Transformer	84/06	20	
Heath HN-31A Cantenna	84/05	42	Dummy Load
Belden Low-Attenuation Coax Cables	84/01	29	
Hustler 6BTV Vertical Antenna	84/01	44	
KLM 15M6	83/12	46	
Austin Omni 2-M Antenna	83/12	47	See 84/01 p 49
Kilo-Tec Weather Boot	83/11	14	
Gorilla Hooks Tower Climbing Accessory	83/11	30	
AEA Hot Rod Antenna for 2-M Handheld	83/11	53	
Viewstar VS1500A Transmatch	83/10	45	See 83/11 p 61
Kilo-Tec Antenna and Dipole Center Connector	83/09	36	
KLM AP-144DII Base Station VHF Antenna	83/09	43	
West Jersey 80-M BN Cage Antenna	83/09	43	
Cushcraft 220B 220-MHz Boomer	83/08	45	
RF Products 5/8-Wave Antennas	83/08	46	For 220 & 450 MHz
Cushcraft 40-2CD 40-M Skywalker Yagi	83/07	43	
Tokyo Hy-Power HC-200 Transmatch	83/05	38	
Cushcraft R3 3-Band Vertical Ant.	83/03	45	
Hy-Gain TH7DX Broadband Super Thunderbird	83/02	50	
Cushcraft A4 Triband Yagi	83/01	41	
Lambda Coaxial Portal Unit	82/11	48	
Cushcraft 617-6B Boomer 6-M Yagi	82/09	41	
Heath SA-2060 Transmatch	82/07	40	
AEA Isopole 220-MHz Vert. Gain Antenna	82/06	49	
Hy-Gain V2 2-M Antenna	82/05	40	
Wilson System 40 Tribander	82/04	49	
McKay-Dymek DA100D Active Antenna	82/03	46	
B & W BNR 2-M Quad Antenna	82/03	48	
Cushcraft A743 40/30-M Add-on Kit	82/03	48	
Alpha Delta Comm. Transi-Trap	82/02	39	See 87/04 p 59
Centurion Tuf-Duck Mini 2-M Antenna	82/02	48	
Cubic Model MMBX Matchbox	82/01	44	
Cushcraft 20-4CD Skywalker 20-M Ant.	81/12	51	
KLM 144-148-13LB Ant. for 2-M Port.	81/10	49	See 85/02 p 41
KLM 50-7LD 6-M Ant.	81/10	49	
Decibel Products PB-702 2-M Ant.	81/09	49	
Vocom Telescoping 5/8 Wave for 2 M	81/09	49	
Hint for Avanti On-Glass Antenna	81/07	44	
KLM KT-34XA Triband Yagi Antenna	81/06	42	
Cushcraft A3 Triband Yagi	81/05	40	
B & W 370-15 Antenna	81/03	50	
Bird 6736 Termaline Wattmeter	81/01	39	
Tandy Wire RG-8/M Coax Cable	80/12	49	
Alliance HD-73 Heavy Duty Rotator	80/12	51	
Cushcraft 32-19 Boomer and 324QK Kit	80/11	48	Includes stacking kit
Bencher ZA1 and ZA2 Baluns	80/10	45	
MSL Digital QSK Kit	80/10	46	
Heath HM-2141 VHF Wattmeter	80/09	41	
Heath SA-7010 Triband Yagi	80/08	45	
Bird 4381 Power Analyst	80/07	43	
Heath SA-1480 Remote Antenna Switch	80/07	43	
Apollo Trans-Systems Tuner 2000X2	80/05	39	
AEA Isopole 2-M Antenna	80/04	51	
TET 3F35DX Triband Antenna	80/04	53	
Tips for Mosley CL33 and 36 Tribanders	80/04	57	
Heath HM-2140 Dual HF Wattmeter	80/02	40	
Dielectric 1000A RF Wattmeter	79/12	53	
Avanti AH151.3G Window Mount Antenna	79/12	54	144 to 174 MHz
Tempo (K6FZ) 20-M Loop Antenna	79/09	42	
CDR Rotators and Control Units (Repair)	79/08	43	Capacitor replacement
Mirage MP2 VHF Wattmeter	79/08	46	
Wilson System Three Triband Yagi	79/08	46	
KLM 16 Element 2-M Yagi	79/08	47	
Beam Ant. Color Codes (Mosley)	79/08	51	Mosley trap color codes
Cushcraft ATV Vertical HF Antennas	79/07	47	
Tonna (F9FT) 144/16 2-M Yagi Ant.	79/07	48	
Cushcraft ATB-34 Ant.	79/06	39	
Butternut HF5VII Multiband HF Vert.	79/05	39	
Daiwa CS-201 and CS-401 Coax Switches	79/05	41	

Commercial Products—Reviews and Modifications, continued	Date	Page	Remarks
Improving Heath HD-1250 Dip Response	79/05	48	
Electrospace HV5 80-10 M Dual-Mode Antenna	79/03	40	And HP2 160-M Matching Unit
VHF Radio Propagation	79/03	73	Book review
Pointers on the Gem Quad Antenna	79/02	50	
Daiwa CN-720 SWR and Power Meter	79/01	47	
RIW Products 432-19 Yagi	78/12	34	
Decibel Products Vapor-Bloc Coax Cable	78/11	38	
Hy-Gain 214 2-M Yagi	78/10	32	
Communications Power WM7000 Wattmeter	78/09	34	
Wilson System One Triband Yagi	78/09	34	
Mosley TA33 First Aid	78/08	32	See 77/11 p 44
Spectrum International 1296-MHz Loop Yagi	78/05	33	
Extra Meters for Heath HM-102	78/05	38	
Antenna Specialists HM187	78/04	38	2-M mag. mount
Ideas for Beam Antennas (Cable Mounting)	78/04	39	For Mosley TA33 series
Propagation Products Insulator and Quad Kit	78/03	39	
Cushcraft DX144 Collinear Array	78/02	30	
J. W. Miller TVI Filters	78/01	37	
Gem Quad	78/01	38	3 band quad
KLM 40-Meter Beam	77/11	42	
Mosley TA33 First Aid	77/11	44	See 78/08 p 32
Delayed Brake Release for the Ham-II	77/08	14	See 77/11 p 20
Sinclair D-236H 2-Meter Antenna	77/04	54	
Antenna Specialists HM224	77/04	55	220 MHz
RF Products Travel-Tenna	77/04	71	
Bunker Ramo Solderless Connector	77/03	46	
Omega-T 2000C Beam Steering Combiner	77/03	46	
Add Switch to Heath HM102 SWR Meter	77/02	43	
CDR Rotators Erratic Direction Ind.	77/02	43	
Palomar Engrs RX Noise Bridge	77/01	35	
Alliance U100 for Elevating Satellite Ant.	77/01	45	
PEP Wattmeter—a'la Heath (HM102 Mod)	76/12	30	77/07 p 50, 77/08 p 32, 77/09 p 51
Drake RCS4 Remote Coax Switch	76/12	40	
Antenna Inc. 10043 Power/VSWR Ind.	76/11	38	
Dentron 160V Skyclaw Ant. (160 M)	76/11	38	
Wilson HT Antenna Adapter	76/11	41	
Decibel Products Mobile VHF Ant. (2 M)	76/10	39	
Ham-M Delayed Braking	76/09	40	
Leader LAC895 Antenna Coupler	76/07	38	
Hy-Gain Hy-Tower Loading Coil	76/07	41	Mounting
Penniman-Rasmussen TVI Filters	76/05	34	
Larsen JM Mobile Ant. Mount	76/04	33	
Spectrum International UHF Filters	76/04	33	
Ham-M Brake Modification	76/02	45	
Heath HD-1250 Dip Meter	76/01	38	
Comm. Power Broadband Transformers	76/01	39	
Skylane Products Cubical Quad Antenna	75/11	39	
Al's Antenna Quick Up Spider	75/08	46	
Rush Multiband Antenna Loading Coils	75/07	46	
Decibel Products DB4048 Duplexer	75/06	46	
Amat. Radio Vert. Ant. Handbook	75/03	44	
Wire Antennas	75/03	44	
Directional Ind. for Hy-Gain 400 Rotator	74/06	16	
Wilson DB54 20 and 15-M Duoband Beam	74/06	40	
Adjusting Cushcraft 144-7 & 144-11 2-M Yagis	74/06	88	
Adjusting Hy-Gain 15 Element 2-M Yagi	74/06	88	
Delayed Braking in the Ham-M Rotator	74/05	49	
Constant Impedance Trap Vertical	74/03	29	Mod. for Hy-Gain 14AVQ
Teletron Slinky Dipole	74/02	49	
North Shore RF Tech. Duplexer Kits	74/01	46	
KLM Log Periodic Antennas	74/01	53	
Swan WM-1500 RF Wattmeter	73/12	54	
Heath HM-2103 RF Load-Wattmeter	73/09	46	
Microwave Assoc. Circulator and Isolators	73/09	49	
Automating the TR44 Rotator	73/06	28	
Cushcraft AFM-44D Gain Ant.	73/06	45	
Ant. Specialists 6-M Rooftop Ant.	73/02	25	
Weinschel Engineering System 1 3-Band Yagi	72/12	41	
Murch UT-2000 Ultimate Transmatch	72/12	43	
KW Electronics KW103 SWR/Wattmeter	72/09	55	
KW Electronics KW107 Supermatch	72/07	48	
Bird Ham Mate Directional Wattmeter	72/05	55	
Cable Supports for Beam Antennas	72/04	68	
Cushcraft FM Antennas	72/01	50	
Millen Solid State Dipper	71/10	61	

DELTA LOOP ANTENNAS

	Date	Page	Remarks
Delta Loop—Full Wave at Low Height	84/10	24	
Half Delta Loop Goes Rectangular	84/07	26	
40 M with a Phased Delta Loop	84/05	20	
Half Twin Delta Loop Array	83/04	39	
Two Band Delta Loop Antenna	83/03	36	
Half-Wave Delta Loop—Anal. and Deployment	82/09	28	See 82/07 p 14
Antenna Matching Remotely	82/07	14	For use on half delta loop
Dual Full Wave Loop Antenna	82/03	27	Coaxial 2 band loop
Phantom Stub	79/12	37	Multibanding a delta loop
Mono-Loop Delta Antenna	79/09	33	3 Band delta loop beam
Expanded Tribander	79/01	33	Add 40 & 80 M to tribander
40-Meter Triangle	76/05	31	
Two Band Delta Loop Array for OSCAR	74/11	11	432 & 144 MHz on single boom
Modified 20-M Delta Loop Beam	73/06	24	
Inverted Dipole Delta Loop	73/01	37	For 160/75/40 M
Delta Loop on 15 or 10 Meters	71/03	48	Inside a 20-M delta loop
Triband Delta Loop Beam	69/12	52	See 69/03 p 50 & 69/05 p 49
Delta Loop Beam for 6 M	69/09	15	See 69/12 p 52
Delta Loop Beam on 144 MHz	69/04	34	
The HRH (10 M) Delta Loop Beam	69/01	27	
Delta Loop Beam for 15 M	69/01	29	

DIVERSITY RECEPTION

	Date	Page	Remarks
Simpler RTTY Diversity Combiner	83/06	39	
Diversity Combiner for RTTY Reception	82/09	42	
Polarization Diversity	74/06	88	
Space Diversity Reception	73/04	96	
Dual Polarization DX Antenna	72/03	22	Transmitting & receiving
Diversity is Worth the Effort	66/04	40	For HF RTTY
Diversity Reception Made Easy	00/00	66	
Polarization Diversity Reception at HF	00/00	67	

DOUBLE BAZOOKA ANTENNAS

(Also known as coaxial dipoles)

	Date	Page	Remarks
Search for Broadband 80-M Dipole	83/04	22	See 83/05 p 43 & 83/06 p 42
Broadband Double Bazooka Antenna	76/09	29	How broad is it
The Double Bazooka Antenna	69/07	38	

DUMMY LOADS

	Date	Page	Remarks
Low Cost 1500-W Dummy Load	88/03	43	
Balanced 52, 70, or 200 Ohm Dummy Load	87/11	42	
Microwave Dummy Loads	87/10	59	
The Ultimate Dummy Load (Not Quite)	81/01	35	
Flower Pot Hides Dummy Antenna	80/06	49	
300-Ohm Standard for Transmatches	66/10	22	
Cantenna as an RF Wattmeter	65/12	29	
Aqueous Dummy Loads	65/06	16	
Car Radio (AM Band) Dummy Antenna	63/03	52	

FILTERS/POWER DIVIDERS AND DUPLEXERS

(See also Phasing Systems and Networks)

	Date	Page	Remarks
Antenna Specialists RF Power Dividers	88/03	39	
Directional Couplers	86/04	60	Waveguide type
Microwave Ferrite Devices	85/05	62	
Multi-Cavity Waveguide Filters	84/03	73	
Multitransmitter Filter Systems	83/07	28	
Power Splitters	83/02	82	
Wave Reflections in Attenuators and Filters	81/11	47	See 82/02 p 52
Wave Traps with 3 Components	81/11	47	
Cavity Duplexer Construction Notes	80/08	42	
Improved Program for Low-Pass Filter	80/05	34	
Solve Multi-Transmitter Intermod by Traps	80/04	57	
Curing High Power TVI	80/01	53	
Low-Pass Filters for Ham Transmitter	79/12	44	
Broadband Hybrid Splitters and Summers	79/10	44	
TVI Filter Resonance	79/10	51	
Low Cost PC Board Duplexer	79/04	11	For 144 & 220 MHz
TVI Filter	79/01	50	
$5 Filter Solves OSCAR Mode J Desense	78/11	21	
BC Band Energy Rejection Filter	78/02	22	
Twisted Wire Directional Coupler	78/01	21	
Helical Resonator Design Techniques	76/06	11	
TVI Cure for 6 Meters	76/04	24	
Monolithic Crystal Filters	75/07	27	Front end for repeaters
2 and 4 Way 50-Ohm Power Dividers	73/10	97	For 144/220/432 MHz

Grounds, Grounding And Lighting Protection, continued

	Date	Page	Remarks
Wintertime Static on Antennas	85/07	42	
Radial Systems for Gnd. Mtd. Vertical Ant.	85/06	28	
An Aid for Driving Ground Rods	85/06	40	
Efficient Gnd. Systems for Vertical Antennas	83/02	20	
Lightning Arrestor Antenna End Insulator	82/03	52	
Install Radials and Protect Your Vertical	81/06	38	
Enclosure for Coax Lightning Arrestor	81/04	56	
Lightning Protection for Ham-M	81/04	56	Control box protection
Antennas and Grounds for Apartments	80/12	40	
Making Husky Ground Terminals	80/12	54	
Grounding Guy Wires Eliminates QRM	80/04	57	
Ground Rod Removal	79/09	46	See 68/07 p 43
Lightning Protection and Weatherproofing	79/06	42	
Short Ground Radials for Short Verticals	78/04	30	
Lightning Arrestor Idea	77/09	44	
Radials Installed the Easy Way	77/08	48	
Galvanic Action and Grounds	77/06	46	
Optimum Gnd. System for Vert. Ant.	76/12	13	
Improving Earth Ground Characteristics	76/12	16	
Grounding Strap Substitute	76/12	42	
Installing Ground Rods	76/10	40	
Lightning Arrestor From Spark Plugs	76/05	35	
Is Your Ground Rod Grounded?	71/05	46	
Ground Rod Installation	71/01	46	
Ground Rod Removal	68/07	43	See 79/09 p 46
Grounds	67/12	24	
Ground Systems for 160 M	65/04	65	

HAIRPIN MATCHES

	Date	Page	Remarks
Hairpin Match for Collinear-Coax Array	84/10	39	
Ant. Hint for DXers	80/04	60	

HELICAL ANTENNAS

	Date	Page	Remarks
(See also Satellite Antennas)			
Helical Ant. for Space Shuttle Comm.	84/12	14	
More on 50-Ohm Helix Feed	81/12	54	
Easy 50-Ohm Feed for Helix	81/06	28	See also 81/12 p 54
Weatherproof Helical Antenna—Economical	72/06	47	
Finding Wire Length for Helicals	68/02	56	
The Basic Helical Beam	65/11	20	See 86/04 p 42
A Quadhelix for the 1215-MHz Band	63/08	36	See 72/12 p 88
Helical Antennas (Chapter 7)	00/00	35	Basis for all helicals
Use of the Helical Antenna on ATV	00/00	76	

INVERTED V DIPOLES, SLOPERS AND HALF SLOPERS

	Date	Page	Remarks
Improved Broadband Antenna Efficiency	87/08	38	See 86/04 p 23
Half Sloper Observations	86/11	45	
Truly Broadband Antenna for 80 and 75 M	86/04	23	See also 86/11 p 45
KI6O 160-M Linear Loaded Sloper	86/04	26	See TC, 87/02 p 43
Update on Sloping Wire Antenna	84/09	40	
Inverted V Antennas Over Real Ground	84/08	48	
Inexpensive 30-M Beam Antennas	83/06	39	
Broadband 80-M Inverted V	82/08	45	
More Notes on Half-Sloper Antenna	82/06	51	
More Thoughts on the Sloper	81/10	31	
Vertical V Antenna	81/05	24	
ZS6U Minishack Special (end-fed V)	81/04	32	See 81/09 p 51
Half Sloper Hardware	81/04	55	
WA1AKR 40- and 75-M Slopers	80/08	42	
2-Band Half Sloper Antenna	80/06	32	
Half Sloper—Successful Deployment	80/05	31	
Quarter Wave Sloper for 160 M	79/07	19	
Additional Notes on the Half Sloper	79/07	20	See 79/12 p 49
75-M DX Antenna	79/03	44	Switched half slopers
Long Wire Inverted V Sans Tuner	69/08	30	
Radiation Resistance of Inverted V	68/10	36	
Determining Length of Inverted V Antenna	68/07	42	
80-M Inverted V for Field Day	68/06	16	
Inverted-V Radiation Patterns	65/05	81	

L AND T ANTENNAS

	Date	Page	Remarks
A Remote Switched Inverted L Antenna	85/05	37	
Windom JL Revisited	85/05	46	
The Inverted L Revisited	83/01	20	
Try the TJ Antenna	82/06	18	
Inverted L Antenna	77/04	32	

Matching Techniques (Except Transmatches), continued	Date	Page	Remarks
Microwave Matching Techniques	81/04	73	
4/1 UNUN (Balun)	80/10	41	Unbalanced to unbalanced
T-Network Semi-Automatic Antenna Tuner	80/04	26	
Tuned Feeders are Better	80/04	47	
Matching the Transmitter to the Load	80/02	22	
Matching Coax Cables	80/01	42	
Balanced Dipoles Fed by Coax Cable	79/05	43	
Graphical Look at the L Network	79/03	24	
Matching Network Design	79/03	26	See 79/05 p 31
Series Section Line Impedance Match	78/07	14	
Antenna Network That Covers 160 M	77/12	46	
Slant Wire Feed for Grounded Towers	77/05	22	
Matching to HF Mobile Antennas	77/03	48	
Using 75-Ohm Line in 50 Ohm Systems	76/04	34	
Simple Broadband Matching Networks	76/01	20	Toroidal transformers
HF Wideband Transformers	75/04	38	
3-Band Matching System for 40-M Doublet	75/01	43	
Some Ideas on Antenna Couplers	74/12	48	
Graphic Solution to Z Matching Networks	73/12	39	
Series Section Matching	73/06	47	
Admittance Matching Groundplane to Coax Line	73/04	55	
Simplified Impedance Matching and Mac Chart	72/12	33	
RF Matching Techniques	72/10	24	
Stretcher for End-Fed Multiband Wire	72/07	32	See 72/09 p 58
Ant. Coupling Unit for WWVL Receiver	72/06	47	60-kHz ant. coupler
Switchable Impedance Balun	71/01	44	
5/8 Wave Vert. Ant. with Coaxial Transformer	71/01	45	
Quick and Easy Antenna Matching	69/11	83	
L Matching Network for Mobile Whips	69/09	49	
A Coax Line Matcher for VHF Use	69/07	20	See 68/09 p 23
L Matching Network and Balun	68/12	24	
Impedance Matching with C-Line (VHF)	68/09	23	See 69/07 p 20
The L Match for 2-M Yagi Arrays	67/07	19	
L Networks for Reactive Loads	66/09	30	
Working 15- and 20-M Antennas on 40 and 80 M	64/09	50	
Resistive Matching with Quarter-Wave Lines	63/02	56	
Easy Match for High Impedance Antenna	63/01	47	
Series Line Matching Sections	00/00	18	
Exponential Line Matching	00/00	58	For broadband ant.
Transmission Line Transformers	00/00	75	Classic textbook

MISC./SPECIAL ANTENNAS

	Date	Page	Remarks
On-Ground Low-Noise Receiving Antennas	88/04	30	"Snake" antennas
Horn Antennas for 10 GHz	87/04	80	See also 87/05 p 63
A Truly Broadband Antenna for 80/75 M	86/04	23	
Interferometer Direction Finder	85/11	33	
Balanced Antennas	85/05	47	
Winners ARRL Antenna Design Competition	85/02	44	See 85/05 p 47
High Gain Monoband Directional Antenna	84/10	41	
Antenna System for the New Novice	84/03	37	
Periscope Antenna Systems (for UHF Antennas)	84/01	70	See 84/02 p 68
Extended Element Beams	83/12	35	
Cage Antennas	83/11	61	
Building and Using 30-M Antennas	83/10	27	
Wire Antennas for the Beginner	83/06	33	See 83/09 p 46
Search for Broadband 80-M Dipole	83/04	22	See 83/05 p 43 & 83/06 p 42
K4YF Special Antenna	82/09	26	
Updating the Double Duck DF	82/05	15	See 81/07 p 11
Microwave Horn Antennas	82/02	74	
Double Ducky Direction Finder for 2-M DF	81/07	11	See 82/05 p 15
Maritime Antenna for 2 M	81/06	36	
ZS6U Minishack Special (end-fed V)	81/04	32	See 81/09 p 51
Knock Down—Lock Out DF Boxes for 2 M	81/04	41	
CB Antenna to PC-Board-Tool Conversion	81/02	48	
Broadband 80-M Antenna	80/12	36	
Maverick Trackdown	80/07	22	
Multi-Element Twin Loop Array	80/01	28	
"Z" Antenna for 10 to 160 M	79/12	59	
The Cornwall Collinear	79/07	28	Fixed 48 ele. for 6 M
Zip Cord Antennas—Do They Work?	79/03	31	See 79/08 p 43 & 79/11 p 49
40-M Midget	79/02	33	Short rotatable dipole
Expanded Tribander	79/01	33	Add 40 & 80 M to tribander
Dopplescant (Doppler Scan Ant. for 2 M)	78/05	24	See 78/07 p 13
Low Noise Receiving Antennas	77/12	36	
160-M Monster Antenna	77/09	27	

Misc./Special Antennas, continued

	Date	Page	Remarks
Kytoon Support	77/06	93	
Continuously Loaded Helical Antenna	77/05	49	
Hybrid 20-M Quad	77/01	42	Using linear reflector
Kytoon Support	76/06	71	
Off Center Loaded Dipole Antennas	74/09	28	See 78/01 p 41
Four-Band Whopper	74/04	11	40-M dipole w/tribander
Half Square Antenna	74/03	11	Bobtail antenna
15-M Dipole Made of Conduit	74/03	49	
7-MHz Vertical Parasitic Array	73/11	39	
Zip Cord Special Antenna	72/05	51	
Dual Polarization DX Antenna	72/03	22	Dual diversity scheme
Heli-Rope Antenna	71/06	32	See 74/08 p 31
2-M Eggbeater Antenna (Omnidirectional)	71/04	44	Horizontally polarized
10-M Collinear Yagi Quartet	69/11	11	
Using a Grounded Tower on 160 M	69/05	48	
Novel Antenna for 80 and 40 M	69/02	40	
Multiband Ant.	68/08	46	
A Complete Multiband Antenna System	67/11	26	
A Simple 80- and 10-M Antenna	67/11	49	
Indoor Dipole	67/09	45	
4 Band Rotatable Dipole	67/03	35	
All Band Antenna	67/03	48	
Are You Ready for 15-M Openings	66/03	34	Novice article
The Antalo (2 M)	64/12	24	Halo with parasitic ele.
Stacking Halos for Omni Coverage	64/08	64	
Skew Planar Wheel Antenna	63/11	11	Circular polarized omni
Broadband Ant. Using Coax Sections	00/00	5	Patent notice
Asymmetrical Folded Dipole Beam	00/00	6	Patent notice
Weather Balloon Verticals	00/00	42	
160-M Transmission Line Antenna	00/00	53	

MOBILE ANTENNAS AND MOUNTS—HF

	Date	Page	Remarks
Tuning Wires for Mobile Antennas	84/12	48	
Retune Hustler 20-M Resonator for 30 M	84/07	44	
Converting a Hustler Mobile Ant. for 30 M	83/09	37	
Multiband Mobile Ant. (Modification)	83/07	38	
Retuning Mobile Antennas Fast and Easy	83/07	38	H&K
Mobile Antenna Matching Automatically	82/10	15	See 82/12 p 53
Swan 45 Mobile Antenna Repairs	82/10	42	
Connecticut Shorthorn	81/04	16	
Deluxe RV 5 Band Ant.	80/10	38	
Novel Antenna Installation for Sailboat	77/08	17	
Matching to HF Mobile Antennas	77/03	48	
Place to Store Mobile Antennas	76/10	41	
Inexpensive Mobile Antenna Mount	74/10	38	
Preventing Mobile Antenna Sway	73/01	52	
Mobile Antenna Tuning Trick	72/02	49	
Ant. Adjustment from the Drivers Seat	71/04	51	
Dual-Band Mobile Antenna	69/10	34	80- & 40-M Bandswitching
L Matching Network for Mobile Whips	69/09	49	
Alpha Special Mobile Antenna	69/07	26	
Mobile Whips and Corona	69/05	50	
Ant. for Travel Trailers and Campers	69/03	34	
Mobile Multiband Ant. System	68/11	18	
The MABAL Antenna (Mobile HF)	68/07	11	
Storing Hustler Resonators	68/02	56	
The Connecticut Longhorn	67/08	11	See 67/09 p 76 & 67/12 p 48
Modeling Radiation Patterns of Whips	67/01	31	
Antenna Bumper Mount	63/10	76	
Remotely Tuned Mobile Antennas	63/06	11	
Short Antenna for Mobile Operation	53/09	30	Classic article

MOBILE ANTENNAS AND MOUNTS—VHF

	Date	Page	Remarks
Austin Triband VHF/UHF Antennas	87/11	45	
Horizontally Polarized 2-M Mobile Ant.	87/05	39	
Broadcast Ant. for 2-M Mobile Operation	84/10	40	
Flexible Mobile Antenna Mount	83/09	37	
Build the Timeless J Antenna	82/11	40	See 84/01 p 48
Hint for Avanti On-Glass Antenna	81/07	44	
Maritime Antenna for 2 M	81/06	36	
2-M Fox Hunt Antenna (Mast for Car)	81/04	56	
VHF/UHF 3-Band Mobile Antenna	80/02	16	
Build 5/8-Wave Antenna for 146 MHz	79/06	15	See 81/04 p 54
Corvette Antenna Mount	79/03	45	
2-M Mobile Ant.	78/02	37	

Propagation—HF, continued

Quad Antennas, continued	Date	Page	Remarks
Two 3-Element Quads for 2 M	65/10	46	
Quadwrangle (Quad Mechanical)	65/09	20	
Strong, Lightweight 3-Band Quad	64/06	46	See 64/12 p 71
The Short Quad	64/02	46	
The Multielement Quad	63/05	11	
Interlaced Quad Array for 50 and 144 MHz	63/02	11	

QUAGI ANTENNAS

	Date	Page	Remarks
A T-Boom Quagi	86/10	48	
Reproducible Quagi for 1296 MHz	81/08	11	
Using Quagi on Other Frequencies	78/04	34	See 77/04 p 11 and 78/02 p 20
Long Boom Quagi	78/02	20	See 78/04 p 34
VHF Quagi	77/04	11	See 77/08 p 42
Metal Boom Quagi	00/00	23	

REPEATER (BASE) ANTENNAS

	Date	Page	Remarks
Noise and the Cushcraft 4-Pole Array	84/01	46	
Gain of Vertical Collinear Antenna	82/10	40	
Low Cost PC Board Duplexer	79/04	11	For 144 & 220 MHz
Predicting the Coverage of Repeaters	77/12	33	
Calculating Vert. Pattern of Repeater Ant.	73/04	24	
Determining Height Above Average Terrain	73/01	54	
A Duplexer for 2-M Repeaters	72/07	22	
Beam Tilt	00/00	71	For vertical collinears

RHOMBIC ANTENNAS

	Date	Page	Remarks
Invisible Rhombic	77/11	38	
5 in 1 Rhombic Array (80-10 M)	74/10	33	Multidirectional coverage
2-M Rhombic	67/04	88	LaPort rhombic
Rhombic Antenna Design	00/00	32	Classic text book

ROTATORS AND DIRECTION INDICATORS

	Date	Page	Remarks
Automatic Rotator Controller	86/09	40	
CD Ham-IV Tips (Rotator)	85/10	46	
Tower Thrust Bearing Protection	83/07	38	
Automatic Control for HD73 Rotator	83/01	17	
Brake Protection for CDE Ham-IV	82/04	52	
Lubrication for Slow-Turning Rotator	81/07	45	See 81/09 p 45
Lightning Protection for Ham-M	81/04	56	Control box protection
Rotator Cable Quick Disconnect	80/04	58	
Ant. Elevation Indicator (Surplus)	79/09	24	Using radio compass meter
CDR Rotators and Control Units (Repair)	79/08	43	Capacitor replacement
Low Cost Antenna Elevation System	79/06	19	See 79/09 p 24
Heavy-Duty Gears for Antenna Rotating	77/11	44	
Delayed Brake Release for the Ham-II	77/08	14	See 77/11 p 20
CDR Rotators Erratic Direction Ind.	77/02	43	
Alliance U100 for Elevating Satellite Ant.	77/01	45	
Side Mount Rotator for Large HF Antennas	76/11	17	
2-Wire Control for Prop Pitch Rotators	75/04	39	
Directional Ind. for Hy-Gain 400 Rotator	74/06	16	
Manual Elevation Control for VHF Yagi	74/05	48	
Delayed Braking in the Ham-M Rotator	74/05	49	
Automating the TR44 Rotator	73/06	28	
Ant. Rotator Heater	72/06	47	
Rejuvenating Prop Pitch Rotators	71/08	36	
Prop-Pitch Rotator Control Circuit	71/08	42	
Delayed Action Braking for Rotators	71/05	42	
Ham-M Rotator Modification	71/02	46	
Drive Shaft for Base Mounted Rotator	69/11	43	
Preventing Rotator Freeze-Up	69/04	47	
Preventing Loose Rotator Bolts	67/07	48	
Ant. Rotators and Indicators—Part 2	67/05	31	See 67/04 p 22
Ant. Rotators and Indicators—Part 1	67/04	22	See 67/05 p 31
Using Ham-M Rotator with Long Control Lines	66/09	84	
Polar Mounts for Moon Tracking	65/09	84	
Rotator Operation for the Handicapped	65/06	51	
400 Hz Supply for Selsyn Ind.	64/05	45	
Beam Rotator	63/03	53	

SATELLITE (OSCAR) ANTENNAS

	Date	Page	Remarks
Mode-L Parabolic Antenna and Feed Horn	86/05	24	For OSCAR 10
Adventures in Satellite DXing—Part 2	86/05	28	Incl. antennas
Antennas for Working OSCAR	85/10	72	

Small Beam Antennas, continued   *Date*   *Page Remarks*

	Date	Page	Remarks
Five for Five (Short ZL Beam for 20 and 15 M)	71/01	40	See 71/03 p 46
All Driven 3-Element Mini Beam	69/05	35	For 20 M
Compact End Loaded 2-Element 20-M Yagi	69/02	16	
Compact 40-M Beam	67/06	20	
The Short Quad	64/02	46	

SMITH CHART

	Date	Page	Remarks
Smith Chart in BASIC	84/11	28	Microsoft® BASIC
On Using the Smith Chart	66/06	40	
Smith Chart for the Amateur—Part 2	66/02	30	See 66/03 p 76 and 66/01 p 22
Smith Chart for the Amateur—Part 1	66/01	22	See 66/03 p 76 and 66/06 p 40
Computerized Smith Chart for Noise Bridge	00/00	2	
The Smith Chart	00/00	57	
Smith Chart Fundamentals	00/00	78	Also 87/05 QEX, p 15

STANDING WAVE RATIO (SWR) THEORY

(See also Antenna Theory)

	Date	Page	Remarks
Reflected Power	84/01	48	See 73/02 p 51
More on Reflected Power	83/09	46	
Line Loss and SWR	83/07	41	
A Simple Approach to Antenna Impedances	83/03	16	
The Reality of Reflected Power	83/02	52	See 84/01 p 48
What Your Wattmeter Really Reads	81/02	26	See 83/02 p 52 and 84/01 p 48
What Does Your SWR Cost You	79/01	19	
Another Look at Reflections—Part 7	76/08	15	Multipart series
Another Look at Reflections—Part 6	74/12	11	Multipart series thru 1976
Another Look at Reflections—Part 5	74/04	26	Multipart series thru 1976
Correction Chart for SWR Measurement	73/12	40	Includes line loss
Another Look at Reflections—Part 4	73/10	22	Multipart series thru 1976
Another Look at Reflections—Part 3	73/08	36	Multipart series thru 1976
Reflections and Transmission Lines	73/08	47	
Transmission Line Measurements and Line Loss	73/07	55	See 73/12 p 40
Another Look at Reflections—Part 2	73/06	20	Multipart series thru 1976
Another Look at Reflections—Part 1	73/04	35	Multipart series thru 1976
Reflected Power	73/02	51	See 84/01 p 48
Reflections on Reflected Wire	72/11	46	
SWR—What Does it Mean	71/11	44	

T MATCHES

	Date	Page	Remarks
Balanced Antennas	85/05	47	

TEST EQUIPMENT—DIP METERS

	Date	Page	Remarks
Compact Transistor Dip Oscillator	81/09	43	
Add-Ons for Greater Dipper Versatility	81/02	37	
Capacitance Measurement w/Dip Meter	80/12	23	
A 1980 Dipper	80/03	11	
Improving Heath HD-1250 Dip Response	79/05	48	
Auditory Dip Meter	78/09	25	
MOSFET Grid Dipper	78/02	34	
Extending Grid-Dip Meter Range	77/09	45	
Dual Gate MOSFET Dip Meter	77/01	16	
Hybrid Gate Dip Oscillator	74/06	33	
Grip Dipping Toroidal Wound Inductors	73/04	60	
Anatomy of a Solid State Dipper	72/12	23	
High Accuracy FET Dipper	72/06	46	
Field Effect Transistor Dipper	68/02	24	
The Gate Dip Oscillator	67/09	45	
The Dipper	65/11	26	
Finding VHF Balun Lengths w/a GDO	64/04	56	
Grid Dipper Calibration	63/09	83	
Modernizing a Transistor Dip Meter	63/05	20	
How to Use Grid-Dip Oscillators	00/00	33	Book

TEST EQUIPMENT—MISC. AND MEASUREMENT METHODS

	Date	Page	Remarks
Microwave Dummy Loads	87/10	59	
Coaxial Feed Lines in Parallel	87/02	45	Z measurement
Feed Line Tests	86/10	48	
Beyond the Dipper	86/05	14	
Directional Couplers	86/04	60	Waveguide type
Field Tester for Antennas	86/03	40	
A Handy RF Sampler for Coax Lines	86/03	48	
Learning to Use Field Strength Meters	85/03	26	See 85/05 p 47
Beam Antenna Pattern Measurement	85/03	31	

Test Equipment—SWR/Power Meters, continued

	Date	Page	Remarks
Peak Reading Meter for SSB Transmitters	81/03	31	Bar-graph display
Important QRT (SWR Monitoring)	81/03	56	
1296 MHz Power and SWR Indicator	80/11	69	
Reflectometer for Twin-Lead	80/10	15	See 80/12 p 58
PEP Wattmeter Modification	80/06	48	
Simple and Sensitive Impedance Bridge	80/03	29	See 80/04 p 47
Tune Up Swiftly, Silently and Safely	79/12	42	
Simple Accurate RF Wattmeter	79/11	40	Terminating type
Ant. Accessories for the Beginner	79/02	15	See 79/11 p 49
Extra Meters for Heath HM-102	78/05	38	
PEP Wattmeter Switching Arrangement	77/10	45	
Add Switch to Heath HM102 SWR Meter	77/02	43	
Calorimeter for VHF/UHF Power Measurement	75/12	11	
Correction Chart for SWR Measurement	73/12	40	Includes line loss
Simple Computing SWR Meter	73/07	23	See 77/03 p 43 & 82/07 p 37
QRP Man's RF Power Meter	73/06	13	
Ant. Changeover Sys. and Power Output Ind.	73/05	17	
SWR and Directional Wattmeter Readings	72/10	57	See 72/11 p 33 & 72/12 p 37
A VHF/UHF RF Power Monitor	72/04	21	Line Sampler
Interpreting Peak and Average Power	71/11	15	For SSB linear amplifiers
Power Bridge and SWR Indicator for 2 M	71/07	37	
SWR Indicator Mounting	71/04	50	
In-Line RF Power Metering	69/12	11	See 74/04 p 45
Etched Circuit Monimatch	69/10	29	See 69/12 p 52 and 86/01 p 49
Slotted Line for UHF SWR Checks	69/01	36	
The Millimatch (QRP Monimatch)	67/08	44	
The Economatch	67/07	32	See 69/09 p 51
The Wavebridge (Wavemeter and SWR Bridge)	66/07	43	
The Varimatcher	66/05	11	Variable Z SWR bridge
Cantenna as an RF Wattmeter	65/12	29	
Telematch Revisited	65/09	68	
Mini Mono-Monimatch	65/03	54	Small mobile SWR meter
Monimatch Construction	65/03	68	
Telematch (Tune-Up System with SWR Bridge)	65/02	21	See 65/04 p 64
Accuracy of SWR Measurements	64/11	50	
The Monimatch MK3 and MK4	64/09	20	
The Monimatch and SWR	64/08	54	
Audible SWR Meter (for the Blind)	00/00	7	
Theory, Limits and Adjustment, SWR Ind.	00/00	8	
Errors in SWR Indicators	00/00	26	

TOWERS AND ANTENNA SUPPORTS

	Date	Page	Remarks
Mid-Michigan Skyhook	87/01	15	
Wrap Your Guy Wires	86/10	48	
Some Advice About Tower Anchors	86/09	49	
Bamboo Antenna Supports	85/10	46	
Try This Field Day Antenna Support	85/05	35	
Effects of Supports on Simple Antennas	83/07	41	See 82/12 p 32
Raising Beam Antennas	83/06	40	
Effects of Supports on Simple Antennas	82/12	32	See 83/07 p 41
Indoor Antenna Support	82/08	46	
Assembling Big Ant. on Fixed Towers	82/03	28	
Ant. Towers—A Warning	81/12	54	
Lubrication for Crank-Up Towers	81/09	44	
Combination Flagpole and Tilt-Over Tower	81/07	44	
2-M Fox Hunt Antenna (Mast for Car)	81/04	56	
Kite Supported 160 (and 80) M Ant.	81/03	40	
More on Removing Tower Sections	80/04	58	
Walking Your Tower Up Safely	80/03	32	See 80/07 p 40
Separating Tower Sections (Technique)	79/09	37	
Dismantling Tower Sections	77/06	47	
Kytoon Support	77/06	93	
Aluminum Towers—Things to Watch for	76/10	41	
Tower Shield (Anti Climbing Cover)	76/09	26	
Kytoon Support	76/06	71	
Fold-Over Tower—New Approach	76/05	36	
Disassembly of Tower Sections	76/04	35	
Easy Side Mount Fixture for Antennas	76/03	73	For VHF antennas on tower
Guy Anchor Test	75/02	47	
Tower Guard System	74/12	25	Automatic wind protection
Guy Wire Safety	74/08	45	
Fence Mount for Vertical Antennas	74/07	30	
Poor Man's Electronic Tower Hoist	74/07	38	
Homemade Tip-Over Tower	73/10	33	
$3 Pushup Mast for VHF	72/10	56	

TRANSMISSION LINES AND WAVEGUIDES

	Date	Page	Remarks
Archer Coaxial Cable Improved	88/03	32	
Connections for ½-In. Hardline	87/08	37	
Hardline Coaxial Connectors You Can Make	87/04	32	
Antenna Hardware You Can Build	87/04	36	
Strain Relief for Coaxial Cables	87/04	58	
Coaxial Feed Lines in Parallel	87/02	45	Z measurement
Feed Line Tests	86/10	48	
Circular Waveguide for 2304 MHz	86/08	62	
Coaxial Cables—Feedback	85/06	42	See also 84/11 p 19
Coaxial Cables—Feedback	85/03	46	See also 84/11 p 19
Waveguide Attenuation	85/02	60	
Coaxial Cables—Their Construction and Use	84/11	19	See 85/03 p 46 and 85/06 p 42
Coaxial Cable for Microwave Use	84/04	73	
Connectors for 1½ Inch Hardline	84/03	44	
Waveguide to Coax Transitions	83/12	84	
Line Loss and SWR	83/07	41	
Freeze Damage to Coaxial Cable	82/12	50	
RG-8 and the PL259	82/09	43	
Antenna Feed Lines for Portable Use	82/02	51	
Apparent SWR—What is it	82/02	51	
Coaxial Cable Wire Stripper	82/01	48	
Tuning and Constructing Balanced Lines	81/05	43	
Ant. and Transmission Line Quiz	81/01	43	Answers in 81/02 p 46
Measuring Transmission Line Velocity Factor	79/06	27	
Zip Cord Antennas—Do They Work	79/03	31	See 79/08 p 43 & 79/11 p 49
Coaxial Fittings on Hardline	78/11	39	
Series Section Line Impedance Match	78/07	14	
Stripping Twin-Lead	77/09	44	
My Feed Line Tunes my Antenna	77/04	40	
Easier Installation of Coax Connectors	77/01	45	
Using 75-Ohm Line in 50 Ohm Systems	76/04	34	
Transmission Line Losses	75/12	48	
Putting the G-Line to Work	74/06	11	
Repairing 450-Ohm Open Wire Line	73/09	42	
Reflections and Transmission Lines	73/08	47	
Supporting Open Wire Line	73/08	50	
Source of Feeder Spreaders	73/07	54	
Transmission Line Measurements and Line Loss	73/07	55	See 73/12 p 40
Inexpensive Time Domain Reflectometer	73/03	19	
Wiring Coax Cable to a Wafer Switch	73/03	57	
Coax Shield Separator	72/05	50	
Ant. Feeders and Transmatches; Some Facts	71/05	33	
Line Sections as RF Chokes and Bypasses	69/11	49	
Feed Lines Slide Rule	69/03	24	
Coaxial Cable Guide	67/03	51	
Low Loss Coax	66/12	15	
Transmission Lines as Circuit Elements	66/11	34	
Open Wire Line Spacers	66/03	45	
When is a Feed Line not a Feed Line	65/08	37	
Ant. and Transmission Line Quiz	65/07	19	Answers in 65/08 p 55
Losses in Coax	65/04	52	
The Whys of Transmission Lines—Part 3	65/03	19	See 65/01 p 25 and 65/02 p 24
The Whys of Transmission Lines—Part 2	65/02	24	See 65/01 p 25 and 65/03 p 19
The Whys of Transmission Lines—Part 1	65/01	25	See 65/02 p 24 and 65/03 p 19
Antennas and Feeders—Part 1	63/10	30	See 63/11 p 36 & 63/12 p 53
Simplified Transmission Line Calculations	63/07	17	
Radiation from Open-Wire Line at 420 MHz	63/07	55	
Resistive Matching with Quarter-Wave Lines	63/02	56	
Z of Special Open Wire Line Configurations	00/00	3	
Coax Loss Program for HP97 and TRS80C	00/00	13	
RF Cable for Microwave Applications	00/00	55	
UHF-Type Connectors for ½-In. Heliax	00/00	79	

TRAP ANTENNAS AND TRAP DESIGN

Add 160 to Your Trap Antenna	87/05	38	
Tune Up Your Tribander	86/04	27	
Coaxial Cable Traps	85/08	43	
Multiband Trap and Parallel Dipoles (HF)	85/05	26	See 85/07 p 43
Optimizing Coaxial Cable Traps	84/12	37	
Harmonics and Trap Antennas	84/06	40	
Ant. Traps Versus Bandwidth and Loading	84/05	46	
Coaxial Antenna Trap Design	84/03	46	
Franklin Broadside	84/02	48	Coaxial trap info.
Retuning Traps for the WARC Bands	83/12	43	
Dual Frequency Antenna Traps	83/11	27	
Lightweight Trap Antenna—Some Thoughts	83/06	15	See 84/06 p 43

Vertical and Groundplane Antennas, continued

VHF/UHF ANTENNAS (GENERAL ARTICLES)

W8JK BEAMS

(See also ZL Special Antennas)

1.8 MHz ANTENNAS AND TOPICS

	Date	Page	Remarks
Add 160 to Your Trap Antenna	87/05	38	
How to Build a 160-M Shortie	86/11	26	
KI6O 160-M Linear Loaded Sloper	86/04	26	See TC, 87/02 p 43
HH160RL Antenna	85/10	45	Receiving loop
Build a 4X Array for 160 Meters	85/02	21	
Franklin Broadside	84/02	48	Multiband Ant.
Top-Fed Vertical Ant. for 1.8 MHz—Plus 3	83/09	25	See 83/10 p 48
Shunt-Fed Towers—Practical Applications	82/10	21	
K4YF Special Antenna	82/09	26	
Try the TJ Antenna	82/06	18	
The Lowbanders One-Antenna Farm	82/02	23	
45 Ft DX Vertical for 160/80/40/30 M	81/09	27	See 81/11 p 50
Kite Supported 160 (and 80) M Ant.	81/03	40	
Ant. Hint for DXers	80/04	60	
Frame Receiving Ant.	80/01	43	
"Z" Antenna for 10 to 160 M	79/12	59	
Quarter Wave Sloper for 160 M	79/07	19	
Big Signal from a Small Lot	79/04	32	
BC Band Energy Rejection Filter	78/02	22	
Low Noise Receiving Antennas	77/12	36	
Small Space 160-M Antenna	77/12	45	
Antenna Network that Covers 160 M	77/12	46	
160-M Vertical Antenna	77/10	46	
160-M Monster Antenna	77/09	27	
Beat Noise with a Scoop Loop	77/07	30	160-M receiving loop
Weak Signal Reception on 160 M	77/06	35	Antenna notes
Kytoon Support	77/06	93	
Inverted L Antenna	77/04	32	
Kytoon Support	76/06	71	
160-M DX with a 2-Element Beam	75/10	20	
Shunt Feeding Tower for Low Frequencies	75/10	22	
Method of Shunt Feeding Tower	75/10	25	Gamma match
160-M Receiving Loop	75/04	40	
Minooka Special	74/12	15	160-M fixed & mobile ant.
Some Antenna Ideas for 1.8-MHz Portables	74/04	44	
Backyard 160-M Vertical	74/04	45	
Receiving Loop for 160 Meters	74/03	38	
Inverted Dipole Delta Loop	73/01	37	For 160/75/40 M
Ant. Traps of Spiral Delay Line	72/11	13	
Using a Grounded Tower on 160 M	69/05	48	
The 160-M Transmatch	67/05	38	
Ground Systems for 160 M	65/04	65	
Flagpole Without a Flag	64/11	36	160/80/40 vert.
160-M Transmission Line Antenna	00/00	53	

900 MHz AND ABOVE ANTENNAS

	Date	Page	Remarks
Horn Antennas for 10 GHz	87/04	80	See also 87/05 p 63
Circular Waveguide for 2304 MHz	86/08	62	
Simple 10-GHz Dish Antenna	86/06	62	
Mode L Parabolic Antenna and Feed Horn	86/05	24	For OSCAR 10
A 902-MHz Loop Yagi Antenna	85/11	30	
Periscope Antenna Systems	84/01	70	See 84/02 p 68
Alford Slot Antenna	82/09	67	
1.3-GHz Alford Slot Antenna	82/06	75	
Microwave Horn Antennas	82/02	74	
Reproducible Quagi for 1296 MHz	81/08	11	
Amateur Microwave Antennas	81/06	60	
Easily Constructed Ant. for 1296 MHz	69/06	47	
Antenna Ideas for 3.5, 5.8 and 10.4 GHz	00/00	81	

SPECIAL REFERENCES

[1] QEX #48, Feb 1986, p 10.

[2] QEX #43, Sep 1985, p 7.

[3] QEX #42, Aug 1985, p 2.

[4] QEX #41, Jul 1985, p 8.

[5] QEX #38, Apr 1985, p 8.

[6] QEX #37, Mar 1985, p 7.

[7] QEX #36, Feb 1985, p 3.

[8] QEX #34, Dec 1984, p 3.

[9] QEX #32, Oct 1984, p 9.

[10] QEX #31, Sep 1984, p 12.

[11] QEX #30, Aug 1984, p 4.

[12] QEX #30, Aug 1984, p 8.

[13] QEX #29, Jul 1984, p 8.

[14] QEX #26, Apr 1984, p 2.

[15] QEX #25, Mar 1984, p 7.

[16] QEX #24, Feb 1984, p 5.

[17] QEX #23, Jan 1984, p 2.

[18] QEX #23, Jan 1984, p 4. See correction QEX #25, Mar 1984, p 2. See also computer program for TS1000 in QEX #41, Jul 1985, p 4, as corrected in QEX #44, Oct 1985, p 2.

[19] QEX #22, Dec 1983, p 3.

[20] QEX #21, Nov 1983, p 7.

[21] QEX #19, Sep 1983, p 1. See also QEX #27, May 1984, p 4, and QEX #21, Nov 1983, p 3.

[22] QEX #11, Dec 1982, p 2.

[23] QEX #9, Oct 1982, p 3.

[24] QEX #8, Sep 1982, p 10.

[25] QEX #6, Jul 1982, p 2. See also QEX #8, Sep 1982, p 10.

[26] QEX #5, Jun 1982, p 2.

[27] QEX #2, Feb 1982, p 9. See correction QEX #17, Jul 1983, p 2.

[28] A. W. Lowe, *Reflector Antennas* (New York: IEEE Press, 1978).

[29] K. Davies, *Ionospheric Radio Propagation—National Bureau of Standards Monograph 80* (Washington, DC: U.S. Government Printing Office, April 1, 1965).*

[30] P. H. Lee (K6TS), *The Amateur Radio Vertical Antenna Handbook,* 1st ed. (Port Washington, NY: Cowan Publishing Corp., 1974).**

[31] V. A. Misek (W1WCR), *The Beverage Antenna Handbook* (Hudson, NH: V. A. Misek, 1977).

[32] A. E. Harper, *Rhombic Antenna Design* (Princeton, NJ: D. VanNostrand Co, Inc, 1941).*

[33] R. P. Turner, *How to Use Grid-Dip Oscillators,* 2nd ed. (New York: Hayden Book Co, 1969).*

[34] P. P. Viezbicke, *Yagi Antenna Design, NBS Technical Note 688* (Washington, DC: U.S. Government Printing Office, Dec 1976).*

[35] J. D. Kraus (W8JK) *Antennas,* 2nd ed. (New York: McGraw-Hill Book Co, 1988).

[36] H. Jasik, *Antenna Engineering Handbook* (New York: McGraw-Hill Book Co, 1960).**

[37] R. W. P. King, H. R. Mimno and A. H. Wing, *Transmission Lines, Antennas and Waveguides* (New York: Dover Publications, Inc, 1965).*

[38] P. N. Saveskie, *Radio Propagation Handbook* (Blue Ridge Summit, PA: Tab Books, Inc, 1980).

[39] J. Devoldere (ON4UN), *80-Meter DX Handbook,* Chapter 2 (Greenville, NH: Communications Technology, Inc, 1977).

[40] W. I. Orr (W6SAI), "The Folded Mini-Monopole Antenna," Ham Radio, May 1968, p 32.

[41] D. W. Covington (K4GSX), "Circularly Polarized Ground-plane Antenna for Satellite Communications," *Ham Radio,* Dec 1974, p 28.

[42] R. L. Guard (K4EPI), "Weather Balloon Verticals," 73, Jun 1971, p 32.

[43] J. R. True (W4OQ), "Low-Frequency Loop Antennas," *Ham Radio,* Dec 1976, p 18.

[44] P. A. Scholz (W6PYK) and G. E. Smith (W4AEO), "Log Periodic Antenna Design," *Ham Radio,* Dec 1973, p 34.

[45] P. Bertini (K1ZJH), "Active Antenna Covers 0.5-30 MHz," *Ham Radio,* May 1985, p 37. See also *Ham Radio,* Jul 1985, p 10.

[46] G. H. Brown, "Directional Antennas," *Proc IRE,* Jan 1937, p 88.

[47] R. D. Thrower (WA6PZR), "Electronic Antenna Rotation," *Radio Electronics,* Aug 1967, p 47.

[48] J. Malone (WØPJG), "Can A 7 Foot 40 M Antenna Work—The Small Loop," *73,* Mar 1975, p 33.

[49] R. C. Wilson (WØKGI), "The Incredible 18 Inch All-Band Antenna," *73,* Mar 1975, p 49.

[50] G. E. Smith (W4AEO), "Yes, I've Built Sixteen Log Periodic Antennas," *73* Mar 1975, p 97.

[51] J. Reisert (W1JR), "Reflector Antennas," *Ham Radio,* Feb 1986, p 51.

[52] J. Reisert (W1JR), "Loop Yagis," *Ham Radio,* Sep 1985, p 56.

[53] T. S. Rappaport (N9NB), "160-Meter Transmission Line Antenna," *Ham Radio,* May 1985, p 87.

[54] R. Ross, "A Sloping Terminated Vee Beam," *Ham Radio,* May 1985, p 71.

[55] H. Weinstein (K3HW), "RF Transmission Cable for Microwave Applications," *Ham Radio,* May 1985, p 106.

[56] B. Wolf and A. Anderson, "Memorandum on the Beverage Wave Antenna for Reception of Frequencies in the 550-1500 Kilocycle Band," *Antennas for Reception of Standard Broadcast Signals* (Washington, DC: Federal Communications Commission Report TRR 9.2.1, Apr 1, 1958).

[57] L. A. Moxon (G6XN), "The Smith Chart," *Radio Communication,* Jan 1977, p 22.

[58] H. D. Hooton (W6TYH), "Exponential Potential," *73,* Sep 1985, p 44.

[59] S. Gibilisco (W1GV/4), "Discover the Discone," *73,* May 1985, p 17.

[60] J. A. Swank (W8HXR), "Rotate the Bobtail Curtain," *73,* May 1985, p 48.

[61] J. A. Ryan (WB5LIM), "Ryan's Vertical Ecstasy," *73,* Oct 1984, p 32.

[62] R. W. Burhans, "VLF-HF Active Antennas," *Radio Electronics,* Apr 1983, p 132.

[63] R. W. Burhans, "All About Loop Antennas for VLF-LF," *Radio Electronics,* Jun 1983, p 83.

[64] M. G. Knitter, Ed., *Loop Antennas—Design and Theory* (Cambridge, WI: National Radio Club, 1983).

[65] M. G. Knitter, Ed., *Beverage and Long Wire Antennas—Design and Theory* (Cambridge, WI: National Radio Club, 1983).

[66] P. H. Lee (W3JHR), "Diversity Reception Made Easy," *CQ,* May 1964, p 44.

[67] P. D. Kennedy and J. W. Ames, *Polarization Diversity Reception of High Frequency Signals,* Technical Bulletin No. 4, 2nd ed. (Palo Alto, CA: Granger Associates, 1967).

[68] G. Chaney (W5JTL), "Extended/Expanded Power Dividers," *Ham Radio,* Oct 1984, p 3.

[69] QEX #49, Mar 1986, p 5.

[70] QEX #56, Oct 1986, p 3.

[71] QEX #55, Sep 1986, p 4.

[72] QEX #55, Sep 1986, p 3.

[73] QEX #58, Dec 1986, p 14.

[74] J. L. Lawson (W2PV), *Yagi-Antenna Design,* 1st ed. (Newington: ARRL, 1986).

[75] J. Sevick (W2FMI), *Transmission Line Transformers,* 1st ed. (Newington: ARRL, 1987).

[76] QEX #59, Jan 1987, p 5.

[77] QEX #60, Feb 1987, p 2.

[78] QEX #61, Mar 1987, p 4.

[79] QEX #66, Aug 1987, p 4.

[80] QEX #69, Nov 1987, p 6.

[81] QEX #71, Jan 1988, p 3.

[82] QEX #72, Feb 1988, p 3.

* Out of print. (Try university libraries on books. In the case of government publications, try your local U.S. Government depository library.)
** Newer edition due out. Either edition should be usable.

Chapter 30

Glossary and Abbreviations

This chapter provides a handy list of terms that are used frequently in Amateur Radio conversation and literature about antennas. With each item is a brief definition of the term. Most terms given here are discussed more thoroughly in the text of this book, and may be located by using the index. Following this glossary is a list of abbreviations that are commonly used throughout this book.

Actual ground—The point within the earth's surface where effective ground conductivity exists. The depth for this point varies with frequency and the condition of the soil.

Antenna—An electrical conductor or array of conductors that radiates signal energy (transmitting) or collects signal energy (receiving).

Antenna tuner—See Transmatch.

Aperture, effective—An area enclosing an antenna, on which it is convenient to make calculations of field strength and antenna gain. Sometimes referred to as the "capture area."

Apex—The feed-point region of a V type of antenna.

Apex angle—The included angle between the wires of a V, an inverted V dipole, and similar antennas, or the included angle between the two imaginary lines touching the element tips of a log periodic array.

Balanced line—A symmetrical two-conductor feed line that has uniform voltage and current distribution along its length.

Balun—A device for feeding a balanced load with an unbalanced line, or vice versa. May be a form of choke, or a transformer that provides a specific impedance transformation (including 1:1). Often used in antenna systems to interface a coaxial transmission line to the feed point of a balanced antenna, such as a dipole.

Base loading—A lumped reactance that is inserted at the base (ground end) of a vertical antenna to resonate the antenna.

Bazooka—A transmission-line balancer. It is a quarter-wave conductive sleeve (tubing or flexible shielding) placed at the feed point of a center-fed element and grounded to the shield braid of the coaxial feed line at the end of the sleeve farthest from the feed point. It permits the use of unbalanced feed line with balanced feed antennas.

Beamwidth—Related to directive antennas. The width, in degrees, of the major lobe between the two directions at which the relative radiated power is equal to one half its value at the peak of the lobe (half power = −3 dB).

Beta match—A form of hairpin match. The two conductors straddle the boom of the antenna being matched, and the closed end of the matching-section conductors are strapped to the boom.

Bridge—A circuit with two or more ports that is used in measurements of impedance, resistance or standing waves in an antenna system. When the bridge is adjusted for a balanced condition, the unknown factor can be determined by reading its value on a calibrated scale or meter.

Capacitance hat—A conductor of large surface area that is connected at the high-impedance end of an antenna to effectively increase the electrical length. It is sometimes mounted directly above a loading coil to reduce the required inductance for establishing resonance. It usually takes the form of a series of wheel spokes or a solid circular disc. Sometimes referred to as a "top hat."

Capture area—See aperture.

Center fed—Transmission-line connection at the electrical center of an antenna radiator.

Center loading—A scheme for inserting inductive reactance (coil) at or near the center of an antenna element for the purpose of lowering its resonant frequency. Used with elements that are less than 1/4 wavelength at the operating frequency.

Coax—See coaxial cable.

Coaxial cable—Any of the coaxial transmission lines that have the outer shield (solid or braided) on the same axis as the inner or center conductor. The insulating material can be air, helium or solid-dielectric compounds.

Collinear array—A linear array of radiating elements (usually dipoles) with their axes arranged in a straight line. Popular at VHF and above.

Conductor—A metal body such as tubing, rod or wire that permits current to travel continuously along its length.

Counterpoise—A wire or group of wires mounted close to ground, but insulated from ground, to form a low-impedance, high-capacitance path to ground. Used at MF and HF to provide an RF ground for an antenna. Also see ground plane.

Current loop—A point of current maxima (antinode) on an antenna.

Current node—A point of current minima on an antenna.

Decibel—A logarithmic power ratio, abbreviated dB. May also represent a voltage or current ratio if the voltages or currents are measured across (or through) identical impedances. Suffixes to the abbreviation indicate references: dBi, isotropic radiator; dBic, isotropic radiator circular; dBm, milliwatt; dBW, watt.

Delta loop—A full-wave loop shaped like a triangle or delta.

Delta match—Center-feed technique used with radiators that are not split at the center. The feed line is fanned near the radiator center and connected to the radiator symmetrically. The fanned area is delta shaped.

Dielectrics—Various insulating materials used in antenna systems, such as found in insulators and transmission lines.

Dipole—An antenna that is split at the exact center for connection to a feed line, usually a half wavelength long. Also called a "doublet."

Direct ray—Transmitted signal energy that arrives at the receiving antenna directly rather than being reflected by any object or medium.

Directivity—The property of an antenna that concentrates the radiated energy to form one or more major lobes.

Director—A conductor placed in front of a driven element to cause directivity. Frequently used singly or in multiples with Yagi or cubical-quad beam antennas.

Doublet—See dipole.

Driven array—An array of antenna elements which are all driven or excited by means of a transmission line, usually to achieve directivity.

Driven element—A radiator element of an antenna system to which the transmission line is connected.

Dummy load—Synonymous with dummy antenna. A non-radiating substitute for an antenna.

E layer—The ionospheric layer nearest earth from which radio signals can be reflected to a distant point, generally a maximum of 2000 km (1250 mi).

E plane—Related to a linearly polarized antenna, the plane containing the electric field vector of the antenna and its direction of maximum radiation. For terrestrial antenna systems, the direction of the E plane is also taken as the polarization of the antenna. The E plane is at right angles to the H plane.

Efficiency—The ratio of useful output power to input power, determined in antenna systems by losses in the system, including in nearby objects.

EIRP—Effective isotropic radiated power. The power radiated by an antenna in its favored direction, taking the gain of the antenna into account as referenced to isotropic.

Elements—The conductive parts of an antenna system that determine the antenna characteristics. For example, the reflector, driven element and directors of a Yagi antenna.

End effect—A condition caused by capacitance at the ends of an antenna element. Insulators and related support wires contribute to this capacitance and lower the resonant frequency of the antenna. The effect increases with conductor diameter and must be considered when cutting an antenna element to length.

End fed—An end-fed antenna is one to which power is applied at one end, rather than at some point between the ends.

F layer—The ionospheric layer that lies above the E layer. Radio waves can be refracted from it to provide communications distances of several thousand miles by means of single- or double-hop skip.

Feed line—See feeders.

Feeders—Transmission lines of assorted types that are used to route RF power from a transmitter to an antenna, or from an antenna to a receiver.

Field strength—The intensity of a radio wave as measured at a point some distance from the antenna. This measurement is usually made in microvolts per meter.

Front to back—The ratio of the radiated power off the front and back of a directive antenna. For example, a dipole would have a ratio of 1, which is equivalent to 0 dB.

Front to side—The ratio of radiated power between the major lobe and that 90° off the front of a directive antenna.

Gain—The increase in effective radiated power in the desired direction of the major lobe.

Gamma match—A matching system used with driven antenna elements to effect a match between the transmission line and the feed point of the antenna. It consists of a series capacitor and an arm that is mounted close to the driven element and in parallel with it near the feed point.

Ground plane—A system of conductors placed beneath an elevated antenna to serve as an earth ground. Also see counterpoise.

Ground screen—A wire mesh counterpoise.

Ground wave—Radio waves that travel along the earth's surface.

H plane—Related to a linearly polarized antenna. The plane containing the magnetic field vector of an antenna and its direction of maximum radiation. The H plane is at right angles to the E plane.

HAAT—Height above average terrain. A term used mainly in connection with repeater antennas in determining coverage area.

Hairpin match—A U-shaped conductor that is connected to the two inner ends of a split dipole for the purpose of creating an impedance match to a balanced feeder.

Harmonic antenna—An antenna that will operate on its fundamental frequency and the harmonics of the fundamental frequency for which it is designed. An end-fed half-wave antenna is one example.

Helical—A helically wound antenna, one that consists of a spiral conductor. If it has a very large winding length to diameter ratio it provides broadside radiation. If the length-to-diameter ratio is small, it will operate in the axial mode and radiate off the end opposite the feed point. The polarization will be circular for the axial mode, with left or right circularity, depending on whether the helix is wound clockwise or counterclockwise.

Helical hairpin—"Hairpin" match with a lumped inductor, rather than parallel-conductor line.

Image antenna—The imaginary counterpart of an actual antenna. It is assumed for mathematical purposes to be located below the earth's surface beneath the antenna, and is considered symmetrical with the antenna above ground.

Impedance—The ohmic value of an antenna feed point, matching section or transmission line. An impedance may contain a reactance as well as a resistance component.

Inverted V—A misnomer, as the antenna being referenced does not have the characteristics of a V antenna. See inverted-V dipole.

Inverted-V dipole—A half-wavelength dipole erected in the form of an upside-down V, with the feed point at the apex. Its radiation pattern is similar to that of a horizontal dipole.

Isotropic—An imaginary or hypothetical point-source antenna that radiates equal power in all directions. It is used as a reference for the directive characteristics of actual antennas.

Lambda—Greek symbol (λ) used to represent a wavelength with reference to electrical dimensions in antenna work.

Line loss—The power lost in a transmission line, usually expressed in decibels.

Line of sight—Transmission path of a wave that travels directly from the transmitting antenna to the receiving antenna.

Litz wire—Stranded wire with individual strands insulated; small wire provides a large surface area for current flow, so losses are reduced for the wire size.

Load—The electrical entity to which power is delivered. The antenna system is a load for the transmitter.

Loading—The process of a transferring power from its source to a load. The effect a load has on a power source.

Lobe—A defined field of energy that radiates from a directive antenna.

Log periodic antenna—A broadband directive antenna that has a structural format causing its impedance and radiation characteristics to repeat periodically as the logarithm of frequency.

Long wire—A wire antenna that is one wavelength or greater in electrical length. When two or more wavelengths long it provides gain and a multilobe radiation pattern. When terminated at one end it becomes essentially unidirectional off that end.

Marconi antenna—A shunt-fed monopole operated against ground or a radial system. In modern jargon, the term refers loosely to any type of vertical antenna.

Matching—The process of effecting an impedance match between two electrical circuits of unlike impedance. One example is matching a transmission line to the feed point of an antenna. Maximum power transfer to the load (antenna system) will occur when a matched condition exists.

Monopole—Literally, one pole, such as a vertical radiator operated against the earth or a counterpoise.

Nichrome wire—An alloy of nickel and chromium; not a good conductor; resistance wire. Used in the heating elements of electrical appliances; also as conductors in transmission lines or circuits where attenuation is desired.

Null—A condition during which an electrical unit is at a minimum. The null in an antenna radiation pattern is that point in the 360-degree pattern where a minima in field intensity is observed. An impedance bridge is said to be "nulled" when it has been brought into balance, with a null in the current flowing through the bridge arm.

Octave—A musical term. As related to RF, frequencies having a 2:1 harmonic relationship.

Open-wire line—A type of transmission line that resembles a ladder, sometimes called "ladder line." Consists of parallel, symmetrical wires with insulating spacers at regular intervals to maintain the line spacing. The dielectric is principally air, making it a low-loss type of line.

Parabolic reflector—An antenna reflector that is a portion of a parabolic revolution or curve. Used mainly at UHF and higher to obtain high gain and a relatively narrow beamwidth when excited by one of a variety of driven elements placed in the plane of and perpendicular to the axis of the parabola.

Parasitic array—A directive antenna that has a driven element and at least one independent director or reflector, or a combination of both. The directors and reflectors are not connected to the feed line. Except for VHF and UHF arrays with long booms (electrically), more than one reflector is seldom used. A Yagi antenna is one example of a parasitic array.

Phasing lines—Sections of transmission line that are used to ensure the correct phase relationship between the elements of a driven array, or between bays of an array of antennas. Also used to effect impedance transformations while maintaining the desired phase.

Polarization—The sense of the wave radiated by an antenna. This can be horizontal, vertical, elliptical or circular (left or right hand circularity), depending on the design and application. (See H plane.)

Q section—Term used in reference to transmission-line matching transformers and phasing lines.

Quad—A parasitic array using rectangular or diamond shaped full-wave wire loop elements. Often called the "cubical quad." Another version uses delta-shaped elements, and is called a delta loop beam.

Radiation pattern—The radiation characteristics of an antenna as a function of space coordinates. Normally, the pattern is measured in the far-field region and is represented graphically.

Radiation resistance—The ratio of the power radiated by an antenna to the square of the RMS antenna current, referred to a specific point and assuming no losses. The effective resistance at the antenna feed point.

Radiator—A discrete conductor that radiates RF energy in an antenna system.

Random wire—A random length of wire used as an antenna and fed at one end by means of a Transmatch. Seldom operates as a resonant antenna unless the length happens to be correct.

Reflected ray—A radio wave that is reflected from the earth, ionosphere or a man-made medium, such as a passive reflector.

Reflector—A parasitic antenna element or a metal assembly that is located behind the driven element to enhance forward directivity. Hillsides and large man-made structures such as buildings and towers may act as reflectors.

Refraction—Process by which a radio wave is bent and returned to earth from an ionospheric layer or other medium after striking the medium.

Resonator—In antenna terminology, a loading assembly consisting of a coil and a short radiator section. Used to lower the resonant frequency of an antenna, usually a vertical or a mobile whip.

Rhombic—A rhomboid or diamond-shaped antenna consisting of sides (legs) that are each one or more wavelengths long. The antenna is usually erected parallel to the ground. A rhombic antenna is bidirectional unless terminated by a resistance, which makes it unidirectional. The greater the electrical leg length, the greater the gain, assuming the tilt angle is optimized.

Shunt feed—A method of feeding an antenna driven element with a parallel conductor mounted adjacent to a low-impedance point on the radiator. Frequently used with grounded quarter-wave vertical antennas to provide an impedance match to the feeder. Series feed is used when the base of the vertical is insulated from ground.

Stacking—The process of placing similar directive antennas atop or beside one another, forming a "stacked array." Stacking provides more gain or directivity than a single antenna.

Stub—A section of transmission line used to tune an antenna element to resonance or to aid in obtaining an impedance match.

SWR—Standing-wave ratio on a transmission line in an antenna system. More correctly, VSWR, or *voltage standing-wave ratio*. The ratio of the forward to reflected voltage on the line, and not a power ratio. A VSWR of 1:1 occurs when all parts of the antenna system are matched correctly to one another.

T match—Method for matching a transmission-line to an unbroken driven element. Attached at the electrical center of the driven element in a T-shaped manner. In effect it is a double gamma match.

Tilt angle—Half the angle included between the wires at the sides of a rhombic antenna.

Top hat—See capacitance hat.

Top loading—Addition of a reactance (usually a capacitance hat) at the end of an antenna element opposite the feed point to increase the electrical length of the radiator.

Transmatch—An antenna tuner. A device containing variable reactances (and perhaps a balun). It is connected between the transmitter and the feed point of an antenna system, and adjusted to "tune" or resonate the system to the operating frequency.

Trap—Parallel L-C network inserted in an antenna element to provide multiband operation with a single conductor.

Unipole—See monopole.

Velocity factor—The ratio of the velocity of radio wave propagation in a dielectric medium to that in free space. When cutting a transmission line to a specific electrical length, the velocity factor of the particular line must be taken into account.

VSWR—Voltage standing-wave ratio. See SWR.

Wave—A disturbance or variation that is a function of time or space, or both, transferring energy progressively from point to point. A radio wave, for example.

Wave angle—The angle above the horizon of a radio wave as it is launched from or received by an antenna.

Wave front—A surface that is a locus of all the points having the same phase at a given instant in time.

Yagi—A directive, gain type of antenna that utilizes a number of parasitic directors and a reflector. Named after one of the two Japanese inventors (Yagi and Uda).

Zepp antenna—A half-wave wire antenna that operates on its fundamental and harmonics. It is fed at one end by means of open-wire feeders. The name evolved from its popularity as an antenna on Zeppelins. In modern jargon the term refers loosely to any horizontal antenna.

Abbreviations and acronyms abound in much modern literature, especially if the topic is one of a scientific or technical nature. Amateur Radio has its share of specialized abbreviations, and this statement certainly applies to the field of antennas. Abbreviations and acronyms that are commonly used throughout this book are defined in the list below. Periods are not part of an abbreviation unless the abbreviation otherwise forms a common English word. When appropriate, abbreviations as shown are used in either singular or plural connotation.

-A-

A—ampere
ac—alternating current
AF—audio frequency
AFSK—audio frequency-shift keying
AGC—automatic gain control
AM—amplitude modulation
ANT—antenna
ARRL—American Radio Relay League
ATV—amateur television
AWG—American wire gauge
az-el—azimuth-elevation

-B-

balun—balanced to unbalanced
BC—broadcast
BCI—broadcast interference
BW—bandwidth

-C-

ccw—counterclockwise
cm—centimeter
coax—coaxial cable
CT—center tap
cw—clockwise
CW—continuous wave

-D-

D—diode
dB—decibel
dBd—decibels referenced to a dipole
dBi—decibels referenced to isotropic
dBic—decibels referenced to isotropic, circular
dBm—decibels referenced to one milliwatt
dBW—decibels referenced to one watt
dc—direct current
deg—degree

DF—direction finding
dia—diameter
DPDT—double pole, double throw
DPST—double pole, single throw
DVM—digital voltmeter
DX—long distance communication

-E-

E—ionospheric layer, electric field
ed.—edition
Ed.—editor
EIRP—effective isotropic radiated power
ELF—extremely low frequency
EMC—electromagnetic compatibility
EME—earth-moon-earth
EMF—electromotive force
ERP—effective radiated power
E_s—ionospheric layer (sporadic E)

-F-

f—frequency
F—ionospheric layer, farad
F/B—front to back (ratio)
FM—frequency modulation
FOT—frequency of optimum transmission
ft—foot or feet (unit of length)
F1—ionospheric layer
F2—ionospheric layer

-G-

GDO—grid- or gate-dip oscillator
GHz—gigahertz
GND—ground

-H-

H—magnetic field, henry
HAAT—height above average terrain
HF—high frequency (3-30 MHz)
Hz—hertz (unit of frequency)

-I-

I—current
ID—inside diameter
IEEE—Institute of Electrical and Electronic Engineers
in.—inch
IRE—Institute of Radio Engineers (now IEEE)

-J-

j—vector notation

-K-

kHz—kilohertz
km—kilometer
kW—kilowatt
kΩ—kilohm

-L-

L—inductance
lb—pound (unit of mass)
LF—low frequency (30-300 kHz)
LHCP—left-hand circular polarization
ln—natural logarithm
log—common logarithm
LP—log periodic
LPDA—log periodic dipole array
LPVA—log periodic V array
LUF—lowest usable frequency

-M-

m—meter (unit of length)
m/s—meters per second
mA—milliampere
max—maximum
MF—medium frequency (0.3-3 MHz)
mH—millihenry
MHz—megahertz
mi—mile
min—minimum
mm—millimeter
ms—millisecond
mS—millisiemen
MS—meteor scatter
MUF—maximum usable frequency
mW—milliwatt
MΩ—megohm

-N-

NBS—National Bureau of Standards
NC—no connection, normally closed
NiCd—nickel cadmium
NO—normally open
no.—number

-O-

OD—outside diameter

-P-

p—page (bibliography reference)
P-P—peak to peak
PC—printed circuit
PEP—peak envelope power
pF—picofarad
pot—potentiometer
pp—pages (bibliography reference)
Proc—Proceedings

-Q-

Q—figure of merit

-R-

R—resistance, resistor
RF—radio frequency
RFC—radio frequency choke
RFI—radio frequency interference
RHCP—right-hand circular polarization
RLC—resistance-inductance-capacitance
r/min—revolutions per minute
RMS—root mean square
r/s—revolutions per second
RSGB—Radio Society of Great Britain
RX—receiver

-S-

s—second
S—siemen
S/NR—signal-to-noise ratio
SASE—self-addressed stamped envelope
SINAD—signal-to-noise and distortion
SPDT—single pole, double throw
SPST—single pole, single throw
SWR—standing wave ratio
sync—synchronous

-T-

tpi—turns per inch
TR—transmit-receive
TVI—television interference
TX—transmitter

-U-

UHF—ultra-high frequency (300-3000 MHz)
US—United States
UTC—Universal Time, Coordinated

-V-

V—volt
VF—velocity factor

VHF—very high frequency (30-300 MHz)
VLF—very low frequency (3-30 kHz)
Vol—volume (bibliography reference)
VOM—volt-ohm meter
VSWR—voltage standing-wave ratio
VTVM—vacuum-tube voltmeter

-W-

W—watt
WARC—World Administrative Radio Conference
WPM—words per minute
WVDC—working voltage, direct current

-X-

X—reactance
XCVR—transceiver
XFMR—transformer
XMTR—transmitter

-Z-

Z—impedance

-Other symbols and Greek letters-

°—degrees
λ—wavelength
λ/dia—wavelength to diameter (ratio)
μ—permeability
μF—microfarad
μH—microhenry
μV—microvolt
Ω—ohm
ϕ—angles
π—3.14159
θ—angles

Index

[Editor's note: Except for commonly used antenna terms and abbreviations, topics are indexed by their noun name—"coupling, mutual," rather than "mutual coupling." When noun modifiers exist, the topic can generally be found under any key noun—"ground wave" or "wave, ground"). Numeric listings appear at the end of the alphabetic section. The letters "ff" after a page number indicate coverage of the indexed topic on succeeding pages.

Notes

Notes

Notes

Notes

Please use this form to give us your comments on this book and what you'd like to see in future editions.

Name _____ Call sign _____

Address _____ Daytime Phone ( ) _____

City _____ State/Province _____ ZIP/Postal Code _____